Xianming Chen • Zhensheng Kang
Editors

Stripe Rust

Springer

Editors
Xianming Chen
US Department of Agriculture, Agricultural Research Service, Wheat Health, Genetics and Quality Research Unit
Pullman, WA, USA

Department of Plant Pathology
Washington State University
Pullman, WA, USA

Zhensheng Kang
State Key Laboratory of Crop Stress Biology for Arid Areas and College of Plant Protection
Northwest A&F University
Yangling, Shaanxi, China

ISBN 978-94-024-1491-2 ISBN 978-94-024-1111-9 (eBook)
DOI 10.1007/978-94-024-1111-9

Printed on acid-free paper

This Springer imprint is published by Springer Nature
The registered company is Springer Science+Business Media B.V.
The registered company address is: Van Godewijckstraat 30, 3311 GX Dordrecht, The Netherlands

Preface

Stripe rust (yellow rust) is an old disease of wheat and barley cereals and some grasses. The disease is still an important problem for wheat production throughout the world and even the most important factor limiting wheat production in many regions. Research on stripe rust has been done for more than 130 years since the late nineteenth century, and especially intensive studies have been conducted in the last two to three decades. However, there is no single book focusing on this important disease. This was the incentive for us to start writing such a book specifically on stripe rust about 2 years ago.

We organize the book into seven chapters. Chapter 1 is pretty much an introduction. It briefly addresses the historical aspects of the disease and its research and introduces the distribution and impact of the disease. Chapter 2 focuses on the variations of the pathogen, especially trying to cover both virulence and molecular aspects of pathogen variations. The most recent discovery of the sexual life cycle and its role in the disease epidemic and pathogen variation are discussed. Chapter 3 provides a discussion on the host-pathogen interactions, especially focusing on the recent progress in the molecular mechanisms of the interactions for the stripe rust pathosystem. Chapter 4 presents the epidemiology of the disease. Host, pathogen, and environmental factors determining the pathogen survival and reproduction and the disease development and damage are discussed. Chapter 5 is dedicated to host resistance as it is the most important approach for the control of stripe rust, and tremendous progress has been made in resistance to the disease. In addition to host resistance, nonhost resistance and other sources of resistance are also discussed. Chapter 6 presents the integrated control of the disease from concepts to methods for reducing damage caused by stripe rust. Chapter 7 entitled "Conclusions and Perspectives" summarizes research progress and discusses future perspectives based on our understanding of the existing issues to be addressed and advancing technologies that may help in the research and development of better strategies and techniques for better control of the disease in the future.

There are many thousands of publications on stripe rust. We have tried to include as many publications as possible. However, it is impossible to include all papers and reports on stripe rust, especially those published in languages other than English and Chinese. We would like to express our regret to those whose papers are missed in this book. We would like to especially thank the scientists who made great contributions to the great body of knowledge on stripe rust. We hope the book will be useful for everyone who is interested in stripe rust and similar diseases. We also welcome any comments and criticisms to this book. We wish more young scientists to dedicate their efforts on studying stripe rust and make great contributions toward more effective control of the destructive disease, protecting the highest potential production of wheat, barley, and cultivated grasses, while reducing the use of fungicides and protecting the environment.

Pullman, WA, USA
Yangling, Shaanxi, China
March 2, 2017

Xianming Chen
Zhensheng Kang

Contents

Chapter 1
Introduction: History of Research, Symptoms, Taxonomy of the Pathogen, Host Range, Distribution, and Impact of Stripe Rust

Xianming Chen and Zhensheng Kang

Abstract Stripe rust, caused by *Puccinia striiformis*, is a disease on cereal crops wheat and barley, some cultivated grasses and many wild grasses. It is an old disease, but is still economically important. Stripe rust is one of the most destructive diseases of wheat worldwide, and can cause huge yield losses or cost big expenses for chemical control. Although currently not as damaging as stripe rust on wheat, stripe rust of barley can cause significant yield losses in localized barley growing regions. Based on the host species specialization of the fungal pathogen, the species of *P. striiformis* can be separated into different formae speciales, most notably *P. striiformis* f. sp. *tritici* causing stripe rust mostly on wheat and *P. striiformis* f. sp. *hordei* mostly on barley. The rapid evolution of the pathogen create various races or pathotypes that are more specialized on cultivars of cereal crops with different race-specific resistance genes. This introduction chapter presents general aspects on historical perspectives, taxonomy and host range of the pathogen and distribution and impact of the disease, and the following chapters will focus specific aspects of the pathogen and disease.

Keywords Alternate host • Auxiliary host • Barley • Formae speciales • Grasses • Host range • Primary host • *Puccinia striiformis* • Stripe rust • Wheat • Yellow rust • Yield loss

---

Mention of trade names or commercial products in this publication is solely for the purpose of providing specific information and does not imply recommendation or endorsement by the U. S. Department of Agriculture. USDA is an equal opportunity provider and employer.

X. Chen (✉)
US Department of Agriculture, Agricultural Research Service, Wheat Health, Genetics and Quality Research Unit, Pullman, WA 99164-6430, USA

Department of Plant Pathology, Washington State University, Pullman, WA 99164-6430, USA
e-mail: xianming@wsu.edu

Z. Kang
State Key Laboratory of Crop Stress Biology for Arid Areas and College of Plant Protection, Northwest A&F University, Yangling, Shaanxi 712100, China

X. Chen, Z. Kang (eds.), *Stripe Rust*, DOI 10.1007/978-94-024-1111-9_1

Stripe rust, also called yellow rust, is a disease on cereal crops and grasses. Economically, stripe rust is mostly important on wheat and barley. The disease was recognized several hundred years ago, and it has been intensively studied during the last 120 years. Hassebrauk (1965, 1970) and Hassebrauk and Röbbelen (1974, 1975) wrote a four-part monograph on stripe rust in Germany. Latter, Röbbelen and Sharp (1978) revised the part on host resistance into a monograph in English. Line and Qayoum (1992) wrote a bulletin summarizing virulence, aggressiveness, evolution, and distribution of races of the stripe rust pathogen in North America from 1968–1987. Since then, there has been no any monograph published on stripe rust. So far, there has been no a single book focusing on stripe rust, but covering every major aspects of the research and control of stripe rust. To fill the gap, this book summarizes more than 100-year research, but mostly focusing on the progresses made in the recent decades. The book comprises seven chapters. Chapter 1 covers the introduction on the history, symptomology and effects on plants, distribution and impact of the disease. In this chapter, the taxonomy of the pathogen is also introduced. Chapter 2 focuses on the pathogen variability including variations in virulence, environmental adaptation, and population diversity. This chapter also contains recent findings on genome and functional genomics of the pathogen and discussions on molecular mechanisms of variability. Chapter 3 presents recent findings on host-pathogen interactions using cytological, physiological, and molecular approaches. Interactions to primary and alternate hosts in the context of life cycle of the pathogen are presented in this chapter. The epidemiology of stripe rust is covered in Chap. 4, which also contains how to predict the disease. As resistance is the most important approach to control stripe rust and has been most intensively studied, whole Chap. 5 is dedicated to the topic. The integrated control of stripe rust, including cultural practices and chemical applications, is discussed in Chap. 6. Chapter 7 presents brief conclusions and perspectives on future stripe rust research and management.

1.1 The History

Rusts of cereal crops are an old group of plant diseases that may have occurred since the domestication, which perhaps occurred 12,000 years ago (Haldorsen et al. 2011). Based on archaeological evidence, stem rust existed more than 3300 years ago (Kislev 1982). It is likely that stripe rust also existed that early. Because reports before the late ninenteenth century did not distinguish stripe rust from other rusts (Eriksson and Henning 1896; Chester 1946), many of the early reports of rusts were likely stripe rust or combinations with leaf rust and/or stem rust. The severe rust occurred in Italy in 1766 described by Giovanni Targioni Tozzetti (Tozzetti 1767) as "universal over the whole of Italy, and in all the different levels and exposures of its territory" was most likely stripe rust because "as first appearing, it is a bright yellow, soon becoming orange, and finally after days, becoming black" (Goidanich 1943). According to Eriksson and Henning (1896), stripe rust was first described by Gadd and Bjerkander in Europe in 1777, and the first description and naming of the fungus was done by Schmit before 1819.

In China, the earliest record of stripe rust was in "*Ancient Chinese Agricultural Practices*" (齐民要术), an earliest Chinese book written by JIA Sixie (贾思勰) about 1470 years ago on agriculture, horticulture, husbandry, fishery, winery, food and nutrition. In that book and a late book, "*Agricultural Proverbs*" (马首农谚) in 1836, stripe rust was called "jaundice", like a human disease (黄疸in Chinese) (Li and Zeng 2002). In the former book, stripe rust was related to spring rainy conditions. In India, stripe rust, together with other rusts, was reported in a scientific journal in 1896 (Cunningham and Prain 1896). In Russia, Yachevski (1909) included stripe rust in his book entitled "Rusts of Grain Crops in Russia", containing information on disease distribution, damage caused, environmental conditions favorable for disease development, disease control, and a list of the important rust fungi in Russia.

In the United States, the first clear recognition of stripe rust was in 1915 in Arizona by F Kolpin Ravn, a visiting scientist from Denmark, who was traveling with a United States Department of Agriculture (USDA) crop survey team in the western US (Carleton 1915). Latter, examination of herbarium specimen collected from various locations in the US found that stripe rust, mostly from grasses, occurred at least as early as 1892 in the western US (Humphrey et al. 1924). It is believed that stripe rust very likely occurred on wheat and grasses in California in the 1770s based on old newspapers reporting rusts and the fact that stripe rust is more prevalent than leaf rust and stem rust in California (Smith 1961; Smith and Bowman 1961; Line 2002). In South America, stripe rust was first discovered by Holway on *Hordeum chilense* in Chile in 1919 and on *Agropyron altenuatum* in Ecuador in 1920 (Arthur 1925). The stripe rust was first observed on wheat in 1929 (Rudorf and Job 1931) and on barley for the first time in 1930 (Hirschhorn 1933) in Argentina.

Since the science of plant pathology emerged in the late 1800s and early 1900s, stripe rust has been intensively studied. The first scientific study on stripe rust was done by Jacob Eriksson in Sweden in the early 1890s (Eriksson 1894). This study separated the species of the stripe rust pathogen into five formae speciales based on specialization on different cereal and grass genera. Although Eriksson studied the cytology of rust fungi, he mistakenly considered haustoria as "corpuscules speciaux", a form in which he thought the rust fungus emerged from an invisible "mycoplasm" stage. His concept of "mycoplasma" was challenged by Harry Marshall Ward, who carefully studied cytology of infection by urediniospores, first with *Puccinia dispersa* on *Bromus* spp. (Ward 1904, 1905) and late extended to wheat infected with *P. glumarum* (*P. striiformis*). He provided an extensive set of clear drawings showing appressoria, penetration pegs, substomatal vesicles, infection hyphae, haustoria at several stages of development, and "runner" hyphae beginning to ramify in host tissue. He also showed the number and location of nuclei for each type of fungal cell (Ward 1904). In his study with wheat stripe rust, he included development on an immune cultivar obtained from Biffen. He described "death changes" in the immune host, nowadays more commonly referred as "hypersensitiveness" termed by Stakman (1915). Resistance responses of nonhost species to several rust fungi, including *P. glumarum*, were described histologically by Ward's student, C. M. Gibson (1904), who showed that the fungi usually were able to enter stomata.

The rediscovery of Mendel's laws of segregation and independent assortment of genes stimulated genetic studies on plant disease resistance in the early twentieth century. Working on stripe rust of wheat, Rowland H. Biffen in Cambridge, United Kingdom, was the first to discover the plant resistance to diseases follows Mendel's laws (Biffen 1905). In his first study, Biffen found that stripe rust resistance was controlled by a recessive gene (Biffen 1905). Later, he found that resistance was independent of other plant traits (Biffen 1907). Biffen's results created doubts about the bridging-host theory of Ward (1903). According to the bridging-host theory, a pathogen could gradually adapt to a resistant host by passing through taxonomically intermediate hosts. Within a few years many more examples of Mendelian inheritance of resistance against cereal rusts were found, which eventually brought the debate to the end and scientists generally accepted the concept that Mendel's genetic principles were applicable to resistance against cereal rusts. Biffen's pioneer work with stripe rust led to control of plant diseases, including stripe rust, through breeding for resistant cultivars (Biffen 1931). Since these early studies, intensive research has been conducted on various aspects of stripe rust in various countries to prevent or reduce damage caused by the devastating disease.

1.2 Disease Symptoms and Effects on Host Plants

Stripe rust is also called yellow rust. Both names are descriptive for the disease symptoms and pathogen signs. However, it can sometimes cause confusion when "stripe" and "yellow" are used to diagnose the disease from others at a particular time point. Like other diseases, the disease of stripe rust is a process that takes days to months to the fully diseased stage.

Because the disease symptoms and signs on alternate hosts will be presented in Sect. 3.1.3 on infection process on alternate hosts in Chap. 3, here only symptoms on primary hosts (wheat or barley) are presented, which are similar to those on auxiliary hosts (grasses). The development of disease of stripe rust starts when urediniospores of the stripe rust pathogen germinate and germ tubes grow into host tissues. Although could not be seen with naked eyes, host tissues will start a series of changes which usually need to be observed under a microscope, therefore referred as microscopic symptoms. The histological changes will be described in detail in Sect. 3.2 of Chap. 3. Here, only macroscopic changes that can be observed with naked eyes, referred as macroscopic symptoms, will be discussed.

Under controlled conditions, obvious symptoms will not be seen until 6–7 days after inoculation on wheat leaves. The disease first appears as chlorotic patches that are easier to seen when holding an inoculated leaf against a light especially in the early symptom development. The chlorotic system is like many viral diseases or abiotic stress. Depending upon temperatures, uredinia (commonly referred to sori or pustules appears (the rust pathogen sporulation) from chlorotic areas about 11–14 days after inoculation. Uredinia are tiny, elongated, and yellow to orange in color. A uredinium is a fruiting body of rust fungi, from which thousands of urediniospores are produced. Soon after formation, uredinia erupt to release urediniospores, which cannot be seen with naked eyes, but in mass are yellow to orange powder, which can be collected by tapping off or vacuuming into a tube. Infected leaves can produce urediniospores for many days until the leaves are dried up, or sucked to die. Uredinia can be produced on any parts of wheat plants, including leaf blades and sheathes, young stems, glumes, awns, and young kernels, but most obviously on leaves. Compared to leaf rust and stem rust, the stripe rust fungus is more likely infect glumes, and therefore, the fungus was previously named as *Puccinia glumarum*, and the disease therefore, could be called glume rust in contrast to leaf rust and stem rust. However, to distinguish the three rusts on cereal crops just by plant parts they infect is not always correct as the three fungal species can infect all plant parts above the ground, especially the stripe rust pathogen.

The three rusts of cereal crops can be easily distinguished by the size, shape, color, and arrangement of uredinia. Typical uredinia of both stripe rust and stem rust are elongated and those of leaf rust are round to ovoid. However, stripe rust uredinia are the smallest (0.3–0.5 × 0.5–1.0 mm), those of stem rust are the largest up to several mm in width and several cm in length, and those of leaf rust are in between (up to 1.5 mm in diameter). Uredinia or urediniospore mass of stripe rust are yellow to orange, those of leaf rust are orange red to brown, and those of stem rust are red to red brown. Because uredinia of stripe rust are tiny, uredinial size does not change very much. In contrast, leaf rust and stem rust can have big variations in uredinial sizes depending upon compatible or incompatible interactions between the pathogen and host. Therefore, both stem rust and leaf rust pathogens can produce uredinia as small as or smaller than stripe rust uredinia. As the common name indicates, stripe rust produces uredinia between leaf veins, and therefore, form stripes on adult-plant leaves. In contrast, uredinia of leaf rust and stem rust are randomly distributed and do not form stripes. However, uredinia of stripe rust on seedling leaves are not arranged in stripes, but in clustered patches. On adult-plant leaves, stripes may not be obvious when the leaf surface covered by uredinia. Because infection by the stripe rust fungus is systemic and one-urediniospore infection can produce many uredinia, its uredinia forms stripes on adult-plant and clustered patches on seedlings. In contrast, infection by the leaf rust or stem rust fungus is not systemic, and one-spore infection typically produces one uredinium. Therefore, uredinia are random as infections are random on plant surface. Because of these, it is easy to see infections of one or few uredinia of leaf rust and stem rust, but it is almost always to see a cluster or clusters of uredinia on seedlings and a stripe or stripes of uredinia of stripe rust. Figure 1.1 shows typical systems and signs in a compatible interaction, in which the pathogen is virulent and wheat plants are susceptible.

When host plants are less susceptible or resistant, various symptoms can occur, ranging from tiny flecks without any uredinia to large necrotic patches with or without uredinia on seedlings, or short to long necrotic stripes (up to several cm in length) on adult-plant leaves. When there are no uredinia associated to necrotic patches, the symptoms caused by stripe rust infection can be easily recognized under controlled conditions after inoculation, but can be hard to distinguish from symptoms caused by genetic, physiological or abiotic stresses; or other pathogens

Fig. 1.1 Typical stripe rust symptoms and signs on wheat leaves in fields. Note the patch of uredinia on the seedling (**a**) and stripes of uredinia on adult-plant leaves (**b**)

at the seedling stage. Taking healthy and unhealthy plants in the same field, rusted plants in nearby fields can add to correct diagnosis. Correct recognition of such necrotic symptoms caused by the stripe rust pathogen on seedlings is less important as the infected plants do not produce spores for secondary infections and generally no control measures are needed. However, if the problem of necrotic patches is caused by other spot disease pathogens, fungicide application may be needed. Further examination of disease symptoms, pathogen isolation, or molecular diagnosis can be helpful. On adult plants, necrotic stripes caused by the stripe rust pathogen can be more easily distinguished from other problems even without uredinia, as stripe rust stripes are rectangular, pale white, and usually with a distinct margin. Most similar symptoms are caused by cereal leaf beetles. However, stripes appear hollow as leaf tissues are eaten by the insects and only leaving the leaf surfaces, and such stripes almost can be seen through. Plants with stripes caused by leaf beetle damage usually have beetles on leaves.

Necrotic patches or stripes with uredinia are easily recognized to be caused by the stripe rust pathogen, rather than other biotic or abiotic stresses. If individual patches or stripes are distinct from each other, uredinia are usually on the margins. In contrast, single uredinia are surrounded by chlorotic or necrosis for leaf and stem rusts. Various ratios of necrotic vs. uredinial areas reflect the level of incomplete resistance of the host plants against the stripe rust pathogen. Because this is important for phenotypically studying resistance, the scales to qualify or quantify the reactions and intensities of stripe rust will be presented in Fig. 2.1 in Chap. 2 on variability of the stripe rust pathogen.

The infection of the stripe rust pathogen has various effects on its host plants. Because the pathogen is an obligate biotrophic parasite, its growth and reproduction completely rely on the host plants. The stripe rust fungus gets all water and nutrition from host plants. Except that urediniospores can survive for a limited period of time, the fungal parasite also uses host plant as a shelter for survival. It appears that host plants do not get anything good from the parasite, probably except for infection-triggered defense responses that may prevent or reduce infection by more virulent isolates or other pathogens. In general, the stripe rust fungus is viewed as an invader and damager, and host plants suffer from the pathogen damage.

When the stripe rust fungus is growing within the plant tissue, it obtains water and nutrition, directly from the host. When fighting against pathogen infection, host plants need to relocate energy and related nutritional substances for defense. The reduction of energy, nutrition and water slows down the host growth. Once disease symptoms, such as chlorosis and necrosis are developing, especially when the fungus is sporulating, green leaf surface and tissue are reduced, and as a result, plant photosynthesis is greatly reduced. Macroscopically, stripe rust infection reduces the vigor and height of plants; the number of tillers, heads, and kernels; the weight and quality of seed or grain. Economically, stripe rust reduces yield and quality of grain and forage and adds additional costs for crop management. Agriculturally, because the stripe rust pathogen can cause large-scale disease epidemics, the disease is one of the most important diseases on wheat, barley and some grasses. In some regions of the world, it is the most damaging disease.

1.3 The Taxonomy of the Pathogen

The stripe rust pathogen, *Puccinia striiformis* Westend., is one of about 4000 species in the genus *Puccinia*, belonging to the family Pucciniaceae, order Pucciniales, class Pucciniomycete, division Basidiomycota of the Fungi kingdom (Kirk et al. 2008). According to Van der Merwe et al. (2008), *P. striiformis* is most closely related to *P. poae-nemoralis* Otth and then to *P. poarum* Nielsen, *P. graminis* Pers. and the *P. recondita* Dietel & Holw. complex including *P. triticina* Erikss. based on partial beta-tubulin sequences. Both *P. poae-nemoralis* and *P. poarum* cause rusts on grasses of the Poaceae family, but they have barberry (*Berberis* spp.) and coltsfoot (*Tussilago farfara*) as alternate hosts, respectively (Stubbs 1986; Cagaś and Marková 1988). Both *P. graminis* and *P. triticina* are important pathogens causing stem rust and leaf rust on wheat, respectively.

The Latin name of the stripe rust fungal species had gone changes of several times before its current *P. striiformis*. The uredinial form of the stripe rust fungus was first described as *Uredo glumarum* by Schmidt (1827), probably before 1819 according to Eriksson and Henning (1896). The teleomorph of the fungus was described as *Trichobasis glumarum* by Lév. (1849) in 1848, *Puccinia striaeformis* Westend. by Westendorp in 1854, and later by Eriksson (1894) and Eriksson and Henning (1894) as *Puccinia glumarum* Erikss. in 1894. In 1860, Fuckel named the pathogen *Puccinia straminis* (1860), but whether it was stripe rust or leaf rust is unclear. *P. glumarum* had been mostly used in the literature until the 1950s (Hylander et al. 1953). It was Cummins and Stevenson (1956) who correctly applied the name used by Westendorp, with the orthographic correction to *P. striiformis*. According to Liu and Hambleton (2010), *T. glumarum*, even earlier than *P. striaeformis*, is invalid because the combinations were not formally made based on the International Code of Botanical Nomenclature (McNeill et al. 2006). Since the 1950s, *P. striiformis* has been used as a species of fungi causing stripe rust on cereals and grasses with recent separations of the stripe rust fungi on bluegrass (*Poa pratensis*) and orchard grass (*Dactylis glomerata*) as different species, *Puccinia pseudostriiformis* M. Abbasi, Hedjaroude & M. Scholle and *Puccinia striiformoides* M. Abbasi, Hedjaroude & M. Scholle, respectively (2004; Liu and Hambleton 2010). In addition, Liu and Hambleton (2010) described a new species, *Puccinia gansensis* Liu & Hambleton, from a single sample on *Achnatherum inebrians* (called drunken horse grass in China) from Gansu, China. The separation of these four species were based on some morphological traits, which were inconclusive when used individually but conclusive when used in combination, and partial sequences of the internal transcribed spacer (ITS) region and the beta-tubulin gene (Liu and Hambleton 2010). Because the four species formed a strong monophyletic group distinct from other species within *Puccinia*, Liu and Hambleton (2010) proposed a series, *Puccinia* Series *Striiformis* Liu & Hambleton, to include these species within the genus *Puccinia*. The species in this series are characterized by having serial arrangement of uredinia and telia on stripes, moderately obscure urediniospore germ pores compared to *P. recondita* and *P. hordei* with clearer pores and *P. coronata* with more obscure pores, and telial locules that are most often complete compared to *P. recondita* and *P. coronata* with mostly incomplete locules. In the study by Liu and Hambleton (2010), the species of *P.* Series *Striiformis* were mostly close to *P. poae-nemoralis*, and the next close species was *P. coronata*, followed by *P. graminis*, *P. recondita*, and *P. triticina*, generally in consistent with Van der Merwe et al. (2008). For recognition of stripe rust fungi (*P.* Series *Striiformis*), Liu and Hambleton (2010) pointed out that stripes of uredinia or telia are more conspicuous on wide-leaf hosts and less conspicuous on narrow-leaf hosts, and that some leaf rust fungi on grasses may appear in stripes and have yellow uredinia. The separation of different species within the *P.* Series *Striiformis* and the phylogenetic relationships of these species to others species should be further tested with more samples and rust fungal genes using both morphological and molecular techniques.

1.4 Host Range and Formae Speciales

Although rust fungi are generally considered to be highly specialized on their plant hosts, some species can have a wide host range. For a cereal rust species, the major cereal crop is generally referred as the primary host because of the economic importance; wild grasses as auxiliary hosts because of sharing same spore stages with the primary host; and non-grass hosts as alternate hosts because they are infected by a different spore stage. Taking the wheat stripe rust fungus (*P. striiformis* f. sp. *tritici*) for example, wheat is the primary hosts; barley and various grasses in the Gramineae family are auxiliary hosts; and barberry (*Berberis* spp.) and mahonia (*Mahonia* spp.) of the Berberidaceae family in the order of Ranunculales are alternate hosts. The various spore stages produced on primary/auxiliary and alternate hosts will be presented in Sect. 3.1.1 of Chap. 3. Here, the focus is on the auxiliary host range of the wheat stripe rust fungus.

Host range of a pathogen is directly related to the crop disease epidemiology and control. Important issues need to be determined, such as how wide the host range is, whether auxiliary hosts are the primary or secondary sources of pathogen inoculum, and how important auxiliary hosts are in pathogen survival and diversity. The determination of host range and at least partially addressing of some of the above issues were done by artificially inoculating typical wheat or barley stripe rust isolates on various grasses and isolates from grasses to cereals under controlled conditions, and/or visually observing rusts on plants of grasses under natural infections. Most studies on the host range of the wheat stripe rust fungus were conducted from the late 1890s and the 1960s.

In North America, scientists paid attention to stripe rust on grasses immediately after wheat stripe rust was discovered in 1915. In the same month (May 1915), AG Johnson observed stripe rust on *Hordeum murinum* near Tehachapi, California (Humphrey et al. 1924). In the same year, stripe rust was found widespread in the western US on *Bromus marginatus*, *Elymus canadensis* and *H. murinum* (Carleton 1915). Hungerford (1923) and Hungerford and Owens (1923) inoculated grasses and cereals to determine the host range of the stripe rust pathogen and made observations in the field. Stripe rust was found on wheat in almost all western states, on barley in eastern Washington and Oregon, central California, and western South Dakota; on rye in northern Idaho, western Oregon and eastern Washington, and on *H. murinum* in California. Stripe rust collections from the US Pacific Northwest infected wheat and 59 grass species. Collections from wheat infected wheat, barley, rye and 48 grass species while collections from 12 other grass species did not infected wheat. These studies indicate that wheat stripe rust has a broad range of hosts and not all stripe rust collections were able to infect wheat. Based on these results, Hungerford and Owens (1923) separate the *P. striiformis* species into varieties and use *P. glumarum* var. *tritici* for the collections able to infect wheat. Hungerford also found that *B. marginatus*, *E. canadensis*, *E. glaucus*, *H. nodosum*, and *H. jubatum* harbored dormant mycelium of the pathogen at low elevations in Oregon (Hungerford 1923). In the studies of Humphrey et al. (1924), stripe rust was found occurring naturally on 34 grass species plus barley, rye and emmer, spelt and common wheat. In that publication, they listed the hosts of the stripe rust fungus in both Europe and North America, and specifically noted that stripe rust occurred on *Dactylis glomerata* in Europe but not in North America. Stripe rust on *D. glomerata* was not reported in the US until the 1960s (Tollenaar 1967).

During the late 1910s–1930s, Canadian scientists also reported stripe rust on grasses in addition to wheat. Fraser found stripe rust on *H. jubatum* near Edmonton in 1918, and later the disease was observed on grasses and wheat in Alberta from Edmonton to the southern border with the US, British Columbia and Saskatchewan (Fraser and Conners 1925; Johnson and Newton 1928; Sanford and Broadfoot 1929; Newton and Johnson 1936). The host range included *Agropyron*, *Elymus* and *Hordeum* species of grasses as well as wheat and barley (Newton and Johnson 1936). The disease was found most common on the native grass species. Collections from *H. jubatum* infected wheat and *Agropyron* species.

More species were added to the list of the host range for the stripe rust pathogen in the 1960s. In the US 126 species of grasses in 20 genera were susceptible to wheat stripe rust (Line 2002). The most important genera were *Aegilops*, *Agropyron*, *Bromus*, *Elymus*, *Hordeum*, *Secale* and *Triticum*. Based on collections of 40 years in California, Tollenaar and Houston (1966) pointed out that stripe rust occurred on grasses in the San Francisco area, the coastal mountains and valleys from San Francisco to the Oregon border, and the Sierra Nevadas mountain range. They found stripe rust on wild *Hordeum*, *Elymus* and *Sitanion* species in the Sierra Nevadas at high altitudes during the summer, and logically concluded that winds could carry the urediniospores from grasses in the mountains to the fall-planted wheat fields in the valleys. In the state of Washington, Hendrix and associates did intensive studies on stripe rust on grasses and how grass stripe rust is related to epidemics on wheat. Hendrix et al. (1965) found stripe rust on grass species *Agropyron bakeri*, *A. riparium*, *A. spicatum* (now known as *Pseudoroegneria spicata*), *Bromus carinatus*, *B. pumpellianus*, *B. sitchensis*, *B. marginatus*, *Hordeum jubatum*, *Sitanion hystrix* and *Poa nemoralis* in the mountains surrounding the Columbia Basin of the Pacific Northwest. They concluded that stripe rust on grasses at high elevations was probably a source of inoculum for the wheat at the low elevations. In contrast, Shaner and Powelson (1971, 1973) believed that grasses were not very important as sources of stripe rust on wheat in Oregon based on their epidemiological studies from 1961 to 1968. Similarly, Sharp and Hehn (1963) concluded that grasses were not an important source of stripe rust of wheat in a study of the fungal survival on winter wheat in Montana.

Dietz and Hendrix (1962) inoculated randomly selected 948 grass lines with a wheat stripe rust isolate in a greenhouse at Pullman, Washington. They observed stripe rust on 372 lines representing 105 species in 16 genera. Tu (1967) selected susceptible grasses reported by Dietz and Hendrix (1962) and studied the behavior of stripe rust on grasses during the summer. Selected grasses were artificially inoculated in a greenhouse and transplanted to the field during the summers of 1963 and 1964 at Pullman, Washington. Continuous sporulation was observed throughout the summer on *Agropyron caespitosum*, *A. trachycaulum*, *B. marginatus* and *E. glaucus*.

Sporulation occurred in early and late summer but with a dormant period near midsummer on *Agropyron brachyphyllum*, *A. caespitosum*, *A. cristatum*, *A. dasystachyum*, *A. intermedium*, *A. spicatum*, *A. sussecundum*, *A. trichophorum*, *Bromus carinatus*, *B. scoparius*, *Elymus crinitus*, *Festuca arundinacea*, *F. rubra*, *Hesperochloa kingie* and *Hordeum bulbosum*. Sporulation was observed on *Alopecurus arundinaceus* and *Hesperochloa kingie* in early summer but the fungus succumbed with the advent of hot weather.

In Europe, stripe rust oversummering was reported on *Agropyron caninum* and *A. repens* at high elevations in Germany (Becker and Hart 1939), but grasses are considered not important in northwestern Europe (Zadoks 1961). As stripe rust was introduced to Australia in late 1970s, studies on host range were conducted in the early 1980s. Holmes and Dennis (1985) found successful infection after inoculating wheat from Australia with urediniospores collected from *Bromus mollis*, *B. unioloides*, *Hordeum hystrix*, *H. leporinum*, *H. marinum*, *H. vulgare*, *Phalaris minor*, *P. paradoxa* and *Triticosecale* (Triticale). It is not clear if these grass species are important for wheat stripe rust epidemics. Natural infection of stripe rust occurs on wild barley grasses in Australia, but the rust pathogen is considered as a different forma specialis (Wellings et al. 2000; Wellings 2011).

Based on Gerechter-Amitai's unpublished data, Wahl et al. (1984) indicated that stripe rust appears sporadically on wheat cultivars and wild grasses belonging to *Triticum dicoccoides*, *Aegilops*, *Hordeum*, and other genera in Israel. Some isolates of the fungus have a host range of over 40 species in 17 genera. The stripe rust isolates from *Ae. kotschyi* and *H. spontaneum* in the arid southern region with average annual rainfall below 100 mm were virulent on wheat.

In China, Prof. Lee Ling in 1945 reported that stripe rust commonly occurred on *Roegneria ciliaris* in the Chengdu Plain and surrounding maintains in the province of Sichuan, but the stripe rust on the grass occurred relatively late, adapted to a relatively high range of temperatures, and relatively low virulent on wheat compared to wheat stripe rust (Li and Zeng 2002). Since then, Chinese scientists found a large number of grass species can be infected by stripe rust. They used the following three criteria to determine grass hosts for wheat stripe rust. First, plants of a species have stripe rust under natural conditions either in cultivated fields or of wildly growing, and the stripe rust isolates from the plants should be able to infect and produce uredinia on common wheat cultivars that are susceptible to wheat stripe rust under artificially inoculation conditions. Second, grass plants in production or experimental fields are infected by stripe rust, and the rust isolates can infect and produce uredinia on common wheat cultivars that are susceptible to wheat stripe rust under artificial inoculation conditions. Third, plants of a grass species can be infected and have uredinia after artificially inoculated with isolates of the wheat stripe rust fungus. A species meets one of the criteria is considered as a host for the wheat stripe rust pathogen. Based on these criteria, a total of 88 species in 16 genera, all in the Poaceae (also called Gramineae) family have been identified as hosts of the wheat stripe rust pathogen (Li and Zeng 2002). The 16 genera are *Aegilops* (23 species), *Agropyron* (8 species), *Aneueolepidum* (2 species), *Bromus* (2 species), *Elymus* (8 species), *Elytrigia* (3 species), *Glyceria* (1 species), *Hordeum* (7 species), *Hystrix* (1 species), *Psathyrostachys* (1 species), *Puccinellia* (1 species), *Roegneria* (12 species), *Secale* (6 species), *Taeniatherm* (1 species), *Thinopyrum* (1 species) and *Triticum* (11 species).

Although Liu and Hambleton (2010) questioned that the early reported stripe rust on some grasses might be other rust species, it is no doubt that the wheat stripe rust pathogen can infect a great large number of grass species. Grass species have different strains or ecotypes that may vary in reaction to stripe rust infection. For example, Dugan et al. (2014) reported lines of basin wild rye (*Leymus cinereus*) either resistant or susceptible to wheat stripe rust. It is almost impossible to include all strains or ecotypes of a grass species in an artificial inoculation study for reactions to stripe rust. Also, the studies of host range for stripe rust have been primarily concentrated in Europe, North America, Israel, China, and Australia. Studies in much of other stripe rust occurring regions in the world have been scarce. More species could be identified as auxiliary hosts of the stripe rust fungus of the cereal species or different stripe rust species.

Not all reported species are equally susceptible to stripe rust. Many of them should be considered opportunistic hosts as they are not sporulating hosts under normal conditions, but can be infected by stripe rust fungus and producing uredinia when weather conditions are extremely favorable for the disease and the inoculum level is extremely high. Because of this reason, stripe rust is more commonly observed on grasses when the disease is severe on cereal crops. So far, no report of stripe rust has been reported on common oat (*Avena sativa*) or wild *Avena* spp. However, the author observed infection of stripe rust on volunteer oat plants near an experimental wheat fields at Pullman, Washington, the USA in 2005 when the weather conditions were extremely favorable for the disease. Even though the reaction of large necrotic stripes with only few uredinia is considered resistant, recovered urediniospores were identified as a typical stripe rust race of *P. striiformis* f. sp. *tritici* in the experimental wheat field. Similarly, rice (*Oryzea sativa*) is a non-host for stripe rust, and no macroscopic symptoms can be observed and cell death of small host tissue can be observed microscopically after inoculated with the stripe rust fungus (Kang and associates, unpublished data). In some rice mutants, the stripe rust fungus is able to form relatively large mat of infectious hyphae and haustoria. On the other hand, some grasses can be even more susceptible than susceptible wheat cultivars. The author observed that stripe rust develops earlier and faster on goat grass plants (*Aegilops cylindrica*) than susceptible wheat lines. To let rust develop fast in experimental fields, goat grass plants are often left from weeding. For example, Fig. 1.2 shows severe stripe rust on goat grass plants in a wheat field. Between host and non-host, cereal and grass species have a relatively continuous variation in their ability to interact with the stripe rust fungus. The variability within the stripe rust fungal species at the cultivar level of a host species (such as wheat or barley), reflected by different races or pathotypes, will be discussed in Sect. 2.1.4 of Chap. 2 on pathogen variability. Here, variability within a fungal species at the host genera level, reflected by formae speciales, will be discussed below.

The concept of formae speciales was first proposed by Eriksson (1894) according to specificity on host genera or species. Based on cross-inoculation tests, he

Fig. 1.2 Stripe rust on goat grass plants (*Aegilops cylindrica*) in a field of wheat without rust

reported five formae speciales within the species of *P. glumarum* (now *P. striiformis*), namely *P. striiformis* f. sp. *tritici* on wheat, f. sp. *hordei* on barley, f. sp. *secales* on rye, f. sp. *elymi* on *Elymus* spp. and f. sp. *agropyron* on *Aropyron* spp. (Eriksson 1894). Latter, four more formae speciales were reported: *P. striiformis* f. sp. *dactylidis* on orchard grass (*Dactylis glomerata*) (Manners 1960; Tollenaar 1967; Zadoks 1961); *P. striiformis* f. sp. *poae* on Kentucky bluegrass (*Poa pratensis*) (Britton and Cummins 1956; Murdoch et al. 1973; Tollenaar 1967; Tollenaar and Houston 1967); and *P. striiformis* f. sp. *leymi* on *Leymus secalinus* (Niu et al. 1991). More recently, the stripe rust fungus infecting barley grass *Hordeum* spp. in Australia was named as *P. striiformis* f. sp. *pseudo-hordei* (Wellings et al. 2000; Wellings 2011). Thus, a total of nine formae speciales of *P. striiformis* have been reported (Table 1.1). Worldwide, most important genera are *Aegilops*, *Agropyron*, *Bromus*, *Elymus*, *Hordeum*, *Leymus* and *Triticum*, and in China, *Roegneria* appears to be an important genus for wheat stripe rust.

The separation of formae speciales based on host specificity is a plant pathological concept. Because they are not taxonomic taxa, formae speciales are not governed by the International Code of Botanic Nomenclature (McNeill et al. 2006). Some of the above mentioned formae speciales have morphological and/or physiological differences. Because of smaller urediniospores than the stripe rust fungus on wheat, the stripe rust fungus on orchard grass was considered by Manners (1960) as a variety, a mycological taxon below species. The stripe rust fungus on bluegrass

Table 1.1 Formae speciales reported for *Puccinia striiformis*

P. striiformis f. sp.	Hosts	References
tritici	Primarily wheat; some barley and rye varieties; and various grass species	Eriksson (1894)
hordei	Primarily barley; some wheat varieties; and various grass species	Eriksson (1894)
secales	Primarily rye	Eriksson (1894)
elymi	*Elymus* spp.	Eriksson (1894)
agropyron	*Agropyron* spp.	Eriksson (1894)
dactylidis[a]	*Dactylis glomerata*	Manners (1960)
poae[a]	*Poa* spp.	Tollenaar (1967)
leymi	*Leymus secalinus*	Niu et al. (1991)
pseudo-hordei	Primarily wild barley (*Hordeum* spp.) and some barley varieties	Wellings (2011)

[a]Abbasi et al. (2004) named the stripe rust fungus on orchard grass (*D. glomerata*) as species *P. striiformoides* and that on bluegrass (*P. pratensis*) as *P. pseudostriiformis*. The naming of species was supported by some morphological characteristics and partial sequence data of the ITS and beta-tubulin genes

was considered as a forma specialis of *P. striiformis* because no obvious morphological differences were found (Tollenaar 1967). Chen et al. (1995) was the first to use both virulence and molecular markers to separate some of the formae speciales. By cross-inoculation, they find that isolates of bluegrass stripe rust only infected bluegrass isolates, but not any of the tested wheat, barley, and rye accessions, while wheat stripe rust isolates infected mostly wheat accessions, but some barley accessions and similarly barley stripe rust isolates infected mostly barley accession and few wheat accessions. Random amplified polymorphism DNA (RAPD) markers clearly separate isolates of wheat, barley and bluegrass into three groups with wheat stripe rust and barley stripe rust more closely related to each other than to bluegrass stripe rust. Compared to other rust species, *P. graminis*, *P. triticina* and *P. hordei*, the three formae speciales of the stripe rust fungus were more closely related. More recently, as mentioned in Sect. 1.3 above, the stripe rust fungi on bluegrass and orchard grass were recently promoted to different species within Series *P. Striiformis* based on combination of several morphological characteristics and partial sequences of ITS and beta-tubulin. Using SSR markers, Cheng et al. (2014) also found that the stripe rust fungi on bluegrass and orchard grass were much more different from the wheat or barley stripe rust pathogens. Such separations of either formae speciales or species are supported by molecular, morphological and pathogenic data.

However, separations of other *P. striiformis* formae speciales are either still remaining controversial and/or lack of studies. Overlapping host range has been the major reason for the debate whether some of the above mentioned formae speciales may not be consider as separate formae speciales from others, especially between *P. striiformis* f. sp. *tritici* and *P. striiformis* f. sp. *hordei* (Sydow and Sydow 1904; Straib 1935; Newton and Johnson 1936; Fang 1944). In contrast, Zadoks (1961) and

Stubbs (1985) considered the wheat and barley stripe rust fungi as two different formae speciales based on greenhouse and field data. The separation of *P. striiformis* f. sp. *tritici* and f. sp. *hordei* was supported by isozyme variation and RAPD markers (Chen et al. 1995). Simple sequence repeat (SSR) markers also separate isolates of *P. striiformis* f. sp. *tritici* and *P. striiformis* f. sp. *hordei* as two distinct groups (Cheng and Chen 2014). In the US, barley stripe rust (*P. striiformis* f. sp. *hordei*) was introduced in 1991 (Roelfs et al. 1992; Chen et al. 1995). Before the introduction, wheat stripe rust (*P. striiformis* f. sp. *tritici*) had been occasionally found on barley plants but never caused significant damage, but after 1991, barley stripe rust have caused significant yield losses in the western US (Chen et al. 1995; Chen 2004). In the recent years, epidemics caused severe damage on wheat crops when weather conditions were favorable to stripe rust, but in contrast stripe rust on barley were not significant although major barley cultivars were more susceptible to barley stripe rust than major wheat cultivars to wheat stripe rust, and fields of barley and wheat were next to each other. The differences are due to different levels of pathogen inoculum resulted from different cropping systems, regions and acreages in the US Pacific Northwest, as both winter and spring wheat crops are gown in continuous large acreages and mainly just spring barley is grown in much smaller acreages (Chen 2005; Wan and Chen 2012). The field situations clearly show that stripe rust of wheat and stripe rust of barley are two distinct diseases. The separation of the diseases and distinction of the pathogens as different formae speciales are useful for the management of the diseases and also for setting different systems to identify races within each of the formae speciales.

The separation of other formae speciales, namely *P. striiformis* f. sp. *secales*, f. sp. *elymi*, f. sp. *agropyron* and f. sp. *leymi* does not have adequate data. As Stubbs (1985) pointed out, for rye stripe rust, there are no data to support the use of f. sp. *secalis* because rye can be infected by both the wheat and barley form. Worldwide, there have been no many studies or reports of stripe rust of rye since the 1970s, which may correlated with the substantial decrease in rye growing areas. In 1986, 24 million hectares of rye were harvested; but in 1996, only 17 million hectares harvest (Bushuk 2001). About 80% of the rye production is concentrated in few European countries, such as Germany, Poland and Russia. In the US, rye acreage has continually decreased, from 1,154,251 ha in 1941 to 255,466 ha in 1971 and 97,976 ha in 2011 (http://www.nass.usda.gov/Data_and_Statistics/). Rye is mainly grown in the eastern US, and no much rye is grown in the stripe rust more frequently occurring western US. Wildly grown rye plants are widespread in some southern Washington and northern Oregon along roads and hill foots, the author has never seen stripe rust on such rye plants even in the years like 2003, 2005, 2010 and 2011 when stripe rust was severe in wheat fields and the predominant races were all virulent to *Yr9*, a stripe rust resistance gene from rye. In 2004, two stripe rust samples collected from rye plants grown in a germplasm nursery near Corvallis, Oregon was identified as race PST-21, the least virulent race of *P. striiformis* f. sp. *tritici* based on the tests with the wheat cultivar differentials (Chen et al. 2010). In 2010, two samples from rye plants in an experimental field in California were identified as PSTv-36, a typical race identified in wheat fields (Wan and Chen 2014). Samples from Triticale varieties were identified as *P. striiformis* f. sp. *tritici* races predominant in wheat fields. These observations indicate that stripe rust on rye is either of the same *tritici* forma specialis or the true secales forma specialis reported by Eriksson (1894) and Eriksson and Henning (1896) in Europe does not exist in the US. Hovmøller et al. (2011) reported a new *P. striiformis* race, which became widespread in Denmark and Sweden in 2008 and 2009 and was frequently sampled from cultivated triticale (x Triticosecale). In addition to triticale, the race could also infect several cultivars of spring wheat, barley and rye. The crossing genera host range is common. In the US, stripe rust samples collected from grass species, mainly of genera *Aegilops*, *Agropyron*, *Hordeum*, *Bromus*, *Elymus* and *Leymus*, were mostly identified as *P. striiformis* f. sp. *tritici* and sometimes as *P. striiformis* f. sp. *hordei* races, mostly predominant in wheat or barley fields and sometimes as least virulent races unable to infect wheat cultivars grown in fields. Like barley can be mainly infected by *P. striiformis* f. sp. *hordei* and occasionally infected by *P. striiformis* f. sp. *tritici*, it is possible that some of the grass species may have their mostly adapted formae speciales and also can be sometime infected by less adapted formae speciales. Not all formae speciales exist in the regions where grasses are either native or introduced.

Stripe rust on grasses can contribute to the disease on cereal crops in several ways. (1) Urediniospores from grasses may initiate epidemics on cereal crops. As speculated by Tollenaar and Houston (1967), stripe rust on the high elevations of the Sierra Nevada may initiate the disease on wheat in the fall in the valleys in California. This speculation is challenged by the fact that stripe rust can survive summer in the Sacramento Valley of California on wheat plants grown as cover crops in irrigated orchards (Line 1976). Although it may not be always the case for stripe rust from grasses to initiate the disease on cereal crops, this is possible in some years or some regions in the world. As discussed above, super susceptible grasses, like *Ae. cylindrical*, may provide spores to infect wheat plants in the same field or adjacent fields. (2) Urediniospores from grasses add to the inoculum pressure. When races are able to infect cereal cultivars grown in the region, spores produced on grasses during the disease season definitely increase inoculum load to cereal fields, speeding up the disease development and causing more damage. (3) Grasses may serve as green bridges for the pathogen survival. Susceptible grasses add more possibilities for the stripe rust fungus to survive the hot or cold periods. The role of providing host tissue for the pathogen to survive varies among grass species. As discussed above, when inoculated with wheat stripe rust, some grass species (such as *Elymus glaucus* and *Bromus marginatus*) showed continuous sporulation throughout the summer (Tu 1967). some grasses (such as *Agropyron trichophorum* and *Bromus carinatus*) had sporulation in early and late summer but with a dormant period near mid-summer and some other grass species (such as *Alopecurus arundinaceus* and *Hesperochloa kingie*) had sporulation in early summer but the fungus succumbed with the advent of hot weather (Tu 1967). (4) Grasses provide reservoirs for the stripe rust fungus to maintain diversity and keep mutants. In regions where stripe rust of cereal crops is a production problem, resistant cultivars are grown. Even cultivars that are susceptible to races occurring on wheat crops often have resistance against previous races or new mutants unable to reproduce on wheat cultivars. These races or variants may be able to grown on grasses. As discussed earlier, least virulent races are often collected from grasses. The diverse populations on grasses that may provide new virulent races able to attack cereal cultivars may come from grasses. As previous studies with stripe rust on grasses only deal with very few represent isolates, stripe rust populations on grasses have not been adequately studied. (5) Grasses provide common hosts for somatic hybridization between different formae speciales or isolates. In a recent study using virulence tests, Cheng et al. (2014) found that isolates from grasses, especially wild barley grasses, were able to infect both wheat and barley cultivars. Using SSR markers, these grass isolates were found to be hybrids between wheat and barley isolates.

From the evolutionary standpoint, long before cereals came into existence, stripe rust may have existed on grasses and the rust has become adapted on cereals since crops are grown. Understanding the rust evolution on grasses will help us to understand the mechanisms of the pathogen variation. Based on research results so far for the stripe rust pathogen, it is no doubt that grasses as auxiliary hosts play more roles to the diseases of stripe rust on wheat and barley and the variations of the pathogens than the alternate hosts of *Berberis* spp. and *Mahonia* spp. As grasses are a rich source of resistance, understanding the various interactions among the stripe rust pathogen and its primary and auxiliary hosts will lead to more efficient exploration and utilization of resistance genes from grasses into cereal crops.

1.5 Distribution and Impact of the Disease

1.5.1 Wheat Stripe Rust

Through checking the literature before 2005, this author found that wheat stripe rust was reported in more than 60 countries (Chen 2005). By conducting an international survey, Wellings (2011) compiled a list and map showing that stripe rust occurred also in more than 60 countries in 2000–2009. Although stripe rust varies in frequency of occurrence and significance of damage, almost every country with significant production of wheat has stripe rust as a problem. Based on the survey, wheat stripe rust epidemics can be widespread in all wheat producing areas, very frequent (in 2 or more out of 5 years), and cause 5–10% yield losses in Australia, New Zealand, China, India, Nepal, Pakistan, Uzbekistan, Yemen, Ethiopia, Kenya, the United Kindom, Chile, Peru, Ecuador, Colombia, Mexico and the US (Fig. 1.3).

Because stripe rust can infect wheat plants starting from one-leaf stage throughout the plant growth season, the disease is able to cause 100% yield loss on highly susceptible wheat cultivars. More than 90% yield loss was evident on a susceptible check in an experimental field near Pullman, Washington, the USA in 2011 (Chen 2014). Worldwide, many devastating wheat stripe rust epidemics have been recorded in the literature.

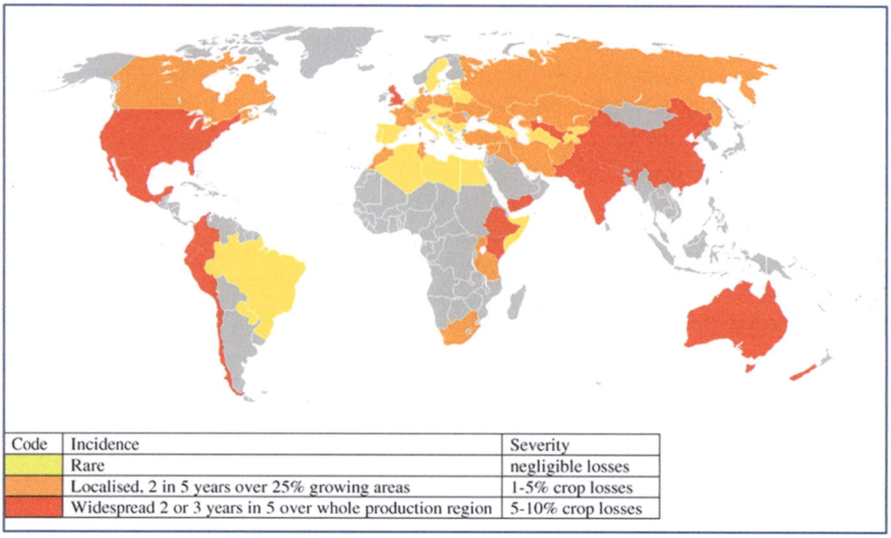

Curtesy of Colin R. Wellings

Fig. 1.3 Wheat stripe rust (*Puccinia striiformis* f. sp. *tritici*) in wheat producing nations for the period 2000–2009

In East Asia, stripe rust has been reported in China, Korea and Japan (Stubbs 1985), but there are not many studies on stripe rust in Korea and Japan. In 2011, the author heard from a colleague of Korea origin that significant wheat rust occurred in South Korea, but it is not clear if the rust was stripe rust. Stripe rust is one of the most important diseases, if not the most important, of wheat in China. The disease is a potential problem throughout wheat growing regions of the country except the far northeast spring wheat growing region. The disease is more frequent and severe in the northwestern and southwestern regions, and often damages the major wheat growing regions along the Yellow and Huai river regions. The 1939–1940 epidemics of stripe rust in Fujian and Sichuan provinces caused 10–15% and 60% of wheat yield losses, respectively (Li and Zeng 2002). In central Shaanxi, severe wheat stripe rust epidemics occurred in 1942, 1946, 1948 and 1949. In the 1950s and 60s, countrywide epidemics occurred in 1950, 1956, 1958, 1960, 1962 and 1964, affecting 222,000–888,667 ha. The extremely widespread epidemics in 1950 and 1964 caused wheat yield losses of 6.0–3.2 million tons, respectively (Li and Zeng 2000). From 1972 to 1983, moderate and severe stripe rust epidemics occurred in 9 out of the 12 years. The 1975 epidemic spread from the northwest and southwest provinces to the east coast, affecting 285,000 ha and causing yield loss of 1.0 million tons. The epidemic in 1990 affected 652,330 ha and caused yield loss of 1.83 million tons (Li and Zeng 2002). In 2002, stripe rust occurred on 6.6 million ha in 11 provinces, and the epidemic caused 1.31 million tons of yield loss and fungicide application on 6.2 million ha (Wan et al. 2004).

In South Asia, stripe rust of wheat is important in the northern areas of India, Pakistan, and Nepal, in southeast Afghanistan, and at the higher elevations in the south (Saari and Prescott 1985). Severe epidemics have occurred about once in every 10 years. Losses of 100% can occur in fields of susceptible cultivars. In the Indian subcontinent, stripe rust oversummers throughout the Himalaya, Hindukush and Sulaiman mountain ranges and in the northwestern Frontier Province, but the largest amounts are in the mountains and valleys of Indus and its tributaries (Nagarajan and Joshi 1985). The disease occurs along the foothills of Himalaya in Jammu and Kashmir, Punjab, Haryana, Himachal Pradesh, and western Uttar Pradesh in early January. In India, primary infections occur near Jammu, Gurdaspur, Pathankot, Ropar, and Jagadhari, where the Tawi, Ravi, Beas, Satluj, and Jamuna rivers enter the plains, respectively. Urediniospores are spread by katabatic wind to the foothills. From such secondary foci, disease spreads gradually during February into the adjoining plains. In Nepal, wheat stripe rust is a dominant disease in the hilly areas, sometimes with losses of 100% (Karki 1980). In Pakistan, wheat is grown in about 8 million ha and stripe rust can attack 70% of the wheat acreage (Anonymous 2006; Aquino et al. 2002). Stripe rust is important in the foothills in the north and also occurs in the central regions and western upland areas (Hassan 1968). The disease is a more serious problem in the Northwest of Pakistan (Chatrath et al. 2007). Since 1948, 13 epidemics have occurred, of which the severe epidemics in 1978, 1997, 1998 and 2005 resulted in losses of 244, 33 and 100 million US dollars, respectively (Hafiz 1986; Duveiller et al. 2007). In Afghanistan, stripe rust is the most important rust disease in wheat, with losses up to 90% in epidemic years (Ghaffor 1970). In Southeast Asia, stripe rust occurs in Bangladesh and Burma, but has not been reported in Thailand, Cambodia, Laos, Vietnam, Sri Lanka, Singapore, Malaysia, Indonesia and the Philippines.

In Central and West Asia, stripe rust of wheat is the most important rust disease especially in Tajikistan, Kyrgyzstan, Kazakhstan, Uzbekistan, Azerbaijan, Iran, Iraq, Syria and Turkey. In Lebanon, Jordan, Israel, Saudi Arabia, Yemen Arab Republic, and Yemen Peoples Democratic Republic, stripe rust regularly appears on wheat and/or barley and can become serious (Stubbs 1985). The most recent epidemic occurred in 2010 throughout Central and West Asia. The 2010 epidemic caused wheat yield loss between 20 and 70% in various countries (http://www.icarda.org/striperust2014/wp-content/uploads/2014/01/Strategies_to_reduce.pdf). The epidemic was so devastating on major wheat cultivars that Syria made an unusual request of help from the US for stripe rust resistant wheat germplasm in that year. In Turkey, stripe rust caused wheat yield losses of 26.5% in 1992, 1.2 million tons (568 million $US) in 1996 and 10 million $US in 2010 (http://www.slideshare.net/bgri/2013-bgrisession31solh). In Iran, stripe rust is important in all wheat-growing regions, particularly along the Caspian Sea where epidemics occur once every 3–4 years (Niemann et al. 1968; Khazra and Bamdadian 1974). Wheat yield loss caused by stripe rust was estimated as 2.5 million tons ($258 million $US) in 1992–1994; in 2007 and 2009, 2 million ha were infected; and in 2010, 650,000 ha were sprayed with fungicides to control stripe rust (http://www.slideshare.net/bgri/2013-bgrisession31solh). In Iraq, stripe rust is a serious wheat disease in the mountainous area in Kurdistan (Stubbs 1985). In Uzbekistan, stripe rust epidemics caused 35% yield loss (worthy 135 million $US) in 1998 and 30% yield loss in 2010.

In Europe, stripe rust epidemics have been reported in varying frequencies in all wheat-growing countries (Hassebrauk 1965; Stubbs 1985; Zadoks and Bouwman 1985). A well-known subarea with a high risk of epidemics comprises the countries of northwest Europe, namely, England, the Netherlands, Belgium, northern France, and northern Germany (Stubbs 1985). Priestley and Bayles (1988) estimated losses in susceptible winter wheats due to stripe rust and leaf rust as £83 million in the United Kingdom in the 1980s. In Italy, stripe rust does not develop to a damage level, but widespread and severe stripe rust occurred on wheat throughout the wheat growing areas in the country in 2014 (M. Maccaferri, personal communication). In eastern Europe, stripe rust sometimes causes significant damage in Poland, Czech Republic, Hungary, Ukraine and Russia. In Russia, stripe rust is one of the major diseases of wheat in the North Caucasus region and its importance has increased in the recent years (S. Sanin, personal communication).

In Africa, stripe rust is a major disease in Ethiopia and Kenya in the east, Morocco and Tunisia in the north, and South Africa in the south. Epidemics sometimes occur and cause damage in Egypt (Mohamed 1963; Abdel-Hak et al. 1972). The disease occasionally occurs in Uganda, Tanzania, Zambia, Malawi and Madagascar (Stubbs 1985). In East Africa, stripe rust infecting wheat and barley in the highlands is a serious problem, particularly in Ethiopia (Wodageneh 1974; Ciccarone 1947) and in Kenya (Martens and Oggema 1972). In Ethiopia, a stripe rust epidemic in wheat occurred in 1983 (Stubbs 1985), and the epidemic in 2010 caused 67–100% yield loss on commercial wheat cultivars (ICARDA 2011). In South Africa, stripe rust of wheat was first reported in 1996 (Pretorius et al. 1997). Since the initial detection in Western Cape, the disease has spread to all wheat growing regions in the country. The 1996 widespread epidemic provided the foothold for stripe rust in South Africa, resulting in epidemics in the central and western Free State in 1997, the eastern Free State in 1998 and all summer rainfall wheat growing areas in 2002. Based on farm and experimental data, stripe rust can reduce yield by more than 50% and hectolitre mass by 14%. The 1998 epidemic on ca. 42,000 ha winter wheat in the eastern Free State resulted in losses of ZAR 12 million (Pretorius 2004).

In Australiasia, stripe rust of wheat was first detected in Victoria, Australia in 1979 and spread to New Zealand in 1980 ((O'Brien et al. 1980; Beresford 1982). Stripe rust is able to cause yield loss of up to 84% in susceptible cultivars (Murray et al. 1995). Severe epidemics occurred in Australia between 1983 and 1984. In 1983, $A8 million was used for chemical application to control the stripe rust epidemic (Wellings and Luig 1984). The disease was restricted in eastern Australia until 2002 when it was first detected and developed quickly in western Australia. During the severe 2003 epidemic in eastern Australia, fungicide control cost in 2003 was estimated in excess of $A40 million (Wellings and Kandel 2004). Brennan and Murray (1988) estimated the potential national cost in susceptible cultivars as $A168 million annually, and control strategies primarily with resistant cultivars at a value of $A139 million. In New Zealand, yield losses in susceptible cultivars are as high as 60% (Beresford 1982).

In South America, stripe rust was first discovered by Holway on *H. chilense* in Chile in 1919 and 1 year later on *A. altenuatum* in Ecuador (Arthur 1925). The disease on wheat was first observed in Argentina in 1929 (Rudorf and Job 1931). In 1930, stripe rust was also observed on barley for the first time (Hirschhorn 1933), but the stripe rust infections on barley as mentioned in literature may have been caused by the form attacking wheat until 1975, when the form specifically infecting barley was found in Colombia (Stubbs 1985). In the Andean zone and in central Chile, stripe rust is a major disease on wheat. In Argentina, Uruguay, and, to a much lesser extent, in Brazil and Paraguay, stripe rust appears regularly in wheat but seldom reaches epidemic levels (Grillo 1937; Marchionatto 1931; Stubbs et al. 1974).

In North America, stripe rust is a major disease of wheat in Mexico and the US, and a less frequent problem in Canada. In Mexico, wheat is grown annually on about 586,000 ha (Huerta-Espino and Singh 2000), mainly under irrigation in the northwest state of Sonora and in the El Bajio region, both lowland areas (http://www.siap.sagarpa.gob.mx). Yield losses of 60% have been recorded in common wheat crops in those areas. In the US, stripe rust was found in states west of the Rocky Mountains, often referred as western US when discuss stripe rust distributions (Line 2002; Chen 2005; Chen et al. 2010; Wan and Chen 2014), including California, Arizona, western New Mexico, western Colorado, Utah, Nevada, western Wyoming, western Montana, Idaho, Oregon and Washington, before the early 1930s. Because stripe rust was not observed in states east of the Rocky Mountains (referred as the eastern US), for more than 25 years, the eastern limit of stripe rust was considered to be the Black Hills of South Dakota in the US, and similarly, Whitewood of Saskatchewan was considered as the eastern limit for stripe rust distribution in Canada (Line 2002). In 1941, McFadden found a small amount of stripe rust in a wheat nursery at College Station, Texas (Humphrey 1941), which was the first time when stripe rust was found in the eastern US. Stripe rust was found as a hot spot at Denton, in 1953 and as a large hot spot at Prairie View in Texas in 1956 (Futrell and Atkins 1954, 1957). Stripe rust became widespread in the Great Plains in 1957. It spread from south and central Texas to Oklahoma, Kansas, Nebraska, South Dakota and Wyoming, but no reports of stripe rust in the spring-wheat region of the northern Great Plains (Daly et al. 1957; Young and Browder 1957; Pady et al. 1957; Pady and Johnston 1958). Stripe rust occurred in the Great Plains again in 1958, spreading from Texas to Oklahoma, Kansas, Nebraska, Wyoming, Minnesota, South Dakota and North Dakota, close to the Canadian border (Bridgmon and Kolp 1958; Davison 1958; Futrell et al. 1959; Gough et al. 1959; Hennen and Komanetsky 1959; Miller and Christensen 1959; Pady and Johnston 1959a, b; Rosen et al. 1959). The epidemic of stripe rust in the Great Plains was attributed to the increase of wheat production and use of new cultivar (Lerma Rojo) in the states of Coahuila, Nuevo Leon and Tamaulipas in Mexico, where stripe rust had occurred in the high elevations for many years (Rodriguez 1946; LeBeau et al. 1956). Although the distribution of stripe rust was much expanded in the 1950s, the disease was not found in eastern states of the US. During the 1960s and 1970s, stripe rust was not a problem in the wheat-growing regions east of the Rocky Mountains, and in the 1980s and 1990s the disease occasionally occurred in the south-central states

(Texas, Louisiana and Arkansas) causing yield losses (Line and Qayoum 1992; Line 2002). However, in 2000, stripe rust occurred in 16 states east of the Rocky Mountains, including Alabama, Arkansas, Colorado, Georgia, Indiana, Kansas, Louisiana, Michigan, Minnesota, Mississippi, North Dakota, Oklahoma, South Dakota, Texas, and Virginia; and caused 7% yield loss in Arkansas (Chen et al. 2002). In 2003, stripe rust occurred much severer in the eastern US, and the disease was first found in Florida and samples were also received from Ontario, Canada (Blount et al. 2005; Chen 2005). In 2005 and 2010, stripe rust was wide-spread, basically covering all wheat growing states in the continental US (Chen 2007; Wan and Chen 2014). In 2005, a leaf sample showing few stripe of necrotic reaction was obtained, but in 2010, stripe rust developed to significant levels in wheat variety trial nurseries in the state of New York. So far, wheat stripe rust has occurred in almost every state, except Alaska, Hawaii, Maine and New Hampshire. However, bluegrass stripe rust was reported in Hawaii (Nurdoch et al. 1973).

In the US, cereal yield losses caused by rusts have been estimated since 1918 (http://www.ars.usda.gov/main/docs.htm?docid=10123). Although stripe rust started to be reported since 1915, estimation of yield losses had not been done until 1958. The lack of yield loss estimates is due to several reasons, including the lack of yield loss studies from the middle 1910s to 1930s, the lack of stripe rust research from the 1930s to the late 1950s, and relatively insignificance of stripe rust due to resistance in diverse cultivars before the late 1950s (Line 2002). The increase of stripe rust in the late 1950s and especially the early 1960s revived stripe rust research. The first yield loss of wheat due to stripe rust was reported in 1958, estimated as 7967 tons, about 4.0% of the total yield in the State of Washington. In 1959, the yield loss due to stripe rust, also just in Washington, was estimated to be 232,466 tons at 10% of the total yield in the state. In 1961, an extreme severe epidemic of wheat stripe rust occurred throughout the entire PNW (Line 2002). Yield loss in the state of Washington was estimated by Dr. J. Walter Hendrix as 7.5 million bushels (204,300 tons, $15 million) in 1960 and 15 million bushels (408,600 tons, $30,000,000) in 1961 (Line 2002). The 1961 epidemic also caused growers in the state of Oregon $15,000,000 (Shaner and Powelson 1971). Dr. George W. Bruehl believed that Hendrix was conservative in his estimates of losses (http://plantpath.wsu.edu/wp-content/uploads/2012/10/History.pdf). According to Bruehl, stripe rust was more as a curiosity than as a real problem in Washington in the middle of 1950s. The release of Omar, a winter club wheat variety of superb quality, smut resistance, and high yield changed everything. By 1960 Omar occupied 67.5% of the wheat acreage of the entire US Pacific Northwest. It was highly susceptible to stripe rust. He witnessed rust spores blew from the earlier areas south and west of the Palouse (around Pullman, Washington and Moscow, Idaho), in a sense a repetition of the smut spore shower phenomenon described by early smut pathologists. Since then, severe stripe rust epidemics has been recorded in the US, but mostly in the Pacific Northwest and California, until 2000, when stripe rust started to cause national scale epidemics (Chen et al. 2002, 2010; Chen 2005, 2007). Because of the huge wheat acreage, yield losses are much bigger whenever stripe rust is severe in the Great Plains. Based on the yield loss data compiled by the USDA-ARS Cereal

Disease Laboratory (http://www.ars.usda.gov/main/docs.htm?docid=10123), from 1958 to 2016, stripe rust caused yield losses of 1,165,823,500 bushels (31,757,032 tons), of which 362,220,900 bushels (9,866,897 tons) were lost from 1958 to 1999 and 803,602,600 bushels (21,890,135 tons) from 2000 to 2016. Table 1.2 lists the severe epidemics caused at least 1.0% yield loss in the nation or at least 5% yield loss in a single state on either winter or spring wheat crop from 1958 to 2016.

1.5.2 Barley Stripe Rust

Barley can be infected by both *P. striiformis* f. sp. *tritici* and *P. striiformis* f. sp. *hordei*, but epidemics are primarily caused by the latter form. Because of this, the distribution and impact of barley stripe rust discussed here are mostly caused by *P. striiformis* f. sp. *hordei*. Barley stripe rust occurs in Africa, Asia, and Europe for a long time. In these continents, barley stripe rust is generally important in countries where wheat stripe rust is important. The barley form had not been reported in the new hemisphere (South and North America) before 1976, and so far has not been reported in Australasia. Compared to wheat stripe rust, stripe rust causes relatively small scale damages, although yield losses can be high in localized regions, due to relatively small barley producing areas, primarily spring barley crops, and relatively short growth season throughout the world. According to the World Food and Agriculture Organization (FAO, http://www.fao.org), wheat is grown in 210–240 million ha, whereas barley is grown only about 56–80 million ha in the world annually over the last several decades. Compared to wheat, barley is grown in similar climates, but usually in the more marginal production environments where limited moisture and poorer soils prevail (Saari and Prescott 1985), which reduces the vulnerability of barley to stripe rust. Therefore, barley stripe rust appears not as important as wheat stripe rust, and as a consequence barley stripe rust has not received as intensive studies as wheat stripe rust.

In China, barley stripe rust is most damaging in the barley growing areas in the southwest such as Yunnan province and the Tibet Autonomous region (Wang 1992; Wang et al. 1989a, b). Although yield loss data are not available, some barley cultivars are highly susceptible in fields for stripe rust in Tibet (Wang 1992; Wang et al. 2002a; T. Wang, personal communication). In a trip to Yunnan in 2011, Dr. Timothy Murray at Washington State University observed very severe stripe rust in barley fields in Xianggelila, Yunnan Province. In contrast, when tested with wheat stripe rust, 101 out of 104 Chinese barley cultivars and lines were highly resistant to all 30 tested races of *P. striiformis* f. sp. *tritici* (Sui et al. 2010). In Japan, stripe rust was known in the 1890s and occurred from Kyushu in the south to Hokkaido in the north (Ito 1909; Hemmi 1934). Epidemics were recorded in 1950–1956 on both wheat and barley (Kajiwara 1964; Kajiwara et al. 1964). Since then, there have been no available reports of stripe rust on either barley or wheat in Japan. Barley stripe rust occurred in India, Burma, Bangladesh, Nepal, Pakistan, Iran and Saudi Arabia (Stubbs 1985).

Table 1.2 Severe epidemics of wheat stripe rust in the United States from 1958 to 2016

	National yield loss		% yield loss in major states[a]
Year	Tons	%	State (%)
1959	232,469	0.84	WA (10.0)
1960	591,127	1.71	WA (25.0)
1961	645,261	1.65	WA (30.0)
1963	120,992	0.43	OR (5.0), WA (4.0)
1967	417,235	1.08	WA (8.0), OR (8.0), MT (1.8)
1974	847,679	1.78	OR (19.0), CA (8.0), MT(2.3), WA(2.0)
1976	1,296,531	2.17	Winter wheat: WA (17.0), ID (10.0), OR (10.0);
			Spring wheat: WA (10.0), ID (10.0), OR (5.0)
1977	270,744	0.49	Winter wheat: OR(5.0), WA (1.0)
1978	349,260	0.71	Winter wheat: WA (5.0), OR (5.0);
			Spring wheat: WA (10.0), OR (10.0), ID (4.0)
1980	825,121	1.26	Winter wheat: WA (5.0), ID (5.0), OR (3.0);
			Spring wheat: OR (20), ID (20), WA (15)
1981	1,008,340	1.31	Winter wheat: WA (10), OR (7.5), ID (5.0), MT (1.0);
			Spring wheat: OR (15), WA (10), ID (10), MT (1.0)
1983	813,002	1.22	Winter wheat: WA (10.0), OR (5.0), ID (2.0), MT(1.0), CA (1.0);
			Spring wheat: WA (8.0), OR (2.0), ID (2.0), MT (1.0)
1984	399,488	0.56	Winter wheat: WA (6.0), OR (3.5), ID (1.0), CA (1.0);
			Spring wheat: WA (1.0), OR (1.0), ID (1.0), MT (1.0)
1987	253,999	0.44	Winter wheat: WA (5.0), LA (5.0), AR (2.5), MS (2.0), ID (1.0), OR (1.0), TN (1.0);
			Spring: WA (2.0)
1990	296,347	0.40	Winter wheat: WA (5.0), ID (2.0), OR (2.0);
			Spring wheat: WA (5.0), OR (2.5)
1999	85,822	0.14	Winter wheat: CA (7.0)
2000	247,026	0.40	Winter wheat: AR (7.0), CA (3.0), WA (1.0), OK (1);
			Spring wheat: OR (1.5); WA (1.0)
2001	1,082,874	1.99	Winter wheat: CO (8.0), KA (7.3), OK (3.0), CA (2.0), TX (1.5), WA (1.0), SD (1.0);
			Spring wheat: WA (2.0), OR (0.2)
2002	217,800	0.49	Winter wheat: CA (6.0), AR (5.0), LA (3.0), WA (1.0), TX (1.0);
			Spring wheat: WA (6.0), OR(1.0)
2003	2,421,121	3.66	Winter wheat: CA (25), KS (10.6), NE (10), LA (5), AR (3), TX (3), MS (3), MO (2.5), SD (2.0), WA(1.5), OK (1.0), CO (1);
			Spring wheat: WA (3.5); Durum wheat: CA(10.0)
2004	333,132	0.56	Winter wheat: CA (5.0), LA (4.5), OR (3.0), UT (2.2), AR(2.0), MS(2.0), WA (1.5), MT (1.0);
			Spring wheat: OR (5.0), WA (3.0); Durum wheat: CA (5.0)

(continued)

Table 1.2 (continued)

	National yield loss		% yield loss in major states[a]
Year	Tons	%	State (%)
2005	2,006,033	3.38	Winter wheat: TX (15.2), KS (8.0), AR (5.0), CA (5.0), LA (5.0), OK (5.0), UT (5.0), NE (4.0), OR (4.0), ID (3.0), MO (2.0), WA (2.0), CO (1.0), Georgia (1.0), MS (1.0);
			Spring wheat: OR (8.0), UT (5.0), WA (4.0), ID (2.0);
			Durum wheat: CA (5.0)
2006	119,396	0.24	Winter wheat: CA (15.0), LA (2.0), AR (1.0);
			Durum wheat: CA (3.0)
2010	2,608,720	4.16	Winter wheat: KS (10.0), TX (10.0), OR (10.0), NE (7.0), OK (5.0), LA (4.0), MT (4.0), CA (3.0), WA (3.0), ND (3.0), CO (2.0), AR(1.0), KY (1.0), MN (1.0), MO (1.0), SD (1.0), WI (1.0);
			Spring wheat: OR (10.0), MT (5.0), ID (3.0), WA (3.0)
2011	960,264	1.73	Winter wheat: MT (10.0), OR (7.0), ID (7.0), CA (6.0), WA (3.5), ND (1.0);
			Spring wheat: OR (7.0), MT (5.0), ID (4.0), WA (2.5), CA (1.0);
			Durum wheat: MT (5.0)
2012	1,492,589	2.36	Winter wheat: KS (5.7), MS (5.0), OK (5.0), CA (4.0), CO (4.0), OR (2.5), AL (2.0), ID (2.0), MN (2.0), NE (2.0), ND (2.0), IN(1.0), MI (1.0), MT (1.0), TX (1.0), WA (1.0);
			Spring wheat: ID (4.0), OR (2.5), CA (2.0), MN (2.0), ND (2.0), WA (1.5), MT (1.0); Durum wheat: MT (1.0)
2013	456,324	0.78	Winter wheat: GA (3.0), IL (3.0), LA (3.0), MS (3.0), MT (3.0), AL (2.0), CA (2.0), OR (2.0), TN (2.0),WI (2.0), ID (1.0), IN (1.0), WA (1.0);
			Spring wheat: ID (5.0), MT (3.0), OR (2.0)
2015	4,509,149	8.07	Winter wheat: CA (3.0), ID (3.0), CO (25.0), IN (2), KS (15.0), MT (2.0), NE (12.0), OK (25.0), SD (15.0), MN (15.0), MO (2.0), WA (1.0), WI (2.0);
			Spring wheat: CO (25.0), ID (6.0), MN (15.0), MT (1.0), ND (5.0), SD (12.0)
2016	3,526,926	5.61	Winter wheat: CO (5.0), DE (1.0), ID (6.0), IN (3.0), KS (9.0), KY (5.0), MI (3.0), MN (3.0), MT (3.0), NE (7.0), NC (1.0), ND (1.0), OK (18.0), SD (7.0), TX (12.0), WA (2.0), WI (8.0);
			Spring wheat: ID (4.0), MN (2.0), MT (1.0), ND (1.0), SD (5.0), WA (2.0); Durum wheat: MT (1.0)

[a]States with yield loss of <1.0% are not listed. *AL* Alabama, *AR* Arkansas, *CA* California, *CO* Colorado, *GA* Georgia, *ID* Idaho, *IL* Illinois, *IN* Indiana, *KS* Kansas, *KY* Kentucky, *LA* Louisiana, *MI* Michigan, *MN* Minnesota, *MS* Mississippi, *MO* Missouri, *MT* Montana, *NE* Nebraska, *ND* North Dakota, *OK* Oklahoma, *OR* Oregon, *SD* South Dakota, *TN* Tennessee, *TX* Texas, *UT* Utah, *WA* Washington, *WI* Wisconsin

In Europe, stripe rust is less common and severe on barley than wheat, because wheat is predominantly winter wheat, which enables the pathogen to survive in wheat plants, whereas barley is grown mainly as spring barley in the high stripe rust risk regions and winter barley is grown in the low risk regions (Stubbs 1985). However, a serious stripe rust epidemic in northwest and central Europe in wheat and barley occurred in 1961 due to very favorable weather conditions for the over-summering and overwintering of both stripe rust forms and for their development in spring (Hassebrauk 1962).

In South America, the barley form, *P. striiformis* f. sp. *hordei*, was detected for the first time near Bogotá, Colombia in 1975, and the pathogen was presumed to be introduced from Europe (Dubin and Stubbs 1986). The disease spread southward as far as Argentina by 1982, covering almost all barley growing areas in South America, except Brazil (Dubin and Stubbs 1986). The yield loss in Colombia in 1975 was estimated as of 70%, equivalent to $US 6 million (Dubin and Stubbs 1986). In Ecuador, barley yield dropped from 0.88 t/h from between 1973 and 1976 to 0.62 t/h until 1980 with a mean yield loss of 30%. In Bolivia, yield losses on various barley cultivars in 1979–1980 were estimated from 47–61%.

Barley stripe rust spread from Colombia northward and reached Mexico by 1987 (Chen et al. 1994). The disease was first observed in a barley breeding nursery near Uvalde, Texas, USA in April, 1991 (Roelfs et al. 1992; Marshall and Sutton 1995). The disease spread to Oklahoma, New Mexico and Colorado by 1992; Arizona and California by spring 1993; southern Idaho and Montana by August 1993; Utah by 1994; and Oregon and Washington by 1995. Since then, the disease has become a major disease of barley in the western US (Chen et al. 1995). Barley stripe rust has also spread to Canada, and localized severe damage has occurred in Alberta (K.Q. Xi personal communication).

Grain yield losses of up to 72% were detected on susceptible barley in experimental fields in Texas in 1992 and 1993 (Marshall and Sutton 1995). In California, barley yield losses caused by stripe rust were estimated as 15% in 1996, 20% in 1997, 15% in 1998, 16% in 1999 and 6% in 2000, and moreover, barley acreage decreased 60% from 1993 to 2000 (Chen 2004). In Oregon, the epidemic in 1997 caused 4% yield loss and in 2000 caused 5% yield loss. In Washington, 3% yield loss was estimated in 1998. In both Oregon and Washington, fungicides were widely used to reduce yield losses. On susceptible cultivars, yield losses were detected as 25% in 2001, 43% in 2002, 28% in 2004 and 27% in 2005 in experimental fields near Pullman, Washington. In the recent years, barley stripe rust levels have generally become lower in the US Pacific Northwest. Even in 2010 and 2011 when the weather conditions were extremely favorable to stripe rust and susceptible wheat varieties had more than 60% (in 2010) and 90% (in 2011) yield losses without fungicide application, barley stripe rust was very low in commercial fields. The decrease of barley stripe rust pressure can be attributed to (1) largely growth of barley cultivars with high-temperature adult-plant resistance, (2) primarily just spring barley crops and (3) decrease of barley acreage due to price disadvantage compared to wheat and other crops. Barley crops could be reversed back to the vulnerability level in the 1990s, if barley is grown in more acreage, winter barley becomes significant, and more susceptible cultivars are grown in the western US. In the recent years, barley stripe rust has not been reported very much in the south-central states like Texas and Oklahoma. Although stripe rust has been detected on barley in the US Midwest (North Dakota and Minnesota), severe epidemics have not occurred in this major barley growing region as the climate is generally not favorable to stripe rust.

References

Abbasi M, Hedjaroude G, Scholler M, Goodwin SB. Taxonomy of *Puccinia striiformis* s.l. in Iran. Rostaniha 5:71–82; 2004, 199–224.

Abdel-Hak T, Stewart DM, Kamel AH. The current stripe rust situation in the near East region. Proc Reg Wheat Workshop, Vol 1. Ford Found, Beirut; 1972

Anonymous. Agricultural statistics of Pakistan. Ministry of Food, Agriculture and Livestock, Government of Pakistan; 2006. 289 pp

Aquino P, Carrion F, Calvo R. Selectd wheat statistics. In: Ekboir J, editor. CIMMYT 2000–2001 world wheat overview and outlook: developing no-till packages for small-scale farmers. Mexico: CIMMYT; 2002. p. 52–62.

Arthur JC. The grass rusts of South America, based on the Holway collection. Proc Am Philos Soc. 1925;64:131–223.

Becker H, Hart H. Das Ayftreteb und die Verbreitung von Gelbrost in Ostharz und den daran angrenzenden Wizenanbaugebieten. Z Pflanzenkrankh. 1939;49:449–81.

Beresford RM. Stripe rust (*Puccinia striiformis*), a new disease of wheat in New Zealand. Cereal Rusts Bulletin. 1982;10:35–41.

Biffen RH. Mendel's law of inheritance and wheat breeding. J Agric Sci. 1905;1:4–48.

Biffen RH. Studies in the inheritance of disease-resistance. J Agric Sci. 1907;2:109–28.

Biffen RH. The cereal rusts and their control. Trans Br Mycol Soc. 1931;16:19–38.

Blount AR, Rizvi S, Barnett RD, Chen XM, Schubert TS, Dankers WH, Momol TM, Dixon WN. First report of stripe rust caused by *Puccinia striiformis* f. sp. *tritici* on wheat (*Triticum aestivum* L.) in Florida. Online. Plant Health Progress;2005. doi:10.1094/PHP-2005-0304-01-HN

Brennan JP, Murray GM. Australian wheat diseases – assessing their economic importance. Agricultural Science New Series. 1988;1:26–35.

Bridgmon GH, Kolp. Stripe rust in Wyoming in 1958. Plant Dis Rep. 1958;43:163–4.

Britton M, Cummins GB. The reaction of species of *Poa* and other grasses to *Puccinia striiformis*. Plant Dis Rep. 1956;40:643–5.

Bushuk W. Rye production and uses worldwide. Cereal Foods World. 2001;46:70–3.

Cagaś B, Marková J. Contributions to the host range of *Puccinia poae-nemoralis* Otth and *Puccinia poarum* Nielsen. Plant Breed. 1988;101:126–31.

Carleton MA. A serious new wheat rust in this country. Science. 1915;42:58–9.

Chatrath R, Mishra B, Ferrara GO, Singh SK, Joshi AK. Challenges to wheat production in South Asia. Euphytica. 2007;157:447–56.

Chen XM. Epidemiology of barley stripe rust and races of *Puccinia striiformis* f. sp. *hordei*: the first decade in the United States. In: abstracts of the 11th Intl cereal rusts and powdery mildews conf, 22–27 August, 2004, Norwich; 2004, pp. A2.8

Chen XM. Epidemiology and control of stripe rust [*Puccinia striiformis* f. sp. *tritici*] on wheat. Can J Plant Pathol. 2005;27:314–37.

Chen XM. Challenges and solutions for stripe rust control in the United States. Aus J Agri Res. 2007;58:648–55.

Chen XM. Integration of cultivar resistance and fungicide application for control of wheat stripe rust. Can J Plant Pathol. 2014;36:311–26.
Chen FQ, Prehn D, Hayes PM. Mapping genes for resistance to barley stripe rust (*Puccinia striiformis* f. sp. *hordei*). Theor Appl Genet. 1994;88:215–9.
Chen XM, Line RF, Leung H. Virulence and polymorphic DNA relationships of *Puccinia striiformis* f. sp. *hordei* to other rusts. Phytopathology. 1995;85:1335–42.
Chen XM, Moore MK, Milus EA, Long DL, Line RF, Marshall D, Jackson L. Wheat stripe rust epidemics and races of *Puccinia striiformis* f. sp. *tritici* in the United States in 2000. Plant Dis. 2002;86:39–46.
Chen XM, Penman L, Wan AM, Cheng P. Virulence races of *Puccinia striiformis* f. sp. *tritici* in 2006 and 2007 and development of wheat stripe rust and distributions, dynamics, and evolutionary relationships of races from 2000 to 2007 in the United States. Can J Plant Pathol. 2010;32:315–33.
Cheng P, Chen XM, See D. Grass hosts harbor more diverse isolates of *Puccinia striiformis* than cereal crops. Phytopathology. 2014;106:362–71.
Ciccarone A. Il problema della ruggine dei grani in Ethiopia. Tre anni di observazioni (1938, 1939, 1940). Riv Agric Subtrop Trop 1947;41:1–42.
Cummins GB, Stevenson JA (1956) A checklist of North American rust fungi (Uredinales). Plant Dis Rep (Suppl. 240):109–193
Cunningham DD, Prain DA. A note on Indian wheat-rusts. Rec Bot Serv India. 1896;1:99–124.
Daly JM, Hennen JG, Lungren EA, Vaughn JR. Sort notes concerning stripe rust in other states. Plant Dis Rep. 1957;41:962–3.
Davison AD. Plant disease of economic crops occurring in Wyoming during 1958. Plant Dis Rep. 1958;42:1409–10.
Dietz SM, Hendrix JW. Reaction of grasses to stripe rust at Pullman, Washington. Phytopathology. 1962;52:730.
Dixon J, Braun HJ, Kosina P, Crouch J, editors. Wheat facts and futures 2009. México: CIMMYT; 2009.
Dubin HJ, Stubbs RW. Epidemic spread of barley stripe rust in South America. Plant Dis. 1986;70:141–4.
Dugan FM, Cashman MJ, Wang MN, Chen XM, Johnson RC. Differential resistance to stripe rust (*Puccinia striiformis*) in collections of basin wild rye (*Leymus cinereus*). Plant Health Progress. 2014;15(2):34–8. doi:10.1094/PHP-RS-14-0002.
Duveiller E, Singh RP, Nicol JM. The challenges of maintaining wheat productivity: pests, diseases, and potential epidemics. Euphytica. 2007;157:417–30.
Eriksson J. Uber die Spezialisierung des Parasitismus bei den Getreiderostpilzen. Ber Dtsch Bot Ges. 1894;12:292–331.
Eriksson J, Henning E. Die Hauptresultate einer neuen Untersuchung über die Getreiderostpilze. Z Pflanzenkr. 1894;4:197–203.
Eriksson J, Henning E. Die Getreideroste. Stockholm: Nortstedt & Söner; 1896.
Fang CT. Physiological specialization of *Puccinia glumarum* Erikss. and Henn. in China. Phytopathology. 1944;34:1020–4.
Fraser WP, Conners IL. The Uredinales of the prairie provinces of Western Canada. Trans R Soc Can. 1925;19:275–308.
Fuckel L. Enumeratio fungorum nassoviae series I. Jahrbücher des Vereins für Naturkunde Herzogthum Nassau. 1860;15:9–10.
Futrell MC, Atkins IM. Diseases of small grains in Texas in 1953. Plant Dis Rep. 1954;38:167–8.
Futrell MC, Atkins IM. Trends in diseases of cereal crops in Texas in recent years. Plant Dis Rep. 1957;41:42–6.
Futrell MC, Atkins IM, Hobbs CD. Unusual occurrence of small-grain diseases in Texas in 1957 and 1958. Plant Dis Rep. 1959;43:777–81.
Ghaffor A (1970) Wheat improvement and production Programme in Afghanistan (1966–69). Proc FAO/Rockefeller Found Wheat Semin, 3rd, 1970, pp 48–52

Gibson CM. Notes of infection experiments with various Uredineae. New Phytol. 1904;3:184–91.
Goidanich G (1943) Life and works of Giovanni Targioni Tozzetti. 3, 184–191. R. Accad. Ital., Stud. Doc. 12, 7–11 (translated from Italian by L. R. Tehon in Phytopathol. Classics No. 9, pp. vii–xi
Gough FJ, Williams ND, Brentzel WE. Occurrence of stripe rust in North Dakota in 1958. Plant Dis Rep. 1959;43:169–71.
Grillo HVS. Lista preliminar dos fungos assignalados en plantas do Brasil. Rodriguésia. 1937;2:39–96.
Hafiz H. Plant diseases. Islamabad: Directorate of Publications, Pakistan Agricultural research Council; 1986. 552 pp
Haldorsen S, Akan H, Celik B, Heun M. The climate of the Younger Dryas as a boundary for Einkorn domestication. Veg Hist Archaeobotany. 2011;20:305–18.
Hassan SF. Cereal rusts situation in Pakistan. In Proc of Eur Mediter Cereal Rusts Conf, Oeiras. 1968;124–125
Hassebrauk K. Die Gelbrostepidemie 1961 in Deutschland. Nachtr. Dtsch. Pflanzenschutzd. Nomenklatur, geographische Verbreitung und Wirtsbereich des Gelbrostes. 1962;14:22–26
Hassebrauk K. Nomenklatur, Geographische Verbreitung und Wirtsbereich des Gelbrostes, *Puccinia striiformis* West. Mittlg. Biol. Bundesanst. Berlin-Dahlem, H. 116, 75 S; 1965.
Hassebrauk K. Der Gelbrost, *Puccinia striiformis* West. II Befallsbid, Morphologie und Biologie der Sporen, Infektion und Weitere Entwicklung, Wirkungen auf die Wirtspflanze. Mittlg. Biol. Bundesanst. Berlin-Dahlem, H. 139, 11 S; 1970.
Hassebrauk K, Röbbelen G. Der Gelbrost, *Puccinia striiformis* West. III. Die Spezialisierung. Mittlg. Biol. Bundesanst. Berlin-Dahlem, H. 156, 150 S; 1974.
Hassebrauk K, Röbbelen G. Der Gelbrost, Puccinia striiformis West. IV. Epidemiologie. – Bekämpfungsmaßnahmen. Mittlg. Biol. Bundesanst. Berlin-Dahlem, H. 164 183 S; 1975
Hemmi T. On the distribution of cereal rusts in Japan and the relation of humidity to germination of urediniospores of some species of *Puccinia*. Proc Pac Sci Congr, 5th, 1933; 1934. pp. 3187–3194.
Hendrix JW, Burleigh JR, Tu J-C. Oversummering of stripe rust at high elevations in the Pacific Northwest – 1963. Plant Dis Rep. 1965;49:275–8.
Hennen J, Komanetsky M. Stripe rust in South Dakota in 1958. Plant Dis Rep. 1959;49:168–9.
Hirschhorn J Dos royas de la cebeda, nuevas para la Argentina. Rev Fac Agron, Univ Nac La Plata 1933;19:390–397
Holmes R, Dennis J. Accessory hosts of wheat stripe rust in Victoria, Australia. Trans Br Mycol Soc. 1985;85:159–60.
Hovmøller MS, Sørensen CK, Walter S, Justesen AF. Diversity of *Puccinia striiformis* on cereals and grasses. Annu Rev Phytopathol. 2011;49:197–217.
Huerta-Espino J, Singh RP. Wheat rust. In: Villaseñor MHE. and Espitia RE (eds) The temporal wheat in México. México: Secretaría de Agricultura, Ganadería y Desarrollo Rural (SAGAR), Instituto Nacional de Investigaciones Forestales, Agrícolas y Pecuarias (INIFAP), Centro de Investigación Regional–Centro (CIR–CENTRO), Campo Experimental Valle de México (CEVAMEX); 2000. pp. 231–249.
Humphrey HB. Brief notes on plant diseases. Plant Dis Rep. 1941;25:337–8.
Humphrey HB, Hungerford CW, Johnson AG. Stripe rust (*Puccinia glumarum*) of cereals and grasses in the United States. J Agric Res. 1924;29:209–27.
Hungerford CW. Studies on the life history of stripe rust, *Puccinia glumarum*. J Agric Res. 1923;24:607–20.
Hungerford CW, Owens CE. Specialized varieties of *Puccinia glumarum* and hosts for variety *tritici*. J Agric Res. 1923;25:363–401.
Hylander N, Jørstad I, Nannfeldt JA. Enumeratio uredinearum scandinavicarum. Opera Botanica. 1953;1:75.

ICARDA. Strategies to reduce the emerging wheat stripe rust disease. Synthesis of a dialog between policy makers and scientists from 31 countries at: International Wheat Stripe Rust Symposium, Aleppo, Syria, April 2011. 2011. http://www.icarda.cgiar.org/publications-and-resources/research-to-action

Ito S. Contributions to the mycological flora of Japan. II. On the Uredineae parasitic on the Japanese Gramineae. J Coll Agric, Tokyo Imp Univ 1909; 3:180–262

Johnson T, Newton M. The occurrence of yellow stripe rust in western Canada. Sci Agric Abstr. 1928;8:464.

Kajiwara T. (1964) Beiträge zur Kenntnis der Physiologischen Spezialisierung des Gerstengelbrostes (*Puccinia striiformis* West. f. sp. *hordei* Erikss.). Nachr Dtsch Pflanzenschutzd Braunschweig

Kajiwara T, Ueda I, Iwata Y. Untersuchungen über die physiologische spezialisierung des gerstengebrostes (*Puccinia striiformis* f. sp. *hordei*) in Japan. Phytopathol Z. 1964;50:313–28.

Karki CB. Report on evaluation of Nepalese wheat and barley varieties in the seedling stage on their resistance to yellow rust. Wageningen: Research Institute of Plant Protion (IPO); 1980

Khazra H, Bamdadian A. The wheat diseases situation in Iran. Proc FAO/Rockefeller Found Wheat Semin, 4th 1973, 1974; pp 292–299.

Kirk PM, Cannon PF, Minter DW, Stalpers JA. Dictionary of the fungi. 10th ed. Wallingford: CAB International; 2008. p. 576.

Kislev ME. Stem rust of wheat 3300 years old found in Israel. Science. 1982;32:833–44.

LeBeau FJ, Sosa Oscar N, Fumagalli A. Yield reduction in wheat by stripe rust. Plant Dis Rep. 1956;40:886.

Li ZQ, Zeng SM. Wheat rusts in China. Beijing: Chinese Agricultural Press; 2002. 379 pp. (In Chinese)

Line RF. Factors contributing to an epidemic of stripe rust on wheat in the Sacramento Valley of California in 1974. Plant Dis Rep. 1976;60:312–6.

Line RF. Stripe rust of wheat and barley in North America: a retrospective historical review. Annu Rev Phytopathol. 2002;40:75–118.

Line RF, Qayoum A. Virulence, aggressiveness, evolution, and distribution of races of *Puccinia striiformis* (The cause of stripe rust of wheat) in North America, 1968–1987. U.S. Department of Agriculture Technical Billetin No. 1788, 44 pp; 1992.

Liu M, Hambleton S. Taxonomic study of stripe rust, *Puccinia striiformis* sensu lato, based on molecular and morphological evidence. Fungal Biol. 2010;114:881–99.

Manners JG. *Puccinia striiformis* Westend. var. *dactylidis* var. nov. Trans Br Mycol Soc. 1960;43:65–8.

Marchionatto JB. La presencia de la roya amarilla, Secc. Prop., Inf., Circ. No. 836, pp. 3–5. Minist. Agric. Nac. Buenos Aires; 1931.

Marshall D, Sutton RL. Epidemiology of stripe rust, virulence of *Puccinia striiformis* f. sp. *hordei*, and yield loss in barley. Plant Dis. 1995;79:732–7.

Martens JW, Oggema MW. The yellow rust situation in Kenya. Proc Reg Wheat Workshop, Vol 1. Ford Found, Beirut; 1972.

McNeill J, Barrie FR, Burdet HM, Demoulin V, Hawksworth DL, Marhold K, Nicolson DH, Prado J, Silva PC, Skog JE, Wiersema JH, Turland NJ, editors. International code of botanical nomenclature (Vienna Code) adopted by the seventeenth international botanical congress Vienna, Austria, July 2005. Ruggell: Gantner Verlag; 2006.

Miller JD, Christensen JJ. Occurrence of stripe rust in Minnesota in 1958. Plant Dis Rep. 1959;43:159.

Mohamed HA. The status of wheat rusts in the U.A.R. (Egypt) in the period 1959–1962. Robigo. 1963;14:13–4.

Murdoch CL, Laemmlen FF, Parvin PE. Response of Kentucky bluegrass and perennial ryegrass cultivars to rust (*Puccinia striiformis* and *P. coronata*) in Hawaii. Plant Dis Rep. 1973;57:215–6.

Murray GM, Ellison PJ, Watson A. Effects of stripe rust on the wheat plant. Aus Plant Pathol. 1995;24:261–70.

Nagarajan S, Joshi LM. Epidemiology in the Indian subcontinent. In: Roelfs AP, Bushnell WR, editors. The cereal rusts, vol. II. Orlando: Academic Press; 1985. p. 362–94.
Newton M, Johnson T. Stripe rust, *Puccinia glumarum*, in Canada. Can J Res. 1936;14:89–108.
Niemann E, Scharif G, Bamdadian A. Die Getreideroste in Iran. Wirtsbereich, Unterscheidung, Bedeutung, Bekämpfung. Entomol Phytopathol Appl. 1968;27:25–41.
Niu YC, Li ZQ, Shang HS. *Puccinia striiformis* West. f. sp. *leymi* and f. sp. *elymi*, two new formae speciales. Acta Univ Agric Boreali-Occident (Suppl). 1991;19:58–62.
O'Brien L, Brown JS, Young RM, Pascoe T. Occurrence an ddistribution of wheat stripe rust in Victoria and susceptibility of commercial wheat cultivar. Aus Plant Pathol. 1980;9:14.
Pady SM, Johnston CO. Cereal rust epidemiology and aerobiology in Kansas in 1957. Plant Dis Rep. 1958;42:726–33.
Pady SM, Johnston CO. Stripe rust in Kansas in 1958. Plant Dis Rep. 1959a;43:159–63.
Pady SM, Johnston CO. Cereal rust aerobiology and epidemiology in Kansas in 1958. Plant Dis Rep. 1959b;43:607–12.
Pady SM, Johnston CO, Rogerson CT. Stripe rust of wheat in Kansas in 1957. Plant Dis Rep. 1957;41:959–61.
Pretorius ZA (2004) The impact of wheat stripe rust in South Africa. In Proc 11th Intl cereal rusts and powdery mildews conf, 22–27. John Innes Centre, Norwich, UK, Eur Medit cereal rust found, Wageningen, Neth. Cereal Rusts and Powdery Mildews Bul, Abstr A1.29. Available from http://www.crpmb.org/icrpmc11/abstracts.htm. Accessed on 3 Aug 2014
Pretorius ZA, Boshoff WHP, Kema GHJ. First report of *Puccinia striiformis* f. sp. *tritici* on wheat in South Africa. Plant Dis. 1997;81:424.
Priestley RH, Bayles RA. The contribution and value of resistant cultivars to disease control in cereals. In: Clifford BC, Lester E, editors. Control of plant diseases, costs and benefits. Oxford: Blackwell Scientific Publications; 1988. p. 53–65.
Röbbelen G, Sharp EL. Mode of inheritance, interaction and application of genes conditioning resistance to yellow rust. Berlin/Hamburg: Verlag Paul Parey; 1978. 88 pp.
Rodriguez VJ. Observations on leaf and stripe rust of wheat in Mexico. Phytopathology. 1946;36:410.
Roelfs AP, Huerta-Espino J, Marshall D. Barley stripe rust in Texas. Plant Dis. 1992;76:538.
Rosen HR, Whitehead MD, Browning JA, Weihing JL, Green GJ, Fischer GW. Short notes concerning stripe rust in other states. Plant Dis Rep. 1959;43:172.
Rudorf W, Job M La existencia de *Puccinia glumarum tritici* (Schmidt) Erikss. et Henn. en los paises del Rio de la Plata. Arch Soc Biol Montevideo. 1931;5(Suppl.):1363–1370.
Saari EE, Prescott JM. World distribution in relation to economic losses. In: Roelfs AP, Bushnell WR, editors. The cereal rusts, vol. II. Orlando: Academic Press; 1985. p. 259–98.
Sanford GB, Broadfoot WC. Stripe rust in Alberta. Sci Agric. 1929;9:337–45.
Schmidt JK (1827) *Uredo glumarum* Schmidt. Allgemeine O¨ konomisch-technische Flora 1:27
Shaner G, Powelson RL. Epidemiology of stripe rust of wheat, 1961–1968. OR Agric Exp Stn Tech Bull. 1971;117:31 pp.
Shaner G, Powelson RL. The oversummering and dispersal of inoculum of *Puccinia striiformis* in Oregon. Phytopathology. 1973;63:13–7.
Sharp EL, Hehn ER. Overwintering of stripe rust in winter wheat in Montana. Phytopathology. 1963;53:1239–40.
Smith RE. Wheat rust as viewed in the early agricultural press in California. Plant Dis Rep. 1961;45:709–12.
Smith RE, Bowman JN. Wheat rust during the Spanish era in California. Plant Dis Rep. 1961;45:632–3.
Stakman EC (1915) Relation between *Puccinia graminis* and plants highly resistant to its attack. J Agric Res 3:193–200 (Washington. D.C.)
Straib W. Infektionsversuche mit biologische Rassen des Gelbrostes (*Puccinia glumarum* (Schm.) Erikss. Et Henn.) im Jahre 1934. Arb Biol Reichsanst Land Forstwirtsch Berl –Dahlem. 1935;21:455–66.

Stubbs RW. Stripe rust. In: Roelfs AP, Bushnell WR, editors. The cereal rusts, vol. II. Orlando: Academic Press; 1985. p. 61–101.
Stubbs FB (eds) (1986) Provisional keys to British plant galls. Pub. Brit Plant Gall Soc.
Stubbs RW, Fuchs E, Vecht H, EJW B. The international survey of factors of virulence of *Puccinia striiformis* Westend. in 1969, 1970 and 1971. Ned Graan-Centrum Tech. 1974;21:1–88.
Sui XX, He ZH, Lu YM, Wang ZL, Xia XC. Molecular mapping of a non-host resistance gene *YrpstY1* in barley (*Hordeum vulgare* L.) for resistance to wheat stripe rust. Hereditas. 2010;147:176–82.
Sydow P, Sydow H. Monographia uredinearum, vol. 1. Leipzig: Fratres Borntraeger; 1904.
Tollenaar H. A comparison of *Puccinia striiformis* f. sp. *poae* on bluegrass with *P. striiformis* f. sp. *tritici* and f. sp. *dactylidis*. Phytopathology. 1967;57:418–20.
Tollenaar H, Houston BR. Effect of temperature during uredospore production and of light in vitro germination of uredospores from *Puccinia striiformis*. Phytopathology. 1966;56:787–9.
Tollenaar H, Houston BR. A study on the epidemiology of stripe rust, *Puccinia striiformis* West in California. Can J Bot. 1967;45:291–307.
Tu JC. The summer biology of *Puccinia striiformis* West. In Southeastern Washington. Dissertation, Washington State University. 1967.
Van der Merwe MM, Walker J, Ericson L, Burdon JJ. Coevolution with higher taxonomic host groups within the *Puccinia/Uromyces* rust lineage obscured by host jumps. Mycol Res. 2008;112:1387–408.
Wahl I, Anikster Y, Manisterski J, Segale A. Evolution at the center of origin. In: Bushnell WR WR, Roelfs AP, editors. The cereal rusts Vol. I, Origins, specificity, structure, and physiology. Orlando: Academy Press; 1984. p. 51–88.
Wan AM, Chen XM. Virulence, frequency, and distribution of races of *Puccinia striiformis* f. sp. *tritici* and *P. striiformis* f. sp. *hordei* identified in the United States in 2008 and 2009. Plant Dis. 2012;96:67–74.
Wan AM, Chen XM. Virulence characterization of *Puccinia striiformis* f. sp. *tritici* using a new set of *Yr* single-gene line differentials in the United States in 2010. Plant Dis. 2014;98:1534–42.
Wan AM, Zhao ZH, Chen XM, He ZH, Jin SL, Jia QZ, Yao G, Yang JX, Wang BT, Li GB, Bi YQ, Yuan ZY. Wheat stripe rust epidemics and virulence of *Puccinia striiformis* f. sp. *tritici* in China in 2002. Plant Dis. 2004;88:896–904.
Wang ZH. On the evolution of the barley-yellow rust pathosystem in Tibet. Acta Phys Sin. 1992;22:151–5.
Wang ZH, Peng YL, Yangjin L, Da Z. Difference of pathogenicity between isolates of barley-attacking and wheat-attacking forms of *Puccinia striiformis* with special reference to the deviation of disease development. Acta Phys Sin. 1989a;19:161–5.
Wang ZH, Yangjin L, Da Z, Peng YL. A preliminary study on the physiologic specialization of *Puccinia striiformis* f. sp. *hordei* in Tibet. Chin J Plant Prot. 1989b;16:186–92.
Ward HM. Further observations on the brown rust of the brome, *Puccinia dispersa* (Erikss.) and its adaptive parasitism. Ann Myol. 1903;1:132–51.
Ward HM. On the histology of *Uredo dispersa*, Eriks., and the mycoplasm hypothesis. Philos Trans R Soc London Ser B. 1904;196:29–46.
Ward HM. Recent researches on the parasitism of fungi. Ann Bot (London). 1905;19:1–14.
Wellings CR. Global status of stripe rust: a review of historical and current threats. Euphytica. 2011;179:129–41.
Wellings CR, Burdon JJ, McIntosh RA, Wallwork H, Raman H, Murray GM (2000). A new variant of *Puccinia striiformis* causing stripe rust on barley and wild *Hordeum* species in Australia. Plant Pathol 49:803–803
Wellings CR, Kandel KR (2004) Pathogen dynamics associated with historic stripe (yellow) rust epidemics in Australia in 2002 and 2003. In: Proc 11th Int cereal rusts and powdery mildews conf 22–27 August 2004, John Innes Centre, Norwich UK. Eur and Midterr cereal rust found, Wageningen. Cereal rusts and Powdery Mildews Bul, Abstr. A2.74. Available from http://www.crpmb.org/icpmc11/abstracts.htm. Accessed 3 Aug 2014

Wellings CR, Luig NH. Wheat rusts, the old and the new. Australian Institute of Agricultural Science, Occasional Publication. 1984;15:5–15.
Wodageneh A. Country reports and programmes: Ethiopia. Proc FAO/Rockefeller Found Wheat Semin, 4th 1973; 1974. pp. 64–68
Yachevski AA. Rusts of grain crops in Russia (in Russian). Dep Zemled Tr Buro Mikol Fitopatol Sib, St. Petersburg, Russia; 1909.
Young Jr HC, Browder LE. Occurrence of stripe rust in Oklahoma in 1957. Plant Dis Rep. 1957;41:958–9.
Zadoks JC. Yellow rust on wheat, studies in epidemiology and physiologic specialization. Neth J Plant Pathol. 1961;67:69–256.
Zadoks JC, Bouwman JJ. Epidemiology in Europe. In: Roelfs AP, Bushnell WR, editors. The cereal rusts, vol. II. Orlando: Academic Press; 1985. p. 321–61.

Chapter 2
Variability of the Stripe Rust Pathogen

Anmin Wan, Xiaojie Wang, Zhensheng Kang, and Xianming Chen

Abstract The stripe rust pathogen, *Puccinia striiformis*, is highly variable. The variation has been studied for about 100 years based on virulence, for about 30 years based on molecular markers, and for about 10 years based on genome sequencing and functional genomics. Virulence characterization using host differential genotypes has identified a large number of races or pathotypes that overcome race-specific resistance genes in cultivars of wheat and barley crops. Such information is essential for developing cultivars with effective resistance to the disease. Molecular characterization of the pathogen using various marker techniques has generated data that are useful in determination of population structure, identification of pathogen introduction and dissemination and understanding of mechanisms of variation. Molecular characterization has conducted to support findings from virulence tests and/or obtain information that is unable to produce through virulence characterization. The high variability of the stripe rust pathogen is due to its high reproductivity, capability of long-distance dissemination, and ability for adapting to various host species and environments. Among the genetic and evolutionary mechanisms for producing variation, mutation is most important in generating new races and genotypes that are under selection of host species, host cultivars and environments. The pathogen also produces new races and genotypes through somatic recombination. Sexual recombination has been recently demonstrated mainly under controlled conditions in generating variation in races and phenotypes. Natural reproduction on alternate hosts, mainly *Berberis* spp., has been so far found in

Mention of trade names or commercial products in this publication is solely for the purpose of providing specific information and does not imply recommendation or endorsement by the U. S. Department of Agriculture. USDA is an equal opportunity provider and employer.

A. Wan
Department of Plant Pathology, Washington State University, Pullman, WA 99164-6430, USA

X. Wang • Z. Kang
State Key Laboratory of Crop Stress Biology for Arid Areas and College of Plant Protection, Northwest A&F University, Yangling, Shaanxi 712100, China

X. Chen (✉)
US Department of Agriculture, Agricultural Research Service, Wheat Health, Genetics and Quality Research Unit, Pullman, WA 99164-6430, USA

Department of Plant Pathology, Washington State University, Pullman, WA 99164-6430, USA
e-mail: xianming@wsu.edu

X. Chen, Z. Kang (eds.), *Stripe Rust*, DOI 10.1007/978-94-024-1111-9_2

China, but its role in generating aeciospores to start stripe rust epidemics on cereal crops and creating new races to attack crop cultivars is still unclear in this country, while ruled out in some other countries. Compared to other fungal pathogens, research on genome sequencing and functional genomics has started relatively late, but recent studies have generated huge data that have helped in understanding the pathogen variation and created useful resources for developing new tools for further studying and monitoring the pathogen variation.

Keywords Adaptation • Aggressiveness • Barley • Clonal population • Fitness • Functional genomics • Genetic drift; Genome • Grasses • Host differentials • Molecular markers • Mutation • Population structure • *Puccinia striiformis* • Races • Sexual recombination • Somatic recombination • Stripe rust • Virulence • Wheat • Yellow rust

In the Introduction part, we discussed the pathogenicity variability of the *Puccinia striiformis* species on different host plant genera or species. Such variability separates the species into different formae speciales. In this chapter, we will focus on the pathogenicity variability of the wheat stripe rust pathogen (*P. striiformis* f. sp. *tritici*) and the barley stripe rust pathogen (*P. striiformis* f. sp. *hordei*). The pathogenicity variability within a forma specialis is mostly determined by the virulence and avirulence of isolates among cultivars. This level of pathogenicity variability separates a forma specialis into races. By definition, a race is a taxon of pathogen particularly characterized by specialization to different cultivars of one host species (Vidhyasekaran 2004). In fact, a race is not a taxonomic term. It is a pathological term as races cannot be generally distinguished morphologically, but can be distinguished by pathogenicity or virulence/avirulence patterns on host cultivars with different race-specific resistance genes. In the literature, races of rust fungi have been traditionally referred as "physiological races" (Bever 1934; Fang 1944; Zadoks 1959; Fuchs 1960; Green 1971; Johnson et al. 1972). However, so far no physiological attributes have been associated with races, and races have been described by virulence/avirulence patterns. Because of this reason, race is preferred to use over physiological race. Equivalently, pathotypes are also used in the same way as races. Any separation of "pathotypes" from "races", such as those based on frequencies used in China (Chen et al. 2009b), is not scientifically sound and sometimes can be practically misleading.

In addition to variations in virulence, *P. striiformis* f. sp. *tritici* strains or populations can also vary in adaptation to different environments. In recent years, significant progress has been made in studying pathogen populations with molecular markers and genomic approaches. In this chapter, we will discuss variations in virulence, environments, populations, and genomics and possible mechanisms of variations.

2.1 Variation in Virulence – Races of *P. striiformis* f. sp. *tritici* and *P. striiformis* f. sp. *hordei*

Eriksson (1894) was the first to use the term forma specialis for rust fungi and described five formae speciales of *Puccinia glumarum*, now known as *Puccinia striiformis*. Later, the formae speciales of *P. striiformis* f. sp. *tritici* (the wheat stripe rust pathogen) and *P. striiformis* f. sp. *hordei* (the barley stripe rust pathogen) are further separated into races based on avirulent/ virulent reactions on cultivars or genetic genotypes of wheat or barley. So far, there are no reports of differentiating races of other formae speciales of *P. striiformis* partially because of less importance of the diseases and less requirement of such information for disease management.

Hungerford and Owens (1923) were the first to report that specialized varieties occurred in *P. striiformis* based on specificity on genera of wheat and grasses in North America. Such varieties were combinations of formae speciales and races. Allison and Isenbeck (1930) were the first to establish the existence of races in *P. striiformis* f. sp. *tritici* in Europe based on specificity on wheat cultivars. A couple of years later, Newton et al. (1933) and Bever (1934) each identified two races of *P. striiformis* f. sp. *tritici* in North America based on different avirulence/virulence patterns on wheat varieties Chinese 166, Red Russian and Heines Kolben. Ten years later, Fang (1944) published the first work of differentiating races of *P. striiformis* f. sp. *tritici* in China.

Races are differentiated by infection types produced by rust pathogen isolates on a set of selected host crop genotypes or single gene lines that are referred to as differentials (Anikster 1984). Throughout the last 80 years, there have been sets of differentials established, modified, and used to differentiate races in many countries (Gassner and Straib 1932; Zadoks 1959; Fuchs 1960; Wang et al. 1963; Line et al. 1970; Johnson et al. 1972; Sharma et al. 1973; Hassebrauk and Röbbelen 1974; Priestley and Doodson 1976; Wellings and McIntosh 1990; Line and Qayoum 1992; Balyes et al. 2000; Hovmøller 2001; Boshoff et al. 2002; Chen et al. 2002, 2009b; Wan et al. 2004; de Vallavieille-Pope 2012; Wan and Chen 2014a). The various sets of wheat differentials used to identify races of *P. striiformis* f. sp. *tritici* were discussed in Chen (2005). In this chapter, we will use the system used in the United States as an example to illustrate the identification of the wheat stripe rust pathogen.

2.1.1 Collection, Handling and Process of Rust Samples and Spore Storage

Obtaining quality samples is the first step in monitoring virulence variation in wheat rust pathogens, especially for the stripe rust pathogen as it is easier than other rust fungi to loss viability. In order to recover and increase urediniospores for virulence testing and race identification, it is important to appropriately collect, handle, ship and store stripe rust samples.

Stripe rust samples can be collected from commercial fields and nurseries (rust screening nurseries, fungicide trial nurseries, yield loss study nurseries and breeding nurseries) of the host crops, auxiliary grass hosts and alternate hosts. Rust samples could be leaves and glumes of cereal crops and grasses or leaves and berries of alternate hosts. The samples should be representative of different locations, regions, cultivars and particular nurseries. Ideally, the collections should be made on a random basis when the major purpose is to determine frequencies and distributions of races and virulences. In fact, it is hard to obtain samples collected randomly due to limited time and in access to sampling sites. Therefore, effort should be made to increase the number of samples to compensate for non-randomness in sampling. Often, it is more important to identify new virulences for specific resistance genes or races with virulences to specific cultivars. Attentions should be paid to observing rust and collecting samples from specific cultivars, breeding lines, and genetic stocks. For example, to identify races virulent to resistance gene *Yr5*, collections should be made from *Yr5*-carring wheat cultivars or lines when stripe rust is observed on such plants. Random sampling is not efficient for identifying new races. It is better to combine random sampling with special sampling.

If plants are not ready for inoculating with the rust samples, the samples should be kept at 4 °C refrigerator until use. Freezers or liquid nitrogen containers with temperatures below 0 °C are not good for keeping fresh- leaf or fresh spore samples, unless samples have been dried. A rust sample is considered as an original isolate. To get enough urediniospores for an isolate to be used in race identification and other studies, urediniospores from sample leaves should be used to inoculate seedlings of a susceptible variety. This process is also called "spore transfer", "recovery" or "initial increase". For spore recovery, transfer spores from the sample leaves onto a "universal" susceptible variety. In the United States, we have been using winter wheat cultivar 'Nugaines' since the 1960s because it is susceptible at the seedling stage to all races of *P. striiformis* f. sp. *tritici* in the United States and samples from other countries (Line and Qayoum 1992; Sharma-Poudyal et al. 2013; Wan and Chen 2014a; Wan et al. 2017). Similarly, winter wheat variety 'Mingxian 169' is commonly used in China for recovering and increasing urediniospores of the same pathogen (Wan et al. 2004; Chen et al. 2009b; Zhan et al. 2016).

Depending upon temperatures used, new urediniospores can be started to harvest about 12 to 16 days after inoculation. Spore collection should be done in a hood or within a plastic booth with a spore collector connecting to a vacuum pump. Wash the hood and hose with 75% ethanol after each sample. Spores can also be collected simply by shaking off onto a filter paper or into a glass tube or plastic bag. In the recent years, cellophane bags are being used in spore increase and collection due to its vapor permeability to reduce moisture in the bags (Sørensen et al. 2016). Different methods can be used for different purposes of storage. For any storage purposes, spores should be dry. Urediniospores can be dried in a desiccator kept at 4 °C for at least 2 days before stored in a sealed glass vial or foil bags. For short-term storage (less than 2 months), urediniospores can be kept at 4 °C. Spores can be stored in a −70 or −80 °C freezer for several months to a couple of years. For long-term storage, urediniospores should be stored in liquid nitrogen. If small foil bags or glass vials for storing spores in liquid nitrogen or a freezer, the bags or vials should be sealed completely.

2.1.2 Selection of Host Genotypes for Use as Differentials

A group of selected crop host lines is used for differentiating races of a rust fungal pathogen. Compared to differentials for identifying races of the wheat leaf rust and stem rust pathogens, there have been no a relatively standardized and internationally accepted set of differentials for the wheat stripe rust pathogen (Wan and Chen 2014a). In considering the local cultivars and resistance genes used, additional differential hosts (mostly local cultivars) were also included in a set or as supplemental differentials. The set of host genotypes chosen as differentials determine the usefulness of race data. In the past, a cultivar or line was added to a set of differentials when it becomes susceptible and is capable of differentiating new races (Line and Qayoum 1992). The addition has immediate usefulness for identifying new races that can be used to avoid using the same sources of resistant germplasm. However, adding every cultivar or line that becomes susceptible has become unpractical because thousands of wheat cultivars or lines are resistant to one or more races of *P. striiformis* f. sp. *tritici* (Chen and associates, unpublished data) and hundreds of genes or quantitative trait loci have been identified for resistance to stripe rust (Chen 2013; Rosewarne et al. 2013; McIntosh et al. 2013; Hou et al. 2015). In order to be practically easy and informative, it is, therefore, important to select genotypes of host crops as differentials for routine race survey.

In selecting host genotypes, investigators give varying weight to different factors. In general, differential hosts should have differentiability which means that they can provide low (resistant) or high (susceptible) reactions to different isolates (Roelfs 1984). The following factors are generally considered for selecting host genotypes as differentials: (1) Historical basis. A historical basis is important because not all races are detected annually. Several years of data are needed to compare race changes. If different genotypes are used to identify races from year to year, the results cannot be compared. (2) Resistance used in commercial cultivars. Resistance genes that are used or under consideration for use in commercial cultivars are important; and genotypes with these genes should be used as differentials. Because differentials should have clear different reactions when tested with different races, genotypes with genes for non-race specific resistance should not be used as differentials. Generally, differentials should not be susceptible or resistant to all isolates evaluated from the rust pathogen populations. (3) Stability of the disease reaction. Ideally, host genotypes selected as differentials must not be severely affected by the range of temperatures, light and inoculum density. (4) Usefulness of information. The differential hosts must provide information useful to the purpose of race analysis and resistance utilization (e.g., resistance discovery, screening and breeding). (5) Single gene differentials. Although it is impossible to be sure whether a host genotype has a single gene for resistance, it is possible to establish genotypes that have only one gene effective against the pathogen population. These single gene lines have advantages as differential hosts. For instances, the pathogen phenotype can be clearly measured. The gene is resistant if there is a low infection type, otherwise the gene is susceptible if there is a high infection type. However when a host genotype that have two or more resistance genes is evaluated, the virulence to one of the gene can be masked by avirulence of the other genes. The resistance reaction on a multigene differential cannot predict which of the genes is effective. Another advantage of single gene line is that it has a smaller range of low infection types, and thus environmental effects are of less importance, and often enables one to predict the infection type of host lines with a known genotype. (6) Seed availability. When a new or single resistance gene is found, sufficient seed is generally not available, especially the single *Yr* gene lines with the same background (e.g., in the same recurrent parent), because it takes time to develop single *Yr* gene lines with the same background (also called near isogenic lines). However, a line with limited seed can be used as a supplemental differential and being tested only with selected isolates. Obviously, maintaining and supply pure seeds of *Yr* single gene lines are very important as well. Meanwhile, some of the available single-gene lines are either winter or spring growth type and may be difficult to grow in some areas of the world. Therefore, maintaining and supply accurate and pure seeds of differential lines continue to be an issue.

As briefly introduced below, there are several sets of differentials that have been used to identify races of *P. striiformis* f. sp. *tritici* in Europe and Australasia (Johnson et al. 1972; Wellings and McIntosh 1990; Hovmøller et al. 2002), China (Fang 1944; Wang et al. 1963; Wan et al. 2004; Chen et al. 2009b), India (Sharma et al. 1973; Nagarajan et al. 1986), and the United States (Line and Qayoum 1992; Line 2002; Chen et al. 2010; Wan and Chen 2014a). The wheat cultivars or genotypes used as differentials in different countries are presented in Table 2.1.

2.1.2.1 The World and European System and Modified Systems

In Europe, the first set of differential genotypes of wheat was established and race analysis of *P. striiformis* f. sp. *tritici* were conducted in Germany in the early 1930s (Gassner and Straib 1932). The set was used by others (Fuchs 1960; Zadoks 1961) for race surveys in Europe, and later for international race surveys by Stubbs (1985). This set included genotypes of wheat (Michigan Amber, Blè Rouge d'Ecosse, Strubes Dickkopf, Webster, Holzapfes Früh, Vilmorin 23, Heines Kolben, Carstens V, Spaldings Prolific, Chinese 166, Rouge Prolifique Barbu and *Triticum dicoccum* var. *tricoccum*), barley (Fong Tien and Heils Franken) and rye (Petkuser Roggen) (Stubbs 1985). Races are named as Race 1, Race 2, etc., including those considered as races of *P. striiformis* f. sp. *hordei* like "Race 24" (Dubin and Stubbs 1986).

Table 2.1 Wheat genotypes for differentiating races of *Puccinia striiformis* f. sp. *tritici* in Europe, China, India, United States and newly established *Yr* single-gene differentials

No.	World (W) and European (E)	China	India	United States	*Yr* single gene line
1	Chinese 166 (*Yr1*) (W)	Trigo Eureka (*Yr6*)	Chinese 166 (*Yr1*)	Lemhi (*Yr21*)	AvSYr1NIL (*Yr1*)
2	Lee (*Yr7*, *Yr22*, *Yr23*) (W)	Fulhard (Unknown)	Lee (*Yr7*, *Yr22*, *Yr23*)	Chinese 166 (*Yr1*)	AvSYr5NIL (*Yr5*)
3	Heines Kolben (*Yr2*, *Yr6*) (W)	Lutescens 128 (Unknown)	Heines Kolben (*Yr2*, *Yr6*)	Heines VII (*Yr2*, *YrHVII*)	AvSYr6NIL (*Yr6*)
4	Vilmorin 23 (*Yr4a*, *YrV23*) (W)	Mentana (Unknown)	Vilmorin 23 (*Yr4a*, *YrV23*)	Moro (*Yr10*, *YrMor*)	AvSYr7NIL (*Yr7*)
5	Moro (*Yr10*, *YrMor*) (W)	Virgilio (*YrVir1*, *YrVir2*)	Moro (*Yr10*, *YrMor*)	Paha (*YrPa1*, *YrPa2*, *YrPa3*)	AvSYr8NIL (*Yr8*)
6	Stubes Dickkopf (*YrSD*, *Yr25*) (W)	Abbondanza (Unknown)	Stubes Dickkopf (*YrSD*, *Yr25*)	Druchamp (*Yr3a*, *YrD*, *YrDru*)	AvSYr9NIL (*Yr9*)
7	Suwon92/Omar (*Yr4*, *YrSu*) (W)	Early Premium (Unknown)	Suwon92/Omar (*Yr4*, *YrSu*)	Riebesel 47/51 (*Yr9*, +); AvSYr5NIL (*Yr5*)[a]	AvSYr10NIL (*Yr10*)
8	Clement (*Yr9*, *YrCle*) (W)	Funo (*YrA*, +)		Produra (*YrPr1*, *YrPr2*)	AvSYr15NIL (*Yr15*)
9	*Triticum spelta* album (*Yr5*) (W)	Danish 1 (*Yr3*)	Hybrid 46(*Yr4b*, *YrH46*)	Yamhill (*Yr2*, *Yr4a*, *YrYam*)	AvSYr17NIL (*Yr17*)
10	Hybrid 46 (*Yr4b*, *YrH46*) (E)	Jubilejina 2 (*YrJu1*, *YrJu2*, *YrJu3*, *YrJu4*)	Heines VII (*Yr2*, *YrHVII*)	Stephens (*Yr3a*, *YrS*, *YrSte*)	AvSYr24NIL (*Yr24*)
11	Reichersberg 42 (*Yr7*, *Yr25*) (E)	Fengchan 3 (*Yr1*)	Compair (*Yr8*, *Yr19*)	Lee (*Yr7*, *Yr22*, *Yr23*)	AvSYr27NIL (*Yr27*)
12	Heines Peko (*Yr2*, *Yr6*, *Yr25*) (E)	Lovrin 13 (*Yr9*, +)	*Triticum spelta* album (*Yr5*)	Fielder (*Yr6*, *Yr20*)	AvSYr32NIL (*Yr32*)
13	Nord Desprez (*Yr3a*, *Yr4a*) (E)	Kangyin 655 (*Yr1*, *YrKy1*, *YrKy2*)	Riebesel 47/51 (*Yr9*, +)	76e (*Yr76*)	AvS/IDO377s (F3-41-1) (*Yr43*)
14	Compair (*Yr8*, *Yr19*) (E)	Suwon 11(*YrSu*)	Sonalika (*YrA*, +)	Tres (*YrTr1*, *YrTr2*)	AvS/Zak (1-1-35-line1) (*Yr44*)
15	Carstens V (*Yr32*, *YrCV2*, *YrCV3*) (E)	Zhong 4 (Unknown)	Kalyansona (*Yr2*)	Hyak (*Yr17*, *Yr76*)	AvS/YrSPNIL (*YrSP*)
16	Spaldings Prolific (*YrSP*) (E)	Lovrin 10 (*Yr9*)		Express (*YrEpx1*, *YrExp2*)	AvSYrTr1NIL (*YrTr1*)
17	Heines VII (*Yr2*, *YrHVII*) (E)	Hybrid46 (*Yr3b*, *Yr4b*)		AvsYr8NIL (*Yr8*)	AvS/Exp (1/1-1 Line 74) (*YrExp2*)

(continued)

Table 2.1 (continued)

No.	World (W) and European (E)	China	India	United States	*Yr* single gene line
18		*Triticum spelta* album (*Yr5*)		AvsYr9NIL (*Yr9*)	AvSYr76NIL (*Yr76*)
19		Guinong 22 (*YrGui1*, *YrGui2*, *YrGui3*)		Clement (*Yr9*, *YrCle*)	
20				Compair (*Yr8*, *Yr19*)	

[a]Riebesel 47/51 was used before 2000 and AvSYr5NIL used after 2000 (Chen et al. 2002)

Johnson et al. (1972) established the "world" and "European" sets of only wheat differentials. The world subset consists of Chinese 166 (*Yr1*), Lee (*Yr7*, *Yr22*, *Yr23*), Heines Kolben (*Yr6*), Vilmorin 23 (*Yr4a*, *YrV23*), Moro (*Yr10*, *YrMor*), Strubes Dickkopf (*YrSD*) and Suwon 92/Omar (*YrSU*); and the European subset consists of Hybrid 46 (*Yr4b*, *YrH46*), Reichersberg 42 (*Yr7*), Heines Peko (*Yr6*), Nord Desprez (*Yr3a*, *YrND*), Compair (*Yr8*, *Yr19*), Carstens V (*Yr32*), Spaldings Prolific (*YrSP*), and Heines VII (*Yr2*, *YrHVII*). They developed a nomenclature system for *P. striiformis* f. sp. *tritici* races by assigning a decanery value to each of the world and European differentials. Chinese 166 and Hybrid 46 have a decanery value 1 ($=2^0$), Lee and Reichersberg 42 2 ($=2^1$), Heines Kolben and Heines Peko 4 ($=2^2$), Vilmorin 24 and Nord Desprez 8 ($=2^3$), Moro and Compair 16 ($=2^4$), Strubes Dickkopf and Carstens V 32 ($=2^5$), Suwon 92 x Omar and Spaldings Prolific 64 ($=2^6$), and Heines VII 128 ($=2^7$). Races are named as two numbers separated by "E" for European; and the number is determined by $\sum X$, where X is the decanery value of a differential with a susceptible reaction. The major advantages of this system are that the differentials include some early genetic stocks used in wheat breeding programs in European countries and possibly in some other countries in other continents. The race names allow derivation of virulence/avirulence patterns on the differentials. The disadvantages of this system include complexity, discontinuous numbers of race names, and difficulties in adding new differential genotypes. Also, the differential cultivars were developed more than 50 years ago and might not be suitable to distinguish current and future races.

The world and European sets of differentials have been used in many other countries. In Pakistan, *P. striiformis* f. sp. *tritici* races were identified using the world and Europe sets of differentials without additions or modifications (Hussain et al. 2012). In Australia and New Zealand, the world and European sets were initially used to identify races of *P. striiformis* f. sp. *tritici* immediately after the fungus was introduced to these countries (Wellings and McIntosh 1990). Later, Clement (*Yr9*,

YrCle; with decanery value 2^7=128) and *Triticum aestivum* subsp. *spelta* Album (*Yr5*; with decanery value 2^8=256) were added to the world set of differentials (Wellings and McIntosh 1990), as proposed by Johnson and Taylor (1976). In addition, three Australian cultivars were used as a group of supplementary differentials including Sonora, W195 or Federation + *Pm4b* as a susceptible control; Oxley (*Yr6*); and Avocet R (previously *YrA* and later identified as complementary *Yr73* and *Yr74*) (Wellings and McIntosh 1990). They use the same nomenclature system as Johns et al. (1972) with addition of "A−" and "A+" to denote avirulence and virulence on Avocet A, and also describe the pathogenicity of pathotypes (races) to resistance genes in the differentials including *Yr1*, *Yr2*, *Yr3*, *Yr4*, *Yr5*, *Yr6*, *Yr7*, *Yr8*, *Yr9*, *Yr10*, *YrSP*, and *YrA* and cultivars Carstens V, Suwon 92/Omar, Strubes Dickkopf and Nord Desprez. Later, Selkirk (*Yr27*) and Trident (*Yr17*), which were important in Australian studies of the stripe rust pathogen, were added to the international set of differentials (Wellings and Kandel 2004; Wellings 2007). The modified international set of differentials and pathotype nomenclature system have been also used in South Africa (Boshoff et al. 2002; Pretorius et al. 2007), Egypt (Shahin et al. 2014) and Syria (AL-Ahmed and Nachit 2014).

In India and Nepal, the world and European sets have been used together with some Indian wheat cultivars (Nagarajan et al. 1984, 1986; Karki 1989; Karki and Sharma 1990; Sharma 1997; Karki et al. 2012). The most recently used set of 24 wheat genotypes are a mix of the world and Indian differentials including Chinese 166, Lee, Heines Kolben, Vilmorin 23, Moro, Strubes Dickkopf, Suwon 92 × Omar, Riebesel 47/51 (*Yr9+*) in Set-A (a modified world set); Hybrid 46, Heines VII, Compair, *T. spelta* album, Tc*6/Lr26 (*Yr9*), Sonalika (*Yr2+*) and Kalyansona (*Yr2*) in Set-B; and WH147, Bilara, WH416, HD2329, HD2667, PBW343, HS240, Anza and A-9-30-1 in Set-0 (Bhardwaj et al. 2014). In Iran, 11 wheat genotypes, *T. spelta* var. album, Anza (*YrA+*), Avocet R, Avocet S, Kalyansona, Federation*4/Kavkaz (*Yr9*), TP981, TP1295 (*Yr25*), Meering + Yr24 (*Yr24*), Chamran (Attila 50) and Bolani (susceptible check), are used as a supplemental set, in addition to the world set (with the addition of Clement (*Yr2*, *Yr9+*) and the European set (Afshari 2012a, b).

Recently, a core set of differentials with resistance to 13 *Yr* genes (*Yr1*, *Yr2*, *Yr3*, *Yr4*, *Yr6*, *Yr7*, *Yr8*, *Yr9*, *Yr10*, *Yr17*, *Yr25*, *Yr32*, and *YrSp*) were used in the Global Rust Reference Center to identify races from wheat stripe rust samples collected from European and other countries (Hovmøller et al. 2016). Of the 13 genes, near-isogenic lines for *Yr1*, *Yr6*, *Yr7*, *Yr8*, *Yr9*, *Yr10*, *Yr17*, *Yr32*, and *YrSp* are available; and cultivars for *Yr2*, *Yr3*, and *Yr4* are likely Heines VII, Vilmorin 23, and Hybrid 46, respectively. These cultivars have other non-permanently named genes. An extended set of wheat differential lines has been used by the Global Rust Reference Center (http://wheatrust.org). This set is a combination of lines from the

'World' and 'European' differential sets and near-isogenic lines in an Avocet background. Currently, this set includes wheat differentials possessing the above 13 genes plus near-isogenic lines or cultivars for 6 more genes or cultivars (*Yr5*, *Yr15*, *Yr24*, *Yr27*, AvS and Amb (http://wheat rust.org). In the Global Rust Reference Center, at least two differential lines are tested for determining avirulence/virulence to commonly used resistance genes like *Yr1*, *Yr2*, *Yr6*, *Yr7*, *Yr8*, *Yr9*, *Yr17*, *Yr25*, *Yr27*, *Yr32* and *YrSp*. Races are presented by '-' for avirulence and *Yr* gene number in the sequential order of the *Yr*-gene differential lines and cultivars. Common race names are given based on relatedness to previous published race names or epidemic sites (shown by countries/years) where a race was first recognized. The set of differentials is similar to the current set used in the United States as discussed below.

2.1.2.2 The Chinese System

In China, the system for identifying *P. striiformis* f. sp. *tritici* races has been evolved since early 1940s. The first set of differentials included seven wheat genotypes (Carstens V, Heines Kolben, Vilmorin 23, 9 H 77, Hybrid 128, Carina, and Michigan Amber) and a barley genotype Heil's Franken, which are of European or U.S origins (Fang 1944). Gassner and Straib's European differentials were also used in China in some studies (Lu et al. 1956). However, these differentials were not suitable for differentiating *P. striiformis* f. sp. *tritici* races in China. Later, seven wheat genotypes were selected as differentials (Wang et al. 1963). Although changes in composition of differentials have been made several times, the major differentials and race nomenclature system have been relatively stable since the early 1960s. The current set of wheat differential includes 19 wheat genotypes: Trigo Eureka (*Yr6*), Fulhard, Lutescens 128, Mentana, Virgilio (*YrVir1*, *YrVir2*), Abbondanza, Early Premium, Funo (*YrA+*), Danish 1 (*Yr3*), Jubilejina 2 (*YrJu1*, *YrJu2*, *YrJu3*, *YrJu4*), Fengchan 3 (*Yr1*), Lovrin 13 (*Yr9+*), Kangyin 655 (*Yr1*, *YrKy1*, *YrKy2*), Suwon 11 (*YrSu*), Zhong 4, Lovrin 10 (*Yr9*), Hybrid 46 (*Yr3b*, *Yr4b*), *Triticum spelta* album (*Yr5*), and Guinong 22 (*YrGui1*, *YrGui2*, *YrGui3*) (Wan et al. 2004, 2015; Chen et al. 2009b; Hu et al. 2012; Liang et al. 2013; Zhan et al. 2016). The most of the wheat cultivars are of foreign origin, many have two or more resistance genes and some cultivars have resistance genes that have not been identified.

2.1.2.3 The US System

In North America, studies of host range of and cultivar resistance to *P. striiformis* f. sp. *tritici* in the early 1920s indicated the existence of races of *P. striiformis* f. sp. *tritici* (Hungerford and Owens 1923; Humphrey et al. 1924). In the early 1930s, Newton et al. (1933) reported two races in Canada and Bever (1934) also reported two races in the United States. In Bever's study, the race from Montana was virulent on wheat cultivars Chinese 166 (*Yr1*) and Red Russian and avirulent on wheat cultivar Heines Kolben (*Yr6*, *YrHK*), whereas the race from Idaho was virulent on

Heines Kolben and avirulent on Chinese 166 and Red Russian. Later, Newton and Johnson (1936) reported that their two races were similar to the two races reported by Bever (1934).

There was no research on race of *P. striiformis* f. sp. *tritici* or any aspect of stripe rust from the middle 1930s until the 1960s as the disease was not a big problem in the United States and Canada (Line and Qayoum 1992; Line 2002). The severe epidemics of stripe rust in the later 1950s and especially the early 1960s resumed stripe rust research, especially the identification of races (Line 2002). A set of five wheat cultivars, Lemhi (*Yr21*), Chinese 166 (*Yr1*), Heines VII (*Yr2*, *YrHVII*), Moro and Suwon 92/Omar were selected as differentials in 1968 (Line et al. 1970). Suwon 92/Omar was replaced by Paha (*YrPa1*, *YrPa2*, *YrPa3*) in 1974 because it was grown in Northwestern United States, had the same parentage and differentiated the same races as Suwon 92/Omar (Line and Qayoum 1992). Later, Druchamp (*Yr3a*, *YrD*, *YrDru*) and Riebesel 47-51 (*Yr9*) were added in 1969; Produra (*YrPr1*, *YrPr2*) and Yamhill (*Yr2*, *Yr4a*, *YrYam*) in 1974, Stephens (*Yr3a*, *YrS*, *YrSte*) in 1976, Lee (*Yr7*, *Yr22*, *Yr23*) in 1977, Fielder (*Yr6*, *Yr20*) in 1980, Tyee (*Yr76* = *YrTye*) in 1983, Tres (*YrTr1*, *YrTr2*) in 1989, Hyak (*Yr17*, *Yr76*) in 1990 and Express (*YrExp1*; *YrExp2*) in 1998 (Line and Qayoum 1992). In 2000, AvSYr8NIL (*Yr8*), AvSYr9NIL (*Yr9*), Clement (*Yr9*, *YrCle*) and Compair (*Yr8*, *Yr19*) were added to the differential set in 2000 (Chen et al. 2002). Also, Riebesel 47/51 was replaced by AvSYr5NIL (*Yr5*) because Riebesel 47/51 did not produce consistent infection with isolates collected in 2000 in the United States and resistance gene *Yr5* was started being used widely in breeding programs (Chen et al. 2002). This set of 20 differentials was used to identify races *P. striiformis* f. sp. *tritici* in the United States from 2000 to 2009 (Chen et al. 2002, 2010; Chen 2005; Wan and Chen 2012a, 2014a, b; Wan et al. 2016). Races were described by virulent/avirulent formula on the sequential numbers of the differentials and named using an open number system based on the order of identification. The advantages of the system include the easiness of naming, communication, counting the number of races and understanding of the identification order; and direct relations to commercial cultivars. However, the system has several disadvantages. For example, virulence to one gene can be masked by avirulence to another gene in the same differential; virulence or avirulence on the old cultivars may not be related to current and future cultivars developed and grown in the same region and those in other regions; and race names do not indicate virulence/avirulence formulae.

2.1.2.4 The *Yr* Single-Gene Differential System

As discussed above, the different systems have been used for identifying races of *P. striiformis* f. sp. *tritici* in different regions and countries of the world. Every system has advantages and drawbacks. The biggest problem is the difficulty to compare virulences of the pathogen populations. There have been very few comparisons of virulences of *Pst* populations from different countries using a set or different sets of differentials (de Vallavieille-Pope and Line 1990; Wellings 2007; Zhan et al. 2012;

Sharma-Poudyal et al. 2013). Such comparative studies are often not feasible because of shipping stripe rust samples from one country to another, repeating tests of same isolates on the same set of differentials and/or on more than one set of differentials, and of course the big time and resource costs. Also as discussed above, every set of differentials used in various regions have cultivars with two or more genes, which makes it difficult to identify virulence to a particular resistance gene. As marker technologies have been developed, it has become possible of identifying and mapping more and more resistance genes. Molecular markers have become available for use in marker-assisted selection for incorporating resistance genes into commercial cultivars. Similar to other diseases, breeding for stripe rust resistance has become more based on genes for resistance, rather than previously based on varieties with unknown genes. Sharing genes for effective resistance and their markers among breeding programs in different countries has become more and more common practices, and breeders have become more eager to know virulences of the pathogen populations to particular genes in the world. This information can be obtained more efficiently by using a set of single resistance gene differentials. Thus, it is important and urgent to establish a set of differentials with single genes for stripe rust resistance.

A great effort was taken by Wellings and colleagues to develop near-isogenic lines for some important wheat genes for resistance to stripe rust (Wellings et al. 2004). They used Australian cultivar Avocet S (AvS) to develop near isogenic lines for resistance genes *Yr1*, *Yr5*, *Yr6*, *Yr7*, *Yr8*, *Yr9*, *Yr10*, *Yr15*, *Yr17*, *Yr18*, *Yr24*, *Yr26* (the same as *Yr24*), *Yr27*, *Yr28*, *Yr29*, *Yr32*, *YrSP* and *YrTr1*. These *Yr*-gene lines have been distributed and used to monitor virulences of the *P. striiformis* f. sp. *tritici* populations throughout the world under field conditions (Wellings et al. 2009). Testing these lines under field conditions can provide information on individual virulences, but often cannot provide information on races, which needs to test individual samples under controlled conditions. Because AvS is a spring wheat variety, these spring type *Yr* single-gene lines may not be conveniently grown under field conditions in areas where only winter wheat is grown. The Plant Protection Institute, Chinese Academy of Agricultural Sciences in Beijing has taken an effort to developing near-isogenic lines of some *Yr* genes in a winter wheat background, which may partially solve the problems of spring wheat lines for some of the *Yr* genes. However, these winter wheat lines have not been widely used. Ideally, it is useful to have a spring wheat line and winter wheat line for each important *Yr* gene.

Although there have been several attempts to use *Yr* near-isogenic lines to monitor virulence changes in *P. striiformis* f. sp. *tritici* populations in different parts of the world by growing them under field conditions (Wellings et al. 2004; 2009), only few studies have tried to use some of the lines for identifying races of the pathogen (Kumar et al. 2012; Holtz et al. 2013; Sharma-Poudyal et al. 2013; Barar and Kutcher 2016; http://wheatrust.org/yellow-rust-tools-maps-and-charts/). However, these studies did not use a standard set of differentials of *Yr* single-gene lines.

To select *Yr* single-gene lines for including in the new differential set, more than 20 single-gene lines were initially used to test *P. striiformis* f. sp. *tritici* (Sharma-Poudyal et al. 2013). Finally, 18 wheat single-gene lines representing 18 *Yr* genes

(*Yr1*, *Yr5*, *Yr6*, *Yr7*, *Yr8*, *Yr9*, *Yr10*, *Yr15*, *Yr17*, *Yr24*, *Yr27*, *Yr32*, *Yr43*, *Yr44*, *YrSP*, *YrTr1*, *YrExp2* and *Yr76*) (Table 2.1) were selected to establish the new single-gene line differentials (Wan and Chen 2014a). The reasons for selecting these 18 *Yr* genes were given by Wan and Chen (2014a, b), and the specificity and suitability of these genes and their carriers are briefly discussed below. With more than 3000 *P. striiformis* f. sp. *tritici* isolates collected from the United States and other countries in 2010–2015, tests with this set of differentials provide more clear information about virulences to specific individual resistance genes (Wan and Chen 2014a, b, 2015; Wan et al. 2016, 2017).

Resistance gene *Yr1* was the first officially named gene in wheat for resistance to stripe rust (Lupton and Macer 1962). Wheat variety Chinese 166, which possesses the single *Yr1* gene (Lupton and Macer 1962; Chen and Line 1992b, c), has been used in the world and US differential sets for identifying races in various parts of the world since 1960s (Line et al. 1970; Johnson et al. 1972; Stubbs 1985; Chen 2005). The variety was used as the donor for developing the *Yr1* near-isogenic line (AvSYr1NIL) in the 'Avocet S' (AvS) background (Wellings et al. 2004). The *Yr1* single-gene line and Chinese 166 produced identical race results (Sharma-Poudyal et al. 2013; Wan and Chen 2014a, b). Race analysis from isolates of 1968–2015 in the United State using the new set of differentials showed that the *Yr1* near-isogenic line can clearly separate *Pst* races into two distinct virulence groups (Wan and Chen 2014a, b; Wan et al. 2016; Liu et al. 2016c). Interestingly, clustering of *Pst* isolates based on molecular markers showed significant correlation to virulence groups based on *Yr1* virulence and avirulence (Chen et al. 1993; Xia et al. 2016a; Chen and associates, unpublished data). These results indicate that *Yr1* has played an important role in shaping the pathogen populations in the United States and other countries. Virulence to *Yr1* exits in many countries at various frequencies (Sharma-Poudyal et al. 2013).

Both *Yr5* and *Yr15* were selected as differentials because of their importance in breeding programs (Wan and Chen 2014a). As *Yr5* was originally in spelt wheat (*Triticum spelta*) (Macer 1966; Kema 1992) and *Yr15* from *Triticium dicoccoided* (Gerechter-Amitai et al. 1989), gene-dragging for undesirable traits could be the major problem to the use of these genes in wheat breeding programs. The development of near-isogenic lines for *Yr5* and *Yr15* in the common wheat AvS background (Wellings et al. 2004) and identification of molecular markers linked to these genes (Chen et al. 2003b; Yan et al. 2003; Murphy et al. 2009) have facilitated and accelerated incorporating these genes into commercial cultivars. For example, spring wheat cultivars Clearwhite 515 (*Yr5*, *Yr15*), Expresso (*Yr15*), Patwin 515 (*Yr15*, *Yr17*) and Seahawk (*Yr5*, *Yr15*) have been widely grown in the western United States. As *Yr5* and *Yr15* are effective against *Pst* throughout the world, they are likely used in wheat breeding in many other countries. Although virulences to these genes have not been identified in the United States and many other countries (Chen et al. 2002; Madariaga et al. 2004; Manninger 2004; Chen 2005; Chen et al. 2010; Rodríguez-García et al. 2010; Zhan et al. 2012; Sharma-Poudyal et al. 2013; Barar and Kutcher 2016; Wan et al. 2016, 2017), the virulence to *Yr5* has been identified in India (Nagarajan et al. 1986) and Australia (Wellings and McIntosh 1990; Wellings 2007);

and the virulence to *Yr15* was reported in Afghanistan and other places (McIntosh et al. 1995). Inclusion of *Yr5* and *Yr15* in the differentials should allow quick identification of races virulent to these genes.

Resistance genes *Yr6*, *Yr7*, *Yr8*, *Yr9*, *Yr10*, *Yr17*, *Yr24*, *Yr27* and *Yr32* are either historically important or have become important throughout the world. Virulences to *Yr6* and *Yr7* were high and common around the world (Wellings et al. 2009; Sharma-Poudyal et al. 2013; Wan et al. 2017). Frequencies of virulences to *Yr6*, *Yr7*, *Yr8*, *Yr9* and *Yr17* are generally high throughout the world; whereas virulences to *Yr10*, *Yr24*, *Yr27* and *Yr32* are low or varied from country to country (Wellings et al. 2009; Sharma-Poudyal et al. 2013; Wan et al. 2017). Including the near isogenic lines of these genes in the differential should allow comparison of the *Pst* races among countries and races in different periods. Most of these near-isogenic lines produce distinct avirulent (IT 1-2) and virulent (IT 8-9) phenotypes, but the *Yr8*, *Yr17* and *Yr27* lines produced intermediate reactions (IT 4-6) when inoculated with some isolates, and their reactions can be affected by different temperatures. The selection of these gene lines was more due to the importance of the genes, rather than distinct reactions (Wan and Chen 2014a). Virulences to *Yr17* and *Yr27* were the major causes for the stripe rust epidemics in many countries in the last 20 years. The sources of *Yr17* resistance, VPM1 and derivatives, have been widely used in wheat breeding programs due to their resistance to multiple diseases and other desirable traits. Similarly, *Yr27* has been also used widely in wheat breeding programs in many countries. A large number of cultivars with *Yr17* and/or *Yr27* have been developed and grown throughout the world. As a result of host selection, the virulences to *Yr17* and *Yr27* have developed in many countries (Bayles et al. 2000; Boshoff et al. 2002; Wellings 2008; Wellings et al. 2009; Rodríguez-García et al. 2010; Sharma-Poudyal et al. 2013; Wan and Chen 2014a; Barar and Kutcher 2016; Wan et al. 2016, 2017). Globally and regionally, it is not doubt that both *Yr17* and *Yr27* have played and will play a very important role in virulence monitoring, understanding the evolution and migration of the *Pst* populations. However, it is important to have well controlled conditions, especially for temperature, in differential tests to obtain reliable reaction data on these single-gene lines. Testing results from XM Chen's program suggest that these lines have not only a single *Yr* gene, but also a linked gene or genes for high-temperature adult-plant resistance that affect the response of these gene lines to stripe rust at high temperature and/or at late growth stages. To produce better differential results and understanding host-pathogen interactions, efforts have been taken to develop a more "near-isogenic" line for *Yr17* and other genes with similar problems.

The inclusion of single-gene lines with *Yr43*, *Yr44*, *YrSP*, *YrTr1*, *YrExp2* and *YrTye* in the differential set was based on their high differential ability and importance mostly in the United States (Wan and Chen 2014a). The *Yr43* and *Yr44* lines were developed from crosses of cultivars IDO377s and Zak with AvS, respectively (Sui et al. 2009; Cheng and Chen 2010). In the United States, virulences to *Yr43* and

Yr44 have become high and presented in major races identified since the year 2000, but either absent (for *Yr43*) or at low frequencies before 2000 (Wan and Chen 2014a; Wan et al. 2016; Liu et al. 2016a, b, c). The near-isogenic line with *YrSP* was selected to be included in the single gene line differentials as it had very clear reactions (either highly resistant or highly susceptible) when tested with US *Pst* isolates (Wan and Chen 2014a) and international isolates (Sharma-Poudyal et al. 2013). *YrSP* was recently mapped to the long arm of chromosome 2B with tightly linked molecular markers, and the gene was found clearly different from other genes, such as *Yr5*, *Yr7*, *Yr43*, and *Yr44*, on the same chromosomal arm (Feng et al. 2015). In the United States, the *YrSP* virulence was very rare before 2000, but has become significant in frequency since 2000 (Wan and Chen 2014a; Liu et al. 2016c; Wan et al. 2016). Virulence to Spaldings Prolific, the *YrSP* donor that has been used in the European set of differentials, was reported in many geographic areas in late 1960s and early 1970s (McIntosh et al. 1995; Stubbs et al. 1974). In a recent study, virulence to *YrSP* was found in isolates from China, Turkey and Uzbekistan, but not in many other countries (Sharma-Poudyal et al. 2013).

The near isogenic line for *YrTr1* was developed from backcrossing of Tres to Avocet S (Wellings et al. 2009). Tres, which has two genes (*YrTr1* and *YrTr2*) for resistance to stipe rust (Chen and Line 1992a), was used in the old US differential set for identifying races (Line and Qayoum 1992; Chen et al. 2002, 2010; Chen 2005; Wan and Chen 2012a). In the United States, the *YrTr1* virulence had a moderate frequency and was present in all epidemiological regions (Wan and Chen 2014a), whereas its frequency varied greatly from country to country (Sharma-Poudyal et al. 2013). The single gene line *YrExp2* is a recombinant inbred line developed from a cross between AvS and Express (Lin and Chen 2008). In the United States, the frequency of virulence to *YrExp2* has been consistently increasing since the 1960s (Wan and Chen 2014a; Liu et al. 2016c; Wan et al. 2016). Although the frequency of the *YrExp2* virulence varied considerably, the virulence presented in all countries with *Pst* isolates tested (Sharma-Poudyal et al. 2013; Wan et al. 2017).

When the *Yr* single-gene differential set was established in 2010 (Wan and Chen 2014a), Tyee was the only winter wheat cultivar included. The cultivar has a single gene named *YrTye* (Chen and Line 1992b, c; Chen et al. 1995a) and has been used in the old US differential set (Line and Qayoum 1992; Chen 2005; Wan and Chen 2012a). *YrTye* is present in many wheat cultivars developed and grown in the U.S. Pacific Northwest (Xiang et al. 2016). The cultivar also differentiated *Pst* races from isolates in many countries (Sharma-Poudyal et al. 2013; Wan et al. 2017). A near-isogenic line in the AvS background was recently developed for *YrTye* and the gene was permanently named *Yr76* (Xiang et al. 2016). The near-isogenic line is used to replace Tyee as the differential for *Yr76*.

Association analysis revealed both significant and insignificant and both positive and negative correlations among virulences to the 18 *Yr* genes (Wan et al. 2016). As *Pst* populations are largely clonal (Cheng and Chen 2014; Wan and Chen 2016; Wang et al. 2016), the significant correlations among some of the virulences are

Table 2.2 Key for designating octal and binary codes of races of *Puccinia striiformis* f. sp. *tritici* differentiated with 18 *Yr* single-gene lines[a]

Differential[b]				Octal code							
Set	*Yr* gene			0	1	2	3	4	5	6	7
1	*Yr1*	*Yr5*	*Yr6*	AAA	AAV[c]	AVA	AVV	VAA	VAV	VVA	VVV
2	*Yr7*	*Yr8*	*Yr9*	AAA	AAV	AVA	AVV	VAA	VAV	VVA	VVV
3	*Yr10*	*Yr15*	*Yr17*	AAA	AAV	AVA	AVV	VAA	VAV	VVA	VVV
4	*Yr24*	*Yr27*	*Yr32*	AAA	AAV	AVA	AVV	VAA	VAV	VVA	VVV
5	*Yr43*	*Yr44*	*YrSP*	AAA	AAV	AVA	AVV	VAA	VAV	VVA	VVV
6	*YrTr1*	*YrExp2*	*Yr76*	AAA	AAV	AVA	AVV	VAA	VAV	VVA	VVV
Binary code				000	001[d]	010	011	100	101	110	111

[a]Reproduced from Wan et al. (2016) for the previous octal numbers
[b]The sets of differentials are expendable by three *Yr* genes in a set
[c]A = avirulent and V = virulent
[d]0 = avirulent and 1 = virulent

expected. None of any pair of the *Yr* single-gene lines identified same *Pst* races in the United States, except those of *Yr5* and *Yr15* as none of the races are virulent to either of the genes (Chen 2005; Chen et al. 2010; Wan and Chen 2014a; Wan et al. 2016). However, an Australian *Pst* isolate was virulent to *Yr5*, but avirulent to *Yr15* (Wellings 2007). Testing *Pst* isolates from 16 other countries using the 18 *Yr* single-gene lines identified different races has proven the differential abilities of these lines and allowed direct comparison of *Pst* races among countries in different continents (Sharma-Poudyal et al. 2013; Wan et al. 2016, 2017).

The new set of 18 *Yr* single-gene differentials is coupled with a nomenclature system for naming *Pst* races (Wan and Chen 2014a; Wan et al. 2016). Races are named with a prefix "PSTv" standing for *Pst* races based on virulence and a number that is given at the time when a new race is identified different from all previously named races. The PSTv prefix distinguishes races with a "PST" prefix identified with the new set of differentials from those identified with the previous set of 20 wheat cultivar differentials (Chen et al. 2002; Chen 2005, 2007; Chen et al. 2010; Wan and Chen 2012a, b). The number system makes it easy to keep tracking the number of *Pst* races identified. The number names of races also indicate relatively when they were identified. Compared with the previous "PST" races, "PSTv" races temporarily lost indication of the sequential order of identification as the system was started to be used in 2010 and the study of historical isolates of the United States with the new set of differentials had not been completed until 2016 (Liu et al. 2016c). From 2016, the year of a new race identified can be indicated by its name.

A PSTv race is also coded using an octal system (Wan and Chen 2014a), and the initial codes have been slightly modified to fit an international binary coding system (Wan et al. 2016). In this system, the 18 *Yr* single-gene differentials are grouped in 6 groups or subsets with 3 differentials in a group (Table 2.2). Lines with *Yr1*, *Yr5* and *Yr6* are classified in the first group; *Yr7*, *Yr8* and *Yr9* in the second, *Yr10*, *Yr15* and *Yr17* in the third; *Yr24*, *Yr27* and *Yr32* in the fourth; *Yr43*, *Yr44* and *YrSP* in the fifth; and *YrTr1*, *YrExp2* and *YrTye* (*Yr76*) in the sixth. Each of all eight possible virulence patterns in each group is assigned to a number from 0 to 7, with 0 for avirulent and 7 for virulent to all three differentials in the group; and 1 for virulence to only the third differential, 2 for virulence to only the second, 3 for virulences to both the second and third, 4 for virulence only to the first, 5 for virulences to the first and third and 6 for virulences to the first and second. In this way, the octal codes (0 to 7) correspond well to binary codes 000, 001, 010, 011, 100, 101, 110 and 111, respectively. Using the octal codes, a race has a six-digit code. For example, race PSTv-1 has a code 400000, indicating that it is virulent only on the first differential (*Yr1*) in the first group, whereas PSTv-51 is coded by 575777, indicating that it is virulent to all 18 *Yr* genes except *Yr5* (the second in the first group) and *Yr15* (the second in the third group) (Table 2.2). The octal code system makes easy for identifying new races, figuring out virulence patterns and determining relationships among races. As octal codes are easy to be expended into binary codes, computer programs can be easily applied to analyze virulence data of races.

So far, more than 280 races of *P. striiformis* f. sp. *tritici* have been identified from isolates from more than 10 countries using the 18 *Yr* single-gene differentials. The virulence/avirulence formulae of these races and their first identified countries are presented in Table 2.3. The virulence spectra range from 0 (PSTv-18) to 16 (PSTv-51) of the virulences to the 18 *Yr* genes. These races and their virulences will serve as the basis for identifying new races.

The new system of differentials and nomenclature is an open system. Additional *Yr* single-gene lines can be added to the set of differentials. Because the octal coding system requires three genes, it will be better to add 3 or 3× lines. This is somehow inconvenient, but is workable.

2.1.3 Issues with Race Identification

None of the past and currently used systems are perfect, and there are several issues related to race identification. Some of the problems associated to the previous race identification systems have been discussed above, which are the reasons for developing the currently used *Yr* single-gene lines for differentiating races. Even for the single-gene line system, there are still some issues that are needed to be considered when use the race information.

2.1.3.1 Race Information Does Not Cover Virulences and Avirulences to All Resistance Genes

The first issue of greatest concern is that races identified with a set of differentials only provide information of virulences and avirulences to the genes in the differentials. Whether a race is virulent or avirulent to other genes require additional tests.

So far, there are 78 permanently named and about 100 temporarily named *Yr* genes for resistance to stripe rust have been identified (McIntosh et al. 2013, 2014;

Table 2.3 Races of *Puccinia striiformis* f. sp. *tritici* and octal codes, virulence formula, the origins of the collections and the year first identified[a]

Race	Octal code	Virulence formula (on *Yr* genes)	Country	Year
PSTv-1	400000	1	USA	2010
PSTv-2	511001	1,6,9,17,76	USA	2010
PSTv-3	531201	1,6,8,9,17,27,76	USA	2010
PSTv-4	511211	1,6,9,17,27,SP,76	USA	2010
PSTv-5	571220	1,6,7,8,9,17,27,44	USA	2010
PSTv-6	570250	1,6,7,8,9,27,43,SP	USA	2010
PSTv-7	531262	1,6,8,9,17,27,43,44,Exp2	USA	2010
PSTv-8	471223	1,7,8,9,17,27,44,Exp2,76	USA	2010
PSTv-9	575221	1,6,7,8,9,10,17,27,44,76	USA	2010
PSTv-10	531253	1,6,8,9,17,27,43,SP,Exp2,76	USA	2010
PSTv-11	571263	1,6,7,8,9,17,27,43,44,Exp2,76	USA	2010
PSTv-12	571067	1,6,7,8,9,17,43,44,Tr1,Exp2,76	USA	2010
PSTv-13	571271	1,6,7,8,9,17,27,43,44,SP,76	USA	2010
PSTv-14	571267	1,6,7,8,9,17,27,43,44,Tr1,Exp2,76	USA	2010
PSTv-15	551273	1,6,7,9,17,27,43,44,SP,Exp2,76	USA	2010
PSTv-16	574566	1,6,7,8,9,10,24,32,43,44,Tr1,Exp2	USA	2010
PSTv-17	571273	1,6,7,8,9,17,27,43,44,SP,Exp2,76	USA	2010
PSTv-18	000000	/	USA	2010
PSTv-19	100000	6	USA	2010
PSTv-20	001000	17	USA	2010
PSTv-21	101000	6,17	USA	2010
PSTv-22	021000	8,17	USA	2010
PSTv-23	004500	10,24,32	USA	2010
PSTv-24	040022	7,44,Exp2	USA	2010
PSTv-25	041022	7,17,44,Exp2	USA	2010
PSTv-26	104700	6,10,24,27,32	USA	2010
PSTv-27	005504	10,17,24,32,Tr1	USA	2010
PSTv-28	151022	6,7,9,17,44,Exp2	USA	2010
PSTv-29	141222	6,7,17,27,44,Exp2	USA	2010
PSTv-30	170026	6,7,8,9,44,Tr1,Exp2	USA	2010
PSTv-31	171026	6,7,8,9,17,44,Tr1,Exp2	USA	2010
PSTv-32	170066	6,7,8,9,43,44,Tr1,Exp2	USA	2010
PSTv-33	170226	6,7,8,9,27,44,Tr1,Exp2	USA	2010
PSTv-34	171226	6,7,8,9,17,27,44,Tr1,Exp2	USA	2010
PSTv-35	171066	6,7,8,9,17,43,44,Tr1,Exp2	USA	2010
PSTv-36	170266	6,7,8,9,27,43,44,Tr1,Exp2	USA	2010
PSTv-37	171266	6,7,8,9,17,27,43,44,Tr1,Exp2	USA	2010
PSTv-38	171666	6,7,8,9,17,24,27,43,44,Tr1,Exp2	USA	2010
PSTv-39	175266	6,7,8,9,10,17,27,43,44,Tr1,Exp2	USA	2010
PSTv-40	174766	6,7,8,9,10,24,27,32,43,44,Tr1,Exp2	USA	2010

(continued)

Table 2.3 (continued)

Race	Octal code	Virulence formula (on *Yr* genes)	Country	Year
PSTv-41	175766	6,7,8,9,10,17,24,27,32,43,44,Tr1,Exp2	USA	2010
PSTv-42	471221	1,7,8,9,17,27,44,76	USA	2011
PSTv-43	470221	1,7,8,9,27,44,76	USA	2011
PSTv-44	470223	1,7,8,9,27,44,Exp2,76	USA	2011
PSTv-45	174226	6,7,8,9,10,27,44,Tr1,Exp2	USA	2011
PSTv-46	511201	1,6,9,17,27,76	USA	2011
PSTv-47	571266	1,6,7,8,9,17,27,43,44,Tr1,Exp2	USA	2011
PSTv-48	510001	1,6,9,76	USA	2011
PSTv-49	575767	1,6,7,8,9,10,17,24,27,32,43,44,Tr1,Exp2,76	USA	2011
PSTv-50	575277	1,6,7,8,9,10,17,27,43,44,SP,Tr1,Exp2,76	USA	2011
PSTv-51	575777	1,6,7,8,9,10,17,24,27,32,43,44,SP,Tr1, Exp2,76	USA	2011
PSTv-52	171262	6,7,8,9,17,27,43,44,Exp2	USA	2012
PSTv-53	510011	1,6,9,SP,76	USA	2012
PSTv-54	101204	6,17,27,Tr1	USA	1976
PSTv-55	541002	1,6,7,17,Exp2	USA	1977
PSTv-56	141204	6,7,17,27,Tr1	USA	1981
PSTv-57	141200	6,7,17,27	USA	1982
PSTv-58	040006	7,Tr1,Exp2	USA	1987
PSTv-59	101005	6,17,Tr1,76	USA	1990
PSTv-60	041206	7,17,27,Tr1,Exp2	USA	1991
PSTv-61	004504	10,24,32,Tr1	USA	1992
PSTv-62	041002	7,17,Exp2	USA	1994
PSTv-63	040002	7,Exp2	USA	1994
PSTv-64	110000	6,9	USA	2000
PSTv-65	104000	6,10	USA	2002
PSTv-66	104500	6,10,24,32	USA	2003
PSTv-67	105500	6,10,17,24,32	USA	2003
PSTv-68	175566	6,7,8,9,10,17,24,32,43,44,Tr1,Exp2	USA	2004
PSTv-69	175362	6,7,8,9,10,17,27,32,43,44,Exp2	USA	2007
PSTv-70	101001	6,17,76	USA	2007
PSTv-71	550263	1,6,7,9,27,43,44,Exp2,76	USA	2012
PSTv-72	175762	6,7,8,9,10,17,24,27,32,43,44,Exp2	USA	2013
PSTv-73	171263	6,7,8,9,17,27,43,44,Exp2,76	USA	2013
PSTv-74	511221	1,6,9,17,27,44,76	USA	2013
PSTv-75	161062	6,7,8,17,43,44,Exp2	USA	2013
PSTv-76	571262	1,6,7,8,9,17,27,43,44,Exp2	USA	2013
PSTv-77	511035	1,6,9,17,44,SP,Tr1,76	USA	2013
PSTv-78	160262	6,7,8,27,43,44,Exp2	USA	2013
PSTv-79	450020	1,7,9,44	USA	2013
PSTv-80	560262	1,6,7,8,27,43,44,Exp2	Nepal	2007
PSTv-81	575667	1,6,7,8,9,10,17,24,27,43,44,Tr1,Exp2,76	Nepal	2007

(continued)

Table 2.3 (continued)

Race	Octal code	Virulence formula (on *Yr* genes)	Country	Year
PSTv-82	575122	1,6,7,8,9,10,17,32,44,Exp2	Chile	2007
PSTv-83	571667	1,6,7,8,9,17,24,27,43,44,Tr1,Exp2,76	Pakistan	2007
PSTv-84	571063	1,6,7,8,9,17,43,44,Exp2,76	Turkey	2010
PSTv-85	570267	1,6,7,8,9,27,43,44,Tr1,Exp2,76	China	2007
PSTv-86	510233	1,6,9,27,44,SP,Exp2,76	China	2006
PSTv-87	101722	6,17,24,27,32,44,Exp2	Turkey	2009
PSTv-88	141462	6,7,17,24,43,44,Exp2	Pakistan	2007
PSTv-89	141226	6,7,17,27,44,Tr1,Exp2	Canada	2007
PSTv-90	141022	6,7,17,44,Exp2	Chile	2007
PSTv-91	161762	6,7,8,17,24,27,32,43,44,Exp2	Turkey	2009
PSTv-92	175526	6,7,8,9,10,17,24,32,44,Tr1,Exp2	Chile	2007
PSTv-93	175022	6,7,8,9,10,17,44,Exp2	Nepal	2008
PSTv-94	160022	6,7,8,44,Exp2	China	2007
PSTv-95	151226	6,7,9,17,27,44,Tr1,Exp2	Pakistan	2007
PSTv-96	151023	6,7,9,17,44,Exp2,76	Pakistan	2007
PSTv-97	561062	1,6,7,8,17,43,44,Exp2	China	2007
PSTv-98	161072	6,7,8,17,43,44,SP,Exp2	China	2007
PSTv-99	161022	6,7,8,17,44,Exp2	China	2007
PSTv-100	575377	1,6,7,8,9,10,17,27,32,43,44,SP,Tr1,Exp2,76	China	2007
PSTv-101	161266	6,7,8,17,27,43,44,Tr1,Exp2	China	2007
PSTv-102	160266	6,7,8,27,43,44,Tr1,Exp2	China	2007
PSTv-103	570237	1,6,7,8,9,27,44,SP,Tr1,Exp2,76	China	2007
PSTv-104	511235	1,6,9,17,27,44,SP,Tr1,76	China	2007
PSTv-105	551242	1,6,7,9,17,27,43,Exp2	Ethiopia	2013
PSTv-106	551262	1,6,7,9,17,27,43,44,Exp2	Ethiopia	2013
PSTv-107	571242	1,6,7,8,9,17,27,43,Exp2	Ethiopia	2013
PSTv-108	555262	1,6,7,9,10,17,27,43,44,Exp2	Ethiopia	2013
PSTv-109	174266	6,7,8,9,10,27,43,44,Tr1,Exp2	Ethiopia	2013
PSTv-110	575762	1,6,7,8,9,10,17,24,27,32,43,44,Exp2	Ethiopia	2013
PSTv-111	574762	1,6,7,8,9,10,24,27,32,43,44,Exp2	Ethiopia	2013
PSTv-112	174762	6,7,8,9,10,24,27,32,43,44,Exp2	Ethiopia	2013
PSTv-113	555722	1,6,7,9,10,17,24,27,32,44,Exp2	Ethiopia	2013
PSTv-114	154722	6,7,9,10,24,27,32,44,Exp2	Ethiopia	2013
PSTv-115	174722	6,7,8,9,10,24,27,32,44,Exp2	Ethiopia	2013
PSTv-116	175722	6,7,8,9,10,17,24,27,32,44,Exp2	Ethiopia	2013
PSTv-117	571246	1,6,7,8,9,17,27,43,Tr1,Exp2	Ethiopia	2013
PSTv-118	004104	10,32,Tr1	Chile	2007
PSTv-119	161262	6,7,8,17,27,43,44,Exp2	USA	2014
PSTv-120	510211	1,6,9,27,SP,76	USA	2014
PSTv-121	450031	1,7,9,44,SP,76	USA	2014
PSTv-122	401021	1,17,44,76	USA	2014

(continued)

Table 2.3 (continued)

Race	Octal code	Virulence formula (on *Yr* genes)	Country	Year
PSTv-123	451023	1,7,9,17,44,Exp2,76	USA	2014
PSTv-124	560170	1,6,7,8,32,43,44,SP	Italy	2014
PSTv-125	550073	1,6,7,9,43,44,SP,Exp2,76	Italy	2014
PSTv-126	551063	1,6,7,9,17,43,44,Exp2,76	Italy	2014
PSTv-127	551072	1,6,7,9,17,43,44,SP,Exp2	Italy	2014
PSTv-128	000002	Exp2	Italy	2014
PSTv-129	411022	1,9,17,44,Exp2	Italy	2014
PSTv-130	000022	44,Exp2	Italy	2014
PSTv-131	160062	6,7,8,43,44,Exp2	Italy	2014
PSTv-132	061462	7,8,17,24,43,44,Exp2	Italy	2014
PSTv-133	061062	7,8,17,43,44,Exp2	Italy	2014
PSTv-134	060062	7,8,43,44,Exp2	Italy	2014
PSTv-135	021022	8,17,44,Exp2	Italy	2014
PSTv-136	020062	8,43,44,Exp2	Italy	2014
PSTv-137	020022	8,44,Exp2	Italy	2014
PSTv-138	030033	8,9,44,SP,Exp2,76	Italy	2014
PSTv-139	170222	6,7,8,9,27,44,Exp2	Ethiopia	2014
PSTv-140	450021	1,7,9,44,76	USA	2015
PSTv-141	550221	1,6,7,9,27,44,76	USA	2015
PSTv-142	150022	6,7,9,44,76	USA	2015
PSTv-143	510015	1,6,9,SP,Tr1,76	USA	2015
PSTv-144	571272	1,6,7,8,9,17,27,43,44,SP,Exp2	USA	2015
PSTv-145	140000	6,7	USA	1968
PSTv-146	000100	32	USA	1968
PSTv-147	440000	1,7	USA	1968
PSTv-148	000020	44	USA	1968
PSTv-149	440022	1,7,44,Exp2	USA	1970
PSTv-150	440002	1,7,Exp2	USA	1973
PSTv-151	100022	6,44,Exp2	USA	1971
PSTv-152	540002	1,6,7,Exp2	USA	1971
PSTv-153	441020	6,7,17,44	USA	1971
PSTv-154	540211	1,6,7,27,SP,76	USA	1972
PSTv-155	000062	43,44,Exp2	USA	1974
PSTv-156	441042	6,7,17,43,Exp2	USA	1974
PSTv-157	000023	44,Exp2,76	USA	1974
PSTv-158	441202	6,7,17,27,Exp2	USA	1974
PSTv-159	101260	6,17,27,43,44	USA	1975
PSTv-160	000400	24	USA	1975
PSTv-161	004570	10,24,32,43,44,SP	USA	1975
PSTv-162	441022	1,7,17,44,Exp2	USA	1975
PSTv-163	141060	6,7,17,43,44	USA	1975

(continued)

Table 2.3 (continued)

Race	Octal code	Virulence formula (on *Yr* genes)	Country	Year
PSTv-164	040214	7,27,SP,Tr1	USA	1975
PSTv-165	100023	6,44,Exp2,76	USA	1975
PSTv-166	004520	10,24,32,44	USA	1975
PSTv-167	140001	6,7,76	USA	1975
PSTv-168	540402	1,6,7,24,Exp2	USA	1975
PSTv-169	100024	6,44,Tr1,	USA	1978
PSTv-170	100224	6,27,44,Tr1	USA	1978
PSTv-171	400031	1,44,SP,76	USA	1981
PSTv-172	104424	6,10,24,44,Tr1	USA	1985
PSTv-173	100001	6,76	USA	1987
PSTv-174	101021	6,17,44,76	USA	1991
PSTv-175	141225	6,7,17,27,44,Tr1,Exp2	USA	1991
PSTv-176	000024	44,Tr1	USA	1992
PSTv-177	000004	Tr1	USA	1993
PSTv-178	100021	6,44,76	USA	1993
PSTv-179	400004	1,Tr1	USA	1993
PSTv-180	541261	1,6,7,17,27,43,44,76	USA	1997
PSTv-181	104524	6,10,24,32,44,Tr1	USA	1997
PSTv-182	140002	6,7.Exp2	USA	1998
PSTv-183	140022	6,7,44,Exp2	USA	1987
PSTv-184	000026	44,Tr1,Exp2	USA	1998
PSTv-185	140020	6,7,44	USA	1998
PSTv-186	540267	1,6,7,27,43,44,Tr1,Exp2,76	USA	1998
PSTv-187	540223	1,6,7,27,44,Exp2,76	USA	1998
PSTv-188	140032	6,7,44,SP,Exp2	USA	1999
PSTv-189	140422	6,7,24,44,Exp2	USA	1999
PSTv-190	160023	6,7,8,44,Exp2,76	USA	2000
PSTv-191	571244	1,6,7,8,9,17,27,43,Tr1	USA	2000
PSTv-192	550277	1,6,7,9,27,43,44,SP,Tr1,Exp2,76	USA	2000
PSTv-193	510020	1,6,9,44	USA	2000
PSTv-194	510021	1,6,9,44,76	USA	2001
PSTv-195	510006	1,6,9,Tr1,Exp2	USA	2001
PSTv-196	450220	1,7,9,27,44,	USA	2001
PSTv-197	161272	6,7,8,17,27,43,44,SP,Exp2	USA	2001
PSTv-198	170262	6,7,8,9,27,43,44,Exp2	USA	2002
PSTv-199	420002	1,8,Exp2	USA	2002
PSTv-200	005100	10,17,32	USA	2002
PSTv-201	020000	8	USA	2002
PSTv-202	551211	1,6,7,9,17,27,SP,76	USA	2003
PSTv-203	511221	1,6,9,17,27,44,76	USA	2003
PSTv-204	164460	6,7,8,10,24,43,44	USA	2003

(continued)

Table 2.3 (continued)

Race	Octal code	Virulence formula (on *Yr* genes)	Country	Year
PSTv-205	450021	1,7,9,44,76	USA	2003
PSTv-206	161526	6,7,8,17,24,32,44,Tr1,Exp2	USA	2003
PSTv-207	050072	7,9,43,44,SP,Exp2	USA	2003
PSTv-208	550241	1,6,7,9,27,43,76	USA	2004
PSTv-209	070022	7,8,9,44,Exp2	USA	2004
PSTv-210	564421	1,6,7,8,10,24,44,76	USA	2004
PSTv-211	570031	1,6,7,8,9,44,SP,76	USA	2005
PSTv-212	511231	1,6,9,17,27,44,SP,76	USA	2005
PSTv-213	534500	1,6,8,9,10,24,32	USA	2005
PSTv-214	470022	1,7,8,9,44,Exp2	USA	2005
PSTv-215	551227	1,6,7,9,17,27,44,Tr1,Exp2,76	USA	2006
PSTv-216	160261	6,7,8,27,43,44,76	USA	2006
PSTv-217	550261	1,6,7,9,27,43,44,76	USA	2007
PSTv-218	174576	6,7,8,9,10,24,32,43,44,SP,Tr1,Exp2	USA	2008
PSTv-219	430022	1,8,9,44,Exp2	USA	2008
PSTv-220	160062	6,7,8,43,44,Exp2	USA	2016
PSTv-221	140262	6,7,27,43,44,Exp2	USA	2016
PSTv-222	154366	6,7,9,10,27,32,43,44,Tr1,Exp2	Mexico	2016
PSTv-223	470020	1,7,8,9,44	Mexico	2016
PSTv-224	165762	6,7,8,10,17,24,27,32,43,44,Exp2	Mexico	2016
PSTv-225	550273	1,6,7,9,27,43,44,SP,Exp2,76	China	2016
PSTv-226	540273	1,6,7,27,43,44,SP,Exp2,76	China	2016
PSTv-227	550272	1,6,7,9,27,43,44,SP,Exp2	China	2016
PSTv-228	550233	1,6,7,9,27,44,SP,Exp2,76	China	2016
PSTv-229	570073	1,6,7,8,9,43,44,SP,Exp2,76	China	2016
PSTv-230	550271	1,6,7,9,43,44,SP,76	China	2016
PSTv-231	550033	1,6,7,9,44,SP,76	China	2016
PSTv-232	550072	1,6,7,9,43,44,SP,Exp2	China	2016
PSTv-233	561262	1,6,7,8,17,27,43,44,Exp2	China	2016
PSTv-234	550003	1,6,7,9,Exp2,76	China	2016
PSTv-235	550031	1,6,7,9,44,SP,76	China	2016
PSTv-236	050233	7,9,27,44,SP,Exp2,76	China	2016
PSTv-237	150223	6,7,9,27,44,Exp2,76	China	2016
PSTv-238	410220	1,9,27,44	China	2016
PSTv-239	571277	1,6,7,8,9,17,27,43,44,SP,Tr1,Exp2,76	China	2016
PSTv-240	530003	1,6,7,Exp2,76	China	2016
PSTv-241	040040	7,43	China	2016
PSTv-242	556272	1,6,7,9,17,27,43,44,SP,Exp2	China	2016
PSTv-243	470270	1,7,8,9,27,43,44,SP	China	2016
PSTv-244	561766	1,6,7,8,17,43,44,Tr1,Exp2	China	2016
PSTv-245	561066	1,6,7,8,17,43,44,Tr1,Exp2	China	2016

(continued)

Table 2.3 (continued)

Race	Octal code	Virulence formula (on *Yr* genes)	Country	Year
PSTv-246	110071	6,9,43,44,SP,76	China	2016
PSTv-247	001051	17,43,SP,76	China	2016
PSTv-248	050040	7,9,43	China	2016
PSTv-249	566262	1,6,7,8,17,27,43,44,Exp2	China	2016
PSTv-250	000230	27,44,SP	China	2016
PSTv-251	010060	9,43,44	China	2016
PSTv-252	510011	1,6,9,SP,76	China	2016
PSTv-253	540043	1,6,7,43,Exp2,76	China	2016
PSTv-254	541273	1,6,7,17,27,43,44,SP,Exp2,76	China	2016
PSTv-255	100025	6,44,Tr1,76	China	2016
PSTv-256	050070	7,9,43,44,SP	China	2016
PSTv-257	150051	6,7,9,43,76	China	2016
PSTv-258	000001	76	China	2016
PSTv-259	170072	6,7,8,9,43,44,SP,Exp2	China	2016
PSTv-260	150071	6,7,9,27,43,44,SP,76	China	2016
PSTv-261	150061	6,7,9,43,44,76	China	2016
PSTv-262	450021	1,7,9,44,76	China	2016
PSTv-263	400040	1,43	China	2016
PSTv-264	450231	1,7,9,27,44,SP,76	China	2016
PSTv-265	450033	1,7,9,44,SP,Exp2,76	China	2016
PSTv-266	540073	1,6,7,17,43,44,SP,Exp2,76	China	2016
PSTv-267	550035	1,6,7,9,44,SP,Tr1,76	China	2016
PSTv-268	551001	1,6,7,9,17,76	China	2016
PSTv-269	550021	1,6,7,9,44,Exp2,76	China	2016
PSTv-270	550053	1,6,7,9,44,SP,Exp2,76	China	2016
PSTv-271	550023	1,6,7,9,44,Exp2,76	China	2016
PSTv-272	550071	1,6,7,9,43,44,SP,76	China	2016
PSTv-273	550075	1,6,7,9,43,44,SP,Tr1,76	China	2016
PSTv-274	551233	1,6,7,9,17,27,44,SP,76	China	2016
PSTv-275	570471	1,6,7,8,9,24,43,44,SP,76	China	2016
PSTv-276	571273	1,6,7,8,9,17,27,43,44,SP,Exp2,76	China	2016
PSTv-277	570273	1,6,7,8,9,27,43,44,SP,Exp2,76	China	2016
PSTv-278	551073	1,6,7,9,17,43,44,SP,Exp2,76	China	2016
PSTv-279	040003	7,Exp2,76	China	2016
PSTv-280	150073	6,7,9,43,44,SP,Exp2,76	China	2016
PSTv-281	510033	1,6,9,44,SP,Exp2,76	China	2016
PSTv-282	100072	6,43,44,SP,Exp2	China	2016
PSTv-283	170262	6,7,8,9,27,43,44,Exp2	China	2016

[a]The data were from Wan and Chen (2014a), Wan et al. (2016, 2017) and XM Chen and associates (unpublished data)

Cheng et al. 2014; Xiang et al. 2016). It is estimated that several hundreds to a couple of thousands of resistance genes existing in wheat germplasm and closely related alien species. It is not practical to use all resistance genes to differentiate races. In the new set of *Yr* single-gene differentials, several well-known genes are not included, such as *Yr2*, *Yr3*, *Yr4*, *Yr21*, *Yr25*, *Yr28*, *Yr31* and *YrA* (=*Yr74*+*Yr75*). These genes are not included either because a single-gene line is not available such as *Yr2*, *Yr3*, *Yr4* and *Yr25* or because a line is susceptible to most isolates in the United States, such as *Yr21*, *Yr25*, *Yr28*, *Yr31* and *YrA*. For the genes without a well-characterized line, additional genes may mask the target virulence and for those largely ineffective, the information of virulence is less important for breeding programs. For the purpose of comparison with races identified with the previous 20 cultivars differentials in the United States and those reported in other countries, type isolates of PSTv races are also tested on the previous differentials and a supplementary set of differentials. The current set of supplementary differentials include Kalyansona (*Yr2*), Vilmorin 23 (*Yr4a*, *YrV23*), Hybrid 46 (*Yr4b*, *YrH46*), Hugenoot (*Yr25*), AvSYr28NIL (*Yr28*), AvSYr31NIL (*Yr31*) and AvSYrANIL (*YrA*). It is worth to note that Vilmorin 23 has been used for *Yr3* and Hybrid 46 for *Yr4* in European and some other countries. However, these cultivars have two or more *Yr* genes, the *Yr3* and *Yr4* loci have multiple alleles, and different alleles have been reported for these cultivars (Lupton and Macer 1962; Chen and Line 1992b, c; Chen et al. 1996).

For wheat cultivars and germplasm lines with resistance of unknown genes, race information may not directly indicate their resistance. It is important to determine and monitor their effectiveness. This can be done in fields under natural *Pst* infection and/or in controlled greenhouse conditions with selected races. The field data indicate the effectiveness in the region and the greenhouse data with individual races from different regions provide the information on race specificity or non-specificity. One of the most important uses of identifying races is to select a group of races for use in screening germplasm. In the United States, four to six races are selected for routine screening of wheat germplasm. Selected races should include predominant races in the recent years, races in different virulence groups, and races with most virulences even in a low frequency.

2.1.3.2 Race Information Does Not Cover Adult-Plant Resistance

So far, race identification of rust pathogens has all been done on seedlings of host plants. There are several reasons for using seedling tests for identifying races. First, seedling testing takes short time and uses small spaces, which allows testing a large number of samples. Testing a large number of samples is critical for identifying new races and predominant races, both of which are the most important objectives of race identification. Second, race identification needs to be conducted under consistent and standardized conditions, which often make the space to do such testing limited. Third, seedling resistance is generally all-stage resistance and race-specific (Chen 2005, 2013). For stripe rust, it is hard to find resistance expressing in seedling stage, but not in adult-plant stage. However, adult-plant resistance expresses when plants are old, is often influenced by temperatures, and generally does not express or expresses at a much low level at seedling stages.

Race-specific adult-plant resistance to stripe rust has been reported in wheat cultivars (Priestley and Doodson 1976; Stubbs 1985; Milus et al. 2015). Milus et al. (2015) attempted to describe virulence patterns using a few *Pst* isolates on some wheat cultivars showing different responses of adult plants. A major issue is that adult-plant resistance of these wheat cultivars have not been genetically characterized, even for the permanently named adult-plant resistance genes such as *Yr11* to *Yr14*. The adult-plant resistances in some of the cultivars are likely controlled by several genes or quantitative trait loci. The sensitivity to temperature makes it hard to well characterize races based on adult-plant resistance. However, the effectiveness or ineffectiveness of adult-plant resistance can be determined using multiple races or isolates in both seedling and adult-plant tests, as did in the study of Milus et al. (2015) and routinely in the senior author's program. Even characterized by race-specific all-stage (seedling resistance), race information is essential to identify adult-plant resistance and determine their specificity or non-specificity.

2.1.4 Testing Isolates on Differentials

The procedures for testing isolates on differentials are basically similar with different sets of differentials with slight variations in temperatures for incubating inoculated plants during the dew period and grown plants after the dew period. Here, the procedures and conditions used in the senior author's program are presented as an example.

Once enough urediniospores (about 0.1–0.2 μg) of an isolate are obtained and differential plants are ready for inoculation, differential testing can be conducted. About 7–10 seeds were planted for each differential, and plants can be grown in a growth chamber or a greenhouse at 10–25 °C with a 14–16 h light period a day. Plants at two-leaf stage (the second leaf is about 1/3 to 1/2 of the length) are dust-inoculated with a mix of fresh urediniospores or those kept dry at 4 °C for less than 2 months with talcum at an about 1:20 ratio. Plants of the two-leaf stage, about 10–12 days after planting, are used because occasionally the primary and second leaves have different reactions. Mixing with talcum allows maximal coverage of all leaves of the differential plants. Instead of talcum, suspension of urediniospores in light mineral oil or in water-tween 20 solution also can be used for mist-inoculation of plants. Plants inoculated with mineral oil need to be kept open for at least 30 min allowing mineral oil to evaporate to avoid plants from toxin damage. The inoculated plants are kept in a dew chamber at 10 °C (or between 7 and 12 °C) without light for 18–24 h (minimum 3 h), and then move to a growth chamber or greenhouse with the temperatures and light programmed at a diurnal cycle gradually changing from 4 °C at 2:00 AM to 20 °C at 2:00 PM and 16 h photo period corresponding to the high temperature period (Chen and Line 1992b, c; Chen et al. 2002). This low temperature range simulates the early growing season of wheat crop in spring in the US Pacific Northwest and many other wheat growing regions. This range can be shortened from both ends for easier setting up programs, for example, 8–16 °C. A relative constant temperature around 12 °C works, but is not ideal as it is hard to find constant temperature under natural conditions in any wheat-growing regions.

About 18–22 days after inoculation, infection type (IT) data can be recorded. There are different scales for recording infection types. Although these scales are interchangeable, the senior author's program has used the 0 to 9 scale described by Line and Qayoum (1992). As shown in Fig. 2.1 for reactions on seedlings (A) and adult plants (B), IT 0 = no visible signs or symptom; 1 = necrotic and/or chlorotic flecks, no sporulation; 2 = necrotic and/or chlorotic blotches or stripes, no sporulation; 3 = necrotic and/or chlorotic blotches or stripes, trace sporulation; 4 = necrotic and/or chlorotic blotches or stripes, light sporulation; 5 = necrotic and/or chlorotic blotches or stripes, intermediate sporulation; 6 = necrotic and/or chlorotic blotches or stripes, moderate sporulation; 7 = necrotic and/or chlorotic blotches or stripes, abundant sporulation; 8 = chlorotic behind sporulation area, abundant sporulation; and 9 = No necrosis or chlorosis, abundant sporulation. An isolate was considered avirulent on a specific differential line when the plants show IT 0-6 or virulent when showing ITs 7-9 (Wan and Chen 2014a; Wan et al. 2016). Intermediate ITs 4-6 on a differential are included into the avirulence class because they may indicate the heterokaryotic state of the avirulence locus. In case of an isolate producing different ITs on the primary and second leaves, the IT of the second leaves, instead of the primary leaves, is recommended to be used for designating the isolate to a race as primary leaves are grow directly from the endosperm (3n = 63) and relatively short lived compared to the secondary leaves (2n = 42) that have the same two copy of the chromosomes. The three copies of the resistance gene in primary cells may provide a higher level of resistance (lower IT) than the two copies in the true leaves. Different from ITs that represent reactions, severity measures percentage of leaf areas are covered by symptoms (such as chlorosis and necrosis) and signs (uredinia) as shown in Fig. 2.1c for comparison with ITs. Severity are generally not used for identifying races, but used for screening host germplasm.

If distinct ITs appear on inoculated leaves of all plants of one or more differentials, the urediniospores from the original sample are likely in a mixture of different isolates of different races. If leaves are overly inoculated with too many urediniospores, different ITs interconnect to each other appearing as intermediate ITs, but the mixture of different races of isolates can be indicated by the intermediate ITs that are different from ITs of previously characterized races. For either of the cases, potential isolates can be separated by a procedure often called sub-isolation. For obtaining the sub-isolates, urediniospores are collected from the differentials that shows the mixture or ITs that have previously not observed with identified races, or two or more differentials with susceptible reactions, but they are not commonly susceptible to any previously identified races. Urediniospores are needed to be purified on such "key" differentials before being tested again on the complete set of differentials.

It is very time consuming to separate different isolates from a sample of mixing races. Such problem can be reduced by collecting a leaf with only one stripe of

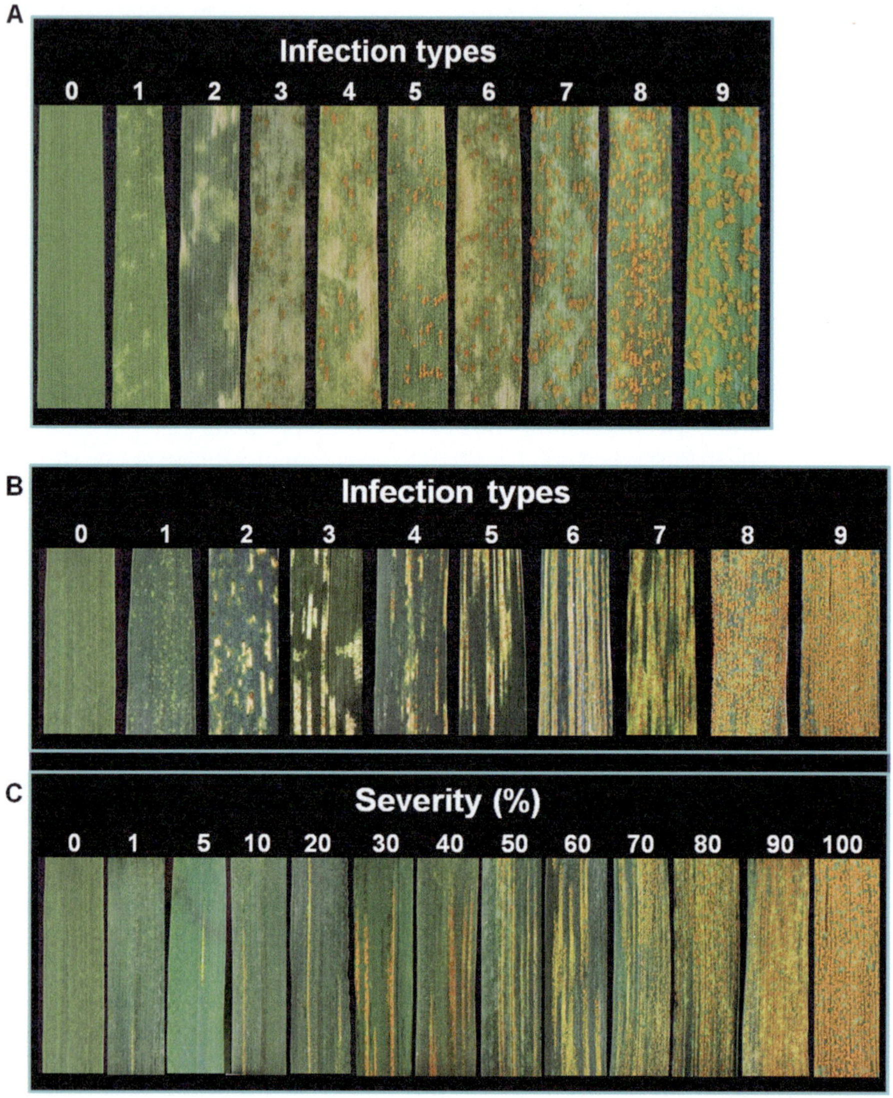

Fig. 2.1 The 0–9 scale of infection types of stripe rust on seedlings (**a**) and adult-plants (**b**), and severity (%) on leaves of adult-plants (**c**)

uredinia, which is produced by infection of a single urediniospore. If an important sample has abundant uredinia, a small piece (1 × 1 cm) of sporulating leaf can be used to reduce the possibility of mixing races. After washing off urediniospores from the piece of leaf, the leaf piece is placed on moist paper in a petri dish and kept at temperatures between 4 and 20 °C for 10 to 20 h. Fresh urediniospores produced on the leaf segment are used to inoculate plants of a susceptible variety to establish a pure isolate.

If an isolate produces a virulence pattern identical or similar to a previous identified race, the race can be assigned to the isolate. If isolate produces a virulence pattern distinct from any previous races, the virulence pattern should be confirmed by testing one or more single-pustule or single-uredinium isolates. In contrast to the leaf rust and stem rust fungi, a single pustule is different from a single uredinium. If few urediniospores are used to inoculated plants, the infection may produce separable pustules, each of them have several clustered uredinia that are produced from infection of a single urediniospores. Keeping one pustule by removing other pustules will producing a good quantity of urediniospores of a pure isolate. If pustules are not separable, picking urediniospores with a fine bush, or needle attached with Vaseline and placing them on leaves for inoculation will also produce relatively good quantity of urediniospores compared with the single-spore isolation method (Chen et al. 1993). Both single-pustule and single-uredinium isolation methods are faster and easier to do than the single-spore method. If the new virulence pattern is confirmed, the isolates and/or sub-isolates can be given a new race name and code, and one of the isolates should be considered as a type isolate for the new race, and used as reference for further race identification, for germplasm testing and/or any other studies.

In additional to the temperature, moisture and light conditions mentioned above, several issues need much attention for conducting successful race identification. (1) Correct and pure seeds of the differential lines should be used. With the morphologically similar plants and seeds of the *Yr* single-gene lines in the AvS background, caution should be taken to increase, keep and use correct seeds. (2) Plants should be grown and kept rust-free and other pathogen or insect-free before inoculation. Plants should be healthy before inoculation and continually healthy (except infected by inoculated *P. striiformis* isolate) before completing the testing. Appropriate fertilization and watering are needed to keep plants healthy and assure successful testing. (3) Viable urediniospores should be used for testing. It is good to use fresh spores or spores kept dry and less than two months (the shorter, the better). Plants should be thoroughly inoculated to avoid any escapes. For increasing spores of isolates kept in liquid nitrogen or freezer, urediniospores should be heat-shocked in a glass vial or foil bag submerged in warm water at 45–50 °C for no more than 2 min. (4) Hands, places and tools (such collectors, dusting bottles and re-useable brushes, glass vials, plastics and booth) should be washed, disinfested or autoclaved between handling different plants and different isolates. (5) Infected plants should be properly disposed to avoid urediniospore escaping from working facilities. The procedures of isolation and autoclave as outlined in a permit issued by the US Department of Agriculture, Animal and Plant Inspection Service (USDA-APHIS) should be strictly followed.

2.1.5 Analyses of Races and Virulences

When stripe rust occurs in a region or on a cultivar, it is important to know which race or races cause the disease. When a race is identified, it is good to know how the distribution and frequency of the race in an area, country, or the world. Such information can be obtained from collecting a large number of samples and identify races of the samples.

When two or more races are identified, it is important to know the relationships between races. This is done by comparing their common and different virulences and avirulences. More virulences shared by two races indicating more relatedness between the two races. Assuming a new virulence arises from a mutation of its avirulence allele, step-wise evolutionary relationships have been found among *Pst* races in Australia (Wellings and McIntosh 1990; Wellings 2007) and in the Unites States (Line and Qayoum 1992; Liu et al. 2016c). In China, the so-called Lovrin and Hybrid 46 "pathotype" series were also based on common virulences (Li and Zeng 2002). Computer software has been developed and used to determine relationships among races for various plant pathogens including *Pst* (Kosman et al. 2008; Schachtel et al. 2012; Wan and Chen 2012a; Kosman 2014; Liu et al. 2016c; Wan et al. 2016). Taking a recent study of *Pst* isolates from 1996 to 2009 as example, the total of 171 races were clustered into two major virulence clusters and further grouped into eight virulence groups (Liu et al. 2016c). The virulence groups differ in the mean number of virulences, common virulences and common avirulences. About 74% of the races appeared to be evolved from previously existing races through single- or double-step mutations, whereas only about 26% of the races might have been introduced and/or produced through somatic recombination. Over the years, races started generally in low frequencies, increased, disappeared and replaced by other races.

During the last several years from testing of more than 5000 isolates from the United States since 1968 and other countries since 2005, more than 200 races, named PSTv-1 to PSTv-219, have been identified (Wan and Chen 2014a; Wan et al. 2016, 2017; Liu et al. 2016c). The octal codes, virulence formulae, countries and years of isolates from which they were first identified are presented in Table 2.3. The number of virulences of these races ranged from 0 (not virulent to any of the 18 *Yr* single-gene lines) to 16. These races show great variations of *Pst* populations in individual countries and the world.

It is important to know the frequencies of individual virulences, as the stripe rust pathogen has the high capability to compile different virulences into a single race, different races have different combinations of virulences and races vary greatly in distributions and frequencies. This information is critical for determination of effective resistance genes in different regions or countries. In the United States, the information of virulence frequencies is reported in every year's race and virulence survey (Chen et al. 2002, 2010; Wan and Chen 2011, 2012a, b, 2013, 2014a, b, 2015, 2016; Wan et al. 2016). Over the years, dynamics of virulences can be determined (Wan et al. 2016; Liu et al. 2016c).

Table 2.4 presents frequency data of virulences to some of the *Yr* single-gene lines that were used in virulence studies with samples from different countries (Sharma-Poudyal et al. 2013; Wan and Chen 2014a; Xi et al. 2015; Wan et al. 2017). In general, virulence to *Yr1* is high in Asian countries and Ethiopia. Virulences to *Yr2*, *Yr6*, *Yr7*, *Yr8*, *Yr9*, *Yr17*, *Yr25*, *Yr28*, *Yr31*, *YrExp2* and *YrA* (= *Yr74*+*Yr75*) are high in very country, except low virulence to *Yr8* in Ecuador. Virulence to *Yr27* is high except Australia and Italy. Virulences to *Yr10*, *Yr24* and *Yr32* were low in most of the countries, except in Ethiopia, Hungary and Pakistan. Virulence to *YrSP* was low except in China, and virulences to *YrTr1* and *YrTye* were low except in Ecuador, Ethiopia and Italy.

Table 2.4 Frequency of virulences in the *Puccinia striiformis* f. sp. *tritici* populations in different countries

Virulence	Frequency (%) of virulence in different countries[a]														
On *Yr* gene	Australia	Canada	Chile	China	Ecuador	Ethiopia	Hungary	Italy	Kenya	Mexico	Nepal	Pakistan	Turkey	Uzbekistan	US
Yr1	40	8	27	42	36	74	50	36	75	9	71	57	39	88	33
Yr2	100	100	100	90	91	100	100	96	100	90	100	89	91	100	–[b]
Yr5	0	0	0	0	0	0	0	0	0	0	0	0	0	0	0
Yr6	60	100	60	93	88	100	100	32	75	94	95	96	96	100	94
Yr7	40	100	80	95	85	100	100	56	100	90	100	91	94	63	90
Yr8	20	99	44	88	6	56	75	60	75	88	86	74	94	63	90
Yr9	40	100	33	50	33	100	75	36	100	91	57	85	74	100	91
Yr10	0	15	33	5	0	35	50	0	25	0	19	11	0	6	6
Yr15	0	0	0	0	0	0	0	0	0	0	0	0	0	0	0
Yr17	100	98	93	68	52	89	100	36	100	66	86	98	93	94	83
Yr24	20	10	27	0	0	27	25	5	0	2	5	28	7	6	6
Yr25	20	–	40	30	100	98	75	96	50	91	38	37	41	75	–
Yr27	0	97	27	67	70	100	75	5	50	85	76	83	83	56	82
Yr28	80	17	87	62	91	75	75	36	25	91	86	78	93	94	–
Yr31	100	98	100	98	100	99	100	72	100	92	100	98	100	100	–
Yr32	0	–	33	2	0	31	50	5	0	2	0	28	7	6	5
YrSP	0	1	0	27	3	0	0	11	0	7	0	0	7	13	3
YrTr1	20	–	33	50	0	10	50	0	25	53	29	46	37	25	67
YrExp2	40	–	80	93	85	100	100	92	100	85	95	93	91	63	89
Yr76	20	–	7	40	9	0	75	24	50	8	43	48	39	75	31
YrA	80	100	60	98	100	83	100	88	50	94	100	70	98	100	–

[a]Data of Sharma-Poudyal et al. (2013) for Australia, Chile, China, Hungary, Kenya, Nepal, Pakistan, Turkey and Uzbekistan; of Xi et al. (2015) for Canada; of Wan and Chen (2014a) and Wan et al. (2016) for the US; of Wan et al. (2017) for Ethiopia; of Wan and Chen (Unpublished) for Ecuador and Italy; and of Figueroa-Lopéz P, Wan and Chen (unpublished)

[b]–, no data

Information on virulence associations is useful for choosing right gene combinations for developing cultivars with resistance lasting longer. For example, using isolates collected from the United States in 2011 and 2012, highly positive associations (>0.60) were found between virulences to *Yr1* and *YrTye*; *Yr7* and *YrExp2*, *Yr44*, *Yr43*, or *Yr8*; *Yr8* and *YrExp2*, *Yr44*, or *Yr43*; *Yr10* and *Yr32*, or *Yr24*; *Yr24* and *Yr32*; *Yr43* and *YrExp2* or *Yr44*; and *Yr44* and *YrExp2* (Wan et al. 2016). Negative associations (<0) were found between virulences to *Yr1* or *YrTye* and *Yr6*, *Yr7*, *Yr8*, *Yr10*, *Yr17*, *Yr24*, *Yr32*, *Yr43*, *Yr44*, *YrTr1*, or *YrExp2*; *YrSP* and *Yr7*, *Yr8*, *Yr17*, *Yr27*, *Yr43*, *Yr44*, *YrExp2*, or *YrTr1*; and *Yr17* and *Yr10*, *Yr24*, or *Yr32*. Most of these positive and negative associations were confirmed in the study with more than 900 isolates collected in the United States from 1968 to 2009 (Liu et al. 2016c). Although the mechanisms of negative associations (disassociations) are not fully understood, the *Yr* genes with negative associations (disassociations) for their virulence are useful for breeding programs. Understanding the mechanisms require more studies for cloning the avirulence genes of the pathogen, resistance genes of the host and their interactions.

2.1.6 Races of the Barley Stripe Rust Pathogen

As barley stripe rust does not cause as widespread as wheat stripe rust in the world, few countries have studied races of the barley stripe rust pathogen, *Puccinia striiformis* f. sp. *hordei* (*Psh*), especially in the past decade. At the 11th International Cereal Rusts and Powdery Mildews Conference in Norwich, UK in 2004, there was only one presentation on barley stripe rust, "Epidemiology of barley stripe rust and races of *Puccinia striiformis* f. sp. *hordei*: the first decade in the United States" given by the senior author (Chen 2004). Similarly, only the senior author made a presentation on barley stripe rust at the 12th International Cereal Rusts and Powdery Mildews Conference in Antalya, Turkey in 2009 (Chen et al. 2009c). There were no any presentations on barley stripe rust in the 13th and 14th International Cereal Rusts and Powdery Mildews Conference in Beijing, China in 2012 and Helsingør, Denmark in 2015, respectively. During the discussion at the 14th conference, there was an indication that barley stripe rust had not been a production problem even in European countries, where most historical studies of barley stripe rust were conducted (Stubbs 1985). In contrast to wheat stripe rust, barley stripe rust may have never been a major disease as Stubbs (1985) stated that stripe rust on barley has not thus far been a major problem in northwestern Europe, even some susceptible spring barleys are being grown. No studies in European countries indicate why stripe rust has become insignificant. However, Stubbs (1985) related the low disease to the lack or low acreage of winter barley as he pointed out that the situation could change if winter barley increases in production. The same factor has been also used to partially explain the relatively low levels of barley stripe rust in the United States (Chen 2004, 2008; Wan and Chen 2012a). Although barley stripe rust has been continually studied by the senior author's program in the United States (Wan and Chen 2011, 2012a, b, 2013, 2014b, 2015, 2016), the damage by barley stripe rust has been decreasing since 2000. The decrease of barley stripe rust is attributed to growth of barley cultivars with race non-specific resistance and reduction of barley acreage in California and the Pacific Northwest, the major regions of barley stripe rust in the United States (Chen 2004, 2008; Wan and Chen 2012a, b). The same factors have also contributed to the stabilization of *Psh* races.

2.1.6.1 The System for Differentiating *Psh* Races in China

Races of *Psh* in China are discussed first, not due to the time of starting race differentiation, but because their differential system is not related much to those in other countries. In China, barley stripe rust has not received as much attention as wheat stripe rust, likely because barley is grown in low acreage and also in remote areas. The situation could change as more barley is grown due to increasing demand of barley for malting, feed and food. Fang (1944) was the first to identify an isolate from barley virulent on 'Heils Franken' in his study with most isolates from wheat. The wheat and barley varieties used in this study were adopted from Gassner and Straib (1932) that will be discussed below. Fang (1944) also found that the barley isolates was able to sporulate on wheat differential Chinese 166. The susceptible or intermediate reaction of Chinese 166 to *Psh* was also found many years later with barley isolates in the United States (Chen et al. 1995b).

In the 1980s and 1990s, a series studies including races were conducted on barley stripe rust, but focusing on Tibet, where barley is an important crop (Wang et al. 1989a, b; Wang 1992). In 1994, the set of differentials was modified and expanded to include six Chinese barley varieties, 'Himalaya 6' (喜马拉6号), 'Uniform Dwarf' (矮秆齐), 'Early Maturity 3' (早熟3), 'Yong 1394' (永1394), 'Yong 802' (永802) and 'Kepin 2' (科品2号) and one Chinese wheat variety 'Hui County Red' (辉县红) (Niu et al. 1994). The wheat variety is susceptible to all *Pst* races identified in China, and resistance genes in the barley differentials have not been studied. Using these differentials, they clearly distinguished *Psh* from *Pst*, and identified four races (barley form 1, 2, 3 and 4) from 40 barley stripe rust samples collected from Tibet, Qinghai, Gansu, Shaanxi, and Henan provinces. Barley forms 1 (moderately virulent) and 2 (low virulent) had a wider distribution, whereas forms 3 and 4 (highly virulent) were identified only in the Tibet samples.

2.1.6.2 The Systems for Differentiating *Psh* Races in Europe, India, Japan and South America

In Europe, *Psh* and *Pst* were not initially separated (Gassner and Straib 1932). Straib (1935) added barley cultivars as differentials. The set of differentials including 12 wheat, 2 barley and 1 rye varieties was continually used until the 1980s (Stubbs 1985). Among the races identified with this set of differentials of mixing crops, races 23, 24 and 48 were considered as *Psh* because these races virulent on one or both of the barley differentials ('Fong Tien' and 'Heils Franken') and largely avirulent to most wheat differentials.

The European set of differentials was also used in India, but races were named with letters instead of numbers (Mehta 1933; Prasada and Lele 1952; Bahl and Bakshi 1963; Bakshi et al. 1964; Bakshi and Bahl 1965). Races of *Psh* have been continually characterized in India and named using a number and letter system (Nagarajan 1983; Nagarajan et al. 1984). Barley cultivars 'Local' and 'Himani' have been used to differentiate *Psh* vs. *Pst* races, but without distinction in naming races (Nagarajan 1983). The presence of race 24 was reported in India (Prasada et al. 1967).

In Japan, the European set of differentials was also used, but supplemented with 'Yechao 35368', a Chinese wheat cultivar, and 'Sakigake', a Japanese barley cultivar, to distinguish four races of both *Pst* and *Psh* (Kajiwara et al. 1964a). Kajiwara (1964) reported a new classification of *Psh* races based on testing 10 European and 7 Japanese isolates of *Psh*. He differentiated seven races with barley cultivars Fong Tien, 'Bavaria', 'Sakigake' and 'Hokkaido Chevalier' plus wheat cultivar Yechao 35368. Stubbs (1985) used these cultivars, but found variable reactions in studies with European isolates.

Stubbs (1985) developed a set of barley cultivar differentials, including 'Cambrinus' (= Heils Franken = 'Astrix'), which is only susceptible to races 24 and 48, 'Mazurka', 'Varunda', 'Emir', 'Abessinian 14', 'Bigo', 'I 5' and 'Topper' for identifying *Psh* races without using wheat varieties. This set of barley differentials has been used in South America for identifying *Psh* races since the barley stripe rust pathogen, mostly race 24, was introduced in Colombia in 1975, and four races were identified and named "race 24" and "race 24 variants" based on additional virulences on Mazurka and Varunda (Dubin and Stubbs 1986). This set of barley varieties provided the basis for the current set of differentials use to identify *Psh* races in the United States and other countries.

2.1.6.3 The Systems for Differentiating *Psh* Races in the United States and Other Countries

Soon after the first observation of barley stripe rust in southern Texas, USA in 1991, scientists conducted studies to differentiate *Psh* races. As the fungus was believed to be introduced from Europe through South America, it was logically to use the Stubbs' set of barley differentials as did by Dubin and Stubbs (1986) for races in South America. By testing *Psh* isolates collected from Texas, New Mexico and Oklahoma, Marshall and Sutton (1995) identified three races, races 23, 24 and TXG, which was similar to race 24 but also virulent on Mazurka, Varunda, Bigo and I 5. Using the same differentials, Chen et al. (1995b) detected three of the four virulence patterns reported in South America. However, they found that the previous differentials used by Stubbs (1985) and Dubin and Stubbs (1986) did not adequately differentiate races, and thus selected 11 barley genotypes for differentiating *Psh* races. This set of differentials included all cultivars used by Dubin and Stubbs (1986), except that Cambrinus was replaced by Heils Franken, plus 'Astrix', 'Hiproly', 'Abed Binder 12' and 'Trumf'. Using the 11 barley cultivar differentials, they identified 14 races, named PSH-1 to PSH-14 from 22 *Psh* isolates collected from Texas, Arizona, Colorado, Idaho and Montana in 1993 and 1994. By 2000, a total of 52 races were identified in the United States (Table 2.5) (Chen and Line 2001).

Table 2.5 Races of *Puccinia striiformis* f. sp. *hordei* identified based on barley differentials

PSH Race	Octal Code[a]	Virulence Description[b]	First year Detected	Country
PSH-1	6000	1,2	1993	US
PSH-2	7000	1,2,3	1994	US
PSH-3	6400	1,2,4	1993	US
PSH-4	6200	1,2,5	1993	US
PSH-5	4140	1,6,7	1994	US
PSH-6	7400	1,2,3,4	1993	US
PSH-7	6600	1,2,4,5	1993	US
PSH-8	7040	1,2,3,7	1994	US
PSH-9	7500	1,2,3,4,6	1993	US
PSH-10	5340	1,3,5,6,7	1994	US
PSH-11	5160	1,3,6,7,8	1993	US
PSH-12	7620	1,2,3,4,5,8	1993	US
PSH-13	7134	1,2,3,6,8,9,10	1994	US
PSH-14	7760	1,2,3,4,5,6,7,8	1993	US
PSH-15	5460	1,3,4,7,8	1995	US
PSH-16	5260	1,3,5,7,8	1995	US
PSH-17	6660	1,2,4,5,7,8	1995	US
PSH-18	5660	1,3,4,5,7,8	1995	US
PSH-19	5360	1,3,5,6,7,8	1995	US
PSH-20	5170	1,3,6,7,8,9	1995	US
PSH-21	5474	1,3,4,7,8,9,10	1995	US
PSH-22	4474	1,4,7,8,9,10	1995	US
PSH-23	7470	1,2,3,4,7,8,9	1995	US
PSH-24	5760	1,3,4,5,6,7,8	1995	US
PSH-25	5670	1,3,4,5,7,8,9	1995	US
PSH-26	7670	1,2,3,4,5,7,8,9	1995	US
PSH-27	5374	1,3,5,6,7,8,9,10	1995	US
PSH-28	7770	1,2,3,4,5,6,7,8,9	1995	US
PSH-29	7764	1,2,3,4,5,6,7,8,10	1995	US
PSH-30	5774	1,3,4,5,6,7,8,9,10	1995	US
PSH-31	7774	1,2,3,4,5,6,7,8,9,10	1995	US
PSH-32	5200	1,3,5	1996	US
PSH-33	4040	1,7	1996	US
PSH-34	4070	1,7,8,9	1996	US
PSH-35	4440	1,4,7	1996	US
PSH-36	4400	1,4	1996	US
PSH-37	6440	1,2,4,7	1996	US
PSH-38	7660	1,2,3,4,5,7,8	1996	US
PSH-39	5060	1,3,7,8	1996	US
PSH-40	7450	1,2,3,4,7,9	1996	US
PSH-41	5550	1,3,4,6,7,9	1996	US

(continued)

Table 2.5 (continued)

PSH Race	Octal Code[a]	Virulence Description[b]	First year Detected	Country
PSH-42	4650	1,4,5,7,9	1996	US
PSH-43	4420	1,4,8	1996	US
PSH-44	5040	1,3,7	1996	US
PSH-45	5560	1,3,4,6,7,8	1996	US
PSH-46	4060	1,7,8	1996	US
PSH-47	5640	1,3,4,5,7	1996	US
PSH-48	4000	1	1997	US
PSH-49	7776	1,2,3,4,5,6,7,8,9,10,11	1997	US
PSH-50	4200	1,5	1998	US
PSH-51	4240	1,5,7	1998	US
PSH-52	4260	1,5,7,8	1998	US
PSH-53	4030	1,8,9	2001	US
PSH-54	4061	1,7,8,12	2001	US
PSH-55	4031	1,8,9,12	2001	US
PSH-56	4261	1,5,7,8,12	2001	US
PSH-57	4074	1,7,8,9,10	2001	US
PSH-58	4065	1,7,8,10,12	2001	US
PSH-59	4066	1,7,8,11,12	2001	US
PSH-60	4275	1,5,7,8,9,10,12	2001	US
PSH-61	5041	1,3,7,12	2002	US
PSH-62	5061	1,3,7,8,12	2002	US
PSH-63	5261	1,3,5,7,8,12	2002	US
PSH-64	4265	1,5,7,8,10,12	2002	US
PSH-65	7461	1,2,3,4,7,8,12	2002	US
PSH-66	5163	1,3,6,7,8,11,12	2002	US
PSH-67	5363	1,3,5,6,7,8,11,12	2002	US
PSH-68	4165	1,6,7,8,10,12	2003	US
PSH-69	4677	1,5,6,7,8,9,10,11,12	2003	US
PSH-70	4045	1,7,10,12	2004	US
PSH-71	5375	1,3,5,6,7,8,9,10,12	2004	US
PSH-72	7777	1,2,3,4,5,6,7,8,9,10,11,12	2004	US
PSH-73	4270	1,5,7,8,9	2005	US
PSH-74	4241	1,5,7,12	2005	US
PSH-75	7775	1,2,3,4,5,6,7,8,9,10,12	2006	US
PSH-76	4375	1,5,6,7,8,9,10,12	2006	US
PSH-77	5275	1,3,5,7,8,9,10,12	2006	US
PSH-78	6721	1,2,4,5,6,7,8,12	2006	US
PSH-79	6675	1,2,4,5,7,8,9,10,12	2006	US
PSH-80	4244	1,5,7,10	2006	US
PSH-81	4044	1,7,10	2006	US
PSH-82	4002	1,11	2008	US

(continued)

Table 2.5 (continued)

PSH Race	Octal Code[a]	Virulence Description[b]	First year Detected	Country
PSH-83	4245	1,5,7,10,12	2014	Iran
PSH-84	4346	1,5,6,7,10,11	2014	Iran
PSH-85	6240	1,2,5,7	2014	Iran
PSH-86	4242	1,5,7,11	2014	Iran
PSH-87	4255	1,5,7,9,10,12	2014	Iran
PSH-88	4221	1,5,8,12	2014	Iran
PSH-89	4205	1,5,10,12	2014	Iran

[a]The octal codes are adapted from Table 2.1.2.4a. The 12 barley differentials (below) are grouped into four sets: Set 1 consists of differentials 1, 2 and 3; Set 2 of 4, 5 and 6; Set 3 of 7, 8 and 9; and set 4 of 10, 11 and 12

[b]Virulence on barley differential cultivars: 1 = Topper (susceptible to all races of *P. striiformis* f. sp. *hordei*), 2 = Heils Franken (*Rps4*, *rpsHF*), 3 = Emir (*rpsEm1*, *rpsEm2*), 4 = Astrix (*Rps4*, *rpsAst*), 5 = Hiproly (*rpsHi1*, *rpsHi2*), 6 = Varunda (*rpsVa1*, *rpsVa2*), 7 = Abed Binder 12 (*rps2*), 8 = Trumpf (*rpsTr1*, *rpsTr2*), 9 = Mazurka (*Rps1.c*), 10 = Bigo (*Rps1.b*), 11 = I 5 (*Rps3*, *rpsI5*), and 12 = Bancroft (*RpsBa*) (Chen et al. 1995b, 2002; Wan and Chen 2012a)

In 2001, barley variety 'Bancroft', which was the first variety purposely developed for resistance to stripe rust in the United States (Wesenberg et al. 2001), was added to the differential set (Chen and Moore 2002). The set of 12 barley cultivars has been used to differentiate *Psh* races in the United States, and by 2008, a total of 82 races have been identified (Table 2.5) (Chen 2004, 2008; Chen et al. 2004; Chen and Penman 2005, 2006; Chen et al. 2007; Wan and Chen 2012a). Since 2009, fewer races of *Psh* have been identified every year; no new races have been identified; and races with narrow virulence spectra have been predominant (Wan and Chen 2011, 2012b, 2013, 2014b, 2015). The barley stripe rust population has appeared stabilized in the United States in the recent years.

In a recent study, Cheng et al. (2016a) tested stripe rust samples from barley, wheat and grasses on both sets of barley and wheat differentials and also with molecular markers. They found that isolates from barley were typical *Psh* races and those from wheat were typical *Pst* isolates. However, isolates from grasses were identified as races of *Psh*, *Pst* or both *Psh* and *Pst* because those isolates were virulent on some differentials of both barley and wheat. The results indicate that grasses host more diverse isolates than cereal crops and that somatic recombination can occur between *Psh* and *Pst* on grasses susceptible to both formae speciales. Because of this, both wheat cultivar 'Nugaines' and barley cultivar 'Steptoe', which are susceptible to all *Pst* races and *Psh* races identified so far in the United States respectively, are used in the senior author's program to recover stripe rust samples collected from grasses. *Psh* isolates from barley are tested on Nugaines together with barley differentials, and if sporulating they are also tested on the wheat differentials. Similarly, *Pst* isolates from wheat are tested on Steptoe together with the wheat differentials, and if sporulating they are also tested on the barley differentials. In the recent years, less than 3% of isolates collected from wheat and barley were able to infect some differentials of the crop that was not the origin crop.

Before the 1980s, barley stripe rust isolates from the Middle East and East Africa analyzed in Wageningen, Netherlands did not shown clear differences in virulence on the differential cultivars supplemented with barley cultivars grown in those regions (Stubbs 1985). However, recent studies in Iran using the 12 barley cultivars, which are used to differentiated races of *Psh* races in the United States (Chen 2004, 2008; Wan and Chen 2012a), clearly showed distinct reactions of the differentials to *Psh* isolates. In Iran, barley stripe rust was first reported in 1947 (Esfandiari 1947). Similar to other countries, the disease is not as important as stripe rust on wheat. However, barley stripe rust is increasing in the northwest and northeast provinces of Iran, and the increase is likely due to cultivation of susceptible cultivars under favorable conditions or changes of virulence factors in pathogen populations (Safavi et al. 2012). From 2007 to 2012, Safavi et al. (2013) planted the 12 barley genotypes, plus an Iranian barley cultivar 'Afzal' as a susceptible check in Ardabil, Sari and Mashhad of Iran for monitoring *Psh* virulences. They found that the *Psh* populations varied in virulence among regions and also varied from year to year in some regions. The rust populations were generally avirulent on differentials Emir (*rpsEm1*, *rpsEm2*), Heils Franken (*rpsHF*, *Rps4*), Varunda (*rpsVa1*, *rpsVa2*) and Astrix (*rpsAst*), but virulent on Abed Binder 12 (*rps2*), Bigo (*Rps1.b*) and I 5 (*Rps3, rpsI5*). From stripe rust samples collected from barley, wheat and foxtail barley in different parts of Iran. Safavi et al. (2013) identified them as *Psh* or *Pst* based on virulence on the 12 barley differentials and Iranian barley cultivars Afzal and 'Torsh' and wheat genotypes AvS, Chinese 166, 'Morocco' and single-gene lines with *YrA*, *Yr1*, *Yr2*, *Yr6*, *Yr7*, *Yr26* and *Yr31*. They found that *Pst* isolates did not infect the 12 barley differentials, but some *Pst* isolates infected barley cultivars Afzal or Torsh. Five pathotypes of *Psh* were virulent on wheat differentials with *Yr1*, *Yr2*, and *Yr7* and also on triticale. Two *Psh* isolates with virulence to *Yr2* or triticale but avirulent on Mazurka and Bigo might be hybrids between *Psh* and *Pst*. The seven *Psh* races were named as PSH-83 to PST-89 and their virulence formulae are presented in Table 2.5.

2.1.6.4 Stripe Rust of Wild Barley Grass in Australia

So far, the barley stripe rust pathogen, *Psh*, has not been detected in Australia, but a stripe rust pathogen form more similar to *Psh* than *Pst* has been reported in this country. In Australia, stripe rust samples were collected from wheat and grasses (principally barley grass *Hordeum* spp.) and occasionally barley during the survey in 1998 (Wellings et al. 2000). Approximately 30% of samples from barley grass were highly avirulent on seedlings of most wheat differentials, partially virulent on the wheat differential Chinese 166, but virulent on a small group of barley cultivars, especially 'Skiff'. For example, a typical wild barley stripe rust isolate (981549) was virulent on barley cultivars Fong Tien, Skill and Sultan and wheat variety Chinese 166, but avirulent to 'Avocet R' that is susceptible to predominant *Pst* races in Australia, indicating that the stripe rust on wild barley grass is not *Pst*. The resistant reactions of most Australian barley cultivars to these isolates primarily from wild barley grass were different from the resistant responses in an earlier test in Mexico with race 24, a typical *Psh* race, indicating that the stripe rust on wild barley grass is not *Psh* race 24. Thus, the "barley grass stripe rust" (BGYR) was considered a new forma specialis introduced into Australia (Wellings et al. 2000), and it was later named as *Puccinia striiformis* f. sp. *pseudohordei* (Wellings 2007). In the senior author's opinion, the wild barley grass stripe rust fungus might be considered as *Psh*. Although it is different from race 24, it could be race 23 based on the Stubbs' differentials (Stubbs 1985). Based on this differential set, race 23 and similar races have been reported in Europe (Stubbs 1985), South America (Dubin and Stubbs 1986), the United States (Chen et al. 1995b; Marshall and Sutton 1995; Wan and Chen 2012a; Cheng et al. 2016a) and Iran (SA Safavi, personal communication). In fact, many *Psh* races listed in Table 2.5 that are virulent on 'Topper' (susceptible to all *Psh* races like Fong Tien) but not virulent to Heils Franken can be considered race 23. More studies are needed to determine the relationship of wild barley stripe rust fungus with *Psh* and *Pst*.

2.2 Variations in Environmental Adaptation, Aggressiveness and Fitness

In addition to variations in pathogenicity to different crop and grass species and virulence to different cultivars of a cereal crop, *P. striiformis* varies greatly in adaptation, aggressiveness and fitness among its formae speciales, isolates or populations. For example, bluegrass stripe rust, caused by *P. striiformis* f. sp. *poae*, which is considered a different species, *P. pseudostriiformis* (Liu and Hambleton 2010), is prevalent in the fall whereas wheat stripe rust (caused by *Pst*) is prevalent in the spring and early summer and barley stripe rust (caused by *Psh*) in the early summer in the US Pacific Northwest. Although the difference in seasons for disease prevalence can be caused by cultural practices, such as mowing bluegrass in laws, the bluegrass stripe rust pathogen may have a better adaptation in the fall seasons than other formae speciales. Similarly, the optimal temperature range (21–24 °C) of urediniospore germination for *P. striiformis* f. sp. *dactylidis* (now *Puccinia striiformoides*), the stripe rust fungus on orchard grass, is much higher than that of *Pst* and *Psh* (Manners 1960). Variations in optimal temperature may exist among different populations of *P. striiformis* f. sp. *dactylidis* as urediniospores of the fungus in New Zealand germinated well between 4 °C (39%) and 25 °C (61%) with the highest germination rate (82%) at 15 °C (Latch 1976). The differences in germination temperatures may indicate variations in adaptation to different environments among different formae speciales and different populations of the same forma specialis.

In this section, we will focus the discussion on the wheat stripe rust fungus as more studies of adaptation, aggressiveness and fitness have been done on this economically more important pathogen.

2.2.1 Adaptation

Adaptation is the evolutionary process whereby an organism becomes better able to live in its habitat or habitats (Dobzhansky 1968). Adaptation refers to both the current state of being adapted (e.g., existing races or pathotypes) and to the dynamic evolutionary process that leads to the adaptation. Adaptations enhance the fitness and survival of individual isolates of the stripe rust fungus. Compared to the stem rust (*P. graminis*) and leaf rust (*P. triticina*) fungi, the stripe rust fungus, especially *Pst*, is adapted to relatively cool environments, and thus the pathogen survives winter much better, but survive summer much less than the other rust fungi. Therefore, *Pst* is more prevalent in the temperate regions and the high elevations of the tropical regions. For a distinct pathogen, this general feature has not changed for *Pst*. However, in the recent years, isolates or populations of *Pst* adapted to warmer environments have been found in different countries. Such isolates that have tolerance to high temperatures infect, reproduce and survive better than other isolates under relatively high temperatures.

Adaptation of *Pst* to warmer temperatures has been studied and discussed since the early 1970s (Macer 1972; Zadoks 1979). Markell and Milus (2008) reported that a novel population of *Pst* has emerged in eastern United States. This population comprises of isolates or races with avirulence to *Yr8* and *Yr9* (Chen et al. 2002, 2010; Chen 2005, 2007). Based on experiments under controlled conditions, these races were found to be more aggressive at 18 °C than old races that were identified before 2000 and avirulent to *Yr8* and *Yr9* (Milus et al. 2006), indicating the new races adapted to the relatively high temperature. It is worthy to note that 18 °C is within the temperature range (from slightly above 0–23 °C) for germination of *Pst* urediniospores, but within the optimal temperature range (10–25 °C) for the pathogen to grow within host tissue and produce urediniospores from infected plants (Rapily 1979; Sharma-Poudyal et al. 2014). Although the more frequent and more damaging stripe rust epidemics were mainly due to the susceptibility of wheat cultivars grown in the eastern United States to the new races (XM Chen and associates, unpublished data), the adaptation to the relatively high temperature also contribute to the increasing importance of stripe rust in the eastern United States. Not only in the eastern United States, races with *Yr9* have also become predominant throughout the United States, including the Pacific Northwest where *Yr9* has not been used as much in wheat cultivars as in the eastern United States. The new races have more virulences than the old races (Liu et al. 2016c), which give them advantage to infect more wheat cultivars, and thus they have a higher cultivar adaptation than old races. The predominant races are adapted to a wider range of environmental conditions, as the high-temperature adapted (or aggressive) strains have been quickly spread throughout the world (Hovmøller et al. 2008).

In Canada, Tran and Kutcher (2015) conducted similar experiments with 10 new isolates (9 from Canada and 1 from the United States) and 4 old isolates (2 from Canada and 2 from the United States) of *Pst*. Instead of using the same temperatures (12 °C for low and 18 °C for high temperature) for the dew period and after the dew period used in Milus et al. (2006), they kept the inoculated plants in a dew chamber for 24 h at 10 °C for all the infection tests, and then grew the plants at 10, 15 and 20 °C. They also did urediniospore germination tests at the three temperatures plus 5 °C. Their results showed that new and old isolates had similar germination rates of urediniospores at 5, 10 and 15 °C; however, new isolates had a higher germination rate at 20 °C than old isolates. On the susceptible cultivar 'AC Bellatrix', there was no significant difference in latent period or AUDPC of new and old isolates at 10°C; however, new isolates had a shorter latent period and a higher AUDPC at both 15 °C and 20°C. They concluded that new isolates have the ability to adapt to a wide range of temperature.

In China, the adaptation to warmer temperature was studied using 126 single-pustule isolates originally collected from 12 provinces spanning three decades from the 1980s to 2010s (Liu et al. 2012c). Medium effect temperature (ET_{50}) was determined as a temperature sensitive parameter for each of the isolates. The values ranged from 18.46 °C to 27.01 °C with a mean of 24.01 °C. Approximately 14.3% isolates had ET_{50} values great than 26 °C and they all were collected after 2010. The isolate with the highest ET_{50} was collected in 2011 from Hubei province. After wheat leaves inoculated at 26 °C, isolates of ET_{50} >24 °C produced extensive chlorotic flecks even with few small uredinia, whereas isolates of ET_{50} <24 °C did not produce any visible symptoms, indicating that high-temperature tolerant isolates are able to germinate and penetrate leaves but sensitive isolates die at 26 °C. This study clearly shows variations in adaptation to high temperatures in the Chinese *Pst* population.

In Australia, the first pathotype introduced from Europe in 1979 was not virulent on *Yr8* and *Yr9*, and the population of this pathotype and derived pathotypes were restricted in the eastern part of the country (Wellings et al. 1987; Wellings and McIntosh 1990; Wellings 2007). In 2002, a new pathotype, with virulences to *Yr8* and *Yr9* and similar virulence/avirulence to pathotype PST-78 that was reported as a foreign incursion in the United States for the first time in 2000, was introduced to the western part of Australia (Wellings et al. 2000, 2003). Several derivative pathotypes were detected in subsequent seasons, and the new pathotype population quickly became predominant in the wheat-growing regions throughout the country, indicating that the new population is more aggressive or has a better adaptability than the old population (Wellings 2007; Loladez et al. 2014). To determine if the new pathotypes are adapted to high temperatures as reported in the United States (Milus et al. 2006), Loladez et al. (2014) tested 16 pathotypes during the 24 h incubation at 10 and 15 °C and during the subsequent post-inoculation at 17 and 23 °C for latent period and infection efficiency on four susceptible wheat cultivars. In addition, they also tested the effect of two contrasting incubation temperatures on urediniospore germination. Their results did not show that pathotypes detected after 2002 are adapted to high temperatures, in contrast to the report by Milus et al. (2006). The contrast results were likely due to the different temperatures used in the different studies. Based on the results, Loladez et al. (2014) thought that other factors, particularly virulence to adult-plant resistance in wheat cultivars, contributed to the observed increased aggressiveness.

2.2.2 Aggressiveness

In a plant-pathosystem, aggressiveness is the relative ability of a pathogen isolate to infect, colonize and cause damage to plants (Shurtleff and Averr III 1997). A more virulent isolate or race is able to cause more sever disease and damage than a less virulent isolate. For rust fungi, a virulent race is able to infect host plants whereas an avirulent race is unable to infect host plants with the corresponding resistance genes. Thus, a virulent race is more aggressive than an avirulent race. Also, a race possessing more virulence genes is more aggressive in comparison with a race possessing fewer virulence genes because the former are able to cause disease on more cultivars. Therefore, virulence is a major component of aggressiveness of rust fungi. Because virulence has been intensively discussed in Sect. 2.1 of this chapter on races and also because as a temperature-adaptation attribute aggressiveness has been discussed above, we focus the discussion in this section on some other attributes of aggressiveness.

As an often quantitative trait, aggressiveness can be measured with different parameters and on different scales. At the epidemic level, aggressiveness is usually evaluated by directly measuring epidemic rates, while at the level of plant-pathogen interactions, aggressiveness can be measured by infection efficiency, latent period, spore production and sporulating period and lesion size. Relative aggressiveness should be measured by comparing different isolates. When compared to an earlier isolate, an isolate has increased aggressiveness if it causes more severe disease and more quickly on susceptible host plants (Hovmøller et al. 2011). Similarly, a new isolate has decreased aggressiveness if it causes less severe disease and more slowly.

Aggressiveness can be measured by germination rates. Volin and Sharp (1973) measured the aggressiveness of *Pst* isolates collected throughout the Pacific Northwest by germinating urediniospores on slides coated with 0.25% ion agar at 5, 10, 15, and 20 °C all with 100% humidity in darkness for 24 h. A germination percentage was calculated daily for a period of 6 days. They found that germination percentages varied significantly among the isolates at all temperatures. The albino (white-urediniospored) isolate had the lowest germination rates at 5, 10 and 15 °C, and did not germinate at 20 °C. Only isolates of two races, both virulent on Chinese 166 (*Yr1*), were able to germinate at the highest temperature; and also had the highest mean germination percentage at the other three temperatures. In this study, races virulent on Chinese 166, such as CDL-1 (PST-1), CDL-17 (PST-17) and CDL-21 (PST-21), were indicated to be more aggressive. In contrast, Line and Qayoum (1992) showed that races virulent on Chinese 166 were not aggressive and do not appear to be well adapted to wheat. These races were slow to sporulate and produced fewer urediniospores than other races on seedlings of susceptible differentials under greenhouse conditions. However, a new group of races with virulence to Chinese 166, represented by PST-127, was relatively aggressive compared to other races identified in the same period of 2007–2009 (Chen et al. 2010). This group of races became predominant in the US Pacific Northwest in 2009–2012. The differences in aggressiveness between the early and late *Yr1*-virulent races could be due to the fact that the late races also have many other virulences. Thus, virulence is a major contributor of aggressiveness, and aggressiveness is also contributed by other traits. Aggressiveness can be determined by measuring the lesion length and spore production. Priestley and Doling (1974) observed significant differences in aggressiveness of four isolates of the same race by measuring the lesion length after point inoculation. Johnson and Taylor (1976) reported different *Pst* isolates of the same race produced different yields of urediniospores. Measuring lesion length was also used in more recent studies (Milus et al. 2006, 2009; Hovmøller et al. 2008).

Aggressiveness can be derived from field inoculation experiments. The low aggressiveness of the albino isolate was also demonstrated by artificial inoculation studies in fields in the 1960s and early 1970s in the Pacific Northwest. The isolate was never detected outside the inoculated fields (Line RF, personal communication), and has never been detected under natural conditions after it was stopped to use in the fields. These observations suggest that albino isolate is less aggressive under field conditions because their white urediniospores are more sensitive to UV light and other environmental conditions than yellow- or orange-urediniospored isolates. Based on the senior author's observations over the last 30 years, stripe rust uredinia have become more orange colored compared yellow colored in the United States. The darker color may have added to the increasing aggressiveness of the current *Pst* population.

A direct way to compared aggressiveness is by co-inoculation of different isolates. Brown and Sharp (1970) inoculated susceptible plants with a mixture of urediniospores of an albino isolate (white-urediniospored) and its yellow-urediniospored isolates under greenhouse conditions and in a one-season field experiment. They did not observe significant differences in survival ability between the isolates by counting the number of uredinia. The two isolates had the same virulences. However, when the albino isolate was mixed with another yellow-urediniospored isolate with a wider virulence spectrum for inoculating susceptible plants in the greenhouse, the more virulent isolate was always predominant in the consecutive 7 generations, indicating that the more virulent isolate was more aggressive. Similar results were observed in a somatic recombination study although the study was not to determine aggressiveness. Under controlled optimal greenhouse conditions for *Pst* infection, susceptible plants inoculated with equal quantity of urediniospores of the white-urediniospored isolate and an orange-urediniospored isolate always produced more orange uredinia than white uredinia (Lei et al. 2016). White-urediniospored isolates derived from co-inoculation could change to orange-urediniospored isolates if contaminated with very few orange urediniospores that could be observed only under a microscope. Interestingly, the albino isolate generally contributed much less than an orange-urediniospored isolate to the genome of somatic recombinant isolates based on molecular markers (Lei et al. 2016).

2.2.3 Fitness

In biology, fitness is the ability of organisms to survive and reproduce in the environment in which they find themselves (Orr 2009). The consequence or impact of fitness is that organisms contribute genes to the next generation. Genes with higher fitness tend to become more common over time. For obligate biotrophic rust fungi, any genetic traits that contribute to the pathogen survival and reproduction are attributes of fitness, including virulence, adaptation, and aggressiveness discussed above. Any host and environment factors that affect the pathogen survival and reproduction will influence the pathogen fitness. Fitness should be a relative term as a pathogen or isolate fits well in one environment, but may not fit in another environment. In general, the stripe rust fungus fits well in low-temperature conditions, the stem rust pathogen in high-temperature conditions and the leaf rust fungus fits a wider range temperature conditions than ether the stripe rust pathogen and the leaf rust pathogen. The three rust pathogens all fit well in the moisture conditions during the host crop growth season.

Fitness can be measured using two methods. One is based on single generation measures of expected contribution of an individual (also called predicated fitness), in which the contribution of an individual is estimated by measuring "fitness components". The above mentioned lesion length and spore production under controlled conditions belong to this method. The other method is based on multi-generation measures of changes in frequency of phenotypes and races. Area under the disease progress curve, which measures disease development, can be used to estimate relative fitness of the pathogen during a host crop season (MacKenzie 1978; Tooley and Fry 1985). Relative fitness also can be estimated for individual races or genotypes by comparison of frequencies over years.

In the senior author's program, frequencies of individual races and virulences of *Pst* have been determined every year in the United States (Chen et al. 2002, 2010; Chen 2005; Wan and Chen 2012a, 2014a; Wan et al. 2016). Comparisons of the frequencies over different years provide relative fitness for different races and virulences. Race with a higher frequency have a better fitness than those with a low frequency. For example, PSTv-37 has been one of the most predominant races and occurred in all epidemiological regions in the United States since 2010 (Wan and Chen 2013, 2014a, b, 2015; Wan et al. 2016). Over the 42 years from 1968 to 2009, virulences to *Yr6* and *Yr7* always had high frequencies whereas virulences to *Yr10*, *Yr24*, *Yr32* and *YrSP* had low frequencies (Liu et al. 2016c), indicating that virulences to *Yr6* and *Yr7* had better fitness than virulences to *Yr10*, *Yr24* and *Yr32*. Similar work has also been done in China (Wan et al. 2004; Chen et al. 2009b).

Relative parasitic fitness has been used to study *Pst* races. In China, a series of parameters were used to determine parasitic fitness, including urediniospore germination rate on leaf surface, number of chlorotic spots, latent period, uredinium length, uredinium density, sporulation period and sporulation intensity (Li and Zeng 2002). Based on data of tests with seven relatively common races in the late 1970s, they determined the contribution rates of the parameter groups. The first component category (mainly sporulation intensity and period) representing reproduction capability explained 45.62%, the second component category (mainly uredinium length and density) representing infection efficiency 26.64%, and the third component category (mainly urediniospore germination rate on leaf surface) representing activity 20.22% of parasitic fitness variation. The seven races were ranked differently in the three categories. Overall, races CYR22 and CYR26 had high; CYR23, CYR25 and

CYR29 had moderate; and CYR27 and CYR28 had low parasitic fitness. They also tested some of these races on five major wheat cultivars in both seedling and adult-plant stages for latent period, sporulation period and uredinium length. Their results showed that races CYR26 and CYR29 had high whereas CYR25 and CYR28 had low parasitic fitness on seedlings, CYR26 had high whereas CYR25 had low parasite fitness on adult plants. Using the low parasite fitness for various races, they explained why races CYR23, CYR25, CYR27 and CYR28 did not develop to major predominant races even under the widely grown susceptible cultivars. Based on testing four predominant races (CYR31, CYR32, SU11-4 and Su11-14) on 12 major wheat cultivars and a susceptible variety, Tian et al. (2008) reported that race CYR32 had the highest parasitic fitness and predicted that SU14-14 would become a predominant race as it had the second highest parasitic fitness. Using the similar approach, Wang et al. (2009a) tested eight races (CYR29, CYR30, CYR31, CYR32, Su11-4, Su11-7, Su-14 and T) and concluded that Su11-4, Su11-14 and CYR32, which have already become the major predominant races, had higher fitness than other tested races. The decrease of CYR31 was due to its low parasitic fitness. They predicted that race T with virulence on 'Zhong 4' would not become a predominant race as it had low parasitic fitness. These two studies generally agreed to each other, but the result about CYR31 were different from a previous report that CYR31 would increase as it had higher parasitic fitness than CYR29 and CYR30 by Qiang et al. (1999). The controversial results were likely due to the different wheat cultivars used in these studies. The cultivars used in Tian et al. (2008) and Wang et al. (2009a) were countrywide whereas the cultivars used in Qiang et al. (1999) were major cultivars grown in Qinghai province. This discussion indicates that parasitic fitness makes sense only for regions the tested cultivars will be continually grown. In addition to these studies, other studies developed theoretical models for estimating parasitic fitness (Sun and Zeng 1995; Zeng 1996, 1998; Zeng and Luo 2008).

As discussed above, most studies for fitness and/or aggressiveness were conducted in a single generation and few studies with a limited number of generations under controlled conditions mostly optimal for stripe rust development. Few studies were conducted in field but just in one growth season. Most studies focused on parasite fitness, but very few studies were on survival fitness. In fact, survival could be more important for the obligate biotrophic stripe rust fungus. Early studies of the stripe rust fungus were conducted at the species or formae species level for its overwintering and oversummering, and host ranges related to crop-off season survival as discussed in Chap. 4 (Epidemiology). No studies have focused on comparison of different races, isolates or genotypes for off-season survival, except few studies on high-temperature adapted isolates as discussed above. Even the high-temperature studies used the temperatures just for urediniospore germination, infection and disease development, but did not use the temperatures for survival. In the future, more studies should be conducted to identify traits and genes involved in the pathogen survival, as well as those involved in virulence and reproduction. Such studies could be important for predicting race or population changes related to environmental changes.

For survival fitness, some traits may increase fitness for one species but reduce fitness for another species, or for an individual species they may increase fitness in one environment but decrease fitness in another environment. For example, telial production is important for *P. graminis* f. sp. *tritici* to survive the hot summer and especially the cold winter in the northern part of the United States, but not important for the southern part of the country as the pathogen can survive the warm winter (Wang and Chen 2015; Wang et al. 2015; Zhao et al. 2016). Because of these, telial production increases survival in the north, but decreases survival by reducing urediniospore production. For *Pst*, which completely reproduces asexually, the pathogen survives winter by infectious hyphae produced by urediniospores in the US Pacific Northwest. Teliospores do not play a role in the pathogen survival in the region as they are covered by host epidermal layer, degraded during the winter and even germinating, cannot catch the period of susceptible barberry leaves (Wang and Chen 2015; Wang et al. 2015). In the US Pacific Northwest, the low temperature at night and dry condition during summer allow *Pst* urediniospores to easily oversummer, and often urediniospores continually infect late spring wheat crops, volunteer wheat plants and susceptible grasses during summer (Dietz and Hendrix 1962; Line 2002; Sharma-Poudyal et al. 2014; Cheng et al. 2016a). Isolates producing more urediniospores and fewer teliospores have more advantages than those producing fewer urediniospores and more teliospores in oversummering and possibly overwintering. In contrast, isolates quickly producing telia have disadvantages in both parasitic fitness and survival fitness. In the US Pacific Northwest, *Pst* generally produces telia very late in the crop season and *Psh* produces telia relatively quick. Under the controlled greenhouse conditions at low temperatures (diurnal cycle of 4–20 °C), *Psh* telia can be readily produced on barley whereas *Pst* does not produce telia. However, a significant change was observed in 2007 when race PST-127 (now PSTv-11) was first identified. This race has the broadest virulence spectrum, virulent on all 20 wheat cultivar differentials except Moro (*Yr10*, *YrMor*), Tres (*YrTr1*, *YrTr2*) and the *Yr5* line (Chen et al. 2010), and particularly readily produces telia in the greenhouses and fields. Several races with similar virulences have been identified in the subsequent years with high telial productivity (Wan and Chen 2012a). PST-127 increased from 1.5% in 2007 to 9.1% and the fifth predominant race in epidemiological region 1 (the eastern Pacific Northwest in 2009. However, the race and its similar races decreased quickly after 2010, and hardly found in 2015 and 2016. Although the race increase from 2007 to 2009 largely due to some of virulences to the wheat cultivars grown the Pacific Northwest, the senior author predicted that the race group should decrease because its quick production of telia. Switching from producing infectious urediniospores to non-infectious teliospores is self-killing for the stripe rust fungus.

Because of variation in telial formation, the senior author's program has conducted telial observation of *Pst* isolates. Percentage of telial formation (proportion changed from uredinia) was examined on susceptible wheat plants 40 days after inoculation with urediniospores under controlled greenhouse conditions. From 1002 isolates collected from 2013 to 2015 throughout the US, 76% of the isolates were able to produce telia, ranging from 1 to 90% with an average of 20% telial productivity (Wan and Chen 2016). Differences in telial formation were observed among race groups, regions of collection, and months of sampling. Telial formation was more related to specific races than the numbers of virulence factors. Generally, isolates from the western United States produced few telia than those from the eastern United States. Isolates collected in earlier season had low telial formation than those collected in the later season. These results could be used for predicting population changes using the telial formation as a negative parameter for fitness.

2.2.4 Trade-Off Among Virulence, Aggressiveness, Fitness and Adaptation

The relationships among virulence, aggressiveness, fitness and adaptation are complex, and one can contributes to another but also can limit another. As discussed above, virulence is a basic component for aggressiveness, fitness and adaptation because the stripe rust pathogen cannot survive and reproduce without a host plant. However, extra virulences may reduce aggressiveness, fitness and/or adaption. In the US, the *Pst* population has generally become more virulent as the recent predominant races have more virulences than those in the 1960s and 1970s (Line and Qayoum 1992; Chen 2005; Chen et al. 2010; Wan and Chen 2012a, 2014a; Wan et al. 2016; Liu et al. 2016c), indicating that more virulent races are more aggressive, more fitting and more adapted to the host crop and environment. However, we also observed the most virulent race (PST-127) based on the 20 wheat cultivar differentials (Chen et al. 2010) quickly disappearing after a couple of year increase, as discussed above. Similarly, race PSTv-51, which is virulent to 16 of the 18 *Yr* genes in the current differentials except *Yr5* and *Yr15*, was detected as 1 or 2 isolates in the United States in 2007, 2009 and 2011, but has not been detected since 2012 (Wan and Chen 2013, 2014b, 2015, Liu et al. 2016c; Wan et al. 2016). These results indicate that the highly virulent race may be less aggressive than predominant races and may not fit the environment. Similarly, highly virulent races of *Psh*, such as PSH-31 (virulent on 10 differentials), PSH-49 (virulent on 11 differentials) and PSH-72 (virulent on all 12 barley differentials) identified in 1995, 1997 and 2004, respectively, have never become significant (Chen et al. 1995b; Chen 2004, 2008; Chen and Line 2001; Chen and Moore 2002; Chen et al. 2003a, b, 2004, 2007; Chen and Penman 2005, 2006; Wan and Chen 2012a, b). In fact, the predominant races have been those with virulences to four or fewer differentials. These results indicate that less virulent races are more aggressive, have a better fit and adapted to the crop and environment, and/or traits other than virulence may affect aggressiveness, fitness and adaptation.

In evolutionary biology, the trade-off effect has been used to study relationships between different but related traits which related to fitness of a parasite under the influence of ecological factors (Poulin 2007; Sacrisan and Garcia-Arenal 2008). There is a natural force providing pressure on the parasite to self-limit virulence. According to this idea, there exists an equilibrium point of virulence, where parasite's fitness is highest. Any movement from the optimal virulence, towards higher or lower virulence, will result in lower fitness for the parasite. Based on this concept, a rust isolate with an extra virulence may trade-off its fitness, reducing the trait(s) for fitness. When the trade-off relationship functions, an optimal number of virulences can exist or can be reached for equilibrium. This issue has not been well studied for rust fungi, especially the stripe rust fungi. However, historical virulence data may provide some information about virulence trade-off with fitness as discussed above, and may give a clue about optimal virulence.

2.3 Molecular Characterization of Variations in the Stripe Rust Pathogen Population

As discussed above, population variations have been studied for the stripe rust pathogen by virulence phenotypes in various parts of the world for more than a century. However, the pathogen variations have been studied using molecular markers for only about 30 years. Although the molecular characterization of the stripe rust pathogen was started later than many other filamentous fungal pathogens, this type of the research has been one of the most active research phases of the stripe rust pathogen in the last two decades.

2.3.1 Isozyme and dsRNA Markers Revealed Differences Between Pst *and* Psh

The earliest molecular markers used to study *P. striiformis* were isozymes and double-stranded (ds) RNA. Newton et al. (1985) determined the activity variations of 29 isozymes and dsRNA markers in the stripe rust pathogen and the leaf rust pathogens of wheat and barley. They found variations in isozymes and dsRNA within and between the rust species *P. striiformis*, *P. triticina* and *P. hordei*. They did not found any differences in isozyme phenotype among 29 diverse *Pst* isolates, and among smaller numbers of isolates of *Psh*, *P. triticina* and of *P. hordei*. They found major differences in isozyme phenotypes among the three species, but found differences in only two out 12 tested isozymes between *Pst* and *Psh*. They detected dsRNA in each species and in all 26 isolates examined. All isolates within each *Pst* and *Psh* had the same or a similar dsRNA phenotype, but each isolate of *P. triticina* and *P. hord*ei had a unique phenotype. Differences in dsRNA phenotypes were detected both among the three species and also between *Pst* and *Psh*. Based on the differences in isozyme and dsRNA phenotype, the authors considered that *Pst* and *Psh* are two discrete populations and should be recognized as different formae speciales.

As they identified different isozyme activities and dsRNA bands, the same authors used the same techniques for studying somatic hybridization between *Pst* and *Psh* isolates (Newton et al. 1986). This study was largely failed as no any somatic hybrid isolates were confirmed by markers. They inoculated barley variety 'Fong Tien' with mixed urediniospores of a white-spored *Pst* isolate and a yellow-spored *Psh* isolates as Fong Tien is susceptible to both isolates. They selected progeny isolates based on virulence markers and non-parental colors, and tested with isozyme and dsRNA markers that clearly differentiated the parental isolates. They obtained two white-spored, non-parental isolates. Except the white color, these isolates resembled the parental *Psh* isolate in virulence, isozyme and dsRNA markers. The authors concluded that these isolates were most likely produced by the color mutation of the *Psh* isolate, rather than somatic hybridization. The major problem of this study is the small number of markers as only 12 isozymes were used and only 5 dsRNA bands were polymorphic between the parental isolates. As the authors pointed out that some of the bands were faint that might affect the results. As discussed later, parental contributions to recombinant progenies are usually uneven, and the less portion of the genome from one parent may not be detectable when a small number of markers are used.

2.3.2 Utilization of DNA Markers for Detecting Stripe Rust Pathogen and Studying the Pathogen Population

As advances in biotechnology have been made, DNA based molecular techniques have been employed to investigate biology, population structure, variations, migration and evolution of *P. striiformis*. Compared to isozymes and dsRNA markers that are under the influence of fungal growth stages and environments for their expression, polymorphic DNA markers are more abundant and reliable.

2.3.2.1 Development of Molecular Markers for Detecting Stripe Rust Pathogen

Although stripe rust is easy to recognize and distinguish from other rusts and other diseases, methods are needed for identifying the pathogen during latent infection in primary leaves, pycnial and aecial stages on alternate hosts, and urediniospores in the air for diagnosis and detection. Such information can be useful for studying biology of the pathogen, epidemiology of the disease and phylogenetics of the pathogen with other rust and fungi and for managing the disease by early prediction. DNA markers are useful for obtaining such information.

2.3.2.1.1 ITS Markers

Ribosomal DNA (rDNA) repeat units have been used for identifying fungal species. The 18S, 5.8S, 28S and 5S rDNA genes are found as multiple copies arranged as head-to-tail repeats separated by noncoding spacers: the internal transcribed spacers (ITSs) between the 18S, 5.8S and 28S RNA sequences and the intergenic spacers

(IGS) residing between each transcription unit (Takamatsu 1998). Zambino and Szabo (1993) distinguished selected cereal and grass rust species including *P. striiformis* and determined their phylogenetic relationships based on ITS sequences. IGS regions exhibit higher inter- and intra-specific variability than ITS regions. Roose-Amsaleg et al. (2002) used the sequences of the intergenic spacer between the 28S and the 5S rDNA genes (IGS 1) for comparing three European isolates representing three *Pst* races from different geographic origins and virulence groups. The sequences were highly conserved with low mutation rates of 0.7 to 1.1% among the isolates studied and the IGS 1-2 was almost identical to the IGS 1-1 except for the 216 bp insertion/deletion. Komjáti et al. (2004) used the IGS sequences to test 12 *Pst* isolates belonging to different races, one *Psh* isolate and one *P. triticina* isolate and found no variation. However, they identified polymorphisms among isolates using sequence related amplified polymorphism (SRAP) markers.

Because ITS and IGS sequences are highly conserved, the sequence information can be used for determining relationships at species level, but is less useful for genetic variations between formae species of *P. striiformis* or among isolates within a forma specialis. This feature makes ITS sequencing very useful for differentiating *P. striiformis* from other *Puccinia* species. Barnes and Szabo (2007) designed primers based on ITS1 sequences of rust fungi and used RT-PCR to detect *P. striiformis*, *P. graminis* and *P. triticina*. Liu and Hambleton (2010) used ITS sequences, together with morphology characteristics to group *Pst* and *Psh* together in to a subclade (Subclade 1 for *P. striiformis*) that is different from other three subclades: Subclade 2 (*P. striiformoides*, previously *P. striiformis* f. sp. *dactylidis*), Subclade 3 (*P. pseudostriiformis*, previously *P. striiformis* f. sp. *poae*) and Subclade 4 (*P. gansensis*). The ITS sequence data also helped them to establish the Series *Striiformis* that includes the four subclades and is distinct from other rust fungi on cereals and grasses including *P. coronate*, *P. poae-nemoralis*, *P. graminis*, *P. recondite*, *P. triticina*, *P. hordei* and *Uromyces dactylidis*.

Zhao et al. (2007a) designed primers PSF (5′-GGATGTTGAGTGCTGCTG TAA-3′ and PSF (5′-TTGAGGTCTTAAGGTTAAAATTG-3′) based on the internal transcribed spacer (ITS) region sequence of *Pst*. The primers amplify a fragment of 169 bp specific for *P. striiformis*. They used the marker to detect *Pst* in wheat leaves. Using the ITS marker and β-tubulin marker of Fraaije et al. (2001), Kumar et al. (2013) successfully detected viable *Pst* mycelia in winter wheat leaves and demonstrated that *Pst* can overwinter in central Alberta, Canada, especially under long-standing snow cover.

As *P. striiformis*, *P. graminis* and some other *Puccinia* spp. can infect the same alternate hosts such as barberry and mahonia (Jin et al. 2010; Wang and Chen 2013; Zhao et al. 2013; Wang et al. 2015), it is important to identify which species on alternate host plants. Currently, no reliable morphological characteristics can be used to distinguish pycnial and aecial stages of different species. Wang et al. (2015) sequenced ITS regions of isolates of *P. striiformis*, *P. graminis*, *P. pseudostriiformis* and *P. striiformoides* and identified differences in sequences among these species. Using the sequences, they designed ITS primer pair, ITS12F (5′-GAATCATCGAATCTTTGAACGC-3′) and ITS12R (5′-GTAGTCCCACCTGATTTGAGG-3′), which amplified a fragment of

326 bp from *P. graminis* f. sp. *tritici*, 314 bp from *Pst*, 316 bp from *P. pseudostriiformis* and 319 bp from *P. striiformoides*. Using the primers to amplify ITS fragments from individual aecia collected from barberry plants, and separating the fragments using capillary electrophoresis, they clearly identified that rust aecia on barberry were mostly *Pgt* and occasionally some other formae speciales of *P. graminis*, but not *P. striiformis*. The ITS markers helped them determine that barberry is essential for stem rust, but does not play any role for stripe rust in the US Pacific Northwest.

2.3.2.1.2 β-Tubulin Gene Markers

Fraaije et al. (2001) designed primers BAF6 (f) (5′-ACCCACAACCGCCAAC ATGCGTGA-3′) and BAF2 (r): (5′-CGTACCGGGCTCGAGATCGACGAG-3′), which amplified fragments of different sizes for different fungi including *P. striiformis*. This primer pair can be used for detecting different fungal species in wheat leaves, but was not very specific. They designed primers YRNT1 (f) (5′- CTTCAAGA TCGGTGGCCTGACCGA-3′) and YRNT 2 (r) (5′- GTGAGCTGTGA AGGGATCGCGGGA-3′). Using YRNT1 and YRNT2, a fragment of 351 bp was specifically amplified from *P. striiformis*. Coram et al. (2008b) used the *Pst* sequences of β-tubulin gene sequence obtained by Ling et al. (2007) to design primers for quantifying *Pst* biomass and growth in their study of genes involved in *Pst*-wheat interactions. The β-tubulin gene was used as a reference for estimating expression of various *Pst* genes during compatible and incompatible interactions (Huang et al. 2011). Using a similar approach, Cao et al. (2008) designed primers YR (f) (5′-CCATGGAAATCGATTCAACC-3′) and YR (r1): (5′-GAGACAA AGAGATTGGGACA-3′) and YR (r2) (5′-TTAAGACTACCGGTGTGAAC-3′. Primers YR (f) and YR (r1) produces a unique band of 160 bp and primers YR (f) and YR (r2) produces a 139-bp fragment from *Pst*. The markers can be detected *Pst* at a limit of 2.20 pg template DNA and in infected wheat leaves.

Yin et al. (2009) designed primers TUB-F (5′-CCGATCAATTCACGGCC ATGTTCA-3′) and TB-R (5′-AACCCTCTTCAACTTCCTCGTCGT-3′) based the β-tubulin gene of *Pst* and used them in qRT-PCR to detected gene expression of *Pst* in inoculated wheat leaves. Pan et al. (2010) also developed a marker based on β-tubulin sequences and used RT-PCR to detect *Pst* during latent infection.

Based on primer pair Bt2a/Bt2b of Glass and Donaldson (1995), Gao et al. (2016) designed two pairs of primers: PSTF117: 5′-CACGCGGGACGTA CTTATTGTC-3′andPSTR363(reverse):5′-CATTTCTGGCGAACATGGAATC-3′; and TF144: 5′-CCTGCGATGGTGTAGACTCA-3′ and TR323: 5′-CGGCGTGT ATGTTCGTGTTG-3′. The first primer pair amplified a 274 bp fragment and the second 180 bp fragment from *Pst*, with a detection limit of 1 pg/μl and 100 fg/μl, respectively. Both markers could detect *Pst* in wheat leaves as soon as 9 h after inoculation. Using the markers in SYBR Green RT-PCR, they were able to detect *Pst* in infected leaves with the detection limit of 1.0 fg DNA from asymptomatic leaf samples of 6 h after inoculation.

2.3.2.1.3 Elongation Factor Markers

Coram et al. (2008b) designed primers EF1-F (5′-TTC GCC GTC CGT GAT ATG AGA CAA-3′ and EF1-R (5′-ATG CGT ATC ATG GTG GTG GAG TGA-3′) based on elongation factor 1-alpha (Ling et al. 2007) and used the gene as a reference to study expression patterns of genes involved in resistance to stripe rust. Similarly, Yin et al. (2009) also used the primers in qRT-PCR to detected gene expression in *Pst*-inoculated wheat leaves. Ma et al. (2016) used these primers in qRT-PCR to determine overwintering of *Pst* in several regions in northwestern China, where the fungus previously was considered not to overwinter.

2.3.2.1.4 Repetitive Sequence Markers

Wang et al. (2008) designed two pairs of primers, Pst1 (5′-ATGCTGGCA GTGTGGTTG-3′) and 5′-ACATTTCCACGAATCTCCTTG) and Pst2 (5′-GTCT GTAAGATGTTAGATGC-3′ and (5′-ATGCTGGCAGTGTGGTTG-3′) based on *Pst* repeat sequences identified by Zheng et al. (2000, 2001). Pst1 primers amplify a fragment of 450 bp and Pst2 primers amplify a 470 bp fragment. The authors found that the sensitivity of Pst1 was o.1 ng and Pst2 0.01 ng DNA template per micro liter solution. Using Pst2 primers, they detected *Pst* in inoculated wheat leaves. The same group further developed a nested PCR method (Wang et al. 2009b). They used the above Pst2 primers as outer primers and designed nester primers: Nesta 5′-GTCGACGAGCCAACTCCTTCT-3′ and Nests 5′-AAGCTCAA CATCATATGCTCGAC-3′. Using the nested PCR methods, they were able to detected *Pst* in 2 fg template DNA, increasing detection sensitive by 1000 times.

2.3.2.1.5 Markers Based on Other Genes for Detecting *P. striiformis*

In addition to the above genes, other genes of *Pst* have been used to detect the pathogen, as a references for studying pathogen genes during *Pst*-host interactions, and distinguishing *P striiformis* from other *Puccinia* species. These genes include histone genes and actin genes (Coram et al. 2008b; Huang et al. 2011). Any genes that are conserved for *P. striiformis* but can separate this species from their host plants and other fungi can be used to develop markers for specific detection of the stripe rust pathogen.

2.3.2.2 Utilization of DNA Markers to Study Populations of *P. striiformis*

Since the early 1990s, various DNA marker techniques have been developed and used to study populations of the stripe rust pathogens. The intensive studies have helped understand population structures, changes, migrations, reproduction, and effects of alternate hosts and grass hosts on disease epidemiology and pathogen evolution.

2.3.2.2.1 RAPD Markers Revealed *Pst* Population Diversity in the US and Differentiated Formae Speciales

After the random amplified polymorphic DNA (RAPD) technique was first developed in 1990 (Williams et al. 1990; Welsh and McClelland 1990), the technique was quickly used to study variations and polymorphisms for many organisms, as the technique is very simple, easy to use, and primers can be universally used for any organisms. For stripe rust, the study of Chen et al. (1993) was the first to use DNA markers to determine the population structure of *Pst*. Using RAPD markers, they characterized 115 single-urediniospore isolates representing 26 *Pst* races in the US. They detected polymorphisms among races and among single-urediniospore isolates within races. Race PST-21 (previously CDL-21) was highly different from other races with only 67% similarity, whereas other races were more similar (about 80% similarity). The 115 isolates were clustered into 5 RAPD groups. They detected a low, but significant correlation coefficient (0.20) between the RAPD data and virulence data. RAPD data clearly separated isolates virulent on resistance gene *Yr1* from other isolate. They found that virulence groups were highly associated with epidemiological regions, whereas the neutral RAPD markers were generally not associated with regions.

The same authors also used RAPD markers to determine the relationships between *Psh* and other rust fungi (Chen et al. 1995b). The marker analysis clearly separated *Psh*, *Pst* and *P. striiformis* f. sp. *poae* (*Psp*., now *P. pseudostriiformis*), and also separate these formae speciales of the stripe rust fungus from the wheat stem rust (*P. graminis* f. sp. *tritici*), wheat leaf rust (*P. triticina*) and barley leaf rust (*P. hordei*) fungi. This study supported the separation of *Psh* and *Pst* and also indicated that *Psp* is more distantly related to *Psh* and *Pst*.

2.3.2.2.2 RFLP Markers Detected *Pst* Population Diversity and Regional Migration in China

Restriction fragment length polymorphism (RFLP) is a relatively old, but widely used marker technique. It requires southern blot hybridization of enzyme-digested DNA fragments with probes (Saiki et al. 1985). Shan et al. (1997) identified a family of dispersed moderately repetitive DNA sequences specific to the genome of *P. striiformis*. Using one of the clones as a probe (PSR331), Shan et al. (1998) determined the genetic variation among 160 isolates of *Pst* collected from six provinces of China. Over 26 putative genetic loci defined by PSR331 and the restriction enzyme *Bgl*II revealed a high level of genetic variation, 97 phenotypes from 160 isolates. They found that diversity, as calculated with the normalized Shannon's index, varied among regions, from 0.37 in Shaanxi to 0.94 in Gansu. They also found a low level of genetic differentiation (GST = 0.0084) among provinces (Gansu, Henan, Shaanxi, Sichuan, and Yunnan provinces) as well as within Gansu and Sichuan provinces. These results provided the molecular evidence for gene flow in the Chinese *Pst* population. Their results support conclusions from virulence surveys that Tianshui of southern Gansu is probably the most important "hotspot" area for generating and maintaining virulence variation. The RFLP analysis also detected additional hotspot areas.

2.3.2.2.3 AFLP Markers Identified *Pst* Population Structures in China and Other Countries

Both RAPD and RFLP techniques have different advantages and disadvantages. RAPD is simple, fast, low cost and does not need sequencing information or probes, but repeatability is a problem and especially for comparing data of different studies. In contrast, RFLP has relatively good repeatability, but it is complex, slow, high cost, requires probes and often uses radioactive materials for Southern blotting. The amplified fragment length polymorphism (AFLP) technique was developed in the 1990s (Vos et al. 1995). AFLP combines the advantages of both RAPD and RFLP and reduces their problems by digesting genomic DNA with restriction enzymes, ligating short sequencings and using selective base pairs in primers. Because of its relative simplicity, time and cost efficiency, without the requirement of sequencing information and no need to use radioactive materials, AFLP quickly gained popularity.

Using a combination of RAPD and amplified fragment length polymorphism (AFLP) markers, Steele et al. (2001) characterized the levels of molecular variation in *Pst* isolates from Australia, New Zealand, UK, Demark and Colombia. Using 60 RAPD primers with more than 300 potentially polymorphic loci on seven Australian isolates representing seven different pathotypes collected between 1979 and 1991, they did not detected polymorphisms. When testing two UK isolates with the same RAPD primers, only 3% of loci showed a polymorphism. They found a similar level of polymorphism between UK isolates using AFLP primers, and 5–15% of fragments polymorphic between an isolate from the UK and an isolate from Denmark, and one from Colombia. They did not found AFLP polymorphisms among 14 Australian and New Zealand isolates tested at over 100 potentially polymorphic loci. The authors concluded that the lack of molecular variation in the Australian and New Zealand collection was consistent with the stepwise mutation of pathotype evolution from a single introduction. Later on, more AFLP markers were identified and used to elucidate genetic variations, determine structures of *Pst* populations in various countries and detect migration and genetic recombination (Zheng et al. 2001; Hovmøller et al. 2002; Justesen et al. 2002; Lorys et al. 2002; Markell et al. 2004; Hovmøller and Justesen 2007; Hovmøller et al. 2008; Duan et al. 2010; Liu et al. 2011b; Liang et al. 2013; Wan et al. 2015; Liang et al. 2016).

Significantly, AFLP markers helped determine the clonality and long distance migration of *Pst* in Europe (Hovmøller et al. 2002; Lorys et al. 2002); rapid spread of two aggressive strains in the world (Hovmøller et al. 2008), high diversity and recombination in the over-summering zone of Gansu (Duan et al. 2010) and both Gansu and Ningxia provinces in China (Liang et al. 2013). Liang et al. (2013) also detected gene flow between the two provinces and the significant effect of the Liupan Mountains on limiting gene flow and population exchange. Using AFLP markers, the same group expanded their population structure research to include Qinhai and Xinjiang (Wan et al. 2015). They found detected frequent genotypic exchange between Xinjiang and Qinhai and between Qinhai and Gansu. The diversity of the *Pst* population in Gansu was much higher than those in both Qinghai and Xinjiang. They detected recombination in the Gansu population and some subpopulations of Qinhai, but not in the Xinjiang population. Surprisingly, they reported an extremely low correlation coefficient values, 0.032 (P = 0.632) for the Xinjiang population and 0.037 (P = 0.659) for the Qinhai population, considering the populations were largely clonal, especially for Xinjiang. Using larger numbers of isolates and the same AFLP technique, the same group compared *Pst* populations in Sichuan and Gansu and detected similar levels of diversities between the populations, asymmetric migration from Gansu to Sichuan, recombination in Gansu and seasonal shift of populations (Liang et al. 2016).

2.3.2.2.4 SSR Markers Widely Used to Study *Pst* Population Structure, Migration and Mode of Reproduction

The stripe rust fungus is dikaryotic in the uredinial stage that is the longest stage during the yearly disease cycle. The two nuclei in each isolate is highly 'heterozygous' (Liu et al. 2012b; Zheng et al. 2013). Dominant markers developed using the above techniques, such as RAPD, AFLP and repetitive sequence RFLP, cannot distinguish between 'heterozygous' and 'homozygous' alleles. Co-dominant markers have advantages to detected heterozygosity of genomic sequences and genes. In the past decade, several studies have been conducted to develop SSR markers for *P. striiformis*, and more than 300 SSR markers are currently available (Enjalbert et al. 2002, 2005; Bahri et al. 2009a; Chen et al. 2009a; Duan et al. 2010; Cheng et al. 2012; Bailey et al. 2013; Luo et al. 2015; Tian et al. 2016).

2.3.2.2.5 Utilization of SSR Markers in Studying *Pst* Population in China

Several studies used SSR markers to study the Chinese population of *Pst*. Using both SSR and AFLP markers in addition to virulence tests, Mboup et al. (2009) assessed the diversity of the *Pst* population using 412 *Pst* isolates from the highlands of Tianshui region in Gansu province of China. They identified remarkably high phenotypic (38%) and genotypic (89%) diversity in the regional population. No any significant genetic structure dependent on the sites sampled (Fst = 0.004) or altitude distribution (Fst = 0.0098) was detected, indicating high gene flow within the region. Based on the genetic recombination between molecular markers, the authors suggested that *Pst* may undergo sexual or parasexual cycles, prior to the discovery of barberry as an alternate host for the stripe rust fungus by Jin et al. (2010).

Expanding from the *Pst* population of Tianshui in southern Gansu, Lu et al. (2011) studied spatial genetic diversity and interregional spread of *Pst* in the Longnan (southern Gansu) and Linxia regions in Gansu and Qinghai provinces, the major over-summering regions for the pathogen and key epidemiological zones in northwestern China. They found that the genetic diversity in the Longnan population was much higher than those in the Linxia and Qinghai populations, confirming that the Longnan region is a center of genetic diversity for *Pst* in northwestern China. The low genetic differentiation and the extensive gene flow were found among the three regions. The authors concluded that the stripe rust inoculum in Qinghai can come from both Longnan and Linxia, but mainly from Longnan directly in the spring.

Further expanded from Gansu and Qinhai provinces, Zhan et al. (2016) studied *Pst* collections from Gansu, Ninxia, Qinhai, Xinjiang and Tibet in western China using 15 SSR markers and virulence tests. They identified more races in Xingjian than other regions. The 35 races identified in Xinjinag comprised of historical (no longer detected in other regions of China) and recently predominant races plus new races unknown to other regions. The SSR marker data also showed that the *Pst* population in Xinjiang had a much higher genetic diversity than those in other regions. These data indicate that Xinjiang is an important region for stripe rust epidemics and the pathogen variation in China. The high diversity in the Xinjiang population was different from the results of Wan et al. (2015). The different results could be due to the different years and areas of sampling, sample sizes, markers and calculation methods. However, for population comparison, it is important to have relatively random samples, cover as much area as possible, and use similar numbers of samples for compared regions. A proper method of calculating diversity should be used.

2.3.2.2.6 Utilization of SSR Markers to Study *Pst* Reproduction in the US

In the US, a large number of *Pst* races are identified every year and predominant races change rapidly. However, whether the high diversity and rapid change of races are caused by sexual reproduction or not was unclear. Before the discovery of barberry and mahonia as alternate hosts for the fungus (Jin et al. 2010; Wang and Chen 2013), a study was initiated to determine the *Pst* reproduction mode. A total of 270 isolates from the US Pacific Northwest where barberry plant existed nearby wheat fields, together with 66 isolates from other regions collected in the same year, were characterized by virulence tests on the wheat cultivar differentials and 20 co-dominant SSR markers (Cheng and Chen 2014). They identified 21 races and 66 multilocus genotypes, of which 15 races and 32 multilocus genotypes were found in the Pacific Northwest region with barberry plants heavily infected by rust. The SSR marker data revealed two genetic groups, which were significantly correlated to the two virulence groups ($r = 0.39$, $P < 0.001$). The analyses of genotype-to-individual ratio, multilocus linkage disequilibrium, and heterozygosity strongly supported asexual reproduction for the pathogen population in the Pacific Northwest and other regions of the US. These results were consistent with the conclusion that barberry is essential for sexual reproduction of the stem rust fungus but not for *Pst* based on the studies with aecial samples from barberry plants (Wang and Chen 2015; Wang et al. 2015).

In the US, *P. striiformis* infects not only cereal crops, but also many grass species. To determine the genetic relationships of stripe rust pathogen on grasses with

Pst and *Psh*, Cheng et al. (2016a) characterized 103 isolates collected from wheat, barley, triticale, rye and grasses using 20 simple sequence repeat (SSR) markers in addition to virulence tests. From the isolates, 44 multi-locus genotypes were identified and the genotypes were clustered into three major molecular groups with one of the major group further classified into three subgroups. Cereal crop isolates were present in one or more of the major or subgroups, but not all, whereas grass isolates were present in all of the major groups and subgroups. This study shows that grasses harbor more diverse isolates of *P. striiformis* than the cereals and play an important role in stripe rust epidemics on cereal crops and also in generating and maintaining high diversity of the pathogen. In this study, the authors utilized a multiplex approach that allowed detecting PCR products amplified with four SSR primer pairs to make the SSR technique more efficient (Cheng et al. 2016a).

To reduce repeating tests of same isolates with molecular markers in different studies, Wan and Chen (2014a) selected 18 co-dominant SSR markers as a set of marker differentials for routine genotyping US isolates of *Puccinia striiformis*. The set of markers identified 137 genotypes from 322 *Pst* isolates collected in the US in 2010. The genotypes were clustered into three major groups. The western US population had more genotypes that were in all three major groups whereas the eastern US population had fewer genotypes that were in only two groups. The western population had much higher genetic diversity than the eastern population. Genetic divergences and distances were much higher within the eastern population than the western population, whereas the gene flow was much higher within the eastern population than the western population. The differences between the western and eastern regions are likely due to the different occurrences and development of stipe rust, number of cultivars and resistance genes in cultivars, climatic conditions for survival of the pathogen, and geographical features for pathogen spreading.

2.3.2.2.7 Utilization of SSR Markers in Comparison of *Pst* Populations in Different Countries

Bahri et al. (2009b) characterized 453 *Pst* isolates collected in 11 countries in the Mediterranean region to assess the geographic distribution of the pathogen population in southern France that is divergent from the northwestern European population. They identified a dominant strain in the western Mediterranean region using 15 SSR markers. Because of the low level of SSR diversity, they used AFLP markers to genotype isolates of the dominant strain and further separated the strain into two south-specific groups, a major group with a broad variability and a wide distribution in both western and eastern Mediterranean countries and a minor group in the western Mediterranean. With the SSR markers, they determined the phylogenetic relationships among the *Pst* population in the Middle East, southern France and northwestern Europe. The authors concluded that migration, local adaptation, the dynamics of clonal lineage diversification and replacement should be considered to define population entities in the strongly clonal pathogen. The same group also used SSR markers to study genetic diversity of the *Pst* population in the North-West

Frontier Province of Pakistan (Bahri et al. 2011). Using 15 SSR markers, the authors identified five groups representing three distinct lineages. They also identified the 49 tested Pakistani isolates as 12 pathotypes using the 15 European and world differentials plus 17 wheat lines with single *Yr* genes or Chinese differentials. The top three predominant pathotypes covered 80% of the isolates and were clustered as the Pakistani lineage, whereas the other isolates were close to either a Mediterranean lineage or a Northern European lineage. The authors concluded that pathogen migration and wheat variety diversity might contribute to and maintained the high genetic diversity of the *Pst* population in Pakistan.

To test the hypothesis that variations in the ability for sexual reproduction could provide an explanation for a geographical gradient in genotypic diversity of *Pst*, Ali et al. (2010) tested 56 isolates representing six *Pst* populations of worldwide geographical origins using 17 SSR markers. The isolates also varied in telial production with those from China, Nepal and Pakistan displaying the highest telial production and those from France and the Mediterranean region having low telial production. The marker data clustered the isolates into five genetic groups corresponding to their geographical origin, with eight inter-group hybrid individuals. The authors discussed the existence of high telial production in genetically diverse populations and its reduction in clonal populations with regard to evolution of sex, *Pst* center of origin and distribution of its alternative host. The same group also studied the origin, migration route and worldwide population structure (Ali et al. 2014a); and contribution of sexual reproduction, migration and off-season survival to the temporal maintenance of *Pst* populations (Ali et al. 2014b) by the means of SSR genotyping.

Zhan et al. (2012) compared Chinese and US populations of *Pst* using virulence tests and SSR markers. From a total of 137 isolates (86 from China and 51 from the US), the genotypic data of 12 SSR markers identified 102 genotypes, of which 71 from China and 31 from the US. They identified a low ($r = 0.34$) but significant ($P = 0.01$) correlation between the virulence and SSR data. Their principal component analyses using either the virulence data or the SSR data separated the isolates into three groups: group a consisting of only Chinese isolates, group b consisting of both Chinese and US isolates and group c consisting of mostly US isolates. The SSR clustering analysis indicated that the *Pst* populations of China and the US in general evolved independently, but with some level of relatedness as suggested by the isolates in group b.

Sharma-Poudyal et al. (2012) genotyped 292 *Pst* isolates from 18 countries using 24 co-dominant SSR markers and identified 273 genotypes. Bayesian analysis discriminated the genotypes into two genetic lineages and one admix group. Genetic lineage 1 consisted of 55% of the isolates from all of the countries; lineage 2 consisted of 35% of the isolates, mostly from China and Uzbekistan; and the admix group consisted of 10% of the isolates, mostly from Asian countries, indicating genetic recombination. Heterozygosities of the isolates ranged from 0 to 75% with an average of 45%. Mean heterozygosities in individual countries ranged from 35 to 60%. Generally, results obtained from Bayesian statistics, principal component analysis and cluster analysis consistently revealed the lack of geographical differentiation among country-wise collections. Identical genotypes were observed between proximity countries as well as between distant countries of different continents, suggesting introduction or migration. Partition of molecular variance at different hierarchies suggested that most variation occurred within countries followed by among international regions. Genetic variation among countries within regions was low but significantly different. Collections from China and Central Asia had significant but low level of genetic differentiation with collections from North American and South America. Either low or non-significant genetic differentiation between other regions may explain the minimum geographic structuring of *Pst* worldwide.

In a study of genome-wide analysis of *Pst*, Luo et al. (2015) identified 4792 SSR loci using the whole genome sequences of six isolates from different regions of the world, with a marker density of one SSR per 22.95 kb. They established a database containing 1113 SSR markers. Through *in silico* comparison, they found that the previously reported SSR markers were mainly in exons, whereas the SSR markers in the database were mostly in intergenic regions. They confirmed 105 SSR markers polymorphic *in silico* among the six isolates. Using the polymorphic markers to test 21 isolates, 82 markers were polymorphic and revealed the genetic relationships among the isolates. Based on the molecular genotypes of the 82 polymorphic SSR markers, the 21 isolates were separated into four groups. Group Ga consisted of two US isolates (PST-1 and PST-43); group Gb three US isolates (PST-17, PST-114, PST-78), an Hungary isolate and four Chinese isolates; group Gc nine Chinese isolates from Xinjiang, Qinghai and Tibet; and group Gd two Chinese isolates from Xinjiang and Gansu. The separation and grouping together of the Chinese and US isolates were consistent with the previous report that the *Pst* populations of China and the US have been generally evolved independently, but with some level of relatedness as discussed above (Zhan et al. 2012).

Using 16 SSR markers, Hovmøller et al. (2016) genotyped a subset of 239 isolates selected from 2605 isolates collected from 12 European countries between 2000 and 2014. Based on virulence tests, they identified at least three races of non-European origin, termed ‘Warrior’, ‘Kranich’ and ‘Triticale aggressive’ in the post-2011 population. The Warrior race that was already present in high frequencies in the first year of detection in most European countries almost replaced the pre-2011 European population, whereas the two other exotic races were localized to certain regions and/or crop type. The marker data detected at least six multilocus genotypes of the Warrior race and five genotypes of the Kranich race in the first year of detection and across large areas. By comparing with reference isolates sampled from six continents, the authors suggested that the Warrior and Kranich races originated from sexually recombining populations in the center of diversity of *Pst* in the near-Himalayan region of Asia. They found that the Triticale aggressive race was most similar to populations in the Middle East/Central Asia.

In South Africa, Visser et al. (2016) used SSR markers to characterize four *Pst* races identified since the first appearance of the pathogen in 1996, including the first detected race (6E16A−) that was proposed to be introduced from Central or Western Asia and subsequently gained additional virulence through step-wise mutations. The genetic relationships of 20 single-pustule isolates representing the four races were determined together with a *P. striiformis* isolate collected from wild rye, two Kenyan *Pst* isolates and a *P. striiformoides* (previously *P. striiformis* f. sp. *dactylidis*) collected from *Dactylis glomerata* using 53 alleles produced with 17 polymorphic SSR markers. The *P. striiformoides* isolate had only 4% similarity with other *Pst* isolates. The South African isolates had 74–100% similarities, but only 48% similarity with the two Kenyan isolates. The wild rye isolates from Southland, South Africa (identified as race 6E16A-) were clearly of *Pst* as they clustered together with other *Pst* isolates. Except the four single-pustule isolates of race 6E16A that were separated into two genotypes with about 84% similarities, all isolates in each of the other three races and the Southland race had an identical genotype, but different races had different genotypes. Because it is not clearly how the four single-pustule isolates were produced, from a single sample or different natural samples, it is hard to assess diversity within each race. Although the authors did not determine the correlation between the phenotypic virulence data and the SSR data, the coefficient value should be very high. Nevertheless, this study suggested that the South African *Pst* population is clonal.

2.3.2.2.8 Development and Utilization of SNP Markers in Studying *Pst* Populations

Compared to SSR markers that utilize insertion and deletion (indels) of short nucleotide repeats (mostly two to four repeats), single-nucleotide polymorphism (SNP) sites are more abundant in the pathogen genome. Even between the two nuclei of a single *Pst* isolate, the number of SNPs is very high. For example, PST-127 has 288,965; PST-78 481,674; CYR32 518,098; PST-1 539,658; and UK94-519 541,789 SNPs (Chen 2014; http://www.broadinstitute.org/annotation/genome/puccinia_group/MultiHome.html). Flanking sequences of SNPs can be used to develop SNP markers to study the pathogen biology, genetics, and population.

Xia et al. (2016a) were the first to develop SNP markers for studying *P. striiformis*. In this study, they developed over a hundred secreted protein gene-derived SNP (SP-SNP) markers and used the markers to study the population structure of the *Pst* population in the US. From 352 isolates collected in 2010 and 2011, 242 multilocus genotypes were identified with 92 SP-SNP markers. They found that the SP-SNP genotypes had a moderate, but significant correlation ($r = 0.34$; $P < 0.0001$) with the virulence phenotype data. Clustering of the multilocus genotypes revealed distinct genetic groups. The marker data detected significant differences between the eastern and western US *Pst* populations. Because SNP markers were codominant, they were able to identify high heterozygosity in the US population and significant differences in heterozygosity among epidemiological regions. The populations of the eastern and western US were highly differentiated, whereas moderate levels of differentiation were observed in populations within the western or eastern US. The western population was more diverse than the eastern population.

Using 97 SP-SNP markers, Xia et al. (2016b) detected SP-SNP markers that are significant associated with avirulence genes *AvYr1*, *AvYr2*, *AvYr6*, *AvYr7*, *AvYr8*, *AvYr44*, *AvYrExp2*, *AvYrSP* and *AvYrTye*. In addition, SP-SNP markers have been used to identified somatic recombinant isolates from an experimental *P. striiformis* population (Lei et al. 2016) and construct linkage maps (Yuan et al. 2016).

2.3.2.2.9 Utilization of SSR and SNP Markers in Studying Long-Term Population Dynamics

The above studies of populations were use *Pst* isolates collected in a year or few years, which cannot directly determine evolutionary changes over a long period. A part of the problem was the lack of historical samples. Historical isolates are useful for comparison of *P. striiformis* populations over years for understanding the pathogen evolution. The 'Stubbs collection' of the stripe rust fungus, which was started in 1956 in the Netherland and lasted until the earlier 1990s (Stubbs 1985, 1988), contains isolates from throughout the world stored in liquid nitrogen. The collection remained untouched until recently. As historical microorganism collections are under the threatening of extinction, exampled by the 'Stubbs collection' of *P. striiformis*, there was an effort in the international scientific community for rescuing historical collections initiated by the American Phytopathological Society in 2005 (APS Collections and Germplasm Committee 2005; Babcock et al. 2007). With the support from the Borlaug Global Rust Initiative, the 'Stubbs collection' was recently moved to the Global Rust Reference Center in Denmark. From the collection, Thach et al. (2015) selected a subset of 231 isolates and recovered 96% of the isolates. Testing 212 isolates from the recovered isolates representing samples collected from five continents between 1958 and 1991 using 19 polymorphic SSR markers. They identified 89 multilocus genotypes that were clustered into seven genetic groups in the past worldwide *Pst* population, which generally corresponded with geographical sample origin except for the Middle East, where six of the seven genetic groups were represented (Thach et al. 2016). They resampled 4 of the 89 genotypes in different geographical regions indicated long-distance migration of the pathogen in the past. When comparing the past isolates with 309 isolates mainly collected between 2001 and 2009, they found the temporal divergence of the populations in all regions except northwestern Europe. This study reported a clear subdivision within the worldwide population structure and migration of *Pst* in the past. Because of a high diversity and recombinant population structure in China and South Asia, and a clonal population structure was in northwestern Europe, East Africa and the Mediterranean region. The authors believed that the high genetic diversity in the Himalayan region supported the region as a putative center of diversity for *P. striiformis*, as previously reported (Ali et al. 2014a, b). However, more genotype groups detected in the Middle East suggested that this region is more likely the center of origin of the pathogen, in consistent with the long-term hypothesis that the Transcaucasia region is the center of origin for *P. striiformis* and his primary grass host (Hassebrauk 1965; Leppik 1970; Stubbs 1985). An alternate hypothesis is that the Middle East is the primary center and the Himalayan region is a secondary center.

Similarly, a *P. striiformis* collection has been stored in liquid nitrogen since the 1960s when stripe rust research re-activated in the US (Line and Qayoum 1992; Line 2002; Chen 2005). After the establishment of the new set of *Yr* single-gene line differentials in 2010 (Wan and Chen 2014a; Wan et al. 2016), isolates of 1968–2009 were sampled and tested on the new differentials (Liu et al. 2016c). A total of 1084 isolates were genotyped using 17 SSR and 92 SP-SNP markers. Both types of markers identified similar number of multilocus genotypes, 568 and 711, respectively, which were clustered into four genetic groups. The four groups includes a group of admix isolates at an average frequency of 3.5% based on the SSR markers and 6.8% based on the SP-SNP markers, indicating a low frequent genetic recombination. The probability of sexual reproduction was extremely low, from 1×10^{-3} to 1×10^{-28}. New multilocus genotypes were identified in every of nine 5-year periods, but much higher in the period of 1983–1987 and 2000–2004, indicating two major introductions in the past over 40 years. The two types of markers also detected similar levels of heterozygous frequencies, 39.0% based on SSR markers and 41.5% based on SP-SNPs. The population structure has become complex since 2000, with only three of the four genotypes prior 2000 and all of the four groups post 2000. Based on the SSR markers, the temporary population changes measured by Nei genetic distance ranged from 0.002 (between the periods of 1978–1982 and 1983–1987) to 0.139 (between the periods of 1978–1982 and 2005–2009). Interestingly, the genetic distance between the periods 1968–1972 and 2005–2009 (0.077) was in between, indicating that some early genotypes re-emerged in the more recent years (TL Liu and XM Chen, unpublished data). The detection of re-emerged genotypes was in consistent with the detection of PSTv-1 and similar races throughout the nine periods and frequency increase of these races in the more recent years (Liu et al. 2016c).

2.3.2.2.10 Utilization of DNA Sequencing to Study *Pst* Populations

Liu et al. (2012b) was the first to use DNA sequencing to determine genetic relationships of *P. striiformis* isolates. They sequenced beta-tubulin (BT), elongation factor (EF) and mitogen-activated protein kinase (MAPK) genes of 21 isolates representing different races from the US and China. They identified 101 polymorphic base pair sites detected in the three genes and 64 (63%) of SNPs had heterokaryotic variations, indicating that heterokaryosis is very common in the population of the stripe rust pathogen. Using the polymorphic base pair sites, they identified 14 genotypes from the 21 isolates. The genotypes were grouped into four sequence lineages. The phylogenetic analysis revealed that mutation is the major evolutionary mechanism to create genetic variations including heterokaryotic variation. This study was the first report of high heterokaryotic variation at the gene sequence level for the stripe rust pathogen. As the conserved BT and EF genes and the highly variable MAPK gene were used in combination, the sequence data should provide a reliable estimation of the genetic distances among the isolates. Based on this study, three US isolates (PST-1, PST-78 and PST-127) and one Chinse isolate (CYR32) were selected for whole genome sequencing (Cuomo et al. 2012; http://www.broadinstitute.org/annotation/genome/puccinia_group/MultiHome.html)

So far, complete and partial sequences of 10 *Pst* isolates have been reported or available in databases (Cantu et al. 2011, 2013; Zheng et al. 2013; Cuomo et al. 2017). Additional 10 US isolates have been sequenced. Using the sequences of all isolates, the genetic relationships of these isolates can be more accurately determines.

Currently, the senior author's program has focused on polymorphic secreted protein genes and their relationship to avirulences (XM Chen and associates, unpublished data). In addition, genomic sequencings of 120 *Pst* isolates produced on barberry have been used to establish genetic linkage maps for the pathogen and used to map virulence genes (Yuan et al. 2016).

2.3.2.2.11 Utilization of RNA Sequencing to Study *Pst* Populations

As discussed above, DNA-based markers distinguish pathogen populations based differences in genomic DNA. DNA markers reflect the genetics of isolates or population, but often cannot be used to study epigenetics effects. Also because *Pst* is a biotrophic fungus, culturing isolates on host plant for DNA extraction is slow and tedious. In the recently years, RNA sequencing (RNAseq) has been emerging as a powerful tool to compare whole genome-wide gene expression and has been used to study population changes.

Hubbard et al. (2015) developed and applied a field pathogenomics approach by transcriptome sequencing infected wheat leaves. This RNAseq approach enabled rapidly gaining insights into the emerging *Pst* population. Using samples collected from fields in 2013, they detected a major shift in the UK *Pst* population in recent years. Through analyzing the population structure, they identified four distinct lineages correlating to the virulence groups. By comparing the genetic diversities of the 2013 samples with historical isolates, they found that the recent population became more diverse. The authors concluded that the recent population shift was likely due to a recent introduction of a diverse set of exotic lineages. This method can be used to directly analyze infected leaf samples without obtaining isolates. However, RNAseq itself does not provide direct information of virulences. Using leaf samples inoculated with isolates with known virulences can partially solve the problem. If combined with the traditional virulence tests, RNAseq can provide quick information for genetic and functional genomic changes without losing virulence information that is essential for developing resistant cultivars for control the disease.

2.4 Evolutionary Perspectives

Any plant pathogen has been and continues to evolve to adapt to the environmental changes, for instance weather conditions, resistance in host plants, fungicide application and so forth. As a species of obligate biotrophic parasite, *P. striiformis* is even under faster evolution due to the consistent and great changes from the hosts by breeding programs. Many of the changes have been discussed above in the sections covering virulence and molecular characterizations of the pathogen. Here, the changes are briefly summarized from an evolutionary perspective.

2.4.1 More Races and More New Races Identified Recently

More and more races or pathotypes have appeared in the recent years. This phenomenon is not just due to the fact that the pathogen has been more intensively studied based on virulence in more countries. It is truly due to the fast evolution of the pathogen as the results of fast changes of wheat cultivars to meet the increasing demand of yield, quality and environmental adaptation, as well as resistance to stripe rust and other pests. Even in a single country where *Pst* races have been characterized for a long time, more races or genotypes have been identified in the recent years than before. For instance, in Australia, one single race identified in 1979 had evolved into 15 races in the first ten years until 1988 (Wellings and McIntosh 1990) and more races were identified until 2006 (Wellings 2007). The similar phenomena have happened in China (Chen et al. 2009b) and Europe (Bayles et al. 2000; De Vallavieille-Pope et al. 2012).

In the US, there were only five races identified before 1970, but 16 new races were identified between 1970 and 1979 and 18 new races between 1980 and 1987 (Line and Qayoum 1992). During the 10 years from 2000 to 2009, a total of 81 new races (named PST-60 to PST-140) were identified, whereas only 59 were identified for over 30 years from the late 1960s to 1999 based on the set of 20 wheat cultivar differentials (Line and Qayoum 1992; Line 2002; Chen et al. 2002, 2010; Wan and Chen 2012a). More races were detected in each year in the 1980s than in the 1960s and 1970s. Using the new set of 18 *Yr* single-gene line nurseries, 171 races were identified from 908 isolates sampled from 1968–2009, the number of races identified in each year ranged from 5 to 28 (average 14) in 1968–1999, whereas from 17 to 34 (average 24) (Liu et al. 2016c). Similarly, the number of new races identified in each year was much higher from 2000–2009 (ranging from 3 to 21 at average of 8.9) than from 1968 to 1999 (ranging from 0 to 12 at average of 3.6).

2.4.2 More Virulences Identified Recently

More and more virulences have been identified in the recent *Pst* populations. As the utilization of more and more resistance *Yr* genes have been used in breeding programs, directional selection pressure has become higher. When a gene conferring race-specific resistance is incorporated into new cultivars, its corresponding virulence will soon or late appears in the pathogen population. For example, virulences to *Yr17* and *Yr32* appeared in Europe a few years after cultivars with the resistance genes were grown (Bayles et al. 2000; Hovmøller et al. 2011). Similarly, the *Yr9* virulence appeared in China in 1981, just few years after cultivars with the resistance gene were grown. In the US, *Yr10* became ineffective in 1968, just three years after the first wheat cultivar with the resistance gene was developed (Line and

Qayoum 1992). Often, virulences can be maintained in the population at a very low level, but their frequency can increase very quickly when the resistance gene is incorporated into cultivars. For example, virulence to *Yr76* (*YrTye*) was low before 1983, but became significantly high after cultivar Tyee with the gene was released in 1980 (Line and Qayoum 1992; Xiang et al. 2016; Liu et al. 2016c).

2.4.3 Races with More Virulences Identified Recently

The virulence spectra of races have s becoming wider. Several studies in Europe, North America and other continents indicated that recent races have more and more virulences (or virulence factors) (Wellings 2007; Bayles et al. 2000; de Vallavieille-Pope et al. 2012; Chen 2005; Chen et al. 2010; Wan and Chen 2012a, 2014a; Wan et al. 2016). In other words, spectra of races have become wider and wider. Taking the virulence changes in the US as an example, from 1968 to 2009, the *Pst* populations changed from simple races with low numbers of virulences to complex races with high numbers of virulences. Based on the tests of 908 isolates of 42 years on the same set of 18 *Yr* single-gene differentials, the average number of virulences of the races detected in 1968 was just 0.8, whereas that of the 2009 races was more than 10.0 (Liu et al. 2016c). The trend of the *Pst* population increased over the years, but more rapid increase occurred during the 2000–2009 periods. When the 42 years were grouped into nine periods, the mean number of virulences in new races was generally significantly higher than those of all races. The differences were greater from 1973–1978 to 1993–1997, but became smaller after 1998–1999. The average number of new races and populations were almost identical in the period of 2005–2009.

2.4.4 Pst *Populations Becoming More Aggressive and More Adaptive to Climate Changes*

As discussed above, several studies and many observations have showed that the *Pst* fungus has produced high-temperature tolerant isolates and high-temperature adapted isolates have become more and more predominant in the pathogen populations (de Vallavieille-Pope et al. 1990; Line and Qayoum 1992; Bayles et al. 2000; Justesen et al. 2002; Chen 2005; Enjalbert et al. 2005; Milus et al. 2006; Milus et al. 2009; Hovmøller et al. 2008; Markell and Milus 2008; Liu et al. 2012c). Environment conditions, breeding programs, cropping systems and other factors have intense and profound impacts on the survival of the stripe rust pathogen and development of the stripe rust disease. The pathogen has to develop mechanisms to become more fit and aggressive to the harsh "living environment".

2.5 Mechanisms of Pathogen Variation

As discussed above, extensive race and population analyses around the world have showed new races and genotypes have occurred in the *P. striiformis* populations over the world and several studies have addressed the mechanisms of the mechanisms of the pathogen variations. It is now known that there are several mechanisms underlining the pathogen evolution. These major mechanisms are briefly discussed below.

2.5.1 *Mutation*

Gassner and Straib (1933) were the first to suggest mutation as a mechanism for formation of new races in *P. striiformis* and Oort (1955) agreed with the suggestion. The mutation theory was proposed mostly based on race analyses. Races of *Pst* produced through mutation can been best seen in Australia. A series of new races have arisen since 1979 in Australia and the best explanation for the production of new races is through stepwise evolution (Wellings and McIntosh 1990; Steele et al. 2001; Wellings 2007). In the US, stepwise race evolution has also been demonstrated (Line and Qayoum 1992). In a recent study of isolates from 1968 to 2009 tested on the 18 *Yr* single-gene line differentials, 127 (74%) of 171 race could be interpreted to arise from previous races (Liu et al. 2016c). These results indicate that the majority of the new races were evolved through mutations.

The majority of mutations have occurred toward increased the number of virulences in new races (de Vallavieille-Pope et al. 1990; Wellings and McIntosh 1990; Line and Qayoum 1992; Bayles et al. 2000; Justesen et al. 2002; Chen 2005; Enjalbert et al. 2005; Wellings 2007; Chen et al. 2010; Wan and Chen 2012a, 2014a; Liu et al. 2016c; Wan et al. 2016), but losses of virulences due to mutations have also been reported (Hovmøller and Justesen 2007; Liu et al. 2016c). When British and Danish isolates collected between 1975 and 2002 was studied, mutations resulting in loss of avirulence were detected 16 times (Hovmøller and Justesen 2007). Similarly, in the study of Liu et al. (2016c) with *Pst* isolates of 1968–2009, 71 (55.9%) of the 127 mutant races added a single virulence, 27 (21.3%) added two virulences, 13 (10.2%) lost one virulence, 12 (9.4%) added one virulence but lost one virulence and 4 (3.1%) lost two virulences.

Hovmøller and Justesen (2007) estimated simultaneous mutation frequency of *Pst* from 1.4×10^{-6} to 4.1×10^{-6} per locus per generation in individual clonal lineages with an assumption of an average of 15 generations per year. They also conducted field experiments for confirming the rate of mutation (Hovmøller and Henriksen 2008).

Through gene sequencing, Liu et al. (2012b) identified 101 polymorphic base pair sites in the sequences of beta-tubulin, elongation factor and mitogen-activated protein kinase genes among 21 *Pst* isolates. Except for few cases of two or three continuous base pair sites, the most of the polymorphic base pairs are not continuous. Therefore, they concluded that point mutation is the major mechanism for the sequence variations because if intragenic recombination had occurred, different sequence blocks should be detected. They noted that only two nucleotides at every base pair site. This result indicates that the mutations are directly specific. This was the first study of *P. striiformis* mutation at the sequence level using specific genes.

Several studied have been conducted to create *Pst* isolates through artificial mutations. Stubbs (1968) was the first to create new races. He found that hydrate urediniospores were more sensitive than dry urediniospores and the fungus in wheat leaves was most sensitive and least sensitive in the reproductive stage to X-ray. He identified the lethal dosages (LD) for 50% spores dead (LD_{50}) and 100% spores dead (LD_{100}) as 125 kR (kilo Röntgen) and 400 kR, respectively. Through selecting urediniospores of three *Pst* races produced on X-ray radiated wheat seedlings, he obtained seven mutant isolates. These mutants had an additional virulence compared to their parental isolates, indicating a single mutation. Of the seven mutant isolates, three were similarly to naturally occurring races while other four were new. One mutant isolate had virulence to *Triticum spelta* album (*Yr5*) that was highly resistant to all races identified from field samples. Later, ultraviolet radiation (Shang et al. 1994; Huang et al. 2005), ethyl methane sulfonate (EMS) (Yao et al. 2006; Li et al. 2016), particle bombardment (Wang et al. 2006; Lu et al. 2009) and GUS gene insertion have been used to develop *Pst* mutants. In all of these studies, mutants with additional virulences were obtained, and some of the studies also used molecular markers to characterize mutants (Huang et al. 2005; Wang et al. 2013; Li et al. 2016). However, some of the techniques, such as EMS, particle bombardment and GUS insertion, need to be improved and mutants need to be better characterized to identify mutated genes or genomic regions.

2.5.2 Somatic Recombination

The role of somatic recombination in the stem rust pathogen was known in the 1950s (Nelson et al. 1955), but unknown to *P. striiformis* until the 1960s. Little and Manners (1969a) conducted first experiment to study *Pst* somatic recombination. They inoculated Strubes Dikkopf with mixed urediniospores of *Pst* races 2B and 8B, both of which are virulent to the wheat variety, and isolated single-urediniospore isolates. From 31 single-urediniospore isolates, they identified three isolates that had reactions different from either the parental isolates tested on a set of differential varieties. The authors postulated that three genes for virulence, at least one of which is dominant, were responsible for the changed reactions on the tested differential varieties. They observed germ tube fusions between races 2B and 8B and cytoplasm from one tube, whose tip was fused to the side of another tube, passed into the second tube (Little and Manners 1969b). They also found higher proportions of urediniospores with three or more nuclei in the somatic recombinant isolates than the parental isolates, suggesting that the nuclear number of somatic recombinant isolates was not stable. Their observations suggested that the reassortment of whole nuclei is a mechanism of production of new races in *Pst* (Manners 1988).

Goddard (1976) conducted a similar experiment with *Pst* races 104E137 and 41E136 and produced three isolates possessing the combined virulence of the two races from 50 single-spore progeny isolates. The isolates were a previously unrecorded race (105E137). The new race inherited virulence on Chinese 166 from 41E136 and virulences on Suwon 92 x Omar and Hybrid 46 from 104E137, as well as the common virulences (Strubes Dickkopf, Vilmorin 23, Heines VII and Nord Desprez) from both parental races. In the same year, Taylor (1976) also reported a similar study of producing somatic recombinant isolates of *Pst*. Following Goddard's study, Wright and Lennard (1980) used a white-spored isolate of race 104E137 and a yellow-spored isolate of race 33E32 for mixed urediniospore inoculation of susceptible wheat variety Sappo. They identified five yellow şpore isolates as race 105E137 that had virulence patterns different form either of the parental isolates. They occasionally observed tri-nucleate urediniospores in the recombinant isolates, but never in the parental isolates. They also concluded that reassortment of whole dikaryotic nuclei was the most likely mechanism responsible for production of the recombinant isolates.

Attempting to produce somatic hybrids between *Pst* and *Psh*, Newton et al. (1986) inoculated barley variety Fong Tien, which is susceptible to both *Pst* and *Psh* isolates used, with a urediniospore mixture of a white-spored *Pst* isolates and a yellow-spored *Psh* isolates, selected isolates for non-parental spore color on selective wheat and barley varieties and tested them for virulence and isozyme and double-stranded RNA (dsRNA) markers. They obtained two white-spored, non-parental isolates that were resembled the parental *Psh* isolate in virulence, isozyme and dsRNA markers. The authors concluded that the recombinant isolates were most likely mutants of the *Psh* isolate.

Chinese scientists have conducted a series of studies on possible role of somatic hybridization. They observed fusions between urediniospore germ tubes of different *Pst* isolates on wheat leaves (Zhao and Wang 1969; Ma et al. 1993), urediniospores and hyphal cells with different numbers of nuclei (Kang et al. 1994a); produced new races by heterokaryosis (Kang et al. 1993) and new isolates with varied number of nuclei (Kang et al. 1994b); determined the nuclear number of major predominant races in China (Li and Zeng 2002).

The above studies on somatic recombination were limited to identification of recombinant isolates by virulence tests and observation of cell fusion and variation in nuclear number. Although molecular markers (isozyme and dsRNA) were used (Newton et al. 1986), the markers did not help in identification of somatic recombinant. Even though the markers could clearly separate the parental *Pst* and *Psh* isolates, the progeny isolates showed only the virulence phenotype and marker genotypes of the parental *Psh* isolate, except white urediniospores. The results made the authors concluded that the progeny isolates were most likely produced by a mutation at spore color and that *Pst* and *Psh* might have genetic barriers to prevent them from recombination.

The advanced marker technologies have made possible to further study somatic recombination. Recently, the senior author's group conducted somatic recombination experiments. From nine combinations of isolates with large differences in virulences, including white-spored and orange-spored *Pst* isolate and orange-spored *Psh* isolates, they obtained more than 90 potential recombinant isolates based on a series of selections on wheat varieties or wheat and barley varieties and virulence tests on wheat differentials (for *Pst* + *Pst* combinations) and both *Pst*+*Psh* combinations) (Cheng et al. 2015a; Lei et al. 2016). Using 51 SSR and 90 SNP markers, Lei et al. (2016) were able to confirm 97% of the tested 68 potential recombinant isolated identified by virulence phenotypes. Mixing inoculations of common susceptible wheat variety with *Pst* and *Psh* isolates were able to produce recombinant isolates although appeared less common than mixes of *Pst* and *Pst* isolates. They found that most recombinant isolates were produced by somatic recombination involving cell fusion, nuclear exchange, and chromosomal reassortment, rather than hybridization without chromosomal reassortment. They were able to identify crossover events in some of the recombinant isolates using markers on single genomic contigs. Moreover, they found that most recombinant isolates have different proportions of markers from their two parental isolates and therefore, concluded that uneven contribution from parental isolates is a general role for somatic recombination, which may be used to distinguish somatic recombination from sexual recombination. The results of experimental isolates support the hypothesis that somatic recombination is one of the major mechanisms for *P. striiformis* variation and evolution and grasses serve as host platforms for somatic recombination to occur, indicated by phenotypic and genotypic characterization of natural populations (Cheng et al. 2016a).

2.5.3 Sexual Recombination

Sexual reproduction of the stripe rust fungus was thought not to occur until the recent discovery of barberry and mahonia as alternate hosts (Jin et al. 2010; Wang et al. 2013; Zhao et al. 2013). Zhao et al. (2016) reviewed the roles of alternate hosts for cereal rusts, particular stripe rust. In general, sexual reproduction on alternate hosts play roles in surviving winter and providing initial inoculum for starting rust epidemic; producing new races with various virulence combinations; and diversifying pathogen populations. The importance of sexual reproduction on alternate hosts varies greatly from pathogen to pathogen and region to region because of different climates; cropping systems; phenology of the primary, alternate and auxiliary hosts; and adaptations of pathogens to various hosts and climates.

For the wheat stripe rust fungus, sexual reproduction was suggested for China, Nepal, Pakistan and other countries in the Himalaya region based on pathogen population diversity and presence of barberry plants (Mboup et al. 2009; Duan et al. 2010; Perveen and Qaiser 2010; Bahri et al. 2011; Ray et al. 2011; Ali et al. 2014a, b). Natural infection of barberry by stripe rust fungi was detected under natural conditions in Minnesota for the bluegrass stripe rust (*P. striiformis* f. sp. *poae*, now *P. pseudostriiformis*) (Jin et al. 2010) and for *Pst* at low frequencies in western China (Zhao et al. 2013; Wang et al. 2016). However, natural infection of barberry by *Pst* has not been found in Europe (Berlin et al. 2012) and the US (Wang et al. 2015).

Wang and Chen (2015) identified factors for why barberry is essential for *Pgt*, but not for *Pst* in the US Pacific Northwest. First, *Pst* telia is covered by an epidermal layer of cereal and grass hosts and teliospores are contained inside of a spherical structure of fungal mycelia, which prevents teliospores from releasing. In contrast, whereas *Pgt* telia are erupted when mature and teliospores are exposed. They observed *Pgt* teliospores germinating naturally on wheat straw, but did not found *Pst* teliospores germinating. Second, *Pst* teliospores do not have or have a very short period of dormancy. If *Pst* teliospores germinate, there are no or few young barberry leaves for basidiospores to infect. Old leaves of barberry are resistant to both *Pst* and *Pgt*. In contrast, *Pgt* teliospores have a very long dormancy. Fluctuation of moisture and temperatures, especially wet and low temperatures such those in the winter, is required for germination. Third, *Pst* teliospores require a minimum 32–40 h of dew formation under the optimum temperature conditions for infect barberry, which is much longer than the dew period required for *Pgt* infection on barberry leaves, and such long dew period hardly occur in the US Pacific Northwest. Fourth, *Pst* teliospores become degraded after wet and cold winter, whereas *Pgt* teliospores become ready for germination after winter. Fifth, barberry new leaves are produced in spring when *Pgt* teliospores reach to the peak of germination and production of basidiospores, but viable *Pst* teliospores are not available.

Based on these factors, regions can be predicted for the importance or unimportance of barberry for stripe rust and stem rust. Using this approach, the authors predicted that stripe rust can infect barberry in northwestern China because *Pst* telia can be produced on late matured spring wheat and some grasses, winter is generally dry, and high and long moist conditions occur in spring. The prediction is consistent with the report of natural infection of barberry by *Pst* in the region (Zhao et al. 2013; Wang et al. 2016). The reported low infection frequencies could be due to low recovery rate of rust samples from barberry, and also could be related to telial structures of *Pst*. For teliospores to germinate, the host epidermal layer and the spherical structure enclosing teliospores need to be broken by wind or other forces.

2.5.4 Migration and Introduction

Mainly through the urediniospores, the stripe rust pathogen is able to migrate or be introduced locally, regionally and continentally. Migration or introduction of different races or genotypes can change the local pathogen populations, either increasing diversity or replacing the old population. Introduced races may increase the chance for mutation and recombination. Long distance and intercontinental migration of the stripe rust fungus either by winds or human activities have been well documented (Zadoks 1961; Stubbs 1985; Nagaranjan and Singh 1990; Wellings and McIntosh 1990; Boshoff et al. 2002; Brown and Hovmøller 2002, Hovmøller et al. 2002; Line 2002; Wellings et al. 2003; Chen 2005; Wellings 2007; Hovmøller et al.

2008; Hovmøller et al. 2010; de Vallavieille-Pope et al. 2012; Hovmøller et al. 2012; Ali et al. 2014a, 2014b).

Natural migration of the stripe rust pathogen in long distance was covered in the disease epidemiology of Chap. 4. There have been several significant cases of introduction of the stripe rust fungus over the world likely due to human activities. For example, *Psh* was introduced from Europe to Colombia in 1975 and the pathogen further spread southward to Chile by 1982 (Dubin and Stubbs 1986) and northward to Mexico by 1987 (Chen et al. 1994). The pathogen was spread to southern Texas in 1991, further spread toward western US and reached Washington in 1995 (Chen et al. 1995b). *Pst* was introduced from Europe to Australia in 1979 and subsequently to New Zealand the following year (O'Brien et al. 1980; Wellings and McIntosh 1990). The introduction was believed by urediniospore on the clothing of international air-travelers (Wellings et al. 1987). Similarly, the first occurrence of *Pst* in South African in 1996 was thought to be introduced from North Africa and the Middle East (Pretorius et al. 2007; Wellings 2011).

There have been reports about the spread of virulences to specific *Yr* genes among countries and regions, for instance, virulences to *Yr8*, *Yr9*, *Yr17* and *Yr27*. Races with virulences to *Yr8* and *Yr9* were more clearly originated from the Middle East and Mediterranean as the frequencies of *Yr8* and *Yr9* virulences were detected in the 1970s and 1980s and high in these regions in 2005–2006 (Bamdadian 1992; Mamluk and El-Naimi 1992; Hakim and Mamluk 1996; Bahri et al. 2009b). The occurrence of the *Yr8* and *Yr9* virulent races in the United States in 2000 was likely from introduction (Chen et al. 2002). The discovery of the similar races in Western Australia in 2002 was also a foreign incursion (Wellings et al. 2003). The spread of *Yr17* virulence to continental Europe was postulated through wind dispersal of urediniospores from a common source, presumably in the United Kingdom (Hovmøller et al. 2011). The virulence to *Yr27* has been spreading quickly in the world. The *Yr27* virulence was recorded in Yemen in 2002, but by 2004 detected in many countries in Central and West Asia and North Africa (Yahyaoui et al. 2005). In 2007, virulence to *Yr27* was also detected in Australia (In 2009–2010, the outbreak of *Yr27* virulent races caused significant yield losses in Azerbaijan, Ethiopia, Iran, Iraq, Kenya, Morocco, Syria, Turkey, Uzbekistan and other countries in the Middle East (ICARDA 2011). *Yr27* virulent races spread to Egypt and caused severe yield losses in 2011–2014 (Shahin and Aly 2015). Virulence to *Yr27* was detected in high frequencies ($\geq$50%) in *Pst* collections of 2005–2010 from Canada, Chile, China, Hungary, Kenya, Nepal, Pakistan, Turkey and Uzbekistan and also detected in Chile at a relatively low frequency (27%) (Sharma-Poudyal et al. 2013).

Compared with the long distance spread by winds, the possibility of spread of urediniospores by humans is increasing due to the significant increasing numbers of travelling around the world. The introduction of wheat stripe rust by human beings is, therefore, continuous a challenge for preventing its' spread, especially from countries to countries, and from continents to continents. It is not possible to stop human moving of rusts through quarantine, but it is definitely helpful to reduce the introduction events by educating people how to avoid inadvertently carrying viable rust spores on their cloth and shoes.

2.5.5 Genetic Drift

Populations of any organisms are constantly changing and adapting to their environment. Drastic changes in environmental conditions can cause drastic changes to the gene pool of the population. For the stripe rust fungus, genetic drift occurs when unfavorable conditions cause decreasing population size and consequently frequencies of alleles in the populations were reduced. The pathogen-unfavorable conditions include no susceptible cultivars for particular races, drought, extreme high temperatures during summer and extremely low temperatures during winter. The population size of *P. striiformis* is greatly reduced during summer after wheat is harvested and during winter, especially when winter weather conditions not good for survival (Chen 2005; Sharma-Poudyal et al. 2014). Because *P. striiformis* is a biotrophic fungus, the absence of hosts and unfavorable weather conditions my lead to low population size after rust season or even pathogen extinction in a region for few years. Under such conditions, genetic drift occurs in the pathogen population, which may eliminate some races or genotypes and make previously unpopular races or genotypes become popular. Since alleles are the genetic variations and driving forces behind the evolution in a population, many alleles are lost and the genetic variation of the population decreases when genetic drift occurs. As the stripe rust fungus is airborne, the effect of genetic drift is dependent on the scope of a region and how much the region relies on the local and outside of the pathogen population for survival and development of the disease. It is obvious that the smaller the population, the greater the impact genetic drift will have, because there are fewer individuals and the gene pool is smaller.

If the effects of genetic drift are strong enough, races or genotypes can be completely removed from the population, reducing the amount of variation in the population's gene pool.

The effects of genetic drift on *Pst* populations in European countries have been well documented and discussed (Zadoks 1961; Stubbs 1985; Hovmøller et al. 2011). The recent replacements of the old races by new races, such as 'Jackie race' and 'Warrior race', are examples of genetic drift (Hovmøller et al. 2016). In many countries of the world, stripe rust does not occur consistently every year (Wellings 2011), as the summer is too hot and/or the winter is too cold.

In the US, *Pst* generally cannot survival in the summer in the south and central parts of the Great Plains due to hot and humid summer and the pathogen generally cannot survive the cold winter in the north part of the Great Plains (Sharma-Poudyal et al. 2014). Genetic drift occurs during the summer or winter in different regions of the Great Plains. Compared to the eastern US, the western US has weather conditions, cropping systems and auxiliary hosts more favorable to the stripe rust pathogen (Chen 2005; Chen et al. 2010). Genetic drift occurs less frequently and to a less extent in the western US than in the eastern US. Because of this, more races and high diversities have been found in the western population than the eastern population (Chen et al. 2010; Wan and Chen 2012a, 2014a; Wan et al. 2014, 2016; Xia et al. 2016a). *Pst* races, such as PSTv-18, which occurred in the 1960s and

1970s, have been continually detected in the recent years (Wan and Chen 2014a; Wan et al. 2016; Liu et al. 2016c). In contrast, races that presented before 2000 has been completely replaced by the *Yr8* and *Yr9* virulent races in the eastern US (Chen et al. 2002, 2010; Chen 2005; Markell and Milus 2008; Milus et al. 2009).

The barley stripe rust pathogen in the US is a good example of genetic drift. In the first few years after the pathogen was introduced to southern Texas in 1991, the pathogen increased its population when spread into the barley producing regions in California and the Pacific Northwest, and 14 races were identified (Chen et al. 1995b) and by 2006, a total of 80 races were identified (Chen 2004, 2008). After 2008, no any new races have been detected and most races presented before 2006 have not been detected (Wan and Chen 2012a). The pathogen has been mostly detected in California and the Pacific Northwest and has not been reported recently in the south-central states like Texas and Oklahoma. Even in California and the Pacific Northwest, the pathogen has been decreasing in significance, the number of races and the number of virulences in races. In addition to stabilized selection of growing barley cultivars with non-race specific HTAP resistance, genetic drift mainly caused by the reduction of barley acreage due to severe stripe rust in the 1990s and early 2000s and low price of grain has played a major role in reducing the pathogen variation.

2.5.6 Founder Effect

The founder effect is the loss or reduction of genetic variation when a new population is established by a very small number of individuals from a larger population. As a result of the reduction of genetic variation, the new population can be different from the parent population from which it is derived. Urediniospores of *Pst* can migrate by winds and leave from the old populations. The migration of urediniospores may establish in a new region a sub population of the parent population and thus reduced the genetic variation compared to the parent population. This phenomenon exists in the populations of *Pst*, but it is not easy to conduct experiments to estimate the effect of the founder effect.

Unlike animal populations or mono-cyclic fungal pathogens, the amount of urediniospores of *Pst* leaving the old populations doesn't have obvious impact on the population size, because *Pst* can reach certain population size very quickly under favorable conditions. However, a founder effect can be seen by comparing the number of races or genotypes of a spore recipient region with that of inoculum source region. In Australia, the number of races in the first few years was much smaller than Europe from which the only one or two races were introduced (Wellings and McIntosh 1990). In China, fewer races of *Pst* usually occur in the spring epidemic regions in the central and eastern provinces than the over-summering regions in the western provinces (Wan et al. 2004; Chen et al. 2009b). In the US, few races are detected every year in the eastern regions, which are among the large number of races detected in the western regions (Chen et al. 2010; Wan and Chen 2012a, 2014a; Wan et al. 2016). Due to the possible founder effect, the *Pst* population has a much lower diversity than the parent population in the western regions especially California (Wan and Chen 2014a; Xia et al. 2016a).

2.5.7 Host Selection

Because the stripe rust fungus is an obligate biotrophic parasite, the pathogen has an intimate relationship with its host plants. Thus, host plants play a major role in selecting particular races and therefore, the prevalence of races or populations is closely associated with the distribution of host plants or cultivars. The highly specialized host preferences and biotrophic nature of *P. striiformis* suggests that host-induced selection can be a major driving force for population dynamics (Hovmøller et al. 2011). The effect of host selection is especially much stronger in modern agricultural systems, in which a single cultivar can be planted in a large area and the crop uniformity is much higher than the old systems. The deployment of wheat cultivars with few race-specific resistance genes in large areas enhances selection in favor of virulence races.

Generally, the frequencies of virulent races are very low in the first several years of the release of the resistant cultivars. However, this directional selection by resistant wheat cultivars definitely increases the frequencies of virulent races which can overcome the resistant cultivars, and as results, the resistant cultivars will become susceptible in a large areas after some years in cultivation. This phenomenon, which is often referred to as the boom and bust cycle, has been reported in numerous cases in the diseases of wheat stripe rust (Bayles et al. 2000; Hovmøller 2001; Chen 2005; Wellings 2007). In general, selection of virulences by growing cultivars with race-specific resistance may play a major role in determining race structure of the pathogen of *Pst* (Line 2002; Chen 2005; Hovmøller et al. 2011). As discussed above, globally using few *Yr* genes, such as *Yr9*, *Yr17* and *Yr27*, have enhanced selection of *Pst* races virulent to those resistance genes (Bayles et al. 2000; McDonald and Linde 2002; Chen 2005; Chen et al. 2009b; Wellings et al. 2012) and may have accelerated the spread and prevalence of such groups of races (Hovmøller et al. 2008, 2016).

2.5.8 Environmental Selection

Environmental conditions are more critical for stripe rust than for other cereal rusts (Stubbs 1985), including temperatures, lights, humidity, air pollution, fungicide application and so forth. As discussed in Chap. 4, the stripe rust pathogen is more sensitive to UV light and air pollution than other cereal rust pathogens (Sharp 1967; Stubbs 1985). A high concentration of large ions in the air due to air pollution decreased the germination rate of urediniospores. It was also showed that the effects of tobacco smoke on germination and infectivity of *Pst* spores (Melching et al. 1974). There were also differences in sensitivity to air pollution between stripe rust cultures from different areas in the world (Stubbs 1985), for example, the northwest European cultures in the 1970s and 1980s were much better adapted to air pollution than those from the Andes Mountains range, East Africa or Asia. A few northwest European cultures were even highly resistant. It was also noticed that inoculations made during days with high air pollution resulted in an increase in the length of the latent period and a shift of infection types toward resistance (Stubbs 1985). Recently, the appearance of temperature adapted isolates or strains have been reported in the US and elsewhere in the world (Markell and Milus 2008; Milus et al. 2009; Wellings 2011; Liu et al. 2012c). Although telial production is genetically controlled, the recent appearance and development of strains of high telial productivity could be selected by environment (Wan and Chen 2016). The stripe rust pathogen appears to have changed from more yellow to more orange. The darker color may have been selected by environment for better adaptation. As climate and agriculture production systems continue changing, the stripe rust pathogen will under more intense selection. Changes in the stripe rust pathogen population are inevitable. However, efforts should be taken to reduce the adverse effects of pathogen changes and utilize the desirable effects of changes. These can be largely done through utilizing diverse and non-race specific resistance genes in breeding programs and cultivars for production to reduce direct selection of highly virulent races and reduce damage caused by the disease in general.

2.6 Genome and Functional Genomics

Studies on genome and functional genomics have improved our understanding the biology and pathogenesis of the stripe rust fungus. Compared to other fungi, genomic and functional genomics research on the stripe rust pathogen started relatively late, but great progress has been achieved in the recent years. This phase of research will continue remaining hot for the next few decades.

2.6.1 The Genome of the Stripe Rust Pathogen

The recent advancement of sequencing technologies has made it possible to sequence rust pathogen. Among the cereal rust fungi, the stem rust fungus was sequenced first due to the global concern about the 'UG99' strain that overcoming the *Sr31* resistance gene that have been widely used in wheat cultivars (Duplessis et al. 2011). Not long after the sequencing of the stem rust fungus, sequencing of the leaf rust and the stripe rust fungi were conducted in different countries. The first study of the stripe rust pathogen was reported by Cantu et al. (2011) on a US *Pst* isolate, and the second of the sequencing study was by Zheng et al. (2013) on a Chinse isolate. In fact, the first study of sequencing stripe rust was initiated in 2008 on five *Pst* isolates, PST-78, PST-1 and PST-127 from the US, CYR32 from China, and UK94-519 from Europe. Although not all data of the five isolates have been officially published, their sequence data are available for use at the Broad Institute *Puccinia* genome website (http://www.broadinstitute.org/annotation/genome/puccinia_group/MultiHome.html). The comparison of the stripe rust fungal genome with those of the leaf rust and stem rust fungi were recently published (Cuomo et al. 2017).

2.6.1.1 The Genome of PST-130

PST-130 is a race of *Pst* that was first identified in Oregon and Washington in 2007 and was characterized using a set of 20 wheat differentials (Chen et al. 2010). This race occurred in the US Pacific Northwest. It was never a predominant race, nor a presentative of an important race group. It was picked for sequencing because of adequate amount of urediniospores for DNA extraction at that time. It was selected also because the sequencing project with well selected three US isolates to present historical population (PST-1), an important group of races since the year 2000 and the most virulent race (PST-127) had started. Although started later, the sequence of PST-130 was published first, but the published sequence was not complete, about 88% of the whole genome (Cantu et al. 2011).

The sequencing of PST-130 was done using the Illumina sequencing technology, commonly known as the "Next Generation" sequencing. Nearly 80 million high quality paired-end reads (about 50× coverage) were obtained and assembled into 29,178 contigs (64.8 Mb) by de novo assembly (Table 2.6). The assembled sequences contained 22,815 putative coding sequences (CDS) encoding predicted proteins with an average length of 277 amino acids. After eliminating 2392 transposable elements (TE) associated proteins, the remaining 20,423 predicted proteins were compared with 458 highly conserved protein families and the *Pgt* genome model. Extensive microsynteny between the *Pst* and *Pgt* genomes and high sequence similarity with annotated *Pgt* genes were identified, and the data supported the quality of the PST-130 contigs. An unexpectedly high level of similarity was found between rust and plant Harbinger elements relative to the long divergence time between their respective two evolutionary lineages (~1500 million years). The authors speculated that an ancient horizontal transfer of Harbinger elements between the rust fungus and plants could result in the high similarity. In addition, they also identified a distant branch of DNA transposons belonging to the MuDR superfamily that encodes proteins containing a GCM (glial cell missing) DNA-binding domain and a DDE-transposase domain in the *Pst* genome.

As secreted protein genes play essential roles in growth and interactions with host plants, analyzing serectome has been a major task for genomics studies of fungi and oomycetes (Torto et al. 2003; Duplessis et al. 2011; Huang et al. 2011). From the sequence of PST-130, Cantus et al. (2011) predicted 1088 sequences to encode secreted proteins with a median length of 165 amino acids (Table 2.6). Among the 1088 putative secreted protein genes, 5 were found to share high similarities with *Uromyces fabae* effector RTP1, and 25 shared similarity with PGTG proteins and were among the 100 most highly up-regulated genes in infected wheat tissue compared to urediniospores. Sixteen largest clusters of the putative secreted protein genes contain the N-terminal [Y/F/W]xC motif, which is associated with secreted proteins in haustorium-forming fungi, such as *Blumeria graminis*, *Pgt* and

Table 2.6 Comparison of *Puccinia striiformis* f. sp. *tritici* genome sequences with those of other sequenced genomes of rust fungi and the wheat powdery mildew fungus

Fungus	Isolate	Size (Mb)	No. of protein coding genes	Secreted proteins	No. of scaffolds	Sequencing approach	TE[a] (%)	References
P. striiformis	PST-130	64.8[b]	18,149	1088	22,815	Illumina Genome Analyser II sequencing	17.8	Cantu et al., (2011)
	CYR32	110.0	25,288	2092	12,833	Fosmid-to-fosmid strategy by Illumina GA paired-end sequencing	48.9	Zheng et al. (2013)
	PST-78	117.3	19,542	2146	9716	Roche 454 FLX and Illumina fosmid-end sequencing	31.5	Cuomo et al. (2017)
P. triticina	1-1	135.3	14,880	1358	14,820	Roche 454 FLX and Sanger fosmid-end and BAC-end sequencing	50.9	Cuomo et al. (2017)
P. graminis	CRL 75-36-700-3	88.6	15,800	1106	392	Sanger sequencing whole-genome shotgun strategy	43.7	Duplessis et al. (2011)
Melampsora larici-populina	98AG31	101.1	16,399	1184	462	Sanger sequencing whole-genome shotgun strategy	0.5	Duplessis et al. (2011)
M. lini	CH5	189.0	16,271	1085	21,130	Illumina sequencing	45.0	
Uromyces fabae	I2	330.0-379.0	23,153	599	59,735	Paired end Illumina sequencing	ND[b]	Link et al. (2014)
Blumeria graminis	96224	97.4	65,400	602	3522	Roche 454 sequencing	90.0	

[a]*TE* Transposable and repetitive elements
[b]This is about 60% of the total genome
[c]*ND* no data

P. triticina (Godfrey et al. 2010). In addition, the three largest clusters of candidate effectors were also characterized by conserved cysteine (C) residues with other eukaryotic *Avr* genes.

2.6.1.2 The Genome of CYR32

By combining different sequencing technologies and pools of large DNA fragments, a *Pst* isolate representing race CYR32, a most predominant race in China, was sequenced using the Illumina sequencing and partially validated using the Sanger sequencing technology (Zheng et al. 2013). In addition, the fosmid-to-fosmid sequencing approach used in this study helped to reach great assembly quality at high sequencing depth. The genome size of CYR32 was estimated to be twice of the partial sequence of the PST-130 isolate. After assembly, the genome of *Pst* was estimated at a much larger size of 110 Mb, and the number of predicted genes was increased to 25,288 (Table 2.6). The *Pst* CYR32 genome size was estimated with Illumina sequencing data of three whole genome shotgun (WGS) pairwise end libraries (with average insert sizes of 224,311 and 515 bp, respectively) based on k-mer (k-base fragment) analysis and the distribution of contig length and depth. Two peaks (16 and 30) were identified in the 17 k-mer depth distribution of WGS Illumina reads (Fig. 2.2), and two peaks of coverage (Cvg) were identified in the distribution of contig length and depth, indicating that the two haploid nuclei of *Pst* urediniospores differ significantly in genome sequences. Repeats and transposable elements were estimated at approximately 48.9% (Table 2.6), of which the class ILTR retro-elements and class II terminal inverted repeat (TIR) DNA transposons are most abundant, accounting for ~27% and ~17% of the *Pst* genome, respectively.

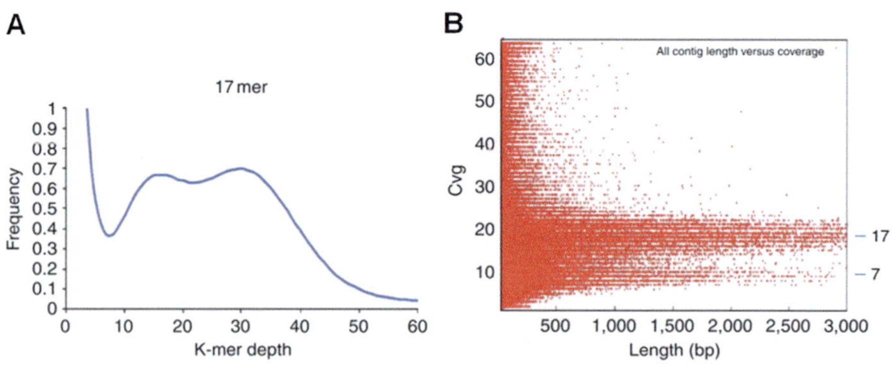

Fig. 2.2 The K-mer distribution and contig frequency analysis of the *Puccinia striiformis* f. sp. *tritici* (*Pst*) genome. (**a**) Seventeen k-mer depth distribution of whole-genome Illumina reads. Two peaks at 16 and 30 were identified. (**b**) Two peaks of Cvg 7 and 17 were identified in the distribution of contig length and depth. These results indicated that the two haploid nuclei of *Pst* urediniospores are highly heterozygous. The haploid genome size can be estimated using a modified Lander–Waterman algorithm based on K-mer analysis (Supplementary Notes 1 and 2). K-mer, a unique sequence of k nucleotides long; Cvg, coverage

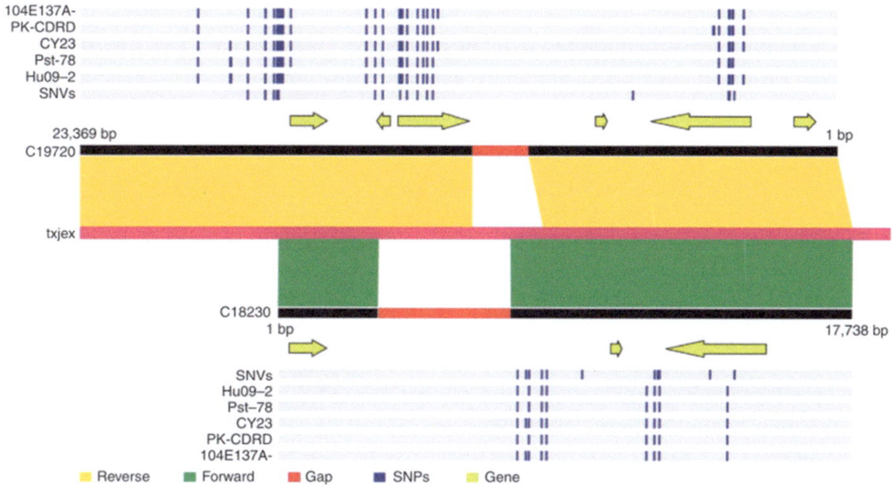

Fig. 2.3 Heterozygous regions in the *Puccinia striiformis* f. sp. *tritici* (*Pst*) genome. Heterozygous regions were identified by aligning the sequence of Txjex, a fosmid fully sequenced by Sanger sequencing with c19720 and c18230, two highly similar scaffolds of the *Pst* race CYR32 assembly generated with Illumina sequences from fosmid pools 7PB_Index52_Index26/20PB_Index47_Index8 and 14PB_Index44_Index29_Index6_Index35, respectively. SNPs between the reference CYR32 genome and those of re-sequenced isolates Hu09-2, PST-78, CY23, PK-CDRD or 104E137A- are marked with blue vertical lines. SNV, heterozygous SNV in the dikaryotic CYR32 genome

High coverage and deep reads (26×) allowed the identification of 83,736 heterozygous single-nucleotide polymorphisms within the sequenced CYR32 genome. The heterozygosity rate (0.68×10^3) was ~1.6 times higher than that of human (0.43×10^3). Therefore, the two nuclei in urediniospores of CYR32 must be highly heterozygous. A total of 2092 proteins encoded by the *Pst* genome (Table 2.6) were predicted by SignalP, TargetP and TMHMM, which account for 8.3% of the total number of proteins, and were grouped into 916 families.

Using the fosmid-to-fosmid strategy, Zheng et al. (2013) independently sequenced and assembled heterozygous regions that were represented in different fosmid pools. This strategy has an advantage for investigating highly heterozygous genomic regions of dikaryotic rust fungi (Fig. 2.3). The strategy greatly helped improve the assembly of the complex genome of *Pst*.

2.6.1.3 The Genomic Sequence of PST-78 and Comparison with Other Fungi

Isolate 2K41-Yr9 (PST-78) collected from the US Great Plains represents a group of races with virulences to resistance genes *Yr8* and *Yr9* first identified in the US in 2000 (Chen et al. 2002). Three similar genomic DNA insert libraries of PST-78 were sequenced using FLX chemistry with a Roche 454 (Cuomo et al. 2017). In addition, paired end Illumina reads were generated for three additional library sizes:

fragment, 3-5 kb insert, and 40 kb Illumina-adapted Fosmids (Fosill library). Three initial assemblies were generated using different algorithms: Life Technologies' Newbler program, the CLC (Qiagen, Hilden, Germany) denovo assembler and ALLPATHS-LG. To provide the most complete representation of the genome, contigs from the ALLPATHS-LG assembly were first selected, and then unique contigs from the CLC assembly were incorporated. Heterozygous positions within the sequenced isolate were identified from Illumina data. Reads were aligned to each assembly using BWA (v0.5.9), and SNP positions were identified with GATK v2.1.9 UnifiedGenotyper, and then filtered by GATK VariantFiltration. Strand-specific libraries were constructed with poly(A) selected RNA samples obtained from haustoria and infected leaf tissues using the dUTP second strand marking method. Gene sets were annotated by incorporating RNA-Seq data and predicted gene structures from multiple de novo predictions. The gene sets of the *Pst*, *Pt* and *Pgt* genomes were compared to each other and those of other fungi: *Melampsora lini*, *Melampsora larici-populina* and *Blumeria graminis* (Table 2.6). Orthologs were identified using OrthoMCL with expectation value 1e-5. The resulting orthologs were input to DAGchainer to identify syntenic blocks, requiring a minimum of 4 genes in the same order and orientation in the compared genomes.

High quality genome assemblies of PST-78 were generated by combining data from multiple sequencing technologies. A range of insert size libraries were sequenced using Roche 454, Illumina and Sanger technologies. The assembly of PST-78 totaled 117.31 Mb and consisted of 9715 scaffolds with N50 length of 519 Kb (Table 2.6). This genome size is slightly larger than that of CYR32 (110 Mb), but both the scaffold and contig total numbers are lower than those reported for CYR32 as presented above for the Chinese *Pst* isolate (Zheng et al. 2013). For comparison, the PST-78 genome size is smaller than that of *Pt* (135.3 Mb), but larger than that of *Pgt* (89 Mb) (Cuomo et al. 2012). The *Pst* genome contains 31.5% repeats, lower than both *Pgt* (36.5%) and *Pt* (50.9%). Of the three sequenced isolates for representing the three *Puccinia* species, the *Pst* isolate has a much higher heterozygosity (473,282 heterozygous SNPs at an average rate of 5.97 SNPs/kb) than those of *Pt* (269,370 heterozygous SNPs at 2.53 SNPs/kb) (Cuomo et al. 2017) and *Pgt* (2.28 SNPs/kb in genic region and 1.72 SNPs/kb in the intergenic region) (Duplessis et al. 2011). The number of species-specific genes is higher in *Pst* (8955) than that of either *Pt* (5443) or *Pgt* (4901) (Cuomo et al. 2017). In general, the rust fungi have a higher number of species-specific genes than other basidiomycete fungi. Based on the genomic sequences, *Pst* diverged earlier than either *Pt* and *Pgt*, and the latter two species are more closely related. However, some features of *Pst* and *Pgt* may be more conserved or have evolved in parallel as both species share the same alternative host.

2.6.1.4 Re-sequencing of Additional *Pst* Isolates

To determine the intra-specific evolution, Zheng et al. (2013) compared the CYR32 genomic sequence with those of five additional *Pst* isolates (PST-78, CYR23, Hu09-2, PK-CDRD, 104E-137A-) from different countries. These isolates were sequenced at 22 - 29× of the genome using the WGS strategy. More than 100,000 SNPs were identified among the *Pst* isolates. The number of SNPs is similar among the different isolates, and heterozygous SNPs occupied more than 80% of SNPs in the genomes. The number of SNPs in gene-coding regions (cSNPs) accounts for ~35% of total SNPs, of which 50% were nonsynonymous mutations that may impact virulence, race specificity and other biological traits and may be responsible for adaptation to host and environmental changes. The phylogenetic tree of the six *Pst* isolates constructed based on total SNPs and cSNPs revealed that the topological features of the cSNP and SNP trees were similar, but the SNP trees differed from the virulence-based tree. Therefore, the evolutionary relations of *Pst* isolates based on SNPs were not significantly correlated with their virulence features.

To identify effector candidates and relate them to their distinct virulence profiles, Cantu et al. (2013) sequenced four *Pst* isolates, PST-21 and PST-43 from the US and PST-87/7 and PST-08/21) from the UK. Using the Illumina WGS sequencing approach, the total number of contigs assembled ranged from 29,178 to 55,502. The completeness of the genome sequences were determined by aligning them to the PST-130 contigs (Cantu et al. 2011) and available *Pst* expressed sequenced tags (ESTs). On average, 20,280 protein-coding genes were identified in the five isolates. As urediniospores constitute asexual dikaryotic spores that contain two independent nuclei, the sequence reads of a particular isolate to the assembled contigs of the same isolate were aligned to assess genetic variation between the two nuclei in the sequenced urediniospores. On average, the heterokaryotic SNPs across isolates were more frequent (5.29 ± 2.23 SNPs/kb) than the homokaryotic SNPs (0.41 ± 0.28 SNPs/kb), indicating the high heterozygosity between the two nuclei. To further characterize the genetic diversity between the five sequenced PST isolates, the number of homokaryotic SNPs within the gene space was assessed. Pair-wise comparison of non-synonymous mutations in these gene sets revealed that more apparent polymorphisms were found in US isolates PST-21 and PST-130 compared to other isolates. The limited genetic diversity found between US isolate PST-43 and the two UK isolates (PST-87/7and PST-08/21) indicated that these isolates belong to the same clonal lineage. This is supported by a diversity study based on 117 amplified fragment length polymorphism (AFLP) fragments which showed that PST isolates collected before 2000 in the US and Europe clustered together, thereby suggesting a common origin. The authors identified proteins that displayed non-synonymous substitutions specifically between the two UK isolates. They identified 5 candidates from 2999 secreted protein genes between the two UK isolates.

Focusing on identification of secreted protein genes and identifying those associated to avirulences, seven *Pst* isolates were sequenced using the Ion Proton technology (Xia, Wang, Cornejoc, Jiwana, See, Chen unpublished data). The 7 isolates, together with previously sequenced 7 isolates present a relatively balanced virulence:avirulence ratio for several loci. The authors reported an unevenly distributed heterozygosity among isolates and among different genomic regions in each isolates, with secreted protein gene regions more variable reflecting different evolutionary forces they suffered during interacting with host genes. Comparison of *Pst* and other rust fungi secretomes indicated a large portion of species specific secreted proteins, suggesting they may have specific roles when interacting with wheat host. Effector candidates were identified by characterizing *Pst* secretomes from diverse aspects including whether they are (a) cysteine-rich, (b) polymorphic among 14 isolates, (c) EffectorP predicted, (d) haustorial differentially expressed and (e) *P. striiformis* species-specific. A class of superoxide dismutase 5 (SOD5)-homologs proteins was identified by searching *Pst* secretome against functionally validated effectors in the PHI-base database, suggesting that antioxidant defense may be a common strategy for *Pst* to suppress host basal immunity. They also identified *Avr* candidates corresponding to five *Yr* genes by correlating polymorphisms to virulence/avirulence profiles. One of the *Avr* candidates showed a presence/absence polymorphism, suggesting that gene loss may be a mechanism responsible for *Pst* virulence changes.

2.6.1.5 Genome Comparison of the Stripe Rust Pathogen with Non-rust Fungi

As more and more organisms have been sequenced, the comparative genomics approach has become a powerful tool to understand phylogenetic relationships of rust fungi like the stripe rust pathogen with other fungi and identify functions of genes referring to other fungi. As discussed above, since *M. larici-populina* and *P. graminis* were first sequenced, several other rust genomes including *Pst* have been sequenced and assembled (Table 2.6). Compared to other basidiomycete fungi including *Pt* and *Pgt*, species-specific gene families are relatively abundant in the *Pst* genome. Because the lack of the specific genes in other fungi, most of these genes are unknown for their function. Among the annotated gene families, helicases, leucine-rich repeat proteins, oligopeptide transporters and different types of glycosyl hydrolases, lipases and peptidases were particularly overrepresented (Zheng et al. 2013). The functional information of *Pst* genes were obtained through bioinformatics comparison to other sequenced fungal species (Ling et al. 2007; Zhang et al. 2008; Yin et al. 2009; Huang et al. 2011; Zheng et al. 2013; Xia et al. 2016a, b).

Compared to the average genome size of sequenced fungi (37.7 Mb) and the average for sequenced basidiomycete fungi (49.9 Mb), the average genome size of rust fungi (225.3 Mb) is much larger (Cuomo et al. 2017). The large genomes of rust fungi are due to the high percentage of repetitive DNA derived from transposable elements that account for 45% of the genome on average as presented above for *Pst*, *Pt* and *Pgt*. The types of transposons that populate the genomes substantially differ among different rust fungi. In *Pgt*, Class I retrotransposons are more predominant than Class II terminal inverted repeat (TIR) transposons, whereas in *Mlp* Class II TIR transposons are more prevalent. In *Pst*, the class ILTR retro-elements and class II TIR DNA transposons are most abundant, accounting for 27% and 17% of the genome, respectively. The heightened retrotransposition activity may result in the increased genome size and loss of functions that are not necessary for life outside its host plant. In an irreversible way, these losses of function eventually lead to the obligate biotrophic life style of rust fungi.

2.6.2 *Identification of Genes and Gene Families Through Genome Sequencing and Functional Genomics Analyses*

Functional genomics of *Pst* has been studied by constructing and sequencing cDNA libraries (Ling et al. 2007; Zhang et al. 2008; Yin et al. 2009), AFLP analyzing cDNA clones (Wang et al. 2009c, 2010), bioinformatics analysis of genomic sequences (Cantu et al. 2011; Cantu et al. 2013), expression of genes using a microarray chip and RT-PCR (Huang et al. 2011), RNAseq (Hubbard et al. 2015) and gene silencing (Cheng et al. 2015b, 2016b, c; Yin et al. 2015; Liu et al. 2016b). These studies developed techniques and/or identified numerous genes in *Pst* underlining the pathogen interactions with the wheat host plants.

2.6.2.1 Secretome of *Pst* and Comparison with Other Rust Fungi

One of the most important objectives of genome and cDNA sequencing is to identify genes encoding proteins and then from the protein genes, secreted protein (SP) genes can be identified using several criteria (Huang et al. 2011; Yin et al. 2009; Cantu et al. 2013; Xia et al. 2016a, b). Huang et al. (2011) identified 276 SP genes from three cDNA libraries of *Pst* constructed from urediniospores (Ling et al. 2007; Zhang et al. 2008; Yin et al. 2009) and utilized the genes and other genes to construct a *Pst* genechip for study gene expression using the microarray technology. These SP genes also have been used to develop SP-SNP markers for studying SP genes associated to virulences and population structure of *Pst* in the US (Xia et al. 2016a, b).

Several studies have reported SP genes of *Pst* by characterizing genes annotated from whole genome sequencing. For instance, Zheng et al. (2013) identified 25,288 putative coding genes in *Pst* race CYR32. This number is similar to the symbiotic basidiomycete *Laccaria bicolor*, but strikingly larger than the corn smut fungi *Ustilago maydis* and *Sporisorium reilianum*, two nonobligate pathogenic biotrophic basidiomycete fungi that encode only approximately 6500 proteins (Kämper et al. 2006; Schirawski et al. 2010). In the genome-wide survey of the rust fungi, candidate effector genes were described by identifying those that encode SPs in the proteome. From the predicted proteome of *Pst* CYR32, a total of 2168 SPs were identified by SignalP, TMHMM and TargetP analyses, account for a large portion of the whole proteome (8.6%) (Zheng et al. 2013). For comparison with other rust fungi, *M. larici-populina* (*Mlp*) was reported to have 16,399 protein genes, of which 74% appear to be species-specific and 1184 (7.2%) are SPs, and In *Pgt*, there are 17,773 proteins, of which 84% are species-specific and 1106 (6.2%) are SPs (Zheng et al. 2013).

Cluster analysis of the *Pst*, *Pgt* and *Mlp* secretomes revealed 2, 271 tribes with at least one member per tribe (Zheng et al. 2013). Secreted proteins of *Pst* are distributed among 993 tribes, of which 474 are unique to *Pst*, 330 are shared with *Pgt*, and

189 are common among all three rust fungi. GenBank searches showed that 1389 of the *Pst* SPs have unknown function. Among them, 101 contain the RXLR-like motif, 137 have nuclear localization signals (NLS) and 217 are repeat-containing proteins (RCPs). In addition, 411 of these putative effector proteins contain up to eight copies of the Y/F/WxC motif.

In PST-78, a total of 19,542 genes encoding proteins were predicted, of which 8955 genes appeared to be specific for *Pst* when compared to *Pt*, *Pgt*, *Mlp* and *Mli* (Table 2.6) (Cuomo et al. 2017). As this study more focused on *Pt* SPs, the authors predicted a total 1358 SP genes for *Pt* and compared the SP genes with *Pst* and *Pgt*. They grouped 914 *Pt* SPs in 385 families or 'tribes' previously assigned to *Pst* or *Pgt*. From the remaining SPs, additional 111 SP contained amino acid sequence similarity to *Pst* and *Pgt* predicted proteins.

Xia CJ and others (unpublished data) re-analyzed PST-78 genome sequence data (http://www.broadinstitute.org/annotation/genome/puccinia_group/MultiHome.html) and identified 20,482 proteins, of which 2146 (10.5%) were SPs, remarkably similar to the number of SPs (2168) reported in CYR32 (Zheng et al. 2013). This percentage is similar to the 9.1% (1425 out of 15,685) in *Pt* and 11.6% (1857 out of 15,979) in *Pgt*. In general, the three *Puccinia* species have a higher percentage of proteins as SP than most other fungi with the percentage of proteins as SPs ranging from 4.0% to 14.0%. The authors analyzed a total of 5428 SPs from *Pst*, *Pgt* and *Pt* and found that *Pst* has the largest proportion of 43.0% (923; in 747 orthologous families), followed by *Pgt* having 35.1% (651; in 523 orthologous families) and *Pt* having 30.8% (439; in 384 orthologous families) of their SPs to be species specific SPs. The number of species-specific SPs in *Pst* is higher than that reported by Cuomo et al. (2017), although the same database was used in both studies. Different numbers could be due to different criteria used to identify SPs. Nevertheless, both studies reported relative large numbers of SPs in the *Puccinia* species particularly *Pst*, compared to other fungi.

2.6.2.2 Carbohydrate-Active Enzymes

In rust pathogen transcriptomes, gene families encoding host-targeted, hydrolytic enzymes acting on plant biopolymers are highly up-regulated in planta, such as proteinases, lipases and several sugar-cleaving enzymes (carbohydrate-active enzymes; CAZymes). Upregulation of these degrading enzymes may aim to cleave the host cell wall by the invading hyphae and thus help penetrate the host cells. However, rust fungi, particularly *Pst*, *Pgt* and *Mlp*, do not have a orthologue of the *srt1* sucrose transporter gene, which was found to be required for virulence of *Ustilago maydis* (Wahl et al. 2010). Thus, the rust fungi may convert sucrose to hexose by secreted invertase, and haustoria then take up hexose from host cells as the major carbohydrate source. More studies are needed to study carbohydrate-active enzymes in the stripe rust pathogen.

2.6.2.3 Expanded Rust Transporters Gene Families Are Expressed During Host Infection

Acquisition of nutrients, including carbohydrates and amino acids, is crucial to the success of rust pathogen biotrophic interactions established by invading hyphae forming haustoria within the host plant. Zheng et al. (2013) identified 512 transporter genes, including 7 sugar and 12 amino acid transporter genes, in *Pst* using Chinese isolate CYR32. Many of these transporter genes are haustorial specific, such as hexose transporter *HXT1*, amino acid transporters *AAT1*, *AAT2* and *AAT3* and H^+-ATPases that have also been reported in *Mlp* and *Pgt*, *Uromyces fabae* (Voegele et al. 2001; Duplessis et al. 2011; Spanu 2012). These haustorial specific transporter genes are known to be highly up-regulated during the interaction with its host plant. In the *Pst* CYR32 genome, two hexose transporters that are closely related to *HXT1* were identified (Zheng et al. 2013). Different from the induced expression during infection of *U. fabae* and *Mlp*, the two *HXT1* homologs in *Pst* were constitutively expressed across the whole infection process.

The detection of 15 amino acid transporter genes in *Pst* by Zheng et al. (2013) is significant. In sequenced basidiomycete genomes, a high number of oligopeptide transporter (OPT) genes was found in *Pgt* (21) and *Mlp* (22) genomes, whereas a low number (5 to 16) of OPTs were found in other basidiomycete fungi. The different numbers of OPTs generally suggest an increased genetic potential for peptide uptake in the rust fungi. OPT genes that are transcriptionally induced in-planta are likely involved in the transport of peptides released by the action of the induced proteinases (e.g. aspartic peptidase, subtilisin) in infected leaf tissues. Although the Major Facilitator Superfamily (MFS) gene family is reduced in the *Mlp* and *Pgt* genomes compared with other basidiomycetes, many MFS transcripts are highly expressed in planta including the two *HXT1* homologs. Combined with the in-planta expression of invertase genes and absence of the sucrose transporter *Srt1 in Mlp* and *Pgt*, it strongly suggested the preferential uptake of host hexoses by invading rust pathogen hyphae. Based on the genomics analysis and in comparison with other fungi, it can be concluded that the uptake of amino acids from host cells by haustoria is essential for protein synthesis in *Pst* and other rust fungi.

2.6.2.4 Nitrate and Sulfate Assimilation Pathway Deficiencies in *Pst* and Other Rust Fungi

Despite of the expanded transporter gene families, *Mlp*, *Pgt* and *Pst* lack genes typically present in saprotrophic basidiomycetes. The nitrate/nitrite porter and the nitritere ductase involved in the $NH4^+$ assimilation pathway are missing in *Pst*, *Pgt* and *Mlp* from the nitrate assimilation gene cluster found in other fungi (Zheng et al. 2013). Meanwhile genes required to perform the primary sulfate assimilation identified in *Mlp* were absent in *Pgt*. In *Mlp*, the transketolase domain of the β-subunits of sulfite reductase (SiR) was missing, while both the α- and β-subunits of SiR were lost in *Pgt* (Duplessis et al. 2011). The apparent incapacitation of nitrate and sulfate assimilation pathways in rust fungi is consistent with their obligate biotrophic lifestyle, as they mainly depend on nitrogen ($NH4^+$ or amino acids) and sulfur acquired from plant cells. These metabolic deficiencies have also been found in other plant pathogens that represent two independent evolutionary lineages of obligate biotrophy in the oomycete (*Hyaloperonospora arabidopsidis*) and ascomycete (*Blumaria graminis*) lineages (Spanu 2012).

2.6.3 Genes Expressing in Different Infection Stages

To date, no stable transformation system has been established for *Pst* and other rust fungi. The obligate biotrophic nature further hampers identification of genes involved in interactions with its host. To overcome this limitation, transcriptomics techniques have been applied to achieve understanding the molecular mechanisms of infection and interactions.

For plant pathogens with little or no history of genetic research, single-pass sequencing of random cDNA clones is a relatively inexpensive and rapid procedure for finding novel genes and information about their expression. To date, extensive expressed sequence tag (EST) databases have been established for various plant pathogenic fungi (Ebbole-Buitrago et al. 2004; Posada-Buitrago and Frederick 2005; Iida et al. 2006). Characterization of ESTs from haustoria of the rust fungus *Uromyces fabae* revealed the major shift in gene expression between urediniospore germination and the biotrophic growth stage (Hahn and Mendgen 1997; Jakupovic et al. 2006). For wheat rust fungi, EST libraries have been established for *Pgt* (Liu et al. 1993; Broeker et al. 2006) and *Pt* (Thara et al. 2003; Hu et al. 2004, 2007), from which numerous genes potentially involved in fungal differentiation and infection have been identified. A total of 25,558 ESTs have been generated from 13 cDNA libraries representing various stages of *Pt*, including urediniospores, germinated urediniospores, appressorium and haustorium formation stages during infection of a susceptible wheat cultivar and a resistant wheat cultivar. Gene sets obtained from these different libraries appeared to be remarkably different, suggesting drastic reprogramming of the transcriptome during these major differentiation processes. Efforts have been made to reveal the transcriptional changes in *Pst* infection and pathogenesis via various approaches, including constructing and sequencing cDNA libraries and analyzing gene expression using microarray techniques.

2.6.3.1 cDNA Libraries of Different Infection Stages of *Pst*

A cDNA library can provide molecular resources for analysis of genes involved in the biology of a plant pathogenic fungus, such as genes responsible for the development, survival, pathogenicity and virulence. In order to initiate studies on the basic genome structure and gene expression of *Pst* with infective state, constructed a full-length stage-specific cDNA library of a predominant race of *Pst* is an irresistible trend. Up to date, a series of cDNA libraries have been developed from different stages of *Pst* development, including urediniospores, germinated urediniospores and haustoria.

2.6.3.1.1 A cDNA Library from Urediniospores

Urediniospore is the most common spore form of *Pst*, which is strictly dependent on living host plants for growth and reproduction. Compared to mycelia, urediniospores are relatively resistant to adverse environmental conditions. Therefore, the urediniospore stage should contain most of the pathogen genes involved in the pathogen development, survival and pathogenicity. Also, urediniospores play the most important role in the disease epidemiology. Such knowledge is critical for understanding the infection mechanisms of this important pathogen and developing better disease management strategies. Thus, revealing the transcription activities of urediniospores can provide useful information in understanding the pathogenicity of *Pst*. A full-length cDNA library from urediniospores was constructed from the US *Pst* race PST-78, consisting of 42,240 clones with an average 1.9 kb cDNA insert (Ling et al. 2007). From 196 sequenced cDNA clones, 73 clones (37.2%) were determined to be of known function, among which 36 clones (18.4%) encode hypothetical proteins, and 37 clones (18.9%) share homology to genes in other fungi. Out of the 73 clones with known functions, 51 genes were involved in amino acid metabolism, cell defense, cell cycle, cell signaling, cell structure and growth, energy cycle, lipid and nucleotide metabolism, protein modification, ribosomal protein complex, sugar metabolism, transcription factor, transport metabolism, and virulence/infection (Ling et al. 2007). The transcript abundance and function annotation analyses revealed the key molecular events in urediniospores, providing important clues in understanding the growth and pathogenicity of *Pst*.

2.6.3.1.2 A cDNA Library from Germinated Urediniospores

Urediniospore germination represents an early stage of the interaction of *Pst* with host plants. As an obligate biotroph, germ tubes of *Pst* cannot progress beyond this stage of development without host plants (Mendgen et al. 2000; Mendgen and Hahn 2002). A cDNA library was constructed from germinated urediniospores of Chinese *Pst* race CYR32 (Zhang et al. 2008). Different from other filamentous fungi, a relatively high level of redundancy was found in this library. Among the total 4798 ESTs sequenced, 3395 of them could be assembled into 1118 unique sequences. A few contigs consist of over 100 EST clones, representing the highly expressed genes during urediniospore germination. The 1118 uniseqs covered a wide range of genes involved in different cellular functions. The most abundant genes during urediniospore germination were those involved in metabolic activities and those responsible for protein biosynthesis, which account for 10.3% and 5.3% of the uniseqs, respectively. A few genes are involved in RNA synthesis, cell signal and communication, cell structure and growth and cell defense, indicating that active metabolism and protein synthesis are important for urediniospore germination and germ tube growth. Several genes that are specifically expressed during urediniospore germination were found to be homologs to genes involved in cell wall modification and calcium-signaling pathway, indicating their role in regulating germ tube emergence and infection structure differentiation. Besides, homologs of fungal pathogenicity factors or effector genes were also identified in this library (Zhang et al. 2008). The data indicated that those genes highly expressed in germinated urediniospores may play important roles in fungal-plant interactions during early infection stages in *Pst*.

2.6.3.1.3 A cDNA Library of Haustoria

As an obligate parasite, *Pst* has adapted to source nutrients from living mesophyll wheat cells through haustoria, similar to other rust fungi (Szabo et al. 2001). A haustorium is thought to be the primary organ for nutrient transfer from the host cell to the fungal vegetative body (Hahn and Mendgen 1997; Dodds et al. 2004). Analyses of haustorial transcripts from other rust fungi (Hahn and Mendgen 1997; Voegele et al. 2001; Dodds et al. 2004; Jakupovic et al. 2006) indicated that a large number of genes are expressed in planta and involved in virulence as well as other aspects of parasitism, like nutrient uptake. A large number of plant-induced and haustorium-specific genes have been identified in the bean rust fungus *Uromyces fabae* (Hahn and Mendgen 1997; Jakupovic et al. 2006). Avirulence genes from *Mli* have been cloned and found to encode small secreted proteins expressed in the fungal haustoria (Dodds et al. 2004; Catanzariti et al. 2006). Thus, identifying genes expressed in haustoria is essential for determine mechanisms of interactions between a rust fungus and its host.

Yin et al. (2009) constructed a cDNA library from haustoria isolated from wheat leaves infected by PST-78. From a total of 5126 EST sequences, 1134 unique sequences were obtained, of which about 10% and 26% were homologous to genes with known functions and hypothetical proteins, respectively, and the remaining 64% showed no significant hits in public databases. The largest group of genes with known functions showed similarities to ribosomal proteins, followed by the group of genes with similarities to proteins involved in primary metabolism and energy production. Similar prevalent proteins with known functions, which were most redundant, have also been reported in other pathosystems during infection (Thara et al. 2003; Broeker et al. 2006; Hu et al. 2007). Overall the less redundant sequences were surprisingly unique especially considering the database searches of the *Pgt* and *Mlp* genome sequences. As virulence proteins from eukaryotic plant pathogens described so far are reported to be diverse in function and share little sequence similarity to proteins of known function. The high percentage of genes expressed in haustoria share little or no sequence similarity to genes in other fungi may indicate that genes expressed in haustoria are more likely involved in virulence, which is probably more specific to *Pst*. In fact, several haustorial transcripts were predicted to encode products homologous to potential virulence related proteins, such as the glutamine synthetase, dihydrolipoyl dehydrogenase and cAMP-dependent protein kinase. In addition, 15 haustorium-specific genes predicted to encode SPs are candidate effectors or virulence factors.

2.6.3.2 Microarray Assays of Compatible/Incompatible Interactions

Microarray analysis is a powerful technique to study global gene expression profiles. It has been used to study gene expression in a large number of plants, but in a relatively small number of fungi. Of about 30 filamentous fungi that have been previously studied, most of the studies were performed under simulated pathogenic conditions *in vitro*, while only a small number were reported for gene expression within their plant hosts (Both et al. 2005a; Both et al. 2005b; Divon et al. 2005; Güldener et al. 2006; Jakupovic et al. 2006; McFadden et al. 2006; Breakspear and Momany 2007), primarily due to the lack of gene sequences, the relatively low amount of transcripts of pathogen genes within the host plant tissue and interference of plant genes. Coram et al. (2008a, b, c, d) were the first to apply the microarray technique in the wheat-*Pst* pathosystem. In a series of studies using the Affymetrix and custom microarrays of wheat, wheat genes involved in race-specific all-stage and nonrace specific high-temperature adult-plant resistance to *Pst* have been identified (Coram et al. 2008a, b, c, d, 2010; Chen et al. 2013).

The above microarray studies all focused on wheat genes as the wheat microarray chip is commercially available. Compared to wheat, a relatively small number of programs work on the stripe rust pathogen, which limits the incentive for developing microarray chip for the stripe rust pathogen. Huang et al. (2011) constructed a custom-made oligonucleotide microarray chip consisting of 442 genes selected from *Pst* cDNA libraries. As the authors pay much attention to SP genes, they included 267 (60%) SPs which largely have unknown functions and the remaining 175 (40%) genes have predicted various functions. The 267 genes included 143 from the germinated urediniospore library (Zhang et al. 2008), 115 from the haustorial library (Yin et al. 2009) and 9 from the urediniospore library (Ling et al. 2007). They determined expression patterns of the genes by hybridizing the microarray with cDNA from urediniospores, germinated urediniospores and infected wheat leaves; and identified 55 transcripts that were differentially expressed during the infection process in a compatible interaction. Some of the 55 genes were identified to have functions related to carbohydrate and lipid metabolism, energy, cell signaling, protein synthesis and cell structure and division. Among the 55 induced genes in various spore and infection stages, 25 (45%) were of SPs. In term of percentage, the highest, 83% (5 out of 6), of the specifically expressed genes was found at the early infection stage, while only 34% of the 35 genes detected at the late infection stage were of SP genes. These results indicate that relatively more proteins are secreted from the pathogen, mainly from haustoria, into host cells to interact with plant gene products.

In an incompatible interaction, the *Yr5* near-isogenic line in the 'Avocet S' (AvS) background inoculated with race PST-78, 17 transcripts of *Pst* were differentially expressed in resistant wheat leaves, ten of which had similar expression patterns to those in the compatible interaction (AvS inoculated with PST-78) (Huang et al. 2011). Several candidates for pathogenicity and virulence/avirulence related genes were also identified. They validated the stage-specific expression patterns of 10 out of 12 selected genes using the qRT-PCR analysis. Six of the total 17 genes were specifically expressed in urediniospores before inoculation and 14 days after inoculation in resistant leaves without sporulation, when *Pst* started to produce urediniospores in the compatible interaction. Except for one gene that was predicted as an RNA polymerase II transcriptional coactivator, all other genes in this group do not have known functions. To the authors' surprisingly, they found that 10 of the 17 genes were highly expressed at 7 and/or 14 days post inoculation (dpi) in the incompatible interaction, which was not expected considering the high resistance of *Yr5* to the inoculated race. The detected gene expression indicates that the pathogen may not be completely dead in such late infection stage. This is supported by the putative functions of the six genes related to signaling, communication, growth and structure of cells, energy and metabolism. Because 7 of the 10 genes were also induced in the compatible interaction, the authors concluded that they have a common role in fungal growth under different environments within the host plant tissue. In a recent study, the senior author's group confirmed the expression of some of the genes in the compatible interaction (Farrakh and Chen, unpublished data). It is also interesting that a gene encoding a plasmid replication initiation protein was detected in all RNA samples, but its expression level was much higher (three- to four-fold) at 12 hours after inoculation and at late infection (7 and 14 dpi) (Huang et al. 2011). Because the gene was not significantly induced in the compatible interaction, the presence or accumulation of the plasmid or a gene carried by the plasmid may play a role in interacting with the host, or possibly play a role in the pathogen's struggle for survival under the adverse environment of host resistance. The study demonstrates that the custom oligonucleotide microarray technology is useful to determine expression patterns of the pathogen genes involved in different types of the host-pathogen interactions and stages of development.

Although it is possible to construct a *Pst* microarray chip containing all possible genes, the complicated process for designing the primers, microarray hybridization and signal detection and relatively high cost have limited its usage. As the RNAseq technology has been advanced and the cost has been tremendously reduced, RNA sequencing has become more and more popular for studying gene expressions.

2.6.3.3 Haustorial Transcriptome Using Next-Generation Sequencing

Similar to the construction of an EST library for haustoria of the US isolate PST-78 (Yin et al. 2009), Garnica et al. (2013) purified haustoria from heavily infected susceptible wheat cultivar Morocco nine days after inoculated with an Australian *Pst* isolate representing race 104E137A-. They used both 454-based transcriptome analysis and Illumina RNAseq technology to sequence and assemble the haustorial transcriptome as well as the transcriptome of germinated uediniospores of the same isolate. They assembled 12,282 transcripts from 454-pyrosequencing data and used them as references for digital gene expression analysis to compare the germinated urediniospores and haustoria transcriptomes based on Illumina RNAseq data. Of the 12,282 transcripts, 4485 had unknown function and the remaining 7797 had various putative functions. More than 400 genes encoding SPs, considered effector candidates, were identified from the haustorial transcriptome, with two thirds of these up-regulated in haustoria compared to germinated urediniospores. They confirmed the expression patterns of 94 effector candidates by RT-PCR analysis. In addition, they detected genes involved in DNA replication and cell cycle, encoding cell wall modification enzymes,

They found that germinated urediniospores were clearly enriched with transcripts classified into the categories of cell cycle, DNA replication and cell division; encoding cell wall modification enzymes; and involving in fatty acid metabolism. Genes highly expressed in haustoria included those for transporting amino acids and oligopeptides, sugars, small molecules and ions, the vitamin co-factor nicotinic acid, ATP binding cassette (ABC), phosphate and nicotinic acid; for energy use – oxidative phosphorylation; associated with the thiamine biosynthesis pathway; S-methylmethionine (SMM) permease for sulfur metabolism; glutamate-cysteine ligase and glutathione synthetase for glutathione metabolism; genes of P450s containing a heme cofactor that catalyse the oxidation of organic substances. Genes encoding enzymes of the pentose phosphate shunt (PPS); enzymes involving in nitrogen metabolism; transcription factors; and enzymes involving in transduction pathways were expressed in both germinated urediniospores and haustoria.

The authors noticed that not all secreted proteins are virulence effectors, and some had clear roles in fungal cell wall modification. The results of this study revealed that urediniospores rely mainly on stored energy reserves for growth and development, whereas haustoria take up host nutrients for massive energy production for biosynthetic pathways and the ultimate production of the next generation of urediniospores.

2.6.3.4 Field Pathogenomics

The above studies on transcriptomics worked with one of two *Pst* isolate to reveal genes involved in different lifecycle stages, such as urediniospores and during infection or in different structures such as urediniospores verse haustoria, with only one study on genes expressed in compatible verse incompatible interactions with a single isolate (Huang et al. 2011). These studies could not identify genes expressed in different isolates interacting with a large number of host cultivars. A recent study by Hubbard et al. (2015) attempted to solve this problem by utilizing the transcriptomics approach to characterize field populations of *Pst*.

In the study by Hubbard et al. (2015), a total 219 single lesion leaf samples of *Pst*-infected wheat and triticale were collected directly from fields in 17 UK counties in 2013, from which 35 wheat samples from wheat varieties covering the resistance spectrum and 4 triticale samples were selected for total RNA extraction, together with selected previously prevalent races of *Pst*. The genomes of 14 UK and 7 French purified PST isolates collected between 1978 and 2011 were sequenced using an Illumina WGS approach. The genome sequence analysis showed that 13 of the 14 historical UK isolates and the 7 French isolates clustered in a single clade with little genetic variation. In contrast, the 2013 field isolates were highly diverse and distantly related to the older isolates. Of the 39 field isolates in 2013, 11 (28%) was genetically similar to the 'Warrior' pathotype identified first in 2011 in the UK. They identified 42 genes that encoded cluster-specific polymorphic proteins that displayed features typical of SPs. The field pathogenomics approach not only captured the pathogen diversity, but also provided opportunity to study host genes responding to *Pst* infection. The authors genotyped the varieties of the field samples by WGS sequencing and using wheat SNP markers. The genotyping data could confirm outbreaks on particular wheat varieties and look for associations between pathogen genotypes and host pedigrees. Based on the results, the authors concluded that the recent *Pst* population was most probably derived from the recent introduction into the UK of diverse assemblage of exotic *Pst* lineages, and that the introduced lineages may have rapidly displaced the previous *Pst* population. The field pathogenomics approach can be integrated into disease survey. Currently, the UK group is collaborating with scientists in many other countries to characterize the global *Pst* populations using this approach.

2.6.4 Characterization of Pathogenicity Related Genes

Due to the lack of transformation and gene disruption methods, it has been difficult to determine functions of genes in rust fungi. The utilization of the virus-induced gene silencing (VIGS) system has made some progress in determining individual *Pst* genes. VIGS uses viral vectors to produce dsRNA of a target gene fragment and trigger RNAi silencing (Becker and Lange 2009). It is a powerful reverse genetics tool that is particularly useful for species where stable transformants are difficult to construct. Based on VIGS, a host-induced gene silencing (HIGS) system was used for transient silencing of pathogen genes, providing a new way for pathogenic research, especially for haustoria-formed rust fungi like *Pst* (Yin et al. 2011). Several genes involving in *Pst* growth and pathogenesis have been characterized, which can be used to develop chemicals for inhibiting the pathogen growth and reproduction, and therefore control stripe rust.

2.6.4.1 Conidiation-Related Gene *PsCon1*

In filamentous fungi, conidiation is a highly effective reproductive mechanism because vast numbers of asexual spores can be produced from a single colony (Adams et al. 1998). The production and germination of spores in fungi play a crucial role in their infection cycles. These processes are regulated by a set of genes, known as conidiation-related genes, which are involved in the production, morphological characteristics, and differentiation of conidia (Shrode et al. 2001). Considering the role of *Pst* urediniospores in pathogen dispersal and disease initiation, dissecting molecular events connected with the production of asexual spores of *Pst* could provide approaches to control stripe rust. The *con* genes for conidiation in *Neurospora crassa* encode proteins containing conserved N-termini and two conserved domains (White and Yanofsky 1993). Guo et al. (2013) characterized a *Pst* conidiation-related gene, *PsCon1*. The PsCon1 protein has two conserved conidiation-specific protein 6 domains. *PsCon1* is relatively conserved among *Pst* races of different virulence with only a single nucleotide difference among five *Pst* races. *PsCon1* was found to express at the highest level in urediniospores, and the level gradually decreased after germination and infection. Based on heterologous expression in *Fusarium graminearum* and subcellular localization assays, *PsCon1* was found to locate in the cytoplasm of conidia. The results indicate that *PsCon1* plays a role in formation or survival of *Pst* urediniospores.

2.6.4.2 Actin of *Pst*

Fungal tip growth is a polarized and dynamic process, involving cytoskeletal changes and interactions mediated by microtubules, microfilaments and motor proteins (Virag and Harris 2006). As an essential component of cellular microfilaments, actin is a ubiquitous and highly conserved protein that plays key roles in several basic functions of eukaryotic organisms, such as cytoskeleton morphology, cell division and motility, signal transduction, cellular interactions and organelle movements (Galletta and Cooper 2009). They are also involved in locomotion, phagocytosis, endocytosis and exocytosis (Peng et al. 2010). Actin proteins are found to be necessary for the integrity and function of the Spitzenkörper (*Spk*), a dynamic structure present at the tips of hyphal cells with a single highly polarized growth site, by treating hyphal tips with F-actin inhibitors, and characterizing actin distribution in actin mutants (Heath et al. 2003). One of the main roles of tip-localized actins is vesicle transport to the hyphal tip (Virag and Griffiths 2004). Liu et al. (2012a) identified a cDNA clone from *Pst* as an actin gene. They observed actin filaments (F-actin) during germ tube formation under a microscope. The most conspicuous actin-containing structures were actin patches, mainly concentrating near the hyphal tips and scattered throughout the cortex. By treating with cytochalasin B, the depolymerization of F-actin was found to greatly reduce the germination rate of urediniospores and disrupt the transport of vesicles to the germ tube tip, indicating that F-actin plays a key role in the tip growth of *Pst* germ tubes.

2.6.4.3 Adenine Nucleotide Translocase *PsANT*

Apoptosis is a form of cell death that plays key roles in fungal development, tissue homeostasis and causing disease (White 1996). Identification of the molecular switches that trigger fungal apoptosis is of paramount importance in generating drug therapies. Mitochondria contribute to apoptosis induction by changing mitochondrial membrane permeability (MMP), and thus an in vitro system for apoptosis induction requires the presence of mitochondria (Newmeyer et al. 1994). An adenine nucleotide translocase (ANT) is a constitutive mitochondrial component that is involved in ADP/ATP exchange and mitochondrion-mediated apoptosis in yeast and mammals.

Tang et al. (2015) identified an ANT gene, designated *PsANT*, from a cDNA library of a compatible wheat-*Pst* interaction. The gene was found to be upregulated during the infection process of *Pst* in host cells. The PsANT protein contains three typical conserved mitochondrion-carrier-protein (mito-carr) domains and shares more than 70% identity with its orthologs from other fungi. Interestingly, different from the multiple ANT members found in other organisms, only one single *ANT* homologue was found in *Pst*. Since the pathogenic fungi mainly depend on their host for nutrient and energy supply, their own energy supply mechanism may have been degenerated during the evolution, and only one ANT member may have remained to supply energy for pathogen growth before the setup of the parasite relationship with its host. Throguh immuno-cytochemical localization experiments, PsANT proteins are found to be concentrated in the mitochondria of infectious hyphae, confirming that PsANT is a member of the ANT family in the mitochondira. More abundant PsANT proteins in germinated urediniospores and invasive hyphae than urediniospores indicated that under the normal condition PsANT may be responsible for supplying energy for the germination of urediniospores and development of the invading *Pst* hyphae through its ATP/ADP translocase activity. Whereas abundant PsANT was also observed in collapsed mitochondria, indicating that *PsANT* is involved in apoptosis-like cell death in *Pst*. Heterologous expression of *PsANT* triggered cell death in tobacco, wheat and fission yeast cells, confirming the cell death promotion role of *PsANT*. Knockdown of *PsANT* using HIGS attenuated the growth and development of virulent *Pst* at the early infection stage but not enough to alter its pathogenicity at the late stage. This may due to that at the late infection stage, haustoria begin to form in great numbers and the expansion of *Pst* mainly depends on the nutrient and energy uptake from host cells via haustoria but not PsANT for energy. This study indicated that *PsANT* contributes to fungal growth and development probably owing to its role of energy supply during the early infection, but is not essential for the pathogen virulence.

2.6.4.4 MAPK Gene *PsMAPK1*

In a variety of eukaryotic organisms, a family of serine/threonine protein kinases known as the mitogen-activated protein kinases (MAPKs) play critical roles in the transduction of a variety of extracellular signals and regulation of various development and differentiation processes (Zhao et al. 2007b). The MAPK cascades are conserved in eukaryotes and have been studied extensively in many organisms. In filamentous fungi, MAPKs mainly fall into three subgroups represented by *Fus3/Kss1*, *Slt2* and *Hog1* of *Saccharomyces cerevisiae* (Kultz 1998; Xu 2000). The *Fus3/Kss1* homolog is more extensively studied than the other two MAPKs in fungal pathogens (Good et al. 2009; Rispail et al. 2009). In the wheat scab fungus *F.*

graminearum, the *map1* deletion mutants are female sterile, non-pathogenic and reduced in conidiation and infectious growth (Jenczmionka et al. 2003; Urban et al. 2003). However, the role of MAPK signal transduction pathways in *Pst* and other rust fungi had not been studied until recently.

Guo et al. (2011) identified a YERK1 MAP kinase homologue, *PsMAPK1*, in *Pst*. The PsMAPK1 protein is more similar to the Fus3/Kss1 orthologues from other basidiomycete than from ascomycete fungi. *PsMAPK1* is most closely related to *Kpp6* of *Ustilago maydis* (Basse and Steinberg 2004; Kahmann and Kamper 2004). Based on the functions of Fus3/Kss1 genes in other fungi (Zhao et al. 2007b; Rispail et al. 2009; Liu et al. 2011a), *PsMAPK1* may play a more important role in the early stages of infection of *Pst* than the other *Fus3/Kss* homologue, which may be involved in stabilization and maintenance of the dikaryotic state. The authors could partially complement the *Magnaporthe oryzae pmk1* mutant by *PsMAPK1* in appressorium formation and plant infection, indicating the functional conservation between *PsMAPK1* and *PMK1*. Expression of *PsMAPK1* was induced at the early infection stage and peaked during haustorium formation, similar to the expression pattern of *PtMAPK1* in *P. triticina*. These observations suggest that the development of highly specialized infection structure such as haustoria in rust fungi is regulated by the well conserved MAPK signaling cascade. Expression of *PsMAPK1* also partially restored the defects of the *F. graminearum map1* mutant in vegetative growth and pathogenicity. The phenotypes of the *map1* and *pmk1* mutants were only partially complemented, indicating that *PsMAPK1* is not fully functional in ascomycetous fungi. This study suggests that *PsMAPK1*may play a role in the regulation of plant penetration and infectious hyphal growth in *Pst*. The authors also demonstrated that *F. graminearum* and *M. oryzae* could be used as surrogate systems for functional analysis of well-conserved *Pst* genes.

2.6.4.5 Protein Kinase Gene *PsSRPKL*

Reversible protein phosphorylation by protein kinases (PKs) and phosphatases is a ubiquitous signaling mechanism in all eukaryotic cells (Miranda-Saavedra and Barton 2007). PKs are generally evolutionarily conserved and proven to be important for pathogenesis in many plant pathogenic fungi (Lee and Kronstad 2002; Zhao et al. 2007b; Nguyen et al. 2008; Wang et al. 2011). Cheng et al. (2015a, b) reported a PK gene, *PsSRPKL* in *Pst*. The gene belongs to a group of PKs that are evolutionarily specific to cereal rust fungi. *PsSRPKL* is highly polymorphic among *Pst* isolates and its proteins are located in nuclei, different from SRPK proteins that are localized in the cytoplasm (Koizumi et al. 1999). The nuclear localization of PsSRPKL may indicate its role in regulating the nuclear localization of SR proteins via phosphorylation. *PsSRPKL* was highly upregulated at 18 hpi prior to haustorium formation. Silencing of *PsSRPKL* using HIGS restricted haustorium formation and development, indicating that *PsSRPKL* contributes to hyphal growth mainly by regulating targets involved in the formation of haustoria (Fig. 2.4). At the late infection stage, a 15–30 fold increase of the *PsSRPKL* transcripts were observed. Along

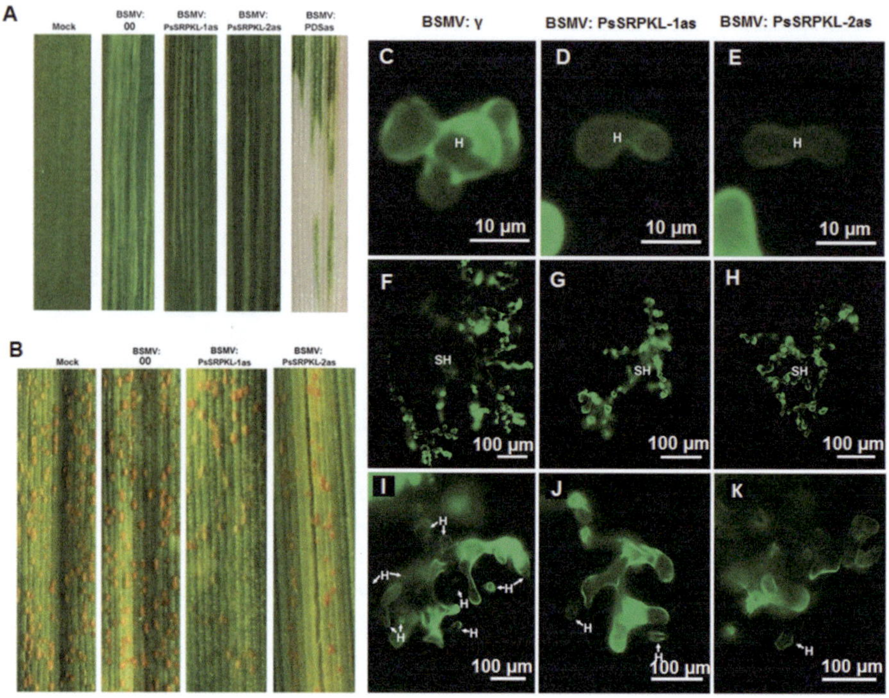

Fig. 2.4 Functional assessment of *PsSRPKL* in the wheat–*Puccinia striiformis* f. sp. *tritici* (*Pst*) interaction by HIGS. (**b**) No change of phenotype was observed in wheat leaves mock-inoculated with FES buffer. Mild chlorotic mosaic symptoms were observed in wheat leaves inoculated with BSMV: γ, BSMV: *PsSRPKL*-1as and *PsSRPKL*-2as. Photobleaching was evident on wheat leaves infected with BSMV: TaPDSas. (**b**) Phenotypes of the fourth leaves of wheat plants inoculated with mock (FES buffer), BSMV: γ, BSMV: PsSRPKL-1as and BSMV: PsSRPKL-2as at 14 dpi with the virulent *Pst* isolate CYR32. (**c–k**) Cytological observation of fungal development in wheat. Leaves inoculated with CYR32 were sampled at 48 and 168 hpi with the CYR32 isolate and examined under an epifluorescence microscope after staining with wheat germ agglutinin conjugated to the fluorophore Alexa-488 (Invitrogen). (**c–e**) Haustoria observed at 48 hpi. (**f–h**) The secondary infection hyphae observed 168 h post inoculation (hpi). (**i–k**) Haustoria observed at 168 hpi. *H* haustorium, *SH* secondary hypha

with the significantly suppressed formation of secondary hyphae and fungal colonization in *PsSRPKL* knockdown plants, *PsSRPKL* appeared to contribute to the further development in late infection stages. Interestingly, *PsSRPKL* was also upregulated during sexual reproduction on barberry, indicating that *PsSRPKL* also plays an important role in barberry infection by regulating haustorial formation and hyphal colonization. In addition, knockdown of *PsSRPKL* by HIGS decreased *Pst* resistance to oxidative stresses. However, overexpression of *PsSRPKL* in fission yeast induces aberrant cell morphology and a decreased resistance to environmental stresses. One likely reason for this difference between *Pst* and *Schizosaccharomyces pombe* may arise from the differences in the origins and levels of H_2O_2 under the experimental conditions. The results indicate that *PsSRPKL* is an important pathogenicity factor responsible for fungal growth and responses to environmental stresses, therefore contributing to *Pst* virulence in wheat. Based on the findings, the authors hypothesized that cereal rust fungi have developed specific PKs as pathogenicity factors for adaptation to their host species.

2.6.4.6 Calcineurin Genes *PsCNA1* and *PsCNB1*

Calcineurin, a serine-threonine-specific calcium/calmodulin dependent protein phosphatase with two subunits (CNA and CNB), regulates a variety of physiological processes, such as growth, morphogenesis, pathogenicity and membrane stress responses through the calcium signaling pathway in eukaryotes (Rusnak and Mertz 2000; Stie and Fox 2008). Calcineurin has been reported to control virulence, hyphal elongation and multiple stress responses in the human pathogens *Candida albicans* and *C. dubliniensis* (Bader et al. 2003; Steinbach et al. 2007; Chen et al. 2011; Juvvadi et al. 2011). Inhibition of calcineurin can decrease fungal growth and arrest tissue invasion (Cowen and Steinbach 2008). This opens possibilities to develop new antifungal agents targeting the calcineurin pathway in fungi (Steinbach et al. 2007).

A few studies reported that the calcium signaling pathway is involved in the initial infection and biotrophic growth of rust fungi (Hu et al. 2007; Zhang et al. 2008). Some homologs involved in calcium signaling, such as CDPK, were identified from *Pst* cDNA libraries (Zhang et al. 2008). Zhang et al. (2012) characterized calcineurin genes *PsCNA1* and *PsCNB1*. Quantitative RT-PCR analyses revealed that both *PsCNA1* and *PsCNB1* expression reached their maximum in the stage of haustorium formation. Suppressed expression of *PsCNA1* and *PsCNB1* in *Pst* using HIGS led to slower extension of fungal hyphae and reduced production of urediniospores. The immune-suppressive drugs cyclosporin A and FK506 markedly reduced the germination rates of urediniospores. These results suggest that the calcineurin signaling pathway participates in stripe rust morphogenetic differentiation, especially the formation of haustoria during the early stage of infection and during the production of urediniospores. Therefore, *PsCNA1* and *PsCNB1* can be considered as important pathogenicity genes involved in the wheat-*Pst* interaction.

2.6.4.7 Two Distinct *Ras* Genes from *Pst*

Ras GTPases are small conserved GTP-binding proteins that act as important regulators of signal transduction pathways (Weeks and Spiegelman 2003). Ras GTPases regulate a variety of cellular processes by interacting with different effector molecules in higher eukaryotes (Mitin et al. 2005). The involvement of Ras genes in the regulation of fungal development and virulence has attracted attention (Lengeler et al. 2000). In several plant pathogenic fungi, such as *M. oryzae*, *U. maydis* and *F. graminearum*, Ras proteins have been reported to act in upstream of MAPK or cyclic adenosine monophosphate-protein kinase A (cAMP-PKA) pathways and appear to play important roles in fungal pathogenesis (Lee and Kronstad 2002; Müller et al. 2003; Park et al. 2006; Bluhm et al. 2007). In addition to regulating fungal pathogenesis, Ras genes have also been shown to be involved in fungal apoptosis (Phillips et al. 2006; Ramsdale 2008). Fungal apoptosis is involved in several important biological processes such as development, aging and stress responses (Phillips et al. 2006; Fröhlich et al. 2007; Fabrizio and Longo 2008; Ramsdale 2008; Sharon et al. 2009).

Based on the available *Pst* genome, Cheng et al. (2016b) identified two Ras genes, *PsRas1* and *PsRas2*, which fall into two distinct phylogenetic clades. Knockdown of *PsRas1* by HIGS caused a significant reduction in the number of haustoria but not in the length of infection hyphae and disease phenotype, indicating that *PsRas1* plays a minor role in rust pathogenicity. However, together with the decreased haustorial number, a decreased hyphal length was clearly evident in *PsRas2*-silenced plants, suggesting that *PsRas2* plays a greater role in *Pst* fungal growth than *PsRas1*. Notably, *PsRas2*-silenced plants exhibited decreased susceptibility to *Pst* infection and formed fewer uredinia, while no significant change in *PsRas1*-silenced plants. Collectively, the results indicate that *PsRas2* but not *PsRas1* significantly contributes to rust pathogenicity. In addition, overexpression of *PsRas1* but not *PsRas2* can trigger cell death in yeast and plants, indicating that *PsRas1* plays an important role in cell death. Cytological observations and molecular characterization showed that *PsRas1*-triggered plant cell death displays similar morphological and molecular characteristics as plant hypersensitive response. Overall, the two distinct *Ras* genes exhibit different functions in the regulation of *Pst* pathogenicity and cell death as *PsRas2* plays a large role in *Pst* fungal growth and virulence, whereas *PsRas1* is likely involved in cell death.

2.6.4.8 Effectors

Effector proteins are key virulence determinants that enhance pathogenesis by suppressing plant defense responses or by means of taking influence on the host's physiology. Verification of effectors may therefore provide new insights into plant defense mechanisms and expand our understanding of a pathogen's adaption to its host. Effectors have been identified in rust fungi *M. lini* and *U. fabae* (Dodds et al. 2004; Catanzariti et al. 2006; Link et al. 2014). Upadhyaya et al. (2014) identified a candidate avirulence effector of *Pgt*. The sequences of various cDNA libraries and whole-genomes of several isolates have provided great opportunities for identifying secreted proteins, many of which could be effectors (Ling et al. 2007; Zhang et al. 2008; Yin et al. 2009; Cantu et al. 2011; Huang et al. 2011; Cantu et al. 2013; Zheng et al. 2013; Cuomo et al. 2017; Xia et al. 2016a, b).

Dong et al. (2011) characterized three putative secreted protein genes, *PstSP2C7*, *PstSP11L10* and *PstSP11P10*, from *Pst* haustoria. The transcripts of these genes were ranged from 769 to 1094 base pairs, encoding predicted proteins of 93 to 203 amino acids without significant similarities to any accessions in the GenBank protein database, but with some homologies to predicted proteins in *Pgt*. *PstSP11L10* had a much higher expression level in infected leaves than in urediniospores and germinated urediospores and therefore should be a good candidate for a gene involved in the wheat-*Pst* interaction.

Gu et al. (2011) reported that the *Ps87* gene from *Pst* encodes a SP that is conserved in diverse fungal pathogens. Ps87 contains an RXLR-like motif and could utilize similar translocation machinery to those of oomycete and fungal pathogens in plant cells. However, no direct suppression activity of plant defense responses was detected for Ps87.

Liu et al. (2016a) identified a *Pst* effector candidate, PEC6. PEC6 was found to compromise pattern-triggered immunity (PTI) in *Nicotiana benthamiana* induced by *Pseudomonas fluorescens* with hampered reactive oxygen species (ROS) accumulation and callose deposition. In Arabidopsis, plants expressing *PEC6* were more susceptible to *Pseudomonas syringae* pv. *tomato* (*Pto*), resulting more bacterial growth. In wheat, PEC6 supressed *P. fluorescens*-elicited PTI as effector-triggered immunity (ETI) by *Pto* was activated. Knocking down of *PEC6* expression by virus mediated HIGS decreased the number of rust pustules, indicating PEC6 as an important pathogenicity factor. A yeast two-hybrid assay showed that PEC6 interacts with both wheat and Arabidopsis adenosine kinases (ADKs). PEC6 and TaADK co-localized and interacted in the nucleus and cytosol. Silencing wheat ADK attenuated leaf growth and enhanced susceptibility to stripe rust, indicating that ADK is important in plant development and defense. ADK has been reported to play essential roles in regulating metabolism, cytokinin interconversion and methyl transfer reactions. Based on these results and facts, the authors proposed a model in which PEC6 may affect one of these processes by targeting ADK to favor fungal growth.

Cantu et al. (2013) re-sequenced genomes of four PST isolates from the US and UK to identify effector candidates and relate them to their distinct virulence profiles using a pipeline integrating genomics, transcriptomics and effector-directed annotation. RNAseq analysis highlighted transcripts encoding SPs that were significantly enriched in haustoria compared to infected tissue. They characterized the expression of 22 candidate effector genes using qRT-PCR. The data revealed distinct temporal expression patterns during infection in wheat. They identified SPs that displayed non-synonymous substitutions specifically between two UK isolates PST-87/7 and PST-08/21, which differ in virulence to two wheat varieties. By focusing on polymorphic variants enriched in haustoria, they identified five polymorphic effector candidates between PST-87/7 and PST-08/21 among 2999 secreted proteins.

Petre et al. (2016) identified and characterized 16 *Pst* candidate effectors which were Pucciniales specific. Nine of the candidate effectors targeted a specific plant subcellular compartment or protein complex. Candidate effector PST02549, accumulated in processing bodies (P-bodies), protein complexes involved in mRNA decapping, degradation and storage. In this study, PST02549 also associated with the P-body-resident enhancer of mRNA decapping protein 4 (EDC4) from *N. benthamiana* and wheat. Co-expression of PST02549 and TaEDC4 increases the size of P-bodies. Thus, the authors proposed that P-bodies are a novel plant cell compartment targeted by pathogen effectors. This study also proved that the pipeline based on an *in planta* protein-protein interaction screening in the heterologous plant *N. benthtlamiana* is a valuable heterologous system for fast-forward effectoromic analysis of plant pathogens. This approach is especially valuable for studying functions of genes in rust fungi.

The availability of *Pst* CYR32 genome sequence has enabled searches for and verification of large sets of effectors. Functional analyzing the 2168 SPs is a challenge. Recently, a study focused on effectors that could suppress programmed cell death (PCD) (Tang et al., unpublished data). Of 150 putative effectors identified based on Cys-rich or up-regulated during plant infection, 14 suppressed PCD that were triggered by *BAX* and INF1 in *N. benthamiana*. The N-terminal regions of the 14 effectors were responsible for the inhibition of PCD although they all have high intra- and inter-species polymorphisms. Delivery of these 14 effectors into wheat plants via a bacterial type-III secretion system efficiently suppressed PTI and ETI, although they individually had only a minor contribution to *Pst* virulence. Interestingly, three of them triggered HR-like PCD in *Yr1* near-isogenic lines (NILs), suggesting that they may function as avirulence genes. Overall, it is proposed that multiple *Pst* effectors function inside plant cells to promote pathogen virulence via suppressing host immunity of both PTI and ETI accumulatively. Effectors are subjects for remarkable temporal and spatial targeting during biotrophic invasion.

From the *Pst* CYR32 genome sequence, Cheng (2016c) identified a candidate effector, PSTha5a23. The *PSTha5a23* gene is *Pst*-specific and shows a low level of polymorphism among *Pst* isolates. The gene protein product has a functional N-terminal signal peptide and is translocated to the host cytoplasm after infection. Overexpression of the gene in tobacco suppressed the PCD triggered by BAX, PAMP-INF1 and two resistance-related mitogen-activated protein kinases (MKK1 and NPK1). Overexpression of *PSTha5a23* in wheat also suppressed PTI-associated callose deposition, and silencing of PSTha5a23 significantly enhanced *Pst* virulence in wheat.

The identification of virulence targets for hundreds of candidate effectors identified by genome sequencing remains a major challenge, partly because functional redundancy seems to be common and because high-throughput cell biological assays are lacking. Stripe rust remains a major food security threat, and incorporation of resistance genes into wheat cultivars is the best way for sustainable control of the disease. High-throughput effectoromics methods for screening the large sets of effector gene candidates available from genome sequencing are already being used to identify new AVR gene–R gene pairs that might be useful to agriculture (Stergiopoulos and de Wit 2009; Vleeshouwers et al. 2011). Understanding *in planta* effector targeting and function is crucial if we are to design durable methods of disease control.

References

Adams TH, Wieser JK, Yu JH. Asexual sporulation in *Aspergillus nidulans*. Microbiol Mol Biol Rev. 1998;62:35–54.

Afshari F. Challenge of a new race of *Puccinia striiformis* f. sp. *tritici* in Iran. In: Yahyaoui A, Rajaram S, editors. Meeting the challenge of yellow rust in cereal crops. Proceedings of the 2nd, 3rd and 4th regional conferences on yellow rust in the Central and West Asia and North Africa (CWANA) Region. ICARDA (International Center for Agricultural Research in the Dry Areas). ISBN: 92-9127-273-6, 39–42. 2012a.

Afshari F. Identification of wheat yellow (stripe) rust pathotypes in Iran. In: Yahyaoui A, Rajaram S, editors. Meeting the challenge of yellow rust in cereal crops. Proceedings of the 2nd, 3rd and 4th Regional Conferences on Yellow Rust in the Central and West Asia and North Africa (CWANA) Region. ICARDA (International Center for Agricultural Research in the Dry Areas), ISBN: 92-9127-273-6, 52–56. 2012b.

Al-Ahmed A, Nachit M. New races of *Puccinia striiformis* f. sp. *tritici* in Syria. In: Poster Index, Borlaug Global Rust Initiative 2014 Technical Workshop 22–25 March, 2014, Universidad La Salle Noroeste Cd. Obregón, Mexico, pp 12. 2014. http://www.globalrust.org/sites/default/files/2014%20BGRI%20Poster%20Abstracts-Monitoring%20the%20Pathogens.pdf. Accessed 24 Sept 2016.

Ali S, Leconte M, Walker AS, Enjalbert J, de Vallavieille-Pope C. Reduction in the sex ability of worldwide clonal populations of *Puccinia striiformis* f.sp. *tritici*. Fungal Genet Biol. 2010;47:828–38.

Ali S, Gladieux P, Leconte M, Gautier A, Justesen AF, Hovmøller MS, Enjalbert J, de Vallavieille-Pope C. Origin, migration routes and worldwide population genetic structure of the wheat yellow rust pathogen *Puccinia striiformis* f.sp. *tritici*. PLoS Pathog. 2014a;10:e1003903.

Ali S, Gladieux P, Rahman H, Saqib MS, Fiaz M, Ahmad H, Leconte M, Gautier A, Justesen AF, Hovmøller MS, Enjalbert J, de Vallavieille-Pope C. Inferring the contribution of sexual reproduction, migration and off-season survival to the temporal maintenance of microbial populations: a case study on the wheat fungal pathogen *Puccinia striiformis* f.sp. *tritici*. Mol Ecol. 2014b;23:603–17.

Allison C, Isenbeck K. Biologische specialisierung von *Puccinia glumarum tritici* Erikss. und Henn. Phytopathologische Zeitschrift. 1930;2:87–98.

Anikster Y The Formae speciales. In: Bushnell RW, Roelfs AP, editors. The cereal rusts, vol. I: Origins, specificity, structure and physiology. New York: Academic; 1984. p. 115–30.

Babcock C, Chen XM, Crous PW, Dugan FM, Goates B, Green PN. Plant germplasm centers and microbial culture collections: A user's guide to key genetic resources for plant pathology. Plant Dis. 2007;91:476–84.

Bader T, Bodendorfer B, Schröppel K, Morschhäuser J. Calcineurin is essential for virulence in *Candida albicans*. Infect Immun. 2003;71:5344–54.

Bahl PN, Bakshi JS. Genetic of rust resistance in barley II. The inheritance of seedling resistance to four races of yellow rust. Indian J Genet Plant Breed. 1963;23:150–4.

Bahri B, Leconte M, de Vallavieille-Pope C, Enjalbert J. Isolation of ten microsatellite loci in an EST library of the phytopathogenic fungus *Puccinia striiformis* f. sp. *tritici*. Conserv Genet. 2009a;10:1425–8.

Bahri B, Leconte M, Ouffroukh A, de Vallavieille-Pope C, Enjalbert J. Geographic limits of a clonal population of wheat yellow rust in the Mediterranean region. Mol Ecol. 2009b;18:4165–79.

Bahri B, Shah SJA, Hussain S, Leconte M, Enjalbert J, de Vallavieille-Pope C. Genetic diversity of the wheat yellow rust population in Pakistan and its relationship with host resistance. Plant Pathol. 2011;60:649–60.

Bailey J, Karaoglu H, Wellings C, Park R. Isolation and characterization of 25 genome-derived simple sequence repeat markers for *Puccinia striiformis* f. sp. *tritici*. Mol Ecol Resour. 2013;13:760–2.

Bakshi JS, Bahl PN. Inheritance of resistance to four races of yellow rust in two varieties of barley. Indian J Genet Plant Breed. 1965;25:239–42.

Bakshi JS, Bahl PN, Kohli SP. Inheritance of seedling resistance to some Indian races of yellow rust in the crosses of rust resistant barley variety E.B. 410. Indian J Genet Plant Breed. 1964;24:72–7.

Bamdadian A. Virulence, evolution and distribution of wheat stripe rust (*Puccinia striiformis* West. f. sp. *tritici*) in Iran. Proc Eur Midditter Cereal Rusts Powdery Mildews Conf. 1992;8:118–20.

Barar GS, Kutcher HR. Race characterization of *Puccinia striiformis* f. sp. *tritici*, the cause of wheat stripe rust, in Saskatchewan and southern Alberta, Canada and virulence comparison with races from the United States. Plant Dis. 2016;100:1744–53.

Barnes CW, Szabo LJ. Detection and identification of four common rust pathogens of cereals and grasses using real-time polymerase chain reaction. Phytopathology. 2007;97:717–27.

Basse CW, Steinberg G. *Ustilago maydis*, model system for analysis of the molecular basis of fungal pathogenicity. Mol Plant Pathol. 2004;5:83–92.

Bayles RA, Flath K, Hovmøller MS, de Vallavieille-Pope C. Breakdown of the *Yr17* resistance to yellow rust of wheat in northern Europe. Agronomie. 2000;20:805–11.

Becker A, Lange M. VIGS–genomics goes functional. Trends Plant Sci. 2009;15:1–4.

Berlin A, Djurle A, Samils B, Yuen J. Genetic variation in *Puccinia graminis* collected from oats, rye, and barberry. Phytopathology. 2012;102:1006–12.

Bever WW. Physiologic specialization in *Puccinia glumerum* in the United States. Phytopathology. 1934;24:686–8.

Bhardwaj SC, Gangwar OP, Prasad P, Khan H. Newsletter "Mehtaensis" (Flowerdale Shimla) of Indian Institute of Wheat and Barley Research. Mehtaensis. 2014;34:1–23. http://dwr.res.in/sites/default/files/Publications/mehtaensisvl34no2.pdf

Bluhm BH, Zhao X, Flaherty JE, Xu JR, Dunkle LD. RAS2 regulates growth and pathogenesis in *Fusarium graminearum*. Mol Plant Microbe Interact. 2007;20:627–36.

Boshoff WHP, Pretprius ZA, van Niekerk BD. Establishment, distribution, and pathogenicity of *Puccinia striiformis* f. sp. *tritici* in South Africa. Plant Dis. 2002;86:485–92.

Both M, Csukai M, Stumpf MPH, Spanu PD. Gene expression profiles of *Blumeria graminis* indicate dynamic changes to primary metabolism during development of an obligate biotrophic pathogen. Plant Cell. 2005a;17:2107–22.

Both M, Eckert SE, Csukai M, Müller E, Dimopoulos G, Spanu PD. Transcript profiles of *Blumeria graminis* development during Infection reveal a cluster of genes that are potential virulence determinants. Mol Plant-Microbe Interact. 2005b;18:125–33.

Breakspear A, Momany M. The first fifty microarray studies in filamentous fungi. Microbiology. 2007;153:7–15.

Broeker K, Bernard F, Moerschbacher BM. An EST library from *Puccinia graminis* f. sp. *tritici* reveals genes potentially involved in fungal differentiation. FEMS Microbiol Lett. 2006;256:273–81.

Brown JKM, Hovmøller MS. Aerial dispersal of pathogens on the global and continental scales and its impact on plant disease. Science. 2002;297:537–41.

Brown JF, Sharp EL. The relative survival ability of pathogenic types of *Puccinia striiformis* in mixtures. Phytopathology. 1970;60:529–33.

Cantu D, Govindarajulu M, Kozik A, Wang MN, Chen XM, Kojima K, Jurka J, Michelmore RW, Dubcovsky J. Next generation sequencing provides rapid access to the genome of wheat stripe rust. PLoS ONE. 2011;8:e24230.

Cantu D, Segovia V, MacLean D, Bayles R, Chen XM, Kamoun S, Dubcovsky J, Saunders DGO, Uauy C. Genome analyses of the wheat yellow (stripe) rust pathogen *Puccinia striiformis* f. sp. *tritici* reveal polymorphic and haustorial expressed secreted proteins as candidate effectors. BMC Genom. 2013;14:270.

Cao LH, Xu SH, Lin RM, Liu TG, Chen WQ. Early molecular diagnosis and detection of *Puccinia striiformis* f. sp. *tritici* in China. Lett Appl Microbiol. 2008;46:501–6.

Catanzariti A-M, Dodds PN, Lawrence GJ, Ayliffe MA, Ellis JG. Haustorially expressed secreted proteins from flax rust are highly enriched for avirulence elicitors. Plant Cell. 2006;18:243–56.

Chen XM. Epidemiology of barley stripe rust and races of *Puccinia striiformis* f. sp. *hordei*: the first decade in the United States. In: Abstracts of the 11th international cereal rusts and powdery mildews conference, 22–27 August, 2004, Norwich, UK, 2004. p. A2.8.

Chen XM. Epidemiology and control of stripe rust *Puccinia striiformis* f. sp. *tritici* on wheat. Can J Plant Pathol. 2005;27:314–37.

Chen XM. Challenges and solutions for stripe rust control in the United States. Aus J Agric Res. 2007;58:648–55.

Chen XM. Races of *Puccinia striiformis* f. sp. *hordei* in the United States from 2004 to 2007. Barley Newsletter 51:WABNL51. 2008. http://wheat.pw.usda.gov/ggpages/BarleyNewsletter/51/WABNL51.htm. Accessed 16 Oct 2016.

Chen XM. High-temperature adult-plant resistance, key for sustainable control of stripe rust. Am J Plant Sci. 2013;4:608–27.

Chen XM. Understanding plant fungal pathogen biology using a genomics approach. In: Conference abstract book of BIT's 4th Annual World Congress of Microbes-2014, June 27–29, 2014, Dalian, China; 2014. p. 116.

Chen XM, Line RF. Genes for resistance to stripe rust in 'Tres' wheat. Crop Sci. 1992a;32:692–6.

Chen XM, Line RF. Inheritance of stripe rust resistance in wheat cultivars used to differentiate races of *Puccinia striiformis* in North America. Phytopathology. 1992b;82:633–7.

Chen XM, Line RF. Identification of stripe rust resistance genes in wheat cultivars used to differentiate North American races of *Puccinia striiformis*. Phytopathology. 1992c;82:1428–34.

Chen XM, Line RF. Races of barley stripe rust in the United States. Barley Newsl 44. 2001. http://wheat.pw.usda.gov/ggpages/BarleyNewsletter/44/WashReport2.html. Accessed 16 Oct 2016.

Chen XM, Moore MK. Epidemics and races of *Puccinia striiformis* in North America in 2001. Phytopathology. 2002;92:S14–5.

Chen XM, Penman L. Races of *Puccinia striiformis* f. sp. *hordei*, the pathogen of barley stripe rust in the United States in 2004. Barley Genet Newsl. 2005;35:23–6. http://wheat.pw.usda.gov/ggpages/bgn/35/Chen.htm (accessed 16 Oct 2016

Chen XM, Penman L. Stripe rust epidemic and races of *Puccinia striiformis* in the United States in 2005. Phytopathology. 2006;96:S23.

Chen XM, Line RF, Leung H. Relationship between virulence variation and DNA polymorphism in *Puccinia striiformis*. Phytopathology. 1993;83:1489–97.

Chen FQ, Prehn D, Hayes PM. Mapping genes for barley stripe rust (*Puccinia striiformis* f. sp. *hordei*). Theor Appl Genet. 1994;88:215–9.

Chen XM, Line RF, Jones SS. Chromosomal location of genes for resistance to *Puccinia striiformis* in winter wheat cultivars Heines VII, Clement, Moro, Tyee, Tres, and Daws. Phytopathology. 1995a;85:1362–7.

Chen XM, Line RF, Leung H. Virulence and polymorphic DNA relationships of *Puccinia striiformis* f. sp. *hordei* to other rusts. Phytopathology. 1995b;85:1335–42.

Chen XM, Jones SS, Line RF. Chromosomal location of genes for resistance to *Puccinia striiformis* in seven wheat cultivars having resistance genes at *Yr3* and *Yr4* loci. Phytopathology. 1996;86:1228–33.

Chen XM, Moore MK, Milus EA, Long DL, Line RF, Marshall D, Jackson L. Wheat stripe rust epidemics and races of *Puccinia striiformis* f. sp. *tritici* in the United States in 2000. Plant Dis. 2002;86:39–46.

Chen XM, Moore MK, Wood DA. Epidemics and control of stripe rust on spring wheat in the Pacific Northwest in 2002. Phytopathology. 2003a;93:S16.

Chen XM, Soria MA, Yan GP, Sun J, Dubcovsky J. Development of sequence tagged site and cleaved amplified polymorphic sequence markers for wheat stripe rust resistance gene *Yr5*. Crop Sci. 2003b;43:2058–64.

Chen XM, Moore MK, Wood DA. Stripe rust epidemics and races of *Puccinia striiformis* in the United States in 2003. Phytopathology. 2004;94:S18.

Chen XM, Penman LN, Richardson KL. Races of *Puccinia striiformis*, the stripe rust pathogen in the United States in 2006. Phytopathology. 2007;97:S21.

Chen CQ, Zheng WM, Buchenauer H, Huang LL, Lu NH, Kang ZS. Isolation of microsatellite loci from expressed sequence tag library of *Puccinia striiformis* f.sp. *tritici*. Mol Ecol Resour. 2009a;9:236–8.

Chen WQ, Wu LR, Liu TG, Xu SC, Jin SL, Peng YL, Wang BT. Race dynamics, diversity, and virulence evolution in *Puccinia striiformis* f. sp. *tritici*, the causal agent of wheat stripe rust in China from 2003 to 2007. Plant Dis. 2009b;93:1093–11.

Chen XM, Penman L, Wan AM, Cheng P. Race change of *Puccinia striiformis* f. sp. *tritici* and *Puccinia striiformis* f. sp. *hordei* in the United States. In: Abstract book of the 12th international cereal rusts and powdery mildews conference, 13–16 October, 2009, Antalya, Turkey, 2009c. p. 27.

Chen XM, Penman L, Wan A, Cheng P. Virulence races of *Puccinia striiformis* f. sp. *tritici* in 2006 and 2007 and development of wheat stripe rust and distributions, dynamics, and evolutionary relationships of races from 2000 to 2007 in the United States. Can J Plant Pathol. 2010;32:315–33.

Chen YL, Brand A, Morrison EL, Silao FGS, Bigol UG, Malbas Jr FF, Nett JE, Andes DR, Solis NV, Filler SG, Averette A, Heitman J. Calcineurin controls drug tolerance, hyphal growth, and virulence in *Candida dubliniensis*. Eukaryot Cell. 2011;10:803–19.

Chen XM, Coram T, Huang XL, Wang MN, Dolezal A. Understanding molecular mechanisms of durable and non-durable resistance to stripe rust in wheat using a transcriptomics approach. Curr Genomics. 2013;14:111–26.

Cheng P, Chen XM. Molecular mapping of a gene for stripe rust resistance in spring wheat cultivar IDO377s. Theor Appl Genet. 2010;121:195–204.

Cheng P, Chen XM. Virulence and molecular analyese support asexual reproduction of *Puccinia striiformis* f. sp. *tritici in the* U.S. Pacific Northwest. Phytopathology. 2014;104:1208–20.

Cheng P, Chen XM, Xu LS, See DR. Development and characterization of expressed sequence tag-derived microsatellite markers for the wheat stripe rust fungus *Puccinia striiformis* f. sp. *tritici*. Mol Ecol Resour. 2012;12:779–81.

Cheng P, Xu LS, Wang MN, See DR, Chen XM. Molecular mapping of genes *Yr64* and *Yr65* for stripe rust resistance in hexaploid derivatives of durum wheat accessions PI 331260 and PI 480016. Theor Appl Genet. 2014;127:2267–77.

Cheng JJ, Kang ZS, Huang LL, Wang MN, Wan AM, Cheng P. Molecular identification of somatic recombination of stripe rust pathogen (*Puccinia striiformis*) in the greenhouse. Mycosystema. 2015a;34:1128–42.

Cheng YL, Wang XJ, Yao JN, Voegele RT, Zhang YR, Wang WM, Huang LL, Kang ZS. Characterization of protein kinase PsSRPKL, a novel pathogenicity factor in the wheat stripe rust fungus. Environ Microbiol. 2015b;17:2601–17.

Cheng P, Chen XM, See D. Grass hosts harbor more diverse isolates of *Puccinia striiformis* than cereal crops. Phytopathology. 2016a;106:362–71.

Cheng YL, Wang WM, Yao JN, Huang LL, Voegele RT, Wang XJ, Kang ZS. Two distinct *Ras* genes from *Puccinia striiformis* exhibit differential roles in rust pathogenicity and cell death. Environ Microbiol. 2016b;18:3910–22.

Cheng YL, Wu K, Yao JN, Li SM, Wang XJ, Huang LL, Kang ZS. PSTha5a23, a candidate effector from the obligate biotrophic pathogen *Puccinia striiformis* f. sp. *tritici*, is involved in plant defense suppression and rust pathogenicity. Environ Microbiol. 2016c. doi:10.1111/1462-2920.13610

Collections APS, Committee G. Is your favorite pathogen collection endangered? Phytopathology. 2005;95:S4.

Coram TE, Brown-Guedira G, Chen XM. Using transcriptomics to understand the wheat genome. CAB Reviews: Perspectives in Agriculture, Veterinary Science, Nutrition and Natural Resource. 2008a;3:1–9.

Coram TE, Settles ML, Chen XM. Transcriptome analysis of high-temperature adult-plant resistance conditioned by *Yr39* during the wheat-*Puccinia striiformis* f. sp. *tritici* interaction. Mol Plant Pathol. 2008b;9:479–93.

Coram TE, Settles ML, Wang MN, Chen XM. Surveying expression level polymorphism and single-feature polymorphism in near-isogenic wheat lines differing for the *Yr5* stripe rust resistance locus. Theor Appl Genet. 2008c;117:401–11.

Coram TE, Wang MN, Chen XM. Transcriptome analysis of the wheat-*Puccinia striiformis* f. sp. *tritici* interaction. Mol Plant Pathol. 2008d;9:157–69.

Coram TE, Huang XL, Zhan GM, Settles ML, Chen XM. Meta-analysis of transcripts associated with race-specific resistance to stripe rust in wheat demonstrates common induction of blue copper binding protein, heat-stress transcription factor, pathogen-induced WIR1A protein, and ent-kaurene synthase transcripts. Funct Integr Genomics. 2010;10:383–92.

Cowen LE, Steinbach WJ. Stress, drugs, and evolution: the role of cellular signaling in fungal drug resistance. Eukaryot Cell. 2008;7:747–64.

Cuomo C, Young S, Wang MN, Yin CT, Hulbert S, Chen XM. Whole genome sequence of *Puccinia striiformis* f. sp. *tritici* and genome size comparisons with *P. graminis* f. sp. *tritici* and *P. triticina*. In: Chen W-Q, editor. Disease risk and food security. Proceedings of the 13th international cereal rusts and powdery mildews conferene, Beijing, China, 28 August – 1 September 2012, China Agricultural Science and Technology Press; 2012. p. 68.

Cuomo CA, Bakkeren G, Khalil HB, Panwar V, Joly D, Linning R, Sakthikumar S, Song X, Adiconis X, Fan L, Goldberg JM, Levin JZ, Young S, Zeng QD, Anikster Y, Bruce M, Wang MN, Yin CT, McCallum B, Szabo LJ, Hulbert S, Chen XM, Fellers JP. Comparative analysis highlights variable genome content of wheat rusts and divergence of the mating loci. 3G: Genes Genomes. Genomics. 2017;7:371–6.

de Vallavieille-Pope C, Line RF. Virulence of North American and European races of *Puccinia striiformis* on North American, world, and European differential wheat cultivars. Plant Dis. 1990;74:739–43.

de Vallavieille-Pope C, Ali S, Leconte M, Enjalbert J, Delos M, Jacques R. Virulence dynamics and regional structuring of *Puccinia striiformis* f. sp. *tritici* in France between 1984 and 2009. Plant Dis. 2012;96:131–40.

Dietz SM, Hendrix JW. Reactions of grasses to stripe rust at Pullman, Washington. Phytopathology. 1962;52:730.

Divon HH, Rothan-Denoyes B, Davydov O, Di Pietro A, Fluhr R. Nitrogen-responsive genes are differentially regulated in planta during *Fusarium oxyspsorum* f. sp. *lycopersici* infection. Mol Plant Pathol. 2005;6:459–70.

Dobzhansky T. On some fundamental concepts of darwinian biology. In: Dobzhansky T, Hecht MK, Steere WC, editors. Evolutionary biology. New York: Appleton-Century-Crofts; 1968. http://link.springer.com/chapter/10.1007%2F978-1-4684-8094-8_1

Dodds PN, Lawrence GJ, Catanzariti A-M, Ayliffe MA, Ellis JG. The *Melampsora lini AvrL567* avirulence genes are expressed in haustoria and their products are recognized inside plant cells. Plant Cell. 2004;16:755–68.

Dong YL, Yin CT, Hulbert S, Chen XM, Kang ZS. Cloning and expression analysis of three secreted protein genes from wheat stripe rust fungus *Puccinia striiformis* f. sp. *tritici*. World J Microbiol Biotechnol. 2011;27:1261–5.

Duan X, Tellier A, Wan A, Leconte M, de Vallavieille-Pope C, Enjalbert J. *Puccinia striiformis* f.sp. *tritici* presents high diversity and recombination in the over-summering zone of Gansu, China. Mycologia. 2010;102:44–53.

Dubin HJ, Stubbs RW. Epidemic spread of barley stripe rust in South America. Plant Dis. 1986;70:141–4.

Duplessis S, Cuomo CA, Lin YC, Aerts A, Tisserant E, Veneault-Fourrey C, et al. Obligate biotrophy features unraveled by the genomic analysis of rust fungi. Proc Nat Acad Sci. 2011;108:9166–71.

Ebbole DJ, Jin Y, Thon M, Pan H, Bhattarai E, Thomas T, Dean R. Gene discovery and gene expression in the rice blast fungus, *Magnaporthe grisea*: analysis of expressed sequence tags. Mol Plant Microbe Interact. 2004;17:1337–47.

Enjalbert J, Duan X, Giraud T, Vaurtin D, de Vallavieille-Pope C, Solignac M. Isolation of 12 microsatellite loci, using an enrichment protocol, in the phytopathogenic fungus *Puccinia striiformis* f. sp. *tritici*. Mol Ecol Note. 2002;2:563–5.

Enjalbert J, Duan X, Leconte M, Hovmøller MS, de Vallavieille-Pope C. Genetic evidence of local adaptation of wheat yellow rust (*Puccinia striiformis* f.sp. *tritici*) within France. Mol Ecol. 2005;14:2065–73.

Eriksson J. Ueber die Specialisierung des Parasitismus bei den Getreiderostpilzen. Ber Dtsch Bot Ges. 1894;12:292–331.

Esfandiari E. Les rouilles de cereales en Iran [Cereal rusts in Iran]. Appl Entomol Phytopathol. 1947;4:67–76.

Fabrizio P, Longo VD. Chronological aging-induced apoptosis in yeast. Biochim Biophys Acta. 2008;1783:1280–5.

Fang CT. Physiological specialization of *Puccinia glumarum* Erikss. and Henn. in China. Phytopathology. 1944;34:1020–2014.

Feng JY, Wang MN, Chen XM, See DR, Zheng YL, Chao SM, Wan AM. Molecular mapping of *YrSP* and its relationship with other genes for stripe rust resistance in wheat chromosome 2BL. Phytopathology. 2015;105:1206–13.

Fraaije BA, Lovell DJ, Coelho JM, Baldwin S, Hollomon DW. PCR-based assays to assess wheat varietal resistance to blotch (*Septoria tritici* and *Stagonospora nodorum*) and rust (*Puccinia striiformis* and *Puccinia recondita*) diseases. Eur J Plant Pathol. 2001;107:905–17.

Fröhlich KU, Fussi H, Ruckenstuhl C. Yeast apoptosis–From genes to pathways. Semin Cancer Biol. 2007;17:112–21.

Fuchs E. Physiologische Rassen bei Gelbrost (*Puccinia glumarum* (Schm) Erikss. & Henn.) auf weizen. Nachrbl Dtsch Pflanzenshutzd Braunschw. 1960;12:49–63.

Galletta BJ, Cooper JA. Actin and endocytosis: mechanisms and phylogeny. Curr Opin Cell Biol. 2009;21:20–7.

Gao L, Yu HX, Shen HM, Li C, Liu TG, Liu B, Kang XH, Chen WQ. Development of SCAR markers and an SYBR green assay to detect *Puccinia striiformis* f. sp. *tritici* in infected wheat leaves. Plant Dis. 2016;100:1840–7.

Garnica DP, Upadhyaya NM, Dodds PN, Rathjen JP. Strategies for wheat stripe rust pathogenicity identified by transcriptome sequencing. PLoS One. 2013;8:e67150.

Gassner G, Straib W. Die Bestimmung der Biologischen Rassen des Weizengelbrostes [*Puccinia glumarum tritici* (Schmidt). Erikss. U. Henn.]. Arb Biol Rechsanst Land Forstwirtsch. 1932;20:141–63.

Gassner G, Straib W. Uber mutationen in einer biologischen rasse von *Puccinia glumarum tritici* (Schmidt) Erikss. und Henn. Mol. Gen. Genet. MGG. 1933;63:154–80.

Gerechter-Amitai ZK, von Silfhout CH, Grama A, Kleitman F. *Yr15* – a new gene for resistance to *Puccinia striiformis* in *Triticium dicoccoided* sel. G-25. Euphytica. 1989;43:187–90.

Glass NL, Donaldson GC. Development of primer sets designed for use with the PCR to amplify conserved genes from filamentous ascomycetes. Appl Environ Microbiol. 1995;61:1323–30.

Goddard MV. The production of a new race, 105E137 of *P. striiformis* in glasshouse experiments. Trans Br Mycol Soc. 1976;67:395–8.

Godfrey D, Böhlenius H, Pedersen C, Zhang Z, Emmersen J, Thordal-Christensen H. Powdery mildew fungal effector candidates share N-terminal Y/F/WxC-motif. BMC Genomics. 2010;11:317.

Good M, Tang G, Singleton J, Remenyi A, Lim WA. The *Ste5* scaffold directs mating signaling by catalytically unlocking the *Fus3* MAP kinase for activation. Cell. 2009;136:1085–97.

Green JC. Physiologic races of wheat stem rust in Canada from 1919 to 1969. Can J Bot. 1971;49:1575–88.

Gu B, Kale SD, Wang Q, Wang D, Pan Q, Cao H, Meng YL, Kang ZS, Tyler BM, Shan WX. Rust secreted protein Ps87 is conserved in diverse fungal pathogens and contains a RXLR-like motif sufficient for translocation into plant cells. PLoS ONE. 2011;6:e27217.

Güldener U, Seong KY, Boddu J, Cho S, Trail F, Xu JR, Adam G, Mewes HW, Muehlbauer GJ, Kistler C. Development of a *Fusarium graminearum* Affymetrix GeneChip for profiling fungal gene expression in vitro and in planta. Fungal Genet Biol. 2006;43:316–25.

Guo J, Dai XW, Xu JR, Wang YL, Bai PF, Liu FR, Duan YH, Zhang H, Huang LL, Kang ZS. Molecular characterization of a Fus3/Kss1 type MAPK from *Puccinia striiformis* f. sp. *tritici*, PsMAPK1. PLoS ONE. 2011;6:e21895.

Guo J, Duan YH, Zhang JS, Shi XX, Chen YY, Zhang H, Huang LL, Kang ZS. A conidiation-related gene is highly expressed at the resting urediniospore stage in *Puccinia striiformis* f. sp. *tritici*. J Basic Microbiol. 2013;53:695–702.

Hahn M, Mendgen K. Characterization of in plant-induced rust genes isolated from a haustorium-specific cDNA library. Mol Plant Microbe Interact. 1997;10:427–37.

Hakim MS, Mamluk OF. Virulence of wheat yellow rust pathogen in Syria and Lebanon. Proc Eur Mediterr Cereal Rusts Powdery Mildew Conf. 1996;9:141.

Hassebrauk K. Nomenklatur, geographische verbreitung und wirtsbereich des gelbrostes, *Puccinia striiformis* West. Mitteilungen aus der Biologischen Bundesanstalt für Land- und Forstwirtschaft, Berlin-Dahlem. 1965;116:1–75.

Hassebrauk K, Röbbelen G. Der Gelbrost *Puccinia striiformis* West. 3. Die Spezialisierung. Mitt Biol Bundesanst Land Forstwirtsch, Berlin-Dahlem. 1974;156:1–150.

Heath IB, Bonham M, Akram A, Gupta GD. The interrelationships of actin and hyphal tip growth in the ascomycete *Geotrichum candidum*. Fungal Genet Biol. 2003;38:85–97.

Holtz MD, Kumar K, Zantinge JL, Xi K. Genetic diversity of *Puccinia striiformis* from cereals in Alberta, Canada. Plant Pathol. 2013;63:415–24.

Hou L, Chen XM, Wang MN, See DR, Chao SM, Bulli P, Jing JX. Mapping a large number of QTL for durable resistance to stripe rust in winter wheat Druchamp using SSR and SNP markers. PLoS ONE. 2015;10:e0126794.

Hovmøller MS. Disease severity and pathotype dynamics of *Puccinia striiformis* f. sp. *tritici* in Denmark. Plant Pathol. 2001;50:181–9.

Hovmøller MS, Henriksen KE. Application of pathogen surveys, disease nurseries and varietal resistance characteristics in a IPM approach for the control of wheat yellow rust. Eur J Plant Pathol. 2008;121:377–85.

Hovmøller MS, Justesen AF. Rates of evolution of avirulence phenotypes and DNA markers in northwest European population of *Puccinia striiformis* f.sp. *tritici*. Mol Ecol. 2007;16:4637–47.

Hovmøller MS, Justesen AF, Brown JKM. Clonality and long-distance migration of *Puccinia striiformis* f.sp. *tritici* in north-west Europe. Plant Pathol. 2002;51:24–32.

Hovmøller MS, Yahyaoui AH, Milus EA, Justesen AF. Rapid global spread of two aggressive strains of a wheat rust fungus. Mol Ecol. 2008;17:3818–26.

Hovmøller MS, Walter S, Justesen AF. Escalating threat of wheat rusts. Science. 2010;329:369.

Hovmøller MS, Sørensen CK, Walter S, Justesen AF. Diversity of *Puccinia striiformis* on cereals and grasses. Annu Rev Phytopathol. 2011;49:197–217.

Hovmøller MS, Walter S, Ali S, Justesen AF, Hansen J, Lassen P, Bayles R, Flath K, de Vallavieille-Pope C. The influence of mutation, recombination and exotic incursions on the recent dynamics of *Puccinia striiformis* in Europe. Page 40. In: Chen WQ, editor. Disease risk and food security, Proceedings of the 13th international cereal rusts and powdery mildews conference. 28 August – 1 September 2012, Beijing. 2012.

Hovmøller MS, Walter S, Bayles RA, Hubbard A, Flath K, Sommerfeldt N, Leconte M, Czembor P, Rodriguez-Algaba J, Thach T, Hansen JG, Lassen P, Justesen AF, Ali S, de Vallavieille-Pope C. Replacement of the European wheat yellow rust population by new races from the centre of diversity in the near-Himalayan region. Plant Pathol. 2016;65:402–11.

Hu GG, Linning R, Kamp A, Joseph C, McCallum B, Banks T, Cloutier S, Butterfield Y, Liu J, Kirkpatrick R, Stott J, Yang G, Smailus D, Jones S, Marra M, Schein J, Pei JM, Westwood T, Bakkeren G. Generation of a wheat leaf rust, *Puccinia triticina*, EST database and microarray from stage-specific cDNA libraries. In: Proceedings of the 11th international cereal rusts and powdery mildew conference 22–27 August 2004, Norwich, England, Abtract. 2004. A1.47

Hu GG, Linning R, Mccallum B, Banks T, Cloutier S, Butterfield Y, Liu J, Kirkpatrick R, Stott J, Yang G, Smailus D, Jones S, Marra M, Schein J, Bakkeren G. Generation of a wheat leaf rust, *Puccinia triticina*, EST database from stage-specific cDNA libraries. Mol Plant Pathol. 2007;8:451–67.

Hu X, Li J, Wang Y, Wang B, Li Q, Kang Z, Yang M, Peng Y, Liu T, Chen W, Xu X. Race composition of *Puccinia striiformis* f. sp. *tritici* in Tibet, China. Plant Dis. 2012;96:1615–20.

Huang LL, Wang XL, Kang ZS, Zhao J. Mutation of pathogenicity induced by ultraviolet radiation in *Puccinia striiformis* f. sp. *tritici* and RAPD analysis of mutants. Mycosystema. 2005;24:400–6.

Huang XL, Chen XM, Coram T, Wang MN, Kang ZS. Gene expression profiling of *Puccinia striiformis* f. sp. *tritici* during development reveals a highly dynamic transcriptome. J Genet Genomics. 2011;38:357–71.

Hubbard A, Lewis CM, Yoshida K, Ramirez-Gonzalez RH, de Vallavieille-Pope C, Thomas J, Kamoun S, Bayles R, Uauy C, Saunders DGO. Field pathogenomics reveals the emergence of a diverse wheat yellow rust population. Genome Biol. 2015;16:23.

Humphrey HB, Hungerford CW, Johnson AG. Stripe rust (*Puccinia glumerum*) of cereals and grasses in the United States. J Agric Res. 1924;29:209–27.

Hungerford CW, Owens CE. Specialized varieties of *Puccinia glumarum* and hosts for variety *tritici*. J Agric Res (Washington, DC). 1923;25:363–401.

Hussain M, Kirmani MAS, Haque E. Agriculture-guided evolution of pathotypes of *Puccinia striiformis* Westend. f. sp. *tritici* in Pakistan. In: Yahyaoui A, Rajaram S, editors. Meeting the challenge of yellow rust in cereal crops. Proceedings of the 2nd, 3rd and 4th regional conferences on yellow rust in the Central and West Asia and North Africa (CWANA) Region. ICARDA (International Center for Agricultural Research in the Dry Areas), ISBN: 92-9127-273-6, 57-63. 2012.

ICARDA. Strategies to Reduce the Emerging Wheat Stripe Rust Disease. 2011. http://www.icarda.org/striperust2014/wp-content/uploads/2014/01/Strategies_to_reduce.pdf

Iida Y, Ohara T, Tsuge T. Identification of genes up-regulated during conidiation of *Fusarium oxysporum* through expressed sequence tag analysis. Fungal Genet Biol. 2006;43:179–89.

Jakupovic M, Heintz M, Reichmann P, Mendgen K, Hahn M. Microarray analysis of expressed sequence tags from haustoria of the rust fungus *Uromyces fabae*. Fungal Genet Biol. 2006;43:8–19.

Jenczmionka NJ, Maier FJ, Losch AP, Schafer W. Mating, conidiation and pathogenicity of *Fusarium graminearum*, the main causal agent of the head blight disease of wheat, are regulated by the MAP kinase GPMK1. Curr Genet. 2003;43:87–95.

Jin Y, Szabo LJ, Carson M. Century-old mystery of *Puccinia striiformis* life history solved with the identification of Berveris as an alternate host. Phytopathology. 2010;100:432–5.

Johnson R, Taylor AJ. Yellow rust of wheat. Cambridge: Plant Breeding Institute, 1975 annual report; 1976. p. 106–9.

Johnson R, Stubbs RW, Fuchs E, Chamberlain NH. Nomenclature for physiologic races of *Puccinia striiformis* infecting wheat. Trans Br Mycol Soc. 1972;58:475–80.

Justesen AF, Ridout CJ, Hovmøller MS. The recent history of *Puccinia striiformis* f. sp. *tritici* in Denmark as revealed by disease incidence and AFLP markers. Plant Pathol. 2002;51:13–23.

Juvvadi PR, Fortwendel JR, Rogg LE, Burns KA, Randell SH, et al. Localization and activity of the calcineurin catalytic and regulatory subunit complex at the septum is essential for hyphal elongation and proper septation in *Aspergillus fumigatus*. Mol Microbiol. 2011;82:1235–59.

Kahmann R, Kamper J. *Ustilago maydis*: how its biology relates to pathogenic development. New Phytol. 2004;164:31–42.

Kajiwara T. Beiträge zur Kenntnis der Physiologischen Spezialisierung der Gerstengelbrostes (*Puccinia striiformis* West. f. sp. *hordei* Erikss.). Nachr Dtsch Pflanzenschutzd Braunschweig. 1964;4:58–9.

Kajiwara T, Ueda I, Iwata Y. Untersuchungen über die Physiologische Spezialisierung des Gerstengebrostes (*Puccinia striiformis* f. sp. *hordei*) in Japan. Phytopathol Z. 1964a;50:313–28.

Kämper J, Kahmann R, Bölker M, Ma LJ, Brefort T, Saville BJ, et al. Insights from the genome of the biotrophic fungal plant pathogen *Ustilago maydis*. Nature. 2006;444:97–101.

Kang ZS, Li ZQ, Shang HS. A new isolate produced by the heterokaryosis of wheat stripe rust. Acta Univ Agric Boreali-Occidentalis. 1993;21:77–9.

Kang ZS, Li ZQ, Shang HS. Nuclear condition of uredinial stage of wheat stripe rust. Acta Phytopathol Sinica. 1994a;24:26–31.

Kang ZS, Li ZQ, Shang HS. On the screening of new heterocaryons of wheat stripe rust and its nuclear dissociation. Acta Phytopathol Sinica. 1994b;24:101–5.

Karki CB. Wheat yellow rust and monitoring virulence of its pathogen, *Puccinia striiformis* f.sp. *tritici*, in Nepal. Regional seminar on microbial research. Kathmandu: Royal Nepal Academy of Science and Technology; 1989.

Karki CB, Sharma S. Wheat disease report 1989–1990. In: Proceedings of the 13th national winter crops seminar. Siddhartha Nagar: NWRP; 1990. p. 189–206.

Karki CB, Sharma S, Duveiller E. Yellow rust of wheat in Nepal: an overview. In: Yahyaoui A, Rajaram S, editors. Meeting the challenge of yellow rust in cereal crops. Proceedings of the 2nd, 3rd and 4th Regional conferences on yellow rust in the Central and West Asia and North Africa (CWANA) Region. ICARDA (International Center for Agricultural Research in the Dry Areas), ISBN: 92-9127-273-6, 13-19. 2012.

Kema GHJ. Resistance in spelta wheat to yellow rust I. Formal analysis and variation for gliadin patterns. Euphytica. 1992;63:207–17.

Koizumi J, Okamoto Y, Onogi H, Mayeda A, Krainer AR, Hagiwara M. The subcellular localization of SF2/ASF is regulated by direct interaction with SR protein kinases (SRPKs). J Biol Chem. 1999;274:11125–31.

Komjáti H, Pasquali M, Hubbard A, Lee D, Bayles R. IGS, SSR and SRAP analysis of *Puccinia striiformis* isolates. In: Proceedings of the 11th international cereal rusts and powdery mildews conference, 22–27 August 2004, John Innes Centre, Norwich, UK. European and Mediterranean Cereal Rust Foundation, Wageningen. Cereal Rusts and Powdery Mildews Bulletin, Abstract A2.33. 2004. Available from http://www.crpmb.org/icrpmc11/abstracts.htm

Kosman E. Measuring diversity: from individuals to populations. Eur J Plant Pathol. 2014;138:467–86.

Kosman E, Dinoor A, Herrmann A, Schachtel GA. Virulence Analysis Tool (VAT). User manual. 2008. Available online http://www.tau.ac.il/lifesci/departments/plant_s/members/kosman/VAT.html

Kultz D. Phylogenetic and functional classification of mitogen- and stress activated protein kinases. J Mol Evol. 1998;46:571–88.

Kumar K, Holtz MD, Xi K, Turkington TK. Virulence of *Puccinia striiformis* on wheat and barley in central Alberta. Can J Plant Pathol. 2012;34:551–61.

Kumar K, Holtz MD, Xi K, Turkington TK. Overwintering potential of the stripe rust pathogen (*Puccinia striiformis*) in central Alberta. Can J Plant Pathol. 2013;35:304–14.

Latch GCM. Stripe rust, *Puccinia striiformis* f.sp. *dactylidis*, on *Dactylis glomerata* in New Zealand. New Zealand J Agric Res. 1976;19:535–6.

Lee N, Kronstad JW. *Ras2* controls morphogenesis, pheromone response, and pathogenicity in the fungal pathogen Ustilago maydis. Eukaryot Cell. 2002;1:954–66.

Lei Y, Wang MN, Wan AM, Xia CJ, See DR, Zhang M, Chen XM. Virulence and molecular characterization of experimental isolates of the stripe rust pathogen (*Puccinia striiformis*) indicate somatic recombination. Phytopathology. 2016;107:329–44.

Lengeler KB, Davidson RC, D'Souza C, Harashima T, Shen WC, Wang P, Pan XW, Waugh M, Heitman J. Signal transduction cascades regulating fungal development and virulence. Microbiol Mol Biol Rev. 2000;64:746–85.

Leppik EE. Gene centres of plants as sources of disease resistance. Annu Rev Phytopathol. 1970;8:324–44.

Li ZQ, Zeng SM. Wheat rusts in China. Beijing: China Agricultural Press; 2002. p. 379.

Li YX, Wang MN, Wan AM, Chen XM. Development of *Puccinia striiformis* f. sp. *tritici* mutants for avirulence characterization. Phytopathology. 2016;S4:20.

Liang JM, Wan Q, Luo Y, Ma ZH. Population genetic structures of *Puccinia striiformis* in Ningxia and Gansu provinces of China. Plant Dis. 2013;97:501–9.

Liang JM, Liu XF, Li Y, Wan Q, Ma ZH, Luo Y. Population genetic structure and the migration of *Puccinia striiformis* f. sp. *tritici* between the Gansu and Sichuan Basin populations of China. Phytopathology. 2016;106:192–201.

Lin F, Chen XM. Molecular mapping of genes for race-specific overall resistance to stripe rust in wheat cultivar Express. Theor Appl Genet. 2008;116:797–806.

Line RF. Stripe rust of wheat and barley in North America: a retrospective historical review. Annual Review of Phytopathology. 2002;40:75–118.

Line RF, Qayoum A. Virulence, aggressiveness, evolution, and distribution of races of *Puccinia striiformis* (the cause of stripe rust of wheat) in North America, 1968–87. US Department of Agriculture, Agriculture Research Services, Technical Bulletin, 1788, 1992. 44 pp.

Line RF, Sharp EL, Powelson RL. A system for differentiating races of *Puccinia striiformis* in the United Sates. Plant Dis Rep. 1970;54:992–4.

Ling P, Wang MN, Chen XM, Campbell KG. Construction and characterization of a full-length cDNA library for the wheat stripe rust pathogen (*Puccinia striiformis* f. sp. *tritici*). BMC Genomics. 2007;8:145–58.

Link T, Seibel C, Voegele RT. Early insights into the genome sequence of *Uromyces fabae*. Front Plant Sci. 2014;5:587.

Little R, Manners JG. Somatic recombination in yellow rust of wheat (*P. striiformis*). I. production and possible origin of 2 new physiologic races. Trans Br Mycol Soc. 1969a;53:251–8.

Little R, Manners JG. Somatic recombination in yellow rust of wheat (*P. striiformis*). I. Germ tube fusion, nuclear number and nuclear size. Trans Br Mycol Soc. 1969b;53:259–67.

Liu M, Hambleton S. Taxonomic study of stripe rust, *Puccinia striiformis* sensu lato, based on molecular and morphological evidence. Fungal Biol. 2010;114:881–99.

Liu Z, Szabo LJ, Bushnell WR. Molecular cloning and analysis of abundant and stage specific mRNAs from *Puccinia graminis*. Mol Plant Microbe Interact. 1993;6:84–91.

Liu WD, Zhou XY, Li GT, Li L, Kong LG, Wang CF, Zhang HF, Xu J-R. Multiple plant surface signals are sensed by different mechanisms in the rice blast fungus for appressorium formation. PLoS Pathog. 2011a;7:e1001261.

Liu X, Huang C, Sun Z, Liang J, Luo Y, Ma Z. Analysis of population structure of *Puccinia striiformis* f. sp. *tritici* in Yunnan Province of China by using AFLP. Eur J Plant Pathol. 2011b;129:43–55.

Liu J, Zhang Q, Chang Q, Zhuang H, Huang LL, Kang ZS. Cloning and characterization of the actin gene from *Puccinia striiformis* f. sp. *tritici*. World J Microbiol Biotechnol. 2012a;28:2331–9.

Liu B, Chen XM, Kang ZS. Gene sequencing reveals heterokaryotic variations and evolutionary mechanisms in *Puccinia striiformis*. Open J Genomics. 2012b;1:1–14. http://rossscience.org/ojgen/articles/2075-9061-1-1.pdf

Liu B, Zhang JQ, Chen WQ, Liu TG, Gao L. Discovery of high temperature tolerant isolates of *Puccinia striiformis* Westend. in China. In: Chen WQ, editor. Disease risk and food security, Proceedings of the 13th international cereal rusts and powdery mildews conference, 26 August – 1 September 2012. Beijing: China Agricultural Science and Technology Press; 2012c. p. 44

Liu CH, Pedersen C, Schultz-Larsen T, Aguilar GB, Madriz-Ordenana K, Hovmøller MS, Thordal-Christensen H. The stripe rust fungal effector PEC6 suppresses pattern- triggered immunity in a host species-independent manner and interacts with adenosine kinases. New Phytologist. 2016a; doi:10.1111/nph.14034.

Liu J, Guan T, Zheng PJ, Chen LY, Yang Y, Huai BY, Li D, Chang Q, Huang LL, Kang ZS. An excellular Zn-only superoxide dismutase from *Puccinia striiformis* confer enhanced resistance to host-derived oxidative stress. Environ Microbiol. 2016b;18:4118–35.

Liu TL, Wan AM, Chen XM. Virulence characterization of *Puccinia striiformis* f. sp. *tritici* in the US for the past 48 years using the *Yr* single-gene differentials. Phytopathology. 2016c;S4:201.

Loladze A, Druml T, Wellings CR. Temperature adaptation in Australasian populations of *Puccinia striiformis* f. sp. *tritici*. Plant Pathol. 2014;63:572–80.

Lorys MMA, Villaréal CL, Claude de Vallavieille-Pope CN. Genetic variability in *Puccinia striiformis* f. sp. *tritici* populations sampled on a local scale during natural epidemics. Appl Environ Microbiol. 2002;68:6138–45.

Lu SI, Fan KF, Shia SM, Mu WT, Kang SL, Yang TM, Wang KN, Lee SP. Studies on stripe rust of wheat. I. Physiologic specialization of *Puccinia glumarum* (Schmidt) Erikss. & Henn. Chin J Plant Pathol. 1956;2:153–66.

Lu LL, Zhang RJ, Wang MN, Wang Y, Jing JX, Li ZQ. Virulent transgenic mutants obtained by particle bombardment transformation of *Puccinia striiformis* f. sp. *tritici* with the GUS gene. Acta Phytopathol Sin. 2009;39:466–75.

Lu NH, Zhan GM, Chen XM, Wang JF, Huang LL, Kang ZS. Spatial genetic diversity and inter-regional spread of *Puccinia striiformis* f. sp. *tritici* in the Northwest China. Eur J Plant Pathol. 2011;131:685–93.

Luo H, Wang X, Zhan G, Wei G, Zhou X, Zhao J, Huang LL, Kang ZS. Genome-wide analysis of simple sequence repeats and efficient development of polymorphic SSR markers based on whole genome re-sequencing of multiple isolates of the wheat stripe rust fungus. PLoS ONE. 2015;10(6):e0130362.

Lupton FGH, Macer RCF. Inheritance of resistance to yellow rust (*Puccinia glumarum* Erikss. & Henn.) in seven varieties of wheat. Trans Br Mycol Soc. 1962;45:21–45.

Ma Q, Kang ZS, Li ZQ. The fusion of urediospore germ tubes in *P. striiformis* West. on wheat leaves. Acta Univ Agric Boreali-Occidentalis. 1993;21:97–8.

Ma LJ, Kong XY, Qiao JX, An F, Hu XP, Xu XM. Overwintering of *Puccinia striiformis* f. *tritici* on winter wheat at varying altitudes in Gansu and Qinghai provinces. Plant Dis. 2016;100:1138–45.

Macer RCF. The formal and monosomic genetic analysis of stripe rust (*Puccinia striiformis*) resistance in wheat. In: MacKey J (ed) Proc 2nd Intl Wheat Genet Symp, Lund, Sweden 1963. Hereditas Suppl. 1966;2:127–42.

Macer RCF. The resistance of cereals to yellow rust and its exploitation in plant breeding. Proc Royal Soc London. 1972;181:281–301.

MacKenzie DR. Estimating parasitic fitness. Phytopathology. 1978;68:9–13.

Madariaga R, Mellado M, Becerra V. Significance of wheat yellow rust (*Yr*) genes in Chile. In: Programme and Abstracts of 11th international cereal rusts powdery mildews conference, 22–27 August, 2004, Norwich: John Innes Center; 2004. p. A2.38

Mamluk OH, El-Naimi M. Occurrence and virulence of wheat yellow rust in Syria. Proc Eur Medditter Cereal Rusts and Mildews Conf. 1992;8:115–7.

Manners J. *Puccinia striiformis* Westend. var. *dactylidis* var. nov. Trans Br Mycol Soc. 1960;43:65–8.

Manners J. *Puccinia striiformis*, yellow rust (stripe rust) of cereals and grasses. Adv Plant Pathol. 1988;6:373–87.

Manninger K. Virulence survey for wheat rusts in Hungary during 2000–2003. In: Proceedings of the 11th international cereal rusts and powdery mildews conference, 22–27 August 2004. Norwich: John Innes Centre; 2004. p. A2.41

Markell SG, Milus EA. Emergence of a novel population of *Puccinia striiformis* f. sp. *tritici* in eastern United States. Phytopathology. 2008;98:632–9.

Markell SG, Milus EA, Chen XM. Genetic diversity of *Puccinia striiformis* f. sp. *tritici* in the United States. In: Proceedings of the 11th international cereal rusts and powdery mildews conference 22–27 August 2004. Norwich: John Innes Centre; 2004. p. A2.43.

Marshall D, Sutton RL. Epidemiology of stripe rust, virulence of *Puccinia striiformis* f. sp. *hordei*, and yield loss in barley. Plant Dis. 1995;79:732–7.

Mboup M, Leconte M, Gautier A, Wan AM, Chen W, de Vallavieille-Pope C, Enjalbert J. Evidence of genetic recombination in wheat yellow rust populations of a Chinese oversummering area. Fungal Genet Biol. 2009;46:299–307.

McDonald BA, Linde C. Pathogen population genetics, evolutionary potential, and durable resistance. Annu Rev Phytopathol. 2002;40:349–79.

McFadden HG, Wilson IW, Chapple RM, Dowd C. Fusarium wilt (*Fusarium oxysporum* f. sp. *vasinfectum*) genes expressed during infection of cotton (*Gossypium hirsutum*). Mol Plant Pathol. 2006;7:87–101.

McIntosh RA, Wellings CR, Park RF. Wheat rusts: an atlas of resistance genes. Melbourne: CSIRO Publications; 1995. 200 pp.

McIntosh RA, Yamazaki Y, Dubcovsky J, Rogers J, Morris C, Appels R, Xia XC. Catalogue of gene symbols for wheat. 12th international wheat genet symposium, 8–13 September 2013, Yokohama, Japan. 2013a. Online: http://www.shigen.nig.ac.jp/wheat/komugi/genes/download.jsp. Accessed 29th Dec 2016.

McIntosh RA, Dubcovsky J, Rogers WJ, Morris C, Appels R, Xia XC. Catalogue of gene symbols for wheat: 2013–2014 supplement. 2014.

Mehta KC. Rusts of wheat and barley in India. Indian J Agric Sci. 1933;3:939–62.

Melching JS, Stanton JR, Koogle DL. Deleterious effects of tobacco smoke on germination and infectivity of spores of *Puccinia graminis tritici* and on germination of spores of *Puccinia striiformis*, *Pyricularia oryzae* and an *Alternaria* species. Phytopathology. 1974;64:1143–7.

Mendgen K, Hahn M. Plant infection and the establishment of fungal biotrophy. Trends Plant Sci. 2002;7:352–6.

Mendgen K, Struck C, Voegele RT, Hahn M. Biotrophy and rust haustoria. Physiol Mol Plant Pathol. 2000;56:141–5.

Milus EA, Seyran E, McNew R. Aggressiveness of *Puccinia striiformis* f. sp. *tritici* isolates in the southcentral United States. Plant Dis. 2006;90:847–52.

Milus EA, Kristensen K, Hovmøller MS. Evidence for increased aggressiveness in a recent widespread strain of *Puccinia striiformis* f. sp. *tritici* causing stripe rust of wheat. Phytopathology. 2009;99:89–94.

Milus EA, Moon DE, Lee KD, Mason RE. Race-specific adult-plant resistance in winter wheat to stripe rust and characterization of pathogen virulence patterns. Phytopathology. 2015;105:1114–22.

Miranda-Saavedra D, Barton GJ. Classification and functional annotation of eukaryotic protein kinases. Proteins. 2007;68:893–914.

Mitin N, Rossman KL, Der CJ. Signaling interplay in Ras superfamily function. Curr Biol. 2005;15:R563–74.

Müller P, Katzenberger JD, Loubradou G, Kahmann R. Guanyl nucleotide exchange factor Sql2 and Ras2 regulate filamentous growth in Ustilago maydis. Eukaryotic Cell. 2003;2:609–17.

Murphy LR, Santra D, Kidwell K, Yan GP, Chen XM, Campbell KG. Linkage maps of wheat stripe rust resistance genes *Yr5* and *Yr15* for use in marker assisted selection. Crop Sci. 2009;49:1786–90.

Nagarajan S. Annual report 1983. Indian Agric Res Inst, Regional Station, Flowerdale, Simla. 1983.

Nagarajan S, Singh DV. Long-distance dispersion of rust pathogens. Annu Rev Phytopathol. 1990;28:139–53.

Nagarajan S, Bahdur P, Nayar SK. Occurrence of a new virulence, 47S102 of *Puccinia striiformis* West., in India during the crop year 1982. Cereal Rusts and Powdery Mildews Bulletin. 1984;12:28–31.

Nagarajan S, Nayar SK, Bahadur P. Race 13 (67S8) of *Puccinia striiformis* virulent on *Triticum spelta* var. *album* in India. Plant Dis. 1986;70:173.

Nelson RR, Wilcoxon RD, Christensen JJ. Heterokaryosis as basis for variation in *Puccinia graminis* var. *tritici*. Phytopathology. 1955;45:639–43.

Newmeyer DD, Farschon DM, Reed JC. Cell-free apoptosis in *Xenopus* egg extracts: inhibition by Bcl-2 and requirement for an organelle fraction enriched in mitochondria. Cell. 1994;79:353–64.

Newton M, Johnson T. Stripe rust, Puccinia glumarum, in Canada. Can J Res. 1936;14:89–109.

Newton M, Johnson T, Brown AM. Stripe rust in Canada. Phytopathology. 1933;23:25–6.

Newton AC, Caten CE, Johnson R. Variation for isozymes and double-stranded RNA among isolates of *Puccinia striiformis* and two other cereal rusts. Plant Pathol. 1985;34:235–47.

Newton AC, Johnson R, Caten CE. Attempted somatic hybridization of *P. striiformis* f. sp. *tritici* and *P. striiformis* f. sp. *hordei*. Plant Pathol. 1986;35:108–13.

Nguyen QB, Kadotani N, Kasahara S, Tosa Y, Mayama S, Nakayashiki H. Systematic functional analysis of calcium-signaling proteins in the genome of the rice-blast fungus, *Magnaporthe oryzae*, using a high-throughput RNA-silencing system. Mol Microbiol. 2008;68:1348–65.

Niu YC, Li ZQ, Shuang HS. On the physiological specialization of barley stripe rust in China. Acta Phytopathol Sinica. 1994;24:285–8.

O'Brien L, Brown JS, Young RM, Pascoe I. Occurrence and distribution of wheat stripe rust in Victoria and susceptibility of commercial wheat cultivars. Aus Plant Pathol. 1980;9:14.

Oort AJP. Verspreiding en verwantshap van physiologische rassen van gele roest (*Puccinia glumarum*) van tarwe in Europa. Tijidschr PlZiekt. 1955;61:202–19.
Orr HA. Fitness and its role in evolutionary genetics. Nat Rev Genet. 2009;10:531–9.
Pan JJ, Luo Y, Huang C, Sun ZY, Zhao L, Yan JH, Ma ZH. Application of real-time PCR to establish the quantitative detection of latent infection of wheat stripe rust quantity method. Acta Phytopathol Sin. 2010;40:504–10.
Park G, Xue C, Zhao X, Kim Y, Orbach M, Xu JR. Multiple upstream signals converge on the adaptor protein Mst50 in *Magnaporthe grisea*. Plant Cell. 2006;18:2822–35.
Peng H, Cheng HY, Yu XW, Shi QH, Zhang H, Li JD, Ma H. Molecular analysis of an actin gene, *CarACT1*, from chickpea (*Cicer arietinum* L.). Mol Biol Rep. 2010;37:1081–8.
Perveen A, Qaiser M. Pollen flora of Pakistan–LXV. Berberidaceae. Pak J Bot. 2010;42:1–6.
Petre B, Saunders DGO, Sklenar J, Lorrain C, Krasileva KV, Win J, Duplessis S, Kamoun S. Heterologous expression screens in *Nicotiana benthamiana* identify a candidate effector of the wheat yellow rust pathogen that associates with processing bodies. PLoS ONE. 2016;11:e0149035.
Phillips AJ, Crowe JD, Ramsdale M. Ras pathway signaling accelerates programmed cell death in the pathogenic fungus *Candida albicans*. Proc Natl Acad Sci USA. 2006;103:726–31.
Posada-Buitrago ML, Frederick RD. Expressed sequence tag analysis of the soybean rust pathogen *Phakopsora pachyrhizi*. Fungal Genet Biol. 2005;42:949–62.
Poulin R. Evolutionaryk. Princeton: Princeton University Press; 2007.
Prasada R, Lele VC. New physiologic races of wheat rusts in India. Indian Phytopathol. 1952;5:128–9.
Prasada R, Joshi LM, Singh SD, Misra DF, Goel LB, Kumari K, Sharma SK, Joshi PC, Ahmad ST. Occurrence of physiologic races of wheat and barley rusts in India during 1962–64 and their sources of resistance. Indian J Agric Sci. 1967;37:273–81.
Pretorius ZA, Pakendorf KW, Marais GF, Prins R, Komen JS. Challenges for sustainable rust control in South Africa. Aus J Agric Res. 2007;58:593–601.
Priestley RH, Doling DA. Aggressiveness of *Puccinia striiformis* isolates on wheat cultivars. Trans Br Mycol Soc. 1974;63:549–57.
Priestley RH, Doodson JK. Physiological specialization of *Puccinia striiformis* to adult plants of winter wheat cultivars in the United Kingdom. In: Proceedings of the fourth European and Mediterranean cereal rusts conference. Interlaken, Switzerland. 1976. p. 87–9.
Qiang ZF, Hou SY, Zhang G. Measurements of parasitic fitness for *Puccinia striiformis* f. sp. *tritici*. J Shangdong Agric University. 1999;30:94–6. (in Chinese)
Ramsdale M. Programmed cell death in pathogenic fungi. Biochim Biophys Acta. 2008;1783:1369–80.
Rapilly F. Yellow rust epidemiology. Annu Rev Phytopathol. 1979;17:59–73.
Ray R, Gururaja KV, Ramchandra TV. Predictive distribution modeling for rare Himalayan medicinal plant *Berberis aristata* DC. J Environ Biol. 2011;32:725–30.
Rispail N, Soanes DM, Ant C, Czajkowski R, Grünler A, Huguet R, Perez-Nadales E, Poli A, Sartorel E, Valiante V, Yang M, Beffa R, Brakhage AA, Gow NAR, Kahmann R, Lebrun M-H, Lenasi H, Perez-Martin J, Talbot NJ, Wendland J, Di Pietro A. Comparative genomics of MAP kinase and calcium-calcineurin signaling components in plant and human pathogenic fungi. Fungal Genet Biol. 2009;46:287–98.
Rodríguez-García MF, Huerta-Espino J, Villaseñor-Mir HE, Sandoval Islas JS, Singh RP. Analysis of wheat (*Triticum aestivum* L.) yellow rust (*Puccinia striiformis* f. sp. *tritici*) virulence in the high valleys of Mexico. Agrociencia. 2010;44:491–502.
Roelfs AP. Race specificity and methods of study. In: Bushnell RW, Roelfs AP, editors. The cereal rusts, vol. I: Origins, specificity, structure and physiology. New York: Academic; 1984. p. 131–64.
Roose-Amsaleg C, de Vallavieille-Pope C, Brygoo Y, Levis C. Characterisation of a length polymorphism in the two intergenic spacers of ribosomal RNA in *Puccinia striiformis* Westend. f. sp. *tritici*, the causal agent of wheat yellow rust. Mycological Research. 2002;106:918–24.

Rosewarne GM, Herrera-Foessel SA, Singh RP, Huerta-Espino J, Lan CX, He ZH. Quantitative trait loci of stripe rust resistance in wheat. Theor Appl Genet. 2013;126:2427–49.

Rusnak F, Mertz P. Calcineurin: form and function. Physiol Rev. 2000;80:1483–521.

Sacrisan S, Garcia-Arenal F. The evolution of virulence and pathogenicity in plant pathogen populations. Mol Plant Pathol. 2008;9:369–48.

Safavi SA, Babai-Ahari A, Afshari F, Arzanlou M. Effect of yellow rust on yield components of barley cultivars with race-specific and slow rusting resistance to yellow rust. Arch Phytopathol Plant Prot. 2012;45:1488–98.

Safavi SA, Jahani-Jelodar Y, Ebrahimnejad S. Virulence patterns of barley yellow rust and effective resistance genes to *Puccinia striiformis* in three parts of Iran. Arch Phytopathol Plant Prot. 2013;46:903–10.

Saiki RK, Scharf S, Faloona F, Mullis KB, Erlich HA, Arnheim N. Enzymatic amplification of beta-globin genomic sequences and restriction site analysis for diagnosis of sickle cell anemia. Science. 1985;230:1350–4.

Schachtel GA, Dinoor A, Herrmann A, Kosman E. Comprehensive evaluation of virulence and resistance data: A new analysis tool. Plant Dis. 2012;96:1060–3.

Schirawski J, Mannhaupt G, Münch K, Brefort T, Schipper K, Doehlemann G, Di Stasio M, Rössel N, Mendoza-Mendoza A, Pester D, Müller O, Winterberg B, Meyer E, Ghareeb H, Wollenberg T, Münsterkötter M, Wong P, Walter M, Stukenbrock E, Güldener U, Kahmann R. Pathogenicity determinants in smut fungi revealed by genome comparison. Science. 2010;330:1546–8.

Shahin A, Aly AAA. Wheat stripe and stem rust situation in Egypt: *Yr27* and *Sr31* virulence. Abstract of 2015 Borlaug Global Rust Initiative Technical Workshop. 2015. http://www.globalrust.org/content/wheat-stripe-and-stem-rust-situation-egypt-yr27-and-sr31-virulence. Accessed 13 Nov 2016.

Shahin AA, Youssef WA, Abu Aly AA, Ashmawy MA. Virulence and diversity of the stripe rust pathogen in Egypt, 2009 to 2013. In: Poster Index, Borlaug Global Rust Initiative 2014 Technical Workshop 22–25 March, 2014, Universidad La Salle Noroeste Cd. Obregón, Mexico. 2014. p. 8. http://www.globalrust.org/sites/default/files/2014%20BGRI%20Poster%20Abstracts-Monitoring%20the%20Pathogens.pdf. Accessed 24 Sept 2016.

Shan WX, Chen SY, Zhang GY, Wu LR, Li ZQ. Identification of a family of dispersed, moderately repetitive DNA sequences in an obligate fungal plant pathogen *Puccinia striiformis*. Sci China (Ser B). 1997;27:241–6.

Shan WX, Chen SY, Kang ZS, Wu LR, Li ZQ. Genetic diversity in *Puccinia striiformis* Westend. f. sp. *tritici* revealed by pathogen genome-specific repetitive sequence. Can J Bot. 1998;76:587–95.

Shang HS, Jing JX, Li ZQ. Mutations induced by ultraviolet radiation affecting virulence in *Puccinia striiformis*. Acta Phytopathol Sin. 1994;25:347–51.

Sharma S. Virulence monitoring and detection of leaf and yellow rust resistance genes in Nepalese wheat varieties. Flowerdale: DWR Regional Station; 1997. 25 pp.

Sharma SK, Singh S, Goel LB. Note on a new record of Sonalika infecting race of yellow rust in India and sources of its resistance. Indian J Agric Sci. 1973;43:964–5.

Sharma-Poudyal D, Chen XM, See D. Genetic diversity of international collections of *Puccinia striiformis* f. sp. *tritici*. Phytopathology. 2012;102(Suppl 4):S4.30.

Sharma-Poudyal D, Chen XM, Wan AM, Zhan GM, Kang ZS, Cao SQ, Jin SL, Morgounov A, Akin B, Mert Z, Shah SJA, Bux H, Ashraf M, Sharma RC, Madariaga R, Puri KD, Wellings CR, Xi KQ, Manninger K, Wanyera R, Ganzalez MI, Koyda M, Sanin S, Patzek LJ. Viruelnce characterization of international collections of the wheat stripe rust pathogen, *Puccinia striiformis* f. sp. *tritici*. Plant Dis. 2013;97:379–86.

Sharma-Poudyal D, Chen XM, Rupp R. Potential oversummering and overwintering regions for the wheat stripe rust pathogen in the contiguous United States. Int J Biometeorol. 2014;58:987–97.

Sharon A, Finkelstein A, Shlezinger N, Hatam I. Fungal apoptosis: function, genes and gene function. FEMS Microbiol Rev. 2009;33:833–54.

Sharp EL. Atmospheric ions and germination of urediospores of *Puccinia striiformis*. Science. 1967;156:1359–60.

Shrode LB, Lewis ZA, White LD, Bell-Pedersen D, Ebbole DJ. *vvd* is required for light adaptation of conidiation-specific genes of *Neurospora crassa*, but not circadian conidiation. Fungal Genet Biol. 2001;32:169–81.
Shurtleff MC, Averre III CW. Glossary of plant-pathological terms. St. Paul: APS Press; 1997. 361 pp.
Sørensen CK, Thach T, Hovmøller MS. Evaluation of spray and point inoculation methods for the phenotyping of *Puccinia striiformis* on wheat. Plant Dis. 2016;100:1064–70.
Spanu PD. The genomics of obligate (and nonobligate) biotrophs. Annu Rev Phytopathol. 2012;50:91–109.
Steele KA, Humphreys E, Wellings CR, Dickinson MJ. Support of s stepwise mutation model for pathogen evolution in Asutralasin *Puccinia striiformis* f. sp. *tritici* by use of molecular markers. Plant Pathol. 2001;50:174–80.
Steinbach WJ, Reedy JL, Cramer Jr RA, Perfect JR, Heitman J. Harnessing calcineurin as a novel anti-infective agent against invasive fungal infections. Nat Rev Microbiol. 2007;5:418–30.
Stergiopoulos I, de Wit PJGM. Fungal effector proteins. Annu Rev Phytopathol. 2009;47:233–63.
Stie J, Fox D. Calcineurin regulation in fungi and beyond. Eukaryot Cell. 2008;7:177–86.
Straib W. Auftreten und Verbreitung biologischer Rassen des Gelbrostes (*Puccinia glumarum* (Schm.) Erikss. et Henn.) im Jahre 1934. Arb Biol Reichsanst Land= Forstwirtsch, Berlin-Dahlem. 1935;21:455–66.
Stubbs RW. Artificial mutation in the study of the relationship between races of yellow rust of wheat. Proc Eur Mediterr Cereal Rusts Conf. 1968;1968:60–2.
Stubbs RW. Stripe rust. In: Roelfs AP, Bushnell WR, editors. The cereal rusts, Disease, distribution, epidemiology, and control, vol. II. New York: Academic; 1985. p. 61–101.
Stubbs RW. Pathogenicity analysis of yellow (stripe) rust of wheat and its significance in a global context. In: Simmonds NW, Rajaram S, editors. Breeding strategies for resistance to the rust of wheat. Mexico: CIMMYT; 1988. p. 23–38.
Stubbs RW, Fuchs E, Vecht H, Basset EJW. The international survey of factors of virulence of *Puccinia striiformis* Westend. in 1969, 1970 and 1971. Technische Bericht Nr. 21, Wageningen: Nederlands Graan-Centrum; 1974.
Sui XX, Wang MN, Chen XM. Molecular mapping of a stripe rust resistance gens in spring wheat cultivar Zak. Phytopathology. 2009;99:1209–15.
Sun P, Zeng SM. Estimation of relative parstitic fitness of pathogens in herterogeneous host populations. Phytopathology. 1995;85:520–1.
Szabo LJ, Bushnell WR, Hidden robbers. The role of fungal haustoria in parasitism of plants. Proc Natl Acad Sci USA. 2001;98:7654–5.
Takamatsu S. PCR applications in fungal phylogeny. In: Bridge PD, Arora DK, Reddy CA, Elander RP, editors. Applications of PCR in mycology. Wallingford: CAB International; 1998. p. 125–52.
Tang CL, Wei JP, Han QM, Liu R, Duan XY, Fu YP, Huang XL, Wang XJ, Kang ZS. *PsANT*, the adenine nucleotide translocase of *Puccinia striiformis*, promotes cell death and fungal growth. Sci Rep. 2015;5:11241.
Taylor EC. The production of behavior of somatic recombinants in Puccinia striiformis. Proc Eur Medit Cereal Rust Conf in Terlaken. 1976;4:36–8.
Thach T, Ali S, Justesen AF, Rodriguez-Algaba J, Hovmøller MS. Recovery and virulence phenotyping of the historic 'Stubbs collection' of the yellow rust fungus *Puccinia striiformis* from wheat. Ann Appl Biol. 2015;167:314–26.
Thach T, Ali S, de Vallavieille-Pope C, Justesen AF, Hovmøller MS. Worldwide population structure of the wheat rust fungus *Puccinia striiformis* in the past. Fungal Genet Biol. 2016;87:1–8.
Thara VK, Fellers JP, Zhou JM. In planta induced genes of *Puccinia triticina*. Mol Plant Pathol. 2003;4:51–6.
Tian HM, Liu TG, Gao L, Chen WQ. Parasitic fitness of four prevalent races of *Puccinia striiformis* f. sp. *tritici* in China. Acta Phytopathol Sin. 2008;38:599–606.

Tian Y, Zhan GM, Chen XM, Tungruentragoon A, Lu X, Zhao J, Huang LL, Kang ZS. Virulence and SSR marker segregation in a *Puccinia striiformis* f. sp. *tritici* population produced by selfing a Chinese isolate on *Berberis shensiana*. Phytopathology. 2016;106:185–91.

Tooley PW, Fry WE. Field assessments of fitness of isolates of *Phytophthora infestans*. Phytopathology. 1985;75:982–8.

Torto TA, Li S, Styer A, Huitema E, Testa A, Gow NA, van West P, Kamoun S. EST mining and functional expression assays identify extracellular effector proteins from the plant pathogen *Phytophthora*. Genome Res. 2003;13:1675–85.

Tran VA, Kutcher HR. Temperature effects on the aggressiveness of *Puccinia striiformis* f. sp. *tritici*, stripe rust of wheat. Proceedings of on-line conference proceedings of soils and crops, University of Saskatchewan. 2015. http://www.usask.ca/soilsncrops/conference-proceedings/2015%20pdf/day-1-presentations/room-1-008-tran.pdf

Upadhyaya NM, Mago R, Staskawicz BJ, Ayliffe MA, Ellis JG, Dodds PN. A bacterial type III secretion assay for delivery of fungal effector proteins into wheat. Mol Plant Microbe Interact. 2014;27:255–64.

Urban M, Mott E, Farley T, Hammond-Kosack K. The *Fusariumgraminearum MAP1* gene is essential for pathogenicity and development of perithecia. Mol Plant Path. 2003;4:347–59.

Vidhyasekara P. Concise encyclopedia of plant pathology. New York: Food Products Press; 2004. 619 pp.

Virag A, Griffiths AJF. A mutation in the *Neurospora crassa* actin gene results in multiple defects in tip growth and branching. Fungal Genet Biol. 2004;41:213–25.

Virag A, Harris SD. The Spitzenkörper: a molecular perspective. Mycol Res. 2006;110:4–13.

Visser B, Herselman L, Pretorius ZA. Microsatellite characterisation of South African *Puccinia striiformis* races. South African J Plant Soil. 2016;33:163–6.

Vleeshouwers VGAA, Raffaele S, Vossen JH, Champouret N, Oliva R, Segretin ME, Rietman H, Cano LM, Lokossou A, Kessel G, Pel MA, Kamoun S. Understanding and exploiting late blight resistance in the age of effectors. Annu Rev Phytopathol. 2011;49:507–31.

Voegele RT, Struck C, Hahn M, Mendgen K. The role of haustoria in sugar supply during infection of broad bean by the rust fungus *Uromyces fabae*. Proc Natl Acad Sci USA. 2001;98:8133–8.

Volin RB, Sharp EL. Physiologic specialization and pathogen aggressiveness in stripe rust. Phytopathology. 1973;63:699–703.

Vos P, Hogers R, Bleeker M, Reijans M, van de Lee T, Hornes M, Friters A, Pot J, Paleman J, Kuiper M, Zabeau M. AFLP: a new technique for DNA fingerprinting. Nucleic Acids Res. 1995;23:4407–14.

Wahl R, Wippel K, Goos S, Kämper J, Sauer N. A novel high-affinity sucrose transporter is required for virulence of the plant pathogen *Ustilago maydis*. PloS Biol. 2010;8(2):e1000303. doi:10.1371/journal.pbio.1000303.

Wan AM, Chen XM (2011) Races and virulences of *Puccinia striiformis* in the United States in 2010. In: Oral Presentations, Poster Abstracts, Participants and Program of BGRI Technical Workshop, June 13-16, 2011, St. Paul, MN, USA, pp 165

Wan AM, Chen XM. Virulence, frequency, and distribution of races of *Puccinia striiformis* f. sp. *tritici* and *P. striiformis* f. sp. *hordei* identified in the United States in 2008 and 2009. Plant Dis. 2012a;96:67–74.

Wan AM, Chen XM. Stripe rust epidemics of wheat and barley and races of *Puccinia striiformis* identified in the United States in 2011. Phytopathology. 2012b;102:S4.130.

Wan AM, Chen XM. Stripe rust epidemics of wheat and barley and races of *Puccinia striiformis* identified in the United States in 2012. Phytopathology. 2013;103(Suppl 2):S2.155.

Wan AM, Chen XM. Virulence characterization of *Puccinia striiformis* f. sp. *tritici* using a new set of *Yr* single-gene line differentials in the United States in 2010. Plant Dis. 2014a;98:1534–42.

Wan AM, Chen XM. Stripe rust epidemics of wheat and barley and races of *Puccinia striiformis* identified in the United States in 2013. Phytopathology. 2014b;104(Suppl 3):S3.125.

Wan AM, Chen XM. Stripe rust epidemics of wheat and barley and races of *Puccinia striiformis* identified in the United States in 2014. Phytopathology. 2015;105(Suppl. 4):S4.144.

Wan AM, Chen XM. Variation of telial formation in the *Puccinia striiformis* f. sp. *tritici* population. Phytopathology. 2016;S4:206–7.

Wan AM, Zhao ZH, Chen XM, He ZH, Jin SL, Jia QZ, Yao G, Yang JX, Wang BT, Li GB, Bi YQ, Yuan ZY. Wheat stripe rust epidemic and virulence of *Puccinia striiformis* f. sp. *tritici* in China in 2002. Plant Dis. 2004;88:896–904.

Wan AM, Wang MN, Chen XM. Regional differences in genetic structure of *Puccinia striiformis* f. sp. *tritici*, the wheat stripe rust pathogen, in the U.S. revealed by SSR markers. Phytopathology. 2014;104(Suppl 3):S3.125.

Wan Q, Liang JM, Luo Y, Ma ZH. Population genetic structure of *Puccinia striiformis* in Northwestern China. Plant Dis. 2015;99:174–1774.

Wan AM, Chen XM, Yuen J. Races of *Puccinia striiformis* f. sp. *tritici* in the United States in 2011 and 2012 and comparison with races in 2010. Plant Dis. 2016;100:966–75.

Wan AM, Muleta KT, Zegeye H, Hundie B, Pumphrey MO, Chen XM. Virulence characterization of wheat stripe rust fungus *Puccinia striiformis* f. sp. *tritici* in Ethiopia. Plant Dis. 2017;101:73–80.

Wang ZH. On the evolution of the barley-yellow rust pathosystem in Tibet. Acta Phytopathol Sinica. 1992;22:151–5.

Wang MN, Chen XM. First report of Oregon grape (*Mahonia aquifolium*) as an alternate host for the wheat stripe rust pathogen (*Puccinia striiformis* f. sp. *tritici*) under artificial inoculation. Plant Dis. 2013;97:839.

Wang MN, Chen XM. Barberry does not function as an alternate host for *Puccinia striiformis* f. sp. *tritici* in the US Pacific Northwest due to teliospore degradation and barberry phenology. Plant Dis. 2015;99:1500–6.

Wang KN, Hong XW, Si QM, Wang JX, Shen JP. Studies of the physiological specialization of stripe rust of wheat in China. Chin J Plant Prot. 1963;2:23–35.

Wang ZH, Peng YL, Yangjin L, Da Z. Difference of pathogenicity between isolates of barley-attacking and wheat-attacking forms of *Puccinia striiformis* with special reference to the deviation of disease development. Acta Phytopathol Sinica. 1989a;19:161–5.

Wang ZH, Yangjin L, Da Z, Peng YL. A preliminary study on the physiologic specialization of *Puccinia striiformis* f. sp. *hordei* in Tibet. Chin J Plant Prot. 1989b;16:186–92.

Wang Y, Wang MN, Zhang RJ, Xu CG, Feng Q, Li ZQ. Transient expression of transformed genes by particle bombardment. Acta Botanica Boreali-Occidentalia. 2006;26:115–8.

Wang XJ, Zheng WM, Buchenauer H, Zhao J, Han QM, Huang LL, Kang ZS. The development of a PCR-based method for detecting *Puccinia striiformis* latent infections in wheat leaves. Eur J Plant Pathol. 2008;120:241–7.

Wang BT, Li GB, Li Q, Wang F, Shi YQ, Liu QR, Kang ZS. Parasitic fitness of the major epidemic strains of *Puccinia striiformis* f. sp. *tritici* in China. Acta Phytopathol Sin. 2009a;103:82–7.

Wang XJ, Chen JL, Buchenauer H, Zhao J, Han QM, Huang LL, Kang ZS. Detection of *Puccinia striiformis* in latently infected wheat leaves by nested polymerase chain reaction. J Phytopathol. 2009b;157:490–3.

Wang XJ, Tang CL, Zhang G, Li YC, Wang CF, Liu B, Qu ZP, Zhao J, Han QM, Huang LL, Chen XM, Kang ZS. cDNA-AFLP analysis reveals differential gene expression in compatible interaction of wheat challenged with *Puccinia striiformis* f. sp. *tritici*. BMC Genom. 2009c;10:289.

Wang XJ, Liu W, Chen XM, Tang CL, Dong YL, Ma JB, Huang XL, Wei GR, Han Q, Huang LL, Kang ZS. Differential gene expression in incompatible interaction between wheat and stripe rust fungus revealed by cDNA-AFLP and comparison to compatible interaction. BMC Plant Biol. 2010;10:9.

Wang CF, Zhang SJ, Hou R, Zhao ZT, Zheng Q, Xu QJ, Zheng DW, Wang GH, Liu HQ, Gao XL, Ma JW, Kistler HC, Kang ZS, Xu JR. Functional analysis of the kinome of the wheat scab fungus *Fusarium graminearum*. PLoS Pathog. 2011;7:e1002460.

Wang Y, Ma DF, Zhang L, Lu LL, Wang MN, Jing JX, Kang ZS. Hereditary stability and pathogenicity of wheat stripe rust mutants with GUS gene insertion. Acta Phytopathol Sin. 2013;43:42–9.

Wang MN, Wan AM, Chen XM. Barberry as alternate host is important for *Puccinia graminis* f. sp. *tritici* but not for *Puccinia striiformis* f. sp. *tritici* in the U. S. Pacific Northwest. Plant Dis. 2015;99:1507–16.

Wang ZY, Zhao J, Chen XM, Lu YZ, Ji JJ, Zhao SL, Huang LL, Kang ZS. Virulence variations of *Puccinia striiformis* f. sp. *tritici* isolates collected from *Berberis* spp. in China. Plant Dis. 2016;100:131–8.

Weeks G, Spiegelman GB. Roles played by Ras subfamily proteins in the cell and developmental biology of microorganisms. Cell Signal. 2003;15:901–9.

Wellings CR. *Puccinia striiformis* in Australia: a review of the incursion, evolution, and adaptation of stripe rust in the period 1979-2006. Aus J Agric Res. 2007;58:567–75.

Wellings CR. Stripe rust pathotype 'Jackie *Yr27*' detected for the first time. Cereal Rust Report Season 2008, Plant Breeding Institute, University of Sydney, vol 6. 2008. p. 1–3 https://sydney.edu.au/agriculture/documents/pbi/cereal_rust_report_2008_vol_6_8.pdf

Wellings CR. Global status of stripe rust: a review of historical and current threats. Euphytica. 2011;179:129–41.

Wellings CR, Kandel KR. Pathogen dynamics associated with historic stripe (yellow) rust epidemics in Australia in 2002 and 2003. In: Proceedings of the 11th international cereal rusts and powdery mildews conference. 22–27 August 2004. John Innes Centre, Norwich, UK. European and Mediterranean Cereal Rust Foundation, Wageningen, Netherlands. Cereal Rusts and Powdery Mildew Bulletin, Abstract A2.74. 2004.

Wellings CR, McIntosh RA. *Puccinia striiformis* f. sp. *tritici* in Australasia: pathogenic changes during the first 10 years. Plant Pathol. 1990;39:316–25.

Wellings CR, McIntosh RA, Walker J. *Puccinia striiformis* f. sp. *tritici* in eastern Australia- possible means of entry and implications for plant quarantine. Plant Pathol. 1987;36:239–41.

Wellings CR, Burdon JJ, McIntosh RA, Wallwork H, Raman H, Murray GM. A new variant of *Puccinia striiformis* causing stripe rust on barley and wild *Hordeum* species in Australia. New Dis Rep. 2000;1:6. http://www.ndrs.org.uk/article.php?id=001006

Wellings CR, Wright DG, Keiper F, Loughman R. First detection of wheat stripe rust in Western Australia: Evidence for a foreign incursion. Aus Plant Pathol. 2003;32:321–2.

Wellings CR, Singh RP, McIntosh RA, Pretorius ZA. The development and application of near isogenic lines for the stripe (yellow) rust pathosystem. In: Proceedings of the 11th international cereal rusts and powdery mildews conference, 22–27 August 2004, John Innes Centre, Norwich, UK. European and Mediterranean Cereal Rust Foundation, Wageningen, Netherlands. Cereal Rusts and Powdery Mildews Bulletin, Abstract A1.39. 2004. Available from http://www.crpmb.org/icrpmc11/abstracts.htm. Accessed 8 July 2005.

Wellings CR, Singh RP, Yahyaoui A, Nazari K, McIntosh RA. The development and application of near-isogenic lines for monitoring cereal rust pathogens. In: McIntosh RA, editor. Proceedings of the Borlaug global rust initiative technical workshop. Mexico: BGRI Cd Obregon; 2009. p. 77–87.

Wellings CR, Boyd LA, Chen XM. Resistance to stripe rust in wheat: Pathogen biology driving resistance breeding. In: Sharma I, editor. Disease resistance in wheat. CAB International 2012. 2012. p. 63–83.

Welsh J, McClelland M. Fingerprinting genome using PCR with arbitrary primers. Nucleic Acids Res. 1990;19:303–6.

Wesenberg DM, Burrup DE, Brown WM, Jr Velasco VR, Hill JP, Whitmore JC, Karow RS, Hayes PM. Registration of 'Bancroft' barley. Crop Sci. 2001;41:265–6.

White E. Life, death, and the pursuit of apoptosis. Genes Dev. 1996;10:1–15.

White BT, Yanofsky C. Structural characterization and expression analysis of the *Neuro spora* conidiation gene *con-6*. Dev Biol. 1993;160:254–64.

Williams JGK, Kubelik AR, Livak KJ, Rafalski JA, Tingey SV. DNA polymorphisms amplified by arbitrary primers are useful as genetic markers. Nucleic Acids Res. 1990;18:6531–5.

Wright RG, Lennard JH. Origin of new race of *Puccinia striiformis*. Trans Brit Mycol Soc. 1980;74:283–7.

Xi K, Kumar K, Holtz MD, Turkington TK, Chapman B. Understanding the development and management of stripe rust in central Alberta. Can J Plant Pathol. 2015;37:21–39.

Xia CJ, Wan AM, Wang MN, Jiwan DA, See DR, Chen XM. Secreted protein gene derived-single nucleotide polymorphisms (SP-SNPs) reveal population diversity and differentiation of *Puccinia striiformis* f. sp. *tritici* in the United States. Fungal Biol. 2016a;120:729–44.

Xia CJ, Wang MN, Wan AM, Jiwan DA, See DR, Chen XM. Association analysis of SP-SNPs and avirulence genes in *Puccinia striiformis* f. sp. *tritici*, the wheat stripe rust pathogen. Am J Plant Sci. 2016b;7:126–37.

Xiang C, Feng JY, Wang MN, Chen XM, See DR, Wan AM, Wang T. Molecular mapping of *Yr76* for resistance to stripe rust in winter club wheat cultivar Tyee. Phytopathology. 2016;106:1186–93.

Xu JR. MAP kinases in fungal pathogens. Fungal Genet Biol. 2000;31:137–52.

Yahyaoui A, Ketata H, Torabi M, Morgounov A, Braun H, Mosaad M, Djumakhanov B, Koichibayev M. Cereal rust monitoring in Central and West Asia and North Africa: current status and future challenges. In: Yahyaoui A, Rajaram S, editors. Meeting the challenge of yellow rust in cereal crops. Proceedings of the 2nd, 3rd and 4th regional conf yellow rust in the Central and West Asia and North Africa (CWANA) Region. 2005. p. 155–70. https://www.ars.usda.gov/ARSUserFiles/50620500/Publications/ICARDAYellowRustProceedings2004-2006-2010.pdf

Yan GP, Chen XM, Line RF, Wellings CR. Resistance gene analog polymorphism markers co-segregating with the *Yr5* gene for resistance to wheat stripe rust. Theor Appl Genet. 2003;106:636–43.

Yao QY, Wang GF, Xu ZB, Wang MN, Wang Y, Jing JX. Virulent mutant in *Puccinia striiformis* induced by ethyl methyl sulfomar (EMS). J Northwest Sci-Tech Univ Agric Forest (Natural Sci Edition). 2006;34:120–3.

Yin CT, Chen XM, Wang XJ, Han QM, Kang ZS, Hulbert SH. Generation and analysis of expression sequence tags from haustoria of the wheat stripe rust fungus *Puccinia striiformis* f. sp. *tritici*. BMC Genom. 2009;10:626.

Yin CT, Jurgenson JE, Hulbert SH. Development of a host-induced RNAi system in the wheat stripe rust fungus *Puccinia striiformis* f. sp. *tritici*. Mol Plant Microbe Interact. 2011;24:554–61.

Yin CT, Downey SI, Klasges-Mundt NL, Ramachandran S, Chen XM, Szabo LJ, Pumphrey M, Hulbert SH. Identification of promising host-induced silencing targets among genes preferentially transcribed in haustoria of *Puccinia*. BMC Genom. 2015;16:579.

Yuan CY, Wang MN, See DR, Chen XM. Towards construction of genetic linkages for mapping virulence genes in *Puccinia striiformis* f. sp. *tritici*, the wheat stripe rust pathogen. Phytopathology. 2016;S4:208.

Zadoks JC. On the formation of physiologic races in plant parasites. Euphytica. 1959;8:104–16.

Zadoks JC. Yellow rust on wheat: studies in epidemiology and physiologic specialization. Tijdschr Plantenziekten. 1961;67:69–256.

Zadoks JC. Strategies in combating cereal diseases in Europe, with special reference to yellow rust of wheat. In: Proceedings of the rice blast workshop. Philippines: Laguna; 1979. p. 183–98.

Zambino PJ, Szabo LJ. Phylogenetic relationships of selected cereal and grass rusts based on rDNA sequence analysis. Mycologia. 1993;85:401–14.

Zeng SM. Studies on measurement of relative pathogenetic fitness of pathogens (an example of wheat stripe rust). Acta Phytopathol Sin. 1996;26:97–104.

Zeng SM. Parasitic fitness of cultivar–race combinations. In: Zeng SM, Zhang SZ, editors. Epidemiological studies in plant breeding for resistance. Beijing: Scientific Press; 1998. p. 41–54.

Zeng SM, Luo Y. Systems analysis of wheat stripe rust epidemics in China. Eur J Plant Pathol. 2008;121:425–38.

Zhan GM, Chen XM, Kang ZS, Huang LL, Wang MN, Wan AM, Cheng P, Cao SQ, Jin SL. Comparative virulence phenotypes and molecular genotypes of *Puccinia striiformis*

f. sp. *tritici*, the wheat stripe rust pathogen in China and the United States. Fungal Biol. 2012;116:643–53.

Zhan GM, Wang FP, Chen XM, Wan CP, Han QM, Huang LL, Kang ZS. Virulence and molecular diversity of the *Puccinia striiformis* f. sp. *tritici* population in Xinjiang in relation to other regions of western China. Plant Dis. 2016;100:99–107.

Zhang YH, Qu ZP, Zheng WM, Liu B, Wang XJ, Xue X, Xu LS, Huang LL, Han QM, Zhao J, Kang ZS. Stage-specific gene expression during urediniospore germination in *Puccinia striiformis* f. sp. *tritici*. BMC Genomics. 2008;9:203.

Zhang H, Guo J, Voegele RT, Zhang JS, Duan YH, Luo HY, Kang ZS. Functional characterization of calcineurin homologs *PsCNA1/PsCNB1* in *Puccinia striiformis* f. sp. *tritici* using a host-Induced RNAi system. PLoS ONE. 2012;7:e49262.

Zhao LS, Wang HR. Preliminary observation of urediniospore germ tube fusion in the wheat stripe rust fungus. Agric Sci China. 1969;1:48–9.

Zhao J, Wang XJ, Chen CQ, Huang LL, Kang ZS. A PCR-based assay for detection of *Puccinia striiformis* f. sp. *tritici* in wheat. Plant Dis. 2007a;91:1669–74.

Zhao XH, Mehrabi R, Xu JR. Mitogen-activated protein kinase pathways and fungal pathogenesis. Eukaryot Cell. 2007b;6:1701–14.

Zhao J, Wang L, Wang Z, Chen XM, Zhang HC, Yao JN, Zhan GM, Chen W, Huang LL, Kang ZS. Identification of eighteen *Berberis* species as alternate hosts of *Puccinia striiformis* f. sp. *tritici* and virulence variation in the pathogen isolates from natural infection of barberry plants in China. Phytopathology. 2013;9:927–34.

Zhao J, Wang MN, Chen XM, Kang ZS. Role of alternate hosts in epidemiology and pathogen variation of cereal rusts. Ann Rev Phytopathol. 2016;54:207–28.

Zheng WM, Chen SY, Kang ZS, Wang Y, Li ZQ, Wu LR. Specificity and stability of PSR (*Puccinia striiformis* Repeat) sequence. Acta Phytopathol Sin. 2000;30:222–5.

Zheng WM, Liu F, Kang Z, Chen SY, Li ZQ, Wu L. AFLP finger-printing of Chinese epidemic strains of *Puccinia striiformis* f. sp. *tritici*. Prog Nat Sci. 2001;11:587–93.

Zheng WM, Huang LL, Huang JQ, Wang XJ, Chen XM, Zhao J, Guo J, Zhuang H, Qiu CZ, Liu J, Liu HQ, Huang XL, Pei GL, Zhan GM, Tang CL, Cheng YL, Liu MJ, Zhang JS, Zhao ZT, Zhang SJ, Han QM, Han DJ, Zhang HC, Zhao J, Gao XN, Wang JF, Ni PX, Dong W, Yang LF, Yang HM, Xu JR, Zhang GY, Kang ZS. High genome heterozygosity and endemic genetic recombination in the wheat stripe rust fungus. Nature Commun. 2013;4:2673.

Chapter 3
Wheat-*Puccinia striiformis* Interactions

Zhensheng Kang, Chunlei Tang, Jie Zhao, Yulin Cheng, Jie Liu, Jun Guo, Xiaojie Wang, and Xianming Chen

Abstract Owing to the scientific and economic importance, the antagonistic wheat-*Puccinia striiformis* f. sp. *tritici* (*Pst*) pathosystem has been a focus of research over the past five decades. With the recent discovery of barberry as alternate hosts, *Pst* has been confirmed to be heteroecious, and the roles of the alternate hosts in disease epidemiology and pathogen variation have been studied. Meanwhile, considerable effort has been taken and significant progress has been made to unravel the wheat-*Pst* interactions. Histological and cytological studies have provided basic information on infection strategies used by the pathogen and defense responses from the host during wheat-*Pst* interactions and identified cellular components involved in the interactions. Physiological studies demonstrated the essential changes in *Pst* infected wheat leaves, unraveling the damage of the pathogen and the countermeasures of the host plants. Transcriptome and genome sequencing has revealed the molecular features and dynamics of the wheat-*Pst* pathosystem. Extensive molecular analyses have led to the identification of major components in the wheat resistance responses. Studies of wheat-*Pst* interactions have now entered a new phase in which cellular and molecular approaches are being used. In this chapter, we present the achievements made in the histological and cytological study of wheat-*Pst* interactions as well as physiological plant pathology. Furthermore, the new insights into wheat immunity provided by the sequence resources and advanced genomic technologies were discussed. Overall, this chapter focuses on the cellular biology of the stripe rust fungus and the wheat-*Pst* interactions, and integrates the emerging data from molecular analyses with the histocytological observations.

Z. Kang (✉) • C. Tang • J. Zhao • Y. Cheng • J. Guo • X. Wang
State Key Laboratory of Crop Stress Biology for Arid Areas and College of Plant Protection, Northwest A&F University, Yangling, Shaanxi 712100, China
e-mail: kangzs@nwafu.edu.cn

J. Liu
State Key Laboratory of Crop Stress Biology for Arid Areas and College of Life Science, Northwest A&F University, Yangling, Shaanxi 712100, China

X. Chen
US Department of Agriculture, Agricultural Research Service, Wheat Health, Genetics and Quality Research Unit, Pullman, WA 99164-6430, USA

Department of Plant Pathology, Washington State University, Pullman, WA 99164-6430, USA

X. Chen, Z. Kang (eds.), *Stripe Rust*, DOI 10.1007/978-94-024-1111-9_3

Keywords Alternate hosts • Cytology • Electron microscopy • Life cycle • Molecular biology • Pathogenicity • Physiology • Plant-pathogen interaction • *Puccinia striiformis* • Ultrastructure • Wheat

As the stripe rust fungus (*Puccinia striiformis*) is a biotrophic obligate parasite, it must depend upon its hosts for growth, reproduction and survival. The pathogen has to be physically in close contact with host cells. Once a spore lands on the surface of a host plant and penetrates the surface, a fighting or interaction between the two organisms takes place, resulting in the susccessful colonization by the pathogen and severe damage on the host, or death or reduced growth of the pathogen and no damage or less disease. During the long coevolution of the host and pathogen, both organisms have developed skills for interacting with its counterpart. In this chapter, we will present research progress and discuss issues in the interactions between host plants and the stripe rust pathogen at various levels, including ecology, population, physiology, cytology, genetics, epigenetics, genome and functional genomics.

3.1 Life Cycle of *P. striiformis*

Among plant pathogenic fungi, rust fungi generally have the most complex life cyle, containing up to five spore stages. All cereal rust fungi have a macrocyclic life style with all five spore stages and require two botanically un-related plants to complete their life cycle. We will discuss the historical perspective and recent discoveries related to the life cycle of the stripe rust fungus in this section.

3.1.1 Alternate Host of P. striiformis

All cereal rust fungi have a macrocyclic heteroecious sexual life cycle with an uredinial stage occurring on cereals and an aecial stage occurring on botanically different plants. However, for centuries, *Puccinia striiformis* Westend., the stripe rust fungus infecting cereal crops and grasses, was known to have an imperfect life cycle consisting of uredinial and telial stages. In 2010, barberry (*Berberis* spp.) was discovered as an alternate (aecial) host for *Puccinia pseudostriiformis* (Syn. *Puccinia striiformis* f. sp. *poae*) under natural conditions in Minnesota, USA and *P. striiformis* f. sp. *tritici* (*Pst*) under controlled conditions (Jin et al. 2010). Thus, *Pst* was determined to be a heteroecious and macrocyclic rust fungus. The complicated life cycle of *Pst* consists of five spore stages (Fig. 3.1): pycniospore (spermatium) (0), aeciospore (I), urediniospore (II), teliospore (III) and basidiospore (IV). Uredinial and telial stages complete on wheat as its primary host or grasses as auxiliary hosts;

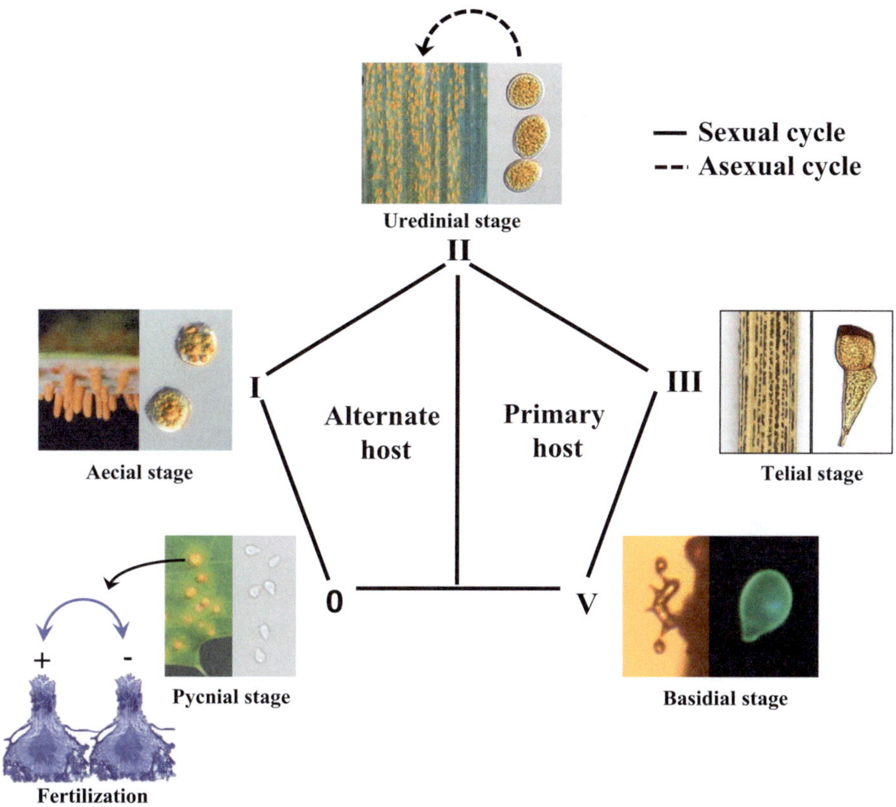

Fig. 3.1 Life cycle for *Puccinia striiformis* f. sp. *tritici*

and basidial, pycnial (spermagonial) and aecial infections occur on alternate host barberry. Urediniospores and aeciospores can infect wheat to produce new urediniospores.

3.1.1.1 History

The discovery of plant species as alternate hosts of *Pst* has taken a quite long time. Historically, many attempts to find alternate hosts for *Pst* were made by mainly European scientists, but all failed. Eriksson (1894) initially tried to look among species of *Boraginaceae* and *Ranunculaceae*. Mains (1933) speculated that *Berberis* and *Mahonia* are most likely to be alternate hosts for *Pst* as it has a close taxonomic relationship to *P. arrhenatheri* Erikss., *P. montanensis* Ellis, and *P. koeleriae* Arth., the aecial infections of which can occur on *Berberis* spp. and *Mahonia* spp. Subsequently, Transzschel (Tranzschel 1934) postulated that *Aecidium valerianella* possibly belongs to *P. striiformis* based on his studies for members of genus

Valerianella. According to Main's and Tranzschel's conjectures, Straib (1937) tested other members of *Valerianella* by inoculation of basidiospores of *P. striiformis* and tried to identify *Pst* from aeciospores derived from *Berberis* spp. and *Mahonia* spp., but did not get success. The later tests on *Berberis*, *Mahonia* and *Valerianella* by Hart and Becker (1939) were unsuccessful either. Until 2009, Jin et al. (2010) identified a stripe rust fungus *Puccinia pseudostriiformis* (syn. *P. striiformis* f. sp. *poae*) by inoculating Kentucky bluegrass with aecium-bearing leaf samples collected from naturally infected plants of *Berberis chinensis* and *Berberis koreana* in an arboretum and ornamental plantings of Emerald Caroudel, an interspecific hybrid between *B. koreana* and *Berberis thunbergii* in Minnesota. They also successfully obtained pycnia (spermagonia) and aecia by inoculating seedlings of *B. chinensis*, *Berberis bolstii*, *B. vulgaris* and limited aecia on *B. koreana* with germinated teliospores of *Pst* collected from a wheat field and then, the mystery of alternate host of *Pst* was finally uncovered. *Berberis vulgaris* was successfully confirmed to be susceptible to *Pst* and *Psh*, with all five spore stages produced using a single isolate under controlled conditions (Wang et al. 2011b). Moreover, *Mahonia aquifolium* was found to be susceptible to the wheat stripe rust pathogen under controlled condition (Wang and Chen 2013). *Berberis* and *Mahonia* have a close taxonomic relationship and were previously known to serve as alternate hosts for *P. graminis* Pers.Pers. f. sp. *tritici* Erikss. & E. Henn. (*Pgt*), the cause of wheat stem rust.

3.1.1.2 Susceptible and Resistant Alternate Host Species

Berberis and *Mahonia* have a large number of species and wide distribution in the globe apart from Australia. Totally, about 500 species of *Berberis* and approximately 100 species of *Mahonia* were recorded, which distribute principally in the northern temperate zone, especially in Asia, except for a few in the southern hemisphere (Africa and South America) (Ying 2001; Whetstone et al. 1989). Approximately a half of the total number of *Berberis* and *Mahonia* are native in China. To date, 33 species of *Berberis* and one species of *Mahonia* are susceptible to *Pst* by inoculation with germinated teliospores of *Pst* (Jin et al. 2010; Wang and Chen 2013; Zhao et al. 2013, 2016) (Table 3.1), and only two species (B. coryi and B. vernalis) were found resistant to *Pst* (Zhao et al. 2013). It is reasonable that more barberry species could be susceptible to *Pst*.

3.1.1.3 Natural Infection of Alternate Host Species by *Pst*

Although natural infection of *Berberis* spp. may occur in some parts of the world, *Pst* infection on barberry under natural conditions has been reported only in China (Zhao et al. 2013; Wang et al. 2016b). In the US Pacific Northwest, no *Pst* was found on barberry, but in contrast, *Pgt* depends upon barberry for initiate stem rust (Wang et al. 2015). Wang and Chen (2015) determined the factors that prevent *Pst* from infecting barberry, including telia underneath of epidermal layers of primary

Table 3.1 Species of *Berberis* and *Mahonia* that have been found susceptible to *Puccinia striiformis* f. sp. *tritici*, the wheat stripe rust pathogen

Species	Country	Infection condition	References
Berberis aggregata	China	Artificial & natural	Zhao et al. (2013) and Wang et al. (2015)
B. aggregata var. integrifolia	China	Artificial	Zhao et al. (2013)
B. atrocarpa	China	Artificial	Zhao et al. (2013)
B. brachypoda	China	Artificial & natural	Zhao et al. (2013)
B. chinensis	US	Artificial	Jin et al. (2010)
B. circumserrata	China	Artificial	Zhao et al. (2013)
B. coryi	China	Artificial	Zhao et al. (2013)
B. dasystachya	China	Artificial	Zhao et al. (2013)
B. davidii	China	Artificial	Zhao et al. (2013)
B. ferdinandi-coburgii	China	Artificial	Zhao et al. (2013)
B. franchetiana var. glabripes	China	Artificial	Zhao et al. (2016)
B. guizhouensis	China	Artificial	Zhao et al. (2013)
B. gyalaica	China	Artificial	Zhao et al. (2016)
B. holstii	US	Artificial	Jin et al. (2010)
B. jaeschkeana var. *bimbilaica*	China	Artificial	Zhao et al. (2016)
B. jamesiana	China	Artificial	Zhao et al. (2013)
B. kongboensis	China	Artificial	Zhao et al. (2016)
B. koreana	US	Artificial	Jin et al. (2010)
B. phanera	China	Artificial	Zhao et al. (2013)
B. platyphylla	China	Artificial	Zhao et al. (2013)
B. poiretii	China	Artificial	Zhao et al. (2013)
B. polyantha	China	Natural	Wang et al. (2015)
B. potaninii	China	Artificial & natural	Zhao et al. (2013)
B. shensiana	China	Artificial & natural	Zhao et al. (2013) and Wang et al. (2015)
B. soulieana	China	Artificial & natural	Zhao et al. (2013)
B. stenostachya	China	Artificial	Zhao et al. (2013)
B. thunbergii	US	Artificial	Jin et al. (2010)
B. vernalis	China	Artificial	Zhao et al. (2013)
B. vulgaris	US	Artificial	Jin et al. (2010) and Wang and Chen (2015)
B. wangii	China	Artificial	Zhao et al. (2013)
B. wilsonae	China	Artificial	Zhao et al. (2016)
B. zanlanscianensis	China	Artificial	Zhao et al. (2016)
Mahonia aquifolium	China	Artificial	Wang and Chen (2013)

host and teliospores enclosed in telia, complete degradation of telia and teliospores after winter, and requirement of at least 32 h dew period for teliospores to germinate and basidiospores to infect barberry, a weather condition hardly occurring during summer when telia form and teliospores are viable. Similar to the situation in the US, no *Pst* was found on barberry plants in Switzerland (Berlin et al. 2012).

In the Himalayan regions including Pakistan, the virulence and pathotype diversities of *Pst* population of *Berberis*-growing zone were higher than that of non-*Berberis*-growing zone, suggesting the determinant role of sexual reproduction in the virulence variation of *Pst* (Ali et al. 2014). However, so far there have been no reports of *Pst* identified from aecia on barberry plants.

In China, natural infection of *Pst* has been found. The identification of *Pst* on barberry was done by inoculating susceptible wheat seedlings with aeciospores collected from six *Berberis* spp. (Zhao et al. 2013; Wang et al. 2016b; Li et al. 2016b). Because the frequency of *Pst* aecia was very low in these studies, how important of alternate host for creating virulent races and providing aeciospores to initiate stripe rust epidemics is still not clear. Urediniospores of *Pst* is able to disseminated over thousands of kilometers by wind and the pathogen is able to survive summer and/or winter in the same or different areas as uredinial stage on primary and auxiliary hosts. It appears that infection or non-infection of alternate hosts does not make big differences for stripe rust epidemics. However, *Pst* infection on barberry may generate new races (Jin 2011; Zhao et al. 2016).

3.1.2 Infection Process on Alternate Hosts

Cytological and ultrastructural studies have been conducted in Dr. Kang's group to illustrate the *Pst* infection process on barberry leaves. Such work has been largely unpublished. We briefly present the results in this section.

Basidiospores are produced from teliospores. When basidiospores are mature, they are instantly released from basidia. After landing on the surface of a barberry leaf, a basidiospore germinates immediately under moist conditions to produce a germ tube, and an obvious appressorium forms at the tip (Fig. 3.2a). Unlike urediniospores, basidiospores cannot survive very long. They will die quickly without water. Thus, dew formation is needed from teliospores starting to germinate to basidiospores to produce germ tubes and grow into barberry leaf tissue. Also, appressoria always form when basidiospores germinate, whereas obvious appressoria form only occasionally when uredinospores germinate (Kang 1996). An infection peg is produced underneath the appressorium and grows directly through the epidermis of the host tissue and extended in the intercellular space in which it produces branches to form slender monokaryotic infectious hyphae (intercellular hyphae) (Fig. 3.2b, c). For comparison, the infection of wheat leaves by urediniospores is through stomata. When an infection hypha produced from basidiospore infection contacts with a host cell wall, it produces a haustorial mother cell separated by a septum (Fig. 3.2d). The haustorial mother cell penetrates directly through

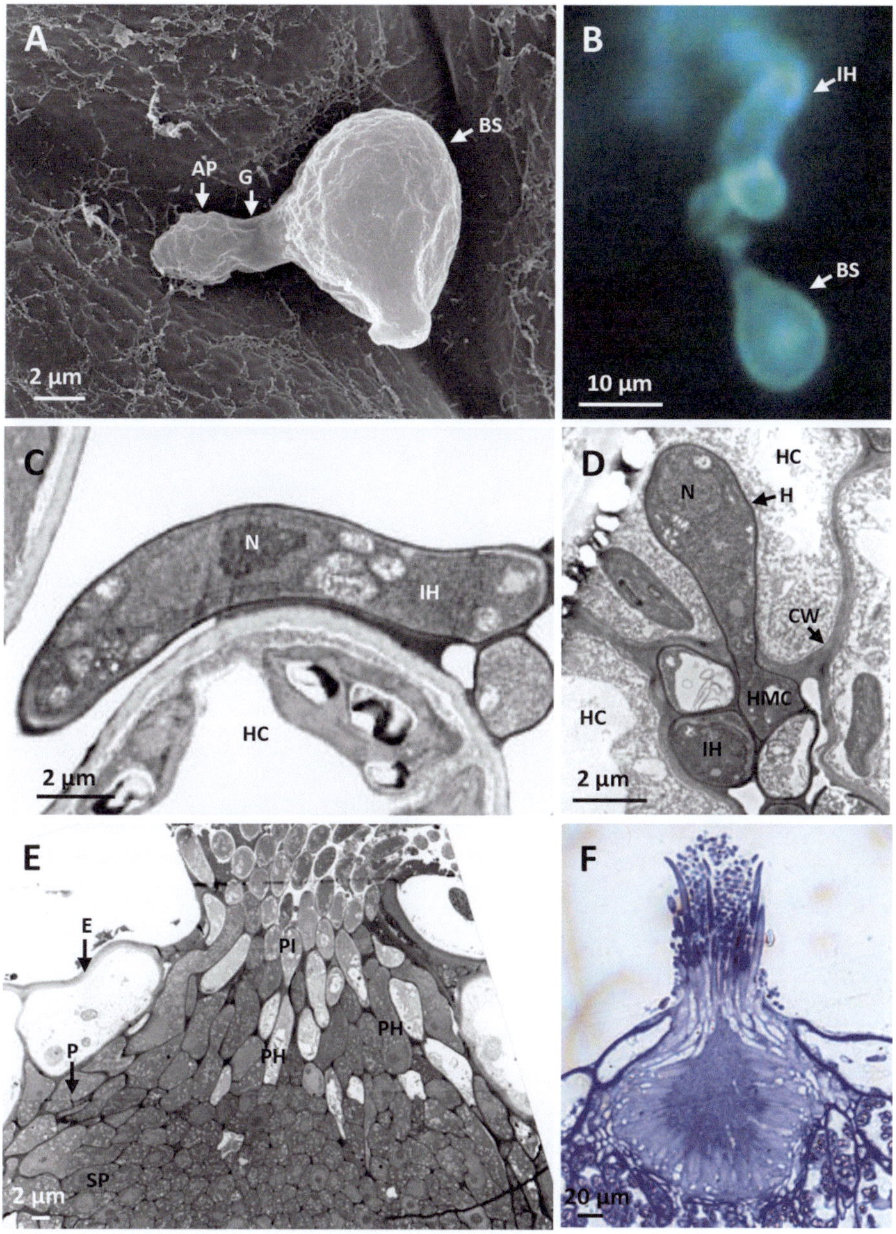

Fig. 3.2 The pycniospore formation process on alternate host barberry. (**a**) A basidiospore (*BS*) germinating to produce germ tube (*G*) and formed an inflated appresorium (*AP*) at the apex of the germ tube, which mechanically penetrated barberry leaf. (**b–c**) The slender monokaryotic (*N*) infection hyphae (*IH*) formed among intercellular space of host cells (*HC*) after penetration of basidiospores. (**d**) Haustorial mother cell (*HMC*) penetrated cell wall (*CW*) of a host cell (*HC*) to form monokaryotic (*N*) haustorium (*H*). (**e–f**) Massive infection hyphae specialized into monokaryotic (*N*) sporogenous cells (*SP*) to form a vase-shaped pycnium (*P*). The specialized hyphae developed pycniosporophores (*PH*) and produced pycniospore initials (*PI*). The cell wall of the pycniospore initial is continuous with the innermost layer of the flared, multilayered collar (*C*) at the peak of the pycniosporophore.

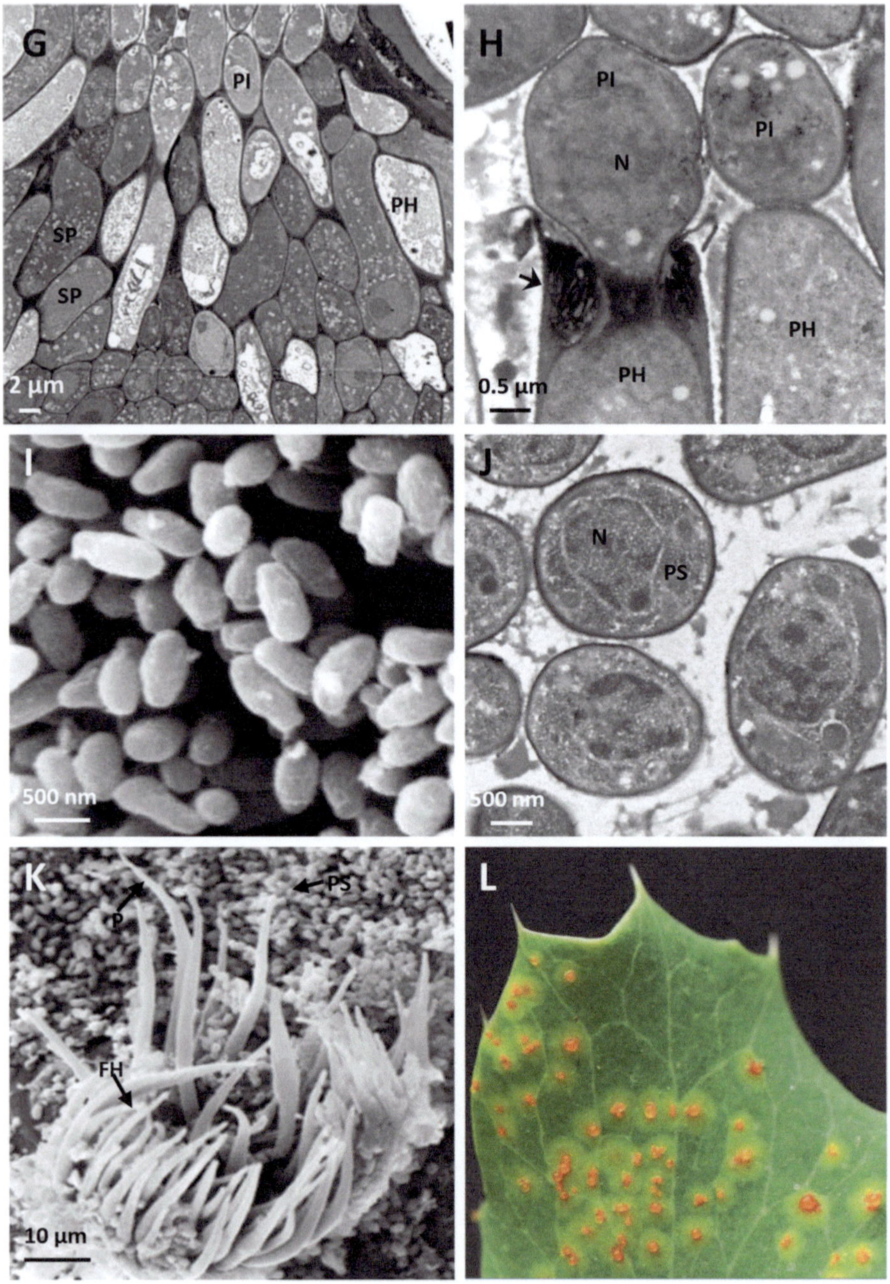

Fig. 3.2 (continued) (**g**) Enlargement of section in (**e**). (**h**) The enlarged pycniosporophores (*PH*) and produce pycniospore initials (*PI*) in (**g**). (**i**) Scanning electron microscopic photograph showing massive oblong pycniospores (*PS*). (**j**) Transmission electron micrograph illustrating monokaryotic pycniospores. *N* = nucleus. (**k**) Scanning electron microscopic (*SEM*) photograph showing accumulation of outgrowing paraphyses (*P*), flexuous hyphae (*FH*), and pycniospores (*PS*) around an ostiole of the pycnium. (**l**) Pycniospores presenting liquid nectar are colorless but generally false light orange at the infection site on the ostiole of the pycnium

the host cell wall, continuously grows into the host cell and forms a balloon-shaped structure called haustorium. Once haustoria are formed, the fungus has established a parasitic relationship with the host, taking water, nutrition and energy from host cell to continually grow within barberry tissue and produce pycnia.

In barberry leaves, intercellular hyphae developed and characterized to form massive monokaryotic sporogenous cells around outer layers of a developing pycnium at the early stage and subsequently developed to form a vase-shaped pycnium (Fig. 3.2e, f). The sporogenous cells continuously develop to form pycniosporophores (Fig. 3.2e, g). As the development of pycniosporophores, pycniospores are produced inside of the apex of the pycniosporophores from which mature pycniospores are released, and a layer of collar was left at the innermost of the pycniosporophores. Pycniospores are produced and released successively from the pycniosporophores, and multi-layered collars are formed (Fig. 3.2e–h). Mature pycniospores are oblong and haploid (Fig. 3.2i, j). The massive pycniospores gathered inside the cavity of the pycnium and are raised up to the ostiole of the pycnium, along with receptive hyphae and paraphyses, which tower the epidermis of the host tissue to form a hump-like bulge (Fig. 3.2k). The cover of the bulge is eventually broken with successively gathering of numerous pycniospores inliquid nectar, colorless but generally false light orange due to reflecting by chlorosis background around the pycnium (Fig. 3.2l).

Because *Pst* is heterothallic, its pycniospores as gametes in one pycnium can fertilize a receptive hypha ofthe opposite mating type in another pycnium. The fertilization (spermatization) can be done through delivering pycniospores in nectar to another pycnium using a brush, a finger or insects. The fertilized receptive hyphae develop toward opposite position of the pycnia and differentiated to form sporogenous cells into which two different nuclei migrate (Fig. 3.3a). The massive dikaryotic sporogenous cells were covered by a layer of peridial cells to form a cylinder-shaped aecium that grows through the epidermis of the opposite surface of the pycnium (Fig. 3.3b). Inside the cylinder-like aecial cups, massive chain-like aeciospores are formed (Fig. 3.3c). During the formation of aeciospores, a septum is formed at the top of the sporogenous cell to produce an aeciospore initial, and two nuclei migrate into the aeciospore initial. Subsequently, an intercalary cell formed with formation of a second septum under the aeciospore initial (Fig. 3.3d, e). The aeciospore initial develops to form a young aeciospore wrapped by primary wall containing immature ornamentals (verrucae) (Fig. 3.3e). A mature aeciospore contains two nuclei and electron-dense cytoplasm with lipid drops, polysaccharide granules, and has germ pores on the surface (Fig. 3.3f). When an aeciospore germinates, a germ tube grows from a germ pore and penetrates a wheat leaf through a stoma without forming an obvious appressorium, similar to a uredinial germ tube penetration. The subsequent infection process of aeciospores in wheat leaf tissue is the same as that of urediniospores (Fig. 3.3g, h).

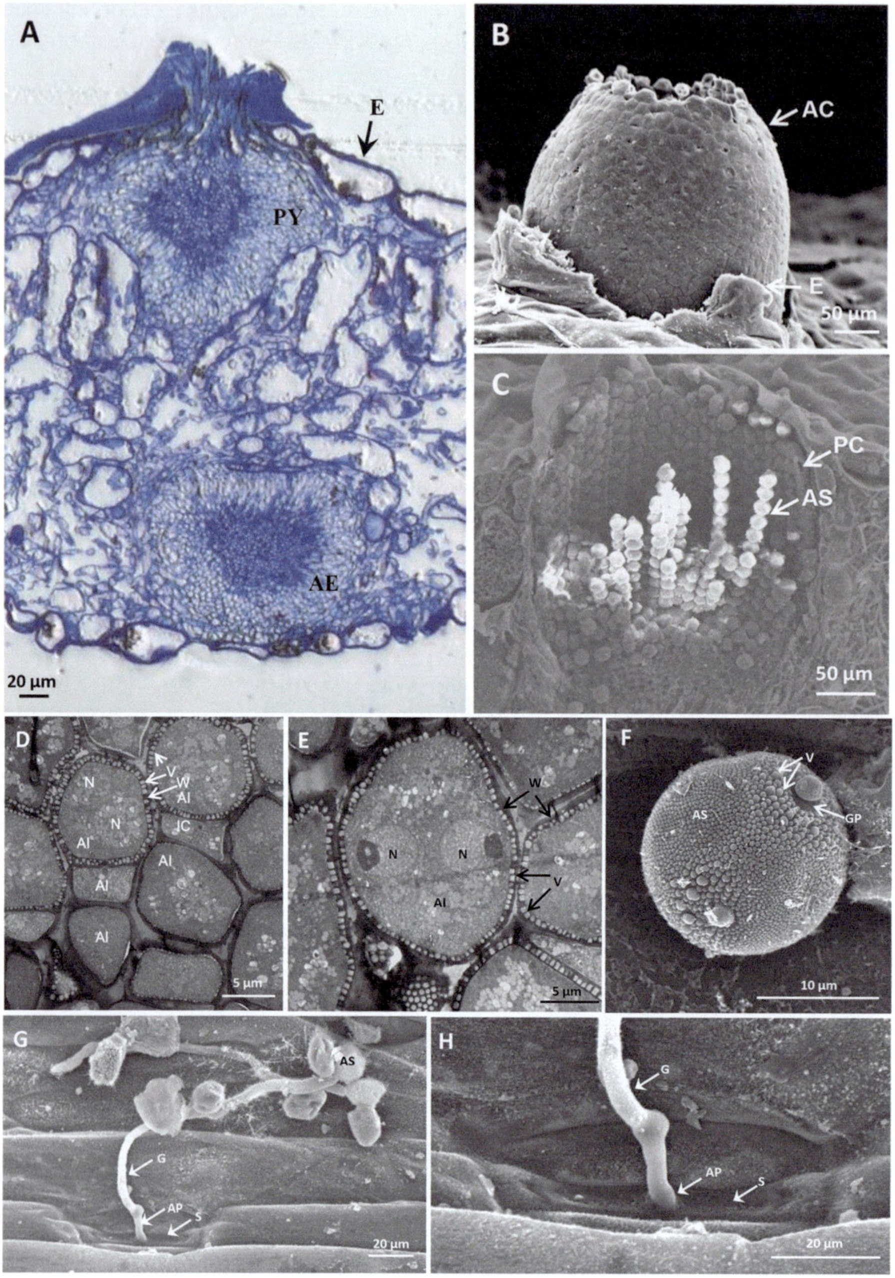

Fig. 3.3 The aeciospore formation process on alternate host barberry leaves. (**a**) Pycniospores as gametes from one pycnium fused a receptive hypha with opposite mating type in another pycnium to form sporogenous cells. (**b**) A cylinder-shaped aecium erupted through epidermis (*E*) of host tissue. (**c**) A cross fracture of an aecial cup showing aeciospores (*AS*) inside the aecial cup in a chain and covered by a layer of peridial cells (*PC*). (**d**) The sporophores are developed to form

3.1.2.1 Asexual Cycle

The asexual cycle of *Pst* is completed on primary cereal hosts or auxiliary grass hosts. As discussed in Chap. 4, *Pst* primarily infects wheat and hundreds of grass species, sometimes infects rye and barley, but generally cannot infect corn, oat, and rice (Stubbs 1985; Chen 2005; Cheng et al. 2016). The conditions for *Pst* infection are also discussed in Chap. 4.

The infection of a single aeciospore or urediniospore can produce an uredinial pustule, which consists of multiple uredinia in a cluster on a seedling leaf or in a stripe on an adult-plant leaf. Uredinia are yellow- to orange-colored and each uredinium contains thousands of urediniospores. Urediniospores are oblong and covered by thorns on the surface. They are mostly dikaryotic (n+n), but occasionally have three or four nuclei (Goddard 1976; Wang et al. 2004).

At the end of wheat plant growth or under unfavorable conditions for urediniospore production, uredinia develop into spherical or elongated sori termed telia, generally underneath epidermis in wheat tissue, which can remain physically intact for several months in dry conditions (Wang and Chen 2015). Telia contain teliospores inside and are surrounded by a peridium of cell layer outside. Telia are dark brown to black and teliospores are yellowish brown, two-celled, diploid when mature. Although teliospores are considered as resting spores, they can germinate as soon as they are mature (Raeder and Bever 1931; Wang and Chen 2015) or have a short dormancy (Wright and Lennard 1978). Viability of teliospores in plant tissue exposed to the air in field decreases after maturity. The duration of teliospore viability depends upon environmental conditions, ranging from a couple weeks under wet conditions to several months under dry conditions. In the US Pacific Northwest, teliospores become degraded after winter (Wang and Chen 2015), whereas viable teliospores still can be found the following April in Gansu, China, where the weather is generally dry and wheat stubbles are often covered by snow.

Fig. 3.3 (continued) binucleate aeciospore initials (*AI*) and interspace cells (*IC*). A dikaryotic aeciospore initial (*AI*) with verrucae (*V*) embedded inside primary wall (*W*) formed on the surface of the aeciospore initial. Aeciospore initials produced successively and separated by an intercalary cell between two aeciospore initials. (**e**) An enlargement of a young aeciospore. (**f**) A mature aeciospore (*AS*) is covered with germ pores (*GP*) and numerous verrucae (*V*) on the surface of the aeciospore. (**g–h**) Scanning electron micrograph showing an aeciospore (*AS*) germinated to produce a germ tube (*G*) which is elongated around stomata (*S*) slit to form an unobvious inflated appresorium (*AP*) penetrating the wheat leaf

3.1.2.2 Sexual Cycle

Teliospores are needed for *Pst* to infect plants of alternate hosts. As *Pst* teliospores do not have or have very short dormancy, they can germinate immediately when mature (Wang and Chen 2015). However, because *Pst* telia are covered by plant epidermis and teliospores are enclosed in telia, it is necessary to break the plant tissue and telia to expose teliospores. The conditions for teliospore germination, basidiospore production and infection of barberry were reported in detail by Wang and Chen (2015) and discussed in Chap. 4 in this book. Cytological and ultrastructural observations were conducted in Dr. Kang's program and are described below.

Teliospores germinate to produce a metabasidium (also called basidium) from each of the two cells accompanying with two meiosis events and the haploid nuclei move into the metabasidium (Fig. 3.4a–d). Each of the two cells of a teliospore usually can germinate (Fig. 3.4e, f, g) to form a four-celled metabasidium, from each of the four cell, a sterigma grows to form a basidiophore and on the top of the basidiophore, a basidiospore is produced (Fig. 3.4h, i). A single nucleus migrates into a basidiospore, but basidiospores with two or three nuclei are also observed. Basidiospores are released from the basidiophore by elastic force within 40 min to 1 h of the formation. Basidiospores are small with average size of 9.4 × 6.4 μm, transparent and oval-shaped. Basidiospores are transported by wind or rain slush, and if land on susceptible leaves (young leaves) of alternate host plants, they may germinate and infect barberry leaves under right conditions. So far, the distance of *Pst* basidiospore dispersal from the source has not been determined because of the lack of data, but it may be speculated to be not more than three hundred meters according to studies on *Pgt* basidiospores (Roelfs 1985). Basidiospores mainly infect young leaves, and sometimes infect young berries and twigs. Leaves of susceptible cultivars become resistant to penetration about 2 weeks after outgrowth of new leaves, and thus phenology of alternate hosts and teliospore survival are essential for infection of barberry by rust fungal basidiospores (Wang and Chen 2015).

The latent period in barberry tissue after basidiospore infection varies under different temperatures. Pycniospores appear in a hyaline nectar that forms at the pycnial ostiole approximately 9–11 days at the 13–16 °C cycle or 12–16 days at the 4–20 °C cycle after inoculation (Zhao et al. 2013; Wang and Chen 2015) (Fig. 3.5a). The nectar on a pycnia is orange-colored and the ostiole of the pycnium is surrounded by chlorosis. Pycniospores are unicellular, haploid (n), and small (average size of 3.1 × 1.5 μm) (Fig. 3.5c).

Functioning as a gamete (+), a pycniospore from one pycnium fuses with a receptive hypha of an opposite gamete (−) from another pycnium to complete fertilization. After fertilization, the color of the pycnium changes from orange to dark brown. Pycniospores can be transferred from one pycnium to another pycnium by using a brush or robbing with a finger under controlled conditions. Under natural conditions, the transfer could be done by insects or rain splash.

Approximately 16–20 days after inoculation, aecia appear on the adaxial surface of the barberry leaf, opposite of the pycnia. One aecium can have one to a large number of aecial cups erupted through the aecium and each contain hundreds of aeciospores in

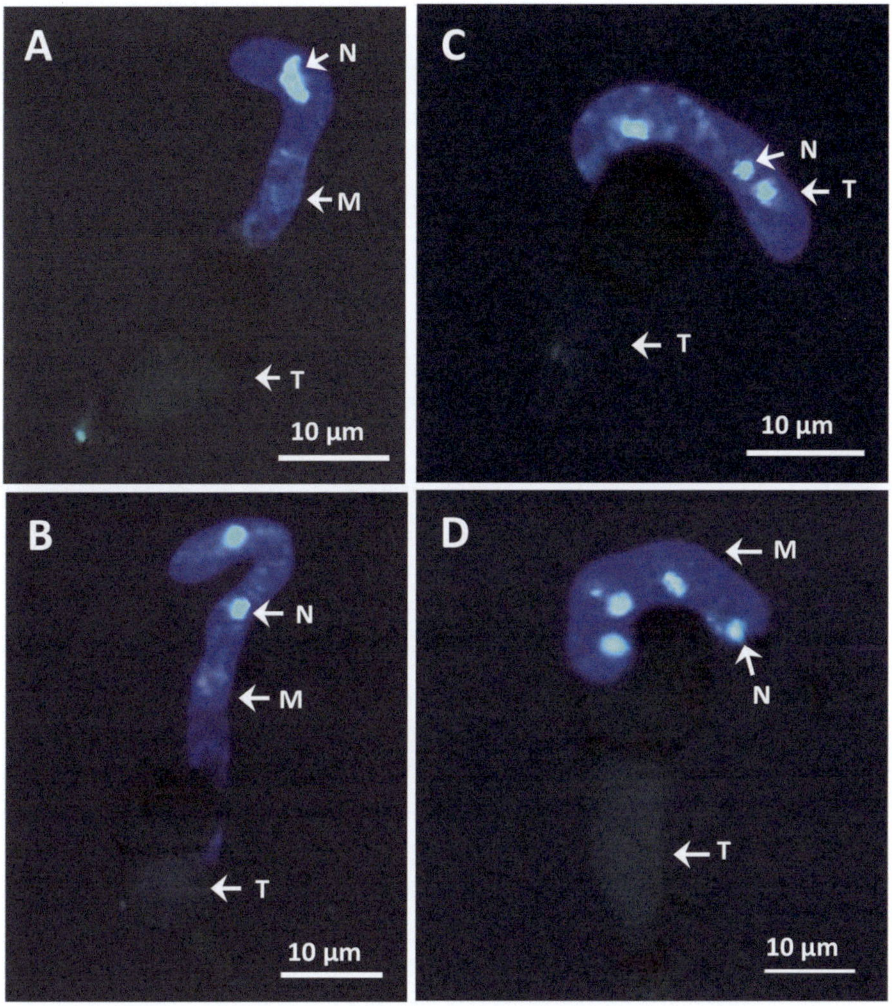

Fig. 3.4 *Puccinia striiformis* f. sp. *tritici* (*Pst*) teliospore germination to form basidiospores. (**a**, **b**) A *Pst* teliospore (*T*) is germinating to produce a metabasidium (*M*) (or promycelium) with nuclei migrating from a cell of the teliospore into the metabasidium, in which the first nuclear division (meiosis I) occurred to produce two daughter nuclei. N = nucleus. (**c**, **d**) The second nuclear division (meiosis II) is occurring in the metabasidium (*M*) after the first nuclear division, producing four daughter nuclei (*N*). (**e**) Each cell of a two-celled teliospore (*T*) is germinated to produce a metabasidium (*M*). (**f**) Production of a cone-shaped basidium (*B*) of *Pst* from a metabasidium (*M*) after teliospore (*T*) germination. (**g**) Four basidiospores (*BS*) produced a metabasidium (*M*) derived from each of a two-celled teliospore (*T*). (**h–i**) Scanning electron micrograph showing development of *Pst* teliospore germinating to form a metabasidium (*M*), basidiophores (*BP*) and basidiospores (*BS*)

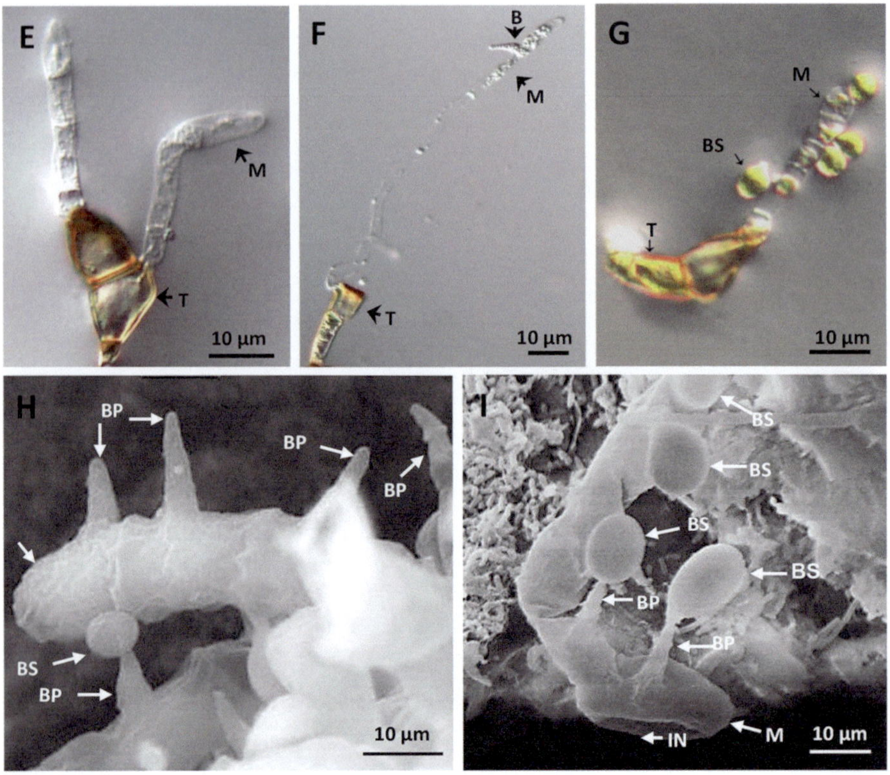

Fig. 3.4 (continued)

chain (Zhao et al. 2013; Wang and Chen 2015) (Fig. 3.5b). Aecial cups are orange in color and cylinder-like, and usually extend to 2–3 mm, and sometimes even up to 5 mm in length (Fig. 3.5d). Aeciospores are dikaryotic (n+n), near round and have an average size of 14.9 × 11.8 µm. Aeciospores are wind-borne, and those of *Pst* can infect wheat and many grasses, and sometimes infect rye and barley as discussed in Chap. 4. Similar to urediniospores, an aeciospore germinates on the surface of a wheat leaf when the moisture and temperature conditions meet the requirement; the germ tube penetrates the host tissue through a stoma; and the infection produces urediniospores.

3.1.3 Infection Process on Cereal Hosts

The most important phase of the *Pst* life cycle is the asexual cycle on wheat, during which urediniospores are produced and recurrently infect wheat plants. Successful *Pst* infection of wheat plants appears as a mass of uredinia arranged in long, narrow stripes mainly on leaves, as well as on leaf sheaths, glumes and awns (Chen et al. 2014). As other rust fungi, the infection process of *Pst* urediniospores on host (wheat) plants (Fig. 3.6) can be divided into three major stages: penetration, parasitic/biotrophic and sporulation stages (Gan et al. 2012).

Fig. 3.5 Pycnial (**a** and **c**) and aecial (**b** and **d**) stages of *Puccinia striiformis* f. sp. *tritici* infecting leaves of *Berberis shensiana* under artificial inoculation conditions

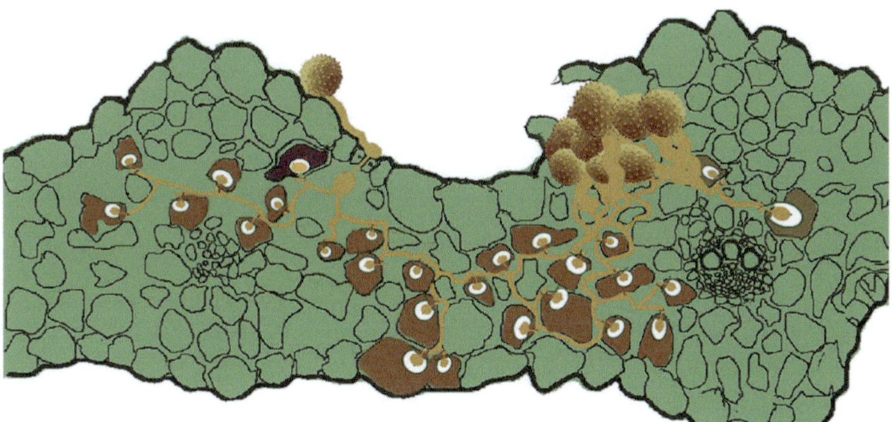

Fig. 3.6 Schematic representation of the whole uredinial infection process of *Puccinia striiformis* f. sp. *tritici*

3.1.3.1 Uredinial Infection Process

Single-celled dikaryotic (n+n) urediniospores (Fig. 3.7a) are deposited on wheat leaf surfaces by wind or raindrops for infection. Under controlled conditions, about 3 h post-inoculation (hpi), *Pst* urediniospore begins to germinate to produce dikaryotic germ tube (Fig. 3.7b) (Zhang et al. 2008a). Interestingly, germ tube extends by tip growth across the leaf surface in a direction perpendicular to the long axis of the

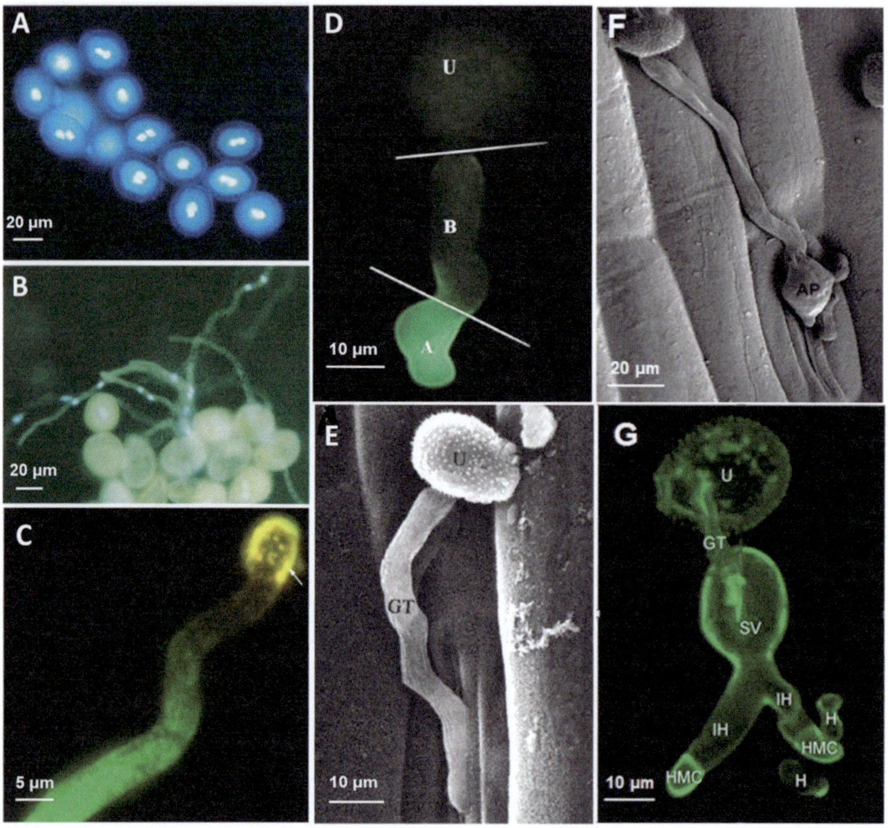

Fig. 3.7 Germination and infection process of *Puccinia striiformis* f. sp. *tritici* (*Pst*) urediniospores on wheat leaves. Dikaryotic urediniospores (**a**) and germ tubes (**b**) of *Pst* stained with hematoxylin and photographed with a fluorescence microscope. (**c**) Visualization of actin (*arrow*) in a germ tube by confocal microscopy. (**d**) Image of a germinating urediniospore after staining with DAF-FMDA to check NO. B, base of germ tube; A, apex of germ tube. (**e** and **f**) A urediniospore (*U*) germinates to produce a germ tube (*GT*) and then enters the leaf through a stoma without an appressorium (*AP*) over a stoma (**e**) or with a spherical appressorium over a stoma (**f**). (**g**) Infection features of *Pst* on wheat observed by an epifluorescence microscope after staining with wheat germ agglutinin conjugated to the fluorophore Alexa 488. A substomatal vesicle (SV) is formed within the stomatal cavity, followed by the formation of intercellular infection hyphae (IH), haustorial mother cells (HMC) and a haustorium (H) at the early infection stages

leaf epidermal cells (Kang 1996; Kang et al. 1997, 2002). Actin filament (F-actin) patterns were visualized microscopically during the germ tube growth (Fig. 3.7c), indicating that actin filament plays a role in the tip growth (Liu et al. 2012b). Nitricoxide (NO) and reactive oxygen species (ROS) as important signaling molecules were proven to regulate the germination of *Pst* urediniospores (Fig. 3.7d) (Yin et al. 2016). Within 6–8 hpi, when encountering a wheat stoma, the germ tube penetrates into the stoma and enters into wheat leaf tissue (Fig. 3.7e, f). Generally *Pst* does not form an appressorium over stoma (Fig. 3.7e), but occasionally can form a spherical appressorium over stoma similar to other rust fungi (Fig. 3.7f) (Moldenhauer et al. 2006; Zhang et al. 2008a). Subsequently, *Pst* forms a substomatal vesicle within the stomatal cavity and infection hypha in the intercellular space (Fig. 3.7g). Up to this point, *Pst* finishes the penetration stage. Noticeably, infection hypha of *Pst* can be divided into two types, typical infection hypha containing septa in hypha branches (Fig. 3.8a) and atypical infection hypha without septa (Fig. 3.8b). The cell wall of infection hypha contains four layers (Fig. 3.8c), and its chemical compositions include polysaccharide, protein, chitin and other substances (Kang et al. 1994). The cell wall of a septum contains three layers, which are connected with the cell wall of infection hypha (Fig. 3.9a) and include proteins, polysaccharides, acetyl shell mannose, glucose and other substances. Septa usually have a septal pore in the middle (Fig. 3.9b–d), and microbodies containing crystalline substances are present at the two sides of septal pores (Fig. 3.9c).

When an infection hypha grows and contacts with a wheat mesophyll cell, the hypha differentiates to form a haustorial mother cell (Fig. 3.7g). A slender neck forms from the haustorial mother cell, which then penetrates the host cell wall and invaginates the cell plasma membrane to form a haustorium inside the host mesophyll cells (Fig. 3.7g) (Kang 1996). Haustoria can be occasionally observed in wheat epidermal cells (Zhang et al. 2008a; Hovmøller et al. 2011). As an infection structure unique to biotrophic pathogens that makes intimate contact with the host cell membrane, haustoria suck nutrient uptake from infected host cells (Voegele and Mendgen 2011). The withdrawing nutrients from the invaded wheat cells by *Pst* haustoria subsequently supports the formation of more infective hyphae and haustoria, finally resulting in extensive colonization and a branching network of fungal mycelium. Transmission electron microscopic observations revealed that multinucleate conditions in infection hyphae, haustorial mother cells and haustoria are very common (Fig. 3.10a–c) (Kang et al. 1994). The fusion of two different infection hyphae was occasionally observed (Fig. 3.10d, e), which offers a prerequisite for the nucleus exchange and the genetic variation.

At the sporulation stage, a spore-bud outgrows from one end of the sporogenous cells, expands to form a septum and then successively develops a urediniospore initial, pedicel cell and primary urediniospore cell (Fig. 3.11a–d). Sporulation (urediniospore production) starts on the surface of wheat leaves 12–14 days after inoculation under favorable conditions (Chen et al. 2014).

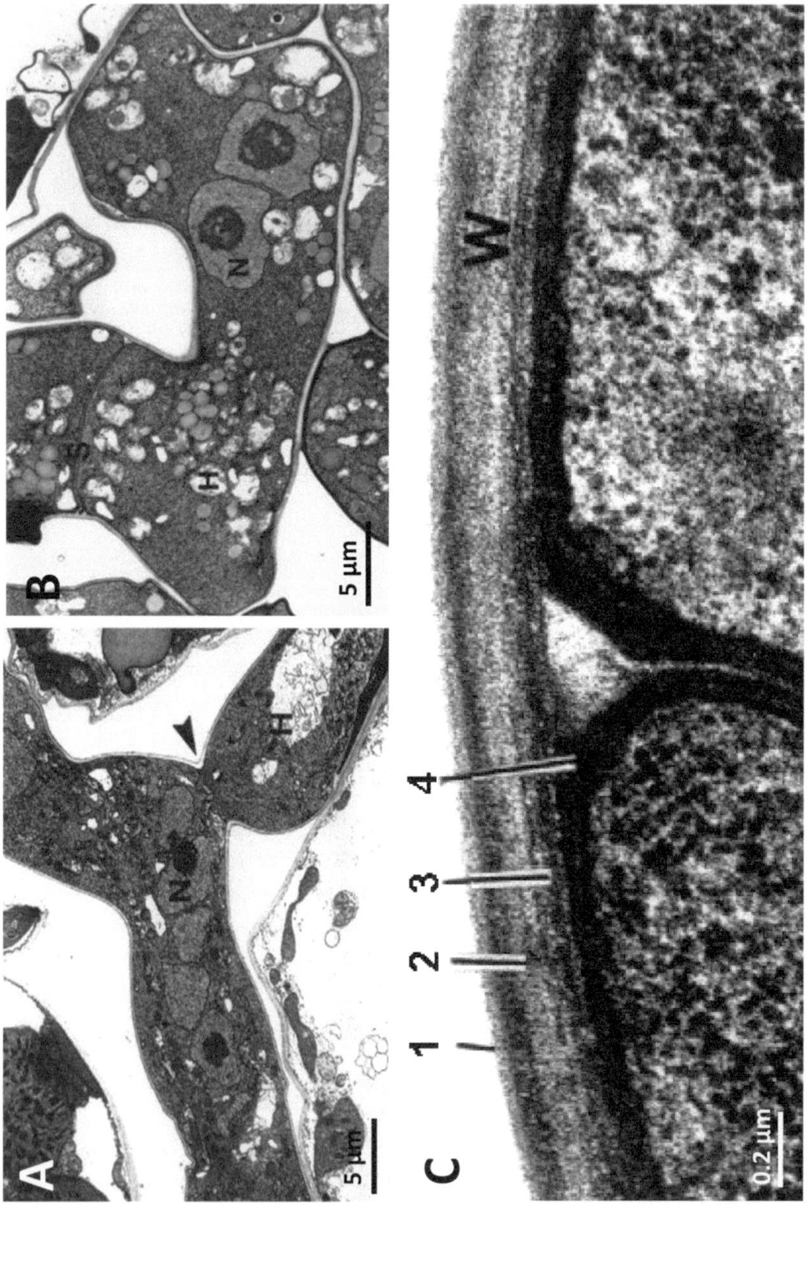

Fig. 3.8 Features of *Puccinia striiformis* f. sp. *tritici* infection hyphae. (**a**) A typical infection hypha containing septa (*arrow*) at hyphal branches. (**b**) An atypical infection hypha without any septa. (**c**) The cell wall of an infectious hypha contains four layers

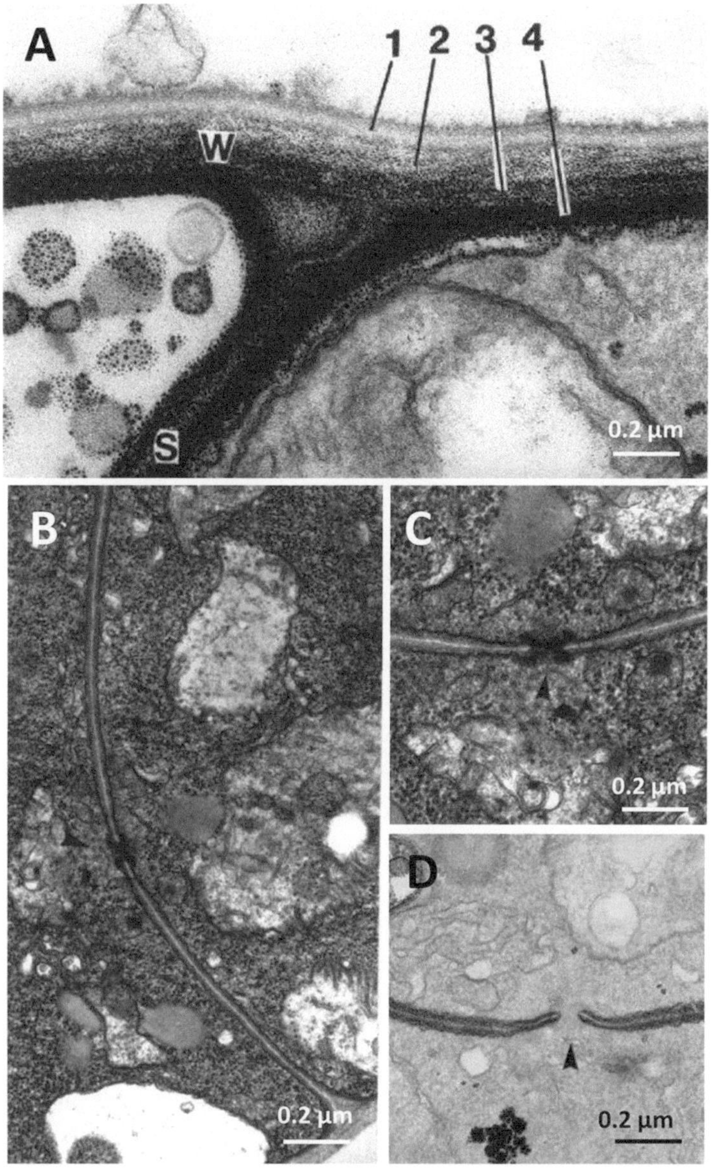

Fig. 3.9 Features of *Puccinia striiformis* f. sp. *tritici* septa in infection hyphae. (**a**) The cell wall of a septum contains three layers and is connected with the cell wall of the infectious hypha. (**b**–**d**) Septa usually have a septal pore in the middle; and (**c**) two sides of septal pores are distributed with microbodies containing a crystalline substance

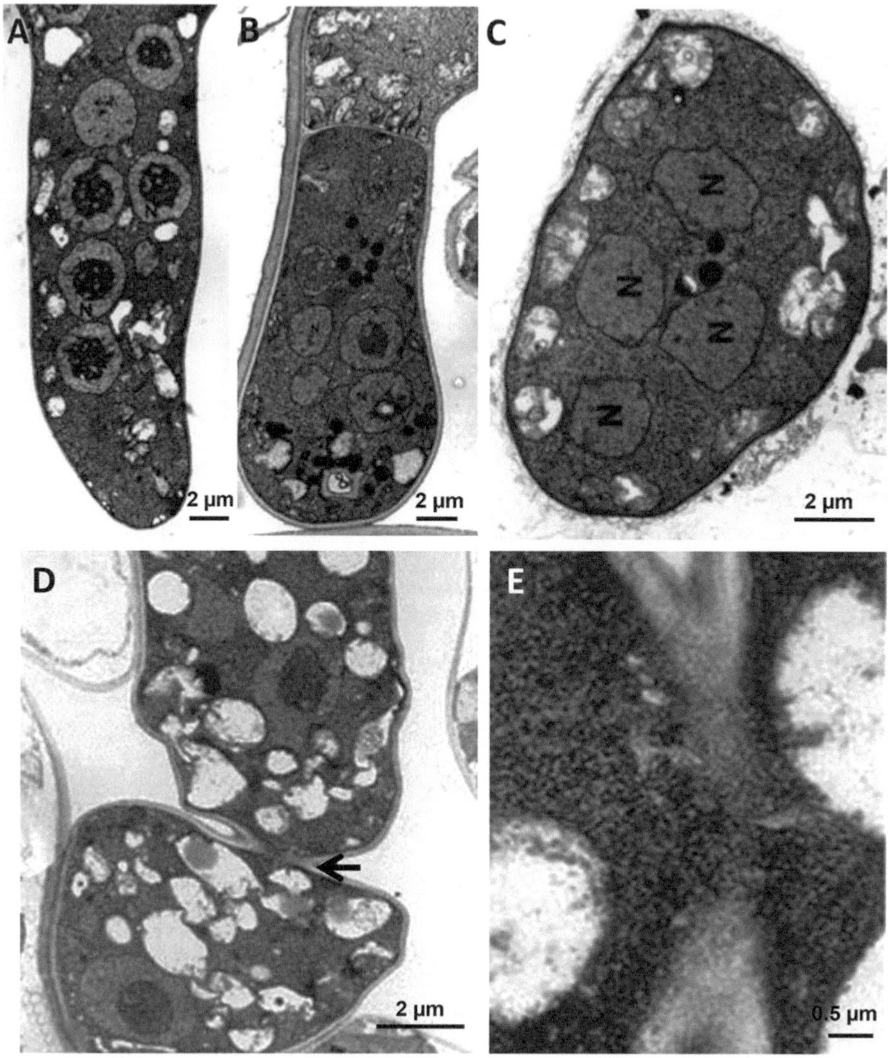

Fig. 3.10 Features of *Puccinia striiformis* f. sp. *tritici* (*Pst*) infection hyphae. (**a**) Multinucleate conditions of infectious hyphae. (**b**) Haustorial mother cells. (**c**) Haustoria, *N*, nucleus. (**d**) The fusion of two different *Pst* infection hyphae. *Arrow* indicates the fusion part. (**e**) Enlarge of the fusion section of **d**

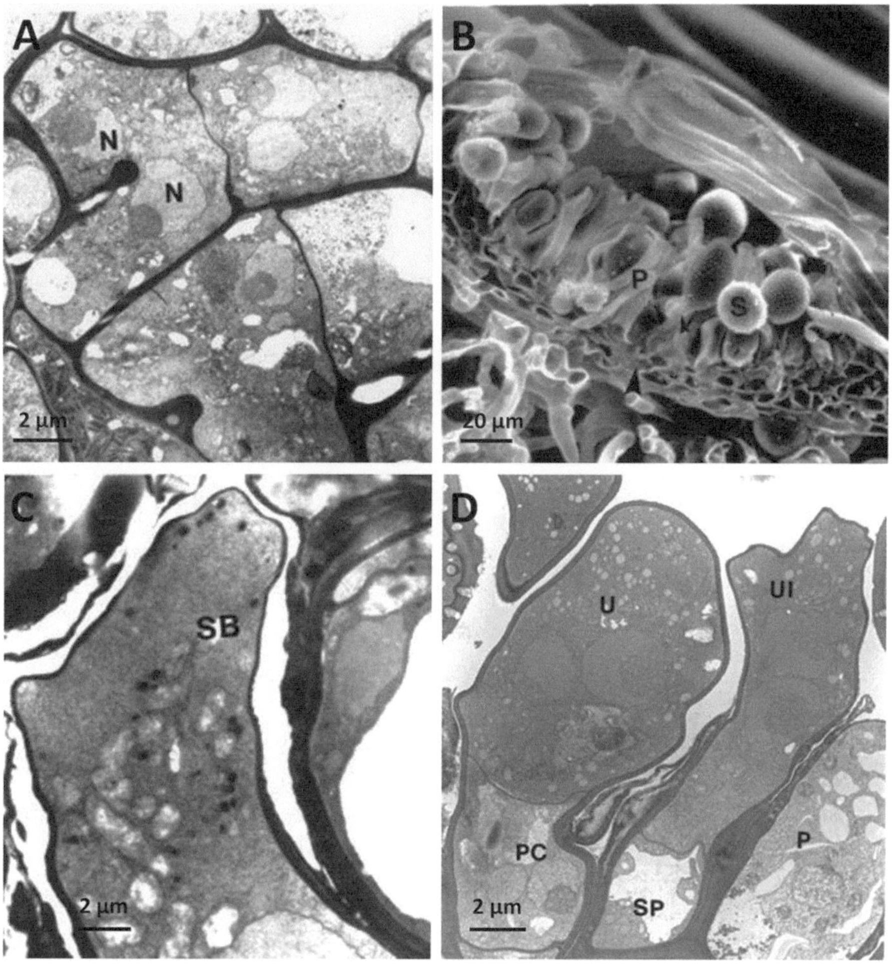

Fig. 3.11 The formation of *Puccinia striiformis* f. sp. *tritici* urediniospores. (**a**) Sporogenous cells containing two nuclei (*N*) gather under the host epidermis. (**b**) A scanning electron microscopic image of the uredinial section. (**c**) Spore buds (*SB*) are formed by producing sporogenous cells. (**d**) The sporogenous part in the sporogenous cells. *UI* urediniospore initial, *PC* pedicel cell, and *U* uredinia

3.1.3.2 Teliospore Formation

When environmental conditions are not suitable for the production of urediniospores, *Pst* will produce teliospores. A teliospore has two-cells separated by a septum, and initially has two nuclei in each cell (n+n) and then become diploid (2n) when mature (Fig. 3.12a, b). Different *Pst* isolates may vary in their ability to produce telia under similar environmental conditions (Chen et al. 2012; Wan and Chen 2016). Histological and ultrastructural observation showed that *Pst* teliospores

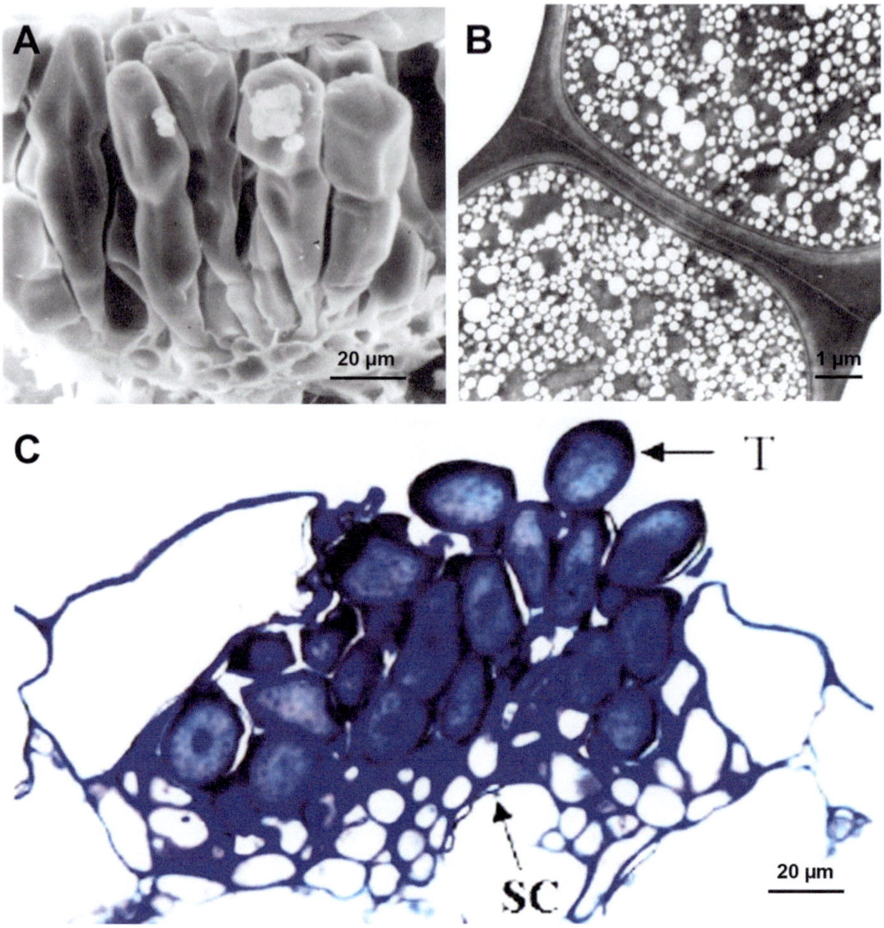

Fig. 3.12 Development of *Puccinia striiformis* f. sp. *tritici* teliospores. (**a**) A telium containing teliospores observed using a scanning electron microscope. (**b**) A section of a teliospore showing the septum separating the two cells. (**c**) A telium containing teliospores rupturing the host epidermis. *SC* sporogenous cell, T teliospore

derived from sporogenous cells lining the base of telia (Fig. 3.12c) (Yao et al. 2012). At the early stage, a spore-bud outgrows from one end of sporogenous cells, expands to form a septum and then the bud differentiates into teliospore initial, pedicel and primary teliospore cell. The primary teliospore cell then develops into a two-celled teliospore with two nuclei in each cell after completion of the primary teliospore nuclear division, and subsequently each cell becomes diploid uninuclear and is separated from the other cell by a septum.

3.1.4 Development and Function of Haustoria

Haustoria are a kind of special structure in rust fungi growing in plant cells when infecting host plants (Szabo 2001). Haustoria of the stripe rust pathogen are morphologically similar to those of other rust fungi, but quite different from the powdery mildew fungus (*Blumaria graminis* (DC.) Speer f. sp. *tritici*) (Szabo 2001). Haustoria penetrate the host plant's cell wall and siphon nutrients from the space between the cell wall and plasma membrane but do not penetrate the plant's cell membrane itself.

3.1.4.1 Development of Haustoria

The development of haustoria can be divided into four major stages, including the formation of haustorial mother cell, the formation of penetration peg, the extension of haustorial neck and the expansion of haustorial body (Kang et al. 2002; Ma and Shang 2004). A haustorial mother cell is located in the tip of an infective hypha and close to the host cell wall (Fig. 3.13a). A haustorial mother cell contains similar organelles as an infective hypha, including nuclei, endoplasmic reticula, ribosomes, microtubules, vacuoles, glycogens and lipid globules. The haustorial mother cell wall consists of six layers with various chemical compositions, such as polysaccharides, proteins, chitin and other substances (Fig. 3.13b), more complicated than that of an infective hypha (Chong et al. 1985; Kang et al. 1993). Then from the fifth and sixth layers that are rich of polysaccharides, an infective structure called penetration peg is produced and it penetrates the host cell wall (Fig. 3.13c). In wheat stem rust and leaf rust fungi, one haustorial mother cell produces only one penetration peg to invade the host cell wall. In contrast, two unusual phenomena of *Pst* penetration peg have been observed: sometimes one haustorial mother cell produces two penetration pegs (Fig. 3.13d) and penetration peg invades its own hypha cell (Fig. 3.13e) (Kang et al. 1994). After penetrating into the host cell wall (Fig. 3.14a), penetration peg continues to extend within the host cell (Fig. 3.14b) and develops a haustorial neck (Fig. 3.14c). Haustorial neck contains common organelles, including nuclei, endoplasmic reticula, ribosomes and mitochondria. Interestingly, the haustorial neck does not pierce the plasma membrane of host cell.

With the development of a haustorial neck, an immature haustorium begins developing (Fig. 3.14d). When a haustorial neck stops to extend, its tip begins to expand and the haustorium becomes mature. The mature haustorium is spherical (Fig. 3.15a) or has multiple prominents (Fig. 3.15b), and is wrapped by a specialized plant plasma membrane called extrahaustorial membrane (EHM) (Fig. 3.15c). EHM is physiologically different from the normal plant plasma membrane, which does not have ATPase activities (Harder and Chong 1991; Hahn and Mendgen 2001). In addition, the cell wall of haustorium is comprised of inner and outer layers (Fig. 3.15c). A special range named extrahaustorial matrix (EM) exits between EHM and haustorial cell wall (Fig. 3.15c) (Harder and Chong 1991), which is rich

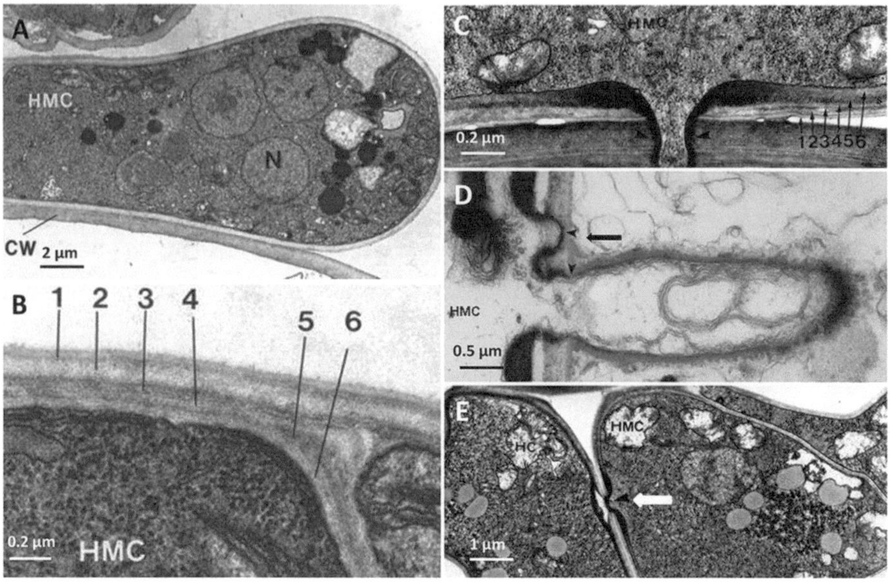

Fig. 3.13 Ultrastructure of *Puccinia striiformis* f. sp. *tritici* haustorial mother cell (*HMC*) and penetration peg. **(a)** A *HMC* is located in the end of an infection hypha and is attaching to the host cell wall (*CW*). (**b**) The cell wall of *HMC* contains six layers. (**c–e**) The features of penetration pegs. Penetration peg (*arrows*) is derived from the fifth and sixth layers of haustorial mother cell (*HMC*) wall (**c**). (**d** and **e**) Two abnormal phenomena of penetration pegs: one haustorial mother cell (*HMC*) can produce two penetration pegs (*arrows*) in **d** and penetration peg invades its own hypha cell (*HC*) in **e**

in fibrous structures (Fig. 3.15d). Usually, haustorium is adjacent to the nucleus or chloroplast of host wheat cells (Fig. 3.16a, b). The host cell protoplasm adjacent to the haustorium also containmany fibrous structures observed in an EM (Fig. 3.16c, d), some of which connect with the host cell plasma membrane (Fig. 3.16e). Thereafter, EM is considered to play an important role in maintaining the biotrophs characteristics and the exchange of material and information between rust fungi and host plants (Heath 1997).

Three dimensional imaging of temporal and spatial of haustoria in wheat plants further revealed the development of haustoria in size and morphology (Sorensen et al. 2012). The average haustorium size significantly increased and their shapes also changed obviously during 22–116 hpi. All observed haustoria were spherical at 22 hpi, but at 44 hpi most haustoria had an irregular structure and at 92 hpi all observed haustoria were branched. Along with the changes of the haustorial body, the haustorial neck changed from being narrow and slender to expand with a rough and invaginated structure.

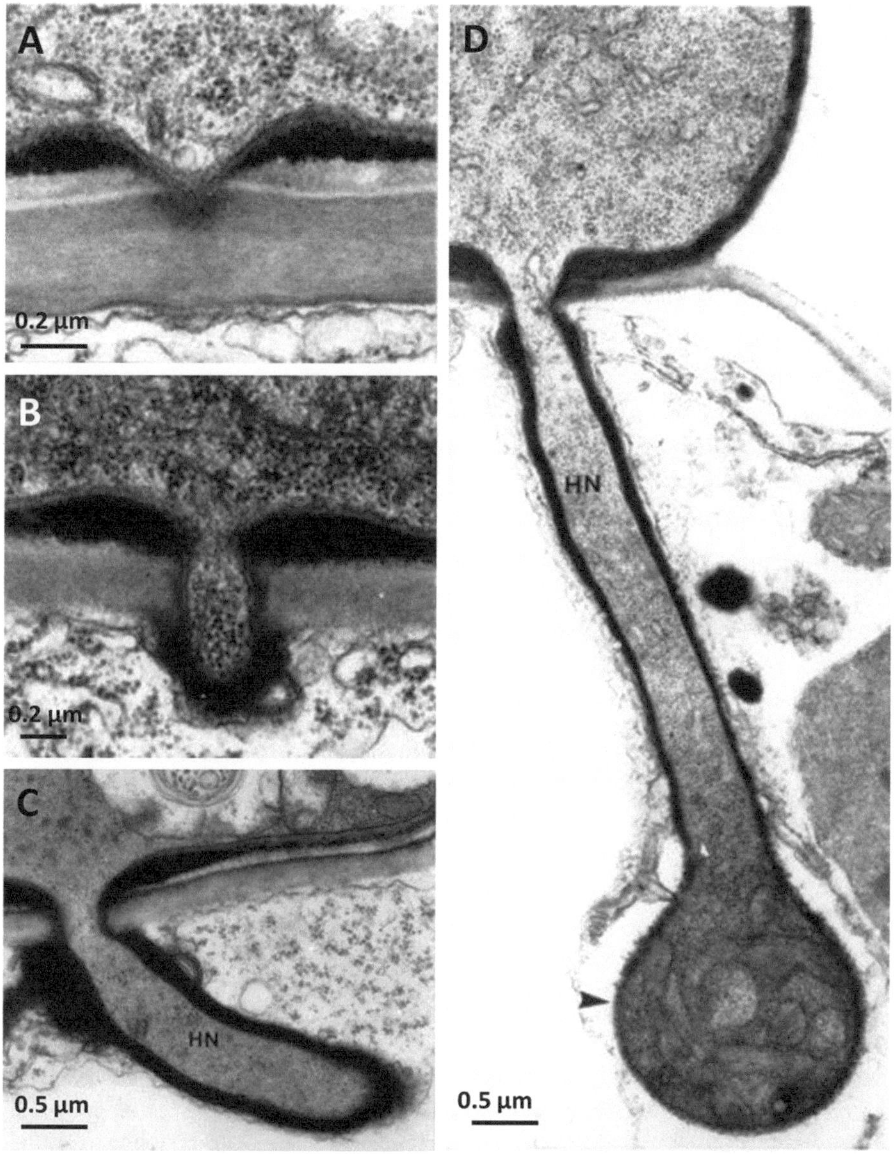

Fig. 3.14 Formation of *Puccinia striiformis* f. sp. *tritici* haustoria. (**a**) A penetration peg begins to penetrate the host cell wall. (**b**) The penetration peg has penetrated host cell wall. (**c**) Extending haustorial neck (*HN*); and (**d**) the development of HN and the development of an immature haustorium (*arrow*)

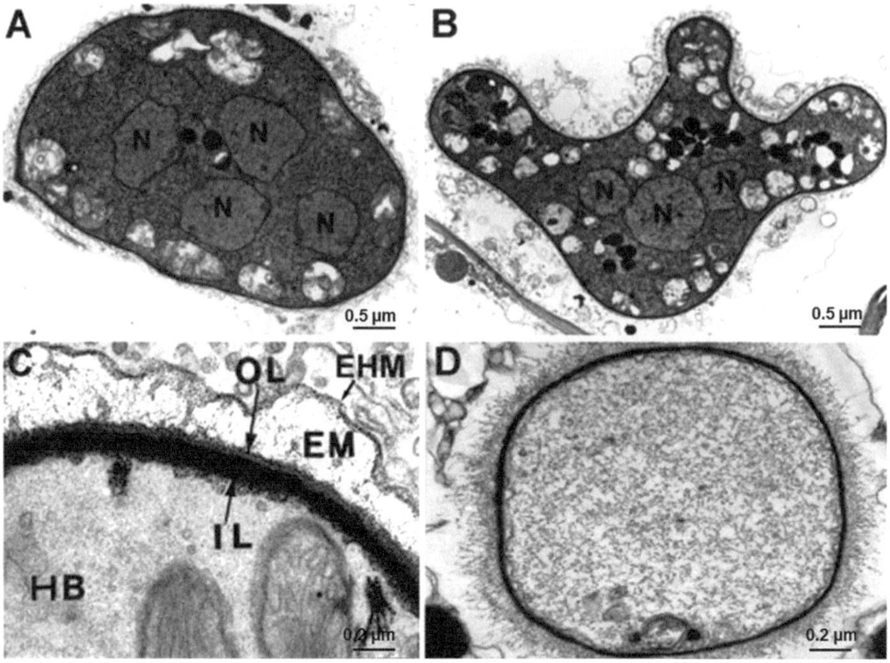

Fig. 3.15 *Puccinia striiformis* f. sp. *tritici* haustoria. (**a** and **b**) A mature haustorium is either spherical (**a**) or has multiple prominents (**b**). (**c**) The structures of a haustorium showing the inner layer (*IL*); outer layer (*OL*), extra-haustorial membrane (*EHM*) and extrahaustorial matrix (*EM*). (**d**) An EM is filled with fibrous structures

3.1.4.2 Functions of Haustoria

As a unique infection structure that invades into host cells, a haustorium makes intimate contact with the host cell membrane. Thus, haustoria are considered as the major exchange place for signals between the rust fungus and host plants, and also the major place for the parasite to draw nutrients and water from the host. Abundant metabolic activities have been observed in haustoria of *U. fabae*, including the amino acid transport, vitamin synthesis and sugar uptaking (Hahn et al. 1997a, b; Hahn 2000; Sohn et al. 2000; Voegele et al. 2001; Voegele and Mendgen 2003), proving the nutrient uptake function of haustoria. In the *Pst* genome, the haustoria specific metabolite-related genes are also identified in the whole genome sequence (Zheng et al. 2013) and haustorial cDNA library (Yin et al. 2009), indicting the similar role of haustoria of *Pst* in nutrient uptaking.

In addition, haustoria are also in charge of exchanges of small RNA molecules and proteins. Although no detailed evidence for small RNA exchanges has been obtained due to the technical difficulties, indirect evidence has been obtained for small RNA exchanges through haustoria. Based on this evidence, the host induced

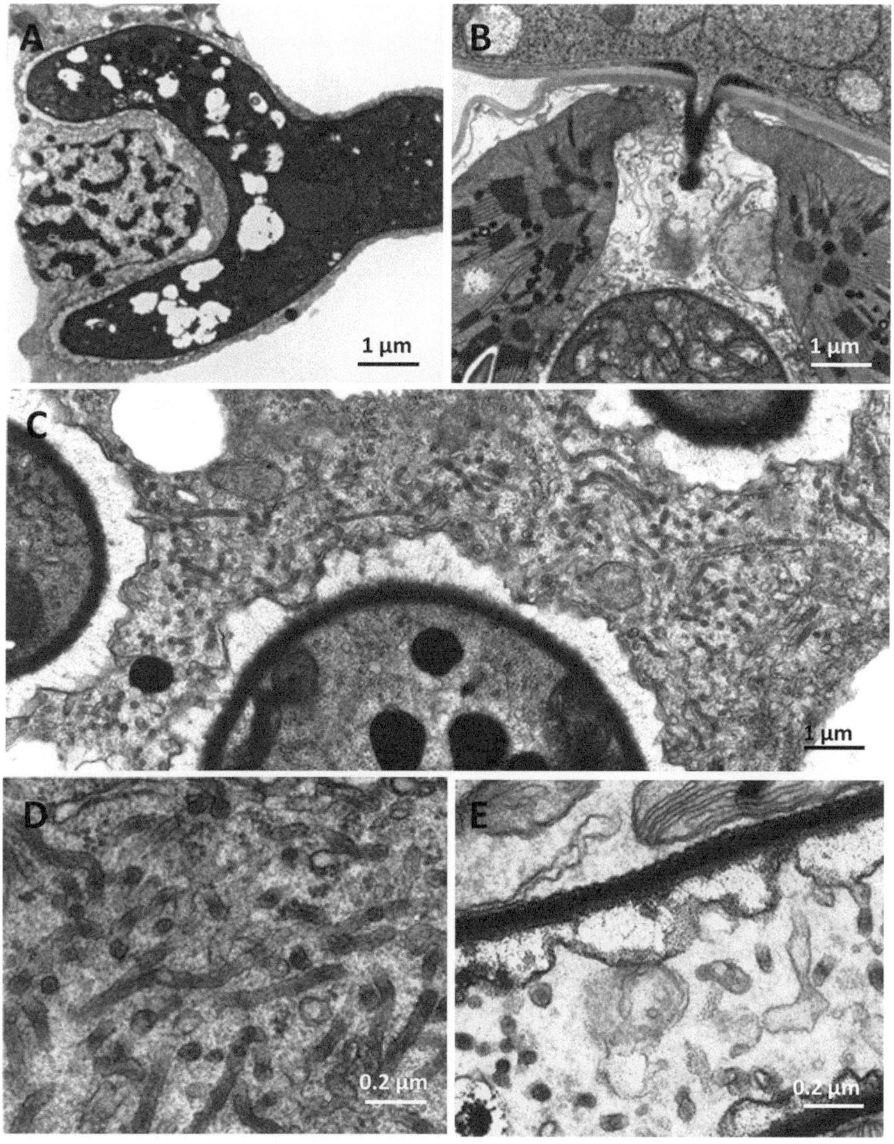

Fig. 3.16 Features of a *Puccinia striiformis* f. sp. *tritici* haustorium. A haustorium is usually adjacent to the nucleus (**a**) or a chloroplast (**b**) of wheat host cells. (**c**) The host cell protoplasm adjacent to a haustorium also contains many fibrous structures. (**d**) The enlargement of fibrous structures. (**e**) Some of fibrous structures connect with the host cell plasma membrane

RNAi system has been established successfully to silence the expression of *Pst* genes during the interactions between *Pst* and wheat (Yin et al. 2011). The host-induced RNAi system is proven to be most efficient for those *Pst* genes that are highly expressed in *Pst* haustoria (Yin et al. 2011). Moreover, haustorium has been reported to be rich in secreted proteins. A large number of haustoria-specific or induced genes have been isolated from the haustoria cDNA library or transcriptome as presented in Chap. 2. It appears that haustoria release a series of secreted proteins into host cells to regulate host immunity and thus promote pathogen virulence. Therefore, haustoria play a key role in determining the interaction between *Pst* and wheat plants.

3.2 Histology and Cytology of Plant Resistance and Defense Against *Pst*

Most plant species are immune to *Pst* infection, and therefore, considered as non-hosts for *Pst*. Within plant species of hosts, ecotypes of wild grass species and cultivars of cereal crops can be resistant to *Pst*. All of them have various ways to defense against *Pst* infection. Even in susceptible plants, there are still some basal levels of defense against the pathogen infection. Resistance will be discussed in Chap. 5, and here, we will focus on histology and cytology of plant resistance and defense against *Pst*.

3.2.1 Host Resistance

Wheat resistance to stripe rust can be broadly categorized as all-stage resistance (also called seedling resistance) and adult-plant resistance (APR) (Chen 2005, 2013). All-stage resistance can be detected at the seedling stage and expressed during all stages of plant growth, but APR is expressed when plants grow old. All-stage resistance is usually of high level, but race specific. All-stage resistance is easily overcome by the rapidly evolved new virulent races (Line and Qayoum 1992; Line and Chen 1995; Chen 2005). In contrast, APR is usually non-race specific and durable, but may not be adequate when the resistance level is low, disease pressure is high and environmental conditions are unfavorable for the expression of APR (Bariana and McIntosh 1995; McIntosh et al. 1995; Chen and Line 1995; Johnson 1988; Chen 2013). Screening of germplasm for stripe rust resistance, genetics, gene identification and mapping and developing resistant cultivars will be presented in Chap. 5. Here, we focus on cytological and ultrastructural studies on plant defense and resistance.

3.2.1.1 All-Stage (Seedling) Resistance

Histological and cytological studies revealed a series of defense responses involved in wheat all-stage resistance against *Pst* infection (Kang et al. 2003c; Wang et al. 2007; Zhang et al. 2008a), including cell wall apposition, papilla formation and accumulation of several antifungal compounds, such as lignin, calloses, plant hydrolases and reactive oxidative species (ROS). Cytochemical studies showed that resistant wheat plants contain higher amounts of lignin (Fig. 3.17), chitinase (Fig. 3.18), thionin (Fig. 3.19) and hydroxyproline-rich glycoprotein (Fig. 3.20) than susceptible wheat plants. Significant hypersensitive response (HR) was observed in the resistant wheat cultivars, which is usually accompanied by oxidative burst. For ROS, H_2O_2 accumulation peaked at 24 hpi (Fig. 3.21a, d) and O_2^- generation was the highest at 12 hpi (Fig. 3.21b, e) (Wang et al. 2007). HR surrounding the infection sites increased continuously from 20 to 96 hpi (Fig. 3.21c, f) (Wang et al. 2007). In addition, electron microscopy observation revealed a series of cytological defense responses in wheat all-stage resistance against *Pst* infection. The host cell wall in contact with intercellular hyphae became dark (Fig. 3.22a), and cell wall apposition occurred between the cell wall and plasmalemma (Fig. 3.22b). Ultimately, along with the necrosis of the host cells, the haustorial bodies became necrotized (Fig. 3.22c). Meanwhile, much more electron-dense materials were deposited in the space between the plasmalemma and the host cell wall in contact with the necrotic host cell (Fig. 3.22d).

The dramatic host resistance response effectively inhibited the fungal growth in all-stage resistant wheat cultivars (Kang et al. 2003b; Zhang et al. 2008a). Similar to a compatible interaction, *Pst* growth was observed in the incompatible interaction during the early infection process. However, the fungal growth was significantly inhibited in the late infection process (after 48 hpi), reducing hyphal growth, hyphal branching, haustorial mother cells, haustoria and colonies. Electron microscopy observations identified a series of abnormal changes of infective hyphae, haustorial mother cells and haustoria in the incompatible interaction. The hyphal cytoplasm contained abundant sacs, electron sparse patches (Fig. 3.23a), vacuoles and lipid drops (Fig. 3.23b). The vacuoles gradually fused into a large vacuole (Fig. 3.23c) and then led to the disintegration and necrosis of the cytoplasm (Fig. 3.23d). The cytoplasm of haustorial mother cells was also vacuolated and the cell wall was irregularly thickened (Fig. 3.23e and f). The haustorial body was surrounded by calloses or electron-dense material (Fig. 3.23g and h), and then the cytoplasm became aggregated and the organelles became disorderly distributed and eventually disintegrated. Meanwhile, the extrahaustorial matrix was irregularly widened, and a great amount of fibril or granular deposits accumulated in the matrix. The extrahaustorial membrane became wrinkled with branches growing into the cytoplasm of host cell and vesicles of different shapes accumulated around the haustorial body (Fig. 3.23i). Finally, the haustorial body became necrotic (Fig. 3.23j).

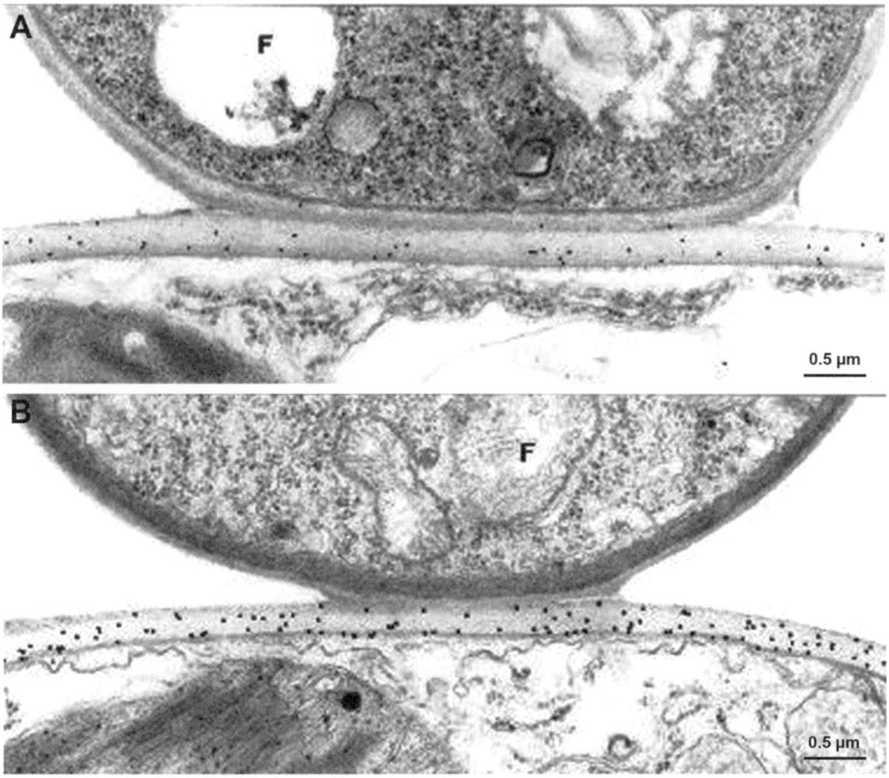

Fig. 3.17 Detection of resistant substance lignin in stripe rust susceptible (**a**) and resistant (**b**) wheat plants using the immune colloidal gold technique

3.2.1.2 Adult-Plant Resistance

As listed in Chap. 5, a good number of permanently named *Yr* genes, such as *Yr11-14*, *Yr16*, *Yr18*, *Yr29*, *Yr30*, *Yr36*, *Yr39*, *Yr46*, *Yr52*, *Yr59* and *Yr62*, and numerous temporarily named *Yr* genes and quantitative trait loci (QTL) have been reported to confer APR to stripe rust (Chen 2013; McIntosh et al. 2013; Zhou et al. 2014; Ren et al. 2012; Lu et al. 2014). However, only few studies on histological and cytological characterization of APR mediated by these adult-plant stripe rust resistance genes have been reported. Mares and Cousen (1977) studied macroscopic and microscopic events associated with the APR to stripe rust in winter wheat cultivars Cappelle Desprez, Holdfast and Maris Widgeon. They found that APR reactions were associated with reduced fungal colony growth rates, the appearance of scattered cells showing hypersensitive necrosis and with more general areas of chlorosis and browning of host tissue. They observed that the differences in resistance

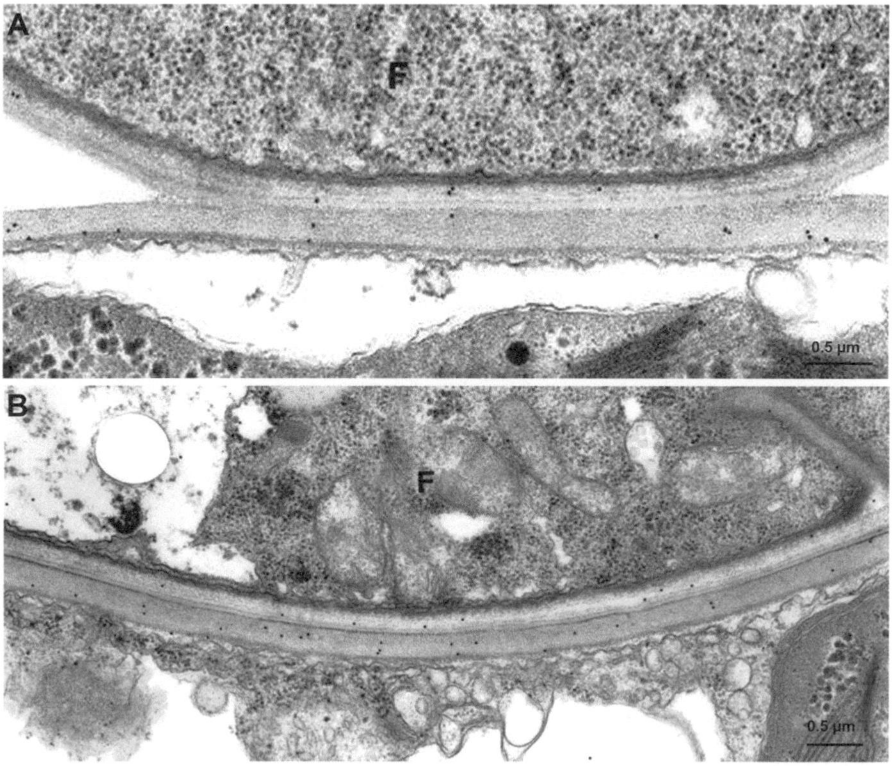

Fig. 3.18 Detection of resistant substance chitinase in stripe rust susceptible (**a**) and resistant (**b**) wheat plants using the immune colloidal gold technique

between cultivars were correlated with different frequencies of necrotic cells. They found that the rate of fungal colony growth during the latter half of the infection process and the development of tissue chlorosis and browning were well correlated with increasing resistance of successive leaves, final reaction type and relative differences in resistance between cultivars.

The *Yr18* and *Yr29* conferring APR resistance are not associated with significant HR (Singh 1992b; Rubiales and Niks 1995; William et al. 2003). *Yr36*, a kinase-START gene, confers temperature-dependent resistance to wheat stripe rust (Fu et al. 2009). To determine if the *Yr36*-mediated resistance is correlated to ROS burst and cell death, wheat near-isogenic lines with *Yr36* (UC1041+*Yr36*) and without *Yr36* (UC1041) were histologically characterized for response to *Pst* infection (Li et al. 2016a). *Yr36* conferred stripe rust resistance at both seedling and adult-plant stages when the gene line was tested with *Pst* race CYR29 at a high-temperature (HT) cycle (12 °C at night and 33 °C during the day). At the HT cycle, the growth of the secondary hyphae was obviously suppressed in both seedlings and adult plants of UC1041+*Yr36* compared to those of UC1041. The percentage of infection

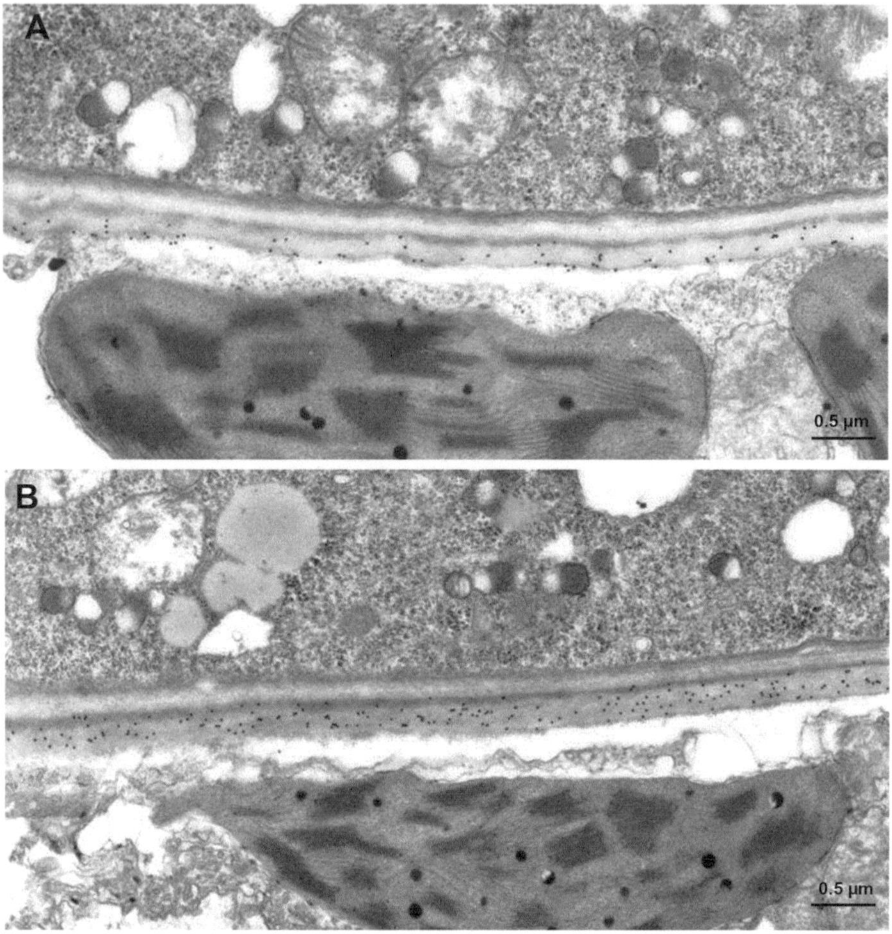

Fig. 3.19 Detection of resistant substance thionin in stripe rust susceptible (**a**) and resistant (**b**) wheat plants using the immune colloidal gold technique

sites with necrotic host cells in UC1041+*Yr36* were significantly higher than that of UC1041 60 hpi at both seedling and adult-plant stages. Mesophyll cell death in the inoculated UC1041+*Yr36* leaves at the HT cycle was stronger than that at a low-temperature (LT) cycle (12 °C at night and 18 °C during the day). At the HT cycle, the level of ROS burst started increasing in the inoculated leaves of UC1041+*Yr36* when *Pst* hyphae started differentiating and extending, and simultaneously, the number of penetration sites with hypersensitive cell death also increased. The results indicate that the *Yr36* product affects the ROS accumulation and cell death of the host in interaction with *Pst*.

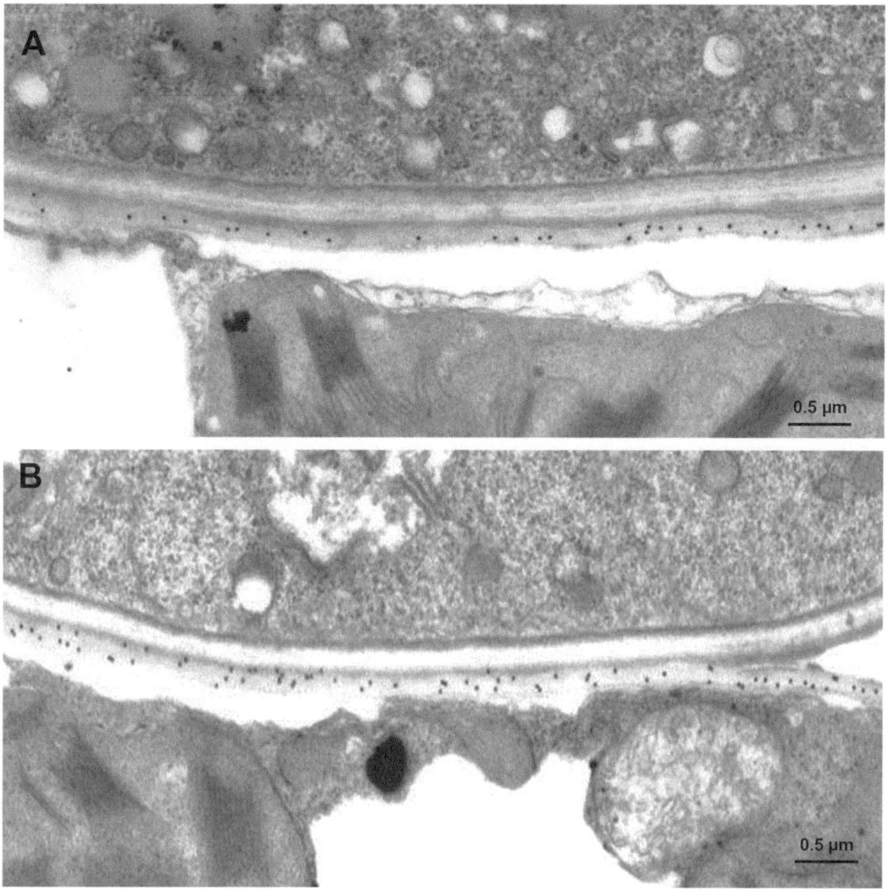

Fig. 3.20 Detection of resistant substance hydroxyproline-rich glycoprotein in stripe rust susceptible (**a**) and resistant (**b**) wheat plants using the immune colloidal gold technique

Wheat cultivar Xingzi 9104 (XZ) is an elite wheat germplasm possessing APR to stripe rust and has been proven to carry at least one APR gene (Liu et al. 2006). The APR of Xingzi 9104 was initiated at the tillering stage, gradually increased as the plant aged and highly expressed after the boot stage. Histological and cytological experiments were conducted to elucidate the mechanisms of APR in XZ (Zhang et al. 2012). The results showed that haustorium formation was retarded in the adult plants and that the differentiation of secondary intercellular hyphae was significantly inhibited, which decreased the development of microcolonies in the adult plants, especially in plants of boot stage. The expression of APR to stripe rust during wheat development was clearly associated with extensive HR with cell death of host

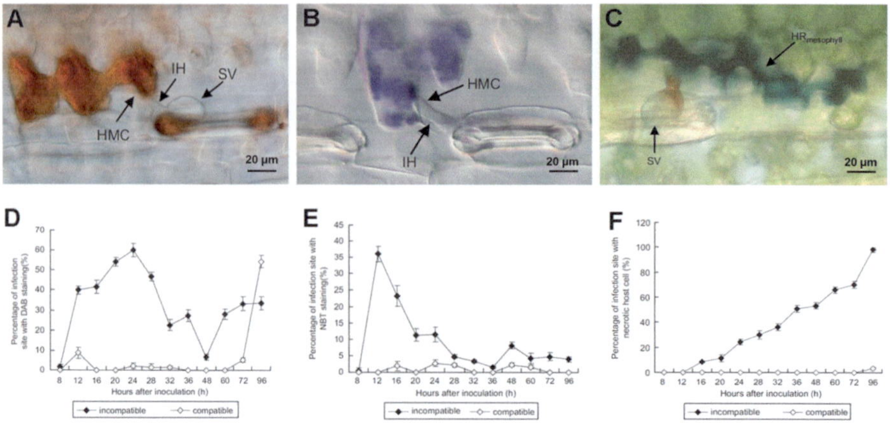

Fig. 3.21 Seeding resistance of wheat cv. 'Suwon 11' against avirulent race CYR23 of *Puccinia striiformis* f. sp. *tritici* (*Pst*). (**a**) Mesophyll cells in contact with haustorial mother cells and haustorial initials showing intensive H_2O_2 accumulation. (**b**) A mesophyll cell in contact with a haustorial mother cell showing O_2^- accumulation. (**c**) Hypersensitive response (*HR*) cells start to lose their original shape. (**d–f**) Percentage of infection sites of mesophyll cells exhibiting H_2O_2, O_2^- and HR accumulation after inoculation with avirulent race CYR23 and virulent race CYR31 of *Pst*. *SV* substomatal vesicle, *IH* infection hyphae, and *HMC* haustorial mother cell

cells and localized production of ROS (Fig. 3.24), which coincided with the restriction of fungal growth in infection sites in adult plants. At the same time, cell wall-related resistance (Fig. 3.25a, b) in adult plants prevented ingression of haustorial mother cells into plant cells. Haustorium encasement (Fig. 3.25c, d) was coincident with malformation or death of haustoria.

Kariega, a South African spring wheat cultivar has a high level of APR to stripe rust (Prins et al. 2005). Confocal laser scanning microscopy observation revealed that *Pst* growth was similar in both resistant (Kariega) and susceptible (cv. Avocet S) cultivars during the first 4 dpi (Moldenhauer et al. 2006). Afterward, the growth of *Pst* colonies increased exponentially in the susceptible cultivar, while fungal hyphal growth was attenuated in the resistant cultivar concomitantly with the earlier and stronger induced cellular lignification and necrosis of host tissue. In Kariega, two major QTL located on chromosomes 7D (*Qyr.sgi-7D*) and 2B (*Qyr.sgi-2B.1*) were identified (Ramburan et al. 2004). In responding to *Pst*, the wheat line carrying *QYr.sgi-2B.1* produced significant lignification and exhibited necrosis but failed to restrict the fungal growth, whereas a doubled haploid (DH) line containing *QYr.sgi-7D* did not exhibit lignified and necrotic host tissue, but limited the fungal growth (Moldenhauer et al. 2008). These results suggested that in addition to lignification and necrosis, the simultaneous and possibly synergistic induction of other resistance reactions may also be involved in the stripe rust APR in Kariega.

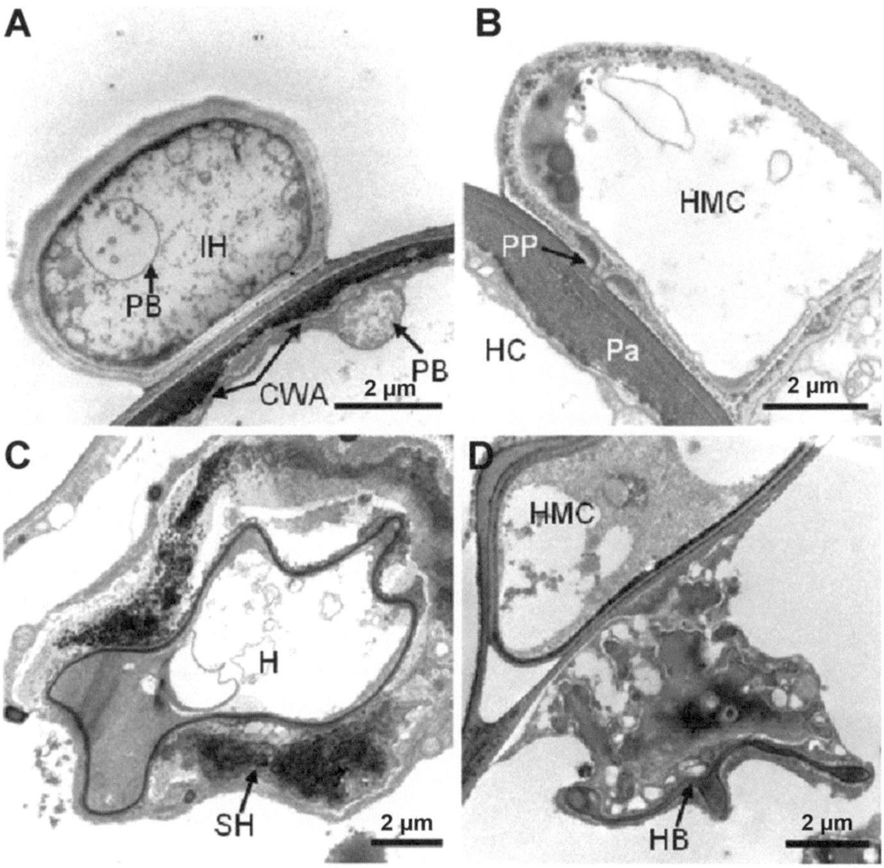

Fig. 3.22 Cytological defense responses in wheat all-stage resistance against *Puccinia striiformis* f. sp. *tritici* infection. (**a**) The cell wall of host wheat is more colored. (**b**) The host cell in contact with intercellular hyphae produces cell wall apposition (*arrow*) between the cell wall and plasmalemma, and the vesicles (*arrow*) come from the cytoplasm of host cell (*HC*) anastomosed with the plasmalemma. (**c**) The host cell (*HC*) is necrotized and a haustorial body (*HB*) inside it is also necrotic. (**d**) Electron-dense material (*arrow*) is deposited between the plasmalemma and the host cell wall in contact with the necrotic host cell (*HC*). *IH* intercellular hyphae, *PB* processing body, *HMC* haustorium mother cell, *SH* secondary hyphae, *H* haustorium, *HB* haustorial body, *Pa* papilla, *CWA* cell wall apposition

Three QTL, *QPst.jic-1B*, *QPst.jic-2D* and *QPst.jic-4B* contributing to the APR of wheat cultivar Guardian were identified by William et al. (2003). *QPst.jic-1B* does not prevent establishment of infection sites, but reduces hyphal growth and delays microcolony formation. The resistance conferred by this QTL is not associated with hydrogen peroxide accumulation or extensive plant cell death (Melichar et al. 2008). In lines carrying both *QPst.jic-2D* and *QPst.jic-4B*, the development of

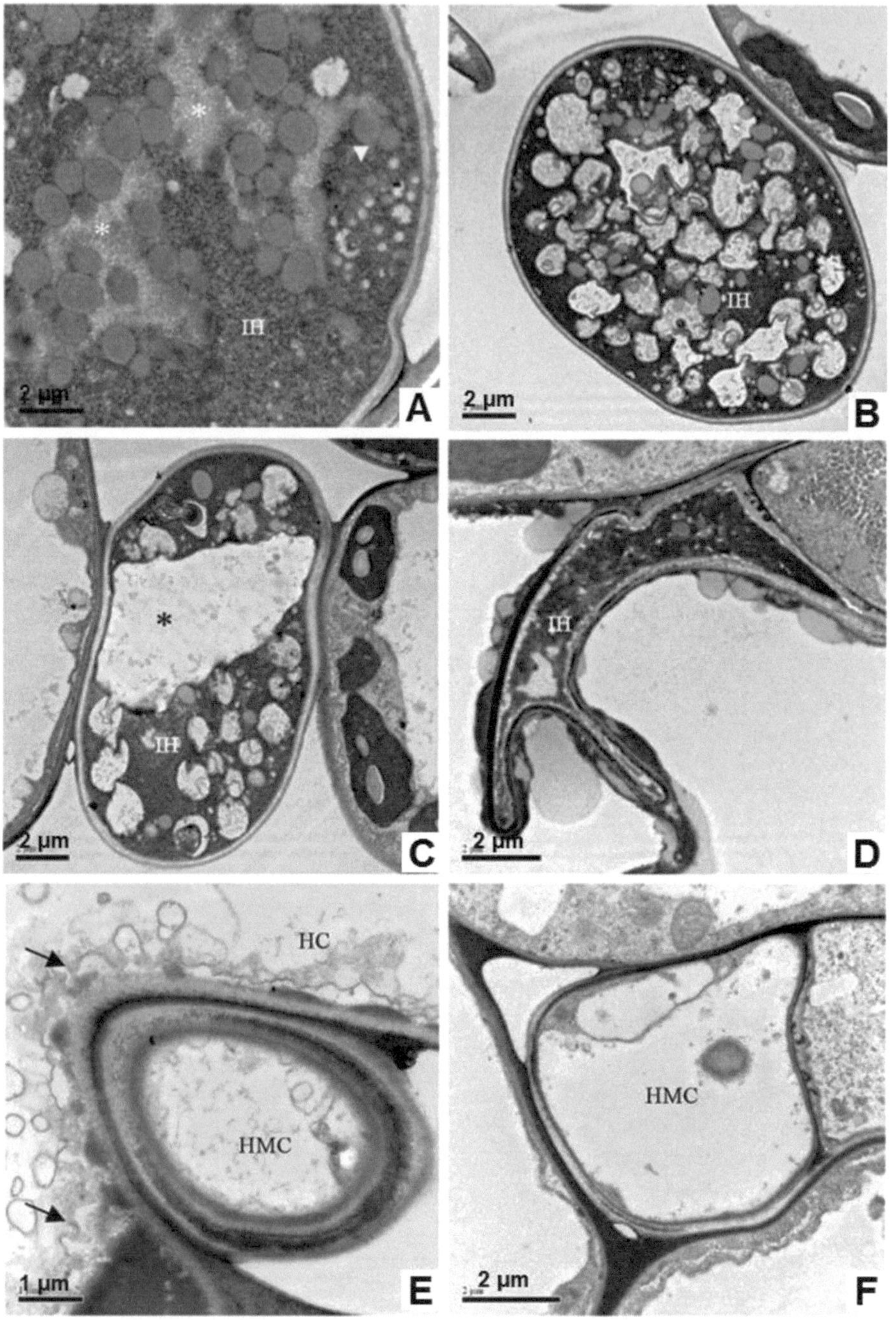

Fig. 3.23 *Puccinia striiformis* f. sp. *tritici* infection hyphae. Abnormal changes occur in infection hyphae (*IH*) (**a–d**), haustorial mother cells (*HMC*) (**e–f**) and haustoria (**g–j**). (**a**) Some sacs (*arrow*) and electron sparse patches (*asterisks*) appear in the hyphal cytoplasm. (**b**) Some vacuoles and lipid drops appear in the hyphal cytoplasm. (**c**) Vacuoles in hyphal cytoplasm gradually fused into a large vacuole (*asterisks*). (**d**) A hyphal cytoplasm begins to disintegrated and necrotized.

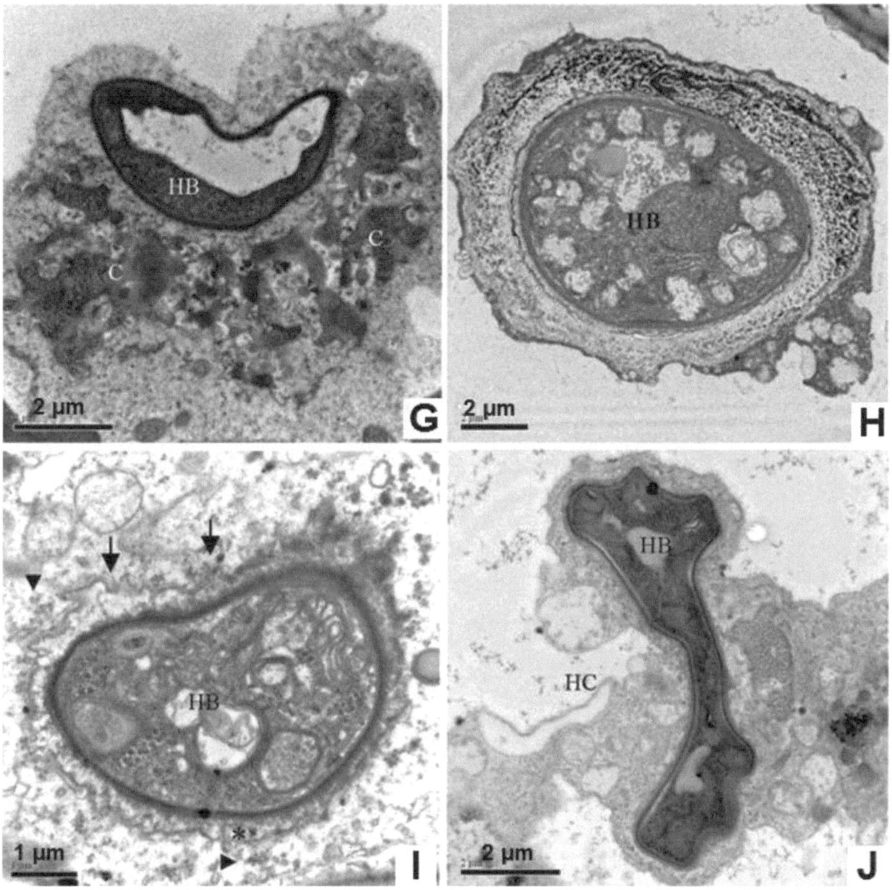

Fig. 3.23 (continued) (**e**) The cytoplasm of a HMC is vacuolated and cell wall of HMC is thickened irregularly and becomes loosened. (**f**) The cytoplasm of a HMC is vacuolated completely and the cell wall of a HMC is thickened irregularly and becomes darker. Much of callose (**g**) or electron-dense substance (**h**) is around a haustorial body (*HB*). (**i**) The cytoplasm of haustorial body becomes aggregating and the organelles are distributed disorderly and disintegrated; the extrahaustorial matrix is widened irregularly and great amount of fibril or granular deposits accumulated in it; the extrahaustorial membrane is wrinkled with branches into cytoplasm of host cell and some vesicles with different shapes accumulated around the haustorial body. (**j**) A HB becomes necrotic

fungal growth was inhibited significantly and no microcolony formation was observed, whereas microcolony formation and extensive cell death was observed in lines carrying only one of the two QTL (Jagger et al. 2011).

Based on the above discussion, different APR genes have different mechanisms to respond to *Pst* infection. APR of wheat to stripe rust is complicated, controlled by diverse genes and resulted from different biochemical pathways.

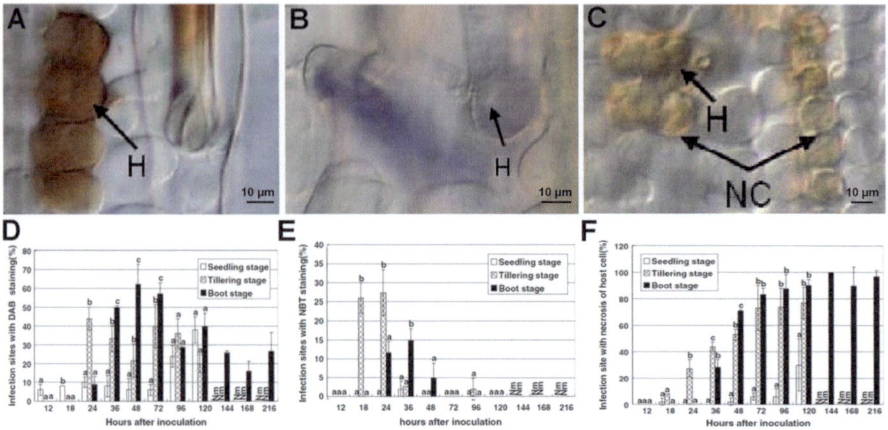

Fig. 3.24 The adult-plant resistance (*APR*) of wheat cv. 'Xingzi 9104' against avirulent race CYR32 of *Puccinia striiformis* f. sp. *tritici*. (**a**) Infected wheat mesophyll cells show intensive H_2O_2 accumulation by DAB staining. (**b**) Infected wheat mesophyll cells show intensive O_2^- accumulation by NBT staining. (**c**) Necrotic cells (*NC*) in infected wheat plants. (**d–f**) The percentages of infection sites of mesophyll cells exhibiting H_2O_2, O_2^- and NC accumulation in Xingzi 9104 leaves of plants at seedling, tillering and boot stages, respectively. *H* haustorium

3.2.2 *Nonhost Resistance*

Nonhost resistance (NHR), which is defined as resistance in all genotypes of a plant species against all isolates or races of a given pathogen species, represents the most robust and durable form of plant resistance in nature (Lipka et al. 2010). Numerous studies have been carried out to illustrate the histological and cytological features of NHR in different nonhost plants (*Arabidopsis*, broad bean, pepper, *Brachypodium distachyon* and rice) against nonpathogen *Pst* (Ayliffe et al. 2011a, b, 2013; Cheng et al. 2012, 2013; Zuo et al. 2013; Yang et al. 2014), and useful information has been obtained for better understanding the mechanisms of NHR to stripe rust.

3.2.2.1 *Arabidopsis* NHR to *Pst*

Arabidopsis is the most important model plants, but it cannot be damaged by any rust fungi. No visible symptoms were observed on *Arabidopsis* leaves inoculated with *Pst* (Cheng et al. 2013). Microscopically, *Pst* urediniospores germinated to produce a germ tube on Arabidopsis leaves, and some urediniospores were able to penetrate *Arabidopsis* leaves through stomata following an infection process similar to that on wheat (Fig. 3.26a, b). However, the frequency of stomatal penetration was significantly reduced in *Arabidopsis* compared with that in wheat. After penetration,

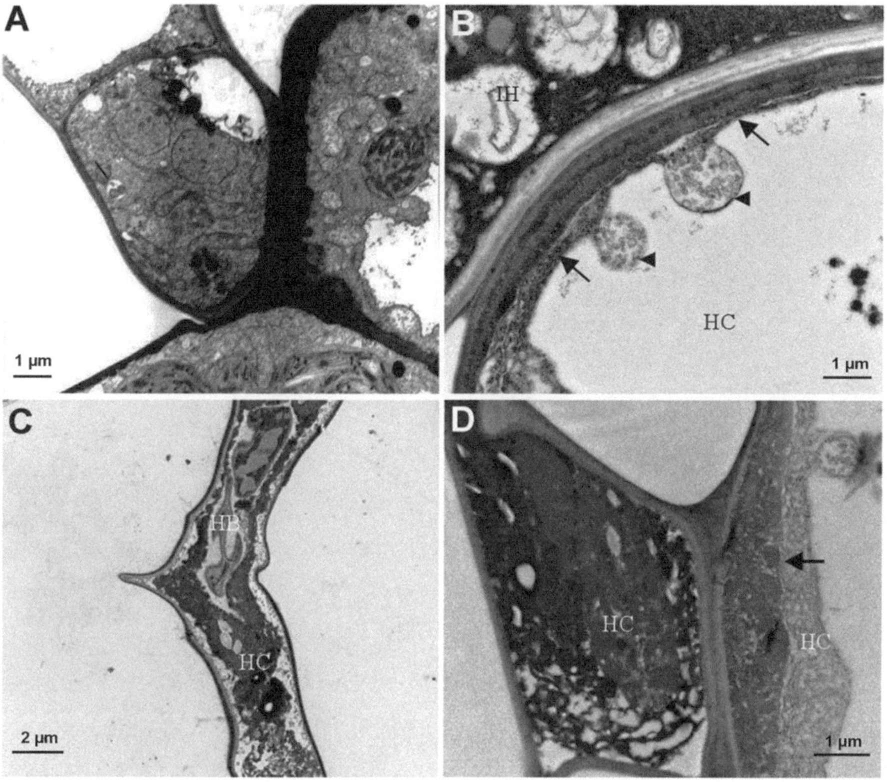

Fig. 3.25 Cytological defense responses in the adult-plant resistance of Xingzi 9104 to *Puccinia striiformis* f. sp. *tritici*. (**a**) The cell wall of a mesophyll cell contacting an infectious hypha (*IH*) shows a higher electron density than unaffected plant cells and exhibited localized cell wall appositions (*CWAs*) between the plant cell wall and plasma membrane. Paramural bodies (*PB*) underlined the CWA in the plant cell. The IH protoplasm was mostly devoid of organelles and cell wall-associated PBs were observed at the periphery of intact cells. (**b**) Attempted penetration by HMCs triggered a dome-shaped papilla (*Pa*) subtending the fungal HMC in the plant cell under attack. The penetration peg (*PP*) appears to puncture the plant cell wall but failed to penetrate the papilla and form a haustorium in the host cell (*HC*). (**c**) A haustorium surrounded by electron-dense sheath (*SH*) substance shows the vacuolation of protoplast. (**d**) Poorly formed haustoria become necrotic and irregular in shape within a live host cell

Pst produced substomatal vesicles within the substomatal chambers in *Arabidopsis* (Fig. 3.26c). The substomatal vesicles usually appeared brown at 24 hpi (Fig. 3.26c) and were detectable by auto-fluorescence (Fig. 3.26d). An active NHR response was involved in suppression of *Pst* growth in *Arabidopsis*. Localized modifications of the surrounding mesophyll cell walls closest to the fungus were observed (Fig. 3.26e). At some points of contacting with substomatal vesicles, the mesophyll cell walls showed an intense bright fluorescence (Fig. 3.26f), likely as a consequence of the accumulation of antimicrobial phenolic compounds. Aniline blue

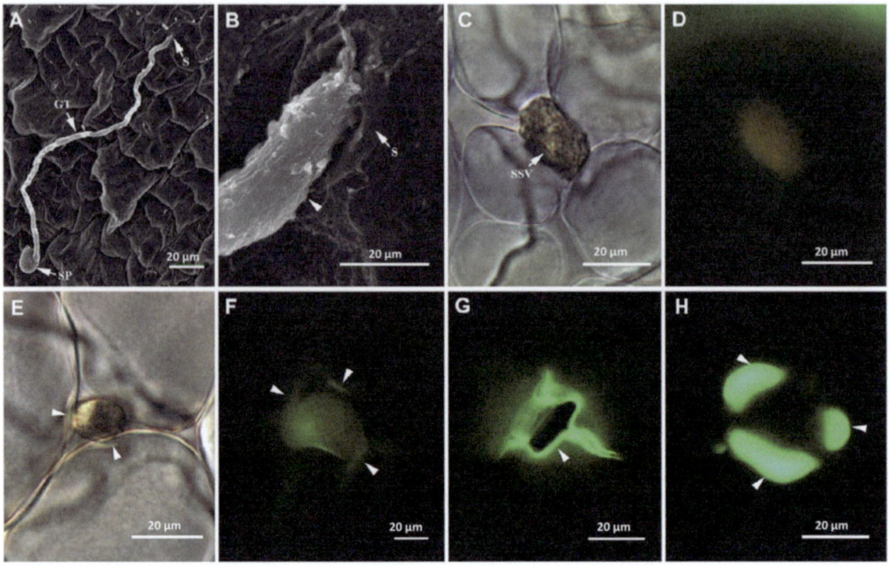

Fig. 3.26 Growth of *Puccinia striiformis* f. sp. *tritici* (*Pst*) and cellular responses in *Arabidopsis thaliana* Col-0 plants. (**a**) Infection of *Pst* in *Arabidopsis* leaves. *SP* urediniospore, *GT* germ tube, *S* stoma. (**b**) A germ tube (*arrow*) penetrating a stoma (*S*). (**c**) A small brown fungal substomatal vesicle (*SSV*) within the substomatal chamber. (**d**) The SSV in (**c**) detected by autofluorescence. (**e**) Localized modifications (*arrows*) of mesophyll walls closest to the fungus. (**f**) Plant cell wall showing bright fluorescence (*arrows*) next to the fungus. (**g–h**) Callose deposition detected in epidermal and mesophyll cells (*arrows*)

staining indicated the presence of callose deposition in both epidermal and mesophyll cells in contact with the fungus (Fig. 3.26g, h). However, there was no generation of the reactive oxygen species H_2O_2 and O_2^- in any of the samples examined by staining with 3,3-diaminobenzidine (DAB) and nitrobluetetrazolium (NBT), respectively.

3.2.2.2 Broad Bean NHR to *Pst*

The stripe rust fungus caused severe disease on wheat crops in the regions where legume crops are often also grown. However, *Pst* does not cause any problem on legumes. To understand the mechanisms of the NPR in legumes to *Pst*, Cheng et al. (2012) inoculated broad bean leaves with *Pst* and examined the inoculated leaves. Inoculation of *Pst* urediniospores caused no visible symptoms on broad bean leaves, successful penetration through stomata and haustoria formation were observed under a microscopy, although the number of haustoria was relatively low compared with broad bean leaves inoculated with *U. fabae* and wheat leaves inoculated with *Pst*. An active NHR response was shown to be involved in the suppression of *Pst* growth on broad bean. A dome-shaped papilla was formed on the inner side of broad bean cell walls at the penetration site adjacent to the HMC (Fig. 3.27a, d). Thickened cell walls were also present in broad bean mesophyll cells in contact with HMC (Fig. 3.27b). Accumulation of H_2O_2 was detected at the sites in direct contact with a substomatal vesicle (SSV), haustorial mother cell (HMC) and in papilla (Fig. 3.27a, b). In the broad bean leaf tissue with haustorium formation, accumulation of H_2O_2 was detected at deposits responsible for haustorium encasement (Fig. 3.27c). Accumulation of H_2O_2 was occasionally observed in the HMC (Fig. 3.27a). The percentage of infection units with accumulation of H_2O_2 increased rapidly from 12 to 24 hpi and then gradually declined. Accumulation of O_2^- was not detected in plant cells of any sample. Aniline blue staining indicated that callose deposition was associated with the *Pst* infection units, and staining at 24 hpi showed the same distribution as H_2O_2 in the plant cell walls at the sites directly contacting with SSV, HMC, papilla and at the encasement of haustoria (Fig. 3.27d–f). Therefore, attempted *Pst* infection of broad bean resulted in rapid production and accumulation of H_2O_2 (but not O_2^-) and callose deposition, presumably related to the generation of subcellular and localized physical barriers. The cell walls of mesophyll cells in contact with infective hyphae and neighboring mesophyll cells showed intense bright fluorescence (Fig. 3.27g, h), which is likely related to the accumulation of antimicrobial phenolic compounds. Hypersensitive response with cell death occurred at infection sites with haustoria formation as evidenced by plant cell autofluorescence at 24 hpi (Fig. 3.27i). Penetrated mesophyll cells began to collapse starting from 48 hpi (Fig. 3.27j). Collectively, attempted infection induced the formation of papillae, cell wall thickening, ROS production, callose deposition and accumulation of phenolic compounds in non-host plant cells. The few haustoria that were formed by *Pst* in broad bean cells were encased in reactive oxygen and callose materials and elicited a post-haustorial HR.

3.2.2.3 Pepper NHR to *Pst*

The NHR of pepper to *Pst* was studied by comparing to the wheat host resistance to *Pst* (Zuo et al. 2013). Similar high germination rates of *Pst* were observed on both non-host pepper and resistant wheat plants. However, the majority of germ tubes failed to form substomatal vesicles and fewer haustoria formed in pepper, and secondary hyphae were not formed. During the NHR of pepper to *Pst*, the accumulation of ROS species was considerably high. In contrast, significantly higher proportion of haustoria, a much lower production of H_2O_2 and secondary hyphae were detectable in the incompatible host resistance deployed by wheat. The incidence of HR was found to be slightly higher in pepper. These results indicate that H_2O_2 and HR play important roles in the NHR of pepper to *Pst*, and NHR may share similar mechanisms with host resistance.

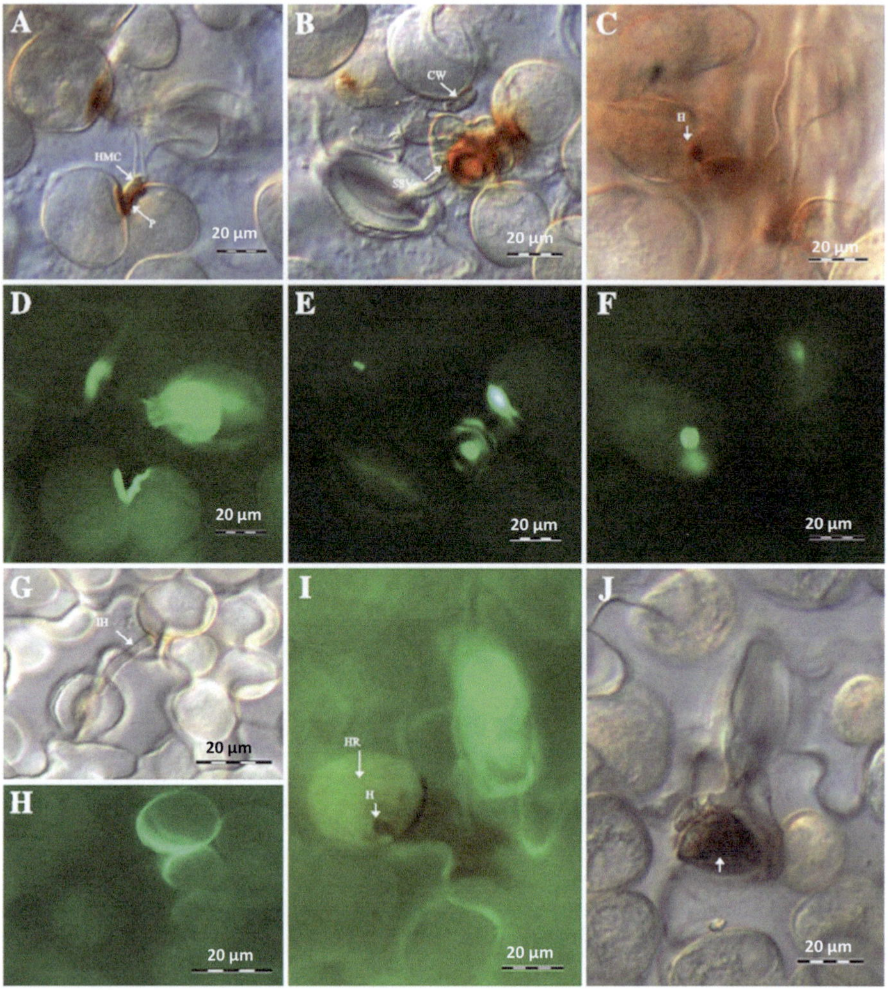

Fig. 3.27 Broad bean responses to attempted *Puccinia striiformis* f. sp. *tritici* (*Pst*) infection. (**a**–**c**) Inoculated broad bean leaves were stained with 3,3-diaminobenzidine (*DAB*) to detect production of H_2O_2 (*reddish brown*) at 24 h post inoculation (*hpi*). A dome-shaped papilla (*P*) (**a**) and thickened cell walls (*CW*) (**b**) are formed in the broad bean cell walls in contact with the fungal haustorial mother cells (*HMC*). DAB staining is visible in the mesophyll cell wall adjacent to a HMC (**a**), mesophyll and epidermal cells in contact with a substomatal vesicle (*SSV*) (**b**), the papilla (**a**), the encasement surrounding a haustorium (*H*) (**c**) and an HMC of *Pst* (**a**). (**d**–**f**) The callose deposition (intensive green fluorescent) in coincident with the H_2O_2 distribution (**a**–**c**) is detected using the aniline blue staining at 24 hpi. (**g**) A mesophyll cell in contact with an infection hypha (*IH*) at 24 hpi. (**h**) Bright fluorescence is observed in the cell walls of a mesophyll cell, and a neighboring mesophyll cell at the same infection unit shown in (**g**). (**i**) In some infection units with the formation of haustoria (**h**), plant cell autofluorescence, indicative of hypersensitive response is detected at 24 hpi. (**j**) Collapsed mesophyll cell (*arrow*) at 72 hpi

3.2.2.4 *Brachypodium distachyon* NHR to *Pst*

Brachypodium distachyon is a model for various studies of temperate grasses. More than 140 *B. distachyon* accessions have been tested for reactions to *Pst* (Ayliffe et al. 2013). The macroscopic symptoms varied from apparent immunity in some lines to the formation of lesions ranging from sparse to common. Callose deposition patterns during infection by *Pst* showed similarity to the wheat basal defense response, while cell death did not appear to be a major component of the defense response. This study indicated that *B. distachyon* is generally a non-host with a small number of accessions that can be infected by *Pst*, which makes the species valuable to study the evolution from nonhost to host.

3.2.2.5 Rice NHR to *Pst*

Rice is one of the most important cereal crops and is also an excellent monocot model for genetic, molecular and genomic studies (Gale and Devos 1998). Unlike other cereals, rice does not have any rust diseases and is considered as non-host to all rust fungi. Histological and cytological studies showed that *Pst* infected rice and produced all infection structures necessary for colonization, including haustoria, but cannot sporulate (Ayliffe et al. 2011a, b). Morphologically abnormal infection structures or aberrant differentiation of *Pst* were also observed in rice (Ayliffe et al. 2011a; Yang et al. 2014). At the same time, an effective NHR was activated, including callose deposition, ROS production and occasional cell death. In addition, *Pst* exhibited different levels of growth and development in different rice varieties, but no rice varieties were able to support rust sporulation (Yang et al. 2014).

Based on the above discussion of NHR in various plant species, we can conclude that NHR comprises of a continuum and multi-layer defenses against *Pst*, including basic incompatibility (*Pst* germ tubes fail to recognize the stomata of non-host plants), structural and chemical strengthening of the cell wall, and occasional HR.

3.3 Physiology of Interactions Between Wheat and *Pst*

Like other rust fungi, the stripe rust fungus is an obligate biotrophic parasite and has evolved a highly intimate and special relationship with plant hosts during the mutual adaption and competition. Different from facultative parasite-host or facultative saprophyte-host interactions, upon *Pst* attack, wheat cells undergo a series of changes but continue to live for a while, meanwhile the rust fungi are able to grow and sporulate. The changes in the host are deemed to benefit the pathogen, but usually in ways that are not entirely clear. Each infection site is a battle field where intense metabolic activities of both host and pathogens take place. Changes in the host tend to be covered up by activities or substances from the pathogen.

Because *Pst* and other rust fungi cannot be artificially cultured and all operations must be performed in living plant tissue, research on plant changes in physiology during the infection process is relatively difficult compared to other plant pathogen systems. Many physiological changes in infected host cells have been described using special experimental approaches, and the physiology between wheat-rust fungi interaction is being gradually understood. Due to its huge harmfulness and relatively intensive studies, wheat stripe rust has received much attention in physiological plant pathology. To date, the essential changes in structure, hormone and metabolic activities in *Pst*-infected wheat leaves have been well studied.

3.3.1 Structural Changes in Pst-Infected Host Tissues

Because *Pst* utilizes water and nutrients from infected plants, diseased plants appear dry even under good moist conditions. If infection occurs in the early growing season, diseased plants will have reduced number of tillers, short plants, small and few spikes or even no spikes, early senescence and reduced number of and small size of kennels. As a result, stripe rust causes various levels of grain yield and quality reduction.

3.3.1.1 Senescence and Related Cellular Changes

Rust infection accelerates the senescence of the host tissues that exhibit loss of water, nucleic acids, proteins, and chlorophylls (Shaw et al. 1958; Shaw and Manocha 1965a). These losses reflect a definite pattern of cytological change, which is characterized by degeneration in the fine structure of subcellular organelles, culminating in the disintegration of the nucleus and the plasma membrane. Approximately 10 days after inoculation with a rust fungus, the cytoplasmic organelles of host cells at pustule centers begin to degenerate and eventually disappear in a slow autolytic digestion process. The endoplasmic reticulum (ER) becomes less abundant and breaks into vesicles. Mitochondria lose their inner membranes (cristae), swell and become vesiculate (Shaw and Manocha 1965b), which may gain small electron-dense inclusions (Reiter et al. 1976).

Besides that another remarked change in rust infected wheat cells is the chloroplast. It has been observed that upon wheat stem rust fungus attack, the volume of host chloroplasts is usually reduced (Allen 1923; Whitney et al. 1962). Although chloroplasts sometimes appear swollen (Ehrlich and Ehrlich 1971) and the volume of chloroplast stroma can increase (Reiter et al. 1976), the chloroplast lamellae may become less compacted and the grana structure is disorganized. The outer chloroplast membrane becomes vesiculate (Shaw and Manocha 1965b) and eventually becomes ruptured (Ehrlich and Ehrlich 1971). Leaf rusts of wheat and barley cause a less detrimental response in which chloroplasts may not shrink (Allen 1928) or show signs of structural degeneration other than loss of chlorophyll (Calonge 1967; Reiter et al. 1976). A light and electron microscopy study of the interaction of *Pst* with a susceptible wheat cultivar revealed electron-dense materials in chloroplasts (Mares 1979), indicating the degradation of chloroplasts. The findings showed that disruption of chloroplast is a common phenomenon in rust infected wheat cells.

As part of the autolytic stage, the host nucleus shrinks rapidly. During wheat-*Pgt* interaction, Allen (1923) showed a decrease in wheat nuclear width and volume 14 dpi, which occurred about 4 days later than chloroplasts beginning to shrink. Similar reduction in host cell nuclear volume as chromatin within the nucleus condensed was observed in wheat cells attacked by wheat leaf rust fungus (Reiter et al. 1976). As autolysis progresses, most host cells remain alive and turgid for as long as 3–4 weeks after inoculation. Vacuoles form to contain residues of degenerated organelles, membranes and electron-dense bodies (Reiter et al. 1976). In advanced stages of autolysis, virtually the only host cell contents remaining are the host nucleus and the fungal haustorium as in corn leaves infected by *Puccinia sorghi* Schw. (Hilu 1965). In wheat leaves infected by *Pst*, the cytoplasmic layer lining the cell wall becomes highly attenuated (Mares 1979).

3.3.1.2 Changes in Host Membranes

In response to rust pathogen attack, the host membranes undergo series of changes and play different roles at the different pathogen infection stages. While penetration pegs of rust fungi invade and extend in wheat cells, host cytoplasm membranes are inward concaved and adhere to the outer wall of the penetration pegs. In the process of haustorial development, as the haustorial body swells inside host cells, the surface area of the plasma membrane around the haustorial body increases, making the host protoplasm separated from haustoria. The increases surface area of host cytoplasm membrane is not attributed to the mechanical expansion effect, but involves in the formation of a new kind of membrane, the extrahaustorial membrane. In fact, the host membrane around haustoria has been significantly changed in morphology and physiology. Harder and Chong (1984) found that the host cell membrane around haustoria of *Pgt* was thicker than that of the healthy host cell. Roland staining of host membrane revealed that the extrahaustorial membrane of flax rust fungi was of shallow color than the normal host cell membrane, which may be due to the changes in structure and constituent (Littlefield and Health 1979). Woods and Gay (1987) assayed the ATPase activity by cytochemistry and found that the extrahaustorial membrane below the neck ring had no ATPase activity. However, the part above the neck ring is of the similar ATPase activity with the healthy host membrane. These results suggest that the extrahaustorial membrane is different from normal host cell membrane in physiology.

Increasing evidence has proven that the cytoplasmic membrane of the host enclosing the haustorial body, the so called extrahaustorial membrane, is modified and therefore no longer resembles a conventional plant plasma membrane.

The extrahaustorial membrane of both bean rust and oat crown rust lacks intramembranous particles, and exhibits a dramatic reduction of sterols (Harder and Mendgen 1982; Harder and Chong 1991). Moreover, the extrahaustorial membrane lacks ATPase for powdery mildew and rust haustoria (Gay et al. 1987; Manners 1989; Baka et al. 1995), implying that there would be no control over solute fluxes from the host cell. Further support for a modification of the extrahaustorial membrane comes from recent studies obtained in the pathosystem *Erysiphe cichoracearum-Arabidopsis thaliana* (Koh et al. 2005), during which eight different plasma membrane markers accumulated in a collar-like formation around the haustorial neck were excluded from the extrahaustorial membrane. This neck region is characterized by electron-dense material, apparently joining the two plasma membranes (Harder and Chong 1984). This "neckband" seals the extrahaustorial matrix against the bulk apoplast (Heath 1976). Based on the sealing by the neckband and the presence of the plant plasma membrane surrounding the whole structure, it was suggested that the extrahaustorial matrix should be considered a symplastic compartment (Heath and Skalamera 1997). However, it might also be regarded as a highly specialized portion of the apoplast, providing conditions different from those present in the bulk apoplast.

With the development of the haustorial body, a zone of separation between the plasma membranes of parasite and host is formed, which is composed of the fungal cell wall and the extrahaustorial matrix (Hahn et al. 1997a). The extrahaustorial matrix resembles an amorphous mixture of mainly carbohydrates and proteins, partly of fungal but primarily of plant origin (Harder and Chong 1991). The initial biotrophic phase of hemibiotrophs is also characterized by the presence of a narrow contact zone between host and parasite (Perfect and Green 2001; Mendgen and Hahn 2002). It seemed that the extrahaustorial matrix represents a formidable trading place for the exchange of nutrients and information and plays an important role in maintaining the biotrophic lifestyle (Heath and Skalamera 1997).

3.3.1.3 Changes in Host Cell Walls and Lignification

Plants attacked by pathogens can evolve a plethora of sophisticated defense mechanisms to resist their potential colonization of microbial pathogens and parasites. In response to the avirulent *Pst*, the host cells usually can form two kinds of defense barrier structure in host cell wall. Cytochemical studies revealed that the first layer is the high electron density material produced between cell wall and plasma membrane, namely cell wall sediments, which contact directly with the hyphae (Kang et al. 2002, 2003c). Besides that, the host cells usually can secrete a lot of calloses at the infection sites where they form a mastoid structure around the haustoria to block the nutrient uptake of haustoria from host cells and ultimately lead to the haustoria abnormality, deformity and death (Kang et al. 1993).

In addition, the synthesis and accumulation of lignin plays an important role in plant disease resistance. Lignin constitutes a mechanical barrier to prevent pathogens from further extension. Lignin accumulation was reported as the main factors of low type disease resistance HR, and the increased levels of lignin are consistent with the host cell death process. Lignin deposition has been assumed to be closely related to the occurrence of the HR in wheat cultivars resistant to stripe rust (Tian et al. 2001). Wang and Shang (1996) found that the lignin levels were obviously increased in avirulent *Pst* infected wheat leaves, so was phenolic acids, the precursor of the lignin synthesis. These results indicate that lignin contributes to wheat resistance to stripe rust.

3.3.2 Hormonal Changes in Pst Infected Host Tissues

Plant hormones, including auxins, cytokinins, gibberellins, ethylene (ET), and abscisic acid (ABA), are important molecules in plant growth, reproduction and survival. They are essential for normal plant growth, but also play important roles in responding to biotic and abiotic stresses, particularly infection by rust fungi.

3.3.2.1 Auxins

Auxins are the earliest discovered endogenous plant hormones that can regulate broad physiological aspects of plant (Lambrecht et al. 2000). Indole-3-acetic acid (IAA) as a classic auxin, is involved in a myriad of developmental and environmental processes, including embryo patterning, cell division and elongation, vascular differentiation, lateral root initiation, gravitropism and phototropism (Zenser et al. 2001). Recent studies reported that auxin is also associated with plant defense and microbial pathogenesis (Kazan and Manners 2009). In plant tissues infected by a rust fungus, auxins apparently increase, particularly in the sites where the rust fungi produce galls and overgrowths of host tissue or at sporulation sites in infected cereal tissues (Pegg 1976). In a compatible host-parasite interaction, a 20-fold increase in the free IAA was observed 10 days after inoculation (Shaw and Hawkins 1958). Before sporulation, only a 2.5-fold increase of free IAA was observed in host cells from 6 to 144 hpi of *Pgt*, and esterified IAA did not increase (Artemenko et al. 1980), but during the sporulation of *Pgt*, the free IAA exhibits a 10- to 50- fold increase and relatively weak increase in bound IAA (Shaw 1963). The increased IAA may be attributed to the synthesized IAA by the rust fungus (Shaw 1963). Of course, it is also possible that a consequence of decreased IAA oxidation by oxidases (decarboxylases) or peroxidases (Daly and Knoche 1976; Pegg 1976). Yin et al. (2014) quantified the auxin level in *Pgt* infected wheat leaves. They found that IAA was low (0.01 ± 0.01 ng/g FW) in inoculated leaves 2 dpi, but increased quickly (3.28 ± 0.09 ng/g FW) at 4 dpi and (4.92 ± 0.66 ng/g FW) 6 dpi. The increased IAA may promote the respiratory or act in concert with other growth hormones, especially cytokinin, to control the metabolic state of the infected host cells in the juvenile stage of host response. Auxin may also have a direct function in growth and sporulation of rust fungi. In addition, auxin could be used by rust fungi as a non-protein effector due to the fact that auxin biosynthesis genes are also present in the rust genome (Spanu 2012) and IAA has been found in urediniospores of *Pgt* (Umnov et al. 1978). Yin et al. (2014) characterized a putative tryptophan 2-monooxygenasegene *Pgt-IaaM* from *Pgt* and found that the haustorially expressed gene is involved in the synthesis of the auxin precursor indole-3-acetamide (IAM) and the rust fungal pathogenicity.

3.3.2.2 Cytokinins

Cytokinin, a well-known plant growth hormone, promotes cell division, nutrient mobilization, and leaf longevity. Many plant pathogens are reported to secrete cytokinins or induce cytokinin production in host plants during host-pathogen interactions (Jameson 2000), and thus cytokinins have been implicated as inducers of the host response to pathogen infection. During wheat-rust fungi interaction, the cytokinin level was significantly increased in the rust fungus infected wheat leaves (Tissera and Ayres 1986). Cytokinins are mainly present in the fungal cytoplasm of the intercellular hyphal cells, haustorial mother cells, haustorial bodies and the extrahaustorial matrix, indicating the fungal origin of cytokinins (Hu and Rijkenberg 1998). However, in bean leaves infected with *U. fabae*, the extracted fraction resembled cytokinin was identified from the host, but not from rust fungal mycelia, indicating the host as the source of the increased cytokithe (Dekhuijzen and Staples 1968; Király et al. 1967). It is presumable that cytokinins can be originated from both rust fungi and host plants, implying its association with the nutrient translocation in the host-fungus interactions. Shaw and Manocha (1965a, b) demonstrated that cytokinin-treated tissues closely resemble tissues infected with *P. graminis* in ultrastructure. Cytokinin application could keep the detached leaves green and delay senescence. Interestingly, the green zones are metabolically active and closely resemble the green islands produced around rust pustules (Bushnell 1967). The cytokinin-induced green zone acts as a sink for substances translocated from the yellow, senescent parts of the detached leaf as do rust-induced green islands. The increases in amount of cytokinin and the similarities in effect of rust and cytokinin provides considerable evidence that cytokinins contribute to maintenance of the normal state in rust fungi infected tissues possibly by maintaining protein synthesis (Stoddart 1981). The stripe rust fungus produces auxins and cytokinins to enhance their pathogenicity and modulate the physiology of host plants (Walters et al. 2008; Choi et al. 2011).

3.3.2.3 Ethylene

Ethylene (ET) is commonly emitted from injured or diseased plant tissues in which cells are killed (Williamson 1950). It is well known that ET produced by plants plays different roles in response to various pathogens. Thomma et al. (1999) proposed that ET serves as a signal for defense against necrotrophic pathogens, whereas it promotes damages when infected with biotrophic pathogens. Rust fungi have been shown to cause the changes of endogenous ET in leaves of their hosts. Emission of ET from wheat leaves can increase tenfold or more as a consequence of virulent rust fungal attack, with the highest rates at the start of sporulation (Daly et al. 1971; Montalbini and Elstner 1977). In an incompatible wheat-rust fungus interaction, ET is sometimes emitted in bursts before or during the HR (Chigrin et al. 1978; Montalbini and Elstner 1977), which may contribute to host cell death. Besides increasing ET in infected host cells, rust fungi also potentiate the neighboring, uninfected tissues to sustain ET production following the normal burst, which may be important to the defense systems of the host (Ayres 1984). In addition, in *U. fabae* infected bean leaves, the increased ET is reported to be responsible for inhibiting stomatal opening, which further blocks the rust fungal penetration (Preeni and Ayres 1986). In wheat-*Pst* interactions, there is no direct evidence for any role of ET in wheat defense response. However, Xu et al. (2010) found that the acireductone-dioxygenase (ARD) gene *TaARD* in *Pst* was likely involved in ET synthesis as the gene expression was upregulated in the compatible interaction but downregulated in the incompatible interaction, indicating that different levels of ET in wheat cells attacked by virulent or avirulent *Pst* isolates. It is likely that ET plays a positive role in disease development caused by the biotrophic stripe rust fungus.

3.3.2.4 Abscisic Acid

Pathogen infection may be involved in changing endogenous abscisic acid (ABA) levels, which might play an indirect role in promoting senescence in healthy leaves and premature yellowing of detached leaves (Stoddart 1981). Application of exogenous ABA markedly reduced the deposition of silica and accumulation of phenolic compounds in the cell walls of bean plants inoculated with the nonpathogenic cowpea rust fungus *Uromyces vignae* (Li and Heath 1990). A significant decrease in ABA levels was found in both leaf tissues and intercellular washing fluids of rust fungus-infected bean leaves (Ryerson et al. 1993). However, ABA levels were increased in wheat leaves infected by *Pgt* (Chigrin et al. 1981). Tang et al. (2016) reported that ABA induced the expression of *TaADF3*, an actin depolymerizing factor gene in *Pst* that negatively regulates wheat resistance to stripe rust in an ROS-dependent manner. It is possible that the ABA changes in host tissues are associated with the invasion of rust fungi.

3.3.3 Metabolism Changes in Host Tissues Infected by Rust Fungi

Infection by rust fungi causes a series of changes in metabolisms in the host tissues, including those related to nucleic acids, proteins and carbohydrates.

3.3.3.1 Nucleic Acids and Proteins

Both RNA and DNA can be changed in cells of plant infected by infection of a rust fungus. The changes are different in plants infected by different rust fungal species and also by different isolates of a single species. Changes in nucleic acids are related to the consequences of resistance to or damaged by the pathogen.

3.3.3.1.1 RNA

Generally, infection by an obligate parasitic fungus has a great effect on RNA synthesis in the host. Increased total RNA content has been observed in compatible wheat-rust fungi interaction. Brown (1936) reported that rust infection markedly decreased the size of the chloroplasts, but increased the size of nucleoli in the infected mesophyll cells, which led to approximately a double RNA content of the nuclei in infected tissue. In *Pgt* infected wheat tissues, accompanied the increased volumes of host cell nuclei and nucleoli, the amount of both nucleolar and extra-nucleolar RNA doubles (Whitney et al. 1962; Bhattacharya et al. 1965). Meanwhile their metabolic activity was enhanced (Shaw 1963), including the doubled incorporation of uridine and cytidine (precursors of RNA) into nuclei and leucine into nuclear proteins. The incorporation of labeled uridine was inhibited by actinomycin D, indicating that incorporation depended on transcription (Heitefuss 1970).

Significant differences in total RNA content have been reported between *Pst*-infected wheat leaves and non-infected leaves. Shang et al. (1994) found that in compatible interactions, the total RNA content of *Pst* infected wheat leaves was significantly higher than that of the non-infected leaves from 5 dpi to sporulation (15 dpi), and the difference was the most apparent at 11 dpi. In contrast, in incompatible interactions, the total RNA content of infected wheat leaves was lower than that of the non-infected wheat leaves (Li and Zeng 2002). Even though *Pst* is growing and probably synthesizing RNA in susceptible wheat leaf tissues, the influence of *Pst* RNA on the total RNA content can be removed by the radioactively labeled method. It appears that *Pst* infection result in induced transcriptional activities in host nuclei, implying a considerable overall enhancement of synthetic activity in the host-parasite complex (Heitefuss 1966a).

3.3.3.1.2 DNA

Slight changes in the host DNA content have been reported in host tissues infected by rust fungi. The amount of nuclear DNA in *Pgt* infected wheat cells does not change until 9 dpi when a slow decline begins (Bhattacharya et al. 1965, 1968). The declined ratio of host DNA content at late infection stage may be due to the fungal DNA synthesis, which probably explains the increased incorporation rate of ^{32}P into a DNA fraction (Heitefuss 1965, 1966b) and the enhanced DNase activity

(Heitefuss and Wolf 1976). In wheat leaves infected by *Pst*, DNA synthesis was low (Li and Zeng 2002). The incorporation rate of synthetic precursor did not reach 1% and decreased gradually as the leaf was aging. In the wheat-*Pst* compatible interactions, the total DNA content of the host changed slightly before sporulation, but reached the peak at the sporulation stage with an increase by 15–19%. At the late stage of sporulation, similar amounts of total DNA were detected in the inoculated and non-inoculated leaves. In the incompatible interactions, there were no detectable changes on DNA content. The content of DNA in a nearly immune wheat cultivar infected by *Pst* was almost identical to the health leaves throughout the infection process, but significant differences were observed in a moderately resistant cultivar at the late stage of sporulation (13 dpi). In comparison, the effect of *Pst* infection on the host DNA content is lower than on the RNA content.

3.3.3.1.3 Proteins

The total amount of proteins in cereal host tissues can be increased by inoculated with a rust fungus. Increases of 20–50% on a fresh-weight basis, paralleling increases in dry weight, at least in the first few days after inoculation were sometimes observed (Shaw and Colotelo 1961; Quick and Shaw 1964). More frequently, the total protein of host and parasite either remains fairly constant (Samborski et al. 1961; Johnson et al. 1968) or declines (Gassner and Franke 1938). There is no doubt that some of the total protein are originated from the developing rust fungus, especially at sporulation and thereafter. In the wheat-*Pgt* pathosystem, early immunochemical and electrophoretic methods revealed that no fungal originated proteins could be detected at 5 dpi (Fric and Heitefuss 1970). After sporulation, two proteins were reduced significantly (Wrigley and Webster 1966). Cheung and Willetts (1973) reported that infection by an avirulent *Pgt* race led to a temporary 29% increase in soluble proteins, including malate dehydrogenase, glucose-6-phosphate dehydrogenase, catalase and amylase from the rust fungus. In the infected leaves, the activity of host glucose-6-phosphate dehydrogenase and catalase changed with the development of the pathogen, while the malate dehydrogenase and amylase of the host were unaffected. In resistant wheat leaves, no obvious changes of any of these enzymes were observed. During the flax-flax rust fungus interaction, no change or only slight decreased leucine incorporation into soluble proteins were observed 6–18 hpi, while increased incorporation into several protein fractions was found in incompatible interactions (von Broembsen and Hadwiger 1972). Similar results were obtained in oat leaves 10–24 hpi by oat crown rust (Tani and Yamamoto 1979). Application of blasticidin S, an inhibitor of protein synthesis, had no effect on crown rust development, suggesting that the protein synthesis in the host was not required for the early fungal growth. Recent studies on the pathogen-regulated genes in susceptible or resistant wheat cultivars revealed the different proteins at different infection stages, including the wall-associated kinase receptor with the leucine-rich repeat domain, putative serine/threonine protein kinase, calcium-mediated signaling protein and 14-3-3 protein (Dmochowska-Boguta et al. 2015).

Studies on the change of total protein of the host infected by *Pst* showed no difference in total protein content from 11 to 15 hpi in a nearly immune wheat cultivar, followed by a subsequent significant decrease in the total protein content (Yang et al. 1991; Zhang et al. 1994). A similar trend of total protein content was observed in a moderate resistant wheat cultivar, which had a larger decline range. In the compatible wheat-*Pst* interactions, the total protein content of the infected host was obviously reduced at the sporulation stage. Furthermore, it has been reported that some specific proteins of wheat were affected due to *Pst* infection. Fullington and Nityagopala (1986) reported that *Pst* infection increased the proportion of albumin-globulin and gliadin proteins but decreased the high Mr glutenins in varieties HD-2009 and WG-377. Studies on the proteome of compatible and incompatible interactions revealed that six proteins were positively regulated, which were involved in photosynthesis, antioxidant defense, signal transduction and metabolism (Li et al. 2011).

Cao et al. (2015) cloned *Pst* heat shock protein 70 (hsp70) and determined its responses in representative isolates with low-and high-temperature sensitivity to high-temperature stress during the heat shock process. Following heat shock at 28 °C, expression of hsp70 transcripts was up-regulated in all isolates with high-and low-temperature sensitivity, and reached the maximum level 2 h after heat treatment. However, the average expression level was increased by 7.12-fold upon heat shock for 2 h in the isolates with low-temperature sensitivity, but only 1.63-fold in the isolates with high-temperature sensitivity.

Using an iTRAQ-based quantitative proteomics approach, Yang et al. (2016) compared proteins in the incompatible *Pst* race CYR23 in infected and non-infected leaves of the wheat cultivar Suwon11. They identified 3475 unique proteins from three key stages of interaction (12, 24 and 48 hpi) and the non-inoculated check, of which 530 proteins were differentially accumulated during the *Pst* infection process, and 10.5% of the 530 proteins were classified as involved in the immune system process and stimulus response. A set of ROS metabolism-related proteins, peptidyl-prolyl cis-trans isomerases (PPIases), RNA-binding proteins (RBPs) and chaperonins were found to be involved in the responses to *Pst* infection. This study was the first to show that PPIases, RBPs and chaperonins participated in the regulation of the plant immune response to a pathogen.

3.3.3.2 Free Amino Acids

During the early responses of plant to pathogen attack, the free amino acids are probably synthesized locally from photosynthates and ammonia, and then translocated from tissues distant from the infection sites to provide the nutrition for fungus to grow. The amount of free amino acids could increase several folds in *Pgt* infected tissues especially in the first 3–6 days after inoculation (Samborski et al. 1961; Shaw and Colotelo 1961). Several amino acids have been identified to increase upon rust fungus infection, such as the glutamine, asparagine, arginine, phenylalanine, leucine or isoleucine and valine (Rohringer 1957; Farkas and Király 1961; Shaw and Colotelo 1961; Siebert 1961; Kim and Rohringer 1969). Growing evidence suggests that the increase of free amino acids was possibly used to meet the nitrogen uptake of pathogens. Mendgen (1979, 1981) reported that the exogenous ^{3}H-labeled amino acids could be up-taken by haustoria of *Uromyces* spp. In addition, putative haustorial secondary amino acid transporters were identified (Hahn and Mendgen 1997; Hahn et al. 1997b). AAT1p (PIG27p) was characterized as a broad specificity amino acid transporter with a main specificity for L-histidine and L-lysine (Struck et al. 2002). Another amino acid transporter, AAT3p, exhibits a substrate preference for L-leucine and the sulphur containing amino acids L-methionine and L-cysteine (Struck et al. 2004). Interestingly, RT-PCR analyses indicate that expression of AAT3 is not restricted to haustoria, indicating that amino acid uptake in *U. fabae* is not limited to haustoria. These specific amino acid transporters imply that nitrogen needed by rust fungi for growth and propagation may come from the host cells.

Infection by a virulent *Pst* isolate has a great influence on free amino acids of wheat plants. In the middle stage of infection, the level of free amino acids in *Pst* infected wheat cells are always higher than that of the healthy cells, and reach the maximum at the beginning of sporulation (Li and Zeng 2002). Similar increases of free amino acids in the sporulation stage were observed in the moderate resistant wheat cultivars. However, no obvious differences in the total amount of free amino acids were detected in near immune wheat cultivars. The increases of the total amount of free amino acids during the *Pst* infection are possibly caused by the increasing activities of proteolytic enzymes or the changes of amino acids de novo anabolism in the infected leaves.

3.3.3.3 Photosynthesis and Photorespiration

Photosynthetic activities are affected by various abiotic and biotic stresses. It is well known that *Pst* infection changes the photosynthetic rate of the host wheat. The photosynthetic rate in the nearly immune wheat cultivar Lovrin 13 inoculated with *Pst* slightly declined in the early infection stage, then decreased significantly by 10.0% when chlorotic blotches started to appear (Li and Zeng 2002). In a resistant wheat cultivar, the net photosynthetic rate is slightly higher than that in non-inoculated healthy leaves before symptom appearance, but was significantly declined after the symptoms and signs develop, especially after sporulation. In compatible interactions, the similar trend was observed with an even greater difference range. The net photosynthetic rate was decreased by 14.6% at 14 dpi and by 57.8% at 18 dpi (Table 3.2), which was possibly attributed to the decline of green leaf areas or damage of chlorophyll. During the late infection when uredinia were breaking through the epidermis of the host leaves, mesophyll cells were largely destroyed and resulted in the decrease of the photosynthetic rate. Based on the study from Doodson et al. (1965), the rate of net photosynthesis in heavily infected leaves by *Pst* is generally reduced to one-third to two-thirds of that in healthy leaves 8–12 dpi, respectively. It was worth noting that net photosynthesis in *Pst* infected leaves can be temporarily stimulated at 0.5% CO_2 in the first few days after inoculation (Doodson et al. 1965). Chen et al. (2015) reported that *Pst* infection caused a more significant

Table 3.2 Net photosynthetic rate of wheat plants inoculated with races of *Puccinia striiformis* f. sp. *tritici* under controlled conditions

Cultivar and treatment[a]	Net photosynthetic rate [$mgCO_2/(dm^2 \cdot h)$] and percentage (%)[b]					
	3d	6d	8d	10d	14d	18d
Lovrin 13						
Non-inoculated	9.99	8.90	8.72	7.30	7.08	6.80
CYR25	9.87	8.18	8.29	6.78	6.47	6.34
	98.80%	91.90%	95.10%	92.90%	91.40%	93.20%
CYR28	10.21	8.90	8.07	6.45	5.30	ND
	102.10%	100.00%	92.50%	89.60%	74.80%	
CYR29	9.70	8.93	7.26	6.48	6.05	2.87
	97.10%	100.70%	100.30%	93.70%	85.40%	42.20%
Huixianhong						
Non-inoculated	10.48	8.42	8.39	7.76	7.31	5.48
CYR25	10.51	8.21	7.82	4.65	3.07	2.83
	100.30%	97.50%	93.20%	59.90%	42.00%	51.60%

[a]Treatments are non-inoculated and inoculated with races (CYR25, CYR28 and CYR29) of *P. striiformis* f. sp. *tritici*

[b]Percentage of the net photosynthetic rate of the inoculated treatment vs. that of the non-inoculated treatment

decrease in the photosynthesis rate in susceptible cultivar Sy95-71 than in resistant cultivar CN19 at 72 hpi. In both the compatible and the incompatible interactions between wheat and *Pst*, the photosynthetic rate exhibited similar increases at 48 hpi and decreases at 120 hpi (Fig. 3.28) (Chang et al. 2013; Liu et al. 2015). The increased photosynthesis at high CO_2 concentration may be attributed to impairment of glycolic acid oxidase and the temporary enhanced ability of chloroplast in synthesizing proteins used in photosynthesis (Tani et al. 1973).

In addition, the net photosynthetic rate of *Pst*-infected wheat plants in the field of mild disease (disease severity 5%), moderate disease (disease severity 40%) and severe disease (disease severity 80%) in the field was monitored (Li and Zeng 2002). The results showed that the daily variation of net photosynthetic rate in mild disease leaves was similar to that of healthy leaves, but showed sensitivity to light intensity change. The net photosynthetic rate was low in the morning, quickly rose with the increasing of light intensity, and reached the maximum value at 10:00 A.M. Then the net photosynthetic rate gradually decreased as the temperature rises until the evening. The daily variation in leaves of mild disease followed an obvious single-peak curve. However, the daily variation of the net photosynthetic rate in moderate disease leaves had a different trend. Moderate disease leaves of susceptible cultivar Huixianhong can keep sensitivity to the increasing light intensity and a relatively high value was maintained from 10:00 A.M. to 12:00 P.M. and a single-peak curve with a much lower peak was observed. For moderate disease leaves of susceptible cultivars Mingxian 169, diseased leaves lost their sensitivity to the light

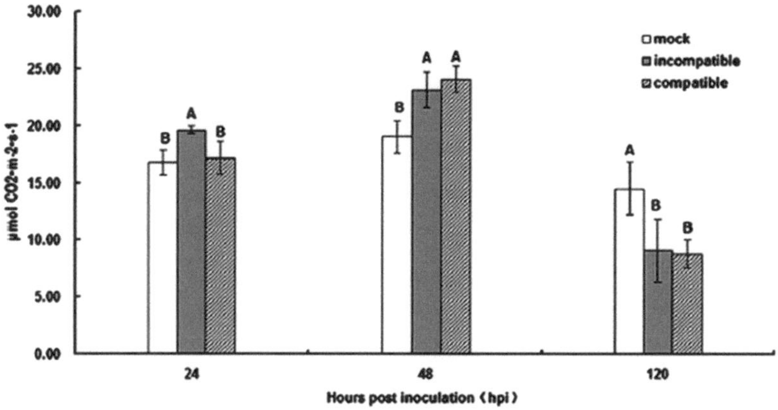

Fig. 3.28 Photosynthetic rates of wheat leaves infected by *Puccinia striiformis* f. sp. *tritici*. At 24 h post inoculation (*hpi*), the photosynthetic rate of the leaves in the incompatible system was higher than those in the other two treatments. The photosynthetic rates of the leaves in both the compatible and the incompatible systems were higher than that of the mock-infected leaves at 48 hpi. The opposite effects were observed at 120 hpi. The means ranked with different letters (*A* and *B*) are significantly different at $P < 0.01$

Table 3.3 Net photosynthetic rate of wheat plants infected by *Puccinia striiformis* f. sp. *tritici* in the field conditions

Cultivar and rust severity (%)	Net photosynthetic rate ((μmol $CO_2/(m^2 \cdot s)$) at different time of day)				
	8:00	10:00	12:00	14:00	16:00
Huixianhong					
0	16.10	21.50	17.50	11.19	9.34
5	17.02	20.02	18.91	11.28	7.39
40	10.28	15.41	14.57	8.31	5.97
80	11.07	9.95	7.49	4.06	2.45
Mingxian 169					
0	13.75	21.64	16.84	9.90	10.43
5	16.30	18.22	12.96	11.84	9.32
40	11.19	11.72	11.49	9.36	4.72
80	10.91	4.99	4.67	3.15	2.12

intensity, although the net photosynthetic rate could remain at a higher value before 2:00 P.M. The net photosynthetic rate of severe disease leaves was different from the health leaves, showing the highest rate at 8:00 A.M., followed by a sharp decline as temperature was increasing, and photosynthesis was kept at a very low level for whole days. The trends of net photosynthetic rate in Huixianhong and Mingxian 169 are basically similar (Table 3.3).

As disease develops into the stage of definite decline in gross photosynthesis, cytological evidence indicates that chloroplasts become degenerated. However, the rate of photosynthesis starts to decline before conspicuous structural changes. To learn what limits the process in the early stages of the decline, the following aspects of photosynthesis have been investigated.

1. **Resistance to diffusion of CO_2 into the leaf**. Diffusion of CO_2 into the leaf was not an important limiting factor for photosynthesis in barley leaf rust fungus infected host cells (Owera et al. 1981). Resistance to diffusion of CO_2 decreased as pustules broke the leaf surface.
2. **Amount of chlorophyll**. Previous studies have indicated that the amount of chlorophyll declines in rust infected cereal tissues (Calonge 1967; Doodson et al. 1965; Mitchell 1979), but it is uncertain whether there is a close correlation between the amounts of chlorophyll and rates of photosynthesis. It is suggested that loss of chlorophyll is a principal limiting factor for photosynthesis in host cells infected by the barley leaf rust pathogen (Owera et al. 1981). In addition, the chlorophyll fluorescence of wheat leaves inoculated with *Pst* was examined by a modulated imaging fluorimeter (Chen et al. 2015). In CN19 (a resistant cultivar), the maximum photochemical efficiency of PSII in the dark-adapted state (F_v/F_m) did not significantly decrease in the *Pst*-infected leaves compared with the healthy leaves. In contrast, Sy95-71 (a susceptible cultivar) decreased significantly in the F_v/F_m value after the inoculation.
3. **Amount of ribulose-1,5-bisphosphate carboxylase**. The decreased ribulose-1,5-bisphosphate carboxylase level was observed in rust infected wheat leaves (Wrigley and Webster 1966). Loss of this enzyme has also been implicated in the decrease in photosynthesis in powdery mildew of sugar beet (Gordon and Duniway 1982). Li et al. (2011) reported that ribulose-1,5-bisphosphate carboxylase/oxygenase (Rubisco) was in response to infection of stripe rust fungi, indicating that chloroplasts and photosynthesis in both the incompatible and compatible interaction were affected by *Pst* infection.
4. **Photophosphorylation**. Although Wynn (1963) detected no change in photophosphorylation in host tissue when infected with the oat crown rust fungus (*P. coronata*), the rate of noncyclic photophosphorylation was reduced by 30% in host tissues infected by the broad bean rust fungus (*Uromyces viciae-fabae*) or the powdery mildew fungus (*Erysiphe betae*) of sugar beets (Montalbini and Buchanan 1974; Magyarosy et al. 1976). The photophosphorylation activity was reduced to 70% of that in uninfected leaves. For powdery mildew, the reduced activity was attributed to a reduction in amount of the cytochromes used for electron transport in noncyclic photophosphorylation (Magyarosy and Malkin 1978). For wheat infected with *Pst*, Shen et al. (2008) found that the rate of electron transport in the susceptible wheat cultivar Mingxian 169 was decreased by 20% compared with the non-infected control.

Overall, the reduction in photosynthetic activity in rust infected leaves seems to be due to loss of chlorophyll and key proteins such as ribulose-l,5-bisphosphate carboxylase or possibly cytochromes. Ribosomal RNA within chloroplasts was reduced in amount with the oat crown rust fungus (Tani et al. 1973) and the electron transport rate was reduced by *Pst* infection (Shen et al. 2008), suggesting that synthesis of chloroplast proteins may be generally depressed.

3.3.3.4 Carbohydrates

Carbohydrates are essential for both host plant and rust fungal growth. During the infection, the rust fungus utilizes carbohydrates from host plant tissue, and as a result, the rust fungal infection reduces plant growth by limiting carbohydrates. Changes in specific carbohydrates are different in different stages of infection and also different between compatible and incompatible interactions between the host and rust pathogen.

3.3.3.4.1 Carbohydrate Requirements of Rust Fungi

The early development of obligate parasitic fungi, such as rust and powdery mildew fungi, is dependent upon the endogenous lipids (18–25% of spore weight) stored in the spores before they establish a parasitic relationship with the host (Malca et al. 1962; Caltrider et al. 1963). Once haustoria are formed, the nutrient supply for the rust fungal parasite completely comes from the host cells. As the endogenous concentrations of soluble carbohydrates (0.01–0.1% of spore weight) in spores are very low (Daly et al. 1967), sugar appears to be the major carbon energy source transferred from the wheat host to the powdery mildew fungus (*Blumaria graminis* f. sp. *tritici* (Mendgen and Nass 1988; Sutton et al. 1999). Glucose fed to the rust infected wheat leaves was utilized by the wheat stem rust fungus without rearrangement of carbons 1 and 6, indicating that the intact glucose molecule was taken up (Pfeiffer et al. 1969). Supply of glucose, fructose, or sucrose are sufficient for rust development on corn or wheat leaves in darkness (Dickson et al. 1959; Silverman 1960), albino corn leaves (Dickson et al. 1959), or on wheat leaves in which photosynthesis is inhibited (Mashaal et al. 1981). Infiltration of sucrose into rust infected leaves specifically inhibited the movement of ^{14}C-labeled sucrose from host to parasite, implicating that sucrose is the principle carbohydrate source for *Puccinia poarum* on leaves of *Poa pratensis*. In addition, in the powdery mildew fungus, feeding experiments with radio-labelled sugars seemed to indicate that glucose not sucrose is the main carbohydrate obtained from epidermal cells. A study investigating the role of haustoria in sugar supply during infection of broad bean (*Vicia faba*) by the biotrophic rust fungus *Uromyces fabae* identified a haustoria specific hexose transporter HXT1 (Voegele et al. 2001). The HXT1 protein was localized exclusively to the haustorial plasma membrane, suggesting that this hexose transporter may have a major role in carbon transfer into the fungus from the extrahaustorial matrix (Voegele et al. 2001). Besides, an invertase from *Uromyces fabae* was identified to be expressed during parasitic growth with increased overall invertase activity to supply substrate for HXT1 (Voegele et al. 2006).

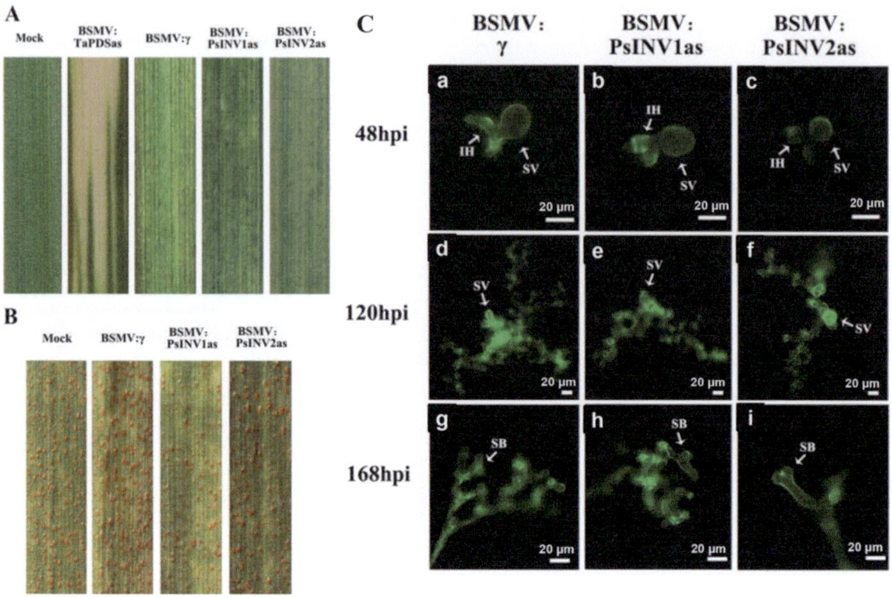

Fig. 3.29 Functional analysis of *PsINV* during the interaction between wheat and *Puccinia striiformis* f. sp. *tritici* (*Pst*) using a BSMV-HIGS system. (**a**) Mild chlorotic mosaic symptoms were observed on the fourth leaves of seedlings at 9 dpi with BSMV, and photobleaching was evident on the fourth leaves of plants infected by BSMV:TaPDS. (**b**) Disease phenotypes of the fourth leaves pre-inoculated with BSMV and then challenged with CYR32. (**c**) Histological observations of *Pst* growth and development in wheat leaves. Wheat leaves pre-inoculated with BSMV:γ, BSMV:PsINV1as or BSMV:PsINV2as were sampled at 48 (*C–E*), 120 (*F–H*) and 168 (*I–K*) hpi with *Pst* CYR32. After staining with wheat germ agglutinin (*WGA*), fungal structure was examined under a fluorescence microscope

Homologue genes of HXT1 have been identified in the genomes of all three wheat rust fungi (*Pst*, *Pgt* and *Pt*) (Duplessis et al. 2011; Zheng et al. 2013), but the distinct orthologue of the *Ustilago maydissrt1* sucrose transporter gene was absent (Wahl et al. 2010). It is possible that the rust fungi may convert sucrose to hexose by secreted invertase, and haustoria then take up hexose from host cells as the major carbohydrate source. An invertase gene, *PsInv*, from *Pst* has been identified. *PsINV* was found to be secreted into host cells and highly up-regulated during the infection of wheat. Furthermore, silencing of *PsINV* by host-induced gene silencing (HIGS) could lead to the reduction of the sporulation of *Pst* (Fig. 3.29a, b), significantly impeded hyphal growth and development (Fig. 3.29c), indicating that *PsINV* plays a pivotal role for sugar absorption during stripe rust fungi infection of wheat (Prof. Kang's group, unpublished data).

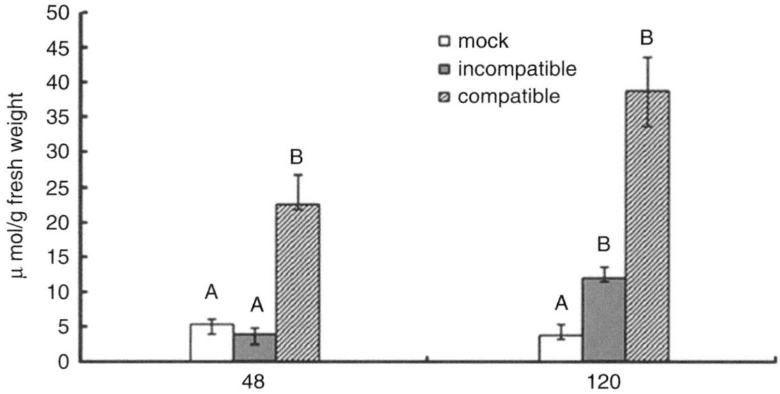

Fig. 3.30 Sucrose levels in wheat leaves inoculated with *Puccinia striiformis* f. sp. *tritici*. The sucrose level of the mock-infected leaves exhibited no significant changes, but the sucrose level was significantly increased in the compatible system at both time points. In the incompatible system, the increase in the sucrose level was mainly observed at 120 hpi. Means ranked with different letters (*A* and *B*) are significantly different at $P < 0.01$

3.3.3.4.2 Carbohydrates in the Host

During the infection of *Pst*, the host carbohydrates change significantly in the early and the late stage. At the early stage of infection, when the photosynthesis of host was strengthened and more sucrose were synthesized in both compatible and incompatible systems, slight changes for all water soluble carbohydrates were observed except for sucrose (Chang et al. 2013) (Fig. 3.30). In the incompatible system, a lot of the products of photosynthesis were transported away from the infected leaves, whereas in the compatible system, the infected leaves were inverted from the source organ into a newly formed sink with higher demands for more carbohydrates with much stronger strength than the other sinks of host, even the grain filling could not compete with it for carbon assimilates (Bancal et al. 2012). At the late stage, infection by *P. triticina* can reduce the green areas of wheat leaves or damage mesophyll cells, leading to the reduction in the photosynthetic rate (Robert et al. 2005). Sucrose, glucose, and fructose often increase several folds in cereal leaves infected by rust fungi as part of the early host response prior to fungus sporulation, and then decline rapidly thereafter (Syamananda and Staples 1963; Lunderstädt 1966; Mitchell et al. 1978).

Starch tends to accumulate along with sugars, so that host tissues within and near pustules could be stained with iodine (Bushnell 1970). Starch in the healthy wheat leaves is very low and constant. In wheat leaves infected by a rust fungus, starch was not affected in the incompatible system, but was depleted steadily during infection in a compatible interaction (Heisterüber et al. 1994). Starch is found to be in chloroplasts, and its amount fluctuates in cereal leaf cells infected by rust fungi (Mirocha and Zaki 1966; MacDonald and Strobel 1970). Host tissues seem to shift rapidly

Table 3.4 Relative water content of wheat leaves inoculated with races of *Puccinia striiformis* f. sp. *tritici* measured at different days after inoculation

		Relative water content (%)					
			Days after inoculation				
Cultivar	Race	Check[a]	4	8	12	16	20
Xiaoyan 6	CY29	96.66	93.55	85.67	92.87	82.67	71.66
	CY29mut3	96.66	95.60	94.53	93.82	85.61	97.58
Lovrin 10	CY29mut3	97.93	93.80	94.03	95.54	96.02	96.10
	CY29	96.60	95.40	93.43	92.46	78.62	74.44
Lovrin 13	CY29mut3	96.60	95.61	94.08	95.57	96.58	96.67

[a]Relative water content of non-inoculated plants measured on the same day of inoculation

between starch synthesis and degradation. Further studies are needed to determine the mechanisms controlling starch synthesis and degradation within host chloroplasts.

3.3.3.5 Water

Water is essential for growth of both host plants and rust fungi. In *Pst*-infected wheat leaves, transpiration was obviously increased, which resulted in wilting, withering and other symptoms of water shortage. Water changes in the diseased wheat plants were not well studied due to difficulties in experimental materials and techniques. Studies in this field have been mainly focused on vascular diseases, but only few studies have been conducted on wheat rust diseases. Smith et al. (1986) studied the effects of stripe rust, irrigation on wheat yield and the leaf temperature and found that the leaf transpiration rate increased 20–40% in the early infection stage but decreased by 10% when lesion spots were formed in the infection period.

Li (1998) studied several parameters of water change in wheat leaves infected by *Pst* using multi-stage inoculation and simultaneous determination. His study showed a series of significant water changes taking place in the *Pst* infected wheat leaves (Table 3.4). These changes are discussed in the following paragraphs.

3.3.3.5.1 Relative Water Content

Obvious differences were found between wheat-*Pst* compatible and incompatible interactions in the relative water content. A high water content was kept in primary leaves in healthy seedlings, while in wheat leaves inoculated with CYR29 (a compatible interaction), relative water content decreased slowly during the incubation period and dramatically dropped after sporulation. In wheat cultivar inoculated with CYR29mut3 (an incompatible interaction), relative water content was relatively stable during the entire infection process, although there was a slight decrease during the early infection and then returned to the normal level as the lesion spots formed.

3.3.3.5.2 Transpiration Rate and Leaves Diffusion Resistance

The trends of transpiration rate changes in compatible and incompatible wheat-*Pst* interactions were different. In compatible interactions, the transpiration of diseased leaves was inhibited during the infection and began to increase after sporulation, with significant increases by 35.3–104.4% after sporulation. In incompatible interactions, the transpiration rate of diseased leaves was also reduced during the early infection process and then returned to the level of healthy leaves after the formation of necrotic spots.

Compatible and incompatible interactions showed significantly differences in leaf diffusion resistance. In compatible interactions, diffusion resistance of diseased leaves was significantly increased before sporulation and gradually decreased after sporulation. When a large number of uredinia were formed, diffusion resistance quickly declined due to the obvious damage of the epidermis and cuticle in diseased leaves. In incompatible interactions, diffusion resistance in diseased leaves was significantly increased during the early infection process and then decreased to the level of healthy leaves after the formation of necrotic spots.

3.3.3.5.3 Water Potential, Osmotic Potential and Pressure Potential

Significant changes in potential of water, osmotic and pressure were observed in *Pst* infected wheat leaves of compatible and incompatible interactions. In compatible interactions, the changes of water potential, osmotic potential and pressure potential in diseased wheat leaves occurred in several periods. Before sporulation, the solute accumulated significantly with decreased potential of water and osmotic, but slight increased pressure potential, possibly due to the nutrient accumulation. At this stage, the disease leaves still had a good water balance, but after sporulation, the accumulation of solute was dramatically increased, accompanied with decreased water potential and a lower pressure potential, which were caused by the increasing transpiration due to the damage of epidermis and cuticle in diseased leaves. After the sporulation peak, osmotic potential was significantly decreased, water potential and pressure potential sharply declined. At this stage, diseased leaves lost the ability to maintain the water balance.

Changes of water potential, osmotic potential and pressure potential in incompatible interactions could be divided into three phases. In the early infection stage, the solute was accumulated significantly in the diseased leaves with decreased water potential and osmotic potential, and slightly increased pressure potential. At this stage the diseased leaves still maintain a good water balance. During the formation of necrotic spots on the diseased leaves, the nutrients were significantly reduced, along with the significantly increased osmotic potential and water potential. However, pressure potential was at a lower level, indicating that the water balance in wheat leaves was slightly affected. After the formation of necrotic spots, the accumulation of solute in diseased leaves was significantly increased, while the osmotic potential significantly decreased, water potential was gradually increased to the control level, and the pressure potential in the diseased leaves was sharply increased. It appeared that more nutrients in stripe rust infected wheat leaves were accumulated in incompatible interactions, which made host leaves possess stronger capability of water absorption and regulation, and led to a better water status.

3.4 Molecular Mechanisms of Interactions Between Wheat and *Pst*

As the molecular and genomic technologies have been developed in the recent years, mechanisms of interactions between wheat and the *Pst* pathogen have become a hot area of stripe rust research. Quite intensive studies have provided useful new knowledge for understanding the host-pathogen interactions, but still leave a lot of interesting questions to be answered in future studies.

3.4.1 Gene Expression in Compatible/Incompatible Interactions

In addition to their economic importance, rust fungi are particularly interesting because of their complex life cycle and the specialized infection structures. A great number of cytological studies have revealed details about spore attachment, germination, appressorium formation over host stomata and host invasion (Hahn 2000). The genetics, histology and pathology of host-rust pathogen interactions have been intensively explored over the last few decades (Kolmer 1996). Recently, the molecular mechanisms of compatible and incompatible interactions between the host and pathogen have been intensively studied. A number of technologies had been applied to study the transcript profiling of compatible and incompatible interactions between wheat and rust fungi, such as differential display reverse transcriptase-PCR (DDRT-PCR), microarrays, cDNA-AFLP, cDNA library and next-generation sequencing (NGS).

3.4.1.1 All-Stage Resistance

All stage resistance, also called seedling resistance against *Pst*, can be highly effective, but it is usually race-specific and can be easily overcome by the rapid evolution of new pathogen races. Due to difficulties of wheat transformation and the lack of a transformation system for *Pst*, the research progress on studying the underlying mechanism of wheat-*Pst* interaction was seriously hindered. Because of these obstacles for cloning resistance genes, transcriptomics has been used to determine the molecular basis of wheat-*Pst* interactions and a lot of progress has been made.

Using the DDRT-PCR method, Bozkurt et al. (2008) identified the expressed genes in the *Yr1* and *Yr10* near-isogenic lines in the 'Avocet S' (AvS) background inoculated with avirulent and virulent *Pst* races. Upon infection, the genes expressed at different time points were evaluated for DDRT-PCR banding profiles. The expression level of 33 differentially expressed genes in DDRT-PCR were further analyzed using quantitative real-time PCR (qRT-PCR), among which 14 gene fragments produced reproducible induction or repression levels. Based on the sequence similarity analysis, all but one were absent in the NCBI database, indicating new gene sequences in wheat. Two genes were identified as cyclophilin like protein (putative antifungal activity) and ubiquitin-conjugating enzyme (E2) or Rad6, respectively. The identified genes may be involved in ubiquitinylation, programmed cell death (apoptosis), putative antifungal activities, disease resistance responses, pathogenesis related (PR) responses, and a few with unknown functions. The findings indicated that genes involved in ubiquitin-mediated protein degradation are regulated in wheat in response to stripe rust avirulent pathogen infection and suggest that ubiquitinylation and protein degradation are significant regulatory mechanisms in wheat stripe rust disease resistance.

To identify transcripts associated with the *Yr5*-mediated incompatible interaction and the *yr5*-compatible interaction, Coram et al. (2008b) was the first to use the Wheat GeneChip to profile the changes in transcriptomics involved in wheat-*Pst* interactions. They first studied wheat genes regulated in wheat near-isogenic lines that differed for the presence of the *Yr5* gene after inoculation with *Pst* races. Their time-course study (6, 12, 24 and 48 h post inoculation) identified 115 transcripts that were induced during the *R*-gene-mediated incompatible interaction, and 73 induced during the compatible interaction. Fifty-four transcripts were induced in both interactions and were considered as basal defense transcripts, whilst 61 transcripts were specific to the incompatible interaction and 19 were specific to the compatible interaction. Annotation of the induced transcripts revealed that the presence of *Yr5* resulted in a rapid and amplified resistance response involving signaling pathways and defense-related transcripts known to occur during *R*-gene-mediated responses, including protein kinase signaling and the production of reactive oxygen species leading to a HR. Basal defense also involved substantial induction of many defense-related transcripts but the lack of *R*-gene signaling resulted in weaker response.

Later, Coram et al. (2010) constructed a custom oligonucleotide microarray to perform a meta-analysis of the transcriptional response involved in race-specific all stage resistance conferred by eight genes, *Yr1*, *Yr5*, *Yr7*, *Yr8*, *Yr9*, *Yr10*, *Yr15* and *Yr17*. By profiling the response of eight resistance genes in a common background, AvS, they identified 28 transcripts significantly involved in the resistance phenotype across all genotypes. The most significant of these were annotated as blue copper-binding protein, heat-stress transcription factor, pathogen-induced WIR1A protein, and ent-kaurene synthase transcripts. Unique transcripts significant in each genotype were also identified, which highlighted some transcriptional events specific to certain genotypes. This approach was effective in narrowing down the list of candidate genes in comparison to studying individual genotypes. They also identified key gene expression events involved in race-specific resistance. Their results confirm the activity of known *R*-gene-mediated pathway race-specific resistance, including an oxidative burst that likely contributes to a HR, as well as PR protein expression and activity of the phenylpropanoid pathway.

The Wheat GeneChip was also used by Bozkurt et al. (2010) to profile the changes occurring in wheat carrying the *Yr1* gene after inoculation with *Pst*. Their microarray study identified a number of defense-related transcripts implicated in *Yr1*-mediated resistance, including classical PR transcripts and genes involved in plant cell defense responses, such as the oxidative burst and cell wall fortification, similar to Coram et al. (2010). A quantitative reverse transcriptase-polymerase chain reaction time course analysis identified a number of defense-related genes, including *PR2*, *PR4*, *PR9*, *PR10* and *WIR1* transcripts, associated with the latter stages of *Yr1*-mediated resistance. A meta-analysis comparison of the *Yr1*-regulated transcriptome with the resistance transcriptomes of the race-specific resistance gene *Yr5* and the race-nonspecific adult plant resistance gene *Yr39* (Coram et al. 2008a, 2010) indicated limited transcript commonality. Common transcripts were largely confined to classic *PR* and defense-related genes (Bozkurt et al. 2010).

Using the cDNA-AFLP technique, the gene expression profiles were investigated over a time course to identify genes involved in race-specific all stage resistance using wheat cultivar Suwon 11 inoculated with *Pst* races CYR31 (compatible interaction) and CYR23 (incompatible interaction (Wang et al. 2009a, 2010a). In the compatible interaction, 54,912 transcript derived fragments (TDFs) were obtained using cDNA-AFLP with 64 primer pairs, and 2306 (4.2%) displayed altered expression patterns after inoculation, of which 966 showed up-regulated and 1340 down-regulated (Wang et al. 2009a). One hundred and eighty six TDFs produced reliable sequences after sequencing of 208 selected TDFs, of which 74 (40%) had known functions through BLAST (Basic Local Alignment Search Tool) searching the GenBank database. The majority of 74 TDFs had predicted gene products involved in energy (13%), signal transduction (5.4%), disease/defense (5.9%) and metabolism. BLAST search of the wheat stem rust fungus genome database identified 18 TDFs possibly from *Pst*, of which 9 were validated of the pathogen origin using PCR-based assays followed by sequencing confirmation. In the incompatible interaction, 52,992 TDFs were generated with the 64 primer pairs and 2437 (4.6%) of them displayed altered expression patterns after inoculation, with 1787 up-regulated and 650 down-regulated (Wang et al. 2010a). Of the 255 reliable TDFs obtained, 113 (44.3%) had putative functions identified. A large group (17.6%) of these genes shared high homology with genes involved in metabolism and photosynthesis; 13.8% to genes with functions related to disease defense and signal transduction; and those in the remaining groups (12.9%) to genes involved in transcription, transport processes, protein metabolism and cell structure. The comparative analysis between the incompatible and compatible interactions by the TBLASTN (translated nucleotide databases using a protein query) searching showed that 161 of the 255 TDFs identified in the incompatible interaction were also induced in the compatible interaction (Wang et al. 2009a), indicating that the 161 TDFs are involved in the basal defense. The remaining 94 TDFs were considered to be expressed more

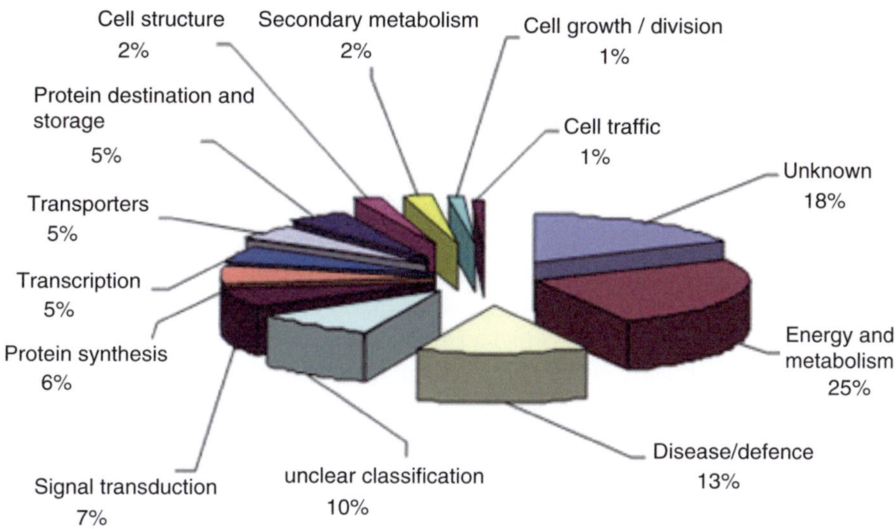

Fig. 3.31 The functional classification of expressed sequence tags (*ESTs*) from a compatible interaction between wheat and *Puccinia striiformis* f. sp. *tritici* based on significant similarities to plant genes in the GenBank database. The homologies of unigenes to plant genes were classified into 13 functional categories based on their cellular roles. About 26% of the ESTs showed significant homologies to enzymes involved in energy and metabolism, representing the largest category. About 18% of the ESTs showed homologies to genes of unknown functions and hypothetic proteins (Reproduced from Ma et al. (2009))

specifically in the incompatible interaction (Wang et al. 2010a). Especially, 11 TDFs expressed specifically in the incompatible interaction observed did not change or were repressed in the compatible interaction through the qRT-PCR validation. These genes should play important roles in the interactions of wheat and *Pst*. Of the 11 TDFs, 6 have unknown functions and encode leucine rich repeat family proteins, CBL-interacting protein kinase, serine/threonine kinase, ET-responsive RNA helicase or protein phosphatase type 2C.

cDNA libraries are useful for identifying the functional genes involved in the wheat-*Pst* interactions. A compatible cDNA library was created from leaves of Suwon11 inoculated with race CYR31, from which 5793 expressed sequence tags (ESTs) were obtained and clustered into 583 contigs and 2160 singletons to give a set of 2743 unisequences (Ma et al. 2009). Of the 2743 unisequences, 52.8% were highly homologous to plant genes; 16.3% to fungal genes and 30% of no-hit. The unigenes homologous to plant genes were classified into 13 functional categories based on the cellular roles of expressed genes. The largest category counted for 26% of the classified ESTs with putative functions as enzymes of energy and metabolism. The second largest was the category of enzymes with unknown and hypothetic proteins, counting for 18% of the classified ESTs. The third category was disease/defense proteins, counting for 13% of the ESTs. Among the 2743 unisequences, 446 showed homologies to genes from model fungal species (Fig. 3.31). Fifteen out of the 446 unisequences showed homologous to other rust fungal genes. Three plant-induced genes (WRIC_348, WRIS_316 and WRIS_1608), as defined by Hahn and Mendgen (1997) were detected. WRIS_2031 and WRIS_3511 were predicted to encode chitinase because of their best homology to the fungal chitinase gene of *P. triticina*. WRIC_312 was predicted to be a rust transferred protein (RTP) because of its similarity to a RTP gene in *Uromyces striatus*. The sequence of WRIS_1741 is similar to the gene encoding a cyclase-associated protein 1 (CAP1) from *Cryptococcus gattii*. WRIC_406 was homologous to the chitin deacetylase, which catalyzes deacetylation of chitin to yield chitosan. WRIS_3617 and WRIS_4517 are two genes homologous to ABC transporters of *Candida albicans* and *Venturia inaequalis*, respectively. WRIS_1157 had a similar sequence to that of a *Cryptococcus neoformans* transketolase gene, which is an enzyme involved in primary carbohydrate metabolism. WRIS_1773 and WRIS_1781, which were predicted to encode actin proteins based on their high sequence similarities to those of *P. graminis*.

To isolate wheat defense-related genes against *Pst*, a suppression subtractive hybridization (SSH) library was constructed for an incompatible interaction with wheat cultivar Suwon 11 inoculated with race CYR23 (Yu et al. 2010). From the SSH library, 652 sequences were determined to be unigenes, of which 31 were determined as genes involved in signal transduction and 77 were predicted to encode defense-related proteins. Defense-related genes were mainly predicted to be involved in three processes including signal recognition and transduction, transportation and expression of resistance. For signal recognition and transductions, four genes were predicted to code for proteins involved in GTP binding, three genes for serine/threonine kinase, five genes for zinc-finger proteins, and 18 other genes involved in signal recognition and transduction. The ET signal process considered to trigger resistance response was predicted to induce genes coding for 1-aminocyc lopropane-1-carboxylate (ACC) and AdoMet synthase. For transport related genes, six ABC transporters and 54 other membrane and transport-related genes were identified. The expression of these genes during the infection process indicated their relationships with ion flux occurring in the early stage of HR. A total of 68 genes were characterized to be related to disease resistance and stress. These included catalase, phenylalanine ammonia-lyase, ascorbate peroxidase, superoxide dismutase considered to be involved in main resistance processes of wheat and hypersensitive-induced reaction protein 1, wound-induced genes, an iron-deficiency-induced gene, and stress-responsive protein genes. The activities of these defense genes work in a sequential and concerted manner to result in a HR.

As shown by various studies discussed above, interactions of *Pst* with wheat plants activate a wide range of host responses. Among different defense responses associated with termination of *Pst* is the expression of particular PR protein genes. Farrakh et al. (2016) conducted a study to identify PR protein genes involved in race-specific all stage resistance controlled by different *Yr* genes in wheat to stripe rust. The expression levels of eight PR protein genes (*PR1*, *PR1.2*, *PR2*, *PR3*, *PR4*, *PR5*, *PR9* and *PR10*) at 0, 1, 2, 7 and 14 days after inoculation (dai) in near isogenic line of wheat with resistance gene *YrTr1*, *Yr76*, *YrSP* and a recombinant inbred line of *YrExp2* inoculated with race PSTv-4 for compatible and PSTv-37 for incompatible interactions were determined. The qRT-PCR analysis revealed that PR protein genes were upregulated and downregulated differently at different time points during the infection process among the wheat lines. However, all of the tested PR protein genes were upregulated at 1 dai in all of the tested *Yr* gene lines. *PR3* and *PR 4* remained upregulated until 14 dai. None of the PR protein genes were induced at 0 dai except for *PR9* (peroxidase) and *PR2* (glucanase). The PR protein genes were also expressed in compatible interactions, but their expression was significantly higher in the incompatible interactions than the compatible interactions. As *PR1*, *PR2* and *PR5* are involved in the salicylic acid (SA) pathway and *PR3*, *PR4* and *PR9* in the jasmonic acid (JA) pathway, the results of this study indicated the presence of both SA and JA pathways in the race-specific all-stage resistance to stripe rust.

3.4.1.2 Adult-Plant Resistance

Adult-plant resistance (APR) does not express or expresses at a low level at the seedling stage, but starts expressing or becomes more resistance during the post-seedling stages, and usually reaches its highest level of effectiveness on flag leaves (Chen 2005). Therefore, APR may be associated with the transition of growth stages. Several studies have been reported on the successful use of cDNA-AFLP and SSH for the isolation of differentially expressed plant genes during wheat-*Pst* interactions. Moldenhauer et al. (2008) detected slightly stronger induction of *PR1*, *PR2* and *PR5* in APR among 'Kariega/Avocet S' wheat derived lines with resistance QTL to stripe rust than a susceptible control. Coram et al. (2008a) identified 99 induced transcripts specific to *Yr39* high-temperature adult-plant (HTAP) resistance with two F_7 recombinant inbred lines (RILs) from an 'AvS/Alpowa' cross that differed at the *Yr39* locus using the wheat Affymetric GeneChip. The results indicated that more than half of the annotated transcripts specifically induced during HTAP resistance were involved in defense and/or signal transduction, including *R* gene homologues and transcripts associated with PR protein production, phenylpropanoid biosynthesis and protein kinase signaling. This study represents the first transcript profiling of HTAP resistance to stripe rust in wheat.

Subsequently, Dr. Chen's group compared genes regulated in wheat lines with different APR or HTAP resistance gene such as *Yr18*, *Yr29*, and *Yr39*; and also compared HTAP resistance –involved genes with genes involved in all stage resistance controlled by different *Yr* genes (*Yr1*, *Yr5*, *Yr7*, *Yr8*, *Yr9*, *Yr10*, *Yr15* and*Yr17*) under low (diurnal cycle of 4-20°C) and high (diurnal cycle of 10–30 °C) temperatures (Chen et al. 2013). They identified that 7 transcripts were significant for *Yr1*, 10 for *Yr5*, 4 for *Yr7*, 31 for *Yr8*, 6 for *Yr9*, 13 for *Yr10*, 6 for *Yr15*, 5 for *Yr17* for all stage resistance; and 40 for *Yr18*, 4 for *Yr29*, 99 for *Yr39* and none for *Yr36* using the custom-constructed wheat gene chips based on their *Yr5* and *Yr39* studies (Coram et al. 2008a, b). In general, transcripts detected in the seedling low-temperature tests were also significant in the adult-plant high-temperature tests for race specific all-stage resistance. After various comparisons, five genes were clearly identified to be involved in race-specific all-stage resistance controlled by *Yr1*, *Yr5*,

Yr7, *Yr9* and *Yr15* and only one gene was commonly expressed in HTAP resistances mediated by different genes (*Yr18*, *Yr29*, *Yr36* and *Yr39*). A greater diversity of genes is involved in HTAP resistance than in all-stage resistance. The genes involved in HTAP resistance are induced more slowly and their expression induction is less dramatic than genes involved in all-stage resistance. The high diversity of genes and less dramatic induction may explain durability and the incomplete expression level of HTAP resistance. The transcript with significant changes across all HTAP resistances is a nonclathrin coat protein. The function of nonclathrin coat protein-related resistance is not clear, but such proteins have been reported to bind to the cytoplasmic dilysine motif of membrane proteins of the early secretory pathway (Harter et al. 1996). The nonclathrin coat protein identified in this study may be involved in transporting antifungal substances across the cell membrane to directly contact *Pst* haustoria or hyphae.

The French wheat variety 'Camp Remy' (CR) possesses a durable, adult plant resistance to stripe rust. To identify genes up- and down-regulated in response to *Pst* inoculation, Mallard et al. (2008) compared the expression of TDFs generated using cDNA-AFLP, in inoculated and mock-inoculated CR and the different CR-derived heterogeneous inbred families (HIFs) that express resistance at different developmental stages. Sequence analysis of the resultant clones revealed several classes of putative genes, including those related to resistance/defense responses, transcription and signal transduction, and primary metabolism. Three of the seven genes validated by real-time PCR were specifically up-regulated during the tillering and/or the two-node stages in the inoculated versus mock-inoculated CR leaves. Two genes, a beclin-1-like and a xip-II gene, were firstly identified as involved in the resistance/defense responses of bread wheat in this study.

Huang et al. (2013) constructed a cDNA library containing genes involved in APR using the SSH approach and characterized selected genes through qRT-PCR of their time-course expression during the interaction between wheat cultivar Xingzi 9104 and *Pst* race CYR32 at the seedling and adult-plant stages. Based on the putative functions of 436 SSH-derived unigene sequences, they identified defense responses triggered during infection and development of adult plant resistance in Xingzi 9104. They found 89 genes involved in defense and signal transductions. Some PR protein genes were found to be kept high expression levels longer in the adult plants than in seedlings after *Pst* infection. This study was the first to report a concanavalin A-like lectin/glucanase involved in plant defense against pathogen infection.

To better understand the mechanism underlying APR, Hao et al. (2016a) performed transcriptome sequencing of seedlings and adult plants of Xingzi 9104 inoculated with a *Pst* race, and obtained 157,689 unigenes, of which 2666 were up- or down regulated at 24 hpi, 783 at 48 hpi, and 2587 at 120 hpi. Among these unigenes, the temporal pattern of the up-regulated unigenes exhibited transient expressions during *Pst* infection, as determined through a gene ontology (GO) enrichment analysis (Fig. 3.32). In addition, the analysis using the Kyoto Encyclopedia of Genes and Genomes (KEGG) pathways showed that many biological processes were involved in APR, including phenylpropanoid biosynthesis, reactive oxygen species,

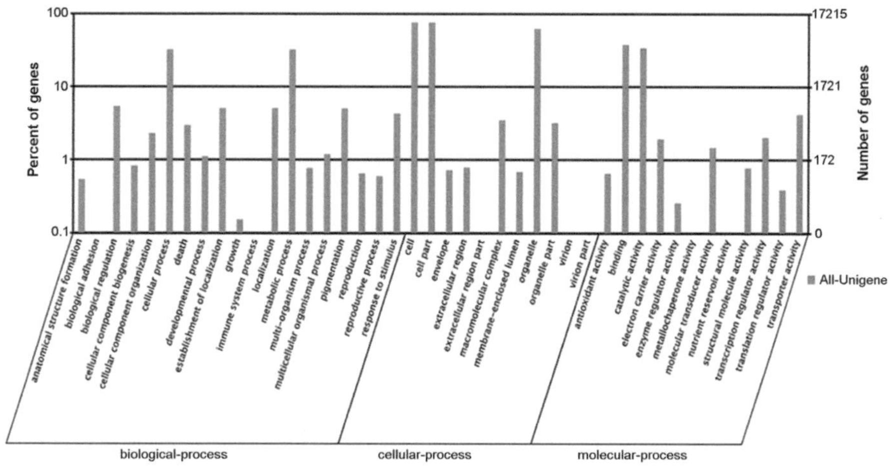

Fig. 3.32 The GO classification of all unigenes sequenced from expressed sequence tags (*ESTs*) of adult-plant resistant wheat cultivar Xingzi 9104 leaves inoculated with *Puccinia striiformis* f. sp. *tritici*. A total of 69,100 unigenes were assigned to GO term annotations using BLAST2GO and then summarized into three main GO categories and 42 sub-categories (functional groups) using WEGO (Reproduced from Hao et al. (2016a)

photosynthesis and thiamine metabolism, which mainly participate in regulation of lignification, reactive oxygen species and sugar, respectively. In particular, the continuous accumulation of reactive oxygen species may potentially contribute to the ability of the adult plant to inhibit fungal growth and development. Based on the GO and KEGG pathway enrichment results, they speculated that six differentially expressed genes might play important roles in the APR of Xingzi 9104 against *Pst* infection, of which two were enriched in the phenylalanine metabolism pathway, two in peroxidase activity and nitric-oxide synthase activity, and two in the chlorophyll biosynthetic process and the photosynthesis process. Phenotypic changes compared to the control were observed on the flag leaves of four candidate gene-knockdown plants (Fig. 3.33). In contrast, no phenotypic changes were observed on plants with the other two genes knockdown plants. This study indicated that the development of *Pst* was markedly inhibited in adult plants at different infection stages by many biological processes, such as phenylpropanoid biosynthesis, reactive oxygen species, photosynthesis, and the thiamine metabolism pathway.

3.4.1.3 Nonhost Resistance

Nonhost resistance is the resistance shown by an entire plant species to all genetic variants of a pathogen (Mysore and Ryu 2004). NHR provides immunity to all members of a plant species against all isolates of a pathogen species that normally infects other plant species (Fan and Doerner 2012). As the most common form of

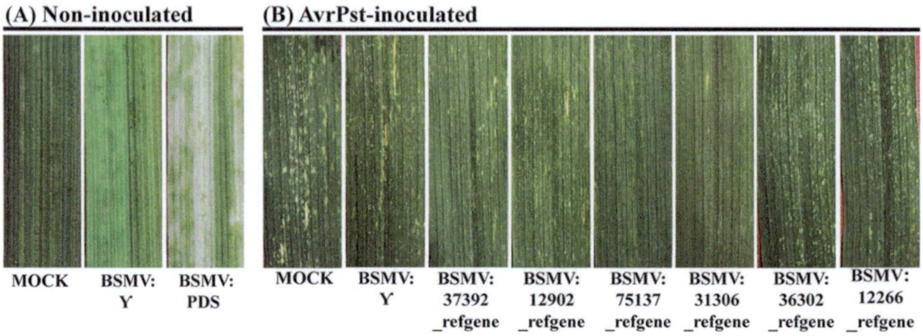

Fig. 3.33 Phenotypes of wheat leaves in six candidate genes-knockdown plants inoculated with *Puccinia striiformis* f. sp. *tritici* (*Pst*). To perform a functional analysis of the candidate genes, virus induced gene silencing (VIGS) was applied to adult plants. (**a**) No change was found in mock-inoculated wheat leaves pre-inoculated with FES buffer; mild chlorotic mosaic symptoms were observed on the flag leaves of wheat at 13 days post inoculation (dpi) with BSMV:γ; photobleaching was evident on the first top leaves of plants 15 dpi with BSMV:PDS. (**b**) Mock-inoculated wheat flag leaves treated with FES buffer and then challenged with *Pst* race CYR32; the flag leaves of knockdown plant (BSMV:γ, BSMV:wheat37392_refgene, BSMV:wheat12902_refgene, BSMV:wheat75137_refgene, BSMV:wheat31306_refgene, BSMV:wheat36302_refgene and BSMV: wheat12266_refgene) challenged with CYR32. Typical leaves were observed and photographed at 15 dpi (Reproduced from Hao et al. (2016a))

plant resistance, NHR is remarkable for its broad-spectrum effectiveness and durability, and has therefore attracted much attention owing to its potential for improving crop resistance (Fan and Doerner 2012). In recent years, the interactions between nonhost plants and rust fungi have been studied at histocytological and genetic levels (Mellersh and Heath 2003; Shafiei et al. 2007; Loehrer et al. 2008; Azinheira et al. 2010). In most cases, plant responses to non-adapted rust fungi were similar to the basic defense responses to the adapted rust fungi. For example, interaction between *Arabidopsis* and non-adapted *P. triticina* (*Pt*) induces production of nitricoxide (NO), SA, camalexin and transient stomatal closure (Shafiei et al. 2007). Genetic studies revealed that multiple quantitative trait loci (QTL) and occasionally resistance genes (*R*) conferred NHR in barley to heterologous rust species. Moreover, most of these genes were effective against only one non-adapted rust species (Niks 2014). These findings reveal that multiple genetic and molecular components contribute to NHR, indicating that a multitude of underlying mechanisms determine the outcome of diverse nonhost interactions.

As a model plant for monocotyledonous crops, rice (*Oryza sativa* L.) is a nonhost species to all known rust fungi. Histocytological studies demonstrated that multiple defense responses, including callose deposition, production of reactive oxygen species and cell death, contributed to rice NHR against several rust species. Rice mutants lacking some defense-related or basic defense genes, did not exhibit increased susceptibility to rust fungi (Ayliffe et al. 2011a, b). These results suggested that the mechanisms underlying rice NHR were different from basic resistance despite the similar responses. To understand the molecular basis of rice NHR to *Pst*, Zhao et al. (2014) identified a set of NHR-related proteins in rice plants inoculated with *Pst* using two-dimensional electrophoresis. The expressed proteins from rice leaves 24 and 48 hpi with *Pst* in comparison with mock-inoculated leaves were identified. Quantitative analysis revealed 27 differentially expressed proteins in response to *Pst* inoculation. Most of these proteins fall into the category "response to stimulus" and are involved in basic resistance processes, such as glycerol-3-phosphate and hydrogen peroxide signaling. A homologue of wheat leaf rust resistance protein Lr10 was also identified, implicating that multiple layers of plant defense are involved in rice NHR to *Pst*. These results demonstrate an intrinsic relationship between host and nonhost resistance.

Recently, *Brachypodium distachyon*, a model grass, is considered as a nonhost for wheat rust diseases. Using microarray techniques and available complete *B. distachyon* genome sequences, 889 genes and 524 genes were found to exhibit differential expression in *B. distachyon* after *Pst* and *Puccinia brachypodii* inoculations, respectively (Prof. Kang's group, unpublished data). GO functional analysis showed that there was overlap between host and nonhost defense mechanisms as the base defense. Besides, the expressions of 669 genes were specially changed in the nonhost interaction. There was a significant difference in JA signaling pathway between host and nonhost interactions. In particular, plant cell wall and secondary metabolites played important roles in resistance to *Pst*. Additionally, defense genes such as *PEN1*, *RPM1*, *PEPR1* and *WRKY* transcription factors which were involved in pathogen-associated molecular pattern (PAMP) were found. These results indicate that powerful basal defense mechanisms participate in the recognition of PAMPs and a high level of accumulation of defense-related gene products may contribute to NHR.

3.4.2 Functional Analysis of Host Defense-Related Genes

In nature, plants exclusively rely on innate immunity to protect cells against pathogen attacks. During the long period host-pathogen co-evolution, plants have evolved a two-layered innate immune system, which are defined as pathogen-associated molecular pattern (PAMP)-triggered immunity (PTI) and effector-triggered immunity (ETI) (Jones and Dangl 2006). PTI, also called basal defense, is governed by the surface pattern recognition receptors (PRRs), which act as initial radars to recognize PAMP and induce weak immune response. PTI is an important barrier that normally prevents non-adapted microbes from successful infection. In turn, certain microbes become adapted to particular host genotypes by evolving virulence factors known as effectors, some of which actively interfere with PTI and dampen the general defense response. The effector-triggered susceptibility drives the host-pathogen co-evolutionary cycle involving reiterating of host defense, in which host species evolve resistance genes that intercept pathogen effectors and induce ETI to stop pathogen growth. Compared to PTI, ETI is rapid and robust, which amplifies PTI transcriptional programs and is often associated with localized plant cell death, referred to as the HR (Jones and Dangl 2006). The local resistance reactions induce longer lasting systemic immunity (systemic acquired resistance), which primes uninfected tissues against subsequent pathogen attack (Fu and Dong 2013).

Activated host immunity triggers an array of immune responses, such as a ROS burst, Ca^{2+} spikes, MAPK cascades, transcriptional reprogramming, callose deposition and localized cell death (Buscaill and Rivas 2014). Previous histological and cytological studies described above have driven our understanding of the cellular defense responses in wheat. Subsequently, the availability of transcriptome and genome sequences for wheat and *Pst* has revealed the molecular process connecting the *Pst* attack to the downstream defense pathway. A picture emerges of a versatile immunity network at the level of pathogen recognition, signal transduction, transcription regulation and coordination of different subcellular defense branches to produce effective and robust resistance.

3.4.2.1 Cloned *Yr* Genes

Resistance genes to stripe rust will be discussed in detail in Chap. 5, here our discussion on resistance genes will be more related to interaction with the pathogen at the molecular level. The most profitable and environmentally friendly strategy for farmers to control stripe rusts is to grow genetically resistant wheat and barley varieties with the deployment of resistance genes, which for stripe rust are known as *Yr* (for yellow rust) genes (Lupton and Macer 1962; Chen 2005, 2013; McIntosh et al. 2013; Maccaferri et al. 2015). Based on different criteria, stripe rust resistance genes can be classified into two categories of several terms (Chen 2013), but based on race specificity, they are generally fallen into two categories: race-specific and race-nonspecific (Chen 2005, 2013; Coram et al. 2008a, b, 2010; Lowe et al. 2011). The race-specific resistance genes, referred to hereafter as *R* genes, detect the presence of specific races of the pathogen and initiate a hypersensitive reaction (Spoel and Dong 2012; Chen 2013). Most of race-specific R genes usually encode NBS-LRR proteins, which include a nucleotide binding site (NBS) and leucine-rich repeat (LRR) domain (Michelmore et al. 2013). In general, resistance controlled by NBS-LRR genes is race-specific and resistance controlled by non NBS-LRR genes is race non-specific (Chen 2013).

In cereals, only a few *R* genes encoding NBS-LRR proteins have been cloned, including three wheat leaf rust resistance genes (*Lr1*, *Lr10* and *Lr21*) (Cloutier et al. 2007; Feuillet et al. 2003; Huang et al. 2003), two stem rust resistance genes (*Sr33* and *Sr35*) (Periyannan et al. 2013; Saintenac et al. 2013), and two powdery mildew resistance genes (*Pm3b* and *Pm8*) (Yahiaoui et al. 2004; Hurni et al. 2013). Among close to 80 permanently named *Yr* genes for resistance to stripe rust in wheat, only one race-specific NBS-LRR gene *Yr10* was cloned (Liu et al. 2014). This gene originates from the Turkish wheat line PI 178383 and is located on chromosome 1B in wheat cultivar Moro (Metzger and Silbaugh 1970; Chen et al. 1995). RAPD analyses revealed a specific DNA fragment from BC_4F_5 lines from a Moro × Fielder

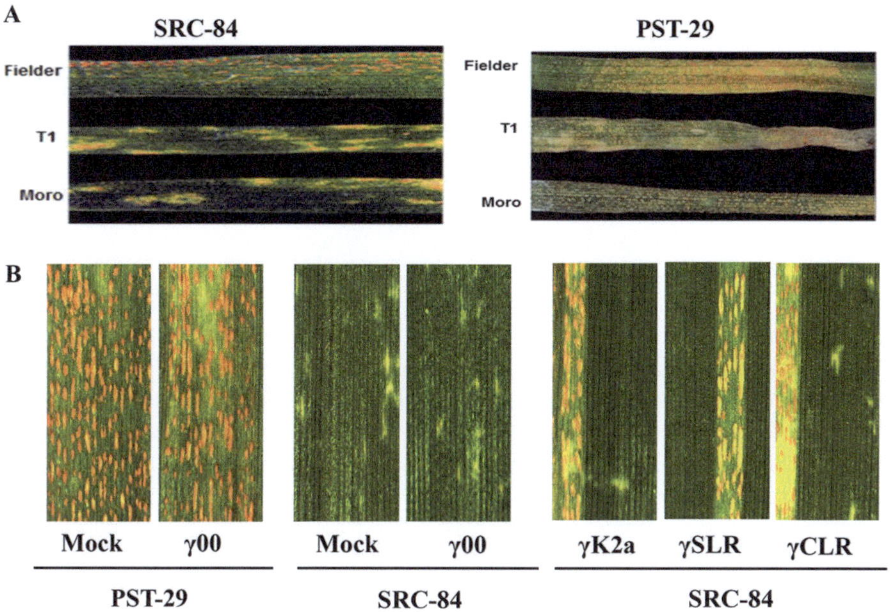

Fig. 3.34 Functional analyses of *Yr10* through transgenic transformation and virus induced gene silencing (VIGS) mediated by BSMV. (**a**) Phenotypic reaction of wheat cultivars Fielder, Moro (*Yr10*), and transformed Fielder (*Yr10*) (T1) after inoculation with races SRC-84 and PST-29 of *Puccinia striiformis* f. sp. *tritici* (*Pst*). (**b**) Stripe rust reaction in silenced Moro leaves inoculated with *Pst* races at 20 dpi. Moro and Moro transfected with BSMV:γ00 were incubated for 8 d, and subsequently inoculated with race PST-29; Moro and Moro transfected with BSMV:γ00 and were subsequently inoculated with race SRC-84 (note necrosis caused by the hypersensitive reaction; and Moro transfected with the γK2a, γSLR, and γCLR and then inoculated with race SRC-84

cross, which was detected in all wheat lines carrying *Yr10* but not in lines without *Yr10*. From a Moro sub-genomic library, hybridization with the specific fragment identified clone 4B and clone 4E containing *Yr10*. Clone 4B of this gene encodes a novel CC-NBS-LRR protein, sharing conserved homologs in the wheat progenitor *Aegilops tauschii* and several host and nonhost monocots of *Pst*. Clone 4E, which shares 84% identity to 4B and does not have a transcription start site (TSS) and putative TATA-box and CAAT-box, is likely a non-expressed pseudogene. Transformation of *Yr10* into the susceptible cultivar Fielder (Su et al. 2003) through bombardment generated one line that developed no pustules and displayed the characteristic HR observed for the resistant cultivar Moro to *Pst* race SRC-84 (avirulent on Moro) (Fig. 3.34a). Silencing of *Yr10* in Moro via the barley stripe mosaic virus (BSMV)-mediated gene silencing (VIGS) resulted in a near fully susceptible phenotype to avirulent *Pst* race SRC-84 (Fig. 3.34b). The results demonstrated that *Yr10* may recognize a corresponding avirulence protein and elicit a HR, and the *Yr10* allele carries all necessary machineries to convert a fully susceptible wheat cultivar to resistance. *Yr10* is the first CC-NBS-LRR gene reported to provide resistance to stripe rust in cereals, which continues to provide effective resistance to stripe rust in most wheat-growing areas but virulent *Pst* races have been reported (Chen et al. 2010).

A candidate of *Yr5* is also a NBS-LRR gene (Wang et al. 2008c; Wang and Chen unpublished data). So far, *Pst* isolates virulent to *Yr5* are extremely rare (Stubbs 1985; Nagarajan et al. 1986). Compared to *Yr10*, *Yr5* is more effective against *Pst* races in the world (Sharma-Poudyal et al. 2013; Wan et al. 2017).

In the wheat-rust pathosystem, the race non-specific resistance is usually associated with a "slow rusting" or "partial resistance" phenotype, of which the pathogen is able to establish itself but is restricted in its growth and sporulation. Two race non-specific resistance genes against wheat rusts have been cloned (Fu et al. 2009; Krattinger et al. 2009). One of the partial resistance genes is *Yr36*, which was first discovered in wild emmer wheat (*Triticum turgidum* ssp. *dicoccoides* accession FA15-3, henceforth DIC). *Yr36* confers HTAP resistance, which is more temperature dependent and less dependent of growth stage (Uauy et al. 2005; Chen 2013). By screening the F_2 population generated from crossing the susceptible durum wheat variety Langdon with the resistant isogenic recombinant substitution lines RSL65 and a RSL65 bacterial artificial chromosome (BAC) library with *Yr36* flanker markers, Fu et al. (2009) initially identified two duplicated genes (WKS1 and WKS2). Subsequent TILLING (Targeting Induced Local Lesions in Genomes) screening of the ethyl methane-sulfonate (EMS)-mutagenized M_2 lines from the common wheat breeding line UC1041+*Yr36* suggested that WKS1 is *Yr36*. WKS1 is composed of a serine/threonine kinase domain and a steroidogenic acute regulatory protein- related lipid transfer (START) domain. Wheat (*T. aestivum*) plants transformed with the complete WKS1 (variant WKS1.1) are resistant to *Pst*, whereas those transformed with an alternative splice variant with a truncated START domain (WKS1.2) are susceptible. WKS1.1 and WKS1.2 preferentially bind to the same lipids mediated by the START domain, but is different in protein-protein interaction. WKS1.1 is targeted to the chloroplast where it phosphorylates the thylakoid-associated ascorbate peroxidase (tAPX) and reduces its ability to detoxify peroxides. Increased expression of WKS1.1 in transgenic wheat plants accelerates leaf senescence in the absence of *Pst*. It is postulated that the phosphorylation of tAPX by WKS1.1 reduces the ability of the cells to detoxify reactive oxygen species and contributes to cell death. This response takes several days longer than typical hypersensitive cell death responses, thus allowing the limited pathogen growth and restricted sporulation. Independent mutations and transgenic complementation confirmed that both of the non-arginine-aspartate (non-RD) kinase domain and START domains are necessary to confer resistance. *Yr36* is present in wild emmer wheat but is absent in modern pasta and bread wheat varieties, except several wheat cultivars recently developed purposely through incorporation of *Yr36* and the linked high protein trait (Uauy et al. 2005; Chen 2013). Later, Gou et al. (2015) determined that the wheat stripe rust resistance protein WKS1 reduces the ability of the thylakoid-associated ascorbate peroxidase to detoxify reactive oxygen species.

The other race non-specific resistance gene is *Lr34/Yr18/Sr57/Pm38* (subsequently referred to as *Yr18*), which supports durable resistance to four biotrophic fungal pathogens, leaf rust, stripe rust, stem rust and powdery mildew (Krattinger et al. 2009). Due to its durability and broad-spectrum effectiveness, *Lr34/Yr18/Sr57/Pm38* has become one of the most important disease resistance genes in wheat breeding. A map-based cloning approach based on three high-resolution populations revealed a gene-rich island containing eight open reading frames on chromosome 7DS, three of which were co-segregating with *Lr34*. However, consistent sequence polymorphism was found only in the adenosine triphosphate (ATP)-binding cassette (ABC) transporter gene between +/− *Lr34* parental lines of the mapping populations. Mutants with interrupted ABC transporter gene were more susceptible to leaf rust, stripe rust, stem rust and powdery mildew, verifying that this gene is responsible for conferring the durable disease resistance based on *Lr34*, *Yr18*, *Sr57* and *Pm38* as well as the leaf tip necrosis. Full-size ABC transporters share a conserved structure consisting of two cytosolic nucleotide binding domains (NBDs) and two transmembrane domains (TMDs). Each TMD is predicted to consist of six membrane-spanning helices (Krattinger et al. 2011), and the TMDs shuttle a specific substrate across a biological membrane. Transgenic wheat plants carrying *Yr18* develop senescence-like processes in the flag leaf tips and edges (Dyck 1991; Singh 1992a, b) and confers resistance on adult wheat plants (Krattinger et al. 2009), revealing that this ABC transporter was sufficient and required for broad-spectrum disease resistance (Risk et al. 2012). *Yr18*-containing germplasm has been a part of wheat improvement since the early twentieth century and wheat cultivars containing *Yr18* occupy more than 26 million ha in various developing countries alone and contribute substantially to yield savings in epidemic years (Marasas et al. 2003). The *Yr18* gene has remained durable, and no evolution of increased virulence toward *Yr18* has been observed for more than 50 years. This is in contrast to many other rust resistance genes, the so-called gene-for-gene class, that provide resistance to some but not all races of a rust fungus (Webb et al. 2002; Cloutier et al. 2007; Ellis et al. 2007; Keller et al. 2007). It should be pointed out that the level of stripe rust resistance provided by *Yr18* alone is very low, lower than many other *Yr* genes and QTL for APR or HTAP resistance, and is often not adequate to protect crops from damage. This gene should be combined with other *Yr* genes to provide an adequate level of resistance.

3.4.2.2 Signal Transduction

Plant cells are equipped with highly efficient signal transduction pathway mechanisms to perceive, transduce and respond to a wide variety of internal and external signals during their growth and development (Sanders et al. 1999). Upon infection by *Pst*, ion exchange takes place from the wheat cell plasma membranes, with release of calciumions, production of a large quantity of reactive oxygen species and phosphorylated proteins, and activation of a series of defense-related genes. All these activities have been shown to be closely interlinked by signal transduction pathways.

3.4.2.2.1 Calcium Ion Influx and Calcium-Binding and Calcium-Sensoring Proteins

As a ubiquitous and versatile messenger, calcium ion is a key factor that is widely used in signal transduction pathways to mediate responses to hormones, biotic/abiotic stress signals and a variety of developmental cues compared to other messengers (Bush 1995; Knight 2000; Reddy and Reddy 2004). Transient changes in the intercellular and cytosolic calcium concentration are primary events of the immunity of plants (Atkinson et al. 1996; Hetherington and Brownlee 2004; Nühse et al. 2007; Ranf et al. 2011). Numerous studies show that calcium plays a role in very early responses of plants to fungi, bacteria, viruses and elicitors in plant innate immunity, including PTI and the HR. Increased calcium in the cytosol is reported to be necessary to induce ROS generation and HR in Arabidopsis (Grant et al. 2000), which is also observed in the incompatible interaction of cowpea and cowpea rust fungus but not in the compatible interaction (Xu and Heath 1998). During potato-*Ralstonia solanacearum* interaction, calcium regulated the induced expression of a non-specific lipid transfer protein (nsLTPs) StLTPa7, which was found to be involved in disease resistance response at the early stages (Kim et al. 2009b; Lee and McNellis 2009).

In wheat challenged with either an avirulent or a virulent isolate of *Pst*, the subcellular localization of calcium was detected by transmission electron microscopy (TEM) as electron dense precipitate after treatment with potassium pyroantimonate (Yin et al. 2015). Cytochemical studies revealed that in the incompatible wheat-*Pst* interaction associated with necrosis of host mesophyll cells, an influx of Ca^{2+} from the intercellular space to the cytoplasm and finally an efflux to the intercellular space again was detected. Calcium precipitates were also observed in mesophyll cells adjacent to necrotic cells. In contrast, calcium flow was not significantly affected in the susceptible wheat plants. The results suggested that the changes of calcium is closely related to HR in the wheat-*Pst* interaction, and Ca^{2+} might function as a secondary messenger in HR induction during wheat resistance to wheat stripe rust.

Changes in the intercellular and cytosolic Ca^{2+} concentration can be sensed by various Ca^{2+} binding proteins, which are characterized with different Ca^{2+} signatures, including EF-hand domains and C2 domains (Nalefski and Falke 1996; Mackrill 1999; Takezawa 2000; Benjamins et al. 2003; Jayasekaran et al. 2006). Through binding to Ca^{2+}, this kind of proteins plays a key role in regulation of the downstream signal transduction cascades that couple signals to cellular processes in plants. EF-hand proteins are characterized by a helix-loop-helix structure, with an interhelical loop of 12 or 14 amino acids that is able to bind a calcium ion. Most of the proteins contain 2-8 EF-hand motifs (Moncrief et al. 1990), which is used as a molecular device to recognize and transduce specific Ca^{2+} signal through binding calcium to change their conformation to interact with downstream proteins (Navazio et al. 1998; Capozzi et al. 2006; Boonburapong and Buaboocha 2007). Several EF-hand calcium binding proteins are known to be involved in defense and responses to various environmental stresses in plants. The EF-hand calcium binding genes in rice EFA27 was induced in response to ABA and osmotic stress (Frandsen et al. 1996), while that of barley was differentially expressed when treated with SA and could strengthen the plant defense to protect barley systematically against the powdery mildew pathogen (Beßer et al. 2001). Feng et al. (2011) showed that a calcium binding EF-hand protein TaCab1 was upregulated in wheat leaves treated with various phytohormones, such as ABA, benzyl adenine, ET, methyl jasmonate (MeJA) and SA, or under different stress stimuli (wounding, low temperature, polyethylene glycol and high salinity). The *TaCab1* gene was shown to be important in regulating wheat basic defense to *Pst* with similar induction and expression pattern in both compatible and incompatible interactions. Their results indicate that *TaCab1* is involved in wheat recognition of *Pst*, and contribute to wheat basal defense to biotic and abiotic stresses mainly through the SA signaling pathway (Feng et al. 2011).

The majority of the Ca^{2+} sensor proteins contain a conserved calcium-dependent membrane targeting module (C2 domain), which could bind phospholipids in a Ca^{2+}-dependent manner (Nalefski and Falke 1996). The five aspartic acid residues form a Ca^{2+}-binding site, which is highly conserved in most C2 domain proteins in plants and animals (Kim et al. 2003a). In plants, C2 domain proteins are involved in mRNA long-distance transport (Xoconostle-Cazares et al. 1999), development (Yang et al. 2008), senescence (Ouelhadj et al. 2005), membrane targeting (Schapire et al. 2008; Yamazaki et al. 2010), abiotic and biotic stress responses (Kim et al. 2000, 2003a, 2004, 2008b; Yang et al. 2006a, 2007; Wang et al. 2009b). From the transcriptomic analysis of wheat APR to stripe rust, a novel C2 domain protein gene *TaERG3* was isolated, which was differentially induced in incompatible and compatible interactions between wheat and *Pst* (Zhang et al. 2013). The TaERG3 protein was sensitive to changes in Ca^{2+} concentration, with a higher expression level in plants treated with 0.5 mM Ca^{2+} than 2.0 mM Ca^{2+}. Besides, *TaERG3* was upregulated by exogenous applied ABA and environmental stimuli including low temperature and high salinity. It was postulated that *TaERG3* might be involved in wheat defense responses against *Pst* and abiotic stresses in an ABA-dependent signaling pathway (Zhang et al. 2013).

In plants, the calcineurin B-like protein (CBL) family represents a unique group of calcium sensors that specifically interacts with and regulates a family of protein kinases, CBL-interacting protein kinases (CIPKs) (Batistič and Kudla 2009). Extensive CBL and CIPK gene families have been identified in the genomes of all higher plants that have been completely sequenced (Kolukisaoglu et al. 2004; Yu et al. 2007). Rice, Arabidopsis and *Populus trichocarpa* each encode 10 CBL calcium sensors and a complement of 26–30 CIPKs. Efforts to determine the functional pairs of CBLs and CIPKs revealed that each CBL interacts with a subset of CIPKs, and each CIPK interacts with one or more CBLs. Some CBLs share common CIPK targets, and some CIPKs share common CBL regulatory subunits. The interaction specificity and overlap among various members of CBL and CIPK family reflect the functional redundancy and specificity, and the complexity of the CBL-CIPK network in plant signal transduction (Luan 2008; Batistič and Kudla 2004). Besides the complicated interaction network, the physiological function of CBL and CIPK has been largely unknown, and only a few of them has been established.

The Arabidopsis CBL calcium sensor AtCBL4 and AtCIPK24 appear to be parts of the calcium regulated signaling pathway that specifically mediates salt stress signaling and adaption (Liu and Zhu 1998; Halfter et al. 2000). CIPK3 and CBL1 function in regulating ABA responses during seed germination and responses to drought and salt stress respectively (Albrecht et al. 2003; Cheong et al. 2003; Kim et al. 2003b). Compared to rice and Arabidopsis, more CBLs and CIPKs loci are identified in the wheat genome, including 23 CBL and 74 CIPK, which indicate a more complicated CBL-CIPK interaction network in wheat. Six CBL and nine CIPK genes have been cloned from cultivar Suwon 11 (Prof. Kang's group, unpublished data). Functional characterization through VIGS showed that silencing of *TaCBL4*, *TaCIPK5*, *TaCIPK10* and *TaCIPK15* renders wheat resistant to *Pst*, while reduced expression of *TaCIPK14* enhanced wheat resistance to *Pst*, suggesting the distinct roles of CBL and CIPK during wheat-*Pst* interactions. Protein-protein interaction assays revealed that TaCIPK10 interacts with a wheat NPR1-like protein (TaNH2) in the cytoplasm of wheat, which plays an important role in wheat defense against *Pst* via the SA signaling pathway. It is postulated that *TaCIPK10* contributes to wheat resistance to *Pst* through regulating the SA -mediated defense pathway. The results imply a crosstalk between the SA and CBL-CIPK pathways to co-regulate wheat immunity. The specific regulation of target proteins is achieved by a complex interplay of mechanisms generating spatial and temporal specificity in the CBL/CIPK signaling network.

By screening a yeast two-hybrid library from wheat leaves infected with *Pst*, Wang et al. (2016a) identified a conserved *Pst* protein that interacts with wheat NPR1 and named it *Puccinia* NPR1 interactor (PNPi). This protein interacts with the wheat NPR1/NIM1-like domain of NPR1 via its C-terminal DPBB_1 domain. Using bimolecular fluorescence complementation assays, they detected the interaction between PNPi and wheat NPR1 in the nucleus of *N. benthamiana* protoplasts. Through a yeast three-hybrid assay, they found that PNPi interaction with NPR1 competes with the interaction between wheat NPR1 and TGA2.2. They also found reduced induction of multiple *PR* genes in the region adjacent to *P. syringae* pv. *tomato* infection in barley transgenic lines over-expressing *PNPi*. The authors postulated that PNPi has a role in manipulating wheat defense response via its interaction with NPR1.

3.4.2.2.2 GTP-Binding Proteins

In plants, during the signal transduction, the small GTP-binding proteins have been found to function as regulators of specific intercellular fundamental biological processes. Small GTP binding proteins constitute a superfamily consisting of at least five families (Ras, Rho, Rab, Sar1/Arf and Ran). The Rab proteins, belonging to the small guanosine triphosphatases (GTPases) superfamily, are thought to act as molecular switches, which play an essential role in endocytic and exocytic traffic in eukaryotic cells. Rab7 and its downstream components are important factors in the pathogenesis of microorganisms (Scianimanico et al. 1999; Harrison et al. 2003;

Vieira et al. 2003; Roberts et al. 2006). In Arabidopsis, overexpression of the Rab7 homolog *RabG3b* leads to expanded HR upon pathogen infection and accelerated leaf senescence (Kwon et al. 2009). A tonoplast-localized rice Rab7 homologue is up-regulated in response to cold, salt and drought stress, suggesting its role in plant adaption to various environmental stimuli (Nahm et al. 2003). During a wheat-*Pst* interaction, a wheat Rab7 homolog *TaRab7* was identified to be specifically induced in the wheat resistant response (Liu et al. 2012a). *TaRab7*was also upregulated in response to salt and drought stresses, indicating that *TaRab7* has a role in plant response to various biotic and abiotic stimuli. Transient expression in onion cells by particle bombardment revealed that TaRab7 was dominantly localized in the cytoplasm, perinuclear area and nucleus. Silencing of *TaRab7* could not block or eliminate the race-specific resistance, but allowed limited fungal growth and uredinial development, indicating the reduced wheat resistance to an avirulent *Pst* isolate. It can be inferred that wheat *TaRab7* GTPase is part of the signaling pathway in responses to *Pst* infection and abiotic stimuli.

3.4.2.2.3 Tyrosine Phosphorylation Proteins

Tyrosine phosphorylation protein A (TypA/BipA) is a member of the ribosome-binding GTPase superfamily. In many bacterial species, TypA acts as a global stress and virulence regulator and also mediates resistance to the antimicrobial peptide bactericidal permeability-increasing protein (Owens et al. 2004). TypA proteins are widely distributed in plants and are present in green algae, mosses, liverworts, gymnosperms and angiosperms (Barak and Trebitsh 2007). TypA genes are reported to play roles in pollen tube growth in Arabidopsis (Lalanne et al. 2004) and in the development of male reproductive organs in cucumber (Barak and Trebitsh 2007). The SsTypA protein has been proposed as a new member of the ROS-scavenging system in the chloroplast of *Suaeda salsa* when subjected to environmental stresses (Wang et al. 2008a). Atkinson (2015) identified a TypA subfamily (mTypA) which is mitochondrially targeted according to transit peptide predictions in some Archaeplastida, Amoebozoa and fungi. TypA proteins have been considered as a novel class of stress tolerance-related proteins in all species ranging from prokaryotes to higher eukaryotes (Wang et al. 2008a). In a wheat-*Pst* interaction, a stress-responsive TypA gene (*TaTypA*) from wheat was isolated and functionally characterized (Liu et al. 2016b). The subcellular localization experiment revealed that TaTypA proteins were mainly localized in chloroplasts of wheat cells. The transcript levels of *TaTypA* were up-regulated in response to the treatment with methyl viologen, which induced ROS in chloroplasts through photoreaction expression. The expression of *TaTypA* was also induced by cold stress and infection of an avirulent isolate of *Pst*. Knockdown of the expression of *TaTypA* through VIGS resulted in reduced ROS accumulation and HR, and increases in *TaCAT* and *TaSOD* expressions, indicating that TaTypA is an important factor in the ROS-scavenging system.

Accompanied with the decreased wheat defense, increases in pathogen hyphal growth and branching were observed. It seems that TaTypA is a new chloroplast-localized defense protein, which contributes to resistance in an ROS-dependent manner (Liu et al. 2016b).

3.4.2.3 Transcriptional Factors

The complex response mechanisms of plants are regulated by a large number of genes that encode important transcription factors (Huang et al. 2005; Yamaguchi-Shinozaki and Shinozaki 2005, 2006; Nakashima and Yamaguchi-Shinozaki 2006; Umezawa et al. 2006). Transcription factors (TFs) bind to specific regulatory sites upstream of genes by direct physical interactions or in combination with other proteins to regulate gene expression in response to a signaling cascade. More than 50 families of TFs have been identified in Arabidopsis and rice. Among these TFs, members of the NAC, WRKY, C2H2-type zinc finger, Apetala2/ET-responsive element binding protein (AP2/EREBP) and MYB families have been well characterized to play key regulatory roles in plant stress and defense responses (Singh et al. 2002). To explore the transcription factors in wheat defense to *Pst*, much attention has been focused on the role of TFs in wheat-*Pst* interactions.

3.4.2.3.1 NAC Proteins

NAC proteins constitute one of the largest families of plant-specific TFs (Olsen et al. 2005), which contains a highly conserved N-terminal DNA-binding domain defined as the NAC domain (for NAM, ATAF1/2, CUC2) by Aida et al. (1997) and a variable C-terminal domain which may serve as a transcriptional activation region (Ren et al. 2000; Xie et al. 2000; Duval et al. 2002). In recent years, increasing evidence has implicated some NAC genes in plant defense responses to pathogen infection and environmental stimuli, particularly the genes in the ATAF subfamily. Overexpression of *OsNAC4* resulted in HR cell death, and in *OsNAC4* knockdown lines, HR cell death was markedly decreased in response to avirulent bacterial strains (*Acidovorax avenae* N1141) (Kaneda et al. 2009). In wheat-*Pst* interaction pathosystem, two novel NAC transcription factors *TaNAC4* and *TaNAC8* were isolated and characterized (Xia et al. 2010a, b). Both *TaNAC4* and *TaNAC8* were upregulated upon infection by an avirulent *Pst* race, suggesting their roles in wheat defense to *Pst* infection. In addition, the expressions of the two transcription factors were induced by exogenously applied MeJA, ET and environmental stimuli including high salinity, wounding and low-temperature. Transcription analysis indicates that these two NAC protein genes function as transcriptional activators to participate in wheat response to biotic and abiotic stresses.

3.4.2.3.2 MYB Transcriptional Factors

MYB TFs are characterized by a MYB domain composed of approximately 52 amino acid residues that adopt a helix-turn-helix conformation to intercalate into the major groove of DNA (Du et al. 2009). Studies revealed that MYB TFs are key factors especially in the regulatory networks of secondary metabolism, cell shape and the enhancement of disease resistance. The first plant MYB gene C1 encoding a c-MYB-like TF was identified in maize (*Zea mays*) and found to be involved in anthocyanin biosynthesis (Paz-Ares et al. 1987). Arabidopsis MYB TF AtMYB30 is involved in an amplification loop or signaling cascade that modulates SA synthesis and thus positively regulates hypersensitive cell death in resistance to pathogens (Vailleau et al. 2002; Raffaele et al. 2006, 2008; Li et al. 2009). Overexpression of rice *OsMYB4* in transgenic plants increased tolerance to chilling, freezing, UV light, drought stresses and diseases (Vannini et al. 2006, 2007; Laura et al. 2010).

In *Pst* infected wheat leaves, one wheat MYB gene *TaMYB4* was identified and shown to be significantly induced in the early stage of infection by an avirulent *Pst* race (Al-Attala et al. 2014). Up-regulated expression of *TaMYB4* was also observed in wheat cells challenged by various plant hormones (SA, ET, ABA and MeJA) or under different stimuli (cold and wound). The varied induced patterns of *TaMYB4* under different hormone treatments implicated the crosstalk between the signaling pathways activated by the SA, ET and ABA hormones in the defenses mediated by TaMYB4. The expression of *TaMYB4* exhibited tissue specificity, which was detected in wheat roots, culms and leaves. Yeast one-hybrid assay verified the transcription activity of TaMYB4 and the C-terminus is necessary for activating the transcription. Coinciding with the transcription activity, TaMYB4 proteins are localized in the nucleus. Furthermore, silencing of *TaMYB4* resulted in significantly reduced transcript levels of *TaPAL* and PR protein genes, which may be responsible for the compromised resistance of wheat cv. Suwon 11 to an avirulent *Pst* race. It can be concluded that *TaMYB4* plays a primary role in conferring resistance to stripe rust, and in wheat responses against different stresses (Al-Attala et al. 2014).

3.4.2.3.3 bZIP Transcriptional Factors

Plant bZIP TFs contain a nuclear localization signal followed by an invariant N-x7--R/K motif that binds to DNA, and a heptad repeat of leucines positioned exactly nine amino acids toward the C-terminus, creating an amphipathic helix involved in protein–protein interactions (Landschulz et al. 1998). bZIP TFs preferentially bind to DNA sequences with an ACGT core, such as the G-box (CACGTG) motif found in the promoters of many stress-responsive genes (Izawa et al. 1993; Kim and Thomas 1998; Martinez-Garcia et al. 1998; Niu et al. 1999; Siberil et al. 2001). Several bZIP TF proteins have been identified to be involved in the regulation of plant defense genes (Rushton and Somssich 1998). Most studies on defense-related bZIP genes were on dicot plants, such as Arabidopsis, tomato, tobacco and potato, but little is known about bZIP genes in defense responses in monocot plants, particularly in wheat. Zhang et al. (2008c) cloned a novel bZIP gene from wheat, designated as *TabZIP1*, and found that the gene transcript was rapidly and strongly induced during an incompatible wheat-*Pst* interaction. Treatments with both MeJA and ET, but not SA, induced the *TabZIP1* expression, indicating that the *TabZIP1* may be involved in defense response to *Pst* infection via the JA/ET-dependent signal transduction pathways (Zhang et al. 2008c).

3.4.2.3.4 EIN3 Proteins

Increasing evidence has implicated ET-insensitive 3 (EIN3) family genes in defense responses to pathogen attack and environmental stimuli. The EIN3 proteins are a family of small plant-specific TFs that contain a special, highly conserved EIN3-binding domain at the N-terminus. Some EIN3/EILs have a DNA binding activity site, A(A/C)G(A/T)A(A/C)CT, which has also been identified in the promoters of many PR protein genes, indicating the potential direct activation of basic PR protein promoters (Kosugi and Ohashi 2000). Tobacco EIN3-like TEIL binds directly to the promoter of *PR1a* gene, functioning as a putative negative transacting factor. EIN3 also participates in a regulatory cascade of PR gene expression by activating ET response factor1 (ERF1) TFs, which act as activators or repressors of PR protein genes by binding to their promoter regions (Ohme-Takagi et al. 2000; Alonso et al. 2003; Chen et al. 2009; Boutrot et al. 2010; Zhang et al. 2011a). EIN3 is a key TF in the ET signaling pathway, which also functions as positive regulators of a subset of jasmonic acid responses, including plant resistance to necrotrophic fungi and root development. EIN3/ EILs participate in the regulation of plant innate immunity not only via the ET signaling pathway, but also via the JA signaling or SA signaling pathway (Chen et al. 2009). EIN3 and EIL1 target directly the promoter of *SID2* which controls SA biosynthesis, negatively regulating *SID2* expression and SA biosynthesis to repress plant immunity (Wildermuth et al. 2001; Chen et al. 2009). Currently, EIN3/EIL are proposed to represent a branching point for the cross-talk between ET, JA and SA signaling pathways and participated in plant defense through these signaling pathways (Chen et al. 2009). Duan et al. (2013a) isolated and characterized a novel wheat *EIN3* homologue, *TaEIL1*, from wheat leaves inoculated with virulent *Pst* race CYR31. The gene was upregulated in wheat leaves during the compatible interaction, indicating the negative regulator role of *TaEIL1* in wheat resistance to *Pst*. The expression of the gene could be reduced under SA treatment. Knockdown of *TaEIL1* partially reduced the susceptibility of wheat to the virulent *Pst* race with decreased disease symptoms and inhibited pathogen growth. Silencing of *TaEIL1* enhanced wheat resistance to an avirulent *Pst* race with increased HR. Moreover, the expression of *TaEIL1* was induced by exogenous application of MeJA and ABA, and was reduced by SA at the early stage, but not affected by ET. In fact, ET has been shown to inhibit EIN3 protelysis and rapidly increase EIN3 protein levels without significant changes in mRNA level (Chao et al. 1997; Tieman et al. 2001; Lee and Kim 2003). The results of Duan et al. (2013a) indicate that *TaEIL1* may serve as a negative regulator in the wheat defense response through cross-talk between SA, JA and ET signaling pathways.

3.4.2.4 Plant Defense

Plant defense is a basic capability for adaptations evolved by plants to improve their survival and reproduction by reducing the impact of biotic and abiotic stresses. There are many mechanisms and genes involved in plant defense. Here, we present those that have been studied in cereals for defense against stripe rust.

3.4.2.4.1 Pathogenesis Related Proteins

Pathogenesis related proteins encode groups of low molecular weight proteins that are activated in the induced defense mechanisms, which are assigned an important role in plants to adapt to stressful environment conditions and to respond especially to pathogenic influences (Bowles 1990; Edreva 2005). Initially, PR proteins are found commonly induced in resistant plants, expressing a HR to viral, fungal and bacterial pathogens. Later, however, it turned out that PR proteins are induced not only in resistant, but also in susceptible plants challenged with plant pathogens, as well as in plants subjected to abiotic stresses (Van Loon 1985). Wang et al. (2010b) isolated and characterized a thaumatin like PR protein gene, *TaPR5*, from wheat leaves (cv. Suwon 11) infected by an avirulent *Pst* race, CYR23. *TaPR5* transcript was significantly induced and upregulated in the incompatible interaction, but at a relative low level in a compatible interaction. Similarly, the expression of *TaGlu*, which is a β-1,3-glucanase PR protein gene was induced during both compatible and incompatible interactions of wheat with *Pst*, but the transcription level was much higher in the incompatible interaction than that in the compatible interaction (Liu et al. 2010). The *TaPR5* gene expression was more effectively activated by the SA signaling pathway, whereas *TaGlu* was more significantly induced by MeJA than SA or ABA. The TaPR5 protein was more concentrated in wheat cell wall, and the *TaGlu* enzyme was localized mainly in the host cell wall and over the extrahaustorial matrix. Commonly, the higher level of transcription and greater abundance of protein labeling in the incompatible wheat-*Pst* interaction than in the compatible interaction further suggested their positive role in wheat resistance to *Pst* (Wang et al. 2010b; Liu et al. 2010). Another PR protein, TaPR10, was shown to participate in the defense responses of wheat APR to *Pst*, due to a much higher expression level in the resistant adult plants than in the susceptible seedling stage (Zhang et al. 2010). Although the roles of PR proteins in plant defense against *Pst* is far from being fully understood, recent results indicate their involvement in wheat resistance to stripe rust. Due to the constitutive role of PR proteins in plant defense, some of the PR proteins have been used as biological markers to evaluate wheat resistance to stripe rust.

3.4.2.4.2 Heat Shock Proteins

Heat shock proteins (HSPs) are a class of ubiquitous and highly conserved proteins which show increased expression in response to an elevated temperature or other environmental stresses (de Maio 1999). In tobacco (*N. benthamiana*), HSP70 and HSP90 have been reported to be essential components in the INF1-mediated HR and function upstream or independent of the MAPK cascade (Kanzaki et al. 2003). HSP70 expression in plants is induced upon infection with various pathogens (Aranda et al. 1996; Ventelon-Debout et al. 2004; Panthee et al. 2007). Duan et al. (2011) isolated a wheat HSP70 gene, *TaHSC70*, from a suppression subtractive hybridization (SSH) cDNA library of wheat leaves inoculated with *Pst*. Following heat shock of wheat seedlings at 40 °C, *TaHSC70* expression increased in inoculated seedlings but remained stable at the same level in non-inoculated seedlings. In the compatible interaction, *TaHSC70* exhibited a transient expression in the early infection process, whereas the gene had a continuous higher level of expression in the incompatible interaction. In addition, a transient increase of the *TaHSC70* expression level was observed in wheat leaves upon treatments with MeJA but SA, ET and ABA had no influence on *TaHSC70* expression. The results indicate that *TaHSC70* plays a role in defense responses elicited by *Pst* infection through a JA-dependent signal transduction pathway (Duan et al. 2011).

Heat shock proteins 90 (Hsp90s), which are ubiquitous molecular chaperones involved in signal transduction, protein folding and protein degradation (Wegele et al. 2004; Taipale et al. 2010), have been found to play an important role in plant immune responses to pathogens. ETI expression in higher plants is known to be mediated by recognition of *R* gene and *Avr* gene products. Many *R* gene proteins have been found to be functionally dependent on cytosolic Hsp90s. The Arabidopsis *R* gene protein RPM1 that mediates HR to *P. syringae* pv. *tomato* was identified as a client of cytosolic Hsp90s (Hubert et al. 2003). Furthermore, cytosolic Hsp90s have been found to interact with two co-chaperones, SGT1 and RAR1, which are major regulatory components of disease resistance triggered by many *R* proteins in both dicotyledonous and monocotyledonous plants (Botër et al. 2007). Wang et al. (2011a) identified and characterized three HSP90 genes, *TaHsp90.1*, *TaHsp90.2* or *TaHsp90.3*, in wheat. Decreasing the expression of *TaHsp90.1* by VIGS caused pronounced inhibition of wheat seedling growth, whereas the suppression of *TaHsp90.2* or *TaHsp90.3* compromised HR of wheat cultivar Suwon 11 to *Pst*, indicating that *TaHsp90.1* plays an important role in the seedling growth of wheat, and *TaHsp90.2* and *TaHsp90.3* are required for the HR resistance of wheat to stripe rust. Their data also showed that the role of *TaHsp90.3* in the *R* gene-mediated resistance to *Pst* was more prominent than that of *TaHsp90.2*. The aggravation of abnormal plant growth owing to *TaHsp90.1* suppression by the simultaneous silencing of *TaHsp90.2* and *TaHsp90.3* suggests that the last two genes may also be involved in the control of wheat seedling growth (Wang et al. 2011a).

3.4.2.4.3 Transport of Defense-Related Materials

SNAREs (soluble N-ethylmaleimide-sensitive factor attachment protein receptors) are key components of vesicle trafficking in eukaryotic cells (Heese et al. 2001). The SNARE complex determines the specificity of intracellular fusion by forming a helix of the SNARE bundle composed of a vesicle membrane-anchored SNARE (v-SNARE) and three target membrane-anchored SNAREs (t-SNAREs) (Antonin et al. 2000; Fukuda et al. 2000). Several SNARE genes in plants have been shown to be involved in plant resistance against various pathogens by mediating trafficking of defense-responsive compounds. HvSNAP34 in barley (*Hordeum vulgare*) participates in callose deposition during nonhost resistance to powdery mildew (Inada and Ueda 2014). NPSN11, a novel plant SNARE (NPSN) gene, has been reported to be involved in the delivery of cell wall precursors to the newly formed cell plate during cytokinesis. AtNPSN11 in Arabidopsis has been reported to be involved in one of the two KNOLLE-containing tetrameric SNARE complexes, which jointly mediate membrane fusion during cytokinesis (El Kasmi et al. 2013). Tobacco (*N. tabacum*) SNARE homologue NbSYP132 has been implicated in plant resistance against bacterial pathogen by mediating the secretion of PR1 protein (Kalde et al. 2007). Another Golgi SNARE, AtMEMB12, was targeted by miR393b* and involved in the accumulation of PR1 protein (Zhang et al. 2011c). SNARE proteins have been reported to be involved in the plant-microbe interactions by mediating the secretory pathway of functional PRR or PR proteins to plasma membrane (Beck et al. 2012; Watanabe et al. 2013).

Wang et al. (2014a) identified three NPSNs (*TaNPSN11*, *TaNPSN12* and *TaNPSN13*) and three plant defense-related SNARE homologs (*TaSYP132*, *TaSNAP34* and *TaMEMB12*) from wheat leaves showing resistant and susceptible responses to *Pst*. Expressions of *TaNPSN11*, *TaNPSN13* and *TaSYP132* were differentially induced by either avirulent or virulent *Pst* races, suggesting a possible involvement of these SNARE homologues in wheat responses to stripe rust. In VIGS assays, resistance of wheat cultivar Xingzi 9104 to *Pst* avirulent race CYR23 was reduced by knocking down *TaNPSN11*, *TaNPSN13* and *TaSYP132*, but not *TaNPSN12*, implying diverse functions of these wheat SNARE genes in reactions to *Pst* infection and hyphal elongation. Silencing of *TaNPSN11* or *TaSYP132* led to reduced resistance of wheat to *Pst* with significant decrease in the accumulation of phenolic auto-fluorescence. TaNPSN11 was primarily localized on vesicle structures near the plasma membrane towards *Pst* infection sites, and subsequently vesicle transport was activated and redirected to the infection sites. TaSYP132 showed a highly specific interaction with TaNPSN11 in both yeast two-hybrid and bimolecular fluorescence complementation (BiFC) assays. Thus, TaNPSN11 seemed to be involved in vesicle trafficking between the Golgi apparatus and plasma membrane, and possibly function in transport of plant defense-related materials through interaction with SYP132. Interestingly, reduced accumulation of ROS was only observed in the *TaNPSN13*- and *TaSYP132*-knockdown plants, indicating that these two SNARE homologues may be involved in the delivery of ROS-related proteins.

Based on all of these results, they concluded that these wheat TaNPSN SNARE genes are involved in vesicle mediated resistance to stripe rust (Wang et al. 2014a).

Q-SNAREs localize to the membrane of target organelles (t-SNAREs), which can be classified as Qa-SNARE motifs (occupy the syntaxin position), Qb-SNARE (SNAP-25 N-terminal) and Qc-SNARE (SNARE-25 C-terminal) (Bock et al. 2001). SYP7 is one of the Qc-SNAREs and contains SYP71, SYP72 and SYP73, among which SYP71 is the most studied. The *Lotus japonicus* SYP71 expressed in vascular tissues has been shown to be involved in symbiotic nitrogen fixation with rhizobia (Hakoyama et al. 2012). SYP71 has been reported to participate in plant defense against various pathogens. For example, overexpression of *OsSYP71* in rice enhanced tolerance to oxidative stress and resistance to rice blast (Bao et al. 2012). Arabidopsis SYP71 is essential for successful viral infection by mediating the fusion of the turnip mosaic potyvirus (*TuMV*)-induced 6K2 vesicles with chloroplasts during *TuMV* infection (Wei et al. 2013). Liu et al. (2016a) identified a homologue of plant-specific Qc-SNARE gene *TaSYP71* in wheat plants inoculated with *Pst*. Subcellular localization via *Agrobacterium*-mediated transient expression revealed that TaSYP71 was localized to the plasma membrane in *N. benthamiana*. The expressions of the three *TaSYP71* homologues in wheat were induced by NaCl, H_2O_2 stress and inoculated with both virulent and avirulent *Pst* isolates. Heterologous expression of *TaSYP71* in *Schizosaccharomyces pombe* elevated tolerance to H_2O_2. Meanwhile, H_2O_2 scavenging gene (*TaCAT*) was down-regulated in the *TaSYP71* silenced plants treated with H_2O_2 compared to that in control plants, which indicated that *TaSYP71* enhanced tolerance to H_2O_2 stress possibly by influencing the expression of *TaCAT* to remove the excessive H_2O_2 accumulation. When *TaSYP71* homologues were all silenced in wheat through VIGS, wheat plants were more susceptible to *Pst*, with larger infection area and increased haustorial formation. Surprisingly, silence of *TaSYP71* homologues also led to increased necrotic area of wheat mesophyll cells. A possible explanation is that the minor contribution of resistance to *Pst* was insufficient to hinder pathogen extension when *TaSYP71*were silenced, and the necrotic area was enlarged accompanied with the pathogen growth. Of course, later cell death could not be excluded. In addition, the expressions of PR protein genes were down-regulated in *TaSYP71* silenced wheat plants. These results together suggest that *TaSYP71* play a positive role in wheat defence against *Pst* (Liu et al. 2016a).

3.4.2.4.4 Cell Death Related Genes

Hypersensitive response, including the death of plant cells localized to the infection sites, is a robust defense response frequently associated with *R* gene-mediated resistance. As a form of programmed cell death (PCD), HR is believed to help plants defend themselves against pathogens by limiting their growth. The development of HR involves a cascade of signaling, during which the earliest changes are accumulation of reactive oxygen intermediates and rapid intracellular ion fluxes (Atkinson et al. 1996; Morel and Dangl 1997; Delledonne et al. 2001). Subsequently, the downstream protein components probably required for HR initiation and execution are activated (Heath 2000). In addition, HR involves expression of defense genes for disease resistance and inducible defense responses (Dangl and Jones 2001).

3.4.2.4.4.1 Hypersensitive-Induced Reaction Genes

Hypersensitive-induced reaction (HIR) genes are a novel group of plant defense genes, which was firstly identified to be able to elicit HR in tobacco challenged by *Tobacco mosaic virus* with induced PR2 and β-glucanase (Karrer et al. 1998). Overexpression of the plasma membrane localized pepper *CaHIR1* and rice *OsHIR1* genes induced spontaneous HR-like lesions and increased expression of defense related genes in transgenic Arabidopsis plants, which show enhanced resistance to bacterial pathogens *P. syringae* and *Xanthomonas campestris*, and oomycete *Hyaloperonospora parasitica* (Jung and Hwang 2007; Jung et al. 2008; Zhou et al. 2010). Arabidopsis HIR proteins form complexes with RPS2, a typical NB-LRR resistance protein, and contribute quantitatively to RPS2-mediated ETI to *Pto* DC3000 *AvrRpt2* (Qi et al. 2011). Increasing evidence shows that HIR genes may act as a positive regulator of cell death involved in plant immunity.

In wheat, two HIR genes have been identified based on the up-regulated expression of *TaHIR1* in wheat seedlings and of *TaHIR3* in adult plants inoculated with *Pst* (Yu et al. 2008; Zhang et al. 2009). The expression of *TaHIR1* and *TaHIR3* was reduced by the environmental stimuli, including low temperature, drought and high salinity stresses. In addition, the two HIR genes were down-regulated upon ET and ABA treatment, but not affected by SA and MeJA. The transcription studies revealed that *TaHIR1* and *TaHIR3* are negatively regulated by various abiotic stimuli. Down-regulation of *TaHIR1* and *TaHIR3* compromised resistance in wheat cv. Suwon 11 against the avirulent *Pst* pathotype CYR23 with reduced area of necrotic cells neighboring the infection sites and altered expression level of the defense-related genes (Duan et al. 2013b). These results suggest that *TaHIR1* and *TaHIR3* function positively in the incompatible interaction of wheat-*Pst*, but have negative transcriptional responses to abiotic stresses.

3.4.2.4.4.2 Negative Regulator of Cell Death

In animal cells, the key proteinases and specific regulatory proteins involved in PCD have been well characterized, including caspases, bcl-2, defender against apoptotic cell death 1 (DAD1) and others (Ruf et al. 1999; Kopitz et al. 2003; O'Donnell et al. 2005; Pozo-Guisado et al. 2005; Mittapalli and Shukle 2008). By virtue of the known PCD mechanism in animals, a series of key components in the cell death pathway of plants have been identified and characterized. DAD proteins and bcl-2 family members are important regulators in mitochondria mediated cell death in animal cells. DAD proteins were well conserved from yeast to mammals and most plants have two closely related paralogous DAD genes. The rice DAD1 orthologue could suppress apoptotic cell death caused by defective dad-1 mutant line of hamster cells (Tanaka et al. 1997), suggesting a negative role in regulation of plant cell death.

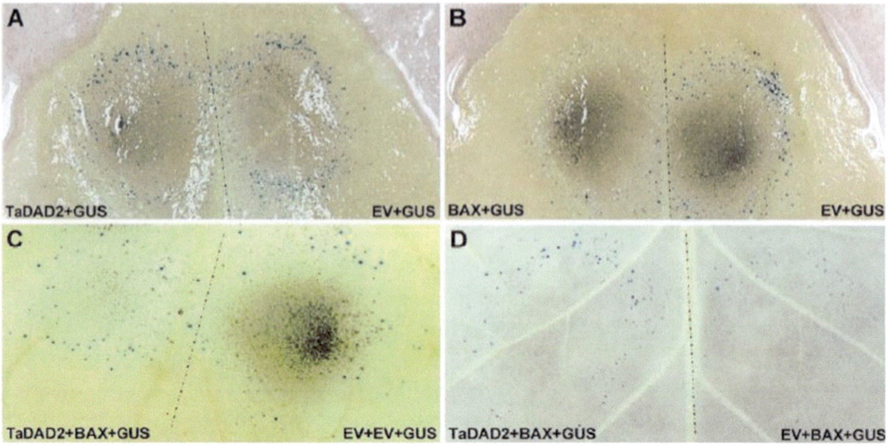

Fig. 3.35 Suppression of Bax-mediated programmed cell death by wheat gene *TaDAD2* in tobacco leaves bombarded with plasmid DNA samples as marked. *EV* empty vector. *Dotted line* marks the position of a divider used to prevent the overlap of two bombardment areas

In wheat, an orthologue of DAD, *TaDAD2*, was isolated (Wang et al. 2011c). Over-expression of *TaDAD2* could inhibit cell death triggered by the mouse pro-apoptotic protein Bax in tobacco cells, suggesting that it may function as a negative regulator of PCD in wheat (Fig. 3.35). The expression level of *TaDAD2* was reduced in the early infection process of *Pst* in both compatible and incompatible interactions, indicating that *TaDAD2* plays a nonspecific response at early stage of wheat-*Pst* interaction. *TaDAD2* expression also was reduced in response to different stresses as early as 1 hpt, indicating the basic function of *TaDAD2* in aiding the survival of wheat cells against biotic or abiotic stresses during the initial reaction. Knocking down *TaDAD2* expression by VIGS weakened hypersensitive cell death in the incompatible interaction with reduced necrotic area around the infection sites. Meanwhile, fungal hyphae were significantly longer. The suppressed host defense response and promoted fungal growth indicated that silencing of *TaDAD2* compromised the resistance of wheat cv. Suwon11 to avirulent race CYR23. Based on the results, *TaDAD2* is postulated to function as a suppressor of cell death in the early stages of wheat-*Pst* interaction. However, it is dispensable for or has an opposite function in HR or cell death triggered by an avirulent race of *Pst* during the late infection process (Wang et al. 2011c).

Members of the Bcl-2 protein family in animals, characterized by the possession of at least one of four conserved Bcl-2 homology domains, have emerged as important regulators of PCD, which act as either activators or suppressors of PCD (Hengartner and Horvitz 1994; Kroemer 1997; Fadeel et al. 1999). Although no Bcl2-associated X protein (BAX) homologues have been identified in plant genomes, homologues of bax inhibitor 1 have been characterized. Overexpression of a rice *BI-1* homologue can suppress the cell death mediated by *Magnaporthe grisea* elicitor (Kawai et al. 1999; Kawai-Yamada et al. 2001, 2004; Matsumura et al. 2003).

Similarly, overexpression of barley *BI-1* reversed the plant resistance conferred by the loss of *MLO* with high hypersusceptibility (Hückelhoven et al. 2003; Eichmann et al. 2006; Babaeizad et al. 2009). Wang et al. (2012b) reported the isolation of a BAX inhibitor 1 homologous gene, *TaBI-1*, from wheat challenged with *Pst*. The transcription of *TaBI-1* was induced in response to both compatible and incompatible interactions between wheat and *Pst*, and also to high salinity. Transient expression of *TaBI-1* blocked BAX-induced cell death in tobacco cells. Silencing of *TaBI-1* converted the resistant wheat genotype to be relatively susceptible to an avirulent *Pst* isolate with increased infection area per infection site, indicating that *TaBI-1*, a negative cell death regulator, contributes to wheat resistance to stripe rust (Wang et al. 2012b).

TaMDAR6, encoding a new monodehydroascorbate reductase (MDAR) from wheat was also reported to be a negative regulator of plant PCD triggered by the *BAX* gene or avirulent *Pst* isolates. Suppression of *TaMDAR6* resulted in significantly induced H_2O_2 production and HR with reduced pathogen development. Interestingly, suppression of *TaMDAR6* during an incompatible interaction increased stomata closure, leading to poor stoma recognition, and as a consequence reduced infection efficiency. In the *TaMDAR6* knocked-down plants, increased infection sites failed to develop substomatal vesicles, presumably due to the abundant ROS accumulation, which might activate other resistance mechanisms that have a negative effect on substomatal vesicle formation. As significant generators of ROS and important elements in PCD regulation, plant chloroplasts have an essential function in plant defense to pathogen attacks (Ishikawa and Shigeoka 2008; Maruta et al. 2012). Thus, the chloroplast localization of TaMDAR6 indicates the involvement of TaMDAR6 in the regulation of PCD (Abou-Attia et al. 2015). *TaMDAR6* can therefore be considered as a negative regulator of plant cell death and participates indirectly in stomatal regulation during the wheat-*Pst* interactions. The MDAR enzyme is involved in the ascorbate–glutathione cycle, and plays an important role in directly reducing monodehydroascorbate (oxidized ascorbate) to ascorbate using NAD(P)H as an electron donor (Apel and Hirt 2004). The role of MDAR have been extensively reported under abiotic stresses such as ozone, salt, polyethylene glycol (PEG) and drought (Sharma and Davis 1997; Sharma and Dubey 2005; Eltayeb et al. 2007), which enhance plant tolerance to abiotic stresses by reducing oxidative damages through maintaining redox status (Eltelib et al. 2011). Few studies have examined the roles of MDAR under biotic stresses. Therefore, the study of Abou-Attia et al. (2015) provides new insights into the role of *TaMDAR* in plant-pathogen interaction.

3.4.2.4.4.3 Metacaspase

Caspases are key components in the animal PCD pathway as the executer of cell death (Kitanaka and Kuchino 1999; Kroemer and Martin 2005). To date, although no functional homologs of animal caspases have been identified in plants, a family of cysteine-dependent proteinases named metacaspases was found in plants, fungi and protozoans (Carmona-Gutierrez et al. 2010; Enoksson and Salvesen 2010). Metacaspases share a complex regulation system with other cell death proteinases, such as vacuolar processing enzyme and subtilisin-like serine proteinases (Woltering 2004). Metacaspases are divided into two types, type I and type II, according to the structural feature of the linker between the P20-like and P10-like domains (Rahman 2010). The specific roles of metacaspases in plant cell death and resistance to pathogens are largely unknown. However, one study shows that two type I Arabidopsis metacaspases (*AtMC*1 and *AtMC3*) are upregulated in plants challenged by various bacterial pathogens and none of the type II metacaspases shows any induction (He et al. 2008). Another study concludes that all three type I metacaspases and two type II metacaspases (*AtMC5* and *AtMC8*) in Arabidopsis are activated upon infection by bacterial pathogens (Watanabe and Lam 2005).

In wheat, a type I metacaspase, *TaMCA1*, and a type II metacaspase, *TaMCA4*, have been identified and they seem to function in an opposite pattern. TaMCA1 exhibited no caspase-1 activity in vitro, but was able to inhibit cell death in tobacco and wheat leaves induced by the mouse *Bax* gene (Hao et al. 2016b). Knockdown of *TaMCA1* via BSMV-VIGS enhanced plant disease resistance to *Pst* with significant hydrogen peroxide (H_2O_2) accumulation. Together with the mitochondria localization of TaMCA1, it was speculated that TaMCA1 function as a negative regulator of cell death mediated by mitochondria, thus contribute to the suppressed HR during the compatible interaction of wheat and *Pst*. Contrast to TaMCA1, the type II metacaspase TaMCA4 failed to induce PCD directly, but were able to enhance cell death triggered by *Bax* or a candidate *Pst* effector (Wang et al. 2012a). In *TaMCA4* knockdown wheat (cv. Suwon 11) seedlings challenged by avirulent *Pst* race CYR23, the necrotic area at per infection site was reduced and the host resistance was dramatically decreased. Considering the significant induction of *TaMAC4* in the early infection stage of avirulent *Pst* before HR, it is reasonable to conclude that *TaMCA4* is activated at the beginning of HR and engaged in ensuring plant defense response to restrict the growth of the initial hyphae and formation of haustorial mother cells. The function of TaMCA4 in resistance to stripe rust is probably related to its special substrates or self-cleavage activity. Vast efforts have been made to identify the substrates or self-processing mechanisms of different metacaspases during the last decade. One study demonstrates that VEIDase TSN (AtSerpin1) is one of the substrates of a type II metacapase in Arabidopsis (Vercammen et al. 2006). Arabidopsis AtMC4 can cleave its inactive form, and in vitro autolysis could also occur under controlled Ca^{2+} conditions (Watanabe and Lam 2011). Although different types of metacaspases have been identified during different wheat-*Pst* interactions, the caspase enzyme activity and their cell death modulation mechanism still need further studies.

3.4.2.4.4.4 Cytoskeleton Dynamics in Cell Death

Actin is one of the most abundant and highly conserved proteins in eukaryotic cells. The dynamic reorganization and rearrangement of the actin cytoskeleton is associated with various cellular processes, and increasing evidence has shown that actin cytoskeleton is precisely regulated to function as a contributing factor to plant immunity against pathogen infection (Shimada et al. 2006; Hardham et al. 2007;

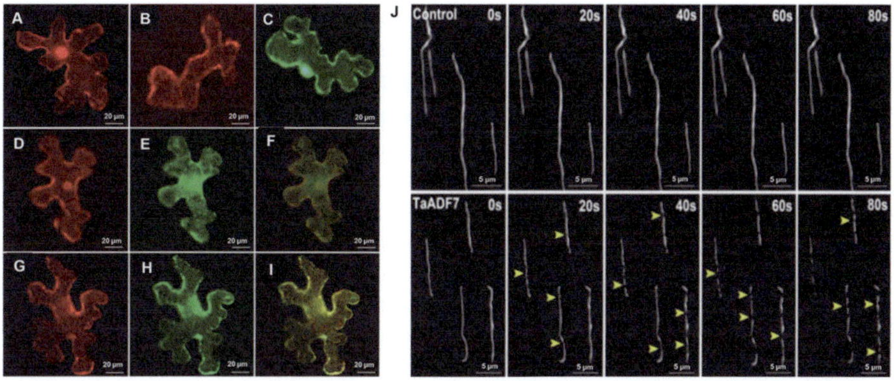

Fig. 3.36 Wheat protein TaADF7 is localized at the actin cytoskeleton and could depolymerize actin filaments in vitro. (**a–h**) Subcellular localization of TaADF7 in *Nicotiana benthamiana*. Laser-scanning confocal micrographs showing the expression of fluorescent proteins in tobacco. DsRed alone (**a**), DsRed-TaADF7 alone (**b**), eGFP alone (**c**), co-expressing DsRed and ABD2-eGFP (**d–f**), co-expressing DsRed-TaADF7 and ABD2-eGFP (**g–i**). The *red channel* shows the localization of DsRed (**a**, **d**) or DsRed-TaADF7 (**b**, **g**). The *green channel* shows the localization of eGFP alone (**c**) or ABD2-eGFP (**e**, **h**). (**f**) An overlay of micrographs (**d**) and (**e**) and (**i**) an overlay of micrographs (**f**) and (**h**). Bars = 20 μm. (**j**) Time-lapse image of polymerized actin filaments every 20 sec in the presence of purified active TaADF7. Numbers in the upper right corner of each frame indicate the time. White arrows mark the locations of new filament-severing events in each frame

Henty-Ridilla et al. 2013). The actin cytoskeleton is modulated by a plethora of actin binding proteins (ABPs), among which the actin-depolymerizing factors (ADFs) are abundant and essential in almost every eukaryotic cell type and are responsible for the high turnover rates if actin filaments *in vivo* (Staiger et al. 1997; Dos Remedios et al. 2003; Van Troys et al. 2008). Corresponding to the regulatory role of the actin cytoskeleton in plants against various environmental stimuli, plant ADFs have been shown to play an important role in response to biological invasion and abiotic stress. The ectopic expression of barley *HvADF3* effectively impedes actin cytoskeleton integrity, thereby enhancing the susceptibility of the *Mlo* genotype to barley powdery mildew and partially breaks down *mlo* resistance with an elevated incidence of fungal entry (Miklis et al. 2007). The Arabidopsis AtADF4 is potentially targeted by the bacterial effector protein AvrPphB under the control of the cognate resistance gene *RPS5*-mediated resistance to *P. syringae* (Porter et al. 2012). Another barley ADF gene, *HvADF3*, was demonstrated to be required for rpg4-mediated resistance to *Puccinia graminis* f. sp. *tritici* by BSMV-VIGS in barley (Wang et al. 2013).

In wheat, an actin-depolymerizing factor TaADF7 was shown to be significantly induced in incompatible wheat-*Pst* interaction (Fu et al. 2014). TaADF7 is located at the cytoskeleton of tobacco cells (Fig. 3.36a–i), and *in vitro* assays revealed that TaADF7 regulated the actin cytoskeletal dynamics by targeting the actin cytoskeleton to execute actin binding/severing activities. Heterologous overexpression of

TaADF7 in fission yeast induced the depolymerization of the actin ring during cell division (Fig. 3.36j). Suppression of *TaADF7* reduced resistance of wheat to avirulent races of *Pst* with reduced ROS accumulation and HR, which could be partially rescued by cytochalasin B. Beside, transient expression of *TaADF7* accelerated the cell death triggered by *Bax* in tobacco cells. It seems that *TaADF7* mainly modulates the dynamics of the actin cytoskeleton to affect ROS accumulation, and then alters subsequent defense responses (Fu et al. 2014). On the contrary, another stress-responsive ADF gene (*TaADF3*) from wheat was inducible by virulent *Pst* isolates but down-regulated in response to avirulent *Pst* isolates (Tang et al. 2016). Knockdown of *TaADF3* enhanced wheat resistance to an avirulent *Pst* isolate and reduced susceptibility to a virulent *Pst* isolate, with increased ROS accumulation and HR. Interestingly, the silencing of *TaADF3* resulted in hindered pathogen penetration and haustorial formation for both avirulent and virulent *Pst* isolates. Moreover, the array and distribution of actin filaments were transformed in *TaADF3*-knockdown epidermal cells, which possibly facilitated attenuating the *Pst* penetration. Thus, the findings suggest that *TaADF3* negatively regulates wheat resistance to *Pst* in an ROS-dependent manner, possibly underlying the mechanism of impeding fungal penetration dependent on the actin architecture dynamics (Tang et al. 2016). It appears that ADFs function in various manners to modulate the dynamic actin cytoskeletal rearrangement, which contribute to the wheat resistance to *Pst*. Except for the different roles between TaADF7 and TaADF3, the two ADF proteins had different subcellular localizations. Transient expression in tobacco cells revealed that TaADF7 was mainly localized at the actin cytoskeleton, while TaADF3 exhibited both cytoplasmic and nuclear localizations in transformed onion cells. AtADF4 has been demonstrated to mediate both ETI and PTI signaling due to its activity in actin rearrangement modulation or translocation of the cytoskeleton into the nucleus through the nuclear localization signal (NLS), where these triggers function as gene expression regulators (Tian et al. 2009; Porter et al. 2012; Henty-Ridilla et al. 2014). The different localizations may determine their distinct roles in cytoskeleton modulation and gene expression. Expanded understanding of the profound regulation and multifaceted functions of ADFs in physiological changes has led to the conclusion that ADF proteins are functional nodes in cell biology (Bernstein and Bamburg 2010). The findings of Fu et al. (2014) and Tang et al. (2016) provide more evidence for the connections of ADFs with host pathogen defense, especially in the wheat-*Pst* interactions.

3.4.2.4.4.5 Lesion-Mimic Mutants

In addition to the HR-induced cell death, a class of lesion-mimic (Les/les) mutants is ubiquitous in plants, which develop cell death lesions resembling those of pathogen induced lesions even in the absence of a pathogen, and thus provide a platform to unravel PCD signaling pathways in plants (Lorrain et al. 2003). Lesion mimics are sometimes associated with enhanced disease resistance, including elevated expression of PR genes and other markers of defense system activation. The Arabidopsis lesion simulating disease 1 (*lsd1*) mutant exhibits a runaway cell death phenotype once initiated by superoxide-dependent signals (Dietrich et al. 1994). The lesion-mimic gene *AtLSD1* contains three zinc-finger motifs, which comprise of a novel plant-specific family of genes that are significantly related to PCD.

In wheat, the homologous gene of *AtLSD1*, *TaLSD1*, encoding an altered zinc finger protein, has been identified to be capable of producing a wheat lesion-simulating disease (Guo et al. 2013). Transient overexpression of *TaLSD1* in *N. benthamiana* partially inhibited *Bax*-triggered PCD, whereas expression of *TaLSD1* alone had no influence on the phenotype of tobacco, indicating the negative role of *TaLSD1* in plant cell death. TaLSD1 was predominantly localized in the nucleus of onion epidermal cells. The mRNA level of *TaLSD1* was up-regulated by oxidative stress generated by methl viologen (MV). Upon *Pst* infection, *TaLSD1* exhibited a similar induced expression pattern at the early infection process, but relatively more significant in the compatible interaction. Silencing of *TaLSD1* enhanced wheat resistance against *Pst* accompanied by enhanced HR, increased in *PR1* gene expression and reduced hyphal growth. The results indicate that *TaLSD1* negatively regulates the plant hypersensitive cell death and is involved in resistance to *Pst* (Guo et al. 2013).

Pheophorbide a oxygenase (PaO) is involved in the oxygenation of pheophorbide a and breaks down chlorophylls. Absence of PaO resulted in the lethal leaf spot (*lls*) mutant of maize, which exhibits a chloroplast mediated cell death phenotype (Gray et al. 1997, 2002) and elevated resistance to fungal pathogens. Wheat *Lls1* gene *TaLls1* encoding PaO was presumed to execute a rheostat role in cell death regulation by controlling a delicate threshold to maintain cell homeostasis in adaption to various stresses (Tang et al. 2013). Silencing of *TaLls1* in wheat led to leaf cell death without pathogen attacks, possibly due to the accumulation of pheophorbide a (upstream substrate of PaO), indicating a suppressor role of *TaLls1*. On the other hand, overexpression of *TaLls1* also triggered cell death in both tobacco and wheat leaves, probably owing to the accumulation of the red chlorophyll catabolite (downstream product of PaO). Taken together, the cell death induction activity of *TaLls1* in either *TaLls1*-knockdown plants or transient overexpressing leaf tissues demonstrated the close relationship of cell death and H_2O_2 accumulation caused by the chlorophyll degradation catabolites resulting from the irregular TaLls1 activity. It appears that there is a threshold for *TaLls1* in maintaining the balance of chlorophyll catabolism and cell homeostasis. TaLls1 seems to serve as a type of rheostat for cell death regulation in plants subjected to various stresses. In the wheat-*Pst* interactions, *TaLls1* was up-regulated with an avirulent *Pst* pathogen infection when the pathogen had not yet formed haustoria but H_2O_2 rapidly increased at infection sites, indicating that *TaLls1* is associated with the generation of H_2O_2 at an early stage of HR initiation. Down-regulation of *TaLls1* in wheat did not change the disease symptoms but enhanced tolerance to *Pst* via a significant increase in H_2O_2 generation, causing elevated cell death and upregulation of PRs (Tang et al. 2013). In conclusion, this study revealed a self-regulation of *TaLls1* in balancing cell death and a negative role in wheat tolerance to *Pst*.

Reactive oxygen species in plants are accumulated in different cellular compartments during the responses to abiotic and biotic stresses. The accumulation of ROS, particularly O_2^- and H_2O_2, has been successively reported in plants infected by various pathogens and is considered as one of the early signs of plant-pathogen interactions. All of the above studies revealed that during wheat-*Pst* interactions, the mitochondria, chloroplast, cytoskeleton and stoma open/closure and other cellular compartments share important roles in ROS generation, by which thereafter to contribute to wheat resistance or susceptibility. These findings roughed out a possible framework of the cell death regulation pathway in wheat challenged by *Pst*.

3.4.2.4.5 Receptor-Like Proteins

Transmembrane receptor-like proteins (RLPs) and receptor-like kinases (RLKs), which contain an extracellular leucine-rich repeat (LRR) domain, a single transmembrane domain and an intracellular kinase domain or a small cytoplasmic tail, comprise a large repertoire of cell surface receptors (Kruijt et al. 2005). Previous studies have identified several RLK genes coding for wheat leaf rust kinases (WLRKs) that are conserved in wheat and related species. The most studied WLRK gene is *LRK10*, which is genetically associated with the *Lr10* locus conferring resistance to leaf rust (Feuillet et al. 1997, 1998, 2001). Despite the close genetic and physical linkages between *Lr10* and *LRK10* (Feuillet et al. 1997, 2003; Stein et al. 2000), there is no direct evidence that *LRK10* is involved in *Lr10*-mediated rust resistance. Moreover, none of the putative WLRKs has been reported to exhibit kinase activity. During the wheat-*Pst* interactions, an important group of light-regulated RLKgenes have been identified, including *TaRLK-R1*, *TaRLK-R2* and *TaRLK-R3* (Zhou et al. 2007). These three new wheat RLK proteins were targeted to the plasma membrane, and the recombinant kinase domain of TaRLK-R3 exhibited auto-phosphorylation activity in *vitro*. Transcription of *TaRLK-R1*, *TaRLK-R2* or *TaRLK-R3* was induced during the HR to *Pst*, and reducing the transcript levels of the three RLK genes individually or simultaneously compromised HR of wheat to the avirulent *Pst* isolate. During the wheat-*Pst* interaction, TaRLK-R1, 2 and 3 may share a basic function in aiding the protection of wheat cells against *Pst* via their positive contribution to wheat HR (Zhou et al. 2007). From an applied point of view, characterization of the three wheat RLKs may provide molecular targets for the development of durable resistance to stripe rust.

Due to the availability of genome sequences, 57 predicted RLP genes were identified that distributed at 33 loci in Arabidopsis, and 90 RLP genes distributed at 38 loci in rice (Wang et al. 2008b; Jiang et al. 2013). Several RLP genes have been identified to be involved in plant-pathogen interactions. The *Cf* genes characterized to be RLP genes mediated HR in tomato to *Cladosporium fulvum* (Jones et al. 1994; Stulemeijer et al. 2007). Besides, the RLP *Ve* genes were involved in plant resistance against vascular wilt, and two *LeEix* genes encodes receptors for the fungal elicitors (Kawchuk et al. 2001; Ron and Avni 2004). In wheat, a novel RLP gene, *TaRLP1.1*, was screened from genes induced by an avirulent *Pst* isolate using a GeneChip (Jiang et al. 2013). Knock-down of *TaRLP1.1* in the stripe rust-resistant plants resulted in increased susceptibility to *Pst* with dramatic decreased HR at the pathogen-host interaction sites. Furthermore, transformation of *TaRLP1.1* into a susceptible wheat variety conferred resistance to *Pst*, accompanied by enhanced HR. It was proposed that *TaRLP1.1* greatly contributed to the HR during the wheat-*Pst* interactions (Jiang et al. 2013).

3.4.2.5 Metabolism

Among the transcriptome dynamics of wheat responses to *Pst*, genes involved in metabolism and energy are strongly represented, indicating the importance of regulation of metabolism in wheat responses to the attack of both avirulent and virulent *Pst* isolates. The roles of metabolism in wheat-*Pst* interactions have been analyzed based on the study of those induced genes participating in different metabolic pathways. *TaAlaAT1* encoding anaminotransferase may function as a regulator of plant carbon and nitrogen metabolism (Duan et al. 2009). As discussed above, the pheophorbide A oxygenase (PaO) homolog TaLls1, a key enzyme during chlorophyll catabolism, contributed to cell death and wheat tolerance to *Pst* by modulation of the chlorophyll degradation metabolism (Tang et al. 2013).

Ethylene production is one of the earliest responses of plants to pathogen attack, which seems to be associated with defense reaction by regulating a wide range of defense related genes (Boller 1991; Deikman 1997; Loon et al. 2006). S-adenosylmethionine (SAM) is one of the precursors for the biosynthesis of ET and polyamines (PAs) (Imaseki 1991; Fluhr and Mattoo 1996; Ravane et al. 1998) and the methionine salvage pathway ensures a high level of ET production even when the pool of free methionine is relatively small. Several enzymes involved in this pathway have been characterized at the biochemical and molecular levels (Burstenbinder et al. 2007; Rzewuski et al. 2007). The acireductone-dioxygenase (ARD) catalyzes the reaction of aci-reductone with dioxygen to produce the immediate precursor of methionine (Dai et al. 1999). Several studies have showed that ARD genes are involved in plant stress responses. The expression of rice *OsRAD1* was increased in the youngest internode of deepwater rice under the submergence, which provides evidence that ET could regulate the methionine cycle (Sauter et al. 2005). In potato, the expression of *StARD* was increased at the wounding surface of potato tubers, suggesting that ARD may play an important role in abiotic stress such as wounding (Kim et al. 2008a). Xu et al. (2010) found that the ARD homologue in wheat *TaARD* was induced in the compatible wheat-*Pst* interaction but suppressed in the incompatible interaction, implying that the methionine cycle may play a positive role in disease symptom development but a negative role in resistance reaction. Exogenous application of various hormones including ABA, ET or SA inhibited the expression of *TaARD*, while *TaARD* was suppressed in the first 6 h but increased afterwards. Besides, the *TaARD* transcript level was suppressed by wounding and environmental stimuli, such as high salinity and low temperature. These findings indicate that the enzyme is involved in the ET synthesis pathway for responding to signals from biotic and abiotic stresses.

Sphingolipids represent the most important and abundant lipids in membranes of animals and plants, beside the glycerophospholipids. Ceramides, consisting of a fatty acid linked to the C-2 amino group of spingosine by an amide bound, represent the metabolic precursors of sphingolipids. The ceramidase catalyzes the conversion of ceramide to sphingosine and sphingosine-a-phosphate, which may function as a second messenger in many cellular events including cell proliferation, differentiation, stress responses and apoptosis (Cuvillier et al. 1996; Mao et al. 2001). Research on the sphinogolipid system in mammalian is more advanced than in plants. In plants, the information of structures, functions and metabolism of sphingolipids is less and their role in plant-pathogen interactions is largely unknown. So far only few ceramidase in plants have been reported to have important roles. Some studies indicated that sphingosine-1-phosphate is related with the closure of guard cells mediated by Ca^{2+}, a sphingosine transfer protein is involved in PCD and plant resistance to fungal toxins is mediated by a plant ortholog of a yeast gene involved in the ceramide synthesis (Ng et al. 2001). From an incompatible SSH cDNA library of wheat leaves infected with *Pst*, a full length wheat ceramidase gene, designated as *Ta-CDase*, was cloned using the RACE (Rapid amplification of cDNA ends) technique (Yu et al. 2011). The ceramidase gene *Ta-CDase* was significantly induced at 48 hpi of an avirulent *Pst* isolate, when haustoria were developing and a series of signal pathways involved in host defense were activated. The up-regulation of *Ta-CDase* in wheat resistance response may suggest the role of sphingolipid compounds in defense reactions.

Glycerol-3-phosphate (G3P) is an obligatory component and precursor of all plant glycerolipid biosynthesis, which also participates in fatty acid biosynthesis. During plant-pathogen interactions, G3P has been proposed to be a regulator of plant defense signaling in basal resistance and systemic acquired resistance (SAR) (Nandi et al. 2004). In plants, G3P is derived via the glycerol kinase (GK)-mediated phosphorylation of glycerol or the NAD-dependent G3P dehydrogenase (G3Pdh)-mediated reduction of dihydroxy-acetone phosphate (DHAP) (Chanda et al. 2011). In Arabidopsis, the *G3Pdh* homologue *GLY1/SFD1* or *GK* homologue *GLI1/NHO1* contributes to resistance against hemibiotrophic fungus (*Colletotrichum higginsianum*) and bacteria (*P. syringae DC 3000*) (Chanda et al. 2008, 2011; Venugopal et al. 2009). Furthermore, *GLI1* is nonspecifically required for nonhost resistance (NHR) to bacterial and fungal pathogens (Lu et al. 2001; Kang et al. 2003a). In the incompatible wheat-*Pst* interaction, increased G3P accumulation accompanied by induced SAR was observed, indicating that G3P-mediated SAR was present in graminaceous plants, and even played an important role in plant resistance to pathogens (Yang et al. 2013). The GLY1-encoded glycerol-3-phosphatede hydrogenase and GLI1-encoded glycerol kinase are two key enzymes engaged in two parallel pathways for the G3P metabolism. Silencing them compromised wheat resistance to *Pst* with suppressed SA accumulation and *PR1* expression (Yang et al. 2013). Considering their regulatory function in the G3P metabolism, the disrupted G3P synthesis may ascribe the affected host resistance.

Due to the ability to maintain high levels of ascorbate in different cell compartments, the ascorbate-glutathione cycle is important in protecting the organism against ROS in many organisms. Monodehydroascorbate reductase (MDHAR) is a key enzyme in the ascorbate-glutathione cycle maintaining the reduced rate of ascorbate. Feng et al. (2014b) found that a wheat peroxisome MDHAR homolog, TaMDHAR, was involved in the early stage of wheat-*Pst* interaction, which is the key stage for the ROS burst. Silencing of *TaMDHAR* in wheat cv. Xingzi 9104 resulted in enhanced wheat resistance to *Pst* by inhibiting sporulation, suppressing *Pst* hyphal growth and increasing necrotic areas in the adult stages. It was concluded that TaMDHAR could contribute to the APR of Xingzi 9104 through the antioxidative and metabolic functions (Feng et al. 2014b). Upon pathogen attack, the interaction between ROS and the AsA-GSH cycle generates compartment-specific redox signals that interfere with other signaling pathways to achieve a complex rearrangement of the primary metabolism from photo-assimilatory pathways into an emergency survival strategy.

The disease resistance and susceptibility of plants is governed by the timely recognition of an invading pathogens and the rapid activation of the host defense responses. The above examples only reveal a small part of the complicated network in wheat defense responses (Table 3.5). Nevertheless, it seems adequate for us to get a glimpse of the integrative resistance/susceptible regulatory network in wheat

Table 3.5 Host defense-related genes identified during wheat interactions with *Puccinia striiformis* f. sp. *tritici*

Category/gene	Function	References
Resistance genes		
Yr10	Race specific resistance	Liu et al. (2014)
Yr18/Lr34	Non race specific resistance	Krattinger et al. (2009)
Yr36	Non race specific resistance	Fu et al. (2009)
Signal transduction		
TaCab1	EF-hand protein	Feng et al. (2011)
TaERG3	C2 domain protein	Zhang et al. (2013)
TaRab7	Small GTP-binding proteins	Liu et al. (2012a)
TaTypA	Small GTP-binding proteins	Liu et al. (2016b)
Transcription regulation		
TaNAC4	NAC transcription factor	Xia et al. (2010a)
TaNAC8	NAC transcription factor	Xia et al. (2010b)
TaMYB4	MYB transcription factor	Al-Attala et al. (2014)
TabZIP1	bZIP transcription factor	Zhang et al. (2008c)
TaEIL1	EIN3 transcription factor	Duan et al. (2013a)
TaPR5	Pathogenesis-related proteins	Wang et al. (2010b)
TaGlu	Pathogenesis-related proteins	Liu et al. (2010)
TaPR10	Pathogenesis-related proteins	Zhang et al. (2010)
TaHSC70	Heat shock protein 70	Duan et al. (2011)

(continued)

Table 3.5 (continued)

Category/gene	Function	References
TaHsp90.1	Heat shock protein 90	Wang et al. (2011a)
TaHsp90.2	Heat shock protein 90	Wang et al. (2011a)
TaHsp90.3	Heat shock protein 90	Wang et al. (2011a)
TaNPSNs	SNAREs	Wang et al. (2014a)
TaSYP71	SNAREs	Liu et al. (2016a)
TaHIR1	Hypersensitive-induced reaction (HIR) genes	Yu et al. (2008), Zhang et al. (2009) and Duan et al. (2013b)
TaHIR3	Hypersensitive-induced reaction (HIR) genes	Yu et al. (2008), Zhang et al. (2009) and Duan et al. (2013b)
Defense		
TaDAD2	Cell death suppressor	Wang et al. (2011c)
TaBI-1	Cell death inhibitor	Wang et al. (2012b)
TaMDAR6	Negative regulator of cell death	Abou-Attia et al. (2015)
TaMCA1	Metacaspase	Hao et al. (2016b)
TaMCA4	Metacaspase	Wang et al. (2012a)
TaADF7	Actin-depolymerizing factor	Fu et al. (2014)
TaADF3	Actin-depolymerizing factor	Tang et al. (2016)
TaLls1	Lethal leaf spot (lls) gene	Tang et al. (2013)
TaLSD1	Lesion simulating disease 1 gene	Guo et al. (2013)
TaRLK-R1	Receptor like kinases	Zhou et al. (2007)
TaRLK-R2	Receptor like kinases	Zhou et al. (2007)
TaRLK-R3	Receptor like kinases	Zhou et al. (2007)
TaRLP1.1	Receptor like protein	Jiang et al. (2013)
Metabolism		
TaARD	Methionine salvage pathway	Xu et al. (2010)
Ta-CDase	Conversion of sphingolipid compounds	Yu et al. (2011)
TaAlaAT1	Carbon and nitrogen metabolism	Duan et al. (2009)
TaGLI1	Glycerol-3-phosphate metabolism	Yang et al. (2013)
TaGLY1	Glycerol-3-phosphate metabolism	Yang et al. (2013)
TaMDHAR	Ascorbate-glutathione cycle	Feng et al. (2014c)

resistance to *Pst*, from the early plant-pathogen recognition, following signal transduction, transcription regulation and the downstream activation of defense responses. The obtained knowledge of the key events makes it potential to improve wheat resistance through silencing essential wheat susceptible genes or overexpressing resistance related genes. This will help to develop new resistant wheat cultivars and provide novel strategies to control stripe rust.

3.4.3 Epigenetics

The complex interaction between wheat and *Pst* can be partly decoded via patterns of gene expression and regulation. Dual RNA-sequencing of pathogen and host is an elegant way to explore the mechanisms in both sides. So far as we know, the full picture of gene regulation during *Pst* infection includes not only protein-coding genes, but also noncoding RNAs. Small RNAs (sRNAs) are short noncoding RNA molecules that fall into two main classes distinguished by their origins, small interfering RNAs (siRNAs) originating from a heteroduplex of two distinct RNA molecules, and microRNAs (miRNAs) transcribed from a single-stranded precursor with self-complementarity. The production and function of small RNAs requires a common set of proteins, the double-stranded RNA (dsRNA)-specific endonucleases such as Dicers, dsRNA-binding proteins and small RNA-binding proteins called Argonaute proteins. Together, the small RNAs and their associated proteins act in distinct but related "RNA silencing" pathways that regulate diverse cellular processes. Identification and characterization of the sRNAs in both wheat and *Pst* should provide new knowledge of the host-pathogen interactions.

3.4.3.1 sRNAs in Wheat

miRNAs are significant regulators of gene expression at the post-transcriptional level by repressing gene translation or degrading target mRNAs in most eukaryotic organisms (Bartel 2004; Chuck et al. 2009). In plants, mature miRNAs, typically 21–24 nucleotides in length, are processed from single strand primary miRNA transcripts (pri-miRNA), which can fold into double-strand RNA (dsRNA) with a stem-loop structure. Once released from the precursor transcript, the miRNA/miRNA duplex is exported to the cytoplasm, where the mature miRNA guide strand is used as a sRNA silencing signal by the RNA induced silencing complex and triggers local transcript cleavage and subsequent degradation or translation repression (Llave et al. 2002). miRNAs have been shown to play essential roles in various biological processes during the plant growth and development (Aukerman and Sakai 2003; Palatnik et al. 2003; Rubio-Somoza and Weigel 2011). They also play critical roles in plant responses to abiotic and biotic stresses (Jin 2008; Padmanabhan et al. 2009; Katiyar-Agarwal and Jin 2010; Zhang et al. 2011b, 2011c). The Arabidopsis miR393 was the first demonstrated miRNA participating in host-pathogen interactions (Navarro et al. 2006). miR393 was induced by a flagellin-derived peptide and negatively regulated messenger RNAs for three F-box auxin receptors to repress the auxin signaling. It also restricts *P. syringae* growth. In Arabidopsis, the miR164 family negatively regulates the NAC transcription factor, which contains a miR164 complementary site to participate in many physiological processes, including the organ enlargement, flower development and age-dependent cell death (Baker et al. 2005; Guo et al. 2005; Sieber et al. 2007; Kim et al. 2009a).

Thus, miR164 was involved in the responses of plants to biotic stresses mainly through regulation of the transcript levels of corresponding genes.

Sequences of miRNAs from various organisms are deposited in the publicly available miRNA database (http://www.mirbase.org). Currently, 192 wheat and 101 barley miRNAs are presented in the miRbase. For comparison, 1500 miRNAs from rice have been deposited. Only 19 miRNAs are presented in the miRbase; and most of them are related to responses to human pathogens and few to plant pathogens. Campo et al. (2013) reported a rice miRNA, osa-miR7695, which negatively regulates an alternatively spliced transcript of OsNramp6 (Natural resistance-associated macrophage protein 6). The overexpression of this miRNA in rice confers resistance to rice blast, illustrating the existence of a novel regulatory network that integrates miRNA function and mRNA processing in plant immunity.

To study if miRNAs of wheat are involved to resistance to stripe rust, Feng et al. (2015, 2016) constructed miRNA libraries from wheat plants (cv. Xingzi 9104) inoculated with *Pst* race CYR32. In the adult plants resistant to CYR32, a total of 596 miRNA candidates were obtained and 420 wheat-specific candidate miRNAs were screened (Feng et al. 2015). Degradome sequencing indicated that most of the annotated target genes of wheat miRNAs were involved in signal transduction, energy metabolism and other functions. Among them RabGAP/TBC domain-containing protein, zinc finger protein and cysteine-rich receptor-like protein kinase 41 might play important role in the adult-plant resistance of Xingzi 9104 to stripe rust. In the seedlings of the cultivars susceptible to stripe rust, a total of 1316 miRNA candidates were obtained, including 145 conserved and 998 wheat-specific miRNAs (Feng et al. 2016). The qRT-PCR analysis showed that some miRNAs were involved in the susceptible reaction to *Pst* infection. For example, tae-miR1432 was highly expressed in compatible interaction, and its predicted corresponding target gene coding for a calcium ion-binding protein, was also upregulated. Simultaneously, PC-3P-7484, a wheat specific miRNA, was highly expressed compared with its dramatical down-regulated target gene ubiquillin. Using microarray-based method, two new miRNAs (PN-tae-miR166a and PN-taemiR168a) in wheat cultivar Suwon 11 were identified to be up-regulated by *Pst* infection at 12 hpi or 24 hpi, exhibiting similar expression patterns with other five known wheat miRNAs (tae-miR156, tae-miR159, tae-miR160, tae-miR164 and tae-miR167) (Feng et al. 2014a). These *Pst*-inducible miRNAs might play important roles in the interactions of wheat with *Pst* by regulating the various types of genes involved in the different signaling pathways.

Based on the established miRNA degradome sequence library of wheat-*Pst* interactions, the candidate target genes of many wheat miRNAs have been detected and several of them have been verified. Basic blue copper protein was also predicted to be the target gene of miR408 in Arabidopsis and wheat (Sunkar and Zhu 2004; Yao et al. 2007). Accumulation of the wheat miRNA tae-miR408 following *Pst* infection leads to negative regulation of plantacyanin-like (basic blue) protein gene *TaCLP* (Feng et al. 2013b). Co-transformation in tobacco leaves verified the interaction between tae-miR408 and *TaCLP*. Overexpression of *TaCLP* in yeast enhances cell tolerance to high salinity and cupric ion stress, while knockdown of *TaCLP* decreased wheat resistance to *Pst*. It seems that miRNA negatively regulated the expression of *TaCLP* gene to benefit the virulence of the pathogen. The wheat miRNAtae-miR159a can cleavage mRNA for *TaMyb3* transcription factor from the MYB subfamily of wheat, which is important in regulation of wheat resistance to *Pst* (Feng et al. 2013a). Another wheat miRNA, tae-miR164, which was decreased in response to avirulent *Pst* isolate, showed contrasting divergent expression patterns with wheat NAC transcription factor *TaNAC21/22* (Feng et al. 2014b). Silencing of *TaNAC21/22* improved wheat resistance to *Pst* with increased necrotic area and blocked fungal growth. Therefore, tae-miR164 appears to positively contribute to wheat resistance by degrading wheat NAC transcription factor genes (Feng et al. 2014a). In addition, two novel wheat miRNAs (1136-P3 and PN-2013) were isolated and identified as cleaving mRNA from monodehydroascorbate reductase gene *TaMDHAR*, an important gene in the ascorbate-glutathione cycle (Feng et al. 2014c). However, only accumulation of PN-2013 downregulated the expression of *TaMDHAR* in Xingzi 9104 and further resulted in enhanced resistance of wheat to *Pst* at adult-plant stages. The specific role of miRNAs in wheat resistance to *Pst* is still largely unknown, but the examples above indicate that wheat possesses a regulatory network that integrates miRNA function to regulate immunity against *Pst*.

In addition, the roles of wheat miRNAs in responses to various abiotic stresses have been studied. Cold, wounding and high-salinity are three common environmental stress stimuli influencing crop growth and development. Based on the identified 31 known miRNAs and 3 novel miRNAs in wheat, qRT-PCR analyses revealed 19 stress-regulated miRNAs (Wang et al. 2014b). Among them, 16 were regulated under cold, 12 under wounding and 8 under high-salinity treatments, respectively. Four of the miRNAs were highly responsive to cold stress in wheat by northern blot, and 6 were regulated by wounding and 3 were regulated by high salinity. Three miRNAs (miR159, miR393 and miR398) were responsive to multiple stress stimuli. The analyses of targets suggested that miR159, miR398 and miR6001 could respond to stress conditions in the regulatory pathways. These results suggest that wheat miRNAs play important roles in response to abiotic stress.

3.4.3.2 sRNAs in *Pst*

Since the first discovery of RNAi in *Neurospora*, functional small RNAs (sRNAs) have been identified from fungi and oomycetes, including many plant pathogens. Because sRNAs can interfere with mRNAs by specifically targeting particular mRNAs for degradation, small interfering RNAs (siRNAs) provide the opportunity to knockdown specific genes. Animal and plant pathogens have siRNAs in their system as arsenal tools to attack their hosts. For example, siRNAs from the necrotrophic fungus *Botrytis cineria* function as virulence factors by silencing plant defense genes (Weiberg et al. 2013). Obligate biotrophic pathogens like rust fungi, which maintain intimate relationships with their hosts both physically and evolutionarily, have particularly adapted to employ sRNA-based effectors. Studies using different approaches have been conducted to understand the functions of sRNAs in pathogenicity. Reduced virulence phenotypes were obtained in the leaf rust fungus (*P. triticina*), stem rust fungus (*P. graminis*) and *Pst* via HIGS, indicating the existence of RNAi mechanism in the rust fungi. Mueth et al. (2015) identified 20,269 sRNAs from leaves of wheat cultivars Louise and 15,490 sRNA from those of Penawawa inoculated with *Pst* race PSTv-37. These sRNAs were identified of *Pst* origin, and accounted only 1.3% or 1.4% of more than 3.5 or 2.9 million non-redundant sRNAs isolated from the infected leaves. The large number of sRNAs from both *Pst* and wheat demonstrate that *Pst* is a well RNAi-equipped fungus capable of producing a large number of specific siRNA sequences and wheat has a much more complex sRNA system for interactions. Interestingly, non-inoculated leaves have more sRNAs than the inoculated leaves. It was not clear whether the reduction of sRNAs in the infected plants was due to *Pst* infection or due to other factors. However, the large number of *Pst* sRNAs isolated from the infected leaves is more interesting. The target predictions of the *Pst* sRNAs revealed numerous putative gene targets, including kinases and small secreted proteins from *Pst*, and some homologs of known plant resistance genes. The diverse predicted target genes in both *Pst* and host cells imply that sRNAs might play a significant role in regulating the complex developmental processes involved in *Pst* pathogenicity. Further studies are needed to identify sRNAs from both the pathogen and host that can be used to create or improve wheat resistance against the pathogen.

References

Abou-Attia MA, Wang X, Al-Attala MN, Xu Q, Zhan G, Kang Z. TaMDAR6 acts as a negative regulator of plant cell death and participates indirectly in stomatal regulation during the wheat-stripe rust fungus interaction. Physiol Plant. 2015;156:262–77.

Aida M, Ishida T, Fukaki H, Fujisawa H, Tasaka M. Genes involved in organ separation in Arabidopsis, analysis of the cup shaped cotyledon mutant. Plant Cell. 1997;9:841–57.

Al-Attala MN, Wang X, Abou-Attia MA, Duan X, Kang Z. A novel TaMYB4 transcription factor involved in the defence response against *Puccinia striiformis* f. sp. *tritici* and abiotic stresses. Plant Mol Biol. 2014;84:589–603.

Albrecht V, Weinl S, Balzevic D, D'Angelo C, Batistic O, Kolukisaouglu Ü, Bock R, Schulz B, Harter K, Kudla J. The calcium sensor CBL1 integrates plant responses to abiotic stresses. Plant J. 2003;36:457–70.

Ali S, Gladieux P, Rahman H, Saqib MS, Fiaz M, Ahmad H, Leconte M, Gautier A, Justesen AF, HovmØller MS, Enjalbert J, Vallavieille-Pope C. Inferring the contribution of sexual reproduction, migration and off-season survival to the temporal maintenance of microbial populations: a case study on the wheat fungal pathogen *Puccinia striiformis* f. sp. *tritici*. Mol Ecol. 2014;23:603–17.

Allen RF. Cytological studies of infection of Baart, Kanred, and Mindum wheats by *Puccinia graminis tritici*. Span J Agric Res. 1923;26:571–604.

Allen RE. A cytological study of *Puccinia glumarum* on *Bromus marginatus* and *Triticum vulgare*. J Agric Res. 1928;36:487–513.

Alonso JM, Stepanova AN, Solano R, Wisman E, Ferrari S, Ausubel FM, Ecker JR. Five components of the ethylene-response pathway identified in a screen for weak ethylene-insensitive mutants in Arabidopsis. Proc Natl Acad Sci U S A. 2003;100:2992–7.

Antonin W, Holroyd C, Fasshauer D, Pabst S, von Mollard GF, Jahn R. A SNARE complex mediating fusion of late endosomes defines conserved properties of SNARE structure and function. EMBO J. 2000;19:6453–64.

Apel K, Hirt H. Reactive oxygen species: metabolism, oxidative stress, and signal transduction. Annu Rev Plant Biol. 2004;55:373–99.

Aranda MA, Escaler M, Wang D, Maule AJ. Induction of *HSP70* and polyubiquitin expression associated with plant virus replication. Proc Natl Sci Acad USA. 1996;93:15289–93.

Artemenko EN, Umnov AM, Chkanikov DI. Changes in the level of indolyl-3-acetic acid and possible paths of it regulation in leaves of wheat infected with stem rust. Sov Plant Physiol (Engl Transl). 1980;27:447–51.

Atkinson GC. The evolutionary and functional diversity of classical and lesser-known cytoplasmic and organellar translational GTPases across the tree of life. BMC Genomics. 2015;16:78.

Atkinson MM, Midland SL, Sims JJ, Keen NT. Syringolide 1 Triggers Ca^{2+} influx, K^+ efflux, and extracellular alkalization in soybean cells carrying the disease-resistance gene *Rpg4*. Plant Physiol. 1996;112:297–302.

Aukerman MJ, Sakai H. Regulation of flowering time and floral organ identity by a MicroRNA and its APETALA2-like target genes. Plant Cell. 2003;15:2730–41.

Ayliffe M, Devilla R, Mago R, White R, Talbot M, Pryor A, Leung H. Nonhost resistance of rice to rust pathogens. Mol Plant-Microbe Interact. 2011a;24:1143–55.

Ayliffe M, Jin Y, Kang ZS, Persson M, Steffenson B, Wang SP, Leung H. Determining the basis of nonhost resistance in rice to cereal rusts. Euphytica. 2011b;179:33–40.

Ayliffe M, Singh D, Park R, Moscou M, Pryor T. Infection of *Brachypodium distachyon* with selected grass rust pathogens. Mol Plant-Microbe Interact. 2013;26:946–57.

Ayres PG. Interactions between environmental stress injury and biotie disease physiology. Annu Rev Phytopathol. 1984;22:53–75.

Azinheira GH, Maria DCS, Talhinhas P, Medeira C, Maia I, Petitot A-S, Fernandez D. Non-host resistance responses of *Arabidopsis thaliana* to the coffee leaf rust fungus (*Hemileia vastatrix*). Botany. 2010;88:621–9.

Babaeizad V, Imani J, Kogel K-H, Eichmann R, Hückelhoven R. Over-expression of the cell death regulator BAX inhibitor-1 in barley confers reduced or enhanced susceptibility to distinct fungalpathogens. Theor Appl Genet. 2009;118:455–63.

Baka ZA, Larous L, Losel DM. Distribution of ATPase activity at the host–pathogen interfaces of rust infections. Physiol Mol Plant Pathol. 1995;47:67–82.

Baker CC, Sieber P, Wellmer F, Meyerowitz EM. The early extra petals1 mutant uncovers a role for microRNA miR164c in regulating petal number in regulating petal number in Arabidopsis. Curr Biol. 2005;15:303–15.

Bancal MO, Hansart A, Sache I, Bancal P. Modeling fungal sink competitiveness with grains for assimilates in wheat infected by a biotrophic pathogen. Ann Bot. 2012;110:113–23.

Bao YM, Sun SJ, Li M, Li L, Cao WL, Luo J, et al. Overexpression of the Qc-SNARE gene *OsSYP71* enhances tolerance to oxidative stress and resistance to rice blast in rice (*Oryza sativa* L.). Gene. 2012;504:238–44.

Barak M, Trebitsh T. A developmentally regulated GTP binding tyrosine phosphorylated protein A-like cDNA in cucumber (*Cucumis sativus* L.). Plant Mol Biol. 2007;65:829–37.

Bariana HS, McIntosh RA. Genetics of adult plant stripe rust resistance in four Australian wheats and the French cultivar 'Hybridede-Bersee'. Plant Breed. 1995;114:485–91.

Bartel DP. MicroRNAs: genomics, biogenesis, mechanism, and function. Cell. 2004;116:281–97.

Batistič O, Kudla J. Integration and channeling of calcium signaling through the CBL calcium sensor/CIPK protein kinase network. Planta. 2004;219:915–24.

Batistič O, Kudla J. Plant calcineurin B-like proteins and their interacting protein kinases. Biochim Biophy Acta (BBA) – Mol Cell Res. 2009;1793:985–92

Beck M, Heard W, Mbengue M, Robatzek S. The INs and OUTs of pattern recognition receptors at the cell surface. Curr Opin Plant Biol. 2012;15:367–74.

Benjamins R, Ampudia CS, Hooykaas PJ, Offringa R. PINOID-mediated signaling involves calcium-binding proteins. Plant Physiol. 2003;132:1623–30.

Berlin A, Djurle A, Samils B, Yuen J. Genetic variation in *Puccinia graminis* collected from oats, rye, and barberry. Phytopathology. 2012;102:1006–12.

Bernstein BW, Bamburg JR. ADF/cofilin: a functional node in cell biology. Trends Cell Biol. 2010;20:187–95.

Beßer K, Jarosch B, Langen G, Kogel KH. Expression analysis of genes induced in barley after chemical activation reveals distinct disease resistance pathways. Mol Plant Pathol. 2001;1:277–86.

Bhattacharya PK, Naylor JM, Shaw M. Nucleic acid and protein changes in wheat leaf nuclei during rust infection. Science. 1965;150:1605–7.

Bhattacharya PK, Shaw M, Naylor JM. The physiology of host–parasite relations. XIX. Further observations on nucleoprotein changes in wheat leaf nuclei during rust infection. Can J Bot. 1968;46:11–6.

Bock JB, Matern HT, Peden AA, Scheller RH. A genomic perspective on membrane compartment organization. Nature. 2001;409:839–41.

Boller T. Ethylene in pathogenesis and disease resistance. In: Mattoo AK, Suttle JC, editors. The plant hormone ethylene. Boca Raton: CRC Press; 1991. p. 293–314.

Boonburapong B, Buaboocha T. Genome-wide identification and analyses of the rice calmodulin and related potential calcium sensor proteins. BMC Plant Biol. 2007;7:4.

Botër M, Amigues B, Peart J, Breuer C, Kadota Y, Casais C, Moore G, Kleanthous C, Ochsenbein F, Shirasu K, Guerois R. Structural and functional analysis of SGT1 reveals that its interaction with HSP90 is required for the accumulation of Rx, an R protein involved in plant immunity. Plant Cell. 2007;19:3791–804.

Boutrot F, Segonzac C, Chang KN, Qiao H, Ecker JR, Zipfel C, Rathjen JP. Direct transcriptional control of the Arabidopsis immune receptor FLS2 by the ethylene-dependent transcription factors EIN3 and EIL1. Proc Natl Acad Sci USA. 2010;107:14502–7

Bowles DJ. Defense-related proteins in higher plants. Annu Rev Biochem. 1990;59:873–907.

Bozkurt O, Unver T, Akkaya MS. Genes associated with resistance to wheat yellow rust disease identified by differential display analysis. Physiol Mol Plant Pathol. 2008;71:251–9.

Bozkurt O, Mcgrann G, MacCormack R, Boyd LA, Akkaya MS. Cellular and transcriptional responses of wheat during compatible and incompatible race-specific interactions with *Puccinia striiformis* f. sp. *tritici*. Mol Plant Pathol. 2010;11:625–40.

Brown W. The physiology of host-parasite relations. Bot Rev. 1936;2:236–81.

Burstenbinder K, Rzewuski G, Wirtz M, Hell R, Sauter M. The role of methionine recycling for ethylene synthesis in Arabidopsis. Plant J. 2007;49:238–49.

Buscaill P, Rivas S. Transcriptional control of plant defence responses. Curr Opin Plant Biol. 2014;20:35–46.

Bush DS. Calcium regulation in plant cells and its role in signaling. Annu Rev Plant Biol. 1995;46:95–122.

Bushnell WR. Symptom development in mildewed and rusted tissues. In: Mirocha CJ, Uritani I, editors. The dynamic role of molecular constituents in plant–parasite interaction. St Paul: American Phytopathological Society Press; 1967. p. 21–39.

Bushnell WR. Patterns in the growth, oxygen uptake, and nitrogen content of single colonies of wheat stem rust on wheat leaves. Phytopathology. 1970;60:92–9.

Calonge FD. Chlorophyll and total nitrogen in barley rust infection. Trans Br Mycol Soc. 1967;50:397–401.

Caltrider PG, Ramachandran S, Gottlieb D. Metabolism during germination and function of glyoxylate enzymes in uredospores of rust fungi. Phytopathology. 1963;53:86–92.

Campo S, Peris-Peris C, Siré C, Moreno AB, Donaire L, Zytnicki M, Notredame C, Llave C, San Segundo B. Identification of a novel microRNA (miRNA) from rice that targets an alternatively spliced transcript of the Nramp6 (Natural resistance-associated macrophage protein 6) gene involved in pathogen resistance. New Phytol. 2013;199:212–27.

Cao HN, Liu B, Liu TG, Gao L, Chen WQ. Cloning of a heat shock protein gene hsp70 of *Puccinia striiformis* f.sp.*tritici* and its expression in response to high-temperature stress. Plant Prot. 2015;41:19–24.

Capozzi F, Casadei F, Luchinat C. EF-hand protein dynamics and evolution of calcium signal transduction: an NMR view. J Biol Inorg Chem. 2006;11:949–62.

Carmona-Gutierrez D, Fröhlich K, Kroemer G, Madeo F. Metacaspases are caspases. Doubt no more. Cell Death Differ. 2010;17:377–8.

Chanda B, Venugopal SC, Kulshrestha S, Navarre DA, Downie B, Vaillancourt L, Kachroo A, Kachroo P. Glycerol-3-phosphate levels are associated with basal resistance to the hemibiotrophic fungus Colletotrichum higginsianum in Arabidopsis. Plant Physiol. 2008;147:2017–29.

Chanda B, Xia Y, Mandal MK, Yu K, Sekine KT, Gao Q-M, Selote D, Hu Y, Stromberg A, Navarre D, Kachroo A, Kachroo P. Glycerol-3- phosphate is a critical mobile inducer of systemic immunity in plants. Nat Genet. 2011;43:421–7.

Chang Q, Liu J, Wang QL, Han LN, Liu J, Li M. Huang LL, Yang JR, Kang ZS. The effect of *Puccinia striiformis* f. sp. *tritici* on the levels of water soluble carbohydrates and the photosynthetic rate in wheat leaves. Physiol Mol Plant Pathol. 2013;84:131–7.

Chao QM, Rothenberg M, Solano R, Roman G, Terzaghi W, Ecker JR. Activation of the ethylene gas response pathway in Arabidopsis by the nuclear protein ETHYLENE-INSENSITIVE3 and related proteins. Cell. 1997;89:1133–44.

Chen XM. Epidemiology and control of stripe rust [*Puccinia striiformis* f. sp. *tritici*] on wheat. Can J Plant Pathol. 2005;27:314–37.

Chen XM. High-temperature adult-plant resistance, key for sustainable control of stripe rust. Am J Plant Sci. 2013;4:608–27.

Chen XM, Line RF. Gene-Action in wheat cultivars for durable, high-temperature, adult-plant resistance and interaction with race-specific, seedling resistance to *Puccinia striiformis*. Phytopathology. 1995;85:567–72.

Chen XM, Line RF, Jones SS. Chromosomal location of genes for resistance to *Puccinia striiformis* in winter wheat cultivars Heines VII, Clement, Moro, Tyee, Tres, and Daws. Phytopathology. 1995;85:1362–7.

Chen H, Xue L, Chintamanani S, Germain H, Lin H, Cui H, Cai R, Zuo J, Tang X, Li X, Guo H, Zhou JM. Ethylene insensitive3 and ethylene insensitive3-like1 repress salicylic acid induction deficient2 expression to negatively regulate plant innate immunity in Arabidopsis. Plant Cell. 2009;21:2527–40.

Chen XM, Penman L, Wan AM, Cheng P. Virulence races of *Puccinia striiformis* f. sp. *tritici* in 2006 and 2007 and development of wheat stripe rust and distributions, dynamics, and evolutionary relationships of races from 2000 to 2007 in the United States. Can J Plant Pathol. 2010;32:315–33.

Chen XM, Wang MN, Wan AM, Cheng P, Cheng JJ. Sexual or asexual reproduction, which one is more important for stripe rust. In: Chen W-Q (ed) Disease risk and food security. Proceedings of the 13th international Cereal Rust and Powdery Mildew Conference. Beijing: China Agricultural Science and Technology Press; 2012. pp 36–37

Chen XM, Coram T, Huang XL, Wang MN, Dolezal A. Understanding molecular mechanisms of durable and non-durable resistance to stripe rust in wheat using a transcriptomics approach. Curr Genomics. 2013;14:111–26.

Chen WQ, Wellings C, Chen XM, Kang ZS, Liu TG. Wheat stripe (yellow) rust caused by *Puccinia striiformis* f. sp. *tritici*. Mol Plant Pathol. 2014;15:433–46.

Chen YE, Cui JM, Su YQ, Yuan S, Yuan M, Zhang HY. Influence of stripe rust infection on the photosynthetic characteristics and antioxidant system of susceptible and resistant wheat cultivars at the adult plant stage. Front Plant Sci. 2015;6:779.

Cheng YL, Zhang HC, Yao JN, Wang XJ, Xu JR, Han QM, Wei GR, Huang LL, Kang ZS. Characterization of non-host resistance in broad bean to the wheat stripe rust pathogen. BMC Plant Biol. 2012;12:96.

Cheng YL, Zhang HC, Yao JN, Han QM, Wang XJ, Huang LL, Kang ZS. Cytological and molecular characterization of non-host resistance in *Arabidopsis thaliana* against wheat stripe rust. Plant Physiol Biochem. 2013;62:11–8.

Cheng P, Chen XM, See D. Grass hosts harbor more diverse isolates of *Puccinia striiformis* than cereal crops. Phytopathology. 2016;106:362–71.

Cheong YH, Kim KN, Pandey GK, Gupta R, Grant JJ, Luan S. CBL1, a calcium sensor that differentially regulates salt, drought, and cold responses in Arabidopsis. Plant Cell. 2003;15:1833–45.

Cheung DSM, Willetts HJ. Changes in proteins and isoenzymes in leaves of wheat when infected by stem rust. Arch Microbiol. 1973;91:365–72.

Chigrin VV, Chumakovskii NN, Zhigalkina TE. Dynamics of ethylene production in the pathogenesis of wheat stem rust. Mikol Fitopatol.1978;12:141–5 (in Russian, transl. by DH Casper)

Choi J, Choi D, Lee S, Ryu CM, Hwang I. Cytokinins and plant immunity: old foes or new friends? Trends Plant Sci. 2011;16:388–94.

Chong J, Harder DE, Rohringer R. Cytochemical studies on *Puccinia graminis* f. sp. *tritici* in a compatible wheat host. I. Walls of intercellular hyphal cells and haustorium mother cells. Can J Bot. 1985;63:1713–24.

Chuck G, Candela H, Hake S. Big impacts by small RNAs in plant development. Curr Opin Plant Biol. 2009;12:81–6.

Cloutier S, McCallum B, Loutre C, Banks T, Wicker T, Feuillet C, Keller B, Jordan M. Leaf rust resistance gene *Lr1*, isolated from bread wheat (*Triticum aestivum* L.) is a member of the large *psr567* gene family. Plant Mol Biol. 2007;65:93–106.

Coram TE, Settles ML, Chen XM. Transcriptome analysis of high-temperature adult-plant resistance conditioned by *Yr39* during the wheat-*Puccinia striiformis* f. sp. *tritici* interaction. Mol Plant Pathol. 2008a;9:479–93.

Coram TE, Wang MN, Chen XM. Transcriptome analysis of the wheat-*Puccinia striiformis* f. sp. *tritici* interaction. Mol Plant Pathol. 2008b;9:157–69.

Coram TE, Huang XL, Zhan GM, Settles ML, Chen XM. Meta-analysis of transcripts associated with race-specific resistance to stripe rust in wheat demonstrates common induction of blue copper-binding protein, heat-stress transcription factor, pathogen-induced WIR1A protein, and ent-kaurene synthase transcripts. Funct Integr Genomics. 2010;10:383–92.

Cuvillier O, Pirianov G, Kleuser B, Vanek PG, Coso OA, Gutkind JS, Spiegel S. Suppression of ceramide-mediated programmed cell death by sphingosine-1-phosphate. Nature. 1996;381:800–3.

Dai Y, Wensink PC, Abeles RH. One protein, two enzymes. J Biol Chem. 1999;274:1193–5.

Daly JM, Knoche HW. Hormonal involvement in metabolism of host–parasite interactions. In: Friend J, Threlfall DR, editors. Biochemical sspects of plant–parasite relationships. New York: Academic Press; 1976. p. 117–33.

Daly JM, Knoche HW, Wiese MV. Carbohydrate and lipid metabolism during Germination of uredospores of *Puccinia graminis* tritici. Plant Physiol. 1967;42:1633–42.

Daly JM, Ludden P, Seevers P. Biochemical comparisons of resistance to wheat stem rust disease controlled by the *Sr6* or *Sr11* alleles. Physiol Plant Pathol. 1971;1:397–407.

Dangl JL, Jones JD. Plant pathogens and integrated defence responses to infection. Nature. 2001;411:826–33.

de Maio A. Heat shock proteins: facts, thoughts, and dreams. Shock. 1999;11:1–12.

Deikman J. Molecular mechanisms of ethylene regulation of gene transcription. Physiol Plant. 1997;100:561–6.

Dekhuijzen HM, Staples RC. Mobilization factors in uredospores and bean leaves infected with bean rust fungus. Contrib Boyce Thompson Inst. 1968;24:39–52.

Delledonne M, Zeier J, Marocco A, Lamb C. Signal interactions between nitric oxide and reactive oxygen intermediates in the plant hypersensitive disease resistance response. Proc Natl Acad Sci U S A. 2001;98:13454–9.

Dickson JG, Syamananda R, Flangas AL. The genetic approach to the physiology of parasitism of the corn rust pathogens. Am J Bot. 1959;46:614–20.

Dietrich RA, Delaney TP, Uknes SJ, Ward ER, Ryals JA, Dangl JL. Arabidopsis mutants simulating disease resistance response. Cell. 1994;77:565–77.

Dmochowska-Boguta M, Alaba S, Yanushevska Y, Piechota U, Lasota E, Nadolska-Orczyk A, Karlowski WM, Orczyk W. Pathogen-regulated genes in wheat isogenic lines differing in resistance to brown rust *Puccinia triticina*. BMC Genomics. 2015;16:742.

Doodson JK, Manners JG, Myers A. Some effects of yellow rust (*Puccinia striiformis*) on ^{14}carbon assimilation and translocation in wheat. J Exp Bot. 1965;16:304–17.

Dos Remedios CG, Chhabra D, Kekic M, Dedova IV, Tsubakihara M, Berry DA, Nosworthy NJ. Actin binding proteins: regulation of cytoskeletal microfilaments. Physiol Rev. 2003;83:433–73.

Du H, Zhang L, Liu L, Tang XF, Yang WJ, Wu YM, Huang YB, Tang YX. Biochemical and molecular characterization of plant MYB transcription factor family. Biochemistry. 2009;74:1–11.

Duan YH, Guo J, Wang SJ, Yu XM, Huang LL, Kang AS. Cloning and expression analysis of alanine amino transferase gene *TaAlaAT*1 in wheat infected with stripe rust fungus. Acta Phys Sin. 2009;39:139–46. (in Chinese)

Duan YH, Guo J, Ding K, Wang SJ, Zhang H, Dai XW, Chen YY, Govers F, Huang LL, Kang ZS. Characterization of a wheat HSP70 gene and its expression in response to stripe rust infection and abiotic stresses. Mol Biol Rep. 2011;38:301–7.

Duan XY, Wang XJ, Fu YP, Tang CL, Li XR, Cheng YL, Feng H, Huang LL, Kang ZS. *TaEIL1*, a wheat homologue of *AtEIN3*, acts as a negative regulator in the wheat-stripe rust fungus interaction. Mol Plant Pathol. 2013a;14:728–39.

Duan YH, Guo J, Shi XX, Guan XN, Liu FR, Bai PF, Huang LL, Kang ZS. Wheat hypersensitive-induced reaction genes *TaHIR1* and *TaHIR3* are involved in response to stripe rust fungus infection and abiotic stresses. Plant Cell Rep. 2013b;32:273–83.

Duplessis S, Cuomo CA, Lin YC, Aerts A, Tisserant E, et al. Obligate biotrophy features unraveled by the genomic analysis of rust fungi. Proc Natl Acad Sci U S A. 2011;108:9166–71.

Duval M, Hsieh T-F, Kim SY, Thomas TL. Molecular characterization of AtNAM: a member of the Arabidopsis NAC domain superfamily. Plant Mol Biol. 2002;50:237–48.

Dyck PL. Genetics of adult-plant leaf rust resistance in "Chinese Spring" and "Sturdy" wheats. Crop Sci. 1991;31:309.

Edreva A. Pathogenesis-related proteins: research progress in the last 15 years. Gen Appl Plant Physiol. 2005;31:105–24.

Ehrlich MA, Ehrlich HG. Fine structure of *Puccinia graminis* and the transfer of C^{14} from uredospores to *Triticum vulgare*. In: Akai S, Ouchi S, editors. Morphological and biochemical events in plant–parasite interaction: Phytopathological Society of Japan; 1971. p. 279–307.

Eichmann R, Dechert C, Kogel K-H, Hückelhoven R. Transient over-expression of barley BAX inhibitor-1 weakens oxidative defence and *MLA12*-mediated resistance to *Blumeria graminis* f. sp. *hordei*. Mol Plant Pathol. 2006;7:543–52.

El Kasmi F, Krause C, Hiller U, Stierhof YD, Mayer U, Conner L, Kong L, Reichardt I, Sanderfoot AA, Jürgens G. SNARE complexes of different composition jointly mediate membrane fusion in Arabidopsis cytokinesis. Mol Biol Cell. 2013;24:1593–601.

Ellis JG, Dodds PN, Lawrence GJ. Flax rust resistance gene specificity is based on direct resistance-avirulence protein interactions. Annu Rev Phytopathol. 2007;45:289–306.

Eltayeb AE, Kawano N, Badawi GH, Kaminaka H, Sanekata T, Shibahara T, Inanaga S, Tanaka K. Overexpression of monodehydroascorbate reductase in transgenic *Nicotiana benthamiana* confers enhanced tolerance to ozone, salt and polyethylene glycol stresses. Planta. 2007;225:1255–64.

Eltelib HA, Badejo AA, Fujikawa Y, Esaka M. Gene expression of monodehydroascorbate reductase and dehydroascorbate reductase during fruit ripening and in response to environmental stresses in acerola (*Malpighia glabra*). J Plant Physiol. 2011;168:619–27.

Enoksson M, Salvesen G. Metacaspases are not caspases – always doubt. Cell Death Differ. 2010;17:1221.

Eriksson J. Über die Spezialisierung des Parasitsimus bei den Getreiderostpilzen. Ber Dtsch Getreiderostpilze Z Pflanzenkr. 1894;4:197–203.

Fadeel B, Zhivotovsky B, Orrenius S. All along the watch tower: on the regulation of apoptosis regulators. FASEB J. 1999;13:1647–57.

Fan J, Doerner P. Genetic and molecular basis of nonhost disease resistance: complex, yes; silver bullet, no. Curr Opin Plant Biol. 2012;15:400–6.

Farkas GL, Király Z. Amide metabolism in wheat leaves infected with stem rust. Physiol Plant. 1961;14:344–53.

Farrakh S, Wang MN, Chen XM. Expression profiling of pathogenesis-related protein genes in wheat resistance to the stripe rust pathogen (*Puccinia striiformis* f. sp. *tritici*). Phytopathology. 2016;S4:197

Feng H, Wang X, Sun Y, Wang X, Chen X, Guo J, Duan Y, Huang L, Kang Z. Cloning and characterization of a calcium binding EF-hand protein gene *TaCab1* from wheat and its expression in response to *Puccinia striiformis* f. sp. *tritici* and abiotic stresses. Mol Biol Rep. 2011;38:3857–66.

Feng H, Zhang Q, Li HY, Wang XJ, Wang XD, Duan XY, Wang B, Kang ZS. vsiRNAs derived from the miRNA-generating sites of pri-taemiR159a based on the BSMV system play positive roles in the wheat response to *Puccinia striiformis* f. sp. *tritici* through the regulation of *TaMyb3* expression. Plant Physiol Biochem. 2013a;68:90–5.

Feng H, Zhang Q, Wang QL, Wang XJ, Liu J, Li M, Huang LL, Kang ZS. Target of tae-miR408, a chemocyanin-like protein gene (*TaCLP1*), plays positive roles in wheat response to high-salinity, heavy cupric stress and stripe rust. Plant Mol Biol. 2013b;83:433–43.

Feng H, Duan XY, Zhang Q, Li XR, Wang B, Huang LL, Wang XJ, Kang ZS. The target gene of tae-miR164, a novel NAC transcription factor from the NAM subfamily, negatively regulates resistance of wheat to stripe rust. Mol Plant Pathol. 2014a;15:284–96.

Feng H, Liu W, Zhang Q, Wang X, Wang X, Duan X, Li F, Huang L, Kang Z. TaMDHAR4, a monodehydroascorbate reductase gene participates in the interactions between wheat and *Puccinia striiformis* f. sp. *tritici*. Plant Physiol Biochem. 2014b;76:7–16.

Feng H, Wang XJ, Zhang Q, Fu YP, Feng CX, Wang B, Huang LL, Kang ZS. Monodehydroascorbate reductase gene, regulated by the wheat PN-2013 miRNA, contributes to adult wheat plant resistance to stripe rust through ROS metabolism. Biochim Biophys Acta (BBA)-Gene Regulat Mech. 2014c;1839:1–12

Feng H, Wang B, Zhang Q, Fu YP, Huang LL, Wang XJ, Kang ZS. Exploration of microRNAs and their targets engaging in the resistance interaction between wheat and stripe rust. Front Plant Sci. 2015;6:469.

Feng H, Wang T, Feng CX, Zhang Q, Zhang XM, Huang LL, Wang XJ, Kang ZS. Identification of microRNAs and their corresponding targets involved in the susceptibility interaction of wheat response to *Puccinia striiformis* f. sp. *tritici*. Physiol Plant. 2016;157:95–107.

Feuillet C, Schachermayr G, Keller B. Molecular cloning of a new receptor-like kinase gene encoded at the Lr10 disease resistance locus of wheat. Plant J. 1997;11:45–52.

Feuillet C, Reuzeau C, Kjellbom P, Keller B. Molecular characterization of a new type of receptor-like kinase (*wlrk*) gene family in wheat. Plant Mol Biol. 1998;37:943–53.

Feuillet C, Penger A, Gellner K, Mast A, Keller B. Molecular evolution of receptor-like kinase genes in hexaploid wheat. Independent evolution of orthologs after polyploidization and mechanisms of local rearrangements at paralogous loci. Plant Physiol. 2001;125:1304–13.

Feuillet C, Travella S, Stein N, Albar L, Nublat A, Keller B. Map-based isolation of the leaf rust disease resistance gene *Lr10* from the hexaploid wheat (*Triticum aestivum* L.) genome. Proc Natl Acad Sci U S A. 2003;100:15253–8.

Fluhr R, Mattoo AK. Critical reviews in plant sciences. In: Conger BV (ed) Ethylene biosynthesis and perception, vol 15. Boca Raton: CRC Press; 1996. pp 479–523

Frandsen G, Müller-Uri F, Nielsen M, Mundy J, Skriver K. Novel plant Ca^{2+}-binding protein expressed in response to abscisic acid and osmotic stress. J Biol Chem. 1996;271:343–8.

Fric F, Heitefuss R. Immunochemische und elektrophoretische Untersuchungen der Proteine von Weizenblättern nach Infektion mit *Puccinia graminis tritici*. Phytopathol Z. 1970;69:236–46.

Fu DL, Uauy C, Distelfeld A, Blechl A, Epstein L, Chen XM, Sela H, Fahima T, Dubcovsky J. A kinase-START gene confers temperature-dependent resistance to wheat stripe rust. Science. 2009;323:1357–60.

Fu ZQ, Dong X. Systemic acquired resistance: turning local infection into global defense. Annu Rev Plant Biol. 2013;64:839–63.

Fu YP, Duan XY, Tang CL, Li XR, Voegele RT, Wang XJ, Wei GR, Kang ZS. *TaADF7*, an actin-depolymerizing factor, contributes to wheat resistance against *Puccinia striiformis* f. sp. *tritici*. Plant J. 2014;78:16–30.

Fukuda R, McNew JA, Weber T, Parlati F, Engel T, Nickel W, Rothman JE, Söllner TH. Functional architecture of an intracellular membrane t-SNARE. Nature. 2000;407:198–202.

Fullington JG, Nityagopala A. Effect of rust infection on the protein components of wheat. Phytochemistry. 1986;25:1289–129.

Gale MD, Devos KM. Plant comparative genetics after 10 years. Science. 1998;282:656–9.

Gan PH, Dodds PN, Hardham AR. Plant infection by biotrophic fungal and oomycete pathogens. In: Silvia P, František B, editors. Signaling and communication in plant symbiosis. New York: Springer; 2012. p. 183–212.

Gassner G, Franke W. Untersuchungen über den Stickstoffhaushalt rostinfizierter Getreideblätter. Phytopathol Z. 1938;11:517–70.

Gay JL, Salzberg A, Woods AM. Dynamic experimental evidence for the plasma membrane ATPase domain hypothesis of haustorial transport and for ionic coupling of the haustorium of *Erysiphe graminis* to the host cell (*Hordeum vulgare*). New Phytol. 1987;107:541–8.

Goddard MV. Cytological studies of *Puccinia striiformis* (yellow rust of wheat). Trans Br Mycol Soc. 1976;66:433–7.

Gordon TR, Duniway JM. Effects of powdery mildew infection on the efficiency of CO_2 fixation and light utilization by sugar beet leaves. Plant Physiol. 1982;69:139–42.

Gou JY, Li K, Wu KT, Wang XD, Lin HQ, Cantu D, Uauy C, Dobon-Alonso A, Midorikawa T, Inoue K, Sáncheza J, Fu DL, Blechl A, Wallington E, Fahima Z, Meeta M, Epstein L, Dubcovsky J. Wheat stripe rust resistance protein WKS1 reduces the ability of the thylakoid-associated ascorbate peroxidase to detoxify reactive oxygen species. Plant Cell. 2015;27:1755–70.

Grant JJ, Yun BW, Loake GJ. Oxidative burst and cognate redox signalling reported by luciferase imaging: identification of a signal network that functions independently of ethylene, SA and Me-JA but is dependent on MAPKK activity. Plant J. 2000;24:569–82.

Gray J, Close PS, Briggs SP, Johal GS. A novel suppressor of cell death in plants encoded by the *Lls1* gene of maize. Cell. 1997;89:25–31.

Gray J, Janick-Buckner D, Buckner B, Close PS, Johal GS. Light-dependent death of maize lls1 cells is mediated by mature chloroplasts. Plant Physiol. 2002;130:1894–907.

Guo HS, Xie Q, Fei JF, Chua NH. MicroRNA directs mRNA cleavage of the transcription factor NAC1 to downregulate auxin signals for Arabidopsis lateral root development. Plant Cell. 2005;17:1376–86.

Guo J, Bai P, Yang Q, Liu F, Wang X, Huang L, Kang Z. Wheat zinc finger protein TaLSD1, a negative regulator of programmed cell death, is involved in wheat resistance against stripe rust fungus. Plant Physiol Biochem. 2013;71:164–72.

Hahn M. The rust fungi: cytology, physiology and molecular biology of infection. In: Kronstadt J, editor. Fungal pathology. Dordrecht: Kluwer Academic Publisher; 2000. p. 267–306.

Hahn M, Mendgen K. Characterization of in planta induced rust genes isolated from a haustorium-specific cDNA library. Mol Plant-Microbe Interact. 1997;10:427–37.

Hahn M, Mendgen K. Signal and nutrient exchange at biotrophic plant-fungus interfaces. Curr Opin Plant Biol. 2001;4:322–7.

Hahn M, Neef U, Struck C, Göttfert M, Mendgen K. A putative amino acid transporter is specifically expressed in haustoria of the rust fungus *Uromyces fabae*. Mol Plant Microbe Interact. 1997b;10:438–45.

Hahn M, Deising H, Struck C, Mendgen K. Fungal morphogenesis and enzyme secretion during pathogenesis. In: Hartleb H, Heitefuss R, Hoppe H-H, editors. Resistance of crop plants against fungi. Jena: Gustav Fischer; 1997a. p. 33–57.
Hakoyama T, Oi R, Hazuma K, Suga E, Adachi Y, Kobayashi M, Akai R, Sato S, Fukai E, Tabata S, Shibata S, Wu G.-J., Hase Y, Tanaka A, Kawaguchi M, Kouchi H, Umehara Y, Suganuma N. The SNARE protein SYP71 expressed in vascular tissues is involved in symbiotic nitrogen fixation in *Lotus japonicus* nodules. Plant Physiol. 2012; 160:897–905
Halfter U, Ishitani M, Zhu JK. The Arabidopsis SOS2 protein kinase physically interacts with and is activated by the calcium-binding protein SOS3. Proc Natl Acad Sci U S A. 2000;97:3735–40.
Hao YB, Wang T, Wang K, Wang XJ, Fu YP, Huang LL, Kang ZS. Transcriptome analysis provides insights into the mechanisms underlying wheat plant resistance to stripe rust at the adult plant stage. PLoS One. 2016a;11:e0150717.
Hao YB, Wang XJ, Wang K, Li HY, Duan XY, Tang CL, Kang ZS. *TaMCA1*, a regulator of cell death, is important for the interaction between wheat and *Puccinia striiformis*. Sci Rep. 2016b;6:26946.
Harder DE, Mendgen K. Filipin-sterol complexes in bean rust- and oat crown rust-fungal/plant interactions: freeze-etch electron microscopy *Uromyces appendiculatus*. Protoplasma. 1982;112:46–54.
Harder D, Chong J. Structure and physiology of haustoria. In: Bushnell WR, Roslfs AP, editors. The Cereal Rusts. New York: Academic Press; 1984. p. 431–76.
Harder DE, Chong J. Rust haustoria. In: Mendgen K, Lesemann D-E, editors. Electron microscopy of plant pathogens. Berlin: Springer; 1991. p. 235–50.
Hardham AR, Jones DA, Takemoto D. Cytoskeleton and cell wall function in penetration resistance. Curr Opin Plant Biol. 2007;10:342–8.
Harrison RE, Bucci C, Vieira OV, Schroer TA, Grinstein S. Phagosomes fuse with late endosomes and/or lysosomes by extension of membrane protrusions along microtubules: role of Rab7 and RILP. Mol Cell Biol. 2003;23:6494–506.
Hart H, Becker H. Beitrage zur Frage des Zwischenwirtes für *Puccinia glumarum*. Z Pflanzenkr (Pflanzenpathol) Pflanzenschutz. 1939;49:559–66.
Harter C, Pavel J, Coccia F, Draken E, Wegehingel S, Tschochner H, Wieland F. Nonclathrin coat protein gamma, a subunit of coatomer, binds to the cytoplasmic dilysine motif of membrane proteins of the early secretory pathway. Proc Natl Acad Sci U S A. 1996;93:1902–6.
He R, Drury GE, Rotari VI, Gordon A, Willer M, Farzaneh T, Woltering EJ, Gallois P. Metacaspase-8 modulates programmed cell death induced by ultraviolet light and H_2O_2 in Arabidopsis. J Biol Chem. 2008;283:774–83.
Heath MC. Ultrastructural and functional similarity of the haustorial neckband of rust fungi and the Casparian strip of vascular plants. Can J Bot. 1976;54:2484–9.
Heath MC. Signalling between pathogenic rust fungi and resistant or susceptible host plants. Ann Bot. 1997;80:713–20.
Heath MC. Hypersensitive response-related death. Plant MolBiol. 2000;44:321–34.
Heath MC, Skalamera D. Cellular interactions between plants and biotrophic fungal parasites. Adv Bot Res. 1997;24:195–225.
Heese M, Gansel X, Sticher L, Wick P, Grebe M, Granier F, Jürgens G. Functional characterization of the KNOLLE-interactingt-SNARE AtSNAP33 and its role in plant cytokinesis. J Cell Biol. 2001;155:239–49.
Heisterüber D, Schulte P, Moerschbacher B. Soluble carbohydrates and invertase activity in stem rust-infected, resistant and susceptible near-isogenic wheat leaves. Physiol Mol Plant Pathol. 1994;45:111–23.
Heitefuss R. Untersuchungen zur Physiologie des temperaturgesteuerten Verträglichkeitsgrades von Weizen und*Puccinia graminis tritici*. I. Veränderungen von Sauerstoffaufnahme und Phosphatstoffwechsel. Phytopathol Z. 1965;54:379–400.
Heitefuss R. Nucleic acid metabolism in obligate parasitism. Annu Rev Phytopathol. 1966a;4:221–42.

Heitefuss R. Untersuchungen zur Physiologie des temperaturgesteuerten Verträglichkeitsgrades von Weizen und *Puccinia graminis tritici*. II. Veränderungen des Nukleinsäurestoffwechsels. Phytopathol Z. 1966b;55:67–85.

Heitefuss R. Der Einfluss von Actinomycin auf *Puccinia graminis tritici* auf Weizen und den Einbau von Orotsäure-C^{14} und Uridin-H^{3} in Wirtspflanze und Parasit. Phytopathol Z. 1970;69:107–14.

Heitefuss R, Wolf G. Nucleic acids in host–parasite interactions. Encycl Plant Physiol, New Ser. 1976;4:480–508.

Hengartner M, Horvitz HR. *C. elegans* cell survival gene ced-9 encodes a functional homologue of the mammalian proto-onco gene *bcl-2*. Cell. 1994;76:665–76.

Henty-Ridilla JL, Shimono M, Li J, Chang JH, Day B, Staiger CJ. The plant actin cytoskeleton responds to signals from microbe-associated molecular patterns. PLoS Pathog. 2013;9:e1003290.

Henty-Ridilla JL, Li J, Day B, Staiger CJ. ACTIN DEPOLYMERIZING FACTOR4 regulates actin dynamics during innate immune signaling in Arabidopsis. Plant Cell. 2014;26:340–52.

Hetherington AM, Brownlee C. The generation of Ca^{2+} signals in plants. Annu Rev Plant Biol. 2004;55:401–27.

Hilu HM. Host-pathogen relationships of *Puccinia sorghi* in nearly isogenic resistant and susceptible seedling corn. Phytopathology. 1965;55:563–9.

Hovmøller MS, Sorensen CK, Walter S, Justesen AF. Diversity of *Puccinia striiformis* on cereals and grasses. Annu Rev Phytopathol. 2011;49:197–217.

Hu GG, Rijkenberg FHJ. Ultrastructural localization of cytokinins in *Puccinia recondite* f. sp. *tritici*-infected wheat leaves. Physiol Mol Plant Pathol. 1998;52:79–94.

Huang L, Brooks SA, Li W, Fellers JP, Trick HN, Gill BS. Map-based cloning of leaf rust resistance gene *Lr21* from the large and polyploid genome of bread wheat. Genetics. 2003;164:655–64.

Huang J, Wang J, Zhang H. Rice *ZFP15* gene encoding fora novel C2H2-type zinc finger protein lacking DLN box, is regulated by spike development but not by abiotic stresses. Mol Biol Rep. 2005;32:177–83.

Huang X, Ma J, Chen X, Wang X, Ding K, Han D, Qu Z, Hang L, Kang Z. Genes involved in adult plant resistance to stripe rust in wheat cultivar Xingzi 9104. Physiol Mol Plant Pathol. 2013;81:26–32.

Hubert DA, Tornero P, Belkhadir Y, Krishna P, Takahashi A, Shirasu K, Dangl JL. Cytosolic HSP90 associates with and modulates the Arabidopsis RPM1 disease resistance protein. EMBO J. 2003;22:5679–89.

Hückelhoven R, Dechert C, Kogel KH. Overexpressionof barley *BAX* inhibitor 1 induces breakdown of mlo-mediatedpenetration resistance to *Blumeria graminis*. Proc Natl Acad Sci U S A. 2003;100:5555–60.

Hurni S, Brunner S, Buchmann G, Herren G, Jordan T, Krukowski P, Wicker T, Yahiaoui N, Mago R, Keller B. Rye *Pm8* and wheat *Pm3* are orthologous genes and show evolutionary conservation of resistance function against powdery mildew. Plant J. 2013;76:957–69.

Imaseki H. The biochemistry of ethylene biosynthesis. In: Mattoo AK, Suttle JC, editors. The plant hormone ethylene. Boca Raton: CRC Press; 1991. p. 1–20.

Inada N, Ueda T. Membrane trafficking pathways and their roles in plant-microbe interactions. Plant Cell Physiol. 2014;55:672–86.

Ishikawa T, Shigeoka S. Recent advances in ascorbate biosynthesis and the physiological significance of ascorbate peroxidase in photosynthesizing organisms. Biosci Biotechnol Biochem. 2008;72:1143–54.

Izawa T, Foster R, Chua NH. Plant bZIP protein DNA binding specificity. J Mol Biol. 1993;230:1131–44.

Jagger LJ, Newell C, Berry ST, MacCormack R, Boyd LA. Histopathology provides a phenotype by which to characterize stripe rust resistance genes in wheat. Plant Pathol. 2011;60:640–8.

Jameson PE. Cytokinins and auxins in plant-pathogen interactions–An overview. Plant Growth Regul. 2000;32:369–80.

Jayasekaran K, Kim KN, Vivekanandan M, Shin JS, Ok SH. Novel calcium-binding GTPase (AtCBG) involved in ABA-mediated salt stress signaling in Arabidopsis. Plant Cell Rep. 2006;25:1255–62.

Jiang Z, Ge S, Xing L, Han D, Kang Z, Zhang G, Wang X, Wang X, Chen P, Cao A. RLP1.1, a novel wheat receptor-like protein gene, is involved in the defence response against *Puccinia striiformis* f. sp. *tritici*. J Exp Bot. 2013;64:3735–46.

Jin H. Endogenous small RNAs and antibacterial immunity in plants. FEBS Lett. 2008;582:2679–84.

Jin Y. Role of *Berberis* spp. as alternate hosts in generating new races of *Puccinia graminis* and *P. striiformis*. Euphytica. 2011;179:105–8.

Jin Y, Szabo LJ, Carson M. Century-old mystery of *Puccinia striiformis* life history solved with the identification of *Berberis* as an alternate host. Phytopathology. 2010;100:432–5.

Johnson R. Durable resistance to yellow (stripe) rust in wheat and its implications in plant breeding. In: Simmonds NW, Rajaram S, editors. Breeding strategies for resistance to the rusts of wheat. Mexico City: CYMMIT; 1988. p. 63–75.

Johnson LB, Brannaman BL, Zscheile FP. Protein and enzyme changes in wheat leaves following infection with *Puccinia recondita*. Phytopathology. 1968;58:578–83.

Jones JD, Dangl JL. The plant immune system. Nature. 2006;444:323–9.

Jones DA, Thomas CM, Hammond-Kosack KE, Balint-Kurti PJ, Jones JD. Isolation of the tomato *Cf-9* gene for resistance to *Cladosporium fulvum* by transposon tagging. Science. 1994;266:789–93.

Jung HW, Hwang BK. The leucine-rich repeat (LRR) protein, CaLRR1, interacts with the hypersensitive induced reaction (HIR) protein, CaHIR1, and suppresses cell death induced by the CaHIR1 protein. Mol Plant Pathol. 2007;8:503–14.

Jung HW, Lim CW, Lee SC, Choi HW, Hwang CH, Hwang BK. Distinct roles of the pepper hypersensitive inducedreaction protein gene *CaHIR1* in disease and osmotic stress, as determined by comparative transcriptome and proteome analyses. Planta. 2008;227:409–25.

Kalde M, Nühse TS, Findlay K, Peck SC. The syntaxinSYP132 contributes to plant resistance against bacteria and secretion of pathogenesis-related protein 1. Proc Natl Acad Sci U S A. 2007;104:11850–5.

Kaneda T, Taga Y, Takai R, Iwano M, Matsui H, Takayama S, Isogai A, Che FS. The transcription factor OsNAC4 is a key positive regulator of plant hypersensitive cell death. EMBO J. 2009;28:926–36.

Kang ZS. Ultrastructure of plant pathogenic fungi. Beijing: China Science &Technology Press; 1996.

Kang ZS, Huang LL, Li ZQ. Ultrastructure of wheat cell responses to invasion of wheat stripe rust. Acta Agric Beoreali-occidentalis Sin. 1993;2:25–8.

Kang ZS, Li ZQ, Shang HS. Nuclear condition of uredinial stage of wheat stripe rust. Acta Phys Sin. 1994;24:26–31.

Kang ZS, Huang LL, Li JY. Morphology of plant pathogenic fungi under scanning electron microscope. Beijing: China Agriculture Press; 1997.

Kang ZS, Huang LL, Buchenauer H. Ultrastructural changes and localization of lignin and callose in compatible and incompatible interactions between wheat and *Puccinia striiformis*. J Plant Dis Prot. 2002;109:25–37.

Kang L, Li J, Zhao T, Xiao F, Tang X, et al. Interplay of the Arabidopsis nonhost resistance gene *NHO1* with bacterial virulence. Proc Natl Acad Sci U S A. 2003a;100:3519–24.

Kang ZS, Huang LL, Buchenauer H. Subcellular localization of chitinase and β-1,3–glucanase in compatible and incompatible interaction between wheat and *Puccinia striiformis* f. sp. *tritici*. J Plant Dis Protect. 2003b;110:170–83.

Kang ZS, Wang Y, Huang LL, Wei GR, Zhao J. Histology and ultrastructure of incompatible combination between *Puccinia striifromis* and wheat cultivars with resistance of low reaction type. Sci Agric Sin. 2003c;36:1026–31.

Kanzaki H, Saitoh H, Ito A, Fujisawa S, Kamoun S, Katou S, Yoshioka H, Terauchi R. Cytosolic HSP90 and HSP70 are essential components of INF1-mediated hypersensitive response and non-host resistance to *Pseudomonas cichorii* in *Nicotiana benthamiana*. Mol Plant Pathol. 2003;4:383–91.

Karrer EE, Beachy RN, Holt CA. Cloning of tobacco genes that elicit the hypersensitive response. Plant Mol Biol. 1998;36:681–90.

Katiyar-Agarwal S, Jin H. Role of small RNAs in host-microbe interactions. Annu Rev Phytopathol. 2010;48:225–46.

Kawai M, Pan L, Reed JC, Uchimiya H. Evolutionallyconserved plant homologue of the *Bax* inhibitor-1 (*BI-1*) gene capableof suppressing Bax-induced cell death in yeast. FEBS Lett. 1999;464:143–7.

Kawai-Yamada M, Jin U, Yoshinaga K, Hirata A, Uchimiya H. Mammalian *Bax*-induced plant cell death can be downregulatedby overexpression of Arabidopsis *Bax* inhibitor-1 (*AtBI-1*). Proc Natl Acad Sci U S A. 2001;98:12295–300.

Kawai-Yamada M, Ohmori Y, Uchimiya H. Dissection of Arabidopsis Bax inhibitor-1 suppressing Bax-, hydrogen peroxide-, and salicylic acid-induced cell death. Plant Cell. 2004;16:21–32.

Kawchuk LM, Hachey J, Lynch DR, Kulcsar F, Gv R, Waterer DR, Robertson A, Kokko E, Byers R, Howard RJ, Fischer R, Prüfer D. Tomato *Ve* disease resistance genes encode cell surface-like receptors. Proc Natl Acad Sci U S A. 2001;98:6511–5.

Kazan K, Manners JM. Linking development to defense: auxin in plant–pathogen interactions. Trends Plant Sci. 2009;14:373–82.

Keller B, Bieri S, Bossolini E, Yahiaoui N. Cloning genes and QTLs for disease resistance in cereals. In: Varshney RK, Tuberosa R, editors. Genomic-assisted crop improvement. Dordrecht: Springer; 2007. p. 103–27.

Kim WK, Rohringer R. Metabolism of aromatic compounds in healthy and rust-infected primary leaves of wheat. III. Studies on the metabolism of tryptophan. Can J Bot. 1969;47:1425–33.

Kim SY, Thomas TL. A family of basic leucine zipper proteins bind to seed specification elements in the carrot Dc3 gene promoter. J Plant Physiol. 1998;152:607–13.

Kim CY, Lee SH, Park HC, Bae CG, Cheong YH, Choi YJ, Han C, Lee SY, Lim CO, Cho MJ. Identification of rice blast fungal elicitor-responsive genes by differential display analysis. Mol Plant-Microbe Interact. 2000;13:470–4.

Kim CY, Koo YD, Jin JB, Moon BC, Kang CH, Kim ST, Park BO, Lee SY, Kim ML, Huang I, Kang KY, Bahk JD, Lee SY, Cho MJ. Rice C2-domain proteins are induced and translocated to the plasma membrane in response to a fungal elicitor. Biochemistry. 2003a;42:11625–31.

Kim KN, Cheong YH, Grant JJ, Pandey GK, Luan S. CIPK3, a calcium sensor-associated protein kinase that regulates abscisic acid and cold signal transduction in Arabidopsis. Plant Cell. 2003b;15:411–23.

Kim YJ, Kim JE, Lee JH, Lee MH, Jun HW, Bahk YY, Hwang BK, Hwang I, Kim WT. The Vr-PLC3 gene encodes a putative plasma membrane-localized phos-phoinositide-specific phospholipase C whose expression is induced by abiotic stress in mung bean (*Vigna radiata* L.). FEBS Lett. 2004;556:127–36.

Kim JH, Kim HS, Lee YH, Kim YS, Oh HW, Joung H, Chae SK, Suh KH, Jeon JH. Polyamine biosynthesis regulated by StARD expression plays an important role in potato wound periderm formation. Plant Cell Physiol. 2008a;49:1627–32.

Kim YC, Kim SY, Choi D, Ryu CM, Park JM. Molecular characterization of a pepper C2 domain-containing SRC2 protein implicated in resistance against host and non-host pathogens and abiotic stresses. Planta. 2008b;227:1169–79.

Kim JH, Woo HR, Kim J, Lim PO, Lee IC, Choi SH, Hwang D, Nam HG. Trifurcate feed-forward regulation of age-dependent cell death involving miR164 in Arabidopsis. Science. 2009a;323:1053–7.

Kim MC, Chung WS, Yun DJ, Cho MJ. Calcium and calmodulin-mediated regulation of gene expression in plants. Mol Plant. 2009b;2:13–21.

Király Z, Elhammad ME, Pozsar BI. Increased cytokinin activity of rust-infected bean and broad bean leaves. Phytopathology. 1967;57:93.

Kitanaka C, Kuchino Y. Caspase-independent programmed cell death with necrotic morphology. Cell Death Differ. 1999;6:508.

Knight H. Calcium signaling during abiotic stress in plants. Int Rev Cytol. 2000;195:269–324.

Koh S, Andre A, Edwards H, Ehrhardt D, Somerville S. *Arabidopsis thaliana* subcellular responses to compatible *Erysiphe cichoracearum* infections. Plant J. 2005;44:516–29.

Kolmer JA. Genetics of resistance to wheat leaf rust. Annu Rev Phytopathol. 1996;34:435–55.

Kolukisaoglu U, Weinl S, Blazevic D, Batistic O, Kudla J. Calcium sensors and their interacting protein kinases: genomics of the Arabidopsis and rice CBL-CIPK signaling networks. Plant Physiol. 2004;134:43–58.

Kopitz J, André S, von Reitzenstein C, Versluis K, Kaltner H, Pieters RJ, Wasano K, Kuwabara I, Liu FT, Cantz M, Heck AJR, Gabius HJ. Homodimeric galectin-7 (p53-induced gene 1) is a negative growth regulator for human neuroblastoma cells. Oncogene. 2003;22:6277–88.

Kosugi S, Ohashi Y. Cloning and DNA-binding properties of a tobacco Ethylene-Insensitive3 (EIN3) homolog. Nucleic Acids Res. 2000;28:960–7.

Krattinger SG, Lagudah ES, Spielmeyer W, Singh RP, Huerta-Espino J, McFadden H, Bossolini E, Selter LL, Keller B. A putative ABC transporter confers durable resistance to multiple fungal pathogens in wheat. Science. 2009;323:1360–3.

Krattinger SG, Lagudah ES, Wicker T, Risk JM, Ashton AR, Selter LL, Matsumoto T, Keller B. *Lr34* multi-pathogen resistance ABC transporter: molecular analysis of homoeologous and orthologous genes in hexaploid wheat and other grass species. Plant J. 2011;65:392–403.

Kroemer G. The proto-oncogene Bcl-2 and its role in regulating apoptosis. Nat Med. 1997;3:614–20.

Kroemer G, Martin SJ. Caspase-independent cell death. Nat Med. 2005;11:725–30.

Kruijt M, DE Kock MJ, de Wit PJ. Receptor-like proteins involved in plant disease resistance. Mol Plant Pathol. 2005;6:85–97.

Kwon SI, Cho HJ, Bae K, Jung JH, Jin HC, Park OK. Role of an Arabidopsis Rab GTPase RabG3b in pathogen response and leaf senescence. J Plant Biol. 2009;52:79–87.

Lalanne E, Michaelidis C, Moore JM, Gagliano W, Johnson A, Patel R, et al. Analysis of transposon insertion mutants highlights the diversity of mechanisms underlying male progamic development in Arabidopsis. Genetics. 2004;167:1975–86.

Lambrecht M, Okon Y, Broek A. V, Vanderleyden J (2000) Indole-3-acetic acid: a reciprocal signalling molecule in bacteria–plant interactions. Trends Microbiol 8:298–300

Landschulz WH, Johnson PF, McKnight SL. The leucine zipper: a hypothetical structure common to a new class of DNA binding proteins. Science. 1998;240:1759–64.

Laura M, Consonni R, Locatelli F, Fumagalli E, Allavena A, Coraggio I, Mattana M. Metabolic response to cold and freezing of *Osteospermum ecklonis* overexpressing *OsMyb4*. Plant Physiol Biochem. 2010;48:764–71.

Lee JH, Kim WT. Molecular and biochemical characterization of VR-EILs encoding mung bean ETHYLENE INSENSITIVE3-LIKE proteins. Plant Physiol. 2003;132:1475–88.

Lee TF, McNellis TW. Evidence that the BONZAI1/COPINE1 protein is a calcium and pathogen-responsive defense suppressor. Plant Mol Biol. 2009;69:155–66.

Li MR. Water relation and photosynthesis in stripe rusted wheat leaves. Dissertation, Department of Plant Protection, Northwest A&F University., Shaanxi, China; 1998

Li A, Heath MC. Effect of plant growth regulators on the interactions between bean plants and rust fungi non-pathogenic on beans. Physiol Mol Plant Pathol. 1990;37:245–54.

Li ZQ, Zeng SM. Wheat stripe rust in China. Beijing: China Agriculture Press; 2002.

Li L, Yu XF, Thompson MG, Yoshida S, Asami T, Chory J, Yin YH. Arabidopsis MYB30 is a direst target of BES1 and cooperates with BED1 to regulate brassinosteroid-induced gene expression. Plant J. 2009;58:275–86.

Li HB, Wei GR, Xu JR, Huang LL, Kang ZS. Identification of wheat proteins with altered expression levels in leaves infected by the stripe rust pathogen. Acta Physiol Plant. 2011;33:2423–35.

Li H, Ren B, Kang ZS, Huang LL. Comparison of cell death and accumulation of reactive oxygen species in wheat lines with or without *Yr36* responding to *Puccinia striiformis* f. sp. *tritici* under low and high temperatures at seedling and adult-plant stages. Protoplasma. 2016a;253:787–802.

Li Q, Qin JF, Zhao YY, Zhao J, Huang LL, Kang ZS. Virulence analysis of sexual progeny of the wheat stripe rust pathogen recovered from wild barberry in Shaanxi and Gansu. Acta Phys Sin. 2016b;46:809–20.

Line RF, Chen XM. Successes in breeding for and managing durable resistance to wheat rusts. Plant Dis. 1995;79:1254–5.

Line RF, Qayoum A. Virulence, aggressiveness, evolution and distribution of races of *Puccinia striiformis* (the cause of stripe rust of wheat) in North America, 1968–1987. United States Department of Agriculture, Agricultural Research Service, Technical Bulletin Number 1788; 1992, 44 pp

Lipka U, Fuchs R, Kuhns C, Petutschnig E, Lipka V. Live and let die- *Arabidopsis* nonhost resistance to powdery mildews. Eur J Cell Biol. 2010;89:194–9.

Littlefield LJ, Health MC. Ultrastructure of rust fungi. New York: Academic Press; 1979.

Liu J, Zhu JK. A calcium sensor homolog required for plant salt tolerance. Science. 1998;280:1943–5.

Liu HM, Liu TG, Xu SC, Liu DQ, Chen WQ. Inheritance of yellow rust resistance in an elite wheat germplasm Xingzi 9104. Acta Agron Sin. 2006;32:1742–5.

Liu B, Xue XD, Cui S, Zhang XY, Han QM, Zhu L, Liang XF, Wang XJ, Huang LL, Chen XM, Kang ZS. Cloning and characterization of a wheat β-1,3-glucanase gene induced by the stripe rust pathogen *Puccinia striiformis* f. sp. *tritici*. Mol Biol Rep. 2010;37:1045–52.

Liu FR, Guo J, Bai PF, Duan YH, Wang XD, Cheng YL, Feng H, Huang LL, Kang ZS. Wheat TaRab7 GTPase is part of the signaling pathway in responses to stripe rust and abiotic stimuli. PLoS One. 2012a;7:e37146.

Liu J, Zhang Q, Chang Q, Zhuang H, Huang LL, Kang ZS. Cloning and characterization of the actin gene from *Puccinia striiformis* f. sp. *tritici*. Microbiol. Biotechnol. 2012b;28:2331.

Liu W, Frick M, Huel R, Nykiforuk CL, Wang X, Gaudet DA, Eudes F, Conner RL, Kuzyk A, Chen Q, Kang Z, Laroche A. The stripe rust resistance gene *Yr10* encodes an evolutionary-conserved and unique CC-NBS-LRR sequence in wheat. Mol Plant. 2014;7:1740–55.

Liu J, Han LN, Huai BY, Zheng PJ, Chang Q, Guan T, Li D, Huang LL, Kang ZS. Down-regulation of a wheat alkaline/neutral invertase correlates with reduced host susceptibility to wheat sripe rust caused by *Puccinia striiformis*. J Exp Bot. 2015;66:7325–38.

Liu MJ, Peng Y, Li HY, Deng L, Wang XJ, Kang ZS. TaSYP71, a Qc-SNARE, contributes to wheat resistance against *Puccinia striiformis* f. sp. *tritici*. Front Plant Sci. 2016a;7:544.

Liu P, Myo T, Ma W, Lan DY, Qi T, Guo J, Song P, Guo J, Kang ZS. TaTypA, a ribosome-binding GTPase protein, positively regulates wheat resistance to the stripe rust fungus. Front Plant Sci. 2016b.; https://doi.org/10.3389/fpls.2016.00873.

Llave C, Kasschau KD, Rector MA, Carrington JC. Endogenous and silencing-associated small RNAs in plants. Plant Cell. 2002;14:1605–19.

Loehrer M, Langenbach C, Goellner K. Characterization of nonhost resistance of Arabidopsis to the Asian soybean rust. Mol Plant-Microbe Interact. 2008;21:1421–30.

Loon LCV, Geraats BPJ, Linthorst HJM. Ethylene as a modulator of disease resistance in plants. Trends Plant Sci. 2006;11:184–91.

Lorrain S, Vailleau F, Balagué C, Roby D. Lesion mimic mutants: keys for deciphering cell death and defense pathways in plants? Trends Plant Sci. 2003;8:263–71.

Lowe I, Cantu D, Dubcovsky J. Durable resistance to the wheat rusts: integrating systems biology and traditional phenotype-based research methods to guide the deployment of resistance genes. Euphytica. 2011;179:69–79.

Lu M, Tang X, Zhou JM. Arabidopsis NHO1 is required for general resistance against Pseudomonas bacteria. Plant Cell. 2001;13:437–47.

Lu Y, Wang MN, Chen XM, See D, Chao SM, Jing JX. Mapping of *Yr62* and a small effect QTL for high-temperature adult-plant resistance to stripe rust in spring wheat PI 192252. Theor Appl Genet. 2014;127:1449–59.

Luan S. The CBL–CIPK network in plant calcium signaling. Cell. 2008;14:37–42.

Lunderstädt J. Effect of rust infection on hexokinase activity and carbohydrate dissimilation in primary leaves of wheat. Can J Bot. 1966;44:1345–64.

Lupton FGH, Macer RCF. Inheritance of resistance to yellow rust (*Puccinia glumarum* Erikss. & Henn.) in seven varieties of wheat. Trans Br Mycol Soc. 1962;45:21–45.

Ma Q, Shang HS. Ultrastructural analysis of the interaction between *Puccinia striiformis* f. sp. *tritici* and wheat after thermal induction of resistance. J Plant Pathol. 2004;86:19–26.

Ma JB, Huang XL, Wang XJ, Chen XM, Qu ZP, Huang LL, Kang ZS. Identification of expressed genes during compatible interaction between stripe rust (*Puccinia striiformis*) and wheat using a cDNA library. BMC Genomics. 2009;10:586.

Maccaferri M, Zhang J, Bulli P, Abate Z, Chao S, Cantu D, Bossolini E, Chen X, Pumphrey M, Dubcovsky J. A genome-wide association study of resistance to stripe rust (*Puccinia striiformis* f. sp. *tritici*) in a worldwide collection of hexa-ploid spring wheat (*Triticum aestivum* L.). G3: Genes Genom Genet. 2015;5:449–65.

MacDonald PW, Strobel GA. Adenosine diphosphate-glucose pyrophosphorylase control of starch accumulation in rust-infected wheat leaves. Plant Physiol. 1970;46:126–35.

Mackrill JJ. Protein-protein interactions in intracellular Ca^{2+} -release channel function. Biochem J. 1999;337:345–61.

Magyarosy AC, Malkin R. Effect of powdery mildew infection of sugar beet on the content of electron carriers in chloroplasts. Physiol Plant Pathol. 1978;13:183–8.

Magyarosy AC, Schurmann P, Buchanan BB. Effect of powdery mildew infection on photosynthesis by leaves and chloroplasts of sugar beets. Plant Physiol. 1976;57:486–9.

Mains EB. Studies concerning heteroecious rusts. Mycologica. 1933;25:407–17.

Malca I, Murray HC, Zscheile FB. Organic conistituents in conidia of *Eripsiplie gramninis tritici*. Phytopathology. 1962;52:891–3.

Mallard S, Nègre S, Pouya S, Gaudet D, Lu ZX, Dedryver F. Adult plant resistance-related gene expression in 'Camp Remy' wheat inoculated with *Puccinia striiformis*. Mol Plant Pathol. 2008;9:213–25.

Manners JM. The host–haustorium interface in powdery mildews. Aust J Plant Physiol. 1989;16:45–52.

Mao C, Xu R, Szulc ZM, Bielawska A, Galadari SH, Obeid LM. Cloning and characterization of a novel human alkaline ceramidase. J Biol Chem. 2001;276:26577–88.

Marasas CN, Smale M, Singh RP. The economic impact of productivity maintenance research: breeding for leaf rust resistance in modern wheat. Agric Econ. 2003;29:253.

Mares DJ. A light and electron microscope study of the interaction of yellow rust (*Puccinia striiformis*) with a susceptible wheat cultivar. Ann Bot. 1979;43:183–9.

Mares DJ, Cousen S. The interaction of yellow rust (*Puccinia striiformis*) with winter wheat cultivars showing adult plant resistance: macroscopic and microscopic events associated with the resistant reaction. Physiol Plant Pathol. 1977;10:257–74.

Martinez-Garcia JF, Moyano E, Alcocer MJ, Martin C. Two bZIP proteins from Antirrhinum flowers preferentially bind a hybrid C-box/G-box motif and help to define a new sub-family of bZIP transcription factors. Plant J. 1998;13:489–505.

Maruta T, Noshi M, Tanouchi A, Tamoi M, Yabuta Y, Yoshimura K, Ishikawa T, Shigeoka S. H_2O_2-triggered retrograde signaling from chloroplasts to nucleus plays specific role in response to stress. J Biol Chem. 2012;287:11717–29.

Mashaal SF, Bama B, Király Z. Effect of photosynthesis inhibitors on wheat stem rust development. Acta Physiol Acad Sci Hung. 1981;16:45–8.

Matsumura H, Nirasawa S, Kiba A, Urasaki N, Saitoh H, Ito M, Kawai-Yamada M, Uchimiya H, Terauchi R. Overexpressionof Bax inhibitor suppresses the fungal elicitor-induced cell death in rice (*Oryza sativa* L.) cells. Plant J. 2003;33:425–34.

McIntosh RA, Wellings CR, Park RF. Wheat rusts: an atlas of resistance genes. Melbourne: CSIRO Publications; 1995.

McIntosh RA, Yamazaki Y, Dubcovsky J, Rogers WJ, Morris CF, Appels R, Xia XC. Catalogue of gene symbols for wheat. 2013. http://wheat.pw.usda.gov/GG2/Triticum/wgc/2013/GeneCatalogueIntroduction.pdf. Accessed 23 Dec 2016.

Melichar JPE, Berry S, Newell C, MacCormack R, Boyd LA. QTL identification and microphenotype characterization of the developmentally regulated yellow rust resistance. Theor Appl Genet. 2008;117:391–9.

Mellersh DG, Heath MC. An investigation into the involvement of defense signaling pathways in components of the nonhost resistance of *Arabidopsis thaliana* to rust fungi also reveals a model system for studying rust fungal compatibility. Mol Plant-Microbe Interact. 2003;16:398–404.

Mendgen K. Microautoradiographic studies on hostparasite interactions. II. The exchange of ^{3}H-lysine between *Uromyces phaseoli* and *Phaseolus vulgaris*. Arch Microbiol. 1979;123:129–35.

Mendgen K. Nutrient uptake in rust fungi. Phytopathology. 1981;71:983–9.

Mendgen K, Hahn M. Plant infection and the establishment of fungal biotrophy. Trends Plant Sci. 2002;7:352–6.

Mendgen K, Nass P. The activity of powdery-mildew haustoria after feeding the host cell with different sugars, as measured with a potentiometric cyanine dye. Planta. 1988;174:283–8.

Metzger RJ, Silbaugh BA. Inheritance of resistance to stripe rust and its association with brown glume color in *Triticum aestivum* L., "P.I. 178383". Crop Sci. 1970;10:567–8.

Michelmore RW, Christopoulou M, Caldwell KS. Impacts of resistance gene genetics, function, and evolution on a durable future. Annu Rev Phytopathol. 2013;51:291–319.

Miklis M, Consonni C, Bhat RA, Lipka V, Schulze-Lefert P, Panstruga R. Barley *MLO* modulates actin-dependent and actin- Independent antifungal defense pathways at the cell periphery. Plant Physiol. 2007;144:1132–43.

Mirocha CJ, Zaki AI. Fluctuation in amount of starch in host plants invaded by rust and mildew fungi. Phytopathology. 1966;56:1220–2.

Mitchell DT. Carbon dioxide exchange by infected first leaf tissues susceptible to wheat stem rust. Trans Br Mycol Soc. 1979;72:63–8.

Mitchell DT, Fung AK, Lewis DH. Changes in the ethanol-soluble carbohydrate composition and acid invertase in infected first leaf tissues susceptible to crown rust of oat and wheat stem rust. New Phytol. 1978;80:381–92.

Mittapalli O, Shukle RH. Molecular characterization and responsive expression of a defender against apoptotic cell death homologue from the Hessian fly, Mayetiola destructor. Comp Biochem Physiol B: Biochem Mol Biol. 2008;149:517–23.

Moldenhauer J, Moerschbacher BM, Van der Westhuizen AJ. Histological investigation of stripe rust (*Puccinia striiformis* f. sp *tritici*) development in resistant and susceptible wheat cultivars. Plant Pathol. 2006;55:469–74.

Moldenhauer J, Pretorius ZA, Moerschbacher BM, Prins R, van der Westhuizen AJ. Histopathology and PR-protein markers provide insight into adult plant resistance to stripe rust of wheat. Mol Plant Pathol. 2008;9:137–45.

Moncrief ND, Kretsinger RH, Goodman M. Evolution of EF-hand calcium-modulated proteins I. Relationships based on amino acid sequences. J Mol Evol. 1990;30:522–62.

Montalbini P, Buchanan BB. Effect of a rust infection on photophosphorylation by isolated chloroplasts. Physiol Plant Pathol. 1974;4:191–6.

Montalbini P, Elstner EF. Ethylene evolution by rust-infected, detached bean (*Phaseolus vulgaris* L.) leaves susceptible and hypersensitive to *Uromyces phaseoli* (Pers.) Wint. Planta. 1977;135:301–6.

Morel JB, Dangl JL. The hypersensitive response and the induction of cell death in plants. Cell Death Differ. 1997;4:671–83.

Mueth NA, Ramachandran SR, Hulbert SH. Small RNAs from the wheat stripe rust fungus (*Puccinia striiformis* f. sp. *tritici*). BMC Genomics. 2015;16:718.

Mysore KS, Ryu CM. Nonhost resistance: how much do we know? Trends Plant Sci. 2004;9:97–104.
Nagarajan S, Nayar SK, Bahadur P. Race 13 (67S8) of *Puccinia striiformis* virulent on *Triticum spelta* var. *album* in India. Plant Dis. 1986;70:173.
Nahm MY, Kim SW, Yun D, Lee SY, Cho MJ, et al. Molecular and biochemical analyses of OsRab7, a rice Rab7 homolog. Plant Cell Physiol. 2003;44:1341–9.
Nakashima K, Yamaguchi-Shinozaki K. Regulons involved in osmotic stress-responsive and cold stress-responsive gene expression in plants. Physiol Plant. 2006;126:62–71.
Nalefski EA, Falke JJ. The C2 domain calcium-binding motif: structural and functional diversity. Protein Sci. 1996;5:2375–90.
Nandi A, Welti R, Shah J. The Arabidopsis thaliana dihydrox-yacetone phosphate reductase gene suppressor of fatty acid desaturase deficiency1 is required for glycerolipid metabolism and for the activation of systemic acquired resistance. Plant Cell. 2004;16:465–77.
Navarro L, Dunoyer P, Jay F, Arnold B, Dharmasiri N, Estelle M, Voinnet O, Jones JDG. A plant miRNA contributes to antibacterial resistance by repressing auxin signaling. Science. 2006;312:436–9.
Navazio L, Sponga L, Dainese P, Fitchette-Laine AC, Faye L, Baldan B, Mariani P. The calcium binding protein calreticulin in pollen of *Liriodendron tulipifera* L. Plant Sci. 1998;131:35–42.
Ng CK, Carr K, McAinsh MR, Powell B, Hetherington AM. Drought-induced guard cell signal transduction involves sphingosine-1-phosphate. Nature. 2001;410:596–9.
Niks RE. How specific is non-hypersensitive host and nonhost resistance of barley to rust and mildew fungi? J Integr Agric. 2014;13:244–54.
Niu XP, Renshaw-Gegg L, Miller L, Guiltinan MJ. Bipartite determinants of DNA-binding specificity of plant basic leucine zipper proteins. Plant Mol Biol. 1999;41:1–13.
Nühse TS, Bottrill AR, Jones AME, Peck SC. Quantitative phosphoproteomic analysis of plasma membrane proteins reveals regulatory mechanisms of plant innate immune responses. Plant J. 2007;51:931–40.
O'Donnell KA, Wentzel EA, Zeller KI, Dang CV, Mendell JT. c-Myc-regulated microRNAs modulate *E2F1* expression. Nature. 2005;435:839–43.
Ohme-Takagi M, Suzuk K, Shinshi H. Regulation of ethylene-induced transcription of defense genes. Plant Cell Physiol. 2000;41:1187–92.
Olsen AN, Ernst HA, Leggio LL, Skriver K. NAC transcription factors: structurally distinct, functionally diverse. Trends Plant Sci. 2005;10:79–87.
Ouelhadj A, Kuschk P, Humbeck K. Heavy metal stress and leaf senescence induce the barley gene *HvC2d1* encoding a calcium-dependent novel C2 domain-like protein. New Phytol. 2005;170:261–73.
Owens GK, Kumar MS, Wamhoff BR. Molecular regulation of vascular smooth muscle cell differentiation in development and disease. Physiol Rev. 2004;84:767–801.
Owera SAP, Farrar JF, Whitbread R. Growth and photosynthesis in barley infected with brown rust. Physiol Plant Pathol. 1981;18:79–90.
Padmanabhan C, Zhang X, Jin H. Host small RNAs are big contributors to plant innate immunity. Curr Opin Plant Biol. 2009;12:465–72.
Palatnik JF, Allen E, Wu X, Schommer C, Schwab R, Carrington JC, Weigel D. Control of leaf morphogenesis by microRNAs. Nature. 2003;425:257–63.
Panthee DR, Yuan JS, Wright DL, Marois JJ, Mailhot D, Stewart Jr CN. Gene expression analysis in soybean in response to the causal agent of Asian soybean rust (*Phakopsora pachyrhizi Sydow*) in an early growth stage. Funct Integr Genomics. 2007;7:291–301.
Paz-Ares J, Ghosal D, Wienand U, Peterson PA, Saedler H. The regulatory c1 locus of *Zea mays* encodes a protein with homology to myb proto-oncogene products and with structural similarities to transcriptional activators. EMBO J. 1987;6:3553–8.
Pegg GF. Endogenous auxins in healthy and diseased plants. Encycl Plant Physiol, New Ser. 1976;4:560–81.
Perfect SE, Green JR. Infection structures of biotrophic and hemibiotrophic fungal plant pathogens. Mol Plant Pathol. 2001;2:101–8.

Periyannan S, Moore J, Ayliffe M, Bansal U, Wang X, Huang L, Deal K, Luo M, Kong X, Bariana H, et al. The gene *Sr33*, an ortholog of barley *Mla* genes, encodes resistance to wheat stem rust race Ug99. Science. 2013;341:786–8.

Pfeiffer E, Jäger K, Reisener HJ. Untersuchungen über Stoffwechselbeziehungen zwischen Parasit und Wirt am Beispiel von *Puccinia graminis* var. *tritici* auf Weizcn. II. Aufnahme von Hexosen aus dem Wirtsgewebe. Planta. 1969;85:194–201.

Porter K, Shimono M, Tian M, Day B. Arabidopsis actin- depolymerizing factor-4 links pathogen perception, defense activation and transcription to cytoskeletal dynamics. PLoS Pathog. 2012;8:e1003006.

Pozo-Guisado E, Merino JM, Mulero-Navarro S, Lorenzo-Benayas MJ, Centeno F, Alvarez-Barrientos A, Fernandez-Salguero PM. Resveratrol-induced apoptosis in MCF-7 human breast cancer cells involves a caspase-independent mechanism with downregulationof *Bcl-2* and *NF-kB*. Int J Cancer. 2005;115:74–84.

Preeni T, Ayres PG. Transpiration and the water relations of faba bean (*Vicia faba*) infected by rust (*Uromyces viciae-fabae*). New Phytol. 1986;102:385–95.

Prins R, Ramburan VP, Pretorius ZA, Boyd LA, Boshoff WHP, Smith PH, Louw JH. Development of a doubled haploid mapping population and linkage map for the bread wheat cross Kariega × Avocet S. South Afr J Plant Soil. 2005;22:1–8.

Qi Y, Tsuda K, Nguyen LV, Wang X, Lin J, Murphy AS, Glazebrook J, Thordal-Christensen H, Katagiri F. Physical association of Arabidopsis hypersensitive induced reaction proteins (HIRs) with the immune receptor RPS2. J Biol Chem. 2011;286:31297–307.

Quick WA, Shaw M. The physiology of host-parasite relations. XIV. The effect of rust infection on the nucleic acid content of wheat leaves. Can J Bot. 1964;42:1531–40.

Raeder JM, Bever WM. Spore germination of *Puccinia glumarum* with notes on related species. Phytopathology. 1931;21:76–789.

Raffaele S, Rivas S, Roby D. An essential role for salicylic acid in AtMYB30-mediated control of the hypersensitive cell death program in Arabidopsis. FEBS Lett. 2006;580:3498–504.

Raffaele S, Vailleau F, Leger A, Joubes J, Mierch O, Huard C, Blee E, Mongrand S, Domergue F, Roby B. A MYB transcription factor regulates very long chain fatty acid biosynthesis for activation of the hypersensitive cell death response in Arabidopsis. Plant Cell. 2008;20:752–67.

Rahman M. Biochemical analysis of type II Metacaspase (mcII-Pa). Masters thesis. Swedish University of Agricultural Sciences, Uppsala, Sweden. 2010

Ramburan VP, Pretorius ZA, Louw JH, Boyd LA, Smith PH, Boshoff WHP, Prins R. A genetic analysis of adult plant resistance to stripe rust in the wheat cultivar Kariega. Theor Appl Genet. 2004;108:1426–33.

Ranf S, Eschen-Lippold L, Pecher P, Lee J, Scheel D. Interplay between calcium signalling and early signalling elements during defence responses to microbe- or damage-associated molecular patterns. Plant J. 2011;68:100–13.

Ravane S, Gakiere B, Job D, Douce R. The specific features of methionine biosynthesis and metabolism in plants. Proc Natl Acad Sci U S A. 1998;95:7805–12.

Reddy VS, Reddy AS. Proteomics of calcium-signaling components in plants. Phytochemistry. 2004;65:1745–76.

Reiter BG, Yudkin LY, Yudnika NB. Submicroscopic structure of *Puccinia recondite* f. sp. *tritici* and leaf cells of *Triticum aestivum*. Mikol Fitopatol. 1976;10:257–60.

Ren T, Qu F, Morris TJ. HRT gene function requires interaction between a NAC protein and viral capsid protein to confer resistance to turnip crinkle virus. Plant Cell. 2000;12:1917–26.

Ren RS, Wang MN, Chen XM, Zhang ZJ. Characterization and molecular mapping of *Yr52* for high-temperature adult-plant resistance to stripe rust in spring wheat germplasm PI 183527. Theor Appl Genet. 2012;125:847–57.

Risk JM, Selter LL, Krattinger SG, Viccars LA, Richardson TM, Buesing G, Herren G, Lagudah ES, Keller B. Functional variability of the *Lr34* durable resistance gene in transgenic wheat. Plant Biotechnol J. 2012;10:477–87.

Robert C, Bancal MO, Ney B, Lannou C. Wheat leaf photosynthesis loss due to leaf rust, with respect to lesion development and leaf nitrogen status. New Phytol. 2005;165:227–41.

Roberts EA, Chua J, Kyei GB, Deretic V. Higher order Rab programming in phagolysosome biogenesis. J Cell Biol. 2006;174:923–9.

Roelfs AP. Wheat and rye stem rust. In: Roelfs AP, Bushnell W, editors. The cereal rusts, vol. II. Orlando: Academic Press; 1985. p. 3–27.

Rohringer R. Untersuchungen zur Biochemie von Weizenkeimpflanzen nach Infektion mit *Puccinia graminis tritici,* Erikss und Henn, phR 126 A. Phytopathol Z. 1957;29:45–64.

Ron M, Avni A. The receptor for the fungal elicitor ethylene inducing xylanase is a member of a resistance-like gene family in tomato. Plant Cell. 2004;16:1604–15.

Rubiales D, Niks RE. Characterization of *Lr34*, a major gene conferring nonhypersensitive resistance to wheat leaf rust. Plant Dis. 1995;79:1208–12.

Rubio-Somoza I, Weigel D. MicroRNA networks and developmental plasticity in plants. Trends Plant Sci. 2011;16:258–64.

Ruf IK, Rhyne PW, Yang H, Borza C, Hutt-Fletcher LM, Cleveland JL, Sample JT. Epstein-Barr virus regulates c-MYC, apoptosis and tumorigenicity in *Burkitt lymphoma*. Mol Cell Biol. 1999;19:1651–60.

Rushton PJ, Somssich IE. Transcriptional control of plant genes responsive to pathogens. Curr Opin Plant Biol. 1998;1:311–5.

Ryerson E, Li A, Young JP, Heath MC. Changes in abscisic acid levels in bean leaves during the initial stages of host and nonhost reactions to rust fungi. Physiol Mol Plant Pathol. 1993;43:265–73.

Rzewuski G, Cornell KA, Rooney L, Burstenbinder K, Wirtz M, Hell R, Sauter M. *OsMTN* encodes a 50-methylthioadenosine nucleosidase that is up-regulated during submergence-induced ethylene synthesis in rice (*Oryza sativa* L.). J Exp Bot. 2007;58:1505–14.

Saintenac C, Zhang W, Salcedo A, Rouse MN, Trick HN, Akhunov E, Dubcovsky J. Identification of wheat gene *Sr35* that confers resistance to Ug99 stem rust race group. Science. 2013;341:783–6.

Samborski DJ, Rohringer R, Person C. Effect of rust-inhibiting compounds on the metabolism of wheat leaves. Can J Bot. 1961;39:1019–27.

Sanders D, Brownlee C, Harper JF. Communicating with calcium. Plant Cell. 1999;11:691–706.

Sauter M, Lorbiecke R, Ouyang B, Pochapsky TC, Rzewuski G. The immediate-early ethylene response gene *OsARD1* encodes an acireductone dioxygenase involved in recycling of the ethylene precursor S-adenosylmethionine. Plant J. 2005;44:718–29.

Schapire AL, Voigt B, Jasik J, Rosado A, Lopez-Cobollo R, Menzel D, Salinas J, Mancuso S, Valpuesta V, Baluska F, Botella MA. Arabidopsis synaptotagmin 1 is required for the maintenance of plasma membrane integrity and cell viability. Plant Cell. 2008;20:3374–88.

Scianimanico S, Desrosiers M, Dermine JF, Méresse S, Descoteaux A, et al. Impaired recruitment of the small GTPase rab7 correlates with the inhibition of phagosome maturation by *Leishmania donovani* promastigotes. Cell Microbiol. 1999;1:19–32.

Shafiei R, Hang C, Kang J-G, Loake GJ. Identification of loci controlling nonhost disease resistance in Arabidopsis against the leaf rust pathogen *Puccinia triticina*. Mol Plant Pathol. 2007;8:773–84.

Shang HS, Zhang H, Li ZQ. RNA synthesis in primary leaves of stripe rusted wheat. Plant Physiol Commun. 1994;30:266–8.

Sharma YK, Davis KR. The effects of ozone on antioxidant responses in plants. Free Radic Biol Med. 1997;23:480–8.

Sharma P, Dubey RS. Drought induces oxidative stress and enhances the activities of antioxidant enzymes in growing rice seedlings. Plant Growth Regul. 2005;46:209–21.

Sharma-Poudyal D, Chen XM, Wan AM, Zhan GM, Kang ZS, Cao SQ, Jin SL, Morgounov A, Akin B, Mert Z, Shah SJA, Bux H, Ashraf M, Sharma RC, Madariaga R, Puri KD, Wellings C, Xi KQ, Wanyera R, Manninger K, Ganzález MI, Koyda M, Sanin S, Patzek LJ. Virulence characterization of international collections of the wheat stripe rust pathogen, *Puccinia striiformis* f. sp. *tritici*. Plant Dis. 2013;97:379–86.

Shaw M. The physiology and host–parasite relations of the rusts. Annu Rev Phytopathol. 1963;1:259–94.

Shaw M, Colotelo N. The physiology of host–parasite relations. VII. The effect of stem rust on the nitrogen and amino acids in wheat leaves. Can J Bot. 1961;39:1351–72.

Shaw M, Hawkins AR. The physiology of host–parasite relations. V. A preliminary examination of the level of free endogenous indoleacetic acid in rusted and mildewed cereal leaves and their ability to decarboxylate exogenously supplied radioactive indoleacetic acid. Can J Bot. 1958;36:1–16.

Shaw M, Manocha MS. Fine structure in detached, senescing wheat leaves. Can J Bot. 1965a;43:747–55.

Shaw M, Manocha MS. The physiology of host–parasite relations. XV. Fine structure in rust-infected wheat leaves. Can J Bot. 1965b;43:1285–92.

Shaw M, Bhattacharya PK, Quick WA. Chloroph, protein and nucleicacid levels in detached, senescing wheat leaves. Can J Bot. 1958;43:739–46.

Shen X, Li HY, Jia QZ, Feng HQ, Li MQ, Liang HG. Influence of wheat stripe rust infection on photosynthetic function and expression protein D1 of wheat leaves. Acta Ecol Sin. 2008;28:669–76.

Shimada C, Lipka V, O'Connell R, Okuno T, Schulze-Lefert P, Takano Y. Nonhost resistance in Arabidopsis-Colletotrichum interactions acts at the cell periphery and requires actin filament function. Mol Plant-Microbe Interact. 2006;19:270–9.

Siberil Y, Doireau P, Gantet P. Plant bZIP G-box binding factors: modular structure and activation mechanisms. FEBS J. 2001;268:5655–66.

Siebert R. Biochemische Untersuchungen zum Wirt-Parasit-Verhältnis am Beispiel von *Puccinia graminis tritici*. Phytopathol Z. 1961;40:221–44.

Sieber P, Wellmer F, Gheyselinck J, Riechmann JL, Meyerowitz EM. Redundancy and specialization among plant microRNAs: role of the MIR164 family in developmental robustness. Development. 2007;134:1051–60.

Silverman W. The development of stem rust on wheat leaves treated with some sugars and sugar alcohols. Phytopathology. 1960;50:114–9.

Singh RP. Association between gene *Lr34* for leaf rust resistance and leaf tip necrosis in wheat. Crop Sci. 1992a;32:874.

Singh RP. Genetic association of leaf rust resistance gene *Lr34* with adult plant resistance to stripe rust in bread wheat. Phytopathology. 1992b;82:835–8.

Singh K, Foley RC, Oñate-Sánchez L. Transcription factors in plant defense and stress responses. Curr Opin Plant Biol. 2002;5:430–6.

Smith RCG, Heritage AD, Stapper M, Barrs MD. Effect of stripe rust (*Puccinia striiformis* West.) and irrigation on the yield and foliage temperature of wheat. Field Crop Res. 1986;14:39–51.

Sohn J, Voegele RT, Mendgen K, Hahn M. High level activation of vitamin B1 biosynthesis genes in haustoria of the rust fungus *Uromyces fabae*. Mol Plant-Microbe Interact. 2000;13:629–36.

Sorensen CK, Justesen AF, Hovmoller MS. 3-D imaging of temporal and spatial development of *Puccinia striiformis* haustoria in wheat. Mycologia. 2012;104:1381–9.

Spanu PD. The genomics of obligate (and nonobligate) biotrophs. Annu Rev Phytopathol. 2012;50:91–109.

Spoel SH, Dong X. How do plants achieve immunity? Defence without specialized immune cells. Nat Rev Immunol. 2012;12:89–100.

Staiger CJ, Gibbon BC, Kovar DR, Zonia LE. Profilin and actin-depolymerizing factor: modulators of actin organization in plants. Trends Plant Sci. 1997;2:275–81.

Stein N, Feuillet C, Wicker T, Schlagenhauf E, Keller B. Subgenome chromosome walking in wheat: a 450-kb physical contig in *Triticum monococcum* L. spans the *Lr10* locus in hexaploid wheat (*Triticum aestivum* L.). Proc Natl Acad Sci U S A. 2000;97:13436–41.

Stoddart JL. Leaf senescence. In: Jeffcoat B (ed) Aspects and prospects of plant growth regulators. Wantage: British Plant Growth Regulatory Group, ARC Letcombe Laboratory; 1981. pp 51–63

Straib W. Untersuchungen über das vorkommen physiologischer rassen des gelbrostes (*Puccinia glumarum*) in den Jahren 1935–1936 und über die Agressivität einiger neuer Formen auf Getreide und Gräsern. Arb. Biol. Reichsanst. Land= Forstwirtsch. Berlin-Dahlem. 1937;22:91–119.

Struck C, Ernst M, Hahn M. Characterization of a developmentally regulated amino acid transporter (AAT1p) of the rust fungus *Uromyces fabae*. Mol Plant Pathol. 2002;3:23–30.

Struck C, Müller E, Martin H, Lohaus G. The *Uromyces fabae* UfAAT3 gene encodes a general amino acid permease that prefers uptake of in planta scarce amino acids. Mol Plant Pathol. 2004;5:183–9.

Stubbs RW. Stripe rust. In: Roelfs AP, Bushnell W, editors. The cereal rusts, vol. II. Orlando: Academic Press; 1985. p. 61–101.

Stulemeijer IJ, Stratmann JW, Joosten MH. Tomato mitogen activated protein kinases LeMPK1, LeMPK2, and LeMPK3are activated during the Cf-4/Avr4-induced hypersensitive response and have distinct phosphorylation specificities. Plant Physiol. 2007;144:1481–94.

Su H, Conner RL, Graf RJ, Kuzyk AD. Virulence of *Puccinia striiformis* f. sp. *tritici*, cause of stripe rust on wheat, in western Canada from 1984 to 2002. Can J Plant Pathol. 2003;25:312–9.

Sunkar R, Zhu J-K. Novel and stress-regulated microRNAs andother small RNAs from Arabidopsis. Plant Cell. 2004;16:2001–19.

Sutton PN, Henry MJ, Hall JL. Glucose, and not sucrose, is transported from wheat to wheat powdery mildew. Planta. 1999;208:426–30.

Syamananda R, Staples RC. The carbohydrate content of rusted corn leaves. Contrib Boyce Thompson Inst. 1963;22:1–8.

Szabo L. Hidden robbers: the role of fungal haustoria in parasitism of plants. Proc Natl Acad Sci. 2001;98:7654–5.

Taipale M, Jarosz DF, Lindquist S. HSP90 at the hub of protein homeostasis: emerging mechanistic insights. Nat Rev Mol Cell Biol. 2010;11:515–28.

Takezawa D. A rapid induction by elicitors of the mRNA encoding CCD-1, a 14 kDa Ca^{2+} -binding protein in wheat cultured cells. Plant Mol Biol. 2000;42:807–17.

Tanaka Y, Makishima T, Sasabe M, Ichinose Y, Shiraishi T, Nishimoto T, Yamada T. *dad-1*, a putative programmed cell death suppressor gene in rice. Plant Cell Physiol. 1997;38:379–83.

Tang CL, Wang XJ, Duan XY, Wang XD, Huang LL, Kang ZS. Functions of the *lethal leaf-spot 1* gene in wheat cell death and disease tolerance to *Puccinia striiformis*. J Exp Bot. 2013;64:2955–69.

Tang CL, Deng L, Chang D, Chen ST, Wang XJ, Kang ZS. TaADF3, an actin-depolymerizing factor, negatively modulates wheat resistance against *Puccinia striiformis*. Front Plant Sci. 2016;6:1214.

Tani T, Yamamoto H. RNA and protein synthesis and enzyme changes during infection. In: Daly JM, Uritani I, editors. Recognition and specificity in plant host–parasite interaction. Baltimore: University Park Press; 1979. p. 273–87.

Tani T, Yoshikawa M, Naito N. Effect of rust infection of oat leaves on cytoplasmic and chloroplast ribosomal ribonucleic acids. Phytopathology. 1973;63:491–4.

Thomma BP, Eggermont K, Tierens KF, Broekaert WF. Requirement of functional ethylene-insensitive 2 gene for efficient resistance of *Arabidopsis* to infection by *Botrytis cinerea*. Plant Physiol. 1999;121:1093–102.

Tian M, Chaudhry F, Ruzicka DR, Meagher RB, Staiger CJ, Day B. Arabidopsis actin-depolymerizing factor AtADF4 mediates defense signal transduction triggered by the *Pseudomonas syringae* effector AvrPphB. Plant Physiol. 2009;150:815–24.

Tian CM, Kang ZS, Li ZQ, Zhap YX. Researches on the histology and cytology of poplar leaf rust. J Northwest Forestry Univ. 2001;16:43–9.

Tieman DM, Ciardi JA, Taylor MG, Klee HJ. Members of the tomato *LeEIL* (EIN3-like) gene family are functionally redundant and regulate ethylene responses throughout plant development. Plant J. 2001;26:47–58.

Tissera P, Ayres PG. Endogenous ethylene affects the behavior of stomata in epidermis isolated from rust infected faba bean (*Vicia faba* L.). New Phytol. 1986;104:53–61.

Tranzschel W. Promežutočnye chozjaeva rzavčiny chlebov i ich der UdSSR. (The alternate hosts of cereal rust fungi and their distribution in the UdSSR). Bull Plant Prot Ser. 1934;2:4–40.

Uauy C, Brevis JC, Chen XM, Khan IA, Jackson LF, Chicaiza O, Distelfeld A, Fahima T, Dubcovsky J. High-temperature adult plant (HTAP) stripe rust resistance gene *Yr36* from *Triticum turgidum* ssp. *dicoccoides* is closely linked to the grain protein content locus *Gpc-B1*. Theor Appl Genet. 2005;112:97–105.
Umezawa T, Fujita M, Fujita Y, Yamaquchi-Shinozaki K, Shinozaki K. Engineering drought tolerance in plants: discovering and tailoring genes to unlock the future. Curr Opin Biotechnol. 2006;17:113–22.
Umnov AM, Artemenko EN, Chkanikov DI. The presence of indolyl-3-acetic acid in uredospores of wheat stem rust. Mikol Fitopatol. 1978;12:222–7.
Vailleau F, Daniel X, Tronchet M, Montillet JL, Triantaphylides C. A R2R3-MYB gene, *AtMYB30*, acts as appositive regulator of the hypersensitive cell death program in plants in response to pathogen attack. Proc Natl Acad Sci U S A. 2002;99:10179–84.
Van Loon L. Pathogenesis-related proteins. Plant Mol Biol. 1985;4:111–6.
Van Troys M, Huyck L, Leyman S, Dhaese S, Vandekerkhove J, Ampe C. Ins and outs of ADF/cofilin activity and regulation. Eur J Cell Biol. 2008;87:649–67.
Vannini C, Iriti M, Bracale M, Locatelli F, Faoro F, Croce P, Pirona R, Di Maro A, Coraggio I, Genga A. The ectopic expression of the rice *OsMyb4* gene in Arabidopsis increases tolerance to abiotic, environmental and biotic stresses. Physiol Mol Plant Pathol. 2006;69:26–42.
Vannini C, Campa M, Iriti M, Genga A, Faoro F, Carravieri S, Rotino GL, Rossoni M, Spinardi A, Bracale M. Evaluation of transgenic tomato plants ectopically expressing the rice *OsMyb4* gene. Plant Sci. 2007;173:231–9.
Ventelon-Debout M, Delalande F, Brizard JP, Diemer H, Van Dorsselaer A, Brugidou C. Proteome analysis of cultivar specific deregulations of *Oryza sativa* indica and *O. sativa* japonica cellular suspensions undergoing rice yellow mottle virus infection. Proteomics. 2004;4:216–25.
Venugopal SC, Chanda B, Vaillancourt L, Kachroo A, Kachroo P. The common metabolite glycerol-3-phosphate is a novel regulator of plant defense signaling. Plant Signal Behav. 2009;4:746–9.
Vercammen D, Belenghi B, Van De Cotte B, Beunens T, Gavigan JA, De Rycke R, Brackenier A, Inzé D, Harris JL, Van Breusegem F. Serpin1 of Arabidopsis thaliana is a suicide inhibitor for metacaspase 9. J Mol Biol. 2006;364:625–36.
Vieira OV, Bucci C, Harrison RE, Trimble WS, Lanzetti L, et al. Modulation of Rab5 and Rab7 recruitment to phagosomes by phosphatidylinositol 3-kinase. Mol Cell Biol. 2003;23:2501–14.
Voegele RT, Mendgen K. Rust haustoria: nutrient uptake and beyond. New Phytol. 2003;159:93–100.
Voegele RT, Mendgen KW. Nutrient uptake in rust fungi: how sweet is parasitic life? Euphytica. 2011;179:41–55.
Voegele RT, Struck C, Hahn M, Mendgen K. The role of haustoria in sugar supply during infection of broad bean by the rust fungus *Uromyces fabae*. Proc Natl Acad Sci U S A. 2001;98:8133–8.
Voegele RT, Wirsel S, Ulla Möll U, Lechner M, Mendgen K. Cloning and characterization of a novel invertase from the obligate biotroph *Uromyces fabae* and analysis of expression patterns of host and pathogen invertases in the course of infection. Mol Plant-Microbe Interact. 2006;19:625–34.
von Broembsen SL, Hadwiger LA. Characterization of disease resistance responses in certain gene-for-gene interactions between flax and *Melampsora lini*. Physiol Plant Pathol. 1972;2:207–15.
Wahl R, Wippel K, Goos S, Kämper J, Sauer N. A novel high-affinity sucrose transporter is required for virulence of the plant pathogen *Ustilago maydis*. PLoS Biol. 2010;8:e1000303.
Walters DR, McRoberts N, Fitt BDL. Are green islands red herrings? Significance of green islands in plant interactions with pathogens and pests. Biol Rev Camb Philos Soc. 2008;83:79–102.
Wan AM, Chen XM. Variation of telial formation in the *Puccinia striiformis* f. sp. *tritici* population. Phytopathology. 2016;S4:206–7.
Wan AM, Muleta KT, Zegeye H, Hundie B, Pumphrey MO, Chen XM. Virulence characterization of wheat stripe rust fungus *Puccinia striiformis* f. sp. *tritici* in Ethiopia and evaluation of Ethiopian wheat germplasm for resistance to races of the pathogen from Ethiopia and the United States. Plant Dis. 2017;101:73–80.

Wang BT, Shang HS. The relation between the high-temperature resistance of wheat to stripe rust and lignin synthesis. Acta Phys Sin. 1996;23:229–34.
Wang MN, Chen XM. First Report of Oregon grape (*Mahonia aquifolium*) as an alternate host for the wheat stripe rust pathogen (*Puccinia striiformis* f. sp. *tritici*) under artificial inoculation. Plant Dis. 2013;97:839.
Wang MN, Chen XM. Barberry does not function as an alternate host for *Puccinia striiformis* f. sp. *tritici* in the U. S. Pacific Northwest due to teliospore degradation and barberry phenology. Plant Dis. 2015;99:1500–6.
Wang Y, Kang ZS, Li ZQ. Study on heterokaryon distribution of *Puccinia striiformis* f. sp. *tritici* West. in the different mountain areas of Tianshui. J Triticeae Crops. 2004;24:110–3.
Wang CF, Huang LL, Buchenauer H, Han QM, Zhang HC, Kang ZS. Histochemical studies on the accumulation of reactive oxygen species (O_2^- and H_2O_2) in the incompatible and compatible interaction of wheat-*Puccinia striiformis* f. sp. *tritici*. Physiol Mol Plant Pathol. 2007;71:230–9.
Wang F, Zhong NQ, Gao P, Wang GL, Wang HY, Xia GX. SsTypA1, a chloroplast-specific TypA/BipA-type GTPase from the halophytic plant *Suaeda salsa*, plays a role in oxidative stress tolerance. Plant Cell Environ. 2008a;31:982–94.
Wang G, Ellendorff U, Kemp B, Mansfield JW, Forsyth A, Mitchell K, Bastas K, Liu CM, Woods-Tör A, Zipfel C, de Wit PJ, Jones JD, Tör M, Thomma BP. A genome-wide functional investigation into the roles of receptor-like proteins in Arabidopsis. Plant Physiol. 2008b;147:503–17.
Wang MN, Coram TE, Ling P, Boyd L, Chen XM. High-resolution genetic and physical mapping of the *Yr5* gene for resistance to stripe rust of wheat. Phytopathology. 2008c;98:S166.
Wang XJ, Tang CL, Zhang G, Li YC, Wang CF, Liu B, Qu ZP, Zhao J, Han QM, Huang LL, Chen XM, Kang ZS. cDNA-AFLP analysis reveals differential gene expression in compatible interaction of wheat challenged with *Puccinia striiformis* f. sp. *tritici*. BMC Genomics. 2009a;10:289.
Wang XY, Li Q, Niu XW, Chen HY, Xu LL, Qi CK. Characterization of a canola C2 domain gene that interacts with PG, an effector of the necrotrophic fungus *Sclerotinia sclerotiorum*. J Exp Bot. 2009b;60:2613–20.
Wang XJ, Liu W, Chen XM, Tang CL, Dong YL, Ma JB, Huang XL, Wei GR, Han QM, Huang LL, Kang ZS. Differential gene expression in incompatible interaction between wheat and stripe rust fungus revealed by cDNA-AFLP and comparison to compatible interaction. BMC Plant Biol. 2010a;10:9.
Wang XJ, Tang CL, Deng L, Cai GL, Liu XY, Liu B, Han QM, Buchenauer H, Wei GR, Han DJ, Huang LL, Kang ZS. Characterization of a pathogenesis-related thaumatin-like protein gene *TaPR5* from wheat induced by stripe rust fungus. Physiol Plant. 2010b;139:27–38.
Wang GF, Wei X, Fan R, Zhou H, Wang X, Yu C, Dong L, Dong Z, Wang X, Kang Z, Ling H, Shen QH, Wang D, Zhang X. Molecular analysis of common wheat genes encoding three types of cytosolic heat shock protein 90 (Hsp90): functional involvement of cytosolic Hsp90s in the control of wheat seedling growth and disease resistance. New Phytol. 2011a;191:418–31.
Wang MN, Wan AM, Chen XM, Evans CK. Barberry is more important as an alternate host for stem rust than for stripe rust in the U.S. Pacific Northwest. In: Oral presentations, poster abstracts, participants and program of BGRI technical workshop, June 13–16, 2011, St. Paul, MN, USA. 2011b, pp 166
Wang XJ, Tang CL, Zhang H, Xu JR, Liu B, Lv J, Han DJ, Huang LL, Kang ZS. *TaDAD2*, a negative regulator of programmed cell death, is important for the interaction between wheat and the stripe rust fungus. Mol Plant-Microbe Interact. 2011c;24:79–90.
Wang XD, Wang XJ, Feng H, Tang CL, Bai PF, Wei GR, Huang LL, Kang ZS. TaMCA4, a novel wheat metacaspase gene functions in programmed cell death induced by the fungal pathogen *Puccinia striiformis* f. sp. *tritici*. Mol Plant-Microbe Interact. 2012a;25:755–64.
Wang XJ, Tang CL, Huang XL, Li FF, Chen XM, Zhang G, Sun YF, Han DJ, Kang ZS. Wheat BAX inhibitor-1 contributes to wheat resistance to *Puccinia striiformis*. J Exp Bot. 2012b;63:4571–84.

Wang X, Richards J, Gross T, Druka A, Kleinhofs A, Steffenson B, Acevedo M, Brueggeman R. The *rpg4*-mediated resistance to wheat stem rust (*Puccinia graminis*) in barley (*Hordeum vulgare*) requires *Rpg5*, a second NBS-LRR gene, and an actin depolymerization factor. Mol Plant-Microbe Interact. 2013;26:407–18.

Wang XD, Wang XJ, Deng L, Chang HT, Dubcovsky J, Feng H, Han QM, Huang LL, Kang ZS. Wheat TaNPSN SNARE homologues are involved in vesicle-mediated resistance to stripe rust (*Puccinia striiformis* f. sp. *tritici*). J Exp Bot. 2014a;65:4807–20.

Wang B, Sun YF, Song N, Wei JP, Wang XJ, Feng H, Yin ZY, Kang ZS. MicroRNAs involving in cold, wounding and salt stresses in *Triticum aestivum* L. Plant Physiol Biochem. 2014b;80:90–6.

Wang MN, Wan AM, Chen XM. Barberry as alternate host is important for *Puccinia graminis* f. sp. *tritici* but not for *Puccinia striiformis* f. sp. *tritici* in the U.S. Pacific Northwest. Plant Dis. 2015;99:1507–16.

Wang XD, Yang BJ, Li K, Kang ZS, Cantu D, Dubcovski J. A conserved *Puccinia striiformis* protein interacts with wheat NPR1 and reduces induction of pathogenesis-related genes in response to pathogens. Mol Plant-Microbe Interact Online: http://dx.doi.org/10.1094/MPMI-10-16-0207-R. 2016a.

Wang ZY, Zhao J, Chen XM, Peng YL, Ji JJ, Zhao SL, Lv YJ, Huang LL, Kang ZS. Virulence variation of *Puccinia striiformis* f. sp. *tritici* isolates collected from *Berberis* spp. in China. Plant Dis. 2016b;100:131–8.

Watanabe N, Lam E. Two Arabidopsis metacaspases AtMCP1b and AtMCP2b are arginine/lysine-specific cysteine proteases and activate apoptosis-like cell death in yeast. J Biol Chem. 2005;280:14691–9.

Watanabe N, Lam E. Calcium-dependent activation and autolysis of Arabidopsis metacaspase 2d. J Biol Chem. 2011;286:10027.

Watanabe S, Shimada TL, Hiruma K, Takano Y. Pathogen infection trial increases the secretion of proteins localized in the endoplasmic reticulum body of Arabidopsis. Plant Physiol. 2013;163:659–64.

Webb CA, Richter TE, Collins NC, Nicolas M, Trick HN, Pryor T, Hulbert SH. Genetic and molecular characterization of the maize *rp3* rust resistance locus. Genetics. 2002;162:381–94.

Wegele H, Muller L, Buchner J. Hsp70 and Hsp90 – a relay team for protein folding. Rev Physiol Bioch P. 2004;151:1–44.

Wei T, Zhang C, Hou X, Sanfaçon H, Wang A. The SNARE protein Syp71 is essential for turnip mosaic virus infection by mediating fusion of virus-induced vesicles with chloroplasts. PLoS Pathog. 2013;9:e1003378.

Weiberg A, Wang M, Lin FM, Zhao H, Zhang Z, Kaloshian I, Huang HD, Jin H. Fungal small RNAs suppress plant immunity by hijacking host RNA interference pathways. Science. 2013;342:118–23.

Whetstone RD, Atkinson TA, Spaulding DD. Berberidaceae. In: Flora of North America, Vol. 3. Website: www.eFloras.org. 1989. Accessed 23 Dec 2016.

Whitney HS, Shaw M, Naylor JM. The physiology of host–parasite relations. XII. A cytophotometric study of the distribution of DNA and RNA in rust-infected leaves. Can J Bot. 1962;40:1533–44.

Wildermuth MC, Dewdney J, Wu G, Ausubel FM. Isochorismate synthase is required to synthesize salicylic acid for plant defence. Nature. 2001;414:562–5.

William HM, Singh RP, Huerta-Espino J, Ortiz-Islas S, Hoisington D. Molecular marker mapping of leaf rust resistance gene *Lr46* and its association with stripe rust resistance gene *Yr29* in wheat. Phytopathology. 2003;93:153–9.

Williamson CE. Ethylene, a metabolic product of diseased or injured plants. Phytopathology. 1950;40:205–8.

Woltering EJ. Death proteases come alive. Trends Plant Sci. 2004;9:469–72.

Woods AM, Gay JL. The interface between haustoria of *Puccinia poarum* (monokaryon) and *Tussilago farfara*. Physiol Mol Plant Pathol. 1987;30:167–85.

Wright RG, Lennard JH. Mitosis in *Puccinia striiformis*. Light Microsc Trans Br Mycol Soc. 1978;70:91–8.
Wrigley CW, Webster HL. The effect of stem rust infection on the soluble proteins of wheat leaves. Aust J Biol Sci. 1966;19:895–901.
Wynn WK. Photosynthetic phosphorylation by chloroplasts isolated from rust-infected oats. Phytopathology. 1963;53:1376–7.
Xia N, Zhang G, Liu XY, Deng L, Cai GL, Zhang Y, Wang XJ, Zhao J, Huang LL, Kang ZS. Characterization of a novel wheat NAC transcription factor gene involved in defense response against stripe rust pathogen infection and abiotic stresses. Mol Biol Rep. 2010a;37:3703–12.
Xia N, Zhang G, Sun YF, Zhu L, Xu LS, Chen XM, Liu B, Yu YT, Wang XJ, Huang LL, Kang ZS. TaNAC8, a novel NAC transcription factor gene in wheat, responds to stripe rust pathogen infection and abiotic stresses. Physiol Mol Plant Pathol. 2010b;74:394–402.
Xie Q, Frugis G, Colgan D, Chua NH. Arabidopsis NAC1 transduces auxin signal downstream of TIR1 to promote lateral root development. Genes Dev. 2000;14:3024–36.
Xoconostle-Cazares B, Xiang Y, Ruiz-Medrano R, Wang HL, Monzer J, Yoo BC, McFarland KC, Franceschi VR, Lucas WJ. Plant paralog to viral movement protein that potentiates transport of mRNA into the phloem. Science. 1999;283:94–8.
Xu H, Heath MC. Role of calcium in signal transduction during the hypersensitive response caused by basidiospore-derived infection of the cowpea rust fungus. Plant Cell. 1998;10:585–98.
Xu LS, Jia JG, Lv J, Liang XF, Han DJ, Huang LL, Kang ZS. Characterization of the expression profile of a wheat aci-reductone-dioxygenase-like gene in response to stripe rust pathogen infection and abiotic stresses. Plant Physiol Biochem. 2010;48:461–8.
Yahiaoui N, Srichumpa P, Dudler R, Keller B. Genome analysis at different ploidy levels allows cloning of the powdery mildew resistance gene *Pm3b* from hexaploid wheat. Plant J. 2004;37:528–38.
Yamaguchi-Shinozaki K, Shinozaki K. Organization of cis-acting regulatory elements in osmotic- and cold-stress responsive promoters. Trends Plant Sci. 2005;10:88–94.
Yamaguchi-Shinozaki K, Shinozaki K. Transcriptional regulatory networks in cellular responses and tolerance to dehydration and cold stresses. Annu Rev Plant Biol. 2006;57:781–803.
Yamazaki T, Takat N, Uemura M, Kawamura Y. Arabidopsis synaptotagmin SYT1, a type I signal-anchor protein, requires tandem C2 domains for delivery to the plasma membrane. J Biol Chem. 2010;285:23165–76.
Yang CY, Shang HS, Li ZQ. Change of nitrogen in wheat leaves infected by stripe rust fungi. Acta Agric Beoreali-occidentalis Sin. 1991;S1:18–22.
Yang H, Li Y, Hua J. The C2 domain protein BAP1 negatively regulates defense responses in Arabidopsis. Plant J. 2006a;48:238–48.
Yang S, Yang H, Grisa WP, Sanchatjate S, Fink GR, Sun Q, Hua J. The BON/CPN gene family represses cell death and promotes cell growth in Arabidopsis. Plant J. 2006b;45:166–79.
Yang H, Yang S, Li Y, Hua J. The Arabidopsis *BAP1* and *BAP2* genes are general inhibitors of programmed cell death. Plant Physiol. 2007;145:135–46.
Yang WQ, Lai Y, Li MN, Xu WY, Xue YB. A novel C2-domain phospholipid-binding protein, OsPBP1, is required for pollen fertility in rice. Mol Plant. 2008;1:770–85.
Yang YH, Zhao J, Liu P, Xing HJ, Li CC, Wei GR, Kang ZS. Glycerol-3-phosphate metabolism in wheat contributes to systemic acquired resistance against *Puccinia striiformis* f. sp. *tritici*. PLoS One. 2013;8:e81756.
Yang YH, Zhao J, Xing HJ, Wang JY, Zhou K, Zhan GM, et al. Different non-host resistance responses of two rice subspecies, japonica and indica, to *Puccinia striiformis* f. sp *tritici*. Plant Cell Rep. 2014;33:423–33.
Yang YH, Yu Y, Bi CW, Kang ZS. Quantitative proteomics reveals the defense response of wheat against *Puccinia striiformis* f. sp. *tritici*. Sci Rep. 2016;6:34261.
Yao Y, Guo G, Ni Z, Sunkar R, Du J, Zhu JK, Sun Q. Cloning and characterization of microRNAs from wheat (*Triticum aestivum* L.). Genome Biol. 2007;8:R96.

Yao JN, Zhang HC, Zhao J, Han QM, Cheng YL, Huang LL, Kang ZS. Histological and ultrastructural observation of teliospore formation in *Puccinia striiformis* f. sp. *tritici*. Mycosystema. 2012;31:560–6.

Yin CT, Chen XM, Wang XJ, Han QM, Kang ZS, Hulbert S. Generation and analysis of expression sequence tags from haustoria of the wheat stripe rust fungus *Puccinia striiformis* f. sp. *tritici*. BMC Genomics. 2009;10:626.

Yin CT, Jurgenson JE, Hulbert SH. Development of a host-induced RNAi system in the wheat stripe rust fungus *Puccinia striiformis* f. sp. *tritici*. Mol Plant-Microbe Interact. 2011;24:554–61.

Yin CT, Park JJ, Gang DR, Hulbert SH. Characterization of traptophan 2-monooxygenase gene from *Puccinia graminis* f. sp. *tritici* involved in auxin biosynthesis and rust pathoenicity. Mol Plant-Microbe Interact. 2014;27:227–35.

Yin SN, Wang CF, Jiao M, Li F, Han QM, Huang LL, Zhang HC, Kang ZS. Subcellular localization of calcium in the incompatible and compatible interactions of wheat and *Puccinia striiformis* f. sp. *tritici*. Protoplasma. 2015;252:103–16.

Yin SN, Gao ZJ, Wang CF, Huang LL, Kang ZS, Zhang HC. Nitric oxide and reactive oxygen species coordinately regulate the germination of *Puccinia striiformis* f. sp. *tritici* urediniospores. Front Microbiol. 2016;7:178.

Ying JS. Berberidaceae. In: The Editorial Committee of Flora of China (ed) Flora of China, Vol. 29. Beijing: Science Press; 2001. pp 54–249.

Yu YH, Xia Xin Li, Yin WL, Zhang HC. Comparative genomics analysis of CIPK gene family in Arabidopsis and Populus. Plant Growth Regul. 2007;52:101–10.

Yu XM, Yu XD, Qu ZP, Huang XJ, Guo J, Han QM, Zhao J, Huang LL, Kang ZS. Cloning of a putative hypersensitive induced reaction gene from wheat infected by stripe rust fungus. Gene. 2008;407:193–8.

Yu X, Wang X, Wang C, Chen X, Qu Z, Yu X, Han Q, Zhao J, Guo J, Huang L, Kang Z. Wheat defense genes in fungal (*Puccinia striiformis*) infection. Funct Integr Genomics. 2010;10:227–39.

Yu XM, Wang XJ, Huang XJ, Buchenauer H, Han QM, Guo J, Zhao J, Qu ZP, Huang LL, Kang ZS. Cloning and characterization of a wheat neutral ceramidase gene *Ta-CDase*. Mol Biol Rep. 2011;38:3447–54.

Zenser N, Ellsmore A, Leasure C, Callis J. Auxin modulates the degradation rate of Aux/IAA proteins. Proc Natl Acad Sci U S A. 2001;98:11795–800.

Zhang H, Shang HS, Li ZQ. Changes in polyribosome and protein synthesis of wheat leaves in the early stage of stripe-rust infection. Acta Phys Sin. 1994;20:339–45.

Zhang HC, Han QM, Wang CF, Huang LL, Zhang QQ, Kang ZS. Histology and ultrastructure of resistant mechanism of a new wheat material Yilipu to *Puccinia striiformis*. Acta Phys Sin. 2008a;38:153–64.

Zhang Y, Qu Z, Zheng W, Liu B, Wang X, Xue X, et al. Stage-specific gene expression during urediniospore germination in *Puccinia striiformis* f. sp. *tritici*. BMC Genomics. 2008b;9:203.

Zhang Y, Zhang G, Xia N, Wang XJ, Huang LL, Kang ZS. Cloning and characterization of a bZIP transcription factor gene in wheat and its expression in response to stripe rust pathogen infection and abiotic stresses. Physiol Mol Plant Pathol. 2008c;73:88–94.

Zhang G, Dong YL, Zhang Y, Li YM, Wang XJ, Han QM, Guo J, Huang LL, Kang ZS. Cloning and characterization of a novel hypersensitive induced-reaction gene from wheat infected by stripe rust pathogen. J Phytopathol. 2009;157:722–8.

Zhang G, Li YM, Zhang Y, Dong YL, Wang XJ, Wei GR, Huang LL, Kang ZS. Cloning and characterization of a pathogenesis related protein gene *TaPR10* from wheat induced by stripe rust pathogen. Agric Sci China. 2010;9:549–56. (in Chinese)

Zhang LX, Li ZF, Quan RD, Li GJ, Wang RG, Huang RF. An AP2 domain-containing gene, ESE1, targeted by the ethylene signaling component EIN3 is important for the salt response in Arabidopsis. Plant Physiol. 2011a;157:854–65.

Zhang W, Gao S, Zhou X, Chellappan P, Chen Z, Zhang X, Fromuth N, Coutino G, Coffey M, Jin H. Bacteria-responsive microRNAs regulate plant innate immunity by modulating plant hormone networks. Plant Mol Biol. 2011b;75:93–105.

Zhang X, Zhao H, Gao S, Wang WC, Katiyar-Agarwal S, Huang HD, Raikhel N. Arabidopsis Argonaute 2 regulates innate immunity via miRNA393*-mediated silencing of a Golgi-localized SNARE gene, MEMB12. Mol Cell. 2011c;42:356–66.

Zhang HC, Wang CF, Chen YL, Chen XM, Han QM, Huang LL, Wei GR, Kang ZS. Histological and cytological characterization of adult plant resistance to wheat stripe rust. Plant Cell Rep. 2012;31:2121–37.

Zhang G, Sun YF, Li YM, Dong YL, Huang XL, Yu YT, Wang JM, Wang XM, Wang XJ, Kang ZS. Characterization of a wheat C2 domain protein encoding gene regulated by stripe rust and abiotic stresses. Biol Plant. 2013;57:701–10.

Zhao J, Wang L, Wang ZY, Chen XM, Zhang HC, Yao JN, Zhan GM, Chen W, Huang LL, Kang ZS. Identification of eighteen *Berberis* species as alternate hosts of *Puccinia striiformis* f. sp. *tritici* and virulence variation in the pathogen isolates from natural infection of barberry plants in China. Phytopathology. 2013;103:927–34.

Zhao J, Yang YH, Kang ZS. Proteomic analysis of rice nonhost resistance to *Puccinia striiformis* f. sp. *tritici* using two-dimensional electrophoresis. Int J Mol Sci. 2014;15:21644–59.

Zhao J, Zhao SL, Peng YL, Qin JF, Huang LL, Kang ZS. Investigation on geographic distribution and identification of six *Berberis* spp. serving as alternate host for *P. striiformis* f. sp. *tritici* in Linzhi, Tibet. Acta Phys Sin. 2016;46:103–11.

Zheng W, Huang L, Huang J, Wang X, Chen X, Zhao J, Guo J, Zhuang H, Qiu C, Liu J, Liu H, Huang X, Pei G, Zhan G, Tang C, Cheng Y, Liu M, Zhang J, Zhao Z, Zhang S, Han Q, Han D, Zhang H, Zhao J, Gao X, Wang J, Ni P, Dong W, Yang L, Yang H, Xu JR, Zhang G, Kang Z. High genome heterozygosity and endemic genetic recombination in the wheat stripe rust fungus. Nat Commun. 2013;4:2673.

Zhou H, Li S, Deng Z, Wang X, Chen T, Zhang J, Chen S, Ling H, Zhang A, Wang D, Zhang X. Molecular analysis of three new receptor-like kinase genes from hexaploid wheat and evidence for their participation in the wheat hypersensitive response to stripe rust fungus infection. Plant J. 2007;52:420–34.

Zhou L, Cheung MY, Li MW, Fu Y, Sun Z, Sun SSM, Lam HM. Rice Hypersensitive Induced Reaction Protein 1(OsHIR1) associates with plasma membrane and triggers hypersensitive cell death. BMC Plant Biol. 2010;10:290.

Zhou XL, Wang MN, Chen XM, Lu Y, Kang ZS, Jing JX. Identification of *Yr59* conferring high-temperature adult-plant resistance to stripe rust in wheat germplasm PI 178759. Theor Appl Genet. 2014;127:935–45.

Zuo H, Wang J, Hao CZ, Zhang B, Ma Q. Histochemical response of nonhost resistance in pepper to the stripe rust fungus (*Puccinia striiformis* f. sp *tritici*). J Plant Pathol. 2013;95:275–83.

Chapter 4
Stripe Rust Epidemiology

Xianming Chen

Abstract Stripe rust, caused by *Puccinia striiformis*, is a polycyclic disease and its epidemic is mainly dependent upon the disease development rate. Although the pathogen has a heteroecious macrocyclic lifecycle that consists of five spore stages, it almost completely reproduces asexually on its primary hosts of cereal crops and auxiliary hosts of wild grasses. The primary inoculum to cause epidemics on cereal crops is mainly from cereal crops, volunteer plants and grasses. Urediniospores can be disseminated by wind for long distance, and also can be carried on clothes and shoes for unintended introduction. The fungus can survive summer and/or winter as mycelium in host tissue for months and/or as viable urediniospores in the air or host surface for different length of time in different regions depending upon environmental conditions. Stripe rust epidemics are affected by various crop and environmental factors, especially host factors such as cultivar susceptibility and cropping systems, and weather factors such as moisture and temperature. Various models for predicting stripe rust have been developed in different regions of the world based on weather factors or the combination of weather factors and cultivar susceptibility. These models have different degrees of usefulness in the disease management.

Keywords Alternate host • Auxiliary host • Disease cycle • Disease epidemic • Disease forecast • Inoculum • Lifecycle • Pathogen survival • Primary host • Prediction model • *Puccinia striiformis* • Spore dissemination • Stripe rust • Yellow rust

---

Mention of trade names or commercial products in this publication is solely for the purpose of providing specific information and does not imply recommendation or endorsement by the U. S. Department of Agriculture. USDA is an equal opportunity provider and employer.

X. Chen (✉)
US Department of Agriculture, Agricultural Research Service, Wheat Health, Genetics and Quality Research Unit, Pullman, WA 99164-6430, USA

Department of Plant Pathology, Washington State University, Pullman, WA 99164-6430, USA
e-mail: xianming@wsu.edu

X. Chen, Z. Kang (eds.), *Stripe Rust*, DOI 10.1007/978-94-024-1111-9_4

For managing a disease, it is essential to understand how the disease develops. Epidemiology is the science that describes the progress of a disease as it becomes epidemic. Depending upon the number of pathogen infection, diseases are broadly classified as simple-interest and compound-interest diseases. In a simple-interest disease, infection occurs only once or in a narrow growth stage during the crop season, and disease severity is highly dependent upon the initial infection rate. Several smut diseases of cereal crops are simple-interest diseases, such as common bunt [caused by *Tilletia caries* (DC.) Tul. & C. Tul. and *T. laevis* J. G. Kühn], dwarf bunt (caused by *T. controversa* J. G. Kühn), flag smut [caused by *Urocystis agropyri* (Preuss) A. A. Fisch. Waldh.], Karnal bunt (caused by *T. indica* Mitra), and loose smut [caused by *Ustilago tritici* (Pers. C. N. Jensen, Kellerm., & Swingle] (Bockus et al. 2010). For a compound-interest disease, infection of a crop by the pathogen occurs many times during a growth season of the crop, and the severity of the disease epidemic depends upon not only the initial infection, but also and often even more on the rate of the disease development. Simple-interest and compound-interest diseases are also referred as single-cycle and multiple-cycle diseases, respectively. Similar to other rusts and many other airborne diseases, stripe rust is a typical compound-interest or multiple-cycle disease. Infection of a cereal crop such as wheat by the stripe rust pathogen, *Puccinia striiformis* f. sp. *tritici* (*Pst*), can occur when a plant just emerges to just before it matures. During a crop season, hundreds of infection cycles can occur, resulting up to 100% grain yield loss.

4.1 Epidemiology of the Disease

As mentioned above, stripe rust is a compound-interest disease and the pathogen can infect host plants whenever plants have green tissue and environmental conditions meet the requirements for infection. Intensive studies on epidemiology of stripe rust were conducted from the beginning of the twentieth century up to 1980s (Hassebrauk 1965, 1970; Hassebrauk and Röbbelen 1975; Rapilly 1979; Stubbs 1985; Li and Zeng 2002; Line 2002; Chen 2005), and revived in the recent years (Jin et al. 2010; Sharma-Poudyal and Chen 2011; Sharma-Poudyal et al. 2014; Wang and Chen 2015; Wang et al. 2015). The epidemiological studies cover the disease and pathogen life cycles, sources of inoculum, pathogen survival and dispersal, cropping and environmental factors on disease development, and disease forecasting.

4.1.1 Pathogen Life Cycle and Disease Cycle

The life cycle, also called life history, for a pathogen is the cyclical progression of stages in the growth and development of the pathogen that occurs between the appearance and reappearance of the same stage of the pathogen (Shurtleff and

Averre III 1997). Similarly, a disease cycle is the sequence of events involved in disease development, including the stages of development of the pathogen and the effect of the disease on the host; the chain of events that occurs between the time of infection and the final expression of disease (Shurtleff and Averre III 1997). For many diseases, the pathogen life cycle and disease cycle can be very similar, but for stripe rust, the disease cycle can be quite different from the pathogen life cycle. In other words, the disease cycle is only a small portion of the pathogen life cycle.

4.1.1.1 Pathogen Lifecycle

For a long time, the stripe rust pathogen was considered to have a hemiform life cycle comprising only the uredinial and telial stages (Stubbs 1985). From the 1890s to the 1930s, several European scientists were searching for alternate hosts of the stripe rust pathogen, without any success. For example, Eriksson and Henning (1894) studied species in Boraginaceae, following the finding of species in Ranunculaceae as the alternate host for the wheat leaf rust pathogen (*P. triticina* Erikss.). Tranzschel (1934) searched species of *Valerianella* and other genera, as they assumed that *Acidium valerianella* could belong to *P. striiformis*. Mains (1933) made a right speculation that *Berberis* spp. and *Mahonia* spp. could be the alternate hosts for the stripe rust pathogen, as *P. striiformis* is closely related to *P. koeleriae* Arth., *P. arrhenatheri* Eriks., and *P. montanensis* Ellis, which have *Berberis* spp. and *Mahonia* spp. as their aecial hosts. Following Tranzschel and Mains' speculations, Straib (1937) inoculated species of *Valerianella* with basidiospores of *P. striiformis*, and did not observe any pycnia. He also tried to isolate *P. striiformis* from aeciospores collected from species of *Berberis* and *Mahonia* without success. Similarly, the attempts of Hart and Becker (1939) with species of *Valerianella*, *Berberis*, and *Mahonia* did not result in positive results. Due to the failure to find any aecial host of the stripe rust pathogen, Hassebrauk (1970) doubted the existence of alternate hosts. Without an alternate host, the stripe rust pathogen could not have its pycinial (spermagonial) and aecial stages in its life cycle. According to Rapilly (1979), the alternate host, if ever found, will escape infection because of the short dormancy of teliospores and readily produced basidiospores (Wright and Lennard 1978). The speculations by Mains and Rapilly have been approved in the recent years.

In 2010, Jin et al. (2010) identified a stripe rust fungus from aecia collected from naturally infected plants of *Berberis chinensis* and *B. koreana* in an arboretum and ornamental plantings of 'Emerald Caroudel', an interspecific hybrid between *B. koreana* and *B. thunbergii* in Minnesota, USA. They inoculated wheat, barley, oat, rye, and Kentucky bluegrass with the aecial samples under controlled conditions. The rust fungus was identified as the bluegrass stripe rust pathogen [*P. pseudostriiformis* (syn. *P. striiformis* f. sp. *poae*)]. They also successfully obtained pycnia (spermagonia) and aecia by inoculating seedlings of *B. chinensis*, *B. holstii*, *B. vulgaris*, and *B. koreana* in the greenhouse with germinated *Pst* teliospores on wheat straw collected from California. Later, *B. vulgaris* was successfully confirmed to be susceptible to *Pst* and the barley stripe rust pathogen (*P. striiformis* f. sp. *hordei*, *Psh*) with all five spore stages produced using a single isolate under controlled conditions (Wang et al. 2011, 2015; Chen et al. 2012; Wang and Chen 2015; X.M. Chen, Y. Lei, M.N. Wang, unpublished data). Wang and Chen (2013) identified *Mahonia aquifolium*, a native plant widely growing in the US Pacific Northwest, to be susceptible to the infection of *Pst* basidiospores under controlled conditions. In China, inoculation of barberry plants with germinated teliospores of *Pst* identified 28 species of *Berberis* susceptible to the pathogen (Zhao et al. 2013; Kang et al. 2015; Wang et al. 2016). Although at low frequencies, *Pst* was identified from aecia collected from naturally infected plants of five species, *B. aggregata*, *B. brachypoda*, *B. polyantha*, *B. shensiana*, and *B. soulieana* (Zhao et al. 2013; Wang et al. 2016). So far, 33 species of *Berberis* and *Mahonia* have been reported susceptible to *Pst*, which are presented in more detail in Chap. 3. These recent studies have proven *P. striiformis* and a related species, *P. pseudostriiformis*, are heteroecious fungi with a macrocyclic life cycle.

The complete life cycle of *Pst* contains a sexual cycle comprises five spore stages (stage 0: pycnia; stage I: aceia; stage II: uredinia; stage III: telia; and stage IV: basidial), which are presented in great detail and illustrated in Chap. 3. Here, these stages are discussed briefly for their relatedness to the disease epidemiology. Stages 0 (pycnia) and I (aecia) occur on alternate hosts (*Berberis* spp. and *Mahonia* spp.); and stages II (uredinia), III (telia), and IV (basidia, also called promycelia) occur on primary hosts of cereals and auxiliary hosts of grasses. Basidiospores released from basidia cannot infect the primary host like wheat, but they can infect an alternate host and on the alternate host, the fungus produces pycnia that contain receptive hyphae and release pycniospores. Both basidiospores and pycniospores (and also receptive hyphae) should be considered as sexual organs of the rust fungus because basidiospores are produced as the results of meiosis that occurs in teliospores and pycniospores from a single pycnium resulted from the infection of a single basidiospore carry the same sexual mating type. Although a high percentage of basidiospores have two nuclei and some basidiospores have three nuclei, in addition to one-nucleus basidiospores (Anikster 1983; Zhao et al. 2016), they are haploid as they have identical nuclei. Pycniospores and receptive hyphal cells are strictly mono-nucleate. The rust fungal sex appears occurring on the alternate host as a pycniospore can fuse with a receptive hypha of opposite mating type. Although this spermatization, often called fertilization, occurs on the alternate host, the fusion of the two nuclei does not occur until very later in teliospores on the primary or an auxiliary host. After the cell fusion, the two nuclei with separate sexual mating types are maintained together in new growing cells within the leaf or other tissue of the alternate host, producing bi-nucleate aeciospores in aecia formed in the other side of the infected leaves. Aeciospores cannot infect the alternate host, but they can infect the primary cereal host or auxiliary grass hosts. The successful infection of wheat by a *Pst* aeciospore produces urediniospores that normally have two nuclei with the identical genotypes as the aeciospores. Urediniospores can infect the same host plants and produce the next generation of urediniopores. This asexual cycle can repeat many times until uredinia change to telia for survival in the environment and host plants which are no longer suitable for infection and production of urediniospores. Teliospores are produced in telia. Different from all other spores that are single-celled, teliospores are two-celled, but the two cells should have the identical genetic make-up, each initially contains two nuclei of different sexual mating types and also other different genes. The real sexual mating occurs in each of the two cells of a teliospore, and thus a mature teliospore has one nucleus but diploid (2n) in each cell. During the germination, a meiosis and mitosis take place in a promycelium from which four basidiospores (n) are produced. Basidiospores are released from the sterigma attaching to the promycelium and carried by wind, rain splash, or insects onto alternate host, and they germinate and infect the alternate host to start another cycle. The sexual life cycle usually takes a year under the natural conditions, and the asexual life cycle takes from 2 weeks under optimal temperature conditions to 5 months at low temperature conditions.

4.1.1.2 Why Is *P. striiformis* Infection on Alternate Hosts Uncommon?

As mentioned above, the attempts from the 1890s to the 1930s all failed to identify alternate hosts of the stripe rust pathogen even *Berberis* and *Mahonia* species were correctly speculated as alternate hosts (Mains 1933), but the early studies were successful for identification of alternate hosts for stem rust and many other cereal rusts (de Bary 1866, 1867; Jackson and Mains 1921; Craigie 1927, 1928; Tranzschel 1914, 1934). Based on the recent studies, it can be concluded that the unsuccessful attempts were mostly due to the fact that infection of *P. striiformis* on alternate hosts is uncommon.

The early studies of searching for alternate hosts of *P. striiformis* were almost exclusively in Europe (Eriksson and Henning 1894; Mains 1933; Tranzschel 1934; Straibs 1937; Hart and Becker 1939). So far, no any report of *P. striiformis* on barberry in Europe. Straibs (1937) did not found *P. striiformis* from aecia samples collected from barberry in Germany. Recently, Berlin et al. (2012) detected the stem rust pathogens of wheat, oat and rye, but did not found any *P. striiformis* from barberry plants in Sweden. These results indicate that infection of *P. striiformis* on barberry does not occur or in very low frequency in these countries. Similarly, Wang et al. (2015) only detected *P. graminis*, but not any *P. striiformis* on barberry plants in the US Pacific Northwest, where stripe rust is the most important disease on wheat. They concluded that barberry is essential for stem rust, but not for stripe rust in this region. So far, the only direct evidence for natural infection of barberry by the stripe rust fungus is from western China for *Pst* (Zhao et al. 2013; Wang et al. 2016) and from Minnesota, USA for *P. pseudostriiformis* (syn. *P. striiformis* f. sp. *poae*) causing stripe rust on bluegrass (Jin et al. 2010). In the first study in China, four *Pst* aecia were isolated from 3703 aecia collected from various *Berberis* spp. (Zhao et al. 2013); and in the second study, 16 *Pst* aecia were isolated from 9297 aecia from various barberry species (Wang et al. 2016), at a frequency of 0.11% and 0.17%, respectively. All of these studies suggest that the stripe rust fungus has some difficulties to infect alternate hosts.

To find the reasons for why *Pst* cannot or have difficulties to infect barberry, Wang and Chen (2015) compared structures of telia and conducted a survey of teliospore viability of both *Pst* and *Pgt* in the US Pacific Northwest from 2010 to 2014, including 2010 and 2011 when wheat stripe rust was extremely severe (Chen 2014). They found that teliospores of *Pst* and *Pgt* differ significantly in structure, dormancy and survival. Telia of *Pst* are submerged under the epidermal layer of wheat tissue and form spherical sori, and teliospores are contained inside the sori and covered by an outside peridial cell layer (Wang and Chen 2015). The two layers from the host plant or the fungus protect *Pst* teliospores from germinating. To germinate teliospores, these layers need to be broken. Although natural factors such as scratching wheat leaves or straws by wind, animals or human activities, the layers should greatly reduce the germination rate of *Pst* teliospores. In comparison, *Pgt* telia form tough and thick pseudoparachemata on the surface of wheat tissue, on which teliospores are readily exposed when mature. The exposed *Pgt* teliospores should easily absorb water to germinate. However, a *Pgt* teliospore could not germinate during summer and fall as it has a long dormancy, whereas *Pst* teliospores do not have or have very short dormancy (Anikster 1983; Wang and Chen 2015). Therefore, the differences in telial structure and teliospore dormancy make *Pst* and *Pgt* teliospores have very different fates. The long dormancy of *Pgt* teliospores protects them from germination in an unfavorable environment and the exposure of teliospores allows them to germinate when the dormancy is broken. In contrast, the closed position of *Pst* teliospores in the telia that are underneath of the host epidemic layer prevents them from germination, and when the layers broken either physically or biochemically, teliospores may germinate when there are no susceptible alternate host tissue or cannot germinate due to degradation (Wang and Chen 2015).

When scratched from wheat plants collected from fields in July when *Pst* telia were just formed, teliospores could be readily germinated under controlled conditions (Wang and Chen 2015). The germination rate gradually decreased before winter, and after winter, *Pst* telia became degraded and no viable teliospores were observed before production of new barberry leaves. In contrast, *Pgt* teliospores could not germinate until the following January, and the germination rate reached the highest in late April and May, catching susceptible new leaves of barberry.

Wang and Chen (2015) speculated that the telial degradation is caused by the wet and fluctuated temperature conditions during the winter time in the US Pacific Northwest, so they determined viability of teliospores under different conditions. They found that the germination rate of *Pst* teliospores kept under conditions of dry and at 22–23 °C dropped gradually from 79% in July when fresh to 1% in following January, while teliospores completely lost viability in 3 weeks if kept wet at 5 °C. The wet condition was found to be detrimental to *Pst* teliospores. Based on these results, they speculated that *Pst* teliospores may function in regions where teliospores form late in the fall, the weather is dry during the winter, and long periods of high moisture conditions are possible in the spring when alternate hosts produce new leaves. Some parts of northwestern China have such conditions.

The prediction has been proven by the finding of low infection of *Pst* on naturally infected barberry (Zhao et al. 2013; Wang et al. 2016).

Bluegrass stripe rust, caused by *P. pseudostriiformis* (syn. *P. striiformis* f. sp. *poae*, *Psp*) (Abbasi et al. 2004; Liu and Hambleton 2010), has a quite different disease cycle compared to wheat stripe rust. Bluegrass stripe rust can occur in spring and summer, but the disease is more prevalence in late fall. The disease is under control by frequent mowing during the summer. Under low temperatures in late fall, bluegrass grows slowly and mowing becomes less frequent so that the stripe rust fungus has enough time to complete its asexual cycle for producing uredinia and later for forming telia. Compared to *Pst* telia, *Psp* telia have a relatively more chance to survive winter and spring to catch young leaves of barberry. This chance should be greatly increased when telia are covered by snow for much of the winter and early spring in regions like Minnesota, USA, where *Psp* aecia were identified from naturally infected barberry plants (Jin et al. 2010). In the US Pacific Northwest, although bluegrass stripe rust occurs mostly in October, *Psp* teliospores have less chance to survive the winter because of the interchanging rain-and-snow conditions.

The stripe rust and stem rust fungi require different temperatures and duration of dew period for infecting barberry. Under saturate moisture conditions, temperatures around 10 °C are optimal for *Pst* teliospore germination and basidiospore formation, while *Pgt* teliospores and basidiospores germinate best at around 20 °C (Wang and Chen 2015). Under their respective optimal temperatures and saturated moisture conditions, *Pgt* basidiospores are produced much faster than the *Pst* basidiospore formation. The minimum and maximum temperatures for *Pst* teliospore and basidiospore germination are slightly below 5 °C and slightly above 22 °C for *Pst*, and those for *Pgt* are slightly below 10 °C and above 30 °C, respectively (Zadoks 1967; Wang and Chen 2015). Wang and Chen (2015) also determined the minimum dew period needed for successful infection of barberry by *Pst*. Under optimal moisture and temperature conditions, no infection was observed for keeping inoculated plants in a dew chamber for less than 32 h. The highest infection was obtained by keeping inoculated plants in the dew chamber for 96 h. The minimum dew period for natural infection should be longer than 32 h because the 32-h dew period was determined under artificial inoculation using teliospore suspension. Such a long period of saturated moisture conditions hardly occur in many regions of the world. It can be speculated that the early unsuccessful attempt of inoculating barberry with *Pst* telia (Hart and Becker 1939) could be due to a short dew period and high temperatures as used for studying stem rust and other rust fungi.

To summarize the reasons why sexual stage or alternate host is not important for the stripe rust pathogen, we can attribute it to the following four factors. (1) The telial structure buried in plant tissue and teliospores enclosed in telial sori prevent teliospores from germination. (2) The lack of dormancy may allow teliospores to germinate under unsuitable environmental and plant phenological conditions for infection of alternate hosts. (3) The requirement of a long dew-period for teliospore germination and basidiospore infection on barberry limits the fungus to perform its teliospore role. (4) Degradation of telia under wet and fluctuating temperature conditions reduces teliospore viability. It is not clear why *Pgt* telia are not degraded under the same wet and fluctuating conditions. The different responses between *Pst* and *Pgt* may be related to the difference in dormancy and the structures of telia and teliospores.

4.1.1.3 Stripe Rust Disease Cycle

After the presence and discussion of *Pst* life cycle, it is easier to present and discuss the stripe rust disease cycle. The disease cycle is the endless repeating the asexual cycle, from infection of wheat plants by urediniospores to production of next generations of urediniospores. However, the stripe rust disease cycle consists of the source of inoculum, spore dispersal and landing on host plants, infection process (including spore germination, penetration, haustorial formation, infectious hypha growth, and reproduction structure differentiation), sporulation, and pathogen survival.

As illustrated in Fig. 4.1, the infection process starts after a viable urediniospore lands on the surface of a wheat leaf and germinates when dew is present and the temperature is within the range from 1 to 23 °C (optimal range of 7–12 °C) (Sharma-Poudyal et al. 2014). The germ tube grows toward a stoma of the wheat leaf. Over the stoma, the germ tube may form an appressorium, from which a penetration peg grows through the stoma into the leaf tissue. Appressoria of *Pst* are usually not as obvious as other cereal rust fungi so that *Pst* is generally thought not to produce

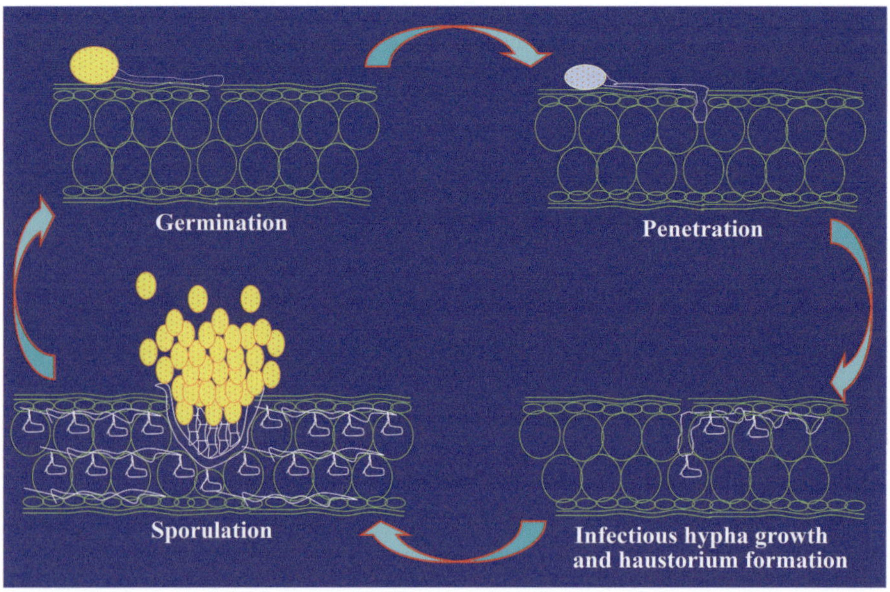

Fig. 4.1 Simplified infection process by a *Puccinia striiformis* urediniospore containing germination, penetration, infectious hypha growth and haustorium formation, and sporulation. Depending upon temperatures, this process takes from about two weeks to several months

appressoria. However, Prof. Zhensheng Kang showed pictures of obvious appressoria produced by *Pst* isolates. Based on his observations, most *Pst* isolates do not form apparent appressoria, but some isolates do when germ tubes grow over stomata. After a penetrate peg grows through a host stoma, it grows swollen to form a sub-vesicle, from which a hypha is growing and branching between host cells. The intercellular hyphae contain numerous nuclei that are produced through mitosis from the initially two nuclei in the urediniospore. When a hypha contacts a cell wall, a haustorial mother cell is formed and divided from the hypha by a septum. The haustorial mother cell grows into the host cell to form a haustorium. The formation of haustoria is the beginning of the fungal parasitic life style. The fungus obtains water and nutrients from the host to grow and reproduce. Most interactions between the fungus and host begin at this point, about 24 h after inoculation (Coram et el. 2008a, b; Wang et al. 2009), although barley response to the stem rust fungus was reported as early as 5 min after inoculation (Nirmala et al. 2007, 2010). Under optimal temperature conditions, disease symptoms, mostly chlorosis, can be observed about 7 days after inoculation. If an artificial inoculation is conducted, this is the earliest time point to find out if the inoculation is successful or not. Infectious hyphae continue growing within the plant tissue, and sporulation (uredinia appear and release urediniospores) start about 12–14 days after inoculation under optimal temperature conditions. Just before sporulation, intercellular hyphae grow tightly connected and differentiate into short cells divided by septa. Each cell contains two nuclei. Tightly interconnected septated hyphae form a structure called uredinium, in which urediniospores are produced. Without mutation, all progeny urediniospores have the identical genotype of the original urediniospore as this asexual reproduction is clonal. Different from the infection of one urediniopsore usually producing one uredinium by the stem rust or leaf rust fungus, the infection of a single urediniospore of *P. striiformis* produces numerous uredinia, each releasing thousands of urediniospores. Because of this, the uredinial infection of the stripe rust pathogen on its primary or auxiliary host is systemic in the host tissue. The mechanism by which the stripe rust fungus keeps the strictly identical two nuclei in the progeny urediniospore cells even through a long period of multi-nucleate hyphal stage has not been studied. However, we can assume that a mechanism is similar to the clam connection that is well demonstrated for other basidiomycete fungi (Alexopoulos and Mims 1979). Without a mechanism to keep the two potentially different nuclei together in the progeny spores, at least three genotypes or races would be obtained from a single urediniospore infection if the two nuclei are different. In fact, the two nuclei in a urediniospore are different in every *Pst* isolates that have been sequenced.

The produced urediniospores have two fates, either die or live. If the spores fall onto the ground or any other non-host substances, they will die. If they are lifted by wind or splashed by raindrops, they will serve as a source of inoculum to infect the same plant, other plants in the same field, and plants in other fields in the same region or far away in other regions.

4.1.2 Sources of Urediniospore Inoculum

Transplanting inoculated plants that have been grown under controlled conditions before sporulation can start stripe rust development and spread the pathogen. This is an efficient method for artificial inoculation of experimental fields (Chen and Line 1995), as the latent period is shortened under optimal condition and the transplanted plants can continuously produce urediniospores to infect plants in the field. In such cases, the inoculum source is the fungal mycelia in the inoculated plants. Under natural conditions, urediniospores are the only form, or major form if aeciospores from alternate hosts play a role, of inoculum for stripe rust to occur in nature. Sources of *Pst* inoculum can be urediniospores produced from the same wheat plants, different wheat plants in the same fields, from different fields in the same region, or from different region, different country and different continent. In addition to wheat crop, urediniospore inoculum also can be from volunteer wheat plants and grasses. Grass species that are susceptible to *Pst* have been reported by many workers (Dietz and Hendrix 1962; Hassebrauk 1965; Tollenaar and Houston 1967; Cheng et al. 2016) and summarized in Chap. 1. Depending upon cropping systems and climatic conditions, major inoculum sources can be different in different regions. In most regions, inoculum from wheat fields is most important.

Any regions where stripe rust occurs are epidemic regions. However, the frequency and level of epidemics can vary among regions depending upon whether the pathogen can survive summer and/or winter in the region and how long the weather conditions are suitable for the pathogen infection and growth. The major inoculum sources are in the regions where the pathogen can survive winter and/or summer. Regions where *Pst* can survive both winter and summer have more frequent epidemics, and these regions are also the sources of inoculum for other regions. Western China; Pacific Coasts of the USA, Canada and Mexico; high-elevation regions in Mexico, Africa, southern, central, and western Asia, and South America; southeastern coast of Australia and New Zealand; and regions in European with cool summer have more stripe rust epidemics (Zadoks 1967; Stubbs 1985; Li and Zeng 2002; Chen 2005; Wellings 2011). In regions where wheat is planted in two different seasons, stripe rust epidemic occurs more frequent than regions where wheat is planted in only one season, such as eastern African where spring wheat is planted in different seasons, and northwestern China and the US Pacific Northwest where both spring and winter wheat are grown. In the US Pacific Northwest, spring wheat and winter wheat are even grown side by side in a single field. Although this cropping system is advantageous for crop rotation, weed management, and reduction of soil erosion, it makes host plants available all year long and expands the stripe rust season (Chen 2005).

4.1.3 Spore Dispersal

Long-distance dispersal of cereal rust pathogens has been well documented by Nagarajan and Singh (1990), Eversmeyer and Kramer (2000) and Brown and Hovmøller (2002). Urediniospores can be disseminated by wind to a wide range of distances from few meters to 8000 km in a crop season (Zadoks 1961; Beresford 1982; Watson and De Sousa 1983; Nagarajan and Singh 1990). The *Puccinia* pathway in North America and similar pathways in other continents (Roelfs 1985; Zadoks and Bouwman 1985; Li and Zeng 2002; Sharma-Poudyal et al. 2014) well demonstrate the ability of rust fungi to move between regions in long distances.

4.1.3.1 Pre-historical Continental Movement of the Stripe Rust Fungus from Its Origin

The center of origin of *P. striiformis* has been assumed to be Transcaucasia, a region lying to the south of the Caucasus Mountains, between the Black Sea and the Caspian Sea, comprising the present-day republics of Georgia, Armenia and Azerbaijan (Stubbs 1985). From this center, the fungus dispersed in all directions to other central and western Asian countries, southern and eastern Asia, northern Africa and Europe. The fungus in North America was assumed from eastern Asia via the Aleutian Islands and Alaska and spread southward into South America along the mountain ranges in western North America and South America (Humphrey et al. 1924; Vallega 1955). These continental dispersals could have occurred thousands or even millions of years ago on grasses long before the domestication of cereal crops. It is difficult to prove or disprove the above overall hypothesis or specific hypothesis. Alternate hypotheses are also possible. For example, instead of the pathogen spreading into North America from eastern Asia via Aleutian Islands and Alaska, the pathogen possibly spread directly from eastern Asia to North America through air movement over the Pacific Ocean. To prove the former hypothesis, evidence of either stripe rust occurring on grasses in Alaska or historical weather data to the fungus able to overwinter is needed. So far, there are no any reports of stripe rust in Alaska. Recently, the Himalayan region has been proposed as a putative center of origin for the wheat stripe rust pathogen based on its high population diversity (Ali et al. 2014a, b; Thach et al. 2016). It was not possible to assess the relative originality of the Himalayan region vs. Transcaucasia as stripe rust samples from the latter region were absent in these studies. Unfortunately, no stripe rust samples from Transcausasia was present in the historical "Stubbs collection" (Thach et al. 2015), probably due to political reasons in the past. Studies including stripe rust samples from Transcausasia are needed to determine which one is the primary center of origin and which one is a secondary center. It is possible that these regions together are the big center of origin of the stripe rust pathogen as these two regions are only few thousands of kilometers away. Urediniospores can easily disseminate from

Transcausasia to the Himalayan region, and thus the stripe rust pathogen might start almost simultaneously in both regions thousands or millions of years ago on grasses.

4.1.3.2 Recent Continental Dispersal of the Stripe Rust Pathogen

Continental dispersal of the stripe rust pathogen has been well demonstrated in recent years, and some of the spread was believed to be through human activities. The stripe rust pathogen was first reported in Australia in 1979 (O'Brien et al. 1980), and the introduction was believed from Europe through urediniospore-contaminated clothing based on the same or similar races (Wellings and McIntosh 1987). The pathogen was further spread to New Zealand in the following year, probably by wind dispersal of urediniospores from eastern Australia (Wellings and McIntosh 1990). Using satellite data, Kim and Beresford (2008) simulated dispersal of *Pst* urediniospores from Australia to New Zealand across the Tasman Sea. The combination of a spectrum model and satellite cloud data allowed them to identify sites at which viable urediniopores were deposited. They modeled urediniospores arrived at sites within 40 km from Otama, New Zealand, where wheat stripe rust was first reported in November 1980. Urediniospores appeared to be transported under an overcast cloud condition, which would increase the chance of successful infection on host crops.

After the wheat stripe rust pathogen was introduced to Australia, the pathogen was occurring only in the east of the country for many years until 2002 when the disease was observed in Western Australia. Based on virulence and molecular analyses (Wellings et al. 2003), the strain responsible for the widespread epidemic in Western Australia in 2002 was believed to be of foreign introduction. As the rust samples from Western Australia had the similar virulence patterns, especially virulent to resistance genes *Yr8* and *Yr9* as reported in the US in 2000 (Chen et al. 2002), the rust was thought to be introduced from the US. Although this is possible, it might be from other countries because the pathogen could be in Western Australia few years before it caused the devastating epidemic in 2002. Also, races virulent to *Yr8* and *Yr9* were detected long before 2000–2002 in many other countries as shown in Fig. 4.2. Races of *Pst* with *Yr9* virulence were identified for the first time in the former Soviet Union in 1973 and the Netherlands in 1974 (Stubbs et al. 1977), China in 1977 (Wang et al. 1986), East Africa (Ethiopia and Kenya) in 1986 (Saari 1996), South America (Ecuador, Colombia, Ecuador, Bolivia, and Chile) and Mexico in 1988 (Stubbs and Yang 1988), the Middle East (Syria and Lebanon) in 1989 (Hakim and Mamluk 1996), and in Australasia in 1991 (Wellings and Burdon 1992). In 1992, 1994, and 1996, the *Yr9* virulence was reported in Iran, Pakistan, and India, respectively (Saari 1996; Nazari and Torabi 2000). In addition, the *Yr9* virulence was reported in Israel, Tanzania, Rwanda, Zambia, and Turkey (Stubbs and Yang 1988; Düsünceli et al. 1996; Manisterski et al. 1996), and France in 1984–2009 (de Vallavieille-Pope et al. 2012). Races with virulence to both *Yr8* and *Yr9* were clearly reported in Kenya in 1986–1989 (Danial et al. 1994) and in India in 1995–2004 (Prashar et al. 2007), and in Australia, Canada, Chile, China, Hungary,

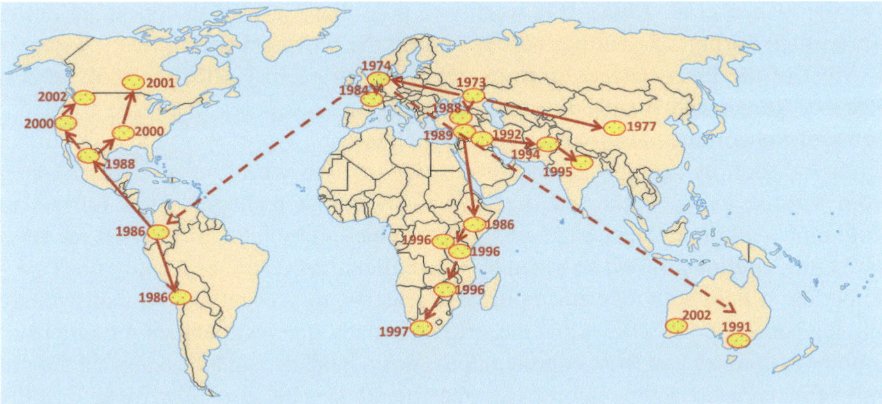

Fig. 4.2 The first years of *Puccinia striiformis* f. sp. *tritici* populations with virulence to *Yr9* (or *Yr8* and *Yr9*) reported in different countries. The possible dispersals through natural dissemination are indicated by the *solid arrows* based on the chronological reports (see the text) and proximities, and by the *dash arrows* through human activities

Kenya, Nepal, Pakistan, Turkey, and Uzbekistan from wheat stripe rust samples collected from 2006 to 2010 (Sharma-Poudyal et al. 2014). The aggressive strain, represented by its virulence to *Yr8* and *Yr9* and adaption to high temperatures, was detected in many countries within a few years (Hovmøller et al. 2008).

In Africa, stripe rust of wheat has occurred in northern and eastern Africa for more than hundreds of years, but the disease was first reported in central Africa (northern Zambia) in 1958 (Angus 1965), and in South Africa in 1996 when it was detected in Western Cape (Pretorius et al. 1997), and from there the pathogen spread to most of South Africa in 1997 (Boshoff et al. 2002).

Stripe rust of barley, caused by *P. striiformis* f. sp. *hordei*, has been less studied than stripe rust of wheat (*P. striiformis* f. sp. *tritici*). The two forms could have been diverged long time ago after barley and wheat were domesticated. Epidemics of barley stripe rust have been reported in northwestern and central European countries (Hassebrauk 1962; Johnson 1968; Stubbs 1985), India (Prasada et al. 1967; Sharma and Gupta 1979; Kumar et al. 1993a, b), Bangladesh (Stubbs 1985), Nepal (Karki 1980), China (Wang et al. 1989a, b; Niu et al. 1991; Wang 1992), and Japan (Narita and Mano 1962; Kajiwara et al. 1964). The barley form had not been reported in the West Hemisphere until 1975 when barley stripe rust was first reported near Bogota, Colombia (Dubin and Stubbs 1986). The rust fungus was postulated to be introduced from Europe based on similar virulence patterns (Dubin and Stubbs 1986). If this postulation is correct, the pathogen must be introduced through human activities. The pathogen dispersed southward to Argentina by 1982 (Dubin and Stubbs 1986) and northward to Mexico by 1987 (Chen et al. 1994), and the US by 1991 (Roelfs et al. 1992; Marshall and Sutton 1995). In the US, the pathogen dispersed from southern Texas in 1991 to Oklahoma, New Mexico, and Colorado by 1992; California, Idaho, and Montana by 1993; Utah by 1994; and Oregon and Washington by 1995 (Chen et al. 1995). It is worthy to note that much of the dispersal in the western US during a 5-year period was against the prevalent winds.

Most of the long distance movements of the stripe rust pathogen took several years or longer and mainly through airborne urediniospores dispersed by wind. Some of the continental-to-continental movements of spores were through human activities, such as the postulated introductions of the wheat stripe rust pathogen from Europe to Australia and the barley stripe rust pathogen from Europe to Colombia. Although the chances of such introduction by human activities are relatively low, cautions should be taken by international travelers, especially those participating in field activities during disease seasons, to prevent from unconsciously moving pathogens. Knowing that the stripe rust pathogen is capable of moving from country to country and from continental to continental, resistance genes that are not only effective against the regional pathogen population but also effective against the populations in other regions or countries should be used in breeding programs.

4.1.3.3 Seasonal Dispersal of the Stripe Rust Pathogen

For stripe rust management on a yearly basis, it is important to understand the dispersal pattern of the stripe rust pathogen within a crop growing season. Urediniospores of the stripe rust fungus can spread by wind for a couple of thousand kilometers in few days. The distance of viable spore spread is a function of many factors. In a spore-producing area, as an area where the pathogen overwintered, the more spores are produced and the longer distance to which spores will spread. Dry spores are spread farther than wet spores. Spores may spread farther under a cloudy weather condition than under a sunny weather, as stronger UV light in a sunny day may kill rust spores. Stronger wind spreads rust spores longer than weaker wind. Spores are spread farther along the wind direction and much less than other directions. For long-distance dispersal, spores are mostly deposited onto crops by precipitation. How large an area in which stripe rust can develop not only depends upon how far spores can be spread, but also depends upon the moisture and temperature conditions. Without favorable moisture and temperature conditions, deposited spores will die.

Seasonal dispersal of rust pathogens has been studied in many countries or among countries. In North America, Stakman (1934) demonstrated the seasonal movement of urediniospores of the wheat stem rust pathogen from northern Mexico/Texas through the Great Plains of the United States to Canada. This urediniospore dispersal route subsequently was defined as the "*Puccinia* Path." Line and Qayoum (1992) documented urediniospore migration from eastern Washington to Montana and Alberta, Canada and to southern Idaho. The stripe rust pathogen also showed similar occurrence patterns from Texas to Canada in the later 1950s and in 2000, 2003, 2005 and 2010 following the same *Puccinia* Path (Pady et al. 1957; Chen et al. 2002, 2010; Chen 2005; Wan and Chen 2014). Based on the *Pst* race and molecular data, migration of *Pst* urediniopsores occurs more from California to south-central, northcentral, and eastern states in the US (Wan and Chen 2010;

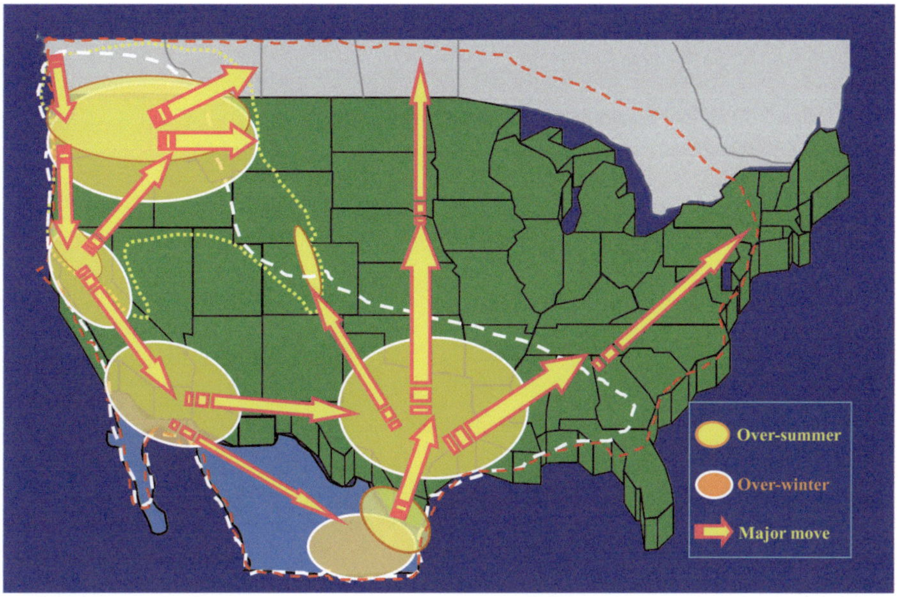

Fig. 4.3 Regions of stripe rust over-summering and over-wintering in North America and movements of stripe rust urediniospores among the regions. Stripe rust can occur in the area within the *red-dash irregular cycle*, over-winter within the *white-dash irregular cycle*, and over-summer within the *yellow-dash irregular cycle*. The major over-summer and over-winter regions are indicated. The major spore movements between regions by wind are indicated by *arrows*. Minor movements between some of the regions by wind and human activities also are possible (Modified from Chen 2005)

Xia et al. 2016a, b; Wan et al. 2016). Based on stripe rust monitoring data and virulence and molecular characterization of the pathogen collections from 2000 to 2015, a generalized pattern, modified from Chen (2005), for major directions of urediniospore movements in North America is shown in Fig. 4.3. The directions of the major spore movements are consistent with the prevailing winds.

In China, interregional urediniospore dispersal of the wheat stripe rust pathogen has been studied since the 1950s (Li and Liu 1956; Chen and Lu 1957; Zeng 1963; Wang and Dai 1965; Yang and Zeng 1992; Sun and Zeng 1994; Li and Zeng 2002; Zeng and Luo 2006). Based on Li and Zeng (2002), there are five oversummering areas. However, it is more reasonable to combine the northern Sichuan over-summering area with the northwestern area covering parts of Gansu, Qinghai, Ningxia and Shaanxi based on the proximity, similar climatic conditions, and recent new race dispersal (Han et al. 2015; Liang et al. 2016). Thus, the four geographical over-summering regions are: (1) the middle region including eastern Qinghai, eastern Gansu, western Shaanxi, and Ningxia; (2) the south region including southeastern Tibet, northern Yunnan, southwestern Guizhou, and southwestern Sichuan; (3) the north region including northern Shanxi, northern Hebei, and southern Inner-Mongolia; and (4) the west region mostly in western Xinjiang (Fig. 4.4).

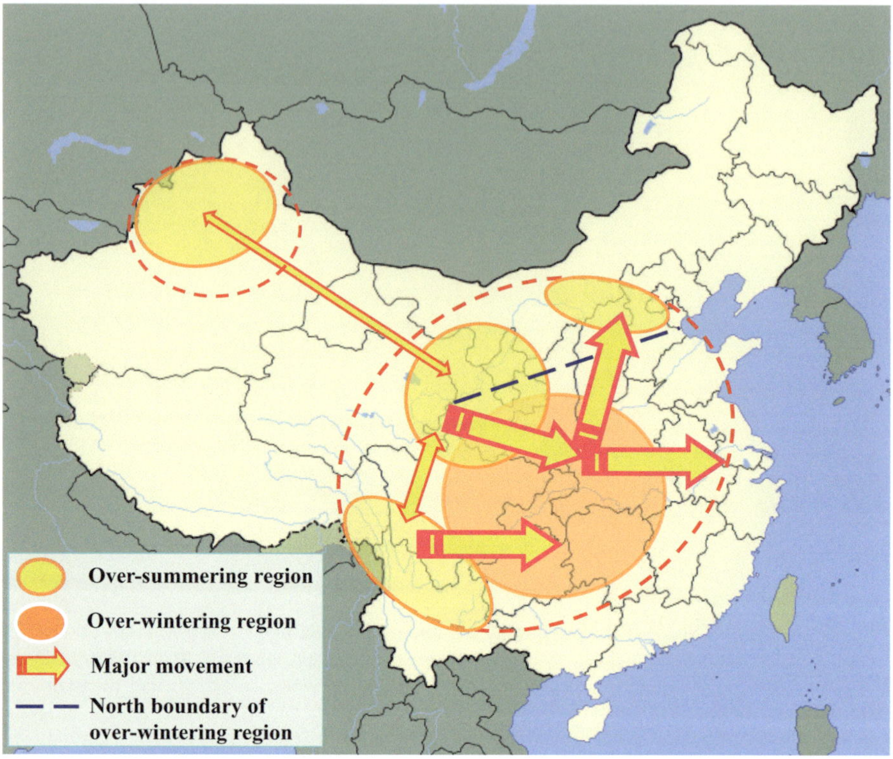

Fig. 4.4 Regions of stripe rust over-summering and over-wintering in China and movements of stripe rust urediniospores among the regions. Stripe rust can occur in the area within the *red-dash cycle*. The major spore movements between regions by wind are indicated by *arrows*

Most over-summering areas are in high elevation mountains. Urediniospores of *P. striiformis* f. sp. *tritici* produced in autumn in the over-summering regions are spread by winds to the nearby basins and farther to the eastern regions of low elevations, where the fungus is able to overwinter in most of wheat growing areas. According to the studies on *Pst* races and molecular genotypes, urediniospore movements occur among the four oversummering regions, although at relative low frequencies compared to the spore movements from the over-summering regions to the overwintering regions (Lu et al. 2011; Wan et al. 2015; Liang et al. 2016; Zhan et al. 2016).

In western and southern Asia, urediniospores of *Pst* occasionally move in waves and leaps from Turkey to India under the influence of "western disturbances." In winter, western disturbances cause widespread rain over the Indo-Pakistani plains and snow in the upper reaches of the Himalayas, and the winter precipitation deposits spores and aids in subsequent infection (Nagarajan and Joshi 1985; Nagarajan and Singh 1990). In Pakistan and India, the stripe rust pathogen oversummers throughout the Himalaya, Hindukush, and Sulaiman mountain ranges and in the northwestern Frontier Province with largest amounts of inoculum are in the mountains and valleys of Indus and its tributaries. In India, Katabatic wind flow carries urediniospores from the primary inoculum sources near Jammu, Gurdaspur, Pathankot, Ropar, and Jagadhari to the foothills in January. Cool night temperatures with dew are common along the foothills and assure successful infection. From such secondary sources, the pathogen spreads into the adjoining plains during February. Similarly, in Pakistan primary areas of stripe rust are along the foothills by January, from which the pathogen spreads to the cooler parts of western Punjab and the northwestern Frontier Area (Hassan 1978). The stripe rust pathogen also survives in the Nilgiri and Palney Hills throughout the year, but does not effectively spread to central and peninsular India as the weather conditions are not favorable for the disease. Thus, urediniospores of the fungus are able to travel for over 1000 km in northern Pakistan and northwestern India, but the dispersion in southern India can be limited by unfavorable weather conditions.

In Europe, Zadoks (1961) reported that stripe rust urediniospores are viable after being wind-transported 800 km. Hermansen and Stapel (1973) suggested that yellow rust epidemics of the wheat variety Kranich in Denmark in 1972 and subsequent years were initiated in 1971 by wind borne urediniospores from the British Isles. However, the experiment in 1975 conducted by Hermansen et al. (1978) did not directly support this hypothesis. In the experiment, they exposed susceptible plants of barley, wheat, and oats at Blavandshuk, the most westerly point of Jutland, Denmark for periods of several hours when air mass movements from the British Isles across the North Sea occurred, and observed leaf rust and powdery mildew on barley and wheat, but did not observed stripe rust. They attributed the failure of observing stripe rust due to the high day temperatures in the greenhouse after the exposure of plants in 1975. Hovmøller et al. (2002) reported a single, clonal population covering the UK, Germany, France and Denmark up to 1700 km apart based on analysis of amplified fragment length polymorphism (AFLP) variation. The pathogen frequently migrates among the four countries and the migration rates between the countries are not equal in both directions. Denmark is the recipient of migrant urediniospores from other countries, predominantly from the UK. They also identified occasions that the pathogen clones were also spread from the UK to Germany and France.

As stripe rust was absent in Australia before 1979, the occurrence of stripe rust covering the area in the first year when the pathogen was introduced well demonstrated the capability of urediniospores. During 1979, the pathogen spread to several regions of Victoria, and also to South Australia and southern and northern New South Wales (Luig 1985). The epidemic area covers more than 1000 km in distance. The occurrence of stripe rust in New Zealand in 1980 demonstrated that urediniospores likely took one step across the Tasman Sea from Australia over 2000 km (Beresford 1982). The incursion of the stripe rust pathogen before or during 2002 in Australia also demonstrates how quick the stripe rust pathogen can spread during a wheat growing season. The pathogen was reported for the first time in the Newdegate shire and the Broomehill shire in the southeastern wheat belt of Western Australia in August 2002 (Wellings et al. 2003). Within 4 weeks infections of 95 fields were confirmed, and by the end of the wheat-growing season in 2002, the disease was present over a wide area encompassing Dongara, Wongan Hills, Esperance, Manjimup, and the major wheat areas in the southeast. The disease occurring region covered areas more than 500 km in distance. It was likely that urediniospores of the pathotype detected in 2002 in Western Australia spread to eastern Australia in the same season as the pathotype also caused severe epidemics in the east region in 2003–2006 (Wellings 2007). Thus, urediniospores of the stripe rust pathogen might travel over 2000 km in that single growing season.

Zadoks and Bouwman (1985) listed criteria for studying long-distance dispersal of rust pathogen, including phenology of the crop and rust and weather conditions in both source and target areas, air trajectories from source to target, spore content of the air between source and target, spore trapping data in the target area, and matching of races in source and target areas. In the past, the long distance movement of the stripe rust pathogen was mainly based on observations of seasonal progression of the disease, spore trapping, and detection of same races. More recently, spore movement has been studied by identifying pathogen genotypes using molecular markers (Hovmøller et al. 2002, 2008; Lu et al. 2011; Wan et al. 2015; Zhan et al. 2016).

4.1.4 Pathogen Survival

4.1.4.1 Survival of Uredinial Stage

Rust pathogens face the challenge of survival during the entire lifecycle and the disease cycle. When urediniospores are produced on leaf surface, they may be splashed by rain drops or blown by wind to leaves of the same plant and nearby plants to cause new infection. They may be blown by wind onto other host plants in long distance as discussed above. Most urediniospores are washed by rain drops or blown by wind to the surface of soil, non-host plants, or other subjects, and die. When urediniopsores are lifted by wind into the air, they may be killed by UV light. Compared to colorless or light colored and thin-walled spores of other fungi, urediniospores of the stripe rust pathogen have relatively high survival ability as they are colored and have relatively thick wall. However, stripe rust urediniospores loss viability more quickly than those of the stem rust and leaf rust pathogens under high temperatures and UV light because of adaptation to relatively low temperatures and light-colored urediniospores compared to the other rust pathogens.

Urediniospores of *P. striiformis* survive better under dry and low-temperature conditions (Rapilly 1979; Chen 2005). For stripe rust research, it is best to dry urediniopores or sporulating leaves quickly under low temperatures (<25 °C) and keep dry spores or leaves as cool as possible. Dry urediniospores can keep viability at 4 °C up to 6 months, at −80 °C for a couple of years, and at −196 °C (in liquid nitrogen) for more than 40 years. When keeping in room temperatures (20–25 °C), stripe rust urediniospores loss viability completely for less than a month. Urediniospores of the stripe rust pathogen can survive on air-dry soil for a month and on dead leaves for 4–6 weeks (Shaner and Powelson 1973). Under the normally dry conditions (with monthly average temperatures of 19 °C in July and August and 15 °C in September and the 3-month total precipitation of 49 mm) during the summer in the inland Pacific Northwest of the US, urediniospores in the air are able to keep viability for about a month, long enough to be carried over from the previous spring wheat crop, which is harvested in August–September, to the winter wheat crop, which is planted in August–October (Chen 2005).

Because the stripe rust fungus is a biotrophic pathogen, it survives better when urediniospores can germinate and infect host plants. Within the host plant tissue, the fungus can be protected from UV lights and higher temperatures than urediniospores can survive. The stripe rust pathogen does not survive temperatures as high as the stem rust and leaf rust pathogens. There were some early indications that the stripe rust pathogen may remain in the dormancy state under relatively high temperatures (Hendrix et al. 1965; Tu 1967). However, no detailed studies were conducted to determine the durations for which the stripe rust fungus can keep alive without sporulation under different temperatures. The study conducted by Shaner and Powelson (1973) in Oregon, USA does not suggest that the stripe rust fungus may survive high temperatures by remaining in dormancy. They observed susceptible reactions on grass species in late spring when temperatures were low and resistant reactions on the same species of grasses in early summer and no rust in late summer. In eastern Washington, USA, highly susceptible reactions are normally observed on susceptible wheat varieties during summer, but occasionally, necrotic stripes with limited uredinia can be observed on these highly susceptible wheat varieties when there are more high-temperature days than normal summers (X. M. Chen, unpublished data).

For many regions of the world, overwintering is a bottleneck for rust pathogen survival. The stripe rust pathogen survives cold temperatures better than the stem rust and leaf rust pathogens. In an area where *P. striiformis* is able to survive winter, the disease usually starts early in spring or even during winter, and epidemics are usually severe. The stripe rust pathogen often survives winter as mycelia in the host plant tissue. The pathogen mycelia remain alive in the plant tissue for several months during the winter and wait for warm temperature for sporulation. Taking the inland US Pacific Northwest for example, where *Pst* is able to oversummer and overwinter, infection of winter wheat occurs in September to November, mostly without sporulation, and sporulation occurs late in March to May. Regions like the US Pacific Northwest and parts of northwestern and southwestern China, where *Pst* is able to survive both summer and winter, stripe rust occurs most often and serves as inoculum sources for other regions (Li and Zeng 2002; Chen 2005, 2014; Wan and Chen 2014). Knowledge of pathogen survival is important for forecast epidemics of the disease (Sharma-Poudyal and Chen 2011; Sharma-Poudyal et al. 2014).

4.1.4.2 Survival of Telial Stage

For many rust fungi, telia are the best structure and stage for survival both hot summer and cold winter, and more important to survive the seasons without primary hosts. In many regions, the telial stage and infection of alternate hosts by basidiospores is essential for epidemics of stem rust of wheat and barley and crown rust of oats (Roelfs 1982; Carson 2008; Wang and Chen 2015; Wang et al. 2015). Teliospores of the stripe rust fungus, however, do not play a role in the pathogen survival and disease epidemics in most epidemic regions of the world (Berlin et al. 2012; Wang and Chen 2015; Wang et al. 2015, 2016). In the US Pacific Northwest, where telia are essential for the stem rust pathogen to over winter and produce basidiospores to infect barberry, telia and alternate hosts of the stripe rust pathogen do not have any role in the pathogen survival and disease epidemics, due to the submerged structure of telia, degradation of teliospores under the wet and fluctuated temperature conditions during winter, and phenology of wheat and barberry plants (Wang and Chen 2015). Natural infection of barberry plants by the wheat stripe rust pathogen has been detected in western China at very low frequencies (Zhao et al. 2013; Wang et al. 2016). Based on the published results so far, the role of telia in survival for the stripe rust pathogen is negligible.

4.1.5 Host Factors

Host is equally important to a disease as the pathogen. Without a host plant, there is no such disease. The scale of susceptible host plants is positively related to the level of epidemics. Rust fungi have three types of hosts: primary, auxiliary, and alternate host plants. The alternate host of the stripe rust pathogen in pathogen survival and disease epidemics is discussed above. Here, we just discuss the primary and auxiliary hosts of the stripe rust pathogen.

4.1.5.1 Primary Hosts

From the evolutionary standpoint, grasses should be the primary hosts for rust fungi. However, crops are considered primary hosts because crops are more important than grasses to humans, and rust diseases are more devastating on large-scale cultivated crops. The species of the stripe rust fungus, *P. striiformis*, is able to infect wheat, barley, rye, and triticale (Eriksson 1894; Chen et al. 1995; Cheng et al. 2016); but has not been reported to cause diseases on any other cultivated cereal crops. As discussed in Chap. 1, the species of the stripe rust fungus can be separated into different formae speciales based on specialization on different crop and grass genera or species, for example, *Pst* on wheat, *Psh* on barley, and *Pss* (*P. striiformis* f. sp. *secalis*) on rye (Eriksson 1894). The primary hosts of *Pst* are all cultivated wheat species including common wheat or bread wheat (*Triticum aestivum* L.; $2n = 6x = 42$), club wheat (*T. aestivum* spp. *compactum*; 2n = 6x = 42), spelta wheat (*T. aestivum* ssp. *spelta*; 2n = 6x = 42), durum wheat [*T. turgidum* spp. *durum* (Desf.) Husn.; 2n = 4x = 28], timopheevi wheat [*T. turgidum* spp. *timopheevi* (Zhuk.); 2n = 4x = 28], emmer wheat (*T. turgidum* spp. dicoccoides (Korn. ex Asch. And Graebn.); 2n = 4x = 28], khorasan wheat (*T. turgidum* ssp. *turanicum*), and einkorn wheat (*T. monococcum* L.; 2n = 2x = 14) (Gustafson et al. 2009). Of these species or subspecies, common wheat and durum wheat are the most cultivated species in the world. Club wheat is significantly and almost exclusively grown in the US Pacific Northwest (Allan 2014). Although cultivated barley varieties can be classified as two-rowed barley vs. six-rowed barley, and hulless vs. hulled barley based on morphological differences, they are easy to cross with each other and are now all considered as one species, *Hordeum vulgare* L. with the genome and chromosome number as 2n = 2x = 14. Cultivated rye is a single species, *Secale cereal* L., with the genome and chromosome number as 2n = 2x = 14. Triticale (× Triticosecale Wittm. ex *A. Camus*) is a hybrid genus between wheat and rye. Although these cereal crops can be damaged by stripe rust, currently the disease is mostly important on wheat and followed by barley. Recently, stripe rust has become more important on triticale in Europe (Hovmøller et al. 2016).

4.1.5.1.1 Host Cropping System

In general, stripe rust on wheat is the most important, followed by on barley, and least on rye, which is related to the areas of cultivation. While *Pst* mostly infects wheat and *Psh* mostly infects barley, the separation of different formae speciales is not a clear cut as they can sometimes or occasionally infect the other crops (Chen et al. 1995). Also *Pst* can be found on rye (Cheng et al. 2016). Nevertheless, the separation of the forms on different crops is very useful for control of different stripe rust diseases, especially in the regions both forms are present. For example, wheat stripe rust was extremely severe under the unusually stripe rust-favorable weather conditions in the US Pacific Northwest in 2010 and 2011 (Chen 2014; Wan and Chen 2014; Wan et al. 2016), but barley stripe rust was not severe. The differences of the two crops related to the difference situations of the stripe rust mainly include the following. Firstly, both winter and spring wheat crops are grown in the region, which provides host plants all year long without crop gaps; in contrast, spring barley is mostly grown, which leaves a 7-month gap between crops, and thus limiting *Psh* survival. Secondly, wheat is grown in much larger areas in the region and also more close to wheat production in other regions, while barley is concentrated in a relatively small area with relatively high rainfall and also far away from barley production in other regions of the US. In addition, the decrease of barley acreage since the late 1990s due to the grain price disadvantage in comparison with wheat is also contributed to the current low level of barley stripe rust in the region.

Because the direct relationship between the host crop area and disease epidemic level and frequency, changing crops have been used to reduce damage caused by major diseases (Li and Zeng 2002; Zhu et al. 2000). In China, reducing wheat acreage has been proposed and practiced to reduce wheat yield loss caused by stripe rust. Fields from reducing wheat acreage are used to grow sunflower, potato and other cash crops. Cao et al. (2015) conducted experiments of different crops interplanted with wheat in southern Gansu in 2007–2009. They found that the relative control efficacies of the intercropping between wheat and maize were 16.7–45.7% for stripe rust and the yield increased by 52.4–140.0%. The relative control efficacies of intercropping between wheat and sunflower were 5.9–28.9% for stripe rust and the yield increased by up to 24.8%. Within a field, intercropping increases the effect of barriers of non-host crop to reduce spore dispersal from plants to plants of the host crops, and may also alter the moisture and/or temperature conditions to reduce infection. The intercropping system is suitable for small fields with intensive labors involved, but may not be practical for large fields with heavy machinery involvement. Cultivar mixture and multiline variety grown in in a single field may solve the intercropping problems. The former has two or more cultivars of a single crop grown in a field and the latter is a cultivar comprising of several lines of morphologically similar but with different resistance genes. Cultivar mixture has been commonly used in the US Pacific Northwest (Chen 2005), and in the same region, two multiline cultivars of club wheat, Crew and Rely, were developed and grown for many years for control of stripe rust on club wheat (Allan et al. 1983, 1993). Mixture of cultivars significantly reduces stripe rust severity and yield losses (Akanda and Mundt 1996). Because of the relative simplicity and beneficial for reducing other problems such as winter kill and frost damage and stabilizing yield under different environments, as well as reduce stripe rust damage, cultivar mixture has reached over 30% of the winter wheat acreage and also become popular for spring wheat crop in the Pacific Northwest. Although multiline cultivars were successfully used to control stripe rust on club wheat in the 1990s, this cropping system is laborious and expensive in keeping and mixing different components. Multiline cultivars have been replaced with highly stripe rust resistant cultivars with multiple resistance genes in each cultivar.

4.1.5.1.2 Host Susceptibility

Susceptibility or resistance of host crop is the most important host factor for stripe rust epidemics. Because the stripe rust pathogen is an obligate parasite, it relies on host plants for reproducing itself and cause disease. In a single field, the disease or damage level depends upon the level of susceptibility or resistance. The fields can be completely disease free if a cultivar with complete resistance is grown even under high disease pressure in the region. If a susceptible cultivar is grown, stripe rust will likely occur, and severity and damage will likely be high. In a region, the scale and severity level of epidemics is largely determined by the acreage of susceptible cultivars. Between complete resistance and complete susceptibility, there is a range of different levels of susceptibility or resistance, often referred to partial or incomplete resistance. Compared to highly susceptible cultivars, a cultivar with partial resistance should have relatively low rust and damage; and compared to highly resistant cultivars, a cultivar with partial resistance should have some level of rust severity. Different levels of susceptibility or resistance are clearer under more severe epidemics. The types of resistance and genes for resistance are presented in more detail in Chap. 5 and utilization of resistance in control of stripe rust are presented in Chap. 6.

4.1.5.1.3 Other Host Factors

Other host plant factors influencing stripe rust epidemics include plant statue, stroma frequency, winter hardness, earliness, shattering, canopy temperature, and possible other traits. These plant characteristics are not direct attributes to susceptibility or resistance to the stripe rust pathogen, but may change micro-environmental conditions favoring or not favoring pathogen infection and disease development. Based on the author's observations (X.M. Chen unpublished data), cultivars with more clustered leaves increase moisture and maintain moisture longer than those with less clustered leaves, and therefore increase pathogen infection and disease severity. Cultivars with broad leaves may have more chances to catch urediniospores than those with narrow leaves. Russell (1975) studied the distribution of urediniospores on leaves, and observed the most efficient catch on the abaxial face in the middle of the lamina. Wheat line W1343 that has erect leaves, had a threefold decrease in the number of urediniospores retained on its leaves, compared with varieties Cappelle-Desprez or Little Joss that have prostrate leaves.

Variation in stoma frequency exists in wheat cultivars, and has been intensively studied for relationships with water use efficiency, photosynthesis efficiency and yield (Bhagwat and Bhatia 1993). Because wheat infection by urediniospores is through stomata, infection efficiency that can be measured by the number of infection in a unit of leaf area may be related to stoma frequency. However, this has not been studied.

Winter hardness related to stripe rust is well known. The stripe rust pathogen survives winter as mycelia in infected leaf tissue. In general, when wheat leaves survive the winter, the stripe rust mycelia in the leaves also survive (Li and Zeng 2002). In a recent study by Ma et al. (2015) conducted under controlled conditions, the survival of *Pst* in cultivars with strong winter-hardiness at 0 and −5 °C was greater than those with weak winter-hardiness. However, such differences among cultivars were negligible at −10, −15, and −20 °C. These results indicate that *Pst* mycelia inside green leaves can be killed by low temperatures rather than through death of green leaves under low temperatures. Nevertheless, survival of wheat leaves is the precondition for survival of rust mycelia.

The earliness of a cultivar is directly related to the duration of stripe rust epidemics and also to inoculum load for the next crop season. An early mature crop or cultivar shortens the period during which the pathogen infect plants, produce urediniospores for increasing disease or as initial inoculum for the following crop, and thus suffer less damage. In the US Pacific Northwest, spring barley and wheat are planted about the same time, but barley matures fast and is harvested about 1–2 weeks earlier, and thus generally has less damage from *Psh* than spring wheat from *Pst* (X.M. Chen, unpublished data). Both barley cultivars Steptoe and Topper are highly susceptible to *Psh*, but final stripe rust severity of Steptoe is often less than Topper as Steptoe matures earlier than Topper. Similar observations have been found among susceptible wheat cultivars, for example, Lemhi often has higher yield loss than Avocet Susceptible, as the latter matures faster.

Shattering can be a harvesting problem, reducing realized grain yield. It also can be related to stripe rust epidemics because shattering often results in more volunteer plants that serve as a green-bridge to carry over the stripe rust pathogen from a previous crop to next, especially when a susceptible cultivar also has a shattering problem. For example, the spring wheat cultivar Zak that became susceptible to a new race of *Pst* appeared in the US Pacific Northwest in 2002 had severe stripe rust in the same year. Due to the mild 2002–2003 winter, *Pst* survived the winter well, so did volunteering Zak plants, which created a significant stripe rust problem in winter wheat in 2003. Spring barley cultivar Steptoe was widely grown in the Pacific Northwest before 2000. It is highly susceptible to every race of *Psh* that reached to the region in 1995 (Chen et al. 1995). The spring cultivar is quite winter hardy and has a shattering problem. The combination of susceptibility, shattering for volunteer plants, and winter hardy contributed to the widespread of barley stripe rust in the region in the late 1990s. The replacement of the most popular cultivar with cultivars with relatively resistant cultivars and less shattering has contributed the reduction of barley stripe rust epidemic (Chen 2004; Wan and Chen 2012).

Crop canopy temperature is a characteristic of wheat cultivars and has been studied in relation to photosynthesis, water use efficiency, adaptation, and yield. Recently, the trait was studied in relation to stripe rust development (Cheng et al. 2015). In the study, five wheat genotypes that have different canopy temperatures and are susceptible to *Pst* at the seedling stage were investigated for resistance to stripe rust at the adult-plant stage in greenhouse and field experiments. In the greenhouse tests, all five varieties were resistant to stripe rust at the adult-plant stage. At the grain-filling stage in the 2007–2010 field experiments at two locations, the two genotypes (NR 9405 and 9430) with higher canopy temperatures had significantly lower disease index and smaller area under the disease progress curve values than the three varieties (Shaan 229, RB 6 and Xiaoyan 6) with lower canopy temperatures. The results indicate that high wheat canopy temperatures provide unfavorable conditions for the development of stripe rust.

In addition to these plant traits, early planting of winter wheat and late planting of spring wheat in some regions like the US Pacific Northwest usually increase stripe rust severity as the former increases stripe rust pathogen infection in autumn and the latter expends the rust season and producing more urediniospores for carrying over to the next crop season. Irrigation and excessive use of fertilizer usually increase stripe rust severity as they create high moisture conditions for infection and lush plant canopy for easy spread of urediniospores from plants to plants.

4.1.5.2 Auxiliary Hosts

Auxiliary hosts are also called accessary hosts. For cereal rust fungi, an auxiliary host is a grass weed host infected by a rust fungal species or forma specialis that causes a major disease on a cereal crop. The interactions of a rust fungus with grass hosts have been much longer than with a cereal crop. The long evolutionary relationship between a rust fungus and grasses has made the pathogen and host at a relatively balance compared to the pathogen and crop system that is under much influence from breeding and growing resistant cultivars.

For rust fungi that have sexual reproduction as an important component in the pathogen lifecycle and disease development, such as stem rust of wheat and barley and crown rust of oat (Roelfs 1982; Carson 2008), the relationships of a pathogen with its primary hosts, alternate hosts, and auxiliary hosts are complex. Primary and auxiliary hosts provide basidiospores to infect alternate hosts, and latter provide aeciospores to infect the former two. Clonal reproduction occurs on primary and auxiliary hosts, while sexual reproduction occurs on alternate hosts. In this way, auxiliary hosts provide not only urediniospores for direct infecting primary hosts, but also through alternate hosts provide new or uncommon virulences to crops. For *Pst*, such complex relationships could exist in western China as *Pst* can be found on barberry plants (Zhao et al. 2013; Wang et al. 2016), and possibly in the Himalaya Mountain regions as speculated based the diversity of *Pst* collections determined using molecular markers (Ali et al. 2014a, b; Thach et al. 2016). However, in the US and Europe, alternate hosts of *Pst* do not play a role in the disease epidemics (Berlin et al. 2012; Wang and Chen 2015; Wang et al. 2015), the relationships of the pathogen are only with primary and auxiliary hosts.

4.1.5.2.1 Species of Auxiliary Hosts

In addition to occurrence on the primary hosts of cereal crops, *P. striiformis* also infects many grass species. The grass species that can harbor the stripe rust fungus were reported in Europe (Eriksson 1894; Becker and Hart 1939; Manners 1950, 1960; Hassebrauk 1962, 1965; Hassebrauk and Röbbelen 1975; Stubbs 1985), North America (Humphrey et al. 1924; Newton et al. 1933; Hendrix et al. 1965; Tollenaar and Houston 1967; Tu 1967; Shaner and Powelson 1973; Chen et al. 1995; Cheng et al. 2016), South America (Arthur 1925), Asia (Abiev et al. 1982; Niu et al. 1991; Li and Zeng 2002), and Australia (Wellings 1982, 2007). In Canada and the US, the following species were susceptible to *Pst* and/or *Psh*, or stripe rust isolates collected from these grasses can infect wheat and/or barley: *Aegilops cylindrica*, *Agropyron cristatum*, *A. bakeri*, *A. dasystachyum*, *A. griffithsi*, *A. riparium*, *A. sibiricum*, *A. spicatum*, *A. trachycaulum* var. *typicum*, *A. trachycaulum* var. *unilaterale*, *Avena fatua*, *Bromus biebersteinii*, *B. carinatus*, *B. ciliatus*, *B. marginatus*, *B. pumpellianus*, *Elymus canadensis*, *E. cinereus*, *E. curvatus*, *E. dahuricus*, *E. fedtschenkoi*, *E. glaucus*, *E. virginicus*, *Hordeum brachyantherum*, *H. jubatum*, *H. spontaneum*, *Hesperochloa kingii*, *Sitanion hystrix*, *Thinopyrum ponticum* (syn. *Agropyron elongatum*, *Elymus elongatus*, *E. obtusiflorus*, *Pascopyrum smithii*, and *Pseudoroegneria spicata*) (Newton and Johnson 1936; Hendrix et al. 1965; Tollenaar 1967; Tu 1967; Shaner and Powelson 1973; Cheng et al. 2016). In the US, stripe rust isolates from *Poae* spp. and *Dactylis* spp. consistently found non-pathogenic to *Pst* and *Psh* (Tollenaar 1967; Shaner and Powelson 1973; Chen et al. 1995; Cheng et al. 2016). In Canada, Newton and Johnson (1936) reported that the wheat stripe rust fungus could infect some native grasses in western Canada. They inoculated native grasses with urediniospores from wheat fields and found that species of *Agropyron* (*A. cristatum*, *A. dasystachyum*, *A. griffithsi*, *A. trachycaulum* var. *typicum*, and *A. trachycaulum* var. *unilaterale*,), *Elymus* (*E. dahuricus* and *E. canadensis*), and *Hordeum* (*H. jubatum*) were highly susceptible.

In China, Li and Liu (1956) collected urediniospores of the stripe rust fungus from 15 wild grass species native in Shaanxi province and used them to inoculate wheat, and used urediniospores from wheat in local fields to inoculate these grasses and 10 cultivated forage grass species from Inner Mongolia. They did not observe susceptible reactions in these tests, indicating that the isolates from grasses were not pathogenic to wheat and wheat isolates were not pathogenic to the grasses. Lu et al. (1958) observed large scale and heavy infections of *P. striiformis* on *Elymus chinensis* [now considered as a synonym of *Leymus chinensis* (Trin.) Tzvelev] in Shanxi Province in northern China. Inoculation of common wheat cultivars with the samples produced limited uredinia. However, after consecutive inoculations of a susceptible wheat, 'Yanda 1885', after five generations, no uredinia could be produced. This grass species was not considered as a host for the wheat stripe rust pathogen and the stripe rust fungus on the grass was not considered as the wheat stripe rust pathogen. They found that *Pst* was able to infect 11 grass species including *Agropyron cristatum*, *Bromus tectorum* and *Elymus sibiricus*. They also found that stripe rust samples from *E. sibiricus* and *Agropyron* spp. were able to infect wheat. Nan (1990) observed stripe rust on cultivated grasses *Elymus breviaristatus* (should be *Bromus breviaristatus* Thurb.), *E. sibiricus*, *Hordeum brevisubulatum*, *Phleum pratense* and *Roegneria semicostata*, but it was not clear whether the rust was caused by the wheat stripe rust fungus or other formae speciales. Niu et al. (1991) inoculated 50 grass species with isolates of four *P. striiformis* formae speciales including *Pst* and *Psh*. They found four grass species, *Elymus excelsus*, *Eremopyrum orientale*, *Hordeum jubatum*, and *H. spontaneum* var. *spontaneum*, were susceptible to both *Pst* and *Psh*; *Agropyron cristatum* and *Thinopyrum ponticum* susceptible to *Pst*; and *Aegilops biuncialis*, *Ae. cylindrica*, *Ae. tauschii*, *Haynaldia villosa*; and *Hordeum agriocrithon* susceptible to *Psh*; and few grass species were susceptible to stripe rust isolates from *Elymus* or *Leymus*. Li and Zeng (2002) summarized the hosts of *P. striiformis* in China and listed 88 species in 16 genera including the primary crops wheat, barley and rye. They classified the studies into three categories: (1) stripe rust found on plants under wild natural habitats and/or fields and inoculation of wheat varieties with the samples produced *Ps* uredinia; (2) stripe rust found on planted grasses and/or experimental plots planted with grasses and inoculation of wheat varieties with the samples produced *Ps* uredinia; and (3) inoculation of the grasses with *Pst* produced uredinia. Of the 88 species, 10 species are in the first, 6 in the second and 72 in the third categories. Although the grass species in the first category can be infected by *Pst* under natural conditions, large scale of infections have not been reported. The distribution of the species in the second category is limited in China. The species in the third category have been only found susceptible to *Pst* under controlled inoculation conditions. Because of these, the authors thought that wild grass hosts may not play an important role in the epidemiology of wheat stripe rust (Li and Zeng 2002).

In South America, *Hordeum chilense* and *Agropyron altenuatum* [this could be *Agropyron attenuatum* (Kunth) Roem. & Schult as no *Agropyron altenuatus* was found in any botany book] are major grass hosts of the stripe rust pathogen (Arthur 1925). In Israel, stripe rust isolates from wild grass *Triticum dicoccoides*, *Aegilops kotschyi* and *Hordeum spontaneum* were virulent on wheat (Wahl et al. 1984).

The actual number of grass species susceptible to stripe rust is difficult to obtain because many species even genera of grasses have been changed names over the years. Based on Hassebrauk (1965), over 320 grass species belonging to 50 genera, but only in family Gramineae, can be infected by the stripe rust fungus, which summarized the early works mostly conducted in Europe and North America. The author (X.M. Chen) summarized 230 species which are recognized as hosts for *P. striiformis* in the US (http://striperust.wsu.edu). Based on these lists, the list of stripe rust hosts in China (Li and Zeng 2002), and the above discussions, a total of 292 species (including subspecies and hybrids) are listed in Table 4.1. These species are in 56 genera, of which *Aegilops* (28 species), *Agropyron* (15 species), *Bromus* (46 species), *Elymus* (38 species), *Hordeum* (21 species) and *Leymus* (11 species) are more important than other genera as they have many species susceptible to stripe rust.

4.1.5.2.2 Possible Roles of Grass Hosts

The roles of grass hosts playing in stripe rust are complicated. They can be important for pathogen survival, providing initial inoculum or increasing inoculum load, spreading, recombination and maintaining diversity. The importance of grass hosts to stripe rust epidemics on cereal crops varies from grass to grass and from region to region. In general, grass hosts can have the following six roles: (1) for pathogen survival and providing initial inoculum; (2) spreading the pathogen inoculum from one region to another region; (3) increasing inoculum load; (4) providing opportunities for somatic and sexual recombination of different races or formae speciales; and (5) maintaining pathogen diversity.

For Pathogen Survival and Providing Initial Inoculum Grass hosts could act as a green bridge between the previous and successive cereal crop seasons for a rust fungus to survive. The stripe rust fungus prefers relatively cool conditions compared to other rust fungi. Hot weather conditions limit the fungus oversummering in the absence of wheat or barley crop in fields. Hendrix et al. (1965) investigated the possible oversummer environments in the US Pacific Northwest and found active stripe rust uredinia on six native grasses in four mountainous areas of 1100 to 3000 m

Table 4.1 Cereal and grass species that have been reported to be infected by *Puccinia striiformis* naturally and/or artificially

Genus/species	Genus/species
Achnatherum (2)	*A. myosuroides* Huds.
A. inebrians (Hance) Keng.[a]	***Aneueolepidum*** (2)
A. nelsonii ssp. *nelsonii* (Scribn.) Barkworth	*A. chinemse* (Trin.) Kitag
Aegilops (28)	*A. dasystachys* (Trin.) Nevsk.
A. aucheri Boiss.[b]	***Arrhenatherum*** (1)
A. bicornis Jaub. & Spach[b]	*A. elatius* (L.) P. Beauv. ex J. Presl. & C. Presl.
A. biuncialis Vis.	***Avena*** (2)
A. caudata L.	*A. fatua* L.
A. caudata L. × *A. umbellulata* Zhuk.	*A. sterilis* L.
A. columnaris Zhuk.	***Beckmannia*** (2)
A. comosa Sibth. Et Sm.[b]	*B. eruciformis* (L.) Host.
A. crassa Boiss.	*B. syzigachne* (Steud.) Fernald
A. cylindrica Host	***Boissiera*** (1)
A. cylindrica var. *globrospica*[b]	*B. squarrosa* (Sol.) Nevski
A. geniculata Roth	***Brachypodium*** (2)
A. juvenlis (Thell.) Eig.[b]	*B. pinnatum* (L.) P. Beauv.
A. kotschyi (Boiss.) Eigen	*B. sylvaticum* (Huds.) P. Beauv.
A. ligustica (Savign.) Coss.[b]	***Briza*** (1)
A. longissima Schweinf. & Muschl.	*B. media* L.
A. neglecta Req. ex Bertol	***Bromus*** (46)
A. ovate L.[b]	*B. anomalus* Rupr. ex E. Fourn.
A. peregrine (Hack.) Maire & Weiller	*B. arvensis* L.
A. searsii Feldmn et Kislev.[b]	*B. biebersteinii* Roem. & Schult.
A. sharonensis Eigen	*B. briziformis* Fisch. & C.A. Mey.
A. sharonensis Eigen. × *A. umbellulata* Zhuk.	*B. carinatus* Hook. et Arn.
A. speltoides Tausch	*B. catharticus* Vahi.
A. tauschii Coss.	*B. ciliates* L.
A. triuncialis L.	*B. commutatus* Schrad.
A. umbellulata Zhuk.	*B. danthoniae* Trin.
A. vatialis Eigen[b]	*B. diandrus* Roth
A. vavilovii (Zhuk.) Chennav.[b]	*B. diandrus* ssp. *diandrus* Roth
A. ventricosa Tausch	*B. erectus* Huds.
Agropyron (15)	*B. frondosus* (Shear) Wooton & Stand.
A. bakeri E. Nels.[c]	*B. hordeaceus* L.
A. brachyphyllum Boiss. & Hausskn.[d]	*B. inermis* Leyss.
A. caespitosum K. Koch[d]	*B. intermedius* Guss.
A. caninum (L) P. Beauv.[e]	*B. japonicas* Thunb.
A. cristatum (L.) Gaertn.	*B. kalmia* A. Gray
A. dasystachyum (Hook.) Scribn.[d]	*B. laevipes* Shear
A. desertorum (Fisch. ex Link) Schult.	*B. lanceolatus* Roth
A. fragile (Roth) P. Candargy	*B. latiglumis* (Shear) Hitchc.

(continued)

Table 4.1 (continued)

Genus/species	Genus/species
A. intermedium (Host) Beauv.[d]	*B. madritensis* L.
A. mongolicum Keng.	*B. marginatus* Nees ex Steud.
A. reparium Scribn. & Smith[c]	*B. mollis* L.[f]
A. repens (L.) Beauv.[d]	*B. oxyodon* Schrenk
A. sibircum (Willd.) Beauv.[b]	*B. pacificus* Shear
A. smithii Rydb.[a]	*B. pectinatus* Thunb.
A. sussecundum (Link) Hitchc.[d]	*B. polyanthus* Scribn. ex Shear
A. trachycaulum (Link) Malte[d]	*B. pumpellianus* Scribn.
A. trichophorum (Link) Richt.[d]	*B. racemosus* L.
Agrostis (1)	*B. ramosus* Huds.
A. capillaris L.	*B. ramosus* Huds. ssp. *benekenii* (Lange) Lindb.
Aira (1)	*B. remotiflorus* (Steud.) Ohwi
A. caryophyllea L.	*B. richardsonii* Link
Alopecurus (2)	*B. rubens* L.
A. arundinaceus Poir.	*B. scoparius* L.
B. secalinus L.	*E. virginicus* L.
B. sitchensis Trin.	*E.* × *brachyphyllus* (Boiss. & Hausskn.) Á.Löve
B. squarrosus L.	*E.* × *mayebaranus* (Honda) S.L. Chen
B. sterilis L.	***Elytrigia*** (6)
B. suksdorfii Vasey	*E. elongate* (Host) Nevski[b]
B. tectorum L.	*E. intermedia* (Host) Nevski[b]
B. tomentellus Boiss.	*E.* × *littorea* (Schumach.) Hyl.
B. unioloides H. B. et K.[f]	*E. lolioides* (P. Candargy) Nevski
B. variegatus M. Bieb.	*E. obtusiuscula* (Lange) Hyl.
B. vulgaris (Hook.) Shear	*E. repens* (L.) Desv.[b]
Calamagrostis (1)	***Eremopyrum*** (4)
C. epigeios (L.) Roth	*E. bonaepartis* (Spreng.) Nevski
Castellia (1)	*E. distans* (K. Koch) Nevski
C. tuberculosa (Moris) Bor.	*E. orientale* Jaub. & Spach
Catabrosa (1)	*E. triticeum* (Gaertn.) Nevski
C. aquatic (L.) P. Beauv.	***Festuca*** (9)
Chloris (1)	*F. arundinacea* Schreb.
C. verticillata Nutt.	*F. capillata* Lam.
Cornucopiae (1)	*F. gautieri* (Hack.) K. Richt
C. cucullatum L.	*F. gigantean* (L.) Vill.
Dactylis (1)	*F. kingie* (*S. Watson*) Cassidy
D. glomerata L.	*F. mairei* St.-Yves
Dasypyrum (1)	*F. pratensis* Huds.
D. villosum (L.) Borbás	*F. rubra* L.
Echinaria (1)	*F. thurberi* Vasey

(continued)

Table 4.1 (continued)

Genus/species	Genus/species
E. capitata (L.) Desf.	***Gaudinia*** (1)
Elymus (38)	*G. fragilis* P. Beauv.
E. albicans (Scribn. & J.G. Sm.) Á. Löve	***Glyceria*** (1)
E. angulatus J. Presl.	*G. ischyroneura* Steud.
E. bakeri (E. Nelson) Á. Löve	***Holcus*** (2)
E. canadensis L.	*H. lanatus* L.
E. caninus L.	*H. mollis* L.
E. ciliaris Tzvelev	***Hordelymus*** (1)
E. ciliaris ssp. *cilaris* (Trin.) Tzvelev	*H. europaeus* (L.) Jexx. ex Harz
E. cordilleranus Davidse & R.W. Pohl	***Hordeum*** (21)
E. dahuricus Turcz. ex. Griseb	*H. bogdanii* Wilensky
E. dentatus (Hook.f.) Tzvelev	*H. brachyantherum* Nevski
E. europaeus L.[b]	*H. brevisubulatum* (Trin.) Link
E. excelsus Turcz. Ex Griseo[b]	*H. bulbosum* L.
E. glabriflorus (Vasey ex L.H. Dewey) Scribn. & C.R. Ball	*H. chilense* Roem. et Schult.
E. glaucus Buckl.	*H. comosum* J. Presl
E. hispidus (Opiz) Melderis	*H. depressum* (Scribn. & J.G.Sm.) Rydb.
E. hystrix L.	*H. distichon* L. (*H. vulgare* var. *nudum* Hook.[b])
E. interruptus Buckl.	*H. euclaston* Steud.
E. lanceolatus (Scribn. & J.G. Sm.) Gould	*H. halophilum* Griseb.
E. lanceolatus ssp. *lanceolatus* (Scribn. & J.G. Sm.) Gould	*H. jubatum* L.
E. longiaristatus (Boiss.) Tzvelev	*H. marinum* Roth ssp. *gussoneanum* (Parl.) Thell.
E. marginatus (H. Lindb.) Á. Löve	*H. marinum* Huds.
E. nipponicus Jaaska	*H. murinum* L. ssp. *leporinum* (Link) Arcang.
E. panormitanus (Parl.) Tzvelev	*H. murinum* L.
E. patagonicus Speg.	*H. procerum* Nevski[b]
E. pycnanthus (Godr.) Melderis	*H. pusillum* Nutt.
E. repens (L.) Gould	*H. secalinum* Schreb.
E. scabriglumis (Hack.) Á. Löve	*H. spontaneum* K. Koch
E. scabrous (R. Br.) Á. Löve	*H. stenostachys* Godr.
E. scribneri (Vasey) M.E. Jones	*H. vulgare* L.
E. semicostatus (Nees ex Steud.) Melderis	***Hystrix*** (3)
E. sibiricus L.	*H. californica* (Bol.) Kuntze
E. smithii (Rydb.) Gould	*H. duthiei* (Stapf) Bor
E. submuticus (Hook.) Smyth	*H. patula* Moench
E. tangutorum (Nevski.) Hand. Mazz[b]	***Koeleria*** (1)
E. trachycaulus (Link) Gould ex Shinners	*K. pyramidata* (Lam.) P. Beauv.
Lagurus (1)	*P. nuttaliana* (Schult.) Hitchc.

(continued)

Table 4.1 (continued)

Genus/species	Genus/species
L. ovatus L.	***Roegneria*** (11)
Lamarckia (1)	*R. canina* Nevsk.[b]
L. aurea (L.) Moench	*R. kamoji* Onwi[b]
Leymus (11)	*R. laxiflora* Keng[b]
L. ambiguous (Vasey & Scribn.) D.R. Dewey	*R. leptoura* Nevski[b]
L. arenarius (L.) Hochst.	*R. multiculmis* Kitag.[b]
L. chinensis (Trin.) Tzvelev	*R. nutans* (Keng) Keng[b]
L. cinereus (Scribn. & Merr.) Á.Löve	*R. parvigluma* Keng[b]
L. condensatus (J. Presl) Á.Löve	*R. pendulina* Nevski[b]
L. flavescens Scribn. & J.G.Sm.	*R. stricata* Keng[b]
L. racemosus (Lam.) Tzvelev	*R. tsukushiense* Honda B. R. Lu, Yen et J. L. Yang[b]
L. ramosus (C. Richt) Tzvelev	*R. turczaninovii* (Drob.) Nevski[b]
L. salinus (M.E.Jones) Á.Löve	***Rostraria*** (3)
L. secalinus (Georgi) Tzvelev	*R. cristata* (L.) Tzvelev
L. triticoides (Buckley) Pilg.	*R. hispida* (Savi) Dogan
Lolium (5)	*R. rohlfsii* (Asch.) Holub
L. multiflorum Lam.	***Secale*** (6)
L. perenne L.	*S. anatolicum* Boiss.[b]
L. remotum Schrank	*S. cereal* L.
L. rigidum Gaudich.	*S. dignoricum* (Vavilov) Roshey.[b]
L. temulentum L.	*S. montanum* Guss.
Milium (1)	*S. sylvestre* Host
M. effusum L.	*S. vavilovii* Grossh[b]
Muhlenbergia (1)	***Sitanion*** (5)
M. huegelii Trin.	*S. hanseni* (Scribn.) J. G. Smith
Parapholis (1)	*S. hystrix* (Nutt.) J. G. Smith
P. incurve (L.) C. E. Hubbard	*S. jubatum* J. G. Smith
Phalaris (4)	*S. longifolium* J. G. Smith
P. arundinacea L.	*S. hanseni* (Scribn.) J. G. Smith
P. canariensis L.	***Taeniatherum*** (1)
P. minor Retz.	*T. caput-medusae* (L.) Nevski
P. paradoxa L.	*T. crinitum* (Schreb.) Nevski
Phleum (1)	***Thinopyrum*** (3)
P. pratense L.	*T. caespitosum* (K. Koch) Zhi W. Liu & R. R.-C. Wang
Poa (8)	*T. elongatum* (Host) D. R. Dewey
P. alopecurus subsp. *prichardii* (Rendle) Giussani & Soreng	*T. podperae* (Nábělek) D. R. Dewey
P. arachnifera Torr.	***Trisetum*** (2)
P. compressa L.	*T. cernuum* Trin.
P. juncifolia Scribn.	*T. flavescens* (L.) P. Beauv.

(continued)

Table 4.1 (continued)

Genus/species	Genus/species
P. litorosa Cheeseman	***Triticum*** (9)
P. nemoralis L.	*T. durum* Desf.
P. pratensis L.	*T. monococcum* L.
P. secunda J. Presl	*T. persicum* Vav.
Psathyrostachys (2)	*T. polonicum* L.
P. huashanica Keng[b]	*T. spelta* L.
P. juncea (Fisch.) Nevski	*T. sphaeroccum* Perc.
Pseudoroegneria (2)	*T. timopheevii* Zhuk.
P. divaricate (Boiss. & Balansa) Á. Löve	*T. turgidum* L.
P. spicata (Pursh) Á. Löve	*T. vavilovii* Jakubz.
Psilurus (1)	***Vulpia*** (6)
P. incuruvus (Gouan) Schinz & Thell.	*V. bormoides* (L.) Gray
Puccinellia (6)	*V. ciliate* Dum.
P. distans (Jacq.) Parl.	*V. delicatula* Dum.
P. hybrid Holmberg	*V. geniculate* Link
P. lemmoni (Vasey) Scribn.	*V. membranacea* (L.) Dumfort.
P. maritime (Huds.) Mutel.	*V. myuros* (L.) C.C. Gmel.
P. nutkaensis (J. Presl.) Fern. et Weath.	

The list of species and subspecies is modified from Hassebrauk (1965) in combination with those reported in later studies as indicated by the following footnotes:
[a]Liu and Hambleton (2010)
[b]Li and Zeng (2002)
[c]Hendrix et al. (1965)
[d]Dietz and Hendrix (1962)
[e]Becker and Hart (1939)
[f]Holmes and Dennis (1985)

elevation around the Columbia Basin in Washington and Oregon states. Wheat plants grown in monitoring plots at Blue Mountains in late summer and early fall were infected by *Pst*. Similarly, grasses have been reported to be important for the survival of the stripe rust pathogen in some Asian and European countries. For example, stripe rust on wild grasses is important for epidemics on wheat crops in the Harz region of Germany (Becker and Hart 1939) and southeastern Kazakhstan (Abiev et al. 1982). In Pakistan, the stripe rust fungus survives summer on grasses (Hassan 1978). In Australia, the stripe rust fungus survives in many local grass species in genera *Agropyron*, *Phalaris* and *Bromus* (Wellings 1982). When wheat or barley crops and volunteer plants are absent off the crop season, the rust pathogen kept by grass hosts can be important to spark epidemics on next crops.

In the literature, there have been controversial conclusions about the significance of grass hosts for rust epidemics on cereal crops. The different conclusions can be due to the different environments, crop systems, rust populations and grass species. They can also be due to the difficulties in testing grasses for susceptibility to rust fungi. For stripe rust, some previous conclusions about whether grasses are important or not were inferred from field investigations or observations, while others were from artificial inoculations under controlled conditions. The latter studies often put the grass-rust fungus interactions under overly saturated rust-favorable conditions, which may seldom occur in the nature. Because of this, cautions should be taken when estimate the importance of grasses based upon data of artificial inoculation. In the US, wild grasses are considered to play an important role for the stripe rust pathogen to oversummer in some regions as discussed above, but other researchers considered wild grasses are not important as inoculum sources to wheat crops based on their observation in fields (Sharp and Hehn 1963; Shaner and Powelson 1973). Based on the author's observations (X. M. Chen), stripe rust appears on grasses usually when the disease is very severe and widespread in crop fields. Occasionally, grass plants, especially *Aegilops* spp., have uredinia before susceptible wheat plants to have stripe rust uredinia. The author also observed severe heavy stripe rust on grasses in or near fields of wheat crops free of rust, which is likely due to the susceptibility of grasses to the stripe rust population, to which the wheat cultivars grown in the fields are resistant. This lets the author speculate another role of grasses, maintaining diverse populations of the pathogen as discussed below. Also, although infection occurs on wheat crops and grasses about the same time, the early sporulation on grasses provides more spores for infecting susceptible wheat crops, and thus, grass hosts increase the amount of inoculum for infecting crops.

Spreading the Pathogen Inoculum from One Region to another Region Grasses act as a bridge to spread rusts from one region to another region, or even from one continent to another continent. In North America, the earliest specimens of stripe rust were collected from *Elymus glaucus*, *Bromus carinatus* and *B. hookerianus* near Seattle and Everett of Washington State in 1892 (Humphrey et al. 1924), more than 20 years before stripe rust was reported on wheat crops (Carleton 1915). Later, Hungerford and Owens (1923) found 39 grass species were naturally infected by the stripe rust fungus, in addition to cultivated wheat and barley. Stripe rust was identified in San Juan Islands and Skagit County in the west coast, and Pullman in the east of Washington from various grass species (*Agropyron spicatum*, *A. violaceum*, *Bromus marginatus*, *B. pacificus*, *B. sitchensis*, *Elymus glaucus*, *Hordeum caespitosum*, *H. jubatum* and *H. nodosum*) (Humphrey et al. 1924). These grasses also grow naturally in Alaska and Vancouver Islands. Some grass species, such as *H. jubatum* and *Agropyron cristatum*, are the most congenial hosts for stripe rust across North America and Siberia. Because of these observations and facts, the stripe rust pathogen was postulated to migrate from Asia to North America and initially established on American grasses before wheat was established as a crop in the continent (Line 2002). It is not unreasonable to accept the postulation that the stripe rust fungus was from Siberia by the way of Kamchatka and Aleutian Islands to North America (Humphrey et al. 1924; Line 2002).

The similar spreading mode was indicated by the stripe rust pathogen in South America, where the earliest reports of stripe rust, caused by *Pst*, were from wild grasses *Hordeum chilense* in Chile and *Agropyron alternuatum* (should be *A. atternuatum*) in Ecuador (Arthur 1925). Grasses have been indicated to play an important role for spreading of the recently introduced barley stripe rust pathogen, caused by *Psh*, in South America. The wild barley grass *Hordeum muticum* in the Andean range can serve as reservoirs for the barley stripe rust fungus, leading it to spread southward and northward in South America and North America (Dubin and Stubbs 1986; Chen et al. 1995; Brown et al. 2001).

Increasing Inoculum Load If stripe rust occurs on grasses and cereal crops about the same time, grasses may not providing initial inoculum, but can provide additional urediniospores to increase inoculum load for epidemics on cereal crops. In California, the stripe rust fungus survives on grasses in the Sierra and Nevada Mountain ranges, which may either provide primary inoculum for fall-sown winter wheat in the regions of low elevation in certain years (Tollenaar and Houston 1967), or add to inoculum generated from volunteer plants or cover crops of wheat in irrigated orchards (Line 1976, 2002). When not serving as a sole source of inoculum, the contribution of grasses has been often less estimated or ignored. The relative contribution of grass hosts to the overall amount of inoculum compared to cereal crops is hard to measure, but should vary from region to region and from season to season, depending upon the distribution of grasses within or outside the crop production region and the relative susceptibility of the primary and auxiliary hosts to the rust populations.

Providing Opportunities for Somatic and Sexual Recombination of Different Races or Formae Speciales Some grass species are susceptible to different formae speciales of a cereal rust species or different races of a same forma specialis, and thus, may provide a platform for genetic exchange. Different but closely related formae speciales infecting same grass plants may produce new races through somatic hybridization. Cheng et al. (2016) studied the virulence phenotypes and SSR genotypes of 42 stripe rust isolates collected from 13 grass species in the US in 2000–2008, and identified isolates capable of infecting some of the wheat and barley differentials. Although these isolates may represent a distinct genetic populations that have a wider host range than either the *Pst* or *Psh* population, some of the isolates could have developed through somatic hybridization between the two forms, as these isolates are highly heterozygous while *Pst* and *Psh* isolates are more homozygous for opposite SSR alleles. Lei et al. (2017) obtained hybrid isolates by co-infection of a common susceptible wheat genotype with a mixture of urediniospores of a *Pst* isolate and a *Psh* isolate. They demonstrated that the hybrid isolates were produced through somatic recombination through both virulence tests and molecular markers. Using the same approach, they also obtained new races through somatic recombination by co-inoculating wheat or barley plants with different races of *Pst* or *Psh*.

The contribution of grasses to rust fungal recombination may also through sexual reproduction on common alternate hosts. This is more likely for the stem rust pathogen than for the stripe rust pathogen. Jin (2011) identified a large number of pathotypes of *P. graminis* in aecial samples collected from *Mahonia* spp. in northeastern Washington, USA. Many of the pathotypes were able to infect some of wheat and rye differentials for distinguishing *Pgt* and *Pgs*, respectively. These pathotypes are likely resulted from *Pgs* and/or *Pgt* produced on wild rye grasses and wheat grasses.

In China, telia were found on *Elymus cylindricus* and *Leymus secalinus* from October, 2014 to March, 2015 (J.F. Qin, Y.Y. Zhao, Q. Li, S.X. Zuo, X.M. Chen, L.L. Huang, J. Zhao, and Z.S. Kang, unpublished data). Inoculated barberry plants under controlled conditions with teliospores produced aecia; and inoculation of a stripe rust susceptible wheat cultivar with aeciospores produced stripe rust. Various infection types (IT) from 1 to 4 using the 0–4 scale were observed on the susceptible wheat cultivar that is highly susceptible (IT 3–4) to *Pst*. The urediniospore isolates were identified as different races, and the isolates produced resistance reaction (IT 1–2) were not typical races found in wheat fields. If teliospores of the stripe rust fungus on grasses can infect barberry under natural conditions, they could produce aeciospores to infect wheat crop. Different races may recombine on common susceptible grass plants and virulences from different isolates produced on grasses may also recombine to produce different races. More studies are needed to test the possibilities.

Maintaining Pathogen Diversity Grass hosts serve as reservoirs to maintain a more diverse population than that occurs on a cultivated cereal crop. As breeders develop and farmers grow resistant cultivars, cereal crops select virulent races of a rust fungus that are able to reproduce. Races that are not able to reproduce on crop cultivars may be able to infect and reproduce on grasses. From stripe rust samples collected from grasses, Cheng et al. (2016) identified *Pst* races that were not found from wheat fields, as well as those commonly found from field fields. Through both pathogenicity and molecular marker tests, they found that the *P. striiformis* population from various grasses was more diverse than either of those from wheat or barley.

Races with a narrow virulence spectrum have been often collected from grasses in the US (Line and Qayoum 1992; Chen et al. 2010). Those with a narrow virulence spectrum may serve as progenitors for evolving into more virulent races. Many races with a new virulence factor often first appear with a narrow virulence spectrum (Line and Qayoum 1992). These races are more likely evolved from races with few virulence factors on grasses than predominant races on wheat crops. These simple races may recombine with other races to form more complex races (Line and Qayoum 1992). Races of *P. graminis* isolated from *Elymus glaucus* also displayed very narrow virulence spectrum (Jin et al. 2014). Therefore, grass hosts can maintain races unable to infect commercially grown cultivars, from which new races capable of infecting crop cultivars may evolve.

4.1.6 Environmental Factors

Stripe rust is more affected by environmental factors than many other plant diseases. Compared to the leaf rust and stem rust fungi, the stripe rust fungus adapts to cool climates, and therefore, stripe rust occurs in the temperate regions, high elevations and cool seasons of hot regions. Because wheat is grown in relatively cool climates, stripe rust can occur in all regions where wheat is grown, but the frequency, severity and scale vary from region to region depending upon the climates in addition to crop systems as discussed above. Even for a single region, the disease varies in severity from year to year depending upon the weather conditions. It is important to understand the environmental factors affecting the pathogen biology and disease development for predicting and managing stripe rust. Most stripe rust studies on environmental factors in the first half of the twentieth century were on survival, spread, and reproduction of the pathogen and development of disease epidemics. Such studies have been well reviewed by several authors (Zadoks 1961; Hassebrauk and Röbbelen 1975; Rapilly 1979; Stubbs 1985; Line 2002; Chen 2005). Among the weather factors affecting stripe rust, moisture, temperature and wind are most important.

4.1.6.1 Moisture

Moisture is the most important factor for stripe rust and also for other rusts, because without adequate moisture for urediniospores to germinate and infect the host crop, there will be no rust diseases. Moisture affects spore germination, infection and survival. Also moisture is more often than any other weather factors to become a limiting factor to stripe rust epidemics.

4.1.6.1.1 Moisture Affecting Spore Germination and Infection

Urediniospores require at least 3 h of continuous moisture (dew formation) on the plant surfaces in order to germinate and infect plants (Rapilly 1979). At suitable temperatures as discussed later, the longer the dew period, the more spore germination and infection of host plants. Dry urediniospores need to absorb water to start growing germ tubes, and germ tubes need high moisture to prevent dried out and to continuous grow to reach host stomata and penetrate. After penetration and especially after haustorial formation, pathogen hyphae can utilize water from host plant tissue, and then free water on the host plants or high moisture around plants are not needed. The relationship of high moisture and rust was recognized by farmers hundreds of years ago. Because of this relationship, farmers in Europe used the cord drawing method with ropes and nets to reduce rusts by shaking off dew on cereal crops in the nineteenth century (Zadoks and Bouwman 1985).

Moist regions with frequent dew formation during the growing season provide conditions conducive for infection by the stripe rust pathogen, and therefore stripe rust is a more problem in high-rainfall regions. For example, western Washington and western Oregon in the Pacific Northwest of the U.S. typically have cool and moist weather patterns in the spring and summer, and therefore, severe stripe rust occurs there almost every year (Chen 2005; Sharma-Poudyal et al. 2014). In the inland Pacific Northwest, east of the Cascade Mountains, more severe damage occurs in the high-rainfall (annual precipitation >500 mm) regions in eastern

Washington, northern Idaho, and northeastern Oregon, and less in the low rainfall (<300 mm) regions in central Washington, central Oregon, and southern Idaho, except fields under irrigation. Because high moisture increases infection, irrigated cereal crops are vulnerable to stripe rust. In rain-fed areas, light rains create the ideal conditions for infection because free water on leaf surfaces allows urediniospores germinating and infecting host plants. After rains, high moisture in the air and soil often create dew for several nights, providing favorable conditions for infection. Rain drops also can disperse spores to spread the pathogen (Rapilly 1979). Stripe rust can be predicted by monitoring precipitation in a region and dew formation in the fields (Sharma-Poudyal and Chen 2011). Although not as immediately affecting stripe rust as during the spring and summer, precipitation as rain or snow during the winter time also increases stripe rust because it increase soil moisture into late spring or even summer. When soil keeps moisture long into the summer, it creates dew formation even when low precipitation occurs. This long effect of precipitation when no new stripe rust infection occurs during the most of winter time is obvious in the inland Pacific Northwest (Chen 2005).

For stripe rust research under controlled conditions, researchers mostly use 16–24 h of dew period at the optimal temperatures (7–12 °C) for inoculation of wheat, barley and grass plants with urediniospores (Chen and Line 1992; Chen et al. 2010). The much longer dew period than the minimal 3-h allows maximal success of inoculation. A dew period as short as 8 h can result in successful tests. A uniform coverage of leaf surfaces with tiny dew is better for infection than sparsely big dew drops. Dry and long-stored urediniospores requires a longer dew-period than fresh spores for germination and infection. A similar period of dew formation is required for infection of wheat and other cereals or grasses with aeciospores (Wang et al. 2015).

Moisture also affects sporulation. Although moisture is not required for the pathogen growing in host tissue during the latent period (the period from penetration to sporulation), the relative humidity needs exceed 50% for sporulation to occur. The number of urediniospores formed per unit of time is related exponentially to the increase in relative humidity, but liquid water stops sporulation (Rapilly 1979).

For infection of barberry and Mahonia plants with teliospores, a much longer period of dew formation is needed. This is because the entire process from teliospore germination to basidiospore infection needs an uninterrupted high moisture condition. A minimum of 32 h of dew formation is required for inoculation of barberry plants with teliospores scratched from wheat straw, and highest infection rates can be obtained when keep inoculated leaves in a dew chamber at 10 °C for 80–96 h (Wang and Chen 2015). Such moisture conditions can be used to determine if alternate hosts can play a role in the pathogen lifecycle and stripe rust epidemic or not during the period of new leaf production in a region.

4.1.6.1.2 Moisture Affecting Pathogen Survival

Moisture promotes stripe rust by favoring spore germination and infection, but adversely affects spore survival. Because urediniospores do not exhibit fungistasis, they can germinate immediately after they are produced if dew is present and the temperature is in the suitable range. Since urediniospores kept under high moisture conditions lose viability more quickly than those kept dry, spores should be kept dry in order to maintain viability during sample shipment and storage. For the purpose of research, sporulating leaf samples are better collected without dew on them, and in the presence of dew, drying leaves for few minutes before putting in a collection bag is needed to keep viable samples. Glassine or paper bags that allow moisture to get out are better than plastic bags for keeping and shipping leaf samples. If plastic bags are used, leaves must be free of dew or high moisture as well as kept cool, and sent quickly. For long term storage, urediniospores should be dry, usually 2–5 days in a desiccator at 4 °C is good, and then can be stored in a refrigerator for 2–6 months, ultra-freezer (−70 to −80 °C) for a couple of years and liquid nitrogen (−197 °C) for several decades. Fresh leaf samples or moist urediniospores should not be stored in a freezer or liquid nitrogen.

Moisture also affects urediniospore dispersal. For stripe rust epidemics, dry urediniospores survive longer than moist spores as dry spores more likely survive during a long-distance dissemination or remaining in the region for infecting next crops. Depending on the relative humidity, urediniospores can spread individually or in clusters (Rapilly 1979). High humidity leads to stronger adhesion of urediniospores to the leaves (Rapilly 1979). Clusters increase in size as humidity increases, and less likely be lifted into air for dispersal. Heavy rain wash urediniospores onto ground, and once stick to soil, they will die quickly. Nevertheless, high humidity generally increases stripe rust by increasing moisture and thus increasing infection frequency during a rust season. However, during the off crop season, dry conditions allow urediniopores to survive longer than moist conditions. The dry weather in late summer and the various stages of wheat crops in the US Pacific Northwest allows urediniospores produced on late harvested spring wheat crop to survive the summer and be able to infect seedlings of winter wheat planted in the fall. This is one of the reasons why stripe rust occurs every year in this region. Similarly, the relatively dry weather conditions in southern and eastern Gansu, China allow urediniopores surviving in late summer, disseminate from high elevation to low elevation fields in the region and beyond to infect fall-planted wheat. Similar dry weather conditions normally occur in central Asia. Thus, stripe rust is more frequent in regions where the weather is dry than moist between crops when the pathogen is mostly in the survive state.

Similarly, teliospores of the stripe rust fungus survive longer under dry conditions than moist conditions. In a recent study by Wang and Chen (2015), the germination rate of *Pst* teliospores kept dry at room temperature gradually decreased from 79% in July when fresh to 1% in following January, whereas teliospores completely lost viability when kept wet at 5 °C in 3 weeks. The high moist condition was detrimental to *Pst* teliospores. They also found that the viability rates of *Pst* teliospores measured by germination were high in July to August (50–90%) when the weather was dry, but decreased to less than 1% after the wet winter and completed dead before May. The decrease of viability was due to the degradation of telia and teliospores during the wet winter (Wang and Chen 2015). Degradation of teliospores by high moisture is one of the factors that prevent *Pst* teliospores from infecting barberry (Wang and Chen 2015; Wang et al. 2015). In contrast, teliospores of the stem rust fungus were not degraded, but their dormancy was broken under the same wet conditions, and as a result, they germinated at highest rates in May when barberry plants produce new leaves (Wang and Chen 2015). The genetic mechanisms for the highly contrasting responses of teliospores to the wet conditions between the two rust species that share common primary and alternate hosts are yet to be studied.

As a special form of precipitation, snow helps the stripe rust pathogen survive cold weather. Snow cover insulates wheat leaves infected by the stripe rust pathogen from air temperatures, so cold temperatures do not eliminate the pathogen (Zadoks 1961; Burleigh 1965). A minimum snow cover of 7.6 cm (3 inches) is sufficient to protect winter wheat from winter kill (Peel and Riveland 1977) and thus, the stripe rust fungus can survive as dormant mycelium as long as infected leaves survive during winter (Sharp and Hehn 1963). During the winter season of 2010–2011, the air temperatures went below −15 °C for several days, but the snow cover during the coldest period protected wheat leaves infected before the winter, resulting in an early re-development and unusually severe stripe rust epidemics in 2011 in the US Pacific Northwest (Chen 2014). In central Alberta, Canada, the cold air temperatures during the winter time, monthly normal average −10 °C in December and −16 °C for January (http://www.accuweather.com/en/ca/lacombe), do not allow the stripe rust pathogen to overwinter. The stripe rust inoculum is often from the US Pacific Northwest (Line 2002; Chen 2005). When the pathogen is from outside of the central Alberta, stripe rust occurrence is usually late in the crop season. However, stripe rust has occurred earlier than or about the same time as in the US Pacific Northwest (KQ Xi, personal communication) in certain years. Kumar et al. (2013) studied *Pst* overwintering at Lacombe, Alberta, Canada from 2008 to 2011, and concluded that persistent snow cover is critical for the survival of overwintering inoculum, as viable pustules and urediniospores rarely survive after snow melt. This study demonstrates the importance of snow cover for the winter survival of the stripe rust pathogen in the regions with cold winter.

4.1.6.2 Temperature

Temperature affects every stages of the pathogen lifecycle and development of stripe rust epidemics. In addition, temperature also influences host resistance, and therefore, influencing the interactions between the pathogen and host.

4.1.6.2.1 Temperature Affecting Spore Germination and Infection

In the literature, there are many studies conducted to determine minimum and maximum temperatures at which urediniospores of the stripe rust pathogen is able to germinate and infect host plants. Sharp (1965) reported that germination occurs from 2 to 15 °C and the optimum temperature for infection is 7–11 °C. Others reported different amounts of germination, infection and disease development in the range of −1 to 11 °C (Naoumova 1937; Zadoks 1961; Burleigh 1965; Dennis 1987; Line 2002). Burleigh (1965) observed infection below −4 °C under field conditions but did not observe infection under laboratory conditions even at −1.5 °C. Schmitt et al. (1964) reported that *Pst* urediniopsores are able to germinate and infect plants at −2.8 to 21.7 °C with the optimum at 9.7 °C. Tu (1967) observed trace infection at 21.1 °C under controlled conditions. By simulating stripe rust epidemics, Zadoks and Rijsdijk (1972) determined daily mean temperatures between 4 and 25 °C for infection. Dennis (1987) reported that infection occurs more at temperatures of 14–15 °C compared to 16–18 °C, but less than at 12–13 °C, whereas de Vallavieille-Pope et al. (1995) reported that infection by *P. striiformis* was restricted to a narrower range 5–12 °C. Park (1990) reported that under controlled conditions, infection decreases from 100% at 15.4 °C to 0.8% at 20.5 °C, and he estimated that no infection occurs at or above 20.8 ± 0.2 °C. The differences in the reported minimum and maximum temperatures in the literature are mainly due to two reasons, (1) the range and number of temperatures used in the studies and (2) different isolates of the stripe rust pathogen. The temperature settings in some of the studies did not allow for the determination of the minimum temperatures for germination and infection. Different isolates can have different optimal temperatures for germination and infection. Milus et al. (2006) reported that *Pst* isolates collected after the year 2000 had higher germination rates at 18 °C than at 12 °C, and had relatively high germination rates and high infection (more aggressive) than those collected before 2000 from the central US. Nevertheless, based on the previous studies, urediniospores of the stripe rust fungus are able to germinate and infect host plants at temperatures from 0 to 22 °C with 7–12 as optimal temperatures for both germination and infection. Sporulation occurs from 4 to 35 °C, but optimal temperatures are in the range from 10 to 25 °C (Tu 1967; Line 2002; Chen 2005).

For research, any constant temperature in the optimal ranges for urediniospore germination, infection and sporulation can be used to allow stripe rust to develop. For inoculation, plants are kept in a dew chamber at temperatures between 8 and 12 °C and then grown between 10 and 20 °C. However, there are no regions with a constant temperature during the wheat or barley growing season. In the stripe rust research program at Pullman, WA, USA, diurnal temperatures cycles of 4–20 °C (low-temperature) and 10–30 °C (high-temperature) are used to simulate the natural temperature ranges in early spring and late spring or early summer, respectively (Chen and Line 1992, 1995; Chen 2013). Different temperature cycles are used for different research purposes. The same temperatures for incubation of inoculated plants in a dew chamber and growing plants after the dew period have been also used for inoculation with aeciospores (Wang and Chen 2015; Wang et al. 2015).

4.1.6.2.2 Temperature Affecting the Pathogen Survival

Hot summer and cold winter are bottlenecks for the stripe rust pathogen. Whether the pathogen can survive the hot and/or cold seasons in a region determines if it can infect fall crop and/or the disease start early in the fields immediately after winter or even develop during the winter. Burleigh (1965) reported that the stripe rust fungus could not survive when 30 day minimum moving average temperatures was −10 °C in field conditions even in the presence of snow cover. However, Li and Zeng (2002) reported that the stripe rust pathogen could successfully overwinter even when the monthly mean temperature below −10 °C if fields is covered with snow during the coldest month. In the absence of sufficient snow cover, a monthly mean temperature of −6 °C or less is enough to kill the pathogen (Li and Zeng 2002). Both these reports indicate that the pathogen cannot overwinter if the mean temperature of the coldest month is below −6 or −7 °C. Based on Rapilly (1979), temperatures between 0 and −10 °C inhibit but do not kill the fungus as long as the temperature does not go below −10 °C. However, this author did not specify how long the temperatures below −10 °C.

Severe cold eliminates the green foliage of host plants and consequently the fungus within leaf tissue is also eliminated. It has been generally thought that the stripe rust fungus can survive cold weather as long as host leaves remain alive. A recent study under controlled conditions showed that the survival of *Pst* in wheat cultivars with strong winter-hardiness at 0 and −5 °C was greater than those with weak winter-hardiness (Ma et al. 2015). However, such differences in the pathogen survival among cultivars were negligible at −10, −15 and −20 °C. These results indicate that *Pst* mycelia inside green leaves can also be killed by low temperatures rather than just through death of green leaves. This study also determined the effects of temperature and duration on *Pst* survival. Overall, the percentage of *Pst* that survived in attached leaves decreased over time at all temperatures tested. Viable fungus were detected after incubation at 0 °C for more than 54 h, at −5 °C for 48 h, at −10 °C for 42 h, and at −15 and −20 °C for 18 h. The percentage of survived *Pst* was 25.8% and 5.5% at 0 °C and −5 °C, respectively after incubation for 54 h in attached leaves. Wheat plants incubated at −20 and −15 °C for 30 h, −10 °C for 42 h, −5 and 0 °C for 54 h were still alive, which were detected by transferring into 4 °C for 1 day and then maintained under normal conditions in a greenhouse for 1 week. These results indicate that *Pst* hyphae can survive cold temperatures as low as −20 °C for a couple of days.

Hot temperatures kill urediniospores of the stripe rust fungus. Temperatures over 30 °C hinder sporulation, and consequently prolong the latent period (Rapilly 1979). Georgievskaja (1966, reviewed by Rapilly 1979) found that once established in the plant the fungus could stand hot temperatures up to 38 °C without being destroyed. However, urediniospores produced at or above 30 °C failed to germinate. Similar to these report, Tu (1967) considered 38 °C is the upper limit for *Pst* survival. In our experience, the stripe rust can tolerate temperatures as high as 35–38 °C for a couple hours every day if temperature drops from these highs, but constantly for a day or longer, these temperatures can kill the fungus. Also, urediniospores kept at room temperature (22–23 °C) for 1 month could not germinate (X.M. Chen unpublished data). Li and Zeng (2002) set 10-day average temperature of 22–23 °C as the upper limit for *Pst* summer survival.

Night temperatures are more important than day-time temperatures for the stripe rust fungus to survive summer (Stubbs 1985; Chen 2005). This is because that both dew formation and low temperatures occur together most frequently at night and therefore, infections are more likely occur at night. Hot nights reduce dew formation and therefore limit new infections and reduce active survival of the pathogen. Because temperature gaps between day-time and night vary greatly from region to region, the use of average and maximum daily temperatures to determine if stripe rust occurs or not can be misleading. For example, if an area has a daily maximum temperature of 28 °C and minimum temperature of 22 °C, while another area has a daily maximum temperature of 34 °C and minimum temperature of 16 °C, and then both areas have an average daily temperature of 25 °C. If we use the daily average temperatures, both regions would be considered no stripe rust. If the maximum temperatures are used, we would conclude that the first area is more likely to have stripe rust than the second area. However, the second area should have more stripe rust as the night low temperature allows new infection and more rapid disease development. In contrast, the lowest temperature in the first area would be marginal for infection to occur. The temperature range in the second area is typical of most areas in the US PNW in the late spring and summer. The cool weather at night allows stripe rust to develop and the pathogen to survive, which makes this region one of the most frequent epidemic regions of stripe rust in the world (Chen 2005; Wellings 2011). Similarly, Park (1990) found that *Pst* infection would not occur if inoculated plants were grown at a constant temperature at or above 20.8 ± 0.2 °C, but high levels of infection occurred when temperatures fluctuated within the range of 19–30 °C in the fields. He conducted an overnight infection experiment over 18 months at Toowoomba, Queensland, Australia, and found that periods of moisture and temperature favoring *Pst* infection occur regularly, even during summer. High-temperature was a limiting factor only in 3% of the nights tested. His study demonstrated that *Pst* is able to over-summer in eastern Australia.

4.1.6.2.3 Effects of Temperatures on Host Resistance

In addition to the effects on the stripe rust pathogen as discussed above, temperature plays an important role on host resistance, especially high-temperature adult-plant resistance (HTAP) resistance. Cereal hosts have two major types of resistance, all-stage resistance and HTAP resistance, which will be discussed in detail in Chap. 5 (Stripe rust Resistance). Temperatures have less influence on all-stage resistance, but have more effect on HTAP resistance (Qayoum and Line 1985; Line 2002; Chen 2005, 2013). In general, HTAP resistance expresses when plants grow old and when weather becomes warm. Some all-stage resistance cannot express at low temperatures, but expresses at high temperatures. As wheat or barley plants are more tolerate high temperatures than the stripe rust pathogen, high temperatures may limit pathogen growth in plant tissue when at relatively high temperatures. However, not all cereal varieties have HTAP resistance. To identify HTAP resistance, it is important to set the right temperatures and have susceptible varieties as checks. In the stripe rust research program at Pullman, WA, USA, we use the two temperature profiles, 4–20 °C (LT) and 10–30 °C (HT), in a four-way test (seedling-LT, seedling-HT, adult-LT, and adult-HT) for identifying HTAP resistance (Chen 2013).

4.1.6.3 Wind

As discussed above, the effects of wind are mostly through disseminating urediniospores, and this helps the pathogen survival by sending spores to areas where host plants are available and weather conditions are suitable for infection and disease development. The effects of wind are so complex that it can either increase or reduce diseases. Wind dries urediniospores quickly and dry spores keep longer viability. Wind blows urediniopores to new plants for infection, but also blows spores to ground where urediniospores will die when stick on soil. Wind reduces moisture, and therefore reduces urediniospore germination and infection. For the purpose of research, wind is hardly included in studies under controlled conditions. However, for field research, wind conditions are considered for inoculation of experimental plots. At the time of dust inoculation of mixing spores with talc, a low speed wind is helpful in distributing inoculum over the field. A windless condition may help dew formation after inoculation. In the US Great Plains, stripe rust occurs early in the season in the southern region and urediniospores are blown by wind to the northern and eastern regions, wind storm events can be used to predict stripe rust in the northern and eastern regions (Chen 2005).

4.1.6.4 Interactions Among Moisture, Temperature and Wind

The above has discussed individual effects of moisture, temperature and wind. Most previous studies usually deal with the variation of one factor (mostly moisture and temperature) at a time by holding other factors at the optimum. However, under the natural conditions, these factors are mixed and interact to each other, and the interactions affect stripe rust differently. For example, under severely cold temperatures, snow cover can reduce cold damage to host plants and the parasitic pathogen. Windchill can increase cold temperature damage to host plants and the pathogen at temperatures under which the fungus can survive if without wind. Wind coupled with rain spreads and deposits urediniopores for developing disease in new areas. Wet and cool weather conditions favor stripe rust development, but humid conditions (although temperatures are not too high) reduce urediniospore survival and therefore reduce disease epidemics (Chen 2005). More studies are needed to determine the effects of environmental factor interactions on stripe rust in order to better understand the disease epidemiology and predict epidemics under changing climates.

4.1.6.5 Other Factors

4.1.6.5.1 Light

Light has been studied for their effects on stripe rust. Papastamati and van den Bosch (2007) determined that light quantity is one of the most important weather factors for the development of stripe rust, only after temperature and dew period. The effect of light on stripe rust was studied by Gassner and Straib (1934), Manners (1950) and Stubbs (1967). In these studies, Carstens V, Holzapfels Früh and Chinese Spring were highly sensitive to changes in light intensity. Low light intensities produced susceptible reactions, whereas high light intensity gave resistant reactions. Modification of host plant reactions to stripe rust by light intensity has also been found by others (Pochard et al. 1962; Georgievskaja 1966; McGregor 1976). Light intensity affects urediniospore production. McGregor and Manners (1985) found that *Pst* urediniospore production is much lower under low light intensity than under high light intensity. The number of pustules per unit area of infected leaves and the daily sporulation rate per pustule increase linearly with increasing light intensity over the range of 10–50 W/m^2. The effects of light on wheat resistance to stripe rust were also studied by Chinese scientists in the 1950s (Lu and Li 1958), and they found that resistances in some wheat cultivars were less sensitive while some cultivars exhibited different reactions under different light intensities.

Because light affects host plant reactions to stripe rust, it is critical to standardize light conditions to obtain consistent and reliable results. Unlike stem rust inoculation that needs few hours of light to have successful infection, inoculation of host plants with the stripe rust pathogen does not need light. The dew period can be completely in dark. However, adequate light is needed during the latent period after the dew period and also during sporulation. Light mainly affects the latent period and urediniospore production. The effects of light on stripe rust are mostly through its effect on host plants. As the stripe rust fungus is an obligate parasite, it likes healthy plants to infect and reproduce. Any light and other conditions good for host plants are also suitable for stripe rust tests. In other words, in order to do a successful rust experiment, it first needs to grow healthy plants. For the purpose of research under controlled conditions, light intensity of 10,000 Lux and light period of 16 h during a daily cycle are commonly used (Stubbs 1985; Chen and Line 1992, 1995). More recently, Pretorius et al. (2007) determined different light settings on stripe rust testing on adult plants of spring wheat. They found that growing plants in a greenhouse under a 14/10-h light/dark cycle produce more vigorous plants than growing under continuous light in a growth chamber. In general, the greenhouse system agreed the field reaction types for the germplasms tested with underestimation of adult-plant reactions for only few entries. They also found that the preconditioning of seedlings less than 24 h of continuous light before inoculation produced unambiguous and consistent infection types. In general 14- to 16-h light at or slightly above 10,000 Lux is adequate for stripe rust testing under controlled conditions.

Not only the light after inoculation affects stripe rust, but also does the preinoculation light. de Vallavieille-Pope et al. (2002) determined preinoculation effects of light quantity on infection efficiency of the stripe rust pathogen. They inoculated wheat seedlings, which had been exposed to various regimes of light duration and intensity for 24 h in either field or controlled conditions, and incubated them in growth chambers set with optimal dew and temperature conditions. They found that infection efficiency increased from 0.4% to 36% depending on the light quantity.

4.1.6.5.2 UV Light and X-Ray

According to Rapilly (1979), ultraviolet (UV) light may reduce the dispersal of urediniospores of the stripe rust pathogen, as the sensitivity of *Pst* urediniospores to UV light is three times higher than that of urediniospores of the stem rust fungus (Maddison and Manners 1972). Stubbs (1985) reported the higher viability of urediniopores collected in the field early in the morning than those collected at noon in the Andes region at 3000 m, and he speculated that the difference in viability is due to different times of exposure to UV light because the high elevation region has the high intensity of UV light. Jing et al. (1993) reported that UV light reduced *Pst* urediniospore germination. The treatment for 1 min reduced germination rate from 100% to below 30% and the treatment for 32 min reduced the germination rate to less than 3%. Different isolates exhibited different levels of sensitivity. They also found that UV-treated urediniospores had low infection efficiency, extended latent period and reduced infection type and severity.

UV light and X-ray cause mutation in the stripe rust fungus. Stubbs (1968) produced races with increased virulence by irradiating urediniospores or infected wheat seedlings with X-rays. The mutants differed from the parent race by only one virulence factor, and one mutant was virulent to *Triticum spelta album* (*Yr5*), generally resistant at that time. Johnson et al. (1978) obtained a similar result with UV light and indicated the high mutability of the genetic locus of virulence to Compair (*Yr8*). By radiating urediniospores of Chinese *Pst* race CYR32 with UV light for 8 min, Chinese scientists found that virulences to different resistant varieties had different mutation rates, ranging from $6.95 \times 10^{-5}\%$ to $1.11 \times 10^{-5}\%$ (Li and Zeng 2002). Each mutant contained one or more mutated virulence/avirulence loci. The mutants containing multiple mutated loci had mutations from both avirulent to virulent and virulent to avirulent.

4.1.6.5.3 Air Pollution

Sharp (1967) demonstrated that the stripe rust fungus is more sensitive to air pollution than other cereal rust fungi. A high concentration of large ions in the air due to air pollution decreased the germination rate of *Pst* urediniospores. Stubbs (1985) indicated differences in sensitivity to air pollution among cultures of the stripe rust fungus from different regions in the world. He found that the northwest European cultures were much better adapted to air pollution than those from the Andes, East Africa and Asia, and a few northwest European cultures even highly resistant to air pollution. He also noticed that inoculations made during days with high air pollution resulted in an increase in the length of the latent period and a shift of infection types toward resistance. He attributed some of the non-reproducibility of infection types even working under strictly controlled conditions partially to the differences in air pollution sensitivity. Stubbs's work indicated that the tolerance of northwest European cultures could have developed under highly polluted air and isolates from other regions have not got a polluted environment at that time. It should be interesting to conduct the similar experiments in countries, such as China, where air pollution has become a serious problem since the 1980s.

Melching et al. (1974) showed the effects of tobacco smoke on germination and infectivity of urediniospores of the stripe rust fungus. They found that urediniospores of the stripe rust fungus, as well as spores of other fungi, could not germinate when 6000 μl of cigarette smoke per litter of air in the incubation chamber. Decreased smoke concentration down to 300 μl inhibited spore germination. Even the exposure of water agar to smoke before placing urediniospores also prevented or inhibited germination depending upon the concentration of smoke and the duration of exposure. When agar plates were placed and wheat plants were inoculated with spores of proven infectivity and incubated in dew chambers in 3000 μl of smoke/liter of air, germination was decreased and essentially no infection occurred. This study has told us that smoking should be prohibited in the places where rust testing is conducted.

4.1.6.5.4 Parasites

Approximately 30 genera of fungi can be hyperparasitic on rust fungi (Littlefield 1981). Among the species in these 30 genera, only five species, *Cladosporium cladosporioides*, *Lecanicillium lecanii*, *Microdochium nivale*, *Typhula idahoensis* and *Verticillium lecanii*, have been reported to infect uredinia and urediniospores of the stripe rust fungus (Littlefield 1981; Mendgen 1981; Zhan et al. 2014). *Verticillium lecanii* can grow on *Pst* uredinia at an relative humidity of higher than 90% (Mendgen 1981). *Cladosporium cladosporioides* reduces the production and viability of urediniospores. Although these hyperparasitic fungi have potential to be used as biocontrol agents for reducing stripe rust, more studies are needed to develop techniques and estimate the efficacy. However, the parasitism is sometimes a problem for rust research. To prevent the problem, it is important to maintain a relatively low relative humidity during the latent and sporulation periods and pre-dry urediniospores before keeping in storage.

In addition to the parasitic fungi, other rust fungi and powdery mildew can be even more serious problems for stripe rust research in a greenhouse. Because stem rust, leaf rust and stripe rust can infect same plants of wheat and barley, leaf samples can carry different rust fungi. It is important to collect stripe rust samples free of other rusts. For leaves with uredinia of two or more rust species, place the leaf samples in a moist petri dish overnight, picking newly produced urediniospores from *Ps* uredinia for inoculation should reduce the possibility of other rust urediniospore contamination. For stripe rust samples of urediniospores contaminated with stem rust urediniospores, growing the inoculated plants after the dew period at low temperature (e. g. less than 10 °C) should reduce stem rust. Contamination by leaf rust urediniospores is more difficult to remove as the leaf rust pathogen had a much wider range optimal temperatures and shorter latent period than the stripe rust fungus (de Vallavieille-Pope et al. 1995, 2002). It is more important to collect leaf rust-free samples of stripe rust. The single-uredinium method and/or selective host varieties for inoculation can be used to purify urediniospores of the stripe rust pathogen. Powdery mildew can be a worse problem for stripe rust research in a greenhouse than other rusts as the fungus is more ubiquitous in greenhouses, does not need free water for germination, and grows on plants much faster. Greenhouse powdery mildew is best controlled by management, such as keeping plants separated and avoiding overlapping plants of different stages as much as possible, appropriate watering and keeping air circulation. In addition, sulfur can be used to control powdery mildew as the chemical highly inhibits the powdery mildew fungus but has a much lower inhibiting effect on the stripe rust pathogen.

4.2 Disease Forecast

For more effectively managing stripe rust, it is important to accurately predict disease epidemics in advance. There have been many studies on prediction of stripe rust (Zadoks 1972, 1981; Zadoks and Rijsdijk 1972; Young et al. 2003; Line 2002; Li and Zeng 2002; Sharma-Poudyal and Chen 2011; Sharma-Poudyal et al. 2014). For developing forecasting models, first, factors affecting the disease need to be identified, and then their effects on disease development need to be quantified.

4.2.1 Identification of Affecting Factors and Developing Forecasting Models

As discussed above, much of stripe rust epidemiological research is to identify factors affecting the disease. Among factors of host, pathogen and environment, environmental factors have been mostly used for predicting stripe rust, as environmental factors are independent of the pathogen and also mostly independent of the host. Because environmental factors are measurable and collectable, most prediction models for stripe rust have been based on environmental factors. As environmental factors are numerous and even the same factor can affect the disease differently in different stages of the disease cycle and crop growth, and also their effects can be different in different regions, it is important to identify affecting factors for the disease in a particular region. These factors should have adequate variation among different crop growth seasons and the variation in disease level is correlated to the environmental factors.

4.2.1.1 Stripe Rust Prediction in Europe

Intensive studies were conducted by European scientists for identifying factors affecting stripe rust and developing models for predicting stripe rust in the 1950s to 1970s (Hassebrauk 1962; Doling and Doodson 1968, Zadoks 1961, 1967, 1971, 1972, 1981; Zadoks and Rijsdijk 1972; Rapilly 1979; Zadoks and Bouwman 1985). Based on field data from 1956 to 1966, Doling and Doodson (1968) observed maximum yield loss of 30% when the foliage is fully infected. They determined empirically that the percentage yield loss was equal to 3 times the square-root of the stripe rust assessment at the soft dough growth stage, but greater yield losses result when ear infection as well as foliar infection occurs. The early studies in Europe were well reviewed by Zadoks and Bouwman (1985). Some studies along this line after the 1990s are discussed.

In France, de Vallavieille-Pope et al. (1995) developed an equation based on five parameters to describe the influence of temperature, dew duration and wetness interruption on stripe rust infection. The influence of temperature on maximal infection (I_{max} (T)) can be described by formula I_{max} $(T) = p\ ((T - T_{min})/(T_{max} - T_{min}))^{n}$ $((T_{max} - T)/(T_{max} - T_{min}))^{m}$. For stripe rust, the minimum (T_{min}) and maximum (T_{max}) temperatures can be fixed at 0 °C and 18 °C, respectively, and the parameter p can be expressed as a function of n and m: $p = (n + m)^{n+m}/(n^{n}m^{m})$. The adjustment can be made by minimizing the mean square error between observed and simulated maximal infection for varying values of n and m. The influence of continuous dew duration on relative infection ($_{Icon}$ (DP_1,T)) can be described as the product of maximal infection $_{Imax}$ (T) and a Richards's function of dew duration with temperature-dependent parameters: $_{Icont}(DP_1, T) = {}_{Imax}\ (T_1\ (1 - \exp\ (-b(DP_1 - DP_{min})))$ for $DP_1 \geq DP_{min}$, where DP_1 is the duration of the initial dew period, DP_{min} is the minimal dew period necessary for infection and b is the initial infection rate, and they are estimated by minimizing the mean square error between measured and simulated values of $_{Icont}(DP_1, T)$ for varying values of b and DP_{min} at each temperature. To describe the influence of wetness interruption, they added a negative exponential term to the second equation:

$_{Iint}(DP_1, T) = {}_{Icont}(DP_1, T) + {}_{Imax}(T)\ \exp.\ (-(DP_1/\tau))$, where $_{Iint}$ (DP_1,T) is the relative infection after 24 h of dew when a 1-h dry period occurs after DP_1 h of dew. This equation consists of two terms: the first term corresponds to infection if a dry period occurs after the minimal wetness duration; the second term corresponds to infection if a dry period occurs before DP_{min}. The second term includes the time constant (τ), which is the time when a dry period leads to an infection equal to maximal infection divided by a factor e (base of natural logarithm). Using the infection data obtained in conditions of interrupted wetness, the parameter τ can be fitted with parameters b and DP_{min} calculated using results of experiments with continuous wetness. Using their observed maximal infection values at different temperatures, they estimated values of m and n as 5 and 7 for stripe rust, respectively. At the highest tested temperature for the stripe rust pathogen, DP_{min} was about 3.5 h, and at the optimal temperatures for infection, it was 5.3 h. The initial infection rate (b) was estimated in the range between 5 and 12 °C. The estimated τ values indicate fast decrease in the response time of infection reduction when a dry period occurs. This information and the models can be used to predict infection of the stripe rust pathogen based on temperature and dew formation, as did by Gladders et al. (2007).

In the United Kingdom, Young et al. (2003) developed a model to predict stripe rust epidemics on the upper canopy of wheat from disease observations on lower leaves. In their study, they used a logistic function, $y = a/(1 + e^{-c-rt})$, to fit the stripe rust severity data, where y is stripe rust percentage severity, a upper asymptote, c horizontal displacement parameter, r rate, and t time. Sporulating lesions on lower leaves were considered to be a measure of inoculum source strength for transfer to upper culm leaves. The time interval t_m was used as a measure of how quickly an epidemic developed on a particular leaf, which is calculated using formula $t_m = a + be^{-k \cdot \text{source strength}}$, where south strength is the percentage area affected by stripe rust lesions. They found that observations of disease two leaves further down the culm were of the greatest predictive value ($R^2 = 86\%$). They also noticed that much unexplained variation, probably due to extraneous variation in upward transfer efficiency of inoculum and rates of epidemic development. This study indicates that integration of disease severity on low leaves with factors affecting upward inoculum transfer and rates of epidemic development is needed to achieve an acceptable predictive precision.

Gladders et al. (2007) analyzed stripe rust survey data from 4475 randomly selected wheat crops from England and Wales during 1985–2000. Using random effects logistic regression (general linear mixed model), they identified factors affecting the incidence of stripe rust, including cultivar resistance ratings, frosts (≤ -5 °C), systemic seed treatment, and foliar fungicides. Their results suggest that the use of risk variables associated with overwintering survival may help adjust fungicide inputs to seasonal risk.

Te Beest et al. (2008) conducted a study to identify key weather factors determining the occurrence and severity of stripe rust epidemics on winter wheat. They used the disease data from field experiments at 12 locations in the UK from 1994 to 2002 with matching data from weather stations within a 5 km range to formulate empirical models to qualitatively predict a damaging epidemic (>5% severity) and quantitatively predict the disease severity. They found the temperature in February to June the most influential factor for a damaging epidemic as well as for disease severity. The best model describing stripe rust severity (=maximum temperature × 0.5 + rain × −4.08 + 2.54) consisted of maximum temperature (accumulation above 16 °C) and rain (number of days ≥8 mm), with an R^2 of 86.4%, but the relationship with rain can be questioned. An alternative model including only maximum temperature (maximum temperature × 0.777–5.73) had an R^2 of 70.2%.

In southern Sweden, stripe rust was usually considered a minor disease, but occurred at a high severity and incidence in some years such as 1990, 1999, 2002 and 2007. Wiik and Ewaldz (2009) analyzed historical data of weather and stripe rust from 1983 to 2007 and found that incidence of stripe rust at soft dough (GS 75) was significantly correlated to temperatures in February and March. Strip rust severity at maximum attack was significantly correlated to precipitation in January and February and temperatures in June. They developed a model ($y = -9.036 + 2.624 \times$ Mar T $- 0.127 \times$ Nov P $+ 1.382 \times$ Oct T) ($R^2 = 0.51$, $P = 0.008$) for predicting stripe rust incidence (y) at GS 75 using March temperature (Mar T), November precipitation (Nov P), and October temperature (Oct T), with $R^2 = 0.51$. The model correctly predicted 4 years (1989, 1990, 1999 and 2007) with high actual incidence of stripe rust. Alternately, stripe rust severity (y) at maximum attack can be predicted by formula $y = -2.304 + 0.022 \times$ Feb P $+ 0.009 \times$ Jan P $+ 0.114 \times$ May T, consisting of independent variables February precipitation (Feb P), January precipitation (Jan P), and May temperature (May T), with $R^2 = 0.46$. This model quite accurately predicted 4 years (1988, 1989, 2002 and 2007) with high actual severity of stripe rust, but the predicted values were higher than the actual levels in some years.

4.2.1.2 Stripe Rust Prediction in China

In China, prediction of wheat stripe rust has been studied since the 1950s, and a series of models for predicting or estimating stripe rust has been developed (Li and Zeng 2002; Zeng and Luo 2008; Kuang et al. 2013). Based on the climatic and stripe rust data, precipitation in April is the most important factor and followed by temperatures in March for determining whether epidemics occur in the Beijing region. Zeng (1962) established a formula for calculating daily disease development rate $r = 1.16\ W + 0.1Y + 0.95Z - 2.6$, where W is the number of rain days, Y amount of precipitation, and Z mean of mean temperatures of 10 days from 11 March to 20 April. Using the r value and initial disease intensity (y_0) in a logistic model ($y_t = y_0 e^{rt}$), disease intensity at time t can be estimated. This model laid the theoretical basis for the establishment of the stripe rust forecasting system in China (Li and Zeng 2002). Late, different models were developed for different regions. Examples are presented below.

For eastern and southern Gansu, a stripe rust "hot-spot" region where the stripe rust can survive both summer and winter, three models have been developed. For Gangu, Gansu, disease prevalence (y) is calculated using formula: $y = -2.07472 + 0.242533x_1 + 0.06744x_2 – 0.0091x_3 + 0.011865x_4 + 0.090152x_5$, where x_1 is average temperature in March, x_2 precipitation in March, x_3 precipitation in April, x_4 precipitation in May, and x_5 susceptible variety area (Fan et al. 2008). For the Longnan mountainous area in southern Gansu, disease prevalence (y) in spring can be estimated using formula $y = -0.547 + 0.025x_1 + 0.03x_2 + 0.29x_3 + 0.081x_4$, where x_1 is percentage of diseased fields in autumn in the preceding year, x_2 proportion of susceptible variety area, x_3 average temperature in January in Huixian, and x_4 mean maximum temperature in March in Wudu (Xiao et al. 2007). For the Pingliang region in eastern Gansu, disease prevalence (y) on autumn wheat seedlings is calculated using $y = -0.1736 + 0.2051x_1 + 0.4345x_2 + 0.0089x_3$, where x_1 is disease prevalence in spring, x_2 amount of oversummering pathogen, and x_3 total precipitation in September and October. The first model is mainly based on climatic factors in spring and a prediction cannot be obtained until May, which leaves limited time for disease management. The second model incorporates disease prevalence in previous fall, and a prediction can be made after March, leaving adequate time for disease control. Different from most predictions, the third model predicts stripe rust prevalence in fall. The usefulness of this model depends upon how frequently the pathogen overwinters in the region. It can be used for control of the disease in autumn if needed, another model for predicting stripe rust epidemics in spring may be more useful.

For the Hanzhong region in Shaanxi Province, where the stripe rust can overwinter but may not be able to oversummer, stripe rust prevalence (y) of a crop season is predicted using formula $y = -4.699158 + 0.199445x_1 + 0.052963x_2 + 0.314299x_3 + 0.014257x_4 + 0.008677x_5$, where x_1 is the amount of pathogen in autumn of the previous year, x_2 amount of pathogen in spring, x_3 average temperature in April, x_4 precipitation in April, and x_5 proportion of susceptible variety area (Hu et al. 2000).

For Shanxi Province, where the stripe rust pathogen generally cannot survive the winter and the disease usually develops late in the crop season, two models have been developed mainly based on spring precipitation. Fan et al. (2007) developed the following model for predicting final disease prevalence (y) usually in June to early July for this province: $y = -0.2766 + 0.0249x_1 + 0.0104x_2 + 0.0132x_3$, where x_1 is precipitation in the middle 10 days of April to the first 10 days of May, x_2 precipitation in the middle 10 days of May and the last 10 days of May, and x_3 average relative humidity in May. Recently, Cheng et al. (2010) developed a model specifically for the Jincheng region in Shanxi Province: $y = 3.7644 + 0.0070x_1 - 0.0097x_2 + 0.0273x_3 + 0.1079x_4 - 0.2145x_5$, where y is final disease prevalence, x_1 precipitation in the last 10-day period of July in the preceding year, x_2 sunlight in the first 10 days of November in the preceding year, x_3 is sunlight in the first 10 days of March in the preceding year, x_4 average temperature in the middle 10 days of April, and x_5 precipitation in the last 10-day period of April. This model is the only one including sunlight variables in predicting stripe rust, probably because sunlight is a limiting factor in this region. The usefulness of this model needs to be determined with stripe rust and climatic data of more years.

As presented above, all of the prediction models developed in China requires stripe rust survey in the early growing season. Because of this, it is important to have a system to collect, store and analyze stripe rust survey data. Zeng and Luo (2008) summarized the studies on stripe rust systems in various levels in China. Kuang et al. (2013) developed a web-based prediction system for wheat stripe rust using the WebGIS technologies. According to the authors, prediction of wheat stripe rust could be conducted using the existing prediction models as presented above and users can also establishing prediction regression models using input historical data.

4.2.1.3 Stripe Rust Prediction in the US

In the US, attempts for predicting stripe rust epidemics can be traced back to the early 1970s. Shrum (1975) conducted research to simulate stripe rust using Epidemic, a flexible plant disease simulator. The simulation approach generally tries to use all factors, including host, pathogen and environmental factors before and throughout the growth stages, to predict disease epidemics. This approach has more value in describing than predicting disease epidemics because when all components are available the disease season gets to the end. For predicting disease epidemics, unrealized scenarios or forecasts have to be used. The simulation approach did not play much role in stripe rust management, except that an Expert System developed later took the basically similar, but more qualitative approach (Line 2002).

A more commonly used approach is to utilize historical data to find correlation between climatic and disease data. In the US Pacific Northwest, stripe rust was not severe before 1958 but has become an important disease since then (Line 2002). Coakley (1978) analyzed the meteorological data, including temperature, precipitation, and relative humidity, of Pendleton, Oregon for 1961–1975. She found that in 1961–1975, the average temperature in January and February was 2 °C higher than for those same months in 1935–1960. In contrast, the average April temperature was over 1.2 °C cooler. Warmer than-normal winter and cooler April temperatures favor the development of stripe rust epidemics. Cool spring temperatures delay high-temperature adult-plant (HTAP) resistance in wheat cultivars such as Gaines and Nugaines. Similarly, the unusually severe stripe rust epidemics in the Pacific Northwest in 2010 and 2011 were explained by the much higher precipitation and lower temperatures in May and/or June and high temperatures in December and January in both years and higher precipitation in September, 2010 than normal (Chen 2011, 2014). The wet and cool weather conditions in 2010 accelerated stripe rust development, reduce HTAP resistance expression, delayed crop maturity, prolonged the rust season, and extended viable urediniospore inoculum into the fall of 2010. The high amount of inoculum and unusual wet conditions in the fall resulted in heavy infection on the winter crop before the winter. The warm winter allowed a high rate of winter survival and an early start of rust development in the early spring, again the wet and cool May and June allowed the development of another and even more severe epidemic in 2011.

Similarly, Coakley and Line (1981) studied the relationships of climatic variation with stripe rust epidemics. Using the historical stripe rust epidemic weather data at Pullman, WA, they found that since 1958 climatic conditions contributed to an increase in the frequency of epidemics. For 1963–1979, stripe rust intensities on susceptible winter wheat cultivar Omar and Gaines that has a low level of high-temperature adult-plant (HTAP) resistance were positively correlated with January temperatures and negatively correlated with April and June temperatures. Spring temperatures were more highly correlated with stripe rust development in the HTAP resistant Gaines than in the highly susceptible cultivar Omar. They also found that frequency of precipitation in June was correlated with stripe rust intensity in both Gaines and Omar. To mathematically relate stripe rust intensity to cumulative temperatures, they calculated negative degree days (NDD) and positive degree days (PDD) by using a 7 °C base, and determined that stripe rust intensity was negatively correlated with NDD and PDD. The best correlations were with NDD that accumulated in both December and January, and PDD that accumulated in April to June. They developed regression equations for disease intensity on Omar and Gains. The equations for the two cultivars had significantly different intercepts but similar slops. Their weather data analysis helped explain why stripe rust was not severe before 1958 and suddenly became severe after 1958. In cultivars differing in resistance, the 20-year period before 1960 was generally colder than normal in the winter and warmer than normal in the spring, and from 1960 to 1979, most winters were warmer and most springs were cooler (Line 2002). More importantly, this study laid the foundation for developing models for predicting stripe rust intensity in the US Pacific Northwest.

Coakley et al. (1982) developed statistic models for different winter wheat cultivars at Pullman, Washington and other four locations in the US Pacific Northwest. In these models, they used standardized NDD (NDDZ) and PDD (PDDZ) by subtracting a long-term mean from each yearly values and dividing by the standard deviation of the mean. Their results indicate that a model for predicting disease intensity by using macro-meteorological conditions can be useful at other locations within a synoptic weather region. Later, they developed two models for predicting stripe rust intensity (Coakley et al. 1983). The first model used NDDZ accumulated during December and January and Julian Day of spring (JDS, defined as the date when 40 or more positive degree days [PDD] had accumulated during the following 14 days) and the second model has PDD for the 80-day period after the JDS. For examples, the prediction model 1 for stripe rust intensity of a susceptible cultivar at the Pullman location is: $y = 2.3400 - 1.8970\text{NDDZ} + 0.0377\text{JDS}$ and model 2 is: $y = 4.5272 - 1.7235\text{NDDZ} + 0.0420\text{JDS} - 0.0054\text{PDD}$. The model 1 is suitable for making timely management decisions, while the model 2 does not leave much time for managerial decisions although it increases the accuracy of prediction. From the location-based models, regional models were developed (Coakley et al. 1984). The regional models ($y = 1.5752 - 1.4174\text{NDDZ} + 0.0253\text{JDS}$ and $y = 3.9390 - 1.2115\text{NDDZ} + 0.0328\text{JDS} - 0.0065\text{PDD}$) have several advantages over location models, such as predictions can be made for each site using only the meteorological data at that site and a variable that helps account for different spring conditions at the various sites is included. The authors noticed incorrect predictions when spring was unusually favorable or unfavorable for stripe rust development. Further improved models were developed for predicting stripe rust on winter wheat in the Pacific Northwest using a computer program WINDOW (Coakley et al. 1988). Meteorological data for 1967–1984 in segments of 21–65 days beginning on 29 July of each year and ending on 24 July of the following year were analyzed. In addition to NDDZ, JDS and PPD, mean maximum temperature (MMAX), total precipitation (TPREC), precipitation frequency, total days maximum temperature greater than 25 °C, and total days average temperatures less than 0 °C (DL0C) were also used in developing models. Some of the models were better than the previous models. However, due to the complexity of these models, the model using NDDZ

[DI = 5.9091 − 1.8317NDDZ, where NDDZ = (NDD-554)/132 and NDD = the absolute value of sum [T(dav) − 7 °C) of 1 December–31 January] (Coakley et al. 1988), was mostly used for predicting stripe rust intensity from the 1980s to 2008 (Line 2002, Chen 2005, Sharma-Poudyal and Chen 2011).

The previous models discussed above predict disease index (DI) using a 0–9 scale at flowering stage (Coakley and Line 1981; Coakley et al. 1983, 1984, 1988). The index is based on converting percent disease intensity (0 = 0% disease, 1 ≤ 1%, 2 = 1–5%, 3 = 6–20%, 4 = 21–40%, 5 = 41–60%, 6 = 61–80%, 7 = 81–95%, 8 = 96–99%, and 9 ≥ 99%). Although DI or disease intensity is correlated positively to yield loss, the latter is also correlated to duration of disease state. The previous models do not concern about the duration of disease. Based on the author's record, the previous models generally predicted DIs similar to the observed values. However, a large variation has been observed in the yield loss data among the years of similar DI or rust severity data (X.M. Chen, unpublished data). By comparison of monthly means of maximum, minimum and average temperatures in November–February of 1993–2007 and 1963–1979 at Pullman, Washington, Sharma-Poudyal and Chen (2011) found that winter has become warmer. To identify weather factors correlated to stripe rust epidemics and develop models for predicting potential yield loss, they conducted correlation and regression analyses using weather parameters and historical yield loss data from 1993 to 2007. They analyzed a total of 1376 weather variables and found 54 significantly correlated to yield loss of winter wheat and 18 to yield loss of spring wheat. Among the seasons, winter temperature variables were more highly correlated to wheat yield loss than the other seasons, and the sum of daily temperatures and accumulated NDDs of February were more highly correlated to winter wheat yield loss than the other monthly winter variables. The number of winter rainfall days was correlated with yield loss. Through regression analyses of the correlated variables, they developed 19 models including 11 single variable models and 8 multiple-variable models for predicting yield loss on susceptible cultivars. The parameters in the single-variable models are the following:

DecANDD(Max): December accumulated NDD based on maximum temperature
FebST(Max): February sum of maximum temperatures
NovDecANDD(Max): November–December accumulated NDD based on maximum temperatures
Jan-Feb20D-MaxMAT(Max): January–February 20-day maximum moving average temperature based on maximum temperatures
DecFeb20D-MaxMAT(Max): December–February 20-day maximum moving average temperatures based on maximum temperatures
DecRFD: December rainfall days
DecJan-RFD: December–January rainfall days
DecFeb-RFD: December–February rainfall days
Apr10D-maxMAT(Mean): April 10-day maximum moving average temperature based on mean temperatures
MarApr10D-MaxMAT(mean): March–April 10-day maximum moving average temperature based on mean temperatures

Jul10D-MinMAT(min): July 10-day minimum moving average temperature based on minimum temperature

Some of the parameters are related to each other. For example, the rainfall days during winter are generally indicate warm temperatures.

The best single-variable models for yield loss are Y = −6.22 + 0.312 [FebST(Max)]; Y = 112–2.56 [DecJan-RFD], and Y = 116–2.01[DecFeb-RFD], as their correlation coefficient (R^2) value is equal or greater than 0.80 (Sharma-Poudyal and Chen 2011). In general, the multiple-variable models, with adjusted R^2 values greater than 90% provide better predictions than the single-variable models, except the two models based on spring weather or summer weather variables. The best model (R^2 = 99.4%) is: $Y = 68.6 + 0.448\ X_{18} - 0.0502\ X_{47} - 0.390\ X_3 - 0.368\ X_7 + 0.265\ X_{43}$, where Y = yield loss (%) for susceptible winter wheat, X_{18} = FebST(Max), X_{47} = NovFebST(Max) (November–February sum temperature based on maximum temperatures), X_3 = DecANDD10D-MAT(Max) (December 10-day accumulated NDD moving average temperature based on maximum temperatures), X_7 = DecST(Max) (December sum of temperatures based on maximum temperature), and X_{43} = DecFebAPDD30D–MAT(Max) (December–February 30-day accumulated PDD moving average temperature based on maximum temperatures). The authors recommended a set of six models for predicting yield loss, which allows prediction as early as after December using model $Y = 13.9 + 6.35\ X_1$, where X_1 = Dec10D-MaxMAT(Mean), and the best model described above. The forecasts made in early January and March allow growers to make decisions for getting ready for possible fungicide application in early spring if severe stripe rust will occur and selecting spring wheat cultivars to planting. Using the models, stripe rust epidemic levels were accurately predicted in the same categories as observed in 5 out of 7 years from 2009 to 2015 (X. M. Chen unpublished data). For the years, the disease was not accurately predicted in March, short-term forecasts were made based on field survey data and weather forecasts from March throughout the crop season for guiding disease control (Chen 2014).

4.2.1.4 Stripe Rust Prediction in Australia

In Australia, Brown and Holmes (1983) determined the infection rate (r) for cultivars commonly grown in Victoria based on disease development data collected from fields using equation $r = [1/(t_2 - t_1)]\ \log_e\ [x_2(1 - x_1)]/[x_1(1 - x_2]$, where x_1 is stripe rust percent attack (PA) at time t_1, and x_2 PA at time t_2. Then, they re-arranged that equation into $x_2 = (0.01_e^{r\Delta t})/(0.99 + 0.01_e^{r\Delta t})$ to calculate the PA at the soft dough growth stage (x_2) using the time change ($\Delta t = t_2 - t_1$) from t_1 when 1% PA is found to the soft dough stage. When the PA at soft dough stage (GS85) is calculated, yield loss (YL) can be calculated using eq. $YL = -0.348 - 0.375\ PA + 6.034\ (PA)^{1/2}PA$, which was developed by 2-year field experiments. This model is used to guide the use of foliar fungicides to control stripe rust. The major advantage of this approach is that it does not need long term historical climatic and disease data. However, the disadvantage is that the *r* value for a cultivar can vary greatly under different weather conditions. Later, Murray et al. (1994) determined the relationship between wheat yield and stripe rust as affected by length of epidemic and temperature at the grain development stage of crop growth. They found that yield was significantly negatively correlated with the proportion of leaf area affected by stripe rust from the end of heading to late milk. The correlation was greatest at the early milk stage of growth where the relationship was logarithmic. They found the length of the epidemic and temperature significantly influenced this relationship. Yield loss increased as the length of the epidemic increased and decreased as temperature increased during grain development. Base on their data, they developed formula $y = 100\ [1 - e^{10^{-4}X\ (393\ +\ 165\ L\ -\ 141\ T)}]$ where y = yield loss (%); *X* = stripe rust (leaf area affected, range 0–99%) at early milk; L = time from 1% disease to early milk (range 0–70 days); and T = mean daily maximum temperature from 7 days before to 14 days after early milk (range 19.8–27.5 °C). This formula can address the issue raised from the previous study (Brown and Holmes 1983).

Dennis (1987) developed a multiple regression equation relating temperature and wet period to stripe rust infection. Using experimental data and the equation, he determined that the optimum temperature range was 7–10 °C, the maximum temperature 18 °C and the minimum temperature less than 0 °C. Minimum wet period ranged from 3 h at 7–10 °C to 6 h at 15 °C. These results were similar to previous reports of scientists and used by others for predicting stripe rust epidemics in other countries (Rapilly 1979; Line 2002; Gladders et al. 2007; Sharma-Poudyal et al. 2014).

4.2.1.5 Stripe Rust Prediction in India and Pakistan

In India, Krishnaa et al. (2014) used hyperspectral reflectance data to detect stripe rust severity. They developed a model, $y = -102.06 - (263.44{*}R_{428nm}) + (399.26{*}R_{672nm}) + (428.42{*}R_{1399nm})$ using reflectance values (R) at 428, 672 and 1399 nm for predicting stripe rust severity. The model had a high correlation scores in the calibration test ($R^2 = 0.89$) and the validation test ($R^2 = 0.93$). Although the model is to indicate real-time stripe rust severity, the technique can be used to measure disease severity in the early stripe rust development and use the early data to predict the epidemic level in the late crop growth season, as the authors showed that the model can detect stripe rust severities from 0% to 100% in a linear relationship.

In Pakistan, Ahmad et al. (2010) conducted a study to "predict" yield losses in various wheat cultivars caused by stripe rust in Faisalabad, an important wheat production region in this country. The field experiment was conducted in only one crop season (2006–2007). They estimated yield loss for each variety based on final disease severity value, which is assumed use the final severity value as the disease ratings were taken on a weekly basis, using Mundy's formula Loss = (0.44 × disease severity) + 3.15 or Loss = (5.06 × the square root of the disease severity) − 17.15 (Munday 1973). Interestingly, the estimated (rather than "predicted") yield losses based on stripe rust severity were remarkably similar to the measured yield loss for each of the 36 wheat cultivars or genotypes including 24 *Lr* single-gene lines with resistance to leaf rust, which have less value for the objective of this study. This is not a truly prediction study, but the yield loss information for individual cultivars can be useful for future disease management. To add the values of prediction, a similar approach as presented by Brown and Holmes (1983) and Murray et al. (1994) as discussed above needs to be taken. The authors also tried to determine the relationships of temperature (maximum and minimum temperatures), relative humidity, rainfall, sun radiation and wind speed to stripe rust severity and grain yield. They did not found significant correlation between stripe rust severity and rainfall, sun radiation or wind speed, but obtained remarkably low but significant correlation between stripe rust severity and maximum temperature (0.0453), minimum temperature (0.0299) and relative humidity (0.0487). The positive correlation between maximum temperature and stripe rust severity is odd. However, as the experiment was conducted only in one crop season, no detail information was provided, and mismatches of the figures and figure captions were in the publication. The correctness of the data analyses and scientific value of this study could not be judged.

4.2.2 Forecasting Based on Climatic Conditions without Historical Disease Data

The methods for stripe rust prediction discussed in the previous section (Sect. 4.2.1) mostly require long-time historical disease data or data from intensive experiments in a relatively short period. Some of the information from previous studies can be directly applied for prediction, while most of them are region-specific and cannot be used directly in another region as different regions can have different limiting factors. It is possible to use climatic data to predict the possibilities of stripe rust based on the biological requirements of the stripe rust pathogen without available historical disease data under the assumptions that the pathogen and host are available.

Sharma-Poudyal et al. (2014) developed a survival index (SI) for summer and winter based on climatic conditions for *Pst* published in the literature (Table 4.2). The 0–10 scale index includes both dormant survival and active survival, so it can be used for predicting the pathogen survival and epidemic development. For summer survival, the 30 day maximum moving average temperature based on daily maximum temperatures [Max30D–MAT(Max)] are used together with 30 day minimum temperature (30D–Min). In contrast, the 30 day minimum moving average temperature based on daily minimum temperatures [Min30D-MAT(Min)] are used together with 30 day maximum temperature (30D-Max). For example, if Max30D-MAT(Max) is equal to or greater than 38 °C, or 30D-Min is greater than 22 °C, the stripe rust pathogen cannot survival and the survival index is 0. In contrast, if Max30D-MAT(Max) is equal to or less than 33 °C and 30D-Min is between 7 and 11 °C, the stripe rust pathogen can actively infect and sporulate, with the highest

Table 4.2 Climatic conditions used for estimation of potential summer and winter survival index (SI) of *Puccinia striiformis* (*Ps*)[a]

Temp (°C); snow cover (cM); other factors		Effect on *Ps*	SI
Summer survival			
Max30D-MAT(Max)[b]	30D-Min		
≥38	Otherwise	Unfavorable for survival	0
Otherwise	>22	Unfavorable for survival	0
37–34	19–22	Dormant survival, putative upper marginal limit for infection	1
37–34	16–18	Dormant survival, marginally favorable for infection	2
37–34	14–15	Dormant survival, moderately favorable for infection	3
37–34	13–12	Dormant survival, favorable for infection	4
37–34	≤11	Dormant survival, highly favorable for infection	5
≤33	19–22	Active survival, putative upper marginal limit for infection	2
≤33	16–18	Active survival, marginally favorable for infection	4
≤33	14–15	Active survival, moderately favorable for infection	6
≤33	13–12	Active survival, favorable for infection	8
≤33	7–11	Active survival, highly favorable for infection	10
Winter survival			
Min30D-MAT(Min); snow cover	30D-Max		
≤−11; ≥7.6	Otherwise	Lower limit for survival	0
≤−6 °C; <7.6	Otherwise	Lower limit for survival	0
Otherwise	≤−5	Lower limit for germination and infection	0
≤−2 to −10; ≥7.6 or ≤−2 to −5; <7.6	≤−3 to −4	Survive as dormant mycelium, putative low marginal limit for germination and infection	1
≤−2 to −10; ≥7.6 or ≤−2 to −5; <7.6	≤−2 to −1	Marginally favorable for infection and infection	2
≤−2 to −10; ≥7.6 or ≤−2 to −5; <7.6	0–1	Moderately favorable for germination infection	3
≤−2 to −10; ≥7.6 or ≤−2 to −5; <7.6	2–11	Favorable for germination and infection	4
≤−2 to −10; ≥7.6 or ≤−2 to −5; <7.6	≥12–30	Favorable disease development	5

(continued)

Table 4.2 (continued)

Temp (°C); snow cover (cM); other factors		Effect on *Ps*	SI
≥−1 to 11	≤−3 to −4	Active survival, putative low marginal limit for germination and infection	2
≥−1 to 11	≤−2 to −1	Marginally favorable for infection and infection	4
≥−1 to 11	0–1	Moderately favorable for germination infection	6
≥−1 to 11	2–11	Favorable for germination and infection	8
≥−1 to 11	≥12–30	Favorable disease development	10
Correction factors for temperature based survival index			
Relative humidity ≥65%; 30D–Min temp ≥20	Unfavorable for survival	0	
Dew point < Min temp; precipitation >30 (mm)	Favorable for infection	NC[c]	
Dew point < Min temp; precipitation 11–30 (mm)	Moderately favorable for infection	−1[d]	
Dew point < Min temp; precipitation 6–10 (mm)	Marginally favorable for infection	−2[d]	
Dew point < Min temp; precipitation ≤5 (mm)	Unfavorable	0	
Dew point ≥ Min temp; precipitation >10 (mm)	Favorable for infection	NC	
Dew point ≥ Min temp; precipitation 6–10 (mm)	Moderately favorable for infection	−1	
Dew point ≥ Min temp; precipitation <5 (mm)	Marginally favorable for infection	−2	

[a]Adapted from Sharma-Poudyal et al. (2014)
[b]*Max* maximum, *MAT* moving average temperature, *Temp* temperature, *Min* minimum, *30D* 30 days
[c]*NC* no correction
[d]The SI value determined based on temperature is extracted by 1 or 2 to get a corrected SI

survival index, 10. Similarly, if [Min30D-MAT(Min)] is equal or less than −6 °C or 30D-Max is equal or less than −5 °C, the pathogen cannot survive. In contrast, if [Min30D-MAT(Min)] is between −1 and 11 °C and 30D-Max is between 12 and 30 °C, the pathogen will actively survive as the survival index is 10. Snow cover (≥7.6 cm) is incorporated for Min30D-MAT(Min) between −2 and −10 °C. Relative humidity, precipitation and the relationship between dew point and minimum temperature are used to make correction for the survival index under certain conditions as specified in the table.

Using the survival index system, Sharma-Poudyal et al. (2014) predicted regions for the summer and/or winter survival using climatic data from 1950 to 2007 using 445 weather stations throughout the contiguous United States. The mapped regions for summer and/or winter survival are strongly supported by observed and reported stripe rust. This system can also be used to predict whether or not stripe rust will occur and the level of potential epidemic, as well as related to summer and/or winter survival, for a particular crop growth season in a specific region with the weather data available.

To a less quantitative extent, specific weather events can be used for predicting stripe rust for a region depending upon outside pathogen inoculum. In the United States, the stripe rust pathogen survives winter in the south-central states such as Texas and Louisiana. The earliness of stripe rust development in this region can be used to predict relative levels of the disease epidemic in this region and further north to the central and northern Great Plains and east to the eastern regions. When the first report of stripe rust (as early as in December and January) in southern Texas is before March, a severe stripe rust epidemic will likely develop in the entire Great Plains and the eastern regions (Chen 2005; Chen et al. 2010; XM Chen unpublished data). If a storm occurs and moves from the southern region to the north and/or to the east during the period stripe rust is actively developing in the south, stripe rust will likely show up in 2–4 weeks in the spore-receiving regions, depending upon temperature and crop stage (Chen 2005; Chen et al. 2010). In the Pacific Northwest, occurrence of stripe rust in the south areas where stripe rust starts early is routinely used to predict the relative time of stripe rust infection in the north areas. The similar approach has been used in New Zealand (Kim and Beresford 2008) and China. In China, stripe rust surveys in the oversummering regions in autumn are used to predict how widely the pathogen can be dispersed by wind and surveys in the overwintering regions in early spring are used to predict epidemic scales and levels (Li and Zeng 2002; Zeng and Luo 2006, 2008).

References

Abbasi M, Hedjaroude G, Scholler M, Goodwin SB. Taxonomy of *Puccinia striiformis* s.l. in Iran. Rostaniha. 2004;5:71–282.

Abiev SA, Zhakhanov A, Kencsarina G, Esengulova BZ. Specialization of yellow rust of wheat in South-Eastern Kazakhstan. Bot Mater Gerbariya, Inst Bot Akad Nauk Kaz SSR. 1982;12:96–8.

Ahmad S, Afzal M, Noorka IR, Iqbal Z, Akhtar N, Iftkhar Y, Kamran M. Prediction of yield losses in wheat (*Triticum aestivum* L.) caused by yellow rust in relation to epidemiological factors in Faisalabad. Pak J Bot. 2010;42:401–7.

Akanda SI, Mundt CC. Effects of two component wheat cultivar mixtures on stripe rust severity. Phytopathology. 1996;86:347–53.

Alexopoulos CJ, Mims CW. Introductory mycology. 3rd ed. New York: Wiley; 1979.

Ali S, Gladieux P, Leconte M, Gautier A, Justesen AF, Hovmøller MS, Enjalbert J, de Vallavieille-Pope C. Origin, migration routes and worldwide population genetic structure of the wheat yellow rust pathogen *Puccinia striiformis* f. sp. *tritici*. PLoS Pathog. 2014a;10:e1003903.

Ali S, Gladieux P, Rahman H, Saqib MS, Fiaz M, Ahmad H, Leconte M, Gautier A, Justesen AF, Hovmøller MS, Enjalbert J, de Vallavieille-Pope C. Inferring the contribution of sexual reproduction, migration and off-season survival to the temporal maintenance of microbial populations: a case study on the wheat fungal pathogen *Puccinia striiformis* f. sp. *tritici*. Mol Ecol. 2014b;23:603–17.

Allan RE. Club wheat. Robert E. Allan; 2014. ISBN 978-1-4951-0549-4 I-V + 354 pp + i-xviii.
Allan RE, Line RF, Peterson CJ, Rubenthaler GL, Morrison KL, Rohde CR. Crew, a multiline wheat cultivar. Crop Sci. 1983;23:1015–6.
Allan RE, Peterson CJ, Line RF, Rubenthaler GL, Morris CF. Registration of "Rely" wheat multiline. Crop Sci. 1993;33:213–4.
Angus A. Annotated list of plant pests and diseases in Zambia. Parts 1–7 and supplements. Chilanga, Zambia: Mount Makulu Research Station; 1965.
Anikster Y. Binucleate basidiospores – a general rule in rust fungi. Trans Br Mycol Soc. 1983;81:624–6.
Arthur JC. The grass rusts of South America; based on the Holway collections. Proc Am Philos Soc. 1925;64:131–223.
Becker H, Hart H. Das Auftreten und die Verbreitung von Gelbrost im Ostharz und der darn angrenzenden Weizenanbaugebieten. Z Pflanzenkr (Pflanzenpathol) Pflanzenschutz. 1939;49:449–81.
Beresford RM. Stripe rust (*Puccinia striiformis*), a new disease of wheal in New Zealand. Cereal Rusts Bull. 1982;10:35–241.
Berlin A, Djurle A, Samils B, Yuen J. Genetic variation in *Puccinia graminis* collected from oats, rye, and barberry. Phytopathology. 2012;102:1006–12.
Bhagwat SG, Bhatia CR. Selection for flag leaf stomatal frequency in bread wheat. Plant Breed. 1993;110:129–36.
Bockus WW, Bowden RL, Hunger RM, Morrill WL, Murray TD, Smiley RW, editors. Compendium of wheat diseases and pests. 3rd ed. APS Press: St. Paul; 2010.
Boshoff WHP, Pretorius ZA, van Niekerk BD. Establishment, distribution, and pathogenicity of *Puccinia striiformis* f. sp. *tritici* in South Africa. Plant Dis. 2002;86:485–92.
Brown JS, Holmes RJ. Guidelines for use of foliar sprays to control stripe rust of wheat in Australia. Plant Dis. 1983;67:485–7.
Brown JKM, Hovmøller MS. Aerial dispersal of pathogens on the global and continental scales and its impact on plant disease. Science (Washington, DC). 2002;297:537–41.
Brown Jr WM, Hill JP, Velasco VR. Barley yellow rust in North America. Annu Rev Phytopathol. 2001;39:367–84.
Burleigh JR. The winter biology of *Puccinia striiformis* West. in the Pacific Northwest. Dissertation, Washington State University; 1965.
Cao SQ, Luo HS, Jin MA, Jin SL, Duan XY, Zhou YL, Chen WQ, Liu TG, Jia QZ, Zhang B, Huang J, Wang XM, Shang XW, Sun ZY. Intercropping influenced the occurrence of stripe rust and powdery mildew in wheat. Crop Prot. 2015;70:40–6.
Carleton MA. A serious new wheat rust in this country. Science. 1915;42:58–9.
Carson ML. Virulence frequencies in oat crown rust in the United States from 2001 through 2009. Plant Dis. 2008;92:381–4.
Chen XM. Epidemiology of barley stripe rust and races of *Puccinia striiformis* f. sp. *hordei*: the first decade in the United States. In: Abstracts of the 11th international cereal rusts and powdery mildews conference, 22–27 August 2004, Norwich; 2004. p. A2.8.
Chen XM. Epidemiology and control of stripe rust [*Puccinia striiformis* f. sp. *tritici*] on wheat. Can J Plant Pathol. 2005;27:314–37.
Chen XM. Stripe rust, why the disease hit so hard in 2010, 2011. Wheat Life Aug/Sep 2011;56–9.
Chen XM. High-temperature adult-plant resistance, key for sustainable control of stripe rust. Am J Plant Sci. 2013;4:608–27.
Chen XM. Integration of cultivar resistance and fungicide application for control of wheat stripe rust. Can J Plant Pathol. 2014;36:311–26.
Chen XM, Line RF. Inheritance of stripe rust resistance in wheat cultivars used to differentiate races of *Puccinia striiformis* in North America. Phytopathology. 1992;82:633–7.

Chen XM, Line RF. Gene action in wheat cultivars for durable high-temperature adult-plant resistance and interactions with race-specific, seedling resistance to stripe rust caused by *Puccinia striiformis*. Phytopathology. 1995;85:567–2572.
Chen SM, Lu SY. Studies on epidemics of stripe rust on winter wheat in north-central China. Acta Phytopathol Sin. 1957;3:63–84.
Chen FQ, Prehn D, Hayes PM. Mapping genes for resistance to barley stripe rust (*Puccinia striiformis* f. sp. *hordei*). Theor Appl Genet. 1994;88:215–9.
Chen XM, Line RF, Leung H. Virulence and polymorphic DNA relationships of *Puccinia striiformis* f. sp. *hordei* to other rusts. Phytopathology. 1995;85:1335–42.
Chen XM, Moore MK, Milus EA, Long DL, Line RF, Marshall D, Jackson L. Wheat stripe rust epidemics and races of *Puccinia striiformis* f. sp. *tritici* in the United States in 2000. Plant Dis. 2002;86:39–46.
Chen XM, Penman L, Wan AM, Cheng P. Virulence races of *Puccinia striiformis* f. sp. *tritici* in 2006 and 2007 and development of wheat stripe rust and distributions, dynamics, and evolutionary relationships of races from 2000 to 2007 in the United States. Can J Plant Pathol. 2010;32:315–33.
Chen XM, Wang MN, Wan AM, Cheng P, Cheng JJ. Sexual or asexual reproduction, which one is more important for stripe rust? In: Chen W-Q, et al., editors. Disease risk and food security, Proceedings of the 13th international cereal rusts and powdery mildews conference, Beijing, 28 Aug–1 Sept 2012. Beijing: China Agricultural Science and Technology Press; 2012. p. 36–7.
Cheng HX, Wang CM, Shuai KJ, Li YF, Song JF, Wang JM. Meteorological forecast models of wheat pests in Jincheng City, Shanxi Province. Jiangsu Agric Sci. 2010;6:159–63. (in Chinese)
Cheng JJ, Li H, Ren B, Zhou CJ, Kang ZS, Huang LL. Effect of canopy temperature on the stripe rust resistance of wheat. New Zealand J Bot. 2015;43:306–15.
Cheng P, Chen XM, See D. Grass hosts harbor more diverse isolates of *Puccinia striiformis* than cereal crops. Phytopathology. 2016;106:362–71.
Coakley SM. The effect of climate variability on stripe rust of wheat in the Pacific Northwest. Phytopathology. 1978;68:207–12.
Coakley SM, Line RF. Quantitative relationships between climatic variables and stripe rust epidemics on winter wheat. Phytopathology. 1981;71:461–7.
Coakley SM, Boyd WS, Line RF. Statistical models for predicting stripe rust on winter wheat in the Pacific Northwest. Phytopathology. 1982;72:1539–42.
Coakley SM, Line RF, Boyd WS. Regional models for predicting stripe rust on winter wheat in the Pacific Northwest. Phytopathology. 1983;73:1382–5.
Coakley SM, Boyd WS, Line RF. Development of regional models that use meteorological variables for predicting stripe rust disease in winter wheat. J Clim Appl Meteorol. 1984;23:1234–40.
Coakley SM, Line RF, McDaniel LR. Predicting stripe rust severity on winter wheat using an improved method for analyzing meteorological and rust data. Phytopathology. 1988;78:543–50.
Coram TE, Settles ML, Chen XM. Transcriptome analysis of high-temperature adult-plant resistance conditioned by *Yr39* during the wheat-*Puccinia striiformis* f. sp. *tritici* interaction. Mol Plant Pathol. 2008a;9:479–93.
Coram T, Wang MN, Chen XM. Transcriptome analysis of the wheat-*Puccinia striiformis* f. sp. *tritici* interaction. Mol Plant Pathol. 2008b;9:157–69.
Craigie JH. Discovery of the function of the pycnia of the rust fungi. Nature. 1927;120:765–7.
Craigie JH. On the occurrence of pycnia and aecia in certain rust fungi. Phytopathology. 1928;18:1005–21015.
Danial DL, Stubbs RW, Parlevliet JE. Evolution of virulence patterns in yellow rust races and its implications for breeding for resistance in wheat in Kenya. Euphytica. 1994;80:165–70.
de Bary A. Neue Untersuchungen über die Uredineen, insbesondere die Entwicklung der *Puccinia graminis* und den Zusammenhang derselben mit Aecidium Berberidis. Monatsber K Preuss Akad Wiss Berlin. 1866.

de Bary A. Neue Untersuchungen iiber Uredineen. Monatsber K Preuss Akad Wiss Berlin. 1867:15–50.
Dennis JI. Temperature and wet-period conditions for infection by *Puccinia striiformis* f. sp. *tritici* race 104E137A +. Trans Br Mycol Soc. 1987;88:119–21.
de Vallavieille-Pope C, Huber L, Leconte M, Goyeau H. Comparative effects of temperature and interrupted wet periods on germination, penetration, and infection of *Puccinia recondita* f. sp. *tritici* and *P. striiformis* on wheat seedlings. Phytopathology. 1995;85:409–15.
de Vallavieille-Pope C, Huber L, Leconte M, Bethenod O. Preinoculation effects of light quantity on infection efficiency of *Puccinia striiformis* and *P. triticina* on wheat seedlings. Phytopathology. 2002;92:1308–14.
de Vallavieille-Pope C, Ali S, Leconte M, Enjalbert J, Delos M, Rouzet J. Virulence dynamics and regional structuring of *Puccinia striiformis* f. sp. *tritici* in France between 1984 and 2009. Plant Dis. 2012;96:131–40.
Dietz SM, Hendrix JW. Reactions of grasses to stripe rust at Pullman, Washington. Phytopathology. 1962;52:730.
Doling DA, Doodson JK. The effect of yellow rust on the yield of spring and winter wheat. Trans Br Mycol Soc. 1968;51:427–34.
Dubin HJ, Stubbs RW. Epidemic spread of barley stripe rust in South America. Plant Dis. 1986;70:141–4.
Düsünceli F, Cetin L, Albustan S, Beniwal SPS. Occurrence and impact of wheat stripe rust (*Puccinia striiformis*) in Turkey in 1994/95 crop season. In: Proceedings of the European and Mediterranean cereal rusts powdery mildew conference, Lunteren, vol. 9; 1996. p. 309.
Eriksson J. Uber die Spezialisierung des Parasitismus bei den Getreiderostpilzen. Ber Dtsch Bot Ges. 1894;12:292–331.
Eriksson J, Henning E. Die Hauptresultate einer neuen Untersuchung über die Getreiderostpilze. Z Pflanzenkr. 1894;4:197–203.
Eversmeyer MG, Kramer CL. Epidemiology of wheat leaf rust and stem rust in the central great plains of the USA. Annu Rev Phytopathol. 2000;38:491–513.
Fan SQ, Xie XS, Li F, Yin QY, Zheng WY. Forecast model for prevalent stripe rust in winter wheat in Shanxi Province. Chin J Eco-Agric. 2007;15:113–5.
Fan ZY, Cao SQ, Luo HS, Jin SL. Prediction model on wheat stripe rust in Gangu County. Gansu Agric Sci Technol. 2008;10:9–21.
Gassner G, Straib W. Weitere Untersuchungen über biologisch Rassen und über die Spezialisierungsverhältnisse des Gelbrostes *Puccinia glumarum* (Schm., Erikss. u. Heen). Ebenda. 1934;21:121–45.
Georgievskaja NA. Quelques lois sur Ie developpement de la rouille Jaune du ble. Tr Vses Nauchno Issled Inst Zabtsh Rost Leningrad. 1966;26:55–63. (In Russian)
Gladders P, Langton SD, Barrie IA, Hardwick NV, Taylor MC, Paveley ND. The importance of weather and agronomic factors for the overwinter survival of yellow rust (*Puccinia striiformis*) and subsequent disease risk in commercial wheat crops in England. Ann Appl Biol. 2007;150:371–82.
Gustafson P, Raskina O, Ma XF, Nevo E. Chapter 1: Wheat evolution, domestication, and improvement. In: Carver BF, editor. Wheat: science and trade. Ames: Wiley-Blackwell; 2009. p. 5–30.
Hakim MS, Mamluk OF. Virulence of wheat yellow rust pathogen in Syria and Lebanon. In: Proceedings of the European and Mediterranean cereal rusts powdery mildew conference, vol. 9; 1996. p. 141.
Han DJ, Wang QL, Chen XM, Zeng QD, Wu JH, Xue WB, Zhan GM, Huang LL, Kang ZS. Emerging *Yr26*-virulent races of *Puccinia striiformis* f. sp. *tritici* are threatening wheat production in the Sichuan Basin, China. Plant Dis. 2015;99:754–60.
Hart H, Becker H. Beitrage zur Frage des Zwischenwirtes für *Puccinia glumarum*. Z Pflanzenkr (Pflanzenpathol) Pflanzenschutz. 1939;49:559–66.
Hassan SF. Rust problems in Pakistan. In: Tahir M, editor. Wheat research and production in Pakistan, vol. I. Islamabad: PARC; 1978. p. 90–3.

Hassebrauk K. Die Gelbrostepidemie 1961 in Deutschland. Nachrichtenbl Dtsch Pflanzenschutzdienst (Berl). 1962;14:22–6.

Hassebrauk K. Nomenklatur, geographische Verbreitung und Wirtsbereich des Gelbrostes *Puccinia striiformis* West. Mitt. Biol. Bundesanst Land=Forstwirtsch, Berlin-Dahlem. 1965;16:1–75.

Hassebrauk K. Der Gelbrost1 *Puccinia striiformis* West. 2 Befallsbild. Morphologie und Biologie der Sporen. Infektion und weitere Entwicklung. Wirkungen auf die Wirtspflanze. 1970;139:1–111.

Hassebrauk K, Röbbelen G. Der Gelbrost, *Puccinia striiformis* West. 4, Epidemiology. Bekämpfungsmassnahmen. Mitt. Biol. Bundesanst. Land= Forstwirtsch. Berlin-Dahlem. 1975;164:1–183.

Hendrix JW, Burleigh JR, Tu JC. Oversummering of stripe rust at high elevations in the Pacific Northwest-1963. Plant Dis Rep. 1965;49:275–8.

Hermansen JE, Stapel C. Notes on the yellow rust epiphytotic in Denmark in 1972. Cereal Rust Bull. 1973;1:5–8.

Hermansen JE, Torp U, Prahm LP. Studies of transport of live spores of cereal mildew and rust fungi across the North Sea. Grana. 1978;17:41–6.

Hovmøller MS, Justesen AF, Brown JKM. Clonality and long-distance migration of *Puccinia striiformis* f. sp. *tritici* in north-west Europe. Plant Pathol. 2002;51:24–32.

Hovmøller MS, Yahyaoui AH, Miles EA, Justesen AF. Rapid global spread of two aggressive strains of a wheat rust fungus. Mol Ecol. 2008;17:3818–26.

Hovmøller MS, Walter S, Bayles R, Hubbard A, Flath K, Sommerfieldt N, Leconte M, Czembor P, Rodriguez-Algaba J, Thach T, Hansen JG, Lassen P, Justesen AF, Ali S, de Vallavieille-Pope C. Replacement of the European wheat yellow rust population by new races from the centre of diversity in the near-Himalayan region. Plant Pathol. 2016;65:402–11.

Hu XP, Yang ZW, Li ZQ, Deng ZY, Ke CH. Studies on the prediction of wheat stripe rust epidemics in Hanzhong District of Shaanxi Province. Acta Univ. 2000;28:18–21.

Humphrey HB, Hungerford CW, Johnson AG. Stripe rust (*Puccinia glumarum*) of cereals and grasses in the United States. J Agric Res (Washington, DC). 1924;29:209–27.

Hungerford CW, Owens CE. Specialized varieties of *Puccinia glumarum* and hosts for variety *tritici*. J Agric Res (Washington, DC). 1923;25:363–401.

Jackson HA, Mains EB. Aecial stage of the orange leaf rust of wheat. J Agric Res. 1921;22:152–72.

Jin Y. Role of *Berberis* spp. as alternate hosts in generating new races of *Puccinia graminis* and *P. striiformis*. Euphytica. 2011;179:105–8.

Jin Y, Szabo LJ, Carson M. Century-old mystery of *Puccinia striiformis* life history solved with the identification of *Berberis* as an alternate host. Phytopathology. 2010;100:432–5.

Jin Y, Rouse M, Groth J. Population diversity of *Puccinia graminis* is sustained through sexual cycle on alternate hosts. J Integr Agric. 2014;13:262–4.

Jing JX, Shang HS, Li ZQ. The biological effects of ultraviolet ray radiation on wheat stripe rust (*Puccinia striiformis* West.). Acta Phytopathol Sin. 1993;24:299–304.

Johnson R. Genetics of resistance to barley yellow rust. In: First international congress of plant pathology abstract 99, London; 1968.

Johnson R, Priestley RH, Taylor EC. Occurence of virulence in Puccinia striiformis for Compair wheat in England. Cereal Rusts Bull. 1978;6:11–3.

Kajiwara T, Ueda I, Iwata Y. Untersuchungen uber die Physiologische Spezialisierung des Gerstengebrostes (*Puccinia striiformis* f. sp. *hordei*) in Japan. Phytopathol Z. 1964;50:313–28.

Kang ZS, Zhao J, Wang ZY, Huang LL. Research progress on the role of sexual hosts for wheat epidemiology in China. In: Abstracts of 14th international cereal rusts and powdery mildews conference, 5–8 July 2015, Helsingør, Denmark; 2015. p. 29.

Karki CB. Report on evaluation of Nepalese wheat and barley varieties in the seedling stage on their resistance to yellow rust. Wageningen: Research Institute of Plant Protection, IPO; 1980.

Kim KS, Beresford RM. Use of a spectrum model and satellite cloud data in the simulation of wheat stripe rust (*Puccinia striiformis*) dispersal across the Tasman Sea in 1980. Agric Forest Meteorol. 2008;148:1374–82.

Krishnaa G, Sahooa RN, Pargalb S, Guptaa VK, Sinhac P, Bhagatd S, Saharane MS, Singha R, Chattopadhyayd C. Assessing wheat yellow rust disease through hyperspectral remote sensing. In: The international archives of the photogrammetry, remote sensing and spatial information sciences, volume XL-8, 2014 ISPRS Technical Commission VIII Symposium, 09–12 December 2014, Hyderabad; 2014. doi:10.5194/isprsarchives-XL-8-1413-2014.

Kuang WG, Liu WC, Ma, ZH, Wang HG. Development of a web-based prediction system for wheat stripe rust. In: Li D, Chen Y, editors. CCTA 2012, Part I, International Federation for Information Processing (IFIP) 2013 AICT 392; 2013. p. 324–35.

Kumar J, Nayar SK, Prashar M, Bhardwaj SC, Bhatnagar R. Virulence survey of *Puccinia striiformis* West in India during 1988–1989. Int Trop Plant Dis. 1993a;11:79–83.

Kumar J, Nayar SK, Prashar M, Bhardwaj SC, Bhatnagar R. Virulence survey of *Puccinia striiformis* in India during 1990-1992. Cereal Rusts Powdery Mildews Bull. 1993b;21:17–24.

Kumar K, Holtz MD, Xi K, Turkington TK. Overwintering potential of the stripe rust pathogen (*Puccinia striiformis*) in central Alberta. Can J Plant Pathol. 2013;35:304–14.

Li ZQ, Liu HW. Preliminary study of development patterns of wheat stripe rust in Shaanxi, Gansu, and Qinghai. Acta Univ Agric Boreali-Occident. 1956;1956(4):1–18.

Lei Y, Wang MN, Wan AM, Xia CJ, See DR, Zhang M, Chen XM. Virulence and molecular characterization of experimental isolates of the stripe rust pathogen (*Puccinia striiformis*) indicate somatic recombination. Phytopathology. 2017;107:329–44.

Li ZQ, Zeng SM. Wheat rusts in China. Beijing: China Agricultural Press; 2002. p. 379.

Liang JM, Liu XF, Li Y, Wan Q, Ma ZH, Luo Y. Population genetic structure and the migration of *Puccinia striiformis* f. sp. *tritici* between the Gansu and Sichuan Basin populations of China. Phytopathology. 2016;106:192–201.

Line RF. Factors contributing to an epidemic of stripe rust on wheat in the Sacramento Valley of California in 1974. Plant Dis Rep. 1976;60:312–6.

Line RF. Stripe rust of wheat and barley in North America: a retrospective historical review. Annu Rev Phytopathol. 2002;40:75–118.

Line RF, Qayoum A. Virulence, aggressiveness, evolution, and distribution of races of *Puccinia striiformis* (the cause of stripe rust of wheat) in North America, 1968–87. US Dep Agric Agric Res Serv Tech Bull. 1992;1788:44pp.

Littlefield LJ. Biology of the plant rust: an introduction. Ames: Iowa State University Press; 1981. 103 pp.

Liu M, Hambleton S. Taxonomic study of stripe rust, *Puccinia striiformis* sensu lato, based on molecular and morphological evidence. Fungal Biol. 2010;114:881–99.

Lu SY, Li GX. Effects of light and temperature on wheat resistance to stripe rust. Acta Phytopathol Sin. 1958;4:129–35.

Lu SY, Yang ZM, Wu WZ, Fan GF, Li WN, Li GX. Study on the stripe rust pathogen on wheat and Gramineae grasses. Acta Phytopathol Sin. 1958;4:137–44.

Lu NH, Zhan GM, Chen XM, Wang JF, Huang LL, Kang ZS. Spatial genetic diversity and interregional spread of *Puccinia striiformis* f. sp. *tritici* in the Northwest China. Eur J Plant Pathol. 2011;131:685–93.

Luig NH. Epidemiology in Australia and New Zealand. In: Roelfs AP, Bushnell WR, editors. The cereal rusts, vol. 2. Diseases, distribution, epidemiology and control. Orlando: Academic; 1985. p. 279–304.

Ma LJ, Qiao JX, Kong XY, Zou YP, Xu XM, Chen XM, Hu XP. Effect of low temperaure and wheat winter-hardness on survival of *Puccinia striiformis* f. sp. *tritici*. PLoS One. 2015;10:e0130691.

Maddison AC, Manners JG. Sunlight and viability of cereal rust uredospores. Trans Br Mycol Soc. 1972;59:429–43.

Mains EB. Studies concerning heteroecious rusts. Mycologia. 1933;25:407–71.

Manisterski J, Kosman E, Eyal Z. Changes in wheat leaf rust and yellow rust populations in Israel. In: Proceedings of the European and Mediterranean cereal rusts & powdery mildews conference, vol. 9; 1996. p.140.

Manners JG. Studies on the physiologic specialization of yellow rust (*Puccinia glumarum* (Schm.) Erikss. et Henn.) in Great Britain. Ann Appl Biol. 1950;37:187–214.

Manners JG. *Puccinia striiformis* Westend. var. *dactylidis* var. nov. Trans Br Mycol Soc. 1960;43:65–8.

Marshall D, Sutton RL. Epidemiology of stripe rust, virulence of *Puccinia striiformis* f. sp. *hordei*, and yield loss in barley. Plant Dis. 1995;79:732–7.

McGregor MJC. Some effects of lights on the growth of yellow rust on wheat. In: Proceedings of the European and Mediterranean cereal rusts conference,. Interlaken; 1976. p. 56–66.

McGregor AJ, Manners JG. The effect of temperature and light intensity on growth and sporulation of *Puccinia striiformis* on wheat. Plant Pathol. 1985;34:263–71.

Melching JS, Stanton JR, Koogle DL. Deleterious effects of tobacco smoke on germination and infectivity of spores of *Puccinia graminis tritici* and on germination of spores of *Puccinia striiformis*, *Pyricularia oryzae* and an *Alternaria* species. Phytopathology. 1974;64:1143–7.

Mendgen K. Growth of *Verticillium lecanii* in pustules of stripe rust (*Puccinia striiformis*). Phytopathol Z. 1981;102:301–9.

Munday EJ. The effect of yellow rust and its control on the yield of Joss Cambier Winter Wheat. Plant Pathol. 1973;22:171–6.

Milu EA, Kristensen K, Hovmoller MS. Evidence for increased agressiveness in a recent widespread strain of *Puccinia striiformis* f. sp. *tritici* causing stripe rust of wheat. Phytopathology. 2009;99:89–94.

Murray G, Ellison M, Watson PJ, Cullis BR. The relationship between wheat yield and stripe rust as effected by length of epidemic and temperature at grain development stage of crop growth. Plant Pathol. 1994;43:397–405.

Nagarajan S, Joshi LM. Epidemiology in the Indian subcontinent. In: Roelfs AP, Bushnell WR, editors. The cereal rusts, vol. 2. Diseases, distribution, epidemiology and control. Orlando: Academic; 1985. p. 371–402.

Nagarajan S, Singh DV. Long-distance dispersion of rust pathogens. Annu Rev Phytopathol. 1990;28:139–53.

Nan ZB. Fungal diseases of cultivated grasses and foreage legumes in loess plateau of eastern Gansu Province. Pratacultural Sci. 1990;7:30–4.

Naoumova NAM. Natural fluctuations of temperature and the duration of the incubation period of *Puccinia glumarum* f. sp. *tritici*. Zaschita Rasteniy. 1937;12:51–66.

Narita T, Mano Y. Researches on the oversummering, overwintering, and source of primary infection of stripe rust in wheat and barley. Spec Rep Plant Dis Insects Forecasting Serv Soc Plant Prot Minist Agric For Jpn. 1962;12:1–36.

Nazari K, Torabi M. Distribution of yellow rust (*Yr*) resistance genes in Iran. Acta Phytopathol Entomol Hung. 2000;35:121–31.

Newton M, Johnson T. Stripe rust, *Puccinia glumarum*, in Canada. Can J Res. 1936;14:89–108.

Newton M, Johnson T, Brown AM. Stripe rust in Canada. Phytopathology. 1933;23:25–6.

Nirmala J, Dahl S, Steffenson B, Kannangara CG, von Wettstein D, Chen XM, Kleinhofs A. Proteolysis of the barley receptor-like protein kinase Rpg1 by a proteasome pathway is required for Rpg1 mediated stem rust resistance. Proc Natl Acad Sci U S A. 2007;104:10276–81.

Nirmala J, Drader T, Chen XM, Steffenson B, Kleinhofs A. Stem rust spores elicit rapid RPG1 phosphorylation. Mol Plant-Microbe Int. 2010;23:1635–42.

Niu YC, Li ZQ, Shang HS. The new discovery on geographic distribution of *Puccinia striiformis* West. f. sp. *hordei* in China. Acta Univ Agric Boreali-Occident (Suppl). 1991;19:63–5.

O'Brien L, Brown JS, Young RM, Pascoe I. Occurrence and distribution of wheat stripe rust in Victoria and susceptibility of commercial wheat cultivars. Aus Plant Pathol. 1980;9:14.

Pady SM, Johnston CO, Rogerson CT. Stripe rust of wheat in Kansas in 1957. Plant Dis Rep. 1957;41:959–61.

Papastamati K, van den Bosch F. The sensitivity of the epidemic growth rate to weather variables, with an application to yellow rust on wheat. Phytopathology. 2007;97:202–10.

Park RF. The role of temperature and rainfall in the epidemiology of *Puccinia striiformis* f. sp. *tritici* in the summer rainfall area of eastern Australia. Plant Pathol. 1990;39:416–23.

Peel MD, Riveland N. Winter wheat production in North Dakota. Ext Bull. 1977;33. http://www.ag.ndsu.edu/pubs/plantsci/smgrains/eb33w.htm

Pochard E, Goujon C, Vergara S. Influence de la température, de l'elairement et du stade de la plante sur l'expression de la sensibilité á la rouille jaune de quelques variétés Francaises de blé. Ann Amelior Plant. 1962;12:45–58.

Prasada R, Joshi LM, Singh SD, Misra DF, Goel LB, Kumari K, Sharma SK, Josh PC, Ahmad ST. Occurrence of physiological races of wheat and barley rusts in India during 1962–64 and their sources of resistance. Ind J Agric Sci. 1967;37:273–81.

Prashar M, Bhardwaj SC, Jain SK, Datta D. Pathotypic evolution in *Puccinia striiformis* in India during 1995–2004. Aus J Agric Res. 2007;58:602–4.

Pretorius ZA, Boshoff WHP, Kema GHJ. First report of *Puccinia striiformis* f. sp. *tritici* on wheat in South Africa. Plant Dis. 1997;81:424.

Pretorius ZA, Pienaar L, Prins R. Greenhouse and field assessment of adult plant resistance in wheat to *Puccinia striiformis* f. sp. *tritici*. Aus Plant Pathol. 2007;36:552–9.

Qayoum A, Line RF. High-temperature adult-plant resistance to stripe rust of wheat. Phytopathology. 1985;75:1121–5.

Rapilly F. Yellow rust epidemiology. Annu Rev Phytopathol. 1979;17:59–73.

Roelfs AP. Effects of barberry eradication on stem rust in the United States. Plant Dis. 1982;66:177–81.

Roelfs AP. Wheat and rye stem rust. In: AP Roelfs, WR Bushnell, editors. The cereal rusts vol. II: Diseases, distribution, epidemiology and control. New York: Academic; 1985. p. 3–37

Roelfs AP, Huerta-Espino J, Marshall D. Barley stripe rust in Texas. Plant Dis. 1992;76:538.

Russell GE. Deposition of *Puccinia striiformis* uredospores on adult wheat plants in laboratory experiments. Cereal Rusts Bull. 1975;3:40–3.

Saari EE. The *Yr9* virulence factor in *Puccinia striiformis* and South Asia: coevolution or migration from a distant place. In: Proceedings of the European and Mediterranean cereal rusts & powdery mildews conference, vol. 9; 1996. p. 142.

Schmitt CG, Hendrix JW, Emge RG, Jones NW. Stripe rust, *Puccinia striiformis* west. Tech. Rept. 43 Plant Sci. Lab., Fort Detrick; 1964. 111pp.

Shaner G, Powelson RL. The oversummering and dispersal of inoculum of *Puccinia striiformis* in Oregon. Phytopathology. 1973;63:13–7.

Sharma KB, Gupta RBL. Control of yellow rust *Puccinia striiformis* of barley by fungicide. Indian J Mycol Plant Pathol. 1979;9:33–5.

Sharma-Poudyal D, Chen XM. Models for predicting potential yield loss of wheat caused by stripe rust in the US Pacific Northwest. Phytopathology. 2011;101:544–54.

Sharma-Poudyal D, Chen XM, Rupp RA. Potential oversummering and overwintering regions for the wheat stripe rust pathogen in the contiguous United States. Int J Biometeorol. 2014;58:987–97.

Sharp EL. Prepenetration and postpenetration environment and development of *Puccinia striiformis* on wheat. Phytopathology. 1965;55:198–203.

Sharp EL. Atmospheric ions and germination of urediospores of *Puccinia striiformis*. Science. 1967;156:1359–60.

Sharp EL, Hehn ER. Overwintering of stripe rust in winter wheat in Montana. Phytopathology. 1963;53:1239–40.

Shrum R. Simulation of wheat stripe rust (*Puccinia striiformis* West.) using EPIDEMIC: a flexible plant disease simulator, Prog. Rep. No. 347. Pennsylvania State University, University Park; 1975.

Shurtleff MC, Averre III CW. Glossary of plant-pathological terms. St. Paul: APS; 1997. p. 361.
Stakman EC. Epidemiology of cereal rusts. In: Proceedings of the 5th Pacific. Science Congress, 1–14 June 1933. Victoria and Vancouver. University of Toronto Press, Toronto, vol. 4; 1934. p. 3177–84.
Straib W. Untersuchungen über das Vorkommen physiologischer Rassen des Gelbrostes (*Puccinia glumarum*) in den Jahren 1935–1936 und über die Agressivität einiger neuer Formen auf Getreide und Gräsern. Arb Biol Reichsanst Land= Forstwirtsch, Berlin-Dahlem. 1937;22:91–119.
Stubbs RW. Influence of light intensity on the reactions of wheat and barley seedlings to *Puccinia striiformis*. Phytopathology. 1967;57:615–7.
Stubbs RW. Artificial mutation in the study of the relationship between races of yellow rust of wheat. In: Proceedings of the 2nd European and Mediterranean cereal rusts conference; 1968. p. 60–2.
Stubbs RW. Stripe rust. In: Roelfs AP, Bushnell WR, editors. The cereal rusts, vol. 2, Diseases, distribution, epidemiology and control. Orlando: Academic; 1985. p. 61–101.
Stubbs R W, Yang HA. Pathogenicity of *Puccinia striiformis* for wheat cultivars with resistance derived from rye. In: Proceedings of the European and Mediterranean cereal rusts conference, vol. 7; 1988. p. 110–2
Stubbs RW, Slovencikova V, Bartos P. Yellow rust resistance of some European wheat cultivars derived from rye. Cereal Rusts Bull. 1977;5:45–7.
Sun P, Zeng SM. Modeling the inter-regional disease spread. Z Pflanzenkrankh Pflanzenschutz. 1994;101:545–9.
Te Beest DE, Paveley ND, Shaw MW, van den Bosch F. Disease–weather relationships for powdery mildew and yellow rust on winter wheat. Phytopathology. 2008;98:609–17.
Thach T, Ali S, Justesen AF, Rodriguez-Algaba J, Hovmøller MS. Recovery and virulence phenotyping of the historic 'Stubbs collection' of the yellow rust fungus *Puccinia striiformis* from wheat. Ann Appl Biol. 2015;167:314–26.
Thach T, Ali S, de Vallavieille-Pope C, Justesen AF, Hovmøller MS. Worldwide population structure of the wheat rust fungus *Puccinia striiformis* in the past. Fungal Genet Biol. 2016;87:1–8.
Tollenaar H. A comparison of *Puccinia striiformis* f. sp. *poae* on bluegrass with *P. striiformis* f. sp. *tritici* and f. sp. *dactylidis*. Phytopathology. 1967;57:418–20.
Tollenaar H, Houston BR. A study on the epidemiology of stripe rust, *Puccinia striiformis* West., in California. Can J Bot. 1967;45:291–307.
Tranzschel W. Kulturversuche mit uredineen in den Jahren 1911–1913. Vorlauf Mitt Mycol CI. 1914;4:70–1.
Tranzschel W. Promežutočnye chozjaeva rzavčiny chlebov i ich der UdSSR. (The alternate hosts of cereal rust fungi and their distribution in the UdSSR). Bull Plant Prot Ser. 1934;2:4–40.
Tu JC. The summer biology of *Puccinia striiformis* West. In Southeastern Washington. Dissertation, Washington State University; 1967.
Vallega J. Wheat rust races in South America. Phytopathology. 1955;45:242–6.
Wahl I, Anikster Y, Manisteriski J, Segal A. Evolution at the center of origin. In: Roelfs AP, Bushnell WR, editors. The cereal rusts, vol 1, Origins, specificity, structure and physiology. Orlando: Academics; 1984. p. 33–77.
Wan AM, Chen XM. Virulence, frequency, and distribution of races of Puccinia striiformis f. sp. tritici and P. striiformis f. sp. hordei identified in the United States in 2008 and 2009. Plant Dis. 2012;96:67–74.
Wan AM, Chen XM. Virulence characterization of *Puccinia striiformis* f. sp. *tritici* using a new set of *Yr* single-gene line differentials in the United States in 2010. Plant Dis. 2014;98:1534–42.
Wan Q, Liang JM, Luo Y, Ma ZH. Population genetic structure of *Puccinia striiformis* f. sp. *tritici* in northwestern China. Plant Dis. 2015;99:1764–74.
Wan AM, Chen XM, Yuen J. Races of *Puccinia striiformis* f. sp. *tritici* in the United States in 2011 and 2012 and comparison with races in 2010. Plant Dis. 2016;100:966–75.

Wang ZH. On the evolution of the barley-yellow rust pathosystem in Tibet. Acta Phytopathol Sin. 1992;22:151–5.
Wang MN, Chen XM. First report of Oregon grape (*Mahonia aquifolium*) as an alternate host for the wheat stripe rust pathogen (*Puccinia striiformis* f. sp. *tritici*) under artificial inoculation. Plant Dis. 2013;97:839.
Wang MN, Chen XM. Barberry does not function as an alternate host for *Puccinia striiformis* f. sp. *tritici* in the US Pacific Northwest due to teliospore degradation and barberry phenology. Plant Dis. 2015;99:1500–6.
Wang JQ, Dai SK. Preliminary study on over-summering patterns of the wheat stripe rust pathogen in southern Gansu. Acta Phytophylacica Sin. 1965;8:1–8.
Wang KN, Hong XW, Wu LR, Xie SX, Meng QY, Chen SM. The analysis of the resistance of varieties in the wheat stripe rust nurseries in 1951–1983. Acta Phytophylacica Sin. 1986;13:112–23.
Wang ZH, Peng YL, Yangjin L, Da Z. Difference of pathogenicity between isolates of barley-attacking and wheat-attacking forms of *Puccinia striiformis* with special reference to the deviation of disease development. Acta Phytopathol Sin. 1989a;19:161–5.
Wang ZH, Yangjin L, Da Z, Peng YL. A preliminary study on the physiologic specialization of *Puccinia striiformis* f. sp. *hordei* in Tibet. Chin J Plant Prot. 1989b;16:186–92.
Wang XJ, Tang CL, Zhang G, Li YC, Wang CF, Liu B, Qu ZP, Zhao J, Han QM, Huang LL, Chen XM, Kang ZS. cDNA-AFLP analysis reveals differential gene expression in compatible reaction of wheat challenged with *Puccinia striiformis* f. sp. *tritici*. BMC Genom. 2009;10:289.
Wang MN, Wan AM, Chen XM, Evans CK. Barberry is more important as an alternate host for stem rust than for stripe rust in the U.S. Pacific Northwest. In: Oral presentations, poster abstracts, participants and program of BGRI technical workshop, 13–16 June 2011, St. Paul; 2011. p. 166.
Wang MN, Wan AM, Chen XM. Barberry as alternate host is important for *Puccinia graminis* f. sp. *tritici* but not for *Puccinia striiformis* f. sp. *tritici* in the U. S. Pacific Northwest. Plant Dis. 2015;99:1507–16.
Wang ZY, Zhao J, Chen XM, Peng YL, Ji JJ, Zhao SL, Lu YJ, Huang LL, Kang ZS. Virulence variations of *Puccinia striiformis* f. sp. *tritici* isolates collected from *Berberis* spp. in China. Plant Dis. 2016;100:131–8.
Watson IA, De Sousa CNA. Long distance transport of spores of *Puccinia graminis tritici* in the Southern Hemisphere. Proc Linn Soc NSW. 1983;106:311–21.
Wellings CR. Annual report 1982. Plant Breed Inst, University of Sydney, Sydney; 1982. p. 24–7.
Wellings CR. *Puccinia striiformis* in Australia: a review of the incursion, evolution, and adaptation of stripe rust in the period 1979–2006. Aust J Agric Res. 2007;58:567–75.
Wellings CR. Global status of stripe rust: a review of historical and current threats. Euphytica. 2011;179:129–41.
Wellings CR, Burdon JJ. Variability in *Puccinia striiformis* f. sp. *tritici* in Australasia. In: Proceedings of the European and Mediterranean cereal rusts & powdery mildews conference, vol. 8; 1992. p. 114.
Wellings CR, McIntosh RA. *Puccinia striiformis* f. sp. *tritici* in Eastern Australia – possible means of entry and implications for plant quarantine. Plant Pathol. 1987;36:239–41.
Wellings CR, McIntosh RA. *Puccinia striiformis* f. sp. *tritici* in Australasia: pathogenic changes during the first 10 years. Plant Pathol. 1990;39:316–25.
Wellings CR, Wright DG, Keiper F, Loughman R. First detection of wheat stripe rust in Western Australia: evidence for a foreign incursion. Aus Plant Pathol. 2003;32:321–2.
Wiik L, Ewaldz T. Impact of temperature and precipitation on yield and plant diseases of winter wheat in southern Sweden 1983–2007. Crop Prot. 2009;28:952–62.
Wright RG, Lennard JH. Mitosis in *Puccinia striiformis* 1. Light microscopy. Trans Br Mycol Soc. 1978;70:91–8.
Xia CJ, Wan AM, Wang MN, Jiwan DA, See DR, Chen XM. Secreted protein gene derived-single nucleotide polymorphisms (SP-SNPs) reveal population diversity and differentiation of *Puccinia striiformis* f. sp. *tritici* in the United States. Fungal Biol. 2016a;120:729–44.

Xia CJ, Wang MN, Wan AM, Jiwan DA, See DR, Chen XM. Association analysis of SP-SNPs and avirulence genes in *Puccinia striiformis* f. sp. *tritici*, the wheat stripe rust pathogen. Am J Plant Sci. 2016b;7:126–37.

Xiao ZQ, Li ZM, Fan M, Zhang Y, Ma SJ. Prediction model on stripe rust influence extent of winter wheat in Longnan mountainous area. Chin J Agrometeorol. 2007;28:350–3.

Yang XB, Zeng SM. Detecting patterns of wheat stripe rust pandemics in time and space. Phytopathology. 1992;82:571–6.

Young CS, Paveley ND, Vaughan TB, Thomas JM, Lockley KD. Predicting epidemics of yellow rust (*Puccinia striiformis*) on the upper canopy of wheat from disease observations on lower leaves. Plant Pathol. 2003;52:338–49.

Zadoks JC. Yellow rust on wheat studies in epidemiology and physiologic specialization. T Pl Ziekten. 1961;67:69–256.

Zadoks JC. Epidemiology of wheat rust in Europe. Pest Articles News Summaries Sect B. Plant Dis Control. 1967;13:29–46.

Zadoks JC. Systems analysis and the dynamics of epidemics. Phytopathology. 1971;61:600–10.

Zadoks JC. Methodology for epidemiological research. Annu Rev Phytopathol. 1972;10:253–76.

Zadoks JC. EPIPRE: a disease and pest management system for winter wheat developed in the Netherlands. OEPP Bull/EPPO Bull. 1981;11:365–9.

Zadoks JC, Bouwman JJ. Epidemiology in Europe. In: Roelfs AP, Bushnell WR, editors. The cereal rusts vol. II: Diseases, distribution, Epidemiology and Control. Orlando: Academic; 1985. p. 329–69.

Zadoks JC, Rijsdijk FH. Epidemiology and forecasting of cereal rust studied by means of a computer simulator named EPISIM. In: Proceedings of the European and Mediterranean cereal rusts conference, vol. 2; 1972. p. 293–6.

Zeng SM. On the mathematic analysis of the epiphytotics of wheat stripe rust: I. Rate of epidemic. Acta Phytophyl Sin. 1962;1:35–48.

Zeng SM. On the mathematical analysis of the epiphytotics of wheat stripe rust: II. Distance of spread. Acta Phytopathol Sin. 1963;6:141–50.

Zeng SM, Luo Y. Long-distance spread and interregional epidemics of wheat stripe rust in China. Plant Dis. 2006;90:980–8.

Zeng SM, Luo Y. Systems analysis of wheat stripe rust epidemics in China. Eur J Plant Pathol. 2008;121:425–38.

Zhan G, Tian Y, Wang F, Chen X, Guo J, Jiao M, Huang LL, Kang ZS. A novel fungal hyperparasite of *Puccinia striiformis* f. sp. *tritici*, the causal agent of wheat stripe rust. PLoS One. 2014;9:e111484.

Zhan GM, Wang FP, Chen XM, Wan CP, Han QM, Huang LL, Kang ZS. Virulence and molecular diversity of the *Puccinia striiformis* f. sp. *tritici* population in Xinjiang in relation to other regions of western China. Plant Dis. 2016;100:99–107.

Zhao J, Wang L, Wang ZY, Chen XM, Zhang HC, Yao JN, Zhan GM, Chen W, Huang LL, Kang ZS. Identification of eighteen *Berberis* species as alternate hosts of *Puccinia striiformis* f. sp. *tritici* and virulence variation in the pathogen isolates from natural infection of barberry plants in China. Phytopathology. 2013;103:935–40.

Zhao J, Wang MN, Chen XM, Kang ZS. Role of alternate hosts in epidemiology and pathogen variation of cereal rusts. Annu Rev Phytopathol. 2016;54:207–28.

Zhu YY, Chen HR, Fan JH, Wang YY, Li Y, Chen JB, Fan JX, Yang SS, Hu LP, Leung H, Mew TW, Teng PS, Wang ZH, Mundt CC. Genetic diversity and disease control in rice. Nature. 2000;406:718–22.

Chapter 5
Stripe Rust Resistance

Meinan Wang and Xianming Chen

Abstract Stripe rust is best controlled by utilizing genetic resistance. Resistance to stripe rust has been studied for more than a century, but most progresses have been made during the last three decades. Two major types of resistance, all-stage resistance (ASR) and adult-plant resistance (APR) or high-temperature adult-plant (HTAP) resistance, have been characterized and used in breeding for resistant cultivars in both wheat and barley. So far, 78 permanently named, 67 temporarily designated and 327 quantitative trait loci (QTL) have been reported in various wheat varieties. The majority of these genes and QTL have been located to wheat chromosomes. Although some of the genes or QTL are the same, these numbers indicate that there are abundant resistance genes in wheat germplasm, which can be used for improving stripe rust resistance in commercially grown cultivars. Although the number is smaller, more than 50 resistance loci have been reported in barley for resistance to stripe rust, and about 30 loci have been mapped to barley chromosomes. Based on the characteristics of different types of resistance and the rich sources of resistance genes and molecular markers, we discuss different strategies for developing resistant cultivars and propose to use the combination of genes for durable APR or HTAP resistance and effective ASR to achieve more effective and sustainable control of stripe rust.

Keywords Barley • Breeding • Chromosomes • Genes • *Hordeum vulgare* • Molecular markers • *Puccinia striiformis* • Resistance • Stripe rust • *Triticum* spp. • Wheat

---

Mention of trade names or commercial products in this publication is solely for the purpose of providing specific information and does not imply recommendation or endorsement by the U. S. Department of Agriculture. USDA is an equal opportunity provider and employer.

M. Wang
Department of Plant Pathology, Washington State University, Pullman, WA 99164-6430, USA

X. Chen (✉)
US Department of Agriculture, Agricultural Research Service, Wheat Health, Genetics and Quality Research Unit, Pullman, WA 99164-6430, USA

Department of Plant Pathology, Washington State University, Pullman, WA 99164-6430, USA
e-mail: xianming@wsu.edu

X. Chen, Z. Kang (eds.), *Stripe Rust*, DOI 10.1007/978-94-024-1111-9_5

Stripe rust is one of the most important diseases of cereal crops in the world. The disease can be reduced by fungicides and some cultural means, but is preferably controlled by utilizing resistance. The resistance approach is the most effective, economic, easy to use and environmentally friendly for control of stripe rust. In this chapter, we will review the concepts and practices of disease resistance with the focus on research progress on stripe rust.

5.1 Resistance to Stripe Rust

Resistance to stripe rust has been studied for more than a century. In this chapter, we will discuss sources, types and genetics of resistance to stripe rust. We will summarize genes and quantitative trait loci (QTL) that have been identified and mapped to chromosomes of wheat and barley. To understand the molecular mechanisms of stripe rust resistance, the cloned genes for resistance to stripe rust will be discussed in comparison with those for resistance to other rusts and powdery mildews of cereals.

5.1.1 Sources of Resistance

Wheat is one of the most important food crops worldwide. It provides food for 40% of the world's population and accounts for more than 20% of the total human calories (http://www.fao.org/faostat/en/#home; http://www.croptrust.org). Wheat is grown on about 17% of all crop areas in the temperate, Mediterranean-type and subtropical parts of the globe. The main wheat production regions distribute in temperate and southern Russia, the Great Plains and Pacific West of the United States, southern Canada, Mediterranean, north-central China, India, Pakistan, Argentina and southern Australia. All these wheat production areas are suffering from stripe rust epidemics.

Barley is also an important cereal crop. Barley grains are used for domesticated animal feet and food and beverage for human consumption. The world production of barley is about 144.8 million tons. Barley is grown in more than 30 countries including Russia, Germany, France, Canada and Spain as the top five producing countries with national production more than 10 million tons (http://www.fao.org). Although epidemics on barley are not as widespread as on wheat, stripe rust of barley can cause significant damage in the major barley producing regions.

5.1.1.1 Origin and Evolution of Wheat and Barley

Sources of stripe rust resistance are closely related to the origin and evolution of cultivated wheat. The modern cultivated wheat cultivars usually refer to two types, hexaploid common wheat [*Triticum aestivum* (2n = 6x = 42, AABBDD)] and tetraploid durum wheat [*Ttiticum turgidum* spp. *durum* (2n = 4x = 28, AABB)]. Common wheat is mostly used for bread, cookie and noodles whereas durum wheat is mainly used for pasta and low-rising bread. Bread wheat accounts for about 95% of world wheat production and the other 5% is durum wheat (http://www.fao.org/faostat/en/#home; http://www.croptrust.org).

Modern wheat has evolved from several species. Wild diploid wheat (*Triticum urartu*, 2n = 2x =14, genome AA) hybridized with goat grass (*Aegilops speltoides*, 2n = 2x =14, genome BB) to produce wild emmer wheat (*Triticum dicoccoides*, 2n = 4x = 28, genome AABB), which was consumed by early human 19,000 years ago (Gustafson et al. 2009). Humans cultivated wild emmer about 10,000 years ago, subconsciously planted and selected to slowly create a cultivated emmer wheat (*Triticum dicoccum*, 2n = 4x =28, genome AABB). The spontaneously hybridization of emmer wheat with another goat grass (*Ae. tauschii*, 2n = 2x = 14, genome DD) about 9000 years ago produced an early spelt wheat (*Triticum spelta*, 2n = 6x = 42, genome AABBDD). Durum wheat (*Triticum parvicoccum*) with free-threshing heads was derived from a natural mutation of emmer wheat about 8500 year ago. *Triticum aestivum* originated from a cross between domesticated hulled tetraploid emmer wheat *T. dicoccum*, the free-threshing hard wheat *T. durum* or the free-threshing *T. parvicoccum*) and the goat grass *Ae. tauschii* about 9000 years ago (Kilian et al. 2009). With the wheat spread from the Fertile Crescent into other places, different wheat varieties were gradually adapted in their growing environments, which formed diverse landraces or traditional varieties under selection by humans. The modern wheat cultivars were derived from continuous selections in landraces or developed by systemic selections from planned crosses in the late of nineteenth century. The natural undomesticated wheat and the earlier cultivated wheat have a higher diversity of genes to abiotic and biotic stresses including diseases. Long-termed selections along with the process of wheat domestication and modern breeding practices have increased wheat yield and resistance, but eliminated genetic diversity of many traits, including resistance to stripe rust.

The cultivated barley is also originated in the Fertile Crescent. Based on the remains of barley (*Hordeum vulgare*) grains found at archaeological sites in the region, the barley crop was domesticated there from its wild relative *Hordeum spontaneum* about 10,000 years ago (Badr et al. 2000). The wild barley populations of (*H. spontaneum*) from Israel-Jordan are more similar than any others to the gene pool of cultivated barley, and thus, the Israel-Jordan area is the region in which barley was brought into culture. Based on the prevalence of diagnostic alleles I and IIIa of the homeobox gene BKn-3 in the *H. spontaneum* populations in Israel and the Himalayas, respectively and their presence in western landraces and modern cultivated varieties, an allelic substitution took place during the migration of barley from the Near East to South Asia. Thus, the Himalayas can be considered a region of domesticated barley diversification.

The ancestral species and their related grass species, as well as cultivated landraces and cultivars of wheat and barley, are the major sources of resistances to stripe rust.

5.1.1.2 Sources of Stripe Rust Resistance

A resistance source (or germplasm) is the foundation of plant breeding without which breeding resistant cultivars is impossible to conduct. Disease resistance germplasms are directly or indirectly used in the breeding programs and subsequently the resistant wheat types are selected. Combing with other traits in the breeding goals, the promising wheat genotypes that are resistant to a disease and adapted to the certain region are released. Stripe rust is one of important diseases in most wheat production regions, and growing resistant wheat cultivars is the best approach to control the disease. With the dynamic rust pathogen populations of the virulent races, when breeders need to improve wheat resistance to stripe rust, they have to find a source of germplasm that would supply the genes needed to undertake the breeding project. Modern cultivars significantly increase the rust resistance level and yield potential, but modern cultivation procedures have narrowed the genetic base of wheat, such as a few resistance genes are intensively used in the most predominant cultivars, which leads to the erosion of genetic resistance along with increasing rust epidemics. Therefore, it is urgent to expand the resistance sources in breeding programs.

The major sources for wheat breeders can be categorized into three broad groups: domesticated wheat, undomesticated wheat and other species or genera. Other categorization of resistance sources can be made based on the concept of gene pools. According to the feasibility of gene transfer or gene flow from those species to wheat, the species of those plant species were sorted as primary, secondary and tertiary gene pools. The primary gene pool consists of biological species that can be intercrossed easily without any problems with fertility of the progeny such as spelt wheat. For example, the spelt wheat variety Album, which carries *Yr5* for resistance to stripe rust (Chen and Line 1992c; Yan et al. 2003), can be easily crossed with common wheat cultivars. The secondary gene pool includes both cultivated and wild relatives of wheat, which are more distantly related with wheat and have some cross-ability problems. Hybrid plants from crosses of wheat with these species may produce seeds but with difficulties. Many *Aegilops* spp. belong to the secondary gene pool for wheat. For example, genes *Yr8* and *Yr17* for resistance to stripe rust were transferred from *Aegilops vertricosa* (Zhuk.) Chennav. and *Aegilops comosa* Sm., respectively into common wheat cultivars (McIntosh et al. 1995). Plant species in the tertiary gene pool are potential genetic resources for wheat improvement. Gene transfer by hybridization between wheat and these species is very problematic, resulting in lethality, sterility and other abnormalities. Special techniques, such as embryo rescue and bridge crossing, may be used to obtain fertile progeny. For example, rye (*Secale cereale* L.) and wheatgrass [*Thinopyrum intermedium* (Host) Barkworth & D.R. Dewey] are valuable resources in this group, as they provided several stripe rust resistance genes, such as *Yr9* and *Yr50*, (Mago et al. 2002; Tang et al. 2006; Yang et al. 2009; Liu et al. 2013a; McIntosh et al. 2013).

Based on materials used in plant breeding, we introduce the following categories as major sources of resistance to stripe rust and other diseases as well: commercial cultivars, landraces, enhanced germplasms, wild relatives, genetic stocks and other sources.

Commercial Cultivars. Effective genes for stripe rust resistance can be taken from commercial cultivars, either currently grown or retired (or historical) cultivars. Commercial cultivars usually possess resistance to the local stripe rust population. Commercial cultivars generally have good resistance and performance, and resistance genes from commercial cultivars also have superior capability of combining with other traits. Some historical cultivars were taken out of wheat production just because new cultivars have other agronomic traits, not because the old cultivars are no longer resistant to stripe rust. The resistance to stripe rust in these retired cultivars can be a good source for current wheat breeding. In many regions, old and current commercial cultivars may have durable resistance that can be continually used in the breeding program. For example, high-temperature adult-plant (HTAP) resistance, one of the important durable types of stripe rust resistance, has been used in the wheat breeding programs over 60 years in the Pacific Northwest of the United States (Line 2002; Chen 2005, 2013). It is still the major resistance type in commercial cultivars in this region.

Sometimes, the resistance is ineffective in one place, but it may be effective to other places due to the virulence composition difference among the pathogen populations, particularly for the regions where wheat just started to grown in recent centuries (like the US) or stripe rust is new introduction (like Australia and South Africa). The rust resistance in early American cultivars was mostly from commercial cultivars in Europe. Similarly, most stripe rust resistance in Australian cultivars was derived from commercial cultivars developed by the International Wheat and Maize Improvement Center (CIMMYT) or other countries. For example, the *Yr9* resistance was not effective in Asia and Europe in late 1980s but it was effective in the US until late 1990s (Chen et al. 2002). *Yr10* is largely ineffective in the western United States, but largely effective in many other countries (Sharma-Poudyal et al. 2013).

Resistance can be recovered by re-selection from heterogeneous cultivars. In a new epidemiological region or a region where stripe rust does not occur frequently, wheat cultivars may have not been well screened for resistance to the disease, and a cultivar may have heterogeneous reactions. Reselection from heterogeneous cultivars may result in cultivars with effective resistance. For example, wheat cultivar Brundage 96, which was reselected from susceptible cultivar Brundage in 1996, has an adequate level of HTAP resistance since its release in the US Pacific Northwest (Chen 2013). Resistance gene *YrA* is heterogeneous in many Australian and international spring wheat cultivars since lack of stripe rust selection until 1979 (Wellings 1986; Dracatos et al. 2016). Many selection lines from Avocet, Banks, Condor and Egret have good resistance and directly used in commercial production (Wellings 1986).

Use of commercial cultivars as a source of disease resistance is a common practice in almost every breeding program as it is easy to do, without much risk of linkage dragging and much sacrifice of adaptation and other agronomic traits. However, the condition for this approach is the presence of genes for effective resistance in commercial cultivars, which may not exist in some regions. Also relying on resistance in local cultivars may put a high pressure on selecting virulence races and increase populations with virulence to such race-specific resistance genes. Usable resistance genes can become fewer and fewer and the diversity of resistance can become lower and lower. To solve such problems, the other sources should be introduced.

Landraces. Wheat landraces are farmer-developed and maintained as traditional cultivars. They are developed over very long periods of time and have complex resistance in the geographical center of wheat origins or long terms growing countries. Landraces of wheat and barley have been collected and maintained in several germplasm centers and also maintained by some breeding programs throughout the world. For examples, the US Department of Agriculture – Agricultural Research Service (USDA-ARS) National Small Grains Collection (NSGC) in Aberdeen, Idaho have collected and maintained thousands of landraces of wheat, barley and other small grains originating throughout the world (https://www.ars.usda.gov/pacific-west-area/aberdeen-id/small-grains-and-potato-germplasm-research/docs/national-small-grains-collection-1/). The germplasm collections have been widely used in breeding programs for many useful traits including resistance to stripe rust.

Wheat landrace collections possess valuable resistance variability in the countries from western, southern and eastern Asia countries: Iran, Iraq, Syria, Turkey, Israel, Pakistan, India and China. These wheat accessions are adapted to the rust populations in the growing regions and have evolved various levels of resistance. For example, stripe rust resistance gene *Yr1* was identified from the Chinese wheat landraces (e.g., Chinese 166) and *Yr10* is from Turkey wheat landrace PI 178383 (Lupton and Macer 1962; Chen and Line 1992c). Gene *Yr18*, which also provides resistance to leaf rust and is named *Lr34*, is derived from Asian wheat landrace and widely used in modern wheat cultivars (McIntosh 1992; Kolmer et al. 2008). *Yr46* is widely distributed in wheat landraces in India and Pakistan before modern wheat cultivars are grown in this region (Moore et al. 2015).

In Europe countries where wheat has a long growing history, some old wheat cultivars have durable resistance and can be used for modern breeding. Van Dijk et al. (1988) evaluated stripe rust resistance in winter wheat cultivars released in the Netherlands before 1930 and the cultivars released from 1930 to 1950. They found that the frequencies of all-stage, adult-plant stage and temperature sensitive resistance were 1.4%, 52% and 54% in the old cultivars and 10%, 62% and 22% in the modern ones, respectively. Some of the landraces and old cultivars with pleiotropic resistance effect (resistance to multiple pathogens, like *Yr18/Lr34/Sr57Pm38* and *Yr46/Lr67/Sr55/Pm46*) are potentially valuable sources for introducing genetic diversity in wheat breeding. This type resistance has been used in CIMMYT programs for improvement of leaf rust resistance.

The USDA–ARS NSGC contains 58,392 accessions of wheat and wild relatives of wheat, among which more than 42,138 accessions are common wheat (Bonman et al. 2015). The majority of the common wheat accessions are landraces collected throughout the world since 1897. Over the last 30 years, the collection has been systematically evaluated for various traits of interest to plant breeders and geneticists. Our program have evaluated all common and durum wheat accessions for responses to stripe rust in fields of Washington under natural infections and in the greenhouse with selected US predominant races of *Pst*. The stripe rust data are kept at the US National Plant Germplasm System (NPGS) Germplasm Resources Information Network (GRIN) database (http://www.ars-grin.gov/). Since 2000, we have selected 4421 spring wheat and 5400 winter wheat accessions with all-stage and/or HTAP resistance to stripe rust (Chen XM and associates, unpublished data). From these sub-collections, we further selected 1000 spring wheat and 1000 winter wheat accessions and tested them for more than 10 years in the fields and in the greenhouse with recent predominant races of *Pst*. Most accessions have continually shown resistance to stripe rust (http://striperust.wsu.edu). We made crosses of more than 200 common and durum wheat landraces with a susceptible common wheat genotype, Avocet S (AvS). From the progenies of these crosses, we developed 70 germplasm lines with stripe rust resistance introgressed from 70 landraces and improved agronomic traits (Wang et al. 2012). Compared to the landraces, these improved germplasm lines are relatively easy for use in breeding for stripe rust resistant cultivars. From these landraces, we have identified and mapped a number of resistance genes, including permanently named genes *Yr45*, *Yr53*, *Yr64* and *Yr65* for effective all-stage resistance and *Yr52*, *Yr59* and *Yr62* for HTAP resistance (Li et al. 2011; Ren et al. 2012a; Cheng et al. 2014; Zhou et al. 2014c; Lu et al. 2014). Use of these *Yr* genes and germplasm lines can greatly increase the diversity of stripe rust resistance genes in commercial wheat cultivars.

From the USDA–ARS NSGC, a core collection of 3836 common wheat accessions (*T. aestivum*) was genotyped with DArT and SNP markers (Bonman et al. 2015). Based on the genotypic data, an informative core (iCore) of 3230 accessions was established for further characterization. Among these accessions, 2326 were resistant to stripe rust based on severity data and 1865 were resistant according to infection type (IT) data. Taking from the collections, Sthapit et al. (2014) studied the stripe rust resistance in 652 wheat landraces from 54 countries and identified 165 accessions showing resistant in fields. Of the 652 accessions, 131 were also resistant to stem rust race Ug99 (Newcomb et al. 2013). They genotyped the 652 landraces with 72 simple sequence repeat (SNP) markers, and 500 of them with 8633 single SNP markers covering the whole wheat genome. The SNP markers grouped the resistant landraces into nine different clusters in an unweighted pair group with arithmetic mean dendrogram and three groups in principal coordinate analyses. The cluster results indicate that the resistant landraces are genetically diverse and are expected to possess diverse resistance genes.

Genome-wide association studies (GWAS) were conducted on the core collection for identifying genes or QTL for resistance to stripe rust. Genotyping 1000 spring wheat accessions with 4585 SNP markers, Maccaferri et al. (2015) found that greater levels of stripe rust resistance were observed in a subpopulation from Southern Asia than in other groups. They identified 97 loci that were significant for stripe rust reaction in at least three environments, including 10 QTL with an experiment-wise adjusted Bonferroni probability <0.10. The 10 QTL explained 15% of the phenotypic variation in IT and all QTL explained 45% of the total phenotypic variation in stripe rust response. The 10 QTL were mapped far from previously identified stripe rust resistance genes and other seven QTL mapped close to known resistance genes. Similarly, Bulli et al. (2016) tested 1175 winter wheat accessions from the core collection using the GWAS approach. They identified 127 stripe rust resistance loci that were effective across at least two environments. Twenty of the loci had significant genome-wide adjusted *P*-values. Five of the 20 QTL are potentially new loci based on relative map positions of previously reported stripe rust resistance genes and QTL. Both studies show that a large number of genes in wheat landraces conferring resistance to stripe rust.

Enhanced Germplasms. Enhanced germplasms are produced by breeding programs or genetic studies. For example, a breeder may release one resistance genotype as a commercial cultivar from a cross, while many resistant genotypes as the intermediate breeding products are retained as enhanced resistance sources to be considered in further research and cultivar development. Enhanced resistant germplasms also can be made from purposely transferring resistance from landraces or other types of wheat to an elite wheat background. As mentioned above, we developed over 70 common wheat germplasms with resistance to stripe rust through crosses between landraces with an elite spring wheat variety, AvS. These germplasms are F_6 or more advance generations with homozygous in stripe rust resistance and improved agronomic traits (Wang et al. 2012). The enhanced germplasms can be used for further studies to map the resistance genes and for use in breeding programs.

Numerous *Yr* genes near-isogenic lines (NILs) were developed in relatively elite wheat backgrounds for genetic studies (Wellings et al. 2004; Xu et al. 2004). Of these NILs, a set of over 20 *Yr* near-isogenic lines in the AvS background developed by the Plant Breeding Institute of University of Sydney have been widely used to monitoring virulence of *Pst* and breeding stripe rust resistant cultivars throughout the world. Almost every genetic and breeding program has enhanced germplasms for use to develop stripe rust resistant cultivars. This source may contain more diverse of resistance genes than commercial cultivars. Like commercial cultivars, they do not have problems in cross-ability, but the breeding process may take little bit longer to get off some adverse trait from the germplasm used in the cross. However, it should be much shorter to incorporate the resistance into new commercial cultivars than wild relatives and related crop species.

Wild Relatives and Related Species. Wheat belongs to *Triticeae* tribe which is consisted of about 350 species (Knüpffer 2009). As we mentioned above, modern wheat evolved from wild relatives. When resistance genes are difficultly found in cultivars of domesticated common wheat (*T. aestivum*), resistance genes in wild relatives can be explored. The species in genus *Aegilops* and *Triticum* and other species in the *Triticeae* tribe, whose genes can be indirectly used or introgressed into wheat gene pool, are valuable resources for resistance breeding. Genus *Triticum* contains three ploidy levels and about 30 species. Most of these species have been investigated as sources of stripe rust resistance and some have been successfully transferred to common wheat. Yildirim et al. (1995) investigated stripe rust resistance in a collection of 279 *Ae. tauschii* (syn. *Triticum tauschii*; *Aegilops squarrosa*) accessions and found 44 (17%) accessions resistant to four races of *Pst* representing all known virulence combinations in the US Pacific Northwest. The resistant accessions were mainly from the Caspian Sea region of Iran and Azerbaijan, with the majority belonging to two subspecies (*strangulate* and *meyeri*) of *Ae. tauschii* (syn.).

Riley et al. (1968) transferred stripe rust resistance from *Ae. comosa* to common wheat cultivar Chinese Spring and developed cultivar Compair. Following his pioneer work, more rust resistance genes were transferred into wheat from wild relative species. Of the 78 permanently named *Yr* genes, 21 were transferred from related species or wild relatives, including *Yr5* from *Triticum spelta*; *Yr8* from *Ae. comosa*; *Yr9* from *S. cereale* ; *Yr15*, *Yr35* and *Yr36* from *T. dicoccoides*; *Yr17* from *Ae. ventricosa*; *Yr24/Yr26*, *Yr53*, *Yr56*, *Yr64*, *Yr65* and *Yr76* from *T. turgidum*; *Yr28* from *Ae. tauschii*; *Yr37* from *Ae. kotschyi*; *Yr38* from *Ae. sharonensis*; *Yr40* from *Ae. geniculate*; *Yr42* from *Ae. neglecta*; *Yr50* from *Thinopyrum intermedium*; and *Yr70* from *Ae. umbellulata* (Chen 2005; McIntosh et al. 2013; Xu et al. 2013; Cheng et al. 2014; Dong et al. 2017). More than 15 temporarily named *Yr* genes also were identified from wild relatives of wheat. Although transferring alien genes from wild relatives is an important means of increasing the genetic diversity of rust resistance in wheat, *Yr* genes from related crop species or wild relatives can have linkage drags. The undesirable botanic or agronomic traits can have adverse impacts on the use of the alien genes in breeding programs. However, several *Yr* genes from wild relatives have been successfully used in commercial wheat cultivars, such as *Yr5*, *Yr9*, *Yr10*, *Yr15*, *Yr17*, *Yr24/Yr26* and *Yr76*.

The development of synthetic wheat is a good way to transfer genes from tetraploid wheat and *Ae. tauschii* into hexaploid wheat. Tetraploid wheat (with AABB genome) is hybridized with *Ae. tauchii* (DD geneome), and the chromosomes of its F_1 hybrid are doubled using the colchicine treatment. The produced fertile synthetic hexaploid (AABBDD) genotypes are fully homologous to common wheat. Their chromosomes will readily recombine with chromosomes of common wheat. Several *Yr* genes were introduced into common wheat through this approach, such as *Yr28* (Singh et al. 2000), *Yr48* (Lowe et al. 2011), *YrAS2388* (Liu et al. 2010) and *YrC142* (Wang et al. 2009). Through this bridge, the stripe rust resistance gene *Yr24* (=*Yr26*) in durum wheat was transferred into common wheat, resulting in cultivar Chuanmai 42 released in China in 2003 (Li et al. 2006a), and this gene is now present in many Chinse wheat cultivars (Zeng et al. 2014; Han et al. 2015).

In addition to the synthetic wheat method, other methods can be used to transfer resistance from alien species to common wheat. The cross ability and gene transfer efficiency vary from species to species, and the variation is related to its relationship to wheat. The genes in the close relatives are easier introgressed into wheat through direct hybridization than those from distantly related species. In transferring stripe rust resistance genes from durum wheat (tetraploid) to common wheat (hexaploid), Chen and his associates combined techniques of cytological identification of progeny plants and making crosses of amphiploid hybrids to obtain stable hexaploid plants within five generations (Wang et al. 2012; Xu et al. 2013; Cheng et al. 2014). Using common wheat plants as recurrent parents in backcrossing with hybrids from common wheat by durum wheat crosses greatly improved the fertility of plants in the BC_2F_1 generation, and stable common wheat families with stripe rust resistance were obtained in the following generations. However, to transfer genes from distant wild relatives into wheat, some genetic barriers can impact the possibility of introgression. This can be caused by the lack of chromosome pairing, preferential transmission of chromosomes harboring gametocidal genes, or hybridization incompatibility. Several of these hurdles have been overcome through the treatment with chemicals for chromosome doubling (such as colchicine or caffeine), cold treatment, bridging crosses, and mutations in chromosome pairing gene *Ph1* and gametocidal genes, irradiation, embryo rescue and other techniques. The chromosome size of an introgression can be highly variable, from whole chromosome to a tiny segment. C-banding and in situ hybridization are powerful cytological techniques allowing the detection of alien chromatin in wheat. C-banding permits identification of the wheat alien chromosomes involved in wheat-alien translocations, whereas genomic in situ hybridization (GISH) allows determination of their size and breakpoint. Often the sites of wheat-alien recombination are unevenly distributed over a chromosome such that the telomeric regions recombine more readily than the pericentromeric region. Friebe et al. (1996) reviewed the introgression of *Aegilops* chromosomes into wheat. Of 57 spontaneous and induced wheat-alien translocations identified 10 had a whole arm translocation with the breakpoint within the centromeric region; 45 had terminal translocations with distal alien segments translocated to wheat chromosome arms. Some cryptic introgressions involve tiny chromosomal segments which cannot be detected using cytogenetic approaches, but molecular markers can be helpful to classify the origin. An excellent example of such a cryptic translocation was the transfer of *Yr40* from *Ae. geneiculata* in to wheat (Kuraparthy et al. 2007).

Although wild relatives of wheat possess valuable resistance against many wheat diseases, some hybridizations between wheat and its relatives did not aim at transferring genes for stripe rust resistance. Depending upon regions, other biotic and abiotic stresses can be top priorities in their breeding programs. For example, a major objective of introgression of alien genes into wheat is to achieve resistance to leaf rust in the US Great Plains. However, some alien chromosome segments carry genes for resistance to multiple diseases, such as *Lr54/Yr37* (Marais et al. 2005a), *Lr56/Yr38* (Marais et al. 2010), *Lr57/Yr40* (Kuraparthy et al. 2007, 2009) and *Lr62/Yr42* (Marais et al. 2009) for resistance to both leaf rust and stripe rust. Stripe rust resistance gene *Yr9* is present in many wheat cultivars developed and grown in the US Great Plains and eastern states, not because it is resistant to stripe rust, but because it is linked with leaf rust resistance gene *Lr26*, stem rust resistance *Sr31* and powdery mildew resistance gene *Pm8* in the 1RS chromatin translocated into wheat (Shi et al. 2001; Chen et al. 2002). In contrast, the widely presence of *Yr9* in Chinese wheat cultivars was because the specific purpose of incorporation of *Yr9* resistance to stripe rust. Some genes can have pleiotropic effect, and the best example is the cloned *Lr34/Yr18* gene, which also provides resistance to stem rust, powdery mildew and other diseases (Krattinger et al. 2009a, 2011, 2013). Therefore, it is necessary to purposely re-evaluate resistance to stripe rust in wheat germplasm derived from wheat-wild relative hybridizations.

Genetic Stocks. Genetic stocks are usually generated from the middle materials in the process of transferring alien genes into wheat. Wheat is allohexaploid species whose genome can buffer various chromosome variations to add or delete a pair of chromosome, single chromosome or part of a chromosome, hence generating the genetic stocks of wheat-alien chromosome addition, substitution, translocation and deletion lines. These lines are the initial requirements of pre-breeding programs aimed at the transfer of alien resistance genes to cultivated wheat cultivars. *Aegilops comosa* is resistant to stripe rust. Through repeated backcrossing with Chinese Spring, resistance on chromosome 2M was transferred into wheat and generated wheat-*Ae. comosa* addition lines (CS-2M), substitution lines (CS 2D/2M) and translocation lines (2DS-2ML, 2AS-2ML) (Riley et al. 1968). Genetic stocks of *Yr36* substitution lines (Uauy et al. 2005) and various deletion lines carrying *Yr9* on translocation chromosome 1BL/1RS (Lukaszewski 2000) were produced and used in wheat resistance breeding.

Other Resistance Sources. Resistance resources also can be produced from somatic clones of wheat tissue culture, mutation, non-host species or transgenic materials. Mutation is an important way to create new resistance sources from susceptible cultivars. Resistant plants can be derived from either spontaneous or induced mutations. The first group of wheat mutants with resistance to stripe rust were developed through irradiation of wheat cultivar Gabo using X-rays at Washington State University in the United States (Konzak 1956). Line et al. (1974) evaluated over 1000 selections from numerous mutagenically treated cultivars at various locations over 3 years and identified at least 12 different recessive, dominant or complementary resistance genes. Cultivar Sujata is a mutant selection from an Indian cultivar C306 (Lan et al. 2015). Wheat line Hobbit 'sib' is susceptible to stripe rust. Mutant lines resistant to strip rust were obtained from Hobbit 'sib' treated with fast neutron irradiation, and it is very likely the suppressor genes were deleted (Smith et al. 2004; Boyd et al. 2006). A mutant wheat line MNR220 displays broad spectrum disease resistance to three rusts in wheat was identified from cultivar Alpowa treated with ethyl-methane sulfonate (EMS) (Campbell et al. 2012). Chinese scientists have developed wheat lines through satellite-carrying and evaluated the lines in fields for resistance to stripe rust. The second author was observed lines with stripe rust resistance in the "Space breeding nurseries" in fields. These lines were presumed to be mutated by space radiation in the satellite. Chinese scientists observed relative uniform reaction of the plants from space-treated seeds and different reactions in the later generations.

Once resistant mutants are developed, the incorporation of the resistance is similar to the use of resistance in cultivars. A possible mutational approach is to directly alter genes for susceptibility in a susceptible plant to develop a resistant genotype. The recently emerging CRISPR (Clustered regularly interspaced short palindromic repeats) technology opens an avenue for using the gene editing approach to improve traits of interest (Ledford 2016). Genes for susceptibility could be removed or altered for providing resistance.

Non-host resistance will be a new source for stripe rust resistance. As a close species to wheat within *Triticeae* tribe, barley (*Hordeum vulgare*) is resistance to most *Pst* races (Chen et al. 1995c). Barley is considered "non-host" or "intermediate host" for the wheat stripe rust pathogen. Pahalawatta and Chen (2005b) identified two genes (one dominant *RpstS1*, and one recessive *rpstS2*) in barley cultivar Steptoe for resistance to *Pst* and mapped the dominant gene *RpstS1* on chromosome 4H. In contrast, Steptoe is susceptible to all race of the barley stripe rust pathogen, *P. striiformis* f. sp. *hordei* (*Psh*), identified so far in the United States (Chen et al. 1995c; Wan and Chen 2012). Another non-host resistance gene (*Rps6* or *YrpstY1*) in barley for resistance to wheat stripe rust was mapped on chromosome 7H (Sui et al. 2010; Dawson et al. 2016; Li et al. 2016a). Similarly, a gene (*RpsLem*) in wheat cultivar Lemhi was identified for resistance to *Psh* (Pahalawatta and Chen 2005a). These studies open a new door for exploring resistance from barley to against wheat stripe rust. More studies are needed to develop techniques from moving the genes from barley to wheat and from wheat to barley through chromosomal manipulation or transformation.

Truly non-host resistance to *Pst* has been studied in Arabidopsis, broad bean, pepper and rice as these plants cannot be infected by the stripe rust fungus (Cheng et al. 2012, 2013; Yang et al. 2014; Zuo et al. 2013; Zhao et al. 2016a). These studies are discussed in detail in Chap. 3. However, these studies focused on histological observations and transcriptome analysis for a general understanding of molecular mechanisms of non-host resistance. No practically workable genes have been discovered for non-host resistance. In the future, key genes for non-host resistance need to be identified to be useful for incorporation into host plants.

Somatic clonal variation can become a source of rust resistance. Two somatic clones, produced from immature embryo-derived calli of wheat cultivar Coker 916, showed improved resistance to leaf rust (Oberthur et al. 1993). During tissue culture, chemical treatments may increase mutation rates. The doubled haploid (DH) technique has become more and more popular for quickly developing wheat populations for mapping genes of interest, including resistance to stripe rust (XM Chen and associates, unpublished data). The DH populations could be used to identify somatic variants providing stripe rust resistance.

The novel biotechnology for gene transformation can easily break biological barriers among species, enabling resistance genes exchanging between different genomes and efficiently creating novel resistance germplasms and cultivars. Plants of barley, rice and corn transformed with the *Yr18/Lr34* gene from wheat displayed resistance to multiple diseases of these crops (Risk et al. 2013; Krattinger et al. 2016; Sucher et al. 2016). Similarly, the *Yr46/Lr67* gene from wheat provided partial resistance to different pathogens of barley when transformed into barley plants (Moore et al. 2015). It is also possible to improve stripe rust resistance in wheat using genes conferring resistance to other diseases from distant species. Once transgenic wheat is publicly accepted, genes from distant plant species for non-host resistance could be a rich source for improving stripe rust resistance in the future.

Genes from plant pathogens could be used for improving disease resistance. This approach has been used to engineer plants for control of various viral diseases and has been recently explored for control fungal diseases. Yin et al. (2015) identified effector genes shared by stripe rust, stem rust and leaf rust fungi, and found that silencing some of the conserved effector genes reduced the development of the three rusts. Expression of an avirulence gene from a pathogen in a susceptible plant with a defeated resistance gene through transformation may cause an incompatible interaction that can be effective against infection by a virulent pathogen. Silencing pathogen pathogenicity genes using pathogen genes or homologs may reduce the pathogen development in plants (Yin et al. 2014). More studies are needed to further demonstrate the principle and develop efficient techniques.

5.1.2 Types of Host Resistance

Disease resistance is a nature of an important biological trait in wheat. Resistance to stripe rust can be characterized as the nature of wheat reducing and/or avoiding damage from *Pst*, hindering the infection and growth of the fungus and/or development of rust. Resistance to stripe rust has been recognized by growers and scientists for a long time. From different viewpoints, various types of resistance to stripe rust have been cataloged (Sharp and Volin 1970; Line et al. 1974; Stubbs 1977; Nelson 1978; Röbbelen and Sharp 1978; Johnson 1981a, b; Parlevliet 1981; Line and Chen 1995; Bariana 2003; Line 2002; Chen 2005, 2013). Terminology dealing with resistance is often far from consistent among researchers. Line and his associates sorted types of resistance into eight groups based on range of ITs on seedlings and adult-plants, race-specificity and pattern of rust development when tested at low and high temperature profiles in controlled greenhouse conditions and in fields (Line 1972; Line et al. 1974; Line et al. 1976; Konzak et al. 1977). Chen (2013) summarized the contrasting types of resistance based on various criteria for separation, particularly emphasized the resistance durability in practice. Table 5.1 summarizes the types of resistance based on Chen (2013) with some modifications.

5.1.2.1 Resistance Types Based on Plant Development and Environment

Based on plant growth stages and environments, stripe rust resistances can be separated into all-stage (also called seedling) resistance (ASR) verses adult-plant resistance (APR); greenhouse resistance and field resistance; temperature insensitive resistance verses temperature sensitive resistance; and non-durable resistance verses durable resistance (Table 5.1). In the early studies, most tests in greenhouses for evaluating wheat varieties for stripe rust response were performed on seedlings and in the fields on adult plants especially for note-taking. Greenhouse testing was usually conducted at relatively low temperatures between 10 to 20 °C and field reactions were usually collected when temperature was warm, but usually not recorded. Thus, these different types of resistance were recognized with terms associated to the testing stage and environments. For stripe rust, there are no clear reports of

Table 5.1 Contrast types of rust resistance based on various criteria

Criterion	Type 1	Definition	Durability	Type 2	Definition	Durability
Based on plant development and environment						
Growth stage	All-stage (seedling) resistance	Resistance that can be detected in the seedling stage, but remains effective throughout all growth stages	Usually not durable	Adult-plant resistance	Plants with adult-plant resistance are susceptible in the seedling stage but become resistant in late stages.	Usually durable
Testing conditions	Greenhouse and field consistent resistance	Resistance that is detected in greenhouse (usually seedlings) is also effective in the field	Usually not durable	Field resistance	Resistance that can be detected in the field (usually in the adult-plant stage), but cannot be detected in greenhouse (usually in seedling stage)	Usually durable
Response to temperature	Temperature insensitive resistance	Resistance that expresses at all ranges of temperatures at which susceptible plants are susceptible	Usually not durable	Temperature sensitive resistance	Resistance that expresses at some (for stripe rust, high-temperature) range of temperature, but not the other range (low-temperature) of the temperatures	Usually durable
Durability	Non-durable resistance	Resistance that is "broken-down" by some races	Not durable	Durable resistance	Resistance that remains effective during its prolonged and widespread use in an environment favorable to the disease	Durable
Based on the degree of resistance						
Level of resistance	Complete resistance (immune)	Resistant plants do not show any visible disease symptom or pathogen sign	Usually not durable	Incomplete (partial) resistance	Resistance that is not complete, showing various degrees of reduction in infection type and/or severity	Usually durable
	Absolute resistance	Resistance is complete	Usually not durable	Relative resistance	Resistance that is incomplete	Depending upon host or non-host resistance
	High-level resistance	Resistance that is relatively high	Depending upon the number of genes	Low-level resistance	Resistance that is relatively low	Usually durable

Speed of symptom and/ or sign development	Fast rusting (no resistance)	Rust develops fast and quickly reaches the highest level.	Not applicable	Slow-rusting	Plants have a susceptible infection type but rust develops slowly on them.	Usually durable
Based on the genetics of resistance						
Number of genes	Monogenic resistance	Resistance that is controlled by a single gene.	Usually not durable	Polygenic resistance	Resistance is controlled by several to many genes	Usually durable
Effects of genes	Major-gene resistance	Resistance that is controlled by one or a few genes with a large effect	Usually not durable	Minor-gene resistance	Resistance that is controlled by one or more genes with minor effects	Usually durable
Inheritance	Qualitative resistance	Resistance that is inherited qualitatively, showing distinct two classes in a segregating population from a cross with a susceptible genotype	Usually not durable	Quantitative resistance	Resistance that is inherited quantitatively, showing continuous variation in a segregating population	Usually durable
Based on the plant-pathogen interaction						
Plant species	Host resistance	Resistance exits in the host species of the rust pathogen	Either durable or not durable	Non-host resistance	Resistance that exits in non-host species	Durability depending upon how close the relationship to the host species
Race-specificity	Race-specific resistance	Resistance that is effective against some races, but not other races	Usually not durable	Non-race specific resistance	Resistance that is effective against all races	Usually durable
	Differential resistance	Resistance that depends upon races	Not durable	General resistance	Resistance that is independent of races	Usually durable
	Vertical resistance	Resistance that is effective against some races, but not other races	Usually not durable	Horizontal resistance	Resistance that is similarly effective against all races	Usually durable

(continued)

Table 5.1 (continued)

Criterion	Type 1	Definition	Durability	Type 2	Definition	Durability
Race-specificity, growth stage and temperature sensitivity	Race-specific all-stage resistance	Resistance that is effective against some races in all growth stages and not sensitive to temperatures	Not durable	Non-race specific high-temperature adult-plant (HTAP) resistance	Resistance that is effective against all races when plants grow old and temperature increases	Durable
Sensitivity to pathogen infection	Hypersensitive resistance	Resistant plants show necrotic tissue result from cell death.	Often not durable	Non-hypersensitive response	Either completely resistant (immune) or reduced severity, but susceptible infection type (e.g. slow-rusting)	Depending upon host or non-host resistance
Relation to hustorial formation	Pre-haustorial resistance	Pathogen enters into plant but growth stops before haustorial formation	Durability Depends upon host or non-host resistance	Post-haustorial resistance	Resistance expresses after pathogen forming hautoria	Either durable or not durable.
Molecular basis	NBS-LRR resistance	Resistance that is controlled by a gene(s) with NBS-LRR domains in gene products	Usually not durable	Non-NBS-LRR resistance	Resistance that is controlled by a gene(s) without NBS-LRR domains in the products	May be durable

resistance only expressing in seedling stage. When plants are resistant in seedling stage to a race or population, they are also resistant to the same race or population in adult-plant stage. Similarly, plants that are resistant in greenhouse seedling tests are also resistant in fields as long as tested with same race or population of the pathogen. However, not all cases of resistance observed on adult-plant stage or in the fields (at adult-plant stage) can be detected in seedling tests or in greenhouses (at seedling stage). From these points, ASR is more preferred than seedling resistance. Similarly, APR is more preferred over field resistance. For temperature sensitivity, both sensitive and insensitive resistances can be observed in seedling tests and adult-plant tests. However, under natural conditions temperature insensitive resistance expresses throughout the growth stages, whereas temperature sensitive resistance usually expresses in late growth stages as temperature become warm, as long as the same race or population of the pathogen is used for the different tests. Therefore, these various terms basically describe two contrast types, ASR and APR, which have been widely used in wheat breeding programs to develop cultivars for resistance to stripe rust (Line 2002; Bariana et al. 2007; Chen 2013; Ellis et al. 2014; Wellings et al. 2012).

5.1.2.1.1 All-Stage Resistance (ASR)

All-stage resistance can be detected in seedling stage and is effective through all growth stages. Identification of ASR is relatively simple as it can be easily observed as low ITs or no rust on seedling testing. The temperature condition does not need to be very specific as long as it allows the stripe rust fungus growing normally. In our laboratory, we inoculate two-leaf seedlings with urediniospores of a particular isolate and keep them at 10 °C (7–12 °C is considered optimal for urediniospores to germinate and penetrate) in a dark dew chamber (100% relative humidity for 16–24 h (Chen and Line 1992a, b). The inoculated plants are then grown in a growth chamber at a diurnal cycle of 4–20 °C and 8-h dark/16-h light and IT data are collected 18–22 days after inoculation. Inoculation of two-leaf seedlings, rather than one-leaf seedlings, is because some wheat genotypes may have different reactions on the primary and second leaves.

The typical characteristic of this type resistance is the resistant plant showing immune or hypersensitive reaction (HR) when challenging with avirulent races. It is usually controlled by single genes conferring high-level resistance, is qualitatively inherited and is easily incorporated into adapted cultivars. This kind of resistance is mostly race-specific and is vulnerably overcome by new virulent races that can rapidly develop in the pathogen population.

5.1.2.1.2 Adult-Plant Resistance (APR)

Plants with APR are susceptible in seedling stage and become more and more resistant when plants grow older and older. There are no a particular stage to switch on APR, but this type of resistance can start expressing in jointing stage, and become more effective in later growth stages. Also, a higher level of resistance tends to appear earlier in the growth stages. APR can be identified in both greenhouse and field conditions. Not all resistances observed in adult-plant stage are APR because ASR also expresses in adult plants. In the literature, some cases of APR were improperly called so because only adult-plant data were used.

Adult-plant resistance is usually quantitatively inherited, and often shows slow rusting. Comparing with ASR, APR may have a relatively long rust latent period, low infection efficiency, small lesion size, low pustule density and long necrotic stripes (Richardson et al. 2006; Guo et al. 2008). Generally APR is non-race specific. However, some APR genes are race-specific, such as *Yr11*, *Yr12*, *Yr13* and *Yr14* (Johnson 1992; McIntosh et al. 1995). Similarly, Australian cultivars Janz and Diamond Bird have moderate and high levels of APR, respectively, to race 110E143A+, but susceptible to race 134E16A+ (Bariana et al. 2007). Ellingboe (1981) suggested that non-race specific resistances are those that are not yet shown to be race specific. This opinion is still arguable.

5.1.2.1.3 High-Temperature Adult-Plant (HTAP) Resistance

High-temperature adult-plant (HTAP) resistance is a type of APR. HTAP resistance considers both plant development and environment. In fact, most cases of APR are HTAP resistance, and they are called APR, but not HTAP resistance, partially because they are identified when temperature effects are not considered. Plants with only HTAP resistance are susceptible in the seedling stage at low temperatures, but gradually become resistant when the plants grow old and the weather becomes warm (Qayoum and Line 1985; Line and Chen 1995; Line 2002; Chen 2005, 2013).

HTAP resistance usually displays in various levels ranging from low to high and is controlled by QTL, depending upon particular QTL and the number of QTL. HTAP resistance can be identified by IT and severity. Because of the various levels of resistance, plants which have ITs and/or severity lower than those of susceptible varieties at late growth stage and relatively warm conditions are considered to have HTAP resistance. Plants have typical HTAP resistance have necrotic stripes with or without uredinia, and generally express resistance when adult plants stage in combination with warm temperatures. However, HTAP resistance can start expressing as early as seedling stage when the weather is warm and also can express to a recognizable level on adult plants when the temperature is relatively low, but expresses to its highest level at the combination of after flag-leaf stage and the highest temperature at which susceptible varieties still are susceptible. The effects of the plant growth stage and temperatures are separable as some genes for HTAP resistance are more sensitive to temperatures, whereas other HTAP resistance genes are more related to the growth stage.

For identifying HTAP resistance, a race or population of the pathogen virulent on seedlings of the testing varieties is required, and tests of both seedlings and adult-plants are needed. These tests can be done in fields as long as infection and sporulation can occur in both seedling and adult-plant stages, but more clear results can be obtained under controlled greenhouse conditions. In our program, we use both greenhouse and field tests to identify HTAP resistance. In the fields, particularly in Mount Vernon, Washington where stripe rust reliably develops in the seedling stage and throughout the remaining stages, we recorded stripe rust data (both IT and severity) at an early stage (e.g. middle jointing, Zadoks 31-32 for winter wheat or barley and early jointing, Zadoks 30 for spring wheat or barley) and an later stage (e.g. boot-flowering, Zadoks 45-60 for both winter and spring wheats or barleys). Lower IT and/or severity in the later stage suggest HTAP resistance. Because the pathogen natural population in fields is often a mixture of several races, field data may not correctly show HTAP resistance. Greenhouse testing is often necessary for clear results of HTAP resistance. For the greenhouse large-scale screening, we first test seedlings with several races at the low temperature cycle (4–20 °C) and then used the seedling-virulent races to test adult-plants at the high temperature cycle (10–30 °C) (Chen and Line 1995a, b). The temperature conditions for the dew-period of both seedling low-temperature and adult-plant high-temperature tests are the same (10 °C) for successful germination and penetration of plants by urediniospores. Because the effects of temperature and growth stage on HTAP resistance can be independent from each other, the above two-way (seedling low-temp and adult high-temp) testing may not be able to separate the two effects, although the testing can produce useful data for selecting germplasm with HTAP resistance. A four-way approach is used for more clear characterization of resistance. In addition to the above two-way tests, wheat and barley varieties are also tested in the seedling stage at the high-temperature cycle and in the adult-plant stage at the low temperature cycle (Chen and Line 1995a, b; Chen 2013). When a variety has HTAP resistance, it most likely shows the highest resistance (lowest IT and severity) in the adult high-temp test and a lower level or levels in the seedling high-temp and/or adult low-temp tests. The four-way tests can produce more clear results for the types of resistance and determine the different effects of temperatures and growth stages. Because the four-way testing approach is more time-consuming, it may be used for further tests of varieties selected from the two-way tests. The determination of temperature and stage effects is important as the information is useful for determining or clarifying race-specificity of resistance and the disease management related to resistance of individual cultivars.

Compared to major-gene controlled ASR, APR or HTAP resistance is relatively difficulty to incorporate into adapted cultivars because the screening needs to be done at the adult-plant stage and mainly because the resistance is mostly a quantitative trait (Milus and Line 1986a, b; Chen and Line 1995a, b). HTAP resistance is usually race non-specific and durable. This type resistance has been widely used in the US Pacific Northwest (PNW) since the early 1960s (Chen 2005; Line 2002;

Chen 2013). For example, wheat cultivars Gaines, Nugaines, Luke and Brevor still show the same level of resistance over 50 years, although their resistance levels are now considered inadequate (Chen 2013). Stephens, released in 1978, and Madsen, released in 1988, are still grown in the PNW and show relatively high levels of HTAP resistance (Chen 2013). In fact, Madsen is considered as a standard of high level of HTAP resistance. Johnson (1981a, b) used term "durable resistance" to describe resistance that remained effective when a cultivar is grown over a large area for a long period of time. The time and frequently severe stripe rust epidemics in the PNW have proven HTAP resistance durable (Chen 2013).

5.1.2.2 Resistance Types Based on the Degree of Resistance

Several terms of resistance have been used based on the level of resistance, such as complete resistance (immune) verses incomplete (partial) resistance, absolute resistance verses relative resistance and high-level resistance verses low-level resistance. Based on the appearance speed of symptoms and signs or disease development, slow rusting has been used in comparison with the fast development of susceptible reactions.

5.1.2.2.1 Complete Resistance and Incomplete (Partial) Resistance

Complete resistance is used to describe the situation in which the pathogen cannot infect the plant or the infection does not produce any observable symptom. This level of resistance is often referred as an immune response. It is equivalent absolute resistance, although the latter term is seldom used. Complete resistance can be included in high-level resistance, but high level is often arbitrary and relative. This type of resistance exits in both non-host resistance and host resistance. In non-host plants, almost all resistances are complete resistance, and their mechanisms are complex and not fully understood. In host plants, complete resistance is usually controlled by a single gene that is often race-specific and not durable. It is worth to note that a gene confers complete resistance in one variety may provide incomplete resistance in another variety, or provide complete resistance to one race but incomplete resistance to another race. For example, *Yr5* in the original genotype *T. spelta* album shows complete resistance (IT 0), but in the near-isogenic line AvSYr5NIL provides incomplete resistance (IT 2) (Chen and Line 1992a; Yan et al. 2003). *Yr9* in Riebesel 47-51 or in the near-isogenic line AvSYr9NIL provides complete resistance (IT 0) or near complete resistance (IT 1) to most races of *Pst* identified before 2000 in the US, but incomplete resistance ITs 2-6 to some avirulent races after 2000 (Chen et al. 2002, 2010).

Incomplete resistance or partial resistance can have a wide range of reactions, ranging from IT 2 to IT 8 in a 0–9 scale (Fig. 2.1 in Chap. 2) and severity of anything from above 0 to less than the percentage of a susceptible check. Incomplete resistance can be controlled by a single major gene or multiple minor genes. This lumped type of resistance can be either race-specific or non-race specific;

not durable or durable. The term is useful only when discuss with complete resistance. Instead of complete and incomplete resistance, high level and low level resistances are more meaningful when compare resistance of different degrees. Both high and low levels of resistance to stripe rust are commonly found in wheat, barley and other related species (Chen 2013, 2014).

5.1.2.2.2 Slow Rusting Verses "Fast Rusting"

The term of slow rusting was first given by Caldwell (1968). It was defined as slow disease development in the filed despite a susceptible infection type and by one or more components, such as longer latent period, smaller uredinium size, low receptivity (infection efficiency) and reduced sporulation. Compared susceptible checks which are "fast rusting" or have a rate of rust development typical for variety without any type of resistance, varieties with slow rusting resistance exhibit relatively low severity and have relatively low yield losses. Slow rusting resistance to stripe rust is often seen on wheat and barley varieties under field conditions and thus, is often referred as APR. However, it also can be detected in seedling stage in greenhouse tests. For example, previous differential varieties Heines VII and Yamhill exhibit slowing rusting, as sporulation starts about two to three days later than other differentials susceptible to the same virulent races. Slow rusting is a partial resistance, similar to other APRs or HTAP resistance. The major difference is its susceptible reaction whereas most cases of APR, HTAP resistance, or temperature sensitive resistance exhibit necrosis with or without sporulation. Slow rusting is quite often used loosely for partial resistance that may have hypersensitive responses. This is acceptable from the practical standard point, but precise characterization is needed for understanding the mechanisms of different types of resistance.

5.1.2.3 Resistance Types Based on the Genetics of Resistance

Plant resistance to various diseases has been studied genetically. Based on genetic characterization, disease resistance can be classified into monogenic resistance verses polygenic resistance based on the number of genes involved; major-gene resistance verses minor-gene resistance; and quantitative resistance verses quantitative resistance based on the mode of inheritance. These terms are parallel, but not identical to each other.

5.1.2.3.1 Monogenic Resistance and Polygenic Resistance

Monogenic resistance is controlled by a single gene. Although it does not necessarily related to whether the single gene has a big effect (a major gene) or small effect (a minor gene) on the phenotype of resistance, monogenic resistance traditionally indicates a gene with a large effect. However, as different genes can be separated into single-gene lines and a minor effect can be quantified, monogenic resistance should not be always considered to be controlled by a major gene. Similarly, resistance controlled by minor-effect genes also should not be always treated as polygenic resistance. The only approach to determine if a resistance in a particular variety is monogenic or polygenic is to conduct a genetic study. If only one gene is involved, the resistance in the variety is monogenic. If two or more genes are involved, the resistance in the parental variety is considered as digenic or polygenic; and monogenic lines can be identified from the progeny. Sometimes, oligogenic (for a few genes) and polygenic (for many genes) resistances are distinguished. The terms of monogenic, oligogenic and polygenic have become less popular than 20 years ago as the number of genes can be determined more easily than before. A specific number of genes for resistance in a particular variety is better than general terms of oligogenic or polygenic resistance.

5.1.2.3.2 Major-Gene Resistance and Minor-Gene Resistance

Major-gene resistance is controlled one or a few genes each with a big effect, whereas resistance in a particular variety is controlled by a gene or more genes each with a small effect is referred as minor-gene resistance. Minor-gene resistance can be determined through quantitative genetics approach, whereas major-gene resistance can be determined by both qualitative and quantitative approaches. From the breeding standpoint, major genes are easy to be incorporated into commercial cultivars, whereas minor genes are relatively difficult to be used. Collectively, several minor genes together can provide high level and potentially durable resistance if they active additively.

5.1.2.3.3 Qualitative Resistance and Quantitative Resistance

The terms of qualitative and quantitative resistances have been widely used. By the terms, qualitative resistance is qualitatively inherited, can be determined by classic genetic methods, and easy to be used in the breeding program. Qualitative resistance is usually controlled by one or a few genes with major effects. In contrast, quantitative resistance is quantitatively inherited and needed to be determined by quantitative genetic approaches. Quantitative resistance in a variety exhibits a continual variation in the F_2 segregating progeny population in a cross with a susceptible variety. Depending upon the number of genes involved and the gene actions and interactions, two (for one gene) or more (for more genes) classes may or may not be distinguished in later segregating populations.

Quantitative resistance is best determined using quantitative trait locus (QTL) mapping, which requires precise phenotyping and large number of molecular markers to cover the genome. Many of the stripe rust resistance genes or QTL, which will be discussed late in this chapter, have been identified through QTL mapping. Confusion should be avoided that not all genes identified by QTL mapping are considered to conferring quantitative resistance because genes for qualitative resistance can also be identified using QTL analysis although they can be more easily identified using qualitative approach. If both qualitative and quantitative resistance genes are present in a variety, QTL approaches should be used to identify them. In general, quantitative resistance is usually non-race specific and more durable than qualitative resistance.

5.1.2.4 Resistance Types Based on Plant-Pathogen Interactions

There are several terms for describing resistance based on plant-pathogen interactions. These terms can be further classified based on plant species, race-specificity, sensitivity to pathogen infection, relation to the infection process and molecular basis.

5.1.2.4.1 Host Resistance and Non-host Resistance

Wheat is considered a host for *Pst* as many varieties are susceptible and also many varieties are resistant. Similarly, barley is a host for *Psh*. The resistance of any wheat varieties to *Pst* and the resistance of any barley variety to *Psh* are of host resistance. Because many grass species are auxiliary hosts to *Pst* and/or *Psh*, some plants or strains of these grasses are resistant to one or both of these formae speciales, and therefore their resistance should be considered host resistance. In contrast, non-host resistance exits in all plants of the species. For example, other cereals such as rice, corn and oats are non-hosts for *P. striiformis* because no any plants of these crops are reported to be susceptible to the fungus. Similarly, the immunity or resistance of any vegetable crops and trees or bushes (except alternate hosts such as barberry and mahonia) is clearly non-host resistance. Similarly, the immunity of wheat or barley to the corn and bean rusts is non-host resistance.

As discussed above, host resistance against stripe rust has been more intensively studied and this type of resistance is easy to be used in breeding programs through crossing. In contrast, non-host resistance is hard or impossible to be used in breeding programs through crossing. In the recent years, non-host resistance of wheat to other diseases and that of other crops, such as Arabidopsis, broad bean, pepper and rice, have been studied using ultrastructural and molecular approaches (Ayliffe et al. 2011a, b, 2013; Cheng et al. 2012, 2013; Zuo et al. 2013; Yang et al. 2014), as discussed in Chap. 3. It will be a long way to go for utilization of non-host resistance for control of stripe rust.

Between the two extreme of host and non-host resistances, there are some intermediates. Wheat is generally resistant to *Psh* and barley generally resistant to *Pst*, but few varieties of wheat or barley are susceptible to *Psh* and *Pst*, respectively (Chen et al. 1995c). These "non-preferred" hosts have been recently defined as "intermediate hosts" and their resistance as "intermediate-host resistance" (Bettgenhaeuser et al. 2014). Intermediate-host resistance has been studied in wheat for resistance to *Psh* (Pahalawatta and Chen 2005a) and in barley for resistance to

Pst (Pahalawatta and Chen 2005a; Dawson et al. 2016; Li et al. 2016a; Sui et al. 2010). Genes and molecular markers of the intermediate-host resistance have been identified. These genes could be transferred to the host crops of the pathogen through transformation or chromosomal manipulation.

5.1.2.4.2 Race-Specific Resistance and Non-race Specific Resistance

Based on race specificity, resistance can be classified as race-specific resistance verses non-race specific resistance; differential resistance and non-differential resistance; or vertical resistance verses horizontal resistance as defined by Vanderplank (1963). As the term describes, race-specific resistance is effective against some races, but not other races. Most permanently named *Yr* genes, such as the *Yr* genes in the single-gene differentials (Wan and Chen 2014), are race-specific resistance genes. Non-race specific resistance is effective against all races of the pathogen in the world. Non-race specificity of resistance is hard to prove as a gene for such resistance can be effective against the current and even the past races of the pathogen, but it is impossible to know whether the gene will be effective against races in the future. Because of this, non-race specific resistance sometimes is referred to the resistance whose specificity has not been found (Ellingboe 1981). However, researchers have used other criteria to determine or suggest non-race specific resistance, as discussed above and below.

5.1.2.4.3 Hypersensitive Resistance and Non-hypersensitive Resistance

Based on sensitivity to pathogen infection, resistance can be classified as hypersensitive resistance and non-hypersensitive resistance. Hypersensitive resistance shows observable necrotic flecks, patches or stripes. For stripe rust, hypersensitive response (HR) varies a lot, ranging from IT 1 to 8 (Fig. 2.1) in Chap. 2. Even IT 0, an immune response showing no obvious symptoms and signs, can indicate different reactions. It can occur in both host and non-host resistances. If observed under a microscope, microscopic cell or tissue death can often be observed for plants showing such so-called immune or non-hypersensitive resistance. Hypersensitivity is useful for describing plant-pathogen interactions, often in cytological and biochemical studies, but often not precise for distinguishing different levels of resistance, especially for stripe rust.

5.1.2.4.4 Pre-haustorial Resistance and Post-haustorial Resistance

The concept of pre-haustorial resistance and post-haustorial resistance is based on the start of expression of resistance during the infection process. As the terms tell, pre-haustorial resistance occurs quickly before haustorial formation, whereas post-haustorial resistance does not express until haustorial formation. As pre-haustorial resistance is quick, it is usually at high level and kills the pathogen. In contrast, post-haustorial resistance takes place late and slowly, it can have various levels of resistance. Because any resistance has many genes involved, different genes may express in different stages during the infection process, it is often hard to distinguish the two types of resistance. These terms are useful in ultrastructural studies for understanding cytological and biochemical events during the interactions between plant and pathogen as discussed in Chap. 3, but less useful from the practical breeding standpoint.

5.1.2.4.5 NBS-LRR Resistance and Non-NBS-LRR Resistance

Cloned plant genes for resistance to diseases generally fall into two categories: gene products containing a nucleotide-binding site (NBS) and a leucine rich repeat (LRR) domain and those containing no NBS-LRR domains. Most cloned plant resistance genes, so called "R" genes, are race or strain specific and contain the NBS and LRR domains. The NBS domain is believed to be involved in signal transduction for regulating defense genes, while the LRR domain determines the specific recognition by the corresponding pathogen virulence gene (Zhang et al. 2003; DeYoung and Innes 2006).

For stripe rust, there are four genes (*Yr10*, *Yr18*, *Yr36* and *Yr46*) that have been cloned so far. *Yr10* is a NBS-LRR gene (Liu et al. 2014a). Although the gene is effective in many countries, *Pst* races virulent to *Yr10* have been identified in the US since 1972 as the gene was widely used in wheat breeding programs in 1960s to 1980s and is still present in many wheat cultivars grown in the US Pacific Northwest (Chen 2013). *Yr18*, the same gene as *Lr34* for resistance to leaf rust and *Pm38* for resistance to powdery mildew, encodes a putative ATP-binding cassette (ABC) transporter (Krattinger et al. 2009a, b). *Yr18* belongs to the pleotropic drug resistance subfamily of ABC transporters and has two cytosolic nucleotide binding domains (NBD) and two hydrophobic transmembrane domains. The pleotropic gene also provides resistance to other disease when transformed into barley, corn, rice and other plant species (Krattinger et al. 2011, 2013, 2016; Risk et al. 2013). The product of *Yr36* contains a predicted kinase and a steroidogenic acute regulatory protein-related lipid transfer (START) domain (Fu et al. 2009). Both the kinase and START domains are necessary for the resistance function. Temperature and *Pst* inoculation consistently up-regulates expression of the resistance alleles, but down-regulates the susceptible alleles. The authors postulated that the START domain has the ability to bind lipids from stripe rust fungus at high temperature and change its conformation, which may cause the kinase domain to initiate a signaling cascade leading to the observed programmed cell death. *Yr46* is also a pleiotropic gene, providing resistance to stripe rust, leaf rust (*Lr67*), stem rust (*Sr55*) and powdery mildew (*Pm46*) (Herrera-Foessel et al. 2011; Chhetri et al. 2016a). The gene encodes a predicted hexose transporter that differs from the susceptible form of the same protein by two amino acids that are conserved in orthologous hexose transporters. The susceptible allele functions as a high-affinity glucose transporter, while the resistance allele has a dominant-negative effect through heterodimerization with functional transporters to reduce glucose uptake. The three APR or HTAP resistance genes, *Yr18*, *Yr36* and *Yr46*, all do not have a LRR domain and belong to the non-NBS-LRR type. They have not shown race specificity. In contrast, *Yr10*, a NBS-LRR gene, has shown race specificity. Our candidate gene for *Yr5*, a gene effective against most *Pst* races but its virulence has been identified (Yan et al. 2003), is also a NBS-LRR gene (Wang and Chen 2016). The cloned genes so far have proven the hypothesis that NBS-LRR genes are race-specific and not durable while non-NBS-LRR genes are non-race specific and durable (Chen 2013; Chen et al. 2013b).

5.1.3 Genetics of Resistance

The genetics of resistance to stripe rust was first reported by Biffen in 1905 (Biffen 1905). Not only for stripe rust, also was his study the first for genetics of plant disease resistance. In crossing the susceptible cultivar Michigan Bronze with the so-called immune cultivar Rivet, Biffen found that the F_2 generation consisted of 195 heavily infected plants and 64 plants without infection, which was segregating as a ratio of 1 resistance to 3 susceptible, suggesting that the resistance in Rivet was controlled by a single recessive gene (Biffen 1905, 1907). This important work demonstrated that the inheritance of resistance to stripe rust follows the Mendel's laws, and resistance was inherited independently of other morphological traits. His study opened a door for genetic studies of resistance to other plant diseases and paved a way for successful breeding for resistance to control plant diseases. Applying the genetics of resistance in his breeding work, Biffen developed the first stripe rust resistant cultivar Little Joss in 1910 (Engledow 1950; Charnley 2009). In 1929, Rudorf first used artificial inoculation and greenhouse conditions to study genetics of resistance (Röbbelen and Sharp 1978), eliminating the variations from environment and pathogen and making the experimental results more reliable.

Unlike other agronomic traits, the resistance to stripe rust involves the interaction of both genes in wheat and rust pathogen. The genotypes in both wheat plant and rust pathogen determine the phenotype of resistance or susceptibility. In 1940s, Flor proposed the gene-for-gene hypothesis based on his genetic studies on the flax and flax rust pathosystem (Flor 1942, 1971). In his series of works, genetic studies of resistance were accomplished through proper testing of segregating populations of host plants for reaction to one or more rust races, and by determining the reaction of a resistant plant to segregating population of fungus cultures (Flor 1942, 1946, 1947, 1955). The gene-for-gene hypothesis states that for each gene conditioning rust reaction in the host there is a specific gene conditioning pathogenicity in the parasite (Flor 1956). Later the gene-for-gene concept is widely adopted and validated in other plant-pathogen systems, including wheat stripe rust. Genetics for race-specific resistance were studied under this theory.

Favret and Vallega (1953) studied the inheritance of stripe rust resistance in wheat variety "Chino 166" (Chinese 166) to Argentine races of *Pst*, and about 10 years later, Lupton and Macer (1962) named the first stripe rust resistance gene, *Yr1*, in Chinese 166. Based on the genetic studies of Chinese 166 and other wheat varieties, Lupton and Macer (1962) first introduced the nomenclature of *Yr* (for Yellow rust) genes for stripe rust resistance in several wheat cultivars based on classic genetics analysis of resistance detected in the seedling stages. With the cytogenetical approaches used in wheat genetics, various genetic stocks enable to study the chromosomal locations of stripe rust resistance genes. The chromosomal locations of several *Yr* genes were obtained through the use of chromosome translocation, substitution and addition lines, and also by means of monosomic and nullisomic analyses (Macer 1966). UK geneticists analyzed the monosomic stocks of wheat cultivar Hobbit Sib and located stripe rust resistance on homoeologous group 5 and translocation 5BS-7BS in several Europe cultivars (Pink et al. 1982; Law et al. 1987). Law and Worland (1991) used fast neutrons to create deletion mutants of susceptible Hobbit Sib and identified chromosomes 4B, 4D, 5B and 5D related to stripe rust resistance. They made substitution lines of Hobbit Sib with chromosomes from Russian wheat cultivar Bezostaya 1 and found that the substitution line of chromosome 7D (carrying *Yr18*) increased stripe rust resistance. Since the middle 1990s, molecular markers have been used to map stripe rust resistance genes on specific positions of chromosomes. The early genetic studies of stripe rust resistance mainly focused on race-specific resistance since the phenotypes in the segregating populations can be clearly categorized into groups according to parental phenotypes and fit the specific ratios of inheritance models. The resistance evaluation can be easily conducted in seedlings in the controlled greenhouse conditions. In the last two decades, the genetics of polygenic or quantitative resistance to stripe rust has been studied in many cultivars partially due to a larger number of molecular markers of wheat have become available, the cost for genotyping mapping populations have been reduced, and the technique platforms for genotyping have been advanced.

5.1.3.1 Analyses of Genes for Stripe Rust Resistance in Wheat Germplasms

Plant genes for resistance to a disease can be identified and characterized using various genetic and non-genetic methods. Each of the methods has its advantages and or limitations. As technology is advancing, issues that could not be addressed before but can be studied now.

5.1.3.1.1 Gene Postulation

Gene postulation applies the gene-for-gene concept to analyzing resistance genes. This approach has been used to identify stripe rust resistance genes in wheat (Sharma et al. 1995; Qamar et al. 2008; Dawit et al. 2012; Singh et al. 2015). Tester cultivars together with a panel of *Yr*-gene lines are inoculated with a set of *Pst* races.

The low infection types will be present only in the interaction between the specific *Yr* gene carriers and its corresponding avirulence *Pst* isolates, while high infection types will occur among other interactions. Then, it is possible to postulate which resistance genes are present in the tester cultivars according to the specific low infection types on the *Yr*-gene lines. The advantage of gene postulation is the information of the possible resistance genes on the tester cultivars can be obtained within several weeks without making crosses. Breeding lines or cultivars with clear pedigree can be evaluated in a short time. For example, Dawit et al. (2012) tested 22 Ethiopian bread wheat cultivars together with 24 *Yr* differential lines with 20 *Pst* races that were collected from Ethiopia, France and Germany. Based on the results, they postulated different combinations of *Yr2*, *Yr3a*, *Yr4a*, *Yr6*, *Yr7*, *Yr8*, *Yr9*, *Yr27*, *Yr32* and *YrSU* presented in the Ethiopian wheat cultivars.

The disadvantages for this method are obvious. First, it can only be used for race-specific genes and cannot be used to analyzing genes for non-race specific resistance. Second, if a cultivar is resistant or susceptible to all the isolates in test, the resistance gene cannot be inferred. Third, if a cultivar carrying more than one resistance genes, this method may not identify all of the resistance genes in the cultivar. Further, the method is limited by the avirulences/virulences, their combinations, and unrecognized avirulences/virulences in the pathogen isolates.

The information of pedigrees and molecular markers can be used together with the gene postulation method to overcome some of the problems. Using the combined methods, Zeng et al. (2014) studied 330 leading cultivars and 164 advanced breeding lines in China, and identified *Yr9*, *Yr17*, *Yr18*, and *Yr26* in 134 (29.4 %), 45 (9.1 %), 10 (2 %) and 15 (3 %) entries, respectively. They also identified *Yr5* in two cultivars and confirmed the absence of *Yr10* and *Yr15* in the entries.

5.1.3.1.2 Genetic Analyses

Following Biffen's pioneer work, a number of genetic studies of resistance to stripe rust were conducted in institutes in Europe, Australia, the North America, India, China and other countries. Genetic analysis can determine the number, gene action and inheritance model of stripe rust resistance genes in wheat cultivars. The procedure for genetic analysis includes the following steps. First, the resistance cultivar being studied is crossed with a susceptible variety. Second, a segregating population is developed. The segregating population can be F_2, F_3, BC_1, BC_1F_2, recombinant inbreed lines (RILs) or doubled haploid (DH) lines. Third, the segregating population is evaluated under the infection of the pathogen, and the reaction of each individual plant or family can be catalogued into resistant, susceptible or segregating groups. Based on the segregating ratio from the population, gene number, action and inheritance mode of the resistance is determined. Majority of studies are conducted using F_2, F_3, BC_1 and BC_1F_2 populations for analysis of race-specific resistance, and RIL and DH populations are used for analysis of quantitative resistance.

5.1.3.1.2.1 Number of Genes for Resistance and Inheritance Modes

The simplest form of inheritance of stripe rust resistance is that it is controlled by a single gene. Resistance to stripe rust can be controlled by either dominant or recessive genes. Single dominant genes for stripe rust resistance were identified in wheat cultivars Comta Klein (Ghosh et al. 1958); Suwon 92 (Allan and Vogel 1961; Helm and Allan 1970); La Prevision (Bakshi and Sawhney 1965), (PI 94349 (Allan and Purdy 1967; Helm and Allan 1970); PI 178383 (Lewellen and Sharp 1968; Helm and Allan 1970); Anza, Bowie and Narino 59 (Omar et al. 1970); Frondoso (Gandhi 1971); Kalyansona and Lerma Rojo (Minhas and Singh 1973); Benno, Orlando and Zorba (Slovencikova 1980); Carstens V (Chen and Line 1993b); Zak (Sui et al. 2009); IDO337s (Cheng and Chen 2010); PI 181434 (Li et al. 2011); PI 480148 (Xu et al. 2013); PI 332610 and PI 480016 (Cheng et al. 2014); and Tyee (Xiang et al. 2016). Recessive single resistance genes were identified in wheat cultivars Frondoso (Nambisan and Kholi 1961); Kans. 587023 (Allan and Purdy 1967); Heines Kolben and Spaldings Prolific (Allan and Purdy 1970); Gelderse Ris (Stubbs et al. 1984); *Yr6* (Wellings 1986); LM168-1 (Feng et al. 2014); Wuhan 2 (Zhou et al. 2014b); AUS91456 (Randhawa et al. 2014); Mianmai 37 (Ren et al. 2015b); Chuanmai 45 (Yang et al. 2016); and Guinong 22 (Li et al. 2016b). Single-gene inheritance for stripe rust resistance was also found in several studies conducted in India in the 1960s (Sikka et al. 1960; Singh and Dhillon 1963; Sawhney and Bakshi 1965). The additional cases of dominant and recessive resistance genes were reviewed by Röbbelen and Sharp (1978) and Wellings (1986).

Although dominant monogenic resistance to stripe rust was detected more frequently in wheat varieties than recessive resistance genes, the latter was reported in many wheat cultivars from the earlier studies. One possible reason is that most stripe rust resistance genes are only partial dominant. The F_1 and the heterozygote F_2 plants phenotypically displayed intermediate infection types. As reviewed by Röbbelen and Sharp (1978), if only the disease free plants were counted as resistant and plants with intermediate and heavy infection were counted as susceptible in F_2, it may draw a conclusion that a recessive gene confers resistance.

Often many wheat cultivars carry multiple resistance genes against different *Pst* races. Chen and Line (1992a, b, c, 1993a, b) systematically studied stripe rust resistance in wheat cultivars which were used to differentiate the stripe rust races in North America or critical commercial cultivars. They found that the gene number, the dominance and recessive and the gene action in certain cultivars were related to the test races and background of crosses. For example, cultivar Tres has one dominant gene for resistance to race PST-1, PST-11, PST-17, PST-20, PST-27 and PST-29, and two dominant genes or one dominant and one recessive gene for resistance to races PST-21 and PST-25 (Chen and Line 1992a). Cultivars Lemhi, Chinese 166, Riebesel 47/51 and Tyee each have a single resistance gene; Heines VII, Moro, Druchamp, Produra, Stephens, Lee and Fielder each have two resistance genes; and Yamhill and Paha each have three resistance genes. The genes in Lemhi, Chinese 166, Heines VII, Moro and Riebesel 47/51, and one of genes in Paha and Lee were dominant. The two genes in Produra, one of genes in Lee, and at least one of the genes in Paha were recessive. The single gene in Tyee and both genes in Druchamp, Fielder and Stephens were dominant or recessive depending on the race, parent in the cross, interaction of race and parent and duration of infection (Chen and Line 1992b). In the studies of Chen and Line (1992a, b, c, 1993a), *Yr1*, *Yr7*, *Yr9*, *Yr10* and *Yr21* were dominant. *YrLee* and *YrPa2* were recessive. *Yr2* and *YrHVII* in Heines VII, *YrMor* in Moro, *YrPa3* in Paha, *Yr3a* and *YrDru* in Druchamp and *Yr4a* in Yamhill, *Yr3a* and *YrSte* in Stephens, *Yr6* and *Yr22* (*YrFie*) in Fielder, and *YrTye* (*Yr76*) in Tyee were either dominant or recessive, depending on races and/or genetic background.

5.1.3.1.2.2 Relationships of Resistance Genes

The relationships of the resistance genes in different cultivars can be determined using allelism analysis by making crosses between two resistant cultivars and analyzing the segregation in F_2 and F_3 generation. Based on the allelism analysis, Chen and Line (1992c) found cultivars Chinese 166, Heines VII, Fielder, Lee, Riebesel 47/51 and Moro have stripe rust resistance genes *Yr1*, *Yr2*, *Yr6*, *Yr7*, *Yr9* and *Yr10*, respectively; Druchamp, Stephens and Yamhill have either *Yr3a* or *Yr4a*; and new genes were identified in cultivars Lemhi, Tyee, Heines VII, Moro, Druchamp, Stephens, Lee and Fielder.

Allelism testing is time consuming and also has other limitations. It is better to have only two genes in question to be tested in a cross. If additional genes involved, it may be hard to get a clear result about the relationship of the two genes. As more and more genes or QTL have been identified, it is almost impossible to make and testing all possible resistant × resistant crosses. However, it is not necessary to make all crosses for determining the relationships of a gene with all other genes. Molecular mapping can greatly reduce the number of crosses to be made. First, the gene in question should be mapped to a chromosomal location. The location is compared to those of the previously reported genes to identify the ones within or near the chromosomal region. Then crosses should be made with lines or cultivars each carrying only one of the previously reported genes within or near the chromosomal region. Progeny populations of F_2 and/or F_3 are tested with a race avirulent to both genes in the cross. The phenotypic data can be used to determine if the two genes from in the parents of the cross are at the same locus or different loci based on the absence or presence of recombinant (susceptible) plants or lines. The proportion of the recombinant plants (lines) can be used to determine the genetic distance between the two loci using the Kosambi function (Kosambi 1944) or other formulae. The lines with genes in comparison can be tested with different races of the pathogen to determine if they have differences in race specificity. Using this combination approach, Chen and associates have identified several new genes or confirmed genes in wheat for resistance to stripe rust (Cheng and Chen 2010; Ren et al. 2012a; Xu et al. 2013; Cheng et al. 2014; Lu et al. 2014; Zhou et al. 2014c; Feng et al. 2015b; Xiang et al. 2016)

5.1.3.1.2.3 *Localization of Resistance Genes to Chromosomes*

Before molecular markers with known chromosomal location were developed for wheat, gene location was determined using cytogenetic methods. Europe wheat cultivar Hybride de Bersee has durable APR to stripe rust. Using ditelosomic lines developed for the cultivar, Johnson and Law (1973, 1975) determined that chromosome 5BS and 7BS are associated to the resistance and found that 5BS was more resistant than 7BS under environments favoring high spore production, but more susceptible under environments favoring low spore production. By crossing resistant cultivars with a set of 21 Chinese Spring monosomic lines, Chen and his colleagues determined chromosomal locations in many wheat cultivars used to differentiate races of *Pst* or as important resistance sources. They found the resistance genes *YrDa1* in chromosome 1A; *Yr3a*, *Yr3c* and *Yr21* in 1B; *YrSte* and *YrV23* in 2B; *YrTr2* in 3A; *YrSte2* in 3B; *YrHVII*, *YrND* and *YrMin* in 4A; *YrCle*, *YrMor* and *YrYam* in 4B; *Yr22* in 4D; *Yr19* and *YrDru* in 5B; *YrDa2* in 5D; *YrH46* and *YrDru2* in 6A; *Yr4a* and *Yr4b* in 6B; *Yr20*, *Yr23*, *YrTye* and *YrTr1* in 6D; and *Yr2* and *Yr6* in 7B (Chen et al. 1995a, b; Chen et al. 1996). The latest reported resistance genes were mapped on chromosomes using combined molecular and cytogenetic approaches.

5.1.3.1.2.4 *Interactions of Resistance Genes*

Various intergenic interactions between stripe rust resistance genes have been reported, including independent, complementary, additive, epistatic and suppression. Independent interactions are commonly found for two or more genes for resistance. This type of interaction is indicated by F_2 ratios of 15(R):1(S) for two dominant genes; 7(R):9(S) for two recessive genes and 13(R):3(S) for one dominant gene and one recessive gene for resistance. For example, the 15(R):1(S) ratio was found in the F_2 populations from cross Druchamp (R) × Chinese 166 (S) when tested with race PST-21 and cross Lemhi (S) × Lee (S) tested with PST-1 (Chen and Line 1992b). Yin et al. (2006) reported two genes, *YrTp1* and *YrTp2*, for resistance to Chinese *Pst* races CYR31 and CYR32 in wheat line A-3 derived from cross of *T. aestivum* × *Th. ponticium.* Both genes originated from *Th. ponticium* accession R431, and they were independent with *YrTp1* dominant and *YrTp2* recessive. Cultivar Xiaoyan 54 carries two independent recessive genes (Zhou et al. 2011). These independent interactions are based on the assumption that the resistance allele at one locus is epistatic to the susceptible allele at the other locus.

Complementary interaction is indicated by F_2 segregating ratios 9(R):7(S) for two dominant genes, 1(R):15(S) for two recessive genes, 3(R):13(S) for one dominant gene and one recessive gene and many other ratios for two or more genes. This type of interaction has been reported for resistance to stripe rust in wheat cultivars Cometa Klein (Nambisan and Kholi 1961); Nord Desprez, Spinkota, Dickson 114 (Allan et al. 1966); Rego (Lewelle and Sharp 1968); E255 (Sandhu and Singh 1970); Alcedo (Walther and Herdan 1984); Robin (Dutlu 1984); and in many cultivars studied by Chen and Line (1992a, b, c, 1993a, b).

The stripe rust resistance in Australian wheat cultivar Avocet is a good example of complementary interactions of two genes. The resistance, controlled by provisionally designated *YrA*, also exists in many other Australian wheat cultivars, such as Banks, Condor, Avocet and Egret, and spring wheat cultivars WW15 (or Anza), Nainari 60, Inia 66, Tanori 71 and Funo in other countries. It was inherited as a pair of dominant complementary genes, which was used to be effective to the initial stripe rust pathogen in Australia (Wellings 1986). The two genes, mapped on chromosome 3DL (*Yr73*) and 5BL (*Yr74*), were determined to confer resistance when both genes are present in the genome (Dracatos et al. 2016). The heterozygote F_1 plants from a cross between each gene displayed intermediate infection types. The susceptible selections of Avocet carried only one of the two resistance genes. The complementary resistance is ineffective against most *Pst* races in the US (Wan and Chen 2014; Wan et al. 2016) and other countries (Sharma-Poudyal et al. 2013).

Additive and epistatic gene interactions play important roles in resistance to stripe rust in some wheat cultivars (Allan et al. 1963; Sharp and Volin 1970; Krupinsky 1977). Krupinsky and Sharp (1978) intercrossed the wheat lines having minor resistance genes and found high level resistance in the F_2 and F_3 progenies. Their study demonstrated the additive gene interaction contributed to improvement of stripe rust resistance. Additive gene interactions were also observed in wild emmer wheat (Grama et al. 1984). Milus and Line (1986a, b) found that stripe rust resistance in Pacific Northwest cultivars Gaines, Nugaines and Luke was partially recessive. They found that epistatic gene action for resistance was significant in Nugaines, but most gene action among loci was additive. They also detected epistatic gene action for susceptibility in cultivar Luke.

Adult-plant resistance is usually inherited as a quantitative trait. Allan et al. (1963) reported that resistance in adult-plant displayed transgressive segregation due to the additive effect of minor genes derived from different levels of resistance in some cultivars. Sharp (1965) found stripe rust resistance involving minor genes and influenced by environments. This temperature-sensitive resistance is unlike the major gene resistance. Cultivars having minor quantitative gene resistance are often resistant to more races than those with major race-specific gene resistance, but each individual minor gene confers lower level resistance. Before the advent of QTL mapping, quantitative resistance was studied through statistical analysis of appropriate experimental populations. Using crosses between resistance and susceptible cultivars, scientists could estimate heritability, the number of relevant genetic loci, the degree of dominance, additivity and the role of gene × gene and gene × environment interactions. These studies focused on phenotypic distributions of populations and correlations in phenotypes among related individuals or lines. By fitting experimental results with predicted genetic models of population means, variances and covariances, it was possible to estimate gene number, gene action and interactions among parameters (Röbbelen and Sharp 1978; Navabi et al. 2004). Chen and Line (1995a) investigated gene actions in resistance cultivars Stephens and Druchamp. They found that additive component for HTAP resistance was significant for both Stephens and Druchamp. When HTAP resistance was effective and ASR resistance was ineffective in the same parent, the dominant component and the additive-additive, additive-dominant and dominant-dominant epistatic interactions contributed significantly to HTAP resistance in Druchamp. When HTAP resistance and seedling resistance were effective in the same parent, the dominant component and additive-additive and dominant-dominant epistatic interaction were significant in both Stephens and Druchamp and additive-dominant epistatic interaction was significant in Stephens. When HTAP resistance was effective in one parent and seedling resistance was effective in the other parent, the dominant component and the additive-additive and dominant-dominant epistatic interactions were significant in the Druchamp crosses but not in the Stephens crosses. A cytoplasm additive gene interaction was significant in the reciprocal crosses if Druchamp with Paha tested with race PST-25. The additive effect of genes in Druchamp was better demonstrated in a more recently study using the QTL mapping approach (Hou et al. 2015).

Interestingly, from the larger number reports the race-specific resistance is more frequently found to be dominant and the HTAP non-race specific durable resistance is usually partially recessive. From a practical point of view, the results of such statistical studies provided critical information about heritability and the prospects for phenotypic progress in crop improvement programs.

The suppressor or inhibitor of resistance genes can influence the resistance gene expression. Johnson and Dyck (1984) proposed that Thatcher (*Yr7*) has a dominant inhibitor of *Yr5* since Thatcher/*T. spelta* album (*Yr5*) was susceptible in F_1 to a race avirulent for *Yr5* and virulent for *Yr7*, and the data for segregating F_3 families supported the inhibitor hypothesis. Kema et al. (1995) reported differential suppression of stripe rust resistance. They studied the expression of resistance in four *Ae. squarrosa* (DD, 2n = 14) and 11 *T. turgidum* subsp. *dicoccoides* (AABB, 2n = 28) accessions by testing 22 synthetic hexaploid (AABBDD, 2n = 42) lines with five *Pst* races in the seedling stage and two races in the adult-plant stage. Resistance in one or both parents was frequently suppressed in the synthetic hexaploid lines, indicating the presence of suppressor genes on the AB and D genomes. The putative suppressor genes were found to be specific as they affecting the expression of specific resistance genes to specific races and in specific growth stages. Wu et al. (2015) identified an inhibitor to *Yr18* in Chinese wheat landrace Sichuan Yonggang 2, which has the *Yr18* allele but was susceptible in adult-plant stage. F_1 plants from cross Sichuan Yonggang 2 and Chinese Spring were susceptible, indicating the inhibitor in Sichuan Yonggang 2 is a dominant gene. The $F_{2:3}$ and BC_1F_2 segregating population confirmed the presence of a dominant inhibitor to *Yr18*.

5.1.3.1.2.5 QTL Mapping and GWAS of Stripe Rust Resistance Genes

With the development of the principal and support techniques of QTL mapping, the quantitative resistance can be explained from a new point of genetics. Based on associations between molecular markers and phenotypes, QTL can be delimited with molecular markers in the specific chromosome regions. The effects of additivity, heritability, and gene actions among QTL can be estimated by analyzing the phenotypic and genotypic data of appropriate progeny populations. Chen and his associates studied the genetics of HTAP resistance using QTL mapping. They found HTAP resistance was mostly partially dominant because the rust infection types on F_1 progenies were either higher than the resistant parent or intermediate between resistant and susceptible parents (Lin and Chen 2007, 2009; Ren et al. 2012a; Lu et al. 2014; Zhou et al. 2014c; Hou et al. 2015). Some cultivars have only one major QTL and explained large proportion of the phenotypic variation, for example, QTL in Alpowa explained 59.1% and 64.2% variation of infection type and rAUDPC, respectively (Lin and Chen 2007). Other cultivars have several QTL, such as three QTL in cultivar Express explaining 49.5– 69.6% variation of rAUDPC (Lin and Chen 2009). In cultivar Druchamp, eight QTL were identified for HTAP resistance, and each QTL explained 1.8–13.9% to 1.7–33.7% variation in multiple-year tests (Hou et al. 2015).

In addition to identification of QTL from biparental cross populations, QTL analysis has recently been used to identify stripe rust resistance genes in various wheat germplasm collections. Through genome-wide association study (GWAS) between molecular markers and phenotypes, more and more QTL for resistance to stripe rust have been identified in recent years (Singh et al. 2013; Zegeye et al. 2014; Jighly et al. 2015; Maccaferri et al. 2015; Naruoka et al. 2015; Bulli et al. 2016; Liu et al. 2017). Even though the GWAS provided valuable information about the chromosome locations of multiple QTL, the appropriate experimental validation is still required to confirm that wheat lines carrying the favorable marker allele actually carry the associated resistance gene. For example, *Yr77* and *Yr78*, which were initially identified as *QYr.ucw-6D* and *QYr.ucw-6B* from GWAS (Maccaferri et al. 2015), were confirmed through analyzing biparental cross populations (Dong et al. 2017).

5.1.3.2 Factors Influencing Resistance Analyses

When conducting genetic analyses of the resistance, some factors can affect the resistance expression. These factors should be considered when studying stripe rust resistance.

5.1.3.2.1 Plant Factors

From the host plant side, ploidy level, genetic background, plant tissue and age are major factors affecting resistance.

Plant Ploidy and Genetic Background. In general, the resistance level can be reduced when the resistance genes transferred from diploid or tetraploid species into hexaploid wheat. For example, Ma et al. (1995) studied stripe rust resistance in 74 synthetic hexaploid wheat lines derived from tetraploid × diploid wheats. Only 15 lines had low ITs similar to the respective donor parents. The remaining 59 lines displayed either intermediate or high ITs. The fact that resistance of donors were not expressed or partially expressed in a synthetic hexaploid background suggests the presence of suppressor genes in the A or B, and D genomes of tetraploid and diploid wheat lines. Xu et al. (2013) reported that *Yr53* in durum wheat PI 180148 produced small necrotic spots in all of the *Pst* race tests, but much larger necrotic blotches or stripes with occasional uredinial pustules were produced when *Yr53* was introgressed into the common wheat Avocet background. Even in the same ploidy level, different genetic backgrounds can give different reactions for a single gene. For example, *Yr8* produced more necrosis in the Avocet than in the Chinese Spring background although both background cultivars were susceptible in the seedling tests (Wellings 1986).

Plant Growth Stage. Growth stage is significantly associated with APR gene expression. Qayoum and Line (1985) studied the effect of plant growth stage and temperature on HTAP resistance. They found that ITs on HTAP resistant cultivars decreased in seedling, jointing, boot and to heading stages from low temperature profiles (2–18 °C or 6–21 °C) to high temperature profiles (13–32 °C, 13–33 °C, 15–35 °C or 15–36 °C). Under high temperature profiles, wheat line Itana/PI 178383-111 showed IT 5–7 (seedling), 5–6 (jointing), 1–3 (boot) and 0 (heading); and Luke showed ITs 7–8 (seedling), 5 (jointing), 1 (boot) and 1–2 (heading), respectively. Even the susceptible cultivars (Omar and Michigan Amber) had high IT 9 in seedlings also displayed relatively low ITs (5, 6–7) in heading stage when temperatures are too high (15–35 °C or 15–36 °C). Because of this reason, the standardized temperature profile for detecting HTAP resistance is set as 10–30 °C, at which susceptible varieties are susceptible, and inoculation is conducted from boot to flowering stages (Chen and Line 1995a, b; Chen 2013).

Similarly, Ma and Singh (1996) evaluated IT, latent period (LP) and area under the disease progress curve (AUDPC) of eight cultivars with APR to stripe rust in different growth stages. Seedling IT and LP for cultivars with APR were similar to those of the susceptible cultivars (IT 6 to 8, LP 12 to 15 days). However, as plant grew, resistance increased and was expressed as lower ITs and longer LPs. Intermediate (IT 4-6) to low (IT 2-3) reactions were observed on cultivars HD2258 and PBW65 as early as the tillering stage and changed to immunity (IT 0) at anthesis. Other six APR cultivars showed intermediate to moderately susceptible response (IT 4-6) at growth stages later than tillering. The lowest IT and longest LP displayed at anthesis in all eight cultivars with APR.

Plant Tissue. Leaf tissue and age also can affect plant reactions to stripe rust. The different ITs can be found on the leaf base and tip of the same plants even using the same inoculum. The example for leaf age affecting IT was reported on Avocet. The IT on second leaf was lower to avirulent races than that on primary leaf (Wellings et al. 1988). In contrast, seedlings carrying other resistance genes such as *Yr6* tend to produce higher ITs on second leaves in response to avirulence races (Wellings 1986). Different reactions of primary and second leaves were also observed in some wheat and barley cultivars to stripe rust in the studies of Chen and Line (1992a, b, c, 1999, 2003). Because of the difference, inoculation of seedlings is typically done at the two-leaf stage in Chen's laboratory.

Plant Condition. As the stripe rust fungus is an obligate biotrophic parasite, plant conditions are important for conducting stripe rust test for reliable results. All practices and conditions to grow healthy plants before inoculation and continually maintain such conditions for normally growing plants are needed for determination of resistance. Any abiotic stresses, such as nutrient deficiency or excessive nutrients, drought, too hot or too cold, inadequate light and toxic soil, and other biotic stresses, such as infection by soilborne pathogens and foliar pathogens like powdery mildew and viruses and infestation of insets, should be prevented in greenhouse tests. Such abiotic and biotic stresses other than stripe rust in fields affect stripe rust development. Field data under such conditions should be carefully analyzed and logically interpreted.

5.1.3.2.2 Pathogen Factors

From the pathogen side, different races or isolates, and the age of spores may affect plant reactions. As discussed above, many *Pst* races identified after 2000 tend to produce intermediate ITs on wheat differential cultivars that used to have either high or low ITs when tested with races before 2000 in the United States.

The effect of heterozygosity of pathogen was observed in resistance to stem rust and other rusts (Roelfs and Bushnell 1985). Although the effect of heterozygosity of the stripe rust pathogen did not well investigate since the sexual life cycle of stripe rust pathogen was just found recently (Jin et al. 2010; Wang et al. 2015), heterozygosity may affect resistance. In our recent studies, we found that many avirulence genes in *Pst* isolates were heterozygous through virulence analysis of progeny isolates produced through sexual reproduction on barberry plants (Chen et al. 2012c; Yuan et al. 2016). For example, *Avr44* in PSTv-4 (isolate 12-368) corresponding to resistance gene *Yr44* was heterozygous. PSTv-4 produced IT 3 on *Yr44*, while some of its selfed progeny isolates produced IT 2 or 8, suggesting that the homozygosity of the locus produce different ITs from its heterozygosity. The similar results were also observed on resistance genes *Yr7*, *Yr43* and *YrExp2* (Y Lei, MN Wang and XM Chen, unpulished data).

Spore viability and amount affect rust severity and also can affect IT data. For reliable results, fresh spores or those kept at 4 °C under dry conditions for less than two months should be used in testing. Plants inoculated with spores kept in liquid nitrogen or freezer tends to sporulate a couple of days later than those inoculated with fresh spores.

5.1.3.2.3 Environmental Factors

Different genes express in different enviroments. The resistance phenomena displays as a results of interactions between resistance genes and pathogen genes under different effects of various environmental factors. As the stripe rust fungus is more sensitive to many environmental factors than other fungi as discussed in Chap. 4, environmental factors strongly affect pathogen infection efficiency and plant reactions. Of various environment factors, temperature, moisture and light are important for studying resistance to stripe rust.

Temperature. Temperature affects spore germination, germ tube growth and penetration, hyphal growth with plant tissue and sporulation as discussed in Chap. 4. Temperature particularly influences plant response to infection, resulting in different reactions under different temperatures, especially for temperature-sensitive resistance. In general, at a higher temperature, plants tend to have a higher level of resistance or lower IT and severity. The resistant ITs produced by cultivars carrying *Yr6* were intermediate at 10 °C and low at 18 °C (Wellings 1986). Qamar et al. (2008) found that ITs on 23 Pakistan cultivars decreased when tested at 15–20 °C compared to at 12–15 °C. For example, cultivar Bowie had IT 3 (susceptible in a 0–4 scale) at the low temperature profile to chlorosis to necrosis (CN) without sporulation at the high temperature profile. Similarly, ITs of cultivar QAL 2000 decreased from IT 3+ to CN and of Stylet from IT 3 to CN1 (with limited sporulation). More examples of temperature effects can be seen in the studies of HTAP resistance (Qayoum and Line 1985; Milus and Line 1986a, b; Chen and Line 1995a, b; Lin and Chen 2007, 2009; Ren et al. 2012a; Lu et al. 2014; Hou et al. 2015; Zhou et al. 2014c).

Moisture. High moisture and adequate duration are essential for infection by the stripe rust fungus. Moisture strongly affects rust severity as it determines infection efficiency and the number of infection cycles under field conditions. For evaluation of plants for resistance under controlled conditions in a greenhouse or growth chamber, in which only one disease cycle occurs, moisture mainly affects severity through determining infection efficiency, but not much on ITs. After the dew period, it is necessary to maintain an adequate level of moisture in soil for normal growth of the plants to ensure good results for plant reactions to pathogen infection.

Light. Light, both intensity and duration, influences ITs and sporulation. In general, high light intensity and adequate length of light period increase sporulation. Under low light intensity and short light duration, plants may become yellow and sporulation may become less, decreasing ITs and severity. However, some stripe rust resistance genes may respond to light differently. For example, wheat lines possessing *YrA* (now the combination of *Yr73* and *Yr74*) produced high ITs under low light intensities when tested with avirulent races (Wellings 1986; Wellings et al. 1988).

5.1.3.2.4 Human Factors

In addition to the above discussed natural factors related to the plant, pathogen and environment, other factors related to techniques, research procedures, and knowledge can influence studies on stripe rust resistance.

The different phenotypic data can be used to study resistance, such as infection type, severity and AUDPC. A continuous phenotype distribution in a population does not necessarily indicate multiple loci controlling the resistance because a resistance gene with low heritability can present a continuous IT or AUDPC distribution in the segregating population. Therefore, the genetic basis of the resistance cannot be determined solely from the observed phenotypic distribution, but must be established by genetic analyses to determine the relationship between the genotypes and phenotypes.

Some analysis results were affected by the researchers' understanding of the resistance, i.e. how they defined the criteria of resistance. Line et al. (1970) considered IT 2 or greater on the 0–4 scale of Gassner and Straib (1932) as susceptible, although many European workers agreed with this, while British scientists regarded IT 3 or greater as susceptible (Johnson et al. 1972). Differences among the interpretation of ITs can influence the understanding of genetic analysis results. Considering the numerous factors influencing resistance, Fuchs (1960) pointed out that the infection type of stripe rust was "more sharply to environmental changes than all other rust species, demands great effort and critical self-restraint".

The 0–4 scale of IT has limitation in describing reactions between IT 0 and IT 4 and especially problematic in the middle. Late, the 0–9 scale, as exampled in Fig. 2.1 in Chap. 2, was adapted by Line and associates (Line and Qayoum 1992; Chen and Line 1992a, b, c) and others. This scale allows use of IT data for QTL analysis. Categorization of ITs into resistant and susceptible or with additional classes of intermediate or moderately resistant and moderately susceptible is still an issue. However, providing ITs when describing reactions by categories is always a good idea to be specific.

5.1.4 Genes of Resistance to Stripe Rust in Wheat

Before the 1960s, studies on the genetics of resistance to stripe rust were reported, but there was no a universal system to distinguish the different resistance genes. For efficient and accurate exchange of information in the rust community and other wheat genetic research groups, two attempts were proposed to develop a model of the genic system for genes conferring resistance to stripe rust. Based on the reactions of 17 wheat varieties to 15 *Pst* races, Zadoks (1961) postulated genes and suggested an alphabetical series of symbols for "genes" determining resistance. The "B" gene symbol was assigned for resistance in Heines Kolben and Heines Peko; "U" for Merlin and Heines VII; "X" for Flamingo; "L" for Chinese 166; "M" for Vilmorin 23, Staring, Nord Desprez and Cappelle Desprez; and "T" for Hope × Timstein. These "genes" were never determined by genetic studies, but genes in some of the wheat varieties have been studied genetically and named using a different system.

Almost simultaneously, Lupton and Macer (1962) proposed the use symbol "*Yr*", standing for Yellow rust, for loci at which alleles determining stripe rust resistance are conditioned. Based on an investigation of intercrosses between seven wheat varieties and using four European *Pst* races, they determined that four independently segregating loci, designated *Yr1* to *Yr4*, were involved. This system is consistent with the proposals of the International Committee on Genic Symbols and Nomenclature, and was quickly and widely adopted by rust researchers. Any qualitative or quantitative resistance locus genetically different from others can be designated a permanent numerical name. Although several of the previously designated *Yr* genes with numerical symbols have never been genetically determined, currently a gene must be shown at a locus clearly different from permanently named genes to have a permanent *Yr* symbol. A tentative name can be used for a resistance locus whose relationship with other genes has not been determined. The following paragraphs will discuss the permanent *Yr* genes, tentative *Yr* genes and quantitative resistance loci.

5.1.4.1 Permanently Designated Genes for Resistance to Wheat Stripe Rust (*Yr1* – *Yr78*)

A total of 78 *Yr* genes have been reported for resistance to stripe rust at the writing time, including 24 for APR or HTAP resistance and 54 for ASR. The chromosomal locations of these genes, and their linked molecular markers and genes are presented in Table 5.2.

Yr1 confers high level all-stage resistance with typically IT 1 to avirulent races. The gene is located on chromosome 2AL, bin 2AL1-0.85-1.00. *Yr1* was identified from winter wheat genotype Chinese 166 with another name 'L' (Zadoks 1961; Lupton and Macer 1962). Its near isogenic lines (NILs) have been developed in several wheat backgrounds, such as Avocet S*6/Chinese 166 (AvSYr1NIL), Aroona*5/*Yr1*, Kite*6/*Yr1*, Warigal*6/*Yr1* and Harrier*2/*Yr1* (McIntosh et al. 1995). Except Chinese 166, cultivars carrying *Yr1* include Chinese wheat cultivars Fengchan 3, Kangyin 655, Bima 1, Xibei 54, Beijing 8, Bainong 3217, Jinmai 33, Yumai 18 (Wan et al. 2007); and other cultivars from other countries such as Dalee, Durin, E2025, E7700, Heines 110, Maris Ranger, Maris Templar, Corin, Odra and Ritmo (Johnson 1981a, b; Chen 2005; McIntosh et al. 2013). Virulence to *Yr1* has been found in *Pst* populations in Australia, Canada, China, Chile, Ecuador, Ethiopia, Hungary, Italy, Kenya, Mexico, Nepal, Pakistan, Turkey, Uzbekistan and the US (Holtz et al. 2013; Sharma-Poudyal et al. 2013; Wan and Chen 2014; Wan et al. 2017). Virulence frequency of *Yr1* particularly was high in East Asia where Chinese 166 originated (Stubbs 1985; Wan et al. 2004; Zhan et al. 2016). Some isolates of *Psh* were virulent to *Yr1* (Chen et al. 1995c; McIntosh et al. 1995; Wan and Chen 2014). Wellings et al. (2009) reported that virulence for *Yr1* was common in China and central Asia, but was rare in the Middle East, and absent in west Asia and east Africa. However, the *Yr1* virulence was found at high frequency in Kenya (50%) (Sharma-Poudyal et al. 2013), Ethiopia (74%) (Wan et al. 2017) and Italy (86%) (Wan and Chen unpublished data). Chinese 166 is used to identify *Pst* races in differential sets of world, Europe and India, while the *Yr1* NIL (AvSYr1NIL) is used

Table 5.2 Permanently named *Yr* genes for resistance to stripe rust in wheat

Gene	Syn.	Type[a]	Source	Germplasm	Chrom.	Chrom. bin	Marker 1 (cM)	Marker 2 (cM)	Linked gene (cM)	References
Yr1	*L*	ASR	Common wheat	AvSYr1NIL, Chinese 166, Corin, Dalee, Durin, E2025, E7700, E8594, Feng-Kang 13, Heines 110, Maris Ranger, maris Templar, Odra, Ritmo	2AL	2AL1-0.85-1.00	*Xstm673acag* (1.1)	ND[b]	*Sr48* (16.5)	Bansal et al. (2009), Lupton and Macer (1962), and Zadoks (1961)
Yr2	*U*	ASR	Common wheat	Derius, Flevina, Hana, HD2329,Heines VII, Kalyansona, Laketch, Leda, Manella, Merlin, Odra, PBW54, PBW120, Slavia, Soissonais Desprez, Yamhill, WG377, WH147, WL711, WL1562	7B	ND	*Xwmc364* (5.6)	ND	ND	Lupton and Macer (1962), Lin et al. (2005), and Chen et al. (1996)
Yr3	*M*	ASR	Common wheat	Enkoy, Minister; Vilmorin 23, Staring	ND	ND	ND	ND	ND	Lupton and Macer (1962), Stubbs (1985), and Zadoks (1961)
Yr3a	No	ASR	Common wheat	Druchamp, Stephens, Nord Desprez	1B	ND	ND	ND	ND	Chen et al. (1996)
Yr3b	No	ASR	Common wheat	Hybrid 46	ND	ND	ND	ND	ND	Lupton and Macer (1962), Chen and Line (1993a), and Chen et al. (1996)

Yr3c	No	ASR	Common wheat	Minister, Maris Beacon	1B	ND	ND	ND	ND	Lupton and Macer (1962) and Chen et al. (1996)
Yr4	*YrRub*	ASR	Common wheat	Avalon, Bolac, EMU S, Kenya Kubangu, Nesser, Rubric AUS33333, Hybrid 46,	3BS	3BS3-0.87-1.00	*Xcfb3530_150bp* (2.4)	ND	ND	Lupton and Macer (1962) and Bansal et al. (2010)
Yr4a	No	ASR	Common wheat	Vilmorin 23, Yamhill, Cappelle-Desprez	6B	ND	ND	ND	ND	Lupton and Macer (1962) and Chen et al. (1996)
Yr4b	No	ASR	Common wheat	Hybrid 46	6B	ND	ND	ND	ND	Lupton and Macer (1962) and Chen et al. (1996)
Yr5	No	ASR	*Triticum spelta* var. album	AvSYr5NIL, By33, E5557, E8510, New Dirkwin, Blanca Grande 515, Clearwhite 515, Patwin 515, Summit 515, Seahawk	2BL	2BL3-C-0.35	*XSTS7/8* (0.3)	*Xbarc349* (0.4), *XS19M93* (co-segregation), *XS23M41* (0.7)	*YrSP* (12.6), *Yr43* (65.5), *Yr44* (41.5), *Yr53* (35.6), presumably allelic to *Yr5* by Zhang et al. (2009)	Macer (1966), Law (1976), Chen et al. (2003), Murphy (2009), Zhang et al. (2009); Feng et al. (2015b), Xu et al. (2013), Cheng and Chen (2010) and Sui et al. (2009)

(continued)

Table 5.2 (continued)

Gene	Syn.	Type[a]	Source	Germplasm	Chrom.	Chrom. bin	Marker 1 (cM)	Marker 2 (cM)	Linked gene (cM)	References
Yr6	*B*	ASR	Common wheat	AvSYr6NIL, Austerlitz, Fielder, Heines Klben, Koga II, Maris Dove, Recital, Takari	7BS	ND	ND	ND	ND	El-Bedewy and Robbelen (1982), Macer (1966) and Zadoks (1961)
Yr7	*T*	ASR	Common wheat	HopexTimstein, AvSYr7NIL, Taichung29*6/Lee, Brock, Lee, Nudif TP257, PBW12, Prinqual, Renard, Talent, Tango, Tommy, WL2265	2BL	ND	*Xgwm526_212bp (5.3)*	ND	*YrSP* (18.6)	Macer (1966), Zadoks (1961), Feng et al. (2015b), Zhang et al. (2009), and Yao et al. (2006)
Yr8	No	ASR	*Aegilops comosa*	AvSYr8NIL, Compair, Chromosome 2D-2M translocations in Hobbit Sib and Maris Widgeon, 2A=2A-2M=T2AS-2M#1L.2M#1S	2A or 2D	2AS.2ML-2MS or 2DS.2ML-2MS	Unpublished	Unpublished	*Sr34*	Reiley and Johnson (1968), Friebe et al. (1996), and McIntosh et al. (1995)
Yr9	No	ASR	*Secale cereale*	AvsYr9NIL, Clement, Petkus, Kavkaz, Lovrin 10, Lovrin 13, Riebesel 47/51	1BS	1B = 1BL.1RS	*Xgwm582* (3.7)	*Xwgp4*, *Xwgp7*, *Xwgp8*, *Xwgp9* (0.0)	*Sr31*, *Lr26*	Macer (1975), Zeller (1973), Shi et al. (2001), Weng et al. (2005) and McIntosh et al. 1995

Yr10	*YrVav*	ASR	Common wheat; *T. vavilovii*; *T. spelta* 415	AvSYr10NIL, Moro, PI178383, QLD709=Janz*2/*T. vavilovii*, *T. spelta* 415, *T. vavilovii* AUS22498	1BS	Telomere of 1BS	*E5_1100* (0.0)	*Xpsp3000* (1.2)	Brown glume (2.03)	Macer (1975), Metzger and Silbaugh (1970), and Liu et al. (2014a)
Yr11	*R11*	APR	Common wheat	Joss Cambier, Heines VII	ND	ND	ND	ND	ND	Johnson and Taylor (1972) and Priestley (1978)
Yr12	*R12*	APR	Common wheat	Fleurus, Frontier, Pride, Mega	ND	ND	ND	ND	ND	Priestley (1978), Priestley and Byford (1979) and Priestley et al. (1982)
Yr13	*R13*	APR	Common wheat	Maris Huntsman, Hustler	ND	ND	ND	ND	ND	Priestley (1978)
Yr14	*R14*	APR	Common wheat	Kador, Hobbbit, Maris Bilbo, Score, Wembley	ND	ND	ND	ND	ND	Priestley (1978) and Priestley and Byford (1979)
Yr15	No	ASR	*T. dicoccoides accession G25*	AvSYr15NIL, Agrestis, Boston, Cortez, Legron, New Dirkwin, Blanca Grande 515, Patwin 515, Summit 515	1BS	ND	*Xwgp34*/ *Xgwm413*/ *Xbrac8* (0.0)	*SC790* (1.0), *SC2700* (0.1), *Xgwm33* (4.5), *UBC199200* (5.6), *Xwmc128*/ *Xgwm273*/ *Xgwm582* (0.4)	*Yr24*, *Yr64* (21.7), *Yr65* (14.1)	Gerechter-Amitai et al. (1989), Mandoulakani et al. (2015), Murphy et al. (2009) and Cheng et al. (2014)

(continued)

Table 5.2 (continued)

Gene	Syn.	Type[a]	Source	Germplasm	Chrom.	Chrom. bin	Marker 1 (cM)	Marker 2 (cM)	Linked gene (cM)	References
Yr16	No	APR	Common wheat	Bersee, Cappelle-Desprez	2DS	ND	ND	*Xgwm102-Xwmc18-Xwmc245-wPt-734322-wPt-664520*	ND	Worland and Law (1986) and Agenbag et al. (2012)
Yr17	No	ASR	*T. ventricosum*	AvSYr17NIL, Apache, Arche, Balthazar, Bill, Brigadier, Caphorn, Clever, Clarus, Cordial, Coraier, Eureka, Hussar, Kris, Pernel, Rapsodia, Renan, Rheia, RL6081, Yamhill, Madsen, Rendezvous, Jagger	2AS	2NS.2AS	*SCY-15*, *Xpsr150-2Nv*, *SC-372*, *SC-385*	ND	*Lr37*, *Sr38*	Bariana and McIntosh (1993), Asakura et al. (1997) and Helguera et al. (2003)
Yr18	No	APR	Common wheat	AvSYr18NIL, Jupateco 73R, Lerma Rojo 64, Nacazari 76, Tesia F 79, Tonichi S 81, Wheaton, Chinese Spring, Chuanmai 18, Frontana, Opata, Anza, Chris, Mentana, Ardito, Forno, Pegaso, Bezostaja, Kavkaz	7DS	ND	*Xgwm120* (0.9), *Xgwm295* (0.7)	*caIND11* (0.0), *caSNP12*, *caSNP4*, *cssfr1* (0.0), *cssfr2* (0.0), *cssfr6* (0.0)	*Lr34*, *Sr57*, *Pm38*, *Ltn*	McIntosh (1992), Lagudah et al. (2009) and Dakouri et al. (2010)
Yr19	*Yrcom*	ASR	Common wheat	Compair	5B	ND	ND	ND	ND	Chen et al. (1995c)
Yr20	*YrFie*	ASR	Common wheat	Fielder	6D	ND	ND	ND	ND	Chen et al. (1995c)

Yr21	*YrLem*	ASR	Common wheat	Lemhi	1B	ND	ND	ND	ND	Chen et al. (1995c) and Pahalawatta and Chen (2005)
Yr22	*YrLe1*	ASR	Common wheat	Lee	4D	ND	ND	ND	ND	Chen et al. (1995c)
Yr23	*YrLe2*	ASR	Common wheat	Lee	6D	ND	ND	ND	ND	Chen et al. (1995c)
Yr24	*YrCH42*, *Yr26*	ASR	*T. turgidum var. durum accession K733*	AvSYr24NIL, Chuanmai 42	1BS	ND	*Xbarc187* (2.3)	*Xgwm498*(1.6)	*Yr15*, *Yr64*, *Yr65*	Li et al. (2006a) and Cheng and Chen (2014)
Yr25	No	ASR	*Common wheat*	Carina, Carstens V, Heines Peko, Heine VII, Hugenout, Heines VII, Heines Peko, Reichersberg 42, TP1295, TP981, Tugela, Spaldings Prolific	1D	ND	ND	ND	ND	Calonnec and Johnson (1998) and Eriksen et al. (2004)
Yr26	*Yr24*	ASR	*T. turgidum* Gamma 80-1, *Haynaldia villosa* L.	AvSYr26NIL, Nannong 9918, Nei 2938, Nei 4221, Neimai 9, 92R137, Yangmai 5	1BL	C-1BL-6-0.32	*CON6*, *CON7*, *CON8*, *CON9*, *CON10*, *CON11* (0.0)	*Xgwm11* (1.9), *Xwe173* (1.4), *Xbarc181* (6.7)	*Yr24*	Ma et al. (2001), Wang et al. (2008) and Zhang et al. (2013)
Yr27	*YrSk*, *YrBjy*	ASR	Common wheat	AvSYr27NIL, Ciano 79, Inquilab 91, Kauz, McMurachy,Opata 85, PWB343, Selkirk, Webster, Attila, Kauz, Nacozari, Buk Buk, Crow	2BS	ND	*Xbcd152*	*Xcdo405*	*Yr31* (recombination value 0.148), *Lr23* (0.131), *Lr13*	McDonald et al. (2004) and Singh et al. (2003)
Yr28	No	ASR	*Ae. tauschii* (accession W-219)	Altar 84/*Ae. tauschii* W-219, Synthetic	4DS	ND	*Xmwg634* (1.0)	*Xbcd265* (~15)	ND	Singh et al. (1998)

(continued)

Table 5.2 (continued)

Gene	Syn.	Type[a]	Source	Germplasm	Chrom.	Chrom. bin	Marker 1 (cM)	Marker 2 (cM)	Linked gene (cM)	References
Yr29	No	APR	Common wheat	Lalbahadur(Parula 1B), Attila, Pavon F76, Quaiu 3	1BL	ND	*XBac24prot* (9.5), *Xwmc44* (2.6)	*XBac17R* (2.9), *Xgwm44* (3.6), *XTG818* (2.1), *Xgwm259* (2.9), *Xgwm140* (9.4)	*Lr46*, *LTN*	William et al. (2003) and Rosewarne et al. (2006)
Yr30	No	APR	Common wheat	Opata 85, Parula, Inia 66, Pavon F76, Quaiu 3	3BS	ND	ND	ND	*Sr2*	McIntosh et al. (2013)
Yr31	No	ASR	Common wheat	Pastor	2BS	ND	ND	ND	*Yr27* (recombination value 0.148), *Lr23* (0.295)	Singh et al. (2003)
Yr32	*YrCv*, *YrCV*	ASR	Common wheat	AvSYr32NIL,Cook*6/ Carstens V, Anouska, Caribo, Consort, Cyrano, Danis, Deben, Hereward, Kraka, Okapi, Oxbow, Senat, Solist, Stakado, Vivant, Wasmo, Carstens V	2AL	ND	*Xwmc198* (2.0); *Xgwm95* (39)	*Xwmc181* (11.0)	*Yr1* (35.0 distal)	Eriksen et al. (2004)
Yr33	No	ASR	Common wheat	Batavia, EGA Gregory, Strezecki	7DL	ND	*Xwgm11*	*Xgwm437*	ND	Zahravi et al. (2003)
Yr34	*YrWA*	ASR	Common wheat	AUS22857, WAWHT2046=AUS91389	5AL	ND	ND	ND	Gene *B1*	Bariana et al. (2006)
Yr35	*YrS8*	ASR	*T. dicoccoides* 479,	98M71=AUS 91388=*T. dicoccoides* 479/7*CS	6BS	ND	*Xgwm191* (18.9)	*Xcfd1* (4.1)	*Lr53* (3.0)	Dadkhodaie et al. (2011) and Marias et al. (2005b)

Yr36	ND	HTAP	*T. dicoccoides* (accession FA-15)	PI 638740, PI 638741, Glupro, UC1041+Yr36, RSL65, Burnside, Glencross, Lilian, Somerset, Farnum	6BS	ND	*Xucw69/ Xbrac101*	*Xucw129, Xucw148*	A high protein gene	Uauy et al. (2005) and Fu et al. (2009)
Yr37	ND	ASR	*Ae. kotschyi* 617 (tetraploid, genome UUSS) Boiss	Line S14, Line 8078	2DL	ND	ND	ND	*Lr54*	Marais et al. (2005a)
Yr38	*YrS12*	ASR	*Ae. sharonensis* (accession 174)	Line 0352-4	6A	6AL-6Lsh.6Ssh	ND	ND	*Lr56*	Marias et al. (2006, 2010)
Yr39	No	HTAP	Common wheat	Alpowa	7BL	No	*Xgwm131* (6.7)	*Xwgp45* (0.8)	*Yr52*, *Yr59*	Lin and Chen (2007)
Yr40	No	ASR	*Ae. geniculata*	TA5602, TA5603	5DS	5DL.5DS-T5MSG	*XLr57/ Yr40-MAS-CAPS16*	ND	*Lr57*	Kuraparthy et al. (2007)
Yr41	*YrCN19*	ASR	Common wheat	AIM5, AIM6, Chuannong 19	2BS	ND	*Xgwm410_391bp* (0.3)	ND	Ciano79 *Yr27* (17.4), Vilmorin23 *Yr3a* (20.2), Pastor *Yr31* (6.2), *YrSP* (12.5)	Luo et al. (2005, 2008)
Yr42	No	ASR	*Ae. neglecta*	Line 03M119-71A	6A	6A=6AL-6LAen.6SAen	ND	ND	*Lr62*	Marais et al. (2009)

(continued)

Table 5.2 (continued)

Gene	Syn.	Type[a]	Source	Germplasm	Chrom.	Chrom. bin	Marker 1 (cM)	Marker 2 (cM)	Linked gene (cM)	References
Yr43	No	ASR	Common wheat	IDO337s=PI 591045, Lolo,	2BL	ND	*Xwgp110* (4.4)	*Xwgp* (5.5)	*Yr44* (13.1), *Yr53* (16.9), *Yr5* (65.5), *YrSP* (15.3)	Cheng and Chen (2010), Xu et al. (2013), and Feng et al. (2015b)
Yr44	*YrZak*	ASR	Common wheat	Zak=PI 607839	2BL	ND	*XpWB5/N1R1* (9.4)	*Xgwp100* (3.9)	*Yr43* (13.1), *Yr5* (41.5)	Sui et al. (2009) and Cheng and Chen (2010)
Yr45	No	ASR	Common wheat	PI 181434, PI 660056	3DL	ND	*Xgwp118* (4.8)	*Xgwp115* (5.8)	ND	Li et al. 2011; Wang et al. 2012
Yr46	No	APR	Common wheat	PI 250413, RL6077 = Thatcher*6/PI 250413	4DL	distal to 0.56	*Xgwm165/ Xgwm192* (0.4)	ND	*Lr67*, *Sr55*	Herrera-Foessel et al. (2011)
Yr47	No	ASR	Common wheat	AUS28183=V336, AUS28187	5BS	5BS-0.71-0.81	*Xgwm234* (10-15)	*Xcfb309* (8-12)	*Lr52* (3-6)	Bansal et al. (2011)
Yr48	No	APR	Synthetic wheat	PI610750=Synthetic 205 (Croc 1/*Ae. tauschii*)/Kauz, UC110/PI610750 RIL4 GSTR 13504 and RIL167 GSTR 13618	5AL	5AL23-0.87-1.00	*Vrn-2A*, *BE495011*, *Xcfa2149*, *Xgpw2181a*, *Xwmc74*, *Xwmc410* (0.0)	*Xwmc727* (4.4), *Xwms291* (0.3)	ND	Lowe et al. (2011)
Yr49	No	APR	Common wheat	Avocet S*3/Chuanmai 18, Chuanmai 18	3DS	3DS-6-0.55-01.00	*Xgps3721* (0.0)	*Xgwm161* (1.0)	ND	McIntosh communication; Spielmeyer et al. unpublished
Yr50	No	ASR	*Thinopyrum intermedium*	CH223, TAI7047	4BL	ND	*Xbarc1096* (8.0)	*Xwmc47* (7.2)	*Yr62* (27.1)	Liu et al. (2013a)
Yr51	No	ASR	Common wheat	Line 5515, AUS 91456, AUS27858	4AL	4AL4-0.80-1.00	*sun104* (2.5)	*Xowm45F3R3* (1.2)	ND	Randhawa et al. (2014)

Yr52	No	HTAP	Common wheat	PI 182527, PI 660057	7BL	7BL3-0.86-1.00	*Xwgp5258* (1.2)	*Xbrac182* (1.1)	*Yr59* (5.4), *Yr39* (31.2), *YrZh84* (12.2)	Ren et al. (2012a)
Yr53	No	ASR	*T. turgidum*	PI 480148, Avocet S/ PI480148 F6 Line128-10; PI 679598	2BL	2BL3-C-0.35	*Xwmc441* (5.6)	*XLRRrev/ NLRRrev_350* (2.7)	linked with *Yr5* (35.6), *YrSP* (13.2), *Yr43*, *Yr44*	Xu et al. (2013), Feng et al. (2015b) and Wang and Chen (unpublished)
Yr54	No	APR	Common wheat	Yr54 RIL GID6032209, Yr54 RIL GID6032334, Quaiu 3	2DL	ND	*Xgwm301* (0.4)	ND	*QYr.tam-2D* in Alcedo	Basnet et al. (2014b)
Yr55	No	APR	Common wheat	Frelon AUS38882	2DL	ND	*Xmag4089* (11.4)	*Xmag3385* (8.4)	ND	McIntosh (communication) and Bariana (unpublished)
Yr56	*YrWol*	ASR	Durum wheat	AUS 91575, Wollaroi = AUS99174,	2AS	ND	*Xsun167* (5.7)	*Xsun168* (7.6), *XwPt4197*, *XwPt9104*	*QYr.sun-2A*	Bansal and Bariana (2014)
Yr57	No	ASR	Common wheat	AUS 91463, AUS 27858	3BS	3BS8-0.78-1.00	*Xgwm389* (2.0)	*XBS00062676* (2.3)	*Yr4* (5.2)	Randhawa et al. (2015)
Yr58	*YrW195*, *QYr. sun-3BS*	HTAP	Common wheat	W195	3BS	ND	*XSun476*, *Xsun533* (0.6)	*X3023704* (1.2)	ND	Chhetri et al. (2016a)
Yr59	No	HTAP	Common wheat	PI 178759, PI 660061	7BL	7BL3-0.86-1.00	*Xwgp52175*	*Xbrac32*	*Yr39* (41.1), *Yr52* (5.4), *YrZh84* (12.1)	Zhou et al. (2014c)
Yr60	No	ASR	Common wheat	Almop	4AL	ND	*Xwmc776* (0.0)	*Xwmc219/ Xwmc313* (0.6)	*Yr51* (10.1)	Herrera-Foessel et al. (2015)
Yr61	No	ASR	Common wheat	Pindong 34	7AS	ND	*STS5467* (1.9)	*XSTS5765* (3.9)	No.	Zhou et al. (2014a)

(continued)

Table 5.2 (continued)

Gene	Syn.	Type[a]	Source	Germplasm	Chrom.	Chrom. bin	Marker 1 (cM)	Marker 2 (cM)	Linked gene (cM)	References
Yr62	No	HTAP	Common wheat	PI 192252, PI 660060	4BL	4BL-0.58-0.70	*Xgwm192_222bp* (2.0)	*Xgwm251_133bp* (0.6)	*Yr50* (27.1)	Lu et al. (2014)
Yr63	No	ASR	Common wheat	AUS 27955	7BS	7BS1-0.27-1.00	*XIWB33120* (0.63)	*XIWB52844* (1.5)	ND	McIntosh communication and Bansal (unpublished)
Yr64	No	ASR	Durum wheat	PI 331260, PI 660064	1BS	1BS9-0.84-1.06	*Xgwm413* (3.5)	*Xgdm33* (2.0)	*Yr65*, *Yr15*, *YrH52*	Cheng et al. (2014)
Yr65	No	ASR	Durum wheat	PI 480016; PI 679621	1BS	1BS10-C-0.50	*Xgwm18* (1.2)	*Xgwm11* (2.1)	*Yr64*	Cheng et al. (2014) and Wang and Chen (unpublished data)
Yr66	*YrVL1*	ASR	Common wheat	AGG91584WHEA = MSP4543.1, VL892 = AGG91586WHEA	3DS	3DS-6-0.55-01.00	*XIWB47165* (3.1)	*XIWB18087/ XIWB56281* (2.9)	ND	McIntosh (communication)
Yr67	*YrVL2*	ASR	Common wheat	C591, C306, Zhongzhi 1, AGG91585WHEA = MSP4543.4, VL892 = AGG91586WHEA	7BL	ND	*Xmag1714* (1.2), *Xbarc182* (0.4), *XIWB62475/ XIWB37096* (1.1), *XIWB71995* (0.6)	ND	*Yr39*, *Yr6*, *Yr52*, *Yr59*	Xu et al. (2014), McIntosh (communication)
Yr68	No	APR	Common wheat	AGG91587WHEA	4BL	4BL1-0.86-1.00	*XIWA4640* (0.0)	*XIWB74301* (0.5), *XIWB28394* (0.5)	ND	McIntosh (communication) and Spielmeyer (unpublished)

Yr69	*YrCH86*	ASR	*Thinopyrum ponticum*	CH7086	2AS	2AS5-0.78-1.00	*X2AS33* (1.9, *Xgwm636* (13.9)	*Xmag3807* (3.1); *Xbarc124* (11.3)	*Yr17*	Hou et al. (2016)
Yr70	*YrUmb*	ASR	*Ae. umbellulata* Pau 3732	IL 393-4, T. durum cv. WH890/*Ae. umbellutata* Pau3732//CS Ph/3/2*WL711	5DS	ND	*Xgwm190* (7.6)	ND	*Yr40*, *Lr76*	McIntosh (communication)
Yr71	*YrSA3*	APR	Common wheat	Sunco	3DL	ND	*Xgwm114b/ KASP_8306/ KASP_12707/ KASP_16434*	ND	*Lr24/Sr24* (1.8)	Bariana (unpublished) and McIntosh (communication)
Yr72	*YrAW4*	ASR	Common wheat	AUS27506, AUS27894	2BL	2BL5-0.59-0.89	*Xsun481* (1.8)	*XIWB12294* (1.5)	*Yr7*, *Yr43*, *Yr44*, *Yr53*	Chhetri (unpublished) and McIntosh (communication)
Yr73	*YrA*	ASR	Common wheat	Teal, Avocet R	3DL	3D-3	1160719|F|0	114046|F|0	ND	Dracatos et al. (2016)
Yr74	*YrA*	ASR	Common wheat	Avocet S, Avocet R	5BL	ND	1091695|F|0	00042510|F|0	ND	Dracatos et al. (2016)
Yr75	No	APR	Common wheat	Axe	7AL	ND	*XIWB34640*	*Xcfa2040*	ND	McIntosh (communication)
Yr76	*YrTye*	ASR	Durum wheat	AvS*4/Tyee, Tyee, ARS-Amber, ARS-Chrystal, Cara, Chukar, Hyak, PI 679624	3AS	3AS4-0.45-1.00	*Xwmc11* (2.6)	*Xwmc532* (3.4)	ND	Xiang et al. (2016)
Yr77	*QYr. ucw-6D*	APR	Common wheat	PI 322118	6DS	ND	*XIWA167*; *Xcfd188* (1.1 to 3.9)	*Xbarc54* (15.2 to 16.8)	ND	Dong et al. (2017)
Yr78	*QYr. ucw-6B,*	APR	Common wheat	PI 519805	6BS	ND	*XIWA7257* (0.6)	*XIWA4408/ Xwmc737* (3.9)	*Yr35* (15.0)	Dong et al. (2017)

[a]Type of resistance: *ASR* all-stage resistance, *APR* adult-plant resistance, *HTAP* high-temperature adult-plant resistance
[b]*ND* no data

as a differential in the North America and China (Chen 2005; Nagarajan et al. 1984; Stubbs 1985; Holtz et al. 2013; Wan and Chen 2014). SSR marker *Xstm673acag* is closely linked with *Yr1* at 1.1 cM proximal and can be used to detect *Yr1* (Bansal et al. 2009). A 120-bp fragment amplified by *Xstm673acag* is present in wheat genotypes carrying *Yr1*, except false positive in Avocet S without *Yr1* but having the 120-bp product. EST-SSR marker *Xbu099658* is closely linked with *Yr1* that generates a 206-bp fragment in *Yr1* carrying lines and no false positive products in Avocet S (Hasancebi et al. 2014).

Yr2 confers intermediate to high level resistance to stripe rust, and it was originally designated to a gene in cultivars Heines VII and Soissonais (Lupton and Macer 1962). Labrum (1980) determined that *Yr2* in cultivar Heines Peko was located on chromosome 7B. Other reference genetic stocks of *Yr2* include B1-B5 Lemhi 53*5/Soissonais (Griffey and Allan 1988); Taichung 29*6/Heines VII (Lin et al. 2005); Kalyansona carrying only *Yr2* (Singh and Johnson 1988); and TP981 and TP1295 that carrying *Yr2* and *Yr25* (Chilosi and Johnson 1990). *Yr2* has another name 'U' (Zodaks 1961). The gene is present in a wide range of wheat cultivars, such as Maris Beacon, Derius, Flevina, Hana, HD2329, Laketch, Leda, Manella, Merlin, Odra, PBW54, PBW120, Slavia, WG377, WH147, WL171, WL1562 and Yamhill (Johnson 1981a, b; Chen and Line 1992c; McIntosh et al. 2013). Virulence for *Yr2* was high (93%, n=235, Sharma-Poudyal et al. 2013) in international *Pst* collections and distributed in Australia, Canada, China, Chile, Ecuador, Ethiopia, Hungary, Kenya, Mexico, Nepal, Pakistan, Turkey, Uzbekistan and the USA (Wan et al. 2004; Sharma-Poudyal et al. 2013; Wan and Chen 2014; Wan et al. 2016, 2017). Cultivars carrying *Yr2* is used as differentials in both the previous European and the North American sets of differentials (Johnson et al. 1972; Line and Qayoum 1992; Chen 2005). SSR marker *Xwmc364* (207-bp) is linked with *Yr2* at 5.6 cM (Lin et al. 2005).

Yr3 and *Yr4* were proposed having multiple alleles. Lupton and Macer (1962) analyzed the inheritance of seven wheat cultivars for resistance to stripe rust and reported Cappelle Desprez having *Yr3a* and *Yr4a*, Hybrid46 *Yr3b* and *Yr4b* and Minister *Yr3c*. However, Chen and Line (1993a) found that a second gene in Hybrid46, presumably *Yr3b*, was not located at the *Yr3* locus, suggesting that the allelism was not strictly demonstrated. Chen et al. (1996) reported *Yr3a* and *Yr3c* on chromosome 1B, *Yr4a* and *Yr4b* on chromosome 6B through monosomic analysis. In addition to cultivars above, *Yr3a* is also present in Stephens, Druchamp, Vilmorin 23 and Nord Desprez (Chen et al. 1996), and Bon Femier, Nudif TP1, Argent, Hobbit, Kinsman, Mardler, Maris Ranger, Top, Viginta and Zadar (McIntosh et al. 2013); *Yr3c* in Cleo and Maris Beacon; *Yr4a* in Vilmorin 23, Druchamp, Hobbit, Huntsman, Kinsman, Maris Ranger, Maris Freeman, Mardler, Nord Desprez, Top, Viginta, Yamhill and Zdar; and *Yr4b* in Avalon, Opal, Staring, Maris Beacon, Nudif TP12 and Stella (McIntosh et al. 2013). Druchamp (*Yr3a*), Stephen (*Yr3a*) and Yamhill (*Yr4a*) were used in the previous set of North American differentials (Line and Qayoum 1992; Chen 2005). Vilmorin 23 (*Yr3a*) is used in the world set of differentials and the Hybrid 46 (*Yr3b*, *Yr4b*) is included in the European and Chinese differential systems (Johnson et al. 1972; Wan et al. 2004; Chen 2005).

The *Yr4* locus was recently proposed in wheat cultivar Rubric (AUS33333) (Bansal et al. 2010). Rubric is an Australian spring wheat cultivar, a selection from the CIMMYT genotype EMU 'S' (AUS17224), which carries a single dominant gene, *YrRub*, displaying a low seedling response against some Australian *Pst* pathotypes. Bansal et al. (2010) suggested that *YrRub* in cultivar Rubric is *Yr4*, but it is located on chromosome 3BS, different from either *Yr4a* or *Yr4b* on chromosome 6B. Markers *Xcfb3530* is closely linked with *YrRub* locus at 2.9 ± 1.3 cM, and placed *YrRub* in the distal of chromosome 3BS, bin 3BS3-0.87-1.00. Cultivars Hybrid 46 and Avalon have the same allele as *YrRub* amplified by the linked markers *Xcfb3530* (150bp) and *Xbarc75* (132bp). Because Hybrid 46 and Avalon carry *Yr4*, Bansal et al. (2010) suggested that *YrRub* is likely to be *Yr4*. Three wheat genotypes Bolac, Nesser and EMU 'S', which produce ITs similar to that of Rubric and have the same alleles as *YrRub* at the *Xcfb3530* and *Xbarc75* loci, were considered carrying *YrRub* (Bansal et al. 2010). Based on these results, the *Yr4* locus is more likely on 3BS as molecular mapping is generally better than monosomic analysis for localizing a gene. However, cultivars previously reported to have stripe rust resistance genes at the *Yr3* and *Yr4* loci need to be further studied, and genes at the two loci and other loci need to be clearly mapped to make their chromosomal locations clear.

Yr5 confers a high level of ASR to stripe rust and is originally from *T. aestivum* subsp. *spelta* cv. Album (TSA). The TSA resistance was first identified by Straib (1939) and the monogenically dominant mode of inheritance of resistance was determined on the segregating populations by Macer (1963), and the gene was eventually designated *Yr5* (Macer 1966). *Yr5* was located on chromosome 2BL through monosomic and disomic analysis (Law 1976). Besides TSA, four European and one Iranian spelt accessions were confirmed to carry the *Yr5* allele through allelism tests (Kema and Lange 1992). The reference genetic stocks of *Yr5* allele include Aroona*6/*Yr5*, Avocet S*6/*Yr5* (AvSYr5NIL), Warigal*6/*Yr5* (McIntosh et al. 1995), Lemhi*8/*Yr5* (Smith et al. 2007) and *T. spelta* var. *album* IPO-DLO collection (Kema and Lange 1992). Other wheat genotypes carrying *Yr5* include By33 (Sun et al. 2002), S2199 (Fang et al. 2008), E5557 and E8570 (Sawhney and Luthra 1970). Due to concomitant repercussions on yield, the resistance of *Yr5* had not been used in commercial wheat cultivars for a long time after it was identified until recently. In the recent years, wheat cultivars carrying *Yr5*, such as Blanca Grande 515, Clearwhite 515, Patwin 515, Summit 515 and Seahawk have been released in the United States since the new molecular technique enables efficient and precise selection of *Yr5* resistance and eliminates genetic drags. Widely used molecular markers of *Yr5* in MAS are sequence-tagged site (STS) STS7/8 marker and *DpnII* CAPS marker developed by Chen et al. (2003), which were converted from *Yr5* co-segregating RGAP markers *Xwgp-17* and *Xwgp-18* (Yan et al. 2003). Beside these two markers, the genetic distance between *Yr5* and the following markers is less than 1.0 cM and can be potentially employed in breeding programs for introgression of *Yr5* into wheat cultivars: STS markers *S19M93*(co-segregating with *Yr5* locus) and *S23M41* (0.7 cM) (Smith et al. 2007), EST-SSR marker *Xdp269* (0.7 cM) (Fang et al. 2008), BAC end STS marker 63H20NBS (0.14 cM) and 151O2 (0.05 cM)

(Wang et al. 2008b), SSR markers *Xbarc349* (0.4 cM) and *Xwmc* 175 (1.4 cM) (Murphy et al. 2009), TaAffx.65234.1.S1_at and Ta.28038 (McGrann et al. 2014), SNP markers IWA6121 and IWA4096 (Naruoka et al. 2016). *Yr5* is used in the China, India, North America, China, and World and European sets of differentials to distinguish *Pst* races (Wellings and McIntosh 1990; Wan and Chen 2014). Virulence for *Yr5* has been rare, and only occasionally recorded in India, Turkey, Australia and Tajikistan (Wellings et al. 2009). The Australian isolate virulent to *Yr5* has a very narrow virulence spectrum (virulent to *Yr2*, *Yr3*, *Yr5* and *Yr21*) based the tests on a set of the single *Yr*-gene lines (Wang and Chen 2016). Virulence analyses of sexually reproduced populations in China and the US suggest that the *Yr5* avirulence locus is largely homozygous in the current *Pst* populations (Tian et al. 2016; Yuan et al. 2016; Y Lei, MN Wang and XM Chen unpublished data).

Yr6, with a synonym B, confers ASR to stripe rust. El-Bewdey and Röbbelen (1982) located *Yr6* in the short arm of chromosome 7B. *Yr6* is common in spring wheat cultivars of South American origin. Many of these cultivars, in particular Frontana and Frontiera, were introduced into the CIMMYT wheat program (Wellings 1986). McIntosh et al. (1995) summarized that *Yr6* is present in cultivars Oxley, Heines Kolben, Heines Peko, Duilio, Latino, Quadruno, Bindawarra, Frontana, Koga II, Maris Dove, Millewa, Récital, Trigo Eureka, etc. It is also present in US spring wheat cultivar Fielder (Chen and Line 1992c; Chen et al. 1995a). The NIL for *Yr6* was developed in the Avocet background (Wellings et al. 2009). Cultivars carrying *Yr6* is used in the China, USA, world and European sets of differentials (Chen 2005). The virulence frequency to *Yr6* was as high as 92% among 235 international *Pst* isolates, and the *Yr6*-virulent isolates distributed in Canada, Chile, China, Hungary, Kenya, Nepal, Pakistan, Turkey, Uzbekistan (Sharma-Poudyal et al. 2013), and also presented at high frequencies in *Pst* isolates from Ecuador, Italy and Mexico (Chen and associates, unpublished data). In 348 isolates of the US *Pst* population collected in 2010, 93.7% isolates were virulent to *Yr6* (Wan and Chen 2014). The *Pst* population in the Indian subcontinent appears to be fixed for virulence of *Yr6* (Kumar et al. 1988). *Yr6* is inherited as a recessive or dominant gene depending on the pathogen culture used in the test (El-Bedewy and Röbbelen 1982). Johnson and Minchin (1992) suggested that the dominance reversal of *Yr6* was related to the pathogenicity for *Yr2* in the test culture. *Yr6* was inherited as dominant gene when *Yr2* was effective; in contrast, *Yr6* appeared to be recessive when *Yr2* was ineffective.

Yr7 confers high level ASR to stripe rust. The gene was originated from durum wheat cultivar Iumillo and was transferred to common wheat cultivar Thatcher. Stubbs (1966) located *Yr7* in chromosome 2B of cultivar Thatcher. In 1976, Gaines confirmed that *Yr7* is located on the long arm of chromosome 2B at 21 cM from centromere (Law 1976). This gene was widely used in cultivars grown in Europe, Australia, New Zealand and North America. The genetic stocks possess *Yr7* including cultivars Thatcher, Lee, near isogenic lines Avocet S*6/*Yr7* (AvSYr7NIL), Chinese Spring*4/Thatcher 2B and Taichung 29*6/*Yr7* (McIntosh et al. 1981; Yao et al. 2006). The virulence frequency to *Yr7* reached to 90% among 235 international *Pst* isolates and the virulent isolates distributed in Canada, Chile, China,

Ecuador, Hungary, Italy, Kenya, Mexico, Nepal, Pakistan, Turkey, Uzbekistan and the US (Sharma-Poudyal et al. 2013; Wan and Chen 2014), and also found in Europe, Australia and New Zealand (McIntosh et al. 1995). These results agree with the occurrence of virulence shortly after commercial deployment of varieties carrying *Yr7* alone or in combination with other *Yr* genes. *Yr7* is included in the *Pst* differential sets used in the US, Europe and Australia (Chen 2005; Wan and Chen 2014). *Yr7* is very closely linked with *Yr5*, *YrSP* and *Sr9* (McIntosh et al. 1981; Zhang et al. 2009; Feng et al. 2015b) and therefore, *Yr7* is also present in many wheat cultivars carrying *Sr9g* for resistance to stem rust. SSR marker *Xgwm526* (212 bp) is linked with *Yr7* at 5.3 cM (Yao et al. 2006). Markers linked with *Yr5* (*STS7/8* and *STS9/10*) or *YrSP* (*IWA2872*, *IWA3973*, *IWA6561*, *IWA1489*, *IWA6317* and *IWA6122*) also can be used to detect *Yr7* (Chen et al. 2003; Zhang et al. 2009; Feng et al. 2015b).

Yr8 confers high level ASR to stripe rust. *Yr8* was originally derived from *Aegilops comosa* Sibth. & Smith (2n=2x=14, MM). The gene was cytogenetically located on chromosome 2M of *Ae. comosa*. Riley et al. (1968) transferred *Yr8* from chromosome 2M of *Ae. comosa* to chromosome 2D of wheat genotype 'Chinese Spring', resulting in a 2D/2M translocation line designating as Compair. Miller et al. (1988) isolated spontaneous non-Robertsonian translocation lines carrying chromosomes 2A/2M or 2D/2M. The translocation chromosomes consisted of the complete short arm, centromere and a large portion of the long arm of chromosome 2M, and segments of the short arms of either chromosomes 2A or 2D of wheat (Riley et al. 1968; Miller et al. 1988). The translocation chromosomes retain the structural rearrangement in chromosome 2M, which confines the further utilization of *Yr8* in wheat resistance breeding (Nasuda et al. 1998). Although not deployed in commercial cultivars, *Yr8* exists in many wheat genetic stocks. Besides wheat line Compair, other *Yr8* translocation lines include CS 2A/2M, Hobbit Sib 2A/2M, Hobbit Sib 2D/2M and Maris Widgeon 2D/2M (Miller et al. 1988). *Yr8* also was introgressed into several wheat genotypes to develop NILs, such as Avocet S*6/*Yr8* (AvSYr8NIL), Harrier*6/*Yr8* and Aroona*6/*Yr8* (McIntosh et al. 1995). The virulence frequency to *Yr8* was 80% among 235 international *Pst* isolates (Sharma-Poudyal et al. 2013). *Yr8* was susceptible to some races in the US, Australia, Canada, Chile, China, Ethiopia, Hungary, Kenya, Nepal, Pakistan, Turkey, Uzbekistan, UK, Israel, German and Netherlands (Riley et al. 1968; Johnson et al. 1978; Wan et al. 2004; Sharma-Poudyal et al. 2013; Wan and Chen 2014; Wan et al. 2017), and in Ecuador, Italy and Mexico (XM Chen and associates, unpublished data). Stubbs (1985) reported that virulence for *Yr8* was common in the center of origin of *Ae. comosa*, likely the Aegean region of Greece and Turkey. Wellings et al. (2009) found *Yr8* virulence in Turkey, Syria, Iran, central Asia (Tajikistan, Kazakhstan), India, New Zealand and South Africa. *Yr8* was included in wheat stripe rust differential sets of the US, Europe and Australia (Chen 2005; Wan and Chen 2014). Fifteen RFLP (Nasuda et al. 1998) and 20 RGAP (XM Chen and associates, unpublished data) markers associated with chromosome 2M can be used to detect *Yr8*. In addition to *Yr8*, an HTAP resistance gene has been identified in translocation chromosome 2M in the *Yr8* NIL (XM Chen and associates, unpublished data).

The translocation 2M segment also carries stem rust resistance gene *Sr34*, which enables this translocation line valuable in rust resistance breeding.

Yr9 confers high level ASR to stripe rust. *Yr9* originated from rye (*S. cereale*) chromosome arm 1RS, which also possesses genes *Lr26*, *Sr31* and *Pm8* for resistance to leaf rust, stem rust and powdery mildew (Bartos et al. 1973; Zeller 1973; Macer 1975; McIntosh 1983). *Yr9* is co-segregating with *Lr26* and *Sr31*, while *Pm8* is 4.4 cM away from *Yr9* locus (Lukaszewski 2000). The donor of chromosome 1RS in the majority of wheat cultivars with 1R chromosome substitution or 1BL.1RS translocation, including the widely used cultivars Kavkaz and Aurora from Russia which can be traced back to the rye variety Petkus, was an important source for rust resistance and yield improvement in many wheat breeding programs (Schlegel and Korzum 1997). The wheat cultivars carrying *Yr9* resistance include Veery, Genaro, Clement, Lovrin 10, Lovrin 13, Rienesel 47/51, Weique and Neuzucht (McIntosh et al. 1995). With extensively deployed in numerous commercially cultivars throughout the world, virulence for *Yr9* was widespread throughout in the 1990s, encompassing regions Africa, Europe, the Middle East (Turkey, Syria, Iraq, Iran, Azerbaijan), central Asia (Tajikistan, Kazakhstan), the Indian subcontinent (India, Pakistan), China, New Zealand and South America (Stubbs 1985; Wellings and Burdon 1992; Wellings et al. 2009). The *Yr9*-virulence was reported in the North America since 2000 and afterward (Chen et al. 2002; Chen 2005; Chen et al. 2010; Wan and Chen 2012, 2014; Wan et al. 2016). *Yr9* is used in the *Pst* differential sets of China and the North America (Wan et al. 2007; Wan and Chen 2014). Many molecular markers have been identified for *Yr9* (Shi et al. 2001). The RFLP probes 1B-159 and RGA39 are co-segregating with *Yr9* locus (Mago et al. 2002). SSR marker *Xgwm582* links with *Yr9* at a genetic distance of 3.7 cM (Weng et al. 2005).

Yr10 is located in chromosome 1BS and confers high level ASR to stripe rust. This gene is initially identified from a Turkey common wheat germplasm PI 178383 (Lewellen et al. 1967; Metzger and Silbaugh 1970), and alternative sources of this gene were found in *T. spelta* accession 415 (Kema and Lange 1992) and *T. vavilovii* accession AUS22498 (Bariana et al. 2002). *Yr10* near-isogenic lines were developed in several backgrounds of wheat genotypes, such as Corrigin, QT3960, 79W93 and Cappelle Desprez. The *Yr10* virulence has been found in low frequencies in the US, Chile, China, Ethiopia, Kenya, Hungary, Pakistan and Uzbekistan (Sharma-Poudyal et al. 2013; Wan et al. 2017), and Italy but not in Ecuador and Mexico (XM Chen and associates, unpublished data). *Yr10* is used in the North America *Pst* differential set (Wan and Chen 2014). *Yr10* is linked with brown glume color (0.2 cM) in wheat genotypes PI 178383 and Moro (Metzger and Silbaugh 1970), while *T. spleta* 415 and *T. vavilovii* AUS22498 have white glumes (Kema and Lange 1992; Bariana et al. 2002). Despite the *Yr10* gene sequence (accession AF149112) was deposited in GenBank in 2000, and the gene function was just confirmed recently (Liu et al. 2014a). The open reading frame (ORF) of *Yr10* comprises of 3630 bp and encodes a highly evolutionary-conserved and unique CC-NBS-LRR, containing the characteristic R gene domain/motifs such as coiled-coil (CC), P-loop, kinase 2a, kinase 3a, conserved domains 2 and 3, Q/EGF and HD. Markers *Xpsp3000* and *Gli-B1* (Bariana et al. 2002) and gene-specific SCAR markers $E5_{1100}$ (Liu et al. 2014a) can be used in marker-assisted selection for *Yr10*. However, the cloned gene AF149112 is about 1.3 cM from the *Yr10* locus (Fu DL, personal communication). More research is needed to confirm this.

Yr11, *Yr12*, *Yr13* and *Yr14* were designated to the presumed APR genes in several European cultivars (Priestley 1978; Priestley and Byford 1979; Priestley et al. 1982). Except *Yr11* (Wellings CR, unpublished data), the other three genes were not confirmed by genetic analysis. The chromosomal locations of these genes are unknown. The reference stocks of *Yr11*, *Yr12*, *Yr13* and *Yr14* are cultivars Joss Cambier, Mega, Maris Huntsman and Hobbit, respectively. Joss Cambier showed slow rusting resistance but was race-specific (Johnson and Taylor 1972). These genes have not been intentionally used in the international breeding programs. The durability of these genes remains unknown (McIntosh et al. 1995). However, European cultivars with these genes in various combinations and also with other genes have been resistance for more than 30 years in our nurseries grown in the US Pacific Northwest. More research is needed to clearly characterize genes in the cultivars that have the so-called *Yr11*, *Yr12*, *Yr13* and/or *Yr14*, including the number of genes, chromosomal locations and race specificity.

Yr15 confers high level ASR to stripe rust. It was originally identified from wild emmer wheat *T. turgidum* var. *dicoccoides* accession G25 because of resistance to all races from the Wageningen culture collection. *Yr15* is also present in other collections of wild emmer wheat (Gerechter-Amitai et al. 1989). *Yr15* was transferred from G25 into a range of cultivated tetraploid and hexaploid wheat backgrounds in many breeding programs of the world. In the test of Gerechter-Amitai et al. (1989), *Yr15* was resistant to 24 *Pst* isolates from 18 countries. In our tests of more than 3000 *Pst* isolates from 17 countries including major wheat production countries, such as Australia, Canada, Chile, China, Pakistan, Ethiopia, Mexico and Italy, no *Yr15*-virulent isolates have been detected (Sharma-Poudyal et al. 2013; Wan and Chen 2014; Wan et al. 2016, 2017). The virulence to *Yr15* was only documented in Afghanistan (virulent race 7E134) (Van Silfhout 1989). *Yr15* is used in the North America *Pst* differential set (Wan and Chen 2014). Despite *Yr15* was introgressed into many wheat germplasm lines in wheat breeding programs in Israel, Australia and the USA, *Yr15* is just recently deployed in a few commercial common wheat cultivars, such as Clearwhite 515, Expresso, Patwin 515, Scarlet 09, Summit 515 and Seahawk in the US. *Yr15* is located on chromosome 1BS near the centromere (Peng et al. 2000a). The genetic distances between *Yr15* and its flanking SSR markers *Xbarc8* and *Xgwm413* are 3.9 and 2.5 cM, respectively. *Xbarc8* and *Xgwm413* have been the most used markers for the incorporation of *Yr15* into wheat cultivars (Murphy et al. 2009; Yaniv et al. 2015). Recently, Mandoulakani et al. (2015) developed two SCAR markers SC790 and SC2700, which were mapped 1.1 and 0.1 cM away from the *Yr15* locus, respectively. These two SCAR markers were distinguishable for 48 wheat genotypes except for lines HSB2408, HSB2955 and Stiletto (Mandoulakani et al. 2015). The two SCAR markers are potentially useful in MAS for transferring *Yr15* gene into wheat cultivars.

Yr16 confers APR to stripe rust. It is located in chromosome 2DS. *Yr16* was identified in French wheat cultivar Cappelle-Desprez (Worland and Law 1986).

Cappelle-Desprez remains durable resistance to stripe rust since it was released in 1946, grown in large acreages in Western Europe until the late 1970s (Boyd 2005). *Yr16* is linked with two important genes on chromosome 2D, semi-dwarfing gene *Rht8* for reducing plant height and photoperiodism gene *Ppd1* for day length insensitivity (Worland et al. 1988). Because of these good agronomic traits, Cappelle-Desprez was used extensively in breeding programs in the UK and Western Europe. A lot of Western European wheat cultivars possess the *Yr16* resistance. The cultivar has also been used and it has contributed HTAP resistance in the US PNW wheat breeding programs (Chen 2013). *Yr16* has been used in South Africa wheat breeding programs since the emergence of stripe rust in the late 1990s. *Yr16* is a quantitative trait locus, explaining about 10% phenotypic variation of both rust IT and severity and is located on an interval between markers *Xgwm102* and *XwPt-664520* (Agenbag et al. 2012).

Yr17 is an ASR gene which is located on a segment of *T. ventricosum* (Tausch) Cess. (synonym *Ae. ventricosa*) (genome $D^vD^vM^vM^v$) chromosome 2NS translocated to the short arm of common wheat chromosome 2A. The segment possessing three rust resistance genes *Yr17*, *Lr37* and *Sr38* for resistance to stripe rust, leaf rust and stem rust, respectively, replaced approximately a half of the short arm of chromosome 2A (distal 25-38 cM). *Yr17* is closely linked with *Lr37* and *Sr38* at 0.5 cM (Bariana and McIntosh 1993). The chromosomal segment was initially introgressed into common winter wheat VPM1, which was widely used in many European wheat breeding programs. *Yr17* is present in cultivars Balthazar, Bowie, Brigadier, Eureka, Folio, Lynx, Pernel, Sunbri, Sunstate, Sunvale, Trident, Tilburi, Madsen, Hyak, Apache, Renan and Jagger (McIntosh et al. 2013). A suppressor of *Yr17/Lr37/Sr38* is present in cultivar Anza, resulting in Anza susceptible to *Yr17* avirulent races (Helguera et al. 2003). The virulent race to *Yr17* was first reported in England in 1994, and spread in Western and Northern Europe counties in the subsequent years. Wellings et al. (2009) reported the virulence to *Yr17* was present in central Asia (Kazakhstan), Middle East (Lebanon, Syria, Iran) and South Africa, but absent in Iraq and Ecuador. Recently, the virulent isolates to *Yr17* were found in the Australia, Canada, Chile, China, Hungary, Kenya, Nepal, Pakistan, Turkey, Uzbekistan and the USA, with frequency of 88% in 235 global isolates (Sharma-Poudyal et al. 2013) and also in Ecuador, Italy and Mexico (XM Chen and associates, unpublished data). A *Yr17* single-gene line (AvSYr17NIL) is used in the North America *Pst* differential set (Wan and Chen 2014). The resistance of *Yr17* is influenced by genetic background, temperature and light intensity. In seedling stage, *Yr17* becomes more susceptible at low temperature and low light intensities. *Yr17* has low level resistance on Thatcher*8/VPM1 with less necrosis/chlorosis and more sporulation. Based on our data, we have hypothesized that the *Yr17* NIL also carries a HTAP resistance gene linked to *Yr17*. Our preliminary results appear proving the hypothesis and we are conducting more studies to separate the genes to clearly prove the hypothesis. Different markers can be used in detecting *Yr17*, such as PCR marker *VENTRIUP/LN2* specific for the 2NS segment, cleaved amplified polymorphic sequence (CAPS) marker *URIC/LN2* specific for the corresponding 2AS segment, and TaqMan real time system with probe *W2NS-55T* (Helguera et al. 2003), and

RFLP marker *Xpsr150-2M*[v], SCAR marker SC-Y15 (0.8 ± 0.7 cM to *Yr17*) (Robert et al. 1999), co-segregated SCAR markers SC-372 and SC-385 (Jia et al. 2011), and several SSR or RGA markers including *cWMG682* and *TaVrga-A1* (Fang et al. 2011).

Yr18 is an APR gene, likely originating from Chinese landraces. The gene was first documented as *Lr34* conferring APR to leaf rust in Canada by Dyck (1987), but *Lr34*-containing germplasm has been a part of wheat improvement since the early twentieth century. The near-isogenic Thatcher line carrying *Lr34* and various other cultivars know to carry *Lr34* also display APR to stripe rust races in Mexico and Australia. Singh (1992) genetically analyzed the association of the APR to stripe rust and leaf rust in *Lr34* carrying lines. He found that the resistance to both rusts were co-segregating and named stripe rust resistance locus as *Yr18*. The further studies found this gene has a pleiotropic effect, conferring resistance to leaf rust (*Lr34*), stem rust (*Sr57*) and powdery mildew (*Pm38*), and inducing leaf tip necrosis (*Ltn1*). Wheat cultivars containing *Lr34/Yr18* occupy more than 26 million ha in various developing countries alone and contribute substantially to yield savings in epidemic years. Although *Yr18* is a partial resistance gene, the resistance of *Yr18* is not adequate under heavy infection in some places. However, this gene has good compatibility with other resistance genes for providing high level resistance. *Yr18* is located on chromosome 7DS and is one of the first cloned stripe rust resistance genes. Sequence of this gene comprises of 24 exons and codes an ATP-binding cassette (ABC) transporter of the pleiotropic drug resistance class (Krattinger et al. 2009a). The level of *Yr18/Lr34*-mediated resistance for rust infection is correlated with the development of leaf tip necrosis, which can be used as a morphological marker in selection. In addition, flag leaf tips of *Yr18/Lr34*-containing cultivars are more resistant than their respective leaf bases. Rubiales and Niks (1995) reported that *Yr18/Lr34* is associated with reduced intercellular hyphal growth but not with a hypersensitive response or papilla formation. The durability of *Yr18* resistance to stripe rust and leaf rust has been observed since early twentieth century. Comparing *Yr18* alleles between resistant and susceptible cultivars in various collections, seven haplotypes were identified. These variations include short deletion/insertion and point mutations among exons 11, 12 and 22, and introns 4 and 6 (Wu et al. 2015). The widely used *Yr18* gene-specific markers *cssfr1 - cssfr7* (Lagudah et al. 2009) and *caSNP*, *caSNP4*, *caSNP12* and *caIND11* (Dakouri et al. 2010) were designed based on a one-bp (A) deletion in exon 10, a 3-bp deletion (TTC) in exon 11 and a G-to-T point mutation in exon 22 between resistant and susceptible haplotypes. With improving leaf rust resistance in many wheat breeding programs, the *Lr34*/*Yr18* resistance has been introgressed into many important germplasms and cultivars including Chinese Spring, Mazhamai and Youzimai in China; Mentana, Ardito and Strampelli in Europe; RL6050 and RL6058 in Canada; Jupateco 73R in CIMMYT, Mexico; Condor in Australia; and Anza in the US. The resistance of *Yr18* is suppressed in some Chinese wheat landraces and a dominant suppressor gene was identified in cultivar Sichuan Yonggang 2 (Wu et al. 2015).

Yr19 (*YrCom*) was identified in cultivar Compair and located on chromosome 5B using monosomic analysis (Chen et al. 1995a). It confers ASR to few US *Pst* races such as PST-21 and PST-45 (formerly CDL-21 and CDL-45). Besides *Yr19*, Compair also has another gene *Yr8* which is derived from *Ae. comosa*. *Yr19* is assumed from *Ae. speltoides* which was used to disrupt the diploid pairing of wheat during the transfer of *Yr8* from *Ae. comosa* chromosome 2M to the wheat chromosome 2D (Chen et al. 1995a). The inheritance of *Yr19* is dominant. Based on the virulence test on Compair, virulence to *Yr19* is very common in the US *Pst* population.

Yr20 (*YrFie*) was identified from a US spring wheat cultivar Fielder and located on chromosome 6D (Chen and Line 1992b, c; Chen et al. 1995a). It confers ASR to few US *Pst* races such as PST-17 (formerly CDL-17). *Yr20* is a dominant gene, but it is not effective in the hemizygous condition according to monosomic analysis (Chen et al. 1995a).

Yr21 (*YrLem*) was identified in the US spring wheat cultivar Lemhi and located on chromosome 1B and, with temporarily name *YrLem* (Chen and Line 1992b, c; Chen et al. 1995a). It confers ASR to PST-21 (formerly CDL-21), which is the only *Pst* race identified so far in the US with avirulence to *Yr21* (Line and Qayoum 1992; Chen 2005; Chen et al. 2010; Wan and Chen 2012). *Yr21* is a dominant gene, but it is not effective in the hemizygous condition according to monosomic analysis (Chen et al. 1995a). Lemhi has been used in the set of wheat cultivars used to differentiate *Pst* races in the US (Line and Qayoum 1992, Chen 2005), and used as a susceptible check in screening wheat germplasm for stripe rust resistance. Although Lemhi is susceptible to all *Pst* races except PST-21, it is highly resistant to barley stripe rust (Chen et al. 1995c). *Yr21* is closely linked (0.3 cM) with another gene *RpsLem*, conferring resistance to *Psh* (Pahalawatta and Chen 2005a). As the two genes mapped so close to each other in Lemhi, they are likely the same gene. If so, the gene that is ineffective against most *Pst* races, but could be useful for control barley stripe rust if introgressed into barley cultivars.

Yr22 (*YrLe1*) was identified in spring wheat cultivar Lee and located on chromosome 4D using monosomic analysis (Chen and Line 1992b, c; Chen et al. 1995a). It confers ASR to few *Pst* races, such PST-35 (formerly CDL-35), in the US. Lee has been used to differentiate *Pst* races in the US (Line and Qayoum 1992; Chen 2005). *Yr22* is a dominant gene. Although allelism tests have not been done, *Yr22* and *Yr20* should be different genes because they have different resistance patterns on a set of *Pst* races and different chromosomal locations. *Yr23* (*YrLe2*) was also identified in Lee, but located on chromosome 6D (Chen and Line 1992b, c; Chen et al. 1995a). It confers ASR to race PST-20 (formerly CDL-20) and some other races in the US. *Yr23* is also dominant (Chen et al. 1995a). In addition to *Yr22* and *Yr23*, Lee also carries another gene *Yr7*. Genes like *Yr6*, *Yr7*, *Yr8*, *Yr9*, *Yr19*, *Yr20*, *Yr21*, *Yr22* and *Yr23* are useful to distinguish *Pst* races and studying the dynamics of the pathogen population, but not very useful for breeding for resistant wheat cultivars as the current populations in the world are highly virulent to these genes.

Yr24 confers ASR to stripe rust and was originally identified in *T. turgidum* var. *durum* accession K733. It is in the same locus as *YrCH42* in Chinese wheat cultivar Chuanmai 42 (Li et al. 2006a). The gene is highly resistant to most *Pst* races with the predominant IT 0; on *Yr24*/3*AvS NIL (AvSYr24NIL) and Chuanmai 42, rarely

IT 0;+, 1+ or 2+ based on the 0–4 scale. *YrCH42* is from *T. turgidum* line Decoy 1, which is the AB genome donor in synthetic wheat Syn 769 (AABBDD). Syn 769 is a parental line for developing Chuanmai 42. *YrCH42* was assigned to chromosome 1BS with SSR markers *Xgwm498* and *Xbarc187* (Li et al. 2006a). Their allelism tests identified *Yr24*, *Yr26* and *YrCh42* in the same locus (see *Yr26* below). The latest work demonstrated that *Yr24/Yr26/YrCH42* locus is on chromosome bin C-1BL-6-0.32 (Wang et al. 2008a). The *Yr24* NIL (AvSYr24NIL) is included in our *Yr* single-gene line differentials (Wan and Chen 2014). Although mostly in low frequencies, races virulent on *Yr24* have been identified in the US, Australia, Chile, China, Hungary, Italy, Nepal, Pakistan, Turkey, Uzbekistan, but not in Ecuador (Wan and Chen 2014; Wan et al. 2016, 2017; XM Chen and associates, unpublished data).

Yr25 was identified in wheat line TP1295 and located on chromosome 1D (Calonnec and Johnson 1998). The gene was assigned to chromosome 1D with Chinese Spring monosomic analysis. It confers ASR to *Pst* race 6E0 (isolate WYR 85-22). The expression of *Yr25* resistance is modified by environment and by at least one minor gene on chromosome 6A. The IT of wheat lines carrying *Yr25* alone ranges from fully resistant (IT ;) to intermediate (IT 2, rarely 3 or 3+) based on the 0–4 IT scale (Calonnec and Johnson 1998). In addition to TP1925, *Yr25* is also present in wheat cultivars/lines TP981, Carina, Hugenout, Strubes Dickkopf, Tugela, Carstens V, Heines Peko, Heines VII, Reichersberg 42 and Spaldings Prolific (Eriksen et al. 2005). We included Hugenoot (*Yr25*) in the set of supplementary differentials (Wan and Chen 2014), and found that *Yr25* is largely ineffective against *Pst* races in the US and other countries (Wan and Chen 2014; Wan et al. 2016, 2017; XM Chen and associates, unpublished data).

Yr26 confers ASR to stripe rust. This gene is possibly originated from Chinese landrace *T. turgidum* Gamma 80-1, a parental line used in cross with *Haynaldia villosa* for developing 6A (6V) substitution and 6AS/6VL translocation lines. Translocation lines R43, R55, R64 and R77 carry *Yr26* resistance (Ma et al. 2001). This gene has been introgressed into many Chinese wheat cultivars including Shanmai 107, Shanmai 175, Mianmai 39, Mianmai 42, Mianmai 96-5, Lantian 17, Zhong G918 and Neimai 8 (Zeng et al. 2014). Although *Yr24/Yr26* has not been intensively deployed in commercial cultivars other than China, the virulence for *Yr24/Yr26* was detected in the US, Ecuador, Chile, Australia, Pakistan and Turkey, with low frequency 11% (Wellings et al. 2009; Sharma-Poudyal et al. 2013). The rapid increase of the *Yr26* virulence is associated to the quick deployment of the gene in wheat cultivars in China (Han et al. 2015). Ma et al. (2001) mapped *Yr26* on chromosome 1BS and found markers *Xgwm11* (193 bp) and *Xgwm18* (184 bp) linked to *Yr26* at 1.9 cM proximal. Later, the gene was located on the long arm of chromosome 1B, bin C-1BL-6-0.32 (Wang et al. 2008a). The fine genetic map delimited *Yr26* locus in a genetic interval of 0.25 cM by markers *CON-4* proximal 0.08 cM and *CON-12* distal 0.17 cM, and six EST-SSR markers (*CON-6*, *CON-7*, *CON-8*, *CON-9*, *CON-10* and *CON-11*) co-segregated with *Yr26* (Zhang et al. 2013). Currently, *Yr26* is generally accepted as the same gene as *Yr24*.

Yr27 confers moderate to high level ASR to stripe rust. It was previously known as *YrBjy*, *YrSk* or 'Selkirk' gene, which is present in many CIMMYT-developed wheat lines, such as Buckbuck (=Ciano 79), Bluejay (=Nacozari 76), Crow, Tesia 79, Opata 85, Kauz (=Bacanora 88, Bakhtawar, WH542, Atrak, Memof) and Attila (=PBW343, MH97, Chamaran, Kubsa, Shirudi) (Rosewarne et al. 2008; McIntosh et al. 2013). The origin of this gene can be traced to cultivar McMurachy, which was used in breeding for resistance to stem rust in Canada (McDonald et al. 2004). *Yr27* was widely deployed in cultivars grown in Ethiopia, India, Iran, Pakistan and Syria. Virulence to *Yr27* was reported from the early 1970s in Near East, eastern Africa and Indian subcontinent (Stubbs 1985), and was common in the recent international *Pst* collections (Sharma-Poudyal et al. 2013). *Yr27* is used in the single-gene differentials (Wan and Chen 2014). *Yr27* was cytogenetically located on the short arm of chromosome 2B proximal to the centromere (McDonald et al. 2004), and it is associated with RFLP markers *Xcdo152-2B* and *Xcdo405-2B* in a QTL analysis (Rosewarne et al. 2008, 2013). The chromosomal region of *Yr27* is also harboring rust resistance genes *Lr13*, *Lr23* and *Lr31*. The recombinant rate between *Yr27* and *Yr31* is 0.148 (Singh et al. 2003).

Yr28 is an ASR gene identified in a synthetic wheat [*T. turgidum* × *Ae. tauschii*] × *T. aestivum* cv. 'Opata 85' cross. The gene is derived from *Ae. tauschii* and confers ITs ranging from 1–2 to 4–5 in the 0–9 scale, depending on the genetic background. The resistance of *Yr28* was reported to express in adult-plant stage with longer day length and relatively warm temperature. *Yr28* was located on the short arm of chromosome 4D based on its linkage to SSR marker *Xmwg634-4DS* (Singh et al. 2000). *Yr28* is largely ineffective. The *Yr28* single-gene line (AvSYr28NIL) is included in the set of supplementary differentials and the gene is ineffective to most *Pst* isolates collected in the US in the recent years (Wan and Chen 2014; Wan et al. 2016). The *Yr28* virulence was detected in isolates from all 10 countries, and the frequencies ranged from 25% in Kenya to 100% Canada with an average of 79% (Sharma-Poudyal et al. 2013).

Yr29 confers moderate level slow-rusting APR to stripe rust. This gene was identified in cultivar Pavon 76. *Yr29* combining with two other *Yr* genes in Pavon remains effective against stripe rust in Mexico, Ethiopia, Australia and other countries. *Yr29* was assigned on chromosome 1BL and confirmed using a F_3 population from a cross between susceptible cultivar Lalbahadur and Lalb (Pavon 1B) substitution line (William et al. 2003). *Yr29* is also present in cultivars Attila, Parula, Saar and Druchamp (Rosewarne et al. 2006; Lillemo et al. 2008; Hou et al. 2015). The leaf rust resistance gene *Lr46* and leaf tip necrosis (*Ltn2*) gene (different from *Ltn* in *Yr18/Lr34* locus) are either closely linked genes or a single gene with pleiotropic effect. The locus was delimited by flanking markers *Xwmc44* 3.6 cM proximal and *XBac17R* 2.1 cM distal, respectively (Roseware et al. 2006).

Yr30 is a minor APR gene introgressed from Yaroslav emmer wheat into bread wheat along with the stem rust resistance gene *Sr2* (Singh et al., 2001). The gene is located on chromosome 3BS and is presented in cultivar Opata 85, Parula, Inia 66 and Pavon F76 (McIntosh et al. 2013). In addition to *Sr2* for stem rust resistance, *Yr30* is also closely linked with *Lr27* for leaf rust resistance (Singh et al. 2000,

2001, 2005). SSR marker *Xgwm389* can be used to detect *Yr30* (Suenaga et al. 2003). As *Sr2* is effective against the UG99 race group of the stem rust pathogen, several markers have been identified and used in MAS for incorporating the gene into wheat cultivars, including SSR marker *Xgwm533* (120 bp) (Spielmeyer et al. 2003), sequence-tagged microsatellites (STM) markers for differentiate different *Xgwm533* variants (Hayden et al. 2004), three SSR markers based on BAC clones (McNeil et al. 2008), CAPS marker *csSr2* (Mago et al. 2011), and KASP marker wMAS000005 developed based on *csSr2* (http://maswheat.ucdavis.edu/protocols/sr2/). These markers for *Sr2* can be potentially used to select *Yr30*.

Yr31 is located on chromosome 2BS, conferring ASR to stripe rust. The gene is present in wheat cultivar Pastor (Singh et al. 2003; Rosewarne et al. 2012) and Chapio (Yang et al. 2013). Significant genotype × environment interactions were detected for *Yr31* on 2BS (Yang et al. 2013). The "environment effect" is likely caused by different *Pst* populations, as the gene was effective in Toluca, Mexico; but a late incursion of a virulent pathotype in 2009 rendered it ineffective and the gene had no effect in China (Yang et al. 2013). *Yr31* suppresses the additive effect of *Yr30* and a 3D gene, but not of *Yr18*. The genetic distance between *Yr31* and other *Yr* genes on 2BS is 6.2 cM for *Yr31-Yr41*, and 12.5 cM for *Yr31-YrSP* (Luo et al. 2008). The recombination value of *Yr31* with other rust resistance genes on 2BS is 0.148 for *Yr31-Yr27* and 0.295 for *Yr31-Lr23* (Singh et al. 2003). *Yr31* is largely ineffective as 99% of the 235 international *Pst* isolates from more than 10 countries were virulent to the gene (Sharma-Poudyal et al. 2013).

Yr32 was identified in near isogeneic line Carstens V/3*Avocet S using monosomic analysis and in an European winter wheat cultivar Senat using a DH population from a cross of Savannah/Senat and QTL analysis (Eriksen et al. 2004). The gene had a tentatively name *YrCV* as it was originally detected from cultivar Carstens V (Stubbs 1985), which is used in the Europe *Pst* differential set (Johnson et al. 1972; Stubbs 1985). Chen and Line (1993b) determined that Carstens V has three genes for resistance to US *Pst* race PST-21, two genes for resistance to races PST-17, PST-20 and PST-29, and one gene for resistance to PST-27, which was also resistance to all other races. Based on the effectiveness of these genes, the gene that was resistant to all tested races should be *Yr32*, as *Yr32* is effective against the majority of the US races since the 1960s (Wan and Chen 2014, Wan et al. 2016, Liu et al. 2016). Similarly, Calonnec et al. (2002) also identified three genes in Carstens V: *Yr25*, *Yr32* and another unknown gene. The virulence isolates was present in Chile, China, Hungary, Pakistan, Turkey, Uzbekistan and the USA (Sharma-Poudyal et al. 2013) at an average frequency of 11%. AFLP marker M62/P19-156 was co-segregating with *Yr32*, and the genetic distances between *Yr32* and two close flanking SSR markers *Xwmc198* and *Xwmc181* were 2.0 and 21 cM, respectively (Eriksen et al. 2004). Another gene *Yr1* on chromosome 2AL is genetically independent from *Yr32* with genetic distance about 35 cM. Eriksen et al. (2004) suggested *Yr32* in cultivar Tres, but this is questionable, because Tres has two genes (*YrTr1* and *YrTr2*) (Chen and Line 1992a), *YrTr2* is ineffective against most US Pst races, and *YrTr1* has very different avirulence/virulence patterns from *Yr32* against US *Pst* races and races in other countries (Sharma-Poudyal et al. 2013; Wan and Chen 2014; Wan et al. 2016, 2017; Liu et al. 2016; XM Chen and associates, unpublished data).

Yr33 confers APR resistance in wheat cultivar Batavia against Australian *Pst* pathotypes 104 E137 A^- and 108 E141 A^- (IT $;1^{-CN}$). It is located on long arm of chromosome 7D through genetic mapping in a DH population from cross Pelsart/Batavia (34 resistant : 36 susceptible) with SSR markers *Xgwm437* and *Xgwm11* (Zahravi et al. 2003). Using F_2 and $F_{2:3}$ populations from cross Avocet S/Batavia and pathotypes 110 E143 A^+ and 134 E16 A^+, Nazari and Wellings (http://ses.library.usyd.edu.au/bitstream/2123/3215/1/P142.pdf) found that Batavia contains two additional genes *YrBat1* and *YrBat2* controlling seedling ITs 12^{-C} and 23^{-C}, respectively. The relationships among *Yr33*, *YrBat1* and *YrBat2* are not clear.

Yr34 confers intermediate resistance (IT 23C of the 0–4 scale) to Australian *Pst* pathotypes in greenhouse and highly resistance (10 R-MR) in the field. It was located on the distal end of the long arm of chromosome 5A in wheat line WAWHT2046. Although suitable linked markers are not available for marker-assisted selection for *Yr34* resistance, the loose linkage of *Yr34* with *B1* (12.2 cM), an awn inhibitor gene, may be helpful for transferring *Yr34* to wheat cultivars (Bariana et al. 2006).

Yr35 confers effective ASR against Australian *Pst* pathotypes. It is derived from wild tetraploid wheat species *T. dicoccoides*. *Yr35* was located on the short arm of chromosome 6B using Chinese Spring monosomic series and telosomic stocks for 6B (Marais et al. 2005b). SSR marker *Xcfd1* at 4.1 cM to *Yr35* can be potentially used in MAS for transferring *Yr35* into wheat cultivars. The translocated alien chromosomal segment also harbors leaf rust resistance gene *Lr53*, at a genetic distance of 3.0 cM to *Yr35* (Dadkhodaie et al. 2011). Virulence to *Yr35* is present in the US PNW (Chen and associates, unpublished data).

Yr36 is derived from *T. turgidum* ssp. *dicoccoides* accession FA15-3 from Israel. It confers high level of HTAP resistance. The *Yr36* HTAP resistance response takes several days longer than typical R gene hypersensitive cell death response, allowing the limited rust pathogen growth and restricted sporulation and resulting in ITs 1 to 3 (based on the 0–9 scale) on adult plants (Fu et al. 2009; Gou et al. 2015). *Yr36* was located to chromosome 6BS co-segregating with markers *Xucw69* and *Xbarc101* and delimited in an interval of 2-cM by proximal to markers *Xucw77* and *Xbarc136* and distal marker *Xucw71* (Uauy et al. 2005). The fine map defines *Yr36* resistance in a 0.02-cM interval between markers *Xucw129* and *Xucw148*. *Yr36* is one of the first cloned stripe rust resistance genes. The gene product has a combination of serine/threonine kinase and steroidogenic acute regulatory protein-related lipid-transfer (START) domains, which targets to the chloroplast and reduces the ability of the thylakoid-associated ascorbate peroxidase to detoxify reactive oxygen species, and both domains are necessary for resistance (Fu et al. 2009; Gou et al. 2015). *Yr36* resistance is present in wild wheat but is absent in modern pasta and bread wheat varieties. The wheat breeding program in University of California introgressed *Yr36* into bread wheat and developed cultivars/germplasms PI 638740, PI 638741, Glupro, UC1041+*Yr36*, RSL65, Burnside, Glencross, Lilian and Somerset. Cultivar Farnum with *Yr36* developed by the wheat breeding programs in Washington

State University has become the most popular hard red winter wheat in the US PNW. As a HTAP resistance gene, *Yr36* appears to be more sensitive to temperature than to plant growth stages. At high temperatures, the resistance tends to express early in the growth stage and low temperatures increase pathogen sporulation (XM Chen and associates, unpublished data). These observations are in consistent with the report Bryant et al. (2014) in the UK. The tight linkage (0.3 cM) between *Yr36* and protein content gene *Gpc-B1* is an advantage for developing wheat cultivars with durable stripe rust and high protein content (Uauy et al. 2005). However, if low protein content is desired for breeding wheat cultivars, both tetraploid (PI 656793) and hexaploid (PI 664549) wheat lines with *Yr36*, but not functional *Gpc-B1*, are available (Hale et al. 2012).

Yr37 is originated from *Ae. kotschyi* Boiss, a tetraploid wild relative of wheat with genome UUSS. It confers ASR to *Pst* races in Canada and South Africa. The resistance is identified on the long arm of homoeologous chromosome group 2 of *Ae. kotschyi* and transferred to wheat chromosome 2D. The genetic stocks carrying *Yr37* include Chinese Spring/*Ae. kotschyi* disomic addition line 8978 and translocation line S14. This alien chromosomal segment also harbors leaf rust resistance gene *Lr54* (Marais et al. 2005a).

Yr38 confers high level ASR to stripe rust. The resistance of *Yr38* was initially identified from *Ae. sharonensis* (Eig) Feldman & Sears (2n=14; $S^{sh}S^{sh}$) with a temporary name *YrS12* (Marais et al. 2003). It is effective against South Africa *Pst* races. It was transferred from chromosome 6S to the distal region of chromosome 6AL of common wheat (Marais et al. 2006). *Yr38* is linked with marker *Xgwm427* (Marais et al. 2010). The translocated alien segment also harbors *Lr56* for leaf rust resistance (Marais et al. 2006, 2010).

Yr39 confers HTAP resistance and was identified from the US spring wheat cultivar Alpowa. The typical symptoms and signs caused by stripe rust infection on Alpowa at later stage of plant growth are necrotic stripes up to several centimeters long with or without sporulation along the edges (Lin and Chen 2007). *Yr39* is located on the long arm of chromosome 7B. The chromosomal region of *Yr39* delimited by two RGAP markers *Xwgp36* and *Xwgp45* explained 64.2% of the phenotypic variance for the averaged rAUDPC and 59.1% for the averaged heading-flowering stage IT (Lin and Chen 2007). The genetic distance of *Yr39* to other 7B *Yr*-genes *Yr52*, *Yr59* and *YrPI182103* is 31.2, 41.1 and 49.45 cM, respectively (Zhou et al. 2014c; Wang and Chen 2016). The HTAP resistance of *Yr39* is generally adequate in the US PNW under the normal weather and stripe rust conditions, but may not completely protect wheat crop from stripe rust damage under unusually long-cool weather conditions and severe stripe rust epidemics (Chen 2013). Wheat lines with various combinations of *Yr39* with other HTAP resistance genes (*Yr52*, *Yr59*, *YrC591* and *YrPI182103*) have been developed and exhibited high level resistance under severe stripe rust epidemics (Wang and Chen 2016). These lines should be more useful for developing wheat cultivars with high level and durable resistance to stripe rust.

Yr40 confers effective ASR to stripe rust and originated from ovate goat grass, *Ae. geniculata* Roth (syn. *Ae. ovata* L.). The stripe rust resistance of *Ae. geniculata* was transferred to wheat through induced homoeologous chromosome pairing between chromosomes $5M^g$ of *Ae. geniculata* and 5D of common wheat (Aghaee-Sarbarzeh et al. 2002), resulting in rust resistant wheat-*Ae. geniculata* translocation lines. The translocation alien fragment also carrying leaf rust resistance gene *Lr57*. *Yr40* and *Lr57* is either closely linked genes or they are the same gene with pleiotropic effect (Kuraparthy et al. 2007). *Yr40/Lr57* was located in chromosome $5M^g$ of *Ae. geniculata* and was translocated to the distal region of chromosome 5DS of common wheat. The resistant wheat-alien translocation line TA5602 has less alien fragment (5%). The completely linked RFLP markers *GSP*, *FBB276* and *BCD873* can be used to detect alien fragment carrying *Yr40/Lr57* gene (Kuraparthy et al. 2007). The diagnostic CAPS marker *XLr57/Yr40-MAS-CAPS16* for the resistance gene(s) is effective for marker-assisted selection of *Yr40/Lr57* in breeding programs (Kuraparthy et al. 2009). Isolates virulent to *Yr40* is common in the US PNW according to the field observations (Wang and Chen 2016).

Yr41 confers ASR to stripe rust and is on chromosome 2BS. It was identified in a Chinese wheat genotype AIM6, with a tentatively name *YrCN19* (Luo et al. 2005). The gene is also present in other Chinese wheat lines AIM 5, Chuannong 19 and Qian 1104. *Yr41* is resistant to Chinese stripe rust races CYR30, CYR31, CYR32, Hybrid 46-4, SY11-4 and SY11-14 (Luo et al. 2005, 2008). SSR marker *Xgwm410* is closely linked with *Yr41* at a distance of 0.3 cM. It amplifies a 319-bp product in *Yr41*-carrying wheat lines (Luo et al. 2008). *Yr41* was reported to link to *Yr27* (wheat genotype Ciano 79), *Yr31* (Pastor), *Yr3a* (Vilmorin 23) and *YrSP* (AvSYrSPNIL) at genetic distances of 17.4, 6.2, 20.2 and 12.5 cM, respectively (Luo et al. 2008).

Yr42 confers ASR and originated from three-awn goat grass *Ae. neglecta* Req. ex Bertol. accession 155 (genomes UUMM). The rust resistance of *Ae. neglecta* was transferred into common wheat through multiple backcrosses with Chinese Spring and Chinese Spring-Short (a short-strawed near-isogenic line) (Marais et al. 2003). A rust resistant line 03M119-71A was selected, and the translocated chromosome composed of a large segment of foreign chromatin replacing the entire 6AS arm and a proximal part of 6AL of wheat (chromosome pattern $6AS^{Aen}.6AL^{Aen}$-6AL) (Marais et al. 2009). *Yr42* was effective against South African *Pst* pathotypes. The translocated fragment also carried leaf rust resistance gene *Lr62* (Marais et al. 2009).

Yr43 confers ASR to stripe rust and is on chromosome 2BL. It was identified in an US spring wheat cultivar IDO337s (PI 591045) (Cheng and Chen 2010). A lot of US PNW wheat cultivars derived from IDO337s were detected to carry *Yr43* based on rust resistance and the two close flanking RGAP markers *Xwgp103* and *Xwgp110*. The genetic distances between *Yr43* and *Xwgp110* and *Xwgp103* are 4.4 and 5.5 cM, respectively. *Yr43* is genetically linked with four *Yr* genes in repulsion on chromosome 2BL. The genetic distances between *Yr43* and other 2B *Yr*-genes *Yr44*, *Yr53*, *YrSP* and *Yr5* were estimated as 13.1, 16.9, 15.3 and 65.5 cM, respectively (Cheng and Chen 2010; Xu et al. 2013; Feng et al. 2015b). An F_3 line 41-1 carrying *Yr43* single-gene and selected from cross Avocet S and IDO337s is included in the *Yr* single-gene set of wheat lines to differentiate *Pst* races (Wan and Chen 2014). Virulence to *Yr43* was rare before 2000, but become common since 2000 in the US

(Cheng and Chen 2010; Wan and Chen 2014; Liu et al. 2016) and also in many other countries including China, Ecuador, Ethiopia, Italy, Mexico, Nepal, Pakistan and Turkey (Wan et al. 2016; XM Chen and associates, unpublished data).

Yr44 confers ASR to stripe rust and was located on chromosome 2BL. It was identified in a US spring wheat cultivar Zak (PI 607839), with a temporary name *YrZak* (Sui et al. 2009). The map distances of two flanking markers *XpWB5/N1R1* and *Xwgp100* to *Yr44* are 9.4 and 3.9 cM, respectively (Sui et al. 2009). The genetic distances between *Yr44* and *Yr43* and *Yr5* are 13.1 and 41.5, respectively (Cheng and Chen 2010; Xu et al. 2015). A F_3 line (1-1-35-1) carrying *Yr44* selected from cross Avocet S × Zak is included in the Yr single-gene set of differentials (Wan and Chen 2014). *Yr44* is effective against most races prior to 2000 in the US Pacific Northwest but became ineffective to the majority of races identified since 2000 (Wan and Chen 2014; Liu et al. 2016). Like the *Yr43* virulence, the *Yr44* virulence has also been detected in many other countries (Wan et al. 2016; XM Chen and associates, unpublished data). In a study of 136 *Pst* isolates collected in 2016 from Yunnan province of China, the frequency of the *Yr44* virulence was 90% (MJ Li, AM Wan and XM Chen, unpublished data).

Yr45 confers ASR to stripe rust and was located on chromosome 3DL. It was identified in a spring wheat landrace PI 181434 originally from Afghanistan (Li et al. 2011). The genetic distances of two flanking markers *Xgwp118* and *Xgwp115* to *Yr45* are 4.8 and 5.8 cM, respectively (Li et al. 2011). The combination of these two markers can be detected *Yr45* from 82% wheat genotypes. A new germplasm line, PI 660056, an F_6 line selected from a cross of Avocet S and PI 181434, carries the single gene *Yr45* (Wang et al. 2012). *Yr45* is resistant to current *Pst* populations in the US and China (Wang et al. 2012; Zhou et al. 2015b).

Yr46 confers APR to stripe rust and is located on the middle of chromosome 4DL. The gene is pleiotropic, conferring resistance to stem rust (*Sr55*), leaf rust (*Lr67*) and powdery mildew (*Pm46*), and also leaf tip necrosis (*Ltn3*) (Herrera-Foessel et al. 2011). *Yr46* is the fourth cloned stripe rust resistance gene (Moore et al. 2015). The genetic stocks of *Yr46* include NIL RL6077 in the Thatcher wheat background, and the Pakistan older wheat collection PI 250413. The gene is common in the older, tall wheat cultivars from Afghanistan, India and Iraq, such as landrace W195 (Chhetri et al. 2016b), but it is rare in the modern wheat cultivars after the 'Green Revolution' of extending semi-dwarf wheat because of *Yr46* susceptible allele linked with a semi-dwarf gene. With the intensely selecting for the semi-dwarf gene *RhtD1b*, which is also located on chromosome 4D, the *Yr46* susceptible allele has been simultaneously fixed in most modern wheat cultivars. *Yr46* was effective in field tests in Pakistan, India, Mexico and Australia. It displays durable resistance. The genetic distance of *Yr46* to a close genetic locus of two co-segregating markers *Xgwm165* and *Xgwm192* is 0.4 cM. *Yr46* is inferred to be located in chromosomal bin 4DL-0.31-0.56, based on the location of the two markers (Herrera-Foessel et al. 2011). This gene was isolated through map-based cloning. The *Yr46* product composes of one 5′ non-translation region, 3 extrons and 2 introns, and one 3′ non-translation region, encoding 514 amino acids, an H+/monosaccharide transporter similar to STP13 of *Arabidopsis thaliana*. Resistant plants differ from susceptible plants by two single nonsynonymous nucleotide substitutions causing Arg144Gly and Leu387Val substitutions. The arginine at position 144 in the predicted fourth transmembrane domain and the leucine at position 387 distinguish the *Yr46* resistant allele from its susceptible allele, which correspond to two SNP positions 514 and 1243, respectively (Moore et al. 2015). Only the susceptible allele was found in the gene pool of *Ae. tauschii*, the diploid D genome progenitor of wheat, indicating that the *Yr46* resistance allele evolved subsequently to bread wheat polyploidization.

Yr47 (*YrW1*) confers ASR to stripe rust and is located on chromosome 5BS, bin 5BS-0.71-0.81. It was identified from two Iranian common wheat landraces AUS28183 (V336) and AUS28187, which have high levels of ASR against Australian pathotypes of *Pst* and the leaf rust pathogen (Bansal et al. 2011). Two flanking SSR markers, *Xgwm234* and *Xcfb309*, assigned the *Yr47* locus on the short arm of chromosome 5B. The genetic distances of *Yr47* to *Xgm234* and *Xcfb309* were estimated about 10–15 and 8–12 cM, respectively. *Yr47* is linked with leaf rust resistance gene *Lr52* with a genetic distance about 3–6 cM (Bansal et al. 2011).

Yr48 is a QTL conferring APR to stripe rust and located on chromosome 5AL, bin 5AL23-0.87-1.00 (Lowe et al. 2011). *Yr48* was derived from wheat genotype PI 610750, a synthetic spring wheat with pedigree Croc1/*Ae. tauschii* (Synthetic 205)// Kauz. The RIL 4 (GSTR1350) and RIL 167 (GSTR 13618) carry *Yr48* alone (Lowe et al. 2011). In the terms of disease severity, the main effect of *Yr48* accounted for 10% of the observed variation in the field and was associated with an average reduction in disease severity of 63%. *Yr48* was determined in an interval of 5.3 cM region, flanked by markers *Xwmc727* at 4.4 cM proximal and *XwPt-9800* at 0.9 cM distal, and the peak of this QTL corresponded to marker *Xcfa2149*. *Yr48* co-segregates with the vernalization locus *Vrn-A2* and markers *XBE495011*, *Xcfa2149*, *Xgpw2181*, *Xwmc74*, *Xwmc410* and *Xgwm291*, and all of these markers mapped on the distal end of chromosome 5AL (Lowe et al. 2011).

Yr49 confers APR to stripe rust and was identified from Chinese wheat cultivar Chuanmai 18. *Yr49* NIL AUS91433 was developed in the Avocet S background, Avocet S*3/Chuanmai 18. It was assigned to the short arm of chromosome 3D, bin 3DS-6-0.5-1.00. *Yr49* co-segregated with marker *Xgps7321*, and at a genetic distance of 1.0 cM to marker *Xgwm161* (McIntosh et al. 2013).

Yr50 confers ASR to stripe rust and was derived from wheatgrass *Th. intermedium* (2n=6x=42; JJJ^{s}J^{s}StSt). It was identified from wheat-*Th. intermedium* translocation line CH223 (Liu et al. 2013a). *Yr50* is resistant to most Chinese *Pst* races including current predominant races CYR32 and CYR33. The gene was mapped on chromosome 4BL with flanking markers *Xbarc1096* at 8.0 cM proximal and *Xwmc47* at 7.2 cM distal (Liu et al. 2013a). *Yr50* is genetically linked with *Yr62* in repulsion at a genetic distance of 27.1 ± 8.6 cM (Lu et al. 2014).

Yr51 confers ASR to stripe rust and is located on chromosome 4BL, bin 4BL4-0.80-1.00. This gene originated from a wheat landrace AUS27858, which possesses two stripe rust resistance genes (*Yr51* and *Yr57*). *Yr51* displaying IT ;1-N to 2C based on the 0–4 scale. A monogenic segregation F_3 population carrying *Yr51* was obtained from cross AUS27858/Westonia and advanced to F_6. *Yr51* is delimited by markers *Xowm45F3R3* at 1.2 cM proximal and *Xsun104* at 2.5 cM distal, respectively, locating *Yr51* on the long arm of chromosome 4B. A reference *Yr51* genetic stock AUS 91456 was developed. The resistance of *Yr51* in AUS 91456 is recessive. It is resistant to Australian *Pst* races 134E16A+, 134E16A+Yr17+, 134E16A+Yr17+Yr27+, 110E143A+ and 108E141A+ and104E137+ (Randhawa et al. 2014). Marker *Xsun104* has a unique 225-bp allele in the *Yr51*-carrying lines and has null allele in a panel of 27 Australian and 13 Indian wheat genotypes, which has been used in MAS for *Yr51* in Australia and India.

Yr52 confers high level HTAP resistance and is located on chromosome 7BL, bin7BL3-0.86-1.00. It was identified from an Indian spring wheat landrace PI 182527 (Ren et al. 2012a). Short necrotic stripes and occasionally with small or un-opened rust pustules (IT 2-3 based on the 0–9 scale) are the typical symptoms of *Yr52* resistance in adult-plant stage. *Yr52* was mapped with the SSR markers and RGAP markers, and the close flanking markers *Xwgp5258* at 1.1 cM proximal and *Xbarc182* at 1.2 cM distal (Ren et al. 2012a). The combination of markers *Xwgp5258* and *Xbarc182* can distinguish the non-resistance allele from 82% in a panel of 74 wheat genotypes. *Yr52* single-gene germplasm line PI 660057 with improved agronomic traits was selected from a single-seed descent F_6 generation from of Avocet S/PI 182527 (Wang et al. 2012). *Yr52* is genetically linked with *Yr39*, *Yr59* and *YrZh84* in repulsion at genetic distances of 31.2 ± 3.7, 5.4 ± 7.6 and 12.2 ± 1.3 cM, respectively (Zhou et al. 2014c). New germplasm lines with combinations of *Yr52* with other *Yr* genes on 7BL, such as PI 679604 (*Yr52* + *YrZh84*) and PI 679605 (*Yr52* + *Yr39*), have been developed and displayed higher levels of resistance (Wang and Chen 2016). These lines are more useful than the single-gene line or the original donor for developing wheat cultivars with durable and high level resistance to stripe rust.

Yr53 confers ASR to stripe rust and is located on chromosome 2BL, bin C-2BL3-0.35 (Xu et al. 2013). It was identified in an Ethiopia durum wheat genotype PI 480148. *Yr53* is resistant to all tested *Pst* races in the US and China (Xu et al. 2013; Zhou et al. 2015b). The gene was mapped in a monogenically segregating F_4 population from cross Avocet S/PI 480148 with SSR and RGAP markers. The genetic distances between *Yr53* and two close flanking markers *Xwmc441* and *XLRRrev/NLRRrev*$_{350}$ is 5.6 and 2.7 cM, respectively (Xu et al. 2013). The flanking markers *Xwmc441*, *STS2F/1R*$_{219}$ and *XLRRrev/NLRRrev*$_{350}$ can be used in MAS to transfer *Yr53* into wheat cultivars. *Yr53* is genetically linked with *Yr5*, *Yr44*, *Yr43* and *YrSP* in repulsion, and the genetic distances from *Yr53* to *Yr5* and *YrSP* were estimated as 35.6 ± 6.7 and 13.2 ± 1.5, respectively (Xu et al. 2013; Feng et al. 2015b). A common spring wheat line (PI 679598) and a line (PI 679599) with a combination of *Yr53* and *Yr5* have been developed (Wang and Chen 2016).

Yr54 is a QTL conferring APR to stripe rust and located on chromosome 2DL. It was identified in a CIMMYT spring wheat line Quaiu 3 (Basnet et al. 2013, 2014b). Quaiu 3 is resistant to both stripe rust and leaf rust. In addition to *Yr54*, Quaiu 3 also carries genes *Yr29/Lr46*, *Yr30*, *Lr42*, and other minor QTL. Through conducting QTL analysis of rust severity and AUDPC in a RIL population from Avocet S/Quaiu 3,

Yr54 was delimited by markers *XwPt-667162* and *XwPt-667054* in an interval of about 3 cM, which explained 49–54% of the phenotypic variation (Basnet et al. 2014b).

Yr55 confers APR and is located in chromosome 2DL in a winter wheat line Frelon (AUS 38882). The genetic distances of *Yr55* to closely linked flanking markers *Xmag4089* and *Xmag3385* were estimated as 11.4 and 8.4 cM, respectively (McIntosh et al. 2013). *Yr55* is well distal to *Yr54* locus.

Yr56 confers APR to stripe rust and is located in chromosome 2AS, bin 2AS5-0.87-1.00. It was initially identified in an Australian durum wheat cultivar Wollaroi (AUS99174). The *Yr56* single-gene reference line is AUS 91465, which shows IT 23N (based on the 0–4 scale) to Australian *Pst* pathotype 150 E16A+ at the 4th-leaf stage under the greenhouse condition. The segregating locus was temporary name *YrWol*, which is coincided with QTL locus *QYr.sun-2A* conferring APR under field conditions. The linked molecular markers assign the *Yr56* locus in the short arm of chromosome 2A, delimited by DArT derived STS markers *Xsun167*-5.7cM-*Yr56*-7.6 cM-*Xsun168* (McIntosh et al. 2013; Bansal and Bariana 2014). Lines carrying *Yr56* singly had 40–50% disease severity, and the gene explained 9.3% of the phenotypic variation (Bansal and Bariana 2014).

Yr57 confers ASR to stripe rust and is located in the distal part of chromosome 3BS. It originated from a common wheat landrace AUS27858. AUS 27858 possesses two stripe rust resistance genes, *Yr51* and *Yr57* (Randhawa et al. 2014, 2015). *Yr57* was identified from a RIL F_6 population developed from a monogenic segregation F_3 family from cross AUS27858/Westonia. *Yr57* is resistant to Australian *Pst* pathotypes 104 E137A+, 108E141A+, 110 E143 A+, 134 E16A+, 134 E16A +Yr17+, 134 E16A+ Yr17+Yr27 and 150 E16A+. Molecular markers SNP *XBS00062676* and SSR *Xgwm389* flank *Yr57* locus at genetic distances of 2.3 and 2.0 cM, proximally and distally, respectively (Randhawa et al. 2015). *Yr57* is genetically linked with *Yr4* in repulsion at 5.2 ± 1.3 cM. Markers *Xgwm389* and *XBS00062676* amplified a 150-bp fragment and the 'A' allele in *Yr57*-carrying lines, respectively, while amplified different alleles in 19 Australian and 4 Indian wheat genotypes. These two markers can be used in MAS (Randhawa et al. 2015).

Yr58 is a HTAP resistance gene and located on chromosome 3BS. This gene was identified in a spring wheat landrace W195 from India (Chhetri et al. 2016a). It had a tentatively name *YrW195* and is corresponding to loci *QYr.sun-3BS* (Chhetri et al. 2016b). The *Yr58* single-gene line is highly susceptible in seedlings, but its resistance increases with the plant growing. It produces IT 23C (based on the 0–4 scale) in the 4th-leaf stage at 21 ± 2 °C. W195 also carries another stripe rust resistance gene *Yr46*, which interacts with *Yr58* and enhances resistance. W195 and the wheat lines carrying both *Yr58* and *Yr46* produce IT ;1C at the 4th-leaf stage. SSR markers *Xsun533* and *Xsun476* mapped 0.6 cM proximal to *Yr58* and SNP marker *X3023704* mapped 1.2 cM distal to *Yr58*. Markers *Xsun533* and *Xsun476* can distinguish *Yr58* among 74 of 76 Australian wheat cultivars and can be used in marker-assisted selection (Chhetri et al. 2016a).

Yr59 is a QTL conferring HTAP resistance to stripe rust and located in chromosome 7BL, bin 7BL-0.86-1.00 (Zhou et al. 2014c). It was identified from an Iraqi spring wheat landrace, PI 178759. This gene showed effective resistance to stripe rust in field over 10 year in the US PNW. The locus was mapped with RGAP and SSR markers in an $F_{2:3}$ population from cross Avocet S × PI 178759. The *Yr59* locus was defined in an interval of 2.1 cM region by flanking markers *Xwgp5175* and *Xbarc32* and explained 31.8 and 54.7 % of phenotypic variation in rAUDPC and IT, respectively. Using the flanking SSR markers *Xwmc557* and *Xbarc32*, a panel of 73 wheat lines with or without *Yr59* could be distinguished, so these markers can be used for MAS (Zhou et al. 2014c). The linkage order of *Yr* genes on 7BL is *Yr39* - 41.1 cM - *Yr52* -5.4 cM - *Yr59* - 6.0 cM - *YrZh84* (Zhou et al. 2014c). A *Yr59* single-gene germplasm line, PI 660059, was developed from an F_7 line of cross AvS × PI 178759 (Wang et al. 2012). *Yr59* has been combined in new germplasm lines with *YrZh84* (PI 679606), *Yr39* (PI 679607) and *QYrPI182103.wgp-7BL* (PI 679615) for developing high-level and durable resistant wheat cultivars (Wang and Chen 2016).

Yr60 (*YrLalb*) confers moderate resistance (ITs 4-5 to 5-6 on the 0–9 scale) to stripe rust in cultivar Almop in both seedling and adult plant stages. It is located on the distal part of chromosome 4AL. *Yr60* co-segregates with SSR marker *Xwmc776* and at a genetic distance of 0.6 cM to two co-segregating markers *Xwmc219* and *Xwmc313* (Herrera-Foessel et al. 2015). *Yr60* is genetically linked with *Yr51* at about 10.1 cM, inferred from the molecular maps of these two genes. A minor QTL for APR to stripe rust was reported in the similar chromosomal region (Prins et al. 2011). Other disease resistance genes have been reported in this chromosomal regions including leaf rust resistance gene *Lr28* from *Ae. speltoides*, linked with *Xwmc313* at 5 cM (Bipinraj et al. 2011), and *Septoria tritici* blotch resistance gene *Stb7* closely linked with *Xwmc313* and *Xwmc219* at 0.3–1.0 cM and 0.5–1.1 cM, respectively (McCartney et al. 2003).

Yr61 confers ASR to stripe rust and is located in chromosome 7AS. This gene was identified from a Chinese wheat cultivar Pindong 34, which is effective against most predominant Chinese *Pst* races from 1980s to 2010, including CYR29, CYR31, CYR32, CYR33, PST-Ch42, PST-Su4 and PST-Su11. *Yr61* is highly resistant with IT 0-1 based on the 0–9 scale (Zhou et al. 2014a). *Yr61* was mapped on chromosome 7AS with RGAP and SSR markers. The genetic distances from *Yr61* to two closely flanking RGAP markers *Xwgp5765b* and *Xwgp5467* are 3.9 and 1.9 cM, respectively. These two RGAP markers have been converted to STS makers, and *Yr61*-carrying lines have a 300-bp allele in *STS5765b* and 175-bp allele in *STS5467*. *STS5467* can be used in MAS for *Yr61* (Zhou et al. 2014a).

Yr62 is a QTL conferring highly effective HTAP resistance to stripe rust and located in chromosome 4BL, bin 0.58-0.70 (Lu et al. 2014). This gene was identified in a Portuguese spring wheat variety PI 192252. This variety is susceptible to predominant US *Pst* races at the seedling stage, but highly resistant at the adult-plant stage under high temperatures. *Yr62* is a major QTL in PI 912252 which explained 40–60% of phenotypic variation for rAUDPC. The genetic distance of *Yr62* and its closely linked SSR markers *Xgwm251* and *Xgwm192* is 2.0 and 0.6 cM, respectively. *Yr62* is genetically linked with *Yr50* in repulsion at a genetic distance 27.1 ± 8.6 cM (Lu et al. 2014). *Yr62* single-gene germplasm PI 660060 was developed from an F_4 line of cross AvS × PI 192252 with improved agronomic trait compared to its donor (Wang et al. 2012). A spring wheat line (PI 679603) with the combination of *Yr62* and *Yr50* displayed high level resistance in both seedling and adult-plant stages (Wang and Chen 2016).

Yr63 confers ASR to stripe rust and is located to chromosome 7BS, bin 7BS1-0.27-1.00 in a Portuguese wheat landrace AUS 27955. The genetic distance of *Yr63* to close markers IWB33120 and IWB52844 is 0.9 and 1.5 cM, respectively (McIntosh et al. 2014).

Yr64 confers ASR to stripe rust and is located in chromosome 1BS, bin 1BS9-0.84-1.06. The resistance was identified in an Ethiopian spring durum wheat landrace PI 331260. Through a cross of Avocet S × PI 331260, the resistance of *Yr64* was transferred from durum wheat into common wheat, and a new germplasm line PI 660064 carrying single gene *Yr64* was selected from an F_8 line of the cross (Cheng et al. 2014; Wang et al. 2012). *Yr64* is highly effective against all tested races in the US with IT 0 or 1 (based the 0–9 scale) in its original durum wheat PI 331260 and 1 or 2 in common wheat PI 660064 (Cheng et al. 2014). The linkage map of *Yr64* and SSR markers on 1BS was constructed using an $F_{6:7}$ monogenically segregating population (Cheng et al. 2014). The flanking markers mapped *Yr64* in the long arm of chromosome 1B near the satellite region, bin 1BS9-0.84-1.06. The genetic distances of *Yr64* to distal marker *Xgw413* and proximal marker *Xgdm33* was estimated as 3.5 and 2.0 cM, respectively. *Yr64* is genetically linked with *Yr15*, *Yr24* and *Yr65* in repulsion at genetic distances of 21.7 ± 0.9, 33.8 ± 3.0 and 24.9 ± 6.2 cM, respectively (Cheng et al. 2014).

Yr65 also confers ASR to stripe rust and is located near the centromere of chromosome 1BS, bin 1BS10-C-0.05 (Cheng et al. 2014). This gene was identified in an Ethiopian durum wheat accession PI 480016. *Yr65* was transferred from durum wheat into the common wheat background through a cross between AvS and PI 480016. Similarly to *Yr64*, *Yr65* is highly effective against all tested US *Pst* races with IT 0 or 1 (on the 1–9 scale) in its original durum wheat PI 480016 and 1 or 2 in its common wheat carrier line 47 (PI 679621), an F_7 line from cross of AvS × PI 480016. The linkage map of *Yr65* and SSR markers on chromosome 1BS was constructed using an $F_{6:7}$ monogenically segregating population. The flanking markers mapped *Yr65* in chromosome 1BS near the centromere, bin C-1BS10-0.5. The genetic distances of *Yr65* to close distal markers *Xgwm18* and proximal marker *Xgwm11* was 1.2 and 2.1 cM, respectively. The estimated genetic distances and order among *Yr65* and other 2BS *Yr* genes are: *Yr24* – (7.1 ± 12.2 cM) – *Yr65* – (14.1 ± 12.2 cM) – *Yr15* – (21.7 ± 0.9 cM) – *Yr64* (Cheng et al. 2014). A common spring wheat line (PI 679623) was developed by pyramiding *Yr64* and *Yr65* and showed IT 0 under the severe stripe rust epidemic in Pullman, Washington in 2016 (Qie et al. 2016).

Yr66 confers ASR to stripe rust and is located in chromosome 3DS, bin 3DS6-0.55-1.00. This gene was identified in wheat line AGG91584WHEA (MSP 4543.1) and also is present in wheat genotype VL892. Line with *Yr66* showed IT 23C (on the 0–4 scale) against race 134E16Yr17+Yr27+ in the seedling stage. The genetic distances between *Yr66* and two flanking marker loci *XIWB47165* and *XIWB18087/XIWB56281* are 3.1 cM and 1.5 cM, respectively. *Yr66* is distal to *Yr49* (McIntosh et al. 2014).

Yr67 confers ASR to stripe rust and is located in the distal part of chromosome 7BL. This gene was identified from an Indian wheat cultivar C591, with a tentative name *YrC591* (Li et al. 2009). *Yr67* shows high level resistance against the predominant *Pst* races in China and race 134 E16Yr17+Yr27+ in Australia. The IT of C591 produced by Chinese *Pst* races is mostly 0; or 1, and that by Australian races is ;1-C to 1C according to the 0–4 scale. *Yr67* is also present in Chinese winter cultivar Zhongzhi 1. The linked molecular markers, including AFLP, SSR and EST markers, assigned *Yr67* in the distal end of long arm of chromosome 7B (Li et al. 2009; Xu et al. 2014). The genetic distances of *Yr67* to close markers *Xmag1714* and *Xbarc182* were estimated as 1.2 and 0.4 cM, respectively. Markers *Xbarc32*, *Xwmc273*, *Xcfa2040*, *Xmag1714* and *XBE425120* can distinguish *Yr67* among different wheat lines and can be used for tracing *Yr67* in MAS (Xu et al. 2014). Based on the comparative maps of other *Yr* genes on 7BL, the order of these *Yr* genes is very likely as *Yr39 - Yr6 - Yr52 - Yr59 - Yr67* (McIntosh et al. 2014; Zhou et al. 2014c).

Yr68 confers APR and is located in chromosome 4BL, bin 4BL1-0.86-1.00. This gene was identified in wheat line AGG91587WHEA1 (csAvYr4BL). The wheat line showed resistance at several international sites. *Yr68* co-segregates with marker *XIWA4640*. The genetic distances of *Yr68* and two flanking markers *XIWB74301* and *XIWB28394* were estimated as 0.5 and 0.5 cM, respectively (McIntosh et al. 2014).

Yr69 confers ASR to stripe rust and is located in chromosome 2AS, bin 2AS5-0.78-1.00. This gene was identified in Chinese wheat genotype CH7086, with a temporary name *YrCH86*. *Yr69* is effective against predominant *Pst* races in China. The linkage map around *Yr69* locus is *Xwmc25 - 2.7 cM - X2AS33 - 1.9 cM - Yr69 - 3.1 cM – Xmag3807*. *Yr69* is different from *Yr17* based on their pedigrees (Hou et al. 2016).

Yr70 confers ASR to stripe rust and is located on chromosome 5DS. This gene was identified from wheat line IL 393-4, a derivative of *T. durum* cv. WH890/*Ae. umbellutata* Pau3732//CS Ph/3/2*WL711. *Ae. umbellulata* Pau 3732 is the original donor of *Yr70*. The genetic distance of *Yr70* to close marker *Xgwm190* is 7.6 cM. *Yr70* behaves as an allele of *Yr40* which is also on 5DS and originated from *Ae. geniculata*. *Yr70* is associated with leaf rust resistance gene *Lr76* (McIntosh 2014, personal communication).

Yr71, temporary named *YrSA3*, confers APR to stripe rust and is located in chromosome 3DL. This gene is present in Australian cultivar Sunco (SUN9E27*4/3Ag14//WW15/3/3*Cook), and wheat lines AGG91588WHEA. Sunco carries three stripe rust resistance genes *Yr18*, *YrCK* and *Yr71*. A *Yr71* single-gene line from cross AvS and Sunco is susceptible at the seedling stage and moderately susceptible in adult-plant stage in the field (Bariana et al. 2016). The genetic map of *Yr71* was constructed using a RIL population from a cross between susceptible and resistant sibs in a BC_1F_2 family from Sunco/2*Avocet S. *Yr71* is mapped at 1.8 cM away from SSR marker *Xgwm114b*. *Xgwm114b* co-segregated with ten KASP markers and *Lr24/Sr24*. *Yr71* is likely distal to *Yr45* based on the positions of their flanking SSR markers (Bariana et al. 2016). Markers *Xgwm114b*, *KASP_16434*, *KASP_12707* and *KASP_20836* were validated for robustness and polymorphism in Australian wheat cultivars and can be used for MAS.

Yr72 is located to chromosome 2BL, bin 2BL5-0.59-0.89. This gene is present in wheat genotypes AUS 27506 and AUS 27894. The linkage map around *Yr72* locus is most likely: *Xsun481-2BL* - 1.8 cM - *Yr72* - 1.2 cM - *XIWB12294* - 1.5 cM - *Xsun482-2BL* - 1.5 cM - *XIWB69000*. Based on the *Yr* gene positions in deletion bin maps of chromosome 2BL, *Yr72* is far distal to previously mapped *Yr5*, *Yr7*, *Yr43*, *Yr44* and *Yr53* (McIntosh 2015, personal communication).

Yr73 and *Yr74* are located in chromosomes 3DL and 5BL, respectively (Dracatos et al. 2016). They are two dominant and complementary genes conferring stripe rust resistance against *Pst* races of A-type in Australia. The resistance was initially found in Australian spring wheat cultivar Avocet R and their derivatives and tentatively named as *YrA* (Wellings 1986). The individual gene of either *Yr73* or *Yr74* is susceptible to *Pst* A- type races. The genotype Avocet Susceptible (AvS) wheat only carries *Yr73* and is susceptible to most *Pst* races due to the terminal region (32% length) of long arm of chromosome 5B is truncated and the *YrA* resistance cannot be expressed. The putative source of *YrA* resistance in Australian cultivars Avocet, Banks, Condor and Egret is likely to have originated from the CIMMYT cultivar Lerma Rojo 64 through WW15 or Anza and Bluebird-derived cultivars (Wellings 1986). DArT markers *X116719[F]0* and *X1091695[F]0* associated with *Yr73* and *Yr74* are mapped at position 36.47 cM on chromosome 3DL and 88.64 cM on 5BL, respectively (Dracatos et al. 2016).

Yr75 confers APR to stripe rust and is located in chromosome 7AL, bin 7AL16-0.86-1.00. The gene was identified from an Australian wheat cultivar Axe. This gene shows R-MR/MR reaction to stripe rust under field conditions and displays IT ;1-2CN at the 4th-leaf stage. The *Yr75* single-gene line RIL#5 was derived from an F_6 line of cross Axe/Nyabing-3. SNP marker *XIWB34640* and SSR marker *Xcfa2040* are closely associated with *Yr75* (McIntosh 2015, personal communication).

Yr76 confers ASR and is located on chromosome 3AS, bin 3AS4-0.45-1.00 (Xiang et al. 2016). This gene was identified in a soft winter club wheat cultivar Tyee (CItr 17773), with a tentative name *YrTye* (Chen and Line 1992c). The resistance in Tyee originated from germplasm line CItr 7805 (selection from PI 58787), a tetraploid wheat introduced from Ethiopia (Allan 2014). *Yr76* displays high level resistance with ITs 1 or 2 on the 0–9 scale. It is highly resistant to *Pst* races in the US prior to 1979. Tyee was added to the North America *Pst* differential set in 1983 shortly after the appearance of a virulent race. Since then 130 races avirulent and 25 races virulent to Tyee have been identified in the US (Line and Qayoum 1992; Chen 2005; Chen et al. 2010; Wan and Chen 2012). Using the new 18 *Yr* single-gene differentials, 35 of 88 races identified in the US and 12 of 55 races in other countries are virulent to *Yr76*. The virulence for *Yr76* were found in Australia, Canada, Chile, China, Ecuador, Hungary, Italy, Kenya, Mexico, Nepal, Pakistan, Russian, Spain, Turkey and Uzbekistan (Sharma-Poudyal et al. 2013; Wan and Chen 2014; Wan et al. 2016, 2017; XM Chen and associates, unpublished data). Linked SSR markers delimited *Yr76* in an interval of 6.0 cM, distal to *Xwmc11* at 2.6 cM and proximal to *Xwmc532* at 3.4 cM. *Yr76* is presented in winter club wheat cultivars ARS-Chrystal,

Cara, Chukar and Hyak, and a soft white common winter wheat cultivar ARS-Amber (PI 665047) (Xiang et al. 2016).

Yr77 confers APR and is located on chromosome 6DS. It was first identified as *QYr.ucw-6D* associated with SNP marker *IWA176* allele 'CC' in a genome-wide associate study (Maccaferri et al. 2015). Dong et al. (2017) validated *QYr.ucw-6D* in six bi-parental cross populations and mapped the peak of the QTL closely linked to SNP marker *XIWA167*, between SSR markers *Xbarc54* (15.2 to 16.8 cM) and *Xcfd188* (1.1 to 3.9 cM). *Yr77* is present in the germplasm lines PI 164377, PI 322118, PI 388095, PI 520350, PI 192463 and PI 623378 (Maccaferri et al. 2015; Dong et al. 2017), and all of these lines were susceptible to the typical US *Pst* races in seedlings and resistance to natural infections in adult-plant stages in fields (Maccafferi et al. 2015). Marker *XIWA167* is not diagnostic for *Yr77* and only validates for the presence of *Yr77* in the six lines. For MAS, the flanking markers *Xbarc54* and *Xcfd188* should be used together with *XIWA167* (Dong et al. 2017).

Yr78 confers APR and is located in chromosome 6BS. Similarly to *Yr77*, *Yr78* resistance was first identified as *QYr.ucw-6B* associated with the SNP marker *XIWA7257* allele 'TT' in GWAS (Maccaferri et al. 2015). *Yr78* was validated in ten populations from bi-parental crosses, and was mapped at 0.6 cM proximal to *XIWA7257* and 3.9 cM distal to *XIWA4408* and *Xwmc737* (Dong et al. 2017). *Yr78* is present in germplasms PI 192493, PI 519805, PI 494101, PI 532116, PI 191351, PI 286543, PI 351878, PI 520108, PI 520265 and PI 520378 (Maccaferri et al. 2015; Dong et al. 2017). *Yr78* is located larger than 15 cM proximal to the seedling resistance gene *Yr35* (Dong et al. 2017). The 'T' allele of *XIWA7257* is a good diagnostic marker to detect the presence of *Yr78*. It is very likely that cultivars Janz (Bariana et al. 2010) and Stephens (Santra et al. 2008) possessed *Yr78* since both cultivars had the 'T' allele of *XIWA7257*. In fact, Santra et al. (2008) mapped two QTL (*QYrst.wgp-6BS.1* and *QYrst.wgp-6BS.2*) for HTAP resistance in Stephens to chromosome 6BS.

5.1.4.2 Temporarily Designated *Yr* Genes for Resistance to Wheat Stripe Rust

A total of 67 genes with provisional names have been reported for resistance to stripe rust, including 61 for ASR and 7 for APR or HTAP resistance (Table 5.3). Most of the genes should be different from each other and different from the permanently designated genes. Many of the genes appear to be effective against stripe rust and are valuable in breeding programs, and some of genes are useful in differentiating *Pst* races.

YrAlp confers ASR to stripe rust and is located in chromosome 1BS. This gene was identified from an US PNW spring wheat cultivar Alpowa. It is resistant to races PST-1 and PST-21, with IT 2 in the 0–9 scale, but susceptible to most *Pst* races in the US. *YrAlp* is a partially dominant gene as F_1 plants from cross AvS × Alpowa had IT 3 (Lin and Chen 2007). *YrAlp* was assigned on chromosome 1BS by SSR markers *Xgwm18* and *Xgwm11* and flanked by two RGAP markers *Xwgp47* at 1.1

Table 5.3 Temporarily named *Yr* genes for resistance to stripe rust in wheat

Gene	Type[a]	Source	Germplasm	Chrom.	Marker 1 (cM)	Marker 2 (cM)	Linked genes (cM)	References
YrAlp	ASR	Common wheat	Alpowa	1BS	*Xwgp47* (1.1)	*Xwgp48* (3.2), *Xgwm18* (13.2)	ND	Lin and Chen (2007)
YrAS2388	ASR	*Aegilops tauschii* ssp. *strangulata*	Syn-SAU-86, Syn-SAU-88	4DS	*Xwmc285* (1.7)	*Xwmc617* (34.6)	ND	Huang et al. (2011)
YrBai	APR	Common wheat	Baidatou	6DS	*Xbarc54* (3.6)	*Xcfd19* (2.1)	ND	Ma et al. (2015a)
YrC142	ASR	Synthetic wheat	CI142	1BS	*Xwmc419* (5.4)	*Xgwm273/Xbarc187* (0.8)	ND	Wang et al. (2009)
YrC51	ASR	Common wheat	C51	2BS	*XC51STS-4* (1.4)	*XC51R1* (5.6), *Xgwm429* (5.1)	ND	Zheng et al. (2014)
yrCH45-1	ASR	Rye	Chuanmai 45	1RS	ND	ND	ND	Yang et al. (2016)
yrCH45-2	ASR	Common wheat	Chuanmai 45	4BL	*Xwmc251* (12.7)	*Xmag1682* (26.0)	ND	Yang et al. (2016)
YrChk	ASR	Common wheat	Chike	1BL	*Xwmc44* (8.3)	*Xgwm259* (9.1)	ND	Liu et al. (2007)
YrCK	ASR	Common wheat	Sumco, Cook	2DS	*Xgdm005* (25)	*Xwmc190* (25)	ND	Bariana et al. (2001)
YrCle	ASR	Common wheat	Clement	4B	ND	ND	ND	Chen et al. (1995b)
YrCN17	ASR	*Secale cereale* L155	Rye L155, Chuanmong 12, Chuannong 17, Chuannong 18	1RS	ND	ND	ND	Tang et al. (2006)
YrDa1	ASR	Common wheat	Daws	1A	ND	ND	ND	Chen et al. (1995b)
YrDa2	ASR	Common wheat	Daws	5D	ND	ND	ND	Chen et al. (1995b)
YrDru	ASR	Common wheat	Druchamp	5B	ND	ND	ND	Chen and Line (1993a); Chen et al. (1996)

YrDru2	ASR	Common wheat	Druchamp	6A	ND	ND	ND	Chen and Line (1993a); Chen et al. (1996)
YrExp1	ASR	Common wheat	Express (PI573003)	1BL	*Xwgp78* (4.2)	*Xwmc631* (3.4)	ND	Lin and Chen (2008)
YrExp2	ASR	Common wheat	Express (PI573003), Expresso, Blanca, Grande, Jeff Pronto, Buck Proto	5BL	*Xwgp81* (1.0)	*Xwgp82* (0.7)	ND	Lin and Chen (2008)
yrGn22	ASR	Unknown	Guinong 22	1B	*Xcfd65*	*Xwe173*	ND	Li et al. (2016b)
YrH122	ASR	*Psathyrostachys huashanica* accession 0503383	H122	1DL	*Xbarc229* (7.7)	*Xwmc339* (4.3)	ND	Tian et al. (2011)
YrH2020 (YrHs)	ASR	*Psa. huashanica* accession 0503383	H9020-1-6-8-3	2DS	*Xgwm445* (5.8)	*Xgwm261* (4.4)	ND	Liu et al. (2008) and Li et al. (2012)
YrH46	ASR	Common wheat	Hybrid 46	6A	ND	ND	ND	Chen et al. (1996)
YrH52	ASR	*Triticum dicoccoide* accession H52	Accession H52	1BS	*Xgwm273a/ Xgwm273b* (2.7)	*Xgwm413/UBC212a/ Nor1* (1.3)	*Yr15* (3.9-8.2)	Peng et al. (2000a)
YrH9020a	ASR	*Psa. huashanica* accession 0503383	H9020-1-6-8-3	6DL	*Xbarc202* (8.3)	*Xbarc96* (4.5)	ND	Liu et al. (2014b)
YrHA	ASR	*Psa. huashanica* accession 0503383	H9014-121-5-5-9	1AL	*Xwmc469* (3.4)	*Xgwm497* (4.6)	ND	Ma et al. (2013)
YrHRMSN-81		Common wheat	HRMSN-81	2DS	*Xwgp_180bp* (3.4)	*Xwmc453* (14.7)	ND	Chen et al. (2013a)
YrHu	ASR	*Psa. huashanica* accession 0503383		3AS	*Xcfd79* (7.2)	*XBG604577* (0.7)	ND	Ma et al. (2016)
YrHua	ASR	*Psa. huashanica* accession 0503383	H9020-17-5	6AL	*XPM14-301* (5.4)	*XPM42-249* (2.7)	ND	Cao et al. (2008)

(continued)

Table 5.3 (continued)

Gene	Type[a]	Source	Germplasm	Chrom.	Marker 1 (cM)	Marker 2 (cM)	Linked genes (cM)	References
YrHVII	ASR	Common wheat	Heines VII	4A	ND	ND	ND	Chen et al. (1995b)
YrJ22	ASR	*Common wheat*	Jimai 22	2AL	*wmc658* (1.0)	*IWA1348* (7.3)	ND	Chen et al. (2016)
YrL693	ASR	*Thinopyrum intermedium*	L693, YU25	1B	*Xgwm273* (0.08)	*Xwmc626* (0.07)	ND	Huang et al. (2014)
yrLM168	APR	Common wheat	LM168, Milan	6A	*XR1* (2.4)	*XR2* (2.4)	ND	Feng et al. (2014)
YrLM168a	ASR	Common wheat	Milan, LM168a, LM168b	6BL	*Xbarc146* (4.6)	*Xwmc756* (4.6)	ND	Feng et al. (2015a)
YrMin	ASR	Common wheat	Minister	4A	ND	ND	*YrND*	Chen et al. (1996)
yrMY37	ASR	Common wheat	Mianmai 37	7BL	*Xgwm297* (0.78)	*Xbarc267* (0.38)	ND	Ren et al. (2015b)
YrMY41	ASR	Common wheat	Mianmai 41	1B	*Xgwm18* (2.0)	*Xwmc206* (4.9)	ND	Ren et al. (2015a)
YrND	ASR	Common wheat	Nord Desprez	4A	ND	ND	ND	Chen et al. (1996)
Yrns-B1	APR	Common wheat	Lgst 79-74	3BS	*Xgwm1329*	*Xgwm533_117bp*	ND	Khlestkina et al. (2007)
YrP138	ASR	Common wheat	PIW138	1BL	*Xbarc61* (6.2)	*Xwmc52* (29.8)	ND	Yue et al. 2010
YrP81	ASR	Common wheat	P81, Xu29	2BS	*Xgwm429* (1.8)	*Xwmc770* (4.1)	ND	Pu et al. (2010)
YrQz	ASR	Common wheat	Qz180	2BL	*P35M48_452bp* (3.4)	*P36M61_163bp* (4.1)	ND	Deng et al. (2004)
YrR61	APR	Common wheat	Pioneer 26R61 (PI 612153), Baldwin, AGS 2035,	2AS (bin 2AS5-0.78-1.00)	*Xbarc124*	*Xgwm359*	ND	Hao et al. (2011)

YrR212	ASR	*Se. cereale* R12	Dwarf rye R12, rye-wheat translocation lines R185, R205, R212	1RS	ND	ND	ND	Luo et al. (2008)
YrS	ASR	Common wheat	Stephens	3B	ND	ND	ND	Chen et al. (1994b)
YrSf	ASR	Common wheat	Suf-Omer	3BL	*Xgwm181* (2.9)	*Xgwm340* (5.1)	ND	Dawit et al. (2012)
YrSM139-1B	ASR	*T. dicocoides* accession AS846	Shaanmai 139	1B	*XSNP43* (3.92)	*XgBRG* (5.45)	*Yr15* (12.7)	Zhang et al. (2016)
YrSM139-2D	ASR	*Th. intermedium*	Shaanmai 139	2DS	ND	ND	ND	Ren et al. (2006) and Zhang et al. (2010)
YrSP	ASR	Common wheat	Spaldings Prolific, AvSYrSPNIL, Cook/YrSP-NIL, Kite/YrSP-NIL, Taichuang/YrSP-NIL	2BL	*Xdp269* (1.6)	*XIWA638* (0.6)	*Yr5* (12.6), *Yr7* (18.6), *Yr43* (15.3), *Yr44* (15.5), *Yr53* (13.2)	Feng et al. (2015b)
YrSph	ASR	*T. sphaerococcum* AS348	wheat line D31	2AS	*Xwmc246* (8.5); *Xgwm372* (16.1)	*F-me15/R-me6* (6.9); *Xwmc149* (12.7)	ND	Chen et al. (2012b)
YrSte	ASR	Common wheat	Stephens	2BL	ND	ND	ND	Chen et al. (1996)
YrSte2	ASR	Common wheat	Stephens	3B	ND	ND	ND	Chen et al. (1996)
YrSuj	ASR	Common wheat	C306	7BL	*Xcfa2040* (2.7)	*Xwmc526* (3.3)	ND	Lan et al. (2015)
YrTp1	ASR	*Th. ponticum* R431	A-3	2BS	*Xwmc477* (0.4)	*Xgwm374* (18.0)	ND	Yin et al. (2006)
YrTp2	ASR	*Th. ponticum* R431	A-3	7BL	*Xgwm297* (10.4)	*Xwmc364* (5.8)	ND	Yin et al. (2006)

(continued)

Table 5.3 (continued)

Gene	Type[a]	Source	Germplasm	Chrom.	Marker 1 (cM)	Marker 2 (cM)	Linked genes (cM)	References
YrTr1	ASR	Common wheat	Tres, AvSYrTr1NIL	1BS	*Xbarc8* (3.2)	*XIWA2872* (4.1)	ND	Feng and Chen (unpublished) and Chen and Line (1992a); Chen et al. (1995b)
YrTr2	ASR	Common wheat	Tres	3A	ND	ND	ND	Chen et al. (1992a, 1995d)
YrTu	ASR	Common wheat	Tussie	7BL	*Xgwm577* (11.9)	ND	ND	Dawit et al. (2012)
YrV3	ASR	*Haynaldia villosa*	Wheat-H. villosa translocation line V3	1B	*Xgwm124* (3.0)	*Xcfa2147* (3.8)	ND	Hou et al. (2013)
YrV23	ASR	Common wheat	Vilmorin 23,	2B	ND	ND	*YrSte*	Chen et al. (1996)
YrV23 (Yr3a)	ASR	Common wheat	Taichuang 29/Vilmorin 23 NIL	2BL	*Xwmc356* (9.4)	ND	ND	Wang et al. (2006)
YrWb	ASR	Common wheat	Wabe	7BL	*Xgwm577* (16.0)	*Xgwm344* (43.7)	ND	Dawit et al. (2012)
Yrwh2	ASR	Common wheat	Wuhan 2	3BS	*Xwmc540* (5.9)	*Xgwm566* (10.0)	ND	Zhou et al. (2014b)
Yrxy1	HTAP	Common wheat	Xiaoyan 54	7AS	*XM8* (3.5)	*XM1* (2.3)	ND	Zhou et al. (2011)

Yrxy2	HTAP	Common wheat	Xiaoyan 54	2AL	*Xgwm794* (4.0); *Xwmc819* (10.4)	*Xbarc5* (6.4)	ND	Zhou et al. (2011)
YrYam	ASR	Common wheat	Yamhill	4B	ND	ND	ND	Chen et al. (1996)
YrYL	ASR	Common wheat	Yilongtuomai	7D	*Xbarc92* (5.35)	*XCLRR for/Pto kin4* (9.86)	ND	Wu et al. (2016)
YrZh84	ASR	Common wheat	Zhou 8425B, Nainari 60, Annong 7959, Zhoumai 11, Zhoumai 12, Zhoumai 16, Aikang 58	7BL, bin 7BL10-0.78-1.00	*Xcfa2040* (1.4)	*Xbarc32* (4.8)	*Yr52* (12.2), *Yr59* (6.0), *YrPI182103* (6.4)	Li et al. (2006a, b) and Zhou et al. (2014c)
Yrzhong12-1	ASR	Common wheat	Zhongliang 12	7AL	*Xcfd20* (3.1)	*Xbarc121* (4.9)	ND	Ma et al. (2015b)
Yrzhong12-2	ASR	Common wheat	Zhongliang 12	1AL	*Xcfa2129* (3.9)	*Xwmc673* (6.3)	ND	Ma et al. (2015b)

[a]Type of resistance: *ASR* all-stage resistance, *APR* adult-plant resistance, *TSR* temperature sensitive resistance, *HTAP* high-temperature adult-plant resistance
[b]*ND* no data

cM and *Xwgp48* at 3.2 cM, respectively. Based on consensus SSR maps, *YrAlp* is likely linked with *Yr15* and *YrH52* in repulsion (Sun et al. 1997; Peng et al. 1999, 2000a, b). Besides *YrAlp*, Alpowa carries an HTAP resistance gene *Yr39*, conferring moderate to high level HTAP resistance as presented above (Lin and Chen 2007).

YrAS2388 confers ASR to stripe rust and is located on chromosome 4DS. This gene was identified from an Iranian *Ae. tauschii* ssp. *strangulata* (2n=2x=14, genome DD) accession AS2388. It has high level ASR to currently prevalent Chinese *Pst* races with IT 0-1 in the 0–9 scale (Liu et al. 2010). Two linked SSR *Xwmc617* and *Xwmc285* markers assigned *YrAS2388* in chromosome 4DS. The genetic distances of *YrAS2388* to *Xwmc617* and *Xwmc285* are 34.6 and 1.7 cM, respectively. The ITs in the synthetic hexaploid wheat lines Syn-SAU-86 (*T. turgidum* ssp. *turgidum* AS2313 × *Ae. tauschii* AS2388) and Syn-SAU-88 (*T. turgidum* ssp. *turgidum* AS2334 × *Ae. tauschii* AS2388) are higher (IT 2-4) than that in the original diploid donor (IT 0-1), indicating that *YrAS2388* only partially expresses in the synthetic hexaploid background (Huang et al. 2011). Trough analysis of 118 collections of *Aegilops* spp., *YrAS2388* was found to be commonly present in accessions of *Ae. tauschii* ssp. *strangulata* (Liu et al. 2013b).

YrBai confers APR and is located on chromosome 6DS. This gene was identified from a Chinese wheat landrace Baidatou. The resistance of *YrBai* is monogenically inherited in F_2 and $F_{2:3}$ populations from cross Mingxian 169 × Baidatou. SSR and SRAP markers mapped *YrBai* in chromosome 6DS. Two closely flanking markers *Xbarc54* and *Xcfd19* are 3.6 and 2.1 cM away from *YrBai*, respectively (Ma et al. 2015a). These two markers can be used in marker-assisted selection for transferring *YrBai* into commercial cultivars.

YrC142 confers ASR to stripe rust and should be located on chromosome 1BS. This gene was identified from synthetic wheat CI142 from cross Garza/Boy//*Ae. squarrosa* 271. *YrC142* is resistant to six Chinese *Pst* races. It was mapped with five SSR markers in a linkage order *Xwmc419* - *YrC142* - *Xwgm273/Xbrac187* - *Xgwm18* - *Xwmc626*, and the genetic distances of *YrC142* to *Xwmc419* and *Xwgm273/Xbrac187* loci are 5.4 and 0.8 cM, respectively. Markers *Xgwm273*$^{-162}$ and *Xgwm*$^{-168}$ can distinguish *YrC142* from *Yr10*, *Yr15* and *Yr24*/*Yr26*. Comparing the SSR maps of *YrC142* and *Yr24/Yr26*, *YrC142* is likely a new gene or new allele at *Yr24/Yr26* locus (Wang et al. 2009). The authors only assigned the gene to chromosome 1B. Based on the linked markers, the gene should be in the long arm of chromosome 1B.

YrC51 confers ASR to stripe rust and is located on chromosome 2BS. This gene was identified in a spring wheat cultivar C51 collected from the International Center for Agricultural Research in the Dry Areas (ICARDA), Syria (Zheng et al. 2014). *YrC51* is resistant to *Pst* races in China. This gene was mapped with SSR and RGAP markers. One closely flanking RGAP marker was genetically located at 1.4 cM to *YrC51* and was converted into STS marker *XC51STS-4* with the 563-bp allele associated with *YrC51*. Other flanking RGAP marker *XCR51R1*, 5.6 cM to *YrC51*, and SSR marker *Xgwm429*, 5.1 cM to *YrC51*, were also validated in a collection of wheat varieties (Zheng et al. 2014).

yrCH45-1 and *yrCH45-2* in Chinese cultivar Chuanmai 45 (CH45) confer ASR to stripe rust (Yang et al. 2016). F_1 plants from crosses CH45 × CH22 and CH45 × CN19 were susceptible and F_2 populations segregated as a ratio of 7 resistant to 9 susceptible, suggesting two recessive genes (*yrCH45-1* and *yrCH45-2*) conferring resistance in cultivar CH45. *yrCH45-2* was mapped in chromosome 4BL with flanking markers *Xwmc251* (12.7 cM) and *Xmag1682* (26.0 cM), respectively. *yrCH45-1* was cytogenetically located on the short arm (1RS) in translocated chromosome 1RS.1BL. Based on the origin and resistance spectrum, *yrH45-1* is different from *Yr9* and other rye originated *Yr* genes (Yang et al. 2016).

YrChk in a Chinese native wheat variety, Chike, is a major gene for conferring ASR to stripe rust (Liu et al. 2007). The gene was located in the long arm of chromosome 1B between SSR markers *Xwmc44* (8.3 cM) and *Xgwm259* (9.1 cM). Based on the origin, resistance type and chromosomal locations of the markers, this gene is likely different from those previously reported in chromosome 1BL.

YrCle confers ASR to stripe rust and is located on chromosome 4B. This gene was identified from cultivar Clement (Chen et al. 1995b). Monosomic analysis assigned this gene on chromosome 4B, but a specific region was not determined. In the hemizygous stage, this gene was resistant to PST-21 (formerly CDL-21) but not effective to PST-29 and PST-45 (formerly CDL-29 and CDL-45) (Chen et al. 1995b).

YrCK confers durable resistance to stripe rust and is located on chromosome 2DS. This gene was identified from Australian wheat cultivar Sunco, a derivative of cultivar Cook (Bariana et al. 2001). *YrCK* is temperature-sensitive, and displayed necrotic flecks to intermediate infection types following incubation of inoculated plants at 20 °C instead of usual temperature conditions for 17 °C. Markers *Xgdm005* and *Xwmc190* flanked *YrCK* at distal and proximal about 25 cM of each, respectively (Bariana et al. 2001). *YrCK* explained 31% of phenotypic variation in F_6 RIL population of cross AvS × Cook. When combined with *Yr18/Lr34*, *YrCK* attributed to a large proportion of pleiotropic effect in resistance to both stripe rust and leaf rust resistance in cultivar Cook (Navabi et al. 2005).

YrCN17 confers ASR to stripe rust. This gene originated from rye *S. cereal* chromosome 1RS. A white rye inbred line L155, selected from rye cultivar Petkus, was crossed with wheat cultivar Jubilar, and their hybrids were multiple backcrossed to wheat cultivars Manyang 11, Xichuang 1378, Fu 16-5 and Tibet 9296 to obtain a wheat-rye translocation 1BL.1RS line R14 (2n=42), which was crossed with wheat cultivar A302 to develop common wheat cultivars Chuannong 12 (CN12), Chuannong 17 (CN17) and Chuannong 18 (CN18) (Ren et al. 2006; Tang et al. 2006). Cultivars CN12, CN17 and CN18 have the same resistance (IT 1 in a 0–4 scale) as the rye donor L155 against current prevalent *Pst* races CYR31 and CYR32 in China (Luo et al. 2008). The resistance of *YrCN17* is a dominant monogenic different from *Yr9* which is susceptible to CYR31 and CYR32 (Ren et al. 2006).

YrDa1 confers ASR to stripe rust and is located in chromosome 1A and *YrDa2* also confers ASR resistance and is located on chromosome 5D. Both *YrDa1* and *YrDa2* were identified from US PNW winter wheat cultivar Daws through monosomic analysis (Chen et al. 1995b). No molecular markers are available for these genes. In addition to the ASR genes, Daws has HTAP resistance to stripe rust (Chen 2013).

YrDru confers ASR to stripe rust and is located on chromosome 5B. *YrDru2* also confers ASR and is located on chromosome 6A. Both *YrDru* and *YrDru2* were identified from a French winter wheat cultivar Druchamp through monosomic analysis (Chen and Line 1993a; Chen et al. 1996). *YrDru* is resistant to US races PST-21 and PST-45 and *YrDru2* is resistant to PST-1, PST-35 and PST-45 (Chen and Line 1993a). In addition to *YrDru* and *YrDru2*, Druchamp also carries *Yr3a*, *Yr29* and other 11 *Yr* genes/QTL for resistance to stripe rust (Hou et al. 2015).

Both *YrExp1* and *YrExp2* confer ASR to stripe rust and identified from a US PNW hard red spring wheat cultivar Express (PI 573003). *YrExp1* is located in chromosome 1BL and *YrExp2* on 5BL (Lin and Chen 2008). *YrExp1* is resistance to US races PST-1 and PST-21, but susceptible to PST-43 and PST-45 and other predominant races. *YrExp2* is resistance to all four races PST-1, PST-21, PST-43 and PST-45 tested. The genetic distances of RGAP marker *Xwgp78* and SSR marker *Xwmc631* flanking *YrExp1* were estimated as 4.2 cM and 3.4 cM respectively. RGAP markers *Xwgp81* and *Xwgp82* flank *YrExp2* about 1.0 cM and 0.7 cM, respectively from the gene. Markers *Xwgp81* and *Xwgp82* can distinguish *YrExp2* from 91% of 44 wheat genotypes tested. *YrExp2* is likely present in cultivars Expresso, Blanca Grande, Jeff Pronto and Buck Proto since the four cultivars and Express have the same alleles at the *Xwgp81* and *Xwgp82* loci. An *YrExp2* single-gene line is included in the *Yr* single-gene set of for identifying *Pst* races (Wan and Chen 2014). Virulence to *YrExp2* has been detected since the 1960s, but its frequency has been much higher since 2000 in the US. The virulence to *YrExp2* also distributes in all other countries, including Canada, Chile, China, Nepal, Pakistan, Turkey, Hungry and Kenya at an average frequency of 89% (Sharma-Poudyal et al. 2013), and Ecuador, Italy and Mexico (XM Chen and associates, unpublished data).

YrF is a QTL and identified from cultivar Francolin#1 (Lan et al. 2014). It is also named as *QYr.cim-2BS* (see the QTL section below).

yrGn22 confers ASR to stripe rust. It was identified from Chinese wheat cultivar Guinong 22, which was presumably developed from a hybridization between *Haynaldia villosa* (L.) Shur (2n = 14, VV) and *T. durum* Desf. cv. Sauwene 20 (2n = 4x =28, AABB) and letting F_1 plants freely pollinated with common wheat varieties. Guinong 22 is used as a differential cultivar for distinguishing *Pst* races in China, and it was highly resistant to most Chinese *Pst* races, but has become susceptible to relatively new races, such as Guinong 22-9, Guinong 22-14 and V26 (Han et al. 2015). In the study of the $F_{2:3}$ population from cross Guinong 22 × Mingxian 169, Li et al. (2016b) identified that *yrGn22* was resistant to races CYR33 and Su11-11 complying with a recessive inheritance model. *yrGn22* was mapped on chromosome 1B near the centromere, flanked by SSR marker *Xcfd65* and STS marker *Xwe173*, which also was linked with *Yr26*. Comparing the resistance patterns of this gene and other *Yr* genes on chromosome 1B to *Pst* races CYR32, CYR33, Guinong 22-9 and V26, Li et al. (2016b) found that *yrGn22* was different from a gene(s) in Vilmorin 23, and also genes *Yr9*, *Yr15* and *Yr21* (Lemhi). Allelism tests showed that *yrGn22* is different from *Yr15* and *Yr10*. *yrGn22* is very likely closely linked with or the same as *Yr26* based on the genetic distance to common linked marker *Xwe173* as 170 F_2 plants from cross of Guinong 22 and AvSYr26NIL were all resistant to CYR33.

YrH46 confers ASR and is located in chromosome 6A. This gene was identified from cultivar Hybrid 46 through monosomic analysis (Chen and Line 1993a; Chen et al. 1996). *YrH46* is resistant to PST-1, but susceptible to PST-21. This resistance character makes *YrH46* different from *YrDru2* (Chen et al. 1996). Besides *YrH46*, Hybrid 46 also carries *Yr3b* and *Yr4b* (Lupton and Macer 1962).

YrH52 confers ASR to stripe rust and is located on chromosome 1BS. This gene was identified in a wild emmer wheat *T. dicoccoides* accession H52 from Mt. Hermon, Israel (Peng et al. 1999). Despite both *Yr15* and *YrH52* are located on chromosome 1BS and originate from wild emmer wheat, *YrH52* is different from *Yr15* in the aspects of resistance spectrum and original carriers. *YrH52* is resistant to all tested races including 7E134 from Afghanistan while *Yr15* is resistant to most races but susceptible to 7E134 (Nevo et al. 1986; The et al. 1993). The two wild emmer wheat accessions, G25 (*Yr15*) and H52 (*YrH52*), are distributed in significantly different eco-geographical populations: *YrH52* is from the Mt. Hermon population (altitude 1300 m) and *Yr15* is from the Rosh Pinna population (altitude 700 m). Eco-geographic factors, e.g., altitude, mean annual temperature, mean day-night temperature difference and mean annual rainfall, are different between these two regions (Nevo et al. 1986; Van Silfhout 1989; Peng et al. 2000b). Molecular markers mapped *YrH52* near the centromere on chromosome 1BS (Peng et al. 1999, 2000a, b). *YrH52* are flanked by *Xgwm273a/Xgwm273d* distal at 2.7 cM and *Xgwm413/UBC212a/Nor1* proximal at 1.3 cM (Peng et al. 2000a, b). The linkage maps of *Yr15* and *YrH52* shared 10 common marker loci. Based on the two maps, *YrH52* is located at distal outwardly to *Yr15* and the map distance between *YrH52* and *Yr15* is 3.9 to 8.2 cM (Peng et al. 2000a).

YrH122, *YrH9020*, *YrH9020a*, *YrHu*, *YrHua* and *YrHA* originate from *Psathyrostachys huashanica* Keng and all confer ASR to stripe rust. *P. huashanica* is a perennial grass species and distributes only in the Huashan Mountain in Shaanxi province of China. The resistance of stripe rust was initially transferred into wheat by an interspecific hybridization between common wheat line 7812 and *P. huashanica* accession 0503383 followed by multiple backcrosses with 7812 (Chen et al. 1991). *P. huashanica* is immune or highly resistant to most *Pst* races in China whereas wheat line 7182 is susceptible. Several wheat-*P. huashanica* translocation lines with stripe rust resistance were obtained. The above six genes were identified and mapped from a series of crosses between wheat-*P. huashanica* resistant translocation lines and susceptible wheat cultivar Mingxian 169 by the stripe rust research team in the Northwest A&F University, Yangling, China. *YrH122* is identified in a translocation line H122, which is highly resistant to races CYR32 and Su11-4. Using SSR markers, *YrH122* was mapped on chromosome 1DL (Tian et al. 2011). The genetic distances of *YrH122* and close markers *Xbrac229* and *Xwmc339* are 7.7 and 4.3 cM, respectively. *YrH9020*, also named *YrHs* (Liu et al. 2008), is identified from a wheat line H9020-1-6-8-3 and is resistant to *Pst* race CYR33 (Tian et al. 2011). *Yr9020* was mapped on chromosome 2DS using SSR markers, with genetic distances to close markers *Xgwm455* and *Xgwm261* at 5.8 and 4.4 cM, respectively (Liu et al. 2008; Li et al. 2012). *YrH9020a* is also identified from wheat line H9020-1-6-8-3. This gene is resistant to CYR32. *YrH9020a* was mapped on chromosome 6DL, flanked by SSR markers *Xbarc202* and *Xbarc96* with genetic distances of 8.3 and 4.5 cM, respectively (Liu et al. 2014b). *YrHA* was identified from a wheat line H9014-121-5-5-9 and effective against race Su11-4. The gene was mapped on chromosome 1AL with flanking SSR markers *Xwmc469* and *Xgwm497* at genetic distances of 3.4 and 4.6 cM, respectively (Ma et al. 2013). *YrHua* was identified from translocation line H9020-17-5 and is resistant to race CYR31. *YrHua* was located on chromosome 6AL by SSR markers and flanked with AFLP markers *XPM14-301bp* (5.4 cM) and *XPM42-249bp* (2.7cM). AFLP marker *XPM14-301bp* was converted into a STS marker *XPM14-296bp*, which could be used in marker-assisted selection for transferring *YrHua* into other wheat cultivars (Cao et al. 2008). *YrHu* was identified from translocation line H9020-17-25-6-4 and resistant to all *Pst* races in China including predominant races CYR32, CYR33 and V26. The gene was mapped in chromosome 3AS with markers *Xcfd79* (7.2 cM) and EST-STS *BG604577* (0.7 cM) (Ma et al. 2016). Based on resistance responses, origins and chromosome locations, these six *Yr* genes are likely novel.

YrHVII confers ASR to stripe rust and is located in chromosome 4A. This gene was identified from European cultivar Heines VII through monosomic analysis (Chen and Line 1992c; Chen et al. 1995b). *YrHVII* is resistant to US *Pst* races PST-6, PST-21 and PST-35. Besides *YrHVII*, Heines VII has also been reported to have *Yr2* and *Yr25* (Chen and Line 1992c; Chen et al. 1995b; McIntosh et al. 2013)

YrJ22 confers ASR to stripe rust. It was identified from wheat cultivar Jimai 22, which is widely grown and also used as an elite parent in wheat breeding programs in China (Chen et al. 2016). *YrJ22* was mapped to the long arm of chromosome 2A with SSR marker *Xwmc658* (1.0 cM proximal) and SNP marker *XIWA1348* (7.3 cM distal). The gene is effective against most current Chinese *Pst* races (CYR31, CYR32, CYR33 and Su-1) but not CYR29. This gene is also effective to *Pst* isolates from Afghanistan, Chile, Ecuador, Egypt, German, Israel and the Netherlands (Chen et al. 2016).

YrL693 confers ASR to stripe rust and is located near the centromere of chromosome 1B. This gene was identified from a Chinese wheat line, L693, a $F_{6:7}$ progeny from cross Mianyang 11 × YU25. On the basis of pedigree analysis, *YrL693* was from YU25 that was derived from *Th. intermedium* by an interspecific cross (Zhang et al. 2011). *YrL693* is resistant to predominant *Pst* races in China. *YrL693* is co-segregating with four SSR markers, *Xgwm18-1B*, *Xgwm11-1B*, *Xbarc187-1B*, *Xbarc187-1B-1* and *Xwmc626-1B-1*, and flanked by *Xgwm273-1B* (0.08 cM) and *Xwmc626-1B-1* (0.07 cM) (Huang et al. 2014).

yrLM168 confers APR to stripe rust and is located on chromosome 6A. This gene was identified in a Chinese wheat line, LM168, a derivative from cross Chuannong 16 and Milan. The resistance of *yrLM168* is recessive. Pedigree analysis found that *yrLM168* originated from the CIMMYT wheat cultivar Milan. SSR markers *Xwmc59* and *Xwmc145* assigned the gene to chromosome 6A and two most close markers were RGAP markers R1 (2.4 cM) and R2 (2.4 cM) flanking the gene (Feng et al. 2014).

YrLM168a is different from *yrLM168*. *YrLM168a* is a dominant gene conferring ASR to stripe rust. This gene was identified from the selected stripe rust resistant lines LM168a and LM168b based on the extreme phenotypes (Feng et al. 2015a). *YrLM168a* was mapped in chromosome 6BL. The nearest flanking markers were SSRs *Xwmc756* (4.6 cM) and *Xbarc146* (4.6 cM).

yrMY37 confers ASR to stripe rust, originating from Chinese cultivar Mianmai 37 (Ren et al. 2015b). The recessive gene was mapped on chromosome 7BL near the centromere with flanking markers *Xgwm297* (0.78 cM proximal) and *Xbarc267* (0.38 cM distal) (Ren et al. 2015b).

YrMY41 confers ASR to stripe rust. This gene was identified from a Chinese cultivar Mianmai 41 (Ren et al. 2015a). Linked molecular markers placed *YrMY41* near the centromere of chromosome 1B. SSR markers *Xgwm18* and *Xwmc406* flanking *YrMY41* were 2.0 and 4.9 cM away from the gene, respectively. The resistance spectrum to 29 *Pst* races showed that *YrMY41* is different from *Yr3*, *Yr9*, *Yr10*, *Yr15*, *Yr26* and *YrCH42*. Allelism tests indicated that *YrMY41* is likely a new allele at *Yr26* locus.

YrMin and *YrND* confer ASR to stripe rust. These genes were identified from European cultivars Minister and Nord Desprez, respectively and both located in chromosome 4A using monosomic analysis (Chen and Line 1992c, 1993a; Chen et al. 1996). Both *YrMin* and *YrND* are effective against US *Pst* races PST-21 and PST-45. Because both Minister and Nord Desprez have cultivars Hatif Inversable and Square Head in their pedigrees, it is possible that *YrMin* and *YrND* are the same genes (Chen et al. 1996).

Yrns-B1 confers non-race-specific APR to stripe rust and is located in chromosome 3BS. This gene was identified from a German cultivar Lgst 79-74 with durable resistance to stripe rust (Borner et al. 2000). *Yrns-B1* was mapped using nine SSR markers. Marker *Xgwm1329* was found having the highest LOD score and was most closely linked to *Yrns-B1*. Combining with resistance and pedigree of Lgst 79-74, the presence of marker *Xgwm533* (117-bp) was characteristic for the *Yrns-B1* resistance among the resistant German and UK cultivars which are progenitors of Lgst 79-74. The *Xgwm533* 117-bp allele is not associated with resistance in Russian wheat collections (Khlestkina et al. 2007).

YrP138 is an ASR gene identified from a Pakistan wheat line PIW138 (Yue et al. 2010). This gene is effective against predominant *Pst* races in China. *YrP138* was mapped in chromosome 1BL, flanked by markers *Xbarc61* (6.2 cM) and *Xwmc52* (29.8 cM) (Yue et al. 2010).

YrP81 confers ASR to stripe rust and is located on chromosome 2BS. This gene was identified from a Chinese spring wheat line, P81. Based on pedigree analysis, *YrP81* is derived from Xu29, a Syria wheat line (Maya S/Sap S) from an ICARDA wheat collection (Pu et al. 2007). *YrP81* is dominant and provides high level resistance to predominant *Pst* races in China. RGAP and SSR markers mapped *YrP81* in chromosome arm 2BS. The map distances between *YrP81* and flanking markers

Xgwm429 (251 bp) and *Xwmc770* (180 bp) were estimated as 1.8 cM (distal) and 4.1 cM (proximal), respectively (Pu et al. 2010).

YrQz confers ASR to stripe rust. This gene was identified in common wheat line Qz180 and mapped to chromosome 2BL. AFLP markers P35M48 (452 bp) and P36M61 (163 bp) were found flanking the gene at the genetic distances of 3.4 cM and 4.1 cM, respectively (Deng et al. 2004).

YrR61 is a QTL conferring APR to stripe rust and located in chromosome 2AS, bin 2AS5-0.78-1.00 (Hao et al. 2011). *YrR61* was identified from a soft red winter wheat cultivar Pioneer 26R61 (XW663, PI 612153, PVPO), a leading cultivar grown in the southern and eastern USA since 2000. Through QTL analysis, *YrR61* was mapped in the distal part of chromosome 2AS and flanked by markers *Xbarc124* and *Xgwm359* in an interval of a genetic distance of 15 cM. *YrR61* explained 56.0% of phenotypic variation for the mean of IT data. Based on the presence of the two flanking markers and field resistance evaluation, *YrR61* is also present in wheat cultivars Baldwin and AGS 2035 and advanced breeding lines GA001138-8E36 and GA021087-9LE33 (Hao et al. 2011).

YrR212 confers ASR to stripe rust and was located on chromosome 1RS of a dwarf rye inbred line R12. A wheat-rye 1RS.1BL translocation line was selected from the backcross progeny of cross Manyang 11*2/R12 and was crossed with cultivar A302 to develop three wheat lines R185, R205 and R212 (Yan et al. 2005; Luo et al. 2008). R185, R205 and R212 displayed immune reaction (IT 0) as its rye donor against predominant *Pst* races in China. The inheritance of resistance in these lines is monogenic, dominant and associated with the 1RS.1BL translocation. The resistance reaction of *YrR212* is different from *Yr9* and *YrCN17*. These genes are either closely linked genes or different alleles at the same locus (Luo et al. 2008).

YrSf confers ASR to stripe rust and was located on chromosome 3BL in Ethiopian common wheat line Suf-Omer (Dawit et al. 2012). This gene is effective against 18 tested *Pst* races from Ethiopia, France and Germany. *YrSf* is flanked by markers *Xgm181* (2.9 cM) and *Xgwm340* (5.1 cM).

YrSM139-1B and *YrSM139-2D* confer ASR to stripe rust in a Chinese wheat cultivar Shaanmai 139. Monosomic analysis of the resistance in Shaanmai 139 assigned these genes in chromosomes 1B and 2D, respectively (Zhang et al. 2010). The resistance of *YrSM139-1B* was derived from *T. dicocoides* accession AS846 and that of *YrSM139-2D* from *Th. intermedium* according to pedigree analysis and chromosome locations (Ren et al. 2006; Zhang et al. 2010). *YrSM139-1B* was mapped on chromosome 1BS flanked by markers *XSNP 43* (3.92 cM) and *XgBRG* (5.45 cM). Allelism tests and resistance spectrum patterns showed that *YrSM139-1B* is different from *Yr10*, *Yr15* and *Yr24/Yr26/YrCH42*. The genetic distance between *YrSM139-1B* and *Yr15* was estimated as 12.86 cM (Zhang et al. 2016).

YrSte and *YrSte2* (*YrS*), conferring ASR to stripe rust, were reported in wheat cultivar Stephens, which has been grown in the US PNW since the late 1970s, partially due to its durable HTAP resistance to stripe rust (Chen 2013). Because it also has race-specific ASR, Stephens was included in the previous set of differentials (Line and Qayoum 1992; Chen et al. 2002). Because two genes were detected early and located to chromosome early in Stephens (Chen and Line 1992b, c; Chen et al.

1994b), one was identified as *Yr3a* and the other was named as *YrS* for the simplicity (Chen et al. 1994b), but a third gene was later identified, and thus the three genes were re-named as *Yr3a*, Y*rSte* and *YrSte2* (*YrS*) and located by monosomic analysis to chromosomes 1B, 2B and 3B, respectively (Chen et al. 1996). *YrSte* is resistant to races PST-6 and PST-21; and *YrSte2* resistant to PST-1 and PST-45. Both *YrSte* and *YrSte2* were dominant, where *Yr3a* was recessive and resistant to PST-21 and PST-29 (Chen et al. 1996). Cultivar Stephens is genetically heterogeneous for several traits. The three genes for ASR resistance were identified from crosses made between Stephens and Michigan Amber (MA) and the set of Chinese Spring monosomic lines using seeds from a single plant of Stephens. A RIL population derived from the same Stephens × MA cross was also used later for mapping QTL for HTAP resistance (Santra et al. 2008), which will be discussed in the QTL section.

YrSP confers ASR to stripe rust and is located in chromosome 2BL, bin 2BL-C-0.5 (Feng et al. 2015b). This gene was identified from wheat cultivar Spaldings Prolific. *YrSP* near-isogenic lines were developed by backcrossing Spaldings Prolific with AvS, Cook and Kite in Australia, and Taichuang in China (McIntosh et al. 1995; Wellings et al. 2009; Guan et al. 2005). Cultivar Spaldings Prolific and AvSYrSPNIL are used in the European and *Yr* single-gene differential sets, respectively (Johnson et al. 1972; Wan and Chen 2014). *YrSP* has high level resistance (IT 0 to 2, severity 0 to 5%) in most trial locations of the US before the virulent races overcome its resistance in 2007. The frequencies of virulence to *YrSP* ranged from 3.4% in 2010 to 18.2% in 2014 in the US (Wan and Chen 2014; Chen et al. 2016; Wan and Chen unpublished data). Among 84 races identified in the United States with the new set of 18 *Yr* single-gene differentials, 9 (10.7%) were virulent to *YrSP* (Wan and Chen 2014; Wan et al. 2016); and of 55 races identified in other countries, 8 (14.5%) were virulent (Sharma-Poudyal et al. 2013). *YrSP* was delineated by a high-resolution map, with the flanking markers *Xdp269* being 1.6 cM distal and SNP *XIWA638* being 0.6 cM proximal to *YrSP* (Feng et al. 2015b). *YrSP* is genetically linked with *Yr5*, *Yr7*, *Yr43*, *Yr44* and *Yr53* in repulsion, with genetic distances to the four *Yr* genes being 12.6, 18.6, 15.3, 15.5 and 13.2 cM, respectively. A set of six SNP linked markers (*XIWA2872*, *XIWA3973*, *XIWA6561*, *XIWA1489*, *XIWA6317* and *XIWA6122*) and three SSR markers (*Xmc332*, *Xdp269* and *Xwmc441*) can be used to distinguish *YrSP* with the linked *Yr* genes and other wheat genotypes (Feng et al. 2015b).

YrSuj confers ASR to stripe rust and was identified in a mutant line of a tall Indian wheat cultivar C306. *YrSuj* was resistant to Indian and Mexican *Pst* races, displaying IT 2-3 to race Mex96 according to the 0–9 scale at seedling stage in the greenhouse (Lan et al. 2015). *YrSuj* was located on the distal region of chromosome 7BL and flanked by markers *Xcfa2040* (2.7 cM) and *Xwmc526* (3.3 cM). This gene is overlapping with QTL *QLr.cim-7BL* conferring leaf rust resistance in cultivar C306 (Lan et al. 2015).

YrTp1 and *YrTp2* are ASR genes and effective against *Pst* races in China (Yin et al. 2006). These two genes were identified in wheat line A-3, derived from progeny of cross *T. aestivum* × *Th. ponticium*. The donor of these genes is *Th. ponticium* accession R431. *YrTp1* is dominant and *YrTp2* recessive. *YrTp1* was mapped in chromosome 2BS, flanked by *Xwmc477* (0.4 cM) and *Xgwm374* (18.0 cM). *YrTp1* was mapped in chromosome 7BL, flanked by *Xwmc364* (5.8 cM) and *Xgwm297* (10.4 cM) (Yin et al. 2006).

YrTr1 and *YrTr2* both confer ASR to stripe rust and were identified in winter club wheat cultivar Tres (Chen and Line 1992a). Through monosomic analysis, *YrTr1* and *YrTr2* were located in chromosomes 6D and 3A, respectively (Chen et al. 1995b). These genes displayed independently dominant inheritance to race PST-35, and complementary dominant interaction to PST-21 (Chen et al. 1995b). A *YrTr1* near-isogenic line (AvSYrTr1NIL) was developed by backcrossing Tres and AvS by Wellings et al. (2009). *YrTr1* was mapped on chromosome 1BS with numerous SSR and SNP markers, and the genetic distances between *YrTr1* and its flanking markers *Xbarc8* and *XIWA2872* were estimated as 3.2 cM (distal) and 4.1 cM (proximal), respectively (Chen and associates, unpublished data). The recent molecular mapping study corrected the chromosomal location previously reported using monosomic analysis.

YrTu confers ASR to stripe rust. The gene was located in chromosome 7BL in Ethiopia common wheat line Tussie (Dawit et al. 2012). This gene is effective against 20 *Pst* races from Ethiopia, France and Germany. A linkage map was constructed with SSR markers using an F_2 population, showing the linkage of this gene with *Xgwm577* at 11.9 cM.

YrV3 confers ASR to stripe rust. The gene was identified from *T. aestivum-Hayaldia villosa* translocation line V3 and mapped to chromosome 1BL (Hou et al. 2013). SSR markers *Xgwm124* and *Xcfa2147* were closely linked to *YrV3* with genetic distances of 3.0 and 3.8 cM, respectively. Based on the chromosomal location, the reaction patterns and pedigree analysis, *YrV3* should be a novel gene for resistance to stripe rust in wheat. Besides *YrV3*, the translocation line was determined to have two independent dominant genes for resistance to race CYR29, a single dominant gene to CYR31, two complementary dominant genes to CYR32-6 and Sun11-4 and two independently dominant genes to CYR33, in addition to a single dominant gene (*YrV3*) for resistance to Sun11-11 (Hou et al. 2013).

YrV23 was identified in European wheat cultivar Vilmorin 23 using US *Pst* races (Chen and Line 1992c, 1993a; Chen et al. 1996). The *YrV23* symbol was used to designate a gene different from another gene (*Yr4a*) in Vilmorin 23 (Chen et al. 1996). This gene confers all-stage resistance to stripe rust and was located on chromosome 2B through monosomic analysis (Chen et al. 1996). *YrV23* is allelic or closely linked with *YrSte* when tested with PST-21 (Chen and Line 1992c, 1993a). Chinese scientists developed a near-isogenic line by backcrossing Vilmorin 23 for a gene, *YrV23* (*Yr3a*), with Chinese wheat cultivar Taichung 29, and identified SSR marker *Xwmc356* linked with the gene at 9.4 cM. The marker assigned this gene to chromosome 2BL (Wang et al. 2006). Although the gene was also referred as *Yr3a*, it is likely the same as *YrV23* identified in the previous studies by Chen and Lines (1992c, 1993a) and Chen et al. (1996) because of the same chromosome.

YrWb confers ASR to stripe rust and was located on chromosome 7BL in Ethiopian common wheat line Tussie (Dawit et al. 2012). This gene is resistant to 20 *Pst* races from Ethiopia, France and Germany. The linkage map with SSR marker based on an F_2 population shows the gene was flanked by markers *Xgwm577* at 11.9 cM and *Xgwm344* at 43.7 cM, respectively.

Yrwh2 confers ASR to stripe rust and is located on chromosome 3BS. This gene was identified in a Chinese winter wheat cultivar Wuhan 2. The resistance of *Yrwh2* is recessive in cross of Mingxian 169 and Wuhan 2. *Yrwh2* was mapped with SSR and RGAP markers, and flanked by SSR markers *Xwmc540* and *Xgwm566* at 5.9 and 10.0 cM, respectively (Zhou et al. 2014b).

Yrxy1 and *Yrxy2* confer HTAP resistance to stripe rust and were identified from a Chinese winter wheat cultivar Xiaoyan 54. Using SSR and RGAP markers, *Yrxy1* and *Yrxy2* were mapped in chromosome 7AS and 2AL, respectively. Both genes are genetically independent and recessive, but the resistance of *Yrxy1* is more effective than *Yrxy2*. *Yrxy1* is flanked by RGAP markers *XM8* and *XM1* at 3.5 cM distal and 2.3 cM proximal, respectively. *Yrxy2* is flanked by SSR markers *Xgwm794* (4.0 cM proximal) and *Xbarc5* (6.4 cM distal) (Zhou et al. 2011).

YrYam confers ASR to stripe rust. This gene was identified in US PNW winter wheat cultivar Yamhill and located in chromosome 4B using monosomic analysis (Chen and Line 1992b, c; Chen et al. 1996). *YrYam* is resistant to PST-1, but susceptible to PST-35. Besides *YrYam*, Yamhill also has *Yr2* and *Yr4a*, these two genes are resistant to PST-1 and PTS-35.

YrYL confers ASR to most Chinese *Pst* races. The gene was identified from Yilongtuomai, a wheat landrace collected from Yilong County, Sichuan province. The resistance of *YrYL* is dominant. A linkage group of 6 SSRs, 2 RGAPs and 1 SRAP was constructed for *YrYL* and mapped the gene in 7DS flanked by SSR marker *Xbarc92* (5.35 cM) and RGAP marker *CLRRfor/Ptokin4* (9.86 cM). Based on pedigrees and race reactions, *YrYL* should be a new gene for resistance to stripe rust.

YrZh84 confers ASR to stripe rust and is in chromosome 7BL, bin 7BL10-0.78-1.00. This gene was identified from a Chinese wheat cultivar, Zhou 8425B. *YrZh84* is effective against 19 out of tested 25 isolates from 15 countries, showing resistant to intermediate resistant reactions (IT 1+ to 2, in the 0–4 scale) to *Pst* isolate 86094 from Kenya and 86107 from Ethiopia, and susceptible to two isolates from Chile and Netherlands (Li et al. 2006b). Molecular markers mapped *YrZh84* in the distal part of chromosome 7BL. Two flanking markers *Xcfa2040* and *Xbarc32* were identified at 1.4 cM proximal and 4.8 cM distal, respectively (Li et al. 2006b). Based on the information of its resistance, pedigree and linked markers, *YrZh84* is derived from cultivar Nainari 60. *YrZh84* is also present in cultivars Annong 7959, Zhoumai 11, Zhoumai 12, Zhoumai 16 and Aikang 58. *YrZh84* is genetically linked with *Yr52*, *Yr59* and *YrPI182103* in chromosome 7BL in repulsion, with genetic distances of 12.2, 6.0 and 6.37 cM, respectively (Zhou et al. 2014c, Wang and Chen 2016).

Yrzhong12-1 and *Yrzhong12-2* are two ASR genes and were identified in a Chinese winter wheat cultivar Zhongliang 12. Based on the pedigree and stripe rust reactions, the resistance in Zhongliang 12 is derived from cultivar Kangyin 655. *Yrzhong12-1* is resistant to race CYR30 and *Yrzhong12-2* is resistant to race CYR31. The resistance of each gene is dominant and independent. *Yrzhong12-1* is located in chromosome 7AL, and flanked by SSR markers *Xcfd20* (3.1 cM) and *Xbarc121* (4.9 cM). *Yrzhong12-2* is located in chromosome 1AL, and flanked by markers *Xcfa2129* (3.9 cM) and *Xwmc673* (6.3 cM) (Ma et al. 2015b).

5.1.4.3 QTL for Resistance to Wheat Stripe Rust

Different from the permanently and temporarily named *Yr* genes discussed above, which were identified and mapped mostly using qualitative genetic approaches, QTL for stripe rust resistance were identified and mapped through QTL mapping although some of the QTL with major effects could be studied qualitatively. These QTL were either given quite long symbols or just given a name as the molecular markers mostly associated to the QTL. So far, a total of 327 QTL with different names (although some may be the same QTL) have been reported for resistance to stripe rust in wheat. To facilitate easy presentation and discussion, we re-name the QTL using numbers with a prefix "Q" followed by the chromosome. The 327 QTL and their wheat germplasm or cultivar names, chromosomes, flanking markers, genetic ranges (in cM), LOD (log10 likelihood ratio) values, phenotypic variation (PEV) explained, frequency of detection in tests, and associated genes for other trait, if available are presented in Table 5.4. Their relative chromosome location and relationships to permanently and temporarily named *Yr* genes are shown in Fig. 5.1.

The 327 QTL distribute on every chromosomes, ranging from 4 in chromosome 6D to 36 in chromosome 2B. The A genome (chromosomes 1A to 7A) has 109, the B genome (chromosomes 1B-7B) 165 and the D genome (chromosomes 1D to 7D) 53 QTL. Regarding to chromosome groups, Group 1 (chromosomes 1A, 1B and 1D) has 47, Group 2 (2A, 2B and 2D) 75, Group 3 (3A, 3B and 3D) 44, Group 4 (4A, 4B and 4D) 39, Group 5 (5A, 5B and 5D) 45, Group 6 (6A, 6B and 6D) 40 and Group 7 (7A, 7B and 7D) 37 QTL. The following discussion of these QTL will be arranged by the chromosome groups and then by chromosomes according to the sequential order in Table 5.4.

5.1.4.3.1 QTL in Group 1 Chromosomes

Fourteen QTL (Q1A.1 – Q1A.14) for resistance to stripe rust have been reported in chromosome 1A and distributed on both the short arm and long arm (Table 5.4; Fig. 5.1). Five QTL, *QYr.cim-1AS* from cultivar Sujata (Lan et al. 2015), *QYrid.ui-1A* from IDO444 (Chen et al. 2012a), *QYr.sig-1A.1* from Kariega (Prins et al. 2011), *QYr.tam-1AS* from TAM111 (Basnet et al. 2014a) and *QYr.sun-1A* from Janz (Bariana et al. 2010), were reported on the short arm. Of these QTL, *QYrid.ui-1A* and *QYr.cim-1AS* are located in the distal region. Nine QTL were reported in the long arm, including *QYr.inra-1A* from Renan (Dedryver et al. 2009), *QYr.tam-1AL* from TAM112 (Basnet et al. 2014a), *QYr.caas-1AL* from Naxos (Ren et al. 2012a), *QRYr1A.1* from Pastor (Jighly et al. 2015), *QYr.cim-1AL* from Pastor (Rosewarne et al. 2012) and *QYrst.orr-1AL* from Stephens (Vazquenz et al. 2012),

Table 5.4 Quantitative trait loci (QTL) for resistance to stripe rust in wheat

Code	Gene	Wheat variety	Chrom.	Marker 1	Marker 2	Genetic interval (cM)	LOD[a]	PEV[b]	Freq.[c]	Note[d]	References
Group 1 chromosomes											
Q1A.1	*QYrid.ui-1A*	IDO444	1AS	*X377889*	*XLMW1*	6.1	3.0	9.0	1/2	NI	Chen et al. (2012a, b, c)
Q1A.2	*QYr.cim-1AS*	Sujata	1AS	*XwPt-9752*	*Xgdm33*	6.3	4.9	13.8		NI	Lan et al. (2015)
Q1A.3	*QYr.sgi-1A.1*	Kariega	1AS	*Xs15m19D*	*XwPt-3698*	ND[e]	5.0	5.5	1/2	NI	Prins et al. (2011)
Q1A.4	*QYr.tam-1AS*	TAM111	1AS	*XTSM120*	*XwPt-6015*	ND	5.1-5.7	17.1-26.4	2/5	NI	Basnet et al. (2014a)
Q1A.5	*QYr.sun-1A*	Janz	1AS	*Xgwm164*	*Xwmc312*	15	3.0–3.3	6.5-7.0	4/6	NI	Bariana et al. (2010)
Q1A.6	*QYr.inra-1A*	Renan	1AL	*Xfba118b*	ND	ND	6.5	9.4	1/1	NI	Dedryver et al. (2009)
Q1A.7	*QYr-1A_IWA5754_APR*	Several[f]	1A	*XIWA5754*	ND	ND	ND	5	1/2	GWAS	Zegeye et al. (2014)
Q1A.8	*QYr.wsu-1A.1_IWA5505*	Several	1A	*XIWA5505*	ND	ND	ND	4.89	4/4	GWAS	Bulli et al. (2016)
Q1A.9	*QYr.tam-1AL*	TAM112	1AL	*XwPt-5167, XwPt-666616*	*XwPt-8016, XwPt-1011*	ND	4.9-6.1	15.7-18.0	2/5	NI	Basnet et al. 2014a
Q1A.10	*QYr.caas-1AL*	Naxos	1AL	*Xwmc59*	*Xbarc213*	ND	3.8	8.2	1/3	NI	Ren et al. (2012a)
Q1A.11	*QRYr1A.1*	Several	1AL	*XwPt-3712, XwPt-0497, XwPt-6754*	ND	0	ND	9.6	1/1	GWAS	Jighly et al. (2015)

(continued)

Table 5.4 (continued)

Code	Gene	Wheat variety	Chrom.	Marker 1	Marker 2	Genetic interval (cM)	LOD[a]	PEV[b]	Freq.[c]	Note[d]	References
Q1A.12	*QYr.cim-1AL*	Pastor	1AL	*XwPt-6005*	*XwPt-4709*	ND	4.9	3.6-4.1	2/4	NI	Rosewarne et al. (2012)
Q1A.13	*QYr.wsu-1A.2_IWA3215*	Several	1A	*XIWA3215*	ND	ND		2.20	4/4	GWAS	Bulli et al. (2016)
Q1A.14	*QYrst.orr-1AL*	Stephens	1AL	*XwPt-4399*	ND	ND	6.8	14	1/2	NI	Vazquez et al. (2012)
Q1B.1	*QYr.cau-1BS_AQ24788-53*	AQ24788-83	1BS	*Xgwm374*	*Xbarc194*	14.4	2.4-5.6	6.9-15.2	2/2	Major QTL	Quan et al. (2013)
Q1B.2	*QYrco.wpg-1B.1_Coda*	Coda	1BS, C-1BS10-0.50	*XIWA4754*	*XIWA2547*	9	38.6-90.3	32-83	9/9	Major QTL	Case et al. (2014)
Q1B.3	*IWA6489_APR*	Several	1B	*XIWA6489*	ND	ND	ND	7	1/2	GWAS	Zegeye et al. (2014)
Q1B.4	*QYr.wpg-1B.1_IWA4715*	Several	1B	*XIWA4715*	ND	ND	ND	5.0	1/3	GWAS	Naruoka et al. (2015)
Q1B.5	*QYr.wsu-1B.1_IWA5963*	Several	1B	*XIWA5963*	ND	ND	ND	1.60	4/4	GWAS	Bulli et al. (2016)
Q1B.6	*QYr.wpg-1B.2_IWA63*	Several	1B	*XIWA63*	ND	ND	ND	5.0-8.0	3/3	GWAS	Naruoka et al. (2015)
Q1B.7	*QYr.caas-1BL.1RS*	SHA3/CBRD	1RS	*Xiag95*	*Xcfa21*	6.7	5.3	7.9	4/4	NI	Ren et al. (2012a)

Q1B.8	*QYrco.wpg-1B.2*	Coda	1BS	*Xbrac119*	*Xbarc137*	5.0	19.0-37.0	1-2	3/9	NI	Case et al. (2014)
Q1B.9	*QYr.wpg-1B.3_IWA7578*	Several	1B	*XIWA7578*	ND	ND	ND	6.0	1/3	GWAS	Naruoka et al. (2015)
Q1B.10	*IWA6891_Seedling*	Several	1BS	*XIWA6891*	ND	ND	ND	5.0-16.1	2/2	GWAS	Zegeye et al. (2014)
Q1B.11	*QYr.cim-1BSb*	Pastor	1BS	*XwPt-6240*	*wPt-8168*	ND	6.6	5.0-5.1	2/3	NI	Rosewarne et al. (2012)
Q1B.12	*QYrdr.wgp-1BL.1_gwm131*	Druchamp	1BL	*Xgwm131*	ND	ND	2.90	5.23	3/6	NI	Hou et al. (2015)
Q1B.13	*QYr-1B_Sachem*	Sachem	1B	*XwPt-3579*	ND	ND	ND	8.7-16.5	1/3	NI	Singh et al. (2013)
Q1B.14	*QYr.wsu-1B.2_IWA5915*	Several	1B	*XIWA5915*	ND	ND	ND	1.86	4/4	GWAS	Bulli et al. (2016)
Q1B.15	*QYrex.wgp-1BL_Express*	Express	1BL	*Xgwm631*	*Xgwm268*	ND	3.1-4.1	4.5-5.2	2/3	NI	Lin and Chen (2009)
Q1B.16	*QPst.jic-1B*	Brigadier	1B	*Xwmc735*	ND	ND	3.23-4.50	7.9-13.1	2/2	Minor QTL	Jagger et al. (2011)
Q1B.17	*QYr.sun-1B_CPI133872*	CPI133872	1B	*XwPt-1313*	ND	ND	2.8-4.8	10-17	3/4	At *Yr29* locus	Zwart et al. (2010)
Q1B.18	*QYr.sun-1B_Kukri*	Kukri	1B	*XwPt-0202*	*Xbrac80*	90	2.5-3.1	4.0-6.0	2/6	NI	Bariana et al. (2010)
Q1B.19	*QYr.sun-1B_Wollaroi*	Wollaroi	1B	*XwPt-1770*	*XwPt-4361*	ND	1.97	6	1/3	NI	Bansal et al. 2014
Q1B.20	*QYr.ucw-1B_IWA3892*	Several	1B	*XIWA3892*	ND	ND	ND	0.7	6/6	GWAS	Maccaferri et al. (2015)

(continued)

Table 5.4 (continued)

Code	Gene	Wheat variety	Chrom.	Marker 1	Marker 2	Genetic interval (cM)	LOD[a]	PEV[b]	Freq.[c]	Note[d]	References
Q1B.21	*QYr.jic-1B_Guardian*	Guardian	1B	*Xgwm259*	*Xgwm818*	ND	5.7-11.7	15-45	2/2	At *Yr29* locus	Melichar et al. (2008)
Q1B.22	*QYr.tam-1B_Quaiu*	Quaiu3	1BL	*XwPt-668027*	*XcSLV46G22*	ND	7.6	6.5	2/2	At *Yr29* locus	Basnet et al. (2014b)
Q1B.23	*QYr.cim-1BL_Francolin#1*	Francolin#1	1BL	*XcsLV46*	*XwPt-9028*	ND	9.3	12.4	3/3	At *Yr29* locus	Lan et al. (2014)
Q1B.24	*QYrdr.wgp-1BL.2_IWA8581*	Druchamp	1BL	*Xwmc830*	*Xgwm259*	6.6	12.21	28.58	6/6	NI	Hou et al. (2015)
Q1B.25	*QYr-1B_Saar*	Saar	1BL	*Xwmc719*	*Xhbe248*	6.8	3.8	17.4	4/4	At *Yr29* locus	Lillemo et al. (2008)
Q1B.26	*QYr.cim-1BL*	Pastor	1BL	*XcsLV46*	*Xgwm818*		23	15.1-22.1	4/4	At *Yr29* locus	Rosewarne et al. (2012)
Q1B.27	*QYr.1BL-Attila*	Attila	1BL	*LTN2*	*XP35/M55*	8.8	13-24	33-64	3/3	At *Yr29* locus	Rosewarne et al. (2008)
Q1D.1	*QYr.caas-1DS_Naxos*	Naxos	1DS	*XUgwm353*	*Xgdm33b*	10.1	3.1	2.1-5.8	3/3	NI	Ren et al. (2012a)
Q1D.2	*QYrst.orr-1DS_Stephens*	Stephens	1DS	*X379337*	ND	ND	3.3	11	1/8	NI	Vazquez et al. (2012)
Q1D.3	*QYr.sun-1D_CPI133872*	CPI133872	1DS	*Xwmc147*	*XP45M53a*	3.2	2.8-4.8	1.0-17.1	2/3	NI	Zwart et al. (2010)
Q1D.4	*QYrdr.wgp-1DS_IWA2268*	Druchamp	1DS	*XIWA2268*	*XIWA7797*	0.1	2.18	3.48	5/6	NI	Hou et al. 2015

Q1D.5	*QYr. wpg-1D.1_IWA6960*	Several	1DS	*XIWA6960*	ND	ND	ND	6.0-18.0	3/3	GWAS	Naruoka et al. (2015)
Q1D.6	*QYr. ucw-1D_IWA980*	Several	1D	*XIWA980*	ND	ND	ND	0.9	3/6	GWAS	Maccaferri et al. (2015)
Group 2 chromosomes											
Q2A.1	*QYr.tam-2AS_TAM 111*	TAM111	2AS	*XwPt-3896*	*XwPt-6431*	15	4.4-5.9	17.0-17.9	2/5	NI	Basnet et al. (2014a)
Q2A.2	*QYr. uga-2AS_26R61*	Pioneer 26R61	2AS	*Xbarc124*	*Xgwm359*	15	25.6	56	2/3	NI	Hao et al. (2011)
Q2A.3	*QYr.ufs-2A*	Cappelle-Desprez	2A	*Xgwm636*	*XwPt-0003*	28	10.6-18.8	28.8-47.8	2/2	Major QTL	Agenbag et al. (2012)
Q2A.4	*QYr. sun-2A_Wollaroi*	Wollaroi	2AS	*XwPt-7285*	*XwPt-1722*	1.4	2.0.-3.37	9.3	3/3	NI	Bansal et al. (2014)
Q2A.5	*QYrtm.pau-2A*	*T. monococcum* accession Pau14087	2AS	*Xwmc407*	*Xwmc170*	3.6	7.8-15	7-12	3/3	NI	Chhuneja et al. (2008)
Q2A.6	*QYr. inra_2AS.1_Recital*	Recital	2AS	*XDupw210*	*Xcfd36*	ND	6.45	9.35	1/1	NI	Dedryver et al. (2009)
Q2A.7	*QYr. ucw-2A.2_IWA422*	Several	2A	*XIWA422*	ND	ND	ND	0.8	5/6	GWAS	Maccaferri et al. (2015)
Q2A.8	*QYr. wpg-2A.2_IWA8274*	Several	2A	*XIWA8274*	ND	ND	ND	5.0-7.0	6/8	GWAS	Naruoka et al. (2015)
Q2A.9	*QYrva. vt-2AS_VA00W-38*	VA00W-38	2AS	*Xgwm296b*	*XwPt-7721*	19.0	14.1-17.3	30.8-66.7	2/4	NI	Christopher et al. (2013)

(continued)

Table 5.4 (continued)

Code	Gene	Wheat variety	Chrom.	Marker 1	Marker 2	Genetic interval (cM)	LOD[a]	PEV[b]	Freq.[c]	Note[d]	References
Q2A.10	*QYr.ucw-2AS_PI610750*	PI610750	2AS	*XwPt-3896*	*Xwmc177*	34.3	3.8	2.2	2/4	NI	Lowe et al. (2011)
Q2A.11	*QYrtb.orz-2AS_cfd36*	Tubbs	2AS	*Xcfd36*	ND	ND	17.9	57.7	7/7	Major QTL	Vazquez et al. 2015
Q2A.12	*QYr.wpg-2A.3_IWA3047*	Several	2A	*XIWA3047*	ND	ND	ND	4.0	3/8	GWAS	Naruoka et al. (2015)
Q2A.13	*QYr.inra-2AS_CampRemy*	Camp Remy	2AS	*Xwgm382a*	*Xgwm359*	29	5.37-6.08	20-40	4/4	NI	Mallard et al. (2005)
Q2A.14	*QYrst.orr-2AS_Stephens*	Stephens	2AS	*XwPt-0003*	ND	ND	7.8	19	6/8	NI	Vazquez et al. (2012)
Q2A.15	*QYr.wpg-2A.1_gwm359*	Several	2A	*Xgwm359*	ND	ND	ND	5.0	2/8	GWAS	Naruoka et al. (2015)
Q2A.16	*QYr.wpg-2A.4_IWA7410*	Several	2A	*XIWA7410*	ND	ND	ND	3.0	1/8	GWAS	Naruoka et al. (2015)
Q2A.17	*QYr.wsu-2A.1_IWA2526*	Several	2A	*XIWA2526*	ND	ND	ND	4.34	4/4	GWAS	Bulli et al. (2016)
Q2A.18	*QYr.sun-2A*	Kukri	2AS	*XwPt-0003*	*XwPt-3114*	50	3.7-4.1	12.9-15	2/6	NI	Bariana et al. (2010)
Q2A.19	*QYr.wsu-2A.2_IWA5824*	Several	2A	*XIWA5824*	ND	ND	ND	2.62	4/4	GWAS	Bulli et al. (2016)

Q2A.20	*QYr. ucw-2A.3_IWA424*	Several	2A	*XIWA424*	ND	ND	ND	0.3	4/6	GWAS	Maccaferri et al. 2015
Q2A.21	*IWA7339_APR*	Several	2A	*XIWA7339*	ND	ND	ND	10	1/2	GWAS	Zegeye et al. 2014
Q2A.22	*IWA544*	Several	2A	*XIWA544*	ND	ND	ND	7.1	4/4	GWAS	Jighly et al. (2015)
Q2A.23	*QYrns. orz-2AL_wPt-7011*	NSA-98-0995	2AL	*XwPt-7011*	ND	ND	12.6	16.7	5/7	NI	Vazquez et al. (2015)
Q2A.24	*QYr. caas-2AL_IWB11764*	Zhong892	2AL	*XBS00092550_51*	*Xwsnp_Ex_c59095_60108097*	8.1	3.0	5.1	2/5	NI	Liu et al. (2015)
Q2A.25	*QYr. wpg-2A.5_IWA5855*	Several	2A	*XIWA5855*	ND	ND	ND	4.0	1/8	GWAS	Naruoka et al. (2015)
Q2A.26	*QYr. wpg-2A.6_IWA966*	Several	2A	*XIWA966*	ND	ND	ND	5.0	1/8	GWAS	Naruoka et al. (2015)
Q2A.27	*QYR2*	Camp Remy	2AL	*Xgwm356*	*Xgwm382*	20.0	1.70	10.7	2/2	NI	Boukhatem et al. (2002)
Q2B.1	*wPt-6271*	Several	2B	*XwPt-6271*	ND	ND	ND	6.6	1/4	GWAS	Jighly et al. (2015)
Q2B.2	*QYr. inra-2BS_Renan*	Renan	2BS	*Xgwm210a*	*Xfcd36*	ND	7.50	8.56	1/1	NI	Dedryver et al. (2009)
Q2B.3	*QYrst. orr-2B.2_Stephens*	Stephens	2BS	*XwPt-0408*	ND	ND	4.1	11	6/8	NI	Vazquez et al. (2012)
Q2B.4	*QYr. wpg-2B.1_IWA7370*	Several	2B	*XIWA7370*	ND	ND	ND	5.0-6.0	4/8	GWAS	Naruoka et al. (2015)
Q2B.5	*QYrlu. cau-2BS1_Luke*	Luke	2BS	*Xwmc154*	*Xgwm148*	38.2	5.1-7.7	31.8-36.6	4/4	NI	Guo et al. (2008)

(continued)

Table 5.4 (continued)

Code	Gene	Wheat variety	Chrom.	Marker 1	Marker 2	Genetic interval (cM)	LOD[a]	PEV[b]	Freq.[c]	Note[d]	References
Q2B.6	*QYr.wpg-2B.2_IWA7697*	Several	2B	*XIWA7697*	ND	ND	ND	4.0	1/8	GWAS	Naruoka et al. (2015)
Q2B.7	*IWA2572*	Several	2B	*XIWA2572*	ND	ND	ND	3.5	1/1	GWAS	Jighly et al. (2015)
Q2B.8	*QYrid.ui-2B.1*	IDO444	2BS	*wPt-9668*	*Xgwm429*	22	12.3	25	2/4	NI	Chen et al. (2012a)
Q2B.9	*QYr.sgi-2B.1_Kariega*	Kariega	2BS, bin 2BS1-0.53-0.75	*Xbarc200*	*XwPt-6278*	25	4.8-29.4	17.7-45.7	2/2	Major QTL	Prins et al. (2011) and Agenbag et al. (2014)
Q2B.10	*QYrid.ui-2B.2*	IDO444	2BS	*Xgwm429*	*Xbarc91*	40	16.5	31.0	2/4	NI	Chen et al. (2012a)
Q2B.11	*QYr.sgi-2B.1a/QYr.sgi-2B.1_Kariega*	Kariega	2BS	*Xgwm148*	*Xs12m60A*	25	28.6-55.2	17-30		NI	Ramburan et al. (2004) and Prins et al. (2011)
Q2B.12	*QYr.cim-2BS_Yr31_Chapio*	Chapio	2BS	*XwPt-7026*	*XwPt-0079*	15.2	3.1-6.6	4.4-11.8	3/8	At *Yr31* locus	Yang et al. (2013) and Rosewarne et al. (2012)
Q2B.13	*QYrlo.wpg-2BS_Louise*	Louise	2BS	*Xwmc474*	*Xwmc148*	16.9	5.5-30.0	11-57	8/8	NI	Carter et al. (2009)
Q2B.14	*QYrst.orr-2BS.2_Stephens*	Stephens	2BS	*XwPt-0408*	ND	ND	4.1	11	5/6	NI	Vazquez et al. 2012

Q2B.15	*QYrlu.cau-2BS2_Luke*	Luke	2BS	*Xgwm148*	*Xbarc167*	23.8	5.5-8.8	32.5-41.5	4/4	NI	Guo et al. (2008)
Q2B.16	*QYr.caas-2BS_Pingyuan50*	Pingyuan 50	2BS	*Xbarc13*	*Xbarc230*	14.2	1.7-1.9	5.1-5.7	2/4	NI	Lan et al. (2010a)
Q2B.17	*QYr3-2B_Opata85*	Opata85	2BS	*Xcdo405*	*Xbcd152*	7.4	7.39	30.7	ND	NI	Boukhatem et al. (2002)
Q2B.18	*QYr.tam-2BL_TAM111*	TAM111	2B	*XwPt-6242*	*XwPt-6471*	5.8	7.2	30.5	5/5	NI	Basnet et al. (2014a)
Q2B.19	*QYr.ucw-2B_UC1110*	UC1110	2BS	*Xgwm429*	*Xbarc230*	34.0	3.3-7.5	4.5	2/4	NI	Lowe et al. (2011)
Q2B.20	*QYr.inra-2B.1_CampRemy*	Camp Remy	2BS	*Xgpw3032*	*Xcfd50a*	30.1	5.96-17.92	22-70		NI	Mallard et al. (2005)
Q2B.21	*QYr.cim-2BS_Francolin#1*	Francolin#1	2BS	*Xwmc474*	*xgwm374*	3.8	12.8	22.7	3/3	Race-specific, *YrF*	Lan et al. (2014)
Q2B.22	*QYr.2BS-Attila*	Attila	2BS	*XP32/M62*	*XP88/M64*	20	ND	4.7-9.3	1/3	At *Yr27* locus	Rosewarne et al. (2008)
Q2B.23	*IWA1488_APR*	Several	2B	*XIWA1488*	ND	ND	ND	6	1/2	GWAS	Zegeye et al. (2014)
Q2B.24	*QRYr2B.2*	Several	2B	*XIWA6093*	ND	ND	ND	6.6	1/4	GWAS	Jighly et al. (2015)
Q2B.25	*QYr.inra-2B.2_CampRemy*	Camp Remy	2BL	*Xbarc101*	*Xwmc175*	7.7	21.70-35.5	42-61	4/4	NI	Mallard et al. (2005)
Q2B.26	*QYr.caas-2BL_Naxos*	Naxos	2BL	*XwPt-8460*	*XwPt-3755*	5.5	4.9	12.2	1/3	NI	Ren et al. (2012b)

(continued)

Table 5.4 (continued)

Code	Gene	Wheat variety	Chrom.	Marker 1	Marker 2	Genetic interval (cM)	LOD[a]	PEV[b]	Freq.[c]	Note[d]	References
Q2B.27	*QYraq.cau-2BL_Aquileja*	Aquileja	2BL	*Xwmc175*	*Xwmc332*	22.3	7.8-13.9	47.0-61.5	4/4	NI	Guo et al. (2008)
Q2B.28	*QYr.caas-2BL.2*	Lumai 21	2BL	*Xcfd73*	*Xgwm47*	13.7	ND		1/3	NI	Ren Y et al. (2015c)
Q2B.29	*QYr.nafu-2BL*	P9897	2BL	*Xcfd73*	*Xgpw5770*	8.8	5.0	10.8	2/2	NI	Zhou et al. (2015a)
Q2B.30	*QYR1*	Camp Remy	2BL	*Xgwm501*	*Xgwm47*	3.5	11.84	45.8	2/2	NI	Boukhatem et al. (2002)
Q2B.31	*QYrdr.wgp-2BL_IWA6286*	Druchamp	2BL	*XIWA6286*	*XIWA7583*	0.3	5.27	10.31	5/6	NI	Hou et al. (2015)
Q2B.32	*QYrns.orz-2BL_wPt-0950*	NSA-98-0995	2BL	*XwPt-0950*	ND	ND	12.2	21.7	5/7	NI	Vazquez et al. (2015)
Q2B.33	*QYr.caas-2BL.3_IWB36313*	Linmai 2	2BL	*XIACX8602*	*XIAAV1798*	6.3	3.4	5.9	2/5	NI	Liu et al. (2015)
Q2B.34	*QYr-2B-Sachem*	Sachem	2B	*XwPt-3632*	ND	ND	ND	26.3	1/3	NI	Singh et al. (2013)
Q2B.35	*QYr.2BL-AvS*	Avocet S	2BL	*Xgwm619*	*Xgwm1027*	12	3.6	6.3	2/3	NI	Rosewarne et al. (2008)
Q2B.36	*QYr.caas-2BS.2*	Lumai 21	2BS	*Xbarc156*	*Xbarc1147*	20.7	2.9	7.4	0/3	NI	Ren et al. (2015c)
Q2D.1	*QYr.caas-2DS_Libellula*	Libellula	2DS	*Xcfd51*	*Xgwm261*	21.5	8.00	9.9	2/4	Minor QTL	Lu et al. (2009)

Q2D.2	*QYr.wpg-2D.1_IWA1939*	Several	2D	*XIWA1939*	ND	ND	ND	7.0-12.0	1/8	GWAS	Naruoka et al. (2015)
Q2D.3	*QYr.cim-2DS*	Avocet	2DS	*XwPt-9780*	*XwPt-3728*	11.1	4.5-5.0	5.3-8.0	2/3	NI	Lan et al. (2014)
Q2D.4	*QYr.caas-2DS.2_Lumai21*	Lumai 21	2DS	*Xbarc168*	*Xwmc18*	26.6	5.5	13.8	3/3	Major QTL	Ren et al. (2015c)
Q2D.5	*QYr.inra-2DS_CampRemy*	Camp Remy	2DS	*Xgwm102*	*Xgwm539*	25.4	5.70-10.9	24-69	4/4	NI	Mallard et al. (2005)
Q2D.6	*QYr.ufs-2DS_Cappelle-Desprez*	Cappelle-Desprez	2DS	*Xgwm102*	*XwPt-664520*	30	6-7	4.7-8.4	1/2	At *Yr16* locus	Agenbag et al. (2012)
Q2D.7	*QYr.caas-2DL_Naxos*	Naxos	2DL	*XwPt-6752*	*Xcfa47*	1.3	2.1-5.6	2.9-10.2	2/3	NI	Ren et al. (2012a, b, c)
Q2D.8	*QYr.wpg-2D.2_IWA6851*	Several	2D	*XIWA6851*	ND	ND	ND	4.0-5.0	2/8	GWAS	Naruoka et al. (2015)
Q2D.9	*QYr.jic-2D_Guardian*	Guardian	2DL	*Xgwm539*	*Xgwm349*	ND	2.1-4.4	8-11	ND	NI	Melichar et al. (2008)
Q2D.10	*QYr-2DL*	Fukuho-komugi	2DL	*Xgwm349*	ND	ND	14.58	6.5-9.6	3/3	NI	Suenaga et al. (2003)
Q2D.11	*QPst.jic-2D*	Alcedo	2DL	*Xgpw8086b*	*Xgwm320*	3	14.36-18.43	30.1-36.2	2/2	GWAS, major QTL	Jagger et al. (2011)
Q2D.12	*QYr.tam_2D_Quaiu*	Quaiu3	2DL	*XwPt-667162*	*XwPt-667054*		43.8	53.5	2/2	At *Yr54* locus	Basnet et al. (2014b)

(continued)

Table 5.4 (continued)

Code	Gene	Wheat variety	Chrom.	Marker 1	Marker 2	Genetic interval (cM)	LOD[a]	PEV[b]	Freq.[c]	Note[d]	References
Group 3 chromosomes											
Q3A.1	*QYr-3A_Saar*	Saar	3AS	*Xstm844tcac*	*Xbarc310*	2.3	5.5	40	1/4	Minor QTL	Lillemo et al. (2008)
Q3A.2	*QYr.caas-3AS_IWB48450*	Zhong892	3AS	*Xkukri_c96747_274*	*Xkukri_rep_c102131_891*	3.7	10.0	15.8	5/5	Major QTL	Liu et al. (2015)
Q3A.3	*QYrst.orr-3AL_Stephens*	Stephens	3AL	*XwPt-1652*	ND	ND	6.8	14	1/2	NI	Vazquez et al. (2012)
Q3A.4	*QYr.cau-3AL_AQ24788-83*	AQ24788-83	3AL	*Xwmc627*	*Xwmc624*	>20	2.6-2.9	4.7-6.8	1/1	Minor QTL	Quan et al. (2013)
Q3A.5	*IWA8630_APR*	Several	3A	*XIWA8630*	ND	ND	ND	11.9	1/2	GWAS	Zegeye et al. (2014)
Q3A.6	*IWA5849*	Several	3A	*XIWA5849*	ND	ND	ND	4.5-5.5	3/4	GWAS	Jighly et al. (2015)
Q3A.7	*QYrdr.wgp-3AL_IWA6834*	Druchamp	3AL	*XIWA6834*	ND	ND	1.09	1.78	3/6	NI	Hou et al. (2015)
Q3A.8	*QYr.wsu-3A_IWA3401*		3A	*XIWA3401*	ND	ND		5.84	4/4	GWAS	Bulli et al. (2016)
Q3A.9	*QYr.cim-3A_Avocet*	Avocet	3AL	*XwPt-6422*	*XwPt-7890*	ND	11.0	2.4-3.2	2/4	NI	Rosewarne et al. (2012)
Q3B.1	*QYr.3BS-Pavon*	Pavon76	3BS	*XpAGGmCAA5*	*XpAATmCAC2*	2.3	2.03	3.3-5.9	3/3	NI	William et al. (2006)

Q3B.2	*QYr.cim-3BS_Yr30_Chapio*	Chapio	3BS	*XPbc*	*Xgwm533*	7.5	3.5-19.2	4.7-18.0	7/8	NI	Yang et al. (2013)
Q3B.3	*QYrrb.ui-3B.1*	Rio Blanco	3BS	*X345897*	*xwPt-3921*	10.6	3.6	10	1/4	NI	Chen et al. (2012a)
Q3B.4	*QYrAlt.syau-3BS*	Alturas	3BS	*Xgwm389*	*Xbarc238*	3.1	17.77	34.28	2/2	NI	Zhao et al. (2012)
Q3B.5	*QYr.nafu-3BS*	P9897	3BS	*Xgwm533.2*	*Xbarc133*	3.3	5.2	13.4	2/2	NI	Zhou et al. (2015a)
Q3B.6	*QYr-3B_Oligoculm*	Oligoculm	3BS	*Xgwm389*	ND	ND	23.16	0.2-4.9	5/5	NI	Suenaga et al. (2003)
Q3B.7	*QYr.inra-3BS_Renan*	Renan	3BS	*Xgwm533*	ND	ND	7.70	10.93	1/1	NI	Dedryver et al. (2009)
Q3B.8	*QYr.tam-3B_Quaiu*	Quaiu3	3BS	*Xgwm533*	*XwPt-7984*	ND	7.1	5.6	2/2	At *Yr30* locus	Basnet et al. (2014b)
Q3B.9	*QYr.nafu-3BS*	Caoxuan 5	3BS	*Xgwm533.2*	*Xgwm493*	10.7	6.5-8.6	17.3-20.1	4/4	NI	Zhou et al. (2015c)
Q3B.10	*QYr.cim-3BS.2_Francolin#1*	Francolin#1	3BS	*Xgwm533.2*	*Xbarc133*	1.3	3.7	5.2	3/3	NI	Lan et al. 2014
Q3B.11	*QYr.ucw-3BS_UC1110*	UC1110	3BS	*XBE446462*	*Xbrac133*	2.5	20-22	24.9	3/4	NI	Lowe et al. (2011)
Q3B.12	*QYrco.wpg-3BS.1_Brundage*	Brundage	3BS, bin C-3BL2-0.22	*XIWA195*	*XIWA7212*	15	4.7-33.5	2-6	7/9	NI	Case et al. (2014)
Q3B.13	*QYr.ucw-3B.2_IWA5202*	Several	3B	*XIWA5202*	ND	ND	ND	1.3	2/6	GWAS	Maccaferri et al. (2015)

(continued)

Table 5.4 (continued)

Code	Gene	Wheat variety	Chrom.	Marker 1	Marker 2	Genetic interval (cM)	LOD[a]	PEV[b]	Freq.[c]	Note[d]	References
Q3B.14	*QYr.uga-3BS.1_AGS2000*	AGS2000	3BS	*XwPt-2557*	*Xbarc133*	1.4	3.8	5.0	1/3	NI	Hao et al. (2011)
Q3B.15	*wPt-800213-IWA6510*	Several	3B	*XwPt-800213*	*XIWA6510*			3.4-6.2	4/4	GWAS	Jighly et al. (2015)
Q3B.16	*QYr.sun-3B*	Kukri	3B	*XwPt-8238*	*XwPt-4209*	29	2.6	5.0	1/6	NI	Bariana et al. (2010)
Q3B.17	*QYr.caas-3BS_IWB35069*	Zhong892	3BS	*XIAAV5662*	*Xexcalibur_c30762_514*	3.3	6.8	10.5	5/5	Major QTL	Liu et al. (2015)
Q3B.18	*QYr.uga-3BS.2*	AGS2000	3BS	*XwPt-730063*	*XwPt-9579*	0.6	3.6	7.0	1/3	NI	Hao et al. (2011)
Q3B.19	*QYr.uga-3BS.3*	AGS2000	3BS	*XwPt-1612*	*XwPt-7486*	1.6	3.1-3.2	4.0-5.0	2/3	NI	Hao et al. (2011)
Q3B.20	*QYrco.wpg-3BS.2_Brundage*	Brundage	3BS	*XwPt-9432*	*Xgdm8*	6.3	3.4-36.1	1-2	6/9	NI	Case et al. (2014)
Q3B.21	*QYr.cim-3B_Pastor*	Pastor	3BL	*XwPt-2458*	*XwPt-0036*	ND	16.2	3.8-5.8	3/4	NI	Rosewarne et al. (2012)
Q3B.22	*QRYr3B.2*	Several	3B	*XIWA6510*	ND	ND	ND	3.2-3.7	2/4	GWAS	Jighly et al. (2015)
Q3B.23	*QYr.inra-3Bcentr_Renan*	Renan	3BL	*Xgwm131b*	*Xbcd131*	ND	4.38	6.31	1/1	NI	Dedryver et al. (2009)

Q3B.24	*QYrpi.vt-3BL*	Pioneer Brand 26R46	3BL	*Xwmc787*	ND	2.0	2.7	6.6	1/4	NI	Christopher et al. 2013
Q3B.25	*IWA3834_APR*	Several	3B	*XIWA3834*	ND	ND	ND	7	1/2	GWAS	Zegeye et al. (2014)
Q3B.26	*QYrns. orz-3BL_wPt-3107*	NSA-98-0995	3BL	*XwPt-3107*	ND	ND	5.3	8.5	3/7	NI	Vazquez et al. (2015)
Q3B.27	*QYr. sun-3B_Wollaroi*	Wollaroi	3B	*XwPt-7454*	*XwPt-2698*	15.4	5.19-9.74	26.7	3/3	NI	Bansal et al. (2014)
Q3B.28	*QYrid.ui-3B.2*	IDO444	3B	*X379646*	*Xgwm299*	3.2	16.5	31.0	1/4	Minor QTL	Chen et al. (2012a)
Q3B.29	*QYr. wpg-3B.1_ IWA6930*	Several	3B	*XIWA6930*	ND	ND	ND	4.0	1/8	GWAS	Naruoka et al. (2015)
Q3B.30	*QYrex. wgp-3BL_Express*	Express	3BL	*Xgwm299*	*Xwgp66*	5.6	9.9	16.4	3/3	NI	Lin and Chen (2009)
Q3D.1	*QYr. tam-3D_Quaiu*	Quaiu3	3DS	*XwPt-672034*	*Xbarc125*	ND	5.3	5.2	2/2	NI	Basnet et al. (2014b)
Q3D.2	*QYr-3DS*	Optata85	3DS	*Xfba241*	*Xfba91*		3.5	14	1/4	NI	Singh et al. (2000)
Q3D.3	*QYR4*	Optata85	3D	*Xcdo407*	*XksuA6*	5.7	2.76	11.7		NI	Boukhatem et al. (2002)
Q3D.4	*QYr.cim-3D*	Chapio	3DL	*Xgdm8*	*Xgdm128*	<23.8	3.8-9.0	3.6-13.5	2/8	NI	Yang et al. (2013)
Q3D.5	*QYr.inra-3DS*	Recital	3DS	*Xgwm456a*	*Xbarc125*	ND	7.29	11.97	1/1	NI	Dedryver et al. (2009)

(continued)

Table 5.4 (continued)

Code	Gene	Wheat variety	Chrom.	Marker 1	Marker 2	Genetic interval (cM)	LOD[a]	PEV[b]	Freq.[c]	Note[d]	References
Group 4 chromosomes											
Q4A.1	*QYr.wpg-4A.1_IWA1940*	Several	4A	*XIWA1940*	ND	ND	ND	5.0	1/8	GWAS	Naruoka et al. (2015)
Q4A.2	*QYr.wsu-4A.1_IWA3981*	Several	4A	*XIWA3981*	ND	ND	ND	3.21	2/4	GWAS	Bulli et al. (2016)
Q4A.3	*IWA3756_APR*	Several	4A	*XIWA3756*	ND	ND	ND	7.0	1/2	GWAS	Zegeye et al. 2014
Q4A.4	*QYr.wsu-4A.2_IWA3774*	Several	4A	*XIWA3774*	ND	ND	ND	1.09	4/4	GWAS	Bulli et al. (2016)
Q4A.5	*IWA8475_APR*	Several	4A	*XIWA8475*	ND	ND	ND	8	1/2	GWAS	Zegeye et al. (2014)
Q4A.6	*QYr-4A_Sachem*	Sachem	4A		ND	ND	ND	1.2	1/3	NI	Singh et al. (2013)
Q4A.7	*QYr.orr-4AL_Stephens*	Stephens	4AL	*XwPt-9901*	ND	ND	3.0	7	1/2	NI	Vazquez et al. (2012)
Q4A.8	*QYr.sgi-4A.2_Kariega*	Kariega	4A	*Xs13m94B*	*Xs12m53B*	ND	24	18	ND	NI	Ramburan et al. (2004)
Q4A.9	*QYr.sig-4A.2*	Avocet S	4AL	*XwPt-0538*	*XwPt-1961*	25	10.4	19.7	1/2	NI	Prins et al. (2011)
Q4A.10	*QYrns.orz-4AL_wPt-6440*	NSA-98-0995	4AL	*XwPt-6440*			3.9	4.9	3/7	NI	Vazquez et al. (2015)

Q4A.11	*QYr.sgi-4A.1*	Kariega	4AL, bin 4AL4-0.80-1.00	*Xgwm219*	*Xbarc78*	16.3	85.2-96.9	20.6-26.5	2/2	Major QTL	Prins et al. (2011) and Agenbag et al. (2014)
Q4A.12	*QYrid.ui-4A*	IDO444	4A	*XwPt-6440*	*XwPt-2151*	41	8.1-10.8	22	2/4	NI	Chen et al. (2012a)
Q4A.13	*QYr.ucw-4A_IWA1034*	Several	4A	*XIWA1034*	ND	ND	ND	1.2	5/6	GWAS	Maccaferri et al. (2015)
Q4A.14	*QYr.wsu-4A.3_IWA6697*	Several	4A	*XIWA6697*	ND	ND	ND	1.75	4/4	GWAS	Bulli et al. (2016)
Q4A.15	*QYr.wsu-4A.4_IWA4651*	Several	4A	*XIWA4651*	ND	ND	ND	1.53	4/4	GWAS	Bulli et al. (2016)
Q4A.16	*QYr.wsu-4A.5_IWA3422*	Several	4A	*XIWA3422*	ND	ND	ND	2.32	2/4	GWAS	Bulli et al. (2016)
Q4A.17	*QYrtb.orz-4AL_wPt-1007*	Tubbs	4AL	*XwPt-1007*	ND	ND	3.2	4.3	1/7	Minor QTL	Vazquez et al. (2015)
Q4B.1	*QYr-4B_Sachem*	Sachem	4B	*XwPt-0872*	ND	ND	ND	1.3-1.9	1/3	NI	Singh et al. (2013)
Q4B.2	*QYr.ufs-4B_Palmiet*	Palmiet	4B	*Xgwm165*	*Xgwm495*	21	4.7-7.6	7.0-11.7	1/2	Minor QTL	Agenbag et al. (2012)
Q4B.3	*QYrrb.ui-4B*	Rio Blanco	4B	*Xgwm165*	*Xgwm495*	4.2	3.0	7.0	1/4	NI	Chen et al. (2012a)
Q4B.4	*QYr.sun-4B_Janz*	Janz	4B	*XwPt-8543*	*Xwmc238*	4.3	3.8-5.2	9.4-16.8	4/5	NI	Zwart et al. (2010)
Q4B.5	*QYr.4BL-AvS*	Avocet S	4BL	*Xgwm149*	*Xgwm495*	4	4.51	7.7-13.1	3/3	NI	William et al. (2006)

(continued)

Table 5.4 (continued)

Code	Gene	Wheat variety	Chrom.	Marker 1	Marker 2	Genetic interval (cM)	LOD[a]	PEV[b]	Freq.[c]	Note[d]	References
Q4B.6	*QYr.caas-4BL_Libellula*	Libellula	4BL	*Xgwm165*	*Xgwm149*	26.2	3.19	3.1	2/4	NI	Lu et al. (2009)
Q4B.7	*QPst.jic-4B*	Alcedo	4B	*Xstm535*	*Xwmc692*	4	11.10-13.89	23.80-28.90	2/2	Major QTL	Jagger et al. (2011)
Q4B.8	*QYr.wpg-4B.1_IWA4348*	Several	4B	*XIWA4348*	ND	ND	ND	5.0	1/8	GWAS	Naruoka et al. (2015)
Q4B.9	*QYrva.vt-4BL_VA00W-38*	VA00W-38	4BL	*Xbarc163*	*Xwmc692*	1.6	3.0-7.2	7.4-16.7	2/4	Major QTL	Christopher et al. (2013)
Q4B.10	*QYr.jic-4B_Guardian*	Guardian	4B	*Xwmc652*	*Xwmc692*	ND	3.2-4.6	8-12		NI	Melichar et al. (2008)
Q4B.11	*QYrst.orr-4BL*	Platte	4BL	*X312980*	ND	ND	4.4	9	6/8	NI	Vazquez et al. (2012)
Q4B.12	*QYr-4B_Oligoculm*	Oligoculm	4BL	*Xfwm538*	ND	ND	24.68	1.8-12.3	5/5	NI	Suenaga et al. (2003)
Q4B.13	*QYr.wpg-4B.2_IWA3994*	Several	4B	*XIWA3994*	ND	ND	ND	6.0	1/8	GWAS	Naruoka et al. (2015)
Q4B.14	*QYrst.orr-4BS*	Stephens	4BS	*XwPt-5265*		ND	2.8	6	2/8	NI	Vazquez et al. (2012)
Q4D.1	*QYr-4DS*	Synthetic	4DS	*Xbcd265*	*Xmwg634*	ND	2.3-4.4	9-17	2/4	At *Yr28* locus	Singh et al. (2000)
Q4D.2	*QYr.cim-4DS_Pastor*	Pastor	4DS	*XwPt-4572*	*XwPt-6880*	ND	11.8	2.8-4.9	2/4	NI	Rosewarne et al. 2012

Q4D.3	*QYr.wsu-4D_IWA5381*	Several	4D	*XIWA5381*	ND	ND	ND	5.0	4/4	GWAS	Bulli et al. (2016)
Q4D.4	*QYr.caas-4DL.2_Lumai21*	Lumai 21	4DL	*Xbarc98*	*Xbarc1148*	9.4	3.7	7.6	1/3		Ren et al. (2015c)
Q4D.5	*QYr.caas-4DL_Bainong64*	Bainong 64	4DL	*Xwmc331*	*Xgwm165*	4.0	3.3	8.0	1/2		Ren et al. (2012b)
Q4D.6	*QYr.ucw-4D_IWA5375*	Several	4D	*XIWA5375*	ND	ND	ND	0.6	6/6	GWAS	Maccaferri et al. (2015)
Q4D.7	*Yr46/Lr67*	RL6077	4D	*Xgwm192*	*Xgwm165*	ND	ND	ND	ND	At *Yr46* locus	Herrera-Foessel et al. (2011)
Q4D.8	*QYr-4D_Oligoculm*	Oligoculm	4DL	*Xwmc399*	ND	ND	31.96	2.5-8.0	5/5	NI	Suenaga et al. (2003)
Group 5 chromosomes											
Q5A.1	*IWA4767_APR*	Several	5A	*XIWA4767*	ND	ND	ND	5.0-8.6	2/2	GWAS	Zegeye et al. (2014)
Q5A.2	*QYr.cau-5AS*	AQ24788-83	5AS	*Xcfa2250*	*Xwmc705*	3.6	11.1-13.5	24.1-35.3	2/2	Major QTL	Quan et al. (2013)
Q5A.3	*QYr.cim-5AL*	Avocet	5AL	*XwPt-733645*	*XwPt-4262*	32.2	3.4-6.0	6.2-7.4	3/3	NI	Lan et al. (2014)
Q5A.4	*IWA6949_APR*	Several	5A	*XIWA6949*	ND	ND	ND	7.0-7.7	2/2	GWAS	Zegeye et al. (2014)
Q5A.5	*QYrtb.pau-5A*	*T. boeoticum* accession pau5088	5AL	*Xbarc151*	*Xcfd12*	8.9	19-21	17-19	3/3	NI	Chhuneja et al. (2008)
Q5A.6	*QYR5-5A*	Optata85	5A	*Xfbb209*	*Xabg391*	18.7	2.80	15.0	ND	NI	Boukhatem et al. (2002)

(continued)

Table 5.4 (continued)

Code	Gene	Wheat variety	Chrom.	Marker 1	Marker 2	Genetic interval (cM)	LOD[a]	PEV[b]	Freq.[c]	Note[d]	References
Q5A.7	*QYr.cim-5AL*	Pastor	5AL	*XwPt-0373*	*XwPt-5231*	ND	13.2	3.9-6.6	2/4	NI	Rosewarne et al. (2012)
Q5A.8	*QYr.caas-5AL.2*	SHA3/ CBRD	5AL	*XwPt-1903*	*XwPt-3334*	ND	2.1-2.3	3.2-3.7	2/3	NI	Ren et al. (2012a)
Q5A.9	*QYr.caas-5AL*	Pingyuan 50	5AL	*Xwmc410*	*Xbarc261*	15.4	4.8	16.5	3/4	NI	Lan et al. (2010a)
Q5A.10	*QYr. wsu-5A_IWA5002*	Several	5A	*XIWA5002*				2.94	4/4	GWAS	Bulli et al. (2016)
Q5A.11	*QYr. ucw-5AL_ PI610750*	PI610750	5A	*Xwmc727*	*XwPt-9800*	5.3	20-21	16.8	2/4	*Yr48*	Lowe et al. (2011)
Q5A.12	*QYr. ucw-5A.1_ IWA6988*	Several	5A	*XIWA6988*	ND	ND	ND	1.0	3/6	GWAS	Maccaferri et al. 2015
Q5A.13	*QYrns. orz-5AL_gwm291*	NSA-98-0995	5AL	*Xgwm291*	ND	ND	5.5	6.6	4/7	NI	Vazquez et al. (2015)
Q5A.14	*QYrdr. wgp-5AL_ IWA2646*	Druchamp	5A	*XIWA2646*	*XIWA2558*		2.02	3.24	5/6	NI	Hou et al. (2015)
Q5B.1	*QYr. uga-5B_AGS2000*	AGS2000	5B	*XwPt-665267*	*Xgdm152*	1.2	3.3	5.0	1/3	NI	Hao et al. (2011)
Q5B.2	*QYr. cim-5BL_Chapio*	Chapio	5BL	*Xgwm46*	*Xbarc74*	5.3	4.1-11.1	2.7-6.9	4/8	NI	Yang et al. (2013)
Q5B.3	*QYrPI192252. wgp-5BS*	PI192252	5BS	*XIWA6910*	*Xgwm335*	10.8	3.6	22.1-27.4	5/5	NI	Lu et al. (2014)

Q5B.4	*QYr. wsu-5B_IWA5166*	Several	5B	*XIWA5166*				1.58	4/4	GWAS	Bulli et al. (2016)
Q5B.5	*QYr. ufs-5B_Cappelle-Desprez*	Cappelle-Desprez	5B	*XwPt-7114*	*Xbarc74*	12	2.8-3.4	4.3-4.9	1/2	Minor QTL	Agenbag et al. (2012)
Q5B.6	*QYr. tem-5B.1_Flinor*	Flinor	5B	*Xbarc89*	*Xwms67*	12	4.44	37.33	1/1	NI	Feng et al. 2011
Q5B.7	*QYr. inra-5BL.1_ CampRemy*	Camp Remy	5BL	*Xgwm639a*	*Xgwm639c*	6.9	6.50-9.70	18-26	4/4	NI	Mallard et al. (2005)
Q5B.8	*QYrdr. wgp-5BL.2_ IWA6867*	Druchamp	5BL	*XIWA6867*	*XIWA6383*	ND	4.99	9.48	5/6	NI	Hou et al. (2015)
Q5B.9	*QYrco. wpg-5B_Coda*	Coda	5BL, bin 5BL1-0.55-0.75	*XIWA5488*	*XwPt-6135*	6.9	7.7-36.8	1-4	7/9	NI	Case et al. (2014)
Q5B.10	*QYr. caas-5BL.1_ Libellula*	Libellula	5BL	*Xwmc415*	*Xwmc537*	4.3	8.19	7.2	3/4	NI	Lu et al. (2009)
Q5B.11	*QYr.jirc-5BL*	Oligoculm	5BL	*Xwmc145*	ND	ND	47.81	2.4-16.1	5/5	NI	Suenaga et al. (2003)
Q5B.12	*QYrns. orz-5BL_wPt-8285*	NSA-98-0095	5BL	*XwPt-8285*	ND	ND	4.0	4.2	3/7	NI	Vazquez et al. (2015)
Q5B.13	*QYrtb. orz-5BL_wPt-6105*	Tubbs	5BL	*XwPt-6105*	ND	ND	3.2	3.7	1/7	Minor QTL	Vazquez et al. (2015)
Q5B.14	*IWA3002_ Seedling*	Several	5B	*XIWA3002*	ND	ND	ND	5.4	1/1	GWAS	Zegeye et al. 2014

(continued)

Table 5.4 (continued)

Code	Gene	Wheat variety	Chrom.	Marker 1	Marker 2	Genetic interval (cM)	LOD[a]	PEV[b]	Freq.[c]	Note[d]	References
Q5B.15	*QRYr5B.1_ IWA2565*	Several	5B	*XIWA2565*	ND	ND	ND	5.3	2/4	GWAS	Jighly et al. (2015)
Q5B.16	*QYr.sun-5B*	Janz	5B	*XwPt-2707*	*XwPt-1733*	11	2.5-3.5	5.8-8.0	5/6	NI	Bariana et al. (2010)
Q5B.17	*QYr.caas-5BL.3_ SHA3/CBRD*	SHA3/ CBRD	5BL	*XwPt-8106*	*Xbarc275*	2.8	1.6-5.2	2.5-7.8	3/3	NI	Ren et al. (2012a)
Q5B.18	*QYr. tem-5B.2_Flinor*	Flinor	5B	*Xwmc235*	*Xwms604*	6.7	4.99	33.12	1/1	NI	Feng et al. (2011)
Q5B.19	*QYr. caas-5BL.2_ Libellula*	Libellula	5BL	*Xbarc142*	*Xgwm604*	19.9	2.53	2.6	1/4	NI	Lu et al. (2009)
Q5B.20	*QYr. inra-5BL.2_ CampRemy*	Camp Remy	5BL	*Xgwm234*	*XDuPw115a*	4.3	9.66-11.17	29-34	4/4	NI	Mallard et al. (2005)
Q5B.21	*QYr. sun-5B_Wollaroi*	Wollaroi	5B	*Xbarc142*	*XwPt-7804*	ND	2.50-2.71	8.7		NI	Bansal et al. (2014)
Q5B.22	*QYrid.ui-5B*	IDO444	5B	*X63541*	*Xbarc59*	ND	3.3	7.0	1/4	NI	Chen et al. (2012a)
Q5B.23	*QYrdr. wgp-5BL.1_ IWA6271*	Druchamp	5BL	*XIWA6271*	ND	ND	6.15, 12.89	30.57, 36.04	PST-45, PST-29	Race-specific, should be *Yr3a*	Hou et al. (2015)
Q5D.1	*QYr. caas-5DS_ Jingshuang16*	Jingshuang 16	5DS	*Xgwm190*	*Xgwm182*	30.7	3.9	9.2	3/3	Major QTL	Ren et al. (2015c)

Q5D.2	*QYr-5DS*	Opata 85	5DS	*Xfba114*	*Xfbb238*	1.8	0.08	1/4		NI	Singh et al. (2000)
Q5D.3	*QYr.nsw-5DL*	Otane	5DL	*Xgwm583a*				6	1/3	NI	Imtiaz et al. (2004)
Q5D.4	*QYr.caas-5DL_IWA4087_Zhong892*	Zhong892	5DL	*Xwsnp_Ex_c508_1008029*	*XBS00110475_51*	7.7	2.9	4.7	2/5	NI	Liu et al. (2015)
Q5D.5	*QYr.jirc-5DL*	Oligoculm	5DL	*Xwmc215*	ND	ND	3.9	5.9	1/5	Minor QTL	Suenaga et al. (2003)
Q5D.6	*QYrdr.wgp-5DL_IWA8331*	Druchamp	5DL	*XIWA8331*	ND	ND	3.27,5.14	9.27, 11.94	PST-35, PST-45, PST-35	Race-specific QTL	Hou et al. (2015)
Q5D.7	*QYrco.wpg-5D_Brundage*	Brundage	5DL, bin C-5DL1-0.60	*Xgwm765*	*Xbarc144*	7.0	7.9	12	PST-114	Race-specific QTL	Case et al. (2014)
Q5D.8	*QYrtb.orz-5DL*	Tubbs	5DL	*XwPt-5870*	ND	ND	3.2	4.5	1/7		Vazquez et al. (2015)
Group 6 chromosomes											
Q6A.1	*QYr.uga-6AS_26R61*	Pioneer 26R61	6AS	*XwPt-671561*	*XwPt-7840*	7.8	4.4	6.0	3/3	NI	Hao et al. (2011)
Q6A.2	*QYr.wgp-6AS_Express*	Express	6AS	*Xgwm334*	*Xwgp56*	3.8	10.5	19.1	3/3	NI	Lin and Chen (2009)
Q6A.3	*QYr.cim-6A_Avocet*	Avocet	6AS	*XwPt-2573*	*XwPt-0959*	ND	7.4	2.0-7.1	2/4	NI	Rosewarne et al. (2012)
Q6A.4	*QYr.nafu-6AL*	Caoxuan 5	6AL	*Xwgp5471*	*Xwgp5576*	4.6	5.3-6.0	12.9-13.0	4/4	NI	Zhou et al. (2015c)

(continued)

Table 5.4 (continued)

Code	Gene	Wheat variety	Chrom.	Marker 1	Marker 2	Genetic interval (cM)	LOD[a]	PEV[b]	Freq.[c]	Note[d]	References
Q6A.5	*QYr.caas-6AL_IWB39473_Zhong892*	Zhong892	6AL	*Xku_c45494_267*	*XBS00040166_51*	8.1	3.0	4.9	2/5	NI	Liu et al. (2015)
Q6A.6	*QRYr6A.1*	Several	6A	*XIWA4949*	ND	ND	ND	11.3	1/4	GWAS	Jighly et al. 2015
Q6A.7	*QYr.wpg-6A.1_IWA3023*	Several	6A	*XIWA3023*	ND	ND	ND	4.0	1/8	GWAS	Naruoka et al. (2015)
Q6A.8	*QYr.cim-6AL*	Avocet	6AL	*XwPt-733679*	*xwPt-733476*	ND	6.4	8.2	2/3	NI	Lan et al. (2014)
Q6A.9	*QYr.ufs-6A*	Avocet S	6A	*XwPt-5550*	*Xpsps3152*		5.6	3.2	1/2	Minor QTL	Prins et al. (2011)
Q6A.10	*QYr-6A*	Avocet	6A	*Xbarc3*	*XwPt-7063*	1.9	5	14.4	1/4	NI	Lillemo et al. (2008)
Q6A.11	*IWA7994_APR*	Several	6A	*XIWA7994*	ND	ND	ND	8	1/2	GWAS	Zegeye et al. (2014)
Q6A.12	*QYr.6AL*	Avocet	6AL	*Xgwm617*	ND	ND	3.09	5.8	3/3	NI	William et al. (2006)
Q6A.13	*QYrpl.orr-6AL*	Platte	6AL	*X378849*	ND	ND	3.1	6	1/2	NI	Vazquez et al. (2012)
Q6A.14	*QYr.wsu-6A_IWA8595*	Several	6A	*XIWA8595*	ND	ND		1.08	4/4	GWAS	Bulli et al. (2016)

Q6A.15	*QYrtb. orz-6AL_wPt-4229*	Tubbs	6AL	*XwPt-4229*	ND	ND	25.9	57.3	7/7	Major QTL	Vazquez et al. (2015)
	QYrns.orz-6BS	NSA-98-0995	6BS	*XwPt-2726*	ND	ND	4.1	4.8	1/7	Minor QTL	Vazquez et al. (2015)
Q6B.1	*QYrdr. wgp-6BL.1_ IWA3297*	Druchamp	6BL	*XIWA3297*	ND	ND	5.57-7.21	13.07-20.36	PST35, PST100, PST114	Race-specific QTL	Hou et al. (2015)
Q6B.2	*QYr. ufs-6B_Kariega*	Kariega	6B	*Xjs14m69E*	*XwPt-7745*	ND	10.0	7.5	1/2	NI	Prins et al. (2011)
Q6B.3	*QYr. tam-6BS_TAM111*	TAM111	6BS	*XwPt-3733*	*XwPt-4742*	10	4.5	14.6	1/5	NI	Basnet et al. (2014a)
Q6B.4	*QYr. caas-6BS.2_Naxos*	Naxos	6BS	*Xwmc104*	*XwPt-0259*	10.6	1.5-2.9	2.9-5.7	2/3	NI	Ren et al. (2012a)
Q6B.5	*QYr.caas-6BS.3*	Bainong 64	6BS	*Xwmc487*	*Xfcd13*	7.0	2.6	6.1	2/2	NI	Ren et al. (2012b)
Q6B.6	*QYr. wgp-6BS.2_ Stephens*	Stephens	6BS	*Xgwm132*	*Xgdm113*	17.8	1.4-8.31	25-43	3/4	NI	Santra et al. (2008)
Q6B.7	*QYr.sun-6B*	Janz	6B	*Xgwm518b*	*wPt-8183, Xgwm644*	12	2.5-2.9	3.9-8.0	4/6	NI	Bariana et al. (2010)
Q6B.8	*QYr-6BSC_ Oligoculm*	Oligoculm	6BS	*Xgwm935.1*	ND	ND	23.00	0.5-3.8	5/5	NI	Suenaga et al. 2003
Q6B.9	*QYr. wgp-6B.1_ Stephens*	Stephens	6BS	*Xbarc1169*	*Xgwm88*	26.6	3.89-6.83	32-45	4/4	NI	Santra et al. (2008)
Q6B.10	*QYr. caas-6BS_ Pingyuan50*	Pingyuan 50	6BS	*Xgwm361*	*Xbarc136*	4.0	2.7	7.6	2/4	NI	Lan et al. (2010a)

(continued)

Table 5.4 (continued)

Code	Gene	Wheat variety	Chrom.	Marker 1	Marker 2	Genetic interval (cM)	LOD[a]	PEV[b]	Freq.[c]	Note[d]	References
Q6B.11	*QYrpi.vt-6BL*	Pioneer Brand 26R46	6BL	*Xwmc756*	*XwPt-7935*	12.4	3.5-5.3	11.1-13.5	2/4	NI	Christopher et al. 2013
Q6B.12	*QYr.wsu-6B.2_IWA4169*	Several	6B	*XIWA4169*	ND	ND	ND	1.14	4/4	GWAS	Bulli et al. (2016)
Q6B.13	*QYr.6BL-Pavon*	Pavon 76	6BL	*Xgwm58*	*Xgwm626*	6.3	7.64	14.0	3/3	NI	William et al. (2006)
Q6B.14	*QYr.inra-6B_Renan*	Renan	6B	*Xcdo270*	*Xgwm193*	ND	4.37	5.59	1/1	NI	Dedryver et al. (2009)
Q6B.15	*QYrdr.wgp-6BL.2_IWA6420*	Druchamp	6BL	*XIWA6420*	ND	ND	1.53	4.08	6/6		Hou ęt al. (2015)
Q6B.16	*QYr.wsu-6B.1_IWA7257*	Several	6B	*XIWA7257*	ND	ND	ND	1.20	1/4	GWAS	Bulli et al. (2016)
Q6B.17	*QYr.ucw-6B_IWA7257*	Several	6B	*XIWA7257*	ND	ND	ND	1.4	6/6	GWAS	Maccaferri et al. (2015)
Q6B.18	*QYr.wpg-6B.1_IWA7257*	Several	6B	*XIWA7257*	ND	ND	ND	6.0	1/8	GWAS	Naruoka et al. (2015)
Q6B.19	*QYr.wsu-6B.3_IWA349*	Several	6B	*XIWA349*	ND	ND	ND	3.38	4/4	GWAS	Bulli et al. (2016)
Q6B.20	*QYr.cim-6BL_Pastor*	Pastor	6BL	*XwPt-6329*	*XwPt-5176*	ND	5.6	2.3-4.4	3/4		Rosewarne et al. (2012)

Q6B.21	*QYr.wpg-6B.2_IWA3222*	Several	6B	*XIWA3222*	ND	ND	ND	4.0	1/8	GWAS	Naruoka et al. 2015
Q6D.1	*QYr.ufs-6D_Cappelle-Desprez*	Cappelle-Desprez	6D	*Xgwm325*	*Xbarc175*	24	2.8-4.2	4.5-7.6	1/2	Minor QTL	Agenbag et al. (2012)
Q6D.2	*QYr.ucw-6D_IWA167*	Several	6D	*XIWA167*	ND	ND	ND	0.7	6/6	GWAS	Maccaferri et al. (2015)
Q6D.3	*QYr-6D_W-7984*	W-7984	6D	*XBcd1510*	*XksuD27*	21.8	2.43	13.1	ND	NI	Boukhatem et al. (2002)
Q6D.4	*IWA4455_Seedling*	Several	6D	*XIWA4455*	ND	ND	ND	6.3	1/1	GWAS	Zegeye et al. (2014)
Group 7 chromosomes											
Q7A.1	*QYr.cim-7AS_Avocet*	Avocet	7AS	*XwPt-4172*	*XwPt-8149*	ND	5.9	0.3–3.3	4/4	NI	Rosewarne et al. (2012)
Q7A.2	*QYr.inra-7A_Recital*	Recital	7A	*Xbcd129b*	*Xfba127c*	ND	3.57	8.77	ND	NI	Dedryver et al. 2009
Q7A.3	*QYr.caas-7A_Jingshuan16*	Jingshuang 16	7AS	*Xbarc127*	*Xbarc174*	28.2	2.4	6.1	1/2	NI	Ren et al. (2015c)
Q7A.4	*QYr.sun-7A_CPI133872*	CPI133872	7A	*XwPt-7214*	*XwPt-4345*	10.0	4.4	10.6	1/4	NI	Zwart et al. (2010)
Q7A.5	*IWA2710_APR*	Several	7A	*XIWA2710*	ND	ND	ND	6	1/2	GWAS	Zegeye et al. (2014)
Q7A.6	*IWA4996_APR*	Several	7A	*XIWA4996*	ND	ND	ND	6	1/2	GWAS	Zegeye et al. 2014
Q7A.7	*IWA4621_APR*	Several	7A	*XIWA4621*	ND	ND	ND	6	1/2	GWAS	Zegeye et al. (2014)

(continued)

Table 5.4 (continued)

Code	Gene	Wheat variety	Chrom.	Marker 1	Marker 2	Genetic interval (cM)	LOD[a]	PEV[b]	Freq.[c]	Note[d]	References
Q7A.8	*IWA4187_ Seedling*	Several	7A	*XIWA4187*	ND	ND	ND	6	1/1	GWAS	Zegeye et al. (2014)
Q7A.9	*QYr. caas-7AL_ IWB24300*	Zhong892	7AL	*XExcalibur_ c25335_306*	*XExcalibur_ c1142_724*	0.1	3.1	6.6	3/5	NI	Liu et al. (2015)
Q7A.10	*QYr. orr-7A_Stephens*	Stephens	7AS	*XwPt-4319*	ND	ND	8.4	15	5/8	NI	Vazquez et al. (2012)
Q7A.11	*QYr. cim-7AL_Avocet*	Avocet	7AL	*XwPt-2260*	*XwPt-2501*	ND	9.1	3.2-6.3	3/4	NI	Rosewarne et al. (2012)
Q7A.12	*QYr. sgi-7A.1_Kariega*	Kariega	7A	*XwPt-3226*	*XwPt-8302*	ND	3.9	2.1	1/2	Minor QTL	Prins et al. 2011
Q7A.13	*QYr.sgi-7A.2*	Kariega	7A	*XwPt-3226*	*Xgwm332*	ND	3.0	2.9	1/2	Minor QTL	Prins et al. (2011)
Q7B.1	*QYr-7B_ Oligoculm*	Oligoculm	7BS	*Xgwm935.3*	ND	ND	30.29	1.0-5.2	5/5	NI	Suenaga et al. 2003
Q7B.2	*QYr.sun-7B*	Kukri	7B	*XwPt-1553*	*XwPt-2356*	39	2.6-3.0	8.0-9.0	6/6	NI	Bariana et al. (2010)
Q7B.3	*IWA8469_APR*	Several	7B	*XIWA8469*	ND	ND	ND	7	1/2	GWAS	Zegeye et al. 2014
Q7B.4	*QYr.caas-7BL.1_ SHA3/CBRD*	SHA3/ CBRD	7BL	*Xbarc176*	*XwPt-8106*	10.4	1.7-2.6	3.7-8.2	3/3	NI	Ren et al. (2012a)
Q7B.5	*QHtap. wsu-7BL_Alpowa*	Alpowa	7BL	*Xwgp36*	*Xwgp45*	0.8	30.4	64.2	3/3	*Yr39*	Lin and Chen (2007)

Q7B.6	*QYr.orr-7BS_Stephens*	Stephens	7BS	*XwPt-7653*	ND	ND	3.3	11	1/2	NI	Vazquez et al. 2012
Q7B.7	*IWA4309_APR*	Several	7B	*XIWA4309*	ND	ND	ND	5	1/2	GWAS	Zegeye et al. (2014)
Q7B.8	*QYr-7B_Tiritea*	Tiritea	7B	*Xgwm611*	ND	ND	ND	37	3/3	NI	Imtiaz et al. (2004)
Q7B.9	*QYrtb.orz-7BL_wPt-2356*	Tubbs	7BL	*XwPt-2356*	ND	ND	3.8	4.9	2/7	NI	Vazquez et al. (2015)
Q7B.10	*QYr.caas-7BL.2_SHA3/CBRD*	SHA3/CBRD	7BL	*Xgwm577*	*XwPt-4300*	17.6	6.4	14.7	1/3	NI	Ren et al. 2012a
Q7B.11	*QYr.cim-7BL_Pastor*	Pastor	7BL	*XwPt-3190*	*XwPt-1475*	ND	25.8	4.2-7.8	4/4	NI	Rosewarne et al. (2012)
Q7B.12	*QYr-7BL_Ologoculm*	Oligoculm	7BL	*Xwmc166*	ND	ND	15.36	2.5-9.0	3/3	NI	Suenaga et al. 2003
Q7B.13	*QYr-7BL_Strongfield*	Strongfield	7BL	*Xgwm146*	ND	ND	ND	18	1/3	NI	Singh et al. (2013)
Q7B.14	*QYr.7BL-Attila*	Attila	7BL	*Xgwm344*	*XP32/M59*	12	ND	3.1	3/3	NI	Rosewarne et al. (2008)
Q7D.1	*wPt-668026*	Several	7D	*XwPt-668026*	ND	ND	ND	5.1	3/4	GWAS	Jighly et al. 2015
Q7D.2	*QYr.caas-7DS.1_IWB56317*	Zhong892	7DS	*XRAC875_c29314_291*	*Xtplb0024a09_2369*	12.3	4.8	9.1	3/5	NI	Liu et al. (2015)
Q7D.3	*QYr.sun-7D*	Janz	7D	*XwPt-3328*	*XwPt-1100, Xgwm295*	5.9	4.6-12.2	14.7-41.6	4/4	At *Yr18* locus	Bariana et al. 2010
Q7D.4	*Yr18/Lr34_csLV23*	Chapio	7DS	*XcsLV34*	*XwPt-1100, Xgwm295*	ND	5.0-50.3	7.5-50.7	8/8	At *Yr18* locus	Yang et al. (2013)

(continued)

Table 5.4 (continued)

Code	Gene	Wheat variety	Chrom.	Marker 1	Marker 2	Genetic interval (cM)	LOD[a]	PEV[b]	Freq.[c]	Note[d]	References
Q7D.5	*QYr.caas-7DS*	Libellula	7DS	*XcsLV34*	*Xgwm295*	15.4	16.16	32.2	4/4	At *Yr18* locus	Lu et al. (2009)
Q7D.6	*QYr-7DS*	Saar	7DS	*XcsLV34*	*XwPt-3328*	10.5	10	40	1/4	At *Yr18* locus	Lillemo et al. (2008)
Q7D.7	*QYr-7DS_Opata85*	Opata 85	7DS	*Xbcd1438*	*Xwg834*		3.0-9.7	12-36	3/4	At *Yr18* locus	Singh et al. (2000)
Q7D.8	*QYr-7D_Otane*	Otane	7D	*Xgwm44*	*Xgwm332c*	21.2	17	13-23	3/3	NI	Imtiaz et al. (2004)
Q7D.9	*QYr.tam-7D*	TAM111	7DL	*XwPt-730455*	*XwPt-730043*	17	3.8	14.1	1/5	NI	Basnet et al. (2014a)
Q7D.10	*QYR4*	Opata85	7D	*XEg834*	*Xbcd1438*	6.9	3.36	13.6	ND	NI	Boukhatem et al. (2002)

[a]*LOD* logarithm of the odds
[b]*PEV* position effect variegation
[c]The frequency ratio indicates that the number of testing environments in which the QTL was effective out of the total number of testing environments
[d]NI denotes no information for the relevant aspects; and GWAS denotes that the QTL was identified using the genome-wide association study (GWAS) approach
[e]*ND* no data
[f]Several denotes several to many wheat varieties possessing the QTL as it was identified through GWAS

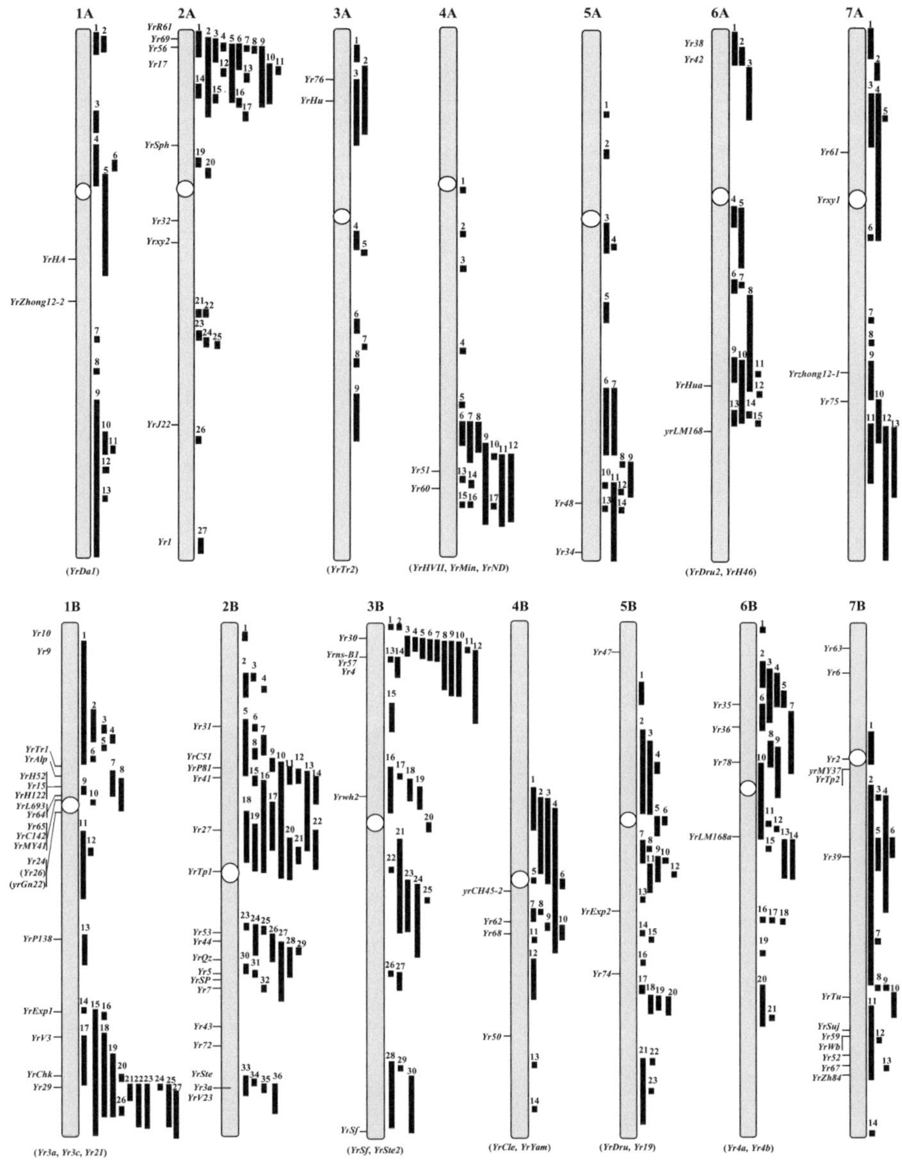

Fig. 5.1 Chromosomal locations of wheat genes or quantitative trait loci (QTL) for resistance to stripe rust. Resistance genes with permanent and provisional *Yr* symbols are shown on the left and QTL on the right of each chromosome. QTL are coded by numbers and can be referred to Table 5.4 for details. Genes that have been reported to a chromosome but no specific locations are given below the chromosomes

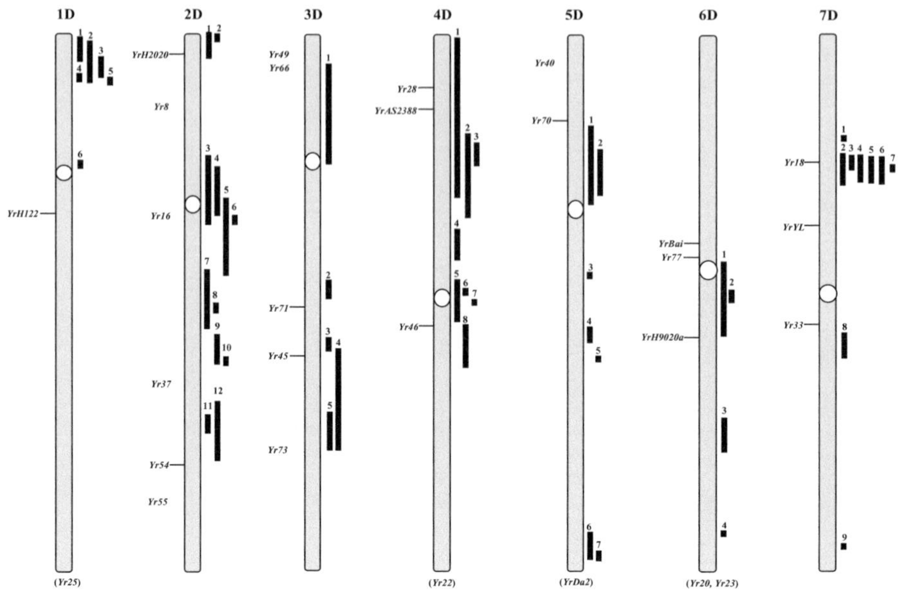

Fig. 5.1 (continued)

QYr-1A_IWA5754 (Zegeye et al. 2014), *QYr.wsu-1A.1* and *QYr.wsu-1A.2* (Bulli et al. 2016). Except the markers of *QYrst.orr-1AL* that could not be located in the integrated map of chromosome 1AL, *QYr.sun-1A* is proximal to the centromere, *QYr-1A_IWA5754* and *QYr.wsu-1A.1_IWA5505* in the middle region, *QYr.cim-1AL* spans a large region from middle to telomere and the others are in distal region of long arm. All QTL in chromosome 1A have similar LOD scores, and most of them were detected in half or more of experiments in the studies, except *QYr.tam-1AS* and *QYr.tam-1AL* were detected in 2 out of 5 the environments. The phenotype explained value (PEV) scores ranged from 3.6 to 26.4%, and four of which were larger than 10%. Ten QTL were identified from bi-parental cross populations and four QTL (*QYr-1A_IWA5754*, *QYr.wsu-1A.1*, *QYr.wsu.1A.2* and *QRYr1A.1*) from genome-wide association analysis (GWAS).

Chromosome 1B has 27 QTL (Q1B.1 – Q1B.27, of which 10 distributed in the short arm and 16 in the long arm (Table 5.4; Fig. 5.1). Several important genes were located in 1B, such as *Yr9* and *Yr10* in the distal of the short arm, *Yr15*, *Yr24/Yr26* near the centromere and *Yr29* in the distal of the long arm. Nineteen QTL were identified from populations of bi-parental crosses and 8 from germplasm panels by GWAS. In the short arm, *QYr.cau-1BS* from wheat line AQ 24788-83 (Quan et al. 2013) and *Yrco.wpg-1B.1* from US PNW winter club cultivar Coda (Case et al. 2014) are located in the distal region; *QYr.caas-1BL.1RS* from wheat-rye translocation line SHA3/CBRD (Ren et al. 2012a) and *QYrco.wpg-1B.2* from Coda (Case et al. 2014) in the middle; and *QYr.cim-1BSb* from Pastor (Rosewarne et al. 2012) in the proximal region. *QYrco.wpg-1B.1* is important and confers high consistent resistance in all test environments (Case et al. 2014). In the long arm, *QYrdr.wgp-1BL.1* from Druchamp (Hou et al. 2015) and *QYr.cim-1BSb* from Pastor (Rosewarne et al. 2012) are located in the proximal region; and *QYr-1B_Sachem* from Sachem (Singh et al. 2013), *QYrex.wgp-1BL* from Express (Lin and Chen 2009) and *QYr.wsu-1B.2* (Bulli et al. 2016) in the middle of 1BL. Twelve QTL are located in the distal region around the *Yr29* locus. Despite a wide range of the LOD and PEV scores of the *Yr29/Lr46* region, seven QTL associated with the locus on the distal region have a significant effect on reducing rust severity across all experiments, including *QYr.sun-1B* from CPI 133872 (Zwart et al. 2010), *QYr.jic-1B* from Guardian (Melichar et al. 2008), *QYr.tam-1B* from Quaiu 3 (Basnet et al. 2014b), *QYr.cim-1BL* from Francolin #1 (Lan et al. 2014), *QYr-1B* from Saar (Lillemo et al. 2008), *QYr.cim-1BL* from Pastor (Rosewarne et al. 2012), and *QYr.1BL* from Attila (Rosewarne et al. 2008). The locus of *QYrex.wgp-1BL* from Express (Lin and Chen 2009) is different from *Yr29* and the other QTL above, likely located on the middle part of the long arm based on the close SSR marker. *QYrPst.jic-1B* from cultivar Brigadier has a small effect on resistance and has some distance from *Yr29* (Jagger et al. 2011). In addition to the above QTL, eight QTL were identified on chromosome 1B through GWAS, of which six were assigned to the long arm and two in the short arm (Zegeye et al. 2014; Maccaferri et al. 2015; Naruoka et al. 2015; Bulli et al. 2016).

Chromosome 1D has six QTL (Q1D.1 – Q1D.6) conferring resistance to stripe rust (Table 5.4; Fig. 5.1). All of them are in the short arm. Five of the QTL, *QYr.caas-1DS* from Naxos (Ren et al. 2012a), *QYrdr.wgp-1DS* from Druchamp (Hou et al. 2015), *QYrst.orr-1DS* from Stephens (Vazquez et al. 2012) and *QYr.sun-1D* from CPI 133872 (Zwart et al. 2010), and *QYr.wpg-1D.1* (Naruoka et al. 2015), are in the telomeric region, while *QYr.ucw-1D* is in the proximal region (Maccaferri et al. 2015). All of these QTL have small effects on stripe rust resistance; and only *QYr.caas-1DS* was detected across all experiments.

5.1.4.3.2 QTL in Group 2 Chromosomes

Chromosome 2A harbors 27 QTL (Q2A.1 – Q2A.27), including 20 in the short arm and 7 in the long arm. This chromosome also has two important ASR genes, *Yr1* in the distal region of the long arm and *Yr17* in the distal region of the short arm (Table 5.4; Fig. 5.1). In the short arm, most QTL are in the distal region, of which *QYr.ufs-2A* from cultivar Cappelle-Desprez (Agenbag et al. 2012), *QYr.sun-2A* from Wollaroi (Bansal et al. 2014), *QYrtb.orz-2AS* from Tubbs (Vazquez et al. 2015) and *QYr.inra-2AS* from Camp Remy (Mallard et al. 2005) are major QTL and effective across all experimental conditions. *QYr.ufs-2A* and *QYr.uga-2AS* have high scores of LOD and PEV (Hao et al. 2011; Agenbag et al. 2012). *QYrtm.pau-2A* from *T. monococcus* accession Pau14087 confers resistance in both seedling and adult-plant stage (Chhuneja et al. 2008). *QYr.wsu-2A.1* and *QYr.wsu-2A.2* were consistently effective against stripe rust in multi-environment conditions in both seedling and adult-plant stages through GWAS (Bulli et al. 2016). In the long arm, *QYR2* from

Camp Remy (Boukhatem et al. 2002) in the distal region and *QYrns.orz-2AL* from NSA-98-0995 (Vazquez et al. 2015) in the proximal region have high scores of PEV, while *QYr.caas-2AL* from Zhong 892 (Liu et al. 2015) has low scores of PEV. In addition to the above QTL, four QTL associated with the marker loci *Xiwa7339*, *Xiwa544* and *Xiwa5855* (*QYr.wpg-2A.5*) in the proximal region and *Xiwa966* (*QYr.wpg-2A.6*) in the middle region were identified by GWAS and also located in the long arm of chromosome 2A (Zegeye et al. 2014; Jighly et al. 2015; Naruoka et al. 2015).

Chromosome 2B carries 36 QTL (Q2B.1 – Q2B.36) for resistance to stripe rust and some important all-stage resistance genes, such as *Yr5*, *Yr7*, *Yr27* and others as discussed above (Table 5.4; Fig. 5.1). The 36 QTL can be sorted into at least five regions with three regions in the short arm and two in the long arm. The first region with one QTL (*wPt-6271*) is close to the telomere, and this QTL were identified through GWAS by Jighly et al. (2015). The second region has three QTL including *QYr.inra-2BS* from Renan (Dedryver et al. 2009), *QYrst.orr-2B.2* from Stephens (Vazquez et al. 2012) and *QYr.wgp-2B.1* (Naruoka et al. 2015). The third region, which spans a large proportion of 2BL, is defined for the QTL identified from 16 studies. The region also includes *Yr27* and *Yr31*. Rosewarne et al. (2008) indicated that Attila had *Yr27* and Yang et al. (2013) identified *Yr31* in Chapio, both falling in this region. Pastor also has *Yr31* (Rosewarne et al. 2012). Guo et al. (2008) identified two loci from a PNW cultivar Luke in this region for HTAP resistances. Other PNW cultivars IDO444, Louise and Stephens also have QTL in this region (Chen et al. 2012a; Carter et al. 2009). Boukhatem et al. (2002) identified a major QTL (*QYr3-2B_Opata85*) from Opata 85 in this region using IT data, and Ramburan et al. (2004) showed a QTL (*QYr.sgi-2B.1a/QYr.sgi-2B.1_Kariega*) Kariega having a major effect on APR using IT, final disease severity and seedling data. *QYr.sgi-2B.1* has a larger effect on IT, showing consistent hypersensitive reaction (Prins et al. 2011; Agenbag et al. 2014). *QYrlu.cau-2BS1* and *QYrlu.cau-2BS2* (Guo et al. 2008), *QYrid.ui-2B.1* and *QYrid.ui-2B.2* (Chen et al. 2012a), *QYr.tam-2BL* (Basnet et al. 2014a) and *QYr3-2B* (Boukhatem et al. 2002) were reported to have high PEV values. These studies show that this region is a gene-rich region which contains a number of ASR, APR or HTAP resistance genes. Furthermore, most QTL in this region were effective across most environments with *QYr.tam-2B* and *QYr.cim-2BS* (*YrF*) effective across all tested experimental conditions. The fourth region is in the long arm and close to the centromere, containing 10 QTL. *QYr.inra-2B.2* and *QYR1* from Camp Remy (Boukhatem et al. 2002; Mallard et al. 2005), *QYraq.cau-2BL* from Aquileja (Guo et al. 2008) and *QYr.nafu-2BL* from P9897 (Zhou et al. 2015a) had high scores of both LOD and PEV and were effective across all tested experimental conditions. The fifth region is in the distal region of the long arm of chromosome 2B. In this region, QTL identified from Avocet, Linmai 2 and Schem were effective in some environments, but not others; and had intermediate LOD and PEV values (Rosewarne et al. 2008; Singh et al. 2013; Liu et al. 2015). A small-effect QTL (*QYr.caas-2BS.2*) from Lumai 21 reported by Ren et al. (2015c) likely physically locates in the telomere of the long arm based on the position of its flanking

SSR marker *Xbarc1147* assigned in Chinese Spring deletion bin map (Sourdille et al. 2004; Song et al. 2005).

Chromosome 2D contains 12 QTL (Q2D.1 – Q2D.12) conferring APR to stripe rust and also two permanently named loci *Yr16* and *Yr54* (Table 5.4; Fig. 5.1). *QYr.caas-2DS* from Libellula (Lu et al. 2009) and *QYr.wpg-2D.1* associated with SNP marker *Xiwa1939* (Naruoka et al. 2015) are located in the telomere of the short arm. *QYr.caas-2DS* had intermediate LOD and PEV scores and was significant in two of the four environments tested (Lu et al. 2009). *QYr.caas-2DS.2* from Lumai 21 (Ren et al. 2015c) and *QYr.inra-2DS* from Camp Remy (Mallard et al. 2005) proximal to the centromere were identified as major QTL and effective in all tested environments. *QYr.ufs-2DS* from Cappelle-Desprez defined in an interval of 30 cM with the peak at *Xwmc18* was believed the *Yr16* locus (Agenbag et al. 2012). A large effect QTL in the long arm of chromosome 2D explaining 49–54% of phenotypic variation in Quaiu 3 was permanently designated as *Yr54* (Basnet et al. 2014b). *QYr.caas-2DL* from Naxos (Ren et al. 2012b) and *QYr.jic-2D* from Guardian (Melichar et al. 2008) were in the proximal region of the long arm and had intermediate LOD and PEV scores. Two QTL in the middle region from cultivars Fukuho-komugi (*QYr-2DL*) and Alcedo (*QPst.jic-2D*) had high LOD scores and PEV and were effective across all tested conditions (Suenaga et al. 2003; Jagger et al. 2011).

5.1.4.3.3 QTL in Group 3 Chromosomes

Chromosome 3A has nine QTL (Q3A.1 – Q3A.9) conferring resistance to stripe rust (Table 5.4; Fig. 5.1). Lillemo et al. (2008) identified a QTL (*QYr-3A_Saar*) in the telomere region of the short arm of chromosome 3A in cultivar Saar, which was only effective in one of four environments, indicating a low value in resistance breeding. *QYr.caas-3AS* from Zhong 892 had high LOD and PEV scores and were effective across all experimental conditions (Liu et al. 2015). *QYrst.orr-3AL* from Stephens had high LOD and PEV scores, but was effective in one of two tested conditions (Vazquez et al. 2012), indicating that this QTL is either race-specific or sensitive to temperature. In the long arm of chromosome 3A, three QTL, *QYr.cau-3AL*, *QYrdr.wgp-3AL* and *QYr.cim-3A*, were identified from cultivars AQ24788-83, Druchamp and Avocet, respectively (Rosewarne et al. 2012; Quan et al. 2013; Hou et al. 2015). The remaining three QTL, *IWA8630_APR* (Zegeye et al. 2014), *IWA5849* (Jighly et al. 2015) and *QYr.wsu-3A_IWA3401* (Bulli et al. 2016), were identified by GWAS and had relatively low PEV values.

Chromosome 3B has 30 QTL (Q3B.1 – Q3B.30) in at least five regions associated with stripe rust resistance (Table 5.4; Fig. 5.1). Consensus mapping in this chromosome was difficult as a group of SSR markers are multi-allelic and each of the markers produced up to three bands located in different chromosome positions. The first region in the distal of the short arm is known to be important as it is the location of *Yr30*, a partial APR gene that is very tightly linked, or pleiotropic, to the slow-stem rusting locus *Sr2* and the phenotypic marker of pseudo-black chaff (*Pbc*). *Yr30* is a valuable gene that has shown to work well in combination with other genes such as *Yr18* (Yang et al. 2013) and gives an intermediate effect in most environments. It is expressed only in the adult plant stage since there is no any evidence of resistance at this locus in seedling tests (William et al. 2006). There are 15 QTL reported around this region. Most of the QTL in this region are consistent across different environments. SSR markers *Xgwm533*, *Xgwm389* and *Xbarc133* are associated with the most QTL in this region, making them valuable diagnostic markers. For example marker, *Xgwm533* is associated with *Yr30* and related to QTL from Rio Blanco (Chen et al. 2012a). This SSR marker has also been associated with *Yr30* in other studies (Börner et al. 2000; Spielmeyer et al. 2005; William et al. 2006; Dedryver et al. 2009; Yang et al. 2013; Basnet et al. 2014b; Lan et al. 2014; Zhou et al. 2015a). In contrast, a major QTL (*QYrAlt.syau-3BS*) identified in a US PNW spring wheat cultivars, Alturas, was more close to *Xgwm389*, but more than 9 cM away from *Xgwm533* (Zhao et al. 2012). These data indicate that the HTAP QTL is likely different from Yr30. The second region associated with five stripe rust resistance QTL was proximal in the short arm. Liu et al. (2015) identified a QTL (*QYr.caas-3BS*) derived from Zhong 892, which showed consistent resistance to stripe rust across all environments, and SNP markers *XIAAV5662* and *XBS00056257_51* are closely linked to this QTL. The molecular marker for *Sr2* was not present in Zhong 892 indicating the absence of *Sr2/Yr30* in Zhong 892 and the QTL is different from *Yr30*. Other three QTL in this region have minor effect with lower PEV and LOD scores. The third region is proximal in the long arm. Cultivars Renan (Dedryver et al. 2009), Pastor (Rosewarne et al. 2012) and Pioneer Brand 26 (Christopher et al. 2013) contributed with intermediate to small effect QTL which were not consistently detected across environments. The fourth region is in the middle of the long arm and harboring two QTL. *QYrrns.orz-3BL_wPt-3107* (Vazquez et al. 2015) has a small effect and another QTL (*IWA3834_APR*) identified by GWAS (Zegeye et al. 2014) needs to be validated. The fifth region is in the telomere of 3BL, containing HTAP resistance *QYrex.wgp-3BL* (Lin and Chen 2009) and *QYrid.ui-3B.2* (Chen et al. 2012a) from the US PNW cultivars Express and IDO444, respectively. *QYrex.wgp-3BL* was consistently identified across environments, had an intermediate level of effect and contributed significantly to lowering IT in adult plants, while *QYrid.ui-3B.2* was a minor QTL and only detected in one of four tested environments, although it had high PEV score.

Chromosome 3D has five QTL (Q3D.1 – Q3D.5) in three regions associated with stripe rust resistance (Table 5.4; Fig. 5.1). Basnet et al. (2014b) reported *QYr.tam-3D* from Quaiu 3, defined by the proximally located markers *XwPt-672034* and *Xbarc125*. This QTL spans a large portion on the short am based on the integrated map (Maccaferri et al. 2015; Bulli et al. 2016). Singh et al. (2000) and Boukhatem et al. (2002) reported stripe rust resistance QTL linked to markers *XksuA6* and *Xbcd1532* that explained 12–14% of the variation in the cultivar Opata 85/Synthetic population. This region is in the middle of the long arm based on the consensus map (Maccaferri et al. 2015; Bulli et al. 2016). Dedryver et al. (2009) reported a small QTL, *QYr.inra-3DS*, flanked by markers *Xbarc125* and *Xgwm456*, responsible for slow rusting resistance. The QTL is derived from cultivar Recital, explained up to 12% of the phenotypic variation. Yang et al. (2013) identified a QTL, *QYr.cim-3D*, flanked by markers *Xdgm8* and *Xgdm128*, derived from cultivar Chapio, and explained 3.6–14% of the phenotypic variation. These three QTL overlap in an interval about 15 cM in the short arm of chromosome 3B based on the integrated genetic and physical wheat maps (Somer et al. 2004; Sourdille et al. 2004; Song et al. 2005). However, the latest consensus map placed the markers on the distal region of the 3B long arm (Maccaferri et al. 2015; Bulli et al. 2016). The accurate position of these QTL in chromosome 3D need to be further investigated.

5.1.4.3.4 QTL in Group 4 Chromosomes

Seventeen QTL (Q4A.1 – Q4A.17) have been identified in chromosome 4A and 12 of them are in the telomere region of the long arm (Table 5.4; Fig. 5.1). Seven QTL in the telomere region were identified in populations from bi-parental crosses. *QYr.sgi-4A.1* and *QYr.sgi-4A.2* were detected from Kariega with large effects and high LOD and PEV values (Ramburan et al. 2004). Later, the single-gene line for *QYr.sgi-4A.1* was developed and shown to be consistently effective across the tested environments and was identified as a major QTL (Prins et al. 2011; Agenbag et al. 2014). The flanking markers narrowed this QTL to a 16.3 cM interval of chromosomal bin 4AL4-0.80-1.00. Flanking markers of *QYr.orr-4AL*, *QYrid.ui-4A*, *QYrtb.orz-4AL* and *QYrns.orz-4AL* also fall into this chromosomal bin (Chen et al. 2012a; Vazquez et al. 2012, 2015). *Yr51* is close to *QYr.sgi-4A.1* because *Xgwm160*, linked to *QYr.sgi-4A.1*, was only 4.3 cM away from *Yr51* (Randhawa et al. 2014). It is very likely that these reported QTL either closely linked or the same gene based on the consensus map and major effects (Maccaferri et al. 2015; Bulli et al. 2016). In addition to the above QTL from analysis of bi-parental crosses, ten QTL were identified through GWAS, of which four QTL were located in the same chromosomal bin in the telomere region of the long arm, while other six QTL were in the proximal region. *QYr.wsu-4A.2* and *QYr.wsu-4A.4* were effective to four *Pst* races in seedlings and also to natural infections in adult plants under field conditions (Bulli et al. 2016). Other QTL detected in the GWAS were also effective in a half of the tested experiments.

Chromosome 4B has 14 QTL (Q4B.1 – Q4B. 14) conferring resistance to stripe rust (Table 5.4; Fig. 5.1). Three QTL are located in the distal region of the long arm, while 11 QTL are around the centromeric region, including three major QTL, *QPst.jic-4B* from Alcedo (Jagger et al. 2011), *QYrva.vt-4BL* from VA00W-38 (Christopher et al. 2013), and *Yr62* from wheat line PI 192252 (Lu et al. 2014). The locus *QPst.jic-4B* in Alcedo is flanked by AFLP marker *S24M37_130* and SSR marker *Xcfd39* in chromosome 4BL with SSR *Xgwm692* as the closest marker to the QTL (Jagger et al. 2011). *Xgwm692* was also close to *QYrva.vt-4BL* (Christopher et al. 2013), but the marker was not polymorphic between PI 192252 and AvS, and *Xcfd39* was 16.5 cM distal to *Yr62* (Lu et al. 2014). Base on the consensus map (Maccaferri et al. 2015), *QPst.jic-4B* and *Yr62* are located in the same region. In addition to these major-effect genes, several small-effect QTL in chromosome 4B were reported. Suenaga et al. (2003) detected a QTL associated with stripe rust severity in cultivar

Oligoculm; William et al. (2006) identified a QTL in Avocet; Melichar et al. (2008) identified a QTL in Guardian; Lu et al. (2009) found QTL in Libellula and Strampelli; Zwart et al. (2010) reported a QTL in Janz; Chen et al. (2012a) found QTL in Rio Blanco; and Agenbag et al. (2012) detected QTL in Palmiet. These small-effect QTL explained only 4–17% of the phenotypic variance and some of them were not detected across different environments. Besides these QTL reported from bi-parental crosses, three QTL were detected through GWAS, however they were only effective in a single year or some conditions of multiple tested locations (Singh et al. 2013; Naruoka et al. 2015). Their effects need to be further confirmed.

Chromosome 4D has eight QTL (Q4D.1 – Q4D.8) and all of these QTL confer moderate resistance to stripe rust (Table 5.4; Fig. 5.1). The recently cloned slow-rusting resistance gene *Yr46* is also in this chromosome (Moore et al. 2015). Seven QTL are located in the short arm. Singh et al. (2000) reported *Yr28* which was derived from *Ae. tauschii*, flanked by markers *Xmwg634* and *Xbcd265*. Rosewarne et al. (2012) identified a minor QTL *QYr.cim-4DS* from cultivar Pastor in 4DS, which is different from *Yr28* based on the resistance type and origin. As discussed above, *Yr46* is an pleotropic resistance gene, conferring partial resistance to stripe rust, leaf rust, stem rust and powdery mildew (*Yr46/Lr67/Sr55/Pm46*). The gene was located in 4DL in the Thatcher derivative RL6077 and closely linked to markers *Xgwm165* and *Xgwm192* (Herrera-Foessel et al. 2011). Ren et al. (2012b) identified a moderate QTL, *QYr.caas-4DL*, from Bainong 64, which is also within the marker interval carrying *Yr46* based on the consensus map (Somers et al. 2004). Ren et al. (2015c) reported a minor QTL, *QYr.caas-4DL.2*, from Lumai 21 flanked by markers *Xbarc98* and *Xbarc1148*, which was near centromere based on consensus map (Song et al. 2005; Maccaferri et al. 2015). Although the locus in 'Lumai 21' conferred resistance to both stripe rust and leaf rust, there was no significant effect in reducing the severity of powdery mildew (Lan et al. 2010b). Herrera-Foessel et al. (2011) tested 'Lumai 21' and 'Jingshuang 16' using the SNP marker *csSNP856* that was co-segregating with the *Lr67/Yr46* locus but did not find any polymorphisms between the two parents, indicating that *QYr.caas-4DL.2* is different from *Yr46*. Suenaga et al. (2003) reported a QTL (*QYr-4D_Oligoculm*) for stripe rust resistance on chromosome 4DL in the Israeli wheat Oligoculm. This QTL was closely linked to *Xwmc399* and contributed only a low level of resistance (variance range 2.5–8.0%). The QTL was approximately 24 cM from *QYr.caas-4DL.2* (He et al. 2011), indicating it is different from *QYr.caas-4DL.2*. In addition to these QTL detected from bi-parental crosses, two QTL, *QYr.wsu-4D* and *QYr.ucw-4D*, were identified from GWAS and effective to 4-5 *Pst* races in seedling tests and resistant in all field tests (Maccaferri et al. 2015; Bulli et al. 2016).

5.1.4.3.5 QTL in Group 5 Chromosomes

Chromosome 5A has 14 QTL (Q5A.1 – Q5A.14) associated with stripe rust resistance (Table 5.4; Fig. 5.1). Two QTL are in the short arm. *QYr.cau-5AS* was derived from wheat line AQ24788-83 and explained for 24.1% to 35.3% phenotypic variation of latent period in two year field trials (Quan et al. 2013). Another QTL associated with SNP marker *IWA4767* was identified through GWAS (Zegeye et al. 2014). Of the 12 QTL in the long arm, four QTL, *QYrtb.pau-5A*, *QYR5-5A*, *QYr.caas-5AL* and *Yr48* (*QYr.ucw-5AL*), were reported to have moderate levels of effect with 15–19% PEV (Boukhatem et al. 2002; Chhuneja et al. 2008; Lan et al. 2010a; Lowe et al. 2011). Chhuneja et al. (2008) mapped HTAP resistance QTL, *QYrtb.pau-5A*, from *T. boeoticum* accession Pau5088, on 5AL in an 8.9 cM interval between *Xbarc151 and Xcfd12*. This QTL is proximal to the centromere based on the consensus map (Somers et al. 2004; Maccaferri et al. 2015) and is clearly different from other QTL on 5AL. *QYR5* in Opata 85 was mapped in 5AL within the interval of RFLP markers *XFbb209 - Xabg391* and SSR marker *Xgwm126* (Boukhatem et al. 2002). *QYr.caas-5AL*, from Chinese landrace Pingyuan 50, was found to be flanked by SSR markers *Xwmc410* and *Xbarc261* close to the centromere (Lan et al. 2010a). *Yr48*, a gene underlying the partial stripe rust resistance in synthetic derivative wheat PI 610750, was mapped in the distal region of 5AL close to marker *Xcfa2149* (Lowe et al. 2011). Ren et al. (2012a) mapped APR QTL *QYr.caas-5AL.2* from wheat line Shanghai 3/Catbird in the marker interval between *XwPt1903-5AL* and *Xwmc727-5AL*, the same chromosomal region as *QYr.caas-5AL* reported by Lan et al. (2010a). Hou et al. (2015) mapped HTAP resistance *QTL QYrdr.wgp-5AL*, from wheat cultivar Druchamp, with closely linked SNP markers *IWA2646* and *IWA2558*, assigning the QTL in the telomere region of chromosome 5AL based on the SNP consensus map (Wang et al. 2014a; Maccaferri et al. 2015). *QYrns.orz-5AL* from NSA-98-0995 (Vazquez et al. 2015) is also located in the same region in according to the consensus map.

Chromosome 5B contains 23 QTL (Q5B.1 – Q5B.23) conferring resistance to stripe rust (Table 5.4; Fig.5.1). Six large-effect QTL (*QYrPI192252-5BS*, *QYr.tem-5B.1*, *QYr.tem-5B.2*, *QYr.inra-5BL.1*, *QYr.inra-5BL.2* and *QYrdr.wgp-5BL*) were mapped in 5B with each QTL explaining over 20% of phenotypic variation. Lu et al. (2014) mapped a QTL from spring wheat line PI 192252 in the short arm of 5B, which explained for 22.1-27.4% phenotypic variation across two locations in 3 years. The parental line and progeny lines with this QTL and *Yr62*, another HTAP resistance gene in PI 192252, confer high level resistance to stripe rust in the US PNW. *QYr.tem-5B.1* and *QYr.tem-5B.2* were detected in Flinor (Feng et al. 2011). *QYr.inra-5BL.1* and *QYr.inra-5BL.2* from cultivar Camp Remy were mapped in in intervals of *Xgwm639a-Xgwm639c* and *Xgwm234a-XDuPw115a*, respectively, and they were effective across all experiments (Mallard et al. 2005). *QYrdr.wgp-5BL.2* was mapped with SNP markers *IWA6867* and *IWA6383* in the telomere region of the long arm in cultivar Druchamp (Hou et al. 2015). Other QTL have small or moderate effects. Hao et al. (2011) identified a minor QTL, *QYr.uga-5B*, from wheat line AGS 2000 that was used as the susceptible parent in a mapping population from a cross with resistant cultivar Pioneer 26R61. This minor QTL was only detected in one location of multiple experimental environments. Yang et al. (2013) reported a minor QTL, *QYr.cim-5BL*, from Chapio, and it was only effective in Chinese experimental environments and not in the cooler highland environments in Mexico, indicating that the QTL is race-specific and/or sensitive to temperature. *QYr.jirc-5BL* in the wheat cultivar Oligoculm is linked to SSR marker *Xwmc415* (Suenaga et al. 2003). *Yr.caas-5BL.1* and *QYr.caas-5BL.2* in Libellula were mapped in the *Xwmc415-Xwmc537* and *Xbarc142-Xgwm604* marker intervals, respectively (Lu et al. 2009). Bariana et al. (2010) identified *QYr.sun-5B* from Janz flanked by markers *XwPt1733* and *XwPt2707*. Ren et al. (2012a) reported *QYr.caas-5BL.3* in Shanghai 3/Catbird in 5BL between *Xwmc75* and *Xbarc275*, a position similar to that of *QYr.caas-5BL.2* and *QYr.sun-5B*. Case et al. (2014) identified *QYrco.wpg-5B* from club wheat Coda in chromosomal bin 5BL1-0.55-0.75, defined by markers *IWA5488* and *XwPt-6135*. Based on the consensus SSR and SNP maps (Somers et al. 2004; Wang et al. 2014a; Maccaferri et al. 2015), *QYrdr.wgp-5BL.2*, *QYrco.wpg-5B*, *QYr.jirc-5BL*, *QYr.caas-5BL.1* and *QYr.inra-5BL.1* should be in the same region. Because these QTL all confer APR and their wheat carriers are all of European origin, they are likely the same gene. Both *QYr.caas-5BL.2* and *QYr.caas-5BL.3* should be distal to *QYrdr.wgp-5BL.2* according to the consensus map (Somers et al. 2004). Three QTL (*QYr.wsu-5B_IWA5166*, *IWA3002_Seedling* and *QRYr5B.1_IWA2565*) were detected through GWAS (Zegeye et al. 2014; Jighly et al. 2015; Bulli et al. 2016). *QY.wsu-5B* associated with marker *XIWA5166* was effective against four *Pst* races in seedlings and natural infections in four field tests (Bulli et al. 2016).

Chromosome 5D has 7 QTL (Q5D.1 – Q5D.7) conferring resistance to stripe rust, 2 in the short arm and 5 in the long arm (Table 5.4; Fig. 5.1). All of QTL in 5D have small to moderate effect. Ren et al. (2015c) identified *QYr.caas-5DS* in Chinese cultivar 'Jingshuang 16', which was close to the centromere and explained 5.1–18.0% phenotypic variance across all environments. Singh et al. (2000) identified a QTL (*QYr-5DS*) for APR to stripe rust in Mexican cultivar Opata 85 in chromosome 5DS, flanked by *Xfbb238* and *Xfba114* and located in the same region as *QYr.caas-5DS* based on the consensus map (Somers et al. 2004; Song et al. 2005). However, the QTL explained 8.0% of the phenotypic variance in only one environment, showing a lower effect and stability than *QYr.caas-5DS*. Suenaga et al. (2003) identified *QYr.jirc-5DL* in chromosome 5DL closely linked to SSR locus *Xwmc215*; and the QTL explained 3.9% of the phenotypic variance. Imtiaz et al. (2004) found *QYr.nsw-5DL* in chromosome 5DL, closely linked to *Xgwm583*, explaining 6.1% of the phenotypic variance. Liu et al. (2015) identified *QYr.caas-5DL* from Zhong 892 flanked by SNP markers *XBS00110475* and *Xwsnp_Ex_c22984_32207214*. As markers *Xwmc215*, *Xgwm583* and *Xwsnp_Ex_c22984_32207214* belong to the same chromosome bin (5DL5-0.76-1.00), these three QTL may be the same, allelic or closely linked. Case et al. (2014) identified an ASR QTL (*QYrbr.wpg-5D*) against race PST-114 from cultivar Brundage. This QTL is flanked by *Xbarc144* and *Xwmc765* in the telomere region of 5DL according to the consensus map (Somer et al. 2004; Song et al. 2005; Maccaferri et al. 2015). Hou et al. (2015) identified an ASR QTL (*QYrdr.wgp-5DL_IWA8331*) from Druchamp, conferring resistance to US *Pst* races PST-35 and PST-45. The QTL is associated with SNP marker *XIWA8331*. Because *XIWA8331* is only 3.0 cM away from *Xbarc144*, and the two QTL confer the same type of resistance, these QTL could be the same gene. Vazquez et al. (2015) reported a minor QTL (*QYrtb.orz-5DL*) from winter wheat cultivar Tubbs, associated with DArt marker *XwPt-5870*. The relationship of this QTL with others is not clear due to the lack of consensus map of *XwPt-5870* and other markers.

5.1.4.3.6 QTL in Group 6 Chromosomes

Chromosome 6A has 15 QTL (Q6A.1 – Q6A.15) associated with stripe rust resistance and they can be sorted into four regions (Table 5.4; Fig. 5.1). The first region is in the short arm, including three QTL, *QYr.uga-6AS* from US winter wheat cultivar Pioneer 26R61 (Hao et al. 2011), *QYr.wgp-6AS* from US spring wheat cultivar Express (Lin and Chen 2009) and *QYr.cim-6A* from Australian spring wheat Avocet (Rosewarne et al. 2012). *QYr.wgp-6AS* contributed to both low disease severity and IT across all tested experiments. *QYr.uga-6AS* has moderate LOD and PEV values. Both QTL confer HTAP resistance and share the same flanking marker *Xgwm334*, indicating that they are likely the same gene. The second region is around the centromere. Two QTL, *QYr.caas-6AL* (Liu et al. 2015) and *QYr.nafu-6AL* (Zhou et al. 2015c), were assigned to this region based on the consensus maps (Somers et al. 2004; Song et al. 2005; Wang et al. 2014a; Maccaferri et al. 2015). *QYr.caas-6AL* and *QYr.nafu-6AL* from Chinese wheat cultivars Zhong 892 and Caoxuan 5, respectively; and both QTL had moderate LOD and PEV scores (Liu et al. 2015; Zhou et al. 2015c); indicating the possibility of the same gene. The third region has two QTL, *QRYr6A.1* and *QYr.wpg-6A.1* identified through GWAS (Jighly et al. 2015; Naruoka et al. 2015). The fourth region is located in the middle of the long arm. In this region, *QYrtb.orz-6AL* is a major QTL from US PNW winter wheat Tubbs and displayed high LOD and PEV scores in all field experiments (Vazquez et al. 2015). The epistatic interaction between *QYrtb.orz-6AL* and *QYrtb.orz-2AS* determined the resistance response in Tubbs. A minor QTL (*QYrpl.orr-6AL*) from Platte (Vazquez et al. 2012) is located in the same region as *QYrtb.orz-6AL*. Four QTL were report in cultivar Avocet, including *QYr.6AL* (William et al. 2006), *QYr.cim-6AL* (Lan et al. 2014), *QYr.ufs-6A* (Prins et al. 2011) and *QYr-6A* (Lillemo et al. 2008). Avocet contains the *Thinopyrum elongatum* translocation in 6AL, which confers stem rust resistance gene *Sr26* (McIntosh et al. 1995). Translocations often carry multiple resistance genes and the positioning of a near-centromeric location for this QTL region is consistent with it being on the *Sr26* translocation (Rao 1996).

Chromosome 6B harbors 22 QTL (Q6B.1 – Q6B.22) for stripe rust resistance, and it also contains important HTAP resistance gene *Yr36* (Table 5.4; Fig. 5.1). The relative position of 11 QTL were assigned to the short arm and 10 QTL to the long arm based on the consensus map (Maccaferri et al. 2015; Bulli et al. 2016), except that of *QYrns.orz-6BS* could not be determined. In the short arm, *QYrdr.wgp-6BL* from Druchamp is a race-specific QTL located in the telomere region (Hou et al. 2015), while the others are non-race specific QTL and proximal in the short arm. *QYr.wgp-6BS.1* and *QYr.wgp-6BS.2* from cultivar Stephens confers HTAP resistance and had high PEV values across field experiments (Santra et al. 2008). The remaining eight QTL in the short arm had low LOD and PEV scores, including *QYrns.orz-6BS* from NSA-98-0995 (Vazquez et al. 2015), *QYr.ufs-6BS* from Kariega (Prins et al. 2011), *QYr.tam-6BS* from TAM111 (Basnet et al. 2014a), *QYr.caas-6BS.2* from Naxos (Ren et al. 2012a), *QYr.caas-6BS.3* from Bainong 64 (Ren et al. 2012b), *QYr.sun-6B* from Janz (Bariana et al. 2010), *QYr-6BSC* from Oligoculm (Suenaga et al. 2003) and *QYr.caas-6BS* from Pingyuan 50 (Lan et al. 2010a). The second region is in the long arm near the centromere, including five QTL, *QYrpi.vt-6BL* from cultivars Pioneer Brand 26R46 (Christopher et al. 2013), *QYr.6BL* from Pavon 76 (William et al. 2006), *QYr.inra-6B* from Renan (Dedryver et al. 2009), *QYrdr.wgp-6BL.2* from Druchamp (Hou et al. 2015) and *QYr.wsu-6B.2* associated with SNP IWA4169 for race-specific resistance in several varieties (Bulli et al. 2016). *QYrpi.vt-6BL* and *QYr.6BL* had high PEV scores, while *QYr.inra-6B* and *QYrdr.wgp-6BL.2* had low PEV values in the tests. However, *QYrdr.wgp-6BL.2* was effective in all six experiments. It should be different from other QTL in this region. The third region has three QTL, *QYr.wsu-6B.1*, *QYr.ucw-6B* and *QYr.wgp-6B.1*. All of them are near the centromere, associated with SNP *IWA7257* and determined in three independent GWAS (Maccaferri et al. 2015; Naruoka et al. 2015; Bulli et al. 2016). They are race-specific QTL and are very likely the same QTL. The fourth region has three QTL, *QYr.cim-6BL* derived from cultivar Pastor (Rosewarne et al. 2012), and *QYr.wsu-6B.3* and *QYr.wgp-6B.2* from GWAS (Naruoka et al. 2015; Bulli et al. 2016). *QYr.cim-6BL* had a small effect with low LOD and PEV values in field tests (Rosewarne et al. 2012), while *QYr.wsu-6B.3* is race-specific (Bulli et al. 2016).

Only four QTL (Q6D.1 – Q6D.4) have been reported in chromosome 6D and all of them are located in the long arm (Table 5.4; Fig. 5.1). Agenbag et al. (2012) identified a minor QTL, *QYr.ufs-6D*, proximal in 6DL in wheat line Yr16DH70. The QTL was believed originating from Cappelle-Desprez. Boukhatem et al. (2002) found an ASR QTL in the middle of the long arm, and the QTL explained 13.1% phenotypic variation, although the author stated that this was only suggestive and not significant with low LOD score. Other two QTL (*QYr.ucw-6D* and *IWA4455_*seedling) were identified though GWAS and both of them were race-specific (Zegeye et al. 2014; Maccaferri et al. 2015). Because these studies were conducted in relatively few environments or with low PEV scores, more work is needed to determine their breeding value.

5.1.4.3.7 QTL in Group 7 Chromosomes

Chromosome 7A has 13 QTL (Q7A.1 – Q7A.13) associated with stripe rust resistance (Table 5.4; Fig. 5.1). All of these QTL have relative minor effect. Five QTL are located in the short arm. *QYr.cim-7AS* from Avocet is in the far distal region of the short arm (Rosewarne et al. 2012). Dedryver et al. (2009) reported that *QYr.inra-7A* from wheat variety Recital was close to *QYr.cim-7AS*, defined by flanking AFLP markers *Xbcd129b* and *Xfba127c*. Three QTL, *QYr.caas-7A* from cultivar Jingshuang 16 (Ren et al. 2015c), *QYr.sun-7A* from CPI 133872 (Zwart et al. 2010) and a QTL associated with SNP *IWA2710* (Zegeye et al. 2014), are likely in the same region, near the centromere in short arm, because the bracket SSR markers of *QYr.sun-7A* were covered the SSR loci interval of *QYr.caas-7A* based on the consensus map (Somers et al. 2004; Maccaferri et al. 2015). Eight QTL are reported in the long arm, of which three from GWAS (Zegeye et al. 2014) are more proximal and the others from bi-parental crosses are more distal. *QYr.caas-7AL* from Zhong 892 was flanked by SNP markers and delimited in a narrow region of a genetic interval of 0.1 cM (Liu et al. 2015). Vazquez et al. (2012) reported *QYr.orr-7A* from Stephens associated with DArT marker *XwPt-4319*. *QYr.cim-7AL* was another QTL from Avocet identified by Rosewarne et al. (2012). Prins et al. (2011) reported two QTL, *QYr.sig-7A.1* and *QYr.sig-7A.2*, both from cultivar Kariega in the far distal region of the long arm. Of all QTL in chromosome 7A, *QYr.sun-7A* and *QYr.orr-7A* have moderate PEV values, and the breeding values of other QTL need to be further determined.

Chromosome 7B has 14 QTL (Q7B.1 – Q7B14) and several important QTL conferring high level HTAP resistance to stripe rust, including *Yr39*, *Yr52*, *Yr59* and *Yr67* (Table 5.4; Fig. 5.1). Except a QTL in cultivar Oligoculm in the short arm (Suenaga et al. 2003), all QTL in 7B are assigned in the long arm according to the consensus map (Maccaferri et al. 2015; Bulli et al. 2016). All of these QTL have small or moderate effect, except HTAP resistance genes such as *Yr39* (*QHtap.wsu-7BL*). *Yr39* (*QHtap.wsu-7BL*) from Alpowa explained 64.2% phenotypic variation with an average LOD value of 30.4 (Lin and Chen 2007). The gene is located in the region proximal to the centromere. Five QTL were assigned to this region according to the consensus map (Maccaferri et al. 2015). *QYr.orr-7BS* from Stephens has a moderate effect, accounting for 11% phenotypic variation (Vazquez et al. 2012). *QYr.sun-7B* from cultivar Kukri spanning 40 cM explained 8–9% phenotypic variation (Bariana et al. 2010). *QYr.caas-7BL.1* from SHA3/CBRD accounted for 3.7–8.2% phenotypic variation (Ren et al. 2012a). Two QTL from GWAS are also in this region (Zegeye et al. 2014). Another important region is in the telomere of the long arm bin 7BL3-0.86-1.00, containing several major genes/QTL, such as *Yr52* (Ren et al. 2012a), *Yr59* (Zhou et al. 2014c), *Yr67* (Xu et al. 2014) and *YrZh84* (Li et al. 2006a, b) as presented above in the section of permanently named *Yr* genes. At least other seven QTL can be assigned to this region based on their flanking markers in the consensus map (Somers et al. 2004; Song et al. 2005; Maccaferri et al. 2015). These QTL include *QYr-7B* from cultivar Titriea (Imtiaz et al. 2004), *QYrtb.orz-7BL* from Tubbs (Vazquez et al. 2015), *QYr.caas-7BL* from Pastor (Rosewarne et al.

2012), *QYr.caas-7BL.2* from SHA3/CBRD (Ren et al. 2012a), *QYr-7BL* from Ologoculm (Suenaga et al. 2003), *QYr-7BL* from Strongfield (Singh et al. 2013) and *QYr.7BL* from Attila (Rosewarne et al. 2008). All of these seven QTL have low or moderate LOD and PEV scores. They are likely different from the major QTL genes, *Yr52*, *Yr59*, *Yr67* (*YrC591*) and *YrZh84*.

Chromosome 7D contains 10 QTL (Q7D.1 – 7D.10) associated with stripe rust resistance (Table 5.4; Fig. 5.1). This chromosome also has *Yr18*, an important gene providing durable stripe rust resistance in many old Asian wheat cultivars and landraces and the modern cultivars derived from CIMMYT wheat germplasm. QTL around the *Yr18* locus have been reported in many studies and they were shown to have stable effect in most environments (Singh et al. 2000; Lillemo et al. 2008; Lu et al. 2009; Bariana et al. 2010; Yang et al. 2013). Most of the QTL in this region should be *Yr18*. In addition to the *Yr18* locus, one QTL associated with marker *XwPt-668026* (Jighly et al. 2015) was detected in GWAS also in the short arm, but it is distal to the *Yr18* locus about 12 cM based on the consensus map (Maccaferri et al. 2015). Two QTL, *QYr.tam-7DL* from wheat cultivar TAM 111 (Basnet et al. 2014a) and *QYr-7D_Otane* from cultivars Otane (Imtiaz et al. 2004), were located in the long arm of chromosome 7D. Both QTL had moderate effects, explaining for over 13% of the phenotypic variation. One QTL from Opata 85 (Boukhatem et al. 2002) explained for 13.6% phenotypic variation, but the relationship between this QTL and others cannot be determined based on the consensus map since its linked markers are absent from the consensus map.

5.1.5 Genetics and Genes of Resistance to Stripe Rust in Barley

Compared to wheat, resistance to stripe rust in barley have been less studied, partially due to the fact that barley is grown in relatively less areas and barley stripe rust is globally less important than wheat stripe rust. There were limited studies on barley for resistance to stripe rust before the barley stripe rust pathogen was introduced to the US. Stripe rust on barley is caused mostly by *Psh* and occasionally by *Pst* (Stubbs 1985; Chen et al. 1995c; Chen and Line 1999, 2003). *Psh* was introduced into South America in 1975 (Dubin and Stubbs 1986) and then found on barley in the USA in 1991 (Roelfs et al. 1992) and became an important disease on barley in 1990s (Line 2002). Chen et al. (1995c) identified seven barley genotypes resistant to all US *Psh* isolates and 11 genotypes resistant to some isolates. They selected eleven of the genotypes, most of them with the European origin, as differentials for identifying *Psh* race, and later added US barley cultivar Bancroft in the differential set (Chen and Line 2001a; Chen 2004, 2008). They also studied genetics of resistance in these genotypes.

5.1.5.1 Modes of Inheritance and Genes Identified for Resistance to Stripe Rust in Barley

The earliest study on the genetics of barley stripe rust resistance was reported in 1942 in India by Murty (1942). In his study, resistance in the American variety Alpha was controlled by two dominant genes. In the 1960s, genetic studies of resistance to stripe rust in nine barley varieties were reported (Bahl and Backshi 1963, Bakshi et al. 1964; Jain and Agrawal 1964; Srinivasan and Padmanabhan 1964; Bakshi and Bahl 1965; Upadhyaya et al. 1965; Bakshi and Luthra 1970). Bakshi and Luthra (1970) reported eight loci for resistance to different Indian races in three barley genotypes, E.B. 145, E.B. 410 and E. B. 438; and named genes *Ps1* to *Ps8* with multiple alleles at the *Ps1* and *Ps3* loci. They found that among the 11 resistant alleles at the eight loci, eight were dominant and three recessive. In these series of studies, different types of epistasis, such as duplicate dominant, independent recessive, complementary recessive, and inhibitory genes, were reported (Bahl and Backshi 1963, Bakshi et al. 1964; Jain and Agrawal 1964; Bakshi and Bahl 1965; Upadhyaya et al. 1965; Bakshi and Luthra 1970).

In Europe, Nover and Scholz (1969) studied resistance in Bigo, Abyssinian 14, BBA 2890, Abed Binder 12 and I 5 to the European race 24 of *Psh*. They reported that the resistance of all five genotypes was controlled by single recessive genes. Based on allelism tests, they suggested that Bigo, Abyssinian 14 and BBA 2890 had resistant alleles at the same locus and was designated *yr*. They designated a gene in Abed Binder 12 as *yr2* and a gene in I 5 as *yr3*. Johnson (1968) reported another gene in barley varieties Europa, Cambrinus and Deba Abed, designated *Yr4*.

In a study on stripe rust resistance of barley varieties to Indian races including an Indian culture considered to be race 24, a typical *Psh* race in Europe (Stubbs 1985), Upadhyay and Prakash (1977) reported that Bigo was susceptible to the Indian culture of race 24 and Abyssinian 14 had one or more additional genes for resistance. In contrast to the study of Nover and Scholz (1969), it is not clear that the races used in India were *Psh* or *Pst* and that the Indian race 24 was the same as the European race 24 or not. Most of races used in the Indian barley studies were virulent on wheat genotype Michigan Amber, which is resistant to race 24 in Europe and all *Psh* races detected in the United States (Stubbs 1985; Chen et al. 1995c; Line and Chen 1996).

After *Psh* was introduced from Europe to South America in 1975 (Dubin and Stubbs 1986), US scientists concerned the potential of spreading of *Psh* into North America and studied resistance of barley cultivars. The first genetic study of barley stripe rust resistance in North America was conducted by Chen et al. (1994a) at Oregon State University in cooperation with the International Wheat and Maize Research Center (CIMMYT) in Mexico. Stripe rust response of a DH barley population was evaluated in Mexico. Using molecular markers and quantitative trait loci (QTL) mapping approaches, they mapped two QTL conferring resistance in an elite six-row spring feed barley to the M arms of chromosomes 7 and 4.

Chen et al. (1995c) tested numerous barley varieties to *Psh* races identified in the US. They found that barley genotypes PI 548708, PI 548734, PI 548747, BBA2890,

Abyssinian 14, Grannelose Zweizeilige, and Stauffers Obersulzer were resistant to all *Psh* races detected in the United States by that time, and that Heils Franken, Emir, Astrix, Hiproly, Varunda, Trumpf, Mazurka, Bigo, I5, Cambrinus and BBA 809 were resistant to some PSH races, but not all. Chen and Line (1999) studied the genetics of 18 barley genotypes. They tested seedlings of parents and F_1, F_2 and F_3 progeny from crosses of the resistant genotypes with susceptible variety Topper or Steptoe with US *Psh* races. They detected one recessive gene in BBA 809, BBA2890, Bigo, Hiproly and Grannelose Zweizeilige; two recessive genes in Emir, I 5, PI 548708, PI 548734, PI 548747, Varunda and Astrix; and one dominant gene and one recessive gene in Abyssinian 14 and Stauffers Obersulzer with PSH-1. When used race PSH-4, they found one recessive gene in BBA 809; two recessive genes in Trumpf; and two partially recessive genes in Astrix. In tests with PSH-13, BBA 2890, Grannelose Zweizeilige, I 5 and PI 548708 had one recessive gene; and Abyssinian 14, Hiproly, PI 548734 and PI 548747 had two recessive genes. In the tests with PSH-20, Bigo had one recessive gene and Heils Franken had a recessive gene and a dominant gene. A gene in Cambrinus for resistance to PSH-10 and a gene in Mazurka for resistance to PSH-20 were dominant based on IT on the first seedling leaf but recessive according to the second leaf IT. In that study, they detected various types of epistasis. The most interesting finding of their study is that most genes for resistance to stripe rust in barley are recessive, very different from the genetics of wheat resistance to stripe rust. The reasons for such difference are not clear, and more studies on the evolution and interactions between the plant hosts and pathogen formae speciales may lead to the mechanisms underlying the different inheritance of resistance to the different forms by different crop species.

Chen and Line (2003) further studied the genes for stripe rust resistance in the 18 barley genotypes through allelism tests. They identified 30 resistance genes and at least 26 of them were different from each other. The single genes in genotypes Bigo, Trumpf, Mazurka and BBA 2890 and one of the genes in PI 548708, PI 548734, PI 548747 and Abyssinian 14 were linked. One of the two genes in PI 548734 was linked to one of the two genes in I 5. The single gene in Cambrinus and one of the genes in Heils Franken and Astrix were the same or closely linked. The single genes in Grannelose Zweizeilige and BBA 809 and the two genes in Stauffers Obersulzer, Emir, Hiproly and Varunda were different from all other genes and different from each other. Based on the results, they named the genes with a prefix "*rps*" for recessive genes and "*Rps*" for dominant genes. These genes, together with those identified by others, are presented in (Table 5.5). Among the genes in the list, only a few genes have been mapped to barley chromosomes, which will be discussed below. More studies are needed to map these genes to barley chromosomes and identify markers useful for use in MSA for developing barley cultivars with resistance to stripe rust.

5.1.5.2 Molecular Mapping of Genes or QTL for Resistance to Stripe Rust in Barley

Similar to wheat, stripe rust resistance in barley also can be separated into ASR and APR or HTAP resistance (Chen and Line 1999, 2003; Chen and Moore 2003; Yan and Chen 2008). In addition to the genes for ASR to stripe rust in barley discussed above and presented in Table 5.5, genes or QTL have been identified and mapped to barley chromosomes using molecular mapping. So far, a total of 28 genes or QTL have been mapped, and these genes are presented in Table 5.6.

Table 5.5 *Rps* genes for resistance to *Puccinia striiformis* f. sp. *hordei* identified in barley varieties

Gene[a]	Barley variety	Races detecting the gene	References
rps1.a	BBA 2890, PI 548734	PSH-1, PSH-13	Chen and Line (2001a, 2003) and Yan and Chen (2007)
rps1.b	Bigo	PSH-1, PSH-10	Chen and Line (2001a, 2003)
rps1.c	Mazurka	PSH-1, PSH-10	Chen and Line (2001a, 2003)
rps2	Abed Binder 12	No data	Johnson (1968) and Chen and Line (2001a, 2003)
rps3	I 5	PSH-1, PSH-13	Chen and Line (2001a, b, 2003)
Rps4	Cambrinus, Heils Franken, Astrix	PSH-10, Europe race 23	Chen and Line (2001a, 2003) and von Wettstein-Knowles (1992)
rpsHF	Heils Franken	PSH-10, PSH-20	Chen and Line (2001a, 2003)
rpsEm1	Emir	PSH-1	Chen and Line (2001a, 2003)
rpsEm2	Emir	PSH-1	Chen and Line (2001a, 2003)
rpsAst	Astrix	PSH-1, PSH-10	Chen and Line (2001a, 2003)
rpsHi1	Hiproly	PSH-1, PSH-13	Chen and Line (2001a, 2003)
rpsHi2	Hiproly	PSH-1	Chen and Line (2001a, 2003)
rpsVa1	Varunda	PSH-1	Chen and Line (2001a, 2003)
rpsVa2	Varunda	PSH-1	Chen and Line (2001a, 2003)
rpsTr1	Trumpf	PSH-1, PSH-10	Chen and Line (2001a, 2003)
rpsTr2	Trumpf	PSH-1, PSH-10	Chen and Line (2001a, 2003)
rpsBBA809	BBA 809	PSH-1, PSH-10	Chen and Line (2001a, 2003)
rpsPI548708-1	PI 548708	PSH-1, PSH-13	Chen and Line (2001a, 2003)
rpsPI548708-2	PI 548708	PSH-1	Chen and Line (2001a, 2003)
rpsPI548734	PI 548734	PSH-1, PSH-13	Chen and Line (2001a, 2003)
rpsPI548747-1	PI 548747	PSH-1, PSH-13	Chen and Line (2001a, 2003)
rpsPI548747-2	PI 548747	PSH-1, PSH-13	Chen and Line (2001a, 2003)
rpsA14-1	Abyssinian 14	PSH-1, PSH-13	Chen and Line (2001a, 2003)
RpsA14-2	Abyssinian 14	PSH-1, PSH-13	Chen and Line (2001a, 2003)
rpsGZ	Grannelose Zweizeilige	PSH-1, PSH-13	Chen and Line (2001a, 2003) and Yan and Chen (2006)
rpsI5	I 5	PSH-1, PSH-13	Chen and Line (2001a, 2003)

(continued)

Table 5.5 (continued)

Gene[a]	Barley variety	Races detecting the gene	References
rpsSO-1	Stauffers Obersulzer	PSH-1, PSH-13	Chen and Line (2001a, 2003)
RpsSO-2	Stauffers Obersulzer	PSH-1, PSH-13	Chen and Line (2001a, 2003)
Rpsh	L94	Non-host resistance	Niks et al. (2015)
Rpsx	CI10587 (PI 243183)	Field *Psh* population at Toluka, Mexico	Castro et al. (2003a)
Rps6 (*YrpstY1*)	PI 466050, Abed Binder 12	Non-host resistance	Li et al. (2016a) and Dawson et al. (2016) and Sui et al. (2010)
RpsLem	PI 478214 (wheat)	Non-host resistance	Pahalawatta and Chen (2005a)
RpstS1	Steptoe	Non-host resistance	Pahalawatta and Chen (2005b)
rpstS2	Steptoe	Non-host resistance	Pahalawatta and Chen (2005b)

[a]Resistance genes *rps1*, *rps2*, *rps3* and *Rps4* were previously named as *yr1*, *yr2*, *yr3* and *yr4*, respectively (Nover and Scholz 1969; Johnson 1968)

5.1.5.2.1 Genes/QTL in Chromosome 1H

One major gene (*Rps4*) and two QTL (*Qpsh-Sh-1Ha* and *Qpsh-Sh-1Hb*) have been reported in chromosome 1H (chromosome 5) (Table 5.6). Thomas et al. (1995) mapped a QTL in the short arm of chromosome 1H using a DH population from cross Blenheim × E224/3. The resistance allele of this QTL is from Blenheim. As Deba Abed, an ancestor of Blenheim, carries *Rps4* (von Wettstei-Knowles 1992), this major QTL is likely *Rps4*. Toojinda et al. (2000) mapped a QTL (*Qpsh-Sh-1Ha*) in chromosome 1H from a DH population from cross Shyri × Galena based on field data of adult plants. The QTL was flanked by markers *act8* and *BMAC213*, and explained 28–50% of phenotypic variation. The allele from Shyri contributed the resistance in this QTL. This QTL is in the same region as *Rps4* (von Wettstei-Knowles 1992; Thomas et al. 1995). Castro et al. (2002a) identified a QTL, *Qpsh-Sh-1Hb* (QTL5), in chromosome 1H using the seedling data of the same DH population tested with *Psh* races PSH-1, PSH-13 and PSH-14. This QTL is in an interval spanning 23.3 cM between markers *LM637/8-3* and *Bmac213*, and was coincident with the QTL identified by Toojinda et al. (2000) using field adult-plant data. Thus, the seedling QTL and adult-plant is the same QTL and likely the same gene, allelic genes or closely linked genes. The QTL from cultivar Shyri was also confirmed in breeding populations developed from Shyri crosses with other cultivars (Castro et al. 2003a). This QTL and a QTL on chromosome 6H were necessary for full resistance in the Shyri × Galena population (Castro et al. 2002a). The resistance of this QTL was introgressed into and validated with a set of barley lines through MAS (Richardson et al. 2006).

Table 5.6 Genes or QTL for resistance to stripe rust mapped to chromosomes in barley

Chrom.	Gene/QTL	Mapping population	Resistance allele	*Peak marker*	*Marker 1*	*Marker 2*	Genetic interval (cM)	LOD	PEV (%)	Freq.	References
1H	*Rps4*	Blenheim × E224/3	Blenheim	*BCD249(5)*	*OPF1-H600*	*BCD98(5)*	ND[a]	ND	73–79	2/2	Thomas et al. (1995)
1H	*Qpsh-Sh-1Ha*	Shyri × Galena	Shyri	*MWG837*	*act8*	*BMAC213*	8.8	30–60	28–50	5/5	Toojinda et al. (2000)
1H	*Qpsh-Sh-1Hb*	Shyri × Galena	Shyri	*LM637/8-3 – BMAC213*	*LM637/8-3*	*BMAC213*	23.3	7–10	ND	3/3	Castro et al. (2002a)
2H	*Qpsh-Ga-2H*	Shyri × Galena	Galena	*E42M47e - EBMAC525*	*E42M47e*	*EBMAC525*	8.7	12	9–12	2/5	Toojinda et al. (2000)
2H	*Qpsh-Bar-2H*	BCD47 × Baronesse	Baronesse	*Bmac093*	*EBmac684*	*Bmag125*	25.3	10	5	1/2	Vales et al. (2005)
3H	*Qpsh-Sh-3H*	Shyri × Galena	Shyri	*ABG004 - T22M3li*	*ABG004*	*T22M3li*	15.2	12	12	1/5	Toojinda et al. (2000)
)3H	*rps1.a*	Steptoe × BBA2890	BBA2890	*B11-B13*	*B11*	*B13*	8.9	NA[b]	NA	NA	Yan and Chen (2007)
3H	*3Ha*	BCD47 × Baronesse	BCD47	*Mnag225 - Bmag606*	*Mnag225*	*Bmag606*	24.0	18	14	1/2	Vales et al. (2005)
3H	*3Hb*	BCD47 × Baronesse	BCD47	*Bmag606 - Bmag013*	*Bmag606*	*Bmag013*	21.0	18	13	1/2	Vales et al. (2006)
3H	*Qpsh-Ban-3H*	Harrington × Bancroft	Bancroft	*H2-H3*	*H2*	*H3*	1.7	35–60	>70	3/3	Yan and Chen (2008)
4H	*Qpsh-Ca-4H*	Calicuchima-sib × Bowman	Calicuchima-sib	*ABG472 - Bmy*	*ABG472*	*Bmy*	40	2.5	10	2/2	Chen et al. (1994a, b)
4H	*Qpsh-BCD-4H*	Calicuchima-sib×Bowman	Calicuchima-sib	*ABG472 - Bmy-1*	*ABG472*	*Bmy-1*	54.6	5	4–26	2/5	Castro et al. (2002b)

(continued)

Table 5.6 (continued)

Chrom.	Gene/QTL	Mapping population	Resistance allele	*Peak marker*	*Marker 1*	*Marker 2*	Genetic interval (cM)	LOD	PEV (%)	Freq.	References
4H	*Qpsh-BCD-4H*	BCD47 × Baronesse	BCD47	*EBmac788 - HvMLO3*	*EBmac788*	*HvMLO3*	3.3–7	19–50	14–34	2/2	Vales et al. (2005)
4H	*rpsGZ*	Steptoe × Grannelose Zweizeilige	GZ	*M7-M8*	*M5*	*M6*	2.9	NA	NA	NA	Yan and Chen (2006)
4H	*Qpsh4Hb*	Lenetah×GZ	GZ	SCRI_RS_10818	*SCRI_RS_188827*	SCRI_RS_157611	7.3-8.5	4–16	12–38	3/3	Klos et al. (2016)
4H	*Rpsh-L94-4H*	Vada×L94	L94	*EBmac0701, E40M32-660*	*EBmac0701*	*E40M32-660*		15-17	46–49	2/2	Niks et al. (2015)
4H	*Qpsh4Ha*	Lenetah×GZ	GZ	*BOPA_12_30839*	*SCRI_RS_190401*	*SCRI_RS_168603*	24.6	5	15	1/3	Klos et al. (2016)
5H	*Qpsh-Ca-5H*	Calicuchima-sib×Bowman	Calicuchima-sib	*CD057 - BG123*	*CD057*	*BG123*	21.8	15	57	2/2	Chen et al. (1994a, b)
5H	*5Ha*	BCD47×Baronesse	BCD47	*Bmag337 -Bmag005*	*Bmag337*	*Bmag005*	3.0-6.3	4–6	3	2/2	Vales et al. (2005)
5H	*5Hb*	BCD47×Baronesse	Baronesse	*Bmag222 - GMS001*	*Bmag222*	*GMS001*	32.6	4	2	1/2	Vales et al. (2005)
6H	*Qpsh-Sh-6H*	Shyri × Galena	Shyri	*KsuA1H2 Linka*	*KsuA1H2*	*Linka*	14.3	11	15–23	2/5	Toojinda et al. (2000)
6H	*Qpsh-Sh-6H*	Shyri × Galena	Shyri	*MWG652A - Linka*	*MWG652A*	*Linka*	37.0	2-3		ND	Castro et al. (2002a)
6H	*Qpsh-BCD-6H*	BCD47 × Baronesse	BCD47	*Bmac316 -Bmag500*	*Bmac316*	*Bmag500*	18.0–23.0	9	5.4–7.8	2/2	Vales et al. (2005)

6H	*Qpsh-GZ-6H*	Lenetah × GZ	GZ	*SCRI_RS_189878*	*BOPA1_1240-844*	*BOPA2_12_30358*	22.07–24.55	4–7	12–18	3/3	Klos et al. (2016)
6H	*Qpsh-Ca-6H*	Calicuchima-sib×Bowman	Calicuchima-sib	*ABG387b*	*ABG466*	*ABG458*	40.0	2–5	9.7–18.2	3/3	Castro et al. (2002b)
7H	*Rpsx*	CI10587 × Galena; AJ, OP, BU	CI10587	*Ris44 - ABG461*	*Ris44*	*ABG461*	22.2	7–12	ND	3/3	Castro et al. (2003a)
7H	*7Ha*	BCD47 × Baronesse	Baronesse	*HvWaxy4a - EBmac603*	*HvWaxy4a*	*EBmac603*	26.4	8	6.5	1/2	Vales et al. (2005)
7H	*7Hb*	BCD47 × Baronesse	Baronesse	*HVCMA - Bmag507*	*HVCMA*	*Bmag507*	26.2–39.2	7–10	4.7–7.2	2/2	Vales et al. (2005)
7H	*Qpsh-GZ-7H*	Lenetah × GZ	GZ	*SCRI_RS_235584*	*BOPA1_478-1291*	*BOPA1_1272-459*	6.17–16.89	2–6	6–18	1/3	Klos et al. (2016)

[a] *ND* no data
[b] *NA* not applicable

5.1.5.2.2 Genes/QTL in Chromosome 2H

Two QTL have been reported on chromosome 2H (chromosome 2) (Table 5.6). Toojinda et al. (2000) mapped a QTL (*Qpsh-Ga-2H*) on chromosome 2H in a DH population from cross Shyri × Galena. The QTL was flanked by markers *E42M47e* and *EBMAC525*, explained 9–12% of phenotypic variation. The allele from Shyri contributed the resistance in this QTL. Vales et al. (2005) mapped a QTL (*Qpsh-Bar-2H*) on 2H using a DH population from cross BCD47 × Baronesse, flanked by markers *EBmac684* and *Bmac093* spanning 25.3 cM. This QTL explained 5.0% phenotypic variation and the resistance allele is from Baronesse, a two-row spring barley cultivar widely grown in the US PNW. These two QTL in chromosome 2H may be in the same chromosome region.

5.1.5.2.3 Genes/QTL in Chromosome 3H

One major gene and three QTL have been reported in chromosome 3H (chromosome 3) (Table 5.6). *rps1.a* was identified in barley cultivar BBA 2890 (Nover and Scholz 1969; Chen and Line 1999, 2003). The name *yr1.a* was given to this gene by Nover and Scholz (1969). It is a recessive gene for ASR effective against all *Psh* races identified by that time in the US (Chen and Line 1999). Yan and Chen (2007) mapped this gene in the long arm of chromosome 3H in a RIL population from cross of Steptoe (susceptible) with BBA 2890 using RGAP and SSR markers. The closest flanking markers *B13* and *B11* to the *rps1.a* locus were 2.1 and 6.8 cM, respectively. Two RGAP markers closely linked with *rps1.a* in coupling have polymorphism in 21 out of 24 tested barley genotypes using in commercial production and *Psh* race differentiation.

Bancroft is the first spring wheat cultivar developed purposely for resistance to stripe rust in the US. It was released in 2000 (Wesenberg et al. 2001), but stripe rust was found on the cultivar in 2001 and thus it was added to the set of barley differentials (Chen and Line 2001b; Chen 2004). However, Bancroft was later found to have HTAP resistance (Chen and Moore 2003). Yan and Chen (2008) identified a dominant gene for resistance using F_1 and F_2 progeny from a cross between cultivar Harrington (susceptible) and Bancroft (resistant) in the adult-plant stage under the controlled high-temperature (10–30 °C cycle) conditions in the greenhouse and F_4 and F_5 generations in the fields under natural infections. The HTAP resistance QTL was mapped in the long arm of chromosome 3H using $F_{5:6}$ RILs with RGAP and SSR makers. The QTL explained over 70% phenotypic variation. The two markers, RGAP marker H3 and SSR marker *Bmag0013*, closely linked with QTL detected 100% polymorphism in a panel of 25 barely genotypes.

In addition to the QTL for HTAP resistance, other two QTL were reported in chromosome 3H (Toojinda et al. 2000; Vales et al. 2005). Vales et al. (2005) mapped two QTL on chromosome 3H in a DH population from cross BCD47 × Baronesse. The resistances of both QTL were contributed from the BCD47 alleles. These two QTL were only detected in one of two experimental locations. The first QTL, *3Ha*, on chromosome 3H spanning 12 cM explained 14% of the phenotypic variation. The second QTL, *3Hb*, spanning 21 cM explained 12.8% of phenotypic variation. Toojinda et al. (2000) mapped a QTL on chromosome 3HL in a DH population from cross Shyri × Galena. The QTL was flanked by RFLP marker *ABG004* and AFLP marker *T22M3li*. It explained 12% of the phenotypic variation, and the allele from Shyri contributed the resistance in this QTL. This QTL is located in the same region as *3Hb* in BCD47 × Baronesse population and should be the same gene (Vales et al. 2005).

5.1.5.2.4 Genes/QTL in Chromosome 4H

Seven QTL is listed for chromosome 4H (chromosome 4), they are really four QTL or genes (Table 5.6). Chen et al. (1994a) mapped a QTL in chromosome 4H using a DH population from a cross between Calicuchima-sib and Bowman. The QTL spanned 40 cM and Calicuchima-sib contributed the resistance allele. Using the same population, Castro et al. (2002b) found this QTL conferring resistance in both seedling and adult plant stages, which flanking by markers *hvm68* and *bmy-1*. The QTL explained 25.9% and 4.2–7.4% phenotypic variation in seedling and adult-plant stages, respectively (Castro et al. 2002b). With the increased marker density, Castro et al. (2002a) separated this QTL into two QTL (*4Ha* and *4Hb*), *4Ha* conferring resistance to adult plant stage and *4Hb* to seedlings. QTL *4Ha* and another QTL on chromosome 6H are necessary for resistance at the seedling stage in the Calicuchima-sib × Bowman mapping population, indicating that these two QTL have complementary interaction. QTL *4Ha* was confirmed in populations AJ, BU and OP, and QTL *4Hb* in populations AJ and BU (Castro et al. 2003a). The 4H QTL is linked with QTL determining barley heading date, which may be related to adult plant resistance (Castro et al. 2002b). Vales et al. (2005) also detected the *4Ha* QTL in the BCD47 × Baronesse population as BCD47 was developed from a cross with Calicuchima-sib. The resistance of 4H QTL was introgressed into and validated in a set of barley lines using MAS (Richardson et al. 2006).

Chen and Line (1999, 2003) studied the F_2 and F_3 populations from a cross between Steptoe (susceptible) and Grannelose Zweizeilige (GZ, resistant), and identified one major recessive gene (*rpsGZ*) conferring resistance to tested *Psh* races in GZ. This gene was subsequently mapped to chromosome 4H using the RIL population derived from the Steptoe × GZ cross tested with races PSH-14 and PSH-54 (Yan and Chen 2006) with RGAP and SSR markers using bulk segregant analysis. The two RGAP markers (*M7* and *M8*) were co-segregating with *rpsGZ*, and other two markers (*M5* and *M6*) were closely linked with *rpsGZ* at a genetic distance of 0.9 and 2.0 cM, respectively. Niks et al. (2015) mapped a major QTL, *Rpsh-L94-4H*, on chromosome 4H. This QTL explained 47% of the phenotypic variation, using RIL from a cross between the Ethiopian line L94 (resistant) and Vada (susceptible). Klos et al. (2016) performed QTL analysis using a RIL population from cross Lenetah × GZ, and identified four QTL conferring resistance in GZ using genome-wide SNP markers. They evaluated these QTL in a population from cross 95SR316A × GZ, and found *Qpsh4Hb* linked with three SNPs (*SCRI_RS_188827*, *SCRI_RS_10818*, and *SCRI_RS_157611*) in GZ. This QTL explained 12–38% phenotypic variation, was consistently present in all tests and in both populations. Given the QTL position on the barley consensus map, the large effect size and the origin of the resistant barley lines, *rpsGZ* and *Qpsh4Hb* should be the same as both from GZ and also likely the same as *Rpsh* (Klos et al. 2016).

Klos et al. (2016) also mapped another QTL, *Qpsh4Ha*, in chromosome 4H (at 72.7 cM). The QTL explained 15% phenotypic variation in seedlings. The effect of this QTL was confirmed in the 95SR316A × GZ population. However, *Qpsh4Ha* was not detected in adult plants in field tests.

5.1.5.2.5 Genes/QTL in Chromosome 5H

Two QTL have been reported in chromosome 5H (chromosome 7) (Table 5.6). The first QTL, *Qpsh-Ca-5H* or *5Ha*, was identified in three mapping populations: Blenheim × E224/3 (Thomas et al. 1995), BSR41 × Steptoe (Toojinda et al. 1998), Calicuchima × Bowman (Chen et al. 1994a; Castro et al. 2002b) and BCD × Baronesse (Vales et al. 2005). This QTL was flanked by SSR markers *Bmag337* and *Bmag005* spanning 6.3 cM, and explained 2.6% of the phenotypic variation in the BCD47 × Baronesse population (Vales et al. 2005). The resistance allele from Blenheim accounted for over 84.2 to 87.2% phenotypic variation in the Blenheim × E224/3 population (Thomas et al. 1995). QTL *5Ha* was also confirmed in populations BU, AJ and OP (Castro et al. 2003a). The second QTL, *5Hb*, spanned 37 cM, flanked by markers *Bmag222* and *GMS001*, and explained 2.2% of the phenotypic variation in the BCD47 × Baronesse population (Vales et al. 2005). The resistances of *5Ha* and *5Hb* are from alleles in BCD47, originating from Calicuchima. These two QTL were introgressed into and validated in a set of barley lines (Richardson et al. 2006).

5.1.5.2.6 Genes/QTL in Chromosome 6H

Five QTL is listed chromosome 6H (chromosome 6) (Table 5.6), but they are at the most three or even two genes. Toojinda et al. (2000) mapped a QTL on chromosome 6H using a DH population from Shyri × Galena. This QTL was defined in an interval between markers *KsuA1H2* and *Linka*, explaining 15–23% phenotypic variation in the adult-plant stage in fields. Using the same population Castro et al. (2002a) detected the QTL in the seedling stage inoculated with *Psh* races PSH-1, PSH-13 and PSH-14. They mapped this QTL in a region spanning 37.0 cM between markers *MWG652A* and *Linka*. The ASR controlled by this gene was contributed by Shyri.

Castro et al. (2002b) mapped a QTL in chromosome 6H using a DH population from cross Calicuchima-sib × Bowman. This QTL was flanked by RFLP markers *ABG466* and *ABG458* (spanning 52 cM), and explained 16.6% and 9.7–18.2% of phenotypic variation in seedlings and adult plant stages, respectively. This QTL is a significant determinant of resistance in seedling stage in two mapping populations

(Castro et al. 2002a, b, 2003b). The resistance alleles originated from cultivar Orea (Castro et al. 2003b). Vales et al. (2005) mapped a QTL in the similar chromosome region in the BCD47 × Baronesse population. Klos et al. (2016) suggested that these three QTL were likely located on the similar chromosome region.

Klos et al. (2016) mapped a QTL (*Qpsh6H*) linked with three SNPs (*BOPA1_1240-844*, *SCRI_RS_189878* and *BOPA2_12_30358*) in chromosome 6H (at 27.19 to 34.46 cM) in the Lenetha × GZ RIL population. *Qpsh6H* explained 12-18% phenotypic variation in seedlings and adult plant stage. The GZ alleles contributed to the resistance. The effect of this QTL was confirmed in the 95SR316A × GZ population.

5.1.5.2.7 Genes/QTL in Chromosome 7H

One major gene and three QTL are listed for chromosome 7H (chromosome 1) (Table 5.6). Castro et al. (2003a) mapped a qualitative resistance gene, *Rpsx*, in the accession CI10587 (PI 243183) in the long arm of chromosome 7H using a DH population from a cross between CI10587 and Galena. *Rpsx* was flanked by RFLP markers *Ris44* and *ABG461* with a genetic distance of 15.5 cM and 6.7 cM, respectively. This gene was also confirmed in three populations (AJ, BU and OP) and combined with other QTL for resistance to stripe rust (Castro et al. 2003a).

Three QTL (*7Ha*, *7Hb* and *Qpsh-GZ-7H*) have been reported in chromosome 7H. The QTL *7Hb* was mapped in three populations: Calicuchima-sib × Bowman (Castro et al. 2002b), and Blenheim × E224/3 (Thomas et al. 1995) and BCD47×Baronesse (Vales et al. 2005). QTL *7Hb* spanned 26.2 to 39.2 cM associated with markers *HvCMA*, *HvACL3* and *Bmag507*, and explained 4.7–7.2% of phenotypic variation (Vales et al. 2005). Resistance alleles were from Bowman and Baronesse. Vales et al. (2005) mapped another minor QTL, *7Ha*, in the BCD47 × Baronesse population, which was detected in one of two experimental locations. *7Ha* spanned 26.4 cM flanking by markers *HvWaxy4a* and *EBmac603*, and explained 6.5% of phenotypic variation. Klos et al. (2016) mapped a QTL *Qpsh-GZ-7H* linked with three SNPs (*BOPA1_478-1291*, *SCRI_RS_235584*, and *BOPA1_1272-459*) in chromosome 7H (at 71.07 to 82.06 cM) in the Lenetha × GZ RIL population. This QTL from GZ explained 6–18% phenotypic variation, but the effect of this QTL did not detected in the 95SR316A × GZ population. *Qpsh-GZ-7H* is closely linked with hulllessness (at 83.79 cM) in GZ and may be difficult to transfer to elite barley germplasm without transferring the hull-less trait.

5.1.6 Genes for "Non-host" Resistance to Stripe Rust in Barley and Wheat

In general, *Psh* does not infect wheat and *Pst* rarely infects barley (Chen et al. 1995c). The resistance of wheat to *Psh* or barley to *Pst* is roughly considered as "non-host" resistance. Four non-host resistance genes were reported (Table 5.5).

Bettgenhaeuser et al. (2014) proposed that interactions involving inappropriate *formae speciales* and non-host plant genera are intermediate host system. The resistance of intermediate host may be durable. Several genes for intermediate host resistance to stripe rust were reported. Although this proposal of intermediate host is acceptable, here we treated barley resistance to *Pst* and wheat resistance to *Psh* as "non-host" for convenient discussion.

Barley cultivar Steptoe is susceptible to all races of *Psh* and resistant to most *Pst* races, while cultivar Russell is susceptible to some *Pst* races and all *Psh* races (Chen et al. 1995c). Pahalawatta and Chen (2005b) studied the resistance of Steptoe to *Pst*. Segregating analysis of a cross between Steptoe × Russell showed that Steptoe has one dominant gene *RpstS1* and one recessive gene *rpstS2* for resistance to *Pst* races PST-41 and PST-45. *RpstS1* was mapped in chromosome 4H with a genetic distance to the closest RGAP markers *M1* and *M2* at 7.7 and 3.8 cM, respectively. SSR marker *HVM68* has a 204-bp allele linked with *RspS1* at 11.6 cM in coupling.

Barley line Y12 is resistant and Y16 is susceptible to all Chinese *Pst* races. Sui et al. (2010) analyzed the F_1 and BC_1 populations from a cross between Y12 and Y18 and found one dominant gene, named *YrpstY1*, conferring resistance to *Pst* in barley Y12. Using SSR markers they mapped *YrpstY1* in chromosome 7H with a genetic distance to the close markers *XAWBMS0022* and *XEBmac0755* at 28.3 cM and 12.2 cM, respectively. Li et al. (2016a) mapped a gene *Rps6* in chromosome 7H in barley accessions PI 466050 and PI 466186, which are resistant to *Pst* races from China and the US. The allelism test between cultivars Y12 (*YrpstY1*) and PI 466050 (*Rps6*) and deletion of the same locus in the cultivated barley Tamlpais suggested that *Rps6* is the same gene as *YrpstY1*, and it may be a frequent component of barley non-host resistance to *Pst*. Using high-density mapping population (10,888 gametes), *Rps6* was delimited in an interval of 0.14 cM region (about 500 kb contig), co-segregating with markers *MlOC_37425* and *MLOC_65262* and flanked by *Mx_ctg_58199* (0.06 cM) and *MLOC_52532* (0.08 cM). This gene was also mapped using Abed Binder 12 × Russell population with QTL analysis (Dawson et al. 2016), one allele in Abed Binder 12 in chromosome 7H accounted for 57.7% and 69.4% of the phenotypic variation for chlorosis and percentage of colonization, respectively. SNP marker *U32_7356_p1* was strongly linked with the peak of QTL. Segregating analysis of 2894 gametes in the Abed Binder 12 × Russell population defined *Rps6* in a 0.1 cM region (about 267 kb) flanking with markers *K_361382* and *K_37596* and co-segregating with *K_57421* and *K_49978*. Both studies mapped *Rps6* in the barley genomic sequence PFC contig_320 from the International Barley Genome Sequencing Consortium (2012).

Niks et al. (2015) studied the QTL in barley for resistance against *Pst*. They mapped one QTL in chromosome 1HS and two QTL in 7HL using a Vada × SusPtrit RIL population, account for 11.2, 13.3 and 14.8% of phenotypic variation (the number of uredinia), respectively. Vada contributed all three resistant alleles.

Wheat genotype PI 478214 is susceptible to all *Psh* races and Lemhi is resistant to most *Psh* races (Chen et al. 1995c). Pahalawatta and Chen (2005a) mapped resistance gene *RpsLem* in Lemhi to *Psh* using a population from a cross between PI 478214 × Lemhi. *RpsLem* is linked with *Yr21*, also in Lemhi for resistance to a *Pst* race, on chromosome 1B and flanked by RGAP marker *M1* in other side with a genetic distance of 2.9 cM. As *RpsLem* and *Yr21* were estimated about 0.3 cM away, they are likely the same gene.

5.1.7 Molecular Markers for Resistance Genes

Molecular markers are extremely useful in modern studies of wheat resistance and breeding for resistance to stripe rust. Earliest used markers are morphological markers and later molecular techniques have been developed. Various DNA-based marker techniques have been developed in the last 30 years and have been intensely applied in resistance gene mapping, diagnosis, marker-assisted selection and phylogenetic analysis. These marker techniques have contributed greatly to our current understanding of genetic variation and genome organization of resistance genes. Widely used marker techniques in rust resistance gene studies included restriction fragment length polymorphism (RFLP), random amplification of polymorphic DNA (RAPD), amplified fragment length polymorphism (AFLP), resistance gene analog polymorphism (RAGP), sequence-related amplified polymorphism (SRAP), target region amplification polymorphism (TRAP), cleaved amplified polymorphic sequences (CAPS), sequence characterized amplification region (SCAR) or sequence-tagged site (STS), single sequence repeat (SSR) or microsatellite, single nucleotide polymorphism (SNP) and Kompetitive allele specific PCR (KASP), diversity arrays technology (DArT) and genotype-by-sequencing (GBS). Every marker technique has advantages and disadvantages and selection of techniques depends upon the purpose and resources.

5.1.7.1 Morphological Markers

Morphological markers were the early type of markers used in genetic studies and breeding of disease resistance. This type of markers is based on associations of a visible trait to disease resistance. Such an association can be determined by long-term observation and/or statistical analysis. Once the association is determined, the marker is relatively easy to use without additional cost. The best example is the brown chaff color associated with the *Yr10* resistance (Metzger and Silbaugh 1970). This trait has been used in selection of wheat cultivars with *Yr10* in the US PNW. Wheat lines with *Yr10* and white chaff have been observed and developed. Similarly, pseudo-black chaff has been found to be associated with the *Yr30* resistance to stripe rust and *Sr2* resistance to stem rust (McIntosh et al. 2013). Leaf-tip necrosis is associated to *Yr18/Lr34* and *Yr29/Lr46* and has been widely used in MAS for the resistances (Rosewarne et al. 2006, Krattinger et al. 2009a). Cautions need to be taken as some of the morphological marker traits are influenced by growth stage and environment. The major disadvantage of morphological traits is the limited number of markers.

5.1.7.2 Isozyme and Protein Markers

Isozymes are the earliest molecular markers developed for finding associations with interested traits such as disease resistance. Isozymes were first described by Hunter and Markert (1957) who defined them as different variants of the same enzyme having identical functions and present in the same individual. Isozymes markers are among the quickest and cheapest to develop and use, but the major limitation is the low polymorphism. Isozyme markers were widely used in the 1970s and early 1980s, and have become less popular as DNA marker techniques have been developed. For resistance to stripe rust resistance, isozyme markers were the first type of molecular markers used. Dr. Robert Allan, the USDA-ARS wheat geneticist and breeder at Pullman, Washington, is probably the first to use isozyme markers to incorporate *Yr17* for resistance to stripe rust and *Pch1* for resistance to eyespot (*Tapesia yallundae* Wallwork & Spooner), originally from *Ae. ventricosa*, in wheat cultivars, releasing common soft white winter wheat cultivar Madsen and club wheat Hyak in 1988 (Allan et al. 1989, 1990). In addition to isozymes, some protein related genes are associated to disease resistance. As discussed above. *Yr36* is tightly linked to high protein content gene *Gpc-B1*, selection for *Gpc-B1* will also likely select the *Yr36* resistance (Uauy et al. 2005; Fu et al. 2009).

5.1.7.3 DNA-Based Markers

Molecular markers based on DNA polymorphisms are currently the most used markers. Various techniques have been developed for identifying and using DNA-based markers. In the following sections, we will briefly discuss the application of these techniques that have been used to identify and map stripe rust resistance genes and develop markers for MSA.

5.1.7.3.1 RFLP

RFLP is one of the earliest techniques for DNA profiling. Its cost is relatively low. The technique was widely used in studies on wheat genetics in the 1980s until the early 2000s. The technique is to detect the DNA variation between samples of homologous in the restriction enzyme sites. The DNA samples were processed in a series steps including digested with restriction enzyme, separated by length through agarose gel electrophoresis, and transferred to a membrane via the Southern blot procedure. Hybridization of the membrane to a labeled DNA probe determines the length of the fragments which are complementary to the probe. Chao et al. (1989) developed RFLP-based genetic maps of wheat homoeologous groups 7 chromosomes and located five disease resistance genes on the map. Singh et al. (2000) used RFLP markers to map *Yr28* and *Yr18* in a synthetic wheat population ([*Triticum turgidum* × *Aegilops tauschii*] × *T. aestivum* cv. 'Opata 85'). Huang and Gill (2001) identified three RFLP markers closely linked with *Lr21*, later found one of the

RFLP markers is the partial sequence of the gene and subsequently cloned *Lr21* (Huang et al. 2003). RFLP markers were intensively used to map genes for resistance to stripe rust in barley varieties (Toojinda et al. 1998, 2000; Castro et al. 2002a, b, 2003a, b). In general, RFLP markers have excellent repeatability. However, the RFLP system is challenged by the high quantity of DNA, complicated hybridization, radioactivity and time consuming, and also limited by the number of available probes.

5.1.7.3.2 RAPD

RAPD markers are DNA fragments from PCR amplification of random segments of genomic DNA with single primers of arbitrary nucleotide sequence (Williams et al. 1990). This technique does not require any specific knowledge of DNA sequences and a set of arbitrary primers can be used for any organisms. The presence or absence of the target fragment depends on whether target region are complementary to the primer sequence. Chague et al. (1999) mapped *Yr15* about 5 cM to RAPD marker *UBC199*^{700}. An early marker of *Yr10* was a RAPD marker co-segregating with the gene, eventually resulting in the cloned gene (Liu et al. 2014a). This technique has been widely used for assessing genetic diversity of large germplasm collections. The major weakness is the low repeatability, and therefore, the technique is no longer used.

5.1.7.3.3 AFLP

AFLP uses restriction enzyme to digest genomic DNA, followed by ligation of adaptors to the sticky ends of the restriction fragments. A subset of the restriction fragments is then selected to be amplified (Vos et al. 1995). Compared to RAPD, AFLP markers have better repeatability, but the technique requires more DNA (although less than RFLP) for enzyme digestion. For wheat and many other organisms with a large-gnome size, selective second PCR amplification is needed. The technique has become popular since its invention in the middle 1990s and is still used, but not as widely as before. Several genes for stripe rust resistance were mapped with AFLP markers, such as *Yr15* (Peng et al. 2000a), *Yr10* (Shao et al. 2001), *YrQz* (Deng et al. 2004), *Yr5* (Smith et al. 2007), *YrHua* (Cao et al. 2008) and *YrC591* (Li et al. 2009).

5.1.7.3.4 RGAP

Analysis of the cloned R genes found that they shared some conserved domains. Based on the sequences of these domains, degenerated primers can be designed and used to identify polymorphic markers from PCR products. Chen et al. (1998) improved the resistance gene analog (RGA) technique by using high-resolution polyacrylamide gel electrophoresis and sensitive silver staining to analyze the amplified products, which is referred as RGAP. Like AFLP primers, RGA primers can be used for any plant species. RGAP markers do not directly indicate chromosomal locations, but they can be used to amplify genetic stocks for such information. Using the RGAP technique and genetic stocks of Chinese Spring nullitetrasomic, ditelosomic and deletion lines, Chen and associates developed a systemic procedure for identification and mapping stripe rust resistance gene and QTL in wheat and barley, and used the procedure in mapping numerous genes such as *Yr9* (Shi et al. 2001), *Yr5* (Yan et al. 2003), *Yr39* (Lin and Chen 2007), *Yr44* (*YrZak*) (Sui et al. 2009), *Yr43* (Cheng and Chen 2010), *Yr45* (Li et al. 2011), *Yr52* (Ren et al. 2012a), *Yr53* (Xu et al. 2013), *Yr59* (Zhou et al. 2014c) and the ASR and HTAP resistance genes/QTL in wheat cultivar Express (Lin and Chen 2008, 2009) and several genes in barley for resistance to stripe rust (Toojinda et al. 2000; Yan and Chen 2006, 2007, 2008). The usefulness of this technique is further evidenced for mapping genes for stripe rust resistance in many other labs (Pu et al. 2010; Chen et al. 2013a; Hou et al. 2013; Feng et al. 2014; Zheng et al. 2014; Zhou et al. 2014a, b; Wu et al. 2015).

5.1.7.3.5 SRAP

Similar to RGAP, SRAP is a PCR-based marker technique, which amplifies open reading frames and produce co-dominant or dominant markers separated on polyacrylamide gel (Li and Quiros 2001). The length of SRAP primers usually is 17–18 bp, and each primer consists of three motifs as filler (10–11 nt), core (4 nt) and select nucleotides (3 nt). The core sequence contains either nucleotides ‘CCGG’ or ‘AATT’ in forward and reverse primers, respectively. The PCR amplification runs at 35 °C for first 5 cycles and followed by 35 cycles at 50 °C. The PCR products are detected on polyacrylamide gel and visualized using silver staining or florescent or radiation labeled primers. Although this marker technique was first developed for mapping genes in *Brasica* spp., it is proven to be useful for mapping genes in wheat. Several stripe rust resistance gene were mapped using this technique, such as *YrSph* (Chen et al. 2012b) and *YrYL* (Wu et al. 2015).

5.1.7.3.6 TRAP

TRAP is another PCR-based marker technique and developed by Hu and Vick (2003). This technique uses two primers of 18 nucleotides to generate markers. One of them is a fixed primer designed from the targeted EST sequence in the database available; the second primer is an arbitrary sequence with either an AT- or GC-rich core to anneal with an intron or exon, respectively. Similar to SRAP, PCR amplification of TRAP products performs at annealing temperature 35 °C for the first 5 cycles, followed by 35 cycles with an annealing temperature at 50 °C. One PCR amplification usually can produce 30–50 scorable fragments ranging from 50–900 bp when separated on a 6.5% polyacrylamide gel. This techniques was used for mapping stripe rust resistance genes *YrSph* (Chen et al. 2012b), *yrLM168* (Feng et al. 2014) and *YrYL* (Wu et al. 2015).

5.1.7.3.7 STS and SCAR

A STS marker is a DNA sequence that has a single occurrence in the genome and whose location and DNA composition are known. STS markers can be developed from sequences of other types of markers such as RAPD, AFLP, RGAP, EST or any other sequences interested. For example, (Chen et al. 2003) converted a RGAP marker co-segregating with *Yr5* into a co-dominant STS marker *Yr5STS/7/8*, which has been widely used in MSA for incorporating *Yr5* into wheat cultivars. Similarly, Smith et al. (2007) converted an AFLP marker tightly linked to *Yr5* into a STS marker. EST-STS markers were used for mapping *Yr26* (Wang et al. 2008a, Zhang et al. 2013). STS markers have been used for mapping several stripe rust resistance genes such as *YrHua* (Cao et al. 2008); *Yr53* (Xu et al. 2013), *YrC591* (Xu et al. 2014), *YrC51* (Zheng et al. 2014) and *Yr69* (Hou et al. 2016).

SCAR markers are STS markers using the same principle and procedure. Mandoulakani et al. (2015) converted six REMAP and IRAP fragments linked with *Yr15* into SCAR markers, which are more efficient than the original markers to detect *Yr15* presence in the germplasms. Li et al. (2009) converted an AFLP marker into a SCAR marker which is linked to *YrC591*.

5.1.7.3.8 CAPS

The CAPS marker technique is based on the genetic difference of restriction endonuclease restriction sites between individuals, and this difference can be detected from the DNA fragment length when PCR products are digested with one or two restriction endonucleases. Chen et al. (2003) developed a CAPS marker from the STS marker tightly linked to *Yr5*. The CAPS marker is more polymorphic than the STS marker for distinguished wheat cultivars carrying or not carrying *Yr5*. Helguera et al. (2003) designed primers URIC and LN2, which can amplify genomes ‘N’ and ‘A’ alleles associated with locus *Yr17/Lr38/Sr38*, based on the absence and presence of *DpnII* restriction site in the N and A genomes, respectively. The CAPS marker can differentiate if wheat germplasms have or not have the *Yr17/Lr38/Sr38* cluster.

5.1.7.3.9 SSR

An SSR is a small DNA region containing certain short repetitive DNA motifs (2–5 bp). The repetitive numbers of DNA motifs between individuals can be used as markers to identify specific individuals. SSRs are present either in non-coding regions or coding regions. EST-based SSR markers are more useful to target functional genes. SSR markers are expensive for developing as the genome or genes sequences are required, but once primers are designed, they are very easy to use. The major advantages of SSR markers included available genome or chromosomal locations, usually co-dominant and the number of SSR markers for a species is large. Also, SSR markers often can be across-used to related species. Multiplex PCR is commonly used for more efficiently screening for and detecting SSR markers.

Over 2500 wheat SSR markers have been developed, and the consensus maps of SSR markers and their chromosomal locations are published (Röder et al. 1998; Somers et al. 2004; Sourdille et al. 2004; Goyal et al. 2005; Song et al. 2005; Kumar et al. 2013) and available in the GrainGene website (https://wheat.pw.usda.gov/GG2). Since a large proportion of SSR markers is genome-specific, amplifying a defined single product from one of the three wheat genomes, SSR marker is the most widely used technique to map resistance genes in wheat. As discussed above, more than 400 genes/QTL have been mapped in wheat and barley and most of them with SSR markers (Tables 5.2, 5.3, 5.4, and 5.6). Closely linked SSR markers have been also used for gene cloning and MAS in breeding. For examples, the recommended SSR markers for MAS of *Yr* genes were *Xbarc8* for *Yr15* (Peng et al. 2000a), *Xcfa2149* (0.06 cM) for *Yr48* (Lowe et al. 2011); and *Xwmc776* for *Yr60* (Herrera-Foessel et al. 2015).

5.1.7.3.10 SNP

SNPs are the predominant type of genetic variation in a single nucleotide due to the substitution or deletion/insertion in any part of the genome as a result of natural or artificial mutation. Since a SNP as point mutation can be evolutionally neutral and escape the pressure of natural selection if the variation occurs in non-coding regions or does not affect the amino acid sequence of the encoded polypeptides, SNPs are widely distributed in wheat genome. With the development of next-generation sequencing techniques, the cost of sequencing wheat genome has continuously declined and obtaining larger number sequences has become more feasible. However, the identification of SNPs is more complex in polyploid genomes due to the high level of similarity between orthologues present in the homoeologous genomes.

It is estimated that in wheat, a homoeologous sequence variant occurs every 24–61 bp, while the frequency of locus-specific SNPs is between 1 in 233 and 1 in 540 bp (Somers et al. 2003; Baker and Edwards 2009; Rustgi et al. 2009). Despite the difficulty for exploring SNPs in wheat, many prospective results in wheat SNP exploring have been obtained. To target the functional genes, transcripts are the best option for developing SNP markers in wheat. Iehisa et al. (2014) identified 13,347 high-confidence SNP in 4872 contigs through transcriptome sequencing of *Ae. tauschii*. Despite the presence of homoeologous and paralogous genes, Akhunov et al. (2009) demonstrated that the Illumina GoldenGate assay could be used for SNP genotyping of homozygous tetraploid and hexaploid wheat lines. Allen et al.

(2011) developed 1114 SNP markers for UK wheat cultivars and mapped 480 SNPs on an Avalon × Cadenza double haploid population. Cavanagh et al. (2013) developed 9K SNP and mapped 4538 SNP on wheat chromosomes using several segregating populations. Wang et al. (2014a) developed 81,587 SNP and assigned 25,252 (31%) of them each to a specific locus (on A, B or D genome).

Hou et al. (2015) used the 9K SNP chip to map a total of eleven QTL/genes for stripe rust resistance in the Druchamp × Michigan Amber recombinant inbreed line population. The 9K or 90K SNP chips have been used to map numerous QTL associated to stripe rust resistance in various panels of wheat germplasms using the GWAS approach (Sthapit et al. 2014; Zegeye et al. 2014; Jighly et al. 2015; Maccaferri et al. 2015; Naruoka et al. 2015; Bulli et al. 2016; Liu et al. 2017). The polymorphisms of the close SNP markers for *Yr62*, *YrSP* and *Yr76* were determined and their presence or absence was validated in the wheat germplasms (Lu et al. 2014; Feng et al. 2015b; Xiang et al. 2016).

5.1.7.3.11 DArT

ADrT markers are based on microarray hybridizations that detect the presence versus absence of individual fragments that are derived from the total genomic DNA of an organism or a population of an organism (Jaccoud et al. 2001). The technology uses methylation-sensitive enzyme to digest genomic DNA, thereby reducing the genome complexity and enriching the low copy sequences for marker development. DArT offers a low-cost high-throughput, robust system with minimal DNA sample requirement capable of providing comprehensive genome coverage even in organisms without any DNA sequence information (http://www.diversityarrays.com). DArT performs well in hexaploid wheat genome (Akbari et al. 2006; Marone et al. 2012). Combining with chromosome sorting, 879 DArT markers were mapped in wheat chromosome 3B, which greatly increased marker intensity (Wenzl et al. 2010). DArT markers have been used for mapping *Yr51* (Randhawa et al. 2014), *Yr56* (Bansal and Bariana 2014) and *Yr58* (Chhetri et al. 2016a) and GWAS analysis to identify stripe rust resistance QTL (Jighly et al. 2015).

5.1.7.3.12 GBS

The GBS technique uses sequence technologies to explore SNP markers. It utilizes a simple but highly multiplexed system for constructing reduced representation libraries to generate a large number of SNPs for use in genetic analysis and genotyping and offers an ultimate tool for marker identification. This genotyping system is simple, quick, extremely specific and highly reproducible. This technique of marker development has become more and more popular. Using three wheat RIL populations and choosing appropriate restriction enzymes, Li et al. (2015) constructed a consensus map containing 28,644 GBS sequence tags at 3757 loci with the average marker distance 0.88 cM, and three rust resistance genes (*Yr18/Sr57/Lr34*,*Yr29/*

Sr58/Lr46, and *Yr30/Sr2/Lr27*) and 15 QTL were validated with high resolution. Within the next 10–20 years, GBS will be the major technique for identifying molecular markers for genes for resistance to stripe rust and other diseases.

5.1.8 Resistance Gene Cloning

In the past several decades, hundreds of wheat rust resistance loci have been identified. To understand resistance at the molecular level, it is essential to isolate the underlying genes and to study their modes of action in the plant. Genes can be isolated using two different strategies, the reverse genetics, from a gene to the phenotype, and the forward genetics approach, from a phenotype to the gene. In the reverse genetics, the gene of interest is directly used for transformation to study its phenotype, or the gene function is disrupted by CRISPR/Cas9, insertional mutant like T-DNA/transposon-tagging, virus induced gene silencing (VIGS), antisense RNA, co-suppression or RNA inference (RNAi). In the forward genetics, the phenotype for the trait of interest is identified and the gene is genetically mapped with a large number markers. This map is then used for the gene isolation. This forward genetics approach of gene isolation is named as map-based cloning (MBC).

Isolation of a gene from wheat is a big challenge for a long time since wheat has a big and complicate genome. Common wheat is an allohexaploid species carrying three sub-genomes (A, B and D), with lager genome size (16, 000 Mb), and over 80% genome is repetitive DNA sequences. The rate of genetic transformation is very low, and the transformation technique still needs to be improved. Although over hundreds disease resistance gene loci have already been genetically mapped to wheat chromosomes, only a small number of genes for disease resistance have been isolated, and the majority of which are rust resistance genes (*Lr1*, *Lr10*, *Lr21*, *Yr10*, *Yr18/Lr34/Sr57*, *Yr36*, *Yr46/Lr67/Sr55*, *Sr33* and *Sr53*). Three approaches were applied for isolation of these rust resistance genes: map-based cloning, gene homology-based cloning and MutReqSeq. So far, the majority of the resistance genes on wheat were isolated by MBC (Table 5.7). However, with the rapid progressing of the wheat genome sequencing project, the MuReqSeq approach will potentially have advantages over other approaches for isolating resistance genes in wheat. We will introduce these three approaches below.

5.1.8.1 Map-Based Cloning

Map-based gene cloning basically involves three steps: i) initial genetic mapping and fine mapping of the resistance gene locus; ii) physical mapping and sequencing, identification and isolation of the candidate gene; and iii) proof of biological function of the candidate gene (Stein and Graner 2004; Krattinger et al. 2009b).

In the first step, the chromosome position of the gene locus is mapped through segregation analysis using the phenotype and molecular markers. The gene locus

Table 5.7 Cloned wheat genes for resistance to stripe rust (*Yr*), leaf rust (*Lr*), stem rust (*Sr*) and powdery mildew (*Pm*, *Mla*)

Gene	Chrom.	Gene function	Population size (no. of gametes)	Cloning method[a]	No. of recomb. plants/critical recomb. plants	Genetic or physical interval	No. of candidate genes	Candidate gene validation[b]	References
Yr10	1BS	CC-NBS-LRR	1748	MBC	55/1	Co-segregating marker	2 RGAs[c]	Transformation, VIGS	Liu et al. (2014a)
Yr36	6BS	Kinase-START	9000	MBC	121/13	0.14 cM/314 kb	4	Mutation, GES, transformation	Fu et al. (2009)
Yr18/ Lr34	7DS	ATP-binding cassette (ABC) transporter	4032	MBC	Not available	0.15 cM/363 kb	8	ADS, mutation, GES	Krattinger et al. (2009a)
Yr46/ Lr67/	4DL	Hexose transporter	2304	MBC	Not available	0.4 cM	3	Mutation, ADS, transformation, complementation, GES	Moore et al. (2015)
Lr1	5DL	CC-NBS-LRR	~7300	MBC	9 (2 + 7)	0.075 cM/~200 kb	4 RGAs	Transformation, VIGS	Qiu et al. (2007) and Cloutier et al. 2007
Lr10	1AS	CC-NBS-LRR	6240	MBC	96	0.13 cM/300 kb	2 RGA	Mutation, ADS, transformation	Stein et al. (2000) and Feuillet et al. (2003)
Lr21	1DS	NBS-LRR	1040	MBC	1	1.7 cM, Cosmid library	1 RGAs	Transformation	Huang et al. (2003)
Sr22	7AL	CC-NBS-LRR	2300	MRS	Not available	Not available	1	Transformation	Steuernagel et al. (2016)
Sr33	1DL	CC-NBS-LRR	2850	MBC	30/8	0.04 cM	7 RGAs	Mutation, transformation, VIGS, Y2H	Periyannan et al. (2013)

(continued)

Table 5.7 (continued)

Gene	Chrom.	Gene function	Population size (no. of gametes)	Cloning method[a]	No. of recomb. plants/critical recomb. plants	Genetic or physical interval	No. of candidate genes	Candidate gene validation[b]	References
Sr35	3AL	CC-NBS-LRR	4575	MBC	5	0.06 cM / 213 kb	4	ADS, mutation, transformation, GES	Saintenac et al. (2013)
Sr45	1DS	CC-NBS-LRR	2300	MRS	Not available	Not available	1		Steuernagel et al. (2016)
Sr50	1RS	CC-NBS-LRR, Mla-homolog	Not available	HBC	Not available	Not available	7 RGAs	Mutant, transformation, ADS	Mago et al. 2015
Pm3b	1AS	CC-NBS-LRR	2,680	MBC	106/6	0.29 cM / 260 kb	3	Mutation, GES, complementation, TEA	Yahiaoui et al. (2004)
Pm8	1RS	CC-NBS-LRR, Mla-homolog	268	HBC	11	Co-segregating marker	1	TEA, transformation, GES	Hurni et al. (2013)
Pm21	6VS	Stpk-V, serine/threonine protein kinase	Not available	HBC	Not available	Microarray and cytogenetic gene localization	1	TEA, transformation, VIGS, GES	Cao et al. (2011)
TmPm3	*T. monococcum*	CC-NBS-LRR, Mla-homolog	Not available	HBC	Not available	Not available	1	TEA	Zhao et al. (2016b)
TmMla1	*Triticum monococcum*	CC-NBS-LRR, Mla-homolog	Not available	HBC	Not available	Not available	3	GES, TEA, Y2H, domain-swap	Jordan et al. (2011)

[a]*MBC* map-based cloning, *HBC* homolog-based cloning, *MRS* MutRenS (mutagenesis and R gene enrichment sequencing)
[b]*VIGS* virus-induced gene silencing, *GES* gene expression study, *ADS* allele density study, *Y2H* yeast-two-hybridization, *TEA* transient expression analysis
[c]*RGAs* resistance gene analogs

can be defined in a small genetic interval (1–10 cM) with a moderate size population, then a larger number of meiotic events are analyzed and the target region will be further narrowed. The proximal markers to the locus should be close enough for chromosome walking in the following step. The initial mapping population can be F_2, DH, backcross or RIL, while the fine mapping population is generally from a large F_2 population. For example, *Lr10* was initially mapped in an interval of 2.0 cM with 385 F_2 plants, then narrowed down to a 0.13 cM region by analyzed 6240 gametes from a F_2 population (Stein et al. 2000). Because recombinant frequencies vary between different parental combinations of segregating populations, the multiple parental combinations may be used for achieving the high resolution of the target gene region. For cloning *Yr18/Lr34*, the locus was mapped in a 3.6 cM interval on chromosome 7DS between markers *Xgwm1220* and *XcsLv34* using 116 RILs (Spielmeyer et al. 2005). Then, three large populations from different parental combinations, i.e. 1728 plants from cross Arina (*Lr34*) × Forno, 1152 plants from Thatcher × PI 5848 (*Lr34*), and 1152 plants from Avocet × Parula (*Lr34*), were used for segregation analysis. The gene locus was narrowed down to a 0.15 cM target interval flanked by *XSWSNP3/XcsLVA1* and *XcsLVE17* (Krattinger et al. 2009a).

To saturate a linkage map in the target region on a wheat chromosome, new markers can be developed using the synteny of model species such as rice and/or *Brachypodium*. The conservation of gene order collinearity in the resistance gene regions between wheat (*Yr36*, *Sr35*, *Lr67/Yr46*) and syntenous regions in the rice and *Brachypodium* genomes provided orthologs to develop close markers (Fu et al. 2009; Saintenac et al. 2013; Moore et al. 2015). However, the conservation region of some NBS-LRR resistance genes, such as *Lr1*, is exceptional (Cloutier et al. 2007).

The difficulties during chromosome walking in wheat involve huge genome size, large amounts of repetitive DNA and the presences of three homologous genomes. The strategies were used to overcome these difficulties include large chromosome fragment insert libraries, such as bacterial artificial chromosome (BAC) libraries and cosmid libraries, and sub-genomic chromosome walking. A BAC library is popular for resistance gene cloning in wheat. For cloning *Yr18/Lr34/Sr57*, two BAC libraries of wheat cultivars Chinese Spring (CS) and Renan were screened using flanking markers *SWSNP3/csLVA1* and *csLVE17*. Four CS BAC clones and one Renan clone were sequenced and from which the candidate genes were identified (Krattinger et al. 2009a). For cloning *Yr36*, a BAC library of wheat line RSL65 with *Yr36* was screened using closely distal marker *Xucw113*, and six BAC clones were identified. BAC ends were used to rescreen the library and extend the contig by chromosome walking. BAC-end marker *Xucw127* was mapped proximal to *Yr36*, completing the physical map with two BAC clones covering the *Yr36* locus in an interval of 314 kb (Fu et al. 2009). To clone *Lr1* (on chromosome 5DL) and *Lr21* (on 1DS), the BAC and cosmid libraries of *Ae. tauschii* were used to construct a physical map for the gene region (Huang et al. 2003; Ling et al. 2003). The cosmid clone carrying *Lr21* gene was directly obtained by a co-segregating RFLP marker (Huang et al. 2003).

Once a physical map of resistance gene locus is constructed with the BAC or cosmid clones in the region, these clones can be sequenced and analyzed to identify candidate genes. Most of the cloned plant *R* genes can be functionally grouped into five distinct classes based on the presence of specific domains (Bent 1996; van Ooijen et al. 2007): 1) the CNL class comprises resistance genes encoding proteins with at least a coiled-coil domain, a nucleotide binding site and a leucine-rich repeat (CC-NB-LRR); 2) the TNL class includes those with a Toll-interleukin receptor-like domain, a nucleotide binding site and a leucine-rich repeat (TIR-NB-LRR, none of this gene class was reported in wheat); 3) the RLP class, acronym for receptor-like protein, groups those with a receptor serine–threonine kinase-like domain, and an extracellular leucine-rich repeat (ser/thr-LRR); 4) the RLK class contains those with a kinase domain, and an extracellular leucine-rich repeat (Kin-LRR); and 5) the 'Others' class includes all other genes which have been described as conferring resistance through different molecular mechanisms, e.g. ABC-transporter for *Yr18/Lr34* (Krattinger et al. 2009a), Kinase-SMART for *Yr36* (Fu et al. 2009) and hexose transporter for *Yr46/Lr67* (Moore et al. 2015). The majority of cloned disease resistance genes in wheat are the CC-NBS-LRR type (Table 5.7).

The function of a candidate gene can be confirmed with various approaches: haplotype analysis, transgenic plants or overexpression of the candidate gene in a susceptible plant, loss-of-function by mutation, virus induced gene silencing (VIGS) and genome editing with CRISPR/Cas9. Both *Agrobacterium* transformation and biolistic particle delivery system have been used for obtaining transgenic plants in wheat, although the transformation rate using *Agrobacterium* is still very low. The loss-of-function mutants can be produced by chemicals [ethyl methane-sulfonate (EMS), or sodium azide] or irradiation (Co^{60} γ-irradiation, fast neutron). Usually a couple of these approaches are used together to validate resistance gene candidates (Table 5.7). For example, in the *Yr18/Lr34* gene cloning, flanking markers delimitated the *Yr18/Lr34* target region in a 363-kb physical interval (Krattinger et al. 2009a). BAC sequencing revealed eight open-reading frames, two of which were interrupted by repetitive sequences and were excluded as candidates for *Yr18/Lr34*. Molecular markers derived from the coding sequences resembling one of the two lectin receptor kinases (SWDEL3), the ABC transporter (SWDEL2/csLVD2) and the hexose carrier (SWDEL1) were co-segregating with *Yr18/Lr34*. The comparison of haplotypes of coding regions among three pairs of +/-*Yr18/Lr34* parental lines of the mapping populations determined that the consistent sequence polymorphism between the alleles of all parental pairs was only in the putative ABC transporter gene. Eight *Yr18/Lr34* gene-function-loss mutants were generated by sodium azide and γ- irradiation. Then, the authors sequenced locus-specific DNA fragments covering parts of the six candidate genes on the eight mutants for loss-of-function of the *Lr34* resistance. Each mutant showed a sequence alteration in the putative ABC transporter gene, leading to either splice site mutations resulting in strongly reduced splicing efficiency or mis-splicing, amino acid exchanges, frame shifts or premature stop codons. The mutants were more susceptible to leaf rust, stripe rust and powdery mildew, and they did not show leaf tip necrosis. Both mutants and haplotype analysis validated the ABC transporter gene responsible for the *Yr18/Lr34* disease resistance (Krattinger et al. 2009a).

In the *Yr36* locus region, the physical map and sequence results revealed two pairs of duplicated genes (Fu et al. 2009). The first pair was two genes with an 'in between RING finger' domain (*IBR1 and IBR2*), and the second pair was two genes designated '*WHEAT KINASE-START 1* and *2* (*WKS1* and *WKS2*). WKS1, WKS2 and IBR1 were absent in the susceptible parent. Primers specific for *WKS1* and *WKS2* kinases and START domains were used to screen a population of 1536 EMS-mutagenized M_2 lines from *Yr36* carrying line and identified 117 mutants. The authors identified six mutants with the changes in conserved amino acids in *WKS1* and three with premature stop codons in *WKS2*. Of the six *WKS1* mutants, five were susceptible, similar to the susceptible control. In contrast, none of the *WKS2* mutants was susceptible. The genomic fragment containing *WKS1* was used to transform wheat cultivar Bobwhite and two T1 transgenic lines were obtained, and they were resistant to stripe rust. The mutation and transformation evidences confirmed that the candidate gene WKS1 is *Yr36* (Fu et al. 2009).

5.1.8.2 Homolog-Based Cloning

Homolog-based cloning is an approach that uses a known resistance gene sequence as a probe to bait homologous gene sequences in related species to isolate a target resistance gene. Four resistance genes in wheat were cloned based on this approach (Table 5.7). Wheat stem rust resistance gene *Sr50* and powdery mildew resistance genes *TaMla1* and *Pm21* were isolated using the homolog-based cloning approach (Cao et al. 2011; Jordan et al. 2011; Mago et al. 2015), and wheat powdery mildew resistance genes *Pm8* and *TmPm3* were isolated as a homolog of *Pm3* (Zhao et al. 2016b). *Mla* encodes a CC-NBS-LRR type resistance protein, which has many homologous alleles in barley and other cereal crops such as wheat and rye. *Sr50* is located on rye chromosome 1RS and confers resistance to wheat stem rust. A probe from *Mla* detected multiple hybridizing sequences at the *Sr50* locus and *Sr50* EMS-mutants, which contained small deletions within this gene cluster. A clone that was present in *Sr50* lines but absent in mutant lines was isolated from a lambda phage genomic library. This clone contained a CC-NBS-LRR sequence and was considered as a *Sr50* candidate. Using the Mla-LRR probe to screen the 1RS-specific BAC library, 175 positive BAC clones were obtained and aligned into six non-overlapping contigs by fingerprinting. Amplifying these positive BAC clones with the primers from the candidate gene identified the candidate sequence in one BAC clone. Using BAC-end markers to test a *Sr50* mutant, three BAC clones were detected to span the deletion region in the mutant. Sequencing these BAC clones identified seven CC-NBS-LRR genes, six of which were present in the deletion region. Crossing two mutant lines produced no resistant progeny, indicating that they had same mutations in the *Sr50* locus. Sequencing these mutants detected that sequences changed in one but not the other CC-NBS-LRR genes. The genomic DNA fragment corresponding to the changed one was used to transform wheat cultivar Fielder, and resistant transgenic lines were obtained, confirming that the candidate gene is *Sr50* (Mago et al. 2015). Of four homolog-based cloned genes (Table 5.7), *TmMla1* and *TmPm3* were isolated from *T. monoccum* using barley *Mla* and wheat *Pm3* probes; and *Sr50* and *Pm8* were isolated from rye chromosome 1RS using the barley *Mla* and wheat *Pm3* as probes, respectively. Combining mutation and transformation validations, the homolog/ortholog-based allele mining strategy was proven to be feasible for isolating resistance genes not only within the same crop species but also among distantly related species (Jordan et al. 2011; Mago et al. 2015).

5.1.8.3 MutRenSeq

The MutRenSeq approach is a reverse genetic approach for cloning genes. The technique was recently reported by Steuernagel et al. (2016). Three stem rust resistance genes, *Sr22*, *Sr33* and *Sr45*, were cloned using this approach (Table 5.7). It can isolate resistance genes in a short time through efficiently integrating several techniques of chemical mutagenesis, exome enrichment capture, high-throughput sequencing and bioinformatics tools. MutRenSeq includes three steps: 1) creating mutants with EMS from resistant lines and identifying mutants with loss of rust resistance; 2) enriching NBS-LRR genes in wild type and mutants and sequencing these NBS-LRR genes; and 3) comparing these genes in mutants and wild types to identify the exact mutations responsible for the loss of rust resistance. Using this approach, Steuernagel et al. (2016) isolated wheat stem rust resistance genes *Sr22* and *Sr45*. First, the authors treated *Sr22* carrier wheat cultivar Schomburgk with EMS and created six susceptible mutants from 1300 M_2 families. Then they designed 60,000 120-mer RNA probes with high similarities to NBS-LRR genes in cereal crops and capture these genes from short insert libraries (500–700 bp) of wild-type *Sr22* and the six mutants. The enriched libraries were sequenced using a high-throughput sequencing technique. Finally, they assembled the sequences and compared mutants with the wild type. They identified the number of mutations ranging from 44 to 84, and 23 contigs mutated in two mutants, three contigs mutated in three mutants, and a single 3408-bp contig containing independent mutations in five of the six mutants. The 3408-bp contig had homology with the C-terminal region of an *Ae. tauschii* NBS-LRR homology. The 5′ end of the *Ae. tauschii* NBS-LRR gene is homologous with a contig from the *Sr22* wild-type that carries an EMS-induced mutation in the N-terminal region of the same gene in the remaining mutants. The full-length sequence of *Sr22* was PCR amplified from genomic and cDNA. The mutation of GC to AT transition was identified in all six *Sr22*-mutants, which cause either nonsense (two) or missense (four) mutations. A marker from this sequence co-segregated with *Sr22* in 2300 gametes, confirming that this gene is *Sr22*. The MutRenSeq approach is fast, cheap, avoiding fine mapping, and does not require constructing physical contigs across the map interval. This method can be used to isolate resistance genes locating in alien chromosomal fragments in wheat-alien translocation germplasms that have not been used in wheat improvement due to linkage drag.

5.1.9 Induced Resistance

A plant might become resistant to a virulent race when an avirulent race infects the plant prior to the infection of the virulent race. Johnson and Allan (1975) used avirulent *Pst* race 104E9 inoculated first leaves of cultivar Maris Templar, and after five days they inoculated the same leaves with virulent race 41E136. They found that the pre-inoculation with the avirulent race caused stripe rust delayed the onset of sporulation of a virulent race and also decreased the spore mass produced. However, the parallel experiments for inoculation with a mixture of equal mass of virulent/avirulent spores did not change stripe rust development and spore production. When challenged the upper and lower leaf surfaces, respectively, they found induced resistance occurred in the leaves where the inoculations was treated on the upper but not the low leaf surface. Cheung and Barber (1972) found that induced resistance occurred in wheat-wheat stem rust system and reduced 80% rust pustules on wheat. The induced resistance may provide an explanation for some mechanisms for multiline resistance. Although induced resistance was reported in greenhouses, the effect of induced resistance has seldom carried out in fields. Because of abundant genetic resistance sources and genes against rusts in wheat and barley, induced resistance appears to be less important and not as well studied as for other diseases to which genetic resistance is hard to find.

5.1.10 Mechanisms of Resistance

Rust resistance is obviously a complex characteristic. The mechanisms of resistance can be dissected from different levels involving the rust infection process in an individual plant and disease spread in wheat populations or epidemiological aspects. The gene interactions of wheat-*Pst* have been introduced in Chap. 3. Here we discuss the mechanisms mainly from the aspects of histology and molecular genetics.

5.1.10.1 Resistance Mechanisms at the Morphological and Histological Levels

Rust infection can be hampered at any stages of fungal development on or in plant of wheat or barley, including spore deposition, germination, stomata recognition, penetration, haustorium formation, infectious hypha spread and uredinium formation (sporulation). Plant leaf pattern (erect or horizontal attitude) can affect spore deposition. Vertical leaf position reduces the amount of spore landing. Russell (1976a, b) found that *Pst* urediniospore germination had difference in leaf surface and position. Higher percentage of *Pst* urediniospore germinated on the adaxial leaf surface, particularly the distal parts, than on the abaxial surface. Most studies showed that there were no significant differences in urediniospore germination on resistant and susceptible wheat plants.

Some researchers believe that the non-host resistance may be durable and tried to explore the mechanism of non-host resistance. Niks (1981) separated non-host resistance as prehaustorium and posthaustorium resistances based on the formation of haustorium, and pointed out that non-host resistance is generally characterized as pre-haustorium resistance. Rubiales and Niks (1996) found that the presence of prominent wax layer over the guard cells on some accessions of *Hordeum chilense* reduces appressorium differentiation of urediniospores of barley leaf rust (*P. hordei*). The extensive wax layer or hairy on wheat leave surfaces may reduce leaf wettability and affect spore germination. On some plant species closely related to wheat, a rust fungus may not be able to locate stomata efficiently. Niks (1981) found that germ tubes have difficulty in locating and recognizing stomata on non-host plants. Epidermal morphology interferes with germ tube growth towards stomata. Rust fungal infection can be inhibited by using an inhibitor that disrupts sulfhydryl bonds of the enzymes used by germ tubes to ingest and metabolize components of epicuticular waxes (Wilkinson and Roberts 1983). However, recent studies showed that non-host resistance may not be pre-haustorium. The mechanism of non-host (rice) resistance to five cereal rust (*P. graminis*, *P. triticina*, *P. striiformis* and *P. hordei*) were investigated (Ayliffe et al. 2011a, b). All infection structures necessary for colonization, including haustoria, can form in rice but sporulation was never observed. Non-host rice plants responded rust fungal infection with active defense response (callose deposition and production of reactive oxygen species) that likely plays a role in limiting colonization and consequently prevent sporulation. Inoculating *Brachypodium distachyon* with urediniospores of cereal rust fungi resulted in the similar conclusion. However, cell death was relatively rare in most tests suggesting that hypersensitive cell death is not a major component of this defense response (Ayliffe et al. 2013). The immunity of *B. distachyon* to *Pst* is controlled by a single dominant gene in a segregating population from BdTR10h × TEK4 or two dominant genes in BdTR13K × Bd21 (Ayliffe et al. 2013). This is an important difference of resistance mechanisms between non-host resistance and host resistance. The genes related to SA- and JA-mediated defense pathways and phytolalexin synthesis are involved in non-host resistance.

A typical character of race-specific resistance mechanism is hypersensitive response (HR) which occurs after rust germ tube penetration and hautorium formation. Marryat (1907) first found the phenomena of HR in wheat-*P. glumarum* (a synonym for *P. striiformis*) pathosystem. Stakman (1915) studied the response of various interactions between cereal crop and *P. graminis* and defined that HR is a common resistance reaction in plant-rust specific interactions. HR as a race-specific resistance mechanism is evidenced by numerous studies in flax-flax rust system (Flor 1942, 1946, 1947, 1955). HR is elicited by products of a pathogen avirulence gene and plant resistance gene. Two classes (TIR-NB-LRR and CC-NB-LRR) of *R* genes involve in R-Avir interactions (Mur et al. 2008). The complex signal transduction pathways were reported in HR, involving changes in protein phosphorylation, production of reactive oxygen species and modification of ion fluxes (Pontier et al. 1998; Coram et al. 2008a, 2010; Mur et al. 2008; Chen et al. 2013b). Protein synthesis, an intact actin cytoskeleton and salicylic acid are necessary for cell death induction. Heath (2000) suggested that cell death within HR acts more as a signal to the rest of the plant rather than as a direct defense mechanism. However, we believe that HR first restricts rust fungal hyphae at the infection site or slows done hyphal growth and then sends signal to other cells. For stripe rust, HR is not only a characteristic of race-specific ASR, but also an important characteristic in non-race specific HTAP resistance (Chen 2013). Zhang et al. (2012) found that APR expression was associated with hypersensitive cell death and the production of reactive oxygen species in rust infection sites. They found that haustorium formation was retarded and infection hyphae was inhibited in the adult-plants, which decreased the development of *Pst* colonies in leaves. They also observed that cell wall-related resistance prevented ingression of haustorium mother cells into plant cells, and caused hautorium malformation or death.

The mechanism of partial resistance or slow rusting resistance in *Yr18* and *Yr46* is different from that of race-specific resistance. In flag leaf, *Yr18* decreased infection frequency and increased the latent period. The fungal development stopped within the pale (nonhypersensitive) flecks. At the microscopic level, the *Yr18* resistance is based on the reduction of haustorium formation in the early stage of infection, which is due to a low rate of intercellular hyphal development (Rubiales and Niks 1995).

5.1.10.2 Resistance Mechanisms at the Molecular and Genomic Levels

To date, twelve wheat rust resistance genes have been cloned, including four genes for stripe rust resistance. These genes can be sorted into three types: *Yr10* is a race-specific resistance gene; *Yr36* a non-race specific resistance gene; and *Yr18* and *Yr46* are non-race-specific multiple pathogen resistance genes with leaf tip necrosis. *Yr18* and *Yr46* genes also function in transgenic barley and rice against barley leaf rust and rice blast.

The gene organizations are different in the genome regions containing race-specific or non-race specific resistance genes. Race-specific resistance genes, like NB-LRR, trend to form gene clusters (Stein et al. 2000; Feuillet et al. 2003; Yahiaoui et al. 2004; Clouter et al. 2007; Qiu et al. 2007; Periyannan et al. 2013; Saintenac et al. 2013). The *Yr10* locus has two CC-NBS-LRR genes with 84% identity in the intron and the LRR domain, and one of which is active transcript for *Yr10* resistance (Liu et al. 2014a). The inactive pseudogene does not have the transcription start site, TATA-box and CAAT-box. This gene is unique in wheat but its homologs have been found in *Ae. tauschii*, *H. vulgare*, *B. distachyon* and other monocots (Liu et al. 2014a). The kinase 2a, kinase 3a and HD domains play important roles for the *Yr10* resistance expression. The interactions of avirulence effectors from rust pathogens with NB-LRR resistance in other plant species have been reported, although the interaction study of *Yr10* and its corresponding avirulence gene cannot be performed due to the fact that the *Yr10* avirulence gene is not available. It is very likely that *Yr10* resistance uses the similar molecular mechanisms.

The *Yr36* locus harbors two duplicate genes *WKS1* and *WKS1*, which have a similar structure consisting of kinase and START domains (Fu et al. 2009). *WKS1* is the active gene for the *Yr36* resistance. This type of gene structure is only present in the B genome in tetraploid wheat before the wheat domestication. The *WSK1* serine-threonine kinase domain has kinase activity. The START domain proteins in human are known to play roles in lipid trafficking, metabolism and sensing; and their binding with sterols and ceramides results in protein conformational changes. If the *WSK1* START domain has the ability to bind lipids from *Pst* and change its conformation, this may cause the kinase domain to initiate a signaling cascade leading to hypersensitive cell death (Fu et al. 2009).

In both *Yr18* and *Yr46* loci, none of their homologous gene cluster is present (Krattinger et al. 2009a; Moore et al. 2015). The sequence of *Yr18* consists of 24 exons and encodes an ATP-binding cassette transporter containing two cytosolic nucleotide binding domains and two hydrophobic trans-membrane domains (Krattinger et al. 2009a). The functional *Yr18* allele emerged after wheat domestication about 8000 years ago and is the result of two gain-of-function mutations (Dakouri et al. 2010; Krattinger et al. 2011, 2013). The resistance allele of *Yr18* is only present in cultivated bread wheat but not in wild wheat progenitors. The expression of *Yr18* does not depend on the presence of rust pathogen. However, the expression level was higher in flag leaves than seedlings. Because *Yr18* also determines leaf tip necrosis, the flag leaf tip of *Yr18*-carrying lines is more resistant than leaf base (Krattinger et al. 2009a). *Yr18* belongs to pleiotropic drug resistance transporters, and its protein uses the energy derived from ATP hydrolysis to shuttle various substrates across cellular membranes. *Yr18* not only confers resistance to multiple pathogens in wheat, but also confers resistance to other pathogens when transferred into other crop plants. Risk et al. (2013) transferred *Yr18* into barley and found that it conferred partial resistance to barley leaf rust (*P. hordei*) and barley powdery mildew (*Blumeria graminis* f. sp. *hordei*). The recent study demonstrated that *Yr18*-transgenic *japonica* rice cultivar Nipponare plants expressed *Yr18* and showed increased resistance against multiple isolates of rice blast pathogen *Magnaporthe oryzae*. Host cell invasion during the pathogen biotrophic growth phase was delayed (Krattinger et al. 2016). The homolog of *Yr18* in *Arabidopsis* contributes to resistance against non-host pathogens, and also involved in translocating toxic compounds derived from glucosinolates into apoplast (Krattinger et al. 2009a).

Yr46 encodes a sugar transporter which contains 12 predicted transmembrane helices and is similar to the STP13 family of H+/monosaccharide symporters, facilitating the transport of hexoses across the plasma membrane (Moore et al. 2015). The *Yr46* resistance is gain-of-function from two mutations, Arg144 and Leu387. The transporter activity is related to mutation Arg144. The Yr46 transporting activity is only for monosaccharide but not for pentose sugar or the di- and tri-saccharides trehalose and raffinose. The resistance allele of *Yr46* in chromosome 4D has been identified only in wheat landraces from the Punjab region of India and the susceptibility allele is present in the gene pool of *Ae. tauschii* and diploid D genome progenitors, which indicates that the *Yr46* resistance emerged after wheat polyploidization. The homoeologous genes of *Yr46* are present in chromosome 4A and 4B. *Yr46* and its susceptible alleles in chromosome 4D and homoeologous alleles on 4A and 4B all were upregulated by rust pathogen infection. The mechanism of *Yr46* resistance may be due to the blocking of apoplastic hexose retrieval by host cells, thereby increasing the hexose/sucrose ratio in the leaf apoplasm, which in turn induces a sugar-mediate signaling response that results in a more hostile environment for pathogen growth (Moore et al. 2015). When *Yr46* was transferred into barley, the transgenic plants also had resistance to barley leaf rust, indicating that the gene has broad resistance (Moore et al. 2015).

5.2 Development of Resistant Cultivars

Developing resistant cultivars is the most preferred strategy for control stripe rust in all areas where stripe rust epidemic occurs either frequently or just sometimes, and for all countries with the stripe rust problem, including those where fungicides are available and relatively cheap. Considering all possible strategies, growing resistant cultivars is the most effective, economic, easy-to-use and environmentally friendly method for control of stripe rust (Chen 2013, 2014). The importance of developing and growing resistant cultivars in the integrated control of stripe rust will be discussed in Chap. 6. In the following section, we will mainly focus on possible strategies for developing resistant cultivars.

5.2.1 Selection of Types of Resistance

As discussed above, different types of resistance have different advantages or advantages. There is no any type that is perfect. ASR generally provides high level resistance, throughout all growth stages and easy to select in breeding. However, single-gene controlled ASR usually does not last long. In contrast, APR, including HTAP resistance, is generally durable, but may not provide adequate resistance, especially in areas where stripe rust often infects crops in seedling stage and weather is cool and wet during the adult-plant stages (Line and Chen 1995; Line 2002; Chen 2005, 2013, 2014).

There was a long-term debate about choosing between ASR and APR in the past. The debate was mainly caused by limited resources and information about resistance including genes. As the current information and resources are much more abundant than in the past, in addition to technologies have become available and more efficient, such debate has become less. Now, the scientific community has pretty much reached to a consensus concept of using combination of different types and combination of genes. Combining different genes for effective ASR can make ASR last longer. Using this strategy needs to pay attention to effectiveness of each gene. Combining two or more genes, each of them is effective against all races, is better than combining genes, of which some is no longer effective. The virulence profiles of rust populations in large scale need to be well characterized and the effectiveness of individual genes needs to be well determined for selecting genes for ASR to pyramid. For example, *Yr5* and *Yr15* are resistant to all *Pst* races identified so far in the US (Wan and Chen 2014; Wan et al. 2016; Liu et al. 2017), these two genes have been combined into several wheat cultivars as presented above. Similarly, genes for APR can be pyramided to increase the level of the durable type of resistance. It even better to combine both effective ASR and APR genes to take the advantages and overcome the drawbacks of both types of resistance for developing high level and durable resistance. Many wheat cultivars grown in the US PNW were developed using this approach (Chen 2013, 2014). Now more genes and markers are available to combine both types of resistance, and more and more cultivars will be developed in the world by combining different genes conferring effective ASR and durable APR.

5.2.2 *Strategies for Utilization of Genetic Resistance*

In the past several decades, several strategies have been proposed and used for developing resistant cultivars, including gene rotation, gene deployment and gene pyramiding. The different strategies were proposed based on different understandings of resistance and availability of resistance resources and technologies for using resistance in breeding programs. Some of the strategies are still useful and some may need rethinking or modified. In the following section, we will discuss the usefulness of the strategies.

5.2.2.1 Gene Rotation

Gene rotation is a strategy to dynamically deploy cultivars with different genes for resistance to the prevalence races in a region where no exotic inoculum was introduced. If the virulent races are rising up, replace the ineffective genes with corresponding effective resistance genes. Then the pathogen population will undergo a genetic bottlenecks and the frequency of virulence genes will be decline. If the frequency of a virulence gene reduces to a very low level, the previously used resistance genes can be used again. We have observed that the *Yr10* virulence have gone through several frequency up-down cycles since the later 1960s. Races virulent to *Yr10* became predominant in the 1970s and early 1980s, decreased in the 1990s and early 2000, became predominant again from 2004 to 2009 and then decreased afterwards (Line and Qayoum 1992; Chen 2005; Chen et al. 2010; Wan and Chen 2012, 2014; Wan et al. 2016; Liu et al. 2016). The early up-down may be related to planting wheat cultivars with *Yr10* and replaced with cultivars without *Yr10*. However, the recent changes of the frequencies of the *Yr10* virulence were not highly related to cultivar changes. Nevertheless, this phenomenon indicates that some defeated resistance genes may have value in breeding programs as the pathogen population changes the virulence profile.

Gene rotation may work in plant-pathogen systems in which the pathogen is highly localized, its race composition highly depending upon resistance genes in crop cultivars grown and less likely to stack different virulence factors into a single race or population. However, gene rotation cannot be a major strategy, especially for stripe rust because the pathogen can infect many grass species and grasses can harbor more diverse races (Cheng et al. 2016). Moreover, as *Pst* is asexual reproduced and mutation is the main mechanism to produce new races, the pathogen is able to accumulate new virulence genes with previously existing ones in new races (Wan and Chen 2014; Wan et al. 2016; Liu et al. 2016). Furthermore, old cultivars are usually not comparable with currently grown cultivars in yield and other traits, and developing a new cultivar is a long process. It is not feasible to schedule which genes should be used in breeding program. However, keeping or using previously defeated resistance genes in cultivars together with effective ASR genes or APR can be a good idea as the defeated genes may become effective.

5.2.2.2 Regional Gene Deployment

Resistance deployment or diversification of resistance, temporally and spatially, can be a useful strategy for wheat stripe rust (Bayles 2002). In this strategy, different resistance genes can be deployed in cultivars grown in different regions and years. From epidemiological standpoint, this strategy can reduce the disease development rate, amount of inoculum and pathogen directional selection pressure. Consequently, the mutation rate and probabilities of evolution of new races of pathogen may be reduced. Resistance deployment can also increase the longevity of the resistance of cultivars because of low disease pressure and less frequent appearance of new virulence races. For regions where pathogen inoculum is mainly from outside, using resistance genes different from those used in the inoculum source region should reduce rust risk. Also, in regions where stripe rust mainly comes later in the growth season, use APR can be a good idea.

In India, gene deployment was successful to slow down the south-north movement of stripe rust for many years by a broad belt of cultivation with cultivar Kalyansona (*Yr2*) until the resistance was breakdown (Sharma et al. 1972). In China different resistance genes are suggested to be deployed in different epidemiological regions. In southern Gansu where stripe rust can over-summer in highland mountains and over-winter in lowland valleys to provide rust inoculum to other places in spring. It has been suggested that different types of resistance and different genes should be use in the different epidemiological regions (Li et al. 1997; Wan et al. 2007). Regional deployment requires many effective resistance genes and the information of pathogen virulence. In Australia, *Yr15* and *Yr33* are effective (Bariana et al. 2007). In the North America, resistance genes *Yr5*, *Yr15*, *Yr45*, *Yr53*, *Yr64* and *Yr65* are effective to the current *Pst* populations (Li et al. 2011; Xu et al. 2013;

Cheng et al. 2014; Wan and Chen 2014). Although a large number of resistance genes have been identified in the recent years, as presented above, their effectiveness to different races in different regions have not be well determined. More genes are needed to apply this strategy.

Although regional deployment is a promising strategy to keep race-specific resistance last longer, there are no reports to provide examples of real use of this strategy. One of the difficulties is to coordinate breeding programs in different regions or different countries for using and not using certain resistance genes. A practical solution is to identify more genes and improve the diversity of resistance at regional and global levels.

5.2.2.3 Cultivar Mixture

Using mixture of cultivars is one of several host genetic diversification strategies for control rusts. This approach was first proposed by Jensen (1952) for control of oat crown rust. Wolfe (1985) defined cultivar mixture as "mixtures of cultivars that vary for many characters including disease resistance, but have sufficient similarity to be grown together." Cultivar mixture can be a composition of several elite cultivars with different resistance genes in each, or the susceptible wheat with superior agronomic characteristics mixing with several resistant wheat cultivars having agronomically inferior. Cultivar mixture is relatively easy to do. The mechanism of cultivar mixtures for control of diseases is assumed through dilution of pathogen inoculum, induced resistance and barriers for spores dispersal and disease movement (Chin and Wolfe 1984; Wolfe 1985; Calonnec et al. 1996), but another suggestion is that host genetic diversity reducing the number of new pathogen founders is the major reason (Mundt 2002, 2014). According to Calonnec et al. (1996), up to one third of the reduction in infection by stripe rust in wheat mixtures was due to induced resistance. Mundt (1994) found that cultivar mixture reduced stripe rust severity from 16 to 64%. The random mixtures of the two cultivars (Tyee susceptible and Tres resistance) with ratio 1:1 provided better stripe rust control than did alternating stripe planting, and the random mixture increased grain yield relative to the pure stand mean (Mundt et al. 1996). Mixtures yielded between 0 and 5% more than the mean of the pure lines in the absence of disease and increased yield between 8 and 13% in the presence of disease (Finckh and Mundt 1992). Wheat cultivar mixture is common practice for reduction of stripe rust and other problems in the US PNW. In 1998, 10% of soft white winter wheat and 76% of club wheat fields in this region were sown to cultivar mixtures. In the recent years, mixtures of two to four cultivars were grown in 30–35% of the total winter wheat acreage. In addition to reduce risk of stripe rust damage, cultivar mixtures also reduce risks from other diseases and abiotic stresses such as winter kill and frost damage. The disadvantages of cultivar mixtures include incompatibility of the varietal components (such as plant height and maturity, planting date), marketing restrictions and processing quality.

5.2.2.4 Multiline Cultivars

Similar to cultivar mixtures, multiline cultivars reduce disease risk by increasing resistance gene diversity in a single field. Different from cultivar mixtures, a multiline cultivar is composed of several lines in the same or similar backgrounds, but with different resistance genes. Thus, a multiline cultivar is more agronomically uniform than a cultivar mixture. The multiline approach was first proposed by Borlaug for control of stem rust (Borlaug 1953, 1959; Borlaug and Gibler 1953). Borlaug (1953) suggested to develop a 'composite variety' by a modification of conventional backcross methods. When distributed commercially, the cultivar will be a mixture of a number of phenotypically similar lines which are genotypically different for resistance. The 'backcross lines' which are eventually employed in the 'composite variety' are developed by crossing a commercial cultivar to a number of varieties carrying different types of resistance or resistance genes. Each single cross is backcrossed several times to a commercial cultivar. When the 'composite variety' is ready for release, the several lines are multiplied separately, and then mixed mechanically to form the cultivar for release to growers. The composition of the multiline cultivar could be varied from year to year in response to race changes.

The first multiline wheat cultivar for commercial production was Miramar 63, released from the Colombian program (Rockefeller Foundation Program Annual Report 1963) for control of stripe rust in Andes. To develop Miramar 63, the Brazilian wheat Frocor was crossed with some of 600 varieties and lines. Over 1200 lines, similar phenotypically to Frocor but with resistance from the 600 non-recurrent parents, were produced. Miramar 63 was a mechanical mixture of equal parts of ten of the best lines giving resistance to both stripe rust and stem rust. Miramar 63 was widely accepted, and its yield was more than doubled those of older varieties in some areas (Rockefeller Foundation Program Annual Report 1964). Within two years two lines in the composition became susceptible to stem rust, but the losses were less than the theoretical maximum of 20%. The two susceptible and other two lines were replaced with other four resistant lines to form a new multiline cultivar Miramar 65. This illustrates the plasticity inherent in multiline cultivars (Rockefeller Foundation Program Annual Report 1965).

In the US PNW, two multiline wheat cultivars have been released, Crew in 1982 and Rely in 1991 purposely for control of stripe rust on club wheat (Allan et al. 1983, 1993). Crew consisted of 10 closely related wheat lines which had resistant to intermediate resistant reactions to the predominant races at the time of the cultivar release. The components of Crew were second or third backcross derivatives of Omar (highly susceptible) or Omar-type semi-dwarf lines. The stripe rust resistance in Crew was believed to be controlled by a minimum of nine different genes among its 10 components. Rely also consisted of 10 closely related and phenotypically similar lines each having race-specific resistance to one or more *Pst* races. The background cultivar was Tres, a major club wheat cultivar released in 1984 (Allan et al. 1986), which has two genes (*YrTr1*, *YrTr2*) for race-specific ASR to stripe rust (Chen and Line 1992a). Collectively, the components of Rely have 12 or more race-specific genes for resistance to stripe rust. Rely was extensively grown in Washington and Oregon between 1994 and 2003 and the resistance remained durable for more than 17 years (Line 2002; Chen 2005; Allan 2014). The development of Rely changed the stripe rust boom-bust problem in club wheat in the US PNW. More importantly, the effort of incorporating the large number of resistance genes laid the foundation of diverse resistance genes in the club wheat program. These ASR genes, together with genes for HTAP, have been combined in different ways into the recently released club cultivars, such as Bruehl, Coda, Chukar and Cara. As a result, club wheat cultivars as a group has remained highly resistant to stripe rust under the frequently extreme severe stripe rust epidemics in the US PNW since 2000 (Chen 2014). The successful story of Rely demonstrates multiline cultivars as an effective approach for developing cultivars for durable resistance using race specific resistance genes. However, the major difficulty is more labor intensive and time consuming for developing a multiline cultivar than a single cultivar. The multiline approach has been replaced by gene pyramiding.

5.2.2.5 Gene Pyramiding

To improve disease resistance in wheat, gene pyramiding is widely accepted by breeders, plant pathologist and farmers. The pyramid of race-specific resistance genes against stem and leaf rust has been applied in wheat breeding programs and achieved successfully rust control for a long time (Green and Campbell 1979; Samborski 1985; Roelfs and Bushnell 1985; Schafer and Roelfs 1985; Kolmer and Dyck 1994; Kolmer 1996; McIntosh and Brown 1997). The advantages of combination of this type resistance include high level resistance and easily selecting in seedling stage. Cultivars with pyramiding stripe rust resistance genes have been developed in the breeding programs in the US PNW as mentioned above. In addition, *Yr5* and *Yr15* have been combined into several spring wheat cultivars including Blanca Grande 515, Clearwhite 515, Summit 515 and Seahawk. In addition to *Yr5* and *Yr15*, Patwin 515 also carries *Yr17*. Cultivar Expresso, with combination of *Yr15* with race-specific ASR and non-race specific HTAP resistance QTL in Express (Lin and Chen 2008, 2009), has become the most grown hard red spring wheat in the PNW. These genes were purposely incorporated into these cultivars and more cultivars through MAS.

As presented above, over 300 gene/QTL conferring different types of resistance to stripe rust have been identified and every chromosome has some genes/QTL. In the case of close repulsion linkages, recombinant genotypes carrying both genes can be identified in segregating populations, which can be used to develop wheat cultivars with pyramided genes. The authors' group (Wang and Chen 2016; Qie et al. 2016) have recently developed the germplasms carrying the combinations of two genes/QTL in same chromosome with the genes previously identified in chromosomes 1BL (Cheng et al. 2014), 2BL (Xu et al. 2013), 4BL (Lu et al. 2014) and 7BL (Ren et al. 2012a; Zhou et al. 2014c). As two genes in each line are linked on the same chromosome, these germplasms are potentially more useful than lines carrying a single gene for developing high level and durable resistant cultivars.

Pyramiding qualitative genes and/or quantitative resistance loci in barley with MAS has produced numerous stripe rust resistant barley lines (Toojinda et al. 1998, 2000; Castro et al. 2003a, b; Richardson et al. 2006). QTL or genes in chromosome 4H, 6H and 7H provide resistance in both seedling and adult-plant stages. For example, BCD47 is a two-row spring barley line selected from a DH population with pyramided QTL on chromosomes 4H and 5H (Castro et al. 2003b; Richardson et al. 2006; Vales et al. 2005).

The major mechanism for effectiveness and durability of pyramided genes is the reduction of possibility for the arising of a virulent race. If resistance genes have not been previously deployed singly or in less complex combinations, the probability of an asexual pathogen, such as the stripe rust fungus, mutating to virulence against all pyramided resistance genes in a cultivar would be the product of the probability for each gene singly, thus reducing the probability of a virulent race (Wheeler and Diachun 1983; Schafer and Roelfs 1985). There is no stand rules for how many resistance genes to be pyramided can obtain ideal resistance level in a cultivar, although Wheeler and Diachun (1983) suggested that a cultivar with four to five resistance genes which conditioned resistance to all races might remain resistance for years or perhaps centuries if the gene-for-gene concept is valid. However, genetic analyses from durable resistance cultivars provided some clues. Cultivar Stephens has been grown in the US PNW for over 30 years. The durable resistance of Stephens to stripe rust is due to a combination of 3 seedling and over 11 HTAP resistance genes (Chen and Line 1995a, b; Chen 2005; Santra et al. 2008; Vazquez et al. 2012). Similarly, a large number of genes contribute to the durable resistance in cultivar Druchamp (Chen and Line 1992a, 1995a, b; Chen et al. 1996; Hou et al. 2015). However, in other cultivars one or two major QTL confer high level resistance, such as Oxley and Cook (McIntosh and Brown 1997), Alpowa (Lin and Chen 2007), Luke (Guo et al. 2008); Louise (Carter et al. 2009); PI 183527 (Ren et al. 2012a), PI 195252 (Lu et al. 2014) and PI 178759 (Zhou et al. 2014c).

Some QTL/genes appear to have an additive resistance or increased additive effect in appropriate combinations (Chen and Line 1995a, b; Vazquez et al. 2012). Certain resistance gene combinations have been found more durable than others. These favorable combinations of QTL/genes have been used in other disease resistance breeding. For example, the combinations of stem rust resistance gene *Sr2* with other genes has played a crucial role in the international programs of CIMMYT (Rajaram et al. 1988; Singh et al. 2011). Another example is wheat-rye translocation 1BL.1RS, which carries resistance genes *Yr9*, *Sr31*, *Lr26* and *Pm8* having good combination ability in various genotype backgrounds, however, these disease resistance genes are no longer effective as they have been intensively used throughout the world. Genes that contribute to durable pyramids (such as *Sr2*, *Lr13*) and/or pleiotropic multiple pathogen resistance (such as *Lr34/Yr18/Sr57/Pm38*, *Yr46/Lr67/Sr55/Pm46*) should have priority to use in the breeding programs.

Adding one gene at a time to a cultivar with genes that have become ineffective generally does not make the resistance durable. For gene pyramiding to be more effective, it is important to combine two or more genes for effective resistance at the same time. These genes are better to be effective against all races if possible. If such gene combinations are not possible, other genes can be selected based on virulence information. Information on virulence frequency and structure can be useful for selecting genes to be combined. We have been monitoring virulence combinations in the *Pst* population in the US. Some virulence genes are less associated or even negatively correlated in the pathogen population than other virulence combinations (Wan and Chen 2014; Wan et al. 2016; Liu et al. 2016). Thus, resistance genes corresponding to the virulence genes with low or negative correlation coefficients can be selected.

Homozygosity verse heterozygosity of virulence loci also can be used for selecting race-specific resistance genes. The dikaryotic stripe rust fungus is highly heterozygous (Liu et al. 2012; Zheng et al. 2013; Cheng and Chen 2014; Xia et al. 2016a). By analyzing sexually reproduced populations, we found that many virulence loci are heterozygous, but some homozygous (Yuan et al. 2016; Y Lei, MN Wang and XM Chen, unpublished data). Homozygous avirulence genes will take longer than heterozygous avirulence genes to change to homozygous virulence genes in the *Pst* population through natural mutation. The avirulence loci corresponding to *Yr5* and *Yr15* are homozygous, which has also been proven by mutagenesis (Li et al. 2016c), and thus, resistance conferred by these two genes may last longer than some other genes.

In addition to pyramiding effective race-specific ASR genes to achieve relative durable resistance, pyramiding non-race specific durable APR or HTAP resistance genes can improve durability and also increase the level of resistance. This approach has been used in CIMMYT for improving wheat resistance to three rusts (Singh et al. 2001, 2005) and in the US PNW for improving stripe rust resistance in both wheat and barley (Line and Chen 1995; Line 2002; Chen 2005, 2013).

5.2.2.6 Combining Different Types of Resistance

Combining APR or HTAP resistance with effective all-stage resistance is the best approach for developing wheat cultivars with high-level and durable resistance because APR can provide complete control when it is effective, and APR, particular HTAP resistance due to its non-race-specific, can reduce damage when ASR is overcome by new races (Chen 2005, 2013). When combining HTAP resistance and ASR into a cultivar, the HTAP resistance should be incorporated into a cultivar first and then followed by ASR, which can ensure the selection through phenotype. However, when useful markers are available for both types of resistance, incorporation of both types of resistance can be performed simultaneously. Lots of molecular markers have been identified for genes for ASR and HTAP resistance (Tables 5.2, 5.3, 5.4 and 5.6), making MAS possible for combining genes for different types of resistance.

In the US PNW, most wheat cultivars have HTAP resistance and many cultivars also have ASR (Line and Chen 1995; Line 2002; Chen 2005, 2013, 2014). Many of these cultivars have been grown for a long time and have remained resistant since their release. For example, winter wheat cultivars Stephens (released in 1978) and

Madsen (released in 1988) were predominant cultivars soon after release and are still widely grown. These cultivars have combined ASR and APR (Chen and Line 1992c, 1993a, 1995b; Chen et al. 1996; Santra et al. 2008; Vazquez et al. 2012; XM Chen and associates, unpublished data). Madsen is currently considered as the standard for high level of HTAP resistance in breeding programs.

Introgression of multiple QTL into a wheat cultivar is no longer as difficult as previously expected, 3–4 minor genes (QTL) should be adequate for high resistance (Singh et al. 2005). HTAP resistance in some cultivars is controlled by one or two major QTL and explained large portion of phenotypic variation (Lin and Chen 2007, 2009; Lu et al. 2014; Ren et al. 2012a; Zhou et al. 2014c). For HTAP resistance, genes with a large effect should be easier to combine than those with a small effect. Chen and associates determined that the durability of a resistance gene is determined by its resistance type rather than its "major" or "minor effect" (Coram et al. 2008a, b, 2010; Chen et al. 2013b). Combining HTAP resistance QTL of a large effect with genes for effective ASR should provide high-level and durable resistance to stripe rust.

5.2.3 *Phenotypic Selections*

For breeding wheat or barley cultivars with resistance to stripe rust, germplasm and breeding lines can be selected phenotypically or in combination of phenotypes and MAS. Using the combination of phenotypes and MAS has become more and more popular. In this section, we will briefly present our procedures for routine germplasm and breeding line screening for supporting breeding programs to develop resistant cultivars

5.2.3.1 Screening for Stripe Rust Resistance

Selecting the desirable trait is very important in the whole process of breeding. The process of selection can be based on phenotype or genotype. In breeding wheat and barley for resistance to stripe rust, phenotypic selection depends on evaluation of various components of reactions, such as infection type, latent period, sporulation period, spore quantity and severity and possible morphological markers. Among the various measurements, infection type and severity are most important and relatively easy to evaluate, especially for large scale screening.

The senior author's program evaluates more than 40,000 wheat and barley entries from various breeding program nurseries annually for resistance to stripe rust. All of the entries are tested under natural infection conditions in fields at Pullman in eastern Washington and Mount Vernon in western Washington. As stripe rust infection occurs almost all year long at Mount Vernon, testing at this location can provide reliable data for early stage infection and late stage infection. Comparison of the late-stage data with the early-stage data may indicate possible HTAP resistance.

However, diverse races in the region may complicate the phenotypic data especially IT response. As many wheat genotypes with an intermediate level of HTAP resistance which typically showing intermediate ITs, similar to ITs produced by a mixture of an avirulent race and a virulent race, HTAP resistance is then indistinguishable with ASR. Screening with individual races separately can sort out HTAP resistance from ASR. Therefore, we also evaluate some of the nurseries in the greenhouse with selected races in both seedling and adult-plant stages under low-temperature (4–20 °C) and high-temperature (10–30 °C) profiles. Races used in the high-temperature adult-plant tests should be virulent on the seedlings of same cultivars. This two-way testing is used for routine and large-scale screening. For separating effects of temperature verses plant stage, a four-way testing is conducted, including seedling low-temperature, seedling high-temperature, adult-plant low-temperature, and adult-plant high-temperature, as previously described (Chen and Line 1995a, b; Chen 2013)

5.2.3.2 Associated Traits and Multi-disease Resistance

The genetic association of stripe rust resistance with other traits allows indirect or direct selection for resistance in breeding. Multiple pathogen resistance in breeding lines, especially resistance to leaf rust, stem rusts and powdery mildew, are often used by breeders. Resistance to several pathogens may be controlled by a single gene or linked genes, which can be treated as a single genetic locus for selection. Examples include *Yr8/Lr34/Sr34*, *Yr9/Lr26/Sr31*, *Yr17/Lr37/Sr38*, *Yr18/Lr34/Pm38*, *Yr29/Lr46*, *Yr40/Lr57* and *Yr46/Lr67/Sr55* as discussed above. Resistance to some pathogens is close linked, such as *Yr35* and *Lr53* (Marias et al. 2005b) with a genetic distance of 3.0 cM and *Yr9* and *Pm8* with a genetic distance of 4.4 cM (Lukaszewski 2000). Most of these multiple pathogen resistance genes are from alien chromosome segments. These traits will be likely transmitted together in the selection process. Some agronomic traits associated with rust resistance is helpful for selection, such as the brown chaff trait linked with *Yr10* with a 2 cM distance (Metzger and Silbaugh 1970), leaf tip necrosis (LTN) associated with *Yr34* (Singh 1992), *Yr46* (Herrera-Foessel et al. 2011) and *Yr29* (William et al. 2003).

5.2.4 Marker-Assisted Selection

Molecular markers are a very useful tool in resistance breeding. Although molecular markers can be any kind of molecules associated with the resistance, DNA markers are the most widely used markers in breeding due to the fact that they are genetically neutral, molecularly stable and practically easy to use. User-friendly markers should be closely linked with or as part of the resistance gene, showing polymorphism across different genetic backgrounds. The presence of the marker genotype will ensure the presence of resistance phenotype. DNA markers can overcome many limits in resistance phenotypic tests, particularly in lack of pathogen inoculum and/or appropriate facility. The advantage of molecular markers is even more important in pyramiding multiple resistance genes. If pyramiding more than two race-specific resistance genes and the corresponding virulence races unavailable, or pyramiding several QTL for resistance and the phenotypic variation is difficult to measure, molecular markers are the best option. Resistance genes *Yr5* and *Yr15* are highly resistant to all *Pst* races in the US. In most wheat backgrounds, *Yr5* produces necrosis and *Yr15* produces small flecks or no visible symptom. It is difficulty to pyramid these two genes in wheat cultivar just through phenotypic selection. Although it is possible to determine if the two genes are in many selected lines through crossing the selected lines with a susceptible line and analyze the progeny populations for two-gene segregation, this procedure is very time consuming. The breeding programs have used *Yr5* marker *STS7/8* and *Yr15* markers *Xbarc8* and *Xgwm273* to incorporate these genes in several wheat cultivars. The molecular markers of most *Yr* genes/QTL identified in the past decades are available, as shown in Tables 5.2, 5.3, 5.4, 5.5 and 5.6. In the recent years, more and more high-throughput genotype platforms (Li-Cor and ABI Genetic Analyzer, Sequenom Mass Array, Illumina Infinite Wheat Gene Chip, Diversity Array Technology) have been used in MSA, which have greatly increased the selection efficiency and accuracy.

The US Department of Agriculture Agricultural Research Service (USDA-ARS) has four genotype labs, located in Raleigh, North Carolina; Fargo, North Dakota; Manhattan, Kansas; and Pullman, Washington, for genotyping screening for various traits including disease resistance using molecular markers. More than 60% of genotyping screening of breeding lines in the Genotype Laboratory of the Wheat Health, Genetics and Quality Research Unit at Pullman, WA is for stripe rust resistance.

5.2.5 New Techniques for Resistance Breeding

The traditional methods for introgression of major resistance genes into wheat cultivars include backcross and/or convergent backcross (Knott 1989), for introgression of minor QTL through double and 4-way crosses (Dubin and Rajaram 1996) or single backcross (Bariana et al. 2004). It takes about 6–12 years to develop a new cultivar. New techniques may shorten the breeding process.

Gene transformation can directly introduce new resistance genes into elite commercial cultivars to improve their resistance to wheat rusts. Gene transformation can overcome the gene boundary among the species. The wheat genes *Yr46* and *Yr18* when transferred into rice and barley conditioned broaden resistance to the pathogens of rice and barley (Risk et al. 2013; Krattinger et al. 2016; Moore et al. 2015). Similarly, genes from other species may be used for improving wheat resistance to stripe rust. Since multiple *Yr*-genes haven been cloned (Fu et al. 2009; Krattinger et al. 2009a; Liu et al. 2014a; Moore et al. 2015), gene cassettes containing multiple resistance genes can be constructed and directly transformed into wheat, which will enhance the resistance durability.

New techniques in genomics, molecular genetics and genetic engineering open a new door for resistance breeding. Next-generation sequencing greatly accelerates genome sequencing of wheat and barley (Chapman et al. 2015; International Wheat Genome Sequencing Consortium 2014) and rust genome sequencing (Cuomo et al. 2017). Deep sequencing and genome annotation have resulted in more markers (Cavanagh et al. 2013; Wang et al. 2014a; Xia et al. 2016a), which will facilitate identification and isolation of resistance genes in wheat and avirulence genes in rust pathogens. The large number of SNP exploration and genome-wide association study (GWAS) provide a powerful tool to identify new resistance and avirulence genes and QTL (Maccaferri et al. 2015; Bulli et al. 2016; Xia et al. 2016b). The large number of markers in wheat also ensures wheat resistance improvement through whole genome selection (Randhawa et al. 2009). Strategies to obtain plants with broad-spectrum disease resistance by transformation with gene cassettes containing both avirulence genes and the corresponding resistance genes will be used in wheat. Recently, the genome editing technique has been applied in hexaploid wheat and simultaneously manipulated the three homoeoalleles to generate broad resistance plants to powdery mildew (Wang et al. 2014b), this technique will profoundly impact the wheat resistance breeding in the future.

References

Agenbag GM, Pretorius ZA, Boyd LA, Bender CM, Prins R. Identification of adult plant resistance to stripe rust in the cultivar Cappelle-Desprez. Theor Appl Genet. 2012;125:109–20.

Agenbag GM, Pretorius ZA, Boyd LA, Bender CM, MacCormack R, Prins R. High-resolution mapping and new marker development for adult plant stripe rust resistance QTL in the wheat cultivar Kariega. Mol Breed. 2014;34:2005–20.

Aghaee-Sarbarzeh M, Ferrahi M, Singh S, Harjit-Singh FB, Gill BS, Dhaliwal HS. *Ph*^{I} induced transfer of leaf and stripe rust resistance genes from *Aegilops triuncialis* and *Ae. geniculata* to bread wheat. Euphytica. 2002;127:377–82.

Akbari M, Wenzl P, Caig V, Carling J, Xia L, Yang S, Uszynski G, Mohler V, Lehmensiek A, Kuchel H, Hayden MJ, Howes N, Sharp P, Vaughan P, Rathmell B, Huttner E, Kilian A. Diversity arrays technology (DArT) for high-throughput profiling of the hexaploid wheat genome. Theor Appl Genet. 2006;113:1409–20.

Akhunov E, Nicolet C, Dvorak J. Single nucleotide polymorphism genotyping in polyploid wheat with the Illumina GoldenGate assay. Theor Appl Genet. 2009;119:507–17.

Allan RE. Club wheat. Robert E. Allan. 2014; ISBN 978-1-4951-0549-4 I–V + 354 pp + i–xviii

Allan RE, Purdy LH. Single gene control of stripe rust resistance in wheat. Plant Dis Report. 1967;51:1041–3.

Allan RE, Purdy LH. Reaction of F_2 seedlings of several crosses of susceptible and resistant wheat selections to *Puccinia striiformis*. Phytopathology. 1970;60:1368–72.

Allan RE, Vogel OA. Stripe rust resistance of Suwon 92 and its relationship to several morphological characteristics in wheat. Plant Dis Report. 1961;45:778–9.

Allan RE, Vogel OA, Purdy LH. Influence of stripe rust upon yields and test weights of closely related lines of wheat. Crop Sci. 1963;3:564–5.

Allan RE, Purdy LH, Vogel OA. Inheritance of seedling and adult reaction of wheat to stripe rust. Crop Sci. 1966;6:242–5.

Allan RE, Line RF, Peterson CJ, Rubenthaler GL, Morrison KL, Rohde CR. Crew, a multiline wheat cultivar. Crop Sci. 1983;23:1015–6.

Allan RE, Peterson CJ, Rubenthaler GL, Line RF, Morrison KJ. Registration of 'Tres' wheat. Crop Sci. 1986;26:203-4.

Allan RE, Peterson CJ, Rubenthaler GL, Line RF, Roberts DE. Registration of 'Madsen' wheat. Crop Sci. 1989;29:1575–6.

Allan RE, Peterson CJ, Rubenthaler GL, Line RF, Roberts DE. Registration of 'Hyak' wheat. Crop Sci. 1990;29:1575–6.

Allan RE, Peterson CJ, Line RF, Rubenthaler GL, Morris CF. Registration of "Rely" wheat multiline. Crop Sci. 1993;33:213–4.

Allen AM, Barker GL, Berry ST, Coghill JA, Gwilliam R, Kirby S, Robinson P, Brenchley RC, D'Amore R, McKenzie N, Waite D, Hall A, Bevan M, Hall N, Edwards KJ. Transcript-specific, single-nucleotide polymorphism discovery and linkage analysis in hexaploid bread wheat (*Triticum aestivum* L.). Plant Biotechnol J. 2011;9:1086–99.

Asakura N, Nakamura C, Ohtsuka I, RAPD markers linked to the nuclear gene from *Triticum timopheevii* that confers compatibility with *Aegilops squarrosa* cytoplasm on alloplasmic durum wheat. Genome. 1997;40:201-10.

Ayliffe M, Devilla R, Mago R, White R, Talbot M, Pryor A, Leung H. Nonhost resistance of rice to rust pathogens. Mol Plant Microbe Interact. 2011a;24:1143–55.

Ayliffe M, Jin Y, Kang ZS, Persson M, Steffenson B, Wang SP, Leung H. Determining the basis of nonhost resistance in rice to cereal rusts. Euphytica. 2011b;179:33–40.

Ayliffe M, Singh D, Park R, Moscou M, Pryor T. Infection of *Brachypodium distachyon* with selected grass rust pathogens. Mol Plant Microbe Interact. 2013;26:946–57.

Badr A, Müller K, Schäfer-Pregl R, El Rabey H, Effgen S, Ibrahim HH, Pozzi C, Rohde W, Salamini F. On the origin and domestication history of barley (*Hordeum vulgare*). Mol Biol Evol. 2000;17:499–510.

Bahl PN, Backshi JS. Genetics of rust resistance in barley – II. The inheritance of seedling resistance to four races of yellow rust. Indian J Genet Plant Breed. 1963;23:150–4.

Baker GLA, Edwards KJ. A genome-wide analysis of single nucleotide polymorphism diversity in the world's major cereal crops. Plant Biotechnol J. 2009;7:318–25.

Bakshi JS, Bahl PN. Inheritance of resistance to four races of yellow rust in two varieties of barley. Indian J Genet Plant Breed. 1965;25:239–42.

Bakshi JS, Luthra JK. Inheritance of resistance to stripe rust (*Puccinia striiformis* West.) in barley. In: Proceedings of the 2nd international Barley genetic symposium. 1970. p. 478–83.

Bakshi JS, Sawhney RN. Inheritance of resistance of La Prevision to three races of yellow rust. Indian J Genet Plant Breed. 1965;25:227–30.

Bakshi JS, Bahl PN, Kohli SP. Inheritance of seedling resistance to some Indian races of yellow rust in the crosses of rust resistant barley variety E.B. 410. Indian J Genet Plant Breed. 1964;24:72–7.

Bansal UK, Bariana H. Mapping of stripe rust resistance gene *Yr56* in durum wheat cultivar Wollaroi. BGRI 2014 Technical workshop, 22–25 March, Obregon, Mexico; 2014.

Bansal UK, Hayden MJ, Keller B, Wellings CR, Park RF, Bariana HS. Relationship between wheat rust resistance genes *Yr1* and *Sr48* and a microsatellite marker. Plant Pathol. 2009;58:1039–43.

Bansal UK, Hayden MJ, Bariana HS. Chromosomal location of an uncharacterized stripe rust resistance gene in wheat. Euphytica. 2010;171:121–7.

Bansal UK, Forrest KL, Hayden MJ, Miah H, Singh D, Bariana HS. Characterisation of a new stripe rust resistance gene *Yr47* and its genetic association with the leaf rust resistance gene *Lr52*. Theor Appl Genet. 2011;122:1461–6.

Bansal UK, Kazi AG, Singh B, Hare RA, Basiana HS. Mapping of durable stripe rust resistance in a durum wheat cultivar Wollaroi. Mol Breed. 2014;33:51–9.

Bariana HS. Breeding for disease resistance. In: Thomas B, Murphy DJ, Murray BG, editors. Encyclopedia of Applied Plant Sciences. Harcourt: Academic Press; 2003. p. 244–53.

Bariana HS, McIntosh RA. Cytogenetic studies in wheat XV. Location of rust resistance genes in VPM1 and their genetic linkage with other disease resistance genes in chromosome 2A. Genome. 1993;36:476–82.

Bariana HS, Hayden MJ, Ahmed NU, Bell JA, Sharp PJ, McIntosh RA. Mapping of durable adult plant and seedling resistances to stripe rust and stem rust diseases in wheat. Australian J Agric Res. 2001;52:1247–55.

Bariana HS, Brown GN, Ahmed NU, Khatkar S, Conner RL, Wellings CR, Haley S, Sharp PJ, Laroche A. Characterisation of *Triticum vavilovii*-derived stripe rust resistance using genetic, cytogenetic and molecular analyses and its marker-assisted selection. Theor Appl Genet. 2002;104:315–20.

Bariana HS, Willey NJ, Venkata BP, Lehmenseik A, Standen GE, Lu M. Breeding methodology to achieve durability for rust resistance in wheat. In: Proceedings of the 54th Aus Cereal Chemistry Conf and 11th Wheat Breeders Assembly. Canberra, ACT. (Eds Black CK, Panozzo JF, Rebetzke GJ) pp 8–12. (Wheat Breeding Society of Australia). 2004

Bariana HS, Parry N, Barclay IR, Loughman R, McLean RJ, Shankar M, Wilson RE, Willey NJ, Francki M. Identification and characterization of stripe rust resistance gene *Yr34* in common wheat. Theor Appl Genet. 2006;112:1143–8.

Bariana HS, Brown GN, Bansal UK, Miah H, Standen GE, Lu M. Breeding tripe rust resistance wheat cultivar for Australia using conventional and marker-assisted selection technologies. Aus J Agric Res. 2007;58:576–87.

Bariana HS, Bansal UK, Schmidt A, Lehmensiek A, Kaur J, Miah H, Howes N, McIntyre CL. Molecular mapping of adult plant stripe rust resistance in wheat and identification of pyramided QTL genotypes. Euphytica. 2010;176:251–60.

Bariana HS, Forrest K, Qureshi N, Miah H, Hayden M, Bansal UK. Adult plant stripe rust resistance gene *Yr71* maps close to *Lr24* in chromosome 3D of common wheat. Mol Breed. 2016;36:98.

Bartos P, Valkoun J, Kosner J, Slovencikova V. On the genetics of rust resistance on the wheat cultivar Kavkaz. Cereal Rusts Bulletin. 1973;1:27.

Basnet BR, Singh RP, Herrera-Foessel SA, Ibrahim AMH, Huerta-Espino J, Calvo-Salazar V, Rudd JC. Genetic analysis of adult plant resistance to yellow rust and leaf rust in common spring wheat Quaiu 3. Plant Dis. 2013;97:728–36.

Basnet BR, Ibrahim AMH, Chen XM, Singh RP, Mason ER, Bowden RL, Liu SY, Hays DB, Devkota RN, Subramanian NK, Rudd JV. Molecular mapping of stripe rust resistance in hard red winter wheat TAM 111 adapted to the U.S. high plains. Crop Sci. 2014a;54:1361–73.

Basnet BR, Singh RP, Ibrahim AMH, Herrera-Foessel SA, Huerta-Espino J, Lan CX, Rudd JC. characterization of *Yr54* and other genes associated with adult plant resistance to yellow rust and leaf rust in common wheat Quaiu 3. Mol Breed. 2014b;33:385–99.

Bayles RA. Variety diversification scheme to reduce spread of yellow rust in wheat. United Kingdom Cereal Pathogen Virulence Survey 2001 Annual Report, 2002. p. 94.

Bent AF. Plant disease resistance genes: function meets structure. Plant Cell. 1996;18:1757–71.

Bettgenhaeuser J, Gilbert B, Ayliffe M, Moscou MJ. Nonhost resistance to rust pathogens-a continuation of continua. Front Plant Sci. 2014;5:664.

Biffen RH. Mendel's laws of inheritance and wheat breeding. J Agric Sci. 1905;1:4–48.

Biffen RH. Studies in the inheritance of disease resistance. J Agric Sci. 1907;2:109–27.

Bipinraj A, Honrao B, Prashar M, Bhardwaj S, Rao S, Tamhankar S. Validation and identification of molecular markers linked to the leaf rust resistance gene *Lr28* in wheat. J Appl Genet. 2011;52:171–5.

Bonman JM, Babiker EM, Cuesta-Marcos A, Esvelt-Klos K, Brown-Guedira G, Chao SM, See D, Chen JL, Akhunov E, Zhang JL, Bockelman HE, Gordon TC. Genetic diversity among wheat accessions from the USDA National Small Grains Collection. Crop Sci. 2015;55:1243–53.

Borlaug NE. New approach to the breeding of wheat varieties resistant to *Puccinia graminis tritici*. Phytopathology. 1953;43:467.

Borlaug NE. The use of multiline or composite varieties to control airborne epidemic disease of self-pollinated crop plants. In: Proceedings of the international wheat genetic symposium, 1st, Winnipeg, 1958. 1959. p. 12–26.

Borlaug NE, Gibler JW. The use of flexible composite wheat varieties to control the constantly changing stem rust pathogen. Agron Abstr. 1953;81

Börner A, Roder MS, Unger O, Meinel A. The detection and molecular mapping of a major gene for non-specific adult plant disease resistance against stripe rust (*Puccinia striiformis*) in wheat. Theor Appl Genet. 2000;100:1095–9.

Boukhatem N, Baret PV, Mingeot D, Jacquemin JM. Quantitative trait loci for resistance against yellow rust in two wheat-derived recombinant inbred line populations. Theor Appl Genet. 2002;104:111–8.

Boyd LA. Can Robigus defeat an old enemy?—Yellow rust of wheat. J Agri Sci. 2005;124:233–43.

Boyd LA, Smith PH, Hart N. Mutants in wheat showing multipathogen resistance to biotrophic fungal pathogens. Plant Pathol. 2006;55:475–84.

Bryant RRM, McGrann GRD, Mitchell AR, Schoonbeek HJ, Boyd LA, Uauy C, Dorling S, Ridout CJ. A change in temperature modulates defence to yellow (stripe) rust in wheat line UC1041 independently of resistance gene *Yr36*. BMC Plant Biol. 2014;14:10.

Bulli P, Zhang JL, Chao XM, Chen XM, Pumphrey M. Genetic architecture of resistance to stripe rust in a global winter germplasm collection. G3 Genes Genomes. Genetics. 2016;6:2237–53.

Caldwell RM. Breeding for general and/or specific plant resistance. In Finlay KW and Sphepard KW, editors. Proceeding 3rd international wheat genetic symposium, Canberra, Australia, 5–9 August 1968. Australian Academy of Science, Canberra, Australia, 1968. p. 263–72.

Calonnec A, Johnson R. Chromosomal location of genes for resistance to *Puccinia striiformis* in the wheat line TP1295 selected from the cross of Soissonais-Desprez with Lemhi. Eur J Plant Pathol. 1998;104:835–47.

Calonnec A, Goyeau H, de Vallavieille-Pope C. Effects of induced resistance on infection efficiency and sporulation of *Puccinia striiformis* on seedlings in varietal mixtures and on field epidemics in pure stands. Euro J Plant Pathol. 1996;102:733–41.

Calonnec A, Johnson R, de Vallavieille-Pope C. Genetic analyses of resistance of the wheat differential cultivars Carstens V and Spaldings Prolific to two races of *Puccinia striiformis*. Plant Pathol. 2002;51:777–86.

Campbell J, Zhang HT, Giroux MJ, Feiz L, Jin Y, Wang MN, Chen XM, Huang L. A mutagenesis-derived broad spectrum disease resistance locus in wheat. Theor Appl Genet. 2012;125:391–404.

Cao ZJ, Deng ZY, Wang MN, Wang XP, Jing JX, Zhang XQ, Shang HS, Li ZQ. Inheritance and molecular mapping of an alien stripe-rust resistance gene from a wheat-*Psathyrostachys huashanica* translocation line. Plant Sci. 2008;174:544–9.

Cao AZ, Xing LP, Wang XY, Yang XM, Wang W, Sun YL, Qian C, Ni JL, Chen YP, Liu DJ, Wang X, Chen PD. Serine/threonine kinase gene Stpk-V, a key member of powdery mildew resistance gene *Pm21*, confers powdery mildew resistance in wheat. Proc Natl Acad Sci USA. 2011;108:7727–32.

Carter AH, Chen XM, Garland-Campbell K, Kidwell KK. Identifying QTL for high-temperature adult-plant resistance to stripe rust (*Puccinia striiformis* f. sp. *tritici*) in the spring wheat (*Triticum aestivum* L.) cultivar 'Louise'. Theor Appl Genet. 2009;119:1119–28.

Case AJ, Naruoka Y, Chen X, Garland-Campbell KA, Zemetra RS, Carter AH. Mapping stripe rust resistance in a Brundage x Coda winter wheat recombinant inbred line population. PLoS ONE. 2014;9(3):e91758.

Castro AJ, Chen XM, Hayes PM, Knapp SJ, Line RF, Toojinda T, Vivar H. Coincident QTL which determine seedling and adult resistance to stripe rust in barley. Crop Sci. 2002a;42:1701–8.

Castro AJ, Hayes PM, Fillichkin T, Rossi C. Update of barley stripe rust resistance QTL in the Calichima-sib × Bowman mapping population. Barley Genetics Newsl. 2002b;32:1–12.

Castro AJ, Capettini F, Corey AE, Filichkin T, Hayes PM, Kleinhofs A, Kudrna D, Richardson K, Sandoval-Islas S, Rossi C, Vivar H. Mapping and pyramiding of qualitative and quantitative resistance to stripe rust in barley. Theor Appl Genet. 2003a;107:922–30.

Castro AJ, Chen XM, Hayes PM, Johnston M. Pyramiding quantitative trait locus (QTL) alleles determining resistance to barley stripe rust: effects on resistance at the seedling stage. Crop Sci. 2003b;43:651–9.

Cavanagh CR, Chao S, Wang S, Huang BE, Stephen S, Kiani S, Forrest K, Saintenac C, Brown-Guedira GL, Akhunova A, See D, Bai G, Pumphrey M, Tomar L, Wong D, Kong S, Reynolds M, da Silva ML, Bockelman H, Talbert L, Anderson JA, Dreisigacker S, Baenziger S, Carter A, Korzun V, Morrell PL, Dubcovsky J, Morell MK, Sorrells ME, Hayden MJ, Akhunov E. Genome-wide comparative diversity uncovers multiple targets of selection for improvement in hexaploid wheat landraces and cultivars. Proc Natl Acad Sci USA. 2013;110:8057–62.

Chague V, Fahima T, Dahan A, Sun GL, Koro AB, Ronin YI, Grama A, Roder MS, Nevo E. Isolation of microsatellite and RAPD markers flanking the *Yr15* gene of wheat using NILs and bulked segregant analysis. Genome. 1999;42:1050–6.

Chao SM, Sharp PJ, Worland AJ, Warham EJ, Koebner RM, Gale MD. RFLP-based genetic maps of wheat homoeologous group 7 chromosomes. Theor Appl Genet. 1989;78:495–504.

Chapman JA, Mascher M, Buluç A, Barry K, Georganas E, Session A, Strnadova V, Jenkins J, Sehgal S, Oliker L, Schmutz J. A whole-genome shotgun approach for assembling and anchoring the hexaploid bread wheat genome. Genome Biol. 2015;16:1.

Charnley B. Rowland Biffen, Little Joss and Yeoman: looking back at two successes behind the birth of NIAB in 1919. Landmark May 2009, pp 3–4 http://berrischarnley.com/pdfs/charnley-landmark-2009.pdf. (2009). Accessed 18 Feb 2017.

Chen XM. Epidemiology of barley stripe rust and races of *Puccinia striiformis* f. sp. *hordei*: the first decade in the United States. In: Abstracts of the 11th Intl Cereal Rusts and Powdery Mildews Conf, 22–27 August, 2004, Norwich, UK. 2004. p. A2.8.

Chen XM. Epidemiology and control of stripe rust [*Puccinia striiformis* f. sp. *tritici*] on wheat. Can J Plant Pathol. 2005;27:314–37.

Chen XM. Races of *Puccinia striiformis* f. sp. *hordei* in the United States from 2004 to 2007. Barley Newsletter 51:WABNL51. (2008) http://wheat.pw.usda.gov/ggpages/BarleyNewsletter/51/WABNL51.htm. Accessed 25 Feb 2017.

Chen XM. Review article: High-temperature adult-plant resistance, key for sustainable control of stripe rust. Am J Plant Sci. 2013;4:608–27.

Chen XM. Integration of cultivar resistance and fungicide application for control of wheat stripe rust. Can J Plant Pathol. 2014;36:311–26.

Chen XM, Line RF. Genes for resistance to stripe rust in 'Tres' wheat. Crop Sci. 1992a;32:692–6.

Chen XM, Line RF. Inheritance of stripe rust resistance in wheat cultivars used to differentiate races of *Puccinia striiformis* in North America. Phytopathology. 1992b;82:633–7.

Chen XM, Line RF. Identification of stripe rust resistance genes in wheat genotypes used to differentiate North America races of *Puccinia striiformis*. Phytopathology. 1992c;82:1428–34.

Chen XM, Line RF. Inheritance of stripe rust resistance in wheat cultivars postulated to have resistance gene at *Yr3* and *Yr4* loci. Phytopathology. 1993a;83:382–8.

Chen XM, Line RF. Inheritance of stripe rust (yellow rust) resistance in the wheat cultivar Carstens V. Euphytica. 1993b;71:107–13.

Chen XM, Line RF. Gene action in wheat cultivars for durable high-temperature adult-plant resistance and interactions with race-specific, seedling resistance to stripe rust caused by *Puccinia striiformis*. Phytopathology. 1995a;85:567–72.

Chen XM, Line RF. Gene number and heritability of wheat cultivars with durable, high-temperature, adult-plant resistance and race-specific resistance to *Puccinia striiformis*. Phytopathology. 1995b;85:573–8.

Chen XM, Line RF. Recessive genes for resistance to races of *Puccinia striiformis* f. sp. *hordei* in barley. Phytopathology. 1999;89:226–32.

Chen XM, Line RF. Genes for resistance to barley stripe rust (*Puccinia striiformis* f.sp. *hordei*). Barley Genet Newsl 31:31. (2001a). Online at http:// www.grain.jouy.inra.fr/ggpages/BarleyNewsletter/44/WashReport2.html.

Chen XM, Line RF. Races of barley stripe rust in the United States. Barley Newsl 44 http://wheat.pw.usda.gov/ggpages/BarleyNewsletter/44/WashReport2.html. (2001b). Accessed 25 Feb 2017.

Chen XM, Line RF. Identification of genes for resistance to *Puccinia striiformis* f. sp. *hordei* in 18 barley genotypes. Euphytica. 2003;129:127–45.

Chen XM, Moore M. High-temperature, adult-plant resistance in the barley cultivar 'Bancroft' against stripe rust. Barley Newsl. 2003;45. http://wheat.pw.usda.gov/ggpages/BarleyNewsletter/45/Barley_temp_resist.html (accessed 27 Feb 2017

Chen SY, Zhang AJ, Fu J. The hybridization between *Triticum aestivum* and *Psathyrotachys huashanica*. Acta Genet Sin. 1991;18:508–12.

Chen FQ, Prehn D, Hayes PM, Mulrooney D, Corey A, Vivar H. Mapping genes for resistance to barley stripe rust (*Puccinia striiformis* f.sp. *hordei*). Theor Appl Genet. 1994a;88:215–9.

Chen XM, Line RF, Jones SS. Chromosomal location of genes for resistance to *Puccinia striiformis* in wheat cultivars Druchamp, Stephens, and Yamhill. Phytopathology. 1994b;84:1116.

Chen XM, Jones SS, Line RF. Chromosomal location of genes for stripe rust resistance in spring wheat cultivars Compair, Fielder, Lee and Lemhi and interactions of aneuploid wheats with races of *Puccinia striiformis*. Phytopathology. 1995a;85:375–81.

Chen XM, Line RF, Jones SS. Chromosomal location of genes for resistance to *Puccinia striiformis* in winter wheat cultivars Heines VII, Clement, Moro, Tyee, Tres and Daws. Phytopathology. 1995b;85:1362–7.

Chen XM, Line RF, Leung H. Virulence and polymorphic DNA relationships of *Puccinia striiformis* f. sp. *hordei* to other rusts. Phytopathology. 1995c;85:1335–42.

Chen XM, Jones SS, Line RF. Chromosomal location of genes for resistance to *Puccinia striiformis* in seven wheat cultivars with resistance genes at the *Yr3* and *Yr4* loci. Phytopathology. 1996;86:1228–33.

Chen XM, Line RF, Leung H. Genome scanning for resistance-gene analogs in rice, barley, and wheat by high-resolution electrophoresis. Theor Appl Genet. 1998;97:345–55.

Chen XM, Moore MK, Milus EA, Long DL, Line RF, Marshall D, Jackson L. Wheat stripe rust epidemics and races of *Puccinia striiformis* f. sp. *tritici* in the United States in 2000. Plant Dis. 2002;86:39–46.

Chen XM, Soria MA, Yan GP, Sun J, Dubcovsky J. Development of user-friendly PCR markers for wheat stripe rust resistance gene *Yr5*. Crop Sci. 2003;43:2058–64.

Chen XM, Penman L, Wan AM, Cheng P. Virulence races of *Puccinia striiformis* f. sp. *tritici* in 2006 and 2007 and development of wheat stripe rust and distributions, dynamics, and evolutionary relationships of races from 2000 to 2007 in the United States. Can J Plant Pathol. 2010;32:315–33.

Chen JL, Chu C, Souza EJ, Guttieri MJ, Chen XM, Xu S, Hole D, Zemetra R. Genome-wide identification of QTL conferring high-temperature adult-plant (HTAP) resistance to stripe rust (*Puccinia striiformis* f. sp. *tritici*) in wheat. Mol Breed. 2012a;29:791–800.

Chen SS, Chen GY, Chen H, Wei YM, Li W, Liu YX, Liu DC, Lan XJ, Zheng YL. Mapping stripe rust resistance gene *YrSph* derived from *Triticum sphaerococcum* Perc. with SSR, SRAP, and TRAP markers. Euphytica. 2012b;185:19–26.

Chen XM, Wang MN, Wan AM, Cheng P, Cheng JJ. Sexual or asexual reproduction, which one is more important for stripe rust? In: Chen W-Q, et al., editors. Disease Risk and Food Security, Proc. 13th International Cereal Rusts and Powdery Mildews Conference, Beijing, China, 28 Aug. – 1 Sept. 2012. Beijing: China Agricultural Science and Technology Press; 2012c. p. 36–7.

Chen SS, Chen GY, Yang C, Wei YM, Wu WX, He YJ, Liu YX, Li W, Pu ZE, Lan JX, Zheng YL. Identification and mapping of a stripe rust resistance gene in spring wheat germplasm HRMSN-81 from CIMMYT. Crop Pasture Sci. 2013a;64:1–8.

Chen XM, Coram T, Huang XL, Wang MN, Dolezal A. Understanding molecular mechanisms of durable and non-durable resistance to stripe rust in wheat using a transcriptomics approach. Curr Genomics. 2013b;14:111–26.

Chen C, He ZH, Lu JL, Li J, Ren Y, Ma CX, Xia XC. Molecular mapping of stripe rust resistance gene *YrJ22* in Chinese wheat cultivar Jimai 22. Mol Breed. 2016;36:118.

Cheng P, Chen XM. Molecular mapping of a gene for stripe rust resistance in spring wheat cultivar IDO377s. Theor Appl Genet. 2010;121:195–204.

Cheng P, Chen XM. Virulence and molecular analyses support asexual reproduction of *Puccinia striiformis* f. sp. *tritici* in the U.S. Pacific Northwest. Phytopathology. 2014;104:1208–20.

Cheng YL, Zhang HC, Yao JN, Wang XJ, Xu JR, Han QM, Wei GR, Huang LL, Kang ZS. Characterization of non-host resistance in broad bean to the wheat stripe rust pathogen. BMC Plant Biol. 2012;12:96.

Cheng YL, Zhang HC, Yao JN, Han QM, Wang XJ, Huang LL, Kang ZS. Cytological and molecular characterization of non-host resistance in *Arabidopsis thaliana* against wheat stripe rust. Plant Physiol Biochem. 2013;62:11–8.

Cheng P, Xu LS, Wang MN, See D, Chen XM. Molecular mapping of genes *Yr64* and *Yr65* for stripe rust resistance in hexaploid derivatives of durum wheat accession PI 331260 and PI 480016. Theor Appl Genet. 2014;127:2267–77.

Cheng P, Chen XM, See D. Grass hosts harbor more diverse isolates of *Puccinia striiformis* than cereal crops. Phytopathology. 2016;106:362–71.

Cheung DSM, Barber HN. Activation of resistance of wheat to stem rust. Trans Br Mycol Soc. 1972;58:333–6.

Chhetri M, Bariana H, Kandiah P, Bansal U. *Yr58*: A new stripe rust resistance gene and its interaction with *Yr46* for enhanced resistance. Phytopathology. 2016a;106:1530–4.

Chhetri M, Bansal U, Toor A, Lagudah E, Bariana H. Genomic regions conferring resistance to rust diseases of wheat in a W195/BTSS mapping population. Euphytica. 2016b;209:637–49.

Chhuneja P, Kaur S, Garg T, Ghai M, Kaur S, Prashar M, Bains NS, Goel RK, Keller B, Dhaliwal HS, Singh K. Mapping of adult plant stripe rust resistance genes in diploid A genome wheat species and their transfer to bread wheat. Theor Appl Genet. 2008;116:313–24.

Chilosi G, Johnson R. Resistance to races of *Puccinia striiformis* in seedlings of Italian wheats and possible presence of the *Yr6* gene in some durum cultivars. J Genet Breed. 1990;44:13–20.

Chin KM, Wolfe MS. The spread of *Erysiphe graminis* f.sp. *hordei* in mixtures of barley varieties. Plant Pathol. 1984;33:89–100.

Christopher MD, Liu S, Hall MD, Marshall DS, Fountain MO, Johnson JW, Millus EA, Garland-Campbell KA, Chen XM, Griffey CA. Identification and mapping of adult plant stripe rust resistance in soft red winter wheat VA00W-38. Crop Sci. 2013;53:871–9.

Cloutier S, McCallum BD, Loutre C, Banks TW, Wicker T, Feuillet C, Keller B, Jordan MC. Leaf rust resistance gene *Lr1*, isolated from bread wheat (*Triticum aestivum* L.) is a member of the large psr567 gene family. Plant Mol Biol. 2007;65:93–106.

Coram TE, Settles ML, Chen XM. Transcriptome analysis of high-temperature adult-plant resistance conditioned by *Yr39* during the wheat-*Puccinia striiformis* f. sp. *tritici* interaction. Mol Plant Pathol. 2008a;9:479–93.

Coram TE, Wang MN, Chen XM. Transcriptome analysis of the wheat-*Puccinia striiformis* f. sp. *tritici* interaction. Mol Plant Pathol. 2008b;9:157–69.

Coram TE, Huang XL, Zhan GM, Settles ML, Chen XM. Meta-analysis of transcripts associated with race-specific resistance to stripe rust in wheat demonstrates common induction of blue copper-binding protein, heat-stress transcription factor, pathogen-induced WIR1A protein, and ent-kaurene synthase transcripts. Func Integ Genomics. 2010;10:383–92.

Cuomo CA, Bakkeren G, Khalil HB, Panwar V, Joly D, Linning R, Sakthikumar S, Song X, Adiconis X, Fan L, Goldberg JM, Levin JZ, Young S, Zeng QD, Anikster Y, Bruce M, Wang

MN, Yin CT, McCallum B, Szabo LJ, Hulbert S, Chen XM, Feller JP. Comparative analysis highlights variable genome content of wheat rusts and divergence of the mating loci. G3: genes genomes. Genetics. 2017;7:371–6.

Dadkhodaie NA, Karaoglou H, Wellings CR, Park RF. Mapping genes *Lr53* and *Yr35* on the short arm of chromosome 6B of common wheat with microsatellite markers and studies of their association with *Lr36*. Theor Appl Genet. 2011;122:479–87.

Dakouri A, McCallum BD, Walichnowski AZ, Cloutier S. Fine-mapping of the leaf rust *Lr34* locus in *Triticum aestivum* (L.) and characterization of large germplasm collections support the ABC transporter as essential for gene function. Theor Appl Genet. 2010;121:373–84.

Dawit W, Flath K, Weber WE, Schumann E, Roder MS, Chen XM. Postulation and mapping of seedling stripe rust resistance genes in Ethiopian bread wheat cultivars. J Plant Pathol. 2012;94:403–9.

Dawson AM, Ferguson JN, Gardiner M, Green P, Hubbard A, Moscou MJ. Isolation and fine mapping of *Rps6*: an intermediate host resistance gene in barley to wheat stripe rust. Theor Appl Genet. 2016;129:831–43.

Dedryver F, Paillard S, Mallard S, Robert O, Trottet M, Nègre S, Verplancke G, Jahier J. Characterization of genetic components involved in durable resistance to stripe rust in the bread wheat "Renan". Phytopathology. 2009;99:968–73.

Deng ZY, Zhang XQ, Wang XP, Jing JK, Wang DW. Identification and molecular mapping of a stripe rust resistance gene from a common wheat line Qz180. Acta Bot Sin. 2004;46:236–41.

DeYoung BJ, Innes RW. Plant NBS-LRR proteins in pathogen sensing and host defense. Nat Immunol. 2006;7:1243–9.

Dong ZZ, Hegarty JM, Zhang JL, Zhang WJ, Chao SM, Chen XM, Zhou YH, Dubcovsky J. Validation and characterization of two QTL for adult plant resistance to stripe rust on wheat chromosome arms 6DS (*Yr77*) and 6BS (*Yr78*). Theor Appl Genet. 2017. (in press).

Dracatos PM, Zhang P, Park RF, McIntosh RA, Wellings CR. Complementary resistance genes in wheat selection 'Avocet R' confer resistance to stripe rust. Theor Appl Genet. 2016;129:65–75.

Dubin HJ, Rajaram S. Breeding disease resistance wheat for tropical highlands and lowlands. Annu Rev Phytopathol. 1996;34:503–26.

Dubin HJ, Stubbs RW. Epidemic spread of barley stripe rust in South America. Plant Dis. 1986;70:141–4.

Dutlu C. The inheritance of resistance to stripe rust in some bread wheat cultivars. In: Proc Fifth Eur Mediter Cereal Rusts Conf; 1984. p. 47–53.

Dyck PL. The association of a gene for leaf rust resistance with the chromosome 7D suppressor of stem rust resistance in common wheat. Genome. 1987;29:467–9.

El-Bedewy R, Röbbelen G. Chromosomal location and change of dominance of a gene for resistance against yellow rust, *Puccinia striiformis* West., in wheat, *Triticum aestivum* L. Zeitschrift für Pflanzenzuchtung. 1982;89:145–57.

Ellingboe AH. Changing concept of host-pathogen genetics. Annu Rev Phytopathol. 1981;19:125–43.

Ellis JG, Lagudah ES, Spielmeyer W, Nodds PN. The past, present and future of breeding rust resistance wheat. Front in Plant Sci. 2014;5:641.

Engledow FL. Rowland Harry Biffen, 1874-1949. Obituary Notices of Fellows of the Royal society. 1950;7:9–25.

Eriksen L, Afshari F, Christiansen MJ, McIntosh RA, Jahoor A, Wellings CR. *Yr32* for resistance to stripe (yellow) rust present in the wheat cultivar Carstens V. Theor Appl Genet. 2004;108:567–75.

Fang TL, Cheng Y, Li GQ, Xu SC, Xie CJ, You MS, Yang ZM, Sun QX, Liu ZY. Molecular characterization of a stripe rust resistance gene from wheat line S2199 and its allelism with *Yr5*. Acta Agron Sin. 2008;34:355–60.

Fang TL, Campbell KG, Liu ZY, Chen XM, Wan AM, Li S, Liu ZJ, Cao S, Chen Y, Bowden RL, Carver BF, Yan L. Stripe rust resistance in the wheat cultivar Jagger is due to *Yr17* and a novel resistance gene. Crop Sci. 2011;51:2455–65.

Favret EA, Vallega J. Inheritance of resistance to Argentine races of *Puccinia glumarum* in the wheat Chino 166. Phytopathology. 1953;43:471.

Feng J, Zuo LL, Zhang ZY, Lin RM, Cao YY, Xu SC. Quantitative trait loci for temperature-sensitive resistance to *Puccinia striiformis* f. sp. *tritici* in wheat cultivar Flinor. Euphytica. 2011;178:321–9.

Feng JY, Chen GY, Wei YM, Liu YX, Jiang QT, Li W, Pu ZE, Lan XJ, Dai SF, Zheng YL. Identification and genetic mapping of a recessive gene for resistance to stripe rust in wheat line LM168-1. Mol Breed. 2014;33:601–9.

Feng JY, Chen GY, Wei YM, Liu YX, Jiang QT, Li W, Pu ZE, Lan XJ, Dai SF, Zhang M, Zheng YL. Identification and mapping stripe rust resistance gene *YrLM168a* using extreme individuals and recessive phenotype class in a complicate genetic background. Mol Genet Genomics. 2015a;290:2271–8.

Feng JY, Wang MN, Chen XM, See DR, Zhen YL, Chao SM. Molecular mapping of *YrSP* and its relationship with other genes for stripe rust resistance in wheat chromosome 2BL. Phytopathology. 2015b;105:1206–13.

Feuillet C, Travella S, Stein N, Albar L, Nublat A, Keller B. Map-based isolation of the leaf rust disease resistance gene *Lr10* from the hexaploid wheat (*Triticum aestivum* L.) genome. Proc Natl Acad Sci USA. 2003;100:15253–8.

Finckh MR, Mundt CC. Stripe rust, yield, and plant competition in wheat cultivar mixtures. Phytopathology. 1992;82:905–13.

Flor HH. Inheritance of pathogenicity in *Melampsora lini*. Phytopathology. 1942;32:653–69.

Flor HH. Genetics of pathogenicity in *Melampsora lini*. Jour Agr Res. 1946;73:335–57.

Flor HH. Inheritance of reaction to rust in flax. Jour Agr Res. 1947;74:241–62.

Flor HH. Host-parasite interaction in flax rust-its genetics and other complications. Phytopathology. 1955;45:680–5.

Flor HH. The complementary genetic systems in flax and flax rust. Adv Genet. 1956;8:29–54.

Flor HH. Current status of the gene-for-gene concept. Annual Rev Phytopathol. 1971;9:275–96.

Friebe B, Jiang J, Raupp WJ, McIntosh RA, Gill BS. Characterization of wheat-alien translocations conferring resistance to diseases and pests: current status. Euphytica. 1996;91:59–87.

Fu D, Uauy C, Distelfeld A, Blechl A, Epstein L, Chen X, Sela H, Fahima T, Dubcovsky J. A kinase-START gene confers temperature-dependent resistance to wheat stripe rust. Science. 2009;323:1357–60.

Fuchs E. Physiologic races in yellow rust (*Puccinia glumarum* (Schm.) Erikss. et Henn.) of wheat. In: Proceedings of the Second European Yellow Rust Conference, Wageningen, The Netherlands; 1960. pp 9–21.

Gandhi SM. Inheritance of adult-plant resistance to stripe rust in wheat. Indian J Genet Plant Breed. 1971;31:256–74.

Gassner G, Straib W. Die Bestimmung der biologischen Rassen des Weizengelbrostes (*Puccinia glumarum* f. sp. *tritici* (Schmidt.) Erikss. und Henn.). Arbeiten der biologischen Reichsanstalt fur Land und Forstwirtschaft, Berlin. 1932;20:141–63.

Gerechter-Amitai ZK, van Silfhout CH, Grama A, Kleitman F. *Yr15*, a new gene for resistance to *Puccinia striiformis* in *Triticum dicoccoides* sel. G25. Euphytica. 1989;43:187–90.

Ghosh S, Sikka SM, Rao MV. Inheritance studies in wheat. IV. Inheritance of rust resistance and other characters. Indian J Genet Plant Breed. 1958;18:142–62.

Gou JY, Wu K, Wang XD, Lin HQ, Cantu D, Uauy C, Dobon-Alonso A, Midorikawa T, Inoue K, Sanchez J, Fu DL, Blechl A, Wallington E, Fahima T, Meeta M, Epstein L, Dubcovsky J. Wheat stripe rust resistance protein WKS1 reduces the ability of the thylakoid-associated ascorbate peroxidase to detoxify reactive oxygen species. Plant Cell. 2015;27:1755–70.

Goyal A, Bandopadhyay R, Sourdille P, Endo TR, Balyan HA, Gupta PK. Physical molecular maps of wheat chromosomes. Func Integ Genomics. 2005;5:260–3.

Grama A, Gerechter-Amitai ZK, Van Silfhout CH. Additive gene action for resistance to *Puccinia striiformis* f. sp. *tritici* in *Triticum dicoccoides*. Euphytica. 1984;33:281–7.

Green GJ, Campbell AP. Wheat cultivars resistant to *Puccinia graminis tritici* in western Canada: Their development, performance, and economic value. Can J Plant Pathol. 1979;1:13–5.

Griffey CA, Allan RE. Inheritance of stripe rust resistance among near-isogenic lines of spring wheat. Crop Sci. 1988;28:48–54.

Guan HT, Guo YH, Wang YB, Liu TG, Lin RM, Xu SC. Microsatellite marker of the resistance gene YrSpP to wheat stripe rust. Sci Agric Sin. 2005;38:1547–77.

Guo Q, Zhang ZJ, Xu YB, Li GH, Feng J, Zhou Y. Quantitative trait loci for high-temperature adult-plant and slow-rusting resistance to *Puccinia striiformis* f. sp. *tritici* in wheat cultivars. Phytopathology. 2008;98:803–9.

Gustafson P, Raskina O, Ma XF, Nevo E. Wheat evolution, domestication, and improvement. In: Carver BF, editor. Wheat Science and Trade. Ames, Iowa, USA: Wiley-Blackwell; 2009. p. 5–30.

Hale I, Zhang X, Fu D, Dubcovsky J. Registration of wheat lines carrying the partial stripe rust resistance gene *Yr36* without the Gpc-B1 allele for high grain protein content. J Plant Registrations. 2012;7:108–12.

Han DJ, Wang QL, Chen XM, Zeng QD, Wu JH, Xue WB, Zhan GM, Huang LL, Kang ZS. Emerging *Yr26*-virulent races of *Puccinia striiformis* f. sp. *tritici* are threatening wheat production in the Sichuan Basin, China. Plant Dis. 2015;99:754–60.

Hao YF, Chen ZB, Wang YY, Bland D, Buck J, Brown-Guedira G, Johnson J. Characterization of a major QTL for adult plant resistance to stripe rust in US soft red winter wheat. Theor Appl Genet. 2011;123:1301–411.

Hasancebi S, Mert Z, Ertugul F, Akan K, Aydin Y, Senturk-Afirat F, Altinkut-Uncuoglu A. An EST-SSR marker, *bu099658*, and its potential use in breeding for yellow rust resistance in wheat. Czech J Genet Plant Breed. 2014;50:11–8.

Hayden MJ, Kuchel H, Chalmers KJ. Sequence tagged microsatellites for the *Xgwm533* locus provide new diagnostic markers to select for the presence of stem rust resistance gene *Sr2* in bread wheat (*Triticum aestivum* L.). Theor Appl Genet. 2004;109:1641–7.

He ZH, Lan CX, Chen XM, Zou YC, Zhuang QS, Xia XC. Progress and perspective in research of adult-plant resistance to stripe rust and powdery mildew in wheat. Sci Agric Sin. 2011;44:2193–215.

Heath MC. Hypersensitive response-related death. Plant Mol Biol. 2000;44:321–434.

Helguera M, Khan IA, Kolmer J, Lijavetzky D, Zhong-qi L, Dubcovsky J. PCR assays for the *Lr37-Yr17-Sr38* cluster of rust resistance genes and their use to develop isogenic hard red spring wheat lines. Crop Sci. 2003;43:1839–47.

Helm JH, Allan RE. A genetic study of reactions to stripe rust and flag smut in hexaploid wheat. Agronomy Abstracts. 1970. p. 11–12.

Herrera-Foessel SA, Lagudah ES, Huerta-Epino J, Hayden M, Bariana H, Singh D, Singh RP. New slow-rusting leaf rust and stripe rust resistance genes *Lr67* and *Yr46* are pleiotropic or closely linked. Theor Appl Genet. 2011;122:239–49.

Herrera-Foessel SA, Singh RP, Lan CX, Huerta-Epino J, Calvo-Salazar V, Bansal UK, Bariana HS, Lagudah ES. *Yr60*, a gene conferring moderate resistance to stripe rust in wheat. Plant Dis. 2015;99:508–11.

Holtz MD, Kumar K, Xi KQ. Virulence phenotypes of *Puccinia striiformis* in Alberta from 2009-2011. Can J Plant Pathol. 2013;35:241–50.

Hou L, Ma DF, Hu ML, He MM, Lu Y, Jing JX. Genetic analysis and molecular mapping of an all-stage stripe rust resistance gene in *Triticum aestivum-Haynaldia villosa* translocation line V3. J Integ Agric. 2013;12:2197–208.

Hou L, Chen XM, Wang MN, See DR, Chao SM, Bulli P, Jing JX. Mapping a large number of QTL for durable resistance to stripe rust in winter wheat Druchamp using SSR and SNP markers. PLoS One. 2015;10:e0126794.

Hou LY, Jia JQ, Zhang XJ, Li X, Yang ZJ, Ma J, Guo HJ, Zhan HX, Qiao LY, Chang ZJ. Molecular mapping of the stripe rust resistance gene *Yr69* on wheat chromosome 2SA. Plant Dis. 2016;100:1717–24.

Hu J, Vick BA. TRAP (target region amplification polymorphism), a novel marker technique for plant genotyping. Plant Mol Biol Reporter. 2003;21:289–94.

Huang L, Gill BS. An RGA-like marker detects all known *Lr21* leaf rust-resistance gene family members in *Aegilops tauschii* and wheat. Theor Appl Genet. 2001;103:1007–13.

Huang L, Brooks SA, Li W, Fellers JP, Trick HN, Gill BS. Map-based cloning of leaf rust resistance gene *Lr21* from the large and polyploid genome of bread wheat. Genetics. 2003;164:655–64.

Huang L, Zhang LQ, Liu BL, Yan ZH, Zhang B, Zhang YL, Liu DC. Molecular tagging of a stripe rust resistance gene in *Aegilops tauschii*. Euphytica. 2011;179:313–8.

Huang Q, Li X, Chen WQ, Xiang ZP, Zhong SF, Chang ZJ, Zhang M, Zhang HY, Tan FQ, Ren ZL, Luo PG. Genetic mapping of a putative *Thinopyrum intermedium*-derived stripe rust resistance gene on wheat chromosome 1B. Theor Appl Genet. 2014;127:843–53.

Hunter RL, Merkert CL. Histochemical demonstration of enzymes separated by zone electrophoresis in starch gels. Science. 1957;125:1294–5.

Hurni S, Brunner S, Buchmann G, Jordan T, Krukowski P, Wicker T, Yahiaoui N, Mago R, Keller B. Rye *Pm8* and wheat *Pm3* are orthologous genes and show evolutionary conservation of resistance function against powdery mildew. Plant J. 2013;76:957-69.

Iehisa JCM, Ohno R, Kimura T, Enoki H, Nishimura S, Okamoto Y, Nasuda S, Takumi S. A high-density genetic map with array-based markers facilitates structural and quantitative trait locus analyses of common wheat genome. DNA Res. 2014;21:555–67.

Imtiaz M, Ahmad M, Cromey MG, Griffin WB, Hampton JG. Detection of molecular markers linked to the durable adult plant stripe rust resistance gene *Yr18* in bread wheat (*Triticum aestivum* L.). Plant Breed. 2004;123:401–4.

International Barley Genome Sequencing Consortium. A physical, genetic and functional sequence assembly of the barley genome. Nature. 2012;491:711–6.

International Wheat Genome Sequencing Consortium. A chromosome-based draft sequence of the hexaploid bread wheat (*Triticum aestivum*) genome. Science. 2014;345:1251788.

Jaccoud D, Peng KM, Feinstein D, Kilian A. Diversity Aarry: a solid state technology for sequence information independent genotyping. Nucleic Acids Res. 2001;29:e25.

Jagger LJ, Newell C, Berry ST, MacCormack R, Boyd LA. The genetic characterisation of stripe rust resistance in the German wheat cultivar Alcedo. Theor Appl Genet. 2011;122:723–33.

Jain KBL, Agrawal RK. Mature plant resistance of barley varieties to Indian races of stripe rust. Indian J Genet Plant Breed. 1964;24:203–8.

Jensen NF. Intra-varietal diversification in oat breeding. Agron J. 1952;44:50–4.

Jia JQ, Li GR, Liu C, Lei MP, Yang ZJ. Characterization of wheat yellow rust resistance gene *Yr17* using EST-SSR and rice syntenic region. Cereal Res Commun. 2011;39:88–99.

Jighly A, Oyiga BC, Makdis F, Nazari K, Youssef O, Tadesse W, Abdalla O, Ogbonnaya FC. Genome-wide DArT and SNP scan for QTL associated with resistance to strip rust (*Puccinia striiformis* f. sp. *tritici*) in elite ICARDA wheat (*Triticum aestivum* L.) germplasm. Theor Appl Genet. 2015;128:1277–95.

Jin Y, Szabo LJ, Carson M. Century-old mystery of *Puccinia striiformis* life history solved with the identification of Berveris as an alternate host. Phytopathology. 2010;100:432–5.

Johnson R. Genetics of resistance of barley to stripe rust. In Abstracts 1st Intlernational Congress Plant Pathology 1968, University of California Press, Berkeley. 1968. p. 99.

Johnson R. Durable disease resistance. In: Jenkyn JF, editor. Strategies for the Control of Cereal Disease. Oxford: Blackwell; 1981a. p. 55–64.

Johnson R. Durable resistance, definition of genetic control, and attainment in plant breeding. Phytopathology. 1981b;71:567–8.

Johnson R. Past, present and future opportunities in breeding for disease resistance, with examples from wheat. Euphytica. 1992;63:3–22.

Johnson R, Allan JD. Induced resistance to rust diseases and its possible role in the resistance of multiline varieties. Ann Appl Biol. 1975;80:359–63.

Johnson R, Dyck PL. Resistance to yellow rust in *Triticum spelta* var. *album* and bread cultivars Thatcher and Lee. In: Proceedings of the Sixth European Mediterranean Cereal Rusts Conference. 1984. p. 71–74
Johnson R, Law CN. Cytogenetic studies on the resistance of the wheat variety Bersee to *Puccinia striiformis*. Cereal Rusts Bull. 1973;1:38–43.
Johnson R, Law CN. Genetic control of durable resistance to yellow rust (*Puccinia striiformis*) in the wheat cultivar Hybride de Bersee. Annals Appl Biol. 1975;81:385–91.
Johnson R, Minchin PN. Genetics of resistance to yellow (stripe) rust of wheat in some differential cultivars. Vorträge für Pflanzenzüchtung. 1992;24:227–9.
Johnson R, Taylor AJ. Isolates of *Puccinia striiformis* collected in England from wheat varieties 'Maris Beacon' and 'Joss Cambier'. Nature. 1972;238:105–6.
Johnson R, Stubbs RW, Fuchs E, Chamberlain NH. Nomenclature for physiologic races of *Puccinia striiformis* infecting wheat. Trans Br Mycol Soc. 1972;58:475–80.
Johnson R, Priestley RH, Taylor EC. Occurrence of virulence in *Puccinia striiformis* for Compair wheat in England. Cereal Rusts Bulletin. 1978;6:11–3.
Jordan T, Seeholzer S, Schwizer S, Töller A, Somssich IE, Keller B. The wheat *Mla* homologue *TmMla1* exhibits an evolutionarily conserved function against powdery mildew in both wheat and barley. The Plant J. 2011;65:610–21.
Kema GHJ, Lange W. Resistance in spelt wheat to yellow rust. II: Monosomic analysis of the Iranian accession 415. Euphytica. 1992;63:219–24.
Kema GHJ, Lang W, van Silfhout CH. Differential suppression of stripe rust resistance in synthetic wheat hexaploids derived from *Triticum turgidum* subsp. *dicoccoides* and *Aegilops squarrosa*. Phytopathology. 1995;85:425–9.
Khlestkina EK, Roder MS, Unger O, Meinel A, Borner A. More precise map position and origin of a durable non-specific adult plant disease resistance against stripe rust (*Puccinia striiformis*) in wheat. Euphytica. 2007;153:1–10.
Kilian B, Özkan H, Pozzi C, Salamini F. Domestication of the Triticeae in the Fertile Crescent. In: Feuillet C, Muehlbauer GJ (eds) Genetics and Genomics of the Triticeae. Series Plant Genetics and Genomics: Crops and Models. 2009;7:81–120.
Klos KE, Gordon T, Bregitzer P, Hayes P, Chen XM, del Blanco IA, Fisk S, Bonman JM. Barley stripe rust resistance QTL: Development and validation of SNP markers for resistance to *Puccinia striiformis* f. sp. *hordei*. Phytopathology. 2016;106:1344–51.
Knott DR. The wheat rusts-breeding for resistance. Monograph on Theor Appl Genet 12. Springer Verlag: Berlin; 1989.
Knüpffer H. Triticeae genetic resources in ex situ gene bank collections. In: Feuillet C, Muehlbauer GJ (eds) Genetics and Genomics of the Triticeae. Plant Genetics and Genomics: Crops and Models. 2009;7:31–80.
Kolmer JA. Genetics of resistance to wheat leaf rust. Annu Rev Phytopathol. 1996;34:435–55.
Kolmer JA, Dyck PL. Gene expression in the *Triticum aestivum-Puccinia recondita* f. sp. *tritici* gene-for-gene system. Phytopathology. 1994;84:437–40.
Kolmer JA, Singh RP, Garvin DF, Viccars L, William HM, Huerta-Espino J, Ogbonnaya FC, Raman H, Orford S, Bariana HS, Lagudah ES. Analysis of the *Lr34/Yr18* rust resistance region in wheat germplasm. Crop Sci. 2008;48:1841–52.
Konzak CF. Induction of mutations for disease resistance in cereals. In: Genetics in Plant Breeding. Brookhaven Symp Biol. 1956;9:157–71.
Konzak CF, Line RF, Allan RE, Schafer JF. Guidelines for the production, evaluation, and use of induced resistance to stripe rust in wheat. Proc Induced Mutation to Plant Dis, International Atomic Energy, Vienna. 1977;1977:437–60.
Kosambi DD. The estimation of map distances from recombination values. Ann Eugen. 1944;12:172–5.
Krattinger SG, Lagudah ES, Spielmeyer W, Singh RP, Huerta-Espino J, McFadden M, Bossolini E, Selter LL, Keller B. A putative ABC transporter confers durable resistance to multiple fungal pathogens in wheat. Science. 2009a;323:1360–3.

Krattinger SG, Wicker T, Keller B. Map-based cloning of genes in Triticeae (wheat and barley). In: Feuillet C, Muehlbauer GJ, editors. Genetics and Genomics of the Triticeae, Plant Genetics and Genomics: Crops and Models 7. LLC: Springer Science + Business Media; 2009b. p. 339–58.

Krattinger SG, Lagudah ES, Wicker T, Risk JM, Ashton AR, Selter LL, Matsumoto T, Keller B. *Lr34* multi-pathogen resistance ABC transporter: molecular analysis of homoeologous and orthologous genes in hexaploid wheat and other grass species. Plant J. 2011;65:392–403.

Krattinger SG, Jordan DR, Mace ES, Raghavan C, Luo MC, Keller B, Lagudah ES. Recent emergence of the wheat *Lr34* multipathogen resistance: insights from haplotype analysis in wheat, rice, sorghum and *Aegilops tauschii*. Theor Appl Genet. 2013;126:663–72.

Krattinger SG, Sucher J, Selter LL, Chauhan H, Zhou B, Tang M, Upadhyaya NM, Mieulet D, Guiderdoni E, Weidenbach D, Schaffrath U, Lagudah ES, Keller B. The wheat durable, multipathogen resistance gene *Lr34* confers partial blast resistance in rice. Plant Biotechnol J. 2016;14:1261–8.

Krupinsky JM. Development of additive resistance in wheat, *Triticum aestivum* L. to stripe rust, *Puccinia striiformis* West. Thesis for PhD degree in plant Pathology in Montana State University. 1977.

Krupinsky JM, Sharp EL. Additive resistance in wheat to *Puccinia striiformis*. Phytopathology. 1978;68:1795–9.

Kumar J, Nayar SK, Bahadur P, Nagarajan S, Bhardwaj SC, Prashar M, Singh SB. Virulence survey of yellow rust of wheat (Puccinia *striiformis* f.sp. *tritici*) and barley (*P. striiformis* f.sp. *hordei*) during 1985-87. Cereal Rusts Powdery Mildews Bull. 1988;16:30–5.

Kumar S, Goyal A, Mohan A, Balyan HS, Gupta PK. An integrated physical map of simple sequence repeats in bread wheat. Aus J Crop Sci. 2013;7:460–8.

Kuraparthy V, Chunneja P, Dhaliwal HS, Kaur S, Bowden RL, Gill BS. Characterization and mapping of cryptic alien introgression from *Aegilops geniculata* with new leaf rust and stripe rust resistance genes *Lr57* and *Yr40* in wheat. Theor Appl Genet. 2007;114:1379–89.

Kuraparthy V, Shilpa S, See DR, Gill BS. Development of a PCR assay and marker-assisted transfer of leaf rust and stripe rust resistance genes *Lr57* and *Yr40* into hard red winter wheats. Crop Sci. 2009;49:120–6.

Labrum KE. The location of *Yr2* and *Yr6*, genes conferring resistance to yellow rust. Proceedings of the 5th European and Mediterranean Cereal Rusts Conference, Bari, Italy. 1980. p. 41–45.

Lagudah ES, Krattinger SG, Herrera-Foessel S, Singh RP, Huerta-Espino J, Spielmeyer Q, Brown-Guedira G, Selter LL, Beller B. Gene-specific markers for the wheat gene *Lr34/Yr18/Pm38* which confers resistance to multiple fungal disease. Theor Appl Genet. 2009;119:889–98.

Lan CX, Liang SS, Zhou XC, Zhou G, Lu QL, Xia XC, He ZH. Identification of genomic regions controlling adult-plant stripe rust resistance in Chinese landrace Pingyuan 50 through bulked segregant analysis. Phytopathology. 2010a;100:313–8.

Lan CX, Ni XW, Yan J, Zhang Y, Xia XC, Chen XM, He ZH. Quantitative trait loci mapping of adult-plant resistance to powdery mildew in Chinese wheat cultivar Lumai 21. Mol Breed. 2010b;25:615–22.

Lan CX, Rosewarne GM, Singh RP, Herrera-Foessel SA, Huerta-Espino J, Basnet BR, Zhang YL, Yang EN. QTL characterization of resistance to leaf rust and stripe rust in the spring wheat line Francolin#1. Mol Breed. 2014;34:789–803.

Lan CX, Zhang YL, Herrera-Foessel SA, Basnet BR, Huerta-Espino J, Lagudah ES, Singh RP. Identification and characterization of pleiotropic and co-located resistance loci to leaf rust and stripe rust in bread wheat cultivar Sujata. Theor Appl Genet. 2015;128:549–61.

Law CN. Genetic control of yellow rust resistance in *T. spelta album*. Plant Breeding Institute, Cambridge. Annual Report. 1976;1975:108–9.

Law CN, Worland AJ. Improving disease resistance in wheat by inactivating genes promoting disease susceptibility. IAEA Vienna, Mutat Breed Newsl. 1991;38:2–5.

Law CN, Snape JW, Worland AJ. Aneuploid in wheat and its uses in genetic analysis. In: Lupton FGH, editor. Wheat breeding: its scientific basis. London: Chapman and Hall; 1987. p. 71–108.

Ledford H. CRISPR: gene editing is just the beginning. Nature. 2016;531(7593):156–9.

Lewellen RT, Sharp EL, Hehn ER. Major and minor genes in wheat for resistance to *Puccinia striiformis* and their responses to temperature changes. Can J Bot 1967;45:2155-72.

Lewellen RT, Sharp EL. Inheritance of minor gene combinations in wheat to *Puccinia striiformis* at two temperature profiles. Can J Bot. 1968;46:21–6.

Li G, Quiros CF. Sequence-related amplified polymorphism (SRAP), a new marker system based on a simple PCR reaction: its application to mapping and gene tagging in Brassica. Theor Appl Genet. 2001;103:455–61.

Li ZQ, Wang MN, Jia MG, Lu HP, Shang HS, Kang ZS, Yang SF, Sun SM, Shi YC. Study on the epidemiology and control strategy of wheat stripe rust in South Gansu. J Northw Agricult Univ (Nat Sci Edn). 1997;25:1–6.

Li GQ, Li ZF, Yang WY, Zhang Y, He ZH, Xu SC, Singh RP, Qu TT, Xia XC. Molecular mapping of stripe rust resistance gene *YrCH42* in Chinese wheat cultivar Chuanmai 42 and its allelism with *Yr24* and *Yr26*. Theor Appl Genet. 2006a;112:1434–40.

Li ZF, Zheng TC, He ZH, Li GQ, Xu SC, Li XP, Yang GY, Singh RP, Xia XC. Molecular tagging of stripe rust resistance gene *YrZh84* in Chinese wheat line Zhou 8425B. Theor Appl Genet. 2006b;112:1089–103.

Li Y, Niu YC, Chen XM. Mapping a stripe rust resistance gene *YrC591* in wheat variety C591 with SSR and AFLP markers. Theor Appl Genet. 2009;118:339–46.

Li Q, Chen XM, Wang MN, Jing JX. *Yr45*, a new wheat gene for stripe rust resistance on the long arm of chromosome 3D. Theor Appl Genet. 2011;122:189–97.

Li Q, Huang J, Hou L, Liu P, Jing J, Wang B, Kang Z. Genetic and molecular mapping of stripe rust resistance gene in wheat-*Psathyrostachys huashanica* translocation line H9020-1-6-8-3. Plant Dis. 2012;96:1482–7.

Li H, Vikram P, Singh RP, Kilian A, Carling J, Song J, Burgueno-Ferreira JA, Bhavani S, Huerta-Espino J, Payne T, Sehgal D. A high density GBS map of bread wheat and its application for dissecting complex disease resistance traits. BMC Genomics. 2015;16:216.

Li K, Hegarty J, Zhang CH, Wan AM, Wu JJ, Guedira GB, Chen XM, Munoz-Amatriain M, Fu DL, Dubcovsky J. Fine mapping of barley locus *Rps6* conferring resistance to wheat stripe rust. Theor Appl Genet. 2016a;129:845–59.

Li Q, Ma DF, Li Q, Fan Y, Shen XX, Jing JX, Wang BT, Kang ZS. Genetic analysis and molecular mapping of a stripe rust resistance gene in Chinese wheat differential Guinong 22. J Phytopathol. 2016b;164:476–84.

Li YX, Wang MN, Wan AM, Chen XM. Development of *Puccinia striiformis* f. sp. *tritici* mutants for avirulence characterization. Phytopathology. 2016c;S4:20.

Lillemo M, Asalf B, Singh RP, Huerta-Espino J, Chen XM, He ZH, Bjornstad A. The adult plant rust resistance loci *Lr34/Yr18* and *Lr46/Yr29* are important determinants of partial resistance to powdery mildew in bread wheat line Saar. Theor Appl Genet. 2008;116:1155–66.

Lin F, Chen XM. Genetics and molecular mapping of genes for race- specific and all-stage resistance and non-specific high-temperature adult-plant resistance to stripe rust in spring wheat cultivar Alpowa. Theor Appl Genet. 2007;114:1277–87.

Lin F, Chen XM. Molecular mapping of genes for race-specific overall resistance to stripe rust in wheat cultivar Express. Theor Appl Genet. 2008;116:797–806.

Lin F, Chen XM. Quantitative trait loci for non-race-specific, high-temperature adult-plant resistance to stripe rust in wheat cultivar Express. Theor Appl Genet. 2009;118:631–42.

Lin F, Xu SC, Zhang ZJ, Miao Q, Zhai Q, Li N. SSR marker of wheat stripe rust resistance gene *Yr2*. J Triticeae Crops. 2005;25:17–9.

Line RF. Recording and processing data on foliar diseases of cereals. Prague: Proc Eur Medit Cereal Rusts Conf; 1972. p. 175–8.

Line RF. Stripe rust of wheat and barley in North America: a retrospective historical review. Annu Rev Phytopathol. 2002;40:75–118.

Line RF, Chen XM. Successes in breeding for and managing durable resistance to wheat rusts. Plant Dis. 1995;79:1254–5.

Line RF, Chen XM. Wheat and barley stripe rust in North America. In: Kema GHJ, Niks RE, Daamen RA (eds) Proceeidings of the 9th European Mediterranean Cereal Rusts and Powdery Mildews conference, 2–6 September 1996, Lunteren, The Netherlands. 1996. p. 101–4.

Line RF, Qayoum A. Virulence aggressiveness, evolution, and distribution of race *Puccinia striiformis* (the cause of stripe rust of wheat) in North America, 1968-87. USDA-ARS Technical Bulletin 1788. USDA-ARS, Washington, DC. 1992.

Line RF, Sharp EL, Powelson RL. A system for differentiating races of *Puccinia striiformis* in the United States. Plant Dis Report. 1970;54:992–4.

Line RF, Konzak CF, Allan RE. Evaluating resistance to *Puccinia striiformis* in wheat. In:Induced Mutations for Disease Resistance in Crop Plants, Proc Res Co-ord Meeting, FAO/IAEA, Novi Sad 1973. Vienna: IAEA; 1974. p. 125–32.

Line RF, Allan RE, Konzak CF. Identifying and utilizing resistance to *Puccinia striiformis* in wheat. In: Proceedings of the induced mutations for disease resistance, crop plants (1975), FAO/IAEA, Ames. 1976. p. 151–8.

Ling HQ, Zhu Y, Keller B. High-resolution mapping of the leaf rust disease resistance gene *Lr1* in wheat and characterization of BAC clones from the *Lr1* locus. Theor Appl Genet. 2003;106:875–82.

Liu FH, Niu YC, Deng H, Tan GJ. Mapping of a major stripe rust resistance gene in Chinese native wheat variety Chike using microsatellite markers. Journal of Genetics and Genomics. 2007;34:1123–30.

Liu P, Yang MN, Zhou XL, Wu HJ, Jing JX. Genetic analysis and molecular mapping of stripe rust resistance of wheat translocation line H9020-1-6-8-3 derived from *Psathyrostachys huashanica* Keng. Acta Phytopathol Sin. 2008;38:104–7.

Liu DC, Zhang LQ, Yan ZH, Lan XJ, Zheng YL. Stripe rust resistance in *Aegilops tauschii* and its genetic analysis. Genet Resour Crop Evol. 2010;57:325–8.

Liu B, Chen XM, Kang ZS. Gene sequencing reveals heterokaryotic variations and evolutionary mechanisms in *Puccinia striiformis*. Open J Genomics. 2012;1:1–14. http://rossscience.org/ojgen/articles/2075-9061-1-1.pdf

Liu J, Chang ZJ, Zhang XJ, Yang ZJ, Li X, Jia JQ, Zhan HX, Guo HJ, Wang JM. Putative *Thinopyrum intermedium*-derived stripe rust resistance gene *Yr50* maps on wheat chromosome 4BL. Theor Appl Genet. 2013a;126:265–74.

Liu M, Zhang CZ, Yuan CL, Zhang LQ, Huang L, Wu JJ, Wang JR, Zheng YL, Zhang HG, Liu DC, Fu DL. Stripe rust resistance in *Aegilops tauschii* germplasm. Crop Science. 2013b;53:2014–20.

Liu W, Frick M, Huel R, Nykiforuk CL, Wang X, Gaudet DA, Eudes F, Conner RL, Kuzyk A, Chen Q, Kang Z, Laroche A. The stripe rust resistance gene *Yr10* encodes an evolutionary-conserved and unique CC-NBS-LRR sequence in wheat. Mol Plant. 2014a;7:1740–55.

Liu ZG, Yao WY, Shen XX, Chao KX, Fan Y, Li MZ, Wang BT, Li Q, Jing JX. Molecular mapping of a stipe rust resistance gene *YrH9020a* transferred from *Psathyrostachys huashanica* Keng on wheat chromosome 6D. Journal of Integrative Agriculture. 2014b;13:2577–83.

Liu JD, He ZH, Wu L, Bai B, Wen WE, Xie CJ, Xia XC. Genome-wide linkage mapping of QTL for adult-plant resistance to stripe rust in a Chinese wheat population Linmai 2 x Zhong 892. PLoS ONE. 2015;10:e0145462.

Liu TL, Wan AM, Chen XM. Virulence characterization of *Puccinia striiformis* f. sp. *tritici* in the US for the past 48 years using the *Yr* single-gene differentials. Phytopathology. 2016;S4:201.

Liu WZ, Macaferri M, Bulli P, Rynearson S, Tuberosa R, Chen XM, Pumphrey M. Genome-wide association mapping of seedling and field resistance to *Puccinia striiformis* f. sp. *tritici* in elite global durum wheat. Theor Appl Genet. 2017. doi: 10.1007/s00122-016-2841-9.

Lowe I, Jankuloski L, Chao SM, Chen XM, See D, Dubcovsky J. Mapping and validation of QTL which confer partial resistance to broadly virulent post-2000 North American races of stripe rust in hexaploid wheat. Theor Appl Genet. 2011;123:143–57.

Lu YM, Lan CX, Liang SS, Zhou XC, Liu D, Zhou G, Lu QL, Jing JX, Wang MN, Xia XC, He ZH. QTL mapping for adult-plant resistance to stripe rust in Italian common wheat cultivars Libellula and Strampelli. Theor Appl Genet. 2009;119:1349–59.

Lu Y, Wang MN, Chen XM, See D, Chao SM, Jing JX. Mapping of *Yr62* and a small-effect QTL for high-temperature adult-plant resistance to stripe rust in spring wheat PI 192252. Theor Appl Genet. 2014;127:1449–59.

Lukaszewski AJ. Manipulation of the 1RS.1BL translocation in wheat by induced homoeologus recombination. Crop Sci. 2000;40:216–25.

Luo PG, Ren ZL, Zhang HQ, Zhang HY. Identification, chromosome location, and diagnostic markers for a new gene (*YrCN19*) for resistance to wheat stripe rust. Phytopathology. 2005;95:1266–70.

Luo PG, Hu XY, Ren ZL, Zhang HY, Shu K, Yang ZJ. Allelic analysis of stripe rust resistance genes on wheat chromosome 2BS. Genome. 2008;51:922–7.

Lupton FCH, Macer RCF. Inheritance of resistance to yellow rust (*Puccinia glumarum* Erikss. and Henn.) in seven varieties of wheat. Trans Br Mycol Soc. 1962;45:21–45.

Ma H, Singh RP. Expression of adult-plant resistance to stripe rust at different growth stages of wheat. Plant Disease. 1996;80:375–9.

Ma H, Sigh RP, Mujeeb-Kazi A. Suppression/expression of resistance to stripe rust in synthetic hexaploid wheat (*Triticum turgidum* x *T. tauschii*). Euphytica. 1995;83:87–93.

Ma JX, Zhou RH, Dong YS, Wang LF, Wang XM, Jia JZ. Molecular mapping and detection of the yellow rust resistance gene *Yr26* in wheat transferred from *Triticum turgidum* L. using microsatellite. Euphytica. 2001;120:219–26.

Ma DF, Zhou XL, Hou L, Bai YB, Li Q, Wang HG, Tang MS, Jing JX. Genetic analysis and molecular mapping of a stripe rust resistance gene derived from *Psathynrostachys huashanica* Keng in wheat line H9014-121-5-5-9. Mol Breed. 2013;32:365–72.

Ma DF, Li Q, Tang MS, Chao KX, Li JC, Wang BT, Jing JX. Mapping of gene conferring adult-plant resistance to stripe rust in Chinese wheat landrace Baidatou. Mol Breed. 2015a;35:157.

Ma DF, Peng F, Fang ZW, Chao KX, Jing JX, Zhang CQ. Genetic and molecular mapping of stripe rust resistance genes in wheat cultivar Zhongliang 12. J Phytopathol. 2015b;163:98–104.

Ma DF, Fang ZW, Lin JL, Chao KX, Jing JX, Li Q, Wang BT. Molecular mapping of stripe rust resistance gene *YrHu* derived from *Psathyrostachys huashanica*. Mol Breed. 2016;36:64.

Maccaferri M, Zhang J, Bulli P, Abate Z, Chao S, Cantu D, Bossolini E, Chen X, Pumphrey M, Dubcovsky J. A genome-wide association study of resistance to stripe rust (*Puccinia striiformis* f. sp. *tritici*) in a worldwide collection of hexaploid spring wheat (*Triticum aestivum* L.). G3: Genes Genomes. Genetics. 2015;5:449–65.

Macer RCF. The formal and monosomic genetic analysis of stripe rust (*Puccinia striiformis*) resistance in wheat. In: Proc 2nd Intl Wheat Genet Symp. Lund, Hereditas Suppl. 1963;2:127–42.

Macer RCF. The formal and monosomic genetic analysis of stripe rust (*Puccinia striiformis*) resistance in wheat. In: MacKey J (ed) Proc 2nd Intl Wheat Genet Symp, Lund, Sweden 1963. Hereditas Suppl. 1966;2:127–42.

Macer RCF. Plant Pathology in a changing world. Trans Br Mycol Soc. 1975;65:351–74.

Mago R, Spielmeyer W, Lawrence GJ, Lagudah ES, Ellis JG, Pryor A. Identification and mapping of molecular markers linked to rust resistance genes located on chromosome 1RS of rye using wheat-rye translocation lines. Theor Appl Genet. 2002;104:1317–24.

Mago R, Simkova H, Brown-Guedira G, Dreisigacker S, Breen J, Jin Y, Singh R, Appels R, Lagudah ES, Ellis J, Dolezel J, Spielmeyer W. An accurate DNA marker assay for stem rust resistance gene *Sr2* in wheat. Theor Appl Genet. 2011;122:735–44.

Mago R, Zhang P, Vautrin S, Šimková H, Bansal U, Luo MC, Rouse M, Karaoglu H, Periyannan S, Kolmer J, Jin Y. The wheat *Sr50* gene reveals rich diversity at a cereal disease resistance locus. Nature Plants. 2015;1:15186.

Mallard S, Gaudet D, Aldeia A, Abelard C, Besnard AL, Sourdille P, Dedryver F. Genetic analysis of durable resistance to yellow rust in bread wheat. Theor Appl Genet. 2005;110:1401–9.

Mandoulakani BA, Yaniv E, Kalendar R, Raats D, Bariana HS, Bihamta MR, Schulman AH. Development of IRAP-and REMAP-derived SCAR markers for marker-assisted selection of the stripe rust resistance gene *Yr15* derived from wild emmer wheat. Theo Appl Genet. 2015;128:211–9.

Marais GF, Pretorius ZA, Marais AS, Wellings CR. Transfer of rust resistance genes from *Triticum* species to common wheat. South African J Plant Soil. 2003;20:193–8.

Marais GF, McCallum B, Snyman JE, Pretorius ZA, Marais AS. Leaf rust and stripe rust resistance genes *Lr54* and *Yr37* transferred to wheat from *Aegilops kotschyi*. Plant Breed. 2005a;124:538–41.

Marais GF, Pretorius ZA, Wellings CR, McCallum B, Marais AS. Leaf and stripe rust resistance genes transferred to common wheat from *Triticum dicoccoides*. Euphytica. 2005b;143:115–23.

Marais GF, McCallum B, Marais AS. Leaf rust and stripe rust resistance genes derived from *Triticum sharonense*. Euphytica. 2006;149:373–80.

Marais F, Marais A, McCallum B, Pretorius Z. Transfer of leaf rust and stripe rust resistance genes *Lr62* and *Yr42* from *Aegilops neglecta* Req. ex Bertol. to common wheat. Crop Sci. 2009;49:871–9.

Marais GF, Badenhorst PE, Eksteen PZA. Reduction of *Aegilops sharonensis* chromatin associated with resistance genes *Lr56* and *Yr38* in wheat. Euphytica. 2010;171:15–22.

Marone D, Panio G, Ficco DB, Russo MA, De Vita P, Papa R, Rubiales D, Cattivelli L, Mastrangelo AM. Characterization of wheat DArT markers: genetic and functional features. Mol Genet Genomics. 2012;287:741–53.

Marryat DCE. Notes on the infection and histology of two wheats immune to the attack of *Puccinia glumarum*, yellow rust. J Agric Sci. 1907;2:129–38.

McCartney CA, Brule-Babel AL, Lamari L, Somers DJ. Chromosomal location of a race-specific resistance gene to *Mycosphaerella graminicola* in the spring wheat ST6. Theor Appl Genet. 2003;107:181–6.

McDonald D, McIntosh RA, Wellings CR, Singh RP, Nelson JC. Cytogenetical Studies in Wheat XIX. Location and linkage studies on gene *Yr27* for resistance to stripe (yellow) rust. Euphytica. 2004;136:239–48.

McGrann GRD, Smith PH, Burt C, Mateos GR, Chama TN, MacCormack R, Wessels E, Agenbag G, Boyd LA. Genomic and genetic analysis of the wheat race-specific yellow rust resistance gene *Yr5*. J Plant Sci Mol Breed. 2014;3:2.

McIntosh RA. A catalogue of gene symbols for wheat (1983 edition). In: Proceedings of the sixth international wheat genetic symposium, Kyoto, Japan. 1983. p. 1197–254.

McIntosh RA. Close genetic linkage of genes conferring adult-plant resistance to leaf rust and stripe rust in wheat. Plant Pathology. 1992;41:523–7.

McIntosh RA, Brown GN. Anticipatory breeding for resistance to rust diseases in wheat. Annu Rev Phytopathol. 1997;35:311–26.

McIntosh RA, Luig NH, Johnson R, Hare RA. Cytogenetical studies in wheat. XI. *Sr9g* for reaction to *Puccinia graminis tritici*. Z Pflanzenzüchtg. 1981;87:274–89.

McIntosh RA, Wellings CR, Park RF. Wheat rust: an atlas of resistance genes. Melbourne, Australia: CSIRO Publishing; 1995. 200 pp.

McIntosh RA, Yamazaki Y, Dubcovsky J, Rogers J, Morris C, Appels R, Xia XC. Catalogue of gene symbols for wheat. In: 12th International wheat genetic symposium, 8–13 September 2013, Yokohama, Japan. Online. 2013. http://www.shigen.nig.ac.jp/wheat/komugi/genes/download.jsp

McIntosh RA, Dubcovsky J, Rogers WJ, Morris C, Appels R, Xia XC. Catalogue of gene symbols for wheat: 2013–2014 supplement. 2014. http://maswheat.ucdavis.edu/CGSW/2013-2014_Supplement.pdf

McNeil MD, Kota R, Paux E, Dunn D, McLean R, Feuillet C, Li D, Kong X, Lagudah E, Zhang JC, Jia JZ, Spielmeyer W, Bellgard M, Appels R. BAC-derived markers for assaying the stem rust resistance gene, *Sr2*, in wheat breeding programs. Mol Breed. 2008;22:15–24.

Melichar JPE, Berry S, Newell C, MacCormack R, Boyd LA. QTL identification and microphenotype characterisation of the developmentally regulated yellow rust resistance in the UK wheat cultivar Guardian. Theor Appl Genet. 2008;117:391–9.

Metzger RJ, Silbaugh BA. Inheritance of resistance to stripe rust and its association with brown glume colour in *Triticum aestivum* L. PI 178383. Crop Sci. 1970;10:567–8.

Miller TE, Reader SM, Singh D. Spontaneous non-Robertsonian translocations between wheat chromosomes and alien chromosome. In: Miller TE, Koebner RMD, editors. Proceedings of the seventh international wheat genetics symposium. Cambridge: Institute of Plant Science Research; 1988. p. 387–90.

Milus EA, Line RF. Number of genes controlling high-temperature adult-plant resistance to stripe rust in wheat. Phytopathology. 1986a;76:93–6.

Milus EA, Line RF. Gene action for inheritance of durable, high-temperature, adult-plant resistance to stripe rust in wheat. Phytopathology. 1986b;76:435–41.

Minhas AS, Singh S. Inheritance of stripe rust resistance in intervarietal crosses of *Triticum aestivum*. Madras Agric J. 1973;60:45–8.

Moore JW, Herrera-Foessel S, Lan C, Schnippenkoetter W, Ayliffe M, Huerta-Espino J, Lillemo M, Viccars L, Milne R, Periyannan S, Kong XY, Spielmeyer W, Talbot M, Bariana H, Patrick JW, Dodds P, Singh R, Lagudah E. A recently evolved hexose transporter variant confers resistance to multiple pathogens in wheat. Nature Genet. 2015;47:1494–8.

Mundt CC. Use of host genetic diversity to control cereal diseases: implications for rice blast. In: Zeigler RS, Leong SA, Teng PS, editors. Rice blast disease. London: CAB International; 1994. p. 293–308.

Mundt CC. Use of multiline cultivars and cultivar mixtures for disease management. Annu Rev Phytopathol. 2002;40:381–410.

Mundt CC. Durable resistance: a key to susceptible management of pathogens and pests. Infect Genet Evol. 2014;27:446–55.

Mundt CC, Brophy LS, Kolar SC. Effect of genotype unit number and spatial arrangement on severity of yellow rust in wheat cultivar mixtures. Plant Pathol. 1996;45:215–22.

Mur LAJ, Kenton P, Lloyd AJ, Ougham H, Prats E. The hypersensitive response: the centenary is upon us but how much do we know? J Exp Bot. 2008;59:501–20.

Murphy LR, Santra D, Kidwell K, Yan GP, Chen XM, Garland-Campbell KA. Linkage maps of wheat stripe rust resistance genes *Yr5* and *Yr15* for use in marker-assisted selection. Crop Sci. 2009;49:1786–90.

Murty SS. Segregation and correlated inheritance of rust-resistance and epidermal characters in a barley cross. Indian J Genet. 1942;2:73–5.

Nagarajan S, Bahadur P, Nayar K. Occurrence of a new virulence, 47S102 of *Puccinia striiformis* West., in India during crop year, 1982. Cereal Rust Bull. 1984;12:28–31.

Nambisan PNN, Kholi SP. Inheritance of seedling resistance to races 13 and H of *Puccinia glumarum* (Schm.) Erikss. and Henn. in crosses of *Triticum aestivum*. Indian J Genet Plant Breed. 1961;21:15–22.

Naruoka Y, Garland-Campbell KA, Carter AH. Genome-wide association mapping for stripe rust (*Puccinia striiformis* f. sp. *tritici*) in US Pacific Northwest winter wheat (*Triticum aestivum* L.). Theor Appl Genet. 2015;128:1083–110.

Naruoka Y, Ando K, Bulli P, Muleta KT, Rynearson S, Pumphrey MO. Identification and validation of SNP markers linked to the stripe rust resistance gene *Yr5* in wheat. Crop Sci. 2016;56:3055–65.

Nasuda S, Friebe B, Busch W, Kynast RG, Gill BS. Structural rearrangement in chromosome 2M of *Aegilops comosa* has prevented the utilization of the Compare and related wheat-*Ae. comosa* translocations in wheat improvement. Thero Appl Genet. 1998;96:780–5.

Navabi A, Singh RP, Tewari JP, Griggs KG. Inheritance of high levels of adult-plant resistance to stripe rust in five spring wheat genotypes. Crop Sci. 2004;44:1156–62.

Navabi A, Tewari JP, Singh RP, McCallum B, Laroche A, Briggs KG. Inheritance and QTL analysis of durable resistance to stripe and leaf rusts in an Australian cultivar, *Triticum aestivum* 'Cook'. Genome. 2005;48:97–107.

Nazari K, Wellings CR Genetic analysis of seedling stripe rust resistance in Australian wheat cultivar 'Batavia'. 2017. http://ses.library.usyd.edu.au/bitstream/2123/3215/1/P142.pdf. Accessed 2nd Mar 2017.

Nelson RR. Genetics of horizontal resistance to plant disease. Annu Rev Phytopathol. 1978;16:359–78.

Nevo E, Gerechter-Amitai ZK, Beiles A, Golenberg EM. Resistance of wild wheat to stripe rust: Predictive method by ecology and allozyme genotypes. Plant Syst Evol. 1986;153:13–30.

Newcomb MN, Acevedo HE, Bockelman HE, Brown-Guedira G, Boates BJ, Jackson EW, Jin Y, Njau P, Rouse MN, Singh D, Wanyera R, Bonman JM. Field resistance to the Ug99 race group of the stem rust pathogen in spring wheat landraces. Plant Dis. 2013;97:882–90.

Niks RE. Appressorium formation of *Puccinia hordei* on partially resistant barley and two nonhost species. Neth J Plant Pathol. 1981;87:201–7.

Niks RE, Alemu SK, Marcel TC, van Heyzen S. Mapping genes in barley for resistance to *Puccinia coronata* from couch grass and to *P. striiformis* from brome, wheat and barley. Euphytica. 2015;206:487–99.

Nover I, Scholz F. Genetic studies on the resistance of barley to yellow rust (*Puccinia striiformis* West.). Theor Appl Genet. 1969;39:150–5.

Oberthur LE, Harrison SA, Croughan TP, Long DL. Inheritance of improved leaf rust resistance in somaclones of wheat. Cop Sci. 1993;33:444–8.

Omar AM, Selim AKA, Hassanein SH, Khalil OHS. Inheritance of mature plant reaction to stripe rust of wheat in UAR. UAR J Bot. 1970;13:131–42.

Pahalawatta V, Chen XM. Genetic analysis and molecular mapping of wheat genes conferring resistance to the wheat stripe rust and barley stripe rust pathogens. Phytopathology. 2005a;95:427–32.

Pahalawatta V, Chen XM. Inheritance and molecular mapping of barley genes conferring resistance to wheat stripe rust. Phytopathology. 2005b;95:884–9.

Parleviet JE. Disease resistance in plants and its consequences for plant breeding. In: Frey KJ, editor. Plant Breeding II. Ames: Iowa State University Press; 1981. p. 309–64.

Peng JH, Fahima T, Roder MS, Li YC, Dahan A, Grama A, Ronin YI, Korol AB, Nevo E. Microsatellite tagging of the stripe rust resistance gene *YrH52* derived from wild emmer wheat, *Triticum dicoccoides*, and suggestive negative crossover interference on chromosome 1B. Theor Appl Genet. 1999;98:862–72.

Peng JH, Fahima T, Röder MS, Huang OY, Dahan A, Li YC, Grama A, Nevo E. High-density molecular map of chromosome region harboring stripe-rust resistance gene *YrH52* and *Yr15* derived from wild emmer wheat, *Triticum dicoccoides*. Genetica. 2000a;109:199–210.

Peng JH, Fahima T, Röder MS, Li YC, Grama A, Nevo E. Microsatellite high density mapping of the stripe rust resistance gene *YrH52* region on chromosome 1B and evaluation of its marker-assisted selection in the F_2 generation in wild emmer wheat. New Phytol. 2000b;146:141–54.

Periyannan S, Moore J, Ayliffe M, Bansal U, Wang X, Huang L, Deal K, Luo M, Kong X, Bariana H, Mago R. The gene *Sr33*, an ortholog of barley *Mla* genes, encodes resistance to wheat stem rust race Ug99. Science. 2013;341:786–8.

Pink DAC, Bennett FGA, Caten CE, Law CN. Correlated effects of homoeologous groups 5 chromosomes upon infection of wheat by yellow rust and powdery mildew. Z Pflanzenzucht. 1982;91:275–94.

Pontier D, Balague C, Roby D. The hypersensitive response. A programmed cell death associated with plant resistance. C R Acad Sci III. 1998;321:721–34.

Priestley RH. Detection of increased virulence in populations of wheat yellow rust. In: Scott PR, Bainbridge A, editors. Plant Disease Epidemiology. Oxford: Blackwell Scientific Publishers; 1978. p. 63–70.

Priestley RH, Byford P. Yellow rust of wheat. UK cereal pathogen virulence survey. 1978 annual report, National Institute of Agricultural Botany. 1979. p. 14–23.

Priestley RH, Bayles RA, Crofts J. Yellow rust of wheat. UK cereal pathogen virulence survey. 1981 annual report, National Institute of Agricultural Botany. 1982. p. 18–28.

Prins R, Pretorius ZA, Bender CM, Lehmensiek A. QTL mapping of stripe, leaf and stem rust resistance genes in a Kariega x Avocet S doubled haploid wheat population. Mol Breed. 2011;27:259–70.

Pu ZJ, Yan ZH, Wei YM, Yang WY, Zheng YL. Genetic analysis of stripe rust resistance in wheat line P81. Acta Phytophyl Sin. 2007;34:511–4.

Pu ZJ, Chen GY, Wei YM, Han ZH, Zheng YL. Identification and molecular tagging of a stripe rust resistance gene in wheat line P81. Plant Breed. 2010;129:53–7.

Qamar M, Ahmad SD, Shan AH, Wellings CR, Batool F. Postulation of stripe rust resistance gene in some Australia bread wheat cultivars and their response to temperature. Pak J Bot. 2008;40:2573–85.

Qayoum A, Line RF. High-temperature, adult-plant resistance to stripe rust of wheat. Phytopathology. 1985;75:1121–5.

Qie YM, Wang MN, Li X, Chen XM. Developing a wheat germplasm with linked genes Yr64 and Yr65 for resistance to stripe rust. Phytopathology. 2016;106(S4):204.

Qiu JW, Schürch AC, Yahiaoui N, Dong LL, Fan HJ, Zhang ZJ, Keller B, Ling HQ. Physical mapping and identification of a candidate for the leaf rust resistance gene *Lr1* of wheat. Theor Appl Genet. 2007;115:159–68.

Quan W, Hou GL, Chen J, Du ZY, Lin F, Guo Y, Liu S, Zhang ZJ. Mapping of QTL lengthening the latent period of *Puccinia striiformis* in winter wheat at the tillering growth stage. Eur J Plant Pathol. 2013;136:715–27.

Rajaram S, Singh RP, Torres E. Current CIMMYT approaches in breeding wheat for rust resistance. In: Simmonds NW, Rajaram S, editors. Breeding Strategies for Resistance to the Rusts of Wheat. Mexico, D.F.: CIMMYT; 1988. p. 101–8.

Ramburan VP, Pretorius ZA, Louw JH, Boyd LA, Smith PH, Boshoff WHP, Prins R. A genetic analysis of adult plant resistance to stripe rust in the wheat cultivar Kariega. Theor Appl Genet. 2004;108:1426–33.

Randhawa HS, Mutti JS, Kidwell K, Morris C, Chen XM, Gill KS. Rapid and targeted introgression of genes into popular cultivars using marker-assisted background selection. PLoS ONE. 2009;4(6):e5752.

Randhawa MS, Bansal U, Valárik M, Klocová B, Doleźel J, Bariana H. Molecular mapping of stripe rust resistance gene *Yr51* in chromosome 4AL of wheat. Theor Appl Genet. 2014;127:317–24.

Randhawa MS, Bariana HS, Mago R, Bansal UK. Mapping of a new stripe rust resistance locus *Yr57* on chromosome 3BS of wheat. Mol Breed. 2015;35:65.

Rao MVP. Close linkage of the *Agropyron elongatum* gene *Sr26* for stem rust resistance to the centromere of wheat chromosome 6A. Wheat Inf Serv. 1996;82:8–10.

Ren ZL, Zhang H, Wang KF, Wang YJ, Cai DM, Ji WQ, Song YL. Development of wheat germplasm with disease resistance: Yuanfeng 139. China Agric Sci Bull. 2006;22:228–31.

Ren RS, Wang MN, Chen XM, Zhang ZJ. Characterization and molecular mapping of *Yr52* for high-temperature adult-plant resistance to stripe rust in spring wheat germplasm PI 183527. Theor Appl Genet. 2012a;125:847–57.

Ren Y, He ZH, Li J, Lillemo M, Wu L, Bai B, Lu QX, Zhu HZ, Zhou G, Du JY, Lu QL, Xia XC. QTL mapping of adult-plant resistance to stripe rust in a population derived from common wheat cultivars Naxos and Shanghai 3/Catbird. Theor Appl Genet. 2012b;125:1211–21.

Ren Y, Li ZF, He ZH, Wu L, Bai B, Lan CX, Wang CF, Zhou G, Zhu HZ, Xia XC. QTL mapping of adult-plant resistances to stripe rust and leaf rust in Chinese wheat cultivar Bainong 64. Theor Appl Genet. 2012c;125:1253–62.

Ren Y, Li SR, Wei YM, Zhou Q, Du XY, He YJ, Zheng YL. Molecular mapping of a stripe rust resistance gene in Chinese wheat cultivar Mianmai 41. J Integr Agric. 2015a;14:295–304.

Ren Y, Li SR, Xia XC, Zhou Q, He YJ, Wei YM, Zheng YL, He ZH. Molecular mapping a recessive stripe rust resistance gene *yrMY37* in Chinese wheat cultivar Mianmai 37. Mol Breed. 2015b;35:97.

Ren Y, Liu LS, He ZH, Wu L, Bai B, Xia XC. QTL mapping of adult-plant resistance to stripe rust in a 'Lumai 21 × Jingshuang 16' wheat population. Plant Breed. 2015c;134:501–7.

Richardson KL, Vales MI, Kling JG, Mundt CC, Hayes PM. Pyramiding and dissecting disease resistance QTL to barley strip rust. Theor Appl Genet. 2006;113:485–95.

Riley R, Chapman V, Johnson R. Introduction of yellow rust resistance of *Aegilops comosa* into wheat by genetically induced homoeologous recombination. Nature. 1968;217:383–4.

Risk JM, Selter LL, Chauhan H, Krattinger SG, Kumlehn J, Hensel G, Viccars LA, Richardson TM, Buesing G, Troller A, Lagudah ES, Keller B. The wheat *Lr34* gene provides resistance against multiple fungal pathogens in barley. Plant Biotechnol J. 2013;11:847–54.

Röbbelen G, Sharp EL. Mode of inheritance, interaction and application of genes conditioning resistance to yellow rust. In: Hannover HW, Röbbelen GG (eds) Advances in plant breeding, Supplement 9 to J Plant Breed. 1978. p. 1–88.

Robert O, Abelard C, Dedryver F. Identification of molecular markers for the detection of the yellow rust resistance gene *Yr17* in wheat. Mol Breed. 1999;5:167–75.

Rockefeller Foundation. Program in Agr Sci, Ann Rpt, 1962–1963, 310 pp. (1963).

Rockefeller Foundation. Program in Agr Sci, Ann Rpt, 1963–1964, 285 pp. (1964).

Rockefeller Foundation. Program in Agr Sci, Ann Rpt, 1964–1965, 262 pp. (1965).

Röder MS, Korzun V, Wendehake K, Plaschke J, Tixier M-H, Leroy P, Ganal MW. A microsatellite map of wheat. Genetics. 1998;149:2007–23.

Roelfs AP, Bushnell WR. The Cereal Rusts II: Disease, Distribution, Epidemiology, and Control. New York: Academic Press; 1985.

Roelfs AP, Huerta-Espino J, Marshall D. Barley stripe rust in Texas. Plant Dis. 1992;76:538.

Rosewarne GM, Singh RP, Huerta-Espino J, William HM, Bouchet S, Cloutier S, McFadden H, Lagudah ES. Leaf tip necrosis, molecular markers and *beta*1-proteasome subunits associated with the slow rusting resistance genes *Lr46/Yr29*. Theor Appl Genet. 2006;112:500–8.

Rosewarne GM, Singh RP, Huerta-Espino J, Rebetzke GJ. Quantitative trait loci for slow-rusting resistance in wheat to leaf rust and stripe rust identified with multi-environment analysis. Theor Appl Genet. 2008;116:1027–34.

Rosewarne GM, Singh RP, Huerta-Espino J, Herrera-Foessel SA, Forrest KL, Hayden MJ, Rebetzke GJ. Analysis of leaf and stripe rust severities reveals pathotype changes and multiple minor QTLs associated with resistance in an Avocet × Pastor wheat population. Theor Appl Genet. 2012;124:1283–94.

Rosewarne GM, Herrera-Foessel SA, Singh RP, Huerta-Espino J, Lan CX, He ZH. Quantitative trait loci of stripe rust resistance in wheat. Theor Appl Genet. 2013;126:2427–49.

Rubiales D, Niks RE. Characterization of *Lr34*, a major gene conferring nonhypersensitive resistance to wheat leaf rust. Plant Dis. 1995;79:1208–12.

Rubiales D, Niks RE. Avoidance of rust infection by some genotypes of *Hordeum chilense* due to their relative inability to induce the formation of appressoria. Physiol Mol Plant Pathol. 1996;49:89–101.

Russell GE. Characterization of adult plant resistance to yellow rust in wheat. In: Proc 4th Eur Meditt Cereal Rust Conf. Interlaken. 1976a;1976:21–3.

Russell GE. Germination of Puccinia uredospores on leaves of adult winter wheat plants. Ann Appl Biol. 1976b;82:71–8.

Rustgi S, Bandopadhyay R, Balyan HS, Gupta PK. EST-SNPs in bread wheat: discovery, validation, genotyping and haplotype structure. Czech J Plant Breed. 2009;45:106–16.

Saintenac C, Zhang W, Salcedo A, Rouse MN, Trick HN, Akhunov E, Dubcovsky J. Identification of wheat gene *Sr35* that confers resistance to Ug99 stem rust race group. Science. 2013;341:783–6.

Samborski DJ. Wheat leaf rust. In: Roelfs AP, Bushnell WR, editors. The Cereal Rusts, vol. 2. Orlando: Academic Press; 1985. p. 39–59.

Sandhu TS, Singh G. Inheritance of field reaction to yellow and brown rusts in crosses of common wheat. Punjab Agricultural University, J Research. 1970;7:5–9.
Santra DK, Chen XM, Santra M, Garland-Campbell KA, Kidwell KK. Identification and mapping QTL for high-temperature adult-plant resistance to stripe rust in winter wheat (*Triticum aestivum* L.) cultivar "Stephens". Theor Appl Genet. 2008;117:793–802.
Sawhney RN, Bakshi JS. Inheritance to three yellow rust races in a Kenya variety of wheat. India J Genet Plant Breed. 1965;25:231–3.
Sawhney RN, Luthra JK. New resistance genes of wheat to Indian races of stripe rust (*Puccinia striiformis*). SABRAO Newsletter, Mishima. 1970;2:155–6.
Schafer JF, Roelfs AP. Estimated relation between numbers of urediniospores of *Puccinia graminis* f. sp. *tritici* and rates of occurrence of virulence. Phytopathology. 1985;75:749–50.
Schlegel R, Korzun V. About the origin of 1RS.1BL wheat-rye chromosome translocation from German. Plant Breed. 1997;116:537–40.
Shao YT, Niu YC, Zhu LH, Zhai WX, Xu SC, Wu LR. Identification of an AFLP markers linked to the stripe rust resistance gene *Yr10* in wheat. Chin Sci Bull. 2001;46:1466–8.
Sharma SK, Joshi LM, Singh SD, Nagarajan S. New virulence of yellow rust on Kalyansona variety of wheat. Proc Europ Medit Cereal Rust Conf, Prague. 1972;1:263–6.
Sharma S, Louwers JM, Karki CB, Snijders CHA. Postulation of resistance genes to yellow rust in wild emmer wheat derivatives and advanced wheat lines from Nepal. Euphytica. 1995;81:271–7.
Sharma-Poudyal D, Chen XM, Wan AM, Zhan GM, Kang ZS, Cao SQ, Jin SL, Morgounov A, Akin B, Mert Z, Shah SJA, Bux H, Ashraf M, Sharma RC, Madariaga R, Puri KD, Wellings CR, Xi KQ, Manninger K, Wanyera R, Ganzalez MI, Koyda M, Sanin S, Patzek LJ. Viruelnce characterization of international collections of the wheat stripe rust pathogen, *Puccinia striiformis* f. sp. *tritici*. Plant Dis. 2013;97:379–86.
Sharp EL. Prepenetration and postpenetration environment and development of *Puccinia striiformis* on wheat. Phytopathology. 1965;55:198–203.
Sharp EL, Volin RB. Additive genes in wheat conditioning resistance to stripe rust. Phytopathology. 1970;60:1146–7.
Shi ZX, Chen XM, Line RF, Leung H, Wellings CR. Development of resistance gene analog polymorphism markers for the *Yr9* gene resistance to wheat stripe rust. Genome. 2001;44:509–16.
Sikka SM, Rao MV, Ahluwalia M. Inheritance studies in wheat. X. Inheritance of field reaction to rusts and other characters. Indian J Agric Sci. 1960;30:223–32.
Singh RP. Genetic association of leaf rust resistance gene *Lr34* with adult plant resistance to stripe rust in bread wheat. Phytopathology. 1992;82:835–8.
Singh G, Dhillon PS. Inheritance of yellow rust resistance in wheat. Indian J Agr. 1963;7:65–71.
Singh H, Johnson R. Genetics of resistance to yellow rust in Heines VII, Soissonais and Kalyansona. In: Miller TE, Koebner RMD, editors. Proc Seventh Intl Wheat Genetics Symposium. Cambridge: UK; 1988. p. 885–90.
Singh RP, Nelson JC, Sorrells ME. Mapping *Yr28* and other genes for resistance to stripe rust in wheat. Crop Sci. 2000;40:1148–55.
Singh RP, Huerto-Espino J, William MH. Slow rusting gene based resistance to leaf and yellow rusts in wheat: genetics and breeding at CIMMYT. In: Proceedings of the 10th assembly of the Wheat Breeding Society of Australia Inc., Mildura, Australia, 16–21 September, 2001. Australia: Wheat Breeding Society of Australia Inc. 2001. p. 103–8.
Singh RP, William HM, Huerta-Espino J, Crosby M. Identification and mapping of gene *Yr31* for resistance to stripe rust in *Triticum aestivum* cultivar Pastor. In: Pogna NE, Romano N, Pogna EA, Galterio G (eds) Proceedings of the 10th international wheat genetics symposium, Instituto Sperimentale per la Cerealcoltura, Roma, Italy, vol 1. 2003. p. 411–3.
Singh RP, Huerta-Espino J, William HM. Genetics and breeding for durable resistance to leaf and stripe rusts in wheat. Turk J Agric For. 2005;29:121–7.

Singh RP, Hodson DP, Huerta-Espino J, Jin Y, Bhavani S, Njau P, Herrera-Foessel S, Singh PK, Singh S, Govindan V. The emergence of Ug99 races of the stem rust fungus is a threat to world wheat production. Annu Rev Phytopathol. 2011;49:465–81.

Singh A, Pandey MP, Singh AK, Knox RE, Ammar K, Clarke JM, Clarke FR, Singh RP, Pozniak CJ, DePauw RM, McCallum BD, Cuthbert RD, Randhawa HS, Fetch TG. Identification and mapping of leaf, stem and stripe rust resistance quantitative trait loci and their interactions in durum wheat. Mol Breed. 2013;31:405–18.

Singh VK, Mathuria RC, Singh DP, Aggarwal R. Characterization of yellow rust resistance genes by using gene postulation and assessment of adult plant resistance in some Indian wheat genotypes. Res Crops. 2015;16:741–50.

Slovencikova V. Genetics of field resistance of the winter wheats Benno, Zorba, Orlando and Saladin to race 232 E137 ("Clement" race) of *Puccinia striiformis*. Cereal Rusts Bulletin. 1980;8:12–6.

Smith PH, Howie JA, Worland AJ, Stratford R, Boyd LA. Mutations in wheat exhibiting growth-stage-specific resistance to biotrophic fungal pathogens. Mol Plant-Microbe Interact. 2004;17:1242–9.

Smith PH, Hadfield J, Hart NJ, Koebner RMD, Boyd LA. STS markers for the wheat yellow rust resistance gene *Yr5* suggest a NBS-LRR-type resistance gene cluster. Genome. 2007;50:259–65.

Somers DJ, Kirkpatrick R, Moniwa M, Walsh A. Mining single-nucleotide polymorphisms from hexaploid wheat ESTs. Genome. 2003;49:431–7.

Somers DJ, Isaac P, Edwards K. A high-density microsatellite consensus map for bread wheat (*Triticum aestivum* L.). Theor Appl Genet. 2004;109:1105–14.

Song QJ, Shi JR, Singh S, Fickus EW, Costa JM, Lewis J, Gill BS, Ward R, Cregan PB. Development and mapping of microsatellite (SSR) markers in wheat. Theor Appl Genet. 2005;110:550–60.

Sourdille P, Singh S, Cadalen T, Brown-Guedira BL, Gay G, Qi L, Gill BS, Dufour P, Murigneux A, Bernard M. Microsatellite-based deletion bin system for the establishment of genetic-physic map relationship in wheat (*Triticum aestivum* L.). Func Integ Genomics. 2004;4:12–25.

Spielmeyer W, Sharp PJ, Lagudah ES. Identification and validation of markers linked to broad-spectrum stem rust resistance gene *Sr2* in wheat (*Triticum aestivum* L.). Crop Sci. 2003;43:333–6.

Spielmeyer W, McIntosh RA, Kolmer J, Lagudah ES. Powdery mildew resistance and *Lr34/Yr18* genes for durable resistance to leaf and stripe rust cosegregate at a locus on the short arm of chromosome 7D of wheat. Theor Appl Genet. 2005;111:731–5.

Srinivasan VK, Padmanabhan TS. Inheritance of field reaction to yellow rust in barley. Indian J Genet Plant Breed. 1964;24:180–2.

Stakman EC. Relation between *Puccinia graminis* and plants highly resistance to its attack. J. Agric Res. 1915;4:193–9.

Stein N, Graner A. Map-based gene isolation in cereal genomes. In:Cereal genomics. Dordrecht: Springer; 2004. p. 331–60.

Stein N, Feuillet C, Wicker T, et al. Subgenome chromosome walking in wheat: a 450-kb physical contig in *Triticum monococcum* L. spans the *Lr10* resistance locus in hexaploid wheat (*Triticum aestivum* L.). Proc Natl Acad Sci. 2000;97:13436–41.

Steuernagel B, Periyannan SK, Hernández-Pinzón I, Witek K, Rouse MN, Yu G, Hatta A, Ayliffe M, Bariana H, Jones JGD, Lagudah ES, Wulff BBH. Rapid cloning of disease-resistance genes in plants using mutagenesis and sequence capture. Nature Biotechnol. 2016;34:652–5.

Sthapit J, Newcomb M, Bonman JM, Chen XM, See DR. Genetic diversity for stripe rust resistance in wheat landraces and identification of accessions with resistance to stem rust and stripe rust. Crop Sci. 2014;54:2131–9.

Straib W. Untersuchungen zur Genetik der Gelbrostresistenz des Weizens. Pthytopath. Z. 1939;7:427–77.

Stubbs RW. Recent aspects of the physiological specialization of yellow rust in the Netherlands. In: Macer RCF, Wolfe MS, editors. Proc Third Eur Yellow Rust Conf. Cambridge, UK: Plant Breeding Institute; 1966. p. 47–54.

Stubbs RW. Observations on horizontal resistance to yellow rust (*Puccinia striiformis* f. sp. *tritici*). Cereal Rust Bull. 1977;5:27–32.

Stubbs RW. Stripe rust. In: Roelfs AP, Bushnell WR, editors. The cereal rusts II. Orlando: Academic; 1985. p. 61–101.

Stubbs RW, Sanders M, Zeven AC. A recessive resistance gene for yellow rust (*Puccinia striiformis* West.) in bread wheat (*Triticum aestivum* L.). Euphytica. 1984;33:561–2.

Sucher J, Boni R, Yang P, Rogowsky P, Buchner H, Kastner C, Kumlehn J, Krattinger SG, Keller B. The durable wheat disease resistance gene *Lr34* confers common rust and northern corn leaf blight resistance in maize. Plant Biotechnol J pp. 2016:1–8.

Suenaga K, Singh RP, Huerta-Espino J, William HM. Microsatellite markers for genes *Lr34/Yr18* and other quantitative trait loci for leaf rust and stripe rust resistance in bread wheat. Phytopathology. 2003;93:881–90.

Sui XX, Wang MN, Chen XM. Molecular mapping of a stripe rust resistance gene in spring wheat cultivar Zak. Phytopathology. 2009;99:1209–15.

Sui XX, He ZH, Lu YM, Wang ZL, Xia XC. Molecular mapping of a non-host resistance gene *YrpstYr1* in barley (*Hordeum vulgare* L.) for resistance to wheat stripe rust. Hereditas. 2010;147:176–82.

Sun GL, Fahima T, Korol AB, Turpeinen T, Grama A, Ronin YI, Nevo E. Identification of molecular markers linked to the *Yr15* stripe rust resistance gene of wheat originated in wild emmer wheat *Triticum dicoccoides*. Theor Appl Genet. 1997;95:622–8.

Sun Q, Wei Y, Ni C, Xie C, Yang T. Microsatellite marker for yellow rust resistance gene *Yr5* introgressed from spelt wheat. Plant Breed. 2002;121:539–41.

Tang ZX, Ren ZL, Wu F, Fu SL, Wang XX, Zhang HQ. The selection of transgenic recipients from new elite wheat cultivars and study on its plant regeneration system. Agric Sci China. 2006;5:417–24.

The TT, Nevo E, McIntosh RA. Response of Israeli wild emmer to selected Australian pathotypes of Puccinia species. Euphytica. 1993;71:75–81.

Thomas WTB, Powell W, Waugh R, Chalmers KJ, Barua UM, Jack P, Lea V, Forster BP, Swanston JS, Ellis RP, Hanson PR, Lance RCM. Detection of quantitative trait loci for agronomic, yield, grain and disease characters in spring barley (*Hordeum vulgare* L.). Theor Appl Genet. 1995;91:1037–47.

Tian YE, Huang J, Li Q, Li GB, Wang BT. Inheritance and SSR mapping of a stripe rust resistance gene *YrH122* derived from *Psathyrostachys huashanica* Keng. Acta Phytopathol Sin. 2011;41:64–71.

Tian Y, Zhan GM, Chen XM, Tungruentragoon A, Lu X, Zhao J, Huang LL, Kang ZS. Virulence and SSR marker segregation in a *Puccinia striiformis* f. sp. *tritici* population produced by selfing a Chinese isolate on *Berberis shensiana*. Phytopathology. 2016;106:185–91.

Toojinda T, Baird E, Booth A, Broers L, Hayes P, Powell W, Thomas W, Vivar H, Young G. Introgression of quantitative trait loci (QTL) determining stripe rust resistance in barley: an example of marker-assisted line development. Theor Appl Genet. 1998;96:123–31.

Toojinda T, Broers LH, Chen XM, Hayes PM, Kleinhofs A, Korte J, Kudrna D, Leung H, Line RF, Powell W, Ramsay L. Mapping quantitative and qualitative disease resistance genes in a doubled haploid population of barley (*Hordeum vulgare*). Theor Appl Genet. 2000;101:580–9.

Uauy C, Brevis JC, Chen XM, Khan I, Jackson L, Chicaiza O, Distenfeld A, Fahima T, Dubcovsky J. High-temperature adult-plant stripe rust resistance gene *Yr36* from *Triticum turgidum* ssp. *dicoccoides* is closely linked to the grain protein content locus *Gpc-B1*. Theor Appl Genet. 2005;112:97–105.

Upadhyaya YM, Srinivasan VK, Murty BN. Inheritance of yellow rust and leaf blotch resistance and their association with linkage groups in barley. Indian J Genet Plant Breed. 1965;25:208–16.

Vales MI, Schön CC, Capettini F, Chen XM, Corey AE, Mather DE, Mundt CC, Richardson KL, Sandoval-Islas JS, Utz HF, Hayes PM. Effect of population size on the estimation of QTL: a test using resistance to barley stripe rust. Theor Appl Genet. 2005;111:1260–70.

Van Dijik P, Parlevliet JE, Kema G, Stubbs RW. Characterization of the durable resistance to yellow rust in old winter cultivars in the Netherlands. Euphytica. 1988;38:149–58.

Van Ooijen G, van den Burg HA, Cornelissen BJ, Takken FL. Structure and function of resistance proteins in solanaceous plants. Annu Rev Phytopathol. 2007;45:43–72.

Van Silfhout CH. Identification and characterization of resistance to yellow rust and powdery mildew in wild emmer wheat and their transfer to bread wheat. PhD thesis, Research Institute for Plant Protection, Wageningen, The Netherlands. 1989.

Vanderplank JE. Plant Diseases: Epidemics and Control. New York and London: Academic Press; 1963. 349 pp.

Vazquez MD, Peterson CJ, Riera-Lizarazu O, Chen X, Heesacker A, Mundt C. Genetic analysis of adult plant, quantitative resistance to stripe rust in wheat cultivar 'Stephens' in multi-environment trials. Theor Appl Genet. 2012;124:1–11.

Vazquez MD, Zemetra T, Peterson CJ, Chen XM, Heesacker A, Mundt CC. Multi-location wheat stripe rust QTL analysis: genetic background and epistatic interactions. Theor Appl Genet. 2015;128:1307–218.

von Wettstein-Knowles P. Cloned and mapped genes: Current status. In: Shewry PR, editor. Barley: Genetics, Biochemistry, Molecular Biology, and Biotechnology. Wallingord, UK: CAB International; 1992. p. 73–98.

Vos P, Hogers R, Bleeker M, Reijans M, van de Lee T, Hornes M, Frijters A, Pot J, Peleman J, Kuiper M. AFLP: a new technique for DNA fingerprinting. Nucleic Acids Res. 1995;23:4407–14.

Walther U, Herdam H. On the inheritance of field resistance to stripe rust (*Puccinia striiformis* West.) in some wheat cultivars. In: Proceedings of the fifth European Mediterranean cereal rusts conference. 1984. p. 107–10.

Wan AM, Chen XM. Virulence, frequency, and distribution of races of *Puccinia striiformis* f. sp. *tritici* and *P. striiformis* f. sp. *hordei* identified in the United States in 2008 and 2009. Plant Dis. 2012;96:67–74.

Wan AM, Chen XM. Virulence characterization of *Puccinia striiformis* f. sp. *tritici* using a new set of *Yr* single-gene line differentials in the United States in 2010. Plant Dis. 2014;98:1534–42.

Wan AM, Zhao ZH, Chen XM, He ZH, Jin SL, Jia QZ, Yao G, Yang JX, Wang BT, Li GB, Bi YQ, Yuan ZY. Wheat stripe rust epidemic and virulence of *Puccinia striiformis* f. sp. *tritici* in China in 2002. Plant Dis. 2004;88:896–904.

Wan AM, Chen XM, He ZH. Wheat stripe rust in China. Aus J Agric Res. 2007;58:605–19.

Wan AM, Chen XM, Yuen J. Races of *Puccinia striiformis* f. sp. *tritici* in the United States in 2011 and 2012 and comparison with races in 2010. Plant Dis. 2016;100:966–75.

Wan AM, Muleta KT, Zegeye H, Hundie B, Pumphrey MO, Chen XM. Virulence characterization of wheat stripe rust fungus *Puccinia striiformis* f. sp. *tritici* in Ethiopia and evaluation of Ethiopian wheat germplasm for resistance to races of the pathogen from Ethiopia and the United States. Plant Dis. 2017;101:73–80.

Wang MN, Chen XM. Pyramiding stripe rust resistance genes on wheat chromosomes 2B, 4B and 7B. Phytopathology. 2016;S4:207.

Wang YB, Xu SC, Xu Z, Liu TG, Lin RM. A microsatellite marker inked to the stripe rust resistance gene *YrV23* in the wheat variety Vilmorin23. Hereditas (Beijing). 2006;28:306–10.

Wang CM, Zhang YP, Han DJ, Kang ZS, Li GP, Cao AH, Chen PD. SSR and STS markers for wheat stripe rust resistance gene *Yr26*. Euphytica. 2008a;159:359–66.

Wang MN, Coram T, Chen XM, Boyd L, Ling P. High-resolution genetic and physical mapping of the *Yr5* gene for resistance to stripe rust of wheat. Phytopathology. 2008b;98:S166.

Wang LM, Zhang ZY, Liu HJ, He MZ, Liu HX, Veisz O, Xin ZY. Identification, gene postulation and molecular tagging of a stripe rust resistance gene in synthetic wheat CI142. Cereal Res Commun. 2009;37:209–15.

Wang MN, Chen XM, Xu LS, Cheng P, Bockelman HE. Registration of 70 common spring wheat germplasm lines resistance to stripe rust. J Plant Reg. 2012;6:104–10.

Wang SC, Wong D, Forrest K, Allen A, Chao SM, Huang BE, Maccaferri M, Salvi S, Milner SG, Cattivelli L, Mastrangelo AM, Whan A, Stephen S, Barker G, Wieseke R, Plieske J, International Wheat Genome Sequencing Consortium, Lillemo M, Mather D, Appels R, Dolferus R, Brown-Guedira G, Korol A, Akhunova AR, Feuillet C, Salse J, Morgante M, Pozniak C, Luo MC, Dvorak J, Morell M, Dubcovsky J, Ganal M, Tuberosa R, Lawley C, Mikoulitch I, Cavanagh C, Edwards KJ, Hayden M, Akhunov E. Characterization of polyploid wheat genomic diversity using a high-density 90,000 single nucleotide polymorphism array. Plant Biotechnol J. 2014a;12:787–96.

Wang YP, Cheng X, Shan QW, Zhang Y, Liu JX, Gao CX, Qiu JL. Simultaneous editing of three homoeoalleles in hexaploid bread wheat confers heritable resistance to powdery mildew. Nature Biotechnol. 2014b;32:947–51.

Wang MN, Wan AM, Chen XM. Barberry as alternate host is important for *Puccinia graminis* f. sp. *tritici* but not for *Puccinia striiformis* f. sp. *tritici* in the U. S. Pacific Northwest. Plant Dis. 2015;99:1507–16.

Wellings CR. Host: pathogen studies of wheat stripe rust in Australia. PhD thesis, University of Sydney. 1986.

Wellings CR, Burdon JJ. Variability in *Puccinia striiformis* f.sp. *tritici* in Australasia. Vorträge für Pflanzenzüchtung. 1992;24:114.

Wellings CR, McIntosh RA. *Puccinia striiformis* f. sp. *tritici* in Australasia: pathogenic changes during the first 10 years. Plant Pathol. 1990;39:316–25.

Wellings CR, McIntosh RA, Hussain M. A new source of resistance to *Puccinia striiformis* f. sp. *tritici* in spring wheats (*Triticum aestivum*). Plant Breed. 1988;100:88–96.

Wellings CR, Singh RP, McIntosh RA, Pretorius ZA. The development and application of near isogenic lines for the stripe (yellow) rust pathosystem. In: Proc 11th Intl Cereal Rusts and Powdery Mildew Conf. England: Norwich; 2004. p. A1.39.

Wellings CR, Singh RP, Yahyaoui A, Nazari K, McIntosh RA. The development and application of near-isogenic lines for monitoring cereal rust pathogens. In: McIntosh RA, editor. Proc Borlaug Global Rust Initiative Technical Workshop. Mexico: BGRI Cd Obregon; 2009. p. 77–87.

Wellings CR, Boyd LA, Chen XM. Resistance to stripe rust in wheat: pathogen biology driving resistance breeding. In: Sharma I, editor. Disease Resistance in Wheat, CAB International, vol. 2012; 2012. p. 63–83.

Weng DX, Xu SC, Lin RM, Wan AM, Li JP, Wu LR. Microsatellite marker linked with stripe rust resistant gene *Yr9* in wheat. Acta Genet Sin. 2005;32:937–41.

Wenzl P, Suchánková P, Carling J, Šimková H, Huttner E, Kubaláková M, Sourdille P, Paul E, Feuillet C, Kilian A, Doležel J. Isolated chromosomes as a new and efficient source of DArT markers for the saturation of genetic maps. Theor Appl Genet. 2010;121:465–74.

Wesenberg DM, Burrup DE, Brown Jr WM, Velasco VR, Hill JP, Whitmore JC, Karow RS, Hayes PM. Registration of 'Bancroft' barley. Crop Sci. 2001;41:265–6.

Wheeler H, Diachun S. Mechanisms of pathogenesis. In: Kommedahl T, Williams PH, editors. Challenging Problems in Plant Health. St. Paul: American Phytopathological Society; 1983. p. 324–33.

Wilkinson RE, Roberts JJ. Methods for inhibiting rust infections of plants. US Patents US5198011 A. 1983.

William M, Singh RP, Huerta-Espino J, Islas SO, Hoisington D. Molecular marker mapping of leaf rust resistance gene *Lr46* and its association with stripe rust resistance gene *Yr29* in wheat. Phytopathology. 2003;93:153–9.

William HM, Singh RP, Huerta-Espino J, Palacios G, Suenaga K. Characterization of genetic loci conferring adult plant resistance to leaf rust and stripe rust in spring wheat. Genome. 2006;49:977–90.

Williams JG, Kubelik AR, Livak KJ, Rafalski JA, Tingey SV. DNA polymorphisms amplified by arbitrary primers are useful as genetic markers. Nucleic Acids Res. 1990;18:6531–5.

Wolfe MS. The current status and prospects of multiline cultivars and variety mixtures for disease control. Annu Rev Phytopathology. 1985;23:251–73.

Worland AJ, Law CN. Genetic analysis of chromosome 2D of wheat I. The location of genes affecting height, day-length insensitivity, hybrid dwarfism and yellow-rust resistance. Zeitschrift fur Pflanzenzuchtung. 1986;96:331–45.

Worland AJ, Petrovic S, Law CN. Genetic analysis of chromosome 2D of wheat. II. The importance of this chromosome to Yugoslavian varieties. Plant Breed. 1988;100:247–59.

Wu L, Xia XC, Rosewarne GM, Zhu HZ, Li SZ, Zhang ZY, He ZH. Stripe rust resistance gene *Yr18* and its suppressor gene in Chinese wheat landraces. Plant Breed. 2015;134:634–40.

Wu XL, Wang JW, Cheng YK, Ye XL, Li W, Pu ZE, Jiang QT, Wei YM, Deng M, Zheng YL, Chen GY. Inheritance and molecular mapping of an all-stage stripe rust resistance gene derived from the Chinese common wheat land race 'Yilongtuomai'. J Heredity. 2016;107:463–70.

Xia C, Wan A, Wang M, Jiwan DA, See DR, Chen X. Secreted protein gene derived-single nucleotide polymorphisms (SP-SNPs) reveal population diversity and differentiation of *Puccinia striiformis* f. sp. *tritici* in the United States. Fungal Biol. 2016a;120:729–44.

Xia C, Wang M, Wan A, Jiwan DA, See DR, Chen X. Association analysis of SP-SNPs and avirulence Genes in *Puccinia striiformis* f. sp. *tritici*, the wheat stripe rust pathogen. Am J. Plant Sci. 2016b;7:126.

Xiang C, Feng JY, Wang MN, Chen XM, See DR, Wan AM, Wang T. Molecular mapping of stripe rust resistance gene *Yr76* in winter club wheat cultivar Tyee. Phytopathology. 2016;106:1186–93.

Xu SC, Wu LR, Wan AM, Wang FL, Niu YC, Zhao WS. Advances in constructions of isogenic lines for wheat stripe rust. In: Proc 15th Intl Plant Protect Cong. Beijing, China: Foreign Languages Press; 2004. p. 60.

Xu LS, Wang MN, Cheng P, Kang ZS, Hulbert SH, Chen XM. Molecular mapping of *Yr53*, a new gene for stripe rust resistance in durum wheat accession PI480148 and its transfer to common wheat. Theor Appl Genet. 2013;126:523–33.

Xu HX, Zhang J, Zhang P, Qie YM, Niu YC, Li HJ, Ma PT, Xu YF, Diaoguo A. Development and validation of molecular markers closely linked to the wheat stipe rust resistance *YrC591* for marker-assisted selection. Euphytica. 2014;198:317–23.

Yahiaoui N, Srichumpa P, Dudler R, Keller B. Genome analysis at different ploidy levels allows cloning the powdery mildew resistance gene *Pm3* from hexaploid wheat. Plant J. 2004;37:528–38.

Yan GP, Chen XM. Molecular mapping of a recessive gene for resistance to stripe rust in barley. Theor Appl Genet. 2006;113:529–37.

Yan GP, Chen XM. Molecular mapping of the *rps1.a* recessive gene for resistance to stripe rust in BBA 2890 barley. Phytopathology. 2007;97:668–73.

Yan GP, Chen XM. Identification of a quantitative trait locus for high-temperature adult-plant resistance against *Puccinia striiformis* f.sp. *hordei* in 'Bancroft' barley. Phytopathology. 2008;98:120–7.

Yan GP, Chen XM, Line RF, Wellings CR. Resistance gene-analog polymorphism markers co-segregating with the *Yr5* gene for resistance to wheat stripe rust. Theor Appl Genet. 2003;106:636–43.

Yan BJ, Zhang HQ, Ren ZL. Molecular cytogenetic identification of a new 1RS/1BL translocation line with secalin absence. Hereditas (Beijing). 2005;27:513–7.

Yang ZJ, Li GR, Jia JQ, Zeng T, Lei MP, Zeng ZX, Tao Z, Ren ZL. Molecular cytogenetic characterization of wheat-*Secale africanum* amphiploids and derived introgression lines with stripe rust resistance. Euphytica. 2009;167:197–202.

Yang EN, Rosewarne GM, Herrera-Foessel SA, Huerta-Espino J, Tang ZX, Sun CF, Ren ZL, Singh RP. QTL analysis of the spring wheat "Chapio" identifies stable stripe rust resistance despite inter-continental genotype × environment interactions. Theor Appl Genet. 2013;126:1721–32.

Yang YH, Zhao J, Xing HJ, Wang JY, Zhou K, Zhan GM, Zhang HC, Kang ZS. Different non-host resistance responses of two rice subspecies, japonica and indica, to *Puccinia striiformis* f. sp *tritici*. Plant Cell Rep. 2014;33:423–33.

Yang EN, Li GR, Li LP, Zhang ZY, Yang WY, Peng YL, Zhu YQ, Yang ZJ, Rosewarne GM. Characterization of stripe rust resistance genes in the wheat cultivar Chuanmai 45. Int J Mol Sci. 2016;17:601.

Yaniv E, Raats D, Ronin Y, Korol AB, Grama A, Bariana H, Dubcovsky J, Schulman AH, Fahima T. Evaluation of marker-assisted selection for the stripe rust resistance gene *Yr15*, introgressed from wild emmer wheat. Mol Breed. 2015;35:43.

Yao ZJ, Lin RM, Xu SC, Li ZF, Wan AM, Ma ZY. The molecular tagging of the yellow rust resistance gene *Yr7* in wheat transferred from differential host Lee using microsatellite markers. Sci Agric Sin. 2006;39:1146–52.

Yildirim A, Jones SS, Murray TD, Cox TS, Line RF. Resistance to stripe rust and eyespot disease of wheat in *Triticum tauschii*. Plant Dis. 1995;79:1230–6.

Yin XG, Shang XW, Pang BS, Song JR, Cao SQ, Li JC, Zhang XY. Molecular mapping of two novel stripe rust resistance genes *YrTp1* and *YrTp2* in A-3 derived from *Triticum aestivum* × *Thinopyrum ponticum*. Agric Sci China. 2006;5:483–90.

Yin CT, Park JJ, Gang DR, Hulbert SH. Characterization of a tryptophan 2-monooxygenase gene from *Puccinia graminis* f. sp. *tritici* involved in auxin biosynthesis and rust pathogenicity. Mol Plant-Microbe Interact. 2014;27:227–35.

Yin CT, Downey SI, Klages-Mundt NL, Ramachandran S, Chen XM, Szabo LJ, Pumphrey M, Hulbert SH. Identification of promising host-induced silencing targets among genes preferentially transcribed in haustoria of *Puccinia*. BMC Genomics. 2015;16:579.

Yuan CY, Wang MN, See DR, Chen XM. Towards construction of genetic linkages for mapping virulence genes in *Puccinia striiformis* f. sp. *tritici*, the wheat stripe rust pathogen. Phytopathology. 2016;S4:208.

Yue YL, Yao ZJ, Ren XX, Wang L. Molecular mapping of a gene for resistance to stripe rust in wheat variety PIW138. Agric Sci China. 2010;9:1285–91.

Zadoks JC. Yellow rust on wheat. Studies in epidemiology and physiologic specialization. J Pl Ziekten. 1961;67:69–256.

Zahravi M, Bariana HS, Shariflou MR, Balakrishna PV, Banks PM, Ghannadha MR. Bulk segregant analysis of stripe rust resistance in wheat (*Triticum aestivum*) using microsatellite markers. In: Pogna NE, Romano M, Pogna EA, Galterio G, editors. Proc 10th Intl Wheat Genet Symp. Rome: Instituto Sperimentale per Cerealcoltura; 2003. p. 861–3.

Zegeye H, Rasheed A, Makdis F, Badebo A, Ogbonnaya FC. Genome-wide association mapping for seedling and adult plant resistance to stripe rust in synthetic hexaploid wheat. PLoS ONE. 2014;9:e105593.

Zeller FJ. 1B/1R wheat-rye chromosome substitutions and translocations. In:Proc Fourth Intl Wheat Genet Symp. Missouri, USA: Columbia; 1973. p. 209–21.

Zeng QD, Han DJ, Wang QL, Yuan FP, Wu JH, Zhang L, Wang XJ, Huang LL, Chen XM, Kang ZS. Stripe rust resistance and genes in Chinese wheat cultivars and breeding lines. Euphytica. 2014;196:271–84.

Zhan GM, Wang FP, Chen XM, Wan CP, Han QM, Huang LL, Kang ZS. Virulence and molecular diversity of *Puccinia striiformis* f. sp. *tritici* population in Xinjiang in relation to other regions of western China. Plant Dis. 2016;100:99–107.

Zhang L, Meakin H, Dickinson M. Isolation of genes expressed during compatible interactions between leaf rust (*Puccinia triticina*) and wheat using cDNA-AFLP. Mol Plant Pathol. 2003;4:469–77.

Zhang P, McIntosh RA, Hoxha S, Dong CM. Wheat stripe rust resistance genes *Yr5* and *Yr7* are allelic. Theor Appl Genet. 2009;120:25–9.

Zhang H, Ren ZL, Hu YG, Wang CY, Ji WQ. Characterization of wheat stripe rust resistance genes in Shaanmai 139. Acta Agron Sin. 2010;36:109–14.

Zhang L, Chang ZJ, Li X, Zhang HY, Ren ZL, Luo PG. Screen and identification of wheat new resistant germplasms to Fusarium head blight. Acta Phytophyl Sin. 2011;38:569–70.

Zhang HC, Wang CF, Cheng YL, Chen XM, Han QM, Huang LL, Wei GR, Kang ZS. Histological and cytological characterization of adult plant resistance to wheat stripe rust. Plant Cell Rep. 2012;31:2121–37.

Zhang XJ, Han DJ, Zeng QD, Duan YH, Yuan FP, Shi JD, Wang QL, Wu JH, Hung LL, Kang ZS. Fine mapping of wheat stripe rust resistance gene *Yr26* based on collinearity of wheat with *Brachypodium distachyon* and rice. PLoS ONE. 2013;8(3):e57885.

Zhang H, Zhang L, Wang CY, Wang YJ, Zhou XL, Lv SK, Liu XL, Kang ZS, Ji WQ. Molecular mapping and marker development for the *Triticum dicoccoides*-derived stripe rust resistance gene *YrSM139-1B* in bread wheat cv. Shaanmai 139. Theor Appl Genet. 2016;129:369–76.

Zhao L, Feng J, Zhang CY, Xu XD, Chen XM, Sun Q, Miao Q, Xu SC, Lin F. The dissection and SSR mapping of a high-temperature adult-plant stripe rust resistance gene in American spring wheat cultivar Alturas. Eur J Plant Pathol. 2012;134:281–8.

Zhao J, Yang YH, Yang DH, Cheng YL, Jiao M, Zhan GM, Zhang HC, Wang JY, Zhou K, Huang LL, Kang ZS. Characterization and genetic analysis of rice mutant crr1 exhibiting compromised non-host resistance to *Puccinia striiformis* f. sp. *tritici* (*Pst*). Frontiers Plant Sci. 2016a;7:1822.

Zhao CZ, Li YH, Dong HT, Geng MM, Liu WH, Li F, Ni ZF, Wang XJ, Xie CJ, Sun QX. Molecular cloning, functional verification, and evolution of *TmPm3*, the powdery mildew resistance gene of *Triticum monococcum* L. Genet Mol Res. 2016b;15(2):gmr.15028056.

Zheng WM, Huang LL, Huang JQ, Wang XJ, Chen XM, Zhao J, Guo J, Zhuang H, Qiu CZ, Liu J, Liu HQ, Huang XL, Pei GL, Zhan GM, Tang CL, Cheng YL, Liu MJ, Zhang JS, Zhao ZT, Zhang SJ, Han QM, Han DJ, Zhang HC, Zhao J, Gao XN, Wang JF, Ni PX, Dong W, Yang LF, Yang HM, Xu JR, Zhang GY, Kang ZS. High genome heterozygosity and endemic genetic recombination in the wheat stripe rust fungus. Nature Commum. 2013;4:2673.

Zheng JM, Yan ZH, Zhao L, Li SZ, Zhang ZY, Garry R, Yang WY, Pu ZJ. Molecular mapping of a stripe rust resistance gene in wheat line C51. J Genet. 2014;93:443–50.

Zhou XL, Wang WL, Wang LL, Hou DY, Jing JX, Wang Y, Xu ZQ, Yao Q, Yin JL, Ma DF. Genetics and molecular mapping of genes for high-temperature resistance to stripe rust in wheat cultivar Xiaoyan 54. Theor Appl Genet. 2011;123:431–8.

Zhou XL, Han DJ, Chen XM, Gou HL, Guo SJ, Rong L, Wang QL, Huang LL, Kang ZS. Characterization and molecular mapping of stripe rust resistance gene *Yr61* in winter wheat cultivar Pindong 34. Theor Appl Genet. 2014a;127:2349–58.

Zhou XL, Han DJ, Gou HL, Wang QL, Zeng QD, Yuan FP, Zhan GM, Huang LL, Kang ZS. Molecular mapping of a stripe rust resistance gene in wheat cultivar Wuhan 2. Euphytica. 2014b;196:251–9.

Zhou XL, Wang MN, Chen XM, Lu Y, Kang ZS, Jing JX. Identification of *Yr59* conferring high-temperature adult-plant resistance to stripe rust in wheat germplasm PI 178759. Theor Appl Genet. 2014c;127:935–45.

Zhou XL, Han DJ, Chen XM, Hu JM, Xue WB, Zeng QD, Wang QL, Huang LL, Kang ZS. QTL mapping of adult-plant resistance to stripe rust in wheat line P9897. Euphytica. 2015a;205:243–53.

Zhou XL, Zhan GM, Huang LL, Han DJ, Kang ZS. Evaluation of resistance to stripe rust in eighty abroad spring wheat germplasms. Sci Agric Sin. 2015b;48:1518–26.

Zhou XL, Zhang Y, Zeng QD, Chen XM, Han DJ, Huang LL, Kang ZS. Identification of QTL for adult plant resistance to stripe rust in Chinese wheat landrace Caoxuan 5. Euphytica. 2015c;204:627–34.

Zuo H, Wang J, Hao CZ, Zhang B, Ma Q. Histochemical response of nonhost resistance in pepper to the stripe rust fungus (*Puccinia striiformis* f. sp *tritici*). J Plant Pathol. 2013;95:275–83.

Zwart RS, Thompson JP, Milgate AW, Bansal UK, Williamson PM, Raman H, Bariana HS. QTL mapping of multiple foliar disease and root-lesion nematode resistances in wheat. Mol Breed. 2010;26:107–24.

Chapter 6
Integrated Control of Stripe Rust

Xianming Chen and Zhensheng Kang

Abstract Stripe rust, caused by *Puccinia striiformis*, can be controlled by growing resistant cultivars, applying fungicides and through some cultural practices. As developing and growing resistant cultivars is the most preferred approach for control of stripe rust and a great body of knowledge has been obtained for resistance, the topic of resistance is discussed in Chap. 5. This chapter focuses on the information and techniques of fungicides and cultural practices related to effecting stripe rust development and control. The advantages and disadvantages of individual strategies are discussed. Based on the discussion, we propose the integrated control approach that primarily relies on developing and growing cultivars with adequate and durable resistance, secondarily utilizes effective fungicides with timely applications when necessary and uses appropriate cultural practices. The integrated approach allows best control on a yearly basis based on available resistant cultivars of acceptable yield and quality, effective fungicides and cultural practices most suitable for the region. The approach also allows continual improving cultivar resistance and other measures for more effective and sustainable control of the disease.

Keywords Barley • Chemical control • Cultural control • Cropping system • Disease management • Disease resistance • Durable resistance • Fungicide • Integrated management • *Puccinia striiformis* • Stripe rust • Sustainable control • Wheat • Yellow rust

---

Mention of trade names or commercial products in this publication is solely for the purpose of providing specific information and does not imply recommendation or endorsement by the U. S. Department of Agriculture. USDA is an equal opportunity provider and employer.

X. Chen (✉)
US Department of Agriculture, Agricultural Research Service, Wheat Health, Genetics and Quality Research Unit, Pullman, WA 99164-6430, USA

Department of Plant Pathology, Washington State University, Pullman, WA 99164-6430, USA
e-mail: xianming@wsu.edu

Z. Kang
State Key Laboratory of Crop Stress Biology for Arid Areas and College of Plant Protection, Northwest A&F University, Yangling, Shaanxi 712100, China
e-mail: kangzs@nwafu.edu.cn

X. Chen, Z. Kang (eds.), *Stripe Rust*, DOI 10.1007/978-94-024-1111-9_6

Stripe rust can be reduced or controlled by growing resistant cultivars, application of fungicides, and utilization of cultural practices. Because host resistance is the most important approach for control of stripe rust, genetics of resistance and development of resistant cultivars is discussed in Chap. 5. This chapter focuses on control of stripe rust through fungicides and cultural practice, and integration of different approaches to achieve best management of the disease on a yearly basis and toward more effective control while reducing the cost, producing safe grain, maximizing grain production and protecting the environment.

6.1 Fungicides for Stripe Rust Control

Use of fungicides can be ranked as the second approach for control of stripe rust after breeding and growing resistant cultivars. As the technology for screening and developing fungicides has been greatly advancing in the recent decades, more and more effective fungicides with different modes of action have become available. Compared to a decade ago, fungicides are currently much cheap, which makes control of stripe rust and other diseases of cereal crops more profitable using chemical control. Fungicide application has become an important component of integrated control of stripe rust in many parts of the world.

6.1.1 Development of Fungicides

Fungicides are chemicals used to limit initial infection by and progress of plant pathogens and thus prevent or reduce damage to plants. Testing and use of chemicals for control of rusts of cereal crops can be traced back to the late nineteenth century. Numerous tests were conducted using dust and spray applications of sulfur, and Eriksson and Henning (1896) summarized the studies and concluded that partial control of rust was possible with sulfur, but its use was unfeasible. In the 1920s, the success of managing apple scab and codling moth in orchards by applying dusts containing finely ground sulfur and lead arsenate with improved equipment stimulated a re-examination of the use of sulfur to control cereal rusts (Rowell 1968).

Great advances were made on development of organic chemicals and evaluation of them for plant disease control in the 1960s and 1970s (Lukens 1983; Russell 2005). Early research on use of chemicals to control cereal rusts was reviewed by Dickson (1959) and Rowell (1968, 1985). By the early 1980s, a number of chemical compounds, such as calcium sulfamate, fenapanil, inorganic salts of nickel, sodium sulfanilate, oxycarboxin, triarimol, triazbutil and triadimefon were found to be effective and potentially useful for control of cereal rusts, but none of them were registered for use on cereals (Rowell 1985), except triadimefon was emergently labeled for control the devastating stripe rust epidemics in the US Pacific Northwest in 1981 (Rowell 1985; Line 2002).

The Pacific Northwest is the first region in the United States and probably one of the first regions in the world where chemicals were commercially used to control stripe rust. Intensive research on chemical control of stripe rust has been conducted by scientists in this region. According to Line (2002), the devastating epidemic of 1961 in the region appeared without warning, leaving growers with no control options. Although some wheat growers planted moderately resistant cultivars, but most continued to plant susceptible cultivars in the 1960s. When stripe rust became severe, growers sprayed several thousand acres of wheat with chemicals of doubtful effectiveness and at considerable expense in an attempt to control the disease. Although protectant chemicals had been tested for control of cereal rusts in the 1950s and 1960s, but the need for frequent applications made them impractical (Purdy 1964; Powelson and Shaner 1966). The chemicals were even less effective against stripe rust because they did not prevent growth of earlier infections within the leaves. Nickel sulfate and nickel chloride were shown promise for control of stripe rust on wheat and bluegrass (Hardison 1963; Line and Rowell 1973). Oxathiin fungicides were found to be able to control stripe rust on wheat seedlings when used as a seed treatment, and also controlled stripe rust on bluegrass when used as soil treatments and foliar sprays (Hardison 1966, 1975; Powelson and Shaner 1966; Powelson 1969). Line and co-workers showed that a single foliar application of the systemic fungicide oxycarboxin (Plantvax) controlled stripe rust when taken up by roots after a rain or after irrigation (Line et al. 1970; Line 1976). As new systemic fungicides were beginning to be developed, Line and associates conducted annual testing of the new chemicals at several sites in the Pacific Northwest. By 1978, systemic fungicides had been shown to provide economical control of stripe rust and guidelines for their use were developed (Line 1976). In the same year, the market for systemic fungicides developed by European companies opened up in the Pacific Northwest after persuasive consultation. In the fall of 1980, both stripe rust and leaf rust were prevalent throughout the Pacific Northwest and continued to increase and spread throughout the winter due to the much warmer-than-normal weather conditions. Based on the previous testing data and through Line's effort, triadimefon (Bayleton), the most effective fungicide for control of stripe rust at that time, was received emergency registration, and was widely used in the region for control of both stripe rust and leaf rust during the 1980–1981 wheat crop season, thereby preventing multimillion dollar losses. In that year, when rusts were not controlled, susceptible cultivars such as Omar were totally destroyed, and moderately resistant cultivars had 50% yield losses (Line 2002). That was the first effective commercial use of fungicides for control of cereal rusts in North America. Treatment of seed with triadimenol (Batan) that could control early development of stripe rust was registered later (Line and Rakotondradona 1980; Rakotondradona and Line 1984; Scott and Line 1985). Since then, new, highly effective foliar fungicides have subsequently been registered for use on wheat for control of stripe rust and other foliar diseases in the region.

Although fungicides have been successfully used in Europe, North America, Australasia and some other countries, such as China and South Africa, in other continents (Priestley 1981; Rakotondradona and Line 1984; Conner and Kuzyk 1988;

Cromey 1989; Scott 1996; Line 2002; Boshoff et al. 2003; Chen 2005; Chen et al. 2014e), they still have not been used in many countries in Africa, Asia and South America. In North America, the major factors of limiting use of chemicals in control of the cereal crops before the 1980s included the difficulties in toxicological and economic evaluations, concern about the consequences of chemical residues and the economics of small grain production. The successful use of triadimefon to control stripe rust of wheat in the Pacific Northwest in 1981 was partially due to the severe epidemic level and high yield of wheat crop (Line 2002). Similarly, chemicals were used intensively in Western Europe for control of cereal diseases, including rusts in the late 1970s and early 1980s, where the intensive cropping systems, high yield and high level of grain prices permitted economic use of chemical control of rusts at crop losses as low as 2% of yield for winter wheat and barley (Rathnell and Skidmore 1982). Since the 1980s, chemical control of stripe rust and other diseases have become more and more popular. However, the economic issue related to cost of chemicals and application is still a major obstacle for use of chemicals to control stripe rust in many countries of the world with low yield and high cost of chemicals. In many countries, lack of research and extension for fungicide application could be another reason for not use of fungicides. It is a great idea to promote breeding for resistant cultivars, but this cannot be used for ignoring or putting side chemical control. At the same time when developing resistant cultivars, making fungicides and application techniques available is urgently needed for every region to deal with sudden and unexpected disease epidemics. Fungicide application is an option for growers to control stripe rust and other diseases when susceptible cultivars are grown for many reasons and when resistant cultivars suddenly become susceptible.

Any compounds that have antifungal activities could be developed into fungicides (Lukens 1983). The historical development of fungicides before the 1940s was mostly testing various inorganic chemicals for control of plant diseases including stripe rust. Most of the chemicals were applied at a high rate (Morton and Staub 2008). From 1940 to 1970, a number of new chemistry classes were introduced as fungicides (Russell 2005). For stripe rust control, it was important to introduce Benzimidazole chemicals (such as Benlate) in the 1960s and early 1970s as foliar fungicides and seed treatments.

Since 1970, the development of modern fungicides, as well as other pesticides, has become a systemic process from designing compounds based on chemical properties and interactions with fungi; synthesis; various tests for antifungal activities; movement in plant tissues; residual in plants, produces and environment; and toxicity to humans, animals and environment. After obtaining adequate data from the various tests, the compounds must be registered before they can be commercially sold and used for certain regions to control certain diseases of certain crops.

In addition to synthetic chemicals, microbe metabolites have become an important source of fungicides. Several fungicides, such as fenpiclonil, fludioxonil and strobilurins, have been developed from microbial metabolites through chemical modification, yielding additional useful features suitable for agricultural fungicides (Takeuchi et al. 1957; Anke et al. 1977; Nyfeler and Ackermann 1992). For example, strobilurin A was found to be fungicidal metabolites in Basidiomycete fungi

Strobilurus tenacellus (Anke et al. 1977). Since then, a number of structurally related compounds have been reported to have fungicidal activity (Kim and Kwang 2007). New environment-friendly fungicides could be developed using genomics and proteomics analyses for finding the fungicide targets (Acero et al. 2011).

Because of the high cost for developing a fungicide, it is important to have a broad spectrum of antifungal activities for commercialization of a fungicide. Other factors that are also needed to be concerned including low toxicity to humans and animals (such as birds, bees and fish), adequate duration of effectiveness, low residual in plant produce and low environmental risk. In the future, when development cost is greatly reduced, more pathogen-specific fungicides could be developed. As a specific fungicide targets to unique structures or biochemical pathways of a particular pathogen, such as the stripe rust fungus, it can be friendly to humans, animals, beneficial microorganisms and environment.

6.1.2 Testing Fungicides

So far, fungicides that are effective for control of stripe rust have been developed from generic chemicals or formulations showing effectiveness in controlling other diseases through testing with the stripe rust pathogen. There are two approaches for testing compounds for inhibiting or killing fungal pathogens including the stripe rust fungus, *in vitro* and *in vivo* testing.

6.1.2.1 *In Vitro* Testing

In vitro testing is commonly used for quick determination if a compound has any effect on a pathogen. As the testing can be performed on glass slides in Petri dishes or in agar plates, a large number of concentrations can be tested to find EC_{50}, the concentration of a fungicide that gives half-maximal response. For testing fungicides for controlling stripe rust, urediniospore germination testing can be conducted to determine if a compound can inhibit germination and/or germ tube growth. For germinating urediniopsores of the stripe rust fungus, temperatures approximately at 10 °C should be used, and germination rate and germ tube length can be recorded from 2–3 h to 24 h after the treatments. Such tests are often done by chemical companies. The information of EC_{50} and other data is used for designing *in vivo* testing.

6.1.2.2 *In Vivo* Testing

In vivo testing can be in a greenhouse or field. Under controlled greenhouse conditions, seedling testing is more commonly used as it takes only a few weeks and small space. Plants of one-to-two leaves, 7–14 days after planting, can be used for seedling tests. Fungicides can be applied before, on the same day or after inoculation of the plants with urediniospores of a virulent isolate to determine different effects of chemicals. The inoculated plants should be incubated at approximately 10 °C for 12–24 h and then grown at the optimal temperature (10–25 °C) and light (16 light/8 dark) conditions for disease development. Stripe rust symptoms and pathogen signs can be observed and recorded from 7 days to 30 or more days after inoculation. If desired, leaf samples can be collected at different time and microscopic or histological observations can be conducted with various microscopes to determine the effects of chemicals on the different stages of the infection process.

Although not commonly used, adult-plants, especially at flag leaf stage, can be used for testing fungicides under controlled greenhouse conditions. Flag leaves should be tested as they contribute more to grain filling than any other leaves. Greenhouse tests can determine fungicide efficacy, but have limitations in determination of efficacy for controlling multicyclic diseases, such as stripe rust, and the effect of fungicide application on yield (Chen et al. 2013c). More informative data should be obtained from tests conducted in fields.

Field testing is most commonly used to determine fungicide efficacy and yield response. The author's program tests various fungicide treatments for control of stripe rust on winter wheat, spring wheat and spring barley for identifying effective fungicides. As stripe rust is most effectively controlled by fungicides as foliar application, most tests have been conducted with foliar fungicides (Chen and Wood 2002a, b, 2003a, b, 2004, 2005a, b, c, 2006a, b, c, 2007a, b, 2008a, b, 2009a, b, c, 2010a, b, c, 2011a, b; Chen et al. 2012a, b, 2013a, b, d, e, 2014a, b, c, d, 2015a, d, 2016a).

The general procedure for testing foliar fungicides in fields is the following. A highly susceptible variety, which can be a genotype not commercially grown but adapts to the region, is planted at seeding rate and time similar to and managed using the similar practices to commercial crop production. The field can be either planted as individual plots or divided into plots by killing plants between plots using an herbicide in the spring or by cutting alleys. It is better to measure the length and width of each plot as a small difference in area for the plots can be translated into large differences when converting yields of small plots to those of per acre or hectare. Before fungicide application, the percentage of plant stand in each plot is recorded as plots may have uneven number of plants. The percentages of plant stands may be used to normalize the grain yields for comparison. Fungicides can be applied at different timings and at different rates according to rust development and/or plant growth stages depending on the treatments. All experiments are a randomized complete block design with four replications for every treatment. To increase the coverage of chemicals on plants surface, a surfactant is often used when make a fungicide suspension with water. Stripe rust data (infection type and severity) are recorded just before fungicide application and various days after application depending on objectives, rust development and plant growth stage, at least three times. If a susceptible variety is used in the test, it should have a highly susceptible infection type in the non-treated check plots and in all plots before fungicide application if stripe rust has occurred. Application of an effective fungicide often change a high infection type (susceptible) to low infection type (appear like a resistant reaction). Multiple stripe rust data allow to determine the duration of effectiveness of a fungicide treatment and to calculate the area under the disease progress curve (AUDPC) for each plot using formula.

$$\mathrm{AUDPC} = \sum\left[\left(\mathrm{S_i} + \mathrm{S_{(i+1)}}\right) \times \mathrm{D} / 2\right]$$

where S_i is disease severity at time *i*, $S_{(i+1)}$ severity at the next time *i* + 1, and D duration in days from time *i* to time *i* + 1. For ease comparison, AUDPC values can be converted to relative AUDPC (rAUDPC) as percentages of the mean AUDPC value of the non-treated plots. Fungicide efficacy can be determined by comparing the mean AUDPC or mean rAUDPC of the treatment with that of the non-treated check or with other fungicide treatments. The effectiveness of a new fungicide or a new treatment (different application time, rate, or mixture of fungicides) can be determined by comparison with the fungicide or fungicides commonly used in the region.

As the damage of stripe rust is reducing grain yield and quality, it is more meaningful to compare grain yield and test weight. To do so, grain yield and test weight is measured at the time of harvest for each plot. For test weight, it is easier and even better to measure the weight of grain in a unit of volume. In this way, grain yield can be expressed as bushels per acre as used in the United States or tons per hectare as used in many other countries. Thus, the effect of fungicide application can be expressed as yield increase or saved by bushels/tons or percentage per acre/hectare compared with the non-treated check, and its economic value can be easily determined.

When infection of stripe rust occurs in the early crop growth season, seed treatment with fungicides can be beneficial especially in regions where infection occurs commonly in fall for winter crops and/or in spring for spring crops (Chen and Wood 2006d, 2007c, d, e; Chen et al. 2013d, e). For testing seed treatment, seed of a susceptible cultivar are treated with different fungicides or different concentration of a fungicide in a chemical hood before planting. Treated seed are planted in individual plots, and field management, disease data recording, grain measurement and data analyses are similar to the testing of foliar fungicides as discussed above. As some chemicals or high concentrations may cause an emergence problem, it is important to recording data of plant emergence. Seed treatment is relatively easy to do and also costs less. However, many fungicides are not suitable for control of stripe rust as seed treatments because their effectiveness is short lasting. Long-lasting fungicides may cause residue problems in soil. More research is needed to develop environmental friendly and effective fungicides as seed treatment for effective control of stripe rust.

6.1.3 *Classification of Fungicides*

Fungicides can be classified in different ways, based on their mobility in the plant, role in the protection of plants, breadth of metabolic activity, mode of action, chemical group or family and Fungicide Resistance Action Committee (FRAC) code (Mueller et al. 2013).

6.1.3.1 Fungicides Classified by Mobility in the Plant

Based on the mobility in the plant, fungicides are classified as contact fungicide and systemic fungicide. A contact fungicide remains on the surface of the plant where it is applied, but not absorbed into plant tissue. This type of fungicide does not have post-infection activity. Contact fungicides can be washed off by rain, irrigation or degraded by environment factors such as sunshine. Thus, contact fungicides need to be repeatedly applied to protect new growth of leaves and heads of cereal crops. Contact fungicides are effective only when applied just before a spore shower. For control of stripe rust that is a multicyclic disease, contact fungicides may not be effective. Most fungicides used for attempted control of stripe rust before the 1970s were contact fungicides, which did not provide adequate control (Line 2002).

In contrast, a systemic fungicide moves into plant tissue and can offer some post-infection activity. Few fungicides are truly moving freely throughout the plant, but some are upwardly moving up in the plant through xylem tissue, and many are locally systemic, moving into treated leaves and redistributing within the treated portion of the plant. Oxycarboxin (Plantvax), which was widely used for control of stripe rust in the US Pacific Northwest in the 1970s and 1980s, can move upwards in wheat plants. Line and associates found that a single foliar application of Plantvax controlled stripe rust when taken up by roots after a rain or irrigation, but was inadequate when the soil surface was dry because the fungicide could not move into the roots (Line and Rowell 1973; Line 1976). Most of the currently used systemic fungicides for control of stripe rust and other foliar diseases belong to the locally systemic fungicides.

6.1.3.2 Fungicides Classified by the Role in the Protection of Plants

The roles of fungicides in the protection of plants include preventive, early infection and antisporulant activities. Preventive activity, also referred to protective activity, occurs when a fungicide is present on the plant as a barrier before the pathogen arrives or begins to develop, and thus prevents infection from occurring. Most contact fungicides have preventive activities.

Early infection activity, which is also referred as curative or kickback activity, occurs when a fungicide is absorbed by the plant and stops the fungus in its early developmental stages, within 24–72 h after infection occurs. Most fungicides that have early infection activity also have preventive activity, and they are most effective when applied before infection occurs. Many of the fungicides currently used for control of stripe rust, such as Strobilurin fungicides (e.g. Headline and Twinline), have early preventive activity.

A fungicide with antisporulant activity prevents spores being produced. For stripe rust, Propiconazole fungicides (e.g. Tilt) exhibit antisporulant activity, necrotic stripes can develop after fungicide application with no uredinial formation (X.M. Chen, unpublished data). Often, use of roles in the protection of plants could not clearly classify fungicides as many fungicides have two or more of the activities. Also, many new fungicides contain different active ingredients which combine different roles of protection. For example, Quilt and Quilt Xcel, which are widely used for control of stripe rust and other foliar diseases of cereal crops in the United States, contain Azoxystrobin and Propiconazole, which combine preventive, anti-early- infection, and antisporulant activities.

6.1.3.3 Fungicides Classified by the Breadth of Metabolic Activity

A single-site fungicide is active against only one point or function in one of the metabolic pathways of a fungus or against a single critical enzyme or protein needed by the fungus. For example, Strobilurin fungicides (e.g. Headline), which are widely used to control stripe rust and other cereal diseases in the United States, target to only one site, Quinone outside inhibitor (QoI). In contrast, a multisite fungicide targets to two or more points or functions in multiple metabolic pathways of a fungus. Many of the recently developed fungicide formulations, such as Quilt (Azoxystrobin + Propiconazole), Aproach (cyproconazole + picoxystrobin), Topguard (Flutriafol + Propyleneglycol) and Prosaro (Prothioconazole + Tebuconazole) effective for control of stripe rust (Chen et al. 2013a, b, 2014a, b, 2015a, b; Chen et al. 2016a), have two or more targets in their modes of action.

6.1.3.4 Fungicides Classified by the Mode of Action

The mode of action is the means by which a fungicide kills or suppresses a target fungus or affects specific biochemical processes of the target fungus. Fungicides are metabolic inhibitors and their modes of action can be classified into four broad groups: (1) inhibitors of electron transport chain, (2) inhibitors of enzymes, (3) inhibitors of nucleic acid metabolism and protein synthesis, and (4) inhibitors of sterol synthesis (Matheron 2001).

Examples of fungicides inhibiting the electron transport chain are sulfur and Strobilurin. Sulfur disrupts electron transport along the cytochromes. Although sulfur is not effective against stripe rust, it is a valuable chemical for stripe rust research as dusting or fumigating sulfur can control powdery mildew in greenhouses. Strobilurins (e.g., azoxystrobin, kresoxim-methyl, pyraclostrobin and trifloxystrobin) are Quinone outside inhibitors (QoIs). This group of fungicides inhibits mitochondrial respiration, blocking the cytochrome bc1 complex. Their target, cytochrome bc1, is an integral membrane protein complex that is essential for fungal respiration. The activity of QoIs relies on its ability to inhibit mitochondrial respiration by binding to the Qo site of the cytochrome bc1 complex, blocking electron transfer and halting ATP synthesis (Fernández-Ortuño et al. 2008). Of the Strobilurin fungicides, Quadris and Headline have been tested and widely used for control of stripe rust (Chen and Wood 2004, 2005a, b, 2006a, b, c, 2007a, b, 2008a, b).

Only a few fungicides inhibiting enzymes have been tested for control of stripe rust. Chlorothalonil, which is one of active ingredients in fungicide Alto effective for control of stripe rust (Chen et al. 2013a, b), can be classified into this group. Chlorothalonil reduces fungal intracellular glutathione molecules to alternate forms which cannot participate in essential enzymatic reactions, ultimately leading to cell death. Penthiopyrad (e. g. Vertisan) is a succinate dehydrogenase inhibitor (SDHI). SDHI fungicides inhibit fungal respiration. In field tests, Vertisan exhibited effect on reduction of stripe rust, but not as good as other best chemicals at the rate and timing tested (Chen et al. 2012a, b).

Benzimidazole fungicides, such as benomyl (e.g. Benlate), are inhibitors of nucleic acid metabolism and protein synthesis. Their effects are through specific binding of the beta-tubulin subunit of fungal tubulin, which consequently interferes with microtubule assembly. As microtubule assembly is essential for numerous cellular processes, such as mitosis and cytoskeleton formation (Ishii 2004), inhibiting β-tubulin formation and interfering microtubule assembly can stop fungal growth.

Fungicides that are inhibitors of sterol synthesis are among the most successful fungicides used for control of stripe rust and other foliar diseases of cereals. The demethylation inhibitor (DMI) fungicides inhibit demethylation at the 14-α carbon of lanosterol or 24-methylene dihydrolanosterol (eburicol), which are the substrates for the cytochrome P450-dependent 14-α demethylase in the biosynthesis of fungal sterols such as ergosterol (Steffens et al. 1996). Some DMI fungicides, such as pyribencarb (benzylcarbamate-type fungicide), the effects are through inhibiting the electron transport system in fungi. DMI fungicides strongly inhibited the development of *P. striiformis* in host leaf tissues, and also altered the structures of the hyphae and haustoria when fungicide-treated leaves were observed under an electron microscope (Kang et al. 1993, 1996; Han et al. 2006). Changes made by DMI fungicides include increasing vacuolation, irregular thickening of cell walls, necrosis or degeneration of cytoplasm and collapse of hyphae and haustoria.

Of DMI fungicides, triadimenol (e. g. Bayleton) was widely used in control of stripe rust in the US Pacific Northwest in the 1970s and 1980s (Line 2002), and still the mostly used fungicide in China (Wan et al. 2004, 2007). Triazole fungicides, especially propiconazole (e.g. Tilt and Quilt) and tebuconazole (e.g. Forlicur, Prosaro), have been widely used for control of stripe rust and foliar diseases of cereal crops (Chen and Wood 2002a, b, 2003a, b, 2004, 2005a, b, c, 2006a, b, c, 2007a, b, 2008a, b, 2009a, b, c, 2010a, b, c, 2011a, b; Chen et al. 2012a, b, 2013a, b, d, e, 2014a, b, c, d, 2015a, d, 2016a).

6.1.3.5 Fungicides Classified by Chemical Group, Family or FRAC Code

A group or class of chemicals has a common biochemical mode of action and may or may not have a similar structure. Agricultural fungicides are classified into more than 40 chemical groups or families (Pscheidt and Ocamb 2016). Fungicides that have been tested and/or widely used in control of stripe rust belong to four families or groups. The Benzimidazole family has benomyl (Benlate). The Carboxamide family (Group 7) contains carboxin (Vitavax), oxycarboxin (Plantvax) and penthiopyrad (Vertisan). The Triazole family (Group 3) contains flutriatol (Topguard), propiconazole (Tilt, Bunper, Stratego), tebuconazole (Folicur, Raxil, Absolute), triadimofon (Bayleton) and triadimenol (Baytan). The QoI Strobilurin family (Group 11) contains azoxystrobin (Quadris, Quilt), fluoxastrobin (Evito), picoxystrobin (Aproach), pyraclostrobin (Headline, Twinline) and trifloxystrobin (Flint, Abolute, Stratego).

The FRAC is an organization established to address the issue of fungicide resistance. This organization developed a code of numbers and letters for distinguishing fungicide groups based on their modes of action. Currently used fungicides effective for stripe rust control are mostly in FRAC Code 3 [demethylation inhibitors (DMI)], such as propoconazole (Tilt), tebuconazole (Folicur) and triadimefon (Bayleton); followed by FRAC Code 11 [Quinone Outside inhibitor (QoI)], such as azoxystrobin (Quadris, Quilt) and pyraclostrobin (Headline). Some of the old fungicides that were used for control of stripe rust, such as Vitavax and Plantvax, are in FRAC Code 7 [Succinate Dehydrogenase Inhibitor (SDHI)] and FRAC Code 1 [Methyl Benzimidazole Carbamate (MBC)], such as benomyl (Benlate).

6.1.4 Registered Fungicides for Stripe Rust Control

There are more than 20 fungicides or formulations currently registered for control of stripe rust on wheat and barley in the United States. These fungicides and their rate of application and duration of effectiveness are listed in Table 6.1. Detailed information about efficacy of fungicides tested for stripe rust in the United States can be found in *Plant Disease Management Reports* (Chen and Wood 2002a, b, 2003a, b, 2004, 2005a, b, c, 2006a, b, c, 2007a, b, 2008a, b, 2009a, b, c, 2010a, b, c, 2011a, b; Chen et al. 2012a, b, 2013a, b, d, e, 2014a, b, c, d, 2015a, d, 2016a). Fungicides registered for control of stripe rust in other countries can be found from relevant websites and also can be find from the major international agricultural chemical companies, such as BASF (http://www.agro.basf.com/agr/AP-Internet/en/content/solutions/fungicides/index), Bayer (https://www.cropscience.bayer.us/products/fungicides), Dupont (http://www.dupont.com/products-and-services/crop-protection/cereals-protection.html), Cheminova-FMC Corporation (http://www.fmc.com/agsolutions/Home.aspx); Helena (http://www.helenachemical.com/products/helena-products/crop-protection/); Syngenta (http://www.syngenta.com/global/corporate/en/products-and-innovation/product-brands/crop-protection/fungicides/Pages/fungicides.aspx).

Table 6.1 Fungicides currently registered for control of stripe rust in the United States

Trade name	Common name (active ingredient %)	Chemical family	FRAC[a] code	Application rate (fl. oz./A)	Control effect[b]	Effective duration (days)	Total limitation per crop (fl. oz./A)	Application restriction (no later than)[c]
Absolute 500 SC	Tebuconazole 22.6, Trifloxystrobin 22.6	Triazole & Strobilurin	3 & 11	5.0–7.7	VG	35	32	35 dbh
Aframe Plus	Azoxystrobin 13.5, Propiconazole 11.7	Triazole & Strobilurin	3 & 11	10.5–14	E	40	28	Feekes 10.5
Alto 100 SL	Cyproconazole 8.9	Troazole	3	3.0–5.5	VG	30	5.5	30 dbh
Aproach SC	Picoxystrobin 22.5	Strobilurin	11	6.0–12.0	VG	30	36	Feekes 10.5
Aproach Prima SC	Cyproconazole 7.17, Picoxystrobin 17.94	Triazole & Strobilurin	3 & 11	3.4–6.8	E	40	6.8	45 dbh
Avaris 2XS	Propiconazole 11.7, Azoxystrobin 7.0	Triazole & Strobilurin	3 & 11	10.5–14.0	E	40	56	30 dbh
Bumper 41.8 EC	Propiconazole 41.8	Triazole	3	4.0	VG	30	8.0	Feekes 10.5
Caramba 0.75 SL	Metconazole 8.6	Triazole	3	10.0–17.0	E	40	34	30 dbh
Custodia	Tebuconazole 18.3, Azoxystrobin 11.0	Triazole & Strobilurin	3 & 11	6.4–8.6	E	40	8.6	Feekes 10.5
Embrace	Tebuconazole 38.7	Triazole	3	4.0	–[d]	–[d]	4.0	30 dbh
Equation 2.08 SC	Azoxystrobin 22.8	Strobilurin	11	4.0–12.0	VG	30	24.0	Feekes 10.54
Evito 480 SC	Fluoxastrobin 40.3	Strobilurin	11	2.0–4.0	VG	20	8.0	Feekes 10.5 and 40 dbh
Fitness	Propiconazole 41.8	Triazole	3	4.0	VG	30	8.0	Feekes 10.5
Folicur 3.6 F	Tebuconazole 38.7	Triazole	3	4.0	E	30	4.0	30 dbh
Headline SC	Pyraclostrobin 23.6	Strobilurin	11	9.0	VG	20	18.0	Feekes 10.5
Monsoon	Tebuconazole 38.7	Triazole	3	4.0	–	–	4.0	30 dbh

Muscle 3.6 F	Tebuconazole 38.7	Triazole	3	4.0	E	30	4.0	30 dbh
Onset 3.6 L	Tebuconazole 38.7	Triazole	3	4.0	–	–	4.0	30 dbh
Orius 3.6 F	Tebuconazole 38.7	Triazole	3	4.0	–	–	4.0	30 dbh
Priaxor 500 SC	Fluxapyroxad 14.3, Pyraclostrobin 28.6	Carboxamide & Strobilurin	7 & 11	8.0	VG	30	16.0	Feekes 10.5
Proline 480 SC	Prothioconazole 41.0	Triazole	3	4.3–5.0	–	–	9.37	30 dbh
Propiconazole E-AG 41.8 EC	Propiconazole 41.8	Triazole	3	4.0	–	–	8.0	Feekes 10.5
PropiMax 3.6 EC	Propiconazole 41.8	Triazole	3	4.0	–	–	8.0	Feekes 10.5
Prosaro 421 SC	Prothioconazole 19.0, Tebuconazole 19.0	Triazole	3	6.5–8.2	E	30	8.2	30 dbh
Quadris 2.08 SC	Azoxystrobin 22.9	Strobilurin	11	4.0–12.0	E	30	24.0	45 dbh
Quilt 200 SC	Propiconazole 11.7, Azoxystrobin 7.0	Triazole & Strobilurin	3 & 11	10.5–14.0	E	40	28.0	Feekes 10.5
Quilt Xcel 2.2 SE	Propiconazole 11.7, Azoxystrobin 13.5	Triazole & Strobilurin	3 & 11	10.5–14.0	E	40	28.0	Feekes 10.5
Stratego 250 EC	Prothioconazole 11.4, Trifloxystrobin 11.4	Triazole & Strobilurin	3 & 11	10.0	VG	30	20.0	Feekes 10.5 and 35 dbh
Stratego YLD	Prothioconazole 10.8, Trifloxystrobin 32.3	Triazole & Strobilurin	3 & 11	4.0	VG	30	8.0	Feekes 10.5 and 35 dbh
Tebucon 3.6 F	Tebuconazole 38.7	Triazole	3	4.0	–	–	4.0	30 dbh
Tebustar 3.6 F	Tebuconazole 38.7	Triazole	3	4.0	–	–	4.0	30 dbh
Tebuzol 3.6 F	Tebuconazole 38.7	Triazole	3	4.0	–	–	4.0	30 dbh
Tegrol	Tebuconazole 38.7	Triazole	3	4.0	–	–	4.0	30 dbh
Tilt 3.6 EC	Propiconazole 41.8	Triazole	3	4.0	VG	30	8.0	Feekes 10.5
Toledo 3.6 F	Tebuconazole 38.7	Triazole	3	4.0	–	–	4.0	30 dbh

(continued)

Table 6.1 (continued)

Trade name	Common name (active ingredient %)	Chemical family	FRAC[a] code	Application rate (fl. oz./A)	Control effect[b]	Effective duration (days)	Total limitation per crop (fl. oz./A)	Application restriction (no later than)[c]
Topguard SC	Flutriafol 11.8	Triazole	3	10.0–14.0	E	30	28.0	30 dbh
Twinline 1.75 EC	Metconazole 7.4, Pyraclostrobin 12.0	Triazole & Strobilurin	3 & 11	7.0–9.0	E	40	18.0	Feekes 10.5
Vertisan	Penthiopyrad 20.6	Carboxamide	7	24.0	E	40	48	Feekes 10.5.1
Viathon 5.1 SC	Tebuconazole 3.3, Potassium phosphite 49.0	Triazole	3	2.0 pt	E	30	16.5	30 dbh

[a]*FRAC* Fungicide Resistance Action Committee
[b]Control effect is relative: *E* excellent, *VG* very good
[c]Crop growth stage and/or days before harvest (dbh), after which the fungicide cannot be applied
[d]–, no data

6.1.5 *Timing of Fungicide Application*

Timing of fungicide application is important for effective control of stripe rust, which can be different from region to region and also different from season to season in a specific region. It depends upon when and at what crop stage the infection of the stripe rust pathogen occurs and starts sporulation, weather conditions, occurrence of other foliar diseases, cost of fungicide and application, yield potential and grain price. In general, three time points can be critical for effective control of stripe rust on winter crops, seedling before winter, seedling after winter, and flag-leaf stage. Application of any time points should be tested with various fungicides to determine the effectiveness and economic values for a specific region before it can be used in commercial fields.

6.1.5.1 Fungicide Application on Seedlings Before Winter

The stripe rust fungus can infect wheat and barley plants just after emerge if urediniospores are around, plants are susceptible at seedling stage and weather conditions are favorable for infection. These conditions can happen in some areas, such as the U.S. Pacific coast and the south and east of Gansu Province, China. If not controlled, heavy infection in the fall can severely reduce plant vigor, which may reduce the ability of the plants for overwinter. Heavy sporulation may provide urediniospores for wind dispersal to other regions with warmer winter for the pathogen easier to overwinter or propagate. Thus, control of stripe rust in the fall may be necessary in the regions where stripe rust frequently occurs on seedlings of winter wheat. Treating seeds with systemic fungicides with effectiveness of 3 weeks or longer may prevent or reduce fall infection. If there are no effective fungicides for seed treatments, foliar fungicides can be used. For regions where fall infection is low and/or cold winter can reduce the fungus, it is not necessary to do seed treatment at the time of planting or apply foliar fungicides before winter.

6.1.5.2 Fungicide Application on Seedlings After Winter

In regions where stripe rust infection occurs in the fall and the pathogen can overwinter, foliar fungicides may be applied when stripe rust starts sporulating in the early spring. For this early spring application, fungicides are often mixed with herbicides and applied at the timing of herbicide application for weed control. As it can get rid of or reduce overwintered stripe rust fungus, easy to do, and does not add cost for application, this practice has become more and more popular in the United States and many other countries where fungicides have become cheap and herbicides are timely used. However, fungicides should be used only when necessary, and efforts, such as conducting field survey and providing accurate forecast of stripe rust, should be taken to prevent unnecessary or excessive use of fungicides.

6.1.5.3 Fungicide Application at a Late Crop Growth Stage

Depending upon how early stripe rust starts developing, susceptibility of crop cultivars, whether fungicide is used or not and what fungicide is previously used, and weather conditions, timing of fungicide application is more complex for the late crop growth season. It is most critical to protect a crop from stripe rust infection after the flag leaves as flag leaves contribute more than other leaves to grain yield. Thus, fungicide application after flag leaves fully expended is often more important than any early season application. As listed in Table 6.1, most fungicides cannot be applied after flowering (anthesis) stage (Feekes 10.5) and some cannot be used 30–40 days before harvest. Most fungicides also have limitation in times and dosage during a crop season and the duration of effectiveness varies greatly (20–40 days) among fungicides. Therefore the timing and choosing of a fungicide in late crop season need to be determined based on whether, when and what fungicide is previously used during the same crop season, stripe rust pressure, crop stage, cultivars with HTAP resistance or not, and weather conditions.

6.1.6 Methods of Fungicide Application

Fungicides can be applied using different techniques. The easiest application method is seed treatment, but it can be only used for the regions where stripe rust infection occurs in the fall and effective fungicides as seed treatment are available. Applications in the early crop season, including at seedling stage before or after winter, is relatively easy as ground application can be used. Ground application is using a tractor to pull a sprayer (Fig. 6.1a). For fungicide spray in late crop growth season, ground application can cause significant damage from tractor tracking, and sometimes can be limited by wet soil conditions. It is more efficient through air application by dust planes (Fig. 6.1b). Air application generally costs more than ground application. During a widespread stripe rust epidemic, such as during the 2010 and 2011 epidemics in the U.S. Pacific Northwest, the capacity of air application in a region can be a limiting factor. Growers may not be able to book dusty planes for immediate and on-time application. Any delay from the best time of application can significantly reduce the effectiveness and benefits of fungicide application. Air application has become popular in Europe and North America, but is still limited in use in less developed countries and regions. In China and many other countries, backpack sprayers are mostly used (Li and Zeng 2002). Such application of fungicides is labor-intensive.

When cereal crops are grown under irrigation using a sprinkler system, a fungicide can be applied through the irrigation system. This approach is often referred as chemigation or fungigation.

To increase the efficacy of a fungicide, an adjuvant is tank mixed or included in the fungicide formulation to improve coverage on the plants or penetration into plant tissue. An adjuvant compound is also known as spreader-sticker and surfactant.

Fig. 6.1 Equipment for application of fungicides. (**a**) A chemical application system for ground application. The beam of application spans 36.576 m, and the tanker contains 8000 l, enough for spaying more than 40 ha. (**b**) A dust plane is spraying fungicide for control of stripe rust over a wheat field

6.1.7 Side Effects

Fungicides are chemicals that amy be harmful to humans and animals, and may have phytotoxicity when over-exposed, over-used, used at unspecified plant stage, or used on not specified plants. For example, azoxystrobin fungicides (such as Abound and Quadris) can cause irritating to skin, eyes and respiratory tract to humans (http://extension.psu.edu/pests/pesticide-education/applicators/fact-sheets/pesticide-safety/potential-health-effects-of-pesticides). Benomyl fungicides has a very low toxicity in animals, but mildly toxic to fish (Lorgue et al. 1996). Some fungicides can have dual usages (Oruc 2010). For example, thiabendazole, a widely used agricultural fungicide, can be used as drugs against intestinal parasites in humans and animals (Lorgue et al. 1996).

Fungicide toxicity studies often use lab animals. Let's take the Propiconazole fungicide (e.g. Tilt 3.6E) as an example, as it has been widely used for control of stripe rust and tested for effects on a wide range of animals (http://pmep.cce.cornell.edu/profiles/extoxnet/metiram-propoxur/propiconazole-ext.html). Based on tests on rabbit and guinea pigs, the formulated products are considered a moderate irritant or sensitizer (Ciba-Geigy 1997). In a 2-year feeding study in mice, the No-Observed-Effect-Level (NOEL) was established at 100 ppm, and significant increases were noted in the incidence of spontaneous liver tumors observed in male mice at the highest feeding level only. In a 2-year rat feeding study, the no-effect-level was also established at 100 ppm, and there were no tumors in the rat at any feeding level. In a 1-year feeding study in dogs, the NOEL was established at 250 ppm, the highest level tested. In a 2-year period, rats and their offspring were fed a diet containing propiconazole at concentrations up to 2500 ppm, the high dose reduced body weight and increased in liver weight of adults and pups, but did not affect any reproductive, foetal and embryonic parameters (Ciba-Geigy 1997). No evidence of mutagenic potential was observed. Animals can rapidly metabolize propiconazole to a wide variety of compounds. The n-propyl side chain of the dioxolane ring can be oxidized to give a series of metabolites consisting of formic, acetic and propionic acids. Propiconazole was found to have the following LD50s for avian species: mallard duck >2510 mg/kg, Peking duck >3000 mg/kg, Japanese quail 2223 mg/kg and bobwhite quail 2825 mg/kg (2). Propiconazole has the following EC50/LC50 values for freshwater fish species: bluegill 1.3–10.2 ppm, brown trout 3.3 ppm, rainbow trout 0.9–13.2 ppm, carp 6.8–21.0 ppm, catfish 2.0–5.1 ppm and fathead minnow 7.6 ppm. Propiconazole has the following EC50/LC50 values for freshwater invertebrate species: crayfish 42.0 ppm and water flea 3.2–11.5 ppm. The marine species oyster and Mysid shrimp have EC50/LC50 values of 0.3 ppm and 0.5–1.4 ppm, respectively. Propiconazole does not appear to have any adverse effects on soil microbes and earthworms (Ciba-Geigy 1997).

Moser et al. (2001) conducted a developmental exposure study with tebuconazole, also a triazole fungicide, in rats. The study demonstrates the feasibility of detecting relatively specific adverse effects caused by chemical exposure. Specifically, they found that Tebuconazole exposure was associated with impaired performance in the Morris Water Maze learning task. This learning deficit correlated well with neuropathological findings (i.e., pyknotic cells and pyramidal cell loss in the hippocampal fields). They concluded that perinatal exposure to tebuconazole produced neurobehavioral deficits and neuropathology in rats, although did not alter immunological or reproductive functions.

In general, currently used fungicides had been tested for their toxicity and harmfulness before they were registered. It is generally considered that the labeled rate of a fungicide does not cause problems to applicators, animals in surrounding areas and crops when properly used. However, people handling fungicides need to read the Material Safety Data Sheets (MSDS) and labels and to know the harmfulness and follow the instructions for protecting themselves when conducting applications. People may not be allowed and animals may be prevented to enter the fields for a certain period of time after fungicide application. More importantly, fungicides should not be applied on cereal crops more than the labeled rates during a crop season or after the labeled growth stage to avoid excessive residue in grains or other products.

Because the fungicides currently used for control of stripe rust are also effective against other foliar diseases and may also have some effects on reducing diseases caused by soil pathogens, application of fungicides to control stripe rust may also prevent other diseases. For example, wheat leaf rust, caused by *Puccinia triticina*, has not occurred at significant levels in the U.S. Pacific Northwest in the recent decades, which can be at least partially attributed to fungicide application for control of stripe rust.

The plant vigor effect, or growth-promoting effect, of fungicides is a controversial issue. Such effect is stated as application of a fungicide increase plant vigor or growth, and as a results, increase yield. In other words, fungicides can have effect of a fertilizer. This has been often used for promoting use of fungicides. However, there is no convincing evidence to support such effect. Based on over a decade field experiments of testing Tilt on various spring and winter wheat cultivars grown in the U.S. Pacific Northwest, the results did not show that fungicide application increases grain yield without stripe rust (Chen 2014). These studies used Tilt, the most widely used fungicide for control of stripe rust. However, whether any more recently developed fungicides have any fertilizer or growth promoting effect need careful studies.

6.1.8 Fungicide Resistance

Fungicide resistance is defined as "the stable, inheritable adjustment by a fungus to a fungicide, resulting in a less than normal sensitivity to that fungicide" (Dekker 1995). Fungicide resistance has been reported in many plant pathogenic fungi (McGrath 2001; Mavroeidi and Shaw 2006; Brent and Holloman 2007; Deising et al. 2008; Godoy 2012; Beckerman 2013; FRAG-UK 2015). Only a few cases of fungicide resistance or less sensitivity have been reported in rust fungi (Godoy 2012). So far, there have been no reports of fungicide resistance in the stripe rust fungus (FRAG-UK 2015).

In the United States, occasional unsuccessful control of stripe rust was thought to be due to fungicide resistance, but eventually found to be caused by either late application and/or less than the full rate. In a recent study conducted by Kang et al. (2015), historical and recent isolates of the wheat stripe rust pathogen were tested for sensitivity to Tilt (propiconazole) and Headline (pyraclostrobin) by germinating urediniospores and infecting wheat seedlings and adult-plants in a greenhouse at various concentrations. No significant difference was observed among isolates at the concentration rates of 1.00 (the full-labeled rate; 1.64 mg/ml) (100% dead) and 0.04; while significant differences were observed among the isolates at the rates of 0.06–0.20 in the germination test with Tilt. With Headline, significant germination differences were observed among the isolates at the rates of 0.0015–0.0040, but no significant differences at the rates between 0.0075 and 1.0 (the full-labeled rate, 3.7 mg/ml) (100% dead). No sporulation was observed at the rates of 1.0 and 2.0 of Tilt or 0.5 and 1.0 of Headline applied 1 day before; 0.25 and higher rates of Tilt or 0.5 and 1.0 of Headline applied at the same day; and at the rate of 1.0 of Headline applied 3 days after pathogen inoculation, while significant differences were observed in the other rates in the seedling tests. Similar results were obtained in the adult-plant tests. This study indicates that the fungicides are still effective at the full-labeled rates, but different sensitivities exist among pathogen isolates, and the results provide the baseline sensitivity for continual monitoring the pathogen population for fungicide resistance.

Although there is no evidence for fungicide resistance in the stripe rust fungus, more studies are needed for monitoring the pathogen population, and management strategies are needed for preventing or delaying the appearance of fungicide resistant strains. The following approaches should be considered: (1) Use an integrated control approach, especially growing resistant cultivars and use proper cultural practices, to reduce the use of fungicides. (2) Apply a fungicide only when it is necessary and avoid unnecessary fungicide application. (3) Use a fungicide only for its effect for control of stripe rust or other diseases caused by fungal pathogens and avoid using it because of unproved other activities. (4) Alternate fungicides that have different modes of action and/or apply mixtures of fungicides with different modes of action. (5) Apply fungicides on time and avoid late application that is usually less effective and may provide chemical conditions for the fungus to gradually develop tolerance. (6) Use the full rate on the label of a fungicide and avoid using a reduced rate for the same reasons as for (5). (7) Use a surfactant with a fungicide and apply under suitable weather conditions to improve the coverage and efficacy of the fungicides. (8) After fungicide application, monitor the crop for signs of sporulation, which may indicate resistance development.

6.1.9 Economics of Chemical Control

The economics of chemical control depends upon many factors, including yield potential, produce (e.g. grain and hay) price, usages of products, impact on the disease control in the region and other regions, potential yield loss, costs of a fungicide and application, and costs of harvesting and handling the proportion of the product saved by the application. All of these factors can be considered to make decisions whether or not a fungicide application should be used. A formula can be developed for making decisions using these factors. Yield potential and produce price make the output in one side of the formula, and potential loss and all costs make the input for the other side of the formula. When the both sides are in balance or the output is greater than the input, it is reasonable to use a fungicide. Such formula is based on a business or market model, which does not take the product usage and impact into the consideration.

In the state of Washington, it was estimated that fungicide application saved growers about $96 million at the cost of about $27 million in 2010 (Chen 2014). In 2011, the stripe rust potential damage was even greater (more than 90% yield loss on susceptible winter wheat cultivars), but the state wheat growers had a bumper harvest because of widely use of fungicides. In that year, fungicide application saved about $136 million at the cost of about $28 million. The input and output ratio for chemical control was about 1:5 on winter wheat. In the same year, the fungicide control cost about $12.3 million, which saved about $39.4 million, at about a 1:3 ratio of input to output for spring wheat. For some fields grown with quite susceptible wheat cultivars (yield losses estimated up to 45%, about half of the over 90% yield loss on highly susceptible experimental varieties), the output was more than ten times of the input of chemical control. Such successful control of stripe rust by fungicides was achieved by relatively on-time application under the environmental conditions for high yield potential and extremely high stripe rust pressure, and the economic conditions of relatively low cost of fungicides and relatively high wheat grain price in the market.

In the world, not all agriculture products are for sale. In many areas, farmers consume what they produce by themselves on their farms. In this situation, decisions for chemical control should be made based on the necessity of produce, such as wheat grain, not on how much the saved product is worthy in monetary, because the product is their only or major food source. This factor can be considered for a farmer, and also for a region.

It is difficult to quantify or determine the impact of control of stripe rust in a field to the whole region or in one region to another. However, it can be considered from the disease epidemiological standpoint. Regions or fields can be separated into sources and recipients of pathogen inoculum and they differ in time of disease development (Li and Zeng 2002; Chen 2005). Control of stripe rust in a source region can reduce inoculum produced in this region that may be blown by wind to other regions. For example, in the US Pacific Northwest, stripe rust first starts in fields planted early with susceptible winter wheat cultivars before or after winter, and the urediniospores produced in such fields provide inoculum for infection of late planted winter wheat and spring wheat in the same region and regions further north and east (Chen 2005). Early control of stripe rust by fungicides can be important for this region and other influenced regions. Similarly, in the Great Plains, urediniopores produced in the southern part are spread by wind to the central and north states and also to the eastern states. Early control of stripe rust in the southern region is beneficial for other regions. Similarly, in China, the over-summering regions provide urediniopores for fall infection of wheat crops in the overwintering regions to the east which in turn provides inoculum to the regions further east and north in the spring (Li and Zeng 2002). Therefore, control of stripe rust in the oversummering regions is beneficial to other regions even the whole country.

6.2 Cultural Practices for Stripe Rust Reduction

Appropriate cultural practices can play an important role in reduction of stripe rust epidemics. The most important cultural practice is planting resistant cultivars, which is so important that we designate a special chapter to discuss resistance to stripe rust. In this section, I mainly discuss other cultural practices. Many cultural practices have effects on stripe rust, but also have effects on other issues, such as yield, weed control, soil conservation and environment protection. The effects can be different from region to region. Therefore, it is important to determine the effects of cultural practices on stripe rust and other issues to choose appropriate practices.

6.2.1 Cropping Systems

Cropping systems play an important role in occurrence, frequency and severity of stripe rust in a region. In general, stripe rust is more frequent in regions where both winter and spring wheat crops are grown than regions where only winter or spring wheat is grown, because the winter and spring wheat co-growing system allows host plants available all year long. For example, in the U.S. Pacific Northwest, winter and spring wheat cultivars are grown, which makes stripe rust epidemic more frequent than any other region in the United States (Chen 2005; Chen et al. 2010; Wan and Chen 2012, 2014). Before 1994, winter wheat occupies 90% and spring wheat 10% of the total wheat acreage in the state of Washington. In contrast, by 1999, spring wheat increased to 25% and since then has been planted in 25–30% (http://www.nass.usda.gov). Such increase of spring wheat has reduced the gaps between crops and has at least partially contributed to the more frequent severe epidemics since 2000. Although spring wheat has lower yield than winter wheat, the benefits of growing spring wheat in the region include a good rotation crop and reducing weed and soil erosion. Barley stripe rust also occurs in the region, but has decreased in severity since 2000 (Chen 2004; Wan and Chen 2012). The factors contributing to the reduction of barley stripe rust pressure include mainly spring barley grown in the region and the reduction of barley acreage in the past decade. In northwestern China, both winter wheat and spring wheat are also grown in some areas that are considered as "hot-spots" for stripe rust (Li and Zeng 2002). In eastern Africa, two crops of spring wheat are grown annually in some areas, where stripe rust is a serious problem (Negassa et al. 2013).

Changing crop systems can be an approach for reducing stripe rust epidemics, for example, growing only winter or spring wheat, but its consequences on individual farms, the regional or even national economics, food security and the environment need to be considered. If changing crop systems are not feasible or not beneficial, other approaches for control of stripe rust need to be taken.

6.2.2 Reduce Volunteer Plants

For wheat, volunteer plants are those growing from kernels dropped onto the ground before or during harvest. Volunteer plants are green-bridges between crop seasons, and if infected, they can carry the stripe rust pathogen to pass the off-season and provide inoculum for infecting next crop. To reduce volunteer plants, cultivars with a less shattering problem should be developed and grown, harvest grain on time to reduce shattering, and use tillage or chemifallow.

No-till or reduced tillage has become more and more popular in many regions for soil conservation, environmental protection, reduction of farming input and other purposes. This cultural practice generally increases volunteer plants. To use the practice without increase stripe rust pressure, it would be better to grow resistant cultivars. Chemifallow may be used to eliminate volunteer plants.

6.2.3 Appropriate Planting Timing

The time of planting of wheat or barley is related to stripe rust epidemics. For winter wheat, early planting generally increases the chance of fall infection, and urediniospores produced in early planted fields are inoculum sources for other fields in the region and also for other regions. In the U.S. Pacific Northwest, winter wheat is normally planted from middle August to late October and harvested from middle July to late August, and spring wheat planted from March to April and harvested from late August to early October, varying among regions with different amounts of precipitation. Winter wheat is planted earlier in dryland areas than in areas of high precipitation. Summer fallow is practiced more in dryland areas, which allows early planting of winter wheat. As precipitation is low and emergency is a major problem, early planting for utilization of soil moisture for establishing a good crop is generally beneficial for high yield. Stripe rust infection in the fall takes place more and rust re-appears earlier in the early planted fields. However, appropriate delayed planting can greatly reduce fall infection and possibly rust pressure after winter. Differences in stripe rust infection between early and late planted fields can be dramatic. In the fall of 2010, unusually severe stripe rust was observed in many of the wheat fields planted in August in the dryland of central Washington State due to the unusually high precipitation in the September, while fields planted after the middle September did not show stripe rust by the November. In the fall of 2013, the fall was dry as normal, no stripe rust was found in the entire central Washington by the November, except one field in which every plant was severely infected. That field was planted in late July, which was too early.

The general recommendation for winter wheat and winter barley is appropriate late planting under the conditions not significantly reduce yield potential.

The effects of planting time for spring wheat and spring barley on stripe rust can be different from for winter crops. A late planting pushes maturity late, and if weather conditions continue to be favorable (wet and cool) for stripe rust throughout the growth season, stripe rust can continue developing into late summer and provide more urediniospores that have a better chance to survive late summer and infect winter wheat in the fall. On the other hand, if weather conditions are not favorable (dry and hot) for stripe rust in the late growth season, late planting may shorten the period of time during which crops are vulnerable to stripe rust. However, late planted spring wheat and spring barley crops are generally vulnerable to leaf rust and stem rust. In regions where both winter and spring wheats are grown, early planted spring wheat is immediately under the pressure of stripe rust infection if stripe rust is severe on winter crops. Therefore, either early or late plantings of spring crops can be more or less vulnerable to stripe rust, depending upon stripe rust pressure on winter crops and weather conditions especially in the late crop season. Because spring crops have a much short growth season, planting time has a much greater effect on yield than winter wheat. General recommendation is to plant spring crops at the time period most suitable for the region mainly based on yield.

6.2.4 Planting Cultivar Mixtures, Multiline Cultivars or Mixture of Crops

A seed mixture of cultivars can consist of two or more different cultivars. If two or more cultivars have compromised characteristics, such as cultivars of high yielding but susceptible to stripe rust mixed with cultivars of low yielding but resistant to the disease, can reduce stripe rust damage if the disease is severe and increase yield if the disease is not severe. In addition to stripe rust, abiotic stresses, such as emergence, winter hardness and drought tolerance, are also considered to choose suitable combinations of cultivars. Several studies have been conducted to determine the effects of cultivar mixtures on stripe rust (Kampmeijer and Zadoks 1977; Finckh and Mundt 1992, 1996; Dileone and Mundt 1994; Mundt et al. 1995; Akanda and Mundt 1996a, b, 1997; Calonnec et al. 1996; Mundt et al. 1996; Garrett and Mundt 2000; Mundt 2002). Planting mixtures of cultivars is easy to do. In the U.S. Pacific Northwest, growing a mixture of two or three cultivars has become a common practice (Chen 2005). In the recent years, about 30% of winter wheat acreage was planted with mixtures of different cultivars.

A multiline cultivar is composed of several lines with different disease resistance genes in a similar background. In the U.S. Pacific Northwest, two multiline cultivars were developed specifically for control of stripe rust on club wheat. Crew (CItr 17,951), consisting of ten lines with different genes for race-specific resistance to stripe rust, was released in 1981 (Allan et al. 1983). The cultivar was grown for about 10 years, and its resistance was moderately durable (Line 2002). Crew was replaced by Rely. Rely (PI 542401), which also consisted of ten lines, was released in 1991 (Allan et al. 1993). The ten individual lines resembled the common recurrent parent, Tres (CItr 17,917), a popular club wheat cultivar grown in the region in the late 1980s but became susceptible to new races of *Pst* (Line and Qayoum 1992). Rely was the predominant club wheat cultivar in the state of Washington from 1996 to 2000. During that time, Rely's resistance remained effective. Multiline cultivars have helped to sustain club wheat production for at least 18 years (Line 2002). Compared to cultivar mixture, multiline cultivars require more intensive breeding efforts to develop and also additional effort is needed to maintain and mix individual lines. The multiline cultivars have not been widely grown and no new multiline cultivars have been developed because new cultivars have higher yields, better quality, and more importantly have better resistance to stripe rust through combining effective all-stage resistance and durable HTAP resistance.

Another approach to diversify crops in a single field is to grow a mixture of different crops. For example, growing two or more crops simultaneously or overlapping in a single field has been practiced in China for hundreds or thousands of years. Currently, intercropping winter wheat and spring corn is a common practice in many regions. Recently, Cao et al. (2015) studied the effects of intercropping on stripe rust and powdery mildew in wheat. They tested intercropping of three stripe rust susceptible winter wheat varieties, Lantian 13, Lantian 6 and Tian 94–3, with potato, chili, corn, sunflower, benne or soybean on reduction of the diseases. The relative control efficacies of the intercropping between wheat and maize were 16.7–45.7% for stripe rust and 14.7–37.0% for powdery mildew compared to the pure stands of wheat, increasing yield by 52.4–140.0%. The relative control efficacies of intercropping between wheat and sunflower were 5.9–28.9% for stripe rust and 11.7–18.4% for powdery mildew, increasing yield by 1.4–24.8%. In contrast, the intercropping system of wheat with potato or chili did not significantly reduce the diseases, but increased the yield by 150.0% or more. Other crop combinations did not result in a significant disease reduction or yield increase. Growing a mixture of different crops or intercropping is labor intensive and has been limited in regions where the land is limited and labor is less expensive.

All of these practices are to reduce disease risk through increasing diversity in a single field. The mechanisms of reducing diseases, such as stripe rust, by growing a mixture of cultivars include the dilution of pathogen inoculum, compensation of resistant plants for susceptible plants through increasing tiller number and induced resistance (Mundt 2002). Although these approaches can reduce stripe rust, they can be more useful or profitable when they have benefits considering other agronomic traits or agricultural factors.

6.2.5 Avoid Excessive Irrigation

Irrigation is used for wheat production in many regions of the world, as it provides essential moisture for growing wheat and barley or for increasing yield. It also provides moisture essential for infection of crops by the stripe rust pathogen and many other fungal pathogens. In general, wheat and barley crops under irrigation are more vulnerable to stripe rust than under rainfed conditions. Thus, it is more important to choose highly resistant cultivars to grown under irrigation. However, if highly resistant cultivars are not available or resistant cultivars become susceptible or less resistant, it is better to balance the frequency and amount of irrigation water for reducing stripe rust and achieving high yield, in addition to use fungicides if possible. Different methods of irrigation may have different effects on stripe rust infection and development. The methods without wetting or shortening the wet period of plant canopy (e.g. drip/trickle and subsurface) should be better for reducing stripe rust than the methods (e. g. sprinkler) of wetting the canopy. Appropriate timing of irrigation may be chosen for less favorable for stripe rust. For example, irrigating when temperature is and continues to be above 23 °C is better than when temperatures are below because urediniospores of the stripe rust fungus generally do not germinate at the higher temperatures. If chemigation is used, it combines irrigation with fungicide application. The general recommendation is to avoid excessive irrigation.

6.2.6 Avoid Excessive Fertilization

Fertilization is essential for growing wheat and barley crops for high yield. It affects growth of host plants and stripe rust fungus, and also their interactions. The effects of fertilizers on stripe rust have been studied since the 1910s (Comes 1913; Spinks 1913; Hiltner 1914; Stranak 1915).

Nitrogen is the most important fertilizer element as it determines the productivity of wheat and barley, and also the most studied element for its effect on stripe rust. In general, higher nitrogen associated with increased disease severity (Darwinkel 1980; Ash and Brown 1991; Bryson et al. 1995; Danial and Parlevliet 1995; Neumann et al. 2004; Devadas et al. 2014). A major mechanism of high nitrogen increase stripe rust is that the increased plant density creates high moisture that is favorable for stripe rust infection (Darwinkel 1980; Danial and Parlevliet 1995). Another mechanism is that it increases nitrogen content in the host tissue acting as a substrate for pathogen growth (Bryson et al. 1995; Neumann et al. 2004). A recent study by Devadas et al. (2014) indicated that the total amount of nitrogen entering the grain was reduced by stripe rust and the effects of stripe rust on nitrogen yield are most likely associated with reduced uptake of nitrogen during grain filling.

Chloride also affects stripe rust development. There have been several reports that chloride reduces stripe rust (Russell 1978; Christensen et al. 1982; Scheyer et al. 1987). When applied to the soil, both NaCl and KCl reduced stripe rust severity in both greenhouse and field experiments (Russell 1978). Because sodium nitrate and sodium phosphate did not reduce stripe rust, the reduction of stripe rust appeared to be by chloride. Less stripe rust was observed on winter wheat cultivars Stephens and Yamhill when fertilized with NH4Cl than with (NH4)2SO4 at an equivalent rate of nitrogen (Christensen et al. 1982). In the study of Scheyer et al. (1987), the reduction of stripe rust infection rate by chloride on three winter wheat tested, Purplestraw, Yamhill and Rew, was apparent in 1 year, but the increase of infection rate was observed on Rew in another year. The authors concluded that Cl-containing fertilizers can affect stripe rust development, but the effects are small, inconsistent and of questionable practical value.

6.2.7 Planting Non-host Crops

In regions where severe epidemics of stripe rust occur very often and provide inoculum for infecting crops in other regions, growing other non-host crops can be used to reduce susceptible wheat areas. This approach has been proposed by Chinese scientists and used in southern and eastern Gansu, the "hot-pots" of oversummering regions for stripe rust epidemics in China (Li and Zeng 2002). The non-host crops that have been grown in the regions include corn, rice, canola, sunflowers and various vegetables. The change crop approach can have some effects on reduction of stripe rust inoculum and epidemics, and possibly increase grower's income. However, the approach is limited by regions and could not be used everywhere because wheat is the most important staple food crop for many countries or regions.

6.3 Integrated Control of the Disease

Stripe rust can be reduced by various approaches as discussed above. Each of the approaches has advantages and disadvantages, and none of the approaches alone can completely prevent stripe rust epidemic or reduce damage to minimal in a region and in a period of time. An integrated strategy is needed for control of stripe rust on a yearly basis based on resources available for a region and the integrated strategy is continually needed to improve to meet the challenges of increasing grain yield and quality for food security, food safety, changes in cropping systems, adaptation to new technologies, changes in climate, protection of environment and reducing production inputs.

6.3.1 Approaches of Stripe Rust Control

Approaches that can be used for controlling and reducing stripe rust are generally categorized as resistance, fungicide and cultural methods (Chen 2005). Individual methods for these categories have been discussed above. The following mainly focus on the advantages and disadvantages for each of the categories.

6.3.1.1 Resistance

Growing resistant cultivars is considered most preferred strategy for control of stripe rust, as it is effective, economic, easy-to-use and environmentally friendly (Chen 2013, 2014). In a field planted with a cultivar of a high level of resistance, the crop will be completely protected from stripe rust damage without fungicide application. More wheat acreage is planted with resistant cultivars, less damage stripe rust can cause in a region. Seed of resistant cultivars may not cost more to growers. Eliminating the need of fungicide application will save the cost of fungicides and application and also eliminating the potential adverse effects to the environment and humans, without potential problems of chemical residue in grain. Growing resistant cultivars is easy to do as it does not need additional farming techniques, equipment and practices. This approach can be used in any regions and under any crop system conditions. Growing resistant cultivars is the only or major options for some cropping systems, such as organic wheat and barley production or in areas of strict use of chemicals to protect streams, bees and wild life, because under such systems, fungicide application cannot be used or the use is limited.

However, growing resistant cultivars is under the conditions that resistant cultivars are available, and they are comparable to susceptible cultivars in yield, quality and price of grain, adaptation, and suitability to other agricultural technologies such as Clearfield for weed control. Developing resistant cultivars needs extra effort in breeding and takes many years, although marker-assistant selection and doubled-haploid techniques have been developed for speeding up the breeding process. Breeding resistant cultivars needs considerable investment, but the effective control and saving cost for fungicide application make the investment cost effective.

The major challenge is that not all genes can provide long-lasting resistance because new races with a new virulence or combination of previously existing virulences, which may develop in the pathogen population, can circumvent the race-specific resistance provided by individual resistance genes. In general, race-specific resistance has a life span of effectiveness about 3.5 years (Chen 2005). Less race-specific and more durable resistance, such as HTAP resistance, is available in wheat and barley for resistance to stripe rust. Many germplasm with stripe rust resistance and genes or QTL for HTAP resistance have been identified in wheat and barley, and this type of resistance has been widely used in the U.S. Pacific Northwest and many other regions in the world (Chen 2014). Although HTAP resistance is generally not influenced by pathogen races, but it is incomplete and under influence of weather condition, crop stage and inoculum pressure (Chen 2005, 2013). Under severe stripe rust epidemics, cultivars with low or moderate level of HTAP resistance can suffer considerable yield losses (Chen 2014). For cultivars with high levels of HTAP resistance, fungicide application is not needed in the late growth season. However, they can be infected in seedling stage in regions where fall infection is common and the pathogen can easily overwinter and start developing before temperatures become high enough and plants grow old enough to allow HTAP resistance expressing. For these regions, the best strategies for developing wheat and barley cultivars with high level, all-stage and durable resistance is through pyramiding genes for effective all-stage resistance and HTAP resistance into individual cultivars. Use of different resistance genes in different crops (winter wheat vs, spring wheat) in a region and different genes in different epidemic regions can be helpful for reducing epidemics and prolonging cultivar resistance (Chen 2013).

6.3.1.2 Fungicides

Now, fungicides that are effective for control of stripe rust are available. Application of fungicides is a major method for preventing stripe rust epidemics and reducing yield losses. When stripe rust occurs, it is necessary to apply fungicides on susceptible cultivars. When resistant cultivars become susceptible and no resistant cultivars are available, chemical control is the only way to keep growing wheat and barley. However, use of fungicides adds extra cost for grain production. Although the registered fungicides are generally safe when used in the registered growth stages and dosage, they may have some minor effects of chronological effects to humans, animals and environments. Fungicides need to be sprayed on time to have their maximal effects. Under certain circumstances, such as organic farming and regions protected for wild life and natural resources, fungicides cannot be used. Overly use of fungicides may provide selection pressure for fungicide-tolerant strains. Voiding or reducing excessive use of fungicides has become a challenge. To use fungicides only when needed and use on time need accurate prediction, diligent disease monitoring, and careful characterization of pathogen races and cultivar resistance (Chen 2014).

6.3.1.3 Cultural Practices

As discussed above, various cultural practices can reduce stripe rust. When used properly, they can have significant effects on reducing stripe rust, but generally their effects are limited. Some practices need extra activities and others may not be compatible to other cultural practices. This category of the approaches needs to be integrated into the overall cropping system best for a region.

6.3.2 Integrated Control

As discussed above, the three major categories of approaches all have advantages and disadvantages. Thus they need to be integrated to take their advantages and overcome the disadvantages. In a general integrated strategy, resistance should be consider most important and as the first line of defense, chemical control as a second line of defense and cultural practices as a supplementary approach. The ranking can be different from region to region, and also from time to time in a single region depending upon availability of resources and technology development.

In the U.S. Pacific Northwest, an integrated control system has been developed based on disease forecast and monitoring, virulences in the pathogen populations, cultivar resistance and yield loss potential, and efficacy and timing of fungicides (Chen 2014). The major concept of the system is to integrate cultivar resistance with fungicide application to achieve maximal protection of crops from stripe rust damage and also reduce unnecessary use of fungicides. The system requires well characterization of stripe rust resistance in each cultivar (Chen et al. 2015b, c, 2016b) and virulent races of the pathogen (Wan and Chen 2012, 2014), monitoring the start of development of the disease, and effectiveness of fungicides (http://striperust.wsu.edu). Avoiding unnecessary early planting of winter wheat and excessive irrigation of wheat crop in irrigation wheat production areas is recommended. Disease forecast, updates and recommendations are provided to growers through e-mails, reports, websites, social media and extension systems (http://striperust.wsu.edu).

6.3.3 Sustainability and Future Perspectives

Each of individual approaches does not lead to sustainable control of stripe rust. There have been debates about taking which strategies. For examples, when resistance breeding is emphasized, fungicide testing is ignored and not pursued. In fact, fungicides have not been tested and utilized in many countries. When epidemics occur, no options are available for solving emerging disease problems. Such situations of stripe rust and other diseases have happened even in the recent years. This problem has been caused by some incorrect retrospectives that fungicides cannot be used in poor or low-yielding regions as fungicides are expensive. However, fungicide technologies have been continually developing, and cost has been greatly reduced. Old perspectives need to be changed. On the other hand, as more and more effective fungicides become available and some generic fungicides become cheap, pushing for using fungicides and not paying enough attention for developing resistant cultivars have become a problem. Either of these directions would not make stripe rust control sustainable as both resistance and fungicide approaches have different advantages and disadvantages as discussed above.

To achieve sustainable control of stripe rust, more efforts should be taken to develop resistant cultivars. Because there are different types of resistance and they also have different advantages and advantages (Chen 2005, 2013, 2014), there have been also debates on use of which types of resistance. When high-level of resistance is emphasized, people look down and even strongly against partial resistance as reviewed by Line (2002). On the other hand, when non-race specific resistance is emphasized, the non-durability of individual genes for race specific resistance is over looked. In this case, people often just look at the epidemics caused by circumvent of resistance genes by new races of the pathogen, but forget the contribution of the individual genes to the control of the disease in the years when they were effective. Of course, one type of resistance can be better than another in a region or in a specific historical period. However, in general and in a long term, an integrating breeding approach of using the combination of different types of resistance and different resistance genes may lead to an improved and sustainable control of stripe rust. This strategy requires a large number of genes for different types of resistance to stripe rust. In the past decade, great progress has been made in discovering new resistance genes and identifying molecular markers for resistance genes through traditional genetics and molecular mapping and genome wide association mapping studies (Imtiaz et al. 2004; Mallard et al. 2005; Christiansen et al. 2006; Li et al. 2006a, b; Yan and Chen 2006; Kuraparthy et al. 2007; Lin and Chen 2007, 2008; Yan and Chen 2007; Chhuneja et al. 2008; Lillemo et al. 2008; Rosewarne et al. 2008; Santra et al. 2008; Yan and Chen 2008; Carter et al. 2009; Dedryver et al. 2009; Fu et al. 2009; Li et al. 2009; Lin and Chen 2009; Lin et al. 2009; Lu et al. 2009; Murphy et al. 2009; Sui et al. 2009; Yao et al. 2009; Cheng and Chen 2010; Li et al. 2010; Bansal et al. 2011; Fang et al. 2011; Hao et al. 2011; Herrera-Foessel et al. 2011; Lowe et al. 2011; Prins et al. 2011; Zhou et al. 2011; Chen et al. 2012c; Naz et al. 2012; Paillard et al. 2012; Ren et al. 2012a, b, c; Vazquez et al. 2012; Wang et al. 2012; Zhao et al. 2012; Chen 2013; Christopher et al. 2013a, b; Li et al. 2013; McIntosh et al. 2013; Rosewarne et al. 2013; Singh et al. 2013; Xu et al. 2013; Agenbag et al. 2014; Bansal et al. 2014; Basnet et al. 2014a, b; Case et al. 2014; Cheng et al. 2014; Feng et al. 2014; Lan et al. 2014; Lu et al. 2014; Sthapit et al. 2014; Xu et al. 2014; Zhou et al. 2014a, b; Feng et al. 2015; Hou et al. 2015; Ma et al. 2015; Maccaferri et al. 2015; Randhawa et al. 2015; Ren et al. 2015; Vazquez et al. 2015; Zhou et al. 2015a, b; Bulli et al. 2016; Klarquist et al. 2016; Li et al. 2016; Xiang et al. 2016). More genes and their useful markers for stripe rust resistance should be identified. The resistant germplasms, genes and their markers from the above and future studies are useful for diversifying resistance genes for developing new cultivars with resistance to stripe rust, which should make stripe rust control more sustainable.

Current fungicides need to be used wisely in order to control the disease when epidemics occur and also to delay potential resistant strains. At the same time, new fungicides with different modes of action should be developed. When the current widely used fungicides become ineffective, new fungicides are available for use. As organic cereal production has become more and more popular and more and more areas require not to use or limited use of chemicals, new products that can control rusts but do not have adverse effects on environment and natural resources should be developed, in addition to develop highly resistant cultivars for the specific cropping systems.

References

Acero FJF, Carbú M, El-Akhal MR, Garrido C, González-Rodríguez VE, Cantoral JM. Development of proteomics-based fungicides: new strategies for environmentally friendly control of fungal plant diseases. Int J Mol Sci. 2011;12:795–816.

Agenbag GM, Pretorius ZA, Boyd LA, Bender CM, MacCormack R, Prins R. High-resolution mapping and new marker development for adult plant stripe rust resistance QTL in the wheat cultivar Kariega. Mol Breed. 2014;34:2005–20.

Akanda SI, Mundt CC. Effects of two component wheat cultivar mixtures on stripe rust severity. Phytopathology. 1996a;86:347–53.

Akanda SI, Mundt CC. Path coefficient analysis of the effects of stripe rust and cultivar mixtures on yield and yield components of winter wheat. Theor Appl Genet. 1996b;92:666–72.

Akanda SI, Mundt CC. Effect of two-component cultivar mixtures and yellow rust on yield and yield components of wheat. Plant Pathol. 1997;46:566–80.

Allan RE, Line RF, Peterson Jr CJ, Rubenthaler GL, Morrison KJ, Rohde CR. Crew, a multiline wheat cultivar. Crop Sci. 1983;23:1015–6.

Allan RE, Peterson CJ, Line RF, Rubenthaler GL, Morris CF. Registration of Rely wheat multiline. Crop Sci. 1993;33:213–4.

Anke T, Oberwinkker F, Steglich W, Schramm G. The strobilurins new antifungal antibiotics from the Basidiomycete *Strobilurus tenacellus*. J Antibiot. 1977;30:806–7.

Ash GJ, Brown JF. Effect of nitrogen nutrition of the host on the epidemiology of *Puccinia striiformis* f. sp. *tritici* and crop yield in wheat. Australas Plant Pathol. 1991;20:108–14.

Bansal U, Forrest K, Hayden M, Miah H, Singh D, Bariana H. Characterization of a new stripe rust resistance gene *Yr47* and its genetic association with the leaf rust resistance gene *Lr52*. Theor Appl Genet. 2011;122:1461–6.

Bansal UK, Kazi AG, Singh B, Hare RA, Bariana HS. Mapping of durable stripe rust resistance in a durum wheat cultivar Wollaroi. Mol Breed. 2014;33:51–9.

Basnet BR, Ibrahim AMH, Chen XM, Singh RP, Mason ER, Bowden RL, Liu SY, Devkota RN, Subramanian NK, Rudd JC. Molecular mapping of stripe rust resistance QTL in hard red winter wheat TAM 111 adapted in the US high plains. Crop Sci. 2014a;54:1361–73.

Basnet BR, Singh RP, Ibrahim AMH, Herrera-Foessel SA, Huerta-Espino J, Lan C, Rudd JC. Characterization of *Yr54* and other genes associated with adult plant resistance to yellow rust and leaf rust in common wheat Quaiu 3. Mol Breed. 2014b;33:385–99.

Beckerman JL. Detection of fungicide resistance. In: Nita M, editors. Fungicides – showcases of integrated plant disease management from around the world. InTech, ISBN 978-953-51-1130-6; 2013. p. 281–310. http://dx.doi.org/10.5772/55981

Boshoff WHP, Pretorius ZA, van Niekerk BD. Fungicide efficacy and the impact of stripe rust on spring and winter wheat in South Africa. South Afr J Plant Soil. 2003;20:11–7.

Brent KJ, Holloman DW. Fungicide resistance in crop Pathogens: how can it be managed? FRAC Monograph No. 1. 2nd ed. Fungicide Resistance Action Committee. 2007. ISBN 90-72398-07-6, 56 pp.

Bryson RJ, Sylvester-Bradley R, Scott RKS, Paveley ND. Reconciling the effects of yellow rust on yield of winter wheat through measurements of green leaf area and radiation interception. Asp Appl Biol. 1995;42:9–18.

Bulli P, Zhang JL, Chao SM, Chen XM, Pumphrey M. Genetic architecture of resistance to stripe rust in a global winter wheat germplasm collection. G3: genes genomes. Genetics. 2016;6:2237–53.

Calonnec A, Goyeau H, de Vallavieille-Pope C. Effects of induced resistance on infection efficiency and sporulation of *Puccinia striiformis* on seedlings in varietal mixtures and on field epidemics in pure stands. Eur J Plant Pathol. 1996;5:733–41.

Cao SQ, Luo HS, Jin MA, Jin SL, Duan XY, Zhou YL, Chen WQ, Liu TG, Jia QZ, Zhang B, Huang J, Wang XM, Shang XW, Sun ZY. Intercropping influenced the occurrence of stripe rust and powdery mildew in wheat. Crop Prot. 2015;70:40–6.

Carter AH, Chen XM, Garland-Campbell K, Kidwell KK. Identifying QTL for high-temperature adult-plant resistance to stripe rust (*Puccinia striiformis* f. sp. *tritici*) in the spring wheat (*Triticum aestivum* L.) cultivar 'Louise'. Theor Appl Genet. 2009;119:1119–28.

Case A, Naruoka Y, Chen XM, Garland-Campbell KA, Zemetra RS, Carter AH. Mapping stripe rust resistance genes in a Brundage x Coda winter wheat recombinant inbred line population. PLoS One. 2014;9:e91758.

Chen XM. Epidemiology of barley stripe rust and races of *Puccinia striiformis* f. sp. *hordei*: the first decade in the United States. In: Abstracts of the 11th international cereal rusts and powdery mildews conference; 2004 Aug 22–27; Norwich, UK. p. A2.8; 2004.

Chen XM. Epidemiology and control of stripe rust [*Puccinia striiformis* f. sp. *tritici*] on wheat. Can J Plant Pathol. 2005;27:314–37.

Chen XM. High-temperature adult-plant resistance, key for sustainable control of stripe rust. Am J Plant Sci. 2013;4:608–27.

Chen XM. Integration of cultivar resistance and fungicide application for control of wheat stripe rust. Can J Plant Pathol. 2014;36:311–26.

Chen XM, Wood DA. Control of stripe rust of spring barley with foliar fungicides, 2001. F&N Tests. 2002a;57:CF01.

Chen XM, Wood DA. Control of stripe rust of spring wheat with foliar fungicides, 2001. F&N Tests. 2002b;57:CF03.

Chen XM, Wood DA. Control of stripe rust of spring barley with foliar fungicides, 2002. F&N Tests. 2003a;58:CF003.

Chen XM, Wood DA. Control of stripe rust of spring wheat with foliar fungicides, 2002. F&N Tests. 2003b;58:CF004.

Chen XM, Wood DA. Control of stripe rust of spring wheat with foliar fungicides, 2003. F&N Tests. 2004;59:CF022.

Chen XM, Wood DA. Control of stripe rust of winter wheat with foliar fungicides, 2004. F&N Tests. 2005a;60:CF027.

Chen XM, Wood DA. Control of stripe rust of spring wheat with foliar fungicides, 2004. F&N Tests. 2005b;60:CF021.

Chen XM, Wood DA. Control of stripe rust of spring barley with foliar fungicides, 2004. F&N Tests. 2005c;60:CF020.

Chen XM, Wood DA. Control of stripe rust of spring wheat with foliar fungicides, 2005. F&N Tests. 2006a;61:CF022.

Chen XM, Wood DA. Control of stripe rust of spring barley with foliar fungicides, 2005. F&N Tests. 2006b;61:CF024.

Chen XM, Wood DA. Control of stripe rust of winter wheat with foliar fungicides, 2005. F&N Tests. 2006c;61:CF023.

Chen XM, Wood DA. Reduction of stripe rust severities on spring wheat by seed treatment with fungicides, 2005. F&N Tests. 2006d;61:ST010.

Chen XM, Wood DA. Control of stripe rust of winter wheat with foliar fungicides, 2006. Plant Dis Manag Rep. 2007a;1:FC108.

Chen XM, Wood DA. Control of stripe rust of spring wheat with foliar fungicides, 2006. Plant Dis Manag Rep. 2007b;1:FC109.

Chen XM, Wood DA. Fungicide seed treatments for reducing yield losses caused by stripe rust on spring barley, 2006. Plant Dis Manag Rep. 2007c;1:ST005.

Chen XM, Wood DA. Fungicide seed treatments for reducing yield losses caused by stripe rust on spring wheat, 2006. Plant Dis Manag Rep. 2007d;1:ST006.

Chen XM, Wood DA. Fungicide seed treatments for reducing yield losses caused by stripe rust on winter wheat, 2006. Plant Dis Manag Rep. 2007e;1:ST007.

Chen XM, Wood DA. Control of stripe rust of spring wheat with foliar fungicides, 2007. Plant Dis Manag Rep. 2008a;2:CF014.

Chen XM, Wood DA. Control of stripe rust of winter wheat with foliar fungicides, 2007. Plant Dis Manag Rep. 2008b;2:CF015.

Chen XM, Wood DA. Control of stripe rust of winter wheat with foliar fungicides, 2008. Plant Dis Manag Rep. 2009a;3:CF038.
Chen XM, Wood DA. Control of stripe rust of spring wheat with foliar fungicides, 2008. Plant Dis Manag Rep. 2009b;3:CF039.
Chen XM, Wood DA. Testing fungicide Evito for control of stripe rust on spring wheat, 2008. Plant Dis Manag Rep. 2009c;3:CF040.
Chen XM, Wood DA. Control of stripe rust of spring wheat with foliar fungicides, 2009. Plant Dis Manag Rep. 2010a;4:CF002.
Chen XM, Wood DA. Control of stripe rust of winter wheat with foliar fungicides, 2009. Plant Dis Manag Rep. 2010b;4:CF003.
Chen XM, Wood DA. Control of stripe rust of spring barley with foliar fungicides, 2009. Plant Dis Manag Rep. 2010c;4:CF004.
Chen XM, Wood DA. Control of stripe rust of spring wheat with foliar fungicides, 2010. Plant Dis Manag Rep. 2011a;5:CF003.
Chen XM, Wood DA. Control of stripe rust of winter wheat with foliar fungicides, 2010. Plant Dis Manag Rep. 2011b;5:CF004.
Chen XM, Penman L, Wan AM, Cheng P. Virulence races of *Puccinia striiformis* f. sp. *tritici* in 2006 and 2007 and development of wheat stripe rust and distributions, dynamics, and evolutionary relationships of races from 2000 to 2007 in the United States. Can J Plant Pathol. 2010;32:315–33.
Chen JL, Chu CG, Souza EJ, Guttieri MJ, Chen XM, Xu S, Hole D, Zemetra R. Genome-wide identification of QTLs conferring high-temperature adult-plant (HTAP) resistance to stripe rust (*Puccinia striiformis* f. sp. *tritici*) in wheat. Mol Breed. 2012a;29:791–800.
Chen XM, Evans CK, Garner JP. Control of stripe rust of spring wheat with foliar fungicides, 2011. Plant Dis Manag Rep. 2012b;6:CF031.
Chen XM, Evans CK, Garner JP. Control of stripe rust of winter wheat with foliar fungicides, 2011. Plant Dis Manag Rep. 2012c;6:CF032.
Chen XM, Evans CK, Garner JP, Liu YM. Control of stripe rust of spring wheat with foliar fungicides, 2012. Plant Dis Manag Rep. 2013a;7:CF031.
Chen XM, Evans CK, Garner JP, Liu YM. Control of stripe rust of winter wheat with foliar fungicides, 2012. Plant Dis Manag Rep. 2013b;7:CF032.
Chen XM, Evans CK, Garner JP, Liu YM. Evaluation of chemical seed treatments for control of stripe rust in wheat under controlled conditions. Plant Dis Manag Rep. 2013c;7:ST003.
Chen XM, Evans CK, Garner JP, Liu YM. Evaluation of chemical seed treatments for control of stripe rust in spring wheat, 2012. Plant Dis Manag Rep. 2013d;7:ST012.
Chen XM, Evans CK, Garner JP, Liu YM. Evaluation of chemical seed treatments for control of stripe rust in winter wheat, 2012. Plant Dis Manag Rep. 2013e;7:ST013.
Chen WQ, Wellings C, Chen XM, Kang ZS, Liu TG. Wheat stripe (yellow) rust caused by *Puccinia striiformis* f. sp. *tritici*. Mol Plant Pathol. 2014a;15:433–46.
Chen XM, Evans KC, Liu YM. Control of stripe rust of winter wheat with various foliar fungicides. Plant Dis Manag Rep. 2014b;8:CF023.
Chen XM, Evans KC, Liu YM. Control of stripe rust of spring wheat with various foliar fungicides. Plant Dis Manag Rep. 2014c;8:CF034.
Chen XM, Evans KC, Liu YM. Control of stripe rust on winter wheat cultivars with foliar fungicide in 2013. Plant Dis Manag Rep. 2014d;8:CF35.
Chen XM, Evans KC, Liu YM. Control of stripe rust on spring wheat cultivars with foliar fungicide in 2013. Plant Dis Manag Rep. 2014e;8:CF036.
Chen XM, Evans CK, Liu YM, Heath M. Control of stripe rust of spring wheat with various foliar fungicides, 2014. Plant Dis Manag Rep. 2015a;9:CF016.
Chen XM, Evans CK, Liu YM, Heath M. Effects of fungicide application on control of stripe rust on winter wheat cultivars in 2014. Plant Dis Manag Rep. 2015b;9:CF017.
Chen XM, Evans CK, Liu YM, Heath M. Effects of fungicide application on control of stripe rust on spring wheat cultivars in 2014. Plant Dis Manag Rep. 2015c;9:CF018.

Chen XM, Evans CK, Liu YM, Heath M. Control of stripe rust of winter wheat with various foliar fungicides, 2014. Plant Dis Manag Rep. 2015d;9:CF021.

Chen XM, Evans CK, Liu YM. Control of stripe rust of winter wheat with various foliar fungicides, 2015. Plant Dis Manag Rep. 2016a;10:CF022.

Chen XM, Evans CK, Liu YM. Responses of winter wheat cultivars to fungicide application for control of stripe rust in 2015. Plant Dis Manag Rep. 2016b;10:C023.

Cheng P, Chen XM. Molecular mapping of a gene for stripe rust resistance in spring wheat cultivar IDO377s. Theor Appl Genet. 2010;121:195–204.

Cheng P, Xu LS, Wang MN, See DR, Chen XM. Molecular mapping of genes *Yr64* and *Yr65* for stripe rust resistance in hexaploid derivatives of durum wheat accessions PI 331260 and PI 480016. Theor Appl Genet. 2014;127:2267–77.

Chhuneja P, Kaur S, Garg T, Ghai M, Kaur S, Prashar M, Bains NS, Goel RK, Keller B, Dhaliwal HS, Singh K. Mapping of adult plant stripe rust resistance genes in diploid A genome wheat species and their transfer to bread wheat. Theor Appl Genet. 2008;116:313–24.

Christensen NW, Jackson TL, Powelson RL. Suppression of take-all root rot and stripe rust diseases of wheat with chloride fertilizers. Plant Nutrition 1982: 9th proceedings of international plant nutrition colloq, vol 1; 1982. p. 111–16.

Christiansen MJ, Feenstra B, Skovgaard IM, Andersen SB. Genetic analysis of resistance to yellow rust in hexaploid wheat using a mixture model for multiple crosses. Theor Appl Genet. 2006;112:581–91.

Christopher MD, Liu SY, Hall MD, Marshall DS, Fountain MO, Johnson JW, Milus EA, Garland-Campbell KA, Chen XM, Griffey CA. Identification and mapping of adult plant stripe rust resistance in soft red winter wheat cultivar 'USG 3555'. Plant Breed. 2013a;132:53–60.

Christopher MD, Liu SY, Hall MD, Marshall DS, Fountain MO, Johnson JW, Milus EA, Garland-Campbell KA, Chen XM, Griffey CA. Identification and mapping of adult plant stripe rust resistance in soft red winter wheat VA00W-38. Crop Sci. 2013b;53:871–9.

Ciba-Geigy. Technical information bulletin for Propiconazole fungicide. Greensboro: Ciba-Geigy; 1997. 15 pp.

Comes O. Della resistenza dei frumenti alle ruggini stato attuale della quistione e provvedimenti. In: Atti R Ist D'Incor Napoli, Ser 6, 64:419–441. Letteratura e Note; 1913. p. 437–440.

Conner RL, Kuzyk AD. Effectiveness of fungicides in controlling stripe rust, leaf rust, and black point in soft white spring wheat. Can J Plant Pathol. 1988;10:321–6.

Cromey MG. Infection and control of stripe rust in wheat spikes. N Z Crop Hortic Sci. 1989;7:159–64.

Danial DL, Parlevliet JE. Effects of nitrogen fertilization on disease severity and infection type of yellow rust on wheat genotypes varying in quantitative resistance. J Phytopathol. 1995;143:679–81.

Darwinkel A. Grain production of winter wheat in relation to nitrogen and diseases: I. Relationship between nitrogen dressing and yellow rust infection. Z Acker Pflanzenbau. 1980;149:299–308.

Dedryver F, Paillard S, Mallard S, Robert O, Trottet M, Nègre S, Jahier J. Characterization of genetic components involved in durable resistance to stripe rust in the bread wheat 'Renan'. Phytopathology. 2009;99:968–73.

Deising HB, Reimann S, Pascholati SF. Mechanisms and significance of fungicide resistance. Braz J Microbiol. 2008;39:286–95.

Dekker J. Development of resistance to modern fungicides and strategies for its avoidance. In: Modern selective fungicides. Properties, applications, mechanisms of action, 2nd ed. New York: Gustav Fischer Verlag Publ.; 1995. p. 22–38.

Devadas R, Simpfendorfer S, Backhouse D, Lamb DW. Effect of stripe rust on the yield response of wheat to nitrogen. Crop J. 2014;2:201–6.

Dickson JG. Chemical control of cereal rusts. Bot Rev. 1959;25:486–513.

Dileone JA, Mundt CC. Effect of wheat cultivar mixtures on populations of *Puccinia striiformis* races. Plant Pathol. 1994;43:917–30.

Eriksson J, Henning E. Die Gelreideroste ihre Geschichte und N atur sawie Massregen gegen diesel ben. Stockholm: P.A. Norstedt und Soner; 1896. 463 pp.

Fang TL, Campbell KG, Liu ZY, Chen XM, Wan AM, Li S, Liu ZJ, Cao S, Chen Y, Bowden RL, Carver BF, Yan L. Stripe rust resistance in the wheat cultivar Jagger is due to *Yr17* and a novel resistance gene. Crop Sci. 2011;51:2455–65.

Feng JY, Chen GY, Wei YM, Liu YX, Jiang QT, Li W, Pu ZE, Lan XJ, Dai SF, Zheng YL. Identification and genetic mapping of a recessive gene for resistance to stripe rust in wheat line LM168-1. Mol Breed. 2014;33:601–9.

Feng JY, Wang MN, Chen XM, See DR, Zheng YL, Chao SM, Wan AM. Molecular mapping of *YrSP* and its relationship with other genes for stripe rust resistance in wheat chromosome 2BL. Phytopathology. 2015;105:1206–13.

Fernández-Ortuño D, Torés JA, de Vicente A, Pérez-García A. Mechanisms of resistance to Qol fungicides in phytopathogenic fungi. Int Microbiol. 2008;11:1–9.

Finckh MR, Mundt CC. Stripe rust, yield, and plant competition in wheat cultivar mixtures. Phytopathology. 1992;82:905–13.

Finckh MR, Mundt CC. Temporal dynamics of plant competition in genetically diverse wheat populations in the presence and absence of stripe rust. J Appl Ecol. 1996;33:1041–52.

FRAG-UK. Fungicide resistance management in cereals; 2015. p. 28. http://cereals.ahdb.org.uk/media/175998/frag-2015-fungicide-resistance-management-in-cereals-frag-2015-.pdf

Fu DL, Uauy C, Distelfeld A, Blechl A, Epstein L, Chen XM, Sela H, Fahima T, Dubcovsky J. A kinase-START gene confers temperature-dependent resistance to wheat stripe rust. Science. 2009;323:1357–60.

Garrett KA, Mundt CC. Effects of planting density and the composition of wheat cultivar mixtures on stripe rust: an analysis taking into account limits to the replication of controls. Phytopathology. 2000;90:1313–21.

Godoy CV. Fungicide resistance in crop protection: risk and management. Oxfordshire: CABI; 2012. ISBN 9781845939052, doi:10.1079/9781845939052.0087.

Han QM, Kang ZS, Buchenauer H, Huang LL, Zhao J. Cytological and immunocytochemical studies on the effects of the fungicde tebuconazole on the interaction of wheat with stripe rust. J Plant Pathol. 2006;88:263–71.

Hao Y, Chen Z, Wang Y, Bland D, Buck J, Brown-Guedira G, Johnson J. Characterization of a major QTL for adult plant resistance to stripe rust in US soft red winter wheat. Theor Appl Genet. 2011;123:1401–11.

Hardison JR. Commercial control of *Puccinia striiformis* and other rusts in seed crops of *Poa pratensis* by nickel fungicides. Phytopathology. 1963;53:209–16.

Hardison JR. Systemic activity of two derivatives of 1,4-ocathiin against smut and rust diseases of grasses. Plant Dis Rep. 1966;50:624.

Hardison JR. Control of *Puccinia striiformis* by two new systemic fungicides, Bay MEB 6447 and BAS 31702 F. Plant Dis Rep. 1975;59:652–5.

Herrera-Foessel SA, Lagudah ES, Huerta-Espino J, Hayden MJ, Bariana HS, Singh D, Singh RP. New slow-rusting leaf rust and stripe rust resistance genes *Lr67* and *Yr46* in wheat are pleiotropic or closely linked. Theor Appl Genet. 2011;122:239–49.

Hiltner L. Neuere beobachtungen über den rostbefall des wintergetreides. Prakt. Bl. Pflanzenbau u. Pflanzenschutz, Jahrg. 1914;17 (n. F. 12):81–84.

Hou L, Chen XM, Wang MN, See DR, Chao SM, Bulli P, Jing JX. Mapping a large number of QTL for durable resistance to stripe rust in winter wheat Druchamp using SSR and SNP markers. PLoS One. 2015;10:e0126794.

Imtiaz M, Ahmad M, Cromey M, Griffin W, Hampton J. Detection of molecular markers linked to the durable adult plant stripe rust resistance gene *Yr18* in bread wheat (*Triticum aestivum* L.). Plant Breed. 2004;123:401–4.

Ishii H. Studies on fungicide resistance in phytopathogenic fungi. J Gen Plant Pathol. 2004;70:379–81.

Kampmeijer P, Zadoks JC. EPIMUL, a simulator of foci and epidemics in mixtures of resistant and susceptible plants, mosaics and multilines. Wageningen: Centre for Agricultural Publishing and Documentation; 1977.

Kang ZS, Shang HS, Li ZQ. Effects of triadimefon on the development of wheat stripe rust in its host. Acta Northwest Agric Univ. 1993;21(Suppl 2):14–8.

Kang ZS, Shang HS, Jing JX, Wei GR, Li ZQ. Effects of systemic fungicide diniconazole on the development of stripe rust and powdery mildew on their host wheat. Acta Plant Pathol. 1996;26:111–6.

Kang ZH, Li X, Wan AM, Wang MN, Chen XM. Fungicide sensitivity of the wheat stripe rust pathogen (*Puccinia striiformis* f. sp. *tritici*). Phytopathology. 2015;105(Suppl 4):S4.69.

Kim BS, Hwang BK. Microbial fungicides in the control of plant diseases. J Phytopathol. 2007;155:641–53.

Klarquist EF, Chen XM, Carter AH. Novel QTL for stripe rust resistance on chromosomes 4A and 6B in soft white winter wheat cultivars. Agronomy. 2016;6:4. doi:10.3390/agronomy6010004.

Kuraparthy V, Chhuneja P, Dhaliwal HS, Kaur S, Bowden RL, Gill BS. Characterization and mapping of cryptic alien introgression from *Aegilops geniculata* with new leaf rust and stripe rust resistance genes *Lr57* and *Yr40* in wheat. Theor Appl Genet. 2007;114:1379–89.

Lan CX, Rosewarne GM, Singh RP, Herrera-Foessel SA, Huerta-Espino J, Basnet BR, Zhang YL, Yang EN. QTL characterization of resistance to leaf rust and stripe rust in the spring wheat line francolin#1. Mol Breed. 2014;34:789–803.

Li ZQ, Zeng SM. Wheat rusts in China. Beijing: China Agricultural Press; 2002. p. 379.

Li G, Li Z, Yang W, Zhang Y, He Z, Xu S, Singh R, Qu Y, Xia X. Molecular mapping of stripe rust resistance gene *YrCH42* in Chinese wheat cultivar Chuanmai 42 and its allelism with *Yr24* and *Yr26*. Theor Appl Genet. 2006a;112:1434–40.

Li ZF, Zheng TC, He ZH, Li GQ, Xu SC, Li XP, Yang GY, Singh RP, Xia XC. Molecular tagging of stripe rust resistance gene *YrZH84* in Chinese wheat line Zhou 8425B. Theor Appl Genet. 2006b;112:1098–103.

Li Y, Niu Y, Chen XM. Mapping a stripe rust resistance gene *YrC591* in wheat variety C591 with SSR and AFLP markers. Theor Appl Genet. 2009;118:339–46.

Li Q, Chen XM, Wang MN, Jing JX. *Yr45*, a new wheat gene for stripe rust resistance mapped on the long arm of chromosome 3D. Theor Appl Genet. 2010;122:189–97.

Li ZF, Singh S, Singh RP, López-Vera EE, Huerta-Espino J. Genetics of resistance to yellow rust in PBW343 × Kenya Kudu recombinant inbred line population and mapping of a new resistance gene *YrKK*. Mol Breed. 2013;32:821–9.

Li K, Hegarty J, Zhang CZ, Wan AM, Wu JJ, Brown-Guedira GL, Chen XM, Fu DL, Dubcovsky J. Fine mapping of barley locus *Rps6* conferring resistance to wheat stripe rust. Theor Appl Genet. 2016;129:845–59.

Lillemo M, Asalf B, Singh R, Huerta-Espino J, Chen X, He Z, Bjørnstad Å. Adult plant rust resistance loci *Lr34/Yr18* and *Lr46/Yr29* are important determinants of partial resistance to powdery mildew in bread wheat line Saar. Theor Appl Genet. 2008;116:1155–66.

Lin F, Chen XM. Genetics and molecular mapping of genes for race-specific all-stage resistance and non-race-specific high-temperature adult-plant resistance to stripe rust in spring wheat cultivar Alpowa. Theor Appl Genet. 2007;114:1277–87.

Lin F, Chen XM. Molecular mapping of genes for race-specific overall resistance to stripe rust in wheat cultivar express. Theor Appl Genet. 2008;116:797–806.

Lin F, Chen XM. Quantitative trait loci for non-race-specific, high-temperature adult-plant resistance to stripe rust in wheat cultivar express. Theor Appl Genet. 2009;118:631–42.

Lin F, Sun Q, Xu S, Chen XM, Zhang L, Zhang C, Xu Y, Miao Q, Qu B, Li N, Zhai Q. Identification of wheat-*Thinopyrum intermedium* alien disomic addition lines conferring resistance to stripe rust. Can J Plant Sci. 2009;89:569–74.

Line RF. Chemical control of *Puccinia striiformis* and *Puccinia recondite* on wheat in northwestern United States. In: Proceedings of European Mediterranean Cereal Rusts conference, 4th, Interlaken, vol 4; 1976. p. 105–108.

Line RF. Stripe rust of wheat and barley in North America: a retrospective historical review. Annu Rev Phytopathol. 2002;40:75–118.

Line RF, Qayoum A. Virulence, agressiveness, evolution, and distribution of races of Puccinia striiformis (the cause of stripe rust of wheat) in North America, 1968-87. United States Department of Agriculture, Agricultural Research Service, Technical Bulletin No. 1788. 1992, 44 pp.

Line RF, Rakotondradona R. Chemical control of *Puccinia striiformis* and *Puccinia recondita*. In: Proceedings of European Mediterranean Cereal Rusts conference, 5th, vol. 1; Bari/Rome, Italy; 1980. p. 239–241

Line RF, Rowell JB. Systemic fungicides for control of rusts and smuts. In: Proceedings of international congress plant pathology, 2nd, Minneapolis: Symp; 1973. p. 549.

Line RF, Sharp EL, Powelson RL. A system for differentiating races of *Puccinia striiformis* in the United States. Plant Dis Rep. 1970;54:992–4.

Lorgue G, Lechenet J, Rivière A. Clinical veterinary toxicology. London: Blackwell Science; 1996. p. 5–194.

Lowe I, Jankuloski L, Chao SM, Chen XM, See D, Dubcovsky J. Mapping and validation of QTL which confer partial resistance to broadly virulent post-2000 North American races of stripe rust in hexaploid wheat. Theor Appl Genet. 2011;123:143–57.

Lu Y, Lan C, Liang S, Zhou X, Liu D, Zhou G, Lu Q, Jing J, Wang M, Xia X, He Z. QTL mapping for adult-plant resistance to stripe rust in Italian common wheat cultivars Libellula and Strampelli. Theor Appl Genet. 2009;119:1349–59.

Lu Y, Wang MN, Chen XM, See DR, Chao SM, Jing JX. Mapping of *Yr62* and a small-effect QTL for high-temperature adult-plant resistance to stripe rust in spring wheat PI 192252. Theor Appl Genet. 2014;127:1449–59.

Lukens RJ. Antimicrobial agents in crop production. In: Block SS, editor. Disinfection, sterilization and preservation. Philadelphia: Lea & Febiger; 1983. p. 695–713.

Ma DF, Li Q, Tang MS, Chao KX, Li JC, Wang BT, Jing JX. Mapping of gene conferring adult-plant resistance to stripe rust in Chinese wheat landrace Baidatou. Mol Breed. 2015;35:157.

Maccaferri M, Zhang JL, Bulli P, Abate Z, Chao SM, Cantu D, Bossolini E, Chen XM, Pumphrey M, Dubcovsky J. A genome-wide association study of resistance to stripe rust (*Puccinia striiformis* f. sp. *tritici*) in a worldwide collection of hexaploid spring wheat (*Triticum aestivum* L.). 3G: genes genomics. Genetics. 2015;5:449–65.

Mallard S, Gaudet D, Aldeia A, Abelard C, Besnard A, Sourdille P, Dedryver F. Genetic analysis of durable resistance to yellow rust in bread wheat. Theor Appl Genet. 2005;110:1401–9.

Matheron M. Modes of action for plant disease management chemistries. The University of Arizona, Cooperative Extension; 2001. http://cals.arizona.Edu/crop/diseases/papers/dischemistry.Html (Presented December 6, 2001 at the 11th Annual Desert Vegetable Crop Workshop, Yuma, AZ).

Mavroeidi VI, Shaw MW. Effects of fungicide dose and mixtures on selection for triazole resistance in *Mycosphaerella graminicola* under field conditions. Plant Pathol. 2006;55:715–25.

McGrath MT. Fungicide resistance in cucurbit powdery mildew: experiences and challenges. Plant Dis. 2001;85:236–45.

McIntosh RA, Yamazaki Y, Dubcovsky J, Rogers J, Morris C, Appels R, Xia XC. Catalogue of gene symbols for wheat. 12th international wheat genetics symposium, 2013 Sept 8–13, Yokohama, Japan; 2013. Online: http://www.shigen.nig.ac.jp/wheat/komugi/genes/download.jsp

Morton V, Staub T. A short history of fungicides. Online, APSnet Features. 2008. doi:10-1094/APSnetFeature-2008-0308.

Moser VC, Barone S, Smialowicz Jr RJ, Harris MW, Davis BJ, Overstreet D, Mauney M, Chapin RE. The effects of perinatal Tebuconazole exposure on adult neurological, immunological, and reproductive function in rats. Toxicol Sci. 2001;62:339–52.

Mueller DS, Wise KA, Dufault NS, Bradley CA, Chilvers MI. Fungicides for field crops. St. Paul: APS Press; 2013. 112 pp

Mundt CC. Use of multiline cultivars and cultivar mixtures for disease management. Annu Rev Phytopathol. 2002;40:381–410.

Mundt CC, Brophy LS, Schmitt MS. Disease severity and yield of pure-line wheat cultivars and mixtures in the presence of eyespot, yellow rust, and their combination. Plant Pathol. 1995;44:173–82.

Mundt CC, Brophy LS, Kolar SC. Effect of genotype unit number and spatial arrangement on severity of yellow rust in wheat cultivar mixtures. Plant Pathol. 1996;45:215–22.

Murphy LR, Santra D, Kidwell K, Yan GP, Chen XM, Campbell KG. Linkage maps of wheat stripe rust resistance genes *Yr5* and *Yr15* for use in marker assisted selection. Crop Sci. 2009;49:1786–90.

Naz AA, Kunert A, Flath K, Pillen K, León J. Advanced backcross quantitative trait locus analysis in winter wheat: dissection of stripe rust seedling resistance and identification of favorable exotic alleles originated from a primary hexaploid wheat (*Triticum turgidum* ssp. *dicoccoides* × *Aegilops tauschii*). Mol Breed. 2012;30:1219–29.

Negassa A, Shiferaw B, Koo J, Sonder K, Smale M, Braun HJ, Gbegbelegbe S, Guo Z, Hodson D, Wood S, Payne T, Abeyo B. The potential for wheat production in Africa: analysis of biophysical suitability and economic profitability. Mexico: CIMMYT; 2013.

Neumann S, Paveley ND, Beed FD, Sylvester-Bradley R. Nitrogen per unit leaf area affects the upper asymptote of *Puccinia striiformis* f. sp. *tritici* epidemics in winter wheat. Plant Pathol. 2004;53:725–32.

Nyfeler R, Ackermann P. Phenylpyrroles, a new class of agricultural fungicides related to the natural antibiotic pyrrolinitrin. In: Baker DR, Fenyes JG, Steffens JJ, editors. Synthesis and chemistry of agrochemicals III, ACS symposium series 504. Washington, DC: American Chemical Society; 1992. p. 395–404.

Oruc HH. Fungicides and their effects on animals. In: Carisse O, editor. Fungicides, ISBN: 978–953–307-266-1. InTech; 2010. p. 349–62. Available from: http://www.intechopen.com/books/fungicides/fungicides-and-theireffects-on-animals

Paillard S, Trotoux-Verplancke G, Perretant M-R, Mohamadi F, Leconte M, Coëdel S, de Vallavieille-Pope C, Dedryver F. Durable resistance to stripe rust is due to three specific resistance genes in French bread wheat cultivar Apache. Theor Appl Genet. 2012;125:955–65.

Powelson RL. Epidemiology of stripe rust (*Puccinia striiformis* West.) of wheat. Contract no. DA 18-064-AMC2208A; 1969. Fort Detrick, Frederick, MD, Final Tech Rep, Feb 1, 1964–Dec 31, 1968.

Powelson RL, Shaner GE. An effective chemical seed treatment for systemic control of seedling infection of wheat by stripe rust (*Puccinia striiformis*). Plant Dis Rep. 1966;50:806–7.

Priestley RH. Fungicide treatment increases yield of cereal cultivars by reducing disease and delaying senescence. EPPO Bull. 1981;11:357–63.

Prins R, Pretorius ZA, Bender CM, Lehmensiek A. QTL mapping of stripe, leaf and stem rust resistance genes in a Kariega × Avocet S doubled haploid wheat population. Mol Breed. 2011;27:259–70.

Pscheidt JW, Ocamb CM (Senior editors). Pacific Northwest plant disease management handbook. Oregon State University; 2016. http://pnwhandbooks.org/plantdisease/node/1835/print

Purdy LH. Inhibition of wheat stripe rust by the antibiotic Phleomgein in the greenhouse. Plant Dis Report. 1964;48:159–61.

Rakotondradona R, Line RF. Control of stripe rust and leaf rust of wheat with seed treatments and effects of treatments on the host. Plant Dis. 1984;68:112–7.

Randhawa MS, Bariana HS, Mago R, Bansal UK. Mapping of a new stripe rust resistance locus *Yr57* on chromosome 3BS of wheat. Mol Breed. 2015;35:65.

Rathmell WG, Skidmore AM. Recent advances in the chemical control of cereal rust diseases. Outlook Agric. 1982;11:37–43.

Ren RS, Wang MN, Chen XM, Zhang ZJ. Characterization and molecular mapping of *Yr52* for high-temperature adult-plant resistance to stripe rust in spring wheat germplasm PI 183527. Theor Appl Genet. 2012a;125:847–57.

Ren Y, He Z, Li J, Lillemo M, Wu L, Bai B, Lu Q, Zhu H, Zhou G, Du J. QTL mapping of adult-plant resistance to stripe rust in a population derived from common wheat cultivars Naxos and Shanghai 3/Catbird. Theor Appl Genet. 2012b;125:1211–21.

Ren Y, Li Z, He Z, Wu L, Bai B, Lan C, Wang C, Zhou G, Zhu H, Xia X. QTL mapping of adult-plant resistances to stripe rust and leaf rust in Chinese wheat cultivar Bainong 64. Theor Appl Genet. 2012c;125:1253–62.

Ren Y, Li S, Xia X, Zhou Q, He Y, Wei Y, Zheng Y, He Z. Molecular mapping of a recessive stripe rust resistance gene *yrMY37* in Chinese wheat cultivar Mianmai 37. Mol Breed. 2015;35:97.

Rosewarne GM, Singh RP, Huerta-Espino J, Rebetzke G. Quantitative trait loci for slow-rusting resistance in wheat to leaf rust and stripe rust identified with multi-environment analysis. Theor Appl Genet. 2008;116:1027–34.

Rosewarne GM, Herrera-Foessel SA, Singh RP, Huerta-Espino J, Lan CX, He ZH. Quantitative trait loci of stripe rust resistance in wheat. Theor Appl Genet. 2013;126:2427–49.

Rowell JB. Chemical control of the cereal rust. Annu Rev Phytopathol. 1968;6:243–62.

Rowell JB. Evaluation of chemicals for rust control. In: Roelfs AP, Bushnell WR, editors. The cereal rusts, Vol 2: Diseases, distribution, epidemiology and control. Orlando: Academic; 1985. p. 538–64.

Russell GE. Some effects of applied sodium and potassium chlorides on yellow rust in winter wheat. Ann Appl Biol. 1978;90:163–8.

Russell PE. A century of fungicide evolution. J Agric Sci. 2005;143:11–25.

Santra D, Chen X, Santra M, Campbell K, Kidwell K. Identification and mapping QTL for high-temperature adult-plant resistance to stripe rust in winter wheat (*Triticum aestivum* L.) cultivar 'Stephens'. Theor Appl Genet. 2008;117:793–802.

Scott DB. Disease control and yield improvement obtained with fungicides applications to spring wheat in South Africa. Crop Prot. 1996;15:167–70.

Scott RB, Line RF. Control of stripe rust and leaf rust of wheat with triadimenol seed treatments and triadimefon foliar sprays. Phytopathology. 1985;75:1383–4.

Scheyer JM, Christensen NW, Powelson RL. Chloride fertilizer effects on stripe rust development and grain yield of winter wheat. Plant Dis. 1987;71:54–7.

Singh A, Pandey MP, Singh AK, Knox RE, Ammar K, Clarke JM, Clarke FR, Singh RP, Pozniak CJ, DePauw RM, McCallum BG, Cuthbert RD, Randhawa HS, Fetch Jr TG. Identification and mapping of leaf, stem and stripe rust resistance quantitative trait loci and their interactions in durum wheat. Mol Breed. 2013;31:405–18.

Spinks GT. Factors affecting susceptibility to disease in plants. J Agric Sci. 1913;5:231–47.

Steffens JJ, Pell EJ, Tien M. Mechanisms of fungicide resistance in phytopathogenic fungi. Curr Opin Biotechnol. 1996;7:348–55.

Sthapit J, Newcomb M, Bonman JM, Chen XM, See D. Genetic diversity for stripe rust resistance in wheat landraces and identification of dual resistance to stem rust and stripe rust. Crop Sci. 2014;54:2131–9.

Stranak F. Zur frage der bekämpfung des gelbrostes, vol. 42. Jahrg: Deut Landw Press; 1915. p. 379.

Sui XX, Wang MN, Chen XM. Molecular mapping of a stripe rust resistance gene in spring wheat cultivar Zak. Phytopathology. 2009;99:1209–15.

Takeuchi S, Hirayama K, Ueda K, Sakai H, Yonehara H. Blasticidin S, a new antibiotic. J Antibiot Ser A. 1957;11:1–5.

Vazquez MD, Peterson CJ, Riera-Lizarazu O, Chen XM, Heesacker A, Ammar K, Crossa J, Mundt CC. Genetic analysis of adult plant, quantitative resistance to stripe rust in wheat cultivar Stephens in multi-environment trials. Theor Appl Genet. 2012;124:1–11.

Vazquez MD, Zemetra R, Peterson CJ, Chen XM, Heesacker A, Mundt CC. Multi-location wheat stripe rust QTL analysis: genetic background and epistatic interactions. Theor Appl Genet. 2015;128:1307–18.

Wan AM, Chen XM. Virulence, frequency, and distribution of races of *Puccinia striiformis* f. sp. *tritici* and *P. striiformis* f. sp. *hordei* identified in the United States in 2008 and 2009. Plant Dis. 2012;96:67–74.

Wan AM, Chen XM. Virulence characterization of *Puccinia striiformis* f. sp. *tritici* using a new set of *Yr* single-gene line differentials in the United States in 2010. Plant Dis. 2014;98:1534–42.

Wan AM, Zhao ZH, Chen XM, He ZH, Jin SL, Jia QZ, Yao G, Yang JX, Wang BT, Li GB, Bi YQ, Yuan ZY. Wheat stripe rust epidemics and virulence of *Puccinia striiformis* f. sp. *tritici* in China in 2002. Plant Dis. 2004;88:896–904.

Wan AM, Chen XM, He ZH. Wheat stripe rust in China. Aust J Agric Res. 2007;58:605–19.

Wang MN, Chen XM, Xu LS, Cheng P, Bockelman H. Registration of 70 common spring wheat germplasm lines resistant to stripe rust. J Plant Reg. 2012;6:104–10.

Xiang C, Feng JY, Wang MN, Chen XM, See DR, Wan AM, Wang T. Molecular mapping of *Yr76* for resistance to stripe rust in winter club wheat cultivar Tyee. Phytopathology. 2016;106:1186–93.

Xu LS, Wang MN, Cheng P, Kang ZS, Hulbert S, Chen XM. Molecular mapping of *Yr53*, a new gene for stripe rust resistance in durum wheat accession PI 480148 and its transfer to common wheat. Theor Appl Genet. 2013;126:523–33.

Xu HX, Zhang J, Zhang P, Qie YM, Niu YC, Li HJ, Ma PT, Xu YF, An JG. Development and validation of molecular markers closely linked to the wheat stripe rust resistance gene *YrC591* for marker-assisted selection. Euphytica. 2014;198:317–23.

Yan GP, Chen XM. Molecular mapping of a recessive barley gene for resistance to stripe rust. Theor Appl Genet. 2006;113:529–37.

Yan GP, Chen XM. Molecular mapping of the *rps1.a* recessive gene for resistance to stripe rust in BBA 2890 barley. Phytopathology. 2007;97:668–73.

Yan GP, Chen XM. Identification of a quantitative trait locus for high-temperature adult-plant resistance against *Puccinia striiformis* f. sp. *hordei* in 'Bancroft' barley. Phytopathology. 2008;98:120–7.

Yao Q, Song YX, Zhou RH, Fu TH, Jia JZ. Quantitative trait loci for adult-plant resistance against yellow rust in a wheat-derived recombinant inbred line population. Sci Agric Sin. 2009;42:4234–41.

Zhao L, Feng J, Zhang CY, Xu XD, Chen XM, Su Q, Miao Q, Xu SC, Lin F. The dissection and SSR mapping of a high-temperature adult-plant stripe rust resistance gene in American spring wheat cultivar Alturas. Eur J Plant Pathol. 2012;134:281–8.

Zhou X, Wang W, Wang L, Hou D, Jing J, Wang Y, Xu Z, Yao Q, Yin J, Ma D. Genetics and molecular mapping of genes for high-temperature resistance to stripe rust in wheat cultivar Xiaoyan 54. Theor Appl Genet. 2011;123:431–8.

Zhou XL, Han DJ, Chen XM, Gou HL, Guo SJ, Rong L, Wang QL, Huang LL, Kang ZS. Characterization and molecular mapping of stripe rust resistance gene *Yr61* in winter wheat cultivar Pindong 34. Theor Appl Genet. 2014a;127:2349–58.

Zhou XL, Wang MN, Chen XM, Lu Y, Kang ZS, Jing JX. Identification of *Yr59* conferring high-temperature adult-plant resistance to stripe rust in wheat germplasm PI 178759. Theor Appl Genet. 2014b;127:935–45.

Zhou XL, Han DJ, Chen XM, Mu JM, Xue WB, Zeng QD, Wang QL, Huang LL, Kang ZS. QTL mapping of adult-plant resistance to stripe rust in wheat line P9897. Euphytica. 2015a;205:243–53.

Zhou XL, Zhang Y, Zeng QD, Chen XM, Han DJ, Huang LL, Kang ZS. Identification of QTL for adult plant resistance to stripe rust in Chinese wheat landrace Caoxuan 5. Euphytica. 2015b;204:627–34.

Chapter 7
Stripe Rust Research and Control: Conclusions and Perspectives

Xianming Chen and Zhensheng Kang

Abstract Stripe rust is an old problem in world production of wheat and barley. Great progress has been made in understanding the pathogen, the hosts and the disease, especially the pathogen biology, host resistance and epidemiology of the disease. Now, it is clear that *Puccinia striiformis* has a macrocyclic lifecycle, but mainly reproduces through the asexual cycle. A large number of cereal germplasms and genes for different types of resistance to stripe rust have been identified and used in breeding programs for developing resistant cultivars. At the same time, more fungicides with different modes of action have become available for use to control stripe rust. Nowadays, growers are able to reduce potentially very high yield losses to minimal with an integrated control approach combining cultivar resistance, appropriate use of fungicide and suitable cultural practices. However, there are still many challenges for control of stripe rust. For example, the role of alternate hosts and auxiliary grass hosts in developing stripe rust epidemics and generating pathogen variation is not clear in many regions of the world. Stripe rust resistance has not received adequate efforts in breeding programs in many countries or regions due to various reasons. Chemical control has not become a practical option for growers in many developing countries. Although tremendous progress has been made in research on the genome, functional genomics and marker development for monitoring the pathogen populations, none of the *P. striiformis* avirulence genes have been cloned and molecularly characterized. Putting these challenges into consideration and research, more progress will likely be made in these areas in the next 10–20 years. Ultimately, the devastating disease can be put under control more effectively.

X. Chen (✉)
US Department of Agriculture, Agricultural Research Service, Wheat Health, Genetics and Quality Research Unit, Pullman, WA 99164-6430, USA

Department of Plant Pathology, Washington State University, Pullman, WA 99164-6430, USA
e-mail: xianming@wsu.edu

Z. Kang
State Key Laboratory of Crop Stress Biology for Arid Areas and College of Plant Protection, Northwest A&F University, Yangling, Shaanxi 712100, China
e-mail: kangzs@nwafu.edu.cn

X. Chen, Z. Kang (eds.), *Stripe Rust*, DOI 10.1007/978-94-024-1111-9_7

Keywords Avirulence • Barley • Disease control • Disease resistance • Functional genomics • Genome • *Puccinia striiformis* • Stripe rust • Wheat • Yellow rust

Various aspects of research on stripe rust and practice of the disease control have been covered in substantial details in the previous chapters. We are attempted to briefly summarize research progress and control technology development in this chapter. We are trying to point out issues for stripe rust workers to concern in their research and technology development. As technology will continue being advanced, scientists will take such opportunities to make more and more achievements in stripe rust research, which will help better understand the pathosystem and improve the control strategies and practices.

7.1 Conclusions

Stripe rust is an old disease that has damaged cereal crops for hundreds or even thousands of years. Scientific research on stripe rust was started about 240 years ago. Milestones in research have been reached for a great body of knowledge in the pathogen, hosts, their interactions and control of the disease.

7.1.1 The Pathogens

Since 1777 when Gadd first described the disease (Eriksson and Henning 1896), stripe rust has been studied for 240 years. During the first 120 years, the studies of stripe rust by only few scientists in Europe were mainly on the nomenclature of the pathogen. Several species names were given to the stripe rust fungus, but the name of *Puccinia striaeformis* given by Westendorp (1854) has been used to the present, with an interruption of *Puccinia glumerum* given by Eriksson (1894) for more than 60 years. The current name, *Puccinia striiformis* Westend., was restored based on *P. striaeformis* by Hylander et al. (1953) and followed by Cummins and Stevenson (1956).

Intensive research has been conducted in the second 120 years since Eriksson (1894) studied the host specificity of the stripe rust fungal species and identified formae speciales. So far nine formae speciales have been reported for *P. striiformis* based on specificities of the pathogen subpopulations adapted to different cereal crops and genera of grasses (Eriksson 1894; Manners 1960; Tollenaar 1967; Niu et al. 1991; Wellings et al. 2000), with formae speciales *P. striiformis* f. sp. *poae* and *P. striiformis* f. sp. *dactylidis* recently considered as different species, *Puccinia pseudostriiformis* and *Puccinia striiformoides*, respectively (Abbasi et al. 2004; Liu and Hambleton 2010). As a species of fungus, *P. striiformis* has a much wider host range than its closely related species, *Puccinia graminis* and *Puccinia triticina*, the causal agents of stem rust and leaf rust, respectively. Even the wheat stripe rust form, *P. striiformis* f. sp. *tritici*, alone can infect more than 290 grass species (Cheng et al. 2016).

The high specificity of the stripe rust fungus is also reflected by its specific interactions with their crop host at the cultivar level, which separates a forma specialis further into different races. Races are identified based on reactions on a set of selected crop host cultivars or specific resistance gene lines. There are several sets of differentials that have been used in different countries or regions for identifying races of the wheat stripe rust pathogen (*P. striiformis* f. sp. *tritici*) and the barley stripe rust pathogen (*P. striiformis* f. sp. *hordei*). A trend that has started in the recent years is to identify races using a common set of differentials, which allows comparison of virulences and races in different parts of the world (Sharma-Poudyal et al. 2013; Wan and Chen 2014; Wan et al. 2016).

Puccinia striiformis is an obligate and biotrophic basidiniomycete fungus, requiring living host plants for growth and reproduction. Like other rust fungi, the stripe rust fungus is still difficult to culture on artificial media, which makes genetic manipulation of the fungus difficult.

The stripe rust pathogen has been recently shown to be heteroecious, requiring a cereal or grass plant and barberry or mahonia plant for completing its sexual lifecycle. Its lifecycle is macrocyclic, consisting of all five spore stages (uredinial, telial, basidial, pycinial and aecial stages) (Jin et al. 2010; Wang and Chen 2013). Of these types of spores, the urediniospore stage is the most important for the pathogen persistence and for causing stripe rust disease. The role of sexual reproduction on alternate hosts in the pathogen variation and disease epidemics is limited as *P. striiformis* has not been found in the stripe rust region of the Pacific Northwest in the US (Wang et al. 2015d) and Europe (Berlin et al. 2012), and at very low frequencies in western China (Zhao et al. 2015; Wang et al. 2016). The function of *P. striiformis* teliospores is limited by enclosure in telia covered by the host epidermal layer, lack of dormancy and degradation under moisture conditions (Wang and Chen 2015). Nevertheless, its capability of sexual reproduction on alternate hosts under controlled conditions allows genetic studies of *P. striiformis*, which was previously impossible (Zhao et al. 2016).

Although *P. striiformis* does not have, or has very low frequency of sexual reproduction, the fungal pathogen is highly variable. The variability of the pathogen has been studied intensively by virulence testing on differentials of cereal crops since the 1930s (Newton et al. 1933; Bever 1934; Straib 1937; Prasada et al. 1967; Stubbs 1985; Wellings and McIntosh 1990; Line and Qayoum 1992; Kumar et al. 1993; Line 2002; Li and Zeng 2002; Chen 2005; Chen et al. 2009a, b; Wan and Chen 2014). Along the development of molecular techniques since the early 1990s, *P. striiformis* have been studied intensively using various molecular markers (Chen et al. 1993, 1995, 2009a, b; Steele et al. 2001; Enjalbert et al. 2002, 2005; Hovmøller et al. 2002, 2008, 2016; Hovmøller and Justesen 2007; Bahri et al. 2009a, b; Mboup et al. 2009; Duan et al. 2010; Liu and Hambleton 2010; Lu et al. 2011; Liu et al. 2012b; Zhan et al. 2012, 2016; Ali et al. 2014a, b; Cheng et al. 2012, 2016; Cheng and Chen 2014; Holtz et al. 2014; Liang et al. 2016; Xia et al. 2016a, b). The molecular characterization studies have greatly improved the understanding of the genetic relationships of *P. striiformis* to related rust fungal species, among formae speciales, among populations of the wheat stripe rust pathogen in different regions, reproduction modes, tagging particular strains and evolutionary mechanisms.

In the past 10 years, genomic sequencing has been done for several isolates of *P. striiformis*. The first published partial genome sequence of a US isolate, PST-130, using an Illumina Genome Analyzer II contained 64.8 Mb assembled into 29,178 contigs and 22,815 putative genes, covering about 88% of the genome (Cantu et al. 2011), which indicates an estimated genome size of 74 Mb. Using the same approach, two more US isolates (PST-21 and PST-43) and two UK isolates (PST-87/7 and PST-08/21) were also partially sequenced (Cantu et al. 2013). The assembled sequence of PST-21 consisted of 73 Mb in 43,106 contigs with 20,653 protein genes, PST-43 71 Mb in 49,784 contigs with 21,036 protein genes, PST-87/7 53 Mb in 55,502 contigs with 20,688 protein genes, and PST-08/21 56 Mb in 50,898 contigs with 20,875 protein genes. The first published whole-genome sequence was from a Chinese isolate using a 'fosmid-to-fosmid' approach. This assembled sequence consisted of 130.7 Mb in 12,833 contigs assembled into 4283 scaffolds with 25,288 protein-coding genes (Zheng et al. 2013). This study estimated the *P. striiformis* genome size as 110 Mb. The second complete sequence was from US isolate 2K41-Yr9 (race PST-78) using FLX chemistry with a Roche 454 (Cuomo et al. 2017). The size of assembled sequences was 117.31 Mb in 17,295 contigs combined into 9715 scaffolds with 19,542 genes. Compared with *P. graminis* and *P. triticina*, *P. striiformis* has a genome size larger than *P. graminis* (88.64 Mb), smaller than *P. triticina* (135.34 Mb), but its number of genes is higher than either *P. graminis* (15,800) or *P. triticina* (14,880) (Cuomo et al. 2017). Using the sequence data, secreted protein genes and mating type genes were identified, and SSR and SNP markers were developed and used to study the genetic structures of the *P. striiformis* populations (Cantu et al. 2013; Garnica et al. 2013; Tian et al. 2016; Xia et al. 2016a, b). Now we have more advanced technologies and rich resources for studying the stripe rust pathogen than before, which should lead to a much better understanding of the pathogen biology, genetics and its pathogenicity to cause stripe rust epidemics.

7.1.2 The Hosts

The *P. striiformis* species has a wide host range for its uredinial stage, with more than 290 species in 56 genera serving as auxiliary hosts for the wheat stripe rust pathogen, *P. striiformis* f. sp. *tritici* (see Chap. 1). Urediniospores on grass auxiliary hosts can initiate stripe rust epidemics on cereal crops and increase inoculum pressure. Grass hosts also can serve as green bridges for the stripe rust fungus to survive between cereal crop seasons, provide reservoirs for the pathogen to maintain diversity and keep mutants, and provide common hosts for somatic recombination to occur between different isolates of same forma specialis or between different formae speciales (Cheng et al. 2016). In addition, grass hosts can provide resistance genes useful in developing cereal cultivars with resistance to stripe rust. Such genes can be incorporated into wheat, barley and triticale through traditional genetic manipulation as done in the past, or through transformation highly likely in the future.

Use of host resistance has been a major approach for control of stripe rust. Different types of resistance have been identified in both wheat and barley for resistance to their stripe rust pathogens. Although there are some variations, these resistances can be separated into two types with various terms. Chen (2013) contrasted the two types based on host growth stages, testing condition, specificity, degree of resistance, sensitivity to pathogen infection, speed of phenotype development, response to temperature, inheritance, the effect of genes, the number of genes, molecular basis and combination of some of these aspects. All-stage resistance and adult-plant resistance are commonly used to describe these two types of resistance. All-stage resistance can be detected in seedling stage and is effective throughout the growth stages. The level of all stage resistance is usually high and it provides complete control when effective. It is usually controlled by a single gene and easy to use in breeding. However, all-stage resistance is usually race specific and not durable. In contrast, adult-plant resistance starts express when plants pass seedling stages and the resistance level increases as plant growth stage progresses. Adult-plant resistance is usually not complete and the level of resistance is usually also affected by temperature, and thus referred as high-temperature adult-plant (HTAP) resistance (Chen 2005, 2013). The effects of growth stage and temperature can be separable as some resistances are more influenced by temperature than by plant growth stage and vice versa. It is important to recognize the effect of temperatures on adult-plant resistance as this is related to how to more effectively use such type of resistance in various regions with different climates, and related to how to distinguish temperature effects from race effects. Adult-plant or HTAP resistance is usually race non-specific and more durable than all-stage resistance.

Research on genetics of host resistance to stipe rust was started in the early 1900s (Biffen 1905) and became intensive after the 1960s (Lupton and Macer 1962). During the last decade, more studies of stripe rust resistance were conducted and more resistance genes were reported than any previous decades. So far, there are 78 permanently named *Yr* genes and more than 300 temporarily named genes or quantitative trait loci (QTL) reported for resistance to stripe rust in wheat (Chen 2013; Christopher et al. 2013a, b; Li et al. 2013, 2015; Liu et al. 2013; Ma et al. 2013, 2015; McIntosh et al. 2013; Rosewarne et al. 2013; Singh et al. 2013, 2014; Xu et al. 2013, 2014; Zhang et al. 2013, 2016; Agenbag et al. 2014; Bansal et al. 2014; Basnet et al. 2014a, b; Buerstmayr et al. 2014; Case et al. 2014; Cheng et al. 2014; Feng et al. 2014, 2015; Herrera-Foessel et al. 2014; Huang et al. 2014; Lan et al. 2014, 2015; Lu et al. 2014; Randhawa et al. 2014, 2015; Sthapit et al. 2014; Tadesse et al. 2014; Zegeye et al. 2014; Zheng et al. 2014; Zhou et al. 2014a, b, 2015a, b; Calvo-Salazar et al. 2015; Hou et al. 2015, 2016; Jighly et al. 2015; Kertho et al. 2015; Maccaferri et al. 2015; Naruoka et al. 2015; Ren et al. 2015a, b; Vazquez et al. 2015; Wang et al. 2015a; Bulli et al. 2016; Dracatos et al. 2016; Klarquist et al. 2016; Xiang et al. 2016). Compared to the number of resistance genes reported in wheat, a relatively small number of genes for resistance to stripe rust in barley has been identified and mapped (Toojinda et al. 2000; Chen and Line 1999, 2003; Castro et al. 2002, 2003; Yan and Chen 2006, 2007, 2008; Klos et al. 2016).

In addition to host resistance genes, genes in wheat for resistance to the barley stripe rust pathogen and in barley to the wheat stripe rust pathogen have also been identified and mapped (Pahalawatta and Chen 2005a, b; Sui et al. 2010; Dawson et al. 2016; Li et al. 2016a). Barley genes conferring resistance to the barley grass stripe rust pathogen (*P. striiformis* f. sp. *pseudo-hordei*) were also identified and mapped (Derevnina et al. 2015). These genes could be useful in the future for introgression from "non-hosts" or intermediate hosts into host cultivars for controlling stripe rust.

Intensive research has been conducted in the recent years to understand the molecular mechanisms of resistance to stripe rust. Transcriptomics studies identified large numbers of defense-related genes and putative biochemical pathways involved in resistance controlled by different *Yr* genes and molecular differences between all-stage resistance and HTAP resistance (Hulbert et al. 2007; Coram et al. 2008a, b, c, 2009, 2010; Chen et al. 2013). Using histological, cytological, and molecular approaches, genes and related biochemical pathways contributing to incompatible reactions in seedling stage and adult-plant resistance were identified (Xia et al. 2010; Wang et al. 2010a, 2011, 2012a, b, 2013, 2014, 2015b, c; Feng et al. 2011,2013a, b; Zhang et al. 2011, 2012, 2015; Liu et al. 2012a; Risk et al. 2012; Duan et al. 2013; Huang et al. 2013; Jiang et al. 2013; Tang et al. 2013, 2015, 2016; Yang et al. 2013; Fu et al. 2014; Gou et al. 2015; Herrera-Foessel et al. 2015; Pei et al. 2015; Yin et al. 2015b; Huang et al. 2016; Li et al. 2016a, b). So far, four genes, *Yr10*, *Yr18*, *Yr36* and *Yr46*, for resistance to stripe rust in wheat have been cloned. *Yr18* is a putative adenosine triphosphate-binding cassette (ABC) transporter, and the gene is present in many wheat cultivars and also confers resistance to leaf rust (as *Lr34*) and other diseases (Krattinger et al. 2009). *Yr36*, which was first identified in wild emmer wheat (*Triticum turgidum* spp. *dicoccoides*) and has been transferred to durum and hexaploid wheat cultivars, is a kinase-START (steroidogenic acute regulatory protein-related lipid transfer) gene (Fu et al. 2009). *Yr10* encodes an evolutionary-conserved and unique CC–NBS–LRR (coiled-coil – nucleotide-binding site – leucine-rich repeat) sequence in wheat (Liu et al. 2014). Encoding a sugar transporter containing 12 transmembrane helices, *Yr46* functions in transporting hexoses across the plasma membrane (Moore et al. 2015). These genes show that stripe rust resistance is controlled by genes with different biochemical functions. In addition, both *Yr18* and *Yr46* also confer resistance to multiple diseases.

7.1.3 The Disease

Stripe rusts occurring on wheat, barley, rye, triticale and cultivated grasses are considered as different diseases because they are caused by different formae speciales or species. Because triticale is more related to wheat, wheat stripe rust can cause damage on triticale. Isolates from triticale have been identified as races of *P. striiformis* f. sp. *tritici* (Cheng et al. 2016). As rye is currently not widely grown and information of rye stripe rust is limited. How different *P. striiformis* f. sp. *secales* is from *P. striiformis* f. sp. *tritici* is not clear. However, stripe rust occurring on wheat generally does not cause significant damage on barley and barley stripe rust doses not damage wheat (Stubbs 1985; Chen et al. 1995; Line 2002; Chen 2005). In the US Pacific Northwest, where both wheat and barley are grown, severe epidemics of wheat stripe rust occurred in eight of the last 16 years, whereas barley stripe rust did not occurred at significant levels (Chen 2014). The difference is not due to resistance of barley to barley stripe rust and wheat resistance to wheat stripe rust, and also not due to the environment, as barley and wheat are grown in the same environment. The low barley stripe rust is dues to the low inoculum of the barley stripe rust pathogen and resistance of barley to wheat stripe rust, in additions to the difference of cropping systems of barley and wheat in the region (Chen 2005; Wan and Chen 2012). As the barley stripe rust pathogen was introduced in the 1990s to the United States (Chen et al. 1995), resistance of barley cultivars to barley stripe rust is generally not as good as resistance of wheat cultivars to wheat stripe rust. Under natural conditions, stripe rust pressure is much lower for barley than for wheat as spring barley is mainly grown verses both spring wheat and winter wheat are grown, and also because of much smaller acreage of barley. Similarly, in Europe, wheat stripe rust has caused severe damage on wheat and triticale, but not on barley in the recent years. It is interesting that in the International Cereal Rusts and Powdery Mildews Conference held in Denmark in 2015, all presentations on stripe rust were about stripe rust on wheat and triticale, but none on barley stripe rust. More importantly, control of these individual diseases has been, and will be in a foreseeable future, through development of resistant cultivars of individual cereals. Treating stripe rusts on different cereals as different diseases is not only scientifically sound, but also practically important for control of the different diseases.

As discussed in Chap. 6, stripe rust can be controlled by host resistance, fungicide and cultural practices. Cultural practices can make differences in delaying and reducing stripe rust epidemics, but generally have a relatively limited role in control of the disease compared to resistance and fungicides. Some cultural practices that can be useful for high yield, regional adaption, and improving the overall agriculture and environment in the region may not be good for control of stripe rust. For example, increasing spring wheat in the US Pacific Northwest has increased stripe rust pressure. Also, quickly releasing wheat cultivars with high yield potential, good quality, and/or herbicide resistance, but reduced stripe rust resistance has contributed to stripe rust epidemics in the region. Therefore, appropriate cultural practices need to be considered in the integrated stripe rust management, but they should be largely adopted to improve productivity and protect environment.

The development of systemic fungicides during the last 40 years is one of the most achievement in control of stripe rust and also many other diseases. The availability of the effective fungicides provides growers a great secondary defense to combat stripe rust when a disease epidemic starts developing. This has ended the situation of hundreds of years that there was no any way to stop damaging diseases and saved million tons of food every year. Because of effective fungicides, there was for a while a trend to rely on chemical control for stripe rust and other diseases in some regions of the world. Compared to host resistance, fungicide application has many drawbacks, such as adding extra cost to food production; potential harmfulness to humans, animals and environment; less or not effective when not applied on time; and potential chemical-tolerant strains. Disease forecasting and monitoring can greatly improve timely control of stripe rust, but prediction models are usually highly specific to particular regions and disease monitoring is also needed for particular regions.

There is less debate about the important role of developing and growing resistant cultivars for control of stripe rust. There are still quite bit debate in how much effort should be used and how to develop resistant cultivars. Because of different climates in different regions, different regions have different vulnerability to stripe rust and other problems. Currently, not all regions take developing stripe rust resistance as a top priority. In the recent decade, stripe rust caused much more damages in regions where stripe rust does not frequently occur at an epidemic level, mainly because not enough resistant cultivars had been developed and also because not adequate management methods are available. As urediniospores are capable of traveling hundreds and thousands of kilometers, spores from other regions can spark disease epidemic on susceptible cultivars when the environment is favorable to the disease. Therefore, stripe rust resistance is needed in cultivars developed for regions where severe disease does not occur every year.

It was a long debate about what types of resistance should be used. Any types of resistance or any levels of resistance are useful compared to no resistance. However, choosing more appropriate types of resistance or resistance genes can make a breeding program more efficient and make resistant cultivars last longer. Breeding for resistant cultivars is largely depends upon the availability of genetic resources and technologies. As discussed above, different types of resistance have their advantages and disadvantages. Because of high level resistance and single-gene inheritance, genes for all-stage resistance have been widely used in breeding programs. Examples of such genes are *Yr9*, *Yr17* and *Yr27* used worldwide; *Yr26* in China; and *Yr10*, *Yr76* (*YrTye*) and *YrTr1* in the US Pacific Northwest. Races to each of these genes have been identified and they have made these genes largely ineffective. These genes have made great contributions to reduce the frequency of stripe rust epidemics and protected wheat crops for many years in many parts of the world when they were effective.

Criticisms have often been given to the use of such all-stage resistance genes because of their race specificity, and they were attributed to the epidemics after becoming ineffective. These genes have not received adequate appreciation for the protection during the years when they were effective. For example, *Yr9* in 1BL/1RS wheat-rye translocation and 1B/1R substitution lines were developed in Germany by Salzmunder and Wiehenstephan in 1920–1930s, and have been widely used in wheat breeding programs in many countries since the 1970s (Rabinovich 1998). The *Yr9*-virulence was reported for the first time in the former Soviet Union in

1973, in the Netherlands in 1974 (Stubbs et al. 1977), in China in 1977 (Wang et al. 1986), in East Africa (Ethiopia and Kenya) in 1986 (Saari 1996), in the Middle East (Syria and Lebanon) in 1989 (Hakim and Mamluk 1996), in South America (Ecuador, Colombia, Ecuador, Bolivia, and Chile) and Mexico in 1988 (Stubbs and Yang 1988), in Australasia in 1991 (Wellings and Burdon 1992), and in the US in 2000 (Chen et al. 2002). In the US, numerous wheat cultivars with 1BL/1RS wheat-rye translocation were developed in the 1980s and 1990s (Rabinovich 1998), but the *Yr9*-virulent races were not detected until 2000 (Chen et al. 2002). Thus, the resistance was effective for about 20 years. As Line and Qayoum (1992) pointed out, stripe rust has become more prevalent in the eastern US, especially in the South–Central states, since 1984. Even so, the disease severity and scale levels in the 1980s and 1990s were much lower than the epidemics caused by *Yr9*-virulent races since 2000 (Chen et al. 2002, 2010; Chen 2005, 2007; Wan and Chen 2012, 2014; Wan et al. 2016). The control of stripe rust in the 1980s and 1990s in this region is at least partially contributed by the *Yr9* resistance. The stem rust resistance provided by *Sr31* from rye still protects wheat crops in the US and throughout the world, except in Africa and few countries of central and western Asia where the UG99 race group has occurred (Singh et al. 2011; Patpour et al. 2016). *Yr10* is a race-specific resistance gene from wheat germplasm PI 178383, and the first wheat cultivar, Moro, with this gene was developed in the US Pacific Northwest in 1965 (Chen and Line 1992). Many club wheat and hard red winter wheat cultivars developed in this region have this resistance gene. A race virulent on Moro was detected in the region in 1968, and several races virulent on *Yr10* have been detected (Line and Qayoum 1992; Chen 2005; Chen et al. 2010; Wan and Chen 2012, 2014; Wan et al. 2016). However, the frequency of *Yr10*-virulent races fluctuated greatly over the last four decades, and wheat cultivars with *Yr10* recovered the high-level resistance when *Yr10*-virulent races were absent or in a low frequency (XM Chen and associates, unpublished data). These examples indicate that not all race-specific all-stage resistance genes are short lived, and not all are resistance genes with virulent races no longer useful. In the past, when limited effective resistance sources were available, every breeding program used a single source. However, as discussed above, many genes are now available, appropriate utilization of effective all-stage resistance through pyramiding and/or deployment in different cultivars or regions can expend the effectiveness of the resistance genes in wheat or barley cultivars.

In contrast to race-specific all-stage resistance, HTAP resistance is more durable. The utilization of HTAP resistance in the US started in the early 1960s when wheat cultivars Gaines and Nugaines were released purposely for reduce damage by stripe rust (Line 2002). Although the levels of these early cultivars are no longer considered adequate, the resistance is still recognizable with the current races of *P. striiformis* f. sp. *tritici* (Chen 2013). The levels of HTAP resistance in current cultivars in the region have been greatly increased over the past 50 years. However, not all cultivars have an equally high level of such durable type of resistance. When stripe rust starts early and quickly develop under low-temperature conditions, HTAP resistance may not expressed to the potential level and cultivars with low to moderate levels of resistance may suffer considerable yield losses, even though much lower than susceptible varieties (Chen 2005, 2014). Under such situations, fungicide applications are needed.

The level of HTAP resistance can be increased by combining different genes. As discussed above, many genes or QTL for HTAP resistance have been identified and can be used for developing cultivars with high level of durable resistance. It is even greater to pyramid effective genes for all-stage resistance and those for HTAP resistance. When all-stage resistance genes are effective, crops are completely protected; and when they become ineffective, HTAP resistance can protect crops from severe yield losses. Molecular markers of the identified genes can facilitate gene combinations. However, challenges still exist to combine high level and durable resistance to stripe rust with high yield, high quality and many other traits including resistance to other biotic and abiotic stresses. With the current genetic resources and more resistance genes identified in the future, developing such cultivars will become more and more efficient as new technologies will be developed.

7.2 Perspectives

Because the stripe rust fungus has a wide range of auxiliary grass hosts and capable of migration by wind for thousands of kilometers, it is impossible to eradicate the pathogen. However, it is possible to prevent huge yield losses under high epidemic pressure and convert stripe rust from a major disease to a minor disease. Developing more effective methods for control of stripe rust is an endless task, which requires continual research and technology development by scientists of many generations. In the next couple of decades, breakthrough progresses will be likely achieved in the biology, genomics and population genetics of the pathogen; resistance of crops; epidemiology of the disease and strategies and technologies in the management of stripe rust.

7.2.1 The Biology of the Pathogen and Epidemiology of Stripe Rust

Through the last 240 years of research, the biology of the stripe rust fungus is generally quite clear. However, there are still many questions to be answered. For example, the stripe rust fungus is now known to have a complete sexual lifecycle with the sexual stage on barberry and mahonia (Jin et al. 2010; Wang and Chen 2013; Zhao et al. 2013), but different results have been reported for the role of alternate hosts under natural conditions in China, Europe and the US (Berlin et al. 2012; Zhao et al. 2013, 2016; Wang et al. 2015d, 2016). The role of alternate hosts for the pathogen variation and disease epidemiology needs to be determined in many other regions in the world, especially in the regions where stripe rust frequently occurs. For the regions where natural infection of alternate hosts by the stripe rust pathogen has been detected, the frequency of infection, conditions allowing infection, and the impact of alternate host infection to the pathogen variation and disease epidemics need to be determined.

In addition to facilitate karyogamy, teliospores function for a rust fungus to survive unfavorable host and environmental conditions. More evidence has indicated that teliospores of the stripe rust fungus do not play a role in its sexual lifecycle under natural conditions (Wang and Chen 2015; Wang et al. 2015d), and therefore do not help the fungus survive. In fact, when the fungus stops producing urediniospores and instead forms teliospores, the fungus basically kills itself. The stripe rust fungus usually forms teliospores in the late host growth stage when the weather is hot, which is not favorable for infection and formation of urediniospores. Recently, we have found that isolates of the stripe rust pathogen vary greatly in their ability of producing telia and some isolates are able to form telia under conditions favorable for urediniospore production (Wan and Chen 2016). It is important to understand the genetics of telial formation and determine biological and ecological factors controlling the life process. Determination of the ability of telial formation of different races or isolates is valuable in predicting the fate of specific races in the pathogen population as telial formation is directly related to the fungal aggressiveness and fitness. Identification of genes and biochemical pathways involved in the transition from infectious uredinial stage to non-infectious telial stage may lead to the development of special fungicides that can put the fungus into a no-wake-up sleep.

As the stripe rust fungus is an obligate biotrophic parasite, it completely depends upon its plant hosts. In addition to the cereal hosts, the stripe rust pathogen is able to infect many grass hosts as discussed in Chap. 2. It is no doubt that host grasses play an important role in the pathogen survival, generating races, changing the pathogen population and developing disease epidemics on cereal crops (Cheng et al. 2016). However, the importance of grass hosts can vary from species to species and also vary from region to region. The role of grasses is not clear in many parts of the world. Monitoring of stripe rust and the pathogen population need to pay attention to various grasses. Not all species that have been reported as hosts of the stripe rust pathogens are equally susceptible. They are in the whole range from highly susceptible host to nonhost. Determine the position of each grass species or biotype in the range is important for the disease management through utilization of resistance genes from grasses and control or not control of rust on grasses. Taking jointed goatgrass (*Aegilops cylindrica* L.) for example, it is a noxious weed often present in wheat fields. This grass is highly susceptible to the wheat stripe rust pathogen. On a beneficial side, races of the stripe rust pathogen that are avirulent on wheat cultivars but virulent on goatgrass can reduce the weed problem. However, on the other hand, rust occurring on the grasses can provide inoculum for infecting wheat crops and may serve as a cradle for developing new races capable of overcoming resistance in wheat cultivars. It is important to survey stripe rust on this grass and other grasses for providing useful information for stripe rust management.

Determination of pathogen survival is critical for understanding the epidemiology of stripe rust, especially for developing models for forecasting disease epidemics (Line 2002; Sharma-Poudyal and Chen 2011; Sharma-Poudyal et al. 2014). Whether the stripe rust pathogen is able to survive winter and/or summer in a particular region often is not a clear cut for an absolute yes or no. It is important to determine the frequency or degree of pathogen survival and disease epidemics. In a scale of 0–10 survival index developed by Sharma-Poudyal et al. (2014), even a region with an index 1 for winter survival based on the historical climatic data, the stripe rust pathogen can survive some winters, although in a low frequency. It is important to check for each winter survival against the index in a region as this has an immediate impact on preparation of early stripe rust control. Because climate changing is a common phenomenon, regions used to be unfavorable can become favorable to stripe rust, and the pathogen can also become adapted to an environment (Line and Qayoum 1992; Milus and Seyran 2006; Milus et al. 2009). As prediction models for stripe rust epidemics are specific for a region as Sharma-Poudyal and Chen (2011) did for the US Pacific Northwest, models are needed for every specific epidemic region which can be defined by their climatic conditions, cropping systems and relationships to other regions.

Early detection of stripe rust is useful for implementing appropriate control measures on time for maximum protection and also for avoiding unnecessary use of chemicals. This can be done by disease survey in the early crop season. However, disease survey requires considerable time of experts or trained personnel especially during the very early disease development. It is impossible to check every field unless to have every grower involved in the survey of their fields. Even the latter is not possible in many parts of the world. Planting trap nurseries including susceptible and commercial cultivars is an efficient way for monitoring stripe rust development if planted in right time and suitable locations. The stripe rust information in the trap nurseries can be used to guide disease management in the environmentally similar regions. Time is generally enough from the first spotting stripe rust pustules to the best time of fungicide application. The results are even better to use disease data from both trapping nurseries and commercial fields. This is a routine practice in the disease management in many countries. In the regions where this has not become a routine practice, the measure should be taken without waiting until stripe rust can be seen easily.

In the recent years, several studies have been reported for developing molecular markers for distinguishing *Puccinia* species and early detection of the stripe rust pathogen before sporulation (Barnes and Szabo 2007; Zhao et al. 2007; Wang et al. 2010b; Yan et al. 2012; Gao et al. 2015; Ma et al. 2016). As stripe rust is relatively easy to be distinguished from other rust on cereal crops and grasses, molecular markers have less value for diagnosis in crop fields, but can be very useful in detection of the pathogen spores from air or rain drop samples for detecting spores in the air, and also useful for distinguish *Puccinia* species on barberry plants (Wang et al. 2015d). Molecular markers are also useful for early detection in cereal crops as the fungus can be killed in plant tissue before stripe rust develops. However, so far, such technology has not been practically used. Because of the molecular detection techniques in the above studies were developed either using inoculated plants or in a stripe rust hot-spot region, the efficiency of the technique, similar to visual disease survey, also face a sampling issue. In many regions, the early infection is at very low incidence, which makes the detection of low infection is difficult. Such problem can be partially solved by planting trap nurseries as discussed above and improving sampling methods. Pooling samples coupled with improved sensitivity will be helpful to detect the pathogen at a very low frequency of infection. Distinguishing viable verses dead pathogen in plant tissue using RNA samples can be more helpful when how much fungal parasite is killed during a winter is considered. Such a technique may also help in determining resistance or susceptibility levels of host cultivars. As illustrated in the study of Yan et al. (2012), accurately identifying "hot-spots" of infection may allow precise application of fungicides, which can avoid unnecessary use of chemicals.

7.2.2 The Genomics and Population Genetics of the Pathogen

In the next couple of decades, many significant accomplishments can be made in the genomics and population genetics of the stripe rust pathogen. Although traditional methods of identification of races based on reactions on host differentials and populations using selected molecular markers will be continually used, it will become easier for comparison of populations in different countries or regions and in different regions over different years. Genome-wide marker (for example SNPs) chips could be developed and used for characterization of pathogen isolates and populations as currently used for humans.

As the sequencing technology is advancing and the cost for sequencing becomes lower and lower, more and more isolates can be sequenced. Historical isolates represent different races and different genetic groups can be sequenced and used as references for identifying new genotypes. Isolates sampled from a particular region in each year can be sequenced and new genotypes can be identified by comparing with the reference genotypes. The sequences of reference isolates from different countries will become available online for easy comparison of the pathogen populations without shipping isolates for comparison studies. Sequence comparison may immediately reveal the origins and genetic mechanisms of changes in new isolates and populations. Tagging new races or strains of the pathogen with specific markers or genome sequences may provide information on distribution, spreading and adaptation, which will allow using deployment of cultivars with effective genes to reduce the damage by new races or populations.

In addition to genomic sequencing, RNA sequencing (RNA-seq) will become more and more popular. Different from genomic sequences that provide genetic codes, RNA sequences can provide genes under regulation at different life stages and processes of interactions with different hosts (Saunders et al. 2013; Hubbard et al. 2015; Dobon et al. 2016). RNA-seq allows identification of molecular patterns in both the host and pathogen during different interactions in a whole-genomic scale. Individual genes of the pathogen can be selected from RNA-seq data for further studies of their functions. Selected genes could be used as targets for developing new fungicides. It is also possible to use pathogen-derived genes for developing resistance by interfering pathogen virulence as used for controlling virus diseases (Gottula and Fuchs 2009) and recently proposed as a new strategy for control of rusts (Ayliffe et al. 2008; Hulbert and Pumphrey 2014).

So far, no virulence genes of the stripe rust pathogen have been cloned and molecularly characterized. A recent study by Liu et al. (2016) identified a *P. striiformis* effector candidate (PEC6) with pattern-triggered immunity suppression function and its corresponding host target. Knocking down of PEC6 expression by virus-mediated host-induced gene silencing decreased the number of rust pustules, suggesting PEC6 as a pathogenicity factor, but not a virulence factor. PEC6 interacts with wheat adenosine kinases (ADKs). Knocking down wheat ADK expression reduced leaf growth and enhanced the number of rust pustules, indicating that ADK is involved in plant development and defense. In the next couple of decades, more genes in the stripe rust pathogen involved in pathogenicity and with a clear function for avirulence corresponding to individual resistance genes will be identified. Cloning avirulence genes may lead to perfect molecular markers for tagging individual virulence genes in the pathogen population. In the future, virulence-tagging markers can be used to identify and monitoring races of the stripe rust pathogen.

7.2.3 Development of Resistant Cultivars

Mapping resistance genes has been the most active field of stripe research in the past decade, and this trend will continue into the next couple of decades. Wheat and barley germplasm especially landraces maintained in the major germplasm collection centers and individual breeding programs is a rich source of resistance to stripe rust. Mapping resistance genes has become easier as more efficient marker techniques have become available. Simple sequence repeat (SSR) and single-nucleotide polymorphism (SNP) markers will continually be used, but genotyping by sequencing (GBS) will become widely used. On the other hand, identification of different genes has become more and more difficult as more and more genes have been mapped. Determination of relationships among genes, especially on a same chromosome, can be more difficult as more genes are involved and more crosses need to be made for allelism tests. Although the relationship of genes is not absolutely necessary for use them in breeding programs, the information is useful for cataloguing genes and deploying different resistance genes in cultivars.

Identifying different resistance genes from the germplasm collections is a huge task. Traditionally, this is done by making crosses to study one source after another. New technologies have become available. For example, genome-wide association studies (GWAS), as conducted recently for studying stripe rust resistance (Sthapit et al. 2014; Zegeye et al. 2014; Kertho et al. 2015; Maccaferri et al. 2015; Naruoka et al. 2015; Bulli et al. 2016), are effective to quickly identify genes shared by a considerable accessions among a large number of germplasm accessions. This approach will be continually used for testing germplasm collections. However, genes present at a low frequency in a panel of germplasm tested could not be identified and mapped. These genes can be more useful for the purpose of diversifying resistance genes in breeding programs because they are more likely different from those that are already present in cultivars. To solve this problem, resistant accessions should be selected after a GWAS study for developing bi-parental populations to map the resistance genes.

In addition to the germplasm of crop species, resistance genes to stripe rust and other diseases have been transferred from related species or subspecies into common wheat. For example, *Yr5* is from *Triticum spelta* L.; *Yr9* from rye (*Secale cereale* L.); *Yr8* from *Aegilops comosa* Sm.; *Yr15*, *Yr35*, *Yr36* and *YrH52* from wild emmer wheat (*T. turgidum* L. var. *dicoccoides*); *Yr17* from *Ae. vertricosa* Tausch.; *Yr24*/*Yr26* from *T. turgidum* L.; *Yr28* from *Ae. tauschii* Coss.; *Yr37* from *Ae. kotschyi* Boiss.; *Yr38* from *Ae. sharonensis* Eig.; *Yr40* from *Ae. geniculata* Roth.; *Yr42* from *Ae. neglecta* Req.; *Yr50* from *Thinopyrum intermedium* (Host) Barkworth & D.R. Dewey; and *Yr53*, *Yr64* and *Yr65* from durum wheat [*Triticum turgidurum* L. subsp. *durum* (Desf) Husn.] (McIntosh et al. 2013; Cheng et al. 2014). As alien species have been used intensively in wheat improvement, more genes for stripe rust resistance will be identified in wheat varieties with alien species in their pedigrees. Usefulness of genes from alien species can be different and is related to other traits. Linkage drag is a common phenomenon for wheat with an alien chromatin. It can be beneficial if a stripe rust resistance gene is linked with resistance to other diseases and desirable traits (Zeven et al. 1983). Resistance genes from alien species also can be deleterious if it is tightly linked to genes for unwanted traits. Undesirable linkage drag can be reduced by shortening the length of alien chromatin through developing near-isogenic translocation lines, and the adverse effect of linkage drag can be compensated by selecting suitable wheat varieties for developing near-isogenic translocation lines (Klindworth et al. 2013). As the wheat stripe rust fungus has many grass hosts, resistance controlled by genes from grass species should not be automatically considered durable just because they are from a different host (Cheng et al. 2016). Their durability, as well as genes from cereal crops, should be determined by their characteristics of resistance, cloned genes and ultimately by long-term utilization in crop production.

As several wheat chromosomes have two or more genes for resistance to stripe rust, these genes linked in repulsion can be converted to be linked in coupling by combining into new genotypes. The linked genes will be more likely incorporated together into new cultivars. The combined genes should make the resistance last longer than each of the individual genes if they are race-specific. The author and associates have identified several genes on chromosomes 1B, 2B and 7B (Yan et al. 2003; Lin and Chen 2007; Ren et al. 2012; Cheng et al. 2014; Zhou et al. 2014b;

XM Chen and associates, unpublished data). They are currently conducting studies to pyramid these genes on same chromosomes in different combinations. New germplasm lines with linked genes will be useful for developing long-lasting resistance cultivars.

Genes with pleiotropic resistance to different diseases have been reported. The best example is the cloned same gene *Yr18* for resistance to stripe rust and *Lr34* to leaf rust (Krattinger et al. 2009). The gene has also been found to confer resistance to stem rust and powdery mildew, and a more recent study has shown that the gene can confer partial resistance to blast when transformed into rice plants (Krattinger et al. 2016). Similarly, *Yr29* (=*Lr46*/*Sr58*/*Pm39*) and *Yr46* (=*Lr67*/*Sr55*/*Pm46*) have been also reported to confer resistance to leaf rust, stem rust and powdery mildew (Ellis et al. 2014). Because of resistance against multiple diseases, these genes have been widely used in wheat breeding programs. In the next couple of decades, more and more genes for resistance to stripe rust and also to other diseases will be identified and utilized.

Pathogen-derived resistance has been successfully used in control of plant virus diseases. The similar approach is expected to be developed for control of stripe rust and other fungal diseases in the near future. As discussed above on the pathogen, effectors or other types of genes from the stripe rust fungus could be used for developing resistance in wheat and barley. A recent study identified effector genes shared by stripe rust, stem rust and leaf rust fungi, and silencing some of the genes reduced the development of the three rusts (Yin et al. 2015a). The results indicate that it may be possible to engineer resistance with a single gene in transgenic wheat plants for sustainable control of the three rusts. If this technology is completely developed, we basically can turn every virulence effector of rust pathogens into plant resistance genes.

In addition to the possible resistance sources in other plants and pathogens, a susceptible host plant can become a rich source of resistance as relevant genes can be modified to provide resistance. Although just the beginning, gene editing has opened a wide avenue for modifying the function of a gene by changing its sequence. CRISPR (clustered regularly interspaced short palindromic repeats), which utilizes the latest synthetic endonuclease CRISPR-associated protein 9 (Cas9), makes it feasible to knockout genes at an unprecedented efficiency (Barakate and Stephens 2016). The CRISPR–Cas9 technology has revolutionized gene manipulation capabilities in many species including crops. CRISPR–Cas9 can be used in plants to knockout all or single multigene family members (Endo et al. 2015) and even multiple unrelated genes (Lowder et al. 2015). As resistance suppressing genes have been reported in wheat to rusts (Ma et al. 1995; Talajoor et al. 2015), knockout of a gene or genes suppressing resistance can make plants resistant. In barley, many stripe rust resistance genes are recessive (Chen and Line 1999, 2003), and it would be possible to knockout the dominant alleles at resistance loci in susceptible varieties to make them resistant. It may be possible to change the specific domain of a resistance gene to be re-recognizable by avirulence effectors for making a defeated gene re-effective using the gene editing function of the CRISPR technology. It is also possible to use the system to unmask the master regulators of disease resistance for improving host defense mechanisms.

Developing cultivars with adequate resistance to stripe rust will continue remaining as a top priority for the wheat and barley breeding programs in the regions where stripe rust is a major problem. Regions where stripe rust is a problem but not frequent, breeding programs should increase effort in developing cultivars with resistance to stripe rust. As more and more resistance genes and useful markers are available, breeding for stripe rust resistant cultivars will become more efficient and easier to incorporate into breeding processes for other priorities. The major change will be more utilization of well characterized resistance genes by marker-assistant selection. Transgenic wheat and barley may eventually become acceptable by consumers, and thus transformation may become a common breeding method. Once the door is open, stripe rust resistance genes can be easily incorporated into elite cereal cultivars through transformation. A large number of cloned genes from host cereals, alien species, non-host plants, pathogens and artificially designed or modified DNA sequences that can provide or enhance resistance should be obtained. These genes can be used to develop wheat and barley cultivars with well-designed combinations of resistance genes by stacking genes into various cassettes for high-level and durable resistance for sustainable control of stripe rust and other diseases.

7.2.4 Control of Stripe Rust

The ultimate goal of any research on stripe rust is to control the disease. The best control can be achieved using the available best resources and technologies on a yearly basis, and in a long run by developing new resources and technologies superior to available ones. As developing resistant cultivars is the most preferred strategy, breeders will continue working together with pathologists, extension specialists and growers to develop more resistant cultivars. In a particular region, it may be useful to set a standard for stripe rust resistance levels based on the epidemic frequency; requirements of yield, quality and resistance to other biotic and abiotic stresses; and other desirable agronomic traits. The standard can be gradually increased and improved for the level and durability of stripe rust resistance. Whenever a cultivar becomes susceptible or below the standard, it should be quickly withdrawn from production. This can be done through the corporation of breeding, pathology, extension programs together with seed suppliers and growers in a region. Of course, this requires accurate disease monitoring and information on pathogen virulence and resistance genes in cultivars. More importantly, new cultivars with equal or superior yield and other traits should be available to replace cultivars to be withdrawn.

Fungicides are needed for control of stripe rust in the foreseeable future. Although obvious fungicide-resistant strains of the stripe rust pathogen have not been reported, new fungicides with diverse modes of action should be developed to delay or reduce the possibility of resistant strain developing. Fungicide sensitivity of the pathogen population needs to be monitored. As application can be more costly than fungicide itself, fungicides tend to be applied together with other chemicals to control weeds and insets and providing fertilizer during the growth season. Compatibilities with other chemicals and surfactants need to be clearly determined. Effective biofungicides could be developed for use in fields under organic production and in areas with restrictions to chemicals for protecting the environment. Timing of fungicide application need to be clearly determined based on disease development patterns. Efforts need to be taken to precisely apply fungicides when needed, and avoid unnecessary use of fungicides. This can be achieved by clearly determining cultivar resistance, accurate disease forecast and timely disease monitoring.

Determining the effects of cultural practices on stripe rust is useful for integrated control of stripe rust. As the effects of similar practices can be different in different regions, individual practices should be studied for their effects or consequent impacts on stripe rust and other diseases. Approaches need to be developed for reducing the undesirable effects and promoting desirable effects. A simple practice of growing more diverse cultivars, rather than one cultivar, in a region or by a farmer can greatly reduce the risk of entire crop failure. Integration of growing resistant cultivars, proper cultural practices and appropriate application of fungicides when needed should put stripe rust under control.

References

Abbasi M, Hedjaroude G, Scholler M, Goodwin SB. Taxonomy of *Puccinia striiformis* s.l. in Iran. Rostaniha. 2004;5:71–82, 199–224.

Agenbag GM, Pretorius ZA, Boyd LA, Bender CM, MacCormack R, Prins R. High-resolution mapping and new marker development for adult plant stripe rust resistance QTL in the wheat cultivar Kariega. Mol Breed. 2014;34:2005–20.

Ali S, Gladieux P, Leconte M, Gautier A, Justesen AF, Hovmøller MS, Enjalbert J, de Vallavieille-Pope C. Origin, migration routes and worldwide population genetic structure of the wheat yellow rust pathogen *Puccinia striiformis* f. sp. *tritici*. PLoS Pathog. 2014a;10:e1003903.

Ali S, Gladieux P, Rahman H, Saqib MS, Fiaz M, Ahmed H, Leconte M, Gautier A, Justesen AF, Hovmøller MS, Enjalbert J, de Vallavieille-Pope C. Inferring the contribution of sexual reproduction, migration and off-season survival to the temporal maintenance of microbial populations: a case study on the wheat fungal pathogen *Puccinia striiformis* f. sp. *tritici*. Mol Ecol. 2014b;23:603–17.

Ayliffe M, Singh R, Lagudah E. Durable resistance to wheat stem rust needed. Curr Opin Plant Biol. 2008;11:187–92.

Bahri B, Leconte M, de Vallavieille-Pope C, Enjalbert J. Isolation of ten microsatellite loci in an EST library of the phytopathogenic fungus *Puccinia striiformis* f.sp. *tritici*. Conserv Genet. 2009a;10:1425–8.

Bahri B, Leconte M, Ouffroukh A, de Vallavieille-Pope C, Enjalbert J. Geographic limits of a clonal population of wheat yellow rust in the Mediterranean region. Mol Ecol. 2009b;18:4165–79.

Bansal UK, Kazi AG, Singh B, Hare RA, Bariana HS. Mapping of durable stripe rust resistance in a durum wheat cultivar Wollaroi. Mol Breed. 2014;33:51–9.

Barakate A, Stephens J. An overview of CRISPR-based tools and their improvements: new opportunities in understanding plant–pathogen interactions for better crop protection. Front Plant Sci. 2016;7:765.

Barnes CW, Szabo LJ. Detection and identification of four common rust pathogens of cereals and grasses using real-time polymerase chain reaction. Phytopathology. 2007;97:717–27.

Basnet BR, Ibrahim AMH, Chen XM, Singh RP, Mason ER, Bowden RL, Liu SY, Devkota RN, Subramanian NK, Rudd JC. Molecular mapping of stripe rust resistance QTL in hard red winter wheat TAM 111 adapted in the US high plains. Crop Sci. 2014a;54:1361–73.

Basnet BR, Singh RP, Ibrahim AMH, Herrera-Foessel SA, Huerta-Espino J, Lan C, Rudd JC. Characterization of *Yr54* and other genes associated with adult plant resistance to yellow rust and leaf rust in common wheat Quaiu 3. Mol Breed. 2014b;33:385–99.

Berlin A, Djurle A, Samils B, Yuen J. Genetic variation in *Puccinia graminis* collected from oats, rye, and barberry. Phytopathology. 2012;102:1006–12.

Bever WM. Physiologic specialization in *Puccinia glumerum* in the United States. Phytopathology. 1934;24:686–8.

Biffen RH. Mendel's law of inheritance and wheat breeding. J Agric Sci. 1905;1:4–48.

Buerstmayr M, Matiasch L, Mascher F, Vida G, Ittu M, Robert O, Holdgate S, Flath K, Neumayer A, Buerstmayr H. Mapping of quantitative adult plant field resistance to leaf rust and stripe rust in two European winter wheat populations reveals co-location of three QTL conferring resistance to both rust pathogens. Theor Appl Genet. 2014;127:2011–28.

Bulli P, Zhang JL, Chao SM, Chen XM, Pumphrey M. Genetic architecture of resistance to stripe rust in a global winter wheat germplasm collection. G3: Genes. Genomes Genet. 2016;6:2237–53.

Calvo-Salazar V, Singh RP, Huerta-Espino J, Cruz-Izquierdo S, Lobato-Ortiz R, Sandoval-Islas S, Vargas-Hernández M, German S, Silva P, Basnet BR, Lan CX, Herrera-Foessel SA. Genetic analysis of resistance to leaf rust and yellow rust in spring wheat cultivar Kenya Kongoni. Plant Dis. 2015;99:1153–60.

Cantu D, Govindarajulu M, Kozik A, Wang MN, Chen XM, Kojima K, Jurka J, Michelmore RW, Dubcovsky J. Next generation sequencing provides rapid access to the genome of wheat stripe rust. PLoS One. 2011;8:e24230.

Cantu D, Segovia V, MacLean D, Bayles R, Chen XM, Kamoun S, Dubcovsky J, Saunders DGO, Uauy C. Genome analyses of the wheat yellow (stripe) rust pathogen *Puccinia striiformis* f. sp. *tritici* reveal polymorphic and haustorial expressed secreted proteins as candidate effectors. BMC Genomics. 2013:14:270

Case A, Naruoka Y, Chen XM, Garland-Campbell KA, Zemetra RS, Carter AH. Mapping stripe rust resistance genes in a Brundage x Coda winter wheat recombinant inbred line population. PLoS One. 2014;9:e91758.

Castro AJ, Chen XM, Hayes PM, Knapp SJ, Line RF, Toojinda T, Vivar H. Coincident QTL which determine seedling and adult plant resistance to stripe rust in barley. Crop Sci. 2002;42:1701–8.

Castro AJ, Chen XM, Corey A, Filichkina T, Hayes PM, Mundt C, Richardson K, Sandoval-Islas S, Vivar H. Pyramiding and validation of quantitative trait locus (QTL) alleles determining resistance to barley stripe rust: Effects on adult plant resistance. Crop Sci. 2003;43:2234–9.

Chen XM. Epidemiology and control of stripe rust [*Puccinia striiformis* f. sp. *tritici*] on wheat. Can J Plant Pathol. 2005;27:314–37.

Chen XM. Challenges and solutions for stripe rust control in the United States. Aust J Agric Res. 2007;58:648–55.

Chen XM. High-temperature adult-plant resistance, key for sustainable control of stripe rust. Am J Plant Sci. 2013;4:608–27.

Chen XM. Integration of cultivar resistance and fungicide application for control of wheat stripe rust. Can J Plant Pathol. 2014;36:311–26.

Chen XM, Line RF. Identification of stripe rust resistance genes in wheat cultivars used to differentiate North American races of *Puccinia striiformis*. Phytopathology. 1992;82:1428–34.

Chen XM, Line RF. Recessive genes for resistance to races of *Puccinia striiformis* f. sp. *hordei* in barley. Phytopathology. 1999;89:226–32.
Chen XM, Line RF. Identification of genes for resistance to *Puccinia striiformis* f. sp. *hordei* in 18 barley genotypes. Euphytica. 2003;129:127–45.
Chen XM, Line RF, Leung H. Relationship between virulence variation and DNA polymorphism in *Puccinia striiformis*. Phytopathology. 1993;83:1489–97.
Chen XM, Line RF, Leung H. Virulence and polymorphic DNA relationships of *Puccinia striiformis* f. sp. *hordei* to other rusts. Phytopathology. 1995;85:1335–42.
Chen XM, Moore MK, Milus EA, Long DL, Line RF, Marshall D, Jackson L. Wheat stripe rust epidemics and races of *Puccinia striiformis* f. sp. *tritici* in the United States in 2000. Plant Dis. 2002;86:39–46.
Chen CQ, Zheng WM, Buchenauer H, Huang LL, Lu NH, Kang ZS. Isolation of microsatellite loci from expressed sequence tag library of *Puccinia striiformis* f. sp. *tritici*. Mol Ecol Resour. 2009a;9:236–8.
Chen WQ, Wu LR, Liu TG, Xu SC, Jin SL, Peng YL, Wang BT. Pathotype dynamics, diversity and virulence evolution in *Puccinia striiformis* f. sp. *tritici*, the causal agent of wheat stripe (yellow) rust in China from 2003 to 2007. Plant Dis. 2009b;93:1093–101.
Chen XM, Penman L, Wan AM, Cheng P. Virulence races of *Puccinia striiformis* f. sp. *tritici* in 2006 and 2007 and development of wheat stripe rust and distributions, dynamics, and evolutionary relationships of races from 2000 to 2007 in the United States. Can J Plant Pathol. 2010;32:315–23.
Chen XM, Coram T, Huang XL, Wang MN, Dolezal A. Understanding molecular mechanisms of durable and non-durable resistance to stripe rust in wheat using a transcriptomics approach. Curr Genomics. 2013;14:111–26.
Cheng P, Chen XM. Virulence and molecular analyses support asexual reproduction of *Puccinia striiformis* f. sp. *tritici* in the U.S. Pacific Northwest. Phytopathology. 2014;104:1208–20.
Cheng P, Chen XM, Xu LS, See DR. Development and characterization of expressed sequence tag-derived microsatellite markers for the wheat stripe rust fungus *Puccinia striiformis* f. sp. *tritici*. Mol Ecol Resour. 2012;12:779–81.
Cheng P, Xu LS, Wang MN, See DR, Chen XM. Molecular mapping of genes *Yr64* and *Yr65* for stripe rust resistance in hexaploid derivatives of durum wheat accessions PI 331260 and PI 480016. Theor Appl Genet. 2014;127:2267–77.
Cheng P, Chen XM, See D. Grass hosts harbor more diverse isolates of *Puccinia striiformis* than cereal crops. Phytopathology. 2016;106:362–71.
Christopher MD, Liu SY, Hall MD, Marshall DS, Fountain MO, Johnson JW, Milus EA, Garland-Campbell KA, Chen XM, Griffey CA. Identification and mapping of adult plant stripe rust resistance in soft red winter wheat cultivar 'USG 3555'. Plant Breed. 2013a;132:53–60.
Christopher MD, Liu SY, Hall MD, Marshall DS, Fountain MO, Johnson JW, Milus EA, Garland-Campbell KA, Chen XM, Griffey CA. Identification and mapping of adult plant stripe rust resistance in soft red winter wheat VA00W-38. Crop Sci. 2013b;53:871–9.
Coram TE, Brown-Guedira G, Chen XM. Using transcriptomics to understand the wheat genome. CAB Reviews: Perspectives in Agriculture, Veterinary Science, Nutrition and Natural Resources. 2008a;3. No. 083
Coram TE, Settles ML, Chen XM Transcriptome analysis of high-temperature adult-plant resistance conditioned by *Yr39* during the wheat-*Puccinia striiformis* f. sp. *tritici* interaction. Mol Plant Pathol. 2008b;9:479–493
Coram TE, Settles ML, Wang MN, Chen XM. Surveying expression level polymorphism and single-feature polymorphism in near-isogenic wheat lines differing for the *Yr5* stripe rust resistance locus. Theor Appl Genet. 2008c;117:401–11.
Coram TE, Settles ML, Chen XM. Large-scale analysis of antisense transcription in wheat using the Affymetrix GeneChip wheat genome array. BMC Genomics. 2009;10:253–64.
Coram TE, Huang XL, Zhan GM, Settles ML, Chen XM. Meta-analysis of transcripts associated with race-specific resistance to stripe rust in wheat demonstrates common induction of blue

copper-binding protein, heat-stress transcription factor, pathogen-induced WIR1A protein, and ent-kaurene synthase transcripts. Funct Integ Genome. 2010;10:383–92.

Cummins GB, Stevenson JA. A check list of North American rust fungi (Uredinales). Plant Dis Rep Suppl. 1956;240:109–93.

Cuomo CA, Bakkeren G, Khalil HB, Panwar V, Joly D, Linning R, Sakthikumar S, Song X, Adiconis X, Fan L, Goldberg JM, Levin JZ, Young S, Zeng QD, Anikster Y, Bruce M, Wang MN, Yin CT, McCallum B, Szabo LJ, Hulbert S, Chen XM, Fellers JP. Comparative analysis highlights variable genome content of wheat rusts and divergence of the mating loci. 3G: Genes Genomes. Genomics. 2017;7:371–6.

Dawson AM, Ferguson JN, Gardiner M, Green P, Hubbard A, Moscou MJ. Isolation and fine mapping of *Rps6*: an intermediate host resistance gene in barley to wheat stripe rust. Theor Appl Genet. 2016;129:831–43.

Derevnina L, Zhou MX, Singh D, Wellings CR, Park RF. The genetic basis of resistance to barley grass yellow rust (*Puccinia striiformis* f. sp. *pseudo-hordei*) in Australian barley cultivars. Theor Appl Genet. 2015;128:187–97.

Dobon A, Bunting DCE, Cabrera-Quio LE, Uauy C, Saunders DGO. The host-pathogen interaction between wheat and yellow rust induces temporally coordinated waves of gene expression. BMC Genomics. 2016;17:380.

Dracatos PM, Zhang P, Park RF, McIntosh RA, Wellings CR. Complementary resistance genes in wheat selection 'Avocet R' confer resistance to stripe rust. Theor Appl Genet. 2016;129:65–76.

Duan X, Tellier A, Wan A, Leconte M, de Vallavieille-Pope C, Enjalbert J. *Puccinia striiformis* f.sp. *tritici* presents high diversity and recombination in the over-summering zone of Gansu, China. Mycologia. 2010;102:44–53.

Duan XY, Wang XJ, Fu YP, Tang CL, Li XR, Cheng YL, Feng H, Huang LL, Kang ZS. *Ta*EIL1, a wheat homologue of *At*EIN3, acts as a negative regulator in the wheat–stripe rust fungus interaction. Mol Plant Pathol. 2013;14:728–39.

Ellis JG, Lagudah ES, Spielmeyer W, Dodds PN. The past, present and future of breeding rust resistant wheat. Front Plant Sci. 2014;5:641.

Endo M, Mikami M, Toki S. Multigene knockout utilizing off-target mutations of the CRISPR/Cas9 system in rice. Plant Cell Physiol. 2015;56:41–7.

Enjalbert J, Duan X, Giraud T, Vautrin D, de Vallavieille-Pope C, Solignac M. Isolation of twelve microsatellite loci, using an enrichment protocol, in the phytopathogenic fungus *Puccinia striiformis* f.sp. *tritici*. Mol Ecol Notes. 2002;2:563–5.

Enjalbert J, Duan X, Leconte M, Hovmøller MS, de Vallavieille-Pope C. Genetic evidence of local adaptation of wheat yellow rust (*Puccinia striiformis* f. sp. *tritici*) within France. Mol Ecol. 2005;14:2065–73.

Eriksson J. Über die Spezialisierung des Parasitsimus bei den Getreiderostpilzen. Ber Dtsch Bot Ges. 1894;12:292–331.

Eriksson J, Henning E. Die getreideroste. Stockholm: Norstedt & Söner; 1896.

Feng H, Wang XM, Sun YF, Wang XJ, Chen XM, Guo J, Duan YH, Huang LL, Kang ZS. Cloning and characterization of a calcium binding EF-hand protein gene TaCab1 from wheat and its expression in response to *Puccinia striiformis* f. sp. *tritici* and abiotic stresses. Mol Biol Rep. 2011;38:3857–66.

Feng H, Zhang Q, Li HY, Wang XJ, Wang XD, Duan XY, Wang B, Kang ZS. vsiRNAs derived from the miRNA-generating sites of pri-tae-miR159a based on the BSMV system play positive roles in the wheat response to *Puccinia striiformis* f. sp. *tritici* through the regulation of *tae-Myb3* expression. Plant Physiol Biochem. 2013a;68:90–5.

Feng H, Zhang Q, Wang QL, Wang XJ, Liu J, Li M, Huang LL, Kang ZS. Target of tae-miR408, a chemocyanin-like protein gene (*TaCLP1*), plays positive roles in wheat response to high-salinity, heavy cupric stress and stripe rust. Plant Mol Biol. 2013b;83:433–43.

Feng JY, Chen GY, Wei YM, Liu YX, Jiang QT, Li W, Pu ZE, Lan XJ, Dai SF, Zheng YL. Identification and genetic mapping of a recessive gene for resistance to stripe rust in wheat line LM168-1. Mol Breed. 2014;33:601–9.

Feng JY, Wang MN, Chen XM, See DR, Zheng YL, Chao SM, Wan AM. Molecular mapping of *YrSP* and its relationship with other genes for stripe rust resistance in wheat chromosome 2BL. Phytopathology. 2015;105:1206–13.

Fu DL, Uauy C, Distelfeld A, Blechl A, Epstein L, Chen XM, Sela H, Fahima T, Dubcovsky J. A kinase-START gene confers temperature-dependent resistance to wheat stripe rust. Science. 2009;323:1357–60.

Fu YP, Duan XY, Tang CL, Li XR, Voegele RT, Wang XJ, Wei GR, Kang ZS. TaADF7, an actin-depolymerizing factor, contributes to wheat resistance against *Puccinia striiformis* f. sp. *tritici*. Plant J. 2014;78:16–30.

Gao L, Yu HX, Shen HM, Li C, Liu TG, Liu B, Kang XH, Chen WQ. Development of SCAR markers and an SYBR green assay to detect *Puccinia striiformis* f. sp. *tritici* in infected wheat leaves. Plant Dis. 2015;100:1840–7.

Garnica DP, Upadhyaya NM, Dodds PN, Rathjen JP. Strategies for wheat stripe rust pathogenicity identified by transcriptome sequencing. PLoS One. 2013;8:e67150.

Gottula J, Fuchs M. Toward a quarter century of pathogen-derived resistance and practical approaches to plant virus disease control. Adv Virus Res. 2009;75:161–83.

Gou JY, Li K, Wu KT, Wang XD, Lin HQ, Cantu D, Uauy C, Dobon-Alonso A, Midorikawa T, Inoue K, Sánchez J, Fu DL, Blechl A, Wallington E, Fahima T, Meeta M, Epstein L, Dubcovsky J. Wheat stripe rust resistance protein wks1 reduces the ability of the thylakoid-associated ascorbate peroxidase to detoxify reactive oxygen species. Plant Cell. 2015;27:1755–70.

Hakim MS, Mamluk OF. Virulence of wheat yellow rust pathogen in Syria and Lebanon. In: Proceedings of European mediterranean cereal rusts powdery mildews conference 9:141. 1996.

Herrera-Foessel SA, Singh RP, Lillemo M, Huerta-Espino J, Bhavani S, Singh S, Lan CX, Calvo-Salazar V, Lagudah ES. *Lr67/Yr46* confers adult plant resistance to stem rust and powdery mildew in wheat. Theor Appl Genet. 2014;127:781–9.

Herrera-Foessel SA, Singh RP, Lan CX, Huerta-Espino J, Calvo-Salazar V, Bansal UK, Bariana HS, Lagudah ES. *Yr60*, a gene conferring moderate resistance to stripe rust in wheat. Plant Dis. 2015;99:508–11.

Holtz MD, Kumar K, Zantinge JL, Xi K. Genetic diversity of *Puccinia striiformis* from cereals in Alberta. Can Plant Pathol. 2014;63:415–24.

Hou L, Chen XM, Wang MN, See DR, Chao SM, Bulli P, Jing JX. Mapping a large number of QTL for durable resistance to stripe rust in winter wheat Druchamp using SSR and SNP markers. PLoS One. 2015;10:e0126794.

Hou LY, Jia JQ, Zhang XJ, Li X, Yang ZJ, Ma J, Guo HJ, Zhan HX, Qiao LY, Chang ZJ. Molecular mapping of the stripe rust resistance gene *Yr69* on wheat chromosome 2AS. Plant Dis. 2016;100:1717–24.

Hovmøller MS, Justesen AF. Rates of evolution of avirulence phenotypes and DNA markers in a northwest European population of *Puccinia striiformis* f. sp. *tritici*. Mol Ecol. 2007;16:4637–47.

Hovmøller MS, Justesen AF, Brown JKM. Clonality and long-distance migration of *Puccinia striiformis* f. sp. *tritici* in north-west Europe. Plant Pathol. 2002;51:24–32.

Hovmøller MS, Yahyaoui AH, Miles EA, Justesen AF. Rapid global spread of two aggressive strains of a wheat rust fungus. Mol Ecol. 2008;17:3818–26.

Hovmøller MS, Walter S, Bayles R, Hubbard A, Flath K, Sommerfieldt N, Leconte M, Czembor P, Rodriguez-Algaba J, Thach T, Hansen JG, Lassen P, Justesen AF, Ali S, de Vallavieille-Pope C. Replacement of the European wheat yellow rust population by new races from the centre of diversity in the near-Himalayan region. Plant Pathol. 2016;65:402–11.

Huang XL, Ma JB, Chen XM, Wang XJ, Ding K, Han DJ, Qu ZP, Huang LL, Kang ZS. Genes involved in adult plant resistance to stripe rust in wheat cultivar Xingzi 9104. Physiol Mol Plant Pathol. 2013;81:26–32.

Huang Q, Li X, Chen WQ, Xiang ZP, Zhong SF, Chang ZJ, Zhang M, Zhang HY, Tan FQ, Ren ZL, Luo PG. Genetic mapping of a putative *Thinopyrum intermedium*-derived stripe rust resistance gene on wheat chromosome 1B. Theor Appl Genet. 2014;127:843–53.

Huang L, Sela H, Feng LH, Chen QJ, Krugman T, Yan J, Dubcovsky J, Fahima T. Distribution and haplotype diversity of WKS resistance genes in wild emmer wheat natural populations. Theor Appl Genet. 2016;129:921–34.

Hubbard A, Lewis CM, Yoshida K, Ramirez-Gonzalez RH, de Vallavieille-Pope C, Thomas J, Kamoun S, Bayles R, Uauy C, Saunders DGO. Field pathogenomics reveals the emergence of a diverse wheat yellow rust population. Genome Biol. 2015;16:23.

Hulbert SH, Pumphrey M. A time for more booms and less busts? Unraveling cereal-rust interactions. MPMI. 2014;27:207–14.

Hulbert SH, Bai J, Fellers JP, Pacheco MG, Bowden RL. Gene expression patterns in near isogenic lines for wheat rust resistance gene *Lr34/Yr18*. Phytopathology. 2007;97:1083–93.

Hylander N, Jørstad I, Nannfeldt JA. Enumeratio uredionearum Scandinavicarum. Opera Bot. 1953;1:1–102.

Jiang ZN, Ge S, Xing LP, Han DJ, Kang ZS, Zhang GQ, Wang XJ, Wang X, Chen PD, Cao AZ. *RLP1.1*, a novel wheat receptor-like protein gene, is involved in the defence response against *Puccinia striiformis* f. sp. *tritici*. J Exp Bot. 2013;64:3735–46.

Jighly A, Oyiga BC, Makdis F, Nazari K, Youssef O, Tadesse W, Abdalla O, Ogbonnaya FC. Genome-wide DArT and SNP scan for QTL associated with resistance to stripe rust (*Puccinia striiformis* f. sp. *tritici*) in elite ICARDA wheat (*Triticum aestivum* L.) germplasm. Theor Appl Genet. 2015;128:1277–95.

Jin Y, Szabo LJ, Carson M. Century-old mystery of *Puccinia striiformis* life history solved with the identification of *Berberis* as an alternate host. Phytopathology. 2010;100:432–5.

Kertho A, Mamidi S, Bonman JM, McClean PE, Acevedo M. Genome-wide association mapping for resistance to leaf and stripe rust in winter-habit hexaploid wheat landraces. PLoS One. 2015;10:e0129580.

Klarquist EF, Chen XM, Carter AH. Novel QTL for stripe rust resistance on chromosomes 4A and 6B in soft white winter wheat cultivars. Agronomy. 2016;6:4.

Klindworth DL, Hareland GA, Elias EM, Xu SS. Attempted compensation for linkage drag affecting agronomic characteristics of durum wheat 1AS/1DL translocation lines. Crop Sci. 2013;53:422–9.

Klos KE, Gordon T, Bregitzer P, Hayes P, Chen XM, del Blanco IA, Fisk S, Bonman JM. Barley stripe rust resistance QTL: development and validation of SNP markers for resistance to *Puccinia striiformis* f. sp. *hordei*. Phytopathology. 2016;106:1344–51.

Krattinger SG, Lagudah ES, Spielmeyer W, Singh RP, Huerta-Espino J, McFadden H, Bossolini E, Selter LL, Keller B. A putative ABC transporter confers durable resistance to multiple fungal pathogens in wheat. Science. 2009;323:1360–3.

Krattinger SG, Sucher J, Selter LL, Chauhan H, Zhou B, Tang M, Upadhyaya NM, Mieulet D, Guiderdoni E, Weidenbach D, Schaffrath U, Lagudah ES, Keller B. The wheat durable, multipathogen resistance gene *Lr34* confers partial blast resistance in rice. Plant Biotechnol J. 2016;14:1261–8.

Kumar J, Nayar SK, Prashar M, Bhardwaj SC, Bhatnagar R. Virulence survey of *Puccinia striiformis* West in India during 1988-1989. Int Trop Plant Dis. 1993;11:79–83.

Lan CX, Rosewarne GM, Singh RP, Herrera-Foessel SA, Huerta-Espino J, Basnet BR, Zhang YL, Yang EN. QTL characterization of resistance to leaf rust and stripe rust in the spring wheat line Francolin#1. Mol Breed. 2014;34:789–803.

Lan CX, Zhang YL, Herrera-Foessel SA, Basnet BR, Huerta-Espino J, Lagudah ES, Singh RP. Identification and characterization of pleiotropic and co-located resistance loci to leaf rust and stripe rust in bread wheat cultivar Sujata. Theor Appl Genet. 2015;128:549–61.

Li ZQ, Zeng SM. Wheat rusts in China. Beijing: China Agricultural Press; 2002. p. 379.

Li ZF, Singh S, Singh RP, López-Vera EE, Huerta-Espino J. Genetics of resistance to yellow rust in PBW343 × Kenya Kudu recombinant inbred line population and mapping of a new resistance gene *YrKK*. Mol Breed. 2013;32:821–9.

Li HH, Vikram P, Singh RP, Kilian A, Carling J, Song J, Burgueno-Ferreira JA, Bhavani S, Huerta-Espino J, Payne T, Sehgal D, Wenzl P, Singh S. A high density GBS map of bread wheat and its application for dissecting complex disease resistance traits. BMC Genomics. 2015;16:216.

Li K, Hegarty J, Zhang CZ, Wan AM, Wu JJ, Brown-Guedira GL, Chen XM, Fu DL, Dubcovsky J. Fine mapping of barley locus *Rps6* conferring resistance to wheat stripe rust. Theor Appl Genet. 2016a;129:845–59.

Li H, Ren B, Kang ZS, Huang LL. Comparison of cell death and accumulation of reactive oxygen species in wheat lines with or without *Yr36* responding to *Puccinia striiformis* f. sp. *tritici* under low and high temperatures at seedling and adult-plant stages. Protoplasms. 2016b;253:787–802.

Liang JM, Liu XF, Li Y, Wan Q, Ma ZH, Luo Y. Population genetic structure and the migration of *Puccinia striiformis* f. sp. *tritici* between the Gansu and Sichuan Basin populations of China. Phytopathology. 2016;106:192–201.

Lin F, Chen XM. Genetics and molecular mapping of genes for race-specific all stage resistance and non-race-specific high-temperature adult-plant resistance to stripe rust in spring wheat cultivar Alpowa. Theor Appl Genet. 2007;114:1277–87.

Line RF. Stripe rust of wheat and barley in North America: a retrospective historical review. Annu Rev Phytopathol. 2002;40:75–118.

Line RF, Qayoum A. Virulence, aggressiveness, evolution, and distribution of races of *Puccinia striiformis* (the cause of stripe rust of wheat) in North America, 1968–87. US Dep Agric Agric Res Serv Tech Bull. 1992;1788:44 pp.

Liu M, Hambleton S. Taxonomic study of stripe rust, *Puccinia striiformis* sensu lato, based on molecular and morphological evidence. Fungal Biol. 2010;114:881–99.

Liu F, Guo J, Bai P, Duan Y, Wang X, Cheng Y, Feng H, Huang LL, Kang ZS. Wheat *TaRab7* GTPase is part of the signaling pathway in responses to stripe rust and abiotic stimuli. PLoS One. 2012a;7:e37146.

Liu B, Chen XM, Kang ZS. Gene sequences reveal heterokaryotic variations and evolutionary mechanisms in *Puccinia striiformis*, the stripe rust pathogen. Open J Genom. 2012b;1:1–14.

Liu J, Chang ZJ, Zhang XJ, Yang ZJ, Li X, Jia JQ, Zhan HX, Guo HJ, Wang JM. Putative *Thinopyrum intermedium*-derived stripe rust resistance gene *Yr50* maps on wheat chromosome arm 4BL. Theor Appl Genet. 2013;126:265–74.

Liu W, Frickb M, Huel R, Nykiforuk CL, Wang XM, Gaudet GA, Eudes F, Conner RL, Kuzyk A, Chen Q, Kang ZS, Laroche A. The stripe rust resistance gene *Yr10* encodes an evolutionary-conserved and unique CC–NBS–LRR sequence in wheat. Mol Plant. 2014;7:1740–55.

Liu CH, Pedersen C, Schultz-Larsen T, Aguilar GB, Madriz-Ordeñana K, Hovmøller MS, Thordal-Christensen H. The stripe rust fungal effector PEC6 suppresses pattern-triggered immunity in a host species-independent manner and interacts with adenosine kinases. New Phytol. 2016; doi:10.1111/nph.14034.

Lowder LG, Zhang D, Baltes NJ, Paul JW 3rd, Tang X, Zheng X, Voytas DF, Hsieh TF, Zhang Y, Qi Y. A CRISPR/Cas9 toolbox for multiplexed plant genome editing and transcriptional regulation. Plant Physiol. 2015;169:971–85.

Lu NH, Zhan GM, Chen XM, Wang JF, Huang LL, Kang ZS. Spatial genetic diversity and interregional spread of *Puccinia striiformis* f. sp. *tritici* in the Northwest China. Eur J Plant Pathol. 2011;131:685–93.

Lu Y, Wang MN, Chen XM, See DR, Chao SM, Jing JX. Mapping of *Yr62* and a small-effect QTL for high-temperature adult-plant resistance to stripe rust in spring wheat PI 192252. Theor Appl Genet. 2014;127:1449–59.

Lupton FGH, Macer RCF. Inheritance of resistance to yellow rust (*Puccinia glumarum* Erikss. & Henn.) in seven varieties of wheat. Trans Br Mycol Soc. 1962;45:21–45.

Ma H, Singh RP, Mujeebkazi A. Suppression expression of resistance to stripe rust in hexaploid wheat (*Triticum turgidum* × *T. tauschii*). Euphytica. 1995;83:87–93.

Ma DF, Zhou XL, Hou L, Bai YB, Li Q, Wang HG, Tang MS, Jing JX. Genetic analysis and molecular mapping of a stripe rust resistance gene derived from *Psathynrostachys huashanica* Keng in wheat line H9014-121-5-5-9. Mol Breed. 2013;32:365–72.

Ma DF, Li Q, Tang MS, Chao KX, Li JC, Wang BT, Jing JX. Mapping of gene conferring adult-plant resistance to stripe rust in Chinese wheat landrace Baidatou. Mol Breed. 2015;35:157.

Ma LJ, Kong XY, Qiao JX, An F, Hu XP, Xu XM. Overwintering of *Puccinia striiformis* f. sp. *tritici* on winter wheat at varying altitudes in Gansu and Qinghai Provinces. Plant Dis. 2016;100:1138–45.

Maccaferri M, Zhang JL, Bulli P, Abate Z, Chao SM, Cantu D, Bossolini E, Chen XM, Pumphrey M, Dubcovsky J. A genome-wide association study of resistance to stripe rust (*Puccinia striiformis* f. sp. *tritici*) in a worldwide collection of hexaploid spring wheat (*Triticum aestivum* L.). 3G: Genes Genomics. Genetics. 2015;5:449–65.

Manners JG. *Puccinia striiformis* Westend. var. *dactylidis* var. nov. Trans Br Mycol Soc. 1960;43:65–8.

Mboup M, Leconte M, Gautier A, Wan AM, Chen W, de Vallavieille-Pope C, Enjalbert J. Evidence of genetic recombination in wheat yellow rust populations of a Chinese oversummering area. Fungal Genet Biol. 2009;46:299–307.

McIntosh RA, Yamazaki Y, Dubcovsky J, Rogers J, Morris C, Appels R, Xia XC. Catalogue of gene symbols for wheat. In: 12th international wheat genetics symposium, 8–13 Sep 2013, Yokohama, Japan. 2013. Online: http://www.shigen.nig.ac.jp/wheat/komugi/genes/download.jsp

Milus EA, Seyran E. Aggressiveness of *Puccinia striiformis* f. sp. *tritici* isolates in the South-Central United States. Plant Dis. 2006;90:847–52.

Milus EA, Kristensen K, Hovmøller MS. Evidence for increased aggressiveness in a recent widespread strain of *Puccinia striiformis* f. sp. *tritici* causing stripe rust of wheat. Phytopathology. 2009;99:89–94.

Moore JW, Herrera-Foessel S, Lan C, Schnippenkoetter W, Ayliffe M, Huerta-Espino J, Lillemo M, Viccars L, Milne R, Periyannan S, Kong XY, Spielmeyer W, Talbot M, Bariana H, Patrick JW, Dodds P, Singh R, Lagudah E. A recently evolved hexose transporter variant confers resistance to multiple pathogens in wheat. Nature Genet. 2015;47:1494–8.

Naruoka Y, Garland-Campbell KA, Carter AH. Genome-wide association mapping for stripe rust (*Puccinia striiformis* F. sp. *tritici*) in US Pacific Northwest winter wheat (*Triticum aestivum* L.). Theor Appl Genet. 2015;128:1083–101.

Newton M, Johnson T, Brown AM. Stripe rust in Canada. Phytopathology. 1933;23:25–6.

Niu YC, Li ZQ, Shang HS. *Puccinia striiformis* West. f. sp. *leymi* and f. sp. *elymi*, two new formae speciales. Acta Univ Agric Boreali-Occident. 1991;19:58–62.

Pahalawatta V, Chen XM. Genetic analysis and molecular mapping of wheat genes conferring resistance to the wheat stripe rust and barley stripe rust pathogens. Phytopathology. 2005a;95:427–32.

Pahalawatta V, Chen XM. Inheritance and molecular mapping of barley genes conferring resistance to wheat stripe rust. Phytopathology. 2005b;95:884–9.

Patpour M, Hovmøller MS, Justesen AF, Newcomb M, Olivera P, Jin Y, Szabo LJ, Hodson D, Shahin AA, Wanyera R, Habarurema I, Wobibi S. Emergence of virulence to *SrTmp* in the Ug99 race group of wheat stem rust, *Puccinia graminis* f. sp. *tritici*, in Africa. Plant Dis. 2016;100:522.

Pei HC, Sun QX, Hao QQ, Lv B, Wu JJ, Fu DL. The HSP90-RAR1-SGT1 based protein interactome in barley and stripe rust. Physiol Mol Plant Pathol. 2015;91:11–9.

Prasada R, Joshi LM, Singh SD, Misra DF, Goel LB, Kumari K, Sharma SK, Josh PC, Ahmad ST. Occurrence of physiological races of wheat and barley rusts in India during 1962-64 and their sources of resistance. Indian J Agric Sci. 1967;37:273–81.

Rabinovich SV. Importance of wheat-rye translocations for breeding modern cultivars of *Triticum aestivum* L. Euphytica. 1998;100:323–40.

Randhawa M, Bansal U, Valárik M, Klocová B, Doležel J, Bariana H. Molecular mapping of stripe rust resistance gene *Yr51* in chromosome 4AL of wheat. Theor Appl Genet. 2014;127:317–24.

Randhawa MS, Bariana HS, Mago R, Bansal UK. Mapping of a new stripe rust resistance locus *Yr57* on chromosome 3BS of wheat. Mol Breed. 2015;35:65.

Ren RS, Wang MN, Chen XM, Zhang ZJ. Characterization and molecular mapping of *Yr52* for high-temperature adult-plant resistance to stripe rust in spring wheat germplasm PI 183527. Theor Appl Genet. 2012;125:847–57.

Ren Y, Li S, Xia X, Zhou Q, He Y, Wei Y, Zheng Y, He Z. Molecular mapping of a recessive stripe rust resistance gene *yrMY37* in Chinese wheat cultivar Mianmai 37. Mol Breed. 2015a;35:97.

Ren Y, Li SR, Wei YM, Zhou Q, Du XY, He YJ, Zheng YL. Molecular mapping of a stripe rust resistance gene in Chinese wheat cultivar Mianmai 41. J Integr Agric. 2015b;14:295–304.

Risk JM, Selter LL, Krattinger SG, Viccars LA, Richardson TM, Buesing G, Herren G, Lagudah ES, Keller B. Functional variability of the *Lr34* durable resistance gene in transgenic wheat. Plant Biotech J. 2012;10:477–87.

Rosewarne GM, Herrera-Foessel SA, Singh RP, Huerta-Espino J, Lan CX, He ZH. Quantitative trait loci of stripe rust resistance in wheat. Theor Appl Genet. 2013;126:2427–49.

Saari EE.The *Yr9* virulence factor in *Puccinia striiformis* and South Asia: coevolution or migration from a distant place. In: Proceedings of European mediterranean cereal rusts powdery mildews conference 9:142. 1996.

Saunders DGO, Cantu D, Segovia V, MacLean D, Bayles R, Chen XM, Kamoun S, Dubcovsky J, Uauy C. Genome analyses of the wheat stripe (yellow) rust pathogen *Puccinia striiformis* f. sp. *tritici* reveal polymorphic and haustorial expressed secreted proteins as candidate effectors. Page 114 in Abstracts of BGRI 2013 technical workshop, 19–22 Aug, New Delhi, India. 2013.

Sharma-Poudyal D, Chen XM. Models for predicting potential yield loss of wheat caused by stripe rust in the US Pacific Northwest. Phytopathology. 2011;101:544–54.

Sharma-Poudyal D, Chen XM, Wan AM, Zhan GM, Kang ZS, Cao SQ, Jin SL, Morgounov A, Akin B, Mert Z, Shah SJA, Bux H, Ashraf M, Sharma RC, Madariaga R, Puri KD, Wellings C, Xi KQ, Wanyera R, Manninger K, Ganzález MI, Koyda M, Sanin S, Patzek LJ. Virulence characterization of international collections of the wheat stripe rust pathogen, *Puccinia striiformis* f. sp. *tritici*. Plant Dis. 2013;97:379–86.

Sharma-Poudyal D, Chen XM, Rupp RA. Potential oversummering and overwintering regions for the wheat stripe rust pathogen in the contiguous United States. Int J Biometeorol. 2014;58:987–97.

Singh RP, Hodson DP, Huerta-Espino J, Jin Y, Bhavani S, Njau P, Herrera-Foessel S, Singh PK, Singh S, Govindan V. The emergence of Ug99 races of the stem rust fungus is a threat to world wheat production. Annu Rev Phytopathol. 2011;49:465–81.

Singh A, Pandey MP, Singh AK, Knox RE, Ammar K, Clarke JM, Clarke FR, Singh RP, Pozniak CJ, DePauw RM, McCallum BG, Cuthbert RD, Randhawa HS, Fetch Jr TG. Identification and mapping of leaf, stem and stripe rust resistance quantitative trait loci and their interactions in durum wheat. Mol Breed. 2013;31:405–18.

Singh A, Knox RE, DePauw RM, Singh AK, Cuthbert RD, Campbell HL, Shorter S, Bhavani S. Stripe rust and leaf rust resistance QTL mapping, epistatic interactions, and co-localization with stem rust resistance loci in spring wheat evaluated over three continents. Theor Appl Genet. 2014;127:2465–77.

Steele KA, Humphreys E, Wellings CR, Dickinson MJ. Support for a stepwise mutation model for pathogen evolution in Australasian *Puccinia striiformis* f. sp. *tritici* by use of molecular markers. Plant Pathol. 2001;50:174–80.

Sthapit J, Newcomb M, Bonman JM, Chen XM, See D. Genetic diversity for stripe rust resistance in wheat landraces and identification of dual resistance to stem rust and stripe rust. Crop Sci. 2014;54:2131–9.

Straib W. Untersuchungen über das Vorkommen physiologischer Rassen des Gelbrostes (*Puccinia glumarum*) in den Jahren 1935–1936 und über die Agressivität einiger neuer Formen auf Getreide und Gräsern. Arb Biol Reichsanst Land=Forstwirtsch, Berlin-Dahlem. 1937;22:91–119.

Stubbs RW. Stripe rust. In: Roelfs AP, Bushnell WR, editors. The cereal rusts, Diseases, distribution, epidemiology and control, vol. 2. Orlando: Academic; 1985. p. 61–101.

Stubbs RW, Yang HA. Pathogenicity of *Puccinia striiformis* for wheat cultivars with resistance derived from rye. Proc Eur Mediterr Cereal Rusts Conf. 1988;7:110–2.

Stubbs RW, Slovencikova V, Bartos P. Yellow rust resistance of some European wheat cultivars derived from rye. Cereal Rusts Bull. 1977;5:45–7.

Sui X, He Z, Lu Y, Wang Z, Xia X. Molecular mapping of a non-host resistance gene *YrpstY1* in barley (*Hordeum vulgare* L.) for resistance to wheat stripe rust. Hereditas. 2010;147:176–82.

Tadesse W, Ogbonnaya FC, Jighly A, Nazari K, Rajaram S, Baum M. Association mapping of resistance to yellow rust in winter wheat cultivars and elite genotypes. Crop Sci. 2014;54:607–16.

Talajoor M, Jin Y, Wan AM, Chen XM, Bhavani S, Tabe L, Lagudah E, Huang L. Specificity of a rust resistance suppressor on 7DL in the spring wheat cultivar Canthatch. Phytopathology. 2015;105:477–81.

Tang CL, Wang XJ, Duan XY, Wang XD, Huang LL, Kang ZS. Functions of the lethal leaf-spot 1 gene in wheat cell death and disease tolerance to *Puccinia striiformis*. J Exp Bot. 2013;64:2955–69.

Tang CL, Wang XJ, Cheng YL, Liu MJ, Zhao MX, Wei JP, Kang ZS. New insights in the battle between wheat and *Puccinia striiformis*. Front Agr Sci Eng. 2015;2:101–14.

Tang CL, Deng L, Chang D, Chen ST, Wang XJ, Kang ZS. TaADF3, an actin-depolymerizing factor, negatively modulates wheat resistance against *Puccinia striiformis*. Front Plant Sci. 2016;6:1214.

Tian Y, Zhan GM, Chen XM, Tungruentragoon A, Lu X, Zhao J, Huang LL, Kang ZS. Virulence and SSR marker segregation in a *Puccinia striiformis* f. sp. *tritici* population produced by selfing a Chinese isolate on *Berberis shensiana*. Phytopathology. 2016;106:185–91.

Tollenaar H. A comparison of *Puccinia striiformis* f. sp. *poae* on bluegrass with *P. striiformis* f. sp. *tritici* and f. sp. *dactylidis*. Phytopathology. 1967;57:418–20.

Toojinda T, Broers LH, Chen XM, Hayes PM, Kleinhofs A, Korte J, Kudrna D, Leung H, Line RF, Powell W, Ramsay L, Vivar H, Waugh R. Mapping quantitative and qualitative disease resistance genes in a doubled haploid population of barley (*Hordeum vulgare*). Theor Appl Genet. 2000;101:580–9.

Vazquez MD, Zemetra R, Peterson CJ, Chen XM, Heesacker A, Mundt CC. Multi-location wheat stripe rust QTL analysis: genetic background and epistatic interactions. Theor Appl Genet. 2015;128:1307–18.

Wan AM, Chen XM. Virulence, frequency, and distribution of races of *Puccinia striiformis* f. sp. *tritici* and *P. striiformis* f. sp. *hordei* identified in the United States in 2008 and 2009. Plant Dis. 2012;96:67–74.

Wan AM, Chen XM. Virulence characterization of *Puccinia striiformis* f. sp. *tritici* using a new set of *Yr* single-gene line differentials in the United States in 2010. Plant Dis. 2014;98:1534–42.

Wan AM, Chen XM. Variation of telial formation in the *Puccinia striiformis* f. sp. *tritici* population. Phytopathology. 2016;S4:206–7.

Wan AM, Chen XM, Yuen J. Races of *Puccinia striiformis* f. sp. *tritici* in the United States in 2011 and 2012 and comparison with races in 2010. Plant Dis. 2016;100:966–75.

Wang MN, Chen XM. First report of Oregon grape (*Mahonia aquifolium*) as an alternate host for the wheat stripe rust pathogen (*Puccinia striiformis* f. sp. *tritici*) under artificial inoculation. Plant Dis. 2013;97:839.

Wang MN, Chen XM. Barberry does not function as an alternate host for *Puccinia striiformis* f. sp. *tritici* in the US Pacific Northwest due to teliospore degradation and barberry phenology. Plant Dis. 2015;99:1500–6.

Wang KN, Hong XW, Wu LR, Xie SX, Meng QY, Chen SM. The analysis of the resistance of varieties in the wheat stripe rust nurseries in 1951–1983. Acta Phytophylacica Sinica. 1986;13:112–23.

Wang XJ, Tang CL, Deng L, Cai GL, Liu XY, Liu B, Han QM, Buchenauer H, Wei GR, Han DJ, Huang LL, Kang ZS. Characterization of a pathogenesis-related thaumatin-like protein gene *TaPR5* from wheat induced by stripe rust fungus. Physiol Plant. 2010a;139:27–38.

Wang B, Hu X, Li Q, Hao B, Zhang B, Li G, Kang Z. Development of race specific SCAR markers for detection of Chinese races CYR32 and CYR33 of *Puccinia striiformis* f. sp. *tritici*. Plant Dis. 2010b;94:221–8.

Wang XJ, Tang CL, Zhang HC, Xu JR, Liu B, Lv J, Han DJ, Huang LL, Kang ZS. *TaDAD2*, a negative regulator of programmed cell death, is important for the interaction between wheat and the stripe rust fungus. MPMI. 2011;24:79–90.

Wang XJ, Tang CL, Huang XL, Li FF, Chen XM, Zhang G, Sun YF, Han DJ, Huang LL, Kang ZS. Wheat BAX inhibitor-1 contributes to wheat resistance to *Puccinia striiformis*. J Exp Bot. 2012a;63:4571–84.

Wang XD, Wang XJ, Feng H, Tang CL, Bai PF, Wei GR, Huang LL, Kang ZS. *TaMCA4*, a novel wheat metacaspase gene functions in programmed cell death induced by the fungal pathogen *Puccinia striiformis* f. sp. *tritici*. MPMI. 2012b;25:755–64.

Wang X, Wang X, Duan Y, Yin S, Zhang H, Huang L, Kang ZS. *TaAbc1*, a member of Abc1-like family involved in hypersensitive response against the stripe rust fungal pathogen in wheat. PLoS One. 2013;8:e58969.

Wang XM, Gaudet DA, Liu W, Frick M, Puchalski B, Lu ZX, Leggett F, Kang ZS, Laroche A. Defence responses including hypersensitive cell death, oxidative burst and defence-gene expression in 'Moro' wheat inoculated with *Puccinia striiformis*. Can J Plant Pathol. 2014;36:202–15.

Wang YJ, Wang CY, Zhang H, Ali M, Xue F, Liu XL, Chen CH, Ji WQ. Molecular mapping of stripe rust resistance gene *Yrse5756* in synthetic hexaploid wheat and its transfer to common wheat. Pak J Bot. 2015a;47:1183–9.

Wang FT, Lin RL, Feng J, Chen WQ, Qiu DW, Xu SC. TaNAC1 acts as a negative regulator of stripe rust resistance in wheat, enhances susceptibility to *Pseudomonas syringae*, and promotes lateral root development in transgenic *Arabidopsis thaliana*. Front Plant Sci. 2015b;6:108.

Wang GF, Fan RC, Wang XP, Wang DW, Zhang XQ. TaRAR1 and TaSGT1 associate with TaHsp90 to function in bread wheat (*Triticum aestivum* L.) seedling growth and stripe rust resistance. Plant Mol Biol. 2015c;87:577–89.

Wang MN, Wan AM, Chen XM. Barberry as alternate host is important for *Puccinia graminis* f. sp. *tritici* but not for *Puccinia striiformis* f. sp. *tritici* in the U. S. Pacific Northwest. Plant Dis. 2015d;99:1507–16.

Wang ZY, Zhao J, Chen XM, Peng YL, Ji JJ, Zhao SL, Lu YJ, Huang LL, Kang ZS. Virulence variations of *Puccinia striiformis* f. sp. *tritici* isolates collected from *Berberis* spp. in China. Plant Dis. 2016;100:131–8.

Wellings CR, Burdon JJ. Variability in *Puccinia striiformis* f. sp. *tritici* in Australasia. In: Proceedings of European mediterranean cereal rusts powdery mildews conference 8:114. 1992.

Wellings CR, McIntosh RA. *Puccinia striiformis* f. sp. *tritici* in Australasia: pathogenic changes during the first 10 years. Plant Pathol. 1990;39:316–25.

Wellings CR, Burdon JJ, McIntosh RA, Wallwork H, Raman H, Murray GM. A new variant of *Puccinia striiformis* causing stripe rust on barley and wild *Hordeum* species in Australia. Plant Pathol. 2000;49:803.

Westendorp GD. Quatrième notice sur quelques Cryptogames récemment découvertes en Belgique. Bull Acad R Sci Belg. 1854;21:229–46.

Xia N, Zhang G, Sun YF, Zhu L, Xu LS, Chen XM, Liu B, Yu YT, Wang XJ, Huang LL, Kang ZS. *TaNAC8*, a novel NAC transcription factor gene in wheat, responds to stripe rust pathogen infection and abiotic stresses. Physiol Mol Plant Pathol. 2010;74:394–402.

Xia CJ, Wan AM, Wang MN, Jiwan DA, See DR, Chen XM. Secreted protein gene derived-single nucleotide polymorphisms (SP-SNPs) reveal population diversity and differentiation of *Puccinia striiformis* f. sp. *tritici* in the United States. Fungal Biol. 2016a;120:729–44.

Xia CJ, Wang MN, Wan AM, Jiwan DA, See DR, Chen XM. Association analysis of SP-SNPs and avirulence genes in *Puccinia striiformis* f. sp. *tritici*, the wheat stripe rust pathogen. Am J Plant Sci. 2016b;7:126–37.

Xiang C, Feng JY, Wang MN, Chen XM, See DR, Wan AM, Wang T. Molecular mapping of *Yr76* for resistance to stripe rust in winter club wheat cultivar Tyee. Phytopathology. 2016;106:1186–93.

Xu LS, Wang MN, Cheng P, Kang ZS, Hulbert S, Chen XM. Molecular mapping of *Yr53*, a new gene for stripe rust resistance in durum wheat accession PI 480148 and its transfer to common wheat. Theor Appl Genet. 2013;126:523–33.

Xu HX, Zhang J, Zhang P, Qie YM, Niu YC, Li HJ, Ma PT, Xu YF, An JG. Development and validation of molecular markers closely linked to the wheat stripe rust resistance gene *YrC591* for marker-assisted selection. Euphytica. 2014;198:317–23.

Yan GP, Chen XM. Molecular mapping of a recessive barley gene for resistance to stripe rust. Theor Appl Genet. 2006;113:529–37.

Yan GP, Chen XM. Molecular mapping of the *rps1.a* recessive gene for resistance to stripe rust in BBA 2890 barley. Phytopathology. 2007;97:668–73.

Yan GP, Chen XM. Identification of a quantitative trait locus for high-temperature adult-plant resistance against *Puccinia striiformis* f. sp. *hordei* in 'Bancroft' barley. Phytopathology. 2008;98:120–7.

Yan GP, Chen XM, Line RF, Wellings CR. Resistance gene analog polymorphism markers co-segregating with the *Yr5* gene for resistance to wheat stripe rust. Theor Appl Genet. 2003;106:636–43.

Yan J-H, Luo Y, Chen T-T, Huang C, Ma Z-H. Field distribution of wheat stripe rust latent infection using real-time PCR. Plant Dis. 2012;96:544–51.

Yang Y, Zhao J, Liu P, Xing H, Li C, Wei G, Kang ZS. Glycerol-3-phosphate metabolism in wheat contributes to systemic acquired resistance against *Puccinia striiformis* f. sp. *tritici*. PLoS ONE. 2013;8:e81756.

Yin CT, Downey SI, Klages-Mundt NL, Ramachandran S, Chen XM, Szabo LJ, Pumphrey M, Hulbert SH. Identification of promising host-induced silencing targets among genes preferentially transcribed in haustoria of *Puccinia*. BMC Genomics. 2015a;16:579.

Yin SN, Wang CF, Jiao M, Li F, Han QM, Huang LL, Zhang HC, Kang ZS. Subcellular localization of calcium in the incompatible and compatible interactions of wheat and *Puccinia striiformis* f. sp. *tritici*. Protoplasma. 2015b;252:103–16.

Zegeye H, Rasheed A, Makdis F, Badebo A, Ogbonnaya FC. Genome-wide association mapping for seedling and adult plant resistance to stripe rust in synthetic hexaploid wheat. PLoS One. 2014;9:e105593.

Zeven AC, Knott DR, Johnson R. Investigation of linkage drag in near isogenic lines of wheat by testing for seedling reaction to races of stem rust, leaf rust and yellow rust. Euphytica. 1983;32:319–27.

Zhan GM, Chen XM, Kang ZS, Huang LL, Wang MN, Wan AM, Cheng P, Cao SQ, Jin SL. Virulence and molecular comparison of Puccinia striiformis f. sp. tritici populations in China and the United States. Fungal Biol. 2012;116:643–53.

Zhan GM, Wang FP, Chen XM, Wan CP, Han QM, Huang LL, Kang ZS. Virulence and molecular diversity of the *Puccinia striiformis* f. sp. *tritici* population in Xinjiang in relation to other regions of western China. Plant Dis. 2016;100:99–107.

Zhang G, Li YM, Sun YF, Wang JM, Liu B, Zhao J, Guo J, Huang LL, Chen XM, Kang ZS. Molecular characterization of a gene induced during wheat hypersensitive reaction to stripe rust. Biol Plant. 2011;55:696–702.

Zhang HC, Wang CF, Cheng YL, Chen XM, Han QM, Huang LL, Wei GR, Kang ZS. Histological and cytological analyses of adult plant resistance to wheat stripe rust. Plant Cell Rep. 2012;31:2121–37.

Zhang X, Han D, Zeng Q, Duan Y, Yuan F, et al. Fine mapping of wheat stripe rust resistance gene *Yr26* based on collinearity of wheat with *Brachypodium distachyon* and rice. PLoS One. 2013;8:e57885.

Zhang ZJ, Chen JM, Su YY, Liu HM, Chen YE, Luo PG, Du XG, Wang D, Zhang HY. *TaLHY*, a 1R-MYB transcription factor, plays an important role in disease resistance against stripe rust fungus and ear heading in wheat. PLoS One. 2015;10:e0127723.

Zhang H, Zhang L, Wang CY, Wang YJ, Zhou XL, Lv SK, Liu XL, Kang ZS, Ji WQ. Molecular mapping and marker development for the *Triticum dicoccoides*-derived stripe rust resistance gene *YrSM139-1B* in bread wheat cv. Shaanmai 139. Theor Appl Genet. 2016;129:369–76.

Zhao J, Wang XJ, Chen CQ, Huang LL, Kang ZS. A PCR-based assay for detection of *Puccinia striiformis* f. sp. *tritici* in wheat. Plant Dis. 2007;91:1669–74.

Zhao J, Wang L, Wang ZY, Chen XM, Zhang HC, Yao JN, Zhan GM, Chen W, Huang LL, Kang ZS. Identification of eighteen *Berberis* species as alternate hosts of *Puccinia striiformis* f. sp. *tritici* and virulence variation in the pathogen isolates from natural infection of barberry plants in China. Phytopathology. 2013;103:935–40.

Zhao J, Zhao SL, Chen XM, Wang ZY, Wang L, Yao JN, Chen W, Huang LL, Kang ZS. Determination of the role of *Berberis* spp. in wheat stem rust in China. Plant Dis. 2015;99:1113–7.

Zhao J, Wang MN, Chen XM, Kang ZS. Role of alternate hosts in epidemiology and pathogen variation of cereal rusts. Annu Rev Phytopathol. 2016;54:207–28.

Zheng WM, Huang LL, Huang JQ, Wang XJ, Chen XM, Zhao J, Guo J, Zhuang H, Qiu CZ, Liu J, Liu HQ, Huang XL, Pei GL, Zhan GM, Tang CL, Cheng YL, Liu MJ, Zhang JS, Zhao ZT, Zhang SJ, Han QM, Han DJ, Zhang HC, Zhao J, Gao XN, Wang JF, Ni PX, Dong W, Yang LF, Yang HM, Xu JR, Zhang GY, Kang ZS. High genome heterozygosity and endemic genetic recombination in the wheat stripe rust fungus. Nat Commun. 2013;4:2673.

Zheng J, Yan Z, Zhao L, Li S, Zhang Z, Garry R, Yang W, Pu Z. Molecular mapping of a stripe rust resistance gene in wheat line C15. J Genet. 2014;93:443–50.

Zhou XL, Han DJ, Chen XM, Gou HL, Guo SJ, Rong L, Wang QL, Huang LL, Kang ZS. Characterization and molecular mapping of stripe rust resistance gene *Yr61* in winter wheat cultivar Pindong 34. Theor Appl Genet. 2014a;127:2349–58.

Zhou XL, Wang MN, Chen XM, Lu Y, Kang ZS, Jing JX. Identification of *Yr59* conferring high-temperature adult-plant resistance to stripe rust in wheat germplasm PI 178759. Theor Appl Genet. 2014b;127:935–45.

Zhou XL, Han DJ, Chen XM, Mu JM, Xue WB, Zeng QD, Wang QL, Huang LL, Kang ZS. QTL mapping of adult-plant resistance to stripe rust in wheat line P9897. Euphytica. 2015a;205:243–53.

Zhou XL, Zhang Y, Zeng QD, Chen XM, Han DJ, Huang LL, Kang ZS. Identification of QTL for adult plant resistance to stripe rust in Chinese wheat landrace Caoxuan 5. Euphytica. 2015b;204:627–34.

Glossary

A

Abiotic Of or pertaining to nonliving.

Abscisic acid (ABA) Plant hormone involved in abscission, dormancy, stomatal closure, growth inhibition or promotion and other plant response.

Accessary host Auxiliary host; alternative host; wild grasses attacked by the same species of pathogen attacking a crop. For a rust pathogen of a cereal crop, a susceptible grass species is considered as accessary host.

Acquired resistance (syn. Induced resistance or acquired immunity) A noninherited resistance response developed by a normally susceptible host following a predisposing treatment, such as inoculation with a bacterium, fungus, virus, or treatment with certain chemicals.

Active ingredient The active component in a formulated product (e.g. fungicide).

Adaptation (adj. adaptive) From an evolutionary standpoint, a characteristics of a living organism that improves its chances for survival in the environment of its habitat; change brought about in a population of an organism as a result of exposure to a particular set of environmental conditions; the change in structural and functional development of an individual or population enabling the organism to adjust to the environmental conditions in question.

Additive effect (of genes) Gene action in which the effects on a genetic trait are enhanced by each additional gene, either an allele at the same locus or genes at different loci.

Adenine One of the purine bases found in DNA, RNA, nucleosides and nucleotides. In DNA, it pairs with thymine.

Adenosine A mononucleotide consisting of adenine and D-ribose that is produced in the hydrolysis of adenosine monophosphate (AMP).

Adenosine 5′-triphosphate (ATP) A compound of one molecule each of adenine and D-ribose with three molecules of phosphoric acid; the phosphorylated condition of ADP. It conveys energy needed for metabolic reactions, and then loses one phosphate group to become ADP.

X. Chen, Z. Kang (eds.), *Stripe Rust*, DOI 10.1007/978-94-024-1111-9

Adenosine diphosphate (ADP) A nucleotide composed of adenine and D-ribose with two phosphate groups attached; ADP and ATP participate in metabolic reactions. These molecules, through the process of being phosphorylated or dephosphorylated, transfer energy within cells to drive metabolic processes.

Adenosine monophosphate (AMP) or adenylate The nucleotide having a phosphate in ribose-O-phosphate ester linkage at position 5 of the ribose moiety.

Adjuvant (Of pesticides) any substance added to a formulation to improve its physical or chemical properties. The term includes such materials as wetting agents, spreaders, stickers, emulsifiers, dispersing agents, foam suppressants, penetrates, correctives, etc.

Admix (Admixture) An isolate or population resulted from interbreeding between two or more previously isolated isolates or populations within a species.

ADP *See* Adenosine diphosphate.

Adult-plant resistance Resistance not expressed at the seedling stage; it increases with plant maturity.

Aeciospore A binucleate rust spore produced in an aecium.

Aecium (pl. aecia) A cup-shaped fruiting body of the rust fungi which produces aeciospores.

Aegilops A genus of grasses including several species of breeding and genetic interest. Some of them contributed to the evolution of bread wheat. They are also used as donors of disease resistance genes and other traits for the cropped wheat.

AFLP *See* Amplified fragment length polymorphism.

Aggressiveness Relative ability to colonize and cause damage to plants. *See also* Virulence.

Agropyron A genus in the family of Poaceae, commonly known as wheatgrass.

Airborne Transported from place to place through the air.

Alien chromosome A chromosome from a more or less related species transferred to a crop plant.

All-stage resistance Resistance that expresses throughout the growth stages of a plant. It is generally controlled by one or a few major genes. It is usually of high level, but often race-specific and not durable.

Allele One of the two or more alternate forms of a gene occupying the same locus on a particular chromosome; a form of gene. *See* Gene.

Allelic Of two or more genes that are derived from the same gene by mutation and are alternative occupants of the same chromosomal locus.

Allelism test genetic or molecular test to determine if genes in question from different genotypes.

Alternate host A plant, different from the principal (primary) host, on which a pathogen (e.g., a heteroecious rust fungus) must develop to complete its life cycle. Do not confuse with secondary, alternative or auxiliary host. *See* Accessary host, Alternative host or Auxiliary host.

Alternative host One of a parasite's several or many hosts; alternative hosts are not required for completion of the pathogen complete life cycle. *See* Accessary host, Alternate host or Auxiliary host.

Amino acids Organic, nitrogenous acids from which protein molecules are constructed. There are 20 common amino acids in living organisms, each with the basic formula NH_2–CHR–COOH.

Amino acids sequence The sequence of amino acid residues in a polypeptide chain that represents the primary structure of a protein. The sequence is unique for each protein and influences the protein structure and function.

Amplified fragment length polymorphism (AFLP) A molecular marker technique that utilizes a pair of restriction enzymes to digest DNA, PCR procedure to selectively amplify and electrophoresis to detect polymorphic DNA fragments.

Analog (analogue; adj. analogous) A resemblance in form, structure or function that is not considered as evidence of evolutionary relatedness.

Aneuploid Having a chromosome number that is not a multiple of the haploid set.

Annealing Forming double-stranded nucleic acid molecules from single-stranded molecules with complementary base sequences.

Annual plant A plant that completes its life cycle from seed and dies within one year.

Anthesis Developmental stage in flowering at which anthers rupture and shed pollen; a state of full bloom.

Antisporulant (Of fungicides) inhibiting sporulation.

Apoptosis A form of cell death in which a programmed sequence of events leads to the elimination of cells without releasing harmful substances into the surrounding area.

Appressorium The swollen tip of a germ tube that facilitates attachment and penetration of the host by the fungus.

Area under the disease progress curve (AUDPC) A useful quantitative summary of disease intensity over time, for comparison across cultivars, time, locations or management tactics. The most commonly used method for estimating the AUDPC, the trapezoidal method, is to discretize the time variable (hours, days, weeks, months or years) and calculate the average disease intensity between each pair of adjacent time points.

Asexual reproduction Any type of reproduction not involving the union of gametes or meiosis.

Assimilation In plant physiology, the production of an organic substance from non-organic elements and compounds via photosynthesis.

ATP *See* Adenosine 5′-triphosphate.

Autoecious fungus A parasitic fungus that can complete its entire life cycle on the same host.

Auxiliary host Alternative host of a parasite. Different from an alternate host that is infected by a different type of spores, an auxiliary host is infected by the same type of spores that infect the primary (principal) host. For example, wheat is the primary host for the wheat stripe rust pathogen and many wild grass species are considered as auxiliary hosts for the same pathogen. *See* Accessary host, Alternative host, Alternate host and Primary host.

Auxin A plant growth-regulating substance controlling cell elongation.

Avirulence (adj. avirulent) Non-pathogenic; lacking virulence; unable to cause disease. In rust fungi, an avirulent isolate produces a resistant reaction on a host plant, which is an incompatible interaction.

Axenic culture A culture in which only a single species, variety, or isolate of an organism is present and entirely free of all other contaminating organisms; for rust fungi, axenic culture is to grow an isolate on an artificial medium.

B

BAC Bacterial Artificial Chromosome.

Backcross A cross of an F_1 hybrid or heterozygote with an individual of genotype identical to that on one or the other of the two parental individuals as the recurrent parent. Backcrosses can be used to determine mode of inheritance, number of genes, linkage, cross-over values and breeding for incorporation of specific traits.

Bacterial Artificial Chromosomal (BAC) library A DNA construct, based on a functional fertility plasmid (or F-plasmid), used for transforming and cloning in bacteria, usually *E. coli* F-plasmids, play a crucial role because they contain partition genes that promote the even distribution of plasmids after bacterial cell division. The bacterial artificial chromosome's usual insert size is 150–350 kbp. A similar cloning vector called a PAC has also been produced from the bacterial P1-plasmid. BACs are often used to sequence the genome of organisms in genome projects. A short piece of the organism's DNA is amplified as an insert in BACs, and then sequenced. Finally, the sequenced parts are rearranged in silico, resulting in the genomic sequence of the organism. BACs were replaced with faster and less laborious sequencing methods like whole genome shotgun sequencing and now more recently next-gen sequencing.

Barberry Deciduous and evergreen shrubs in the genus of *Berberis*, numerous species of *Berberis* serve as alternate hosts for the stem rust fungus under natural conditions and for the stripe rust fungus under controlled conditions.

Barley *Hordeum vulgare*; a commonly diploid cereal crop; its grain is used as feed for animals, malt and human food.

Basal immunity The first level of plant immunity induced by the interaction of the plant with a pathogen.

Base pair A pair of nucleotides (nitrogen bases) held together by hydrogen bonding and found in double-stranded nucleic acids.

Basic Local Alignment Search Tool (BLAST) A bioinformatics program for finding regions of similarity between biological sequences. The program compares nucleotide or protein sequences to sequence databases and calculates the statistical significance.

Basidiomycetes The class of fungi producing their sexual spores, basidiospores, on basidia.

Basidiospore A sexually produced spore borne on a basidium.

Basidium A club-shaped structure on which basidiospores are borne.

Berberidaceae A family of several genera of flowering plants commonly called barberry plants; many species in this family can be infected by basidiospores of the stem rust and stripe rust fungi.

Berberis A genus in the family Berberidaceae commonly known as barberry, the genus consisting of more than 500 species in the world and many of them are alternate hosts for the stem rust fungus under natural conditions and for the stripe rust fungus under natural and/or controlled conditions.

Binary system A system involving only two elements, as 0 and 1 or yes and no; a system of naming in which binomial nomenclature is used for species.

Binomial nomenclature or system The scientific method of naming organisms. *See* Latin binomial.

Binucleate Containing two nuclei.

Bio-fungicide Fungicide that contains one or more microorganisms (usually bacterium or fungi) as the active ingredient(s), which either directly inhibit the infection and growth of one or more plant pathogens or induce plant resistance to the pathogens.

Bioinformatics The science of collecting and analyzing complex biological data such as genetic codes. As an interdisciplinary field of science, bioinformatics combines computer science, statistics, mathematics and engineering to analyze and interpret biological data.

Biological control An approach or method for pest control through counterbalance of microorganisms and other natural components of the environment; control of diseases or pests by means of living predators, parasites, disease-producing organisms, competitive microorganisms and decomposing plant material, which reduce the population of the pathogen or pest.

Biolistic particle delivery A gene engineering system that injects DNA-coated particles directly into cells, cultures and tissues of organisms including plants and fungi.

Biotechnology The use of genetically engineered microorganisms and/or modern techniques and process with biological systems for industrial production; also, technology concerned with machines in relation to human needs.

Biotroph (adj. biotrophic) A parasite that obtains its energy from living host cells, commonly an obligate parasite. *See* Nectrotroph.

BLAST The Basic Local Alignment Search Tool (BLAST) for finding regions of local similarity between sequences.

Breeding The propagation and genetic modification of organisms for the purpose of selecting improved offspring; various techniques of hybridization and selection are applied.

C

Callose Hard or thick, sometimes rough; a carbohydrate component of plant cell walls often forming over sieve plates and in calcified cell walls; a glucan formed in response to injure or infection.

Canidiation The generation and development of asexual spores, such as conidia, in fungi.

Canopy Expanded leafy top of a plant or plants.

CAPS *See* Cleaved amplified polymorphic sequence (CAPS).

Carbohydrate Substance composed of carbon, hydrogen and oxygen (CH_2O) with the last two in a 2 to 1 ratio, as in water (H_2O).

Catalase An enzyme that catalyzes the degradation of hydrogen peroxide to water and the oxidation by hydrogen peroxide of alcohols to aldehydes during seed germination.

Causal agent (or organism) The organism or agent that incites a given disease or injury.

C-banding A selective chromosome banding technique that uses Giemsa stain after most of the DNA is denatured or extracted by treatment with alkali, acid, salt or heat; only heterochromatic regions close to the centromeres and rich in satellite DNA are stained; the banding pattern is used to differentiate chromosomes.

cDNA Abbreviation for complementary DNA that is synthesized from a messenger RNA template. *See* complementary DNA.

Cell The structural and functional unit of all animal, microorganism and plant life. The living organism may have from one cell to hundreds of billions of cells. The essential feature of a cell is its living protoplasm limited by a membrane. It is surrounded in plant and fungal cells by a wall.

Cell division The process by which a cell divides to form two daughter cells, each of which contains the same genetic material as the original cell and roughly half of its cytoplasm.

Cell membrane The outer membrane of the cytoplasm, next to the cell wall, that regulates the flow of material into and out of the cell.

Cell wall Protective, resistant, but permeable polysaccharide structure secreted externally to the cell membrane in plants and certain other organisms; the barrier that develops between nuclei during mitosis. Bacteria and fungi typically have cell walls with other components.

Centromere The region of the chromosome to which the spindle fiber is attached during cell division (both mitosis and meiosis). The centromere is the constricted point at which the two chromatids forming the chromosome are joined together.

Centimorgan (cM) A unit of measurement of recombination frequency; one centimorgan is equal to a 0.01 chance that a genetic locus will be separated from a marker because of recombination in a single generation.

Cereal Grass grown primarily for its edible seed.

Chemical control An approach or method of control of plant diseases or pests using chemicals as active ingredients.

Chemfallow A cultural practice mainly for control of weeds using chemicals followed by fallowing. It may reduce fertilizer input costs, conserve moisture and reduce and reduce some soil-borne plants pathogens.

Chemigation The application of chemicals, such as pesticides or system maintenance compounds, through an irrigation system.

Chitin A complex, nitrogen-containing polysaccharide in many fungal cell walls.

Chitinase An enzyme that degrades chitin. Chitinases are generally found in organisms that either needs to reshape their own chitin or dissolve and digest the chitin of fungi or animals.

Chlorophyll (adj. chlorophyllous) The green, light-sensitive pigments found chiefly in the chloroplasts of leaves and other green parts of higher plants, that absorbs the light energy used in the process called photosynthesis.

Chloroplast Specialized cytoplasmic organelle (plastid) in plant cells that contains chlorophyll and is the site of photosynthesis.

Chlorosis Yellowing of normally green tissue due to chlorophyll destruction or failure of chlorophyll formation.

Chromatin Deeply staining nuclear material of which hereditary determiners are composed.

Chromosome A DNA-histone protein thread usually associated with RNA, occurring in the nucleus of a cell. It bears the genes, which constitute the hereditary material. Generally each species has a constant number of chromosomes.

Chromosome arm One part of a chromosome apart from the centromere.

Cleaved amplified polymorphic sequence (CAPS) A molecular marker technique that utilizes a restriction enzyme(s) to generate length-polymorphic DNA fragments from PCR amplicons.

Climate The variations of cold and heat, dryness and moisture, calm and wind in a given region or country; in general, the combined result of all meteorological phenomena of any region.

Clone (adj. clonal) The aggregate of individual organisms produced asexually from one individual; a population of cells, spores, or organisms of identical genotype; use of in vitro recombination techniques to insert a particular DNA sequence into a vector.

Cluster analysis Mathematic treatment of character comparisons used in the identification of taxonomic groups from similarity values.

Coding sequence The process by which nucleotides within a certain area of RNA or DNA determine the sequence of amino acids in the synthesis of a particular protein.

Co-dominance (adj. co-dominant) The expression of both alleles in the heterozygote.

Co-evolution (of fungi) Evolution of parasitic or symbiotic fungi together with their hosts.

Collinearity The correspondence between the order of nucleotides in a section of DNA (cistron) and the order of amino acids in the polypeptide that the cistron specifies.

Common name A generic name given to the active ingredient of a fungicide, nematicide, or other pesticide by a recognized committee (e.g., American Standard Association (ASA), Sectional Committee on Common Names for Pest Control Chemicals, the United States Standards Organization (USSO), the International Standards Organization (ISO), etc.).

Comparative mapping Localization (mapping) of a common set of DNA markers or genes onto linkage maps of different species. This approach shows synteny of markers among related species or genera.

Compartmentalization Isolation of a specific tissue area by host barrier tissue.

Compatible Describes an interaction between a plant and a pathogen resulting in disease. *See* Incompatible.

Complementary (Of gene interactions) two non-allelic genes are required for the phenotypical expression of a specific trait, such as disease resistance.

Complementary DNA (cDNA) DNA synthesized by reverse transcription from an RNA template; a single-stranded DNA molecule that is complementary in base sequence to the single strand from which it was transcribed. If the cDNA is made double-stranded and clones, it is described as a cDNA clone.

Complementation test A test to determine whether mutations in two strains are in different genes. Complementation will not occur if the mutations are in the same gene.

Compound-interest disease A disease in which the inoculum increases in successive "generations" of infection in a season, similar to the capital that is increased by the amount of "interest".

Contact pesticide A pesticide that kills in contact.

Contig A set of overlapping DNA segments that together represent a consensus region of DNA. In bottom-up sequencing projects, a contig refers to overlapping sequence data (reads). In top-down sequencing projects, contig refers to the overlapping clones that form a physical map of the genome that is used to guide sequencing and assembly. Contigs can thus refer both to overlapping DNA sequence and to overlapping physical segments (fragments) contained in clones.

Control of plant diseases Prevention, retardation or alleviation of diseases. There are four principal methods: *Exclusion* – Keeping the pathogen away from a disease-free area through quarantines, embargoes, and disinfection of plants and plant parts; *Eradication* – Destruction (roguing) of infected plants or plant parts or killing of the pathogen or agents on or in the host; *Protection* – Application of a chemical or physical barrier to prevent entrance of the pathogen; and *Immunization* – Production of genetically resistant or immune plant cultivars, chemotherapy or other treatment to inactivate or nullify the effect of the pathogen within the plant. Some scientists recognize a fifth method, Avoidance, which involves choosing cultural practices (e.g., sanitation, rotation, choice of planting site and date, propagating and planting only disease-free material, etc.) that avoid disease. Methods of control can be generally classified as regulatory, cultural, biological, physical, and chemical, depending upon the nature of the agents employed to control the disease.

Correlation coefficient A number that quantifies some type of correlation and dependence, meaning statistical relationships between two or more random variables or observed data values.

Cosegregating Describing closely linked genes or markers that are inherited together.

Cosmid A synthetic word derived from the designations **cos** and plas**mid**. A comid is a plasmid (e.g., pBR322) with so-called cos-sites of the DNA. They offer the chance of incorporation of alien DNA fragments of sizes between 32–45 kb.

Coupling Linkage in which both interested alleles at different loci are from one parent.

Covariance A statistic measure of the joint variability of two random variables.

Cover crop A crop grown between orchard trees or on fields between the cropping seasons of a main crop, to protect the soil against erosion and leaching and for improvement of soil. Wheat is often planted as a cover crop in orchards.

CRISPR Clustered regularly interspaced short palindromic repeats (CRISPR) are segments of prokaryotic DNA containing short, repetitive base sequences. Each repetition is followed by short segments of spacer DNA from previous exposures to foreign DNA (e.g., a virus or plasmid). The technique can be used for gene editing.

Crown rust A disease of oats, other cereals and grasses caused by *Puccinia coronata*.

Cultivar A cultivated variety; assemblage of closely related plants of common origin within a species that differ from other cultivars in certain minor details (e.g., form, color, flower, fruit, height) that when reproduced retain their distinguishing features.

Cultivar mixture Growing a mixture of two or more cultivars in a single field. *See* Blend.

Cultural control Control or reduction of a plant disease using agricultural practices normally used for crop production.

Cultural practice An agricultural practice for growing crops, which may also affect disease development.

Cytogenetics A branch of genetics that is concerned with the study of the structure and function of the cell, especially the chromosomes.

Cytokinins A group of hormones important in many growth responses of plants, including nucleic acid and protein metabolism, cell division and enlargement, bud and root dormancy, flower photoperiod, and delaying senescence.

Cytology Science or study of cells and their components, structure, organic processes and functions.

Cytoplasm All of the living substances of a cell outside of the nucleus.

Cytoplasmic inheritance A non-Mendelian (extra-chromosomal) inheritance via genes in cytoplasmic organelles (mitochondria, plastids).

Cytoplasmic membrane Membranes on the surface of cytoplasm and between cytoplasm and vacuoles

Cytosine One of the nitrogenous bases found in DNA and RNA.

Cytoskeleton A complex network of interlinking filaments and tubules that extend throughout the cytoplasm, from the nucleus to the plasma membrane.

D

DArT *See* Diversity arrays technology (DArT).

Database Repository of information.

Degeneration The state or process of being or becoming degenerate, decline or deterioration; deterioration and loss of function in the cells of a tissue or organ.

Degradation The act or process of damaging or ruining something; breaking, deterioration and loss of viability of a living organism.

Degradome sequencing A modified version of 5′-Rapid Amplification of cDNA Ends (RACE) using high-throughput, deep sequencing method using. Degradome sequencing provides a comprehensive means of analyzing patterns of RNA degradation; it is also referred as parallel analysis of RNA ends (PARE).

Deletion mutant A mutant generated by the loss of one or more nucleotides from a coding sequence.

Deletion line A line of plants with one or more chromosomes removed.

Demethylation The chemical process resulting in the removal of a methyl group (CH_3) from a molecule. A common way of demethylation is the replacement of a methyl group by a hydrogen atom, resulting in a net loss of one carbon and two hydrogen atoms. The counterpart of demethylation is methylation.

Deoxyribonucleic acid (DNA) The gene bearing material of each organism cell; the principal constituent of chromatin. *See* DNA.

Depolymerization The process of converting a polymer into a monomer or a mixture of monomers.

DH lines *See* Doubled-haploid.

Diagnosis (pl. diagnoses) Identification of the nature and cause of a disease problem; a shortened, Latin version of taxonomic description of a species or other taxon.

Dicot An abbreviated name for dicotyledon, which refers to plants having two seed leaves, e.g., barberry. *See* Monocot.

Dicotyledons Plant species having two cotyledons; flower parts arranged in fours or fives or multiples thereof; net-veined leaves and vascular bundles in the stem, arranged in a ring.

Differential hosts (varieties) The specific varieties or genotypes of host plants whose reactions are used for determining races or pathotypes.

Differential display reverse transcriptase-PCR (DDRT-PCR) A technique for comparing and identifying changes in gene expression at the mRNA level between two or more samples. It was the most commonly used method to compare expression profiles of two eukaryotic cell samples in the 1990s, and by 2000, and it was superseded by DNA microarray approaches.

Differentiation Development of an organism from one cell to many cells together with modification of the new cells, tissues or organs for the performance of definite functions.

Dikaryon A cell or strain of fungal cells with two genetically distinct haploid nuclei (n + n); a specialized binucleate condition common in the Basidiomycetes and Ascomycetes.

Dikaryotic Mycelium or spores containing two nuclei in each cell. Most urediniospores of a rust fungus are dikaryotic.

Diploid Applied to the 2n condition, the chromosomes being twice as many in the nucleus of a cell as in the haploid condition (n); e.g., a matured teliospore of the stripe rust fungus is diploid.

Directional selection Selection resulting in a shift in the population mean in the direction desired by the breeder. In population genetics, directional selection is a mode of natural selection in which an extreme phenotype is favored over other phenotypes, causing the allele frequency to shift over time in the direction of that phenotype.

Disease of plant Any disturbance of a plant that interferes with its normal structure, function, or economic value.

Disease cycle The chain of events involved in disease development, including the stages of development of the pathogen and effect of the disease on the host.

Disease resistance The ability to resist disease or the agent of disease and to remain healthy.

Disease tolerance Capability of a plant to maintain fairly normal vigor, without excessive injure or loss in yield, even though a pathogen is established within the plant.

Disome (adj. disomic) A cell or an individual showing the two homologous sets of chromosomes.

Dispersal The transfer (spread) or transport of infectious material (inoculum) to healthy plant tissue by wind, water, insects, animal, machinery, humans or other means. *See* dissemination.

Dissemination Transfer of inoculum from its source to healthy plants or from the place of production to another place.

Distal Remote from the point of attachment or origin; away from the center of the body; terminal; often used to describe the position of a marker or gene away from the centromere or toward the telomere of a chromosome. *See* Proximal.

Distribution Spread of a pathogen to areas outside of its previous geographical range.

Ditelosomic A cell or an individual that has two telocentrics of one chromosome arm.

Diurnal Occurring daily.

Divergence (of biology) The process by which a species or population evolves into two or more descendant or different forms or subpopulations; also, the process of tracing two or more species (populations) back to a common ancestor.

Diversity The state of being diverse; consisting of different individuals or genotypes.

Diversity array technology (DArT) A molecular marker technology used to develop sequence markers for genotyping and other genetic analysis. DArT is based on microarray hybridizations that detect the presence versus absence of individual fragments in genomic representations. The technology has significant advantages over other array based single-nucleotide polymorphism detection technologies in the analysis of polyploid plants.

Division A taxonomic category made up of closely related classes, used in classification of plants and fungi.

DNA (deoxyribonucleic acid) A molecule occurring in nuclei of living organism cells and composed of repeating subunits of nucleotides containing deoxyribose (a five-carbon sugar), phosphoric acid and one of four nitrogenous bases (adenine, cytosine, guanine and thymine). Every inherited characteristic has its origin in the code of an individual's complement of DNA.

DNA polymerase Any enzyme that catalyzes the synthesis of DNA strands from deoxyribonucleoside triphosphates using single-stranded DNA as a template.

DNA polymorphism One of two or more alternate forms (alleles) of a chromosomal locus that differ in nucleotide sequence or have variable numbers of repeated nucleotide units.

DNA sequencing Determining the relative order of nucleotide pairs in a stretch of DNA, a gene, a chromosome, or an entire genome.

DNase An enzyme that breaks down DNA by hydrolysis.

Domain (of protein) A protein domain is a conserved part of a given protein sequence and (tertiary) structure that can evolve, function and exist independently of the rest of the protein chain. Each domain forms a compact three-dimensional structure and often can be independently stable and folded.

Domestication The process by which plants and animals are genetically modified over time by humans for traits that are more advantageous or desirable for humans.

Dominance (adj. dominant) A genetic principle, established by Mendel, that when two contrasting characters are brought together in a hybrid as a result of a cross, one character (dominant) may mask the other (recessive).

Dominant gene A gene that is fully expressed in the phenotype of the heterozygote; refers to the gene (or the expression of the character it influences) that, when present in a hybrid with a contrasting gene, completely dominates in the development of the character. *See* Recessive gene.

Donor An individual, line, population or variety from which pollen or genetic material is used for transfer to another.

Dormancy A period in an organism's life cycle when growth and development are temporarily stopped. This minimizes metabolic activity and therefore helps an organism to conserve energy and to survive the period in which environmental conditions are unfavorable.

Dormant (n. dormancy) Being in a state of reduced physiological activity.

Dosage (dose) Quantity of a substance applied per unit of plant, soil, or other surface; also called *rate*.

Double-stranded RNA (dsRNA) RNA with two complementary strands, similar to the DNA found in all cells; dsRNA can trigger RNA interference in eukaryotes.

Doubled haploid A diploid plant that results from spontaneous or induced chromosome doubling of a haploid cell or plant, usually after anther or microspore culture by using different means.

Doubled haploid method A method used to speed up the production of homozygotes and to decrease the population size for selection or genetic and molecular studies. In other words, generating haploid plants by parthenogenesis or by anther culture followed by doubling of the number of their chromosomes (spontaneously or induced).

Doubled haploid lines (DH lines) Homozygous lines derived from haploidization and doubling again the chromosome number.

Durable resistance Resistance that remains effective and stable for a long period of time during which the resistant cultivars are widely grown under continuous epidemics of the disease.

Durum wheat *Triticum durum*; a cultivated tetraploid (2n = 4× = 28) wheat producing hard, flinty kernels. Its flour is used to make bread, other bakery products and pasta items.

E

EC_{50} The concentration or duration of a treatment that reduces an organism by half.

Effector (biology) A molecule that binds to a protein and thereby alters the activity of that protein; a molecule capable of responding to a stimulus.

Effectoromics Study of effectors; a high-throughput functional genomics approach that uses effectors of a pathogen to probe plant germplasm to detect disease resistance genes in plants.

Effector-triggered immunity (ETI) A property of the innate immune system. The innate immune system is common to all multicellular organisms and forms the first line of defense against pathogens.

Electron microscopy Use of an electron microscope, in which a focused beam of electrons produces a greatly enlarged image of minute objects, in the same way as light forms an image in a compound microscope.

Electrophoresis The differential migration of charged molecules and macromolecular ions (e.g., fragments of nucleic acid or proteins) in a free solution or through a porous medium in an electric field. The porous supporting medium may be filter paper, cellulose acetate.

Elicitor A chemical compound or physical process produced by the host (or pathogen) that induces a response by the pathogen (or host).

Elongation factor Protein-forming part of the ribosomal binding complex that promotes elongation of a polypeptide chain.

Embryo An organism in the early stages of development; the part of the seed that has developed from the zygote.

Embryo rescue a method of inducing the artificial growth of embryos by excising the young embryos under septic conditions and placing them on suitable nutrient media. The method is often applied when postgamous incompatibility exits (e.g., in wide crosses).

Emergence Appearance of the shoot above the soil surface.

Emmer wheat *Triticum turgidum* subsp. *dicoccum* and *T. turgidum* conv. *durum*; two-grained, hulled wheat with awn; tetraploid (2n = 4× = 28); one of the first crops domesticated in the Near East, and grown in limited areas in Asia and Europe.

EMS *See* Ethyl methane sulfonate.

Endonuclease An enzyme that cleaves the phosphodiester bond within a polynucleotide chain.

Endosperm Nutritive tissue formed within the seeds of angiosperms and on which the embryo feeds while developing. It develops from the sexual fusion of a sperm cell with two polar nuclei of the embryo sac and is thus triploid (3n).

Enhancer A modifier gene that may enhance the action or effect of another gene.

Environment The soil, water, the atmosphere, plants, animals, other organisms and human beings that surround the object or organism being discussed.

Enzyme A protein produced by living cells that can catalyze a specific organic reaction.

Epidemic A widespread and severe outbreak of a disease.

Epidemiology (adj. epidemiological) (Of plant) the branch of plant pathology concerned with disease in plant populations; study of the factors influencing the initiation, development and spread of infectious causes of disease; the study of epidemics (epiphytotics).

Epidermis The superficial layer of cells occurring on all plants parts.

Epigenetics The study of the mechanisms by which genes bring about their phenotypic effects.

Epistasis (adj. epistatic) The nonreciprocal interaction of nonallelic genes; the gene interaction in which one gene interferes with the phenotypic expression of another nonallelic gene so that the phenotype is determined effectively by the former and the latter is described as a hypostatic gene.

Equilibrium The state of stable conditions in which all significant factors remain more or less constant over a period, and there is little or no inherent tendency for change.

EST(s) Expressed sequence tag(s).

Ethylene Unsaturated, colorless, flammable gas ($CH_2{=}CH_2$) with plant growth regulator properties commonly produced by some living plants, injured plants, fruits and during incomplete combustion of fuels. Ethylene is involved in various aspects of vegetative growth, fruit ripening and abscission or senescence of plant parts.

Ethyl methane sulfonate (EMS) A chemical compound that acts as a mutagen in plants and fungi.

Ethylene A volatile plant hormone; synthesis is promoted by auxin or damage in seedlings, in shoot apex and various organs. It is known as a ripening hormone. It stimulates flowering and fruit ripening. It inhibits elongation of stem, root and leaves.

Eukaryote (adj. eukaryotic) An organism having membrane-limited organelles and nuclei that divide by mitosis; all organisms except bacteria and blue-green algae. *See* Prokaryote.

Evaluation The recording of those characters whose expression is often influenced by environmental factors.

Evolution The development of a species or other group of organisms over a long period of time.

Evolutionary divergence The mode of evolutionary change whereby an ancestral population is split into different genotypic populations or phyletic lines.

Exome The part of the genome formed by exons, the sequences which when transcribed remain within the mature RNA after introns are removed by RNA splicing.

Expressed sequence tags (ESTs) DNA sequences derived by sequencing an end of random cDNA clone from a library of interest. Usually, tens of thousands of such ESTs are generated as part of a given genome project. These ESTs provide a rapid way of identifying cDNAs of interest, based on their sequence "tag".

F

f. sp. *See* Forma specialis.

F_1 The first generation progeny of a cross between any two parents.

F_2, F_3, etc. The second, third, etc. generations following a cross.

Fallow Pertains to cropland not planted for one or more seasons; maintained without plant growth.

Family A taxonomic grouping of organism genera according to resemblance and similarity; denotes natural relationships; a category above a genus and below an order.

Fatty acid Organic compound containing carbon, hydrogen and oxygen that combines with glycerol to make fat.

Feekes scale A scale from 0–11 for description of growth stages of cereal crops.

Fertilization The sexual union of two protoplasts resulting in doubling of chromosome numbers.

Fertilizer Any organic or inorganic material of natural or synthetic origin containing nutrients (elements) essential to the growth of plants.

Field resistance Resistance observed under natural field conditions but usually not detected under experimental conditions.

Filial generation (F_1) The offspring resulting from first experimental crossing of the plants; the parental generation with the genetic experiment starts is referred to as P_1.

Fitness The relative ability of an organism to survive and transmit its genes to the next generation.

Flag leaf Leaf originating from the first culm note below the rachis; uppermost leaf on cereals and grasses.

Flag smut A disease caused by *Urocystis agropyri* on wheat and grasses.

Flanking markers DNA markers tightly linked to a gene or QTL of interest at both side.

Flanking region The DNA sequences extending on either side of a specific locus or gene.

Fleck A small or minute white to tan lesion, often translucent and visible through a leaf; a hypersensitive reaction by an avirulent isolate of a rust fungus.

Fluorescence (adj. fluorescent) Emission of radiation (light) when a substance or organism is placed in ultraviolet (UV) or other radiation.

Foliar Pertaining to leaves.

Foliar fungicide A fungicide that is applied onto leaves or canopy.

Forage Green or dehydrated vegetation used as food by livestock, e.g., hay, pasture and silage.

Forma specialis (pl. formae speciales, abbr. f. sp.) Special form; a biotype (or group of biotypes) of a species of pathogen that differ from others in the ability to infect selected genera or species of susceptible plants; an intraspecific taxon characterized from a pathological standpoint (especially host adaptation) but scarcely or not at all from a morphological standpoint)

Formulation (in relation to pesticide) A specific mixture containing active ingredients, the carrier and other additions (adjuvants) required to make it ready for sale (as an emulsifiable concentrate (EC), dust (D), flowable (F), gradule (G), wettable powder (WP), oil solution, etc.).

Fosmid Fosmids are similar to cosmids but are based on the bacterial F-plasmid. Fosmids can hold DNA inserts of up to 40 kb in size. A fosmid library is prepared by extracting the genomic DNA from the target organism and cloning it into the fosmid vector. The ligation mix is then packaged into phage particles and the DNA is transfected into the bacterial host. Bacterial clones propagate the fosmid library. Fosmids are useful for constructing stable libraries from complex genomes for genome sequencing and gene cloning.

Founder effect Genetic drift due to the founding of a population by a small number of individuals.

Frame shift A mutation caused by the insertion or deletion of one or more nucleotides that changes the reading frame of a codon, giving a changed amino acid sequence starting at the mutated codon; also a change from the reading frame of one cistron to that of an overlapping cistron.

Fungicide A compound toxic to fungi.

Fungigation Application of fungicide through an irrigation system.

Fungistasis Effect against fungi.

Fungus A single- or many-celled eukaryotic organism that lacks chlorophyll, usually with a chitinous wall, reproducing by sexual and/or asexual spores, and commonly producing mycelium.

G

Gamete A male or female reproductive cell or the nuclei within a gametangium.

GBS *See* Genotyping by sequencing.

Gel A jellylike colloidal mass often used for separating DNA, RNA and proteins.

Gel electrophoresis In gel electrophoresis, DNA, RNA or proteins migrate in an electric field through a porous matrix (gel), typically composed of polyacrylamide or starch. Mobilities are directly related to charge and inversely related to size.

Gene The base triplets of the DNA molecule in a chromosome that determines or conditions one or more hereditary characters; the smallest functioning unit of genetic material on a chromosome; the fundamental unit of inheritance; an ordered sequence of nucleotides that specify the manufacture of a single type of protein.

Gene action The expression of the gene based on the transcription into complementary RNA sequences, the subsequent translation of mRNA into polypeptides, which may form a specific protein.

Gene cloning The isolation and multiplication of an individual gene sequence by its insertion, usually into a bacterium, where it can multiply.

Gene deployment A system of assigning specific resistance genes to a specific geographic area to control pests. Gene deployment is sometimes used for incorporating different genes into different cultivars.

Gene editing Changing the function of a gene through insertion, deletion or replacement of DNA at a specific site of the gene of an organism or cell.

Gene expression The expression of the genetic material of an organism as specific traits; the transcription of mRNA from the DNA sequence of a gene and the subsequent translation of the mRNA to give the protein product.

Gene family A group of similar or identical genes, usually along the same chromosome, that originate by gene duplication of a single original gene. Some members of the family may work in concert while others may be silenced and become pseudogenes. In genomics, a family or genes have the same or similar functions.

Gene flow In population genetics, gene flow (also known as gene migration) is the transfer of alleles or genes from one population to another.

Gene interaction Modification of gene action by nonallelic gene or genes, generally the interaction between products of nonallelic genes.

Gene location Determination of physical or relative distance of a gene on a particular chromosome.

Gene map A graphic presentation of the linear arrangement of a chromosome or segment; it shows the relative distance between loci gained in linkage experiments.

Gene pool All the genes of an interbreeding program.

Gene pyramiding A method aimed at combining two or more desirable genes from multiple parents into a single genotype for a specific trait.

Gene silencing Inactivation of gene expression, usually with the implication that inactivation is not due to mutation, but rather to epigenetic modification such as methylation or change in chromatin structure.

Gene stacking The insertion of two or more genes into the genome.

Gene symbol Designating a gene, usually by an abbreviation of the name or description of a given gene.

Gene transfer The physical transfer of a gene by crossing, chromosomal manipulation and molecular means.

Gene-coding region The coding region of a gene, also known as the coding sequence or CDS (coding DNA sequence), is the portion of a gene's DNA or RNA, composed of exons, that codes for protein. The region is bounded nearer the 5′ end by a start codon and nearer the 3′ end with a stop codon.

Gene-for-gene hypothesis The concept that corresponding genes for resistance and virulence exist in the host and pathogen, respectively.

Genera *See* Genus.

General linear mixed model An extension to the generalized linear model in which the linear predictor contains random effects in addition to the usual fixed effects. Such models also extend the idea of linear mixed models to non-normal data. They provide a more flexible approach for analyzing nonnormal data when random effects are present.

General resistance Resistance against all strains or races of a pathogen; nonspecific host resistance.

Genetic Relating to heredity; describing heritable characteristics as influenced by germplasm.

Genetic background The remaining genetic constitution when a particular locus or allele of a given individual or taxon is studied.

Genetic distance A measure of gene differences between individuals or populations based on the differences of several characters. Such distance may be based on phenotypic traits, allele frequencies or DNA sequences. When allele frequencies are estimated for many loci, the genetic distance is calculated by averaging over these loci.

Genetic drift The random fluctuations of gene frequencies in a population such that the genes among offspring are not a perfectly representative sampling of the parental genes.

Genetic erosion The loss of genetic materials that occur when highly adaptable cultivars are developed and threaten the survival of their more locally adapted ancestors, which form the genetic base of the crop.

Genetic interaction The interaction between genes resulting in different phenotypic expression.

Genetic map A graphic representation of the linear arrangement of genes on a chromosome or genome.

Genetic marker A gene or DNA sequence with a known location on a chromosome that can be used to identify individuals or species. It can be described as a variation (which may arise due to mutation or alteration in the genomic loci) that can be observed. *See* Molecular marker.

Genetic recombination The process of forming new allelic combination in offspring by exchanges between genetic materials (as exchange of DNA sequences between DNA molecules). This process is a natural process, such as the crossing over between homologous chromosomes during meiosis.

Genetic resistance Resistance against pathogens or other pests due to specific or general gene action.

Genetic resources The gene pool in natural and cultivated stocks of organisms that are available for human exploitation.

Genetic stock A variety or strain known to carry specific traits.

Genetic variance (V_g) A portion of phenotypic variance that results from the varying genotypes of the individuals in a population. Together with the environmental variance (V_e), it adds up the total phenotypic variance observed among individuals in a population. It is divided into additive (resulting from differences between homozygotes, V_{add}) and dominance variance (resulting from specific effects of various alleles in heterozygotes, V_{dom}).

Genetically modified organism (GMO) A term that designates crops that carry new traits which have been inserted through genetic engineering methods, rather than traditional breeding methods.

Genetics The science or study of patterns of inheritance of specific traits.

Genome The complete genetic complement of an organism. Size is usually denoted in base pairs of DNA or RNA for some viruses.

Genome editing A type of genetic engineering in which DNA is inserted, deleted or replaced in the genome of a living organism using engineered nucleases, or "molecular scissors." *See* Gene editing.

Genome-wide association study (GWAS) In genetics, a genome-wide association study (GWA study, or GWAS), also known as whole genome association study (WGA study, or WGAS). It is an examination of a genome-wide set of genetic variants in different individuals to see if any variant is associated with a trait. GWASs typically focus on associations between single-nucleotide polymorphisms (SNPs) and traits like disease resistance.

Genomic in situ hybridization (GISH) A fluorescence in situ hybridization technique that uses total genomic DNA of a given species as a probe and total genomic DNA of another species as a blocking DNA. It is a useful method to detect interspecific or intergeneric genome differentiation, chromosome rearrangements (translocations) and substitutions or additions.

Genomic library A collection of clones made up of a set of overlapping DNA fragments representing the entire genome.

Genomic selection (GS) A method in which the number of polymorphic bands resembling the recurrent parent is used as the selection criterion.

Genomics The study of genes and their role in a structure, growth, and other aspects of an organism.

Genotype The aggregate of genes in an organism that influences the phenotype; class or group of individuals sharing a specific genetic makeup; the total genetic constitution of an organism; a group of individuals identified by a set of common molecular markers.

Genotyping by sequencing A method to discover single nucleotide polymorphisms (SNPs) in order to perform genotyping studies. It uses restriction enzymes to reduce genome complexity and genotype multiple DNA samples. After digestion, PCR is performed to increase fragments pool and then the libraries are sequenced using next generation sequencing technologies, usually resulting in about 100 bp single-end reads.

Genotypic Phenomena and processes associated with the genotype.

Genus (pl. genera) A taxonomic category that includes a group of closely related (structurally or phylogenetically) species. The genus or generic name is the first name in a Latin binomial.

Germ pore A small, round, thin area or hallow within a spore wall (especially in rust fungi) through which a germ tube may emerge.

Germ tube Hyphal thread resulting from an outgrowth of the spore wall and/or cytoplasm during germination of a fungal spore.

Germinate (n. germination) To begin growth of a seed, spore, or other reproductive body starting with absorption of water.

Germplasm (Germ plasm) Material capable of transmitting heritable characteristics sexually or asexually; the total genetic variability available to an organism represented by the pool of germ cells or seed.

GISH *See* Genomic in situ hybridization.

Glumes The outermost pair of bract-like structures of each spikelet, a chaff-like bract.

GMO *See* Genetically modified organism.

Golgi apparatus A system of flattened, smooth-surfaced, membranaceous cisternae, arranged in parallel 20–30 nm apart and surrounded by numerous vesicles; a feature of all eukaryotic cells. This structure is involved in the packaging of many products of cell metabolism.

Grain The simple, indehiscent fruit of a grass, particularly a cereal or grain crop, whose ovary walls are fused to form the seed.

Grain yield Yield of grains for cereal crops, e.g., tons per hectare, bushels per acre, etc.

Grain test weight Test weight of grains per unit of volume or number, e.g., grams of 1000 kernels, pounds per bushel, etc.

Gramineae Former taxonomic family name for grasses; now the Poaceae; any monocotyledonous, mostly herbaceous plants, having joined stems, slender sheathing leaves and flowers borne in spikelets of bracts; including cereals.

Gramineous Of or relating to a grass; grassy.

Greenhouse A building, room or area, usually of glass, in which the temperature is maintained within a desired range; used for cultivating plants.

Growth habit The mode of growth of a plant; crops may be classified as (1) annuals, biennials or perennials; cereals can be classified as spring or winter crops.

Growth inhibitor A substance that inhibits the growth of a plant or microorganism.

Growth regulator A synthetic or natural substance that in low concentrations regulates the enlargement, division or activation of plant cells.

Growth season The period between commencing of growth in the spring and cessation of growth in the fall; the period between planting and harvesting.

Growth stage The discrete portion of the life cycle of a plant, such as vegetative growth, reproduction or senescence. There are systems to various crops in order to subdivide the broad physiological and/or morphological stages.

Guard cells Paired, specialized epidermal cells that contain chloroplasts and surround a stoma.

H

Haploid A cell or an organism whose nuclei have a single complete set of chromosomes.

Haplotype A combination of alleles of closely linked loci found in a single chromosome; sometimes, a combination of particular nucleotide variants within a given DNA sequence.

Haustorium (pl. haustoria) A projection of a hypha into host cells which acts as a penetration and absorbing organ.

Heading Emerging spikes.

Heat shock protein When certain organisms are exposed to high temperature, heat shock proteins are synthesized. They provide thermal protection to subsequent heat stress.

Hectare (ha) A land area in the metric system equal to 2.471 acres = 10,000 m^2 = 0.01 km^2.

Hemibiotroph A parasite that initially forms an association with living cells of the host, much like a biotroph, and then in the later stages of infection it becomes necrotrophic, actively killing host cells.

Hemizygous An individual, generally diploid, having a given gene and/or allele present once (e.g., in monosomic or haploids).

Herbaceous plant A higher plant that does not develop woody tissue.

Herbicide Any chemical or physical agent used for killing or inhibiting the growth of plants where they are not wanted; a weed or grass killer; a type pesticide.

Heritability A measure of the degree to which a phenotype is genetically influenced and can be modified by selection; it is presented by the symbol h^2, equals V_g/V_p, where V_g is the variance due to genes with additive effects and V_p is the phenotypic variance.

Heteroecious Requiring two different kinds of hosts to complete its life cycle; pertaining particularly to rust fungi.

Heterogeneous Differing in structure, tissues, qualities or nature. *See* Homogeneous.

Heterokaryon Cell or mycelium with two or more nuclei of different constitution; also used to denote a fungal strain with this characteristic; a hybrid cell formed by the fusion of two cells from different species. *See* Homokaryon.

Heterokaryosis The condition in which a mycelium contains two genetically different nuclei per cell.

Heterologous expression The expression of a gene or part of a gene in a host organism, which does not naturally have this gene or gene fragment. Insertion of the gene in the heterologous host is performed by recombinant DNA technology.

Heterothallic fungus A fungus producing compatible male and female gametes on physiologically distinct mycelia.

Heterozygosity The presence of different alleles of a given gene at a particular gene locus.

Heterozygote A zygote that possesses two different alleles in a single gene pair.

Heterozygous Having mixed hereditary factors; not a pure line; having heterokaryosis resulting from the fusion of unlike gametes; having different alleles at various loci. *See* Homozygous.

Hexaploid Having six sets of chromosomes.

High-temperature adult-plant (HTAP) resistance Resistance expressed or increasingly expressed when temperature is higher and when the plant grows older. This type of resistance is commonly found in wheat and barley for resistance to stripe rust.

HIGS *See* Host-induced gene silencing.

Histology The science or study that deals with the microscopic structures of organisms.

Histones The five main classes of basic water-soluble proteins in cell nuclei in close association with DNA forming chromatin. They contain tyrosine but little tryptophan.

Homeostasis The tendency toward a relatively stable equilibrium between interdependent elements, especially as maintained by physiological processes.

Homoeologous chromosomes Two or more chromosomes that are partially homologous and have originated from ancestral homologous chromosomes, e.g., wheat chromosomes 1A, 1B and 1D are homoeologous chromosomes.

Homokaryon (adj. homokaryotic) An individual (cell or mycelium) whose nuclei are genetically alike (of one genotype); hybrid cell formed by the fusion of two cells of the same species. *See* Heterokaryon.

Homologous, homology The same with respect to type or species; (of fungi) showing a resemblance in form or structure, but not necessarily function. Homology is considered to be evidence of evolution relatedness.

Homologous chromosomes The two members of a chromosome pair.

Homothallism (adj. homothallic) Self-fertile; the condition in which sexual reproduction can occur without the interaction of two different thalli; compatible male and female gametes on the same mycelium. A dikaryotic mycelium develops from a single basidiospore and can eventually give rise to basidia. Three types have been recognized: (1) primary homothallism, as occurs in a homokaryotic individual; (2) secondary homothallism, also called pseudo- or facultative heterothallism; and (3) amphithallism, as in a thallus derived from a heterokaryotic spore containing nuclei of compatible mating type. *See* Heterothallism.

Homozygosity The presence of identical alleles at one or more loci in homologous chromosomal segments.

Homozygous Having paired identical genes present in the same cell or organism; without allelic differences; a pure breed or line. *See* Heterozygous.

Horizontal resistance Partial resistance that is effective against most or all races (genetic variants of a pathogen).

Horizontal transfer The movement of genetic material between different organisms other than via vertical transmission (the transmission of DNA from parent to offspring.).

Hormone A chemical substance produced in one part of plant and used in minute quantity to induce a growth response in another part.

Host A living organism (e.g. plant) harboring or invaded by a parasite and from which the parasite obtains part or all of its nourishment.

Host range The complete range of plants that may be attacked by a given pathogen.

Host-induced gene silencing (HIGS) An RNAi-based process where small RNAs made in the plant silence genes of pathogens or pests that attack the plant.

Host-pathogen interaction Defined as how plant pathogens sustain themselves within host organisms on a molecular, cellular, organismal or population level.

Hot spot (in genetics) A region usually within a gene or genome in which mutations occurs at an unusually high frequency; (in rust and other diseases) a focus or small area with plants severely infected, visually distinguishable from the rest of the field.

HTAP resistance *See* High-temperature adult-plant (HTAP) resistance.

Humidity, relative The weight of water vapor in a given quantity of air as compared to the total weight of water vapor that the air is capable of holding at a given temperature and expressed as a percentage.

Hybridization The cross of two individuals that differ in one or more heritable characteristics. The joining of two complementary single strands of DNA or of RNA and DNA, usually from different sources, to form a double-stranded molecule.

Hydrolysis (v. hydrolyze) Enzymatic disintegration of a large molecule into smaller component ones; to undergo or cause to undergo a chemical reaction with ions of water. Digestion is a process of hydrolysis.

Hyperparasite An organism parasitic on another parasite.

Hypersensitive response (HR) A mechanism, used by plants, to prevent the spread of infection by microbial pathogens. The HR is characterized by the rapid death of host cells in the local region surrounding an infection. The HR serves to restrict the growth and spread of pathogens to other parts of the plant.

Hyperspectral reflectance Collecting and processing information from across the electromagnetic spectrum; the goal of hyperspectral reflectance (imaging) is to obtain the spectrum for each pixel in the image of a scene, with the purpose of finding objects, identifying materials or detecting processes.

Hypersensitivity Excessive sensitivity of plant tissue to certain pathogen. Affected cells are killed quickly, blocking the advance of obligate parasites.

Hypha (pl. hyphae, adj. hyphal) The basic vegetative unit of structure and function of most fungi; a largely microscopic tubular filament that increases in length by growth at its tip. New hyphae arise as lateral branches. Hyphae constitute the body (mycelium) of a fungus. They may be divided into cells by cell walls (septa) or be one long cell. Some hyphae are specialized for producing spores, penetrating host tissues, absorbing food and overwintering.

Hyphal fusion Joining of fungal hyphae, usually with some exchange of cell contents; anastomosis.

I

IAA Indole-acetic acid
Identification The study of characters of an organism to determine its name.
IGS *See* Intergenic spacer.
Illumina sequencing A technique used to determine the series of base pairs in DNA. The reversible terminated chemistry concept was invented by Bruno Canard and Simon Sarfati at the Pasteur Institute in Paris. This sequencing method is based on reversible dye-terminators that enable the identification of single bases as they are introduced into DNA strands. It can also be used for whole-genome and region sequencing, transcriptome analysis, metagenomics, small RNA discovery, methylation profiling and genome-wide protein-nucleic acid interaction analysis.
Immune (n. immunity) Not affected by or responsive to disease; exempt from infection due to its inherent properties (e.g. tough outer wall, hairiness, nature of natural openings, waxy coating, thick cuticle etc.).
Inbred line A pure line usually originating by self-pollination and selection.
Inbreeding The formation of heterokaryons from plasmogamy between compatible homokaryons from the same paternal source; breeding of closely related individuals or races; used in plant breeding to obtain a homozygous condition.
Incompatible Describes an interaction between a plant and a pathogen not resulting in disease, usually a hypersensitive reaction occurs. *See* Compatible.
Incomplete dominance A condition that occurs when a dominant gene is only partially expressed in the phenotype of the heterozygote; a condition in heritance in which neither member of a pair of contrasting characters masks the other.
Incomplete resistance A type of resistance that is not complete; partial resistance.
Incubate Hold in a steady, environmental state to await development; to allow a microorganism to grow undisturbed under a given set of conditions.
Incubation period (or stage) The elapsed time between penetration of a host by a pathogen and the first appearance of disease symptoms and/or signs on the host; the time period during which microorganisms inoculated into a medium are allowed to grow; a period of rest of a pathogen prior to infection; maintaining inoculated plants or pathogens in an environment favorable for disease development. *See* Latent period.
Indel A molecular biology term for the insertion or the deletion of bases in the DNA of an organism.
Independent variable If two random variables "a" and "b" are independent, then the probability of any given value of "a" is unchanged by the value of "b".
Indoleacetic acid (IAA) A natural or synthetic organic compound belonging to the auxin class of growth-regulating substances.
Induced resistance *See* Acquired resistance.
Infection (v. infect) Process or act of a pathogen entering (invasion, penetration) and establishing a parasitic (often pathogenic) relationship with a host plant.

Infection type Gross appearance of symptoms and/or signs, as from infection by a rust fungus.

Infectious Capable of infection and spreading disease (a pathogen) from plant to plant.

Infectious disease A disease caused by a pathogen that can multiply and spread from a diseased to healthy plant.

Infective (infectivity) (Of a pathogen) able to infect a living organism.

Ingredient statement The part of the label on a pesticide container that gives the name and amount of each active ingredient and the amount of inactive material(s) in the mixture.

Inheritance The acquisition of characters or qualities by transmission from parent(s) to offspring.

Inhibition Prevention of growth or multiplication of microorganisms.

Inoculate (n. inoculation) To artificially introduce a pathogen at the site of infection of a host to induce disease or into a culture medium. Loosely, and incorrectly, used to describe the transfer of living cells of a microorganism to any place in which they will grow and develop.

Inoculum (pl. inocula) The pathogen or its parts (e.g., fungus spores, mycelium, bacterial cell, nematodes, virus particle, etc.) used for inoculating plants to producing disease.

Insertion Foreign DNA cloned into a bacterial plasmid, virus or other gene vector.

Integrated control (management) An attempt to use all available methods (biological, chemical, cultural, genetic, legal, and physical) to control the diseases and pests of a crop plant for best control results but with the least cost and the least damage to the environment.

Intercellular Between or among (host) cells.

Intergenic spacer A region of non-coding DNA between genes. The term is used particularly for the spacer DNA between the many tandemly repeated copies of the ribosomal RNA genes.

Internal transcribed spacer (ITS) The spacer DNA situated between the small-subunit ribosomal RNA (rRNA) and large-subunit rRNA genes in the chromosome or the corresponding transcribed region in the polycistronic rRNA precursor transcript; often used to identify different species.

Inter-simple sequence repeats (ISSRs) A genome region between microsatellite loci; the complementary sequences to two neighboring microsatellites are used as PCR primers; the variable region between them gets amplified; the limited length of amplification cycles during PCR prevents excessive replication of overly long contiguous DNA sequences, so the result will be a mix of a variety of amplified DNA strands which are generally short but vary in length.

Interspecific cross A natural or intentional cross between two species.

Intracellular Inside or through a cell or cells.

Introgression The incorporation of genes of one species into the gene pool of another.

Introns The DNA sequences that interrupt the protein-coding sequences of a gene. Introns are transcribed into mRNA but are eliminated from the message before it is translated into protein. They are the region of DNA transcribed into RNA initially but then removed by splicing, in the formation of functional RNA.

Irradiation Exposure of pollen, seed or other plant parts, or fungal spores or mycelium, to radiant energy of various types (X-rays or other short wavelength (gamma) radiations) to increase mutation rates.

Irrigation Applying water to soil, other than natural rainfall.

Isolate (Verb) to separate a culturable microorganism from an infected plant and grow it in the absence of other organisms; to separate a rust fungus from an infected plant and grow it on plants as a pure culture or separate different races from a plant infected by two or more races; (noun) a sample (e.g., a rust fungus) from a defined source.

Isolation Prevention of crossing among plant population due to distance or geographic barriers; prevention of a pathogen spreading among regions or from across contamination; the process of getting an organism into pure culture or the culture itself.

Isozyme (isoenzyme) A species of enzyme that exists in two or more structural forms, which are easily identified by electrophoretic methods.

ISSRs *See* Inter-simple sequence repeats.

J

Jasmonates A group of cyclopentanone derivatives, originate biosynthetically from linolenic acid via and inducible octadecanoid pathway consisting of at least seven enzymatic steps; the end product is (+)-7-iso-jasmonic acid, a physiologically active substance that is rapidly converted to its stereoisomer, stable (-)-jasmonic acid.

Jasmonic acid A substance in higher plants synthesized from linolenic acid via the octadecanoic pathway with an important role of operating as a "master switch" responsible for the activation of signal transduction pathways in response to predation and pathogen attack.

K

Karyogamy The fusion of two sex nuclei after cell fusion, i.e. after plasmogamy.

kb Kilobases, An abbreviation for 1000 base pairs of DNA.

Kilobase (pairs) (kb, kbp) One thousand nucleotides in a nucleotide chain; a measure of the size of nucleic acid molecule. Kb refers to single-stranded nucleic acid, with kbp referring to double-stranded nucleic acid.

Kinase Any enzyme that catalyzes phosphorylation reaction.

Kosami function Recombination fractions and map distances correspond only over relatively short recombinational distance. As genetic distance increases, the probability of a second (and correcting) recombination also increases, hence the measured recombination for two loci is less than would be apparent if a third intervening locus were present. Various mapping functions have been suggested to permit single recombination fractions to be converted to map distances. the Kasami function is the simplest and probably most general, which is given as x = Log_n [(1+2y)/(1−2y)]; x is the map distance (cM) corresponding to the recombination fraction, y.

L

Label (of pesticides) Pesticide product labels provide critical information about how to safely and legally handle and use pesticide products. Unlike most other types of product labels, pesticide labels are legally enforceable, and all of them carry the statement: "It is a violation of Federal law to use this product in a manner inconsistent with its labeling." In other words, the label is the law.

Landraces A set of populations or clones of a crop species produced and maintained by farmers; inbreeding, a mixture of a great number of different genotypes, which are well adapted to the environmental conditions of its habitat. In countries of highly developed agriculture, landraces have been superseded by highly advanced varieties; however, for selection landraces are a suitable material in which a great diversity of useful genotypes may be found.

Latent infection The state in which a host is infected with a pathogen but does not show any symptom.

Latent period (phase) The period between inoculation of a plant with a pathogen and appearance of symptoms and/or signs.

Latin binomial The scientific name of an organism. It is composed of two names, the first designating the genus, and the second, the species.

Leaf rust The disease caused by *Puccinia hordei* on barley and by *P. triticina* or *P. recondita* of wheat, rye and many grasses.

Lesion Well-marked, localized, often sunken area of diseased or disordered tissue; a wound.

Lethal Capable of causing death.

Library (of biotechnology and molecular genetics) A collection of cells, usually bacteria or yeast, that have been transformed with recombinant vectors carrying DNA inserts from a single species (e.g. BAC, cDNA), expression, or genomic library.

Life cycle Cyclical progression of stages in the growth and development of an organism that occurs between the appearance and reappearance of the same stage of the organism.

Ligase *See* DNA and RNA ligase.

Lignification, lignified Hardening of mainly xylem tissue from wall deposition of lignins.

Lignin Complex organic material derived from phenylpropane and distinct from carbohydrates that, in part, constitutes wood; imparts rigidity and strength, especially to woody tissues; associated with cellulose in secondary cell wall, especially xylem of many plants.

Limiting factor A limiting factor of a system is a variable such that a small change in value would cause a non-negligible change in an output or other measure of the system. The identification of a factor as limiting is of use only in distinction to one or more other factors which is/are non-limiting.

Line An inbred homozygous strain; a type of cultivar; an inbred group of progeny derived from a same plant.

Linkage Tendency of certain genes (markers) to remain together in inheritance because of their location on the same chromosome; the proximity of two or more markers on a chromosome. The closer together the markers are, the lower the possibility that they will be separated during meiosis. This gives an idea of the probability that they will be inherited together.

Linkage disequilibrium In population genetics, linkage disequilibrium is the non-random association of alleles at different loci. Loci are said to be in linkage disequilibrium when the frequency of association of their different alleles is higher or lower than what would be expected if the loci were independent and associated randomly. Linkage disequilibrium is influenced by many factors, including selection, rates of recombination and mutation, genetic drift, system of mating, population structure and genetic linkage.

Linkage drag The transfer of undesirable traits along with the genes or traits of interest due to their linkage.

Lipase An enzyme that hydrolyzes fats into glycerin (glycerol) and fatty acids.

Lipids Fatty compounds, substance whose molecule consists of glycerin and fatty acids and sometimes certain additional types of compounds; generic term for oils, fats, waxes and related products found in living tissues.

Locus (pl. loci) (Of genetics) the fixed position of a gene in a chromosome.

Logistic regression In statistics, logistic regression, or logit regression (model), is a regression model where the dependent variable is categorical. The binary logistic model is used to estimate the probability of a binary response based on one or more predictor (or independent) variables. It allows one to say that the presence of a risk factor increases the probability of a given outcome by a specific percentage.

M

Macrocyclic A life cycle of a rust fungus comprised of aecia, uredinia, telia and usually pycnia (spermogonia); they may be autoecious or heteroecious. *See* Microcyclic.

Macroscopic Visible to the unaided eye.

Major gene A gene that has a large observable effect upon the phenotype.

Map To determine the relative or physical position of a gene, marker or chromosome segment; a linkage to present the position and relative distances of genes or markers on a chromosome or chromosome fragment.

Map distance The distance between any two genes or markers on a genetic map, based on the percentage of cross-over; the minimum distance between linked genes is 1% and maximum 50%.

Map-based cloning The isolation of genes of interest by cloning the gene in question on the basis of molecular maps.

Mapping The process and the result of determination of map distances within or between linkage groups; sometimes it refers to the localization of genes or chromosome segments.

Marker An identifiable physical location on a chromosome whose inheritance can be monitored. Markers can be morphological or molecular.

Marker-assisted selection (MAS) Indirect selection exploiting the association between markers and desirable traits. It is a strategy permitting plant selection at the juvenile stage from early generations. The essential requirements for MAS in plant breeding are: (1) marker(s) should cosegregate or be closely linked with the desired trait; (2) an efficient means of screening large populations for molecular markers should be available; and (3) the screening technique should have high reproducibility across laboratories, to be economical to use, and should be user friendly.

Mating type Compatible strains, usually designated + and – or A and B, necessary for sexual reproduction in heterothallic fungi.

Maturity The state of ripeness. It usually refers to that stage of development that results in maximum quality.

Maxam and Gilbert method A technique for sequencing DNA using chemical base-specific modification and cleavage.

Mean The sum of an array of quantity divided by the number of quantities in the group.

Median effective time (ET_{50}) The amount of time for a substance to produce a response in 50% of the hosts (subjects) to whom it is given.

Membrane A lipid bilayer that separates the internal contents of a cell or organelle from its surroundings; also used for nitrocellulose or nylon fibers used in blotting.

Mendel's laws A set of three laws on the inheritance of chromosomal genes on the basis of chromosome theory of heredity, formulated by Gregor J. Mendel (1822–1884), an Austrian biologist. The laws are: (1) law of dominance or of uniformity of hybrids; (2) law of segregation; and (3) law of independent assortment.

Meosis Reduction division; a process that involves two nuclear divisions, during which both reductional division (separation of homologs) and equational division (separation of chromatids) occur; process by means of which the chromosome numbers are reduced by half; process by which a zygote (2n) divides twice and produces four haploid (n) gametes. Meiosis compensates for the chromosome-doubling effect of fertilization.

Mesophyll Chlorophyllous parenchyma tissue in leaves between the two epidermal layers.

Messenger RNA (mRNA) A chain of ribonucleotides that codes for a specific protein. mRNA moves from the nucleus to the ribosomes, where protein is synthesized in the cytoplasm.

Metabasidium (pl. metabasidia) Often defied as the part of the basidium in which meiosis takes place; also called the promycelium in certain basidiomycetes.

Metabolic pathway A sequential series of enzymatic reactions involving the synthesis, degradation or transformation of a metabolite. The pathway can be linear, branched or cyclic, and directly or indirectly reversible.

Metabolism The sum total of the chemical processes (physiological activities) occurring in the body of a living organism; the process by which organisms utilize nutrients to build structural components and break down cellular material to obtain energy and simple substances for special functions.

Metabolite Any chemical component of a reaction series or process; a product of metabolism.

Methylation The addition of methyl groups to nucleic acids or to other compounds.

Microarray A multiplex lab-on-a-chip. It is a 2D array on a solid substrate (usually a glass slide or silicon thin-film cell) that assays large amounts of biological material using high-throughput screening miniaturized, multiplexed and parallel processing and detection methods.

Microcyclic A life cycle of a rust fungus comprised of spermagonia and telia, or of telia only. *See* Macrocyclic.

Microorganism Microscopic organism (includes all unicellular prokaryotes and eukaryotes and some multicellular eukaryotes such as bacteria, fungi, nematodes, viruses, microscopic algae and protozoa. Same as Microbe.

Microsatellite marker Microsatellite or simple sequence repeats are a type of molecular markers; which consist of tandem repeats of 1–6 nucleotide motifs.

Microscope An optical instrument having a magnifying lens or a combination of lenses for observing objects too small to be seen distinctly by the un-aided eye.

Microscopic Too small to be seen except with the aid of a microscope.

Microtubules Any of the minute tubular structures of an eukaryotic cell that are widely distributed in protoplasm and made up of longitudinal fibrils.

Migration The movement of individuals or their propagules from one area to another.

Minor gene A gene that has a small observable effect upon the phenotype.

Mitochondrion (pl. mitochondria) A minute, membrane-bound, long or round cellular organlle found outside the nucleus of an eukaryotic cell. It contains the enzyme systems required for the citric acid cycle, electron transport and oxidative phosphorylation; possesses DNA, messenger RNA and small ribosomes; is capable of protein synthesis; and produces intracellular energy for the cell through respiration and is in rich in fats, proteins and enzymes.

Mitosis Usual process of nuclear cell division in which each chromosome duplicates, pulls apart and produces two daughter nuclei of the same genetic constitution as the parent nucleus. Mitosis is usually, but not always, followed by cellular division (cytokinesis).

Modulation The functional and morphologic fluctuation of cells in response to changing environmental conditions; the regulation of the rate of translation of mRNA by a modulating codon; a change in the kinetics of an enzyme or metabolic pathway; and systematic variation in a characteristic of a sustained oscillation to code additional information.

Moisture A basis for representing moisture content of a product. It is calculated from the net weight of water lost by drying, divided by dried weight of the material. The answer is multiplied by 100 to give percent.

Molecular biology A field of biology concerned with the interaction of biochemistry and genetics in the life of an organism. Molecular techniques in use include DNA-DNA hybridization, DNA fingerprinting, PCR (polymerase chain reaction), DNA sequencing and various molecular marker techniques. Most of these techniques are based on DNA that first must be extracted.

Molecular cloning DNA segments of different sizes of prokaryotic or eukaryotic origin are identically multiplied as a part of bacterial plasmids or phages. The alien DNA is incorporated into the host DNA ring molecule, and by the rapid division of the host cells the alien DNA segments are simultaneously cloned. Subsequently, those cloned DNA segments can be excised, separated and purified for further utilization.

Molecular genetics A branch of genetics that deals with molecular aspects of genetic mechanisms.

Molecular marker Particular DNA sequences and/or segments that are closely linked to a gene locus and/or a morphological or other characters of a plant; those segments can be detected and visualized by molecular techniques; roughly, three groups of markers can be classified as the following: (1) hybridization-based DNA markers such as restriction fragment length polymorphisms (RFLPs) and oligonucleotide fingerprinting; (2) PCR-based DNA markers such as random amplified polymorphic DNAs (RAPDs), which can also be converted into sequence characterized amplified regions (SCARs), simple sequence repeats (SSRs) or microsatellites, sequence-tagged sites (STS), amplified fragment length polymorphisms (AFLPs), inter-simple sequence repeat amplification (ISA), cleaved amplified polymorphic sequences (CAPS) and amplicon length polymorphisms (ALPs); and (3) DNA chip and sequencing-based DNA markers such as single nucleotide polymorphisms (SNPs).

Molecular weight (MW) The weight of a molecule expressed as the sum of the atomic weights of its constituents atoms.

Molecule A unit of matter; the smallest portion of an element or a compound composed of one or more atoms that retains the chemical identity with the substance in mass. A molecule usually consists of two or more atoms; some organic molecules contain hundreds to a million or more atoms.

Monocot An abbreviation of monocotyledon, referring to plants having single-seed leaves; flower parts arranged in threes or multiples thereof, parallel-veined leaves, closed vascular bundles arranged randomly in the stem tissue, e.g. wheat. *See* Dicot.

Monocotyledoneae (adj. monocotyledonous) The subclass of flowering plants, including the cereals and grasses, having one cotyledon (seed leaf) in the embryo; characterized by parallel-veined leaves and fibrous roots.

Monocyclic Having one cycle per growing season; no second infections.

Monoecism (adj. monoecious) Having male and female sex structures (flowers, gametangia) on the same plant; a fungus (particularly a rust) that has all stages of life cycle on a single species of plant (autoecious); also, the stamens and pistils in separate flowers, but borne on the same individual (as in maize). *See* Dioecious, heteroecious and homothallism.

Monogenic Containing or conditioned by one gene.

Monogenic resistance Resistance determined by a single gene.

Monokaryon A fungal mycelium or hypha in which each cell contains a single nucleus and a mononuclear spore or cell of a fungus that produces a dikaryon in its life cycle.

Monokaryotic Having a single nucleus in each cell.

Monosomic A genome that lacks a homologue in a diploid or higher ploidy organism.

Monosomic analysis A method for gene mapping (e.g., in hexaploid wheat). When genes determining phenotypes of interest for which an aneuploid series is not available, crosses can be made to a monosomic series in a variety with a contrasting phenotypic pattern. The monosomics are used as female parents in order to ensure the majority of progeny (about 72%; the transmission of n – 1 gametes is about 75% by the egg cell, but only about 4 % by the pollen) will become monosomic. If the gene involved is both recessive and hemizygous-effective (an uncommon situation), direct phenotypic observations on F_1 monosomic progenies enable the researcher to locate the gene on a particular chromosome; and only monosomic individuals in the critical cross will exhibit the phenotype, whereas disomic sibs and monosomic (or disomic) progenies from all other crosses will display the dominant phenotype. If the plants of the monosomic series carry the dominant allele of a gene of interest, monosomic individuals in the critical cross exhibit the recessive phenotype, whereas disomic sibs display the dominant phenotype.

Morphology Study or science of the form, structure and development of organisms.

mRNA Messenger ribonucleic acid.

Multigenic A trait that is controlled by many genes, as opposite to monogenic.

Multiline A cultivar or variety that is composed by many more or less defined lines.

Multiline cultivar (variety) A composite (blended) population of several genetically related lines of a self-pollinated crop, but bearing different genes (e.g., for resistance to a pathogen).

Multilocus genotypes A unique combination of alleles across two or more loci. For organisms such as plant pathogens that often reproduce clonally, multilocus genotypes are invaluable for identifying the mode and spread of an organism.

Multinucleate Describes cells that have more than one nucleus; usually for cells having more than two nuclei (as binucleate is for cells having two nuclei).
Multiple alleles The existence of several known allelic forms of a gene.
Multiple-cycle disease A disease caused by a pathogen with the capacity to be disseminated to other suscepts and have secondary infection cycles during the current growing season; a polycyclic disease.
Mutagenesis The production of mutations, often caused by chemical and physical agents.
Mutant An individual, strain or isolate of an organism bearing a mutant gene that expresses itself in the phenotype.
Mutation A change in the structure or amount of the genetic material of an organism; in cytogenetics, a gene or a chromosome set that has undergone a structural change.
Mycelium (pl. mycelia) A mass of hyphae that form the body of a fungus.

N

Natural infection Infection caused by naturally occurring pathogen spores, rather than artificial inoculation.
Natural logarithm A number that is its logarithm to the base of the mathematical constant *e*, where *e* is an irrational and transcendental number approximately equal to 2.718281828459. The natural logarithm of x is generally written as ln(x) or $\log_e(x)$.
Natural openings Stomata, lenticels, hydathodes and nectarthodes in plants.
Natural selection The process in nature by which the best-fitted types of organisms survived and produce offspring, the less competitive being eliminated; an important feature of Darwin's theory of the mechanisms of evolution.
NBS-LRR protein A protein containing a nucleotide-binding site (NBS) and leucine-rich repeats (LRRs). Products of many plant disease resistance genes are NBS-LRR proteins.
Near-isogenic lines Not fully isogenic. Near-isogenic lines are developed through multiple sequential backcrosses of genotypes carrying interested genes with a common recurrent parental genotype. For example, in wheat, near-isogenic lines have been produced for different *Rht* (reduced height) genes causing different straw length, and different genes for resistance to different rusts.
Necrosis (pl. necroses, adj. necrotic) Localized or general death and disintegration of plant cells or plant parts, usually resulting in tissue turning brown or black due to oxidation of phenolics or pale as commonly seen on leaves showing resistant reactions to infection by a rust fungus; hypersensitive reaction.
Necrotroph A parasite that typically kills and obtains its energy from dead host cells. *See* Biotrophy.
Nectar The stick, often sweet, secretion, (especially spermogonia (pycnia) of rust fungi) in which pycniospores (spermatia) may be freed, and that has an attraction

for insects; a sweet fluid secreted by the nectaries of plant and the chief raw material of honey.

Negative association The association of two traits or parameters that have a negative correlation coefficient value; an increase of one parameter results in a decrease of the other parameter.

Negative degree days (NDD) The temperatures of the days with average daily temperatures negatively departing from a base temperature value. As the low end of the optimal temperature range for the infection of the stripe rust pathogen is 7 °C, 7 °C is used as the base value for calculating accumulated NDD values for various periods during the winter and spring time for determining the effects of temperature on the survival of the stripe rust pathogen and development of the disease. *See* Positive degree days.

Next-generation sequencing Also known as high-throughput sequencing, it is the catch-all term used to describe a number of different modern sequencing technologies including Illumina (Solexa) sequencing, Roche 454 sequencing, Ion torrent:Proton/PGM sequencing and SOLiD sequencing.

Nitrogen (N) An element that is essential to all plants. Its chemical properties are especially important in the structure of proteins and nucleic acids, and its deficiency causes chlorosis and etiolation.

Nomenclature The scheme (belied to be a system) by which names are attached to objects, including microorganisms. The naming of organisms is governed by the International Code of Botanical Nomenclature. The Code is composed of principles, rules (termed articles) and recommendations.

Noncoding DNA A certain portion of DNA that obviously does not determine a gene product, such as a protein and/or character.

Nonhost resistance Inability of a pathogen to infect a plant because the plant is not a host of the pathogen due to lack of something in the plant that the pathogen needs or to the presence of substances incompatible with the pathogen.

Nonsense mutation A mutation that alters a gene so that a nonsense codon is inserted. Such a codon is the one for which no normal tRNA molecule exists, therefore it does not code for an amino acid. Usually, nonsense codons cause the termination of translation. Several nonsense codons are recognized (e.g., amber, ochre, opal).

Non-synonymous mutation A nucleotide mutation that alters the amino acid sequence of a protein. It is contrasted with synonymous substitution which does not alter amino acid sequences. As nonsynonymous substitutions result in a biological change in the organism, they are subject to natural selection.

No-till (stubble culture) A cultural system most often used with annual crops, in which the new crop is seeded or planted directly in a field on which the preceding crop plants were cut down, had the tops harvested, or were destroyed by a nonselective herbicide, but the old crop is not removed or incorporated into the soil as is common in preparing a plant bed.

Nontill rotation A method of planting crops that involves no seedbed preparation in the rotation other than opening small areas in the soil for placing seed at the intended depth. Moreover, there is no cultivation during crop production, despite chemicals that are used for vegetation control.

Nucleic acid A compound of high molecular weight that consists of pentose (ribose or deoxyribose), phosphate and nitrogen bases (purines and pyrimidines) joined in a long chain of repeating units (nucleotide complex). Nucleic acids are presented in all living cells.

Nucleotide(s) The building blocks of DNA and RNA consisting of a phosphate group joined to a five-carbon sugar (ribose or deoxyribose) that, in turn, is joined to a base; a phosphate ester of a nucleoside. Thousands of nucleotides are linked to form a DNA or RNA molecule.

Nucleus (pl. nuclei) The double-membrane-bound organelle containing the chromosomes that is found in most nondividing eukaryotic cells. It disappears temporarily during cell division. Within the nucleus several independent approaches point to the compartmentalization of particular activities such as transcription, RNA processing and replication. Chromosomes are revealed to occupy defined domains and to represent highly differentiated structures. The numerous activities that use DNA and RNA as template occur with a defined spatial and temporal relationships (e.g., compartmentalization of nuclear functions is particularly seen with replication). DNA moves through a fixed architecture containing the molecular machines directing replication.

Nullisomic analysis In nullisomic analysis, observations are made for phenotypic or other differences between the nullisomic for each chromosome and the disomic condition within the same variety. The method is applied for localization of genes within a given genome and can only be used in polyploids (e.g., hexaploid wheat), while diploids commonly do not tolerate the loss of both homologous chromosomes.

Nulli-tetrasomic A cell or individual, usually allopolyploid (e.g., hexaploid wheat), that lacks a pair of chromosomes, but has an additional pair of another, usually homoeolous, chromosome. The effect of the missing pair of chromosomes can be partially compensated by the tetrasomic (four-fold dosage) chromosomes. A whole series of nulli-tetrasomic lines was produced in hexaploid wheat and successfully used in numerous genetic and molecular studies.

Nutrient Essential elements available for organisms to utilize for metabolisms and growth.

O

Obligate parasite A parasite that can grow and multiply in nature only on or in living tissue and that cannot be cultured on an artificial medium.

Oligonucleotide A short-chain nucleic acid molecule, often used as polymerase chain reaction (PCR) primers.

Oligopeptide A short chain of amino acids.

Oomycetes Fungus-like eukaryotic microorganisms. They are filamentous, microscopic, absorptive organisms that reproduce both sexually and asexually. Oomycetes occupy both saprophytic and pathogenic lifestyles, and include some of the most notorious pathogens of plants. The oomycetes are often referred to as water molds, although the water-preferring nature is not true for most species.

Optimum Most favorable, e.g., optimum temperature is that most favorable for growth and reproduction of a culture, spore germination, invasion of a host, etc.

Order Taxonomic category ranking between a family and a class. An order is composed of related families.

Organelle Membrane-delimited structure of body within a cell (e.g., Golgi apparatus, mitochrondria) having a specialized function.

Ornamental plant A plant grown for accent, attraction, beautification, color, pleasure, specimen and other aesthetic reasons.

Orthologous genes Homologous genes that have become differentiated in different species derived from a common ancestral species, as opposed paralogous genes.

Osmotic potential The potential energy of water per unit volume relative to pure water in reference conditions; water potential quantifies the tendency of water to move from one area to another due to osmosis, gravity, mechanical pressure or matrix effects such as capillary action. The concept of water potential has proved useful in understanding and computing water movement within plants, animals and soil. Water potential is typically expressed in potential energy per unit volume and very often is represented by the Greek letter ψ.

Ostiole Small, more or less circular, schizogenous, paraphysis-lined pore or cavity in the papilla or neck of a spermagonium (pycnium), perithecium or pycnidium through which spores are discharged after secession from conidiogenous cells.

Oversummer To survive the winter.

Overwinter To survive the winter.

Oxidase Enzyme of qualitative significance in some organisms; accelerates the oxidation of some substance.

Oxidation A chemical reaction in which oxygen combines with another substance or in which hydrogen atoms or electrons are removed from a substance.

Oxidative burst The production of reactive oxygen species (ROS) by certain cells following the infection by a pathogen. ROS are generated irrespective of the type of pathogen. ROS generation leads to the killing of pathogens. *See* Reactive oxygen species.

Oxidative phosphorylation Utilization of energy released by the oxidative reactions of respiration to form high-energy ATP bonds.

Oxidative respiration The chemical decomposition of food (glucose, fats and proteins) requiring oxygen as a terminal electron acceptor and yielding carbon dioxide, water and energy. The energy is commonly in ATP.

Ozone (O_3) A highly reactive form of oxygen (a photochemical oxidant) that in relatively high concentrations may injure plants, humans, animals and microbes.

P

Parameter (of statistics) A numerical quantity that specifies a population in respect to some characteristics); (in microbiology) points on or near the limits of a boundary. These variable points or boundaries present known values.

Paraphysis (paraphyses) Sterile, upward growing, basally attached hyphal filament or cell in a hymenium, especially in the Ascomycotina and certain Basidiomycotina, where they are generally clavate or filiform in shape, branched or unbranched and free at the apex. Paraphyses are often produced among fertile conidiophores or conidiogenous cells, and the free ends frequently make an epithecium over the aci.

Parasexual cycle A mechanism in filamentous fungi in which a recombination of hereditary properties is based on the mitotic cycle. Its essential features are: production of diploid nuclei in a heterokaryotic haploid mycelium; multiplication of diploid nuclei along with haploid nuclei in a heterokaryotic mycelium; the sorting out of a diploid homokaryon; segregation by crossing-over at mitosis; and haploidization of the diploid nuclei.

Parasite An organism (fungus, bacterium, nematode, insect, etc.), virus or viroid living with, in or on another living organism (host) and obtaining food from it; may benefit the host in return, but more frequently causes disease in the host; a biotroph or necrotroph.

Parasitism A non-mutual relationship between species, where one species, the parasite, benefits at the expense of the other, the host. Traditionally a parasite is referred primarily to an organism visible to the naked eye, or macroparasites (such as helminths and parasitic plants), but they can be microparasites, which are typically smaller, such as bacteria, fungi and viruses.

Partial resistance Resistance that is expressed by the slower development of fewer pustules or lesions, compared with normal susceptible varieties.

Particle bombardment A method for gene transfer using a biolistic approach.

Pathogen An organism or agent (e.g., fungus, bacterium, nematode, virus or viroid) capable of causing disease in a particular host or range of hosts. Most pathogens are parasites, but there are few exceptions.

Pathogen-associated molecular patterns (PAMPs) Molecules that are associated with groups of pathogens, which are recognized by the innate immune system of host cells. These molecules can be referred to as small molecular motifs conserved within a class of microbes. They are recognized by Toll-like receptors (TLRs) and other pattern recognition receptors (PRRs) in both plants and animals. A vast array of different types of molecules can serve as PAMPs, including glycans and glycoconjugates.

Pathogen-regulated genes Plant genes that are regulated by the infection of a pathogen and involved in interactions with the pathogen.

Pathogenesis The sequence of processes in disease development from the time of infection to the final reaction in the host; production and development of disease.

Pathogenesis related (PR) proteins Proteins that accumulate within cells in plant tissue as a reaction to the hypersensitive response to fungal or viral infection or to certain chemical treatments.

Pathogenicity The ability of a pathogen to cause (incite) disease; the state or condition of being pathogenic.

Pathosystem A subsystem of an ecosystem defined by the phenomenon of parasitism. A plant pathosystem is one in which the host species is a plant and the parasite is any species in which the individual spends a significant part of its lifespan inhabiting one host individual and obtaining nutrients from it.

Pathotype A subspecific classification of a pathogen distinguished from others of the species by its pathogenicity (virulence) on a specific host.

Pattern-triggered immunity (PTI) Basal immunity trigged by pathogen-associated molecular patterns.

PCR *See* Polymerase chain reaction.

Pedigree A record of ancestry; an ancestral line or chart depicting the lineage or descent of an individual; a diagram showing the lineage or genealogy of an individual and all the direct ancestors, usually to analyze or follow the inheritance of trait.

Penetration The initial invasion (entrance) of a host by pathogen.

Penetration peg A thin hypha strand developing on the underside of an appressorium and penetrating between closed guard cells into the substomatal cavity or through the cuticle and epidermal cell wall into an epidermal cell; infection peg.

Pentose phosphate shunt A metabolic pathway parallel to glycolysis. It generates NADPH and pentoses (5-carbon sugars) as well as ribose 5-phosphate, the last precursor for the synthesis of nucleotides.

Peptide A compound consisting of two or several amino acids jointed by peptide bonds.

Pericentromeric region Describing a chromosome region situated near, or on each side of, the centromere of a chromosome.

Peridial cells (especially of aecia) Cells of the peridium.

Peridium (pl. peridia) The outer cellular structure (wall) or limiting membrane of a sporangium, perithecium, rust sorus or other fruitbody.

Permeability (adj. permeable) The quality or condition allowing a fluid or substance in a fluid to pass or diffuse through a membrane.

Peroxidase Enzyme that catalyzes the decomposition of peroxides; used as a reporter molecule in diagnostic tests such as ELISA.

Pest Any organism (e.g., some bacteria, fungi, insects, mites, nematodes, rodents, weeds) injuring or detrimental to a beneficial plant, plant product, animal, or human (or human property or that annoys humans) or the environment).

Pesticide Any chemical or physical agent that destroys, prevents, mitigates, repels or attracts pests (e.g., acaricide or miticide, bactericide, fungicide, herbicide, insecticide, nematicide, rodenticide, etc.).

Phenotype External, visible physical characteristics of an organism determined by its genotype and modified by the environment.

Phenotypic variance The total variance observed in a character; it includes experimental error, genotype × environment interaction and the genotypic variance.

Phenylpropanoid biosynthesis A group of plant secondary metabolites derived from phenylalanine and having a wide variety of functions both as structural and signaling molecules. Phenylalanine is first converted to cinnamic acid by deamination. It is followed by hydroxylation and frequent methylation to generate coumaric acid and other acids with a phenylpropane (C6-C3) unit. Reduction of the CoA-activated carboxyl groups of these acids results in the corresponding aldehydes and alcohols. The alcohols are called monolignols, the starting compounds for biosynthesis of lignin.

Phosphatase An enzyme that catalyzes the hydrolysis and synthesis of phosphoric acid esters and the transfer of phosphate groups from phosphoric acid to other compounds; an enzyme that splits phosphate from its organic compound.

Phosphorylation The addition of a phosphate group, for example H_2PO_4, to a compound.

Photoperiodism The physiological reaction of organisms to the length of day or night. It occurs in plants and animals. Photoperiodism can also be defined as the developmental responses of plants to the relative lengths of light and dark periods.

Photophosphorylation The production of ATP by the addition of a phosphate group to ADP using the energy of light-excited electrons produced in the light reactions of photosynthesis. Photo = light; phosphorylation = adding phosphorus.

Photorespiration A process in plant metabolism where the enzyme RuBisCO oxygenates RuBP, causing some of the energy produced by photosynthesis to be wasted.

Photosynthesis The series of metabolic reactions that occur in certain autotrophs, whereby organic compounds are synthesized by the reduction of carbon dioxide using energy absorbed by chlorophyll from light.

Phylogenetic Indicating natural evolutionary relationships.

Phylogenetic tree a diagram showing evolutionary lineages of organisms.

Phylogeny History of the evolution of an organism group in relation to other groups; the study of "family trees."

Physic map Representation of a DNA molecule showing the location of restriction sites or gene; map showing relative distances of genes and/or markers based on their locations on a chromosome, often determined by deletions of the chromosomes and/or genome sequence.

Physiological stress A stress (disease or disorder) produced by some unfavorable genetic, physical or environmental factors (e.g., excess of deficiency of light, water, soil nutrients, chemical, physical or other injure, etc.).

Physiological race Subdivision within a species of fungi or other organism: members of a race are alike in morphology but differ from other races in virulence, symptom expression, biochemical and physiological properties, or host range.

Physiological specialization The occurrence of physiological races within a species, variety or forma specialis of a pathogen.

Phytotoxicity (adj. phytotoxic) Poisonous, injurious or lethal to plants; usually describing a chemical.

Phytotoxin A substance (toxin) injurious to plants; or a toxin produced by plant.

Plant density The rate at which seed or vegetative propagules are placed in a field or experimental planting.

Plant disease *See* Disease of plant.

Plant hormone A compound that is synthesized by a plant, but is not a nutrient, coenzyme or detoxification product, and which regulates growth, differentiation or other specific physiological processes.

Plant pathology The science or study of diseases, their nature and effects on plants, causes and their control.

Plant physiology The science or study of the functions and activities of living plants.

Plasmalemma Outer three-ply membrane bounding the protoplast next to the cell wall; cell, cytoplasmic or plasma membrane composed of phospholipids and proteins that regulate the exchange of materials between the cell and its environment.

Plasmid Generally a small, covalently closed, circular piece of nonchromosomal, double-stranded DNA, found in certain bacteria and fungi that carries generally nonessential genetic information and is self-replicating. Some plasmids are linear molecules. Plasmids are used in recombinant DNA experiments as acceptors of foreign DNA.

Pleitropic An allele or gene that affects several traits at the same time.

Ploidy The number of complete chromosome sets in the cell nucleus (e.g., diploid, tetraploid, hexaploid, etc.).

Poaceae A large and nearly ubiquitous family of monocotyledonous flowering plants known as grasses. Poaceae includes the cereal grasses, bamboos and the grasses of natural grassland and cultivated lawns (turf) and pasture. Grasses have stems that are hollow except at the nodes and narrow alternate leaves borne in two ranks. The lower part of each leaf encloses the stem, forming a leaf-sheath. With about 780 genera and 12,000 species, Poaceae is the fifth-largest plant family.

Pollen Minute yellow bodies produced within the anthers of flowering plants or the cones of other seed plants; the male (micro) gametophyte.

Pollen tube A tube-like structure containing the tube nucleus in the microspore that helps guide the sperm from the pollen grain through the stigma and style of a flower to the embryo sac. A pollen grain and a mature tube are a male (micro-) gametophyte.

Pollination The transfer of pollen grain from a stamen (or staminate cone) to a stigma or seed cone.

Polyacrylamide gel electrophoresis (PAGE) Electrophoresis in a gel composed of polyacrylamide made by cross-linking acrylamide, usually with N, N′-methylene-bis-acrylamide. It is used for the separation of nucleic acid or protein molecules according to their molecular size and thus in estimating relative molecular weights.

Polycyclic disease A disease of which many cycles occur in one growing season, resulting in many secondary infections.

Polymerase An enzyme that catalyzes the formation of RNA or DNA by the addition of ribonucleotide or deoxyribonucleotide triphosphates.

Polymerase chain reaction (PCR) A technique for continuous amplification of DNA and/or DNA fragments in vitro with oligonucleotide primers that are complementary to the extremes of the fragment(s) that are to be amplified; heat stable DNA polymerase (e.g., from *Thermus aquaticus*) is used for DNA synthesis.

Polymorphic Having different forms; pleomorphic.

Polymorphism The existence of two or more forms that are genetically distinct from one another but contained within the same interbreeding population.

Polypeptide A chain of amino acids linked together by peptide bonds from the synthesis or partial hydrolysis of a protein.

Polyploid, polyploidy Having more than two sets of 2n chromosomes per nucleus.

Polyploidization Hybridization that leads to polyploidy.

Polysaccharide A large, long-chain, organic molecule (e.g., cellulose, glycogen, starch) consisting of many monosaccharide units (simple sugars).

Population A community of individuals, which share a common gene pool; in statistics, a hypothetical (often infinitely large) series of potential observations among which observations actually made constitute a sample.

Population dynamics Studies on the size and age composition of populations as dynamical systems, and the biological and environmental processes driving them (such as birth and death rates, and by immigration and emigration).

Population genetics The study of inherited variation in populations and its modulation in time and space. It relates the heritable changes in populations to the underlying individual processes of inheritance and development.

Population size The number of individuals in a population that are included in reproduction during a certain generation.

Population structure The composition of a given population, which is broken down into categories. Population structure can be used to categorize populations into many subsections based on different categories, levels or scales.

Positive association (correlation) The phenomenon or relationship in which the values of one variable tend to increase as the values of the other variable increase. A perfect positive association means that a relationship appears to exist between two variables, and that relationship is positive 100% of the time.

Positive degree days (PDD) The temperatures of the days with average daily temperatures positively departing from a base temperature value. As the low end of the optimal temperature range for the infection of the stripe rust pathogen is 7 °C, 7 °C is used as the base value for calculating accumulated PDD values for various periods during the wheat growth season for determining the effects of temperature on the survival of the stripe rust pathogen and development of the disease. *See* Negative degree days.

Post-transcription regulation The control of gene expression at the RNA level, therefore between the transcription and the translation of the gene.

Potassium (K) An element that is required for plant growth; deficiency leads to reduced growth and to dark or blue-green coloration in the leaves.

Powdery mildew Fungus (or disease) that forms a superficial white coating on the surface of leaves, stem, fruits, buds and flowers. It generally refers to members of Erysiphales or a type of disease caused by these fungi.

Precipitation Any product of the condensation of atmospheric water vapor that falls under gravity. The main forms of precipitation include drizzle, rain, sleet, snow, graupel and hail. Precipitation occurs when a portion of the atmosphere becomes saturated with water vapor, so that the water condenses and "precipitates".

Prediction A statement about an uncertain event, also referred as forecast. It is often, but not always, based on experience or knowledge. Although guaranteed accurate information about the future is in many cases impossible, prediction can be useful to assist in making plans about possible developments.

Predominant being most frequent or common; having superior strength, influence, or authority; prevailing.

Prevalence The observed frequency of a trait or disease in a population, often at a particular age or time.

Primary host (Of cereal rusts) the most important cereal host. *See* Alternate host, Auxiliary host.

Primary infection The first infection by a pathogen after going through a resting or dormant period.

Primary inoculum Propagules or vegetative structures of a pathogen, usually from an overwintering source, that causes initial rather than secondary outbreaks of disease.

Primary leaf One of the first or first pair of leaves to emerge above the cotyledon during the development of a seedling; it is often morphologically distinct from subsequent leaves.

Primary mycelium (of basidiomycetes) Haploid mycelium formed directly by the germinating basidiospores.

Primer Small fragment of nucleic acid with a free 3′-hydroxyl group necessary for the initiation of DNA and, sometimes, RNA synthesis.

Probability The measure of the likelihood that an event will occur. Probability is quantified as a number between 0 and 1 (where 0 indicates impossibility and 1 indicates certainty).

Probasidium (pl. probasinia) The portion of the basdium in which karyogamy takes place.

Probe A radioactive nucleic acid used to detect the presence of a complementary strand by hybridization; a specific sequence of DNA or RNA used to detect complementary sequence by hybridization.

Progenitor The original, ancestral or parental cell, individual or species.

Progeny The young or offspring.

Programed cell death (PCD) The death of a cell mediated by an intracellular program. PCD is carried out in a biological process, which usually confers advantage during an organism's life-cycle. PCD serves fundamental functions during both plant and animal tissue development. The forms of PCD include apoptosis, autophagy and especially necrosis that occurs as a result of pathogen infection or injury.

Prokaryote (adj. prokaryotic) An organism lacking membrane-limited nuclei or organelles and not exhibiting mitosis (e.g., bacteria and blue-green algae); organisms with a genome of a simple, circular, double-stranded DNA free in the cytoplasm and not in organelles. *See* Eukaryote.

Promoter A region of DNA or RNA that is recognized by RNA polymerase in initiation of transcription.

Promycelium (pl. promycelia) Basidium of the rust and smut fungi; initial short and short-lived hypha produced upon teliospore germination from which promycelial spores (now called sporidia) are produced.

Protectant (of pesticides) An agent, usually chemical, that tends to prevent or inhibit infection by a pathogen.

Protection Placement of a chemical or physical barrier between the pathogen and the host that prevents infection.

Protective fungicide Fungicide used to protect an organism against infection by a fungal pathogen.

Protein Complex, high molecular weight, organic, nitrogenous substance (polymer compound), built up of amino acids joined by peptide bonds.

Protein kinase Enzyme that catalyzes the phosphorylation of proteins usually in the presence of cyclic AMP or cyclic GMP.

Protein-protein interaction The physical contacts of high specificity established between two or more protein molecules as a result of biochemical events steered by electrostatic forces including the hydrophobic effect.

Proteinase An enzyme that hydrolyzes proteins to polypeptides.

Proteomics The large-scale study of proteins.

Protoplasm Living material within a cell in which all vital functions of nutrition, secretion, growth and reproduction depend; essential semifluid, viscous, translucent colloid of all plant and animal cells.

Protoplast The organized living unit of cell exclusive of its wall; the cytoplasmic membrane and the cytoplasm, nucleus and cytoplasmic organelles are a part of it.

Proximal Nearest, next to or nearest the point of attachment or origin; opposite of distal; often used to describe the position of a marker or gene close to the centromere of a chromosome.

Pustule A small, blisterlike or pimplelike, frequently erumpent elevation from which erupts a fruiting structure of a fungus that produces spores; a cluster of uredinia produced from one urediniospore infection.

Pycniospore Hapoid spores produced in a pycnium; also called a spermatium.

Pycniosporophore A hypha-like structure at the ostiole of a pycnium producing pycniospores.

Pycnium (pl. pycnia) Flask-shaped fruitbody of the rust fungi (Uredinales) that contains pycniospores (spermatia) and filaments (receptive hyphae) that extend through the ostiole; designated by spore state 0; also called spermogonium.

Q

QTL *See* Quantitative trait locus.
Qualitative trait A trait in which variation is discontinuous.
Qualitative inheritance An inheritance of a character that differs markedly in its expression amongst individuals of a species. Variation is discontinuous. Such characters are usually under the control of one or a few major genes.
Qualitative resistance Resistance that has a qualitative inheritance.
Quantitative trait A character in which variation is continuous so that classification into discrete categories is arbitrary.
Quantitative genetics A branch of genetics that deals with the inheritance of quantitative traits. Sometimes it is also called biometrical or statistical genetics.
Quantitative inheritance An inheritance of a character that depends upon the cumulative action of many genes, each of which produces only a small effect. The character shows continuous variation.
Quantitative resistance Resistance that has a quantitative inheritance.
Quantitative trait locus (QTL) A section of DNA (the locus) that correlates with variation in a phenotype of a quantitative trait. Usually the QTL is linked to, or contains, the genes that control that phenotype.

R

Race A subgroup or biotype of pathogens within a species, variety or forma specialis distinguished by virulence but not by morphology; different in virulence to cultivars (varieties) of the same host species.
RACE *See* Rapid amplification of cDNA ends
Race nonspecific resistance Host plant resistance that is effective against all races of a pathogen species. It can be variable, sensitive to environmental changes and usually quantitatively inherited and polygenetically controlled.
Race-specific resistance Host plant resistance that is effective against one or more, but not all, races of a pathogen species; of generally high level; and usually qualitatively inherited and controlled by one or a few genes.
Random amplified polymorphic DNA (RAPD) A molecular marker technique that uses short (commonly 10 nucleotides) primers to amplify DNA fragments with complementary sequences of the primers. As primers are designed randomly, no sequence information is needed for using the technique, and the same set of primers can be used for any organisms. The technique is simple and easy to use, but repeatability can be a problem. Markers identified using RNAP may be converted to more repeatable sequence-tagged site (STS) or sequence characterized amplified region (SCAR) markers.
Random genetic drift Variation in gene frequency from one generation to another due to chance fluctuations.

Random mating For a giving population, an individual of one sex has an equal probability of mating with any individual of the opposite sex.

Random sample A sample of a population selected so that all items in the population are equally likely to be included in the sample.

Randomized complete block design (RCB) An experimental design in which each block contains a plot for each candidate to be tested; in that case the classification of the data according to the blocks and the classification according to the candidates are orthogonal.

RAPD *See* Random amplified polymorphic DNA.

Rapid amplification of cDNA ends (RACE) A molecular technique used to obtain the full length sequence of an RNA transcript found within a cell. RACE results in the production of a cDNA copy of the RNA sequence of interest, produced through reverse transcription, followed by PCR amplification of the cDNA copies. The amplified cDNA copies are then sequenced and, if long enough, should map to a unique genomic region.

Rate The amount of active ingredient of a pesticide applied to a given surface; also called dose or dosage.

rDNA Recombinant DNA. A new DNA molecule produced by enzymatic insertion of one piece into another.

Reactive oxygen species (ROS) Chemically reactive chemical species containing oxygen; examples include peroxides, superoxide, hydroxyl radical and singlet oxygen.

Real-time PCR Also known as quantitative polymerase chain reaction (qPCR). It is a molecular technique based on the polymerase chain reaction (PCR). It monitors the amplification of a targeted DNA molecule during the PCR, i.e. in real-time, and not at its end, as in conventional PCR. Real-time PCR can be used quantitatively (quantitative real-time PCR), and semi-quantitatively, i.e. above/below a certain amount of DNA molecules (semi quantitative real-time PCR).

Rearrangement The action or process of changing the positions of genes, chromosomal parts or genomic sequences.

Reassortment The formation of individuals with combinations of genes in order different from the parents. Several different processes contribute to reassortment, including assortment of chromosomes and chromosomal crossover.

Receptive hypha Specialized hypha that protrudes from the top of a spermogonium (pycnium) and functions in sexual reproduction; flexuous hypha, trichogyne and possibly other similar structures.

Receptor A molecule, structure, cell or organ that selectively receives signals and initiates responses.

Recessive One of a pair of contrasting characters that is masked, when both are present, by the other (dominant) character. It also refers genes determining such characters.

Reciprocal crosses A pair of crosses in which the two parental genotypes are used as female and male parents; often used to determine maternal or cytoplasmic effects or inheritance.

Recombinant, recombination A new line or strain of an organism that occurs as a result of the breakage and renewal of covalent links in a nucleic acid chain, so that genes are swapped or replaced with different alleles; a cell or clone or cells resulting from recombination; formation in daughter cells of gene combinations not present in either parents; the mixing of genotypes that results from sexual reproduction.

Recombinant frequency The number of recombinants divided by the total number of progeny, expressed as a percentage of fraction. Such frequencies indicate relative distances between loci on a genetic map.

Recombinant inbred lines (RILs) A collection of lines that are developed through single-seed descent from the F_2 generation of a cross. The collection can be used for trait mapping and analysis.

Recurrent parent The parent to which a hybrid is crossed in a backcross. It replaces the dragged alleles step by step with the alleles of the original variety.

Reduced tillage Also called conservation tillage. It is a practice of minimizing soil disturbance and allowing crop residue or stubble to remain on the ground instead of being thrown away or incorporated into the soil.

Regression A statistic measure of the relation between the mean value of one variable and corresponding values of other variables.

Regression analysis A statistical process for estimating the relationships among variables. It includes many techniques for modeling and analyzing several variables, when the focus is on the relationship between a dependent variable and one or more independent variables. More specifically, regression analysis helps one understand how the typical value of the dependent variable changes when any one of the independent variables is varied, while the other independent variables are held fixed.

Relative humidity (RH) The ratio of the quantity of water vapor pressure in the atmosphere to the quantity that would saturate it at the same temperature.

Repetitive sequence A type of DNA sequence that constitutes a significant fraction of the total DNA. It is characterized by its large number of copies of repeated nucleotide sequences; some may be redundant DNA of unknown function, but mRNA, tRNA, 5S-RNA and histones are coded by repetitive sequences.

Repression Decrease in the rate of synthesis of an enzyme specifically caused by a small molecule, usually the end product of a biosynthesis pathway, e.g. an amino acid or nucleotide.

Repressor A protein that prevents RNA polymerase from starting RNA synthesis by binding to a specific DNA sequence upstream of the transcription initiation site.

Reproduction, sexual development of new plants by seeds (except in apomixis); the sexual cycle of a fungus or other organism.

Reproduction, vegetative Reproduction by other than sexually produced seed; including grafting, cutting, etc., as well as apomixis.

Repulsion Linkage in which two interested alleles at different loci are from different parents.

Residual Pertaining to or constituting a pesticide residue that persists after application in amounts sufficient to kill pests for several days to several weeks or longer.

Residue (in relation to pesticides) Amount of chemical that remains on or in a harvested crop. It is illegal to have a pesticide residue on a crop at harvest above the set tolerance or on a crop that is not registered.

Resistance The inherent ability of an organism (host plant) to overcome or retard, completely or to some degree, the activity (infection) of a pathogen or other damaging factor; (of a pest) able to withstand exposure to certain pesticides. Acquired resistance is a noninherited resistance response in a normally susceptible host following a predisposing treatment.

Resistance breeding Special crossing and selection methods in order to improve the inherent capability of a crop plant to prevent or retard the development of an infectious disease or other pests.

Resistance gene analog (RGA) A DNA sequence or gene that contains domains (e.g. nucleotide-binging site (NBS) and leucine rich repeat (LRR) domains) that are commonly found in disease resistance genes.

Resistance gene analog polymorphism (RGAP) A molecular marker technique that uses primers designed based on sequences of cloned disease resistance genes to amplify polymorphic DNA sequences analogous to resistance genes.

Resistant Possessing qualities that hinder the development of a particular pathogen or the effects of other damaging factors. A plant may be slightly, moderately or highly resistant.

Respiration A series of enzymatic reactions (oxidation processes) in which a living organism produces energy for cellular activities, usually by oxidizing carbohydrate and releasing carbon dioxide; the oxidation of food by plants and animals to yield energy for cellular activities.

Restriction endonucleases A group of enzymes, usually bacterial, that hydrolyze DNA at highly specific sites. In recombinant DNA experiments, restriction enzymes are used to cut dsDNA from a given organism into defined fragments of characteristic size by breaking internal bonds before it is recombined with a vector. Type I enzymes bind to the recognition site but usually cut the DNA at random site. Type II enzymes bind at the recognition site and cut either at or close to that site. They are used extensively in recombinant DNA technology.

Restriction enzyme Restriction endonuclease.

Restriction fragment A poly nucleotide fragment produced by cutting DNA with a restriction endonuclease.

Restriction fragment length polymorphism (RFLP) A molecular marker technique that utilizes restriction enzymes to digest DNA molecules, probes and Southern blotting to detect polymorphic fragments; a technique that uses PCR to amplify a gene or DNA fragment, digested with a specific restriction enzyme(s), and then separate polymorphic length through gel electrophoresis.

Retrotransposon Retrotransposons are a ubiquitous and major component of plant genomes; those with long terminal DNA repeats are widely distributed over the chromosomes of many plant species.

Retrotransposon-microsatellite amplified polymorphism (REMAP) A molecular marker technique involving amplification of fragments which lie between a retrotransposon insertion site and a microsatellite site.

Reverse genetics An approach to discover the function of a gene by analyzing the phenotypic effects of specific engineered gene sequences. This investigative process proceeds in the opposite direction of so-called forward genetic screens of classical genetics.

Reverse transcriptase An enzyme coded by certain viruses that makes a DNA copy of a primed RNA molecule. The enzyme can also synthesize DNA from a DNA template.

Reverse transcription Copying of the genetic information from RNA into DNA.

Revolutions per minute (rpm) A measure of the speed of a centrifuge.

RFLP *See* Restriction fragment length polymorphism.

RGAP *See* Resistance gene analog polymorphism.

RH *See* Relative humidity.

Ribonucleases (RNases) A group of enzymes with different activities that hydrolyze RNA. They can be used in the characterization and sequencing of RNA, and also DNA purification by digesting RNA.

Ribonucleic acid (RNA) Any of a number of nucleic acid containing ribose, uracil, guanine, cytosine and adenine occurring in cell cytoplasm and the nucleus; RNA is associated with control of cellular chemical activities. RNA is the nucleic acid type of most plant viruses.

Rice blast A disease of rice caused by *Magnaporthe oryzae* (formerly *Magnaporthe grisea*).

RNA interference (RNAi) A biological process in which RNA molecules inhibit gene expression or translation, by neutralizing targeted mRNA molecules.

RNA polymerase An enzyme synthesizing RNA from a DNA or RNA template.

RNA sequencing (RNAseq) Also called whole transcriptome shotgun sequencing (WTSS). It uses next-generation sequencing (NGS) to reveal the presence and quantity of RNA in a biological sample at a given moment in time. RNA-Seq is used to analyze the continually changing cellular transcriptome. Specifically, RNA-Seq facilitates the ability to look at alternative gene spliced transcripts, post-transcriptional modifications, gene fusion, mutations/SNPs and changes in gene expression. In addition to mRNA transcripts, RNA-Seq can look at different populations of RNA to include total RNA, small RNA, such as miRNA, tRNA and ribosomal profiling. RNA-Seq can also be used to determine exon/intron boundaries and verify or amend previously annotated 5′ and 3′ gene boundaries.

Rotation Sequence and duration of crop cultivation in a given area; deployment of resistance genes in commercial cultivars in different time periods.

rRNA Ribosomal RNA.

Roche 454 sequencing A DNA sequencing technology that uses a large-scale parallel pyrosequencing system run on the Genome Sequencer FLX with GS FLX Titanium series reagents. The system relies on fixing nebulized and adapter-ligated DNA fragments to small DNA-capture beads in a water-in-oil emulsion. The DNA fixed to these beads was then amplified by PCR. Each DNA-bound bead was placed into a ~29 μm well on a PicoTiterPlate, a fiber optic chip. A mix of enzymes such as DNA polymerase, ATP sulfurylase and luciferase is also packed into the well. The PicoTiterPlate is then placed into the GS FLX System for sequencing.

RT-PCR *See* real-time PCR.

Rust A disease caused by one of the rust fungi (Uredinales) in the Basidiomycotina; a disease giving a "rusty" appearance to a plant. The lifecycle of a rust fungus may involve up to five different types of spores. A rust fungus may parasitize one species of plant during their life cycle (autoecious) or two nonrelated species (heteroecious).

S

Salicylic acid (SA) A monohydroxybenzoic acid, a type of phenolic acid and a beta hydroxy acid. It has the formula $C_7H_6O_3$. This colorless crystalline organic acid is widely used in organic synthesis and functions as a plant hormone.

Sanger sequencing A method of DNA sequencing first commercialized by Applied Biosystems, based on the selective incorporation of chain-terminating dideoxynucleotides by DNA polymerase during in vitro DNA replication. The technique was developed by Frederick Sanger and colleagues in 1977 and has been the most widely used sequencing method. More recently, higher volume Sanger sequencing has been supplanted by "Next-Gen" sequencing methods, especially for large-scale, automated genome analyses. However, the Sanger method remains in wide use for validation of Next-Gen results and for obtaining especially long contiguous DNA sequence reads (>500 nucleotides).

Scaffold In genomic sequencing, it is the first complete sequence assembled, to which new reads are aligned. Once a scaffold is obtained, *de novo* assembly may not be needed anymore so that one can find a region to which any given new read is most similar and subsequently look for genetic variation.

Scanning electron microscopy (SEM) A technique for examining the surface features at high magnification by coating specimens with a thin layer of gold, palladium, or aluminum and studying the images produced by a scanning electron beam.

SCAR *See* Sequence characterized amplified region.

Secreted protein Any protein, whether it is endocrine or exocrine, which is secreted by a cell. Secretory proteins include many hormones, enzymes, toxins and antimicrobial peptides. Secretory proteins are synthesized in the endoplasmic reticulum.

Secretome The entirety of product types secreted by a cell, parasite or other entity.

Seed treatment A chemical or nonchemical treatment applied to seeds for control of a disease or other pests.

Seedling A young plant grown from seed.

Segregation A genetic principle, established by Mendel, that the factors (or genes) of a pair are separated from each other in reduction division.

Selection The process of isolating and preserving certain individuals or characters from a group of individuals or characters.

Selection pressure The effectiveness of natural selection in altering the genetic composition of a population over a series of generations.

Selective advantage An advantage of survival of a genotype in a population and for production of viable progeny as compared to other genotypes, which may show a selective disadvantage with respect to fitness and viability.

Selective disadvantage Inferior fitness of one genotype compared to others in the population.

Selective neutrality The situation in which different alleles of a certain gene confer equal fitness.

Selfing When a pistil is fertilized with pollen from the same plant that bears the pistil; also applied to seed resulting from such fertilization. *See* Self-pollination.

Self-pollination Transfer of pollen from the stamen(s) to the stigma of the same flower, of another flower on the same plant or other plant of the same genotype.

SEM *See* Scanning electron microscopy.

Senescence The gradual deterioration of function characteristic of most complex lifeforms that on the level of the organism increases mortality after maturation. Senescence can be at the cellular, tissue or the whole plant.

Sensitivity (adj. sensitive) A diseased organism reacting with severe symptoms to the attack of a given pathogen; a pathogen reacting (reducing or losing viability) to a chemical or environmental factors.

Septate (n. septation) With more or less regularly occurring cross walls; divided by partitions or septa; having septa.

Septum (pl. septa) A dividing cell wall or partition in a hypha or spore.

Sequence The order of nucleotides in RNA or DNA or of amino acids in a polypeptide.

Sequence characterized amplified region (SCAR) marker A molecular marker based on characterized sequence of a gene or DNA fragment; also is referred as a sequence-tagged-site (STS) marker.

Sequence-related amplified polymorphism (SRAP) A molecular marker technique for detecting genetic variation in the open reading frames (ORFs) of genomes of plants and related organisms.

Sequence-tagged-site (STS) marker A molecular marker that is detected with a pair of primers designed based on the unique sequence of interested; also referred as sequence characterized amplified region (SCAR) marker.

Sequencing The determination of a sequence of a nucleic acid or protein.

Sexual cycle The life cycle or stage in which the binucleate fungal mycelium eventually forms the telia and/or basidia in which nuclear fusion and meiosis take place prior to the formation of the basidiospores. Fungi usually reproduce both sexually and asexually. The asexual cycle produces mitospores (asexual spores), and the sexual cycle produces meiospores (sexual spores).

Sexual population A population developed through sexual reproduction.

Sexual recombination Recombination resulted from sexual reproduction.

Sexual reproduction Reproduction requiring the union of two compatible nuclei or gametes; reproduction involving nuclear fusion and meiosis.

Shattering The opening or disintegrating of the seed coat, fruit or husk before harvesting; the consequence is loss of seeds in crop plant and potential growth of volunteer plants.

Shotgun method A way for preparing DNA for genetic studies; random-cut fragments of DNA are cloned into a vector. The fragments resemble a clone library to which several molecular techniques are applicable, such as DNA sequencing and transformation.

Sign Any indication of disease on a host plant from direct visibility of the pathogen or its parts and products (spores, mycelium, exudate, fruitbodies or rhizomorphs).

Signal A molecule initiating an interaction when combined with a specific receptor.

Signal transduction The biochemical events that conduct the signal of a hormone or growth factor from the cell exterior, through the cell membrane and into the cytoplasm. It involves a number of molecules, including receptors, proteins and messengers.

Silver staining Using silver nitrate ($AgNO_3$) and an appropriate cell pretreatment interphase nuclei can be made visible without using a phase contrast microscope. Silver staining also is used to stain DNA fragments in polyacrylamide gels.

Simple-interest disease Denotes a plant disease with a pattern of increase mathematically analogous to simple interest in money. There is only one generation of pathogen reproduction during a plant growth season or a disease cycle. Examples of simple-interest diseases are many smuts and bunts of cereal crops and grasses.

Simple sequence repeat (SSR) markers Molecular markers developed using primers flanking DNA fragments containing variable numbers of microsatellite sequences (simple sequence repeats).

Simulation The imitation of the operation of a real-world process or system over time. The act of simulating something first requires that a model be developed, which represents the key characteristics or behaviors/functions of the selected physical or abstract system or process. The model represents the system itself, whereas the simulation represents the operation of the system over time.

Single-cycle disease A disease that occurs only once during a growing cycle. *See* Monocyclic, Simple-interest disease.

Single-nucleotide polymorphisms (SNPs) Variations in individual nucleotides that occur within DNA at the rate of approximately one in every 1300 bp in most organism. SNPs usually occur in the same genomic location in different individuals and are widely used as molecular markers for genotyping.

Single-seed descent Derivation of plants by a selection procedure in which F_2 plants and their progeny are advanced by single seeds until genetic purity is achieved.

Slow rusting A type of rust resistance with which plants exhibit slow rust development. The rust severity may never reach or take much longer to reach the highest than a susceptible genotype.

SNAREs An acronym derived from "SNAP (Soluble NSF Attachment Protein) REceptors"; a large protein superfamily consisting of at least 24 members in yeasts and more than 60 members in mammalian cells. The primary role of SNARE proteins is to mediate vesicle fusion, that is, the fusion of vesicles with their target membrane bound compartments (such as a lysosome).

Small grains Small-grain cereals that are characterized by their relative small grains, compared to maize, and specific morphological traits such as relatively short plants, spikelets, scutellum, etc. There are six cultivated plants: wheat, barley, rye, triticale, oat and rice.

Smut A disease caused by a smut fungus (Ustilaginales) in the Basidiomycotina, characterized by masses of dark brown or black, dusty to greasy masses of teliospores that generally accumulate in black, powdery sori.

SNP *See* single-nucleotide polymorphism.

Somatic cell A cell other than a sex or germ cell.

Somatic hybridization Production of hybrid cell by fusion of two protoplasts differing in genetic makeup.

Somatic recombination Production of recombinants through somatic hybridization.

Sorus (pl. sori) Compact fruiting structure, especially the erumpent spores mass in the rust (Uredinales) and smut (Ustilaginales) fungi.

Source of inoculum The place or object on or in which inoculum is produced.

sp. (pl. spp.) Abbreviation for species; one kind of organism life subordinate to a genus but above a race, strain or variety. A genus name followed by "sp." means that the particular species is undetermined, while followed by "spp." means that several species are grouped together without being named individually. *See* Subspecies.

Species – (pl. species) Any one kind of life subordinate to a genus but above a race, strain or variety; usually the smallest unit in the classification of organisms; a group of closely related individuals of the same ancestry, resembling one another in certain inherited characteristics of structure and behavior and relative stability in nature; the individuals of a species ordinarily interbreed freely and maintain themselves and their characteristics in nature.

Specimen An individual taken as a representative of a large group, e.g., a single plant; also, a portion or sample of material. From such a specimen, a culture can often be made and one or more strains (isolates) isolated.

Spermatium (pl. spermatia) Nonmotile, uninucleate, haploid (+ or –) "sex" cell (haploid male gamete), variously termed as pycniospore or microcondinium; nonmotile, uninucleate, sporelike male structure that empties its contents into a receptive female structure during plasmogamy. Spermatia are variously regarded as gametes or gametangia.

Spermatization Placing of spermatia or microcondinia on structures, such as receptive hyphae, for diploidization.

Spermogonium (pl. spermogonia) Flask-shaped, walled fungal structure in which spermatia are produced, as in ascomycetes; a pycnium of a rust fungus.

Spontaneous mutation A mutant that occurs naturally.

Spore A one- to many-celled, microscopic reproductive body in bacteria, fungi and some lower plants that becomes free and can develop into a new plant. Spores may be sexually or asexually produced and of a wide variety of sizes, shapes, colors and origins.

Sporegenous Pertaining to a cell that generates spores; producing, having or supporting spores.

Sporulation Actively producing spores.

Spreader Material added to a spray preparation to make the spray droplets spread out, increase the area covered and improve contact between the chemical and the plant or other treated surface; susceptible genotypes planted for quickly spreading inoculum of the pathogen and developing the disease in a screening field.

Spring wheat A growth habit of wheat that does not require vernalization or overwintering.

ssp. Subspecies.

SRAP *See* Sequence related amplified polymorphism.

SSH *See* Suppression subtractive hybridization.

SSR *See* Simple sequence repeat.

Stabilizing selection A type of selection that removes individuals from a population (i.e., from both ends of a phenotypic distribution divided by deviation of a sample of means).

Stage A phase in the life cycle; frequently equals state. For fungi, it is better reserved for circumstances in which the succession of two or more sates is regular.

START domain A START (StAR-related lipid-transfer) domain is a lipid-binding domain in StAR, HD-ZIP and signaling proteins. The archetypical domain is found in StAR (Steroidogenic acute regulatory protein), a mitochondrial protein that is synthesized in steroid-producing cells.

Standard deviation A measure of dispersion or spread of data values, given in the same unit as the mean and the original observations. It is of great value in the intuitive interpretation of the data and is often used in the estimation of confidence levels and significance tests.

Stem rust Rust on cereals and grasses caused by *Puccinia graminis*, characterized by large and elongated red to brown colored uredinial and black telial sori on stems and leaves. The fungus has a heteroecious macrocyclic life cycle with basidiospores infecting plants in family Berberidaceae.

Stigma Part of a pistil, usually the apex, that receives pollen and upon which pollen grains germinate.

Stimulus (pl. stimuli) A chemical, physical or environmental factor or change that induces a reaction in a living organism.

Stoma or stomate (pl. stomata or stomates) Regulated microscopic opening (pore) in the epidermis of leaves and stems, controlled by two guard cells, for passage of gases and water vapor; sometimes serves as a point of entry for pathogens.

Strain Biotype; race; form; isolate; an organism or group of similar organisms that differ in minor aspects from other organisms of the same species or variety; descendants of a single isolation in pure culture.

Stripe rust (yellow rust) of cereals and grasses Disease caused by *Puccinia striiformis*, a basidiomycete fungus that typically forms yellow- to orange-colored uredinial sori arranging in stripes on leaves and leaf sheaths of susceptible adult plants.

Stroma (pl. stromata, adj. stromatic, stromatal) Compact mass or matrix of specialized vegetative hyphae (with or without host tissue or substratum), sometimes sclerotium-like, in or on which fruitbodis (reproductive structures) and/or spores are usually produced; often a cushion-like mass of fungal cells or closely interwoven hyphae.

Stubble Stem bases and crowns of harvested plants still rooted in soil.

Subspecies (ssp.) An intraspecific population defined on the basis of more than one sharply defined character (morphologic for most organisms), and often possessing a distinct habitat, that distinguish its members from typical representatives of the species.

Substitution A mutation that exchanges one base for another (i.e., switching an A to a G). Such a substitution could change a codon to one that encodes a different amino acid and cause a small change in the protein produced. Also a plant or line with one or more chromosomes replaced by chromosomes from a different but related species.

Substomatal Just below a stomatal opening.

Summer fallow The tillage of uncropped land during the summer to control weeds and other pests and store moisture in the soil for the growth of a later crop.

Suppression The act of suppressing, or the state of being suppressed.

Suppression subtractive hybridization (SSH) A technology that allows for PCR-based amplification of only cDNA fragments that differ between a control (driver) and experimental transcriptome. The technique relies on the removal of dsDNA formed by hybridization between a control and test sample, thus eliminating cDNAs or gDNAs of similar abundance, and retaining differentially expressed, or variable in sequence, transcripts or genomic sequences.

Suppressor A gene that overcomes or suppresses the effects of mutations in other unlinked genes.

Surfactant Monomolecular compound that decreases surface tension between two unlinked materials (e.g., oil and water) and increases the emulsifying, dispersing, spreading, wetting or other surface-modifying properties of a pesticide formulation. A spreader, emulsifier, dispersing or sticking agent, wetting agent, etc., used to increase coverage (wettability) of the surface being sprayed.

Susceptibility The inability of a plant to resist the effect of a pathogen or other damaging factor.

Susceptible Lacking resistance or having poor resistance; prone to infection.

Sustainable agriculture The appropriate use of crop and livestock systems, and the agricultural inputs supporting their activities, that maintain economic, environmental and social viability while preserving the high productivity and quality of the land.

Symptom Indication of disease by reactions of the host; the visible external or internal effects produced in or on a plant by the presence of a pathogen. *See* sign.

Synonym (in biology) Another name for an organism (species o group), especially a later or illegitimate name; a rejected scientific name, other than a homonym.

Synteny Chromosomal association of genes (linkage) established in somatic cell structure; in biotechnology, the conservation of gene order on the chromosomes, which allows understanding of the evolutionary relationships between species.

Synthetic variety A variety developed by selecting a number of inbred lines with good general combining ability, intercrossing them in all possible combinations and mixing the seeds of all the crosses conducted, in equal quantity, as used in maize breeding. Synthetic hexaploid wheats are developed by crossing the wild goatgrass wheat ancestor *Aegilops tauschii* and various durum wheats to increase the genetic diversity of cultivated wheats.

Systemic Applied to chemicals and pathogens (or single infections) that generally spread internally throughout the plant body as opposed to remaining localized; an infection that spreads throughout a plant.

Systemic fungicide Fungicide or fungistat taken up systemically by plants.

T

Tag polymerase A thermostable DNA polymerase named after the thermophilic bacterium *Thermus aquaticus* from which it was originally isolated. It is frequently used in the polymerase chain reaction (PCR), a method for greatly amplifying the quantity of short segments of DNA.

Target (of pesticides) The plant, area, buildings, animals or pests intended to be treated with a pesticide application.

Target region amplification polymorphism (TRAP) A molecular marker technique that uses a fixed primer coupled with arbitrary primers to generate polymorphic DNA fragments. A fixed primer of about 18 nucleotides is designed from the target sequence of an expressed sequence tag (EST) or gene, and an arbitrary primer of about the same length is designed with either an AT- or GC-rich core to anneal with an intron or exon, respectively. Thus, TRAP markers can be easily connected with target candidate ESTs or genes.

TATA box A canonic DNA sequence; part of a plant promoter. It promotes the transcription of DNA.

Taxon (pl. taxa) A taxonomic group of any rank such as order, family, genus, species, etc.

Taxonomy Science or study of the systematic describing, naming and classification of organisms based on their natural relationships.

TBLASTN A bioinformatics method for identifying database sequences encoding proteins similar to the query. TBLASTN consists of sequences from databases in the International Nucleotide Sequence Database Collaboration and RefSeq sequences but excludes expressed sequence tags (EST), sequence-tagged sites, whole-genome shotgun sequences, transcriptome shotgun assembly sequences and patent sequences as well a phase 0, 1 and 2 high-throughput genomic sequences.

Teleomorph The sexual or perfect state of a fungus, i.e., the form involved in producing meiotic spores.

Teliospore (teleutospores) Thick-walled resting or overwintering spore produced by the rust (Uredinales) and smuts (Ustilaginales) fungi in which karyogamy occurs; it germinates to form a promycelium (basidium) in which meiosis occurs).

Telium (pl. telia) Sorus of a rust fungus producing teliospores, i.e., the terminal soru (designated by the number III), of the dikaryophase; the site of karyogamy.

Telomere One of the two terminal chromomeres of a chromosome; telomeres have certain nucleotide sequences that promote the replication of the DNA double strand. A telomere also plays a role in the spatial orientation of chromosomes within the interphase nucleus.

Terminal inverted repeat (TIR) transposon A transposon that contains inverted repeat sequences at the ends of the transposon.

Tetrasomic Having one or more chromosomes of a complement represented four times in each nucleus.

Tetraploid Having four sets of chromosomes (4n) per nucleus, twice the usual diploid number.

Threshold A critical value on an underlying scale of liability above which individuals manifest a trait or disease.

Tillage The preparation of soil for seeding. It includes maturing, ploughing, harrowing and rolling land, or whatever is done to bring it to a proper state.

Tillering stage The phase of cereal plant growth in which lateral shoot, culm or stalk arising from a crown bud.

Tissue culture An in vitro technique of cultivating (propagating) cells, tissues or plants in a sterile, synthetic medium. The technique is often used to eliminate viruses and other pathogens from vegetatively propagated plants and used to develop doubled haploid wheat plants.

Tolerance (adj. tolerant) Ability of a plant to endure an infectious or noninfectious disease, adverse conditions or chemical injure without serious damage or yield loss; ability of a fungal pathogen to endure the rate of fungicide, at which nontolerant strains (populations) will be killed; (of pesticides) the amount of chemical residue legally permitted on an agricultural product entering commercial channels and usually measured in parts per million (ppm).

Toxin Poisonous secretion produced by a living organism; a non-enzymatic metabolite produced by one organism that is injurious to another.

Trade-off A situation that involves losing one quality or aspect of something in return for gaining another quality or aspect. More colloquially, if one thing increases, some other thing must decrease.

Transcript derived fragments (TDFs) DNA sequences derived from transcripts. TDFs have been used to identify genes that differentially express under different conditions.

Transcription Synthesis of RNA using one strand of DNA as the template; catalyzed by the enzyme DNA-dependent RNA polymerase; copying of a gene into RNA. *See* Reverse transcription.

Transcription factor A protein that is involved in the process of converting, or transcribing, DNA into RNA. Transcription factors include a number of proteins, excluding RNA polymerase, that initiate and regulate the transcription of genes.

Transcriptome The set of all messenger RNA molecules in one cell or a population of cells. It differs from the exome in that it includes only those RNA molecules found in a specified cell population, and usually includes the amount or concentration of each RNA molecule in addition to the molecular identities.

Transduction The process by which DNA is transferred from one bacterium to another by a virus. It also refers to the process whereby foreign DNA is introduced into another cell via a viral vector.

Transformation In general, the heritable modification of the properties of a plant; in biotechnology, the transfer of genetic information to a recipient strain of bacteria by DNA extracted from a donor strain, and recombination of that DNA with the DNA of the recipient; in plant pathology, a plant disease symptom in which a pathogen ramifies and replaces host tissue within an organ of the host, usually a floral organ.

Transgenic plant A plant that contains an alien or modified DNA (gene) introduced by biotechnological means, and which is more or less stable inherited.

Transgressive segregation The segregation of individuals in the F_2 or a later generation of a cross that shows more extreme development of a character than either parent.

Transient expression The expression of genes for a short period of time. Sometimes, the techniques can be used to introduce a gene into a cell but do not result in a stable genetic transformation. Such events can result in foreign gene expression for a short period of time.

Translation (adj. translational) Synthesis of protein on ribosomes as directed by successive triplets of nucleotides moved along an mRNA.

Translocation Segmental exchange between non-paired (non-homologous) chromosomes to form two new chromosomes. Distribution of a chemical from the point of absorption (plant leaves, stems, roots) to other leaves, buds and root tips; movement of water, minerals, food, pathogens, wasters, etc., within or throughout a plant.

Transmembrane A type of integral membrane protein that spans the entirety of the biological membrane to which it is permanently attached. Many transmembrane proteins function as gateways to permit the transport of specific substances across the biological membrane.

Transmission electron microscope (TEM) An electron microscope in which an electric beam passes through a thin specimen or a thin section of tissue mounted on a grid. The image of the specimen is magnified with electrostatic lenses and observed on a fluorescent screen.

Transpiration Diffusion and loss of water vapor from aerial parts of plants, mainly through stomata in the leaves.

Transplanting Relocating or removing to a new growing place; in biotechnology, the cultured tissue or explant, relocated or transferred to a new site *in vitro*.

Transport protein A protein involved in the transport of small molecules around the cell.

Transposon (transposable element) A segment of chromosomal DNA that can insert itself (transposon) into the genome and integrated at different sites on the chromosomes. The ends of the transposon DNA are usually inverted repeats. Transposons are widely used to produce insertional mutations and to study gene function by examining their effect when inserted into different regions of the genome.

TRAP *See* Target region amplification polymorphism.

Trap nursery Sets of plant genotypes are assembled that carry specific resistance to the pathogen in question. It is grown in different geographic locations; it provide information on pathogen populations and may provide resistant genotype for breeding.

Triticale A amphiploid hybrid between wheat (*Triticum*) and rye (*Secale*) species in which wheat is the donor of the cytoplasm; it is a cereal crop created by humans from wheat and rye based on work starting at the end of the nineteenth century. The aim was to combine the quality advantages of wheat with the stress insensitivity of rye with growth habit similar to wheat. It differs from wheat by a greater vigor and large size of spikes and grains. Among the basic ploidy levels so far developed for triticale, the hexaploid type has become the most important for breeding and agriculture.

U

Ubiquitin A single-chain polypeptide found in living cells that plays a role in the degradation of defective and superfluous proteins.

Ubiquitination (ubiquitylation) The addition of ubiquitin to a substrate protein is called ubiquitination or ubiquitylation. Ubiquitination can affect proteins in many ways. It can signal for their degradation via the proteasome, alter their cellular location, affect their activity and promote or prevent protein interactions.

Ultrastructure Submicroscopic structure of a macromolecule, cell or tissue.

Ultraviolet radiation Electromagnetic radiation in the wavelength range of 40–400 nm.

Uninucleate Having one nucleus per cell.

Unisequences DNA or RNA sequences that are different from each other in a cell, individual or a species.

Upstream A term describing the position of the position of a DNA sequence within a DNA or protein molecule. It means that the position of the sequence lies away from the direction of the synthesis of a DNA or protein molecule.

Uredinales The order of rust fungi.

Urediniospore (uredospore, urediospore) Binucleate, dikaryotic (n + n), asexual, one-celled, repeating or summer spore of rust fungi; borne in a uredinium (uredium).

UV Ultraviolet (radiation).

V

Vacuole The relatively clear, bubble-like, membrane-lined body; the cytoplasm of a plant cell that filled with a water solution of sugar and various other plant products and byproducts.

Variability The property or ability of an organism to change its characteristics from one generation to the other as the result of an adaptation to the environment or mutation.

Variance When all values in a population are expressed as plus or minus deviations from the population mean, the variance is the mean of the squared deviations. It is a measure of variation of a population. It can be divided into phenotypic variance, genotypic variance and environmental variance.

Variant An organism showing some variation from the parental strain or population.

Variation Differences in structural or physiological characters in an organism from those typical or common in the species to which the organism belongs; differences among individuals of a population.

Variety (botanical) Group of closely related plants or other organisms of common origin within a species that differ from other varieties in distinct morphological characters; a subgroup within a species given a Latin name according to the rules of the International Code of Botanical Nomenclature. A taxonomic variety is known by the first validly published name applied to it so that nomenclature tends to be stable.

Variety (cultivated) *See* cultivar.

Vector (in genetic engineering) a vector is a self-replicating DNA molecule, such as a plasmid or virus, used to introduce a fragment of foreign DNA into a host cell (as the receptor for DNA molecule); a living organism (e.g., inset, mite, bird, animal, nematode, parasitic plant, human, etc.) able to carry and transmit a pathogen (virus, bacterium, fungus, nematode) and disseminate disease.

Vegetative Asexual; somatic; not forming spores or sex organs; also refers to a significant increase in size; the growth and maintenance (assimilative) stage of an organism in contrast to resting or reproductive stages.

Vernalization Natural or artificial induction of early flowering by exposure of seeds or seedlings to low temperatures for a certain period of time; also the requirement for breaking bud dormancy of certain temperate woody perennials.

Viable (n. viability) State of being alive; capable of germinating, growing or developing; applied usually to seeds and spores.

VIGS *See* Virus-induced gene silencing (VIGS).

Virulence The degree or measure of pathogenicity of a given pathogen; relative capacity to produce disease; (of rust fungi) the capability to cause a susceptible reaction (compatible interaction. *See* Aggressive.

Virus A submicroscopic, filterable agent that causes disease and multiplies only in living cells; a submicroscopic, intracellular, obligate parasite consisting of a core of one or more infectious nucleic acid (either RNA or DNA) molecules usually surrounded by a protein coat. Viruses contain no energy-producing enzyme systems, no functional ribosomes or other cellular organelles, which are supplied by the host cell in which viruses replicate.

Virus-induced gene silencing (VIGS) A technique that utilizes an RNA-mediated antiviral defense mechanism to determine function of a gene of interest or modify their expression. VIGS has been used widely in plants for analysis of gene function and has been adapted for high-throughput functional genomics.

Volunteer plant Self-sown plant; plant seeded by chance; unwanted plants.

W

Weed Any plant growing where it is not wanted.

Wheat A commonly grown cereal crop in the genus *Triticum* spp. (most commonly *T. aestivum*) producing grain for human food and other purposes.

Wild emmer wheat *Triticum turgidum* subsp. *dicoccoides*; tetraploid ($2n = 4\times = 28$), distinguished from cultivated emmer wheat by its shattering seeds.

Wild type The naturally occurring species or taxon as opposed to morphological variants resulting from culture in vitro or biochemical mutants obtained from culture.

Winter crop A crop that is customarily planted in the fall, overwinters as small plants and completes growth, flowering and seed production the following spring or summer before dying.

Winter hardness The ability of a plant to tolerate severe winter conditions.

Winter killing Plants are killed by frost, cold temperature and/or wind during winter.

Winter wheat A growth habit of wheat, which is planted in the fall, overwinters as seedlings, grows and matures in the spring or summer of the following year. Vernalization (treatment at low temperature 2–5 °C) for 1–2 months is generally required to grow winter wheat for seeds.

X

X-ray An electromagnetic wave of high energy and very short wavelength, which is able to pass through many materials opaque to light; a photographic or digital image of the internal composition of something, especially a part of the body, produced by X-rays being passed through it and being absorbed to different degrees by different materials.

Xylem A plant tissue consisting of various types of cells that transports water and dissolved substances toward the leaves.

Y

Yeast A microscopic fungus consisting of single oval cells that reproduce by budding. Yeats are capable of converting sugar into alcohol and carbon dioxide.

Yeast two-hybrid (Y2H) assay A molecular technique that is used to discover protein–protein interactions and protein–DNA interactions by testing for physical interactions (such as binding) between two proteins or a single protein and a DNA molecule, respectively.

Yellow rust Stripe rust, a disease caused by *Puccinia striiformis* on wheat, barley, rye, triticale and other grasses.

Yield Commonly, the aggregate of products resulting from the growth or cultivation of a crop and usually expressed in quantity per area.

***Yr* gene** A gene for resistance to yellow rust (*Yr*) or stripe rust.

Z

Zinc finger (protein) A protein motif involved in the recognition of DNA; the structure contains a complex zinc ion and consists of an antiparallel protein-chain hairpin and loop followed by a helix.

Zone Any continuous area that differs in some respect, or is distinguished for some purpose, from adjoining areas, or within which certain distinctive circumstances exist or are established; an area characterized by a particular set of plants or other organisms, whose presence is determined by environmental conditions.

Index

X. Chen, Z. Kang (eds.), *Stripe Rust*, DOI 10.1007/978-94-024-1111-9

D

H

Q

R

S

T

MIX
Papier aus verantwortungsvollen Quellen
Paper from responsible sources
FSC® C105338

If you have any concerns about our products,
you can contact us on
ProductSafety@springernature.com

In case Publisher is established outside the EU,
the EU authorized representative is:
Springer Nature Customer Service Center GmbH
Europaplatz 3, 69115 Heidelberg, Germany

Printed by Libri Plureos GmbH
in Hamburg, Germany